T0331838

FUNCTIONAL ANALYSIS

This comprehensive introduction to functional analysis covers both the abstract theory and applications to spectral theory, the theory of partial differential equations, and quantum mechanics. It starts with the basic results of the subject and progresses towards a treatment of several advanced topics not commonly found in functional analysis textbooks, including Fredholm theory, form methods, boundary value problems, semigroup theory, trace formulas, and a mathematical treatment of states and observables in quantum mechanics.

The book is accessible to graduate students with basic knowledge of topology, real and complex analysis, and measure theory. With carefully written out proofs, more than 300 problems, and appendices covering the prerequisites, this self-contained volume can be used as a text for various courses at the graduate level and as a reference text for researchers in the field.

Jan van Neerven holds an Antoni van Leeuwenhoek professorship at Delft University of Technology. Author of four books and more than 100 peer-reviewed articles, he is a leading expert in functional analysis and operator theory and their applications in stochastic analysis and the theory of partial differential equations.

Functional Analysis

JAN VAN NEERVEN
Delft University of Technology

CAMBRIDGE
UNIVERSITY PRESS

CAMBRIDGE
UNIVERSITY PRESS

Shaftesbury Road, Cambridge CB2 8EA, United Kingdom

One Liberty Plaza, 20th Floor, New York, NY 10006, USA

477 Williamstown Road, Port Melbourne, VIC 3207, Australia

314–321, 3rd Floor, Plot 3, Splendor Forum, Jasola District Centre, New Delhi – 110025, India

103 Penang Road, #05–06/07, Visioncrest Commercial, Singapore 238467

Cambridge University Press is part of Cambridge University Press & Assessment, a department of the University of Cambridge.

We share the University's mission to contribute to society through the pursuit of education, learning and research at the highest international levels of excellence.

www.cambridge.org
Information on this title: www.cambridge.org/9781009232470

DOI: 10.1017/9781009232487

First published 2022

A catalogue record for this publication is available from the British Library.

ISBN 978-1-009-23247-0 Hardback

Contents

Preface

This book is based on notes compiled during the many years I taught the course "Applied Functional Analysis" in the first year of the master's programme at Delft University of Technology, for students with prior exposure to the basics of Real Analysis and the theory of Lebesgue integration. Starting with the basic results of the subject covered in a typical Functional Analysis course, the text progresses towards a treatment of several advanced topics, including Fredholm theory, boundary value problems, form methods, semigroup theory, trace formulas, and some mathematical aspects of Quantum Mechanics. With a few exceptions in the later chapters, complete and detailed proofs are given throughout. This makes the text ideally suited for students wishing to enter the field.

Great care has been taken to present the various topics in a connected and integrated way, and to illustrate abstract results with concrete (and sometimes nontrivial) applications. For example, after introducing Banach spaces and discussing some of their abstract properties, a substantial chapter is devoted to the study of the classical Banach spaces $C(K)$, $L^p(\Omega)$, $M(\Omega)$, with some emphasis on compactness, density, and approximation techniques. The abstract material in the chapter on duality is complemented by a number of nontrivial applications, such as a characterisation of translation-invariant subspaces of $L^1(\mathbb{R}^d)$ and Prokhorov's theorem about weak convergence of probability measures. The chapter on bounded operators contains a discussion of the Fourier transform and the Hilbert transform, and includes proofs of the Riesz–Thorin and Marcinkiewicz interpolation theorems. After the introduction of the Laplace operator as a closable operator in L^p, its closure Δ is revisited in later chapters from different points of view: as the operator arising from a suitable sesquilinear form, as the operator $-\nabla^\star\nabla$ with its natural domain, and as the generator of the heat semigroup. In parallel, the theory of its Gaussian analogue, the Ornstein–Uhlenbeck operator, is developed and the connection with orthogonal polynomials and the quantum harmonic oscillator is established. The chapter on semigroup theory, besides developing the general theory, includes a detailed treatment of some important examples such as the heat semigroup, the Poisson semigroup, the Schrödinger group, and the wave group. By presenting the material in this integrated manner, it is hoped that the reader will appreciate Functional Analysis as a subject that, besides having its own depth and beauty, is deeply connected with other areas of Mathematics and Mathematical Physics.

In order to contain this already lengthy text within reasonable bounds, some choices had to be made. Relatively abstract subjects such as topological vector spaces, Banach algebras, and C^*-algebras are not covered. Weak topologies are introduced *ad hoc*, the use of distributions in the treatment of weak derivatives is avoided, and the theory of Sobolev spaces is developed only to the extent needed for the treatment of boundary value problems, form methods, and semigroups. The chapter on states and observables in Quantum Mechanics is phrased in the language of Hilbert space operators.

A work like this makes no claim to originality and most of the results presented here belong to the core of the subject. Not just the statements, but often their proofs too, are part of the established canon. Most are taken from, or represent minor variations of, proofs in the many excellent Functional Analysis textbooks in print.

Special thanks go to my students, to whom I dedicate this work. Teaching them has always been a great source of inspiration. Arjan Cornelissen, Bart van Gisbergen, Sigur Gouwens, Tom van Groeningen, Sean Harris, Sasha Ivlev, Rik Ledoux, Yuchen Liao, Eva Maquelin, Garazi Muguruza, Christopher Reichling, Floris Roodenburg, Max Sauerbrey, Cynthia Slotboom, Joop Vermeulen, Matthijs Vernooij, Anouk Wisse, and Timo Wortelboer pointed out many misprints and more serious errors in earlier versions of this manuscript. The responsibility for any remaining ones is of course with me. A list with errata will be maintained on my personal webpage. I thank Emiel Lorist, Lukas Miaskiwskyi, and Ivan Yaroslavtsev for suggesting some interesting problems, Jock Annelle and Jay Kangel for typographical comments, and Francesca Arici, Martijn Caspers, Tom ter Elst, Markus Haase, Bas Janssens, Kristin Kirchner, Klaas Landsman, Ben de Pagter, Pierre Portal, Fedor Sukochev, Walter van Suijlekom, and Mark Veraar for helpful discussions and valuable suggestions.

A significant portion of this book was written in the extraordinary circumstances of the global pandemic. The sudden decrease in overhead and the opportunity of working from home created the time and serenity needed for this project. Paraphrasing the epilogue of W. F. Hermans's novel *Onder Professoren* (Among Professors), the book was written entirely in the hours otherwise spent on departmental meetings, committee meetings, evaluations, accreditations, visitations, midterms, reviews, previews, etcetera, and so forth. All that precious time has been spent in a very useful way by the author.

Delft, April 2022

In the present corrected version we have fixed numerous small misprints, a few misformulations and editing errors, as well as a small number of mathematical oversights. In some proofs, additional details have been written out, and some arguments have been streamlined. I thank Jan Maas for some valuable suggestions in this direction.

Delft, May 2023

Notation and Conventions

We write $\mathbb{N} = \{0, 1, 2, \dots\}$ for the set of nonnegative integers, and $\mathbb{Z}$, $\mathbb{Q}$, $\mathbb{R}$, and $\mathbb{C}$ for the sets of integer, rational, real, and complex numbers. Whenever a statement is valid both over the real and complex scalar field we use the symbol $\mathbb{K}$ to denote either $\mathbb{R}$ or $\mathbb{C}$. Given a complex number $z = a + bi$ with $a, b \in \mathbb{R}$, we denote by $\bar{z} = a - bi$ its complex conjugate and by $\mathrm{Re}\, z = a$ and $\mathrm{Im}\, z = b$ its real and imaginary parts. We use the symbols $\mathbb{D}$ and $\mathbb{T}$ for the open unit disc and the unit circle in the complex plane, respectively. The indicator function of a set A is denoted by $\mathbf{1}_A$. In the context of metric and normed spaces, $B(x; r)$ denotes the open ball with radius r centred at x. The interior and closure of a set S are denoted by S° and $\bar{S}$, respectively. We write $S' \subseteq S$ to express that S' is a subset of S. The complement of a set S is denoted by $\complement S$ when the larger ambient set, of which S is a subset, is understood. We write $|x|$ both for the absolute value of a real number $x \in \mathbb{R}$, the modulus of a complex number $x \in \mathbb{C}$, and the euclidean norm of an element $x = (x_1, \dots, x_d) \in \mathbb{K}^d$. When dealing with functions f defined on some domain f, we write $f \equiv c$ on $S \subseteq D$ if $f(x) = c$ for all $x \in S$. The null space and range of a linear operator A are denoted by $\mathsf{N}(A)$ and $\mathsf{R}(A)$ respectively. When A is unbounded, its domain is denoted by $\mathsf{D}(A)$. A comprehensive list of symbols is contained in the index.

Unless explicitly otherwise stated, the symbols X and Y denote Banach spaces and H and K Hilbert spaces. In order to avoid frequent repetitions in the statements of results, these spaces are always thought of as being given and fixed. Conventions with this regard are usually stated at the beginning of a chapter or, in some cases, at the beginning of a section. The same pertains to the choice of scalar field. In Chapters 1–5, the scalar field $\mathbb{K}$ can be either $\mathbb{R}$ or $\mathbb{C}$, with a small number of exceptions where this is explicitly stated, such as in our treatment of the Hahn–Banach theorem, the Fourier transform, and the Hilbert transform. From Chapter 6 onwards, spectral theory and Fourier transforms are used extensively and the default choice of scalar field is $\mathbb{C}$. In many cases, however, statements not explicitly involving complex numbers or constructions involving them admit counterparts over the real scalars which can be obtained by simple complexification arguments. We leave it to the interested reader to check this in particular instances.

1

Banach Spaces

The foundations of modern Analysis were laid in the early decades of the twentieth century, through the work of Maurice Fréchet, Ivar Fredholm, David Hilbert, Henri Lebesgue, Frigyes Riesz, and many others. These authors realised that it is fruitful to study linear operations in a setting of abstract spaces endowed with further structure to accommodate the notions of convergence and continuity. This led to the introduction of abstract topological and metric spaces and, when combined with linearity, of topological vector spaces, Hilbert spaces, and Banach spaces. Since then, these spaces have played a prominent role in all branches of Analysis.

The main impetus came from the study of ordinary and partial differential equations where linearity is an essential ingredient, as evidenced by the linearity of the main operations involved: point evaluations, integrals, and derivatives. It was discovered that many theorems known at the time, such as existence and uniqueness results for ordinary differential equations and the Fredholm alternative for integral equations, can be conveniently abstracted into general theorems about linear operators in infinite-dimensional spaces of functions.

Stefan Banach, 1898–1945

A second source of inspiration was the discovery, in the 1920s by John von Neumann, that the – at that time brand new – theory of Quantum Mechanics can be put on a solid mathematical foundation by means of the spectral theory of selfadjoint operators on Hilbert spaces. It was not until the 1930s that these two lines of mathematical thinking were brought together in the theory of Banach spaces, named after its creator Stefan Banach (although this class of spaces was also discovered, independently and about the same time, by Norbert Wiener). This theory provides a unified perspective on Hilbert spaces

1

and the various spaces of functions encountered in Analysis, including the spaces $C(K)$ of continuous functions and the spaces $L^p(\Omega)$ of Lebesgue integrable functions.

1.1 Banach Spaces

The aim of the present chapter is to introduce the class of Banach spaces and discuss some elementary properties of these spaces. The main classical examples are only briefly mentioned here; a more detailed treatment is deferred to the next two chapters. Much of the general theory applies to both the real and complex scalar field. Whenever this applies, the symbol $\mathbb{K}$ is used to denote the scalar field, which is $\mathbb{R}$ in the case of real vector spaces and $\mathbb{C}$ in the case of complex vector spaces.

1.1.a Definition and General Properties

Definition 1.1 (Norms). A *normed space* is a pair $(X, \|\cdot\|)$, where X is a vector space over $\mathbb{K}$ and $\|\cdot\| : X \to [0, \infty)$ is a *norm*, that is, a mapping with the following properties:

(i) $\|x\| = 0$ implies $x = 0$;
(ii) $\|cx\| = |c|\,\|x\|$ for all $c \in \mathbb{K}$ and $x \in X$;
(iii) $\|x + x'\| \leqslant \|x\| + \|x'\|$ for all $x, x' \in X$.

When the norm $\|\cdot\|$ is understood we simply write X instead of $(X, \|\cdot\|)$. If we wish to emphasise the role of X we write $\|\cdot\|_X$ instead of $\|\cdot\|$.

The properties (ii) and (iii) are referred to as *scalar homogeneity* and the *triangle inequality*. The triangle inequality implies that every normed space is a metric space, with distance function

$$d(x, y) := \|x - y\|.$$

This observation allows us to introduce notions such as openness, closedness, compactness, denseness, limits, convergence, completeness, and continuity in the context of normed spaces by carrying them over from the theory of metric spaces. For instance, a sequence $(x_n)_{n \geqslant 1}$ in X is said to *converge* if there exists an element $x \in X$ such that $\lim_{n \to \infty} \|x_n - x\| = 0$. This element, if it exists, is unique and is called the *limit* of the sequence $(x_n)_{n \geqslant 1}$. We then write $\lim_{n \to \infty} x_n = x$ or simply '$x_n \to x$ as $n \to \infty$'.

The triangle inequality (ii) implies both $\|x\| - \|x'\| \leqslant \|x - x'\|$ and $\|x'\| - \|x\| \leqslant \|x' - x\|$. Since $\|x' - x\| = \|(-1) \cdot (x - x')\| = \|x - x'\|$ by scalar homogeneity, we obtain the *reverse triangle inequality*

$$\big| \|x\| - \|x'\| \big| \leqslant \|x - x'\|.$$

It shows that taking norms $x \mapsto \|x\|$ is a continuous operation.

If $\lim_{n\to\infty} x_n = x$ and $\lim_{n\to\infty} x'_n = x'$ in X and $c \in \mathbb{K}$ is a scalar, then $\|cx_n - cx\| = \|c(x_n - x)\| = |c| \|x_n - x\|$ implies

$$\lim_{n\to\infty} \|cx_n - cx\| = 0.$$

Likewise, $\|(x_n + x'_n) - (x + x')\| = \|(x_n - x) + (x'_n - x')\| \leqslant \|x_n - x\| + \|x'_n - x'\|$ implies

$$\lim_{n\to\infty} \|(x_n + x'_n) - (x + x')\| = 0.$$

This proves sequential continuity, and hence continuity, of the vector space operations.

Throughout this work we use the notation

$$B(x_0; r) := \{x \in X : \|x - x_0\| < r\}$$

for the *open ball* centred at $x_0 \in X$ with radius $r > 0$, and

$$\overline{B}(x_0; r) := \{x \in X : \|x - x_0\| \leqslant r\}$$

for the corresponding *closed ball*. The *open unit ball* and *closed unit ball* are the balls

$$B_X := B(0; 1) = \{x \in X : \|x\| < 1\}, \quad \overline{B}_X := \overline{B}(0; 1) = \{x \in X : \|x\| \leqslant 1\}.$$

Definition 1.2 (Banach spaces). A *Banach space* is a complete normed space.

Thus a Banach space is a normed space X in which every Cauchy sequence is convergent, that is, $\lim_{m,n\to\infty} \|x_n - x_m\| = 0$ implies the existence of an $x \in X$ such that $\lim_{n\to\infty} \|x_n - x\| = 0$.

The following proposition gives a necessary and sufficient condition for a normed space to be a Banach space. We need the following terminology. Given a sequence $(x_n)_{n\geqslant 1}$ in a normed space X, the sum $\sum_{n\geqslant 1} x_n$ is said to be *convergent* if there exists $x \in X$ such that

$$\lim_{N\to\infty} \left\| x - \sum_{n=1}^{N} x_n \right\| = 0.$$

The sum $\sum_{n\geqslant 1} x_n$ is said to be *absolutely convergent* if $\sum_{n\geqslant 1} \|x_n\| < \infty$.

Proposition 1.3. *A normed space X is a Banach space if and only if every absolutely convergent sum in X converges in X.*

Proof 'Only if': Suppose that X is complete and let $\sum_{n\geqslant 1} x_n$ be absolutely convergent. Then the sequence of partial sums $(\sum_{j=1}^{n} x_j)_{n\geqslant 1}$ is a Cauchy sequence, for if $n > m$ the triangle inequality implies

$$\left\| \sum_{j=1}^{n} x_j - \sum_{j=1}^{m} x_j \right\| = \left\| \sum_{j=m+1}^{n} x_j \right\| \leqslant \sum_{j=m+1}^{n} \|x_j\|,$$

which tends to 0 as $m, n \to \infty$. Hence, by completeness, the sum $\sum_{n\geqslant 1} x_n$ converges.

'If': Suppose that every absolutely convergent sum in X converges in X, and let $(x_n)_{n \geqslant 1}$ be a Cauchy sequence in X. We must prove that $(x_n)_{n \geqslant 1}$ converges in X.

Choose indices $n_1 < n_2 < \ldots$ in such a way that $\|x_i - x_j\| < \frac{1}{2^k}$ for all $i, j \geqslant n_k$, $k = 1, 2, \ldots$ The sum $x_{n_1} + \sum_{k \geqslant 1} (x_{n_{k+1}} - x_{n_k})$ is absolutely convergent since

$$\sum_{k \geqslant 1} \|x_{n_{k+1}} - x_{n_k}\| \leqslant \sum_{k \geqslant 1} \frac{1}{2^k} < \infty.$$

By assumption it converges to some $x \in X$. Then, by cancellation,

$$x = \lim_{m \to \infty} \left(x_{n_1} + \sum_{k=1}^{m} (x_{n_{k+1}} - x_{n_k}) \right) = \lim_{m \to \infty} x_{n_{m+1}},$$

and therefore the subsequence $(x_{n_m})_{m \geqslant 1}$ is convergent, with limit x. To see that $(x_n)_{n \geqslant 1}$ converges to x, we note that

$$\|x_m - x\| \leqslant \|x_m - x_{n_m}\| + \|x_{n_m} - x\| \to 0$$

as $m \to \infty$ (the first term since we started from a Cauchy sequence and the second term by what we just proved). $\qquad\square$

The next theorem asserts that every normed space can be completed to a Banach space. For the rigorous formulation of this result we need the following terminology.

Definition 1.4 (Isometries). A linear mapping T from a normed space X into a normed space Y is said to be an *isometry* if it preserves norms. A normed space X is *isometrically contained* in a normed space Y if there exists an isometry from X into Y.

Theorem 1.5 (Completion). *Let X be a normed space. Then:*

(1) *there exists a Banach space $\overline{X}$ containing X isometrically as a dense subspace;*
(2) *the space $\overline{X}$ is unique up to isometry in the following sense: If X is isometrically contained as a dense subspace in the Banach spaces $\overline{X}$ and $\overline{\overline{X}}$, then the identity mapping on X has a unique extension to an isometry from $\overline{X}$ onto $\overline{\overline{X}}$.*

Proof As a metric space, $X = (X, d)$ has a completion $\overline{X} = (\overline{X}, \overline{d})$ by Theorem D.6. We prove that $\overline{X}$ is a Banach space in a natural way, with a norm $\| \cdot \|_{\overline{X}}$ such that $\overline{d}(x, x') = \|x - x'\|_{\overline{X}}$. The properties (1) and (2) then follow from the corresponding assertions for metric spaces.

Recall that the completion $\overline{X}$ of X, as a metric space, is defined as the set of all equivalence classes of Cauchy sequences in X, declaring the Cauchy sequences $(x_n)_{n \geqslant 1}$ and $(x'_n)_{n \geqslant 1}$ to be equivalent if $\lim_{n \to \infty} d(x_n, x'_n) = \lim_{n \to \infty} \|x_n - x'_n\| = 0$. The space $\overline{X}$ is a vector space under the scalar multiplication

$$c[(x_n)_{n \geqslant 1}] := [c(x_n)_{n \geqslant 1}]$$

and addition

$$[(x_n)_{n\geqslant 1}] + [(x'_n)_{n\geqslant 1}] := [(x_n + x'_n)_{n\geqslant 1}],$$

where the brackets denote the equivalence class.

If $(x_n)_{n\geqslant 1}$ is a Cauchy sequence in X, the reverse triangle inequality implies that the nonnegative sequence $(\|x_n\|)_{n\geqslant 1}$ is Cauchy, and hence convergent by the completeness of the real numbers. We now define a norm on $\overline{X}$ by

$$\|[(x_n)_{n\geqslant 1}]\|_{\overline{X}} := \lim_{n\to\infty}\|x_n\|.$$

Denoting by $\overline{d}$ the metric on $\overline{X}$ given by $\overline{d}(x,x') := \lim_{n\to\infty} d(x_n,x'_n)$, where $x = (x_n)_{n\geqslant 1}$ and $x' = (x'_n)_{n\geqslant 1}$, it is clear that $\overline{d}(x,x') = \|x-x'\|_{\overline{X}}$. $\qquad\square$

1.1.b Subspaces, Quotients, and Direct Sums

Several abstract constructions enable us to create new Banach spaces from given ones. We take a brief look at the three most basic constructions, namely, passing to closed subspaces and quotients and taking direct sums.

Subspaces A subspace Y of a normed space X is a normed space with respect to the norm inherited from X. A subspace Y of a Banach space X is a Banach space with respect to the norm inherited from X if and only if Y is closed in X.

To prove the 'if' part, suppose that $(y_n)_{n\geqslant 1}$ is a Cauchy sequence in the closed subspace Y of a Banach space X. Then it has a limit in X, by the completeness of X, and this limit belongs to Y, by the closedness of Y. The proof of the 'only if' part is equally simple and does not require X to be complete. If $(y_n)_{n\geqslant 1}$ is a sequence in the complete subspace Y such that $y_n \to x$ in X, then $(y_n)_{n\geqslant 1}$ is a Cauchy sequence in X, hence also in Y, and therefore it has a limit y in Y, by the completeness of Y. Since $(y_n)_{n\geqslant 1}$ also converges to y in X, it follows that $y = x$ and therefore $x \in Y$.

Quotients If Y is a closed subspace of a Banach space X, the quotient space X/Y can be endowed with a norm by

$$\|[x]\| := \inf_{y\in Y}\|x-y\|,$$

where for brevity we write $[x] := x+Y$ for the equivalence class of x modulo Y. Let us check that this indeed defines a norm. If $\|[x]\| = 0$, then there is a sequence $(y_n)_{n\geqslant 1}$ in Y such that $\|x-y_n\| < \frac{1}{n}$ for all $n \geqslant 1$. Then

$$\|y_n - y_m\| \leqslant \|y_n - x\| + \|x - y_m\| < \frac{1}{n} + \frac{1}{m},$$

so $(y_n)_{n\geqslant 1}$ is a Cauchy sequence in X. It has a limit $y \in X$ since X is complete, and we have $y \in Y$ since Y is closed. Then $\|x - y\| = \lim_{n\to\infty} \|x - y_n\| = 0$, so $x = y$. This implies that $[x] = [y] = [0]$, the zero element of X/Y. The identity $\|c[x]\| = |c|\,\|[x]\|$ is trivially verified, and so is the triangle inequality.

To see that the normed space X/Y is complete we use the completeness of X and Proposition 1.3. If $\sum_{n\geqslant 1} \|[x_n]\| < \infty$ and the $y_n \in Y$ are such that $\|x_n - y_n\| \leqslant \|[x_n]\| + \frac{1}{n^2}$, the proposition implies that $\sum_{n\geqslant 1}(y_n - x_n)$ converges in X, say to x. Then, for all $N \geqslant 1$,

$$\left\| [x] - \sum_{n=1}^{N} [x_n] \right\| = \left\| \left[x - \sum_{n=1}^{N} x_n \right] \right\| \leqslant \left\| x - \sum_{n=1}^{N} x_n + \sum_{n=1}^{N} y_n \right\| = \left\| x - \left(\sum_{n=1}^{N} x_n - y_n \right) \right\|.$$

As $N \to \infty$, the right-hand side tends to 0 and therefore $\lim_{N\to\infty} \sum_{n=1}^{N} [x_n] = [x]$ in X/Y.

Direct Sums A *product norm* on a finite cartesian product $X = X_1 \times \cdots \times X_N$ of normed spaces is a norm $\|\cdot\|$ satisfying

$$\big\| (0,\ldots,0,\underbrace{x_n}_{n-\text{th}},0,\ldots,0) \big\| = \|x_n\| \leqslant \|(x_1,\ldots,x_N)\|$$

for all $x = (x_1,\ldots,x_N) \in X$ and $n = 1,\ldots,N$. For instance, every norm $|\cdot|$ on $\mathbb{K}^N$ assigning norm one to the standard unit vectors induces a product norm on X by the formula

$$\|(x_1,\ldots,x_N)\| := \big| (\|x_1\|,\ldots,\|x_N\|) \big|. \tag{1.1}$$

As a normed space endowed with a product norm, the cartesian product will be denoted

$$X = X_1 \oplus \cdots \oplus X_N$$

and called a *direct sum* of $X_1,\ldots,X_N$. If every X_n is a Banach space, then the normed space X is a Banach space. Indeed, from

$$\|x\| = \left\| \sum_{n=1}^{N} (0,\ldots,0,x_n,0,\ldots,0) \right\| \leqslant \sum_{n=1}^{N} \|x_n\| \leqslant N\|x\| \tag{1.2}$$

we see that a sequence $(x^{(k)})_{k\geqslant 1}$ in X is Cauchy if and only if all its coordinate sequences $(x_n^{(k)})_{k\geqslant 1}$ are Cauchy. If the spaces X_n are complete, these coordinate sequences have limits x_n in X_n, and these limits serve as the coordinates of an element $x = (x_1,\ldots,x_N)$ in X which is the limit of the sequence $(x^{(k)})_{k\geqslant 1}$.

1.1.c First Examples

The purpose of this brief section is to present a first catalogue of Banach spaces. The presentation is not self-contained; the examples will be revisited in more detail in the next chapter, where the relevant terminology is introduced and proofs are given.

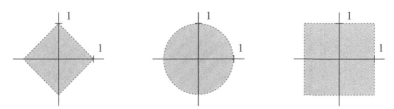

Figure 1.1 The open unit balls of $\mathbb{R}^2$ with respect to the norms $\|\cdot\|_1, \|\cdot\|_2, \|\cdot\|_\infty$.

Example 1.6 (Euclidean spaces). On $\mathbb{K}^d$ we may consider the euclidean norm

$$\|a\|_2 := \left(\sum_{j=1}^{d} |a_j|^2 \right)^{1/2},$$

and more generally the p-norms

$$\|a\|_p := \left(\sum_{j=1}^{d} |a_j|^p \right)^{1/p}, \quad 1 \leqslant p < \infty,$$

as well as the supremum norm

$$\|a\|_\infty := \sup_{1 \leqslant j \leqslant d} |a_j|.$$

It is not immediately obvious that the p-norms are indeed norms; the triangle inequality $\|a+b\|_p \leqslant \|a\|_p + \|b\|_p$ will be proved in the next chapter. It is an easy matter to check that the above norms are all *equivalent* in the sense defined in Section 1.3. *In what follows the euclidean norm of an element $x \in \mathbb{K}^d$ is denoted by $|x|$ instead of the more cumbersome $\|x\|_2$.*

Example 1.7 (Sequence spaces). Thinking of elements of $\mathbb{K}^d$ as finite sequences, the preceding example may be generalised to infinite sequences as follows. For $1 \leqslant p < \infty$ the space ℓ^p is defined as the space of all scalar sequences $a = (a_k)_{k \geqslant 1}$ satisfying

$$\|a\|_p := \left(\sum_{k \geqslant 1} |a_k|^p \right)^{1/p} < \infty.$$

The mapping $a \mapsto \|a\|_p$ is a norm which turns ℓ^p into a Banach space. The space ℓ^∞ of all bounded scalar sequences $a = (a_k)_{k \geqslant 1}$ is a Banach space with respect to the norm

$$\|a\|_\infty := \sup_{k \geqslant 1} |a_k| < \infty.$$

The space c_0 consisting of all bounded scalar sequences $a = (a_k)_{k \geqslant 1}$ satisfying

$$\lim_{k \to \infty} a_k = 0$$

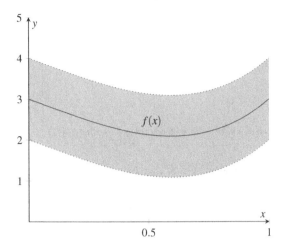

Figure 1.2 The open ball $B(f; 1)$ in $C[0, 1]$ consists of all functions in $C[0, 1]$ whose graph lies inside the shaded area.

is a closed subspace of ℓ^∞. As such it is a Banach space in its own right.

Example 1.8 (Spaces of continuous functions). Let K be a compact topological space. The space $C(K)$ of all continuous functions $f : K \to \mathbb{K}$ is a Banach space with respect to the supremum norm

$$\|f\|_\infty := \sup_{x \in K} |f(x)|.$$

This norm captures the notion of uniform convergence: for functions in $C(K)$ we have $\lim_{n \to \infty} \|f_n - f\|_\infty = 0$ if and only if $\lim_{n \to \infty} f_n = f$ uniformly.

Example 1.9 (Spaces of integrable functions). Let $(\Omega, \mathscr{F}, \mu)$ be a measure space. For $1 \leqslant p < \infty$, the space $L^p(\Omega)$ consisting of all measurable functions $f : \Omega \to \mathbb{K}$ such that

$$\|f\|_p := \left(\int_\Omega |f|^p \, \mathrm{d}\mu \right)^{1/p} < \infty,$$

identifying functions that are equal μ-almost everywhere, is a Banach space with respect to the norm $\| \cdot \|_p$. The space $L^\infty(\Omega)$ consisting of all measurable and μ-essentially bounded functions $f : \Omega \to \mathbb{K}$, identifying functions that are equal μ-almost everywhere, is a Banach space with respect to the norm given by the μ-essential supremum

$$\|f\|_\infty := \mu\text{-}\mathrm{ess}\sup_{\omega \in \Omega} |f(\omega)| := \inf\{r > 0 : |f| \leqslant r \ \mu\text{-almost everywhere}\}.$$

Example 1.10 (Spaces of measures). Let $(\Omega, \mathscr{F})$ be a measurable space. The space

$M(\Omega)$ consisting of all $\mathbb{K}$-valued measures of bounded variation on $(\Omega, \mathscr{F})$ is a Banach space with respect to the variation norm

$$\|\mu\| := |\mu|(\Omega) := \sup_{\mathscr{A} \in \mathbb{F}} \sum_{A \in \mathscr{A}} |\mu(A)|,$$

where $\mathbb{F}$ denotes the set of all finite collections of pairwise disjoint sets in $\mathscr{F}$.

Example 1.11 (Hilbert spaces). A *Hilbert space* is an inner product space $(H, (\cdot|\cdot))$ that is complete with respect to the norm

$$\|h\| := (h|h)^{1/2}.$$

Examples include the spaces $\mathbb{K}^d$ with the euclidean norm, ℓ^2, and the spaces $L^2(\Omega)$. Further examples will be given in later chapters.

1.1.d Separability

Most Banach spaces of interest in Analysis are *infinite-dimensional* in the sense that they do not have a finite spanning set. In this context the following definition is often useful.

Definition 1.12 (Separability). A normed space is called *separable* if it contains a countable set whose linear span is dense.

Proposition 1.13. *A normed space X is separable if and only if X contains a countable dense set.*

Proof The 'if' part is trivial. To prove the 'only if' part, let $(x_n)_{n \geqslant 1}$ have dense span in X. Let Q be a countable dense set in $\mathbb{K}$ (for example, one could take $Q = \mathbb{Q}$ if $\mathbb{K} = \mathbb{R}$ and $Q = \mathbb{Q} + i\mathbb{Q}$ if $K = \mathbb{C}$). Then the set of all Q-linear combinations of the x_n, that is, all linear combinations involving coefficients from Q, is dense in X. □

Finite-dimensional spaces, the sequence spaces c_0 and ℓ^p with $1 \leqslant p < \infty$, the spaces $C(K)$ with K compact metric, and $L^p(D)$ with $1 \leqslant p < \infty$ and $D \subseteq \mathbb{R}^d$ open, are separable. The separability of $C(K)$ and $L^p(D)$ follows from the results proved in the next chapter.

1.2 Bounded Operators

Having introduced normed spaces and Banach spaces, we now introduce a class of linear operators acting between them which interact with the norm in a meaningful way.

1.2.a Definition and General Properties

Let X and Y be normed spaces.

Definition 1.14 (Bounded operators). A linear operator $T : X \to Y$ is *bounded* if there exists a finite constant $C \geqslant 0$ such that

$$\|Tx\| \leqslant C\|x\|, \quad x \in X.$$

Here, and in the rest of this work, we write Tx instead of the more cumbersome $T(x)$. A *bounded operator* is a linear operator that is bounded.

The infimum C_T of all admissible constants C in Definition 1.14 is itself admissible. Thus C_T is the least admissible constant. We claim that it equals the number

$$\|T\| := \sup_{\|x\| \leqslant 1} \|Tx\|.$$

To see this, let C be an admissible constant in Definition 1.14, that is, we assume that $\|Tx\| \leqslant C\|x\|$ for all $x \in X$. Then $\|T\| = \sup_{\|x\| \leqslant 1} \|Tx\| \leqslant C$. This being true for all admissible constants C, it follows that $\|T\| \leqslant C_T$. The opposite inequality $C_T \leqslant \|T\|$ follows by observing that for all $x \in X$ we have

$$\|Tx\| \leqslant \|T\|\|x\|,$$

which means that $\|T\|$ an admissible constant. This inequality is trivial for $x = 0$, and for $x \neq 0$ it follows from scalar homogeneity, the linearity of T and the definition of the number $\|T\|$:

$$\|Tx\| = \left\|\frac{1}{\|x\|}Tx\right\|\|x\| = \left\|T\frac{x}{\|x\|}\right\|\|x\| \leqslant \|T\|\|x\|.$$

Proposition 1.15. *For a linear operator $T : X \to Y$ the following assertions are equivalent:*

(1) *T is bounded;*
(2) *T is continuous;*
(3) *T is continuous at some point $x_0 \in X$.*

Proof The implication (1)$\Rightarrow$(2) follows from

$$\|Tx - Tx'\| = \|T(x - x')\| \leqslant \|T\|\|x - x'\|$$

and the implication (2)$\Rightarrow$(3) is trivial. To prove the implication (3)$\Rightarrow$(1), suppose that T is continuous at x_0. Then there exists a $\delta > 0$ such that $\|x_0 - y\| < \delta$ implies $\|Tx_0 - Ty\| < 1$. Since every $x \in X$ with $\|x\| < \delta$ is of the form $x = x_0 - y$ with $\|x_0 - y\| < \delta$ (take $y = x_0 - x$) and T is linear, it follows that $\|x\| < \delta$ implies $\|Tx\| < 1$. By scalar homogeneity and the linearity of T we may scale both sides with a factor δ, and obtain that $\|x\| < 1$ implies $\|Tx\| < 1/\delta$. From this, and the continuity of $x \mapsto \|x\|$, it follows that $\|x\| \leqslant 1$ implies $\|Tx\| \leqslant 1/\delta$, that is, T is bounded and $\|T\| \leqslant 1/\delta$. $\qquad\square$

Easy manipulations involving the properties of norms and linear operators, such as those used in the above proofs, will henceforth be omitted.

The set of all bounded operators from X to Y is a vector space in a natural way with respect to pointwise scalar multiplication and addition by putting

$$(cT)x := c(Tx), \quad (T + T')x := Tx + Tx'.$$

This vector space will be denoted by $\mathscr{L}(X,Y)$. We further write $\mathscr{L}(X) := \mathscr{L}(X,X)$.

For all $T, T' \in \mathscr{L}(X,Y)$ and $c \in \mathbb{K}$ we have

$$\|cT\| = |c|\|T\|, \quad \|T + T\| \leqslant \|T\| + \|T'\|.$$

Let us prove the second assertion; the proof of the first is similar. For all $x \in X$, the triangle inequality gives

$$\|(T + T)x\| \leqslant \|Tx\| + \|T'x\| \leqslant (\|T\| + \|T'\|)\|x\|,$$

and the result follows by taking the supremum over all $x \in X$ with $\|x\| \leqslant 1$.

Noting that $\|T\| = 0$ implies $T = 0$, it follows that $T \mapsto \|T\|$ is a norm on $\mathscr{L}(X,Y)$. Endowed with this norm, $\mathscr{L}(X,Y)$ is a normed space. If $T : X \to Y$ and $S : Y \to Z$ are bounded, then so is their composition ST and we have

$$\|ST\| \leqslant \|S\|\|T\|.$$

Indeed, for all $x \in X$ we have

$$\|STx\| \leqslant \|S\|\|Tx\| \leqslant \|S\|\|T\|\|x\|$$

and the result follows by taking the supremum over all $x \in X$.

Proposition 1.16. *If Y is complete, then $\mathscr{L}(X,Y)$ is complete.*

Proof Let $(T_n)_{n \geqslant 1}$ be a Cauchy sequence in $\mathscr{L}(X,Y)$. From $\|T_n x - T_m x\| \leqslant \|T_n - T_m\|\|x\|$ we see that $(T_n x)_{n \geqslant 1}$ is a Cauchy sequence in Y for every $x \in X$. Let Tx denote its limit. The linearity of each of the operators T_n implies that the mapping $T : x \mapsto Tx$ is linear and we have $\|Tx\| = \lim_{n \to \infty} \|T_n x\| \leqslant M\|x\|$, where $M := \sup_{n \geqslant 1} \|T_n\|$ is finite since Cauchy sequences in normed spaces are bounded. This shows that the linear operator T is bounded, so it is an element of $\mathscr{L}(X,Y)$. To prove that $\lim_{n \to \infty} \|T_n - T\| = 0$, fix $\varepsilon > 0$ and let $N \geqslant 1$ be so large that $\|T_n - T_m\| < \varepsilon$ for all $m, n \geqslant N$. Then, for $m, n \geqslant N$, from

$$\|T_n x - T_m x\| \leqslant \varepsilon\|x\|$$

it follows, upon letting $m \to \infty$, that

$$\|T_n x - Tx\| \leqslant \varepsilon \|x\|.$$

This being true for all $x \in X$ and $n \geqslant N$, it follows that $\|T_n - T\| \leqslant \varepsilon$ for all $n \geqslant N$. $\quad\square$

The important special case $Y = \mathbb{K}$ leads to the following definition.

Definition 1.17. The *dual space* of a normed space X is the Banach space

$$X^* := \mathscr{L}(X, \mathbb{K}).$$

For $x \in X$ and $x^* \in X^*$ one usually writes $\langle x, x^* \rangle := x^*(x)$. The elements of the dual space X^* are often referred to as *bounded functionals* or simply *functionals*. Duality is a subject in its own right which will be taken up in Chapter 4. In that chapter, explicit representations of duals of several classical Banach spaces are given. For Hilbert spaces this duality takes a particularly simple form, described by the Riesz representation theorem, to be proved in Chapter 3.

It often happens that a linear operator can be shown to be well defined and bounded on a dense subspace. In such cases, a *density argument* can be used to extend the operator to the whole space.

Proposition 1.18 (Density argument – extending operators). *Let X be a normed space and Y be a Banach space, and let X_0 be a dense subspace of X. If $T_0 : X_0 \to Y$ is a bounded operator, there exists a unique bounded operator $T : X \to Y$ extending T_0. The norm of this extension satisfies $\|T\| = \|T_0\|$.*

Proof Fix $x \in X$, and suppose that $\lim_{n \to \infty} x_n = x$ with $x_n \in X_0$ for all $n \geqslant 1$. The boundedness of T_0 implies that $\|T_0 x_n - T_0 x_m\| \leqslant \|T_0\| \|x_n - x_m\| \to 0$ as $m, n \to \infty$, so $(T_0 x_n)_{n \geqslant 1}$ is a Cauchy sequence in Y. Since Y is complete, we have $T_0 x_n \to y$ for some $y \in Y$.

If also $x'_n \to x$, the same argument shows that $T_0 x'_n \to y'$ for some (possibly different) $y' \in Y$. From

$$\|T_0 x'_n - T_0 x_n\| \leqslant \|T_0\| \|x'_n - x_n\| \leqslant \|T_0\| (\|x'_n - x\| + \|x - x_n\|)$$

it follows that

$$\|y' - y\| = \lim_{n \to \infty} \|T_0 x'_n - T_0 x_n\| = 0$$

and therefore $y' = y$.

Denoting the common limit $y = y'$ by Tx, we thus obtain a well-defined mapping $x \mapsto Tx$. It is evident that this mapping extends T_0, for if $x \in X_0$ we may take $x_n = x$ and then $Tx = \lim_{n \to \infty} T_0 x_n = T_0 x$.

It is easily checked that T is linear. To show that it is bounded, with $\|T\| \leqslant \|T_0\|$, we just note that

$$\|Tx\| = \lim_{n \to \infty} \|T_0 x_n\| \leqslant \|T_0\| \lim_{n \to \infty} \|x_n\| = \|T_0\| \|x\|.$$

The converse inequality $\|T\| \geqslant \|T_0\|$ trivially holds since T extends T_0.

Finally, if the bounded operators T and T' both extend T_0, then the bounded operator $T - T'$ equals 0 on the dense subspace X_0 and hence, by continuity, on all of X. □

Under a uniform boundedness assumption, a similar density argument can be used to extend the existence of limits from a dense subspace to the whole space.

Proposition 1.19 (Density argument – extending convergence of operators). *Let X be a normed space and Y a Banach space, and let X_0 be a dense subspace of X. Let $(T_n)_{n \geqslant 1}$ be a sequence of operators in $\mathscr{L}(X,Y)$ satisfying $\sup_{n \geqslant 1} \|T_n\| < \infty$. If the limit $\lim_{n \to \infty} T_n x_0$ exists in Y for all $x_0 \in X_0$, then the limit $Tx := \lim_{n \to \infty} T_n x$ exists in Y for all $x \in X$. Moreover, the operator $T : x \mapsto Tx$ is linear and bounded from X to Y, and*

$$\|T\| \leqslant \liminf_{n \to \infty} \|T_n\|.$$

Proof We will show that the sequence $(T_n x)_{n \geqslant 1}$ is Cauchy for every $x \in X$. Fix arbitrary $x \in X$ and $\varepsilon > 0$ and choose $x_0 \in X_0$ in such a way that $\|x - x_0\| < \varepsilon / M$, where $M := \sup_{n \geqslant 1} \|T_n\|$. Since $(T_n x_0)_{n \geqslant 1}$ is a Cauchy sequence, there is an $N \geqslant 1$ such that $\|T_n x_0 - T_m x_0\| < \varepsilon$ for all $m, n \geqslant N$. Then, for all $m, n \geqslant N$,

$$\|T_n x - T_m x\| \leqslant \|T_n x - T_n x_0\| + \|T_n x_0 - T_m x_0\| + \|T_m x_0 - T_m x\|$$
$$\leqslant M\|x - x_0\| + \varepsilon + M\|x_0 - x\| < 3\varepsilon.$$

The sequence $(T_n x)_{n \geqslant 1}$ is thus Cauchy. Since Y is complete this sequence has a limit, which we denote by Tx. Linearity of $T : x \mapsto Tx$ is clear, and boundedness along with the estimate for the norm follow from

$$\|Tx\| = \lim_{n \to \infty} \|T_n x\| = \liminf_{n \to \infty} \|T_n x\| \leqslant \liminf_{n \to \infty} \|T_n\| \|x\|.$$

□

This proposition should be compared with Proposition 5.3, which provides the following partial converse: if X is a Banach space, Y is a normed space, and $(T_n)_{n \geqslant 1}$ is a sequence in $\mathscr{L}(X,Y)$ such that $Tx := \lim_{n \to \infty} T_n x$ exists in Y for all $x \in X$, then $\sup_{n \geqslant 1} \|T_n\| < \infty$.

Definition 1.20 (Null space and range). The *null space* of a bounded operator $T \in \mathscr{L}(X,Y)$ is the subspace

$$\mathsf{N}(T) := \{x \in X : Tx = 0\}.$$

The *range* of T is the subspace

$$\mathsf{R}(T) := \{Tx : x \in X\}.$$

By linearity, both the null space $\mathsf{N}(T)$ and the range $\mathsf{R}(T)$ are subspaces. By continuity, the null space of a bounded operator is closed. The following result gives a useful sufficient criterion for the range of a bounded operator to be closed.

Proposition 1.21. *Let X be a Banach space and Y be a normed space. If $T \in \mathscr{L}(X,Y)$ satisfies $\|Tx\| \geqslant C\|x\|$ for some $C > 0$ and all $x \in X$, then T is injective and has closed range.*

Proof Injectivity is clear. Suppose that $Tx_n \to y$ in Y; we must prove that $y \in \mathsf{R}(T)$. From $\|x_n - x_m\| \leqslant C^{-1}\|Tx_n - Tx_m\|$ it follows that $(x_n)_{n \geqslant 1}$ is a Cauchy sequence in X and therefore converges to some $x \in X$. Then $y = \lim_{n \to \infty} Tx_n = Tx$. □

We conclude by introducing some terminology that will be used throughout this work. In the next four definitions, X and Y are normed spaces.

Definition 1.22 (Isomorphisms). An *isomorphism* is a bijective operator $T \in \mathscr{L}(X,Y)$ whose inverse is bounded as well. An *isometric isomorphism* is an isomorphism that is also isometric. The spaces X and Y are called (isometrically) isomorphic if there exists an (isometric) isomorphism from X to Y.

Definition 1.23 (Contractions). A *contraction* is an operator $T \in \mathscr{L}(X,Y)$ satisfying $\|T\| \leqslant 1$.

Definition 1.24 (Uniform boundedness). A subset $\mathscr{T}$ of $\mathscr{L}(X,Y)$ is said to be *uniformly bounded* if it is a bounded subset of $\mathscr{L}(X,Y)$, i.e., if $\sup_{T \in \mathscr{T}} \|T\| < \infty$.

Definition 1.25 (Uniform, strong, and weak convergence of operators). A sequence $(T_n)_{n \geqslant 1}$ in $\mathscr{L}(X,Y)$ is said to:

(1) *converge uniformly* to an operator $T \in \mathscr{L}(X,Y)$ if

$$\lim_{n \to \infty} \|T_n - T\| = 0;$$

(2) *converge strongly* to an operator $T \in \mathscr{L}(X,Y)$ if

$$\lim_{n \to \infty} \|T_n x - Tx\| = 0, \quad x \in X;$$

(3) *converge weakly* to an operator $T \in \mathscr{L}(X,Y)$ if

$$\lim_{n \to \infty} \langle T_n x - Tx, y^* \rangle = 0, \quad x \in X, \ y^* \in Y^*,$$

where Y^* is the dual of Y and $\langle y, y^* \rangle := y^*(y)$ for $y \in Y$. In these situations we call T the *uniform limit*, respectively the *strong limit*, respectively the *weak limit*, of the sequence $(T_n)_{n \geqslant 1}$. Uniqueness of weak limits is assured by the Hahn–Banach theorem (see Corollary 4.11).

Uniform convergence implies strong convergence and strong convergence implies weak convergence, but the converses generally fail. For instance, the projections onto the first n coordinates in ℓ^p, $1 \leqslant p < \infty$, converge strongly to the identity operator, but not uniformly; and the operators T^n, where T is the right shift in ℓ^p, $1 < p < \infty$, converges weakly to the zero operator but not strongly (for the case $p = 1$ see Problem 4.33).

1.2.b Subspaces, Quotients, and Direct Sums

Restrictions If T is a bounded operator from a normed space X into a normed space Y, then the restriction of T to a subspace X_0 of X defines a bounded operator $T|_{X_0}$ from X_0 into Y of norm $\|T|_{X_0}\| \leqslant \|T\|$.

Quotients Let Y be a closed subspace of a Banach space X. By the definition of the quotient norm, the *quotient map* $q : x \mapsto x + Y$ is bounded from X to X/Y of norm $\|q\| \leqslant 1$.

Let Z be a normed space and let $T \in \mathscr{L}(X, Z)$ be a bounded operator with the property that Y is contained in the null space $\mathsf{N}(T)$. We claim that

$$T_{/Y}(x + Y) := Tx, \quad x \in X,$$

defines a well-defined and bounded *quotient operator* $T_{/Y} : X/Y \to Z$ of norm $\|T_{/Y}\| = \|T\|$. Well-definedness of $T_{/Y}$ is clear, and for all $x \in X$ and $y \in Y$ we have $\|Tx\| = \|T(x+y)\| \leqslant \|T\| \|x+y\|$. Taking the infimum over all $y \in Y$ gives the bound

$$\|T_{/Y}(x+Y)\| = \|Tx\| \leqslant \|T\| \inf_{y \in Y} \|x+y\| = \|T\| \|x+Y\|.$$

Hence $T_{/Y}$ is bounded and $\|T_{/Y}\| \leqslant \|T\|$. For the converse inequality we note that

$$\|Tx\| = \|T_{/Y}(x+Y)\| \leqslant \|T_{/Y}\| \|x+Y\| = \|T_{/Y}\| \inf_{y \in Y} \|x-y\| \leqslant \|T_{/Y}\| \|x\|.$$

Direct Sums If X_n is a normed space and $T_n \in \mathscr{L}(X_n)$ for $n = 1, \ldots, N$, then the *direct sum operator*

$$T = \bigoplus_{n=1}^{N} T_n : (x_1, \ldots, x_N) \mapsto (T_1 x_1, \ldots, T_N x_N)$$

is bounded on $X = \bigoplus_{n=1}^{N} X_n$ with respect to any product norm; this follows from (1.2). If the product norm is of the form (1.1), then $\|T\| = \max_{1 \leqslant n \leqslant N} \|T_n\|$.

1.2.c First Examples

We revisit the examples of Section 1.1.c and discuss how various natural operations used in Analysis give rise to bounded operators.

Example 1.26 (Matrices). Every $m \times n$ matrix $A = (a_{ij})_{i,j=1}^{m,n}$ defines a bounded operator in $\mathscr{L}(\mathbb{K}^n, \mathbb{K}^m)$ and its norm satisfies

$$\|A\|^2 = \sup_{|x| \leqslant 1} |Ax|^2 = \sup_{|x| \leqslant 1} \sum_{i=1}^{m} \left| \sum_{j=1}^{n} a_{ij} x_j \right|^2 \leqslant \sum_{i=1}^{m} \sum_{j=1}^{n} |a_{ij}|^2, \tag{1.3}$$

where the last step follows from the Cauchy–Schwarz inequality. More generally, every linear operator from a finite-dimensional normed space X into a normed space Y is bounded; this will be shown in Corollary 1.37.

The upper bound (1.3) for the norm of a matrix A is not sharp. An explicit method to determine the operator norm of a matrix is described in Problem 4.14.

Example 1.27 (Point evaluations). Let K be a compact topological space. For each $x_0 \in K$ the point evaluation $E_{x_0} : f \mapsto f(x_0)$ is bounded as an operator from $C(K)$ into $\mathbb{K}$ with norm $\|E_{x_0}\| = 1$. Boundedness with norm $\|E_{x_0}\| \leqslant 1$ follows from

$$|E_{x_0} f| = |f(x_0)| \leqslant \sup_{x \in K} |f(x)| = \|f\|_\infty.$$

By considering $f = \mathbf{1}$, the constant-one function on K, it is seen that $\|E_{x_0}\| = 1$.

Example 1.28 (Integration). Let $(\Omega, \mathscr{F}, \mu)$ be a measure space. The mapping $I_\mu : f \mapsto \int_\Omega f \, d\mu$ is bounded from $L^1(\Omega)$ to $\mathbb{K}$ with norm $\|I_\mu\| = 1$. Boundedness with norm $\|I_\mu\| \leqslant 1$ follows from

$$|I_\mu f| = \left| \int_\Omega f \, d\mu \right| \leqslant \int_\Omega |f| \, d\mu = \|f\|_1.$$

By considering nonnegative functions it is seen that $\|I_\mu\| = 1$.

Example 1.29 (Pointwise multipliers). Let $(\Omega, \mathscr{F}, \mu)$ be a measure space and fix $1 \leqslant p \leqslant \infty$. For any $m \in L^\infty(\Omega)$, the pointwise multiplier $T_m : f \mapsto mf$ defines a bounded operator on $L^p(\Omega)$ with norm $\|T_m\| = \|m\|_\infty$. Indeed, for μ-almost all $\omega \in \Omega$ we have

$$|(mf)(\omega)| = |m(\omega)||f(\omega)| \leqslant \|m\|_\infty |f(\omega)|.$$

For $1 \leqslant p < \infty$, upon integration we obtain

$$\|T_m f\|_p^p = \int_\Omega |mf|^p \, d\mu \leqslant \|m\|_\infty^p \int_\Omega |f|^p \, d\mu = \|m\|_\infty^p \|f\|_p^p,$$

T_m is bounded on $L^p(\Omega)$ and $\|T_m\| \leqslant \|m\|_\infty$. For $p = \infty$ the analogous bound follows by taking essential suprema. Equality $\|T_m\| = \|m\|_\infty$ is obtained by considering, for

$0 < \varepsilon < 1$, functions supported on measurable sets $F_\varepsilon \in \mathscr{F}$ where $|m| \geqslant (1-\varepsilon)\|m\|_\infty$ μ-almost everywhere.

Example 1.30 (Integral operators). Let μ be a finite Borel measure on a compact metric space K. With respect to the *product metric* $d((s,t),(s',t')) := d(s,t) + d(s',t')$, $K \times K$ is a compact metric space (see Proposition D.13). Let $k \in C(K \times K)$ and define, for $f \in C(K)$, the function $Tf : K \to \mathbb{K}$ by

$$Tf(s) := \int_K k(s,t)f(t)\,d\mu(t), \quad s \in K.$$

Using the uniform continuity of k (see Theorem D.12), it is easy to see that Tf is a continuous function. Indeed, given $\varepsilon > 0$, choose $\delta > 0$ so small that $d((s,t),(s',t')) < \delta$ implies $|k(s,t) - k(s',t')| < \varepsilon$. Then $d(s,s') < \delta$ implies

$$|Tf(s) - Tf(s')| \leqslant \varepsilon \int_K |f(t)|\,d\mu(t) \leqslant \varepsilon\mu(K)\|f\|_\infty.$$

As a result, T acts as a linear operator on $C(K)$. To prove boundedness, we estimate

$$|Tf(s)| \leqslant \int_K |k(s,t)||f(t)|\,d\mu(t) \leqslant \mu(K)\|k\|_\infty\|f\|_\infty.$$

Taking the supremum over $s \in K$, this results in

$$\|Tf\|_\infty \leqslant \mu(K)\|k\|_\infty\|f\|_\infty.$$

It follows that T is bounded and $\|T\| \leqslant \mu(K)\|f\|_\infty$.

For kernels $k \in L^\infty(K \times K, \mu \times \mu)$ the same prescription defines a bounded operator on $L^\infty(K,\mu)$ satisfying the same estimate. If one takes $k \in L^2(K \times K, \mu \times \mu)$, this prescription gives a bounded operator T on $L^2(K,\mu)$ satisfying

$$\|T\| \leqslant \|k\|_2. \tag{1.4}$$

Indeed, by the Cauchy–Schwarz inequality (its abstract version for Hilbert spaces will be proved in Chapter 3) and Fubini's theorem we obtain

$$\int_K \left| \int_K k(s,t)f(t)\,d\mu(t) \right|^2 d\mu(s)$$
$$\leqslant \int_K \left(\int_K |k(s,t)|^2\,d\mu(t) \right) \left(\int_K |f(t)|^2\,d\mu(t) \right) d\mu(s) = \|k\|_2^2\|f\|_2^2$$

and the claim follows. This inequality generalises the one of Example 1.26.

Example 1.31 (Volterra operator). For all $f \in L^2(0,1)$, the Cauchy–Schwarz inequality implies that the indefinite integral

$$Tf(s) := \int_0^s f(t)\,dt, \quad s \in [0,1],$$

is well defined and that $|Tf(s) - Tf(s')| \leqslant |s - s'|^{1/2}\|f\|_2$ for all $s,s' \in [0,1]$. From this we infer that $Tf \in C[0,1]$ and, by taking $s' = 0$, that $\|Tf\|_\infty \leqslant \|f\|_2$. This implies that T is bounded from $L^2(0,1)$ into $C[0,1]$ with norm $\|T\| \leqslant 1$.

Composing T with the natural inclusion mapping from $C[0,1]$ into $L^2(0,1)$, the indefinite integral can be viewed as a bounded operator on $L^2(0,1)$ of norm at most 1. A sharper bound is obtained by applying the last part of the preceding example (with $k(s,t) = \mathbf{1}_{(0,s)}(t)$). This gives that T is bounded as an operator on $L^2(0,1)$ with norm

$$\|T\| \leqslant \|k\|_2 = 1/\sqrt{2} \approx 0.7071\ldots.$$

Interestingly, this norm bound is not sharp; it can be shown that the norm of this operator equals

$$\|T\| = 2/\pi \approx 0.6366\ldots$$

This will be proved using the spectral theory of selfadjoint operators in Chapter 8.

As this brief list of examples already shows, operators occurring naturally in Analysis have a tendency to be bounded. This raises the natural question whether linear operators acting between Banach spaces X and Y are always bounded. If one is willing to accept the Axiom of Choice the answer is negative, even for separable Hilbert spaces X and $Y = \mathbb{K}$ (see Problem 3.23). In Zermelo–Fraenkel Set Theory without the Axiom of Choice, it is consistent that every linear operator acting between Banach spaces is bounded. The reader is referred to the Notes to Chapter 3 for a further discussion of this topic.

1.3 Finite-Dimensional Spaces

The aim of this section is to prove that every finite-dimensional normed space is a Banach space. This will be deduced as an easy consequence of the fact that every two norms on a finite-dimensional normed space are equivalent, in the sense made precise in the next definition.

Definition 1.32 (Equivalent norms). Two norms $\|\cdot\|$ and $\|\!|\cdot|\!\|$ on a vector space X are *equivalent* if there exist constants $0 < c \leqslant C < \infty$ such that for all $x \in X$ we have

$$c\|x\| \leqslant \|\!|x|\!\| \leqslant C\|x\|.$$

Example 1.33. Any two product norms on the product $X = X_1 \times \cdots \times X_N$ of normed spaces are equivalent. Indeed, (1.2) shows that every product norm on X is equivalent to the product norm $\|x\|_1 := \sum_{n=1}^N \|x_n\|$ on X.

In the above situation we have the inclusions of open balls

$$B_{\|\cdot\|}(x;r/C) \subseteq B_{\|\!|\cdot|\!\|}(x;r) \subseteq B_{\|\cdot\|}(x;r/c).$$

Hence if two norms on a given vector space are equivalent the resulting normed spaces have the same open sets. This implies that topological notions such as openness, closedness, compactness, convergence, and so forth, are preserved under passing to an equivalent norm.

Theorem 1.34 (Equivalence of norms in finite dimensions). *Every two norms on a finite-dimensional vector space are equivalent.*

Proof Let $(X, \|\cdot\|)$ be a finite-dimensional normed space, say of dimension d, and let $(x_j)_{j=1}^d$ be a basis for X. Relative to this basis, every $x \in X$ admits a unique representation $x = \sum_{j=1}^d c_j x_j$. We may use this to define a norm $\|\cdot\|_2$ on X by

$$\Big\| \sum_{j=1}^d c_j x_j \Big\|_2 := \Big(\sum_{j=1}^d |c_j|^2 \Big)^{1/2}.$$

The theorem follows once we have shown that the norms $\|\cdot\|$ and $\|\cdot\|_2$ are equivalent.

Let $M = \max_{1 \leqslant j \leqslant d} \|x_j\|$. By the triangle inequality and the Cauchy–Schwarz inequality, for all $x = \sum_{j=1}^d c_j x_j$ we have

$$\|x\| \leqslant \sum_{j=1}^d |c_j| \|x_j\| \leqslant M \sum_{j=1}^d |c_j| \leqslant M d^{1/2} \Big(\sum_{j=1}^d |c_j|^2 \Big)^{1/2} = M d^{1/2} \|x\|_2. \tag{1.5}$$

This gives one of the two inequalities in the definition of equivalence of norms.

To prove that a similar inequality holds in the opposite direction, let S_2 denote the unit sphere in $(X, \|\cdot\|_2)$. Since $(c_1, \ldots, c_d) \mapsto \sum_{j=1}^d c_j x_j$ maps the unit sphere of $\mathbb{K}^d$ isometrically (hence continuously) onto S_2, S_2 is compact. Consider the identity mapping $I : x \mapsto x$, viewed as a mapping from $(X, \|\cdot\|_2)$ to $(X, \|\cdot\|)$. The inequality (1.5) implies that I is bounded and therefore continuous. Since taking norms is continuous as well and S_2 is compact, the mapping $x \mapsto \|Ix\|$ is continuous from S_2 to $[0, \infty)$ and takes a minimum at some point $x_0 \in S_2$.

Denoting this minimum by m, we claim that $m > 0$. It is clear that $m \geqslant 0$. Reasoning by contradiction, if we had $m = \|Ix_0\| = 0$, then $Ix_0 = 0$ in X, hence $x_0 = 0$ as an element of S_2. Then $\|x_0\|_2 = 0$, while at the same time $\|x_0\|_2 = 1$ because $x_0 \in S_2$. This contradiction proves the claim.

For any nonzero $x \in X$ we have $\frac{x}{\|x\|_2} \in S_2$ and therefore $\|I \frac{x}{\|x\|_2}\| \geqslant m$. This gives the estimate

$$m \|x\|_2 \leqslant \|Ix\| = \|x\|$$

for nonzero $x \in X$; for trivial reasons it also holds for $x = 0$. $\qquad\square$

Corollary 1.35. *Every d-dimensional normed space is isomorphic to $\mathbb{K}^d$. In particular, every finite-dimensional normed space is a Banach space.*

Proof The first assertion has been proved in the course of the proof of Theorem 1.34, and the second assertion follows from it since $\mathbb{K}^d$ is complete.   □

Corollary 1.36. *Every finite-dimensional subspace of a normed space is closed.*

Proof By Corollary 1.35, every a finite-dimensional subspace of a normed space is complete, and it has been shown in the first paragraph of Section 1.1.b that every complete subspace of a normed space is closed.   □

Corollary 1.37. *Every linear operator from a finite-dimensional normed space X into a normed space Y is bounded.*

Proof Let $(x_j)_{j=1}^d$ be a basis for X. If $T : X \to Y$ is linear, for $x = \sum_{j=1}^d c_j x_j$ we obtain, by the Cauchy–Schwarz inequality,

$$\|Tx\| = \left\| \sum_{j=1}^d c_j T x_j \right\| \leqslant \sum_{j=1}^d |c_j| \|Tx_j\| \leqslant M d^{1/2} \|x\|_2,$$

where $\|x\|_2 := (\sum_{j=1}^d |c_j|^2)^{1/2}$ as in Theorem 1.34 and $M := \max_{1 \leqslant n \leqslant d} \|Tx_n\|$. By Theorem 1.34 there exists a constant $K \geqslant 0$ such that $\|x\|_2 \leqslant K\|x\|$ for all $x \in X$. Combining this with the preceding estimate we obtain

$$\|Tx\| \leqslant M d^{1/2} \|x\|_2 \leqslant K M d^{1/2} \|x\|.$$

This means that T is bounded with norm at most $KMd^{1/2}$.   □

Every bounded subset of a finite-dimensional normed space X is relatively compact; this follows from the corresponding result for $\mathbb{K}^d$ and the fact that X is isomorphic to $\mathbb{K}^d$ for some $d \geqslant 1$ by Corollary 1.35. Conversely, a normed space with the property that every bounded subset is relatively compact is finite-dimensional:

Theorem 1.38 (Finite-dimensional Banach spaces). *The unit ball of a normed space X is relatively compact if and only if X is finite-dimensional.*

The proof depends on the following lemma:

Lemma 1.39 (Riesz). *If Y is a proper closed subspace of a normed space X, then for every $\varepsilon > 0$ there exists a norm one vector $x \in X$ with $d(x, Y) \geqslant 1 - \varepsilon$.*

Here, $d(x, Y) = \inf_{y \in Y} \|x - y\|$ is the distance from x to Y.

Proof Fix any $x_0 \in X \setminus Y$; such x_0 exists since Y is a proper subspace of X. Fix $\varepsilon > 0$ and choose $y_0 \in Y$ such that $\|x_0 - y_0\| \leqslant (1 + \varepsilon) d(x_0, Y)$. The vector $(x_0 - y_0)/\|x_0 - y_0\|$ has norm one, and for all $y \in Y$ we have

$$\left\| \frac{x_0 - y_0}{\|x_0 - y_0\|} - y \right\| = \left\| \frac{x_0 - y_0 - y\|x_0 - y_0\|}{\|x_0 - y_0\|} \right\| \geqslant \frac{d(x_0, Y)}{(1 + \varepsilon)d(x_0, Y)} = \frac{1}{1 + \varepsilon}.$$

It follows that

$$d\left(\frac{x_0 - y_0}{\|x_0 - y_0\|}, Y\right) \geq \frac{1}{1 + \varepsilon}.$$

Since $(1 + \varepsilon)^{-1} \to 1$ as $\varepsilon \downarrow 0$, this completes the proof. $\square$

Proof of Theorem 1.38 It remains to prove the 'only if' part. Suppose that X is infinite-dimensional and pick an arbitrary norm one vector $x_1 \in X$. Proceeding by induction, suppose that norm one vectors $x_1, \ldots, x_n \in X$ have been chosen such that $\|x_k - x_j\| \geq \frac{1}{2}$ for all $1 \leq j \neq k \leq n$. Choose a norm one vector $x_{n+1} \in X$ by applying Riesz's lemma to the proper closed subspace $Y_n = \mathrm{span}\{x_1, \ldots, x_n\}$ and $\varepsilon = \frac{1}{2}$ (that Y_n is closed follows from Corollary 1.36). Then $\|x_{n+1} - x_j\| \geq \frac{1}{2}$ for all $1 \leq j \leq n$.

The resulting sequence $(x_n)_{n \geq 1}$ is contained in the closed unit ball of X and satisfies $\|x_j - x_k\| \geq \frac{1}{2}$ for all $j \neq k \geq 1$, so $(x_n)_{n \geq 1}$ has no convergent subsequence. It follows that the closed unit ball of X is not compact. $\square$

1.4 Compactness

Let X be a normed space. By Theorem 1.38, the collections of bounded subsets of X and relatively compact subsets of X coincide if and only if X is finite-dimensional. Thus, in infinite-dimensional spaces, relative compactness is a stronger property than boundedness. The purpose of the present section is to record some easy but useful general results on compactness that will be frequently used. Compactness in the spaces $C(K)$ and $L^p(\Omega)$ will be studied in the next chapter, and *compact operators*, that is, operators which map bounded sets into relatively compact sets, are studied in Chapter 7.

By a general result in the theory of metric spaces (Theorem D.10), every relatively compact set in a normed space is totally bounded, and the converse holds in Banach spaces. This fact is used in the proof of the following necessary and sufficient condition for compactness. For sets A and B in a vector space V we write

$$A + B := \{u + v : u \in A, v \in B\}.$$

Proposition 1.40. *A subset S of a Banach space X is relatively compact if and only if for all $\varepsilon > 0$ there exists a relatively compact set $K_\varepsilon \subseteq X$ such that $S \subseteq K_\varepsilon + B(0; \varepsilon)$.*

Proof 'If': The existence of the sets K_ε implies that S is totally bounded and hence relatively compact, for if the balls $B(x_{1,\varepsilon}; \varepsilon), \ldots, B(x_{n_\varepsilon, \varepsilon}; \varepsilon)$ cover K_ε, then the balls $B(x_{1,\varepsilon}; 2\varepsilon), \ldots, B(x_{n_\varepsilon, \varepsilon}; 2\varepsilon)$ cover S.

'Only if': This is trivial (take $K_\varepsilon = S$ for all $\varepsilon > 0$). $\square$

The *convex hull* of a subset F of a vector space V is the smallest convex set in V

containing F. This set is denoted by $\mathrm{co}(F)$. When F is a subset of a normed space, the closure of $\mathrm{co}(F)$ is denoted by $\overline{\mathrm{co}}(F)$ and is referred to as the *closed convex hull* of F.

As a first application of Proposition 1.40 we have the following result.

Proposition 1.41. *The closed convex hull of a compact set in a Banach space is compact.*

Proof Let K be a compact subset of the Banach space X. For every $N \geqslant 1$ the set

$$\mathrm{co}_N(K) := \left\{ \sum_{n=1}^{N} \lambda_n x_n : x_n \in K \text{ and } 0 \leqslant \lambda_n \leqslant 1 \text{ for all } n = 1, \dots, N, \ \sum_{n=1}^{N} \lambda_n = 1 \right\}$$

is contained in the image of the compact set $[0,1]^N \times K^N$ under the continuous mapping that sends $((\lambda_1, \dots, \lambda_N), (x_1, \dots, x_N))$ to $\sum_{n=1}^{N} \lambda_n x_n$.

Let $\varepsilon > 0$ be arbitrary, let the open balls $B(\xi_1; \varepsilon), \dots, B(\xi_M; \varepsilon)$ cover K, and consider an element $x \in \mathrm{co}(K)$, say $\sum_{j=1}^{k} \lambda_j x_j$. For each $j = 1, \dots, k$ let $1 \leqslant m_j \leqslant M$ be an index such that

$$\|x_j - \xi_{m_j}\| = \min_{m=1,\dots,M} \|x_j - \xi_m\|.$$

Then

$$\left\| x - \sum_{j=1}^{k} \lambda_j \xi_{m_j} \right\| \leqslant \sum_{j=1}^{k} \lambda_j \|x_j - \xi_{m_j}\| < \sum_{j=1}^{k} \lambda_j \varepsilon = \varepsilon.$$

Since $\sum_{j=1}^{k} \lambda_j \xi_{m_j} = \sum_{m=1}^{M} \left(\sum_{j: m_j = m} \lambda_j \right) \xi_m \in \mathrm{co}_M(K)$, this implies that $x \in \mathrm{co}_M(K) + B(0; \varepsilon)$. This shows that $\mathrm{co}(K) \subseteq \mathrm{co}_M(K) + B(0; \varepsilon)$. It now follows from Proposition 1.40 that $\mathrm{co}(K)$ is relatively compact. $\qquad\square$

The second result asserts that strong convergence implies uniform convergence on relatively compact sets.

Proposition 1.42. *Let X and Y be normed spaces, let the operators $T_n \in \mathscr{L}(X,Y)$, $n \geqslant 1$, be uniformly bounded, and let $T \in \mathscr{L}(X,Y)$. If $\lim_{n\to\infty} T_n = T$ strongly, then for all relatively compact subsets K of X we have*

$$\lim_{n\to\infty} \sup_{x\in K} \|T_n x - T x\| = 0.$$

It will be shown in Proposition 5.3 that if X is a Banach space, strong convergence $T_n \to T$ already implies uniform boundedness of the operators T_n.

Proof Let K be a relatively compact subset of X, let $\varepsilon > 0$ be arbitrary, and select finitely many open balls $B(x_1; \varepsilon), \dots, B(x_k; \varepsilon)$ covering K. Choose $N \geqslant 1$ so large that $\|T_n x_j - T x_j\| < \varepsilon$ for all $n \geqslant N$ and $j = 1, \dots, k$. Let $M := \sup_{n \geqslant 1} \|T_n\|$; this number is finite by assumption. Fixing an arbitrary $x \in K$, choose $1 \leqslant j_0 \leqslant k$ such that $\|x - x_{j_0}\| < \varepsilon$. Then, for $n \geqslant N$,

$$\|T_n x - T x\| \leqslant \|T_n x - T_n x_{j_0}\| + \|T_n x_{j_0} - T x_{j_0}\| + \|T x_{j_0} - T x\|$$
$$\leqslant M\varepsilon + \varepsilon + M\varepsilon = (2M+1)\varepsilon.$$

Taking the supremum over $x \in K$, it follows that if $n \geqslant N$, then

$$\sup_{x \in K} \|T_n x - T x\| \leqslant (2M+1)\varepsilon.$$

Since $\varepsilon > 0$ was arbitrary, this proves the final assertion. $\qquad \square$

1.5 Integration in Banach Spaces

In a variety of circumstances, some of which will be encountered in later chapters, one wishes to integrate X-valued functions, where X is a Banach space. In order to have the tools available when they are needed, we insert a brief discussion of the X-valued counterparts of the Riemann and Lebesgue integrals.

1.5.a The Riemann Integral

Let K be a compact metric space and let μ be a finite Borel measure on K. We will set up the Riemann integral with respect to μ for continuous functions $f : K \to X$. To this end we need the following terminology. A *partition* of K is a finite collection of pairwise disjoint Borel subsets of K whose union equals K. The *mesh* of a partition is the diameter of the largest subset in the partition.

Proposition 1.43 (Riemann integral). *Let μ be a finite Borel measure on a compact metric space K, let X be a Banach space, and let $f : K \to X$ be a continuous function. There exists a unique element in X, denoted by $\int_K f \, d\mu$, with the fol-*

Bernhard Riemann, 1826–1866

lowing property: for every $\varepsilon > 0$ there exists a $\delta > 0$ such that whenever $(K_n)_{n=1}^N$ is a partition of K of mesh less than δ and $(t_n)_{n=1}^N$ is a collection of points in K with $t_n \in K_n$ for all $n = 1, \ldots, N$, then

$$\left\| \int_K f \, d\mu - \sum_{n=1}^N \mu(K_n) f(t_n) \right\| < \varepsilon.$$

The proof of this theorem follows the undergraduate construction of the Riemann integral for continuous functions $f : [0,1] \to \mathbb{K}$ step-by-step and is therefore omitted. The element $\int_K f \, d\mu$ is called the *Riemann integral of f with respect to* μ. Whenever this is convenient we use the more elaborate notation $\int_K f(t) \, d\mu(t)$.

Proposition 1.44. *Let μ be a finite Borel measure on a compact metric space K, let X be a Banach space, and let $f : K \to X$ be a continuous function. Then*

$$\left\| \int_K f \, d\mu \right\| \leq \int_K \|f\| \, d\mu.$$

Proof For any partition $(K_n)_{n=1}^N$ of K and any collection of points $(t_n)_{n=1}^N$ in K with $t_n \in K_n$ for all $n = 1, \ldots, N$ we have

$$\left\| \sum_{n=1}^N \mu(K_n) f(t_n) \right\| \leq \sum_{n=1}^N \mu(K_n) \|f(t_n)\|$$

by the triangle inequality. The result follows by taking the limit along any sequence of partitions whose meshes tend to zero.   □

In the special case where $K = [0,1]$ and μ is the Lebesgue measure, the usual calculus rules apply (defining differentiability of an X-valued function in the obvious way):

Proposition 1.45. *Let X be a Banach space and let $f : [0,1] \to X$ be a function. Then:*

(1) *if f is differentiable at the point $t_0 \in [0,1]$, then f is continuous at t_0;*
(2) *if f is differentiable on $(0,1)$ and $f' \equiv 0$ on $(0,1)$, then f is constant on $(0,1)$;*
(3) *if f is continuously differentiable on $[0,1]$, then*

$$\int_0^1 f'(t) \, dt = f(1) - f(0).$$

Proof (1): Fix an arbitrary $\varepsilon > 0$. The assumption implies there exists $\delta > 0$ such that if $t \in [0,1]$ with $|t - t_0| < \delta$, then

$$\left\| \frac{f(t) - f(t_0)}{t - t_0} - f'(t_0) \right\| < \varepsilon.$$

Then $\|f(t) - f(t_0)\| < (\varepsilon + \|f'(t_0)\|)|t - t_0|$ and continuity at t_0 follows.

(2): The usual calculus proof via Rolle's theorem does not extend to the present setting, as it uses the order structure of the real numbers.

Fix an arbitrary $\varepsilon > 0$. For each $t \in (0,1)$, the assumption $f'(t) = 0$ implies that there exists $h(t) > 0$ such that the interval $I_t := (t - h(t), t + h(t))$ is contained in $(0,1)$ and

$$\|f(t) - f(s)\| \leq \varepsilon |t - s|, \quad s \in I_t.$$

Fix a closed subinterval $[a,b] \subseteq (0,1)$. The intervals I_t, $t \in [a,b]$, cover the compact set

$[a,b]$ and therefore this set is contained in the union of finitely many intervals $I_{t_1}, \ldots, I_{t_N}$. By adding the intervals I_a and I_b and relabelling (and perhaps discarding some of the intervals), we may assume that $a = t_1$, $b = t_N$, and $I_{t_n} \cap I_{t_{n+1}} \neq \varnothing$ for $n = 1, \ldots, N-1$. Choosing $s_n \in I_{t_n} \cap I_{t_{n+1}}$ we have

$$\|f(t_{n+1}) - f(t_n)\| \leqslant \|f(t_{n+1}) - f(s_n)\| + \|f(s_n) - f(t_n)\|$$
$$\leqslant \varepsilon(t_{n+1} - s_n) + \varepsilon(s_n - t_n) = \varepsilon(t_{n+1} - t_n).$$

Now let $t \in [a,b]$, say $t \in I_{t_k}$. Then

$$\|f(t) - f(a)\| \leqslant \|f(t) - f(t_k)\| + \|f(t_k) - f(t_{k-1})\| + \cdots + \|f(t_2) - f(t_1)\|$$
$$\leqslant \varepsilon(t - t_k) + \varepsilon(t_k - t_{k-1}) + \cdots + \varepsilon(t_2 - t_1) = \varepsilon(t - a).$$

This being true for all $\varepsilon > 0$ it follows that $f(t) = f(a)$ for all $t \in [a,b]$. This proves that f is constant on every subinterval $[a,b] \subseteq (0,1)$ and therefore on $(0,1)$.

(3): For the function $g : [0,1] \to X$, $g(t) := f(t) - \int_0^t f'(s)\,ds$, we have

$$\lim_{h \to 0} \frac{1}{h}(g(t+h) - g(t)) = f'(t) - \lim_{h \to 0} \frac{1}{h} \int_t^{t+h} f'(s)\,ds = 0$$

by continuity, and therefore g is continuously differentiable on $[0,1]$ with derivative $g' = 0$. It follows from (2) that g is constant on $(0,1)$, hence on $[0,1]$ by continuity, and then $g(0) = f(0)$ implies

$$f(t) - \int_0^t f'(s)\,ds = g(t) = g(0) = f(0), \quad t \in [0,1].$$

Taking $t = 1$ gives the result. $\qquad\qquad\square$

In Chapter 4 we will sketch a different proof using duality.

1.5.b The Bochner Integral

We turn next to the more delicate problem of generalising the Lebesgue integral to functions taking values in a Banach space X. The results of this section will be needed only in Chapter 13.

In what follows we fix a measure space $(\Omega, \mathscr{F})$. It is a matter of experience that if one attempts to define the measurability of a function $f : \Omega \to X$ by imposing that $f^{-1}(B)$ be in $\mathscr{F}$ for all Borel (equivalently, for all open) subsets of X, one arrives at a notion of measurability that is not very practical, the problem being that it does not connect well with approximation theorems such as the dominated convergence theorem. It turns out that it is better to start from the following necessary and sufficient condition for measurability in the scalar-valued setting: *A scalar-valued function is measurable if and only if it is the pointwise limit of a sequence of simple functions.*

For a function $f : \Omega \to \mathbb{K}$ and $x \in X$ we define $f \otimes x : \Omega \to X$ by

$$(f \otimes x)(\omega) := f(\omega)x. \tag{1.6}$$

Definition 1.46 (Simple functions, strong measurability). A function $f : \Omega \to X$ is called *simple* if it is a finite linear combination of functions of the form $\mathbf{1}_F \otimes x$ with $F \in \mathscr{F}$ and $x \in X$, and *strongly measurable* if it is the pointwise limit of a sequence of simple functions.

A scalar-valued function is strongly measurable if and only if it is measurable, and for such functions we omit the adjective 'strongly'.

Theorem 1.47 (Pettis measurability theorem, first version). *A function $f : \Omega \to X$ is strongly measurable if and only if f takes its values in a separable closed subspace X_0 of X and the nonnegative functions $\|f(\cdot) - x_0\|$ are measurable for all $x_0 \in X_0$.*

A second version of this theorem will be proved in Chapter 4 (see Theorem 4.19).

Proof 'If': Let $(x_n)_{n \geq 1}$ be dense in X_0 and define the functions $\phi_n : X_0 \to \{x_1, \ldots, x_n\}$ as follows. For each $y \in X_0$ let $k(n,y)$ be the least integer $1 \leq k \leq n$ such that

$$\|y - x_k\| = \min_{1 \leq j \leq n} \|y - x_j\|,$$

and put $\phi_n(y) := x_{k(n,y)}$. Since $(x_n)_{n \geq 1}$ is dense in X_0 we have

$$\lim_{n \to \infty} \|\phi_n(y) - y\| = 0, \quad y \in X_0.$$

Now define $\psi_n : \Omega \to X$ by

$$\psi_n(\omega) := \phi_n(f(\omega)), \quad \omega \in \Omega.$$

We have

$$\{\omega \in \Omega : \psi_n(\omega) = x_1\} = \left\{\omega \in \Omega : \|f(\omega) - x_1\| = \min_{1 \leq j \leq n} \|f(\omega) - x_j\|\right\}$$

and, for $2 \leq k \leq n$,

$$\{\omega \in \Omega : \psi_n(\omega) = x_k\}$$
$$= \left\{\omega \in \Omega : \|f(\omega) - x_k\| = \min_{1 \leq j \leq n} \|f(\omega) - x_j\| < \min_{1 \leq j < k-1} \|f(\omega) - x_j\|\right\}.$$

In both identities, the set on the right-hand side is in $\mathscr{F}$. Hence each ψ_n is simple, takes values in X_0, and for all $\omega \in \Omega$ we have

$$\lim_{n \to \infty} \|\psi_n(\omega) - f(\omega)\| = \lim_{n \to \infty} \|\phi_n(f(\omega)) - f(\omega)\| = 0.$$

'Only if': Let $f_n \to f$ pointwise with each f_n simple. Let X_0 be the closed linear span of the ranges of the functions f_n. Then X_0 is separable and f takes its values in X_0. Moreover, $\omega \mapsto \|f(\omega) - x_0\| = \lim_{n \to \infty} \|f_n(\omega) - x_0\|$ is measurable. $\qquad \square$

Corollary 1.48. *If* $\lim_{n\to\infty} f_n = f$ *pointwise, with each* f_n *strongly measurable, then* f *is strongly measurable.*

Proof We check the conditions of the Pettis measurability theorem. Every function $f_n :$ $\Omega \to X$ is the pointwise limit of a sequence of simple functions $f_{nm} : \Omega \to X$, and every f_{nm} takes at most finitely many different values. It follows that f takes its values in the closed linear span of these countably many finite sets, which is a separable subspace of X. The measurability of the functions $\|f_n - x_0\|$ implies that $\|f - x_0\|$ is measurable. $\qquad\square$

Definition 1.49 (μ-Simple functions). A simple function $f = \sum_{n=1}^{N} 1_{F_n} \otimes x_n$ is called μ-*simple* if $\mu(F_n) < \infty$ for all $n = 1, \ldots, N$. For such functions we define

$$\int_\Omega f \, d\mu := \sum_{n=1}^{N} \mu(F_n) x_n.$$

We leave it as a simple exercise to verify that $\int_\Omega f \, d\mu$ is well defined in the sense that it does not depend on the representation of f as a linear combination of functions $1_{F_n} \otimes x_n$ with $\mu(F_n) < \infty$. If f is μ-simple, the triangle inequality implies

$$\left\| \int_\Omega f \, d\mu \right\| \leqslant \int_\Omega \|f\| \, d\mu. \tag{1.7}$$

Definition 1.50 (Bochner integral). A strongly measurable function $f : \Omega \to X$ is said to be *Bochner integrable* with respect to μ if there is a sequence of μ-simple functions $f_n : \Omega \to X$ such that

$$\lim_{n\to\infty} \int_\Omega \|f - f_n\| \, d\mu = 0. \tag{1.8}$$

In that case we define the *Bochner integral* of f by

$$\int_\Omega f \, d\mu := \lim_{n\to\infty} \int_\Omega f_n \, d\mu. \tag{1.9}$$

The nonnegative functions $\|f - f_n\|$ are measurable by the Pettis measurability theorem, so the integral in (1.8) is well defined. The limit in (1.9) exists since the assumption together with (1.7) (applied to $f_n - f_m$) implies that $(\int_\Omega f_n \, d\mu)_{n\geqslant 1}$ is a Cauchy sequence in X. We leave it as another simple exercise to verify that $\int_\Omega f \, d\mu$ is well defined in the sense that it does not depend on the sequence of approximating functions f_n. It is equally elementary to verify that if $\Omega = K$ is a compact metric space and $\mathscr{F}$ is its Borel σ-algebra, then every continuous function $f : K \to X$ is Bochner integrable with respect to μ and the Bochner integral coincides with the Riemann integral.

Proposition 1.51. *A strongly measurable function* $f : \Omega \to X$ *is Bochner integrable with respect to* μ *if and only if*

$$\int_\Omega \|f\| \, d\mu < \infty.$$

In this situation we have

$$\left\| \int_\Omega f \, d\mu \right\| \leqslant \int_\Omega \|f\| \, d\mu.$$

Proof 'If': Let f be a strongly measurable function satisfying $\int_\Omega \|f\| \, d\mu < \infty$. Let g_n be simple functions such that $\lim_{n\to\infty} g_n = f$ pointwise and define

$$f_n := \mathbf{1}_{\{\|g_n\| \leqslant 2\|f\|\}} g_n.$$

Then each f_n is simple and we have $\lim_{n\to\infty} f_n = f$ pointwise. Since $\|f_n\| \leqslant 2\|f\|$ pointwise and $\int_\Omega \|f\| \, d\mu < \infty$, each f_n is μ-simple and by dominated convergence we obtain

$$\lim_{n\to\infty} \int_\Omega \|f - f_n\| \, d\mu = 0.$$

'Only if': If f is Bochner integrable and the μ-simple function $g : \Omega \to X$ is such that $\int_\Omega \|f - g\| \, d\mu \leqslant 1$, then

$$\int_\Omega \|f\| \, d\mu \leqslant 1 + \int_\Omega \|g\| \, d\mu < \infty.$$

The final assertion follows from (1.7) by approximation. □

Problems

1.1　Show that in any normed space X, for all $x_0 \in X$ and $r > 0$ the following assertions hold:

(a) $B(x_0; r) = \{x \in X : \|x - x_0\| < r\}$ is an open set.

(b) $\overline{B}(x_0; r) = \{x \in X : \|x - x_0\| \leqslant r\}$ is a closed set.

(c) $\overline{B}(x_0; r) = \overline{B(x_0; r)}$, that is, $\overline{B}(x_0; r)$ is the closure of $B(x_0; r)$.

1.2　Let X be a normed space.

(a) Show that if $x, y \in X$ satisfy $\|x - y\| < \varepsilon$ with $0 < \varepsilon < \|x\|$, then $y \neq 0$ and

$$\left\| x - \frac{\|x\|}{\|y\|} y \right\| < 2\varepsilon.$$

(b) Show that the constant 2 in part (a) is the best possible.

1.3　Show that a norm $\|\cdot\|$ on the product $X = X_1 \times \cdots \times X_N$ of normed spaces is a product norm if and only if $\|x\|_\infty \leqslant \|x\| \leqslant \|x\|_1$ for all $x = (x_1, \ldots, x_N) \in X$, where

$$\|x\|_\infty := \max_{1 \leqslant n \leqslant N} \|x_n\|, \quad \|x\|_1 := \sum_{n=1}^{N} \|x_n\|.$$

1.4　Show that if $X = X_1 \oplus \cdots \oplus X_N$ is a direct sum of normed spaces, then each summand X_n is closed as a subspace of X.

1.5 Prove that if $T \in \mathscr{L}(X,Y)$ is bounded, then

$$\|T\| = \sup_{\|x\|=1} \|Tx\| = \sup_{\|x\|<1} \|Tx\|.$$

1.6 Let X and Y be normed spaces and let $T \in \mathscr{L}(X,Y)$. Prove that for all $x \in X$ and $r > 0$ we have

$$\sup_{y \in B(x;r)} \|Ty\| \geqslant r\|T\|.$$

1.7 Let $X_0 := C_c^1(0,1)$ be the vector space of all C^1-functions $f : (0,1) \to \mathbb{K}$ with compact support in $(0,1)$.

   (a) Show that $X := \{f \in C[0,1] : f(0) = f(1) = 0\}$ is a Banach space and that X_0 can be naturally identified with a dense subspace in X.

   (b) Show that for each $f \in X_0$ the limit $\lim_{n \to \infty} T_n f$ exists with respect to the norm of X and equals f', where

$$T_n f(t) = \frac{f(t+1/n) - f(t)}{1/n}.$$

   (c) Show that there are functions $f \in X$ for which the limit $\lim_{n \to \infty} T_n f$ does not exist in X.

This example shows that the uniform boundedness assumption cannot be omitted in Proposition 1.19.

1.8 Show that if two norms $\| \cdot \|$ and $\| \cdot \|'$ on a normed space X are equivalent, then the norms of the completions of $(X, \| \cdot \|)$ and $(X, \| \cdot \|')$ are equivalent.

1.9 Let $\| \cdot \|$ and $\| \cdot \|'$ be two norms on a vector space X. Show that the following assertions are equivalent:

   (1) there exists a constant $C \geqslant 0$ such that $\|x\| \leqslant C\|x\|'$ for all $x \in X$;
   (2) every open set in $(X, \| \cdot \|)$ is open in $(X, \| \cdot \|')$;
   (3) every convergent sequence in $(X, \| \cdot \|')$ is convergent in $(X, \| \cdot \|)$;
   (4) every Cauchy sequence in $(X, \| \cdot \|')$ is Cauchy in $(X, \| \cdot \|)$.

1.10 Let X be a Banach space with respect to the norms $\| \cdot \|$ and $\| \cdot \|'$. Suppose that $\| \cdot \|$ and $\| \cdot \|'$ agree on a subspace Y that is dense in X with respect to both norms. We ask whether the norms agree on all of X.

   (a) Comment on the following attempt to prove this: Apply Proposition 1.18 to the identity mapping on Y, viewed as a mapping from the normed space $(Y, \| \cdot \|)$ to the normed $(Y, \| \cdot \|')$ and as a mapping in the opposite direction.

   (b) Comment on the following attempt to prove this: Let $x \in X$ be fixed and, using density, choose a sequence $x_n \to x$ with $x_n \in Y$ for all $n \geqslant 1$. Then $\|x\| = \lim_{n \to \infty} \|x_n\| = \lim_{n \to \infty} \|x_n\|' = \|x\|'$.

   (c) Comment on Problem 2.8 as an attempt to disprove this.

(d) Prove that the answer is affirmative if we make the additional assumption that $\|\cdot\| \leqslant C\|\cdot\|'$ for some constant $0 < C < \infty$.

1.11 Provide the details to the 'if' part of the proof of Proposition 1.13.

1.12 Let X be a normed space.

(a) Show that if X is separable, then the completion of X is separable.

(b) Show that if X is a Banach space and Y is a closed subspace of X, then X is separable if and only if both Y and X/Y are separable.

1.13 Determine whether the following sets are open and/or closed in $C[0, 1]$:

(a) $\{f \in C[0, 1] : f(t) \geqslant 0 \text{ for all } t \in [0, 1]\}$;

(b) $\{f \in C[0, 1] : f(t) > 0 \text{ for all } t \in [0, 1]\}$.

Consider the set $S := \{f \in L^1(0, 1) : f(t) \geqslant 0 \text{ for almost all } t \in (0, 1)\}$.

(c) Determine whether S is a closed subset of $L^1(0, 1)$.

(d) Characterise the functions belonging to the interior S°.

1.14 This problem gives an example of a bounded operator that does not attain its norm. Let X be the space of continuous functions $f : [0, 1] \to \mathbb{K}$ satisfying $f(0) = 0$.

(a) Show that X is a closed subspace of $C[0, 1]$.

Thus, with the norm inherited from $C[0, 1]$, X is a Banach space.

(b) Show that the operator $T : X \to \mathbb{K}$,

$$Tf := \int_0^1 f(t)\, dt,$$

is bounded and has norm $\|T\| = 1$.

(c) Prove that $|Tf| < 1$ for all $f \in X$ with $\|f\|_\infty \leqslant 1$.

1.15 This problem gives an example of a bounded operator whose range is not closed. Consider the linear operator T on $C[0, 1]$ given by the indefinite integral

$$Tf(t) = \int_0^t f(s)\, ds, \quad t \in [0, 1].$$

(a) Show that T is bounded and compute its norm.

(b) Show that $R(T)$ is not closed in $C[0, 1]$.

1.16 Let X be a Banach space and Y be a normed space. Show that if $T : X \to Y$ is a bounded operator satisfying $\|Tx\| \geqslant C\|x\|$ for some $C > 0$ and all $x \in X$, then its range $R(T)$ is complete and T is an isomorphism from X to $R(T)$.

1.17 Let X and Y be finite-dimensional normed spaces. Prove that if $T_n, T \in \mathscr{L}(X, Y)$, then the following assertions are equivalent:

(1) $\lim_{n \to \infty} T_n = T$ uniformly;

(2) $\lim_{n \to \infty} T_n = T$ strongly;

(3) $\lim_{n\to\infty} T_n = T$ weakly.

1.18 Let $1 \leqslant p < \infty$.

    (a) Show that ℓ^p is a dense subspace of c_0.

    (b) Show that the inclusion mapping of ℓ^p into c_0 is bounded.

1.19 Show that a normed space X and its completion $\overline{X}$ have the same dual, that is, the restriction mapping $\overline{x}^* \mapsto \overline{x}^*|_X$ is an isometric isomorphism from $\overline{X}^*$ onto X^*.

1.20 Let X be a real vector space. The product $X \times X$ can be given the structure of a complex vector space by introducing a complex scalar multiplication as follows:

$$(a+ib)(x,y) := (ax - by, bx + ay).$$

The idea is to think of the pair $(x,y) \in X \times X$ as "$x + iy$".

    (a) Check that this formula for the scalar multiplication does indeed turn $X \times X$ into a complex vector space.

The resulting complex vector space is denoted by $X_{\mathbb{C}}$.

    Suppose now that X is a real normed space.

    (b) Prove that the formula

$$\|(x,y)\| := \sup_{\theta \in [0,2\pi]} \|(\cos\theta)x + (\sin\theta)y\|$$

defines a norm on $X_{\mathbb{C}}$ which turns it into a complex normed space. Show that $X_{\mathbb{C}}$ is a Banach space if and only if X is a Banach space.

    (c) Show that this norm on $X_{\mathbb{C}}$ extends the norm of X in the sense that $\|(x,0)\| = \|(0,x)\| = \|x\|$ for all $x \in X$.

    (d) Show that $\|(x,y)\| = \|(x,-y)\|$ for all $x,y \in X$.

    (e) Show that any two norms on $X_{\mathbb{C}}$ which satisfy the identities in parts (c) and (d) are equivalent.

1.21 Let X be a real Banach space and let $X_{\mathbb{C}}$ be the complex Banach space constructed in Problem 1.20.

    (a) Show that if T is a (real-)linear bounded operator on X, then T extends to a bounded (complex-)linear operator $T_{\mathbb{C}}$ on $X_{\mathbb{C}}$ by putting $T_{\mathbb{C}}(x,y) := (Tx, Ty)$.

    (b) Show that $\|T_{\mathbb{C}}\| = \|T\|$.

1.22 As a variation on Proposition 1.40, show that a bounded subset S of a Banach space X is relatively compact if and only if for every $\varepsilon > 0$ there exists a finite-dimensional subspace X_{ε} of X such that $S \subseteq X_{\varepsilon} + B(0;\varepsilon)$.

1.23 Show that a subset K of a Banach space X is relatively compact if and only if K is contained in the closed convex hull of a sequence $(x_n)_{n\geqslant 1}$ in X satisfying $\lim_{n\to\infty} x_n = 0$.

Hint: For the 'only if' part, cover K with finitely many balls of radius 3^{-n} and let C_n be the set of their centres; $n = 1, 2, \ldots$ Let $D_1 := C_1$ and, for $n \geqslant 2$,

$$D_n := \{c_n - c_{n-1} : c_n \in C_n, \ c_{n-1} \in C_{n-1}, \ \|c_n - c_{n-1}\| < 3^{-n+1}\}.$$

Check that each $x \in K$ can be represented as an absolutely convergent sum $x = \sum_{n \geqslant 1} d_n$ with $d_n \in D_n$. Consider the sequence $(x_n)_{n \geqslant 1}$ given by $x_n := 2^n d_n$.

1.24 Let $(\Omega, \mathscr{F})$ be a measurable space. Adapting the proof of Theorem 1.47 show that if $f : \Omega \to X$ is strongly measurable, there are simple functions $f_n : \Omega \to X$ such that $f_n \to f$ and $\|f_n\| \leqslant \|f\|$ pointwise.

1.25 Let K be a compact metric space, let μ a finite Borel measure on K, and let X be a Banach space. Prove that every continuous function $f : K \to X$ is Bochner integrable with respect to μ and that its Bochner integral equals its Riemann integral.

1.26 Let $(\Omega, \mathscr{F}, \mu)$ be a measure space and let X_0 be a closed subspace of the Banach space X. Let $f : \Omega \to X$ satisfy $f(\omega) \in X_0$ for all $\omega \in \Omega$.

(a) Show that if f is strongly measurable as an X-valued function, then f is strongly measurable as an X_0-valued function.

Assume now that $f : \Omega \to X$ satisfies $f(\omega) \in X_0$ for almost all $\omega \in \Omega$.

(b) Show that if f is strongly measurable as an X-valued function, then f is strongly measurable as an X_0-valued function.

(c) Show that if f is Bochner integrable as an X-valued function, then f is Bochner integrable as an X_0-valued function.

1.27 Let $(\Omega, \mathscr{F}, \mu)$ be a measure space. Show that if $T : X \to Y$ is a bounded operator and $f : \Omega \to X$ is Bochner integrable with respect to μ, then $Tf : \Omega \to Y$ is Bochner integrable with respect to μ and

$$T \int_\Omega f \, d\mu = \int_\Omega Tf \, d\mu.$$

1.28 Let $(\Omega, \mathscr{F}, \mu)$ be a measure space and let $(\Omega', \mathscr{F}')$ be a measurable space. Let $\phi : \Omega \to \Omega'$ be measurable and let $f : \Omega' \to X$ be strongly measurable. Let $\nu = \mu \circ \phi^{-1}$ be the image measure of μ under ϕ.

(a) Show that $f \circ \phi$ is strongly measurable.

(b) Show that $f \circ \phi$ is Bochner integrable with respect to μ if and only if f is Bochner integrable with respect to ν, and that in this situation we have

$$\int_\Omega f \circ \phi \, d\mu = \int_{\Omega'} f \, d\nu.$$

1.29 Let $(\Omega, \mathscr{F}, \mu)$ be a probability space. Prove that if $f : \Omega \to X$ is Bochner integrable, then $\int_\Omega f \, d\mu$ is contained in the closed convex hull of $\{f(\omega) : \omega \in \Omega\}$.

2

The Classical Banach Spaces

Before proceeding any further we pause to undertake a detailed study of the classical Banach spaces introduced in the previous chapter.

2.1 Sequence Spaces

Besides the finite-dimensional spaces $\mathbb{K}^d$, perhaps the simplest examples of Banach spaces are provided by the class of sequence spaces. By definition, these are spaces of sequences which, endowed with a suitable norm, turn into Banach spaces. Here we introduce the most important sequence spaces, namely, c_0 and ℓ^p, $1 \leqslant p \leqslant \infty$.

The Spaces c_0 and ℓ^∞ The space c_0 consisting of all scalar sequences $a = (a_k)_{k \geqslant 1}$ satisfying $\lim_{k \to \infty} a_k = 0$ is a Banach space with respect to the supremum norm

$$\|a\|_\infty := \sup_{k \geqslant 1} |a_k|.$$

A justification of this notation is given in the next paragraph. That this is indeed a norm is left as an exercise; the proof of completeness runs as follows. Suppose $(a^{(n)})_{n \geqslant 1}$ is a Cauchy sequence in c_0. Then each coordinate sequence $(a_k^{(n)})_{n \geqslant 1}$ is Cauchy in $\mathbb{K}$ and therefore has a limit which we denote by a_k. We wish to prove that the sequence $a := (a_k)_{k \geqslant 1}$ belongs to c_0 and that $\lim_{n \to \infty} \|a^{(n)} - a\|_\infty = 0$.

Fix $\varepsilon > 0$ and choose N so large that $\|a^{(n)} - a^{(m)}\|_\infty < \varepsilon$ for all $m, n \geqslant N$. Choose N' so large that $|a_k^{(N)}| < \varepsilon$ for all $k \geqslant N'$. Then, for $k \geqslant N'$,

$$|a_k| \leqslant |a_k - a_k^{(N)}| + |a_k^{(N)}| = \lim_{m \to \infty} |a_k^{(m)} - a_k^{(N)}| + |a_k^{(N)}| \leqslant \varepsilon + \varepsilon = 2\varepsilon.$$

It follows that $\lim_{k \to \infty} a_k = 0$, so $a \in c_0$.

Finally, for all $k \geqslant 1$ and $m, n \geqslant N$ we have $|a_k^{(n)} - a_k^{(m)}| < \varepsilon$. Letting $m \to \infty$ while keeping n fixed, for all $k \geqslant 1$ we obtain

$$|a_k^{(n)} - a_k| \leqslant \varepsilon.$$

Taking the supremum over $k \geqslant 1$ we infer that $\|a^{(n)} - a\|_\infty \leqslant \varepsilon$ for all $n \geqslant N$, and the convergence $a^{(n)} \to a$ in c_0 follows.

In the same way one proves that the space ℓ^∞ consisting of all bounded scalar sequences $a = (a_k)_{k \geqslant 1}$ is a Banach space with respect to the supremum norm. This space contains c_0 isometrically as a closed subspace.

The Spaces ℓ^p For $1 \leqslant p < \infty$, the space ℓ^p of scalar sequences $a = (a_k)_{k \geqslant 1}$ satisfying

$$\|a\|_p := \left(\sum_{k \geqslant 1} |a_k|^p \right)^{1/p}$$

is finite is a Banach space with respect to the norm $\| \cdot \|_p$. That this is indeed a norm on ℓ^p is nontrivial; the validity of the triangle inequality $\|a + b\|_p \leqslant \|a\|_p + \|b\|_p$ can be proved by following the line of proof of Proposition 2.19. Completeness of ℓ^p can be proved as in Theorem 2.20. Alternatively, these facts can be deduced as special cases of Proposition 2.19 and Theorem 2.20 by taking $\Omega = \{1, 2, 3, \dots\}$ with the counting measure, that is, the measure which assigns mass 1 to every element of Ω.

It is easy to see (see Problem 2.1) that $1 \leqslant p \leqslant q \leqslant \infty$ implies $\ell^p \subseteq \ell^q$ and

$$\|a\|_q \leqslant \|a\|_p$$

for all $a \in \ell^p$, and that if $a \in \ell^p$ for some $1 \leqslant p < \infty$, then

$$\lim_{\substack{q \to \infty \\ q \geqslant p}} \|a\|_q = \|a\|_\infty.$$

This justifies the notation $\| \cdot \|_\infty$ for the supremum norm.

Remark 2.1. In some applications it is useful to use countable index sets I other than the positive integers. We then define

$$\|a\|_{\ell^p(I)} := \left(\sum_{n \geqslant 1} |a_{i_n}|^p \right)^{1/p},$$

where $(i_n)_{n \geqslant 1}$ is an enumeration of I. This definition is independent of the choice of the enumeration, and the space $\ell^p(I)$ of all mappings $a : I \to \mathbb{K}$ for which this expression is finite is again a Banach space.

2.2 Spaces of Continuous Functions

In this section we study some properties of the space $C(K)$ of continuous functions defined on a compact topological space K.

2.2.a Completeness

It is a standard result in any introductory course in Analysis that the uniform limit of a sequence of continuous functions is continuous. The following theorem recasts this result as a completeness result.

Theorem 2.2 (Completeness). *Let K be a compact topological space. The space $C(K)$ is a Banach space with respect to the* supremum norm

$$\|f\|_\infty := \sup_{x \in K} |f(x)|.$$

The elementary verification that this is indeed a norm is left to the reader. The above supremum is finite (and actually a maximum) since K is compact.

Proof Suppose that $(f_n)_{n \geqslant 1}$ is a Cauchy sequence in $C(K)$. Then for each $x \in K$, $(f_n(x))_{n \geqslant 1}$ is a Cauchy sequence in $\mathbb{K}$ and therefore convergent to some limit in $\mathbb{K}$ which we denote by $f(x)$. We will prove that the function f thus defined is continuous and that $\lim_{n \to \infty} \|f_n - f\|_\infty = 0$.

Fix $\varepsilon > 0$ and choose $N \geqslant 1$ so large that $\|f_n - f_m\|_\infty < \varepsilon$ for all $m, n \geqslant N$. Then in particular for all $m, n \geqslant N$ and all $x \in K$ we have $|f_n(x) - f_m(x)| < \varepsilon$. Passing to the limit $m \to \infty$ while keeping n fixed we obtain

$$|f_n(x) - f(x)| \leqslant \varepsilon. \tag{2.1}$$

Now fix $x \in K$ arbitrary and let $U \subseteq K$ be an open set containing x such that $|f_N(x) - f_N(x')| < \varepsilon$ whenever $x' \in U$. Then, for $x' \in U$,

$$|f(x) - f(x')| \leqslant |f(x) - f_N(x)| + |f_N(x) - f_N(x')| + |f_N(x') - f(x')| \leqslant \varepsilon + \varepsilon + \varepsilon = 3\varepsilon,$$

where we applied (2.1) to $n = N$ and the points x and x'. An argument of this type is called a *3ε-argument*. This proves the continuity of f at the point x. Since $x \in K$ was arbitrary, f is continuous and therefore belongs to $C(K)$. Finally, since (2.1) holds for all $x \in K$ it follows that

$$\|f_n - f\|_\infty = \sup_{x \in K} |f_n(x) - f(x)| \leqslant \varepsilon$$

for all $n \geqslant N$. This proves that $\lim_{n \to \infty} \|f_n - f\|_\infty = 0$. $\qquad\square$

We give three more examples of spaces of functions that are Banach spaces with respect to the supremum norm. The proofs that these spaces are complete are similar to the ones for c_0, ℓ^∞, and $C(K)$, and are left as an exercise.

- The space $B_b(X)$ of bounded Borel measurable functions on a topological space X.
- The space $C_b(X)$ of bounded continuous functions on a topological space X.
- The space $C_0(X)$ of continuous functions on a locally compact topological space X which vanish at infinity (the precise definitions are given in Section 4.1.c).

2.2.b The Stone–Weierstrass Approximation Theorem

The Stone–Weierstrass theorem provides a useful density criterion for the spaces $C(K)$. We begin with the more elementary Weierstrass approximation theorem for $K = [a,b]$.

Theorem 2.3 (Weierstrass approximation theorem). *The polynomials with coefficients in $\mathbb{K}$ are dense in $C[a,b]$.*

Proof By translation and scaling it suffices to prove the theorem for the space $C[0,1]$. Our proof is constructive in that it produces an actual sequence of polynomials approximating a given function. Let $f \in C[0,1]$ be arbitrary and fixed and define the *Bernstein polynomials* associated with f by

$$B_n^{(f)}(x) := \sum_{k=0}^{n} \binom{n}{k} f(\frac{k}{n}) x^k (1-x)^{n-k}, \quad x \in [0,1], \ n \in \mathbb{N}.$$

We will show that $\lim_{n\to\infty} \|B_n^{(f)} - f\|_\infty = 0$. To begin with, the binomial identity

$$\sum_{k=0}^{n} \binom{n}{k} x^k (1-x)^{n-k} = [x + (1-x)]^n = 1 \tag{2.2}$$

implies

$$B_n^{(f)}(x) - f(x) = \sum_{k=0}^{n} \binom{n}{k} x^k (1-x)^{n-k} \left(f(\frac{k}{n}) - f(x) \right).$$

Fix an arbitrary $\varepsilon > 0$. Since f is uniformly continuous there is a real number $0 < \delta < 1$ such that $|f(x) - f(x')| < \varepsilon$ whenever $x, x' \in [0,1]$ satisfy $|x - x'| < \delta$. Fix $x \in [0,1]$ and set $I := \{0 \leqslant k \leqslant n : |\frac{k}{n} - x| < \delta\}$ and $I' = \{0 \leqslant k \leqslant n : k \notin I\}$. The sum over the indices $k \in I$ can be estimated by

$$\sum_{k\in I} \binom{n}{k} x^k (1-x)^{n-k} |f(\frac{k}{n}) - f(x)| \leqslant \varepsilon \sum_{k=0}^{n} \binom{n}{k} x^k (1-x)^{n-k} = \varepsilon,$$

while for $k \in I'$ we have $\delta^2 \leqslant (\frac{k}{n} - x)^2$ and therefore

$$\delta^2 \sum_{k\in I'} \binom{n}{k} x^k (1-x)^{n-k} |f(\frac{k}{n}) - f(x)| \leqslant \sum_{k\in I'} (\frac{k}{n} - x)^2 \binom{n}{k} x^k (1-x)^{n-k} |f(\frac{k}{n}) - f(x)|$$

$$\leqslant 2\|f\|_\infty \sum_{k=0}^{n} (\frac{k}{n} - x)^2 \binom{n}{k} x^k (1-x)^{n-k}$$

$$\overset{(*)}{=} 2\|f\|_\infty \frac{x(1-x)}{n} \leqslant \frac{2}{n} \|f\|_\infty,$$

where in $(*)$ we used the binomial identity (2.2) in combination with the identities

(which are proved by induction on n)

$$\sum_{k=0}^{n} \frac{k}{n} \binom{n}{k} x^k (1-x)^{n-k} = x, \quad \sum_{k=0}^{n} (\frac{k}{n})^2 \binom{n}{k} x^k (1-x)^{n-k} = \frac{n-1}{n} x^2 + \frac{1}{n} x,$$

to see that

$$\sum_{k=0}^{n} (\frac{k}{n} - x)^2 \binom{n}{k} x^k (1-x)^{n-k} = \frac{n-1}{n} x^2 + \frac{1}{n} x - 2x \cdot x + x^2 \cdot 1 = \frac{x(1-x)}{n}.$$

Combining things, we obtain

$$|B_n^{(f)}(x) - f(x)| \leqslant \varepsilon + \frac{2}{\delta^2 n} \|f\|_\infty.$$

Taking the supremum over $x \in [0,1]$ and letting $n \to \infty$, it follows that

$$\limsup_{n \to \infty} \|B_n^{(f)} - f\|_\infty \leqslant \varepsilon.$$

This shows that f can be approximated arbitrarily well by polynomials. $\qquad\square$

Remark 2.4. The same argument shows that if $f : [0,1] \to \mathbb{K}$ is *any* function which is continuous at a point $x_0 \in [0,1]$, then $\lim_{n \to \infty} B_n^{(f)}(x_0) = f(x_0)$.

The proof of Theorem 2.3 has an interesting connection with the law of large numbers. Suppose $\xi_1, \xi_2, \xi_3, \ldots$ are independent identically distributed random variables taking the values 0 and 1 with probability $1 - p$ and p, respectively. Let

$$S_n := \frac{1}{n} \sum_{k=1}^{n} \xi_k.$$

Suppose that $f : [0,1] \to \mathbb{K}$ is continuous at a point $x_0 \in [0,1]$. Denoting expectation and probability by $\mathbb{E}$ and $\mathbb{P}$ respectively,

$$\mathbb{E} f(S_n) = \sum_{k=0}^{n} f(\frac{k}{n}) \cdot \mathbb{P} \left(S_n = \frac{k}{n} \right)$$

and

$$\mathbb{P} \left(S_n = \frac{k}{n} \right) = \binom{n}{k} p^k (1-p)^{n-k}.$$

From Remark 2.4 we therefore obtain

$$\lim_{n \to \infty} \mathbb{E} f(S_n) = \lim_{n \to \infty} \sum_{k=0}^{n} \binom{n}{k} f(\frac{k}{n}) p^k (1-p)^{n-k} = \lim_{n \to \infty} B_n^{(f)}(p) = f(p).$$

In particular we recover the *weak law of large numbers*, which is the assertion that this convergence holds for all $f \in C[0,1]$.

The proof of the Weierstrass theorem using Bernstein polynomials offers little room for generalisation, but the theorem itself does admit a far-reaching generalisation:

Theorem 2.5 (Stone–Weierstrass theorem, algebra version). *Let K be a compact Hausdorff space and suppose that Y is a subspace of* $C(K)$ *with the following properties:*

(i) $1 \in Y$*;*
(ii) $g \in Y$ *implies* $\overline{g} \in Y$*;*
(iii) $g \in Y$ *and* $h \in Y$ *implies* $gh \in Y$*;*
(iv) Y *separates the points of K.*

Then Y is dense in $C(K)$*.*

Karl Weierstrass, 1815–1897

By definition, condition (iv) means that for any two distinct points $x, y \in K$ there exists a function $g \in Y$ such that $g(x) \neq g(y)$.

As a preliminary observation we note that it suffices to prove the theorem for real-valued functions. Indeed, the complex version of the theorem follows from the real version as follows. If $g \in Y$, then the real-valued functions $\operatorname{Re} g = \frac{1}{2}(g + \overline{g})$ and $\operatorname{Im} g = \frac{1}{2i}(g - \overline{g})$ belong to Y. From this it is easy to see that the real-linear space $Y_{\mathbb{R}}$ of all real-valued functions contained in Y satisfies (i)–(iv) again. Now if $f = u + iv \in C(K)$ we may use the real version of the theorem, with Y replaced by $Y_{\mathbb{R}}$, to approximate u and v by functions $u_n, v_n \in Y_{\mathbb{R}}$. Then the functions $u_n + iv_n$ approximate f.

The real version of the theorem will be deduced from its companion where condition (iii) is replaced by closedness under taking pointwise absolute values:

Theorem 2.6 (Stone–Weierstrass theorem, lattice version). *Let K be a compact Hausdorff space and suppose that Y is a subspace of* $C(K)$ *with the following properties:*

(i) $1 \in Y$*;*
(ii) $g \in Y$ *implies* $\overline{g} \in Y$*;*
(iii) $g \in Y$ *implies* $|g| \in Y$*;*
(iv) Y *separates the points of K.*

Then Y is dense in $C(K)$*.*

Proof　Reasoning as before, it suffices to prove the theorem over the real scalars.

For the minimum $a \wedge b := \min\{a, b\}$ and maximum $a \vee b := \max\{a, b\}$ we have the formulas

$$a \wedge b = \frac{1}{2}\big((a+b) - |a-b|\big), \quad a \vee b = \frac{1}{2}\big((a+b) + |a-b|\big).$$

They imply that Y is closed under taking pointwise maxima and minima.

Fix $f \in C(K)$ and $\varepsilon > 0$.

Step 1 – We prove that for each $x \in K$ there exists a function $g_x \in Y$ such that $g_x(x) = f(x)$ and $g_x < f + \varepsilon$ pointwise.

Since Y is a subspace containing the constant functions and separating the points of K, for all $y \in K$ there exists a function $g_{xy} \in Y$ such that $g_{xy}(x) = f(x)$ and $g_{xy}(y) = f(y)$. The set $U_{xy} = \{z \in K : g_{xy}(z) < f(z) + \varepsilon\}$ is open and contains both x and y. Since K is compact, the open cover $\{U_{xy} : y \in K\}$ has a finite subcover, say $\{U_{xy_n} : n = 1, \ldots, N_x\}$. The function $g_x := g_{xy_1} \wedge \cdots \wedge g_{xy_{N_x}}$ has the required properties.

Step 2 – We prove that there exists a function $g \in Y$ such that $f - \varepsilon < g < f + \varepsilon$; this implies $\|f - g\|_\infty \leqslant \varepsilon$ and concludes the proof.

For each $x \in K$ the set $U_x = \{z \in K : f(z) - \varepsilon < g_x(z)\}$ is open and contains x. Since K is compact, the open cover $\{U_x : x \in K\}$ has a finite subcover, say $\{U_{x_n} : n = 1, \ldots, N\}$. The function $g := g_{x_1} \vee \cdots \vee g_{x_N}$ has the required properties. □

Proof of Theorem 2.5 As has already been noted that it suffices to prove the theorem over the real scalar field. Let Y be a subspace of $C(K)$ with the properties (i)–(iv) stated in Theorem 2.5. If we can approximate any $f \in C(K)$ with functions from the closure $\overline{Y}$, we can also approximate with functions from Y. Since $\overline{Y}$ also satisfies the properties (i)–(iv) of Theorem 2.5, we may assume that Y is closed. The strategy of the proof is then to show, under this additional closedness assumption, that Y satisfies the assumptions of Theorem 2.6. For this we need to show that if $f \in Y$, then also $|f| \in Y$.

Fix a function $g \in Y$ and let $\varepsilon > 0$. Since K is compact, the range of g is contained in some compact interval $[a, b]$. By Theorem 2.3 there exists a polynomial $p : [a, b] \to \mathbb{R}$ such that $\|p - q\|_\infty < \varepsilon$, where $q(t) := |t|$ is the absolute value function. Since Y is an algebra containing the constant functions, $p \circ g$ belongs to Y and satisfies $\|p \circ g - |g|\|_{C(K)} < \varepsilon$. Since $\varepsilon > 0$ was arbitrary and Y is closed, it follows that $|g| \in \overline{Y} = Y$. □

Remark 2.7. Theorems 2.5 and 2.6 are stated for compact Hausdorff spaces, but the Hausdorff property was not used in the proofs. Note, however, that the separation-of-points assumptions in these theorems already imply the Hausdorff property.

As a first application of Theorem 2.5 we have the following separability result.

Proposition 2.8. *If K is a compact metric space, then $C(K)$ is separable.*

Proof We must find a countable set in $C(K)$ with dense span. Let $(x_n)_{n \geqslant 1}$ be a countable dense set in K (such a sequence can be realised by covering K, for each integer $k \geqslant 1$, with finitely many open balls of radius $1/k$ using compactness and collecting their centres). We may assume that all points in this sequence are distinct. For all pairs $m \neq n$ the open balls $B_{mn} = B(x_m; \frac{1}{3}d(x_m, x_n))$ and $B_{nm} = B(x_n; \frac{1}{3}d(x_n, x_m))$ have disjoint closures. The collection $\mathscr{B} = \{B_{mn} : m \neq n\}$ is countable and has the property that whenever $x, y \in K$ are two distinct points, they can be separated by two balls contained in $\mathscr{B}$ with disjoint closures. By Urysohn's lemma (Proposition C.11), for any two balls $B_0 := B_{m_0,n_0}$ and $B_1 := B_{m_1,n_1}$ in $\mathscr{B}$ with disjoint closures there exists a function $f \in C(K)$ such that $f \equiv 0$ on B_0 and $f \equiv 1$ on B_1. The subspace Y spanned by the countable set of all finite products of functions of this form and the constant-one function $\mathbf{1}$ satisfies the assumptions of Theorem 2.5 and is therefore dense in $C(K)$.   □

The next two examples give further illustrations of the Stone–Weierstrass theorem.

Example 2.9. The *trigonometric polynomials*, that is, linear combinations of the functions

$$e_n(\theta) := \exp(in\theta), \quad n \in \mathbb{Z},$$

are dense as functions in $C(\mathbb{T})$, where $\mathbb{T}$ denotes the unit circle, which we think of as parametrised with $(-\pi, \pi]$. Indeed, they satisfy the requirements of Theorem 2.5. An explicit procedure to approximate functions in $C(\mathbb{T})$ with trigonometric polynomials is described in Section 3.5.a.

Example 2.10. Let $K_1, \ldots, K_k$ be compact topological spaces. The linear combinations of functions of the form

$$f(x) = f_1(x_1) \cdots f_k(x_k), \quad x = (x_1, \ldots, x_k) \in K_1 \times \cdots \times K_k,$$

with $f_j \in C(K_j)$ for all $j = 1, \ldots, k$, are dense in $C(K_1 \times \cdots \times K_k)$. Indeed, they satisfy the requirements of Theorem 2.5.

2.2.c The Arzelà–Ascoli Compactness Theorem

The next theorem gives a necessary and sufficient condition for relative compactness in $C(K)$. We need the following terminology. A subset $S \subseteq C(K)$ is said to be *equicontinuous* at the point $x \in K$ if for all $\varepsilon > 0$ there exists an open set U in K such that for all $x' \in U$ and $f \in S$ we have $|f(x) - f(x')| < \varepsilon$, and it is said to be *equicontinuous* if it is equicontinuous at every point of K. The set S is said to be *pointwise bounded* if for all $x \in K$ we have $\sup_{f \in S} |f(x)| < \infty$.

Theorem 2.11 (Arzelà–Ascoli). *Let K be a compact topological space. For any subset S of $C(K)$, the following assertions are equivalent:*

(1) *S is relatively compact;*
(2) *S is bounded and equicontinuous;*
(3) *S is pointwise bounded and equicontinuous.*

An equivalent way of formulating the theorem is that a subset of $C(K)$ is compact if and only if it is closed, (pointwise) bounded, and equicontinuous.

Proof (1)⇒(2): Suppose that $S \subseteq C(K)$ is relatively compact. Then obviously S is bounded, so all we need to do is to prove that S is equicontinuous. To this end let $x_0 \in K$ and $\varepsilon > 0$ be arbitrary and fixed. We can cover the compact set $\overline{S}$ with finitely many (say, n) open balls of radius ε. Let $f_1, \ldots, f_n$ be their centres. Using the continuity of these (finitely many) functions we can find an open set U containing x_0 such that $|f_j(x) - f_j(x_0)| < \varepsilon$ for all $x \in U$ and $j = 1, \ldots, n$. Now consider an arbitrary $f \in S$. Choose $j_0 \in \{1, \ldots, n\}$ such that $\|f - f_{j_0}\|_\infty < \varepsilon$; this is possible by the choice of the functions $f_1, \ldots, f_n$. Then for all $x \in U$ we have

$$|f(x) - f(x_0)| \leqslant |f(x) - f_{j_0}(x)| + |f_{j_0}(x) - f_{j_0}(x_0)| + |f_{j_0}(x_0) - f(x_0)| < 3\varepsilon.$$

This verifies the equicontinuity condition.

(2)⇒(3): This implication is trivial.

(3)⇒(1): Let $S \subseteq C(K)$ be pointwise bounded and equicontinuous, and fix $\varepsilon > 0$. By equicontinuity, for every $x \in K$ there is an open set U_x in K such that $|f(x) - f(x')| < \varepsilon$ for all $x' \in U_x$ and $f \in S$. By compactness, finitely many of these open sets cover K, say $U_{x_1}, \ldots, U_{x_k}$. By pointwise boundedness, for each $j = 1, \ldots, k$ the set $\{f(x_j) : f \in S\}$ is bounded. It follows that we can find $c_1, \ldots, c_N \in \mathbb{K}$ such that for all $f \in S$ and $j = 1, \ldots, k$ we have $\min_{1 \leqslant n \leqslant N} |f(x_j) - c_n| < \varepsilon$. Let $\mathcal{N} = \{n = (n_1, \ldots, n_k) : 1 \leqslant n_j \leqslant N$ for all $j = 1, \ldots, k\}$. For $n \in \mathcal{N}$ let

$$B_n = \{f \in S : |f(x_j) - c_{n_j}| < \varepsilon \text{ for all } j = 1, \ldots, k\}.$$

By what we just observed,

$$S = \bigcup_{n \in \mathcal{N}} B_n.$$

Suppose that $f, g \in B_n$ and let $x \in K$ be arbitrary. Then x belongs to at least one of the sets U_{x_j}. Then,

$$|f(x) - g(x)| \leqslant |f(x) - f(x_j)| + |f(x_j) - g(x_j)| + |g(x_j) - g(x)|$$
$$\leqslant \varepsilon + |f(x_j) - c_{n_j}| + |c_{n_j} - g(x_j)| + \varepsilon < 4\varepsilon,$$

where the last inequality holds uniformly with respect to $x \in K$. It follows that $\|f - g\|_\infty < 4\varepsilon$. If, for each $n \in \mathcal{N}$ for which B_n is nonempty, we pick a function $f_n \in B_n$ and consider the open balls $B(f_n; 4\varepsilon)$, we obtain a finite cover of S with 4ε-balls. Since $\varepsilon > 0$ was arbitrary this means that S is totally bounded and hence relatively compact, by Theorem D.10. □

2.2.d Applications to Differential Equations

As an interlude to the main development of the theory, in this section we apply the completeness result of Theorem 2.2 and the compactness result of Theorem 2.11 to study the following initial value problem:

$$\begin{cases} u'(t) = f(t, u(t)), & t \in [0, T], \\ u(0) = u_0, \end{cases} \qquad \text{(IVP)}$$

where $f : [0, T] \times \mathbb{K}^d \to \mathbb{K}^d$ is continuous and $u_0 \in \mathbb{K}^d$ is given.

Global Existence and Uniqueness A *global solution* is continuously differentiable function $u : [0, T] \to \mathbb{K}$ satisfying $u(0) = u_0$ and $u'(t) = f(t, u(t))$ for all $t \in [0, T]$.

Theorem 2.12 (Existence & uniqueness, Picard–Lindelöf). *If $f : [0, T] \times \mathbb{K}^d \to \mathbb{K}^d$ is continuous and there is a constant $L \geqslant 0$ such that for all $t \in [0, T]$ and $x, x' \in \mathbb{K}^d$ we have*

$$|f(t, x) - f(t, x')| \leqslant L|x - x'|,$$

then (IVP) *admits a unique global solution.*

The condition on f is often summarised by saying that f is *Lipschitz continuous in its second variable, uniformly with respect to its first variable.*

The proof of Theorem 2.12 is based on the following abstract fixed point theorem.

Theorem 2.13 (Banach fixed point theorem). *Let X be a complete metric space and let $f : X \to X$ be* uniformly contractive, *that is, there exists a constant $0 \leqslant c < 1$ such that*

$$d(f(x), f(x')) \leqslant c\, d(x, x'), \quad x, x' \in X.$$

Then f has a unique fixed point, that is, there exists a unique element $x \in X$ with the property that $f(x) = x$.

Proof If x and x' are both fixed points, then $d(x, x') = d(f(x), f(x')) \leqslant c\, d(x, x')$, which is only possible if $d(x, x') = 0$, that is, if $x = x'$. It follows that a fixed point, if it exists, is unique.

To prove that a fixed point exists, choose an arbitrary $x_0 \in X$ and define the sequence $(x_n)_{n \geqslant 0}$ by $x_{n+1} := f(x_n)$ for $n \geqslant 0$. We claim that this is a Cauchy sequence. Indeed, for all $n \geqslant 1$ we have

$$d(x_{n+1}, x_n) = d(f(x_n), f(x_{n-1})) \leqslant c\, d(x_n, x_{n-1}),$$

and therefore by induction one sees that $d(x_{n+1}, x_n) \leqslant c^{n-1} d(x_2, x_1)$ for all $n \geqslant 1$. For all $m \geqslant n \geqslant N$ we have

$$d(x_m, x_n) \leqslant d(x_m, x_{m-1}) + \cdots + d(x_{n+1}, x_n)$$

$$\leqslant (c^{m-2} + \cdots + c^{n-1}) \cdot d(x_2, x_1) \leqslant \left(\sum_{k=N-1}^{\infty} c^k \right) \cdot d(x_2, x_1) = \frac{c^{N-1}}{1 - c} \cdot d(x_2, x_1),$$

and the right-hand side can be made small by taking N large. This proves the claim.

Since X was assumed to be complete, the sequence $(x_n)_{n \geqslant 1}$ converges in X. Let x be its limit. Then the continuity of f implies $f(x) = \lim_{n \to \infty} f(x_n) = \lim_{n \to \infty} x_{n+1} = x$, which shows that x is a fixed point for f. $\qquad\square$

By $C([0,T]; \mathbb{K}^d)$ we denote the space of all continuous functions $f : [0,T] \to \mathbb{K}^d$. Endowed with the supremum norm, this space is a Banach space. Indeed, suppose that $(f^{(n)})_{n \geqslant 1}$ is a Cauchy sequence in $C([0,T]; \mathbb{K}^d)$. Then the d sequences of coordinate functions $(f_j^{(n)})_{n \geqslant 1}$ are Cauchy in $C[0,T]$ and therefore converge to limits f_j in $C[0,T]$. This easily implies that the sequence $(f_{n \geqslant 1}^{(n)}$ converges in $C([0,T]; \mathbb{K}^d)$ to the function f with coordinate functions f_j.

We will use the Banach fixed point theorem to prove that the (nonlinear) mapping $I_T : C([0,T]; \mathbb{K}^d) \to C([0,T]; \mathbb{K}^d)$ defined by

$$(I_T u)(t) := u_0 + \int_0^t f(s, u(s)) \, ds, \quad t \in [0,T],$$

where the integral is interpreted as a $\mathbb{K}^d$-valued Riemann integral, has a fixed point. This will prove the theorem in view of the next lemma.

Lemma 2.14. *A function $u \in C([0,T]; \mathbb{K}^d)$ satisfies* (IVP) *for all $t \in [0,T]$ if and only if u is a fixed point of I_T.*

Proof Indeed, u is a fixed point of I_T if and only if $u(t) = u_0 + \int_0^t f(s, u(s)) \, ds$ for all $t \in [0,T]$. By integration, this identity holds if u is a solution, and conversely if the identity holds, then u is continuously differentiable (since the right-hand side is) and by differentiation we obtain that u is a solution. $\qquad\square$

Proof of Theorem 2.12 Let us start with a preliminary estimate that will be refined shortly. For all $u, v \in C([0,T]; \mathbb{K}^d)$ and all $t \in [0,T]$ we have

$$|(I_T(u))(t) - (I_T(v))(t)| = \left| \int_0^t f(s, u(s)) - f(s, v(s)) \, ds \right|$$
$$\leqslant \int_0^t L|u(s) - v(s)| \, ds \leqslant \int_0^t L\|u - v\| \, ds \leqslant LT \|u - v\|.$$

Taking the supremum over $t \in [0,T]$ we find that

$$\|I_T(u) - I_T(v)\| \leqslant LT \|u - v\|.$$

If $LT < 1$, then I_T is uniformly contractive and the Banach fixed point theorem guarantees the existence of a unique fixed point. This proves Theorem 2.12 in the special case that the smallness condition $LT < 1$ is satisfied.

To get around this condition we modify the norm of $C([0,T]; \mathbb{K}^d)$. Fix real number $\lambda > 0$ (in a moment we will see that we need $\lambda > L$), and define

$$\|f\|_\lambda := \sup_{t \in [0,T]} e^{-\lambda t} |f(t)|.$$

It is clear that this defines a norm on $C([0,T];\mathbb{K}^d)$ and we have

$$e^{-\lambda T}\|f\| \leqslant \|f\|_\lambda \leqslant \|f\|.$$

This implies that a sequence in $C([0,T];\mathbb{K}^d)$ is Cauchy with respect to the norm $\|\cdot\|_\lambda$ if and only if it is Cauchy with respect to the norm $\|\cdot\|$, and since $C([0,T];\mathbb{K}^d)$ is complete with respect to the latter, we conclude that $C([0,T];\mathbb{K}^d)$ is a Banach space with respect to the norm $\|\cdot\|_\lambda$. Using this norm, we redo the above computations and find

$$
\begin{aligned}
\|I_T(u) - I_T(v)\|_\lambda &= \sup_{t\in[0,T]} e^{-\lambda t}|(I_T(u))(t) - (I_T(v))(t)| \\
&\leqslant \sup_{t\in[0,T]} e^{-\lambda t} \int_0^t L e^{\lambda s} e^{-\lambda s}|u(s) - v(s)|\,\mathrm{d}s \\
&\leqslant \sup_{t\in[0,T]} e^{-\lambda t}\|u-v\|_\lambda \int_0^t L e^{\lambda s}\,\mathrm{d}s \\
&= \sup_{t\in[0,T]} e^{-\lambda t}\|u-v\|_\lambda \cdot \frac{L}{\lambda}(e^{\lambda t} - 1) \leqslant \frac{L}{\lambda}\|u-v\|_\lambda.
\end{aligned}
$$

Hence if we choose $\lambda > L$, then I_T is uniformly contractive on $C([0,T];\mathbb{K}^d)$ with respect to the norm $\|\cdot\|_\lambda$. Now an application of the Banach fixed point theorem produces a unique fixed point for I_T. $\square$

Remark 2.15. Theorem 2.12 remains true if we replace the interval $[0,T]$ by $[0,\infty)$ and assume that $f:[0,\infty)\times\mathbb{K}^d \to \mathbb{K}^d$ satisfies

$$|f(t,x) - f(t,x')| \leqslant L|x - x'|$$

for all $t \in [0,\infty)$ and $x,x' \in \mathbb{K}^d$. Indeed, the preceding argument produces a solution u_T on every interval $[0,T]$. We may now define $u:[0,\infty) \to \mathbb{K}^d$ by setting $u := u_T$ on the interval $[0,T]$. Since by uniqueness we have $u_T = u_S$ on $[0,S\wedge T]$, this is well defined. The resulting function is continuously differentiable and satisfies (IVP) on every interval $[0,T]$, hence on all of $[0,\infty)$, and is therefore a global solution on $[0,\infty)$.

Remark 2.16. All that has been said extends to the case where $\mathbb{K}^d$ is replaced by a Banach space X. This equally pertains to the results of the next paragraph.

Local Existence As an application of Theorem 2.11 we present a local existence result for differential equations with continuous right-hand side. In contrast to the situation in Theorem 2.12, where a Lipschitz continuity assumption was made, we do not get uniqueness of the solution.

The problem (IVP) *admits a local solution* if there exists $0 < \delta \leqslant T$ and a continuously differentiable function $u:[0,\delta] \to \mathbb{K}^d$ satisfying $u(0) = u_0$ and $u'(t) = f(t,u(t))$ for all $t \in [0,\delta]$.

Theorem 2.17 (Local existence, Peano). *If $f : [0,T] \times \mathbb{K}^d \to \mathbb{K}^d$ is continuous, then the problem* (IVP) *admits a local solution.*

A global solution need not always exist. Indeed, $u(t) = 1/(1-t), t \in [0,\delta]$ with $0 < \delta < 1$, is a local solution of the problem

$$\begin{cases} u'(t) = (u(t))^2, & t \in [0,1], \\ u(0) = 1, \end{cases}$$

but this problem does not have a global solution. This follows from the fact that a local solution on a subinterval $[0,\delta]$, if one exists, is unique. To see this, suppose that u_1 and u_2 are two solutions on $[0,\delta]$. Both u_1 and u_2 are continuous, hence bounded, let us say by the constant M. Consider now the function $\widetilde{\phi}(x) := \min\{x^2, M\}$. This function is globally Lipschitz continuous on $[0,\delta] \times \mathbb{R}$, and therefore the problem

Giuseppe Peano, 1858–1932

$$\begin{cases} u'(t) = \widetilde{\phi}(u(t)), & t \in [0,\delta], \\ u(0) = 1 \end{cases}$$

has a unique solution on $[0,\delta]$, say u. On the other hand u_1 and u_2 are solutions on $[0,\delta]$, because $\widetilde{\phi}(u_1(t)) = (u_1(t))^2$ and $\widetilde{\phi}(u_2(t)) = (u_2(t))^2$ for all $t \in [0,\delta]$, and therefore we must have $u_1 = u = u_2$ on $[0,\delta]$. The upshot of all this is that if a global solution exists, it must be equal to $1/(1-t)$ on every subinterval $[0,\delta]$, hence on the interval $[0,1)$; but the function $1/(1-t)$ cannot be extended to a continuous function on $[0,1]$.

The above uniqueness proof for local solutions made use of the fact that the right-hand side was locally Lipschitz continuous in the second variable in a neighbourhood of the initial value. In general, however, a local solution need not be unique. Indeed, both $u_1(t) = 0$ and $u_2(t) = t^{3/2}$ are solutions of the problem

$$\begin{cases} u'(t) = \dfrac{3}{2}(u(t))^{1/3}, & t \in [0,1], \\ u(0) = 0. \end{cases}$$

The function $\phi(t,x) := \frac{3}{2}x^{1/3}$ fails to be locally Lipschitz continuous in the second variable in a neighbourhood of the initial value 0.

As a first step towards the proof of Theorem 2.17 we show that the problem (IVP) is equivalent to an integrated version of it. For the remainder of this section we assume that f is continuous.

Lemma 2.18. *A function $u \in C([0,\delta];\mathbb{K}^d)$ is a local solution of* (IVP) *if and only if for all $t \in [0,\delta]$ we have*

$$u(t) = u_0 + \int_0^t f(s,u(s))\,\mathrm{d}s. \tag{2.3}$$

The function $s \mapsto f(s,u(s))$ is continuous on $[0,\delta]$, so the integral is well defined as a Riemann integral with values in $\mathbb{K}^d$.

Proof If $u \in C([0,\delta];\mathbb{K}^d)$ is a local solution, then for all $t \in [0,\delta]$ we have

$$u(t) - u_0 = u(t) - u(0) = \int_0^t u'(s)\,\mathrm{d}s = \int_0^t f(s,u(s))\,\mathrm{d}s$$

and (2.3) holds for all $t \in [0,\delta]$. Conversely, if $u \in C([0,\delta];\mathbb{K}^d)$ satisfies (2.3) for all $t \in [0,\delta]$, then u is continuously differentiable on $[0,\delta]$. Using $u(0) = u_0$ and differentiating, we find

$$u'(t) = \frac{\mathrm{d}}{\mathrm{d}t}\int_0^t f(s,u(s))\,\mathrm{d}s = f(t,u(t))$$

for all $t \in [0,\delta]$, that is, (IVP) holds. $\square$

Now we are ready for the proof of Theorem 2.17. It relies on a compactness argument. The idea is to construct, for small enough $\delta \in (0,T]$, a sequence of approximate solutions $(u_n)_{n\geqslant1}$ in $C([0,\delta];\mathbb{K}^d)$ and show its equicontinuity (the definition of which extends to vector-valued functions in the obvious way). An appeal to the Arzelà–Ascoli theorem (which extends to the vector-valued case as well, without change in the proof) then produces a subsequence $(u_{n_k})_{k\geqslant1}$ that converges in $C([0,\delta];\mathbb{K}^d)$. The limit $u \in C([0,\delta];\mathbb{K}^d)$ will be shown to solve (IVP) on the interval $[0,\delta]$.

Proof of Theorem 2.17 Let $M := \sup_{(t,x)\in[0,T]\times\overline{B}(u_0;1)} |f(t,x)|$ and $\delta := \min\{T,1/M\}$. For $n = 1,2,\ldots$ we equipartition the interval $[0,\delta]$ using the partition points $t_{j,n} := jh_n$ for $j = 0,\ldots,2^n$, where $h_n := 2^{-n}\delta$, and define inductively

$$u_n(0) := u_0, \quad u_n(t_{j+1,n}) := u_n(t_{j,n}) + h_n f(t_{j,n},u_n(t_{j,n})), \quad j = 0,\ldots,2^{n-1}. \tag{2.4}$$

For the remaining values of $t \in [0,\delta]$ we define $u_n(t)$ by piecewise linear interpolation. Since each u_n is piecewise continuously differentiable with derivatives bounded by M (by an inductive argument based on (2.4), each $u(t_{j,n})$ belongs to $\overline{B}(u_0;1)$), we have

$$|u_n(t) - u_n(s)| \leqslant M|t-s|, \quad s,t \in [0,\delta].$$

This implies that the functions u_n are equicontinuous. The estimate

$$|u_n(t)| \leqslant |u_n(t) - u_n(0)| + |u_n(0)| \leqslant M\delta + |u_0| \leqslant 1 + |u_0|, \quad t \in [0,\delta],$$

shows that they are also uniformly bounded. By the Arzelà–Ascoli theorem, some subsequence $(u_{n_k})_{k\geqslant1}$ converges to a limiting function u in $C([0,\delta];\mathbb{K}^d)$.

Since f is uniformly continuous on $[0,T] \times \overline{B}(u_0; 1)$ we have $\lim_{n\to\infty} C_n = 0$, where

$$C_n := \sup_{|t-s| \leqslant h_n} \sup_{|x-y| \leqslant h_n M} |f(t,x) - f(s,y)|.$$

Writing

$$u_n(t) = u_0 + \sum_{j=0}^{2^n-1} \int_{t_{j,n}}^{t_{j+1,n}} \mathbf{1}_{[0,t]}(s) f(t_{j,n}, u_n(t_{j,n})) \, \mathrm{d}s, \quad t \in [0,\delta],$$

we see that

$$\left| u_n(t) - \left(u_0 + \int_0^t f(s, u_n(s)) \, \mathrm{d}s \right) \right|$$

$$\leqslant \sum_{j=0}^{2^n-1} \int_{t_{j,n}}^{t_{j+1,n}} \mathbf{1}_{[0,t]}(s) |f(t_{j,n}, u_n(t_{j,n})) - f(s, u_n(s))| \, \mathrm{d}s$$

$$\leqslant C_n \sum_{j=0}^{2^n-1} \int_{t_{j,n}}^{t_{j+1,n}} \mathbf{1}_{[0,t]}(s) \, \mathrm{d}s \leqslant C_n 2^n h_n = C_n \delta.$$

The right-hand side tends to 0 as $n \to \infty$. Taking limits, it follows that

$$u(t) = \lim_{k\to\infty} u_{n_k}(t) = u_0 + \lim_{k\to\infty} \int_0^t f(s, u_{n_k}(s)) \, \mathrm{d}s = u_0 + \int_0^t f(s, u(s)) \, \mathrm{d}s,$$

and therefore u solves the integrated version of (IVP) on $[0,\delta]$. By Lemma 2.18, u then solves (IVP) on the interval $[0,\delta]$. $\qquad\square$

2.3 Spaces of Integrable Functions

Let $(\Omega, \mathscr{F}, \mu)$ be a measure space and fix $1 \leqslant p < \infty$. We define $\mathscr{L}^p(\Omega)$ as the set of all measurable functions $f : \Omega \to \mathbb{K}$ such that

$$\int_\Omega |f|^p \, \mathrm{d}\mu < \infty.$$

For such functions we set

$$\|f\|_p := \left(\int_\Omega |f|^p \, \mathrm{d}\mu \right)^{1/p}.$$

For $p = \infty$ we define $\mathscr{L}^\infty(\Omega)$ as the set of all measurable functions $f : \Omega \to \mathbb{K}$ that are *μ-essentially bounded*, meaning that there is set $N \in \mathscr{F}$ of μ-measure 0 such that f is bounded on $\complement N$. For such functions we define $\|f\|_\infty$ as the *μ-essential supremum* of f,

$$\|f\|_\infty := \mu\text{-}\mathrm{ess\,sup}_{\omega \in \Omega} |f(\omega)| := \inf\{r > 0 : |f| \leqslant r \ \mu\text{-almost everywhere}\}.$$

When there is no risk of confusion, the measure μ is omitted from this notation.

The spaces $\mathscr{L}^p(\Omega)$ are vector spaces:

Proposition 2.19 (Minkowski inequality). *Let*
$1 \leqslant p \leqslant \infty$. *For all functions* $f, g \in \mathscr{L}^p(\Omega)$ *we have* $f + g \in \mathscr{L}^p(\Omega)$ *and*

$$\|f + g\|_p \leqslant \|f\|_p + \|g\|_p.$$

Proof   The result is trivial for $p = \infty$, so we only consider the case $1 \leqslant p < \infty$.

By elementary calculus it is checked that for all nonnegative real numbers a and b one has

Henri Lebesgue, 1875–1941

$$(a + b)^p = \inf_{t \in (0,1)} t^{1-p} a^p + (1-t)^{1-p} b^p.$$

Applying this identity to $|f(\omega)|$ and $|g(\omega)|$ with $\omega \in \Omega$ and integrating with respect to μ, for all fixed $t \in (0, 1)$ we obtain

$$\int_\Omega |f + g|^p \, d\mu \leqslant \int_\Omega (|f| + |g|)^p \, d\mu \leqslant t^{1-p} \int_\Omega |f|^p \, d\mu + (1-t)^{1-p} \int_\Omega |g|^p \, d\mu.$$

Stated differently, this says that

$$\|f + g\|_p^p \leqslant t^{1-p} \|f\|_p^p + (1-t)^{1-p} \|g\|_p^p.$$

Taking the infimum over all $t \in (0, 1)$ gives the result.   □

In spite of this result, $\|\cdot\|_p$ is not a norm on $\mathscr{L}^p(\Omega)$, because $\|f\|_p = 0$ only implies that $f = 0$ μ-almost everywhere. In order to get around this imperfection, we define an equivalence relation $\sim$ on $\mathscr{L}^p(\Omega)$ by

$$f \sim g \Leftrightarrow f = g \ \mu\text{-almost everywhere}.$$

The equivalence class of a function f modulo $\sim$ is denoted by $[f]$. On the quotient space

$$L^p(\Omega) := \mathscr{L}^p(\Omega) / \sim,$$

whose elements are the equivalence classes $[f]$ of functions $f \in \mathscr{L}^p(\Omega)$, we define a scalar multiplication and addition in the natural way:

$$c[f] := [cf], \quad [f] + [g] := [f + g].$$

We leave it as an exercise to check that both operations are well defined. With these operations, $L^p(\Omega)$ is a normed vector space with respect to the norm

$$\|[f]\|_p := \|f\|_p.$$

When we explicitly wish to express the dependence on $\mathscr{F}$ or μ we write $L^p(\Omega, \mathscr{F})$ or $L^p(\Omega, \mu)$. *Following common practice we make no distinction between functions in*

$\mathscr{L}^p(\Omega)$ *and their equivalence classes in* $L^p(\Omega)$, and call the latter "functions" as well. In the same vein, we will not hesitate to talk about the "sets"

$$\{\omega \in \Omega : f(\omega) \in B\}$$

when $B \subseteq \mathbb{K}$ is a Borel set and f is an element of $L^p(\Omega)$. The rigorous interpretation is that it defines an equivalence classes of sets in $\mathscr{F}$, the representatives of which are obtained by selecting pointwise defined measurable representatives for f.

2.3.a Completeness

For the remainder of this section we fix a measure space $(\Omega, \mathscr{F}, \mu)$. The main result of this section is the following completeness result for the spaces $L^p(\Omega)$.

Theorem 2.20 (Completeness). *For all* $1 \leqslant p \leqslant \infty$ *the normed space* $L^p(\Omega)$ *is complete.*

Proof First let $1 \leqslant p < \infty$, and let $(f_n)_{n \geqslant 1}$ be a Cauchy sequence with respect to the norm $\|\cdot\|_p$ of $L^p(\Omega)$. By passing to a subsequence we may assume that

$$\|f_{n+1} - f_n\|_p \leqslant \frac{1}{2^n}, \quad n = 1, 2, \ldots$$

Define the nonnegative measurable functions

$$g_N := \sum_{n=0}^{N-1} |f_{n+1} - f_n|, \quad g := \sum_{n=0}^{\infty} |f_{n+1} - f_n|,$$

with the convention that $f_0 = 0$. By the monotone convergence theorem,

$$\int_{\Omega} g^p \, d\mu = \lim_{N \to \infty} \int_{\Omega} g_N^p \, d\mu.$$

Taking pth roots and using Minkowski's inequality we obtain

$$\|g\|_p = \lim_{N \to \infty} \|g_N\|_p \leqslant \lim_{N \to \infty} \sum_{n=0}^{N-1} \|f_{n+1} - f_n\|_p = \sum_{n=0}^{\infty} \|f_{n+1} - f_n\|_p \leqslant 1 + \|f_1\|_p.$$

If follows that g is finitely valued μ-almost everywhere, which means that the sum defining g converges absolutely μ-almost everywhere. As a result, the sum

$$f := \sum_{n=0}^{\infty} (f_{n+1} - f_n)$$

converges on the set $\{g < \infty\}$. On this set we have

$$f = \lim_{N \to \infty} \sum_{n=0}^{N-1} (f_{n+1} - f_n) = \lim_{N \to \infty} f_N.$$

Defining f to be identically zero on the null set $\{g = \infty\}$, the resulting function f is measurable. From

$$|f_N|^p = \left| \sum_{n=0}^{N-1} (f_{n+1} - f_n) \right|^p \leqslant \left(\sum_{n=0}^{N-1} |f_{n+1} - f_n| \right)^p \leqslant |g|^p$$

it follows $|f|^p \leqslant |g|^p$ and hence

$$|f - f_N|^p \leqslant 2^p (\tfrac{1}{2}(|f| + |f_N|))^p \leqslant 2^p \cdot \tfrac{1}{2}(|f|^p + |f_N|^p) \leqslant 2^p |g|^p,$$

using the convexity of $t \mapsto t^p$ (recall that a function $f : I \to \mathbb{R}$, where I is an interval, is called *convex* if for all $x_0, x_1 \in I$ and $0 \leqslant \lambda \leqslant 1$ we have $f((1 - \lambda)x_0 + \lambda x_1) \leqslant (1 - \lambda)f(x_0) + \lambda f(x_1)$). From the dominated convergence theorem we conclude that

$$\lim_{N \to \infty} \|f - f_N\|_p = 0.$$

We have proved that a subsequence of the original Cauchy sequence converges to f in $L^p(\Omega)$. As is easily verified, this implies that the original Cauchy sequence converges to f as well. This completes the proof for exponents $1 \leqslant p < \infty$.

It remains to establish the result for $p = \infty$. Let $(f_n)_{n \geqslant 1}$ be a Cauchy sequence with respect to the norm $\| \cdot \|_\infty$ of $L^\infty(\Omega)$. By passing to a subsequence we may assume that

$$\|f_{n+1} - f_n\|_\infty \leqslant \frac{1}{2^n}, \quad n = 1, 2, \ldots$$

Choose μ-null sets F_n such that $|f_{n+1}(\omega) - f_n(\omega)| \leqslant \frac{1}{2^n}$ for all $\omega \in \complement F_n$. Defining the functions g_N and g as before, we note that outside the μ-null set $F := \bigcup_{n \geqslant 1} F_n$ we have uniform convergence $g_n \to g$. Defining the function f as before, this implies that $f_N \to f$ uniformly outside F. This, in turn, means that $f_N \to f$ in $L^\infty(\Omega)$. $\qquad\square$

In the course of the proof we obtained the following result:

Corollary 2.21. *Every convergent sequence $(f_n)_{n \geqslant 1}$ in $L^p(\Omega)$, with $1 \leqslant p \leqslant \infty$, has a μ-almost everywhere convergent subsequence $(f_{n_k})_{k \geqslant 1}$, and this subsequence may be chosen to satisfy $|f_{n_k}| \leqslant g$ almost everywhere for some fixed $0 \leqslant g \in L^p(\Omega)$.*

In the majority of applications the first part of this corollary suffices, but the second part is sometimes helpful in setting the stage for an application of the dominated convergence theorem.

Remark 2.22. Except when $p = \infty$, in the setting of Corollary 2.21 it need not be the case that the sequence $(f_n)_{n \geqslant 1}$ itself is μ-almost everywhere convergent to its $L^p(\Omega)$-limit f (see Problems 2.13 and 2.14).

The inequality in the next result is known as *Hölder's inequality*. For $p = q = 2$ and $r = 1$ it reduces to a special case of the Cauchy–Schwarz inequality (see Proposition 3.3).

Proposition 2.23 (Hölder's inequality). *Let* $1 \leqslant p, q, r \leqslant \infty$ *satisfy* $\frac{1}{p} + \frac{1}{q} = \frac{1}{r}$. *If* $f \in L^p(\Omega)$ *and* $g \in L^q(\Omega)$, *then* $fg \in L^r(\Omega)$ *and*

$$\|fg\|_r \leqslant \|f\|_p \|g\|_q.$$

For $r = 1$ the condition on p and q reads $\frac{1}{p} + \frac{1}{q} = 1$; we call such p and q *conjugate exponents*.

Proof It suffices to prove the inequality for $r = 1$; the general case follows by applying this special case to the functions $|f|^r$ and $|g|^r$.

For $p = 1, q = \infty$ and for $p = \infty, q = 1$, the first inequality follows by a direct estimate. Thus we may assume from now on that $1 < p, q < \infty$. The inequality is then proved in the same way as Minkowski's inequality, this time using the identity

$$ab = \inf_{t>0} \left(\frac{t^p a^p}{p} + \frac{b^q}{qt^q} \right).$$

$\square$

Remark 2.24. Let $1 \leqslant p_1, \ldots, p_N, r \leqslant \infty$ satisfy $\frac{1}{p_1} + \cdots + \frac{1}{p_N} = \frac{1}{r}$. If $f_n \in L^{p_n}(\Omega)$ for $n = 1, \ldots, N$, then $\prod_{n=1}^N f_n \in L^r(\Omega)$ and

$$\left\| \prod_{n=1}^N f_n \right\|_r \leqslant \prod_{n=1}^N \|f_n\|_{p_n}.$$

This more general version of Hölder's inequality follows from Proposition 2.23 by an easy induction argument.

As an immediate corollary of Hölder's inequality we have the following result.

Corollary 2.25. *Let* $1 \leqslant p, q, r \leqslant \infty$ *satisfy* $\frac{1}{p} + \frac{1}{q} = \frac{1}{r}$. *Then the mapping*

$$(f, g) \mapsto fg$$

is jointly continuous from $L^p(\Omega) \times L^q(\Omega)$ *into* $L^r(\Omega)$. *In particular, if* $1 \leqslant p, q \leqslant \infty$ *satisfy* $\frac{1}{p} + \frac{1}{q} = 1$ *and* $f_n \to f$ *in* $L^p(\Omega)$, *then for all* $g \in L^q(\Omega)$ *we have*

$$\lim_{n \to \infty} \int_\Omega f_n g \, d\mu = \int_\Omega fg \, d\mu.$$

Proof If $f_n \to f$ in $L^p(\Omega)$ and $g_n \to g$ in $L^q(\Omega)$, then Hölder's inequality implies that $f_n g_n, f_n g, f g_n, fg$ belong to $L^r(\Omega)$ and

$$\|f_n g_n - fg\|_r \leqslant \|(f_n - f)g\|_r + \|f_n(g_n - g)\|_r$$
$$\leqslant \|f_n - f\|_p \|g\|_q + M\|g_n - g\|_q \to 0 \quad \text{as } n \to \infty,$$

where $M := \sup_{n \geqslant 1} \|f_n\|_p$. By Proposition D.8, this proves the asserted continuity. $\square$

A useful special case of Hölder's inequality concerns the case of a finite measure. If $\mu(\Omega) < \infty$ and $1 \leqslant r \leqslant p \leqslant \infty$, then Hölder's inequality implies that if $f \in L^p(\Omega)$, then $f \in L^r(\Omega)$ and

$$\|f\|_r \leqslant \mu(\Omega)^{\frac{1}{r} - \frac{1}{p}} \|f\|_p.$$

In the case of a probability measure μ this takes the simpler form $\|f\|_r \leqslant \|f\|_p$.

The following result provides a converse to Hölder's inequality. We formulate it for exponents $\frac{1}{p} + \frac{1}{q} = 1$; as in the proof of Hölder's inequality, this implies a more general version for exponents $\frac{1}{p} + \frac{1}{q} = \frac{1}{r}$. A further variation will be given in Proposition 5.13.

Proposition 2.26. *Let $1 \leqslant p, q \leqslant \infty$ satisfy $\frac{1}{p} + \frac{1}{q} = 1$. Let (Ω, μ) be a measure space, which is assumed to be σ-finite if $p = \infty$. A measurable function f belongs to $L^p(\Omega)$ if and only if*

$$fg \in L^1(\Omega) \quad \text{and} \quad \|fg\|_1 \leqslant M\|g\|_q$$

for some constant $M \geqslant 0$ and all $g \in Y$, where Y is a dense subspace of $L^q(\Omega)$. In that case we have $\|f\|_p \leqslant M$.

Proof The 'only if' part is immediate from Hölder's inequality. To prove the 'if' part we may assume that f is not identically 0.

Step 1 – By assumption, the mapping $g \mapsto fg$ is bounded, of norm at most M, as a mapping from the dense subspace Y of $L^q(\Omega)$ to $L^1(\Omega)$. Hence by Proposition 1.18 it admits a unique extension to a bounded operator, of norm at most M, from $L^q(\Omega)$ to $L^1(\Omega)$. Denote this operator by T. If $g_n \to g$ in $L^q(\Omega)$ with each g_n in Y, then $Tg = \lim_{n \to \infty} Tg_n = \lim_{n \to \infty} fg_n$ with convergence in $L^1(\Omega)$. Using Corollary 2.21 we may pass to a subsequence such that $g_{n_k} \to g$ and $fg_{n_k} \to Tg$ μ-almost everywhere, and therefore

$$Tg = \lim_{n \to \infty} fg_n = fg \quad \mu\text{-almost everywhere}.$$

This also implies that $fg \in L^1(\Omega)$ for all $g \in L^q(\Omega)$ and

$$\|fg\|_1 \leqslant M\|g\|_q, \quad g \in L^q(\Omega). \tag{2.5}$$

Step 2 – In this step we prove the proposition for $1 \leqslant p < \infty$ by showing that

$$\int_\Omega |f|^p \, d\mu \leqslant M^p. \tag{2.6}$$

To this end let ϕ be a μ-simple function satisfying $0 \leqslant \phi \leqslant |f|$ μ-almost everywhere, say $\phi = \sum_{j=1}^k c_j \mathbf{1}_{F_j}$ with coefficients $c_j \in \mathbb{K}$ and the sets F_j disjoint and of finite measure. We first prove that

$$\int_\Omega |\phi|^p \, d\mu \leqslant M^p. \tag{2.7}$$

If $\int_\Omega |\phi|^p \, d\mu = 0$ this inequality trivially holds, so we may assume that $\int_\Omega |\phi|^p \, d\mu > 0$. To prove (2.7) in this case, set $g := |\phi|^{p-1}$ (with $g := 1$ if $p = 1$). Then

$$\int_\Omega |\phi|^p \, d\mu = \int_\Omega |\phi| g \, d\mu \leqslant \int_\Omega |f| g \, d\mu = \|fg\|_1 \leqslant M \|g\|_q. \tag{2.8}$$

For $p = 1$ we have $\|g\|_q = \|\mathbf{1}\|_\infty = 1$ and (2.7) follows from (2.8). For $1 < p < \infty$ we have $1 < q < \infty$ and

$$\|g\|_q^q = \sum_{j=1}^k |c_j|^{(p-1)q} \mu(F_j) = \sum_{j=1}^k |c_j|^p \mu(F_j) = \int_\Omega |\phi|^p \, d\mu.$$

Taking qth roots on both sides and substituting the result into (2.8), we obtain

$$\int_\Omega |\phi|^p \, d\mu \leqslant M \left(\int_\Omega |\phi|^p \, d\mu \right)^{1/q},$$

which is the same as saying that (2.7) holds.

Now let $0 \leqslant \phi_n \uparrow |f|$ μ-almost everywhere in (2.7), with each ϕ_n a μ-simple function. Applying the previous inequality to ϕ_n, the monotone convergence theorem gives (2.6). This proves that $f \in L^p(\Omega)$ and $\|f\|_p \leqslant M$. This completes the proof for $1 \leqslant p < \infty$.

Step 3 – Suppose next that $p = \infty$ and $(\Omega, \mathscr{F}, \mu)$ is σ-finite. Suppose, for a contradiction, that f does not belong to $L^\infty(\Omega)$. Then for all $n = 1, 2, \ldots$ the set $A_n := \{|f| \geqslant n\}$ has strictly positive measure. If $\Omega = \bigcup_{j \geqslant 1} B_j$ with $\mu(B_j) < \infty$ for all j (such sets exist by σ-finiteness), then for each n there must be an index $j = j_n$ such that $A_n \cap B_{j_n}$ has strictly positive (and finite) measure μ_n. Then $g_n := \frac{1}{\mu_n} \mathbf{1}_{A_n \cap B_{j_n}}$ belongs to $L^1(\Omega)$ and has norm one, and we have

$$\|fg_n\|_1 = \frac{1}{\mu_n} \int_{A_n \cap B_{j_n}} |f| \, d\mu \geqslant n = n \|g_n\|_1.$$

This contradicts (2.5). $\qquad\square$

Remark 2.27. The argument of Step 3 proves, more generally, that if $(\Omega, \mathscr{F}, \mu)$ is a σ-finite measure space and $1 \leqslant p \leqslant \infty$, then a measurable function f belongs to $L^\infty(\Omega)$ if and only if $fg \in L^p(\Omega)$ and $\|fg\|_p \leqslant M \|g\|_p$ for some constant $M \geqslant 0$ and all $g \in Y$, where Y is a dense subspace of $L^p(\Omega)$; in that case we have $\|f\|_\infty \leqslant M$.

2.3.b Approximation by Mollification

It is generally difficult to handle L^p-functions directly. There are two ways of dealing with this problem: by approximation it often suffices to consider functions that are easier to deal with, and by interpolation one can reduce matters to exponents that are easier to deal with. The present section is devoted to approximation techniques; interpolation is treated in Section 5.7.

We begin by proving that the μ-simple functions are dense in $L^p(\Omega)$ for $1 \leqslant p < \infty$. Recall from Definition 1.49 that a *μ-simple function* is a simple function supported on sets of finite μ-measure.

Proposition 2.28 (Approximation by μ-simple functions). *For $1 \leqslant p < \infty$, the μ-simple functions are dense in $L^p(\Omega)$. The same result holds for $L^\infty(\Omega)$ if $\mu(\Omega) < \infty$.*

Proof Fix a function $f \in L^p(\Omega)$.

First let $1 \leqslant p < \infty$. By dominated convergence we have

$$\lim_{n \to \infty} \mathbf{1}_{\{\frac{1}{n} \leqslant |f| \leqslant n\}} f = f$$

in $L^p(\Omega)$. Moreover,

$$\mu\{|f| \geqslant 1/n\} = \int_\Omega \mathbf{1}_{\{|f| \geqslant \frac{1}{n}\}} \, \mathrm{d}\mu \leqslant \int_\Omega \mathbf{1}_{\{|f| \geqslant \frac{1}{n}\}} |nf|^p \, \mathrm{d}\mu \leqslant n^p \|f\|_p^p < \infty.$$

We may therefore assume that f is bounded and μ is a finite measure. By considering real and imaginary parts separately we may also assume that f is real-valued. Under these assumptions we have $f_k \to f$ in $L^p(\Omega)$, where

$$f_k := \sum_{j \in \mathbb{Z}} \mathbf{1}_{\{j2^{-k} \leqslant f < (j+1)2^{-k}\}} j2^{-k}$$

are μ-simple functions (by the boundedness of f these sums have only finitely many nonzero contributions).

If $f \in L^\infty(\Omega)$ with $\mu(\Omega) < \infty$, the functions f_k defined above are μ-simple and approximate f uniformly. $\qquad\square$

More interesting is the fact that if D is an open subset of $\mathbb{R}^d$, then the vector space $C_c^\infty(D)$ of all compactly supported smooth functions $f : D \to \mathbb{K}$ is dense in $L^p(D)$ for every $1 \leqslant p < \infty$. Here, and in what follows, the *support* $\mathrm{supp}(f)$ of a continuous function $f : D \to \mathbb{K}$ is defined as the complement of the largest open set U such that $f \equiv 0$ on $D \cap U$ or, equivalently, as the closure of the set $\{x \in D : f(x) \neq 0\}$.

Proposition 2.29 (Approximation by compactly supported smooth functions). *Let $1 \leqslant p < \infty$ and let $D \subseteq \mathbb{R}^d$ be open. Then $C_c^\infty(D)$ is dense in $L^p(D)$.*

Proof For $f \in L^p(D)$ we have $\lim_{n \to \infty} \|f - \mathbf{1}_{B(0;n)} f\|_p = 0$ by dominated convergence, so there is no loss of generality in assuming that D is bounded. Also, by Proposition 2.28, every $f \in L^p(D)$ can be approximated by simple functions supported on D. Hence it suffices to prove that every simple function supported on a bounded open set D can be approximated in $L^p(D)$ by functions in $C_c^\infty(D)$. By linearity and the triangle inequality, it even suffices to approximate indicator functions of the form $\mathbf{1}_B$ for Borel sets $B \subseteq D$.

Given $\varepsilon > 0$, choose an open set $U \subseteq D$ and a closed set $F \subseteq D$ such that $F \subseteq B \subseteq U$

and $|U \setminus F| < \varepsilon$; this is possible by the regularity of the Lebesgue measure on D (Proposition E.16). Let $\phi \in C_c^\infty(D)$ satisfy $0 \leqslant \phi \leqslant 1$ pointwise, $\phi \equiv 1$ on F, and $\phi \equiv 0$ outside U. As outlined in Problem 2.9, the existence of such functions can be demonstrated by elementary calculus arguments. Then

$$\|\phi - \mathbf{1}_B\|_{L^p(D)}^p = \int_D |\phi|_{U \setminus F}|^p \, dx \leqslant |U \setminus F| < \varepsilon.$$

Since the choice of $\varepsilon > 0$ was arbitrary, this completes the proof. $\qquad\square$

The corresponding result for $p = \infty$ is wrong: if D is nonempty and open, then $C_c(D)$, the vector space of all compactly supported continuous functions $f : D \to \mathbb{K}$, fails to be dense in $L^\infty(D)$. Indeed, if D' is a nonempty open set properly contained in D, then $\|f - \mathbf{1}_{D'}\|_{L^\infty(D)} \geqslant \frac{1}{2}$ for every $f \in C_c(D)$.

The separability of the spaces $C(\overline{B(0;n)})$ implies:

Corollary 2.30. *Let* $1 \leqslant p < \infty$ *and let* $D \subseteq \mathbb{R}^d$ *be open. Then* $L^p(D)$ *is separable.*

Remark 2.31. Since finite Borel measures on metric spaces are regular (Proposition E.16), using Urysohn's lemma (Proposition C.11) the same argument proves that if μ is a finite Borel measure on a compact metric space K, then $C(K)$ is dense in $L^p(K, \mu)$ for all $1 \leqslant p < \infty$.

Combining this observation with Proposition 2.8, as a corollary we obtain that, under these assumptions, $L^p(K, \mu)$ is separable for all $1 \leqslant p < \infty$.

As an application of Proposition 2.29 we prove an L^p-continuity result for translation.

Proposition 2.32 (Continuity of translation). *Let* $f \in L^p(\mathbb{R}^d)$ *with* $1 \leqslant p < \infty$. *Then*

$$\lim_{|h| \to 0} \|f(\cdot + h) - f(\cdot)\|_p = 0.$$

Proof Define, for $h \in \mathbb{R}^d$ and $x \in \mathbb{R}^d$, $(\tau_h f)(x) := f(x + h)$, that is, $\tau_h f$ is the translate of f over h. Clearly, $\|\tau_h f\|_p = \|f\|_p$.

First consider a function $f \in C_c(\mathbb{R}^d)$. Such a function is uniformly continuous, so given an $\varepsilon > 0$, we may choose $\delta > 0$ such that $|x - x'| < \delta$ implies $|f(x) - f(x')| < \varepsilon$. Hence if $|h| < \delta$, then for all $x \in \mathbb{R}^d$ we have $|\tau_h f(x) - f(x)| < \varepsilon$. Choose $r > 0$ large enough such that the support of f is contained in the rectangle $(-r, r)^d$. If $|h| < \delta$ is so small that the support of $\tau_h f$ is also contained in $(-r, r)^d$, then

$$\|\tau_h f - f\|_p^p = \int_{(-r,r)^d} |f(x+h) - f(x)|^p \, dx \leqslant \varepsilon^p \int_{(-r,r)^d} dx = \varepsilon^p (2r)^d.$$

This proves that $\lim_{|h| \to 0} \|\tau_h f - f\|_p = 0$ for all $f \in C_c(\mathbb{R}^d)$.

Now let $f \in L^p(\mathbb{R}^d)$ be arbitrary. Since $C_c(\mathbb{R}^d)$ is dense in $L^p(\mathbb{R}^d)$ by Proposition

2.29, we can find $g \in C_c(\mathbb{R}^d)$ such that $\|f - g\|_p < \varepsilon$. Choose $\eta > 0$ so small that $|h| < \eta$ implies $\|\tau_h g - g\|_p < \varepsilon$; this is possible by what we just proved. Then, for $|h| < \eta$,

$$\|\tau_h f - f\|_p \leqslant \|\tau_h f - \tau_h g\|_p + \|\tau_h g - g\|_p + \|g - f\|_p < \varepsilon + \varepsilon + \varepsilon = 3\varepsilon,$$

noting that $\|\tau_h f - \tau_h g\|_p = \|\tau_h(f - g)\|_p = \|f - g\|_p < \varepsilon$. $\qquad\square$

We now turn to an approximation technique based on convolution. It relies on the following fundamental inequality.

Proposition 2.33 (Young's inequality). *Let* $1 \leqslant p, q, r \leqslant \infty$ *satisfy* $\frac{1}{p} + \frac{1}{q} = 1 + \frac{1}{r}$, *and let* $f \in L^p(\mathbb{R}^d)$ *and* $g \in L^q(\mathbb{R}^d)$. *Then:*

(1) *for almost all* $x \in \mathbb{R}^d$ *the function* $y \mapsto f(x - y)g(y)$ *is integrable;*
(2) *the function* $f * g : \mathbb{R}^d \to \mathbb{K}$, *defined for almost all* $x \in \mathbb{R}^d$ *by*

$$(f * g)(x) := \int_{\mathbb{R}^d} f(x - y)g(y)\,\mathrm{d}y,$$

belongs to $L^r(\mathbb{R}^d)$ *and we have*

$$\|f * g\|_r \leqslant \|f\|_p \|g\|_q.$$

Moreover we have $f * g = g * f$ *in* $L^r(\mathbb{R}^d)$; *and if* $\frac{1}{r} + \frac{1}{s} = 1 + \frac{1}{t}$ *and* $\frac{1}{q} + \frac{1}{s} = \frac{1}{u}$, *then* $\frac{1}{p} + \frac{1}{u} = 1 + \frac{1}{t}$ *and for all* $h \in L^s(\mathbb{R}^d)$ *we have* $(f * g) * h = f * (g * h)$ *in* $L^t(\mathbb{R}^d)$.

The most important special case corresponds to the choices $q = 1$ and $r = p$, for which the proof below simplifies considerably.

Proof The identity $\frac{1}{p} + \frac{1}{q} = 1 + \frac{1}{r}$ implies $\frac{1}{r} + \frac{r-p}{pr} + \frac{r-q}{qr} = 1$ with $r \geqslant p, q$. Hence, by elementary rewriting and Hölder's inequality (for three functions, see Remark 2.24), for all $x \in \mathbb{R}^d$ we have

$$\int_{\mathbb{R}^d} |f(x - y)g(y)|\,\mathrm{d}y$$

$$= \int_{\mathbb{R}^d} (|f(x - y)|^p |g(y)|^q)^{1/r} \cdot |f(x - y)|^{(r-p)/r} \cdot |g(y)|^{(r-q)/r}\,\mathrm{d}y$$

$$\leqslant \left\| (|f(x - \cdot)|^p |g(\cdot)|^q)^{1/r} \right\|_r \left\| |f(x - \cdot)|^{(r-p)/r} \right\|_{\frac{pr}{r-p}} \left\| |g(\cdot)|^{(r-q)/r} \right\|_{\frac{qr}{r-q}}$$

$$= (I) \cdot (II) \cdot (III).$$

Now

$$(I) = \left(\int_{\mathbb{R}^d} |f(x - y)|^p |g(y)|^q\,\mathrm{d}y \right)^{1/r},$$

$$(II) = \left(\int_{\mathbb{R}^d} |f(x - y)|^p\,\mathrm{d}y \right)^{(r-p)/pr} = \|f\|_p^{(r-p)/r},$$

$$(III) = \left(\int_{\mathbb{R}^d} |g(y)|^q \, \mathrm{d}y \right)^{(r-q)/qr} = \|g\|_q^{(r-q)/r}.$$

Putting things together and using Fubini's theorem, it follows that

$$\int_{\mathbb{R}^d} |(f * g)(x)|^r \, \mathrm{d}x \leqslant \|f\|_p^{r-p} \|g\|_p^{r-q} \int_{\mathbb{R}^d} \int_{\mathbb{R}^d} |f(x-y)|^p |g(y)|^q \, \mathrm{d}y \, \mathrm{d}x$$

$$= \|f\|_p^{r-p} \|g\|_q^{r-q} \int_{\mathbb{R}^d} |g(y)|^q \int_{\mathbb{R}^d} |f(x-y)|^p \, \mathrm{d}x \, \mathrm{d}y = \|f\|_p^r \|g\|_q^r.$$

This implies the first assertion as well as the second.

The identity $f * g = g * f$ follows by a change of variables and $(f * g) * h = f * (g * h)$ by Fubini's theorem. $\qquad\square$

The function $f * g$ is called the *convolution* of f and g.

Proposition 2.34 (Approximation by mollification). *Let* $f \in L^p(\mathbb{R}^d)$ *with* $1 \leqslant p < \infty$. *Let* $\phi \in L^1(\mathbb{R}^d)$ *satisfy* $\int_{\mathbb{R}^d} \phi(x) \, \mathrm{d}x = 1$. *For* $\varepsilon > 0$ *define*

$$\phi_\varepsilon(x) := \varepsilon^{-d} \phi(\varepsilon^{-1} x), \quad x \in \mathbb{R}^d.$$

Then

$$\lim_{\varepsilon \downarrow 0} \|\phi_\varepsilon * f - f\|_p = 0.$$

Maurice Fréchet, 1878–1973

Proof We proceed in three steps.

Step 1 – First assume that ϕ and f belong to $C_c(\mathbb{R}^d)$. Since $\int_{\mathbb{R}^d} \phi_\varepsilon(y) \, \mathrm{d}y = 1$, for all $x \in \mathbb{R}^d$ we have

$$\phi_\varepsilon * f(x) - f(x) = \int_{\mathbb{R}^d} \phi_\varepsilon(y)[f(x-y) - f(x)] \, \mathrm{d}y = \int_{\mathbb{R}^d} \phi(y)[f(x-\varepsilon y) - f(x)] \, \mathrm{d}y.$$

Taking $L^p(\mathbb{R}^d)$ norms on both sides and using that the $L^p(\mathbb{R}^d)$-valued function $y \mapsto \phi(y)[f(\cdot - \varepsilon y) - f(\cdot)]$, which is continuous by Proposition 2.32 and compactly supported (and hence supported on some large enough closed cube), is Riemann integrable, by Proposition 1.44 we obtain

$$\|\phi_\varepsilon * f - f\|_p = \left\| \int_{\mathbb{R}^d} \phi(y)[f(\cdot - \varepsilon y) - f(\cdot)] \, \mathrm{d}y \right\|_p \leqslant \int_{\mathbb{R}^d} |\phi(y)| \|f(\cdot - \varepsilon y) - f(\cdot)\|_p \, \mathrm{d}y.$$

Since $\|f(\cdot - \varepsilon y) - f(\cdot)\|_p \leqslant 2\|f\|_p$ uniformly in ε and y, and since $\phi \in L^1(\mathbb{R}^d)$, by dominated convergence it suffices to show that $f(\cdot - \varepsilon y) \to f(\cdot)$ in $L^p(\mathbb{R}^d)$ for each fixed y. This again follows from Proposition 2.32. This completes the proof for functions f and ϕ in $C_c(\mathbb{R}^d)$.

Step 2 – Still assuming that $\phi \in C_c(\mathbb{R}^d)$, we next extend the result to general $f \in$

$L^p(\mathbb{R}^d)$. Fix $\varepsilon > 0$ and choose $g \in C_c(\mathbb{R}^d)$ such that $\|f - g\|_p < \varepsilon$. This is possible by Proposition 2.29. By Young's inequality and the identity $\|\phi_\delta\|_1 = \|\phi\|_1$, for any $\delta > 0$ we have

$$\|\phi_\delta * f - f\|_p \leqslant \|\phi_\delta * f - \phi_\delta * g\|_p + \|\phi_\delta * g - g\|_p + \|g - f\|_p$$
$$\leqslant \|\phi_\delta\|_1 \|f - g\|_p + \|\phi_\delta * g - g\|_p + \varepsilon \leqslant \varepsilon \|\phi\|_1 + \|\phi_\delta * g - g\|_p + \varepsilon.$$

Letting $\delta \downarrow 0$, the result of Step 1 implies that

$$\limsup_{\delta \downarrow 0} \|\phi_\delta * f - f\|_p \leqslant \varepsilon(\|\phi\|_1 + 1).$$

Since $\varepsilon > 0$ was arbitrary, this proves that $\lim_{\delta \downarrow 0} \|\phi_\delta * f - f\|_p = 0$. *Step 3* – We now pass to the general case where $\phi \in L^1(\mathbb{R}^d)$ satisfies $\int_{\mathbb{R}^d} \phi(x) \, dx = 1$. In order to apply the result of the preceding step, choose a function $\psi \in C_c(\mathbb{R}^d)$ such that $\|\phi - \psi\|_1 < \varepsilon$ and $\int_{\mathbb{R}^d} \psi(y) \, dy = 1$. Such a function exists by Proposition 2.29. Then, by Young's inequality and the result of Step 2 applied to ψ,

$$\|\phi_\delta * f - f\|_p \leqslant \|\phi_\delta * f - \psi_\delta * f\|_p + \|\psi_\delta * f - f\|_p$$
$$\leqslant \|\phi_\delta - \psi_\delta\|_1 \|f\|_p + \|\psi_\delta * f - f\|_p \leqslant \varepsilon \|f\|_p + \|\psi_\delta * f - f\|_p.$$

Letting $\delta \downarrow 0$ using the result of Step 2, it follows that

$$\limsup_{\delta \downarrow 0} \|\phi_\delta * f - f\|_p \leqslant \varepsilon \|f\|_p.$$

Since $\varepsilon > 0$ was arbitrary, this proves that $\lim_{\delta \downarrow 0} \|\phi_\delta * f - f\|_p = 0$. $\qquad\square$

2.3.c The Fréchet–Kolmogorov Compactness Theorem

In this section we prove a characterisation of relatively compact sets in $L^p(\mathbb{R}^d)$. It will be used in Chapter 11 to prove the Rellich–Kondrachov theorem on compact embeddings of Sobolev spaces.

Recall the notation $\tau_h f$ for the translate of a function f over h,

$$\tau_h f(x) = f(x + h).$$

Theorem 2.35 (Fréchet–Kolmogorov). *Let $1 \leqslant p < \infty$. A subset S of $L^p(\mathbb{R}^d)$ is relatively compact if and only if it satisfies the following two conditions:*

(i) $\displaystyle \lim_{|h| \to 0} \sup_{f \in S} \|\tau_h f - f\|_p = 0$;

(ii) $\displaystyle \lim_{\rho \to \infty} \sup_{f \in S} \int_{\complement B(0;\rho)} |f(x)|^p \, dx = 0$.

Proof 'If': Let us begin by proving that (i) and (ii) together imply that the set S is bounded in $L^p(\mathbb{R}^d)$. Choose $r > 0$ such that $\sup_{f \in S} \|\tau_h f - f\|_p \leqslant 1$ for all $h \in \mathbb{R}^d$ with $|h| \leqslant r$, and choose $R > 0$ such that $\sup_{f \in S} \int_{\complement B(0;R)} |f(x)|^p \, dx \leqslant 1$. Fix $h \in \mathbb{R}^d$ with $|h| = r$. For all $f \in S$ and $x \in \mathbb{R}^d$ we have

$$\|\mathbf{1}_{B(x;R)} f\|_p \leqslant \|\mathbf{1}_{B(x;R)} (f - \tau_h f)\|_p + \|\mathbf{1}_{B(x;R)} \tau_h f\|_p$$
$$= \|\mathbf{1}_{B(x;R)} (f - \tau_h f)\|_p + \|\mathbf{1}_{B(x+h;R)} f\|_p \leqslant 1 + \|\mathbf{1}_{B(x+h;R)} f\|_p.$$

Hence, by induction,

$$\|\mathbf{1}_{B(0;R)} f\|_p \leqslant N + \|\mathbf{1}_{B(Nh;R)} f\|_p.$$

Choose $N \geqslant 1$ such that $Nr = N|h| > 2R$. Then $B(Nh;R) \subseteq \complement B(0;R)$ and

$$\|f\|_p = \|\mathbf{1}_{B(0;R)} f\|_p + \|\mathbf{1}_{\complement B(0;R)} f\|_p \leqslant N + \|\mathbf{1}_{B(Nh;R)} f\|_p + \|\mathbf{1}_{\complement B(0;R)} f\|_p \leqslant N + 2. \tag{2.9}$$

This proves that S is bounded in $L^p(\mathbb{R}^d)$.

Let us now prove that if S satisfies (i) and (ii), then it is relatively compact. Fix $\varepsilon > 0$ and choose $R_\varepsilon > 0$ such that $\sup_{f \in S} \int_{\complement B(0;R_\varepsilon)} |f(x)|^p \, dx < \varepsilon^p$. Set $B_\varepsilon := B(0;R_\varepsilon)$ and

$$S_\varepsilon := \{\mathbf{1}_{B_\varepsilon} f : f \in S\}.$$

If $f \in S$, then

$$\|f - \mathbf{1}_{B_\varepsilon} f\|_p = \|\mathbf{1}_{\complement B_\varepsilon} f\|_p < \varepsilon \tag{2.10}$$

and therefore $S \subseteq S_\varepsilon + B_p(0;\varepsilon)$, where $B_p(0;\varepsilon)$ is the open ball in $L^p(\mathbb{R}^d)$ with radius ε centred at 0.

Choose $r_\varepsilon > 0$ such that $\sup_{f \in S} \|\tau_h f - f\|_p \leqslant \varepsilon$ for all $f \in S$ and $h \in \mathbb{R}^d$ with $|h| \leqslant r_\varepsilon$. For such h, (2.10) implies that for all $f \in S$ we have

$$\|\tau_h(\mathbf{1}_{B_\varepsilon} f) - \mathbf{1}_{B_\varepsilon} f\|_p \leqslant \|\tau_h(\mathbf{1}_{B_\varepsilon} f - f)\|_p + \|\tau_h f - f\|_p + \|f - \mathbf{1}_{B_\varepsilon} f\|_p \leqslant 3\varepsilon. \tag{2.11}$$

Let $0 \leqslant \phi \in C_c(B(0;r_\varepsilon))$ satisfy $\int_{\mathbb{R}^d} \phi(x) \, dx = 1$ and set

$$S^\varepsilon := \{\phi * g : g \in S_\varepsilon\}.$$

For $g \in S_\varepsilon$, the estimate (2.11) implies

$$\|\phi * g - g\|_p = \left\| \int_{\mathbb{R}^d} \phi(y)(g(\cdot - y) - g(\cdot)) \, dy \right\|_p$$
$$\leqslant \int_{\mathbb{R}^d} \phi(y) \|g(\cdot - y) - g(\cdot)\|_p \, dy = \int_{B(0;r_\varepsilon)} \phi(y) \|g(\cdot - y) - g(\cdot)\|_p \, dy \leqslant 3\varepsilon,$$

where we used Proposition 1.44 (which can be applied in view of Proposition 2.32). This shows that $S_\varepsilon \subseteq S^\varepsilon + B_p(0;3\varepsilon)$ and hence

$$S \subseteq S_\varepsilon + B_p(0;\varepsilon) \subseteq S^\varepsilon + B_p(0;4\varepsilon).$$

If we can prove that S^ε is relatively compact, it follows from Proposition 1.40 that S is relatively compact.

Every $h \in S^\varepsilon$ is supported in $B(0; R_\varepsilon + r_\varepsilon)$. We claim that every $h \in S^\varepsilon$ is continuous and that the set S^ε, as subset of $C(\overline{B}(0; R_\varepsilon + r_\varepsilon))$, is equicontinuous and bounded.

Let $h \in S^\varepsilon$, say $h = \phi * g$ with $g \in S_\varepsilon$, say $g = \mathbf{1}_{B_\varepsilon} f$ with $f \in S$. By uniform continuity, given $\eta > 0$ there exists $0 < \delta < 1$ such that for all $x, x' \in \mathbb{R}^d$ with $|x - x'| < \delta$ we have $|\phi(x) - \phi(x')| < \eta$. Hence, for all $x, x' \in \mathbb{R}^d$ with $|x - x'| < \delta$,

$$
\begin{aligned}
|h(x) - h(x')| &\leqslant \int_{\mathbb{R}^d} |\phi(x - y) - \phi(x' - y)||g(y)|\, dy \\
&= \int_{B_\varepsilon} |\phi(x - y) - \phi(x' - y)||g(y)|\, dy \\
&\leqslant \eta \int_{B_\varepsilon} |g(y)|\, dy = \eta \int_{B_\varepsilon} |f(y)|\, dy \leqslant \eta |B_\varepsilon|^{1/q}(N + 2),
\end{aligned}
$$

applying Hölder's inequality and (2.9) in the last step. This proves the continuity of h. The estimate being uniform with respect to $h \in S^\varepsilon$, it also proves the equicontinuity of S^ε.

Boundedness of S^ε in $C(\overline{B}(0; R_\varepsilon + r_\varepsilon))$ follows from the boundedness of S. Indeed, if $h = \phi * g \in S^\varepsilon$ with $g \in S_\varepsilon$, then by Hölder's inequality with $\frac{1}{p} + \frac{1}{p'} = 1$,

$$
|h(x)| \leqslant \int_{\mathbb{R}^d} \phi(y)|g(x - y)|\, dy \leqslant \|\phi\|_{p'} \|g\|_p.
$$

By the Arzelà–Ascoli theorem, S^ε is relatively compact in $C(\overline{B}(0; R_\varepsilon + r_\varepsilon))$. Since the natural inclusion mapping from $C(\overline{B}(0; R_\varepsilon + r_\varepsilon))$ into $L^p(\mathbb{R}^d)$ is bounded by Hölder's inequality, S^ε is relatively compact as a subset of $L^p(\mathbb{R}^d)$.

'Only if': If S is relatively compact in $L^p(\mathbb{R}^d)$, then (i) and (ii) follows from Proposition 1.42 applied to the operators $f \mapsto \tau_h f$ for $|h| \downarrow 0$ and $f \mapsto \mathbf{1}_{\complement B(0;\rho)} f$ for $\rho \to \infty$, respectively. $\qquad\square$

We have the following immediate corollary for bounded domains.

Corollary 2.36. *Let* $1 \leqslant p < \infty$ *and let* D *be a bounded open subset of* $\mathbb{R}^d$. *A subset* S *of* $L^p(D)$ *is relatively compact if and only if*

$$
\lim_{|h| \to 0} \sup_{f \in S} \|\tau_h f - f\|_p = 0.
$$

Here we identify functions in $L^p(D)$ with their zero extensions in $L^p(\mathbb{R}^d)$.

2.3.d The Lebesgue Differentiation Theorem

By $L^1_{\mathrm{loc}}(\mathbb{R}^d)$ we denote the vector space of functions $f : \mathbb{R}^d \to \mathbb{K}$ that are *locally integrable*, that is, integrable on every compact subset of $\mathbb{R}^d$, identifying two such functions when they are equal almost everywhere. The aim of this section is to prove the

Lebesgue differentiation theorem, which says that if $f \in L^1_{\text{loc}}(\mathbb{R}^d)$, then at almost every point $x \in \mathbb{R}^d$ one has

$$\lim_{\substack{B \ni x \\ |B| \to 0}} \frac{1}{|B|} \int_B |f(y) - f(x)| \, dy = 0,$$

the limit being taken along the balls B in $\mathbb{R}^d$ containing x, letting $|B|$ denote the Lebesgue measure of a measurable set B. The proof of this theorem is based on the following lemma. For balls $B = B(x;r)$ in $\mathbb{R}^d$ and real numbers $\lambda > 0$ we set $\lambda B := B(x; \lambda r)$.

Lemma 2.37 (Vitali covering lemma). *Every finite collection $\mathscr{B}$ of open balls in $\mathbb{R}^d$ has a subcollection $\mathscr{B}_0$ of pairwise disjoint balls such that each ball $B \in \mathscr{B}$ is contained in $3B_0$ for some ball $B_0 \in \mathscr{B}_0$.*

Proof We proceed by induction on the number n of balls in $\mathscr{B}$. For $n = 1$ the lemma is trivial, for we can take $\mathscr{B}_0 = \mathscr{B}$. Suppose the claim has been verified for every collection of n balls, and let $\mathscr{B}$ be a collection of $n+1$ balls. Let $\mathscr{B}' := \mathscr{B} \setminus \{B_0\}$, where B_0 is a ball in $\mathscr{B}$ of minimal radius. By the induction assumption there is a subcollection $\mathscr{B}'_0 \subseteq \mathscr{B}'$ of pairwise disjoint balls such that each ball $B \in \mathscr{B}'$ is contained in $3B'$ for some ball $B' \in \mathscr{B}'_0$. We now distinguish two cases.

Case 1. If B_0 is disjoint from each ball $B' \in \mathscr{B}'_0$, then the subcollection $\mathscr{B}_0 := \mathscr{B}'_0 \cup \{B_0\}$ has the required properties.

Case 2. If B_0 intersects a ball $B' \in \mathscr{B}'_0$, then the radius of B' is at least as large as that of B_0, from which it follows that $B_0 \subseteq 3B'$. The subcollection $\mathscr{B}_0 := \mathscr{B}'_0$ then has the required properties. $\square$

For $f \in L^1_{\text{loc}}(\mathbb{R}^d)$ we define the *Hardy–Littlewood maximal function $Mf : \mathbb{R}^d \to [0, \infty]$* by

$$Mf(x) := \sup_{B \ni x} \frac{1}{|B|} \int_B |f(y)| \, dy,$$

where the supremum is taken over all balls B containing x. Since the supremum in the definition of $Mf(x)$ can be realised by using only a fixed countable collection of balls, Mf is a measurable function.

Theorem 2.38 (Hardy–Littlewood maximal theorem). *For all $f \in L^1(\mathbb{R}^d)$ and $t > 0$ we have the* weak L^1-*bound*

$$t|\{Mf > t\}| \leqslant 3^d \|f\|_1.$$

Moreover, for all $1 < p \leqslant \infty$ there exists a constant $C_{d,p} \geqslant 0$ such that for all $f \in L^p(\mathbb{R}^d)$ we have $Mf \in L^p(\mathbb{R}^d)$ and

$$\|Mf\|_p \leqslant C_{d,p} \|f\|_p.$$

Proof We begin with the proof of the first assertion. By the definition of Mf, for every $x \in \{Mf > t\}$ there exists a ball B containing x such that $\frac{1}{|B|} \int_B |f| > t$. If $K \subseteq \{Mf > t\}$ is a compact subset, it can be covered by a finite collection $\mathscr{B}$ of such balls. Let $\mathscr{B}_0$ be a disjoint subcollection of this cover provided by the Vitali covering lemma. Then,

$$|K| \leqslant \left| \bigcup_{B \in \mathscr{B}} B \right| \leqslant \left| \bigcup_{B \in \mathscr{B}_0} 3B \right| = \sum_{B \in \mathscr{B}_0} 3^d |B| \leqslant \sum_{B \in \mathscr{B}_0} \frac{3^d}{t} \int_B |f(y)| \, dy \leqslant \frac{3^d}{t} \|f\|_1.$$

This being true for all compact sets K contained in the open set $\{Mf > t\}$, the first assertion follows.

For $1 < p < \infty$ the second assertion follows from the first by using the integration by parts identity

$$\int_{\mathbb{R}^d} |g(x)|^p \, dx = p \int_0^\infty t^{p-1} |\{|g| > t\}| \, dt \tag{2.12}$$

for $g \in L^p(\mathbb{R}^d)$ as follows. For any $f \in L^p(\mathbb{R}^d)$ and $t > 0$ the function $f_t(x) := \mathbf{1}_{\{|f| \geqslant t/2\}} f$ belongs to $L^p(\mathbb{R}^d)$ and satisfies the pointwise bound

$$Mf \leqslant \sup_{B \ni x} \frac{1}{|B|} \int_B |f_t(y)| \, dy + \sup_{B \ni x} \frac{1}{|B|} \int_B |\mathbf{1}_{\{|f| < t/2\}} f(y)| \, dy \leqslant Mf_t + t/2,$$

which implies

$$\{Mf > t\} \subseteq \{Mf_t > t/2\}.$$

Hence, by the first part of the theorem,

$$|\{Mf > t\}| \leqslant |\{Mf_t > t/2\}| \leqslant \frac{2 \cdot 3^d}{t} \|f_t\|_1 = \frac{2 \cdot 3^d}{t} \int_{\{|f| \geqslant t/2\}} |f(x)| \, dx. \tag{2.13}$$

By (2.12), (2.13), and Fubini's theorem,

$$\int_{\mathbb{R}^d} |Mf(x)|^p \, dx \leqslant p \int_0^\infty t^{p-1} \left(\frac{2 \cdot 3^d}{t} \int_{\{|f| \geqslant t/2\}} |f(x)| \, dx \right) dt$$

$$= 3^d \cdot 2p \int_{\mathbb{R}^d} |f(x)| \int_0^{2|f(x)|} t^{p-2} \, dt \, dx$$

$$= \frac{3^d \cdot 2p}{p-1} \int_{\mathbb{R}^d} |f(x)| (2|f(x)|)^{p-1} \, dx = C_{d,p}^p \int_{\mathbb{R}^d} |f(x)|^p \, dx,$$

where $C_{d,p} = 2(\frac{3^d p}{p-1})^{1/p}$.

For $p = \infty$ the second assertion follows trivially from the pointwise inequality $Mf \leqslant \|f\|_\infty$, with constant $C_{d,\infty} = 1$. $\square$

Inspection of this proof reveals that the derivation of the L^p-bound for Mf in the second part of the theorem does not use any properties of this function other than the weak L^1-bound contained in the first part of the theorem. This observation lies at the basis of the Marcinkiewicz interpolation theorem in Chapter 5 (see Theorem 5.46).

As a corollary to Theorem 2.38 we have the following fundamental result.

Theorem 2.39 (Lebesgue differentiation theorem). *If $f \in L^1_{\mathrm{loc}}(\mathbb{R}^d)$, then for almost all $x \in \mathbb{R}^d$ we have*

$$\lim_{\substack{B \ni x \\ |B| \to 0}} \frac{1}{|B|} \int_B |f(y) - f(x)|\,dy = 0. \tag{2.14}$$

The correct way of interpreting this theorem is as follows. For every pointwise defined locally integrable function $\widetilde{f}$ on $\mathbb{R}^d$, the limit in (2.14) (with f replaced by $\widetilde{f}$) exists for almost all $x \in \mathbb{R}$, say on a Borel set $\Omega \subseteq \mathbb{R}^d$ such that $|\mathbb{R}^d \setminus \Omega| = 0$. If both $\widetilde{f}_1$ and $\widetilde{f}_2$ are pointwise representatives, the symmetric difference of corresponding sets Ω_1 and Ω_2 has measure 0.

The set of all points $x \in \mathbb{R}^d$ for which (2.14) holds is called the set of *Lebesgue points* of f. As just explained, this set is uniquely determined only up to a set om measure 0 (see also Problem 2.30).

Proof A point $x \in \mathbb{R}^d$ is a Lebesgue point of f if and only if it is a Lebesgue point of $\mathbf{1}_U f$, for any bounded open set U containing x. Hence, upon replacing f by $\mathbf{1}_U f$ if necessary, it suffices to prove the theorem under the stronger assumption $f \in L^1(\mathbb{R}^d)$.

Fixing a pointwise defined representative of f, for all $x \in \mathbb{R}^d$ let

$$Nf(x) := \limsup_{\substack{B \ni x \\ |B| \to 0}} \frac{1}{|B|} \int_B |f(y) - f(x)|\,dy.$$

We wish to prove that $Nf(x) = 0$ for almost all $x \in \mathbb{R}^d$. For this purpose it suffices to show that $|\{Nf > \varepsilon\}| = 0$ for any fixed $\varepsilon > 0$.

For any fixed $\delta > 0$, Proposition 2.29 provides us with a function $g \in C_c(\mathbb{R}^d)$ such that $\|f - g\|_1 < \delta$. Then,

$$Nf \leqslant N(f - g) + Ng \leqslant M(f - g) + |f - g| + 0,$$

using the pointwise inequality $Nh(x) \leqslant Mh(x) + |h(x)|$ and the continuity of g, which implies $Ng(x) = 0$. Therefore, by Theorem 2.38,

$$|\{Nf > \varepsilon\}| \leqslant |\{M(f - g) > \varepsilon/2\}| + |\{|f - g| > \varepsilon/2\}|$$
$$\leqslant \frac{2 \cdot 3^d}{\varepsilon} \|f - g\|_1 + \frac{2}{\varepsilon} \|f - g\|_1 \leqslant \frac{2}{\varepsilon}(3^d + 1)\delta.$$

Since $\delta > 0$ was arbitrary it follows that $|\{Nf > \varepsilon\}| = 0$. $\qquad\square$

2.4 Spaces of Measures

In this section we introduce the space $M(\Omega)$ of $\mathbb{K}$-valued measures on a given measurable space $(\Omega, \mathscr{F})$ and discuss some of its properties. From the functional analytic point of view, the importance of this space resides in the fact that $M(\Omega)$ is a vector space in a natural way by setting

$$(c\mu)(F) := c\mu(F), \quad (\mu + \nu)(F) := \mu(F) + \nu(F), \quad F \in \mathscr{F},$$

and that it is a Banach space with respect to the variation norm introduced in Definition 2.42 (see Theorem 2.44).

2.4.a The Banach Space $M(\Omega)$

In what follows we fix a measurable space $(\Omega, \mathscr{F})$.

Definition 2.40 ($\mathbb{K}$-valued measures). A $\mathbb{K}$-*valued measure* on $(\Omega, \mathscr{F})$ is a mapping $\mu : \mathscr{F} \to \mathbb{K}$ with the following properties:

(i) $\mu(\varnothing) = 0$;
(ii) for all disjoint $\mathscr{F}$-measurable sets $F_1, F_2, \dots$ we have

$$\mu\Big(\bigcup_{n \geqslant 1} F_n\Big) = \sum_{n \geqslant 1} \mu(F_n).$$

Remark 2.41. An ordinary measure (in the sense of Definition E.3) is a real-valued measure if and only if it is finite.

The terms 'real-valued measure' and 'complex-valued measure' are often abbreviated to 'real measure' and 'complex measure'.

Definition 2.42 (Variation). Let μ be a $\mathbb{K}$-valued measure on $(\Omega, \mathscr{F})$. The *variation of μ on the set* $F \in \mathscr{F}$ is defined by

$$|\mu|(F) := \sup_{\mathscr{A} \in \mathbb{F}_F} \sum_{A \in \mathscr{A}} |\mu(A)|,$$

where $\mathbb{F}_F$ denotes the set of all finite collections of pairwise disjoint $\mathscr{F}$-measurable subsets of F.

It is immediate to verify that $|\mu(F)| \leqslant |\mu|(F)$ and $|\mu|(F) \leqslant |\mu|(F')$ whenever $F, F' \in \mathscr{F}$ and $F \subseteq F'$. Also, $|c\mu| = |c||\mu|$ for all $c \in \mathbb{K}$ and $|\mu + \nu|(F) \leqslant |\mu|(F) + |\nu|(F)$ for all $F \in \mathscr{F}$. If μ takes values in $[0, \infty)$, then $|\mu| = \mu$.

Proposition 2.43. *If μ is a $\mathbb{K}$-valued measure, then $|\mu|$ is a finite measure.*

Proof We proceed in two steps.

Step 1 – We prove that $|\mu|$ is a measure. It is clear that $|\mu|(\varnothing) = 0$. Let $(F_n)_{n \geqslant 1}$ be a sequence of pairwise disjoint measurable sets and let F be their union. We must prove that $|\mu|(F) = \sum_{n \geqslant 1} |\mu|(F_n)$.

If $\mathscr{A}_n \in \mathbb{F}_{F_n}$ is a finite collection of pairwise disjoint measurable subsets of F_n, then for every $N \geqslant 1$ the union $\bigcup_{n=1}^{N} \mathscr{A}_n$ is a finite collection of pairwise disjoint measurable subsets of F and therefore

$$\sum_{n=1}^{N} \sum_{A \in \mathscr{A}_n} |\mu(A)| \leqslant |\mu|(F).$$

Taking the supremum over all $\mathscr{A}_n \in \mathbb{F}_{F_n}$, it follows that $\sum_{n=1}^{N} |\mu|(F_n) \leqslant |\mu|(F)$. This being true for all $N \geqslant 1$ we conclude that

$$\sum_{n \geqslant 1} |\mu|(F_n) \leqslant |\mu|(F).$$

In the converse direction suppose that the measurable subsets $A_1, \ldots, A_k$ of F are disjoint. With $F_{j,n} := A_j \cap F_n$ we have

$$\sum_{j=1}^{k} |\mu(A_j)| \leqslant \sum_{j=1}^{k} \sum_{n \geqslant 1} |\mu(F_{j,n})| = \sum_{n \geqslant 1} \sum_{j=1}^{k} |\mu(F_{j,n})| \leqslant \sum_{n \geqslant 1} |\mu|(F_n).$$

Taking the supremum over all finite families of pairwise disjoint measurable subsets of F, we obtain

$$|\mu|(F) \leqslant \sum_{n \geqslant 1} |\mu|(F_n).$$

Step 2 – We prove that the measure $|\mu|$ is finite, that is, $|\mu|(\Omega) < \infty$. By considering real and imaginary parts, it suffices to do this in the real-valued case.

If a finite set $\{r_1, \ldots, r_N\}$ of real numbers is given, then either the positive or the negative numbers in this set (or both) contribute at least half to the sum $\sum_{n=1}^{N} |r_n|$. Enumerating this set (or one of them) as $r_{n_1}, \ldots, r_{n_k}$, we thus have

$$\left| \sum_{j=1}^{k} r_{n_j} \right| \geqslant \frac{1}{2} \sum_{n=1}^{N} |r_n|. \tag{2.15}$$

Suppose, for a contradiction, that some measurable set F satisfies $|\mu|(F) = \infty$. Choose disjoint measurable subsets $F_1, \ldots, F_N$ of F such that

$$\sum_{n=1}^{N} |\mu(F_n)| \geqslant 2(1 + |\mu(F)|).$$

By (2.15) the union F' of a suitable subcollection of the F_j satisfies

$$|\mu(F')| \geqslant \frac{1}{2} \sum_{n=1}^{N} |\mu(F_j)| \geqslant 1 + |\mu(F)|.$$

For $F'' := F \setminus F'$ we then have

$$|\mu(F'')| \geqslant \big||\mu(F)| - |\mu(F')|\big| \geqslant 1.$$

Thus if a set $F \in \mathscr{F}$ satisfies $|\mu|(F) = \infty$, there is a disjoint decomposition $F = F' \cup F''$ with $\mu(F') \geqslant 1$ and $\mu(F'') \geqslant 1$. Since $|\mu|$ is a measure, at least one of the numbers $|\mu|(F')$ and $|\mu|(F'')$ equals ∞. We take one of them and continue applying what we just proved inductively. This produces a sequence of pairwise disjoint measurable sets $G_1, G_2, \ldots$, each of which satisfies $\mu(G_k) \geqslant 1$. Let G be their union. Since μ is a $\mathbb{K}$-valued measure we have $\mu(G) = \sum_{k \geqslant 1} \mu(G_k)$. This sum cannot converge since its terms fail to converge to 0. This is the required contradiction. $\qquad\square$

Theorem 2.44 (Completeness). *Endowed with the variation norm*

$$\|\mu\| := |\mu|(\Omega),$$

$M(\Omega)$ *is a Banach space.*

Proof We leave it as an exercise to prove that $|\mu|(\Omega)$ defines a norm.

To prove completeness, let $(\mu_n)_{n \geqslant 1}$ be a Cauchy sequence in $M(\Omega)$. For all $F \in \mathscr{F}$,

$$|\mu_n(F) - \mu_m(F)| = |(\mu_n - \mu_m)(F)| \leqslant |\mu_n - \mu_m|(F) \leqslant \|\mu_n - \mu_m\|,$$

proving that the sequence $(\mu_n(F))_{n \geqslant 1}$ is Cauchy in $\mathbb{K}$. Let $\mu(F)$ denote its limit. We wish to show that the resulting mapping $\mu : \mathscr{F} \to \mathbb{K}$ is a $\mathbb{K}$-valued measure and that $\lim_{n \to \infty} \mu_n = \mu$ with respect to the norm of $M(\Omega)$.

It is clear that $\mu(\varnothing) = \lim_{n \to \infty} \mu_n(\varnothing) = 0$. Suppose now that $(F_m)_{m \geqslant 1}$ is a sequence of pairwise disjoint measurable sets and let $F := \bigcup_{m \geqslant 1} F_m$. Given $\varepsilon > 0$, choose $N \geqslant 1$ so large that $\|\mu_j - \mu_k\| < \varepsilon$ for all $j, k \geqslant N$. Since μ_N is countably additive we may choose $N' \geqslant 1$ so large that $|\mu_N(F) - \sum_{m=1}^{M} \mu_N(F_m)| < \varepsilon$ for all $M \geqslant N'$. Then, for $M \geqslant N'$,

$$\left|\mu(F) - \sum_{m=1}^{M} \mu(F_m)\right| = \lim_{n \to \infty} \left|\mu_n(F) - \sum_{m=1}^{M} \mu_n(F_m)\right|$$

$$\leqslant \left|\mu_N(F) - \sum_{m=1}^{M} \mu_N(F_m)\right| + \sup_{n \geqslant N+1} \left|(\mu_n - \mu_N)\Big(\bigcup_{m \geqslant M+1} F_m\Big)\right|$$

$$\leqslant \left|\mu_N(F) - \sum_{m=1}^{M} \mu_N(F_m)\right| + \sup_{n \geqslant N+1} \|\mu_n - \mu_N\| \leqslant 2\varepsilon.$$

Since $\varepsilon > 0$ was arbitrary, this proves that $\sum_{m \geqslant 1} \mu(F_m) = \mu(F)$.

Finally, if $F_1, \ldots, F_k$ are disjoint and measurable, then for all $m \geqslant N$ we have

$$\sum_{j=1}^{k} |(\mu - \mu_m)(F_j)| = \lim_{n \to \infty} \sum_{j=1}^{k} |(\mu_n - \mu_m)(F_j)|$$

$$\leqslant \limsup_{n \to \infty} |\mu_n - \mu_m|(\Omega) = \limsup_{n \to \infty} \|\mu_n - \mu_m\| \leqslant \varepsilon.$$

Taking the supremum over finite disjoint families of measurable sets, we find that $\|\mu - \mu_m\| \leqslant \varepsilon$ for all $m \geqslant N$. This proves the required convergence. $\qquad\square$

If μ is a complex measure on $(\Omega, \mathscr{F})$, then

$$(\mathrm{Re}\,\mu)(F) := \mathrm{Re}(\mu(F)), \quad (\mathrm{Im}\,\mu)(F) := \mathrm{Im}(\mu(F)),$$

define real measures on $(\Omega, \mathscr{F})$ and we have $\mu = \mathrm{Re}\,\mu + i\,\mathrm{Im}\,\mu$. The next result shows that real measures allow decompositions into positive and negative parts:

Theorem 2.45 (Hahn–Jordan decomposition). *If μ is a real measure on a measurable space $(\Omega, \mathscr{F})$, then*

$$\mu^+(F) := \sup\{\mu(A) : A \in \mathscr{F}, A \subseteq F\},$$
$$\mu^-(F) := -\inf\{\mu(A) : A \in \mathscr{F}, A \subseteq F\}$$

are finite measures on $(\Omega, \mathscr{F})$ and

$$\mu = \mu^+ - \mu^-, \quad |\mu| = \mu^+ + \mu^-.$$

The measures μ^+ and μ^- are supported on disjoint sets, in the sense that there exists a disjoint decomposition $\Omega = \Omega^+ \cup \Omega^-$ with $\Omega^\pm \in \mathscr{F}$ such that for all $F \in \mathscr{F}$ we have

$$\mu^+(F) = \mu(F \cap \Omega^+), \quad \mu^-(F) = -\mu(F \cap \Omega^-).$$

If ν_1 and ν_2 are finite measures on $(\Omega, \mathscr{F})$ such that $\mu = \nu_1 - \nu_2$, then for all $F \in \mathscr{F}$ we have

$$\nu_1(F) \geqslant \mu^+(F), \quad \nu_2(F) \geqslant \mu^-(F).$$

The decomposition $\mu = \mu^+ - \mu^-$ for real measures μ is called the *Jordan decomposition*; the existence of a corresponding decomposition $\Omega = \Omega^+ \cup \Omega^-$ for their supports is often referred to as the *Hahn decomposition theorem*.

Proof We begin with the construction of the sets Ω^+ and Ω^-. Let us call a set $F \in \mathscr{F}$ *positive* (resp. *negative*) if for all $A \in \mathscr{F}$ with $A \subseteq F$ we have $\mu(A) \geqslant 0$ (resp. $\mu(A) \leqslant 0$). We use the notation $F \geqslant 0$ (resp. $F \leqslant 0$) to express that $F \in \mathscr{F}$ and F is positive (resp. negative).

Finite and countable unions of positive sets are positive. Indeed, suppose that the sets F_n, $n \geqslant 1$, are positive. Set $G_1 := F_1$ and $G_n := F_n \setminus \bigcup_{j=1}^{n-1} F_j$ for $n \geqslant 2$. Then $\bigcup_{n \geqslant 1} F_n =$

$\bigcup_{n\geqslant 1} G_n$. If $A \in \mathscr{F}$ is contained in this union, the countable additivity of μ implies $\mu(A) = \sum_{n\geqslant 1} \mu(G_n \cap A) \geqslant 0$, keeping in mind that $G_n \cap A \subseteq F_n$ and $F_n \geqslant 0$.

Let

$$M := \sup_{F \geqslant 0} \mu(F)$$

and note $M < \infty$. Choose positive sets F_n, $n \geqslant 1$, such that $\lim_{n\to\infty} \mu(F_n) = M$. By the observation just made we may assume that $F_1 \subseteq F_2 \subseteq \ldots$ By the same observation, $\Omega^+ := \bigcup_{n\geqslant 1} F_n$ is positive and therefore $\mu(\Omega^+) \leqslant M$. The positivity of Ω^+ also implies $\mu(F_n) \leqslant \mu(\Omega^+)$, and therefore $M = \lim_{n\to\infty} \mu(F_n) \leqslant \mu(\Omega^+)$. We have shown that $\mu(\Omega^+) = M$.

We show next that $\Omega^- := \complement \Omega^+$ is a negative set. Suppose, for a contradiction, that this is false. Then Ω^- contains subset $A_0 \in \mathscr{F}$ with $\mu(A_0) > 0$. If A_0 were positive, then so would be $\Omega^+ \cup A_0$, but then $\mu(\Omega^+ \cup A_0) = M + \mu(A_0) > M$ contradicts the choice of M. It follows that there exists a smallest integer k_1 with the property that A_0 contains a subset $A_1 \in \mathscr{F}$ with $\mu(A_1) \leqslant -\frac{1}{k_1}$. Since $\mu(A_0 \setminus A_1) = \mu(A_0) - \mu(A_1) > 0$ we can repeat this construction to find the smallest integer k_2 with the property that $A_0 \setminus A_1$ contains a subset $A_2 \in \mathscr{F}$ with $\mu(A_2) \leqslant -\frac{1}{k_2}$. Continuing this way we obtain a sequence of pairwise disjoint sets $(A_n)_{n\geqslant 1}$, all contained in A_0, such that $\mu(A_n) \leqslant -\frac{1}{k_n}$ for all $n \geqslant 1$. We must have $\lim_{n\to\infty} k_n = 0$, since otherwise the union $A = \bigcup_{n\geqslant 1} A_n$ would satisfy $\mu(A) = -\infty$.

Let $B := A_0 \setminus A$. Then $\mu(B) = \mu(A_0) - \mu(A) > 0$ and $B \geqslant 0$: for if we had $C \in \mathscr{F}$ with $C \subseteq B$ and $\mu(C) < 0$, then $\mu(C) < \frac{1}{k}$ for some integer k. The existence of such a set C contradicts the maximality of the k_n for large enough n. The set $\Omega^+ \cup B$ is positive and satisfies $\mu(\Omega^+ \cup B) = M + \mu(B) > M$, contradicting the choice of M. We conclude that Ω^- is negative.

We have shown that for all $F \in \mathscr{F}$ we have

$$\mu(F \cap \Omega^+) \geqslant 0, \quad \mu(F \cap \Omega^-) \leqslant 0.$$

We may thus define measures $\mu_\pm$ by

$$\mu_+(F) := \mu(F \cap \Omega^+), \quad \mu_-(F) := -\mu(F \cap \Omega^-).$$

It is clear that $\mu = \mu_+ - \mu_-$.

Since $\mu(A) = \mu_+(A) - \mu_-(A) \leqslant \mu_+(A) = \mu(A \cap \Omega^+)$ we see that

$$\mu^+(F) = \sup\{\mu(A) : A \in \mathscr{F}, A \subseteq F\} \leqslant \sup\{\mu(A \cap \Omega^+) : A \in \mathscr{F}, A \subseteq F\}.$$

The converse inequality trivially holds, for $A \subseteq F$ implies $A \cap \Omega^+ \subseteq F$. Hence we have equality, and then $\mu_+(A) = \mu(A \cap \Omega^+)$ implies

$$\mu^+(F) = \sup\{\mu(A \cap \Omega^+) : A \in \mathscr{F}, A \subseteq F\}$$
$$= \sup\{\mu_+(A) : A \in \mathscr{F}, A \subseteq F\} = \mu_+(F).$$

The identity $\mu^- = \mu_-$ is proved in the same way. The countable additivity of μ^+ and μ^- is an immediate consequence.

Next, $|\mu|(F) = |\mu^+ - \mu^-|(F) \leqslant |\mu^+|(F) + |\mu^-|(F) = \mu^+(F) + \mu^-(F)$. In the converse direction, write $F = F^+ \cup F^-$ with $F^\pm := F \cap \Omega^\pm$. Then the positivity of F^+ and the negativity of F^- imply

$$|\mu|(F) \geqslant |\mu(F^+)| + |\mu(F^-)| = \mu(F^+) - \mu(F^-) = \mu^+(F) + \mu^-(F).$$

Finally, if ν_1 and ν_2 are finite measures such that $\mu = \nu_1 - \nu_2$,

$$\nu_1(F) \geqslant \nu_1(F \cap \Omega^+) \geqslant \nu_1(F \cap \Omega^+) - \nu_2(F \cap \Omega^+) = \mu(F \cap \Omega^+) = \mu^+(F).$$

The proof that $\nu_2(F) \geqslant \mu^-(F)$ is similar. □

2.4.b The Radon–Nikodým Theorem

If $f : \Omega \to \mathbb{K}$ is integrable with respect to the measure μ, by dominated convergence the formula

$$\nu(F) := \int_F f \, d\mu, \quad F \in \mathscr{F},$$

defines a $\mathbb{K}$-valued measure ν. This measure is *absolutely continuous* with respect to μ, that is, $\mu(F) = 0$ implies $\nu(F) = 0$. The following theorem provides a converse under a σ-finiteness assumption.

Theorem 2.46 (Radon–Nikodým). *Let $(\Omega, \mathscr{F}, \mu)$ be a σ-finite measure space. If the measure $\nu : \mathscr{F} \to \mathbb{K}$ is absolutely continuous with respect to μ, then there exists a unique $g \in L^1(\Omega, \mu)$ such that*

$$\nu(F) = \int_F g \, d\mu, \quad F \in \mathscr{F}.$$

Proof Uniqueness being clear, the proof is devoted to proving existence. By considering real and imaginary parts separately it suffices to consider the case of real scalars. Then, decomposing ν into positive and negative parts via the Hahn–Jordan decomposition, it suffices to consider the case where ν is a finite nonnegative measure.

Consider the set

$$S := \left\{ f \in L^1(\Omega, \mu) : f \geqslant 0, \int_F f \, d\mu \leqslant \nu(F) \text{ for all } F \in \mathscr{F} \right\}.$$

Then $0 \in S$, so S is nonempty. Let

$$M := \sup_{f \in S} \int_\Omega f \, d\mu.$$

For all $f \in S$ we have $\int_\Omega f \, d\mu \leqslant \nu(\Omega)$ and therefore $M \leqslant \nu(\Omega) < \infty$.

Step 1 – In this step we prove that there exists a function $g \in S$ for which the supremum in the definition of M is attained. Let $(f_n)_{n \geqslant 1}$ be a sequence in S with the property that $\lim_{n \to \infty} \int_\Omega f_n \, d\mu = M$. Set $g_n := f_1 \vee \cdots \vee f_n$. Any set $F \in \mathscr{F}$ can be written as a disjoint union of sets $F_1^{(n)}, \ldots, F_n^{(n)} \in \mathscr{F}$ such that $g_j = f_j$ on $F_j^{(n)}$ and therefore

$$\int_F g_n \, d\mu = \sum_{j=1}^n \int_{F_j^{(n)}} f_j \, d\mu \leqslant \sum_{j=1}^n \nu(F_j^{(n)}) = \nu(F).$$

It follows that $g_n \in S$. The sequence $(g_n)_{n \geqslant 1}$ is nondecreasing and therefore its pointwise limit $g := \lim_{n \to \infty} g_n$ is well defined as a $[0, \infty]$-valued function. By the monotone convergence theorem, for all $F \in \mathscr{F}$ we have

$$\int_F g \, d\mu = \lim_{n \to \infty} \int_F g_n \, d\mu \leqslant \nu(F) \tag{2.16}$$

and therefore g takes finite values μ-almost everywhere and belongs to S. Moreover,

$$M = \lim_{n \to \infty} \int_\Omega f_n \, d\mu \leqslant \lim_{n \to \infty} \int_\Omega g_n \, d\mu = \int_\Omega g \, d\mu \leqslant M$$

and therefore equality holds at all places. This proves that the supremum in the definition of M is attained by the function $g \in S$.

Step 2 – Under the additional assumption that μ is a finite measure, we show next that g has the required properties. To this end we must show that $\eta = 0$, where the finite measure η is defined by

$$\eta(F) := \nu(F) - \int_F g \, d\mu, \quad F \in \mathscr{F}.$$

Assume, for a contradiction, that $\eta(\Omega) > 0$. Consider the real-valued measures

$$\eta_n := \eta - \frac{1}{n}\mu, \quad n \geqslant 1.$$

(It is here that we use the assumption that μ is finite; without this assumption η_n would not be a real-valued measure.) For each $n \geqslant 1$ we decompose $\Omega = \Omega_n^+ \cup \Omega_n^-$ with respect to η_n as in Theorem 2.45, and set $\Omega^- := \bigcap_{n \geqslant 1} \Omega_n^-$. If we had $\eta(\Omega^-) > 0$, then for large enough $n \geqslant 1$ we have

$$0 \leqslant \eta(\Omega^-) - \frac{1}{n}\mu(\Omega^-) = \eta_n(\Omega^-) \leqslant 0$$

and therefore, upon letting $n \to \infty$, we obtain $\eta(\Omega^-) = 0$. This contradiction proves that $\eta(\Omega^-) = 0$. If we had $\eta(\Omega_n^+) = 0$ for all $n \geqslant 1$ it would follow that $\eta(\Omega) = \eta(\Omega_n^-)$ for all $n \geqslant 1$, and therefore $\eta(\Omega) = \eta(\Omega^-) = 0$. Having assumed that $\eta(\Omega) > 0$, we conclude that $\eta(\Omega_n^+) > 0$ for some $n \geqslant 1$. If $F \in \mathscr{F}$ is a subset of Ω_n^+, then

$$\eta(F) - \frac{1}{n}\mu(F) = \eta_n(F) = \eta_n(F \cap \Omega_n^+) = \eta_n^+(F) \geqslant 0$$

and therefore $\eta(F) \geqslant \frac{1}{n}\mu(F)$. Letting $h := g + \frac{1}{n}\mathbf{1}_{\Omega_n^+}$ we obtain, for arbitrary $F \in \mathscr{F}$,

$$
\begin{aligned}
\int_F h \, d\mu = \int_F g \, d\mu + \frac{1}{n}\mu(F \cap \Omega_n^+) &\leqslant \int_F g \, d\mu + \eta(F \cap \Omega_n^+) \\
&= \int_{F \cap \Omega_n^-} g \, d\mu + \nu(F \cap \Omega_n^+) \\
&\leqslant \nu(F \cap \Omega_n^-) + \nu(F \cap \Omega_n^+) = \nu(F),
\end{aligned}
$$

using (2.16) in the last inequality. This proves that $h \in S$. Then, by the definition of M,

$$
M \geqslant \int_\Omega h \, d\mu = \int_\Omega g \, d\mu + \frac{1}{n}\mu(\Omega \cap \Omega_n^+) = M + \frac{1}{n}\mu(\Omega_n^+).
$$

Since $M < \infty$, this is only possible if $\mu(\Omega_n^+) = 0$. By (2.16) and absolute continuity this would imply $\int_{\Omega_n^+} g \, d\mu \leqslant \nu(\Omega_n^+) = 0$ and therefore, by the definition of η and the nonnegativity of g,

$$
0 < \eta(\Omega_n^+) = -\int_{\Omega_n^+} g \, d\mu \leqslant 0.
$$

This contradiction concludes the proof that $\eta(\Omega) = 0$.

Step 3 – For finite measures μ the theorem has now been proved. It remains to extend the result to the σ-finite case. Again it suffices to consider the case where ν is a finite nonnegative measure.

Write $\Omega = \bigcup_{n \geqslant 1} \Omega_n$, where the sets $\Omega_n \in \mathscr{F}$ are disjoint and satisfy $\mu(\Omega_n) < \infty$. Define the nonnegative function g on Ω by

$$
g(\omega) := g_n(\omega) \text{ for } \omega \in \Omega_n, \quad n \geqslant 1,
$$

where the nonnegative functions $g_n \in L^1(\Omega_n, \mu|_{\Omega_n})$ are given by Step 2 applied to Ω_n, that is,

$$
\nu(F \cap \Omega_n) = \int_{F \cap \Omega_n} g_n \, d\mu = \int_{F \cap \Omega_n} g \, d\mu, \quad F \in \mathscr{F}.
$$

By additivity, this implies

$$
\nu\left(F \cap \bigcup_{n=1}^N \Omega_n\right) = \int_{F \cap \bigcup_{n=1}^N \Omega_n} g \, d\mu, \quad F \in \mathscr{F}.
$$

Letting $N \to \infty$, by monotone convergence we obtain

$$
\nu(F) = \int_F g \, d\mu, \quad F \in \mathscr{F}.
$$

Taking $F = \Omega$ and using that ν is finite, we see that g is integrable with respect to ν, that is, we have $g \in L^1(\Omega, \nu)$. The function g has the required properties. $\qquad\square$

An alternative proof of the Radon–Nikodým theorem, based on Hilbert space methods, is outlined in Problem 3.22.

Example 2.47. If $f : \Omega \to \mathbb{K}$ is integrable with respect to μ, then

$$\nu(F) := \int_F f \, d\mu, \quad F \in \mathscr{F},$$

defines a $\mathbb{K}$-valued measure and for all $F \in \mathscr{F}$ we have

$$|\nu|(F) = \int_F |f| \, d\mu.$$

If f is real-valued, then ν is a real measure and

$$\nu^{\pm}(F) = \int_F f^{\pm} \, d\mu.$$

To prove the first assertion, let $A_1, \ldots, A_n \in \mathscr{F}$ be disjoint subsets of F. Then

$$\sum_{j=1}^n |\nu(A_j)| \leqslant \sum_{j=1}^n \int_{A_j} |f| \, d\mu \leqslant \int_F |f| \, d\mu,$$

which gives the upper bound '$\leqslant$'. To prove the lower bound '$\geqslant$', we use Proposition F.1 to choose simple functions $g_n : \Omega \to \mathbb{K}$ such that $g_n \to 1_F f$ and $0 \leqslant |g_n| \leqslant 1_F |f|$, say $g_n = \sum_{j=1}^{N_n} c_j^{(n)} 1_{A_j^{(n)}}$ with $A_1^{(n)}, \ldots, A_{N_n}^{(n)} \in \mathscr{F}$ disjoint subsets of F. Then

$$\int_F |f| \, d\mu = \lim_{n \to \infty} \sum_{j=1}^{N_n} |c_j^{(n)}| \mu(A_j^{(n)}) \leqslant \limsup_{n \to \infty} \sum_{j=1}^{N_n} |\nu(A_j^{(n)})| \leqslant |\nu|(F).$$

To prove the second assertion we note that the sets $\Omega^+ = \{f \geqslant 0\}$ and $\Omega^- = \{f < 0\}$ satisfy the requirements of the second part of Theorem 2.45, and the second part of the proof of the theorem shows that the decomposition $\mu = \mu^+ - \mu^-$ is obtained from any such decomposition of Ω. For real scalars this also gives a second proof of the first assertion:

$$|\nu|(F) = \nu^+(F) + \nu^-(F) = \int_F f^+ + f^- \, d\mu = \int_F |f| \, d\mu.$$

Example 2.48. If μ is a $\mathbb{K}$-valued measure, then μ is absolutely continuous with respect to its variation $|\mu|$. By the Radon–Nikodým theorem, there exists $h \in L^1(\Omega, |\mu|)$ such that

$$\mu(F) = \int_F h \, d|\mu|, \quad F \in \mathscr{F}.$$

By the result of Example 2.47,

$$|\mu|(F) = \int_F |h| \, d|\mu|, \quad F \in \mathscr{F},$$

so $|h| = 1$ μ-almost everywhere.

2.4.c Integration with Respect to $\mathbb{K}$-Valued Measures

A measurable function f is said to be *integrable* with respect to a $\mathbb{K}$-valued measure μ if it is integrable with respect to $|\mu|$. The function f is integrable with respect to a real measure μ if and only if it is integrable with respect to the measures μ^+ and μ^-, where $\mu = \mu^+ - \mu^-$ is the Jordan decomposition, and f is integrable with respect to a complex measure μ if and only if f is integrable with respect to the real and imaginary parts of μ.

The integral of an integrable function f with respect to a real measure μ is defined by

$$\int_\Omega f \, \mathrm{d}\mu := \int_\Omega f \, \mathrm{d}\mu^+ - \int_\Omega f \, \mathrm{d}\mu^-,$$

and the integral of an integrable function f with respect to a complex measure μ by

$$\int_\Omega f \, \mathrm{d}\mu := \int_\Omega f \, \mathrm{d}\mathrm{Re}\,\mu + i \int_\Omega f \, \mathrm{d}\mathrm{Im}\,\mu.$$

Proposition 2.49. *If f is integrable with respect to a $\mathbb{K}$-valued measure μ, then*

$$\left| \int_\Omega f \, \mathrm{d}\mu \right| \leqslant \int_\Omega |f| \, \mathrm{d}|\mu|.$$

Proof First let $f = \sum_{n=1}^N c_n \mathbf{1}_{F_n}$ be a simple function, with the sets $F_n \in \mathscr{F}$ disjoint. Then

$$\left| \int_\Omega f \, \mathrm{d}\mu \right| = \left| \sum_{n=1}^N c_n \mu(F_n) \right| \leqslant \sum_{n=1}^N |c_n| |\mu(F_n)| \leqslant \sum_{n=1}^N |c_n| |\mu|(F_n) = \int_\Omega |f| \, \mathrm{d}|\mu|.$$

The general case follows from this by observing that the simple functions are dense in $L^1(\Omega, |\mu|)$ and that $f_n \to f$ in $L^1(\Omega, |\mu|)$ implies $\int_\Omega |f_n - f| \, \mathrm{d}\nu \to 0$ for each of the measures $\nu \in \{\mathrm{Re}\,\mu, \mathrm{Im}\,\mu, \mu^+, \mu^-\}$. □

A more elegant, but less elementary, alternative definition of the integral $\int_\Omega f \, \mathrm{d}\mu$ can be given with the help of the Radon–Nikodým theorem. Indeed, defining $\int_\Omega f \, \mathrm{d}\mu$ as above, by the result of Example 2.48 for functions $f \in L^1(\Omega, |\mu|)$ we have the identity

$$\int_\Omega f \, \mathrm{d}\mu = \int_\Omega f h \, \mathrm{d}|\mu|,$$

where $\mathrm{d}\mu = h \, \mathrm{d}|\mu|$ as in the example (note that $fh \in L^1(\Omega, |\mu|)$ since $|h| = 1$ μ-almost everywhere). This identity could be taken as an alternative definition for the integral $\int_\Omega f \, \mathrm{d}\mu$.

2.5 Banach Lattices

Over the real scalar field, all Banach spaces discussed in this chapter are examples of Banach lattices, a class of Banach spaces that will be briefly discussed in this section.

The main result, Theorem 2.57, shows that any complete norm on a Banach lattice X which is monotone with respect to the partial order of X is equivalent to the given norm of X.

Let $(S, \leqslant)$ be a partially ordered set and let S' be a subset of S. An element $x \in S$ is said to be a *lower bound* for S' if we have $x \leqslant x'$ for all $x' \in S'$. Such an element is called a *greatest lower bound* for S' if $y \leqslant x$ holds for every lower bound y for S'. Similarly an element $x \in S$ is said to be an *upper bound* for S' if we have $x' \leqslant x$ for all $x' \in S'$, and such an element is called a *least upper bound* for S' if $x \leqslant y$ holds for every upper bound y for S'. Greatest lower bounds and least upper bounds, if they exist, are unique.

Definition 2.50 (Lattices). A partially ordered set $(S, \leqslant)$ is called a *lattice* if every pair of elements has a greatest lower bound and a least upper bound.

The greatest lower bound and the least upper bound of the pair $\{x, y\} \subseteq S$ in a partially ordered set S will be denoted by $x \wedge y$ and $x \vee y$, respectively.

Definition 2.51 (Vector lattices). A *vector lattice* is a partially ordered real vector space $(V, \leqslant)$ with the following properties:

 (i) $(V, \leqslant)$ is a lattice;
 (ii) for all $0 \leqslant c \in \mathbb{R}$ and $u, v \in V$ we have $u \leqslant v \Rightarrow cu \leqslant cv$;
 (iii) for all $u, v, w \in V$ we have $u \leqslant v \Rightarrow u + w \leqslant v + w$.

Let $(V, \leqslant)$ be a vector lattice. If $u, u', v'v' \in V$ satisfy $u \leqslant v$ and $u' \leqslant v'$, then $u + u' \leqslant v + u'$ and $v + u' \leqslant v + v'$. Thus,

$$[u \leqslant v \text{ and } u' \leqslant v'] \Rightarrow u + u' \leqslant v + v'$$

by transitivity. Also, if $u \leqslant v$, then $-v = u + (-u - v) \leqslant v + (-u - v) = -u$, and the converse inequality is obtained similarly. Thus,

$$u \leqslant v \Leftrightarrow (-v) \leqslant (-u). \tag{2.17}$$

For $v \in V$ we define

$$v^+ := v \vee 0, \quad v^- := (-v) \vee 0, \quad |v| := v \vee (-v).$$

If $0 \leqslant c \in \mathbb{R}$, then $(cv)^{\pm} = cv^{\pm}$, and if $c \in \mathbb{R}$, then $|cv| = |c||v|$; the easy proofs are left to the reader. Furthermore, from $\pm(u + v) \leqslant |u| + |v|$ it follows that

$$|u + v| \leqslant |u| + |v|. \tag{2.18}$$

The next proposition lists some slightly less trivial identities.

Proposition 2.52. *Let $(V, \leqslant)$ be a vector lattice. Then for all $u, v, w \in V$ we have:*

(1) $(-u) \wedge (-v) = -(u \vee v)$;

(2) $u + (v \vee w) = (u + v) \vee (u + w)$;
(3) $u + v = u \wedge v + u \vee v$;
(4) $v = v^+ - v^-$;
(5) $|v| = v^+ + v^-$.

The representation of v as the difference of two positive elements in (4) is minimal in a sense explained in Problem 2.42.

Proof Let $u, v, w \in V$.

(1): We have $u \leqslant u \vee v$, so $-(u \vee v) \leqslant -u$ by (2.17). In the same way we obtain $-(u \vee v) \leqslant -v$. If follows that $-(u \vee v)$ is a lower bound for $\{-u, -v\}$. To prove that it is the greatest lower bound, we must show that if $w \leqslant -u$ and $w \leqslant -v$, then $w \leqslant -(u \vee v)$. This follows by noting that $u \leqslant -w$ and $v \leqslant -w$, so $-w$ is an upper bound for $\{u, v\}$ and therefore $u \vee v \leqslant -w$. By (2.17) this implies $w \leqslant -(u \vee v)$ as required.

(2): We have $u + v \leqslant u + (v \vee w)$ and $u + w \leqslant u + (v \vee w)$, so $u + (v \vee w)$ is an upper bound for $\{u + v, u + w\}$. To prove that is the least upper bound we must show that if $u + v \leqslant x$ and $u + w \leqslant x$, then $u + (v \vee w) \leqslant x$. But $v \leqslant x - u$ and $w \leqslant x - u$ imply $v \vee w \leqslant x - u$ and therefore $u + (v \vee w) \leqslant x$ as desired.

(3): In view of (1) we must show that $u + v + (-u) \wedge (-v) = u \wedge v$.
We have $u + v + (-u) \wedge (-v) \leqslant u + v + (-v) = u$ and similarly $u + v + (-u) \wedge (-v) \leqslant v$. It follows that $u + v + (-u) \wedge (-v)$ is a lower bound for $\{u, v\}$. To prove that it is the greatest lower bound we must show that if $w \leqslant u$ and $w \leqslant v$, then $w \leqslant u + v + (-u) \wedge (-v)$, or equivalently $w - u - v \leqslant (-u) \wedge (-v)$. By (2.17) and (1), this in turn is equivalent to $u \vee v \leqslant u + v - w$. To prove this inequality we note that $w \leqslant u$ implies $0 \leqslant u - w$ and hence $v \leqslant u + v - w$. In the same way we obtain $u \leqslant u + v - w$, and together these inequalities imply $u \vee v \leqslant u + v - w$ as desired.

(4): Taking $u = 0$ in (3) and using (1) we obtain $v = 0 \wedge v + 0 \vee v = v^+ - (0 \wedge (-v)) = v^+ - v^-$.

(5): By (2), $|v| = v \vee (-v)$ implies $|v| - v = 0 \vee (-2v) = (2v)^- = 2v^-$. It follows that $|v| = v + 2v^- = v^+ - v^1 + 2v^- = v^+ + v^-$. $\qquad\square$

Definition 2.53 (Normed vector lattices). A *normed vector lattice* is a triple $(X, \|\cdot\|, \leqslant)$ with the following properties:

(i) the pair $(X, \|\cdot\|)$ is a real normed space;
(ii) the pair $(X, \leqslant)$ is vector lattice;
(iii) for all $x, y \in X$ we have $|x| \leqslant |y| \Rightarrow \|x\| \leqslant \|y\|$.

In any normed vector lattice, as an immediate consequence of (iii) we have

$$\| |x| \| = \|x\|. \tag{2.19}$$

Moreover, the lattice operations $x \mapsto x^+$, $x \mapsto x^-$, $x \mapsto |x|$ are continuous. Indeed, by (2.18) we have $|x| - |y| \leqslant |x - y|$ and therefore, by (2.19),

$$\big\| \, |x| - |y| \, \big\| \leqslant \big\| \, |x - y| \, \big\| = \|x - y\|.$$

This gives the continuity of $x \mapsto |x|$. The other two assertions follow from this by noting that $x^+ = \frac{1}{2}(x + |x|)$ and $x^- = x^+ - x$. As a consequence we note that the *positive cone*

$$X^+ := \{x \in X : x \geqslant 0\}$$

is closed.

Definition 2.54 (Banach lattices). A *Banach lattice* is a complete normed vector lattice.

The spaces c_0, ℓ^p, $C(K)$, and $L^p(\Omega)$ with $1 \leqslant p \leqslant \infty$, are Banach lattices with respect to their natural pointwise ordering, and $M(\Omega)$ is a Banach lattices with respect to the ordering given by declaring $\mu \leqslant \nu$ if and only if the measure $\nu - \mu$ is nonnegative; the greatest lower bound and least upper bound of μ and ν are given by

$$\mu \wedge \nu = \mu - (\mu - \nu)^+ = \nu - (\nu - \mu)^+,$$
$$\mu \vee \nu = \mu + (\nu - \mu)^+ = \nu + (\mu - \nu)$$

respectively (cf. Problem 2.41); here, $(\nu - \mu)^+$ is defined as in Theorem 2.45. Alternatively one may use the analogues for $M(\Omega)$ of the formulas appearing in Theorem 4.5. The Jordan decomposition of a real measure now becomes a special case of Proposition 2.52(4).

Definition 2.55. Let V and W be vector lattices. A linear operator $T : V \to W$ is said to be *positivity preserving* if $v \geqslant 0$ implies $Tv \geqslant 0$.

If $T : V \to W$ is positivity preserving, then

$$|Tv| \leqslant T|v|, \quad v \in V. \tag{2.20}$$

Indeed, from $v \leqslant |v|$ we have $Tv \leqslant T|v|$ and from $0 \leqslant |v|$ we have $0 = T0 \leqslant T|v|$. Combining these inequalities gives $(Tv)^+ \leqslant T|v|$. In the same way we see that $(Tv)^- \leqslant T|v|$, and the claim follows.

Theorem 2.56. *Let X and Y be Banach lattices. Every positivity preserving linear operator $T : X \to Y$ is bounded.*

Proof Reasoning by contradiction, suppose that T is not bounded. Then for all $n \geqslant 1$ there is a norm one vector $x_n \in X$ such that $\|Tx_n\| \geqslant n^3$. By (2.19) and (2.20), upon replacing x_n by $|x_n|$ we may assume that $x_n \geqslant 0$ for all $n \geqslant 1$. In view of $\sum_{n \geqslant 1} \|x_n\|/n^2 < \infty$ the sum $\sum_{n \geqslant 1} x_n/n^2$ converges in X. For all $1 \leqslant n \leqslant N$ we have $x_n/n^2 \leqslant \sum_{m=1}^{N} x_m/m^2$, and the closedness of the positive cone X^+ implies that for all $n \geqslant 1$ we have $x_n/n^2 \leqslant \sum_{m \geqslant 1} x_m/m^2$. Hence, for all $n \geqslant 1$,

$$n \leqslant \frac{1}{n^2} \|Tx_n\| \leqslant \left\| T \sum_{m \geqslant 1} \frac{x_m}{m^2} \right\|.$$

This contradiction completes the proof. $\qquad\qquad\square$

This theorem has an interesting consequence:

Theorem 2.57. *Any two norms which turn a vector lattice into a Banach lattice are equivalent.*

Proof Suppose the vector lattice $(X, \leqslant)$ is a Banach lattice with respect to the norms $\| \cdot \|$ and $\| \cdot \|'$. Then by Theorem 2.56 the identity mapping from $(X, \| \cdot \|)$ to $(X, \| \cdot \|')$ and its inverse are bounded. $\qquad\qquad\square$

Problems

2.1 Let $1 \leqslant p \leqslant \infty$. Show that if $a \in \ell^p$, then $a \in \ell^q$ for all $p \leqslant q \leqslant \infty$ and

$$\|a\|_\infty \leqslant \|a\|_q \leqslant \|a\|_p, \quad \lim_{q \to \infty} \|a\|_q = \|a\|_\infty.$$

Hint: First show that it suffices to consider sequences $a = (a_n)_{n \geqslant 1}$ satisfying $|a_n| \leqslant 1$ for all $n \geqslant 1$.

2.2 Show that c_0 and ℓ^p with $1 \leqslant p < \infty$ are separable, but ℓ^∞ is not.
Hint: ℓ^∞ contains an uncountable family $(a^{(i)})_{i \in I}$ such that $\|a^{(i)} - a^{(i')}\| = 1$ for all $i \neq i'$. Prove that a normed space X containing such a sequence is nonseparable.

2.3 Prove the completeness assertions at the end of Section 2.2.a.

2.4 Let K be a compact metric space. Our aim is to prove that if $(f_n)_{n \geqslant 1}$ is a sequence in $C(K)$ satisfying $f_1(x) \geqslant f_2(x) \geqslant \cdots \geqslant 0$ and $\lim_{n \to \infty} f_n(x) = 0$ for all $x \in K$, then $\lim_{n \to \infty} f_n = 0$ uniformly on K. This result, known as *Dini's theorem*, provides one of the rare instances where pointwise convergence implies uniform convergence.

 (a) Reasoning by contradiction, show that if the result is false, then there exists an $\varepsilon > 0$, a sequence $(x_n)_{n \geqslant 1}$ in K, and an $x \in K$, such that $\lim_{n \to \infty} x_n = x$ and $f_n(x_n) \geqslant \frac{1}{2}\varepsilon$ for all $n \geqslant 1$.

 (b) Using that also $f_n(x) \downarrow 0$ as $n \to \infty$ and $f_n(x_n) \leqslant f_m(x_n)$ when $n \geqslant m$, show that this leads to a contradiction.

2.5 Find a sequence $(f_n)_{n \geqslant 1}$ in $C_b(0,1)$ such that $0 \leqslant f_{n+1}(x) \leqslant f_n(x) \leqslant 1$ for all $x \in (0,1)$ and $n \geqslant 1$ and $f_n(x) \downarrow 0$ for all $x \in (0,1)$, but $\|f_n\|_\infty \not\to 0$. Compare this with Dini's theorem (Problem 2.4).

2.6   Let K be a compact topological space and X be a Banach space. Prove that the space $C(K;X)$ of all continuous functions $f : K \to X$ is a Banach space with respect to the supremum norm $\|f\|_\infty := \sup_{x \in K} \|f(x)\|$.

2.7   Let D be a bounded open subset of $\mathbb{R}^d$. By $C^k(\overline{D})$ we denote the space of functions $f : \overline{D} \to \mathbb{K}$ that are k times continuously differentiable on D and all of whose partial derivatives $\partial^\alpha f$ extend continuously to $\overline{D}$ for all multi-indices $\alpha = (\alpha_1, \ldots, \alpha_d) \in \mathbb{N}^d$ satisfying $|\alpha| := \alpha_1 + \cdots + \alpha_d \leqslant k$. Here, $\partial^\alpha := \partial_1^{\alpha_1} \circ \cdots \circ \partial_d^{\alpha_d}$, where ∂_j is the partial derivative in the jth direction. Prove that $C^k(\overline{D})$ is a Banach space with respect to the norm

$$\|f\|_{C^k(\overline{D})} := \max_{|\alpha| \leqslant k} \|\partial^\alpha f\|_\infty.$$

2.8   Consider the vector space $P[0,1]$ of all polynomials on $[0,1]$.

   (a) Show that

$$\left\| t \mapsto \sum_{n=0}^{N} c_n t^n \right\| := \max \left\{ \left\| t \mapsto \sum_{k=0}^{\lfloor \frac{N}{2} \rfloor} c_{2k} t^{2k} \right\|_\infty, 2 \left\| t \mapsto \sum_{k=1}^{\lceil \frac{N}{2} \rceil} c_{2k-1} t^{2k-1} \right\|_\infty \right\}$$

   defines a norm on $P[0,1]$. Here, $\lfloor y \rfloor$ is the greatest integer $n \leqslant y$ and $\lceil y \rceil$ is the least integer $n \geqslant y$.

   (b) Show that the functions $1, t^2, t^4, \ldots$ span a subspace of $P[0,1]$ which is dense with respect to the supremum norm.

   (c) Conclude that two different norms on a normed space may agree on a subspace which is dense with respect to one of these norms.

2.9   We prove the existence of smooth functions with various properties.

   (a) Show that the function $f : \mathbb{R} \to [0, \infty)$ defined by

$$f(x) := \begin{cases} \exp\left(-1/x^2\right), & x > 0, \\ 0, & x \leqslant 0, \end{cases}$$

   belongs to $C^\infty(\mathbb{R})$.

   (b) Show that there exists a function $f \in C_c^\infty(0,1)$ such that $f \geqslant 0$ pointwise and $\int_0^1 f(x)\,dx = 1$.

   (c) Show that if $D \subseteq \mathbb{R}^d$ is open and nonempty, there exists a function $f \in C_c^\infty(D)$ such that $f \geqslant 0$ pointwise and $\int_D f(x)\,dx = 1$.

   (d) Show that if $f \in C_c^\infty(\mathbb{R}^d)$ and $g : \mathbb{R}^d \to \mathbb{K}$ is continuous, then the convolution $f * g$ is smooth and

$$\partial^\alpha (f * g) = (\partial^\alpha f) * g$$

   for every multi-index $\alpha \in \mathbb{N}^d$; notation is as in Problem 2.7.

(e) Show that if $K \subseteq D \subseteq \mathbb{R}^d$, where K is nonempty and compact and D is open, then there exists a function $f \in C_c^\infty(D)$ such that $0 \leqslant f \leqslant 1$ pointwise and $f \equiv 1$ on K.

Hint: Let $\delta := d(K, \complement D)$ and put

$$K' := \left\{ x \in D : d(x, K) \leqslant \frac{1}{3}\delta \right\}, \quad D' := \left\{ x \in D : d(x, K) < \frac{2}{3}\delta \right\}.$$

Apply part (c) to select a nonnegative function $f \in C_c^\infty(B(0; \frac{1}{3}\delta))$ satisfying $\int_{B(0;\frac{1}{3}\delta)} f(x)\,dx = 1$ and apply part (d) to the function

$$g(x) := \frac{d(x, \complement D')}{d(x, \complement D') + d(x, K')}, \quad x \in D.$$

2.10 Let

$$F := \left\{ f \in C[0,1] : 0 \leqslant f \leqslant 1, \ f(0) = 0, \ f(1) = 1 \right\}$$

and consider the linear mapping $T : C[0,1] \to C[0,1]$ defined by

$$(Tf)(t) = t f(t), \quad t \in [0,1].$$

(a) Show that F is bounded, convex, and closed in $C[0,1]$.

(b) Show that T maps F into F and satisfies

$$\|Tf - Tg\|_\infty < \|f - g\|_\infty$$

for all $f, g \in F$, $f \neq g$.

(c) Show that T has no fixed point in F.

(d) Compare this result with the Banach fixed point theorem.

2.11 Let $1 \leqslant p \leqslant \infty$.

(a) Show that for $1 \leqslant p < \infty$ the space $L^p(0,1)$ is the completion of $C[0,1]$ with respect to the norm

$$\|f\|_p = \left(\int_0^1 |f(t)|^p \, dt \right)^{1/p}.$$

(b) Show that $C[0,1]$ can be identified in a natural way with a proper closed subspace of $L^\infty(0,1)$.

2.12 Prove the assertion in Remark 2.27. Can the σ-finiteness condition be omitted?

2.13 Let $(\Omega, \mathscr{F}, \mu)$ be a measure space and let $f_n : \Omega \to \mathbb{K}$ ($n \geqslant 1$) and $f : \Omega \to \mathbb{K}$ be bounded measurable functions. Show that if $f_n \to f$ in $L^\infty(\Omega)$, then there is a μ-null set N such that $\lim_{n\to\infty} \sup_{\omega \in \complement N} |f_n(\omega) - f(\omega)| = 0$. Compare this with Corollary 2.21.

2.14 Show that passing to a subsequence is necessary for Corollary 2.21 to be true.
Hint: Consider the unit circle $\mathbb{T} = \{z \in \mathbb{C} : |z| = 1\}$ with its Lebesgue measure. Let $t_n := \sum_{m=1}^{n} \frac{1}{m}$ and consider the indicator functions of the arcs $I_n := \{e^{2\pi i t} : t \in (t_n, t_{n+1})\}$.

2.15 Consider the set

$$S := \{f \in L^1(0,1) : f(t) > 0 \text{ for almost all } t \in (0,1)\}.$$

  (a) Determine whether S is an open subset of $L^1(0,1)$.
  (b) Characterise the functions belonging to the closure $\overline{S}$.

Consider the set

$$S' := \{f \in L^1(0,1) : f(t) \geqslant 0 \text{ for almost all } t \in (0,1)\}.$$

  (c) Determine whether S' is a closed subset of $L^1(0,1)$.
  (d) Characterise the functions belonging to the interior of S'.

2.16 Let $(\Omega, \mathscr{F}, \mu)$ be a measure space and let $\mathscr{G} \subseteq \mathscr{F}$ be a sub-σ-algebra. Show that for all $1 \leqslant p \leqslant \infty$ the set $L^p(\Omega, \mathscr{G})$ consisting of all $f \in L^p(\Omega)$ that are equal μ-almost everywhere to a $\mathscr{G}$-measurable function is a closed subspace of $L^p(\Omega)$.

2.17 For $n \in \mathbb{N}$ and $j \in \{0, 1, \ldots, 2^n - 1\}$ we consider the interval $I_{j,n} := (\frac{j}{2^n}, \frac{j+1}{2^n}) \subseteq (0,1)$. Let $1 \leqslant p < \infty$ and define the operators $E_n : L^p(0,1) \to L^p(0,1)$, $n \geqslant 1$, by

$$f \mapsto E_n f := \sum_{j=0}^{2^n - 1} \mathbf{1}_{I_{j,n}} \cdot \frac{1}{|I_{j,n}|} \int_{I_{j,n}} f(t)\,\mathrm{d}t, \quad f \in L^p(0,1).$$

  (a) Show that each E_n is bounded on $L^p(0,1)$ with norm $\|E_n\| = 1$.
  (b) Show that for all $f \in L^p(0,1)$ we have $\lim_{n \to \infty} E_n f = f$ in $L^p(0,1)$.
    Hint: Consider what happens for functions in the linear span of the set $\{\mathbf{1}_{I_{j,n}} : 0 \leqslant j \leqslant 2^n - 1, n \in \mathbb{N}\}$.

2.18 Using Young's inequality, show that if $f \in L^p(\mathbb{R}^d)$, $g \in L^q(\mathbb{R}^d)$, $h \in L^r(\mathbb{R}^d)$ with $1 \leqslant p, q, r \leqslant \infty$ such that $\frac{1}{p} + \frac{1}{q} + \frac{1}{r} = 2$, then

$$\int_{\mathbb{R}^d} \int_{\mathbb{R}^d} |f(x)g(x-y)h(y)|\,\mathrm{d}x\,\mathrm{d}y \leqslant \|f\|_p \|g\|_q \|h\|_r.$$

2.19 Write out a proof of Corollary 2.30.

2.20 Let $(\Omega, \mathscr{F}, \mu)$ be a finite measure space and let $1 \leqslant p \leqslant \infty$. Prove that

$$\|f\|_p := \left| \int_\Omega f\,\mathrm{d}\mu \right| + \left\| f - \left(\int_\Omega f\,\mathrm{d}\mu \right) \mathbf{1} \right\|_p$$

defines an equivalent norm on $L^p(\Omega)$. Here, $\|\cdot\|_p$ is the usual norm on $L^p(\Omega)$.

2.21 Let $(\Omega, \mathscr{F}, \mu)$ be a measure space and fix $1 \leqslant p, q \leqslant \infty$.

(a) Prove that $L^p(\Omega) \cap L^q(\Omega)$ is a Banach space with respect to the norm

$$\|f\|_{L^p(\Omega)\cap L^q(\Omega)} := \max\{\|f\|_{L^p(\Omega)}, \|f\|_{L^q(\Omega)}\}.$$

(b) Prove that $L^p(\Omega) + L^q(\Omega) = \{g + h : g \in L^p(\Omega), h \in L^q(\Omega)\}$ is a Banach space with respect to the norm

$$\|f\|_{L^p(\Omega)+L^q(\Omega)}$$
$$:= \inf\{\|g\|_{L^p(\Omega)} + \|h\|_{L^q(\Omega)} : f = g + h, \ g \in L^p(\Omega), h \in L^q(\Omega)\}.$$

Hint: For the proof of completeness use Proposition 1.3.

(c) Prove that if $1 \leqslant p \leqslant r \leqslant q \leqslant \infty$, then

$$L^p(\Omega) \cap L^q(\Omega) \subseteq L^r(\Omega) \subseteq L^p(\Omega) + L^q(\Omega)$$

and that the inclusion mappings are continuous.

Hint: Write $f = \mathbf{1}_{\{|f|\leqslant 1\}}f + \mathbf{1}_{\{|f|>1\}}f$.

(d) Prove that if $1 \leqslant p \leqslant r \leqslant q \leqslant \infty$ and $0 \leqslant \theta \leqslant 1$ are such that $\frac{1}{r} = \frac{1-\theta}{p} + \frac{\theta}{q}$, then for all $f \in L^p(\Omega, \mu) \cap L^q(\Omega, \mu)$ we have

$$\|f\|_r \leqslant \|f\|_p^{1-\theta} \|f\|_q^{\theta}.$$

Hint: Use Hölder's inequality with suitable exponents.

2.22 Prove *Lusin's theorem*: If $D \subseteq \mathbb{R}^d$ is open and bounded, and $f : D \to \mathbb{K}$ is measurable, then for every $\varepsilon > 0$ there exists a function $g \in C_c(D)$ such that

$$|\{x \in D : f(x) \neq g(x)\}| < \varepsilon.$$

Hint: Study the proof of Proposition 2.29.

2.23 Let $(\Omega, \mathscr{F}, \mu)$ be a finite measure space, let $1 \leqslant p \leqslant \infty$, and suppose that $(f_n)_{n\geqslant 1}$ is a bounded sequence in $L^p(\Omega)$ converging to a measurable function f μ-almost everywhere.

(a) Using Fatou's lemma show that $f \in L^p(\Omega)$ and $\|f\|_p \leqslant \liminf_{n\to\infty}\|f_n\|_p$.

(b) Show that for $p = 1$ we have $\lim_{n\to\infty} f_n = f$ in $L^1(\Omega)$.

Hint: First show that for every $\varepsilon > 0$ there exists an $r \geqslant 0$ such that

$$\sup_{n\geqslant 1} \int_{\{|f_n|>r\}} |f_n|\,d\mu < \varepsilon.$$

(c) Does the result of (b) extend to exponents $1 < p \leqslant \infty$?

2.24 Let $(\Omega, \mathscr{F}, \mu)$ be a probability space and let $\phi : \mathbb{K} \to \mathbb{R}$ be a convex function. Prove *Jensen's inequality*: If a function $f \in L^1(\Omega)$ has the property that $\phi \circ f$ is integrable, then

$$\phi\left(\int_\Omega f\,d\mu\right) \leqslant \int_\Omega \phi \circ f\,d\mu.$$

Hint: A convex function ϕ is the pointwise supremum of all affine functions $\psi(x) = ax + b$ satisfying $\psi \leqslant \phi$ pointwise.

2.25 On $\mathbb{R}_+ = (0, \infty)$ we consider the Borel measure μ given by

$$\mu(B) := \int_B \frac{1}{t}\, dt, \quad B \in \mathscr{B}(\mathbb{R}_+).$$

(a) Show that for all $B \in \mathscr{B}(\mathbb{R}_+)$ and $s \in \mathbb{R}_+$ we have

$$\mu(B) = \mu(sB), \quad \mu(B) = \mu(B^{-1}),$$

where $sB := \{st : t \in B\}$ and $B^{-1} := \{t^{-1} : t \in B\}$.

(b) For $h \in L^1(\mathbb{R}_+, \frac{dt}{t})$ and $s \in \mathbb{R}_+$ show that

$$\int_0^\infty h(st)\frac{dt}{t} = \int_0^\infty h(t)\frac{dt}{t} = \int_0^\infty h(t^{-1})\frac{dt}{t}.$$

Fix $1 \leqslant p < \infty$. For $f \in L^p(\mathbb{R}_+, \frac{dt}{t})$ and $g \in L^1(\mathbb{R}_+, \frac{dt}{t})$ we define the *multiplicative convolution*

$$f \diamond g(t) := \int_0^\infty f(t/s)g(s)\frac{ds}{s}, \quad t \in \mathbb{R}_+.$$

(c) Show that the multiplicative convolution is well defined for almost all $t \in \mathbb{R}_+$ and that the following analogue of Young's inequality holds:

$$\|f \diamond g\|_{L^p(\mathbb{R}_+, \frac{dt}{t})} \leqslant \|f\|_{L^p(\mathbb{R}_+, \frac{dt}{t})}\|g\|_{L^1(\mathbb{R}_+, \frac{dt}{t})}.$$

(d) Show that $f \diamond g = g \diamond f$.

2.26 Let $k \colon (0,1) \times (0,1) \to \mathbb{K}$ be measurable and suppose that

$$A := \operatorname*{ess\,sup}_{y \in [0,1]} \int_0^1 |k(x,y)|\, dx < \infty,$$

$$B := \operatorname*{ess\,sup}_{x \in [0,1]} \int_0^1 |k(x,y)|\, dy < \infty.$$

Let $1 \leqslant p \leqslant \infty$ and define, for $f \in L^p(0,1)$,

$$T_k f(x) := \int_0^1 k(x,y)f(y)\, dy, \quad x \in (0,1).$$

Show that $T_k \colon L^p(0,1) \to L^p(0,1)$ is a well-defined linear operator which satisfies the so-called *Schur estimate*

$$\|T_k f\|_p \leqslant A^{1/p}B^{1-1/p}\|f\|_p, \quad f \in L^p(0,1).$$

Hint: Use Hölder's inequality.

2.27 Let $(\Omega, \mathscr{F}, \mu)$ be a measure space and let X be a Banach space. For $1 \leqslant p \leqslant \infty$ we denote by $L^p(\Omega; X)$ the space of all (equivalence classes of) strongly measurable functions $f : \Omega \to X$ for which $\omega \mapsto \|f(\omega)\|$ belongs to $L^p(\Omega)$.

(a) Prove that $L^p(\Omega; X)$ is a Banach space with respect to the norm

$$\|f\|_p := \big\| \omega \mapsto \|f(\omega)\| \big\|_{L^p(\Omega)}.$$

By $L^p(\Omega) \otimes X$ we denote the vector space obtained as the linear span in $L^p(\Omega; X)$ of the set of all functions of the form $f \otimes x$ (cf. (1.6)) with $f \in L^p(\Omega)$ and $x \in X$.

(b) Show that if $1 \leqslant p < \infty$, then $L^p(\Omega) \otimes X$ is dense in $L^p(\Omega; X)$.

2.28 Let $(\Omega, \mathscr{F}, \mu)$ be a measure space and $1 \leqslant p < \infty$. Let T be a bounded operator on $L^p(\Omega)$ and let X be a Banach space. Consider the linear mapping from $L^p(\Omega) \otimes X$ into itself defined by

$$(T \otimes I)(f \otimes x) := (Tf) \otimes x.$$

(a) Show that the operator $T \otimes I$ is well defined.
(b) Prove that if T is a positive operator, then $T \otimes I$ admits a unique extension to a bounded operator on $L^p(\Omega; X)$, and that its norm of equals the norm of T.

2.29 Let $(\Omega, \mathscr{F}, \mu)$ and $(\Omega', \mathscr{F}', \mu')$ be measure spaces and let $1 \leqslant p \leqslant q < \infty$.

(a) Show that the identity mapping on linear combinations of functions of the form $(\mathbf{1}_A \otimes \mathbf{1}_B)(\omega, \omega') := \mathbf{1}_A(\omega) \mathbf{1}_B(\omega')$ extends uniquely to a contraction operator from $L^p(\Omega; L^q(\Omega'))$ into $L^q(\Omega'; L^p(\Omega))$.
  Hint: Use (1.7).
(b) Deduce that if $(\Omega, \mathscr{F}, \mu)$ and $(\Omega', \mathscr{F}', \mu')$ are σ-finite and $f : \Omega \times \Omega' \to \mathbb{K}$ is a measurable function, then the *continuous Minkowski inequality* holds:

$$\left(\int_{\Omega'} \left(\int_{\Omega} |f(\omega, \omega')|^p \, d\mu(\omega) \right)^{q/p} \right)^{1/q} \leqslant \left(\int_{\Omega} \left(\int_{\Omega'} |f(\omega, \omega')|^q \, d\mu(\omega') \right)^{p/q} \right)^{1/p}.$$

2.30 Let $f \in L^1_{\mathrm{loc}}(\mathbb{R}^d)$ be given.

(a) Show that for all $c \in \mathbb{K}$ there exists a Borel null set $N_c \subseteq \mathbb{R}^d$ such that for all $x \in \complement N_c$ we have

$$\lim_{\substack{B \ni x \\ |B| \to 0}} \frac{1}{|B|} \int_B |f(y) - c| \, dy = |f(x) - c|. \tag{2.21}$$

(b) Show that there exists a Borel null set $L \subseteq \mathbb{R}^d$ with $|\complement L| = 0$ such that (2.21) holds all $x \in L$ and $c \in \mathbb{K}$.
  Hint: Consider $\bigcup_{n \geqslant 1} N_{c_n}$ with $(c_n)_{n \geqslant 1}$ a dense sequence in $\mathbb{K}$.

A set L with the properties of part (b) is called a *Lebesgue set* for f.

2.31 Prove the following one-sided version of the Lebesgue differentiation theorem for $d = 1$: If x is a Lebesgue point of $f \in L^1_{\mathrm{loc}}(\mathbb{R})$, then

$$\lim_{h \downarrow 0} \frac{1}{h} \int_x^{x+h} |f(y) - f(x)| \, dy = \lim_{h \downarrow 0} \frac{1}{h} \int_{x-h}^x |f(y) - f(x)| \, dy = 0.$$

2.32 Show that a subset K of c_0 is relatively compact if and only if there is a $y \in c_0$ such that for all $x \in K$ and $n \geqslant 1$ we have $|x_n| \leqslant |y_n|$.

2.33 Let $1 \leqslant p < \infty$. Show that a bounded subset S of ℓ^p is relatively compact if and only if for every $\varepsilon > 0$ there exists an index $N \geqslant 1$ such that

$$\sup_{x \in S} \sum_{n \geqslant N} |x_n|^p \leqslant \varepsilon^p.$$

2.34 Let S be a nonempty set. For $1 \leqslant p < \infty$, let $\ell^p(S)$ be the completion of the space of all finitely nonzero functions $f : S \to \mathbb{K}$, that is, functions such that $f(s) \neq 0$ for at most finitely many different $s \in S$, with the norm

$$\|f\|_p := \left(\sum_{s \in S} |f(s)|^p \right)^{1/p},$$

where the sum extends over the finitely many $s \in S$ for which $f(s) \neq 0$.

(a) Show that $\ell^p(S)$ can be isometrically identified with the space of all countably nonzero functions $f : S \to \mathbb{K}$, that is, functions such that $f(s) \neq 0$ for at most countably many different $s \in S$, for which

$$\|f\|_p := \left(\sum_{s \in S} |f(s)|^p \right)^{1/p}$$

is finite. How should this sum be interpreted?

(b) Show that $\ell^p(S)$ is a Banach space in a natural way.

2.35 Show that a $\mathbb{K}$-valued measure ν is absolutely continuous with respect to a measure μ if and only if for every $\varepsilon > 0$ there exists a $\delta > 0$ such that $|\nu(F)| < \varepsilon$ whenever $F \in \mathscr{F}$ satisfies $\mu(F) < \delta$.

2.36 A $\mathbb{K}$-valued measure μ on a topological space X is said to be *regular*, respectively *Radon*, if its variation $|\mu|$ is regular (see Definition E.15), respectively Radon (see Definition E.20). Prove that the sets $M_\mathrm{r}(X)$ and $M_\mathrm{R}(X)$ of all $\mathbb{K}$-valued Borel measures on X that are regular, respectively Radon, are closed subspaces of $M(X)$.

2.37 A function $f : [0,1] \to \mathbb{K}$ is said to be *absolutely continuous* if for every $\varepsilon > 0$ there exists a $\delta > 0$ such that whenever $(a_n)_{n=1}^N$ and $(b_n)_{n=1}^N$ are finite sequences in $[0,1]$ satisfying $\sum_{n \geqslant 1} (b_n - a_n) < \delta$, then

$$\sum_{n=1}^N |f(b_n) - f(a_n)| < \varepsilon.$$

It is said to be of *bounded variation* if

$$\text{var}(f;\pi) := \sum_{n=1}^{N} |f(t_n) - f(t_{n-1})| < \infty$$

where the supremum is taken over all finite partitions $\pi = \{t_0, \ldots, t_n\}$ of $[0,1]$.

(a) Show that a function $f : [0,1] \to \mathbb{K}$ is absolutely continuous and satisfies $f(0) = 0$ if and only if there exists a function $g \in L^1(0,1)$ such that

$$f(t) = \int_0^t g(s)\,ds, \quad t \in [0,1],$$

and that this function g, if it exists, is unique.

Hint: For the 'only if' part use the Radon–Nikodým theorem.

(b) Show that the space $NBV[0,1]$ of functions $f : [0,1] \to \mathbb{K}$ of bounded variation satisfying $f(0) = 0$ is a Banach space with respect to the norm

$$\|f\|_{NBV[0,1]} = \sup_\pi \text{var}(f;\pi).$$

(c) Show that the space of all absolutely continuous functions $f : [0,1] \to \mathbb{K}$ satisfying $f(0) = 0$ is a closed subspace of $NBV[0,1]$ and that

$$\|f\|_{NBV[0,1]} = \|g\|_{L^1(0,1)}, \quad f \in NBV[0,1],$$

where $g \in L^1(0,1)$ is the function of part (a).

2.38 The *disc algebra* $A(\mathbb{D})$ is the closed subspace of the Banach space $C(\overline{\mathbb{D}})$ consisting of those functions that are holomorphic on $\mathbb{D}$. By the maximum principle,

$$\|f\| = \sup_{\theta \in [-\pi,\pi]} |f(e^{i\theta})|.$$

(a) Show that for all $f \in C(\overline{\mathbb{D}})$ and $z_0 \in \mathbb{D}$ we have

$$f(z_0) = \frac{1}{2\pi i} \int_{\mathbb{T}} \frac{f(z)}{z - z_0}\,dz.$$

(b) Show that a function $f \in C(\overline{\mathbb{D}})$ belongs to $A(\mathbb{D})$ if and only if $\widehat{f}(n) = 0$ for all $n \in \mathbb{Z} \setminus \mathbb{N}$, where

$$\widehat{f}(n) := \frac{1}{2\pi} \int_{-\pi}^{\pi} f(e^{i\theta}) e^{-in\theta}\,d\theta, \quad n \in \mathbb{Z}.$$

2.39 Let $\text{Lip}[0,1]$ be the vector space of all functions $f : [0,1] \to \mathbb{K}$ for which

$$\|f\|_{\text{Lip}[0,1]} := |f(0)| + \sup_{\substack{0 \leqslant x,y \leqslant 1 \\ x \neq y}} \left| \frac{f(x) - f(y)}{x - y} \right|$$

is finite. Show that $\text{Lip}[0,1]$ is a Banach space with respect to the norm $\|\cdot\|_{\text{Lip}[0,1]}$.

2.40 A function $f \in L^1_{loc}(\mathbb{R}^d)$ is said to have *bounded mean oscillation* if

$$|f|_{BMO(\mathbb{R}^d)} := \sup_B \frac{1}{|B|} \int_B |f(x) - \text{av}_B(f)| \, dx$$

is finite, where the supremum is taken over all balls B in $\mathbb{R}^d$, $|B|$ is the Lebesgue measure of B, and $\text{av}_B(f) := \frac{1}{|B|} \int_B f(y) \, dy$ is the average of f on B.

(a) Show that if f and g have bounded mean oscillation, then:

(i) cf has bounded mean oscillation and

$$|cf|_{BMO(\mathbb{R}^d)} = |c| |f|_{BMO(\mathbb{R}^d)};$$

(ii) $f + g$ has bounded mean oscillation and

$$|f + g|_{BMO(\mathbb{R}^d)} \leqslant |f|_{BMO(\mathbb{R}^d)} + |g|_{BMO(\mathbb{R}^d)}.$$

(b) Show that $|f|_{BMO(\mathbb{R}^d)} = 0$ if and only if f is almost everywhere constant.

(c) Show that every $f \in L^\infty(\mathbb{R}^d)$ has bounded mean oscillation and

$$|f|_{BMO(\mathbb{R}^d)} \leqslant 2\|f\|_\infty.$$

(d) Show that the unbounded function $x \mapsto \log|x|$ has bounded mean oscillation.

(e) Show that the quotient space $BMO(\mathbb{R}^d) = BMO(\mathbb{R}^d)/\mathbb{K}$, obtained as the quotient modulo the constant functions of the vector space $BMO(\mathbb{R}^d)$ of functions with bounded mean oscillation, is a Banach space in a natural way.

2.41 Show that in a vector lattice $(V, \leqslant)$, the greatest lower bound and the least upper bound of two elements $u, v \in V$ satisfy $u \wedge v = u - (u - v)^+ = v - (v - u)^+$ and $u \vee v = u + (v - u)^+ = v + (u - v)^+$.

2.42 Let V be a vector lattice.

(a) Show that for all $v \in V$ we have $v^+ \wedge v^- = 0$ and $v^+ \vee v^- = |v|$.

(b) Show that if $v, w, w' \in V$ satisfy $v = w - w'$ with $w \geqslant 0$ and $w' \geqslant 0$, then $w \geqslant v^+$ and $w' \geqslant v^-$.

(c) Show that if $v, w, w' \in V$ satisfy $v = w - w'$ with $w \geqslant 0$, $w' \geqslant 0$, and $w \wedge w' = 0$, then $w = v^+$ and $w' = v^-$.

2.43 Prove that if X is a normed vector lattice, the lattice operations $(x, y) \mapsto x \wedge y$ and $(x, y) \mapsto x \vee y$ are continuous from $X \times X$ to X.

2.44 Provide the missing details to the proof, outlined at the end of Section 2.5, that the spaces $M(\Omega)$ studied in Section 2.4 are Banach lattices.

3

Hilbert Spaces

Arguably the most important class of Banach spaces is the class of Hilbert spaces. These spaces play a central role in the theory and in various areas of applications, some of which will be discussed in later chapters. The present introductory chapter develops the basic geometric properties of Hilbert spaces arising from the presence of an inner product generating the norm, such as the orthogonal complementation of closed subspaces, the existence of orthonormal bases, and the selfduality of Hilbert spaces embodied by the Riesz representation theorem.

3.1 Hilbert Spaces

Let V be a vector space. A mapping $\phi : V \times V \to \mathbb{K}$ is called *sesquilinear* if it is linear in the first variable and conjugate-linear in the second variable, that is,

$$\phi(v+v',w) = \phi(v,w) + \phi(v',w), \quad \phi(cv,w) = c\phi(v,w), \tag{3.1}$$

$$\phi(v,w+w') = \phi(v,w) + \phi(v,w'), \quad \phi(v,cw) = \bar{c}\phi(v,w), \tag{3.2}$$

for all $c \in \mathbb{K}$ and $v, v', w, w' \in V$. The complex conjugation in (3.2) is of course redundant when the scalar field is real and sesquilinearity reduces to bilinearity in that case.

Definition 3.1 (Inner products). An *inner product space* is a pair $(H, (\cdot|\cdot))$, where H is a vector space and $(\cdot|\cdot)$ is an *inner product* on $H \times H$, that is, a sesquilinear mapping from $H \times H$ to $\mathbb{K}$ with the following properties:

(i) $(x|x) \geqslant 0$ for all $x \in H$ and $(x|x) = 0 \Rightarrow x = 0$;
(ii) $(x|y) = \overline{(y|x)}$ for all $x, y \in H$.

The conjugation bar in (ii) is again redundant when the scalar field is real. If (ii) holds, then (3.1) implies (3.2).

It will be used frequently without further comment that

$$\text{if } (x|y) = 0 \text{ for all } y \in H, \text{ then } x = 0.$$

Indeed, the hypothesis implies that $(x|x) = 0$, and then $x = 0$ by the definition of an inner product.

When the inner product $(\cdot|\cdot)$ is understood we simply write H instead of $(H, (\cdot|\cdot))$.

Example 3.2. Here are some examples of inner products:

  (i) on $\mathbb{K}^d$ an inner product is given by $(x|y) = \sum_{n=1}^{d} x_n \overline{y_n}$;
 (ii) on ℓ^2 an inner product is given by $(a|b) = \sum_{n \geqslant 1} a_n \overline{b_n}$;
(iii) on $L^2(\Omega, \mu)$ an inner product is given by $(f|g) = \int_\Omega f \overline{g} \, d\mu$.

In order to turn inner product spaces into normed vector spaces we need the following inequality. Its finite-dimensional version has already been used in various places in Chapters 1 and 2.

Proposition 3.3 (Cauchy–Schwarz inequality). *Let H be a vector space and consider a sesquilinear mapping $(\cdot|\cdot) : H \times H \to \mathbb{K}$ with the following properties:*

  (i) $(x|x) \geqslant 0$ *for all $x \in H$;*
 (ii) $(x|y) = \overline{(y|x)}$ *for all $x, y \in H$.*

David Hilbert, 1862–1943

Then for all $x, y \in H$ we have

$$|(x|y)|^2 \leqslant (x|x)(y|y).$$

Proof   We may assume that $y \neq 0$, since otherwise the inequality trivially holds. Fix a scalar $c \in \mathbb{K}$. Then

$$\begin{aligned}
0 \leqslant (x - cy | x - cy) &= (x|x) - (x|cy) - (cy|x) + (cy|cy) \\
&= (x|x) - \overline{c}(x|y) - c\overline{(x|y)} + |c|^2 (y|y) \\
&= (x|x) - 2\,\text{Re}(\overline{c}(x|y)) + |c|^2 (y|y).
\end{aligned}$$

The choice $c = \overline{(x|y)}/(y|y)$ results in the inequality

$$0 \leqslant (x|x) - 2\,\text{Re}\,\frac{|(x|y)|^2}{(y|y)} + \frac{|(x|y)|^2}{(y|y)} = (x|x) - \frac{|(x|y)|^2}{(y|y)}.$$

Multiplying with $(y|y)$ gives the desired result. $\square$

This result is valid without assuming the nondegeneracy assumption in the second part of the defining property (i) of an inner product.

Proposition 3.4. *Every inner product space H can be made into a normed space by defining*

$$\|x\| := (x|x)^{1/2}, \quad x \in H.$$

Proof We must check that $\| \cdot \|$ defines a norm on H. It is immediate that $\|x\| = 0$ implies $x = 0$ and that $\|cx\| = |c| \|x\|$. The triangle inequality follows from the Cauchy–Schwarz inequality:

$$\|x + y\|^2 = \|x\|^2 + 2\operatorname{Re}(x|y) + \|y\|^2 \leqslant \|x\|^2 + 2\|x\| \|y\| + \|y\|^2 = (\|x\| + \|y\|)^2.$$

$\square$

Henceforth it is understood that inner product spaces are always endowed with this norm.

As a corollary to the Cauchy–Schwarz inequality we record:

Corollary 3.5. *Every inner product $(x|y)$ is jointly continuous as a function of x and y.*

Proof It suffices to show that if $x_n \to x$ and $y_n \to y$, then $(x_n|y_n) \to (x|y)$. We have

$$|(x_n|y_n) - (x|y)| \leqslant |(x_n|y_n) - (x_n|y)| + |(x_n|y) - (x|y)|$$
$$= |(x_n|y_n - y)| + |(x_n - x|y)| \leqslant \|x_n\| \|y_n - y\| + \|x_n - x\| \|y\|.$$

Since convergent sequences are bounded, the number $M := \sup_{n \geqslant 1} \|x_n\|$ is finite, and we find

$$|(x_n|y_n) - (x|y)| \leqslant M\|y_n - y\| + \|x_n - x\| \|y\|.$$

Both terms on the right-hand side tend to 0 as $n \to \infty$. $\square$

Proposition 3.6 (Parallelogram identity). *In every inner product space H the parallelogram identity holds: for all $x, y \in H$ we have*

$$2\|x\|^2 + 2\|y\|^2 = \|x + y\|^2 + \|x - y\|^2.$$

Conversely, if X is a normed space with the property that the parallelogram identity holds for all $x, y \in X$, then X is an inner product space, that is, there exists an inner product on X generating the norm of X.

In what follows we only need the first part of the proposition. Its converse is included for reasons of completeness and can be safely skipped upon first reading.

The inequality '$\geqslant$' admits L^p-versions, known as Clarkson's inequalities (see Problem 5.26).

Proof   The proof of the first part is routine:

$$\|x+y\|^2 + \|x-y\|^2 = (x+y|x+y) + (x-y|x-y) = 2(x|x) + 2(y|y) = 2\|x\|^2 + 2\|y\|^2.$$

The proof of the second assertion is quite involved and relies on finding a formula for the inner product in terms of the norm of X. To get an idea what this formula might look like we first *assume* there is such an inner product and denote it by $(\cdot|\cdot)$. Then

$$\|x \pm y\|^2 = \|x\|^2 \pm 2\operatorname{Re}(x|y) + \|y\|^2.$$

From this we get

$$\operatorname{Re}(x|y) = \frac{1}{4}\left(\|x+y\|^2 - \|x-y\|^2\right). \tag{3.3}$$

If the scalar field $\mathbb{K}$ is real, then $\operatorname{Re}(x|y) = (x|y)$ and the above identity expresses the inner product in terms of the norm. If the scalar field $\mathbb{K}$ is complex, by the previous identity we obtain

$$\operatorname{Im}(x|y) = \operatorname{Re}(x|iy) = \frac{1}{4}\left(\|x+iy\|^2 - \|x-iy\|^2\right).$$

This leads to the identity

$$\begin{aligned}(x|y) &= \operatorname{Re}(x|y) + i\operatorname{Im}(x|y) \\ &= \frac{1}{4}\left(\|x+y\|^2 - \|x-y\|^2 + i\|x+iy\|^2 - i\|x-iy\|^2\right).\end{aligned} \tag{3.4}$$

In an arbitrary normed space we could now try to *define* an inner product by (3.3) if $\mathbb{K} = \mathbb{R}$, respectively by (3.4) if $\mathbb{K} = \mathbb{C}$, but this does not always give a norm. If, however, the parallelogram identity holds, then it does. Let us check this for the case $\mathbb{K} = \mathbb{R}$. First,

$$(x|x) = \frac{1}{4}\left(\|x+x\|^2 - \|x-x\|^2\right) = \|x\|^2 \geqslant 0$$

and $(x|x) = 0$ implies $\|x\| = 0$ and hence $x = 0$. Second, the identity $(x|y) = (y|x)$ is immediate. Third,

$$\begin{aligned}(x+x'|y) &= \frac{1}{4}\left(\|(x+x')+y\|^2 - \|(x+x')-y\|^2\right) \\ &= \frac{1}{4}\left(2\|x\|^2 + 2\|x'+y\|^2 - \|x-(x'+y)\|^2\right) \\ &\quad - \frac{1}{4}\left(2\|x\|^2 + 2\|x'-y\|^2 - \|x-(x'-y)\|^2\right) \\ &= \frac{1}{2}\left(\|x'+y\|^2 - \|x'-y\|^2\right) \\ &\quad - \frac{1}{4}\left(\|x-(x'+y)\|^2 - \|x-(x'-y)\|^2\right) \\ &= 2(x'|y) + (x-x'|y).\end{aligned}$$

This proves that

$$(x+x'|y) - (x-x'|y) = 2(x'|y). \tag{3.5}$$

Taking $x = x'$ in (3.5) we obtain $(2x|y) = 2(x|y)$. Now let $x_0, x_1 \in X$ be arbitrary. Applying (3.5) to $x = \frac{1}{2}(x_0 - x_1)$ and $x' = \frac{1}{2}(x_0 + x_1)$ gives

$$(x_0|y) - (-x_1|y) = 2\left(\frac{1}{2}(x_0 + x_1)\Big|y\right),$$

which, in view of the earlier identities, simplifies to

$$(x_0|y) + (x_1|y) = (x_0 + x_1|y).$$

This gives additivity in the first coordinate. Additivity in the second coordinate is proved similarly. Using this inductively, for positive integers k we obtain

$$\begin{aligned}
(kx|y) &= ((k-1)x + x|y) = ((k-1)x|y) + (x|y) \\
&= ((k-2)x + x|y) + (x|y) = ((k-2)x|y) + 2(x|y) \\
&= \cdots = (x|y) + (k-1)(x|y) = k(x|y).
\end{aligned}$$

Applying this to $k^{-1}x$ we also find $k^{-1}(x|y) = (k^{-1}x|y)$. Next let $q = m/n$ with m, n positive integers. By what we just proved,

$$(qx|y) = m(n^{-1}x|y) = mn^{-1}(x|y) = q(x|y).$$

This proves homogeneity with respect to multiplication with the positive rationals. For such rationals we also have $(-qx|y) = -(qx|y) = -q(x|y)$, and therefore homogeneity holds for all rationals. Finally, by the continuity of the norm $\|\cdot\|$, the mapping $q \mapsto (qx|y)$ is continuous. The mapping $q \mapsto q(x|y)$ is continuous for trivial reasons. Therefore, the identity $(cx|y) = c(x|y)$ for arbitrary $c \in \mathbb{R}$ follows by approximation by rationals.

This completes the proof that for $\mathbb{K} = \mathbb{R}$, the formula for $(\cdot|\cdot)$ in (3.3) indeed defines an inner product. We have already seen that $(x|x) = \|x\|^2$, so this inner product generates the norm of X.

For $\mathbb{K} = \mathbb{C}$ it can be verified in a similar manner that the formula for $(\cdot|\cdot)$ in (3.4) defines an inner product and that it generates the norm of X. $\qquad\square$

Definition 3.7 (Hilbert spaces). A *Hilbert space* is a complete inner product space.

Thus, by definition, every Hilbert space is a Banach space.

Example 3.8. By the completeness results in the preceding chapter, all three spaces $\mathbb{K}^d$, ℓ^2, $L^2(\Omega, \mu)$ featuring in Example 3.2 are Hilbert spaces.

Further examples of Hilbert spaces will be given in the problems section.

Proposition 3.9 (Completion). *Let H be an inner product space. On its completion $\overline{H}$ as a normed space, a well-defined inner product is obtained by setting*

$$(x|y) := \lim_{n\to\infty}(x_n|y_n), \quad x,y \in \overline{H},$$

whenever $x_n, y_n \in H$ satisfy $x = \lim_{n\to\infty} x_n$ and $y = \lim_{n\to\infty} y_n$. The norm associated with this inner product coincides with the norm of $\overline{H}$ obtained by completing H.

Proof The proof relies on a few routine verifications, some of which are left as an exercise to the reader.

First of all, it easily follows from Corollary 3.5 that $(x|y)$ is independent of the approximating sequences and agrees with their inner product in H when $x,y \in H$. To see that $(x|y)$ is indeed an inner product, suppose that $(x|x) = 0$ and let $x = \lim_{n\to\infty} x_n$ with each $x_n \in H$. Then $\lim_{n\to\infty}(x_n|x_n) = 0$ and therefore $\lim_{n\to\infty} x_n = 0$ which, by the construction of the completion, means that the Cauchy sequence $(x_n)_{n\geqslant 1}$ defines the zero element of $\overline{H}$. It follows that $x = \lim_{n\to\infty} x_n = 0$ in $\overline{H}$. The remaining properties of an inner product follow trivially by limiting arguments.

Finally, the norm of $\overline{H}$ agrees with the norm generated by its inner product. This follows again by approximation: if $x = \lim_{n\to\infty} x_n$ with each $x_n \in H$, then for the norm $\|\cdot\|$ obtained via the completion procedure in the proof of Theorem 1.5 we have

$$\|x\|^2 = \lim_{n\to\infty} \|x_n\|^2 = \lim_{n\to\infty}(x_n|x_n) = (x|x),$$

the first step being true from the definition of this norm, the second because the norm of H is generated by its inner product, and the third because the inner product $(\cdot|\cdot)$ is jointly continuous. $\square$

3.2 Orthogonal Complements

Throughout this section we fix a Hilbert space H.

Definition 3.10 (Orthogonality). The elements $x, x' \in H$ are said to be *orthogonal*, notation

$$x \perp x',$$

if $(x|x') = 0$. Two subsets A and B of H are called *orthogonal* if $a \perp b$ for all $a \in A$ and $b \in B$.

Orthogonal elements $x \perp x'$ satisfy the Pythagorean identity

$$\|x + x'\|^2 = \|x\|^2 + \|x'\|^2,$$

as is seen by expanding the square norms in terms of inner products.

Definition 3.11 (Orthogonal complement). The *orthogonal complement* of a subset A of H is the set

$$A^\perp := \{x \in H : x \perp a \text{ for all } a \in A\}.$$

The orthogonal complement $A^\perp$ of a subset A is a closed subspace of H. Indeed, it is trivially checked that $A^\perp$ is a vector space. To prove its closedness, let $x_n \to x$ in H with $x_n \in A^\perp$. Then, by the continuity of the inner product, for all $a \in A$ we obtain $(x|a) = \lim_{n\to\infty}(x_n|a) = 0$.

The most important result on orthogonality is certainly the fact that every closed subspace Y of a Hilbert space is orthogonally complemented by $Y^\perp$. This is the content of Theorem 3.13 below. For its proof we need the following approximation theorem for convex closed sets in Hilbert space. Recall that a subset C of a vector space is called *convex* if for all $x_0, x_1 \in C$ we have $(1-\lambda)x_0 + \lambda x_1 \in C$ for all $0 \leqslant \lambda \leqslant 1$.

Theorem 3.12 (Best approximation). *Let C be a nonempty convex closed subset of H. Then for all $x \in H$ there exists a unique $c \in C$ that minimises the distance from x to the points of C:*

$$\|x - c\| = \min_{y \in C} \|x - y\|.$$

Proof Let $(y_n)_{n \geqslant 1}$ be a sequence in C such that

$$\lim_{n \to \infty} \|x - y_n\| = \inf_{y \in C} \|x - y\| =: D.$$

We claim that this sequence is Cauchy. By the parallelogram identity of Proposition 3.6, applied to the vectors $x - y_m$ and $x - y_n$,

$$\|y_n - y_m\|^2 + \|2x - (y_n + y_m)\|^2 = 2\|x - y_m\|^2 + 2\|x - y_n\|^2.$$

As $m, n \to \infty$, the right-hand side tends to $2D^2 + 2D^2 = 4D^2$, whereas from $\frac{1}{2}(y_m + y_n) \in C$ (by convexity) it follows that

$$\|2x - (y_n + y_m)\|^2 = 4\|x - \tfrac{1}{2}(y_n + y_m)\|^2 \geqslant 4D^2.$$

It follows that

$$\limsup_{m,n \to \infty} \|y_n - y_m\|^2 \leqslant 4D^2 - 4D^2 = 0.$$

The limit superior is also nonnegative, and therefore it equals 0. This proves the claim.

Since H is complete we have $\lim_{n\to\infty} y_n = c$ for some $c \in H$, and since C is closed we have $c \in C$. Now $\|x - c\| = \lim_{n\to\infty} \|x - y_n\| = D$, so c minimises the distance to x. $\square$

Both the existence and uniqueness parts of this theorem fail for general Banach spaces; see Problems 3.5 and 3.6.

Theorem 3.13 (Closed subspaces are orthogonally complemented). *If Y is a closed linear subspace of H, then we have an orthogonal direct sum decomposition*

$$H = Y \oplus Y^\perp,$$

that is, we have $Y \cap Y^\perp = \{0\}$, $Y + Y^\perp = H$, and $Y \perp Y^\perp$.

Proof   We have already seen that $Y^\perp$ is a closed subspace. If $y \in Y \cap Y^\perp$, then $y \perp y$, so $(y|y) = 0$ and $y = 0$. It remains to show that $Y + Y^\perp = H$.

Let $x \in H$ be arbitrary and fixed. We must show that $x \in Y + Y^\perp$. Let $\pi_Y : H \to Y$ denote the mapping arising from Theorem 3.12, that is, $\pi_Y x$ is the unique element of Y minimising the distance to x:

$$\|x - \pi_Y x\| = \min_{y \in Y} \|x - y\|.$$

Set $y_0 := \pi_Y x$ and $y_1 := x - y_0$. Then $y_0 \in Y$, and for all $y \in Y$ we have

$$\|y_1\| = \|x - y_0\| \leqslant \|x - \underbrace{(y_0 - y)}_{\in Y}\| = \|y + (x - y_0)\| = \|y + y_1\|. \tag{3.6}$$

We claim that (3.6) implies $y_1 \in Y^\perp$. To see this, fix a nonzero $y \in Y$. For any $c \in \mathbb{K}$ we have, by (3.6),

$$\|y_1\|^2 \leqslant \|cy + y_1\|^2 = |c|^2 \|y\|^2 + 2\operatorname{Re}(cy|y_1) + \|y_1\|^2.$$

Taking $c = -\overline{(y|y_1)}/\|y\|^2$, this gives

$$0 \leqslant \frac{|(y|y_1)|^2}{\|y\|^2} - 2\frac{|(y|y_1)|^2}{\|y\|^2},$$

which is only possible if $(y_1|y) = 0$. Since $0 \neq y \in Y$ was arbitrary, this shows that $y_1 \in Y^\perp$. This proves the claim. It follows that $x = y_0 + y_1$ belongs to $Y + Y^\perp$. $\qquad\square$

Definition 3.14 (Orthogonal projection onto a closed subspace). The projection π_Y onto Y along $Y^\perp$ given by

$$\pi_Y(y + y^\perp) := y, \quad y \in Y, \ y^\perp \in Y^\perp,$$

is called the *orthogonal projection* onto Y.

The Pythagorean inequality implies that $\|\pi_Y\| \leqslant 1$ (with equality if $Y \neq \{0\}$). As was shown in the course of the proof of Theorem 3.13, the projection π_Y coincides with the mapping arising from Theorem 3.12. For general closed convex sets C, the distance minimising mapping of Theorem 3.12 is generally nonlinear.

As a corollary to Theorem 3.13, for closed subspaces Y we get $(Y^\perp)^\perp = Y$, and more generally for any subspace Y we get (since $Y^\perp = (\overline{Y})^\perp$)

$$(Y^\perp)^\perp = \overline{Y}.$$

By way of example, if $(h_n)_{n=1}^N$ is an *orthonormal sequence*, i.e., $(h_j|h_k) = \delta_{jk}$ for all $1 \leqslant j, k \leqslant N$, the orthogonal projection π_N in H onto the span of $h_1, \ldots, h_N$ is given by

$$\pi_N x = \sum_{n=1}^N (x|h_n) h_n,$$

as the reader will have no difficulty checking.

3.3 The Riesz Representation Theorem

Let H be a Hilbert space. As a first application of Theorem 3.13 we prove the *Riesz representation theorem*, which sets up a conjugate-linear identification of a Hilbert space H and its dual $H^* = \mathscr{L}(H, \mathbb{K})$.

By the Cauchy–Schwarz inequality, every $h \in H$ defines a bounded functional $\psi_h : H \to \mathbb{K}$ by taking inner products:

$$\psi_h(x) := (x|h), \quad x \in H.$$

Frigyes Riesz, 1880–1956

Boundedness is evident from $|\psi_h(x)| = |(x|h)| \leqslant \|x\| \|h\|$, which shows that $\|\psi_h\| \leqslant \|h\|$. From $\psi_h(h) = (h|h) = \|h\|^2$ we see that also $\|\psi_h\| \geqslant \|h\|$, so that $\|\psi_h\| = \|h\|$.

All bounded functionals $\phi : H \to \mathbb{K}$ arise in this way:

Theorem 3.15 (Riesz representation theorem). *If $\phi : H \to \mathbb{K}$ is a bounded functional, there exists a unique element $h \in H$ such that $\phi = \psi_h$, that is,*

$$\phi(x) = (x|h), \quad x \in H.$$

Proof If $\phi(x) = 0$ for all $x \in H$, we take $h = 0$. Henceforth we shall assume that $\phi \neq 0$. Then $(\mathsf{N}(\phi))^\perp \neq \{0\}$ by Theorem 3.13, and we can choose a norm one vector $y_0 \in (\mathsf{N}(\phi))^\perp$. Fix an arbitrary $x \in H$. With $c := \phi(x)/\phi(y_0)$ we have $\phi(x - cy_0) = \phi(x) - c\phi(y_0) = 0$. This means that $x - cy_0 \in \mathsf{N}(\phi)$, so $x - cy_0 \perp y_0$ and

$$\phi(x) = c\phi(y_0) = \phi(y_0)(cy_0|y_0) = \phi(y_0)(x|y_0) = (x|\overline{\phi(y_0)}y_0).$$

This proves that $\phi = \psi_h$ with $h := \overline{\phi(y_0)}y_0$.

To prove uniqueness, suppose that $\phi = \psi_h = \psi_{h'}$ for $h, h' \in H$. Then $\|h - h'\|^2 = (h - h'|h - h') = \psi_h(h - h') - \psi_{h'}(h - h') = 0$ and therefore $h' = h$. $\qquad\square$

Proposition 3.16. *For every bounded sequence $(x_n)_{n \geqslant 1}$ in H there exist a subsequence* $(x_{n_k})_{k \geqslant 1}$ *and an* $x \in H$ *such that*

$$\lim_{k \to \infty} (h | x_{n_k}) = (h | x), \quad h \in H.$$

Proof Let H_0 denote the closed linear span of $(x_n)_{n \geqslant 1}$ and let $H_0^\perp$ be its orthogonal complement, so that we have the orthogonal decomposition $H = H_0 \oplus H_0^\perp$.

Step 1 – We begin by proving that there exist a subsequence $(x_{n_k})_{k \geqslant 1}$ and an $x \in H_0$ such that

$$\lim_{k \to \infty} (h | x_{n_k}) = (h | x), \quad h \in H_0.$$

Let $(y_j)_{j \geqslant 1}$ be a sequence whose linear span Y is dense in H_0. The $(x_n)_{n \geqslant 1}$ sequence has a subsequence which, after relabelling, we may call $(x_n^{(1)})_{n \geqslant 1}$, such that the limit $\phi(y_1) := \lim_{n \to \infty} (y_1 | x_n^{(1)})$ exists. This sequence has a further subsequence which, after relabelling, we may call $(x_n^{(2)})_{n \geqslant 1}$, such that the limit $\phi(y_2) := \lim_{n \to \infty} (y_2 | x_n^{(2)})$ exists. Note that we also have $\phi(y_1) = \lim_{n \to \infty} (y_1 | x_n^{(2)})$. Continuing inductively, for every $k \geqslant 1$ we obtain a subsequence $(x_n^{(k)})_{n \geqslant 1}$ with the property that the limit $\phi(y_j) = \lim_{n \to \infty} (y_j | x_n^{(k)})$ exists for $j = 1, \ldots, k$. The 'diagonal subsequence' $(x_n^{(n)})_{n \geqslant 1}$ has the property that the limit $\phi(y_j) = \lim_{n \to \infty} (y_j | x_n^{(n)})$ exists for all $j \geqslant 1$. By linearity, the limit $\phi(y) := \lim_{n \to \infty} (y | x_n^{(n)})$ exists for all $y \in Y$. Clearly $y \mapsto \phi(y)$ is linear and $|\phi(y)| \leqslant M \|y\|$, where $M := \sup_{n \geqslant 1} \|x_n\|$. This shows that ϕ is bounded as a mapping from Y to $\mathbb{K}$. Since Y is dense in H_0, Proposition 1.18 implies that ϕ has a unique bounded extension of the same norm to all of H_0. By the Riesz representation theorem, applied to the Hilbert space H_0, there exists an $x \in H_0$ such that $\phi(h) = (h | x)$ for all $h \in H_0$. This element has the required properties: for all $h \in Y$ we have

$$\lim_{n \to \infty} (h | x_n^{(n)}) = \phi(h) = (h | x).$$

For general $h \in H_0$ the same identity holds by an approximation argument, using that Y is dense in H_0.

Step 2 – We now show that

$$\lim_{n \to \infty} (h | x_n^{(n)}) = (h | x), \quad h \in H,$$

where $(x_n^{(n)})_{n \geqslant 1}$ and $x \in H_0$ are as in Step 1. To this end let $h \in H$ be arbitrary and write $h = h_0 + h_0^\perp$ along the orthogonal decomposition $H = H_0 \oplus H_0^\perp$. Step 1 gives $\lim_{k \to \infty} (h_0 | x_n^{(n)}) = (h_0 | x)$ for $h_0 \in H_0$. Trivially, $\lim_{k \to \infty} (h_0^\perp | x_n^{(n)}) = 0 = (h_0^\perp | x)$ for all $h_0^\perp \in H_0^\perp$. This concludes the proof. $\qquad \square$

The argument is Step 1 is known as a *diagonal argument*.

3.4 Orthonormal Systems

We have the following simple criterion for the convergence of a series whose terms are pairwise orthogonal.

Proposition 3.17. *Let* $(x_n)_{n \geq 1}$ *be a sequence in* H *with* $x_m \perp x_n$ *for all* $m \neq n$. *The following assertions are equivalent:*

(1) $\sum_{n \geq 1} x_n$ *converges in* H;
(2) $\sum_{n \geq 1} \|x_n\|^2 < \infty$.

In this situation,

$$\left\| \sum_{n \geq 1} x_n \right\|^2 = \sum_{n \geq 1} \|x_n\|^2.$$

Proof   Let us first note that if I is any finite set of positive integers, then

$$\left\| \sum_{n \in I} x_n \right\|^2 = \left(\sum_{m \in I} x_m \Big| \sum_{n \in I} x_n \right) = \sum_{m \in I} \sum_{n \in I} (x_m | x_n) = \sum_{n \in I} \|x_n\|^2$$

since $(x_m | x_n) = 0$ if $m \neq n$.

(1)$\Rightarrow$(2): If $\sum_{n \geq 1} x_n$ converges in H, say to x, then $x = \lim_{N \to \infty} \sum_{n=1}^{N} x_n$ in H and therefore

$$\|x\|^2 = \lim_{N \to \infty} \left\| \sum_{n=1}^{N} x_n \right\|^2 = \lim_{N \to \infty} \sum_{n=1}^{N} \|x_n\|^2.$$

It follows that $\sum_{n \geq 1} \|x_n\|^2 = \|x\|^2$. This also proves the final identity in the statement of the proposition, since by definition $\sum_{n \geq 1} x_n = x$.

(2)$\Rightarrow$(1): Suppose, conversely, that $\sum_{n \geq 1} \|x_n\|^2 < \infty$. Then

$$\lim_{\substack{M,N \to \infty \\ N > M}} \left\| \sum_{n=1}^{N} x_n - \sum_{n=1}^{M} x_n \right\|^2 = \lim_{\substack{M,N \to \infty \\ N > M}} \left\| \sum_{n=M+1}^{N} x_n \right\|^2 = \lim_{\substack{M,N \to \infty \\ N > M}} \sum_{n=M+1}^{N} \|x_n\|^2 = 0.$$

It follows that $(\sum_{n=1}^{N} x_n)_{N \geq 1}$ is Cauchy, and hence convergent.   □

As a special case of Proposition 3.17 we record:

Proposition 3.18 (Parseval). *Let* $(h_n)_{n \geq 1}$ *be an orthonormal sequence in* H. *For a scalar sequence* $(c_n)_{n \geq 1}$, *the following assertions are equivalent:*

(1) $\sum_{n \geq 1} c_n h_n$ *converges in* H;
(2) $\sum_{n \geq 1} |c_n|^2 < \infty$.

In this situation the Parseval identity *holds:*

$$\left\| \sum_{n \geq 1} c_n h_n \right\|^2 = \sum_{n \geq 1} |c_n|^2.$$

Definition 3.19 (Orthonormal system, orthonormal basis). Let I be a nonempty set. A family $(h_i)_{i \in I}$ in H is called an *orthonormal system* if

$$(h_i | h_j) = \delta_{ij} := \begin{cases} 1 & \text{if } i = j \\ 0 & \text{otherwise,} \end{cases}$$

for all $i, j \in I$ with $i \neq j$. In the case of a countable set I, an orthonormal system $(h_i)_{i \in I}$ is called an *orthonormal basis* if every $x \in H$ can be represented as a convergent series

$$x = \sum_{i \in I} c_i h_i \tag{3.7}$$

for suitable coefficients $c_i \in \mathbb{K}$.

The convergence of the sum (3.7) is understood in the following sense. We pick an enumeration of the index set, say $I = (i_n)_{n \geq 1}$, and ask for convergence of the sum $\sum_{n \geq 1} c_{i_n} h_{i_n}$. By Parseval's theorem, this sum converges if and only if $\sum_{n \geq 1} |c_{i_n}|^2 < \infty$. Thus, whether or not the sum converges is independent of the enumeration chosen. To see that the sum $x := \sum_{n \geq 1} c_{i_n} h_{i_n}$ is independent of the enumeration, let $(j_m)_{m \geq 1}$ be another enumeration of I and set $y := \sum_{m \geq 1} c_{j_m} h_{j_m}$. Then for any $i \in I$, say $i = i_n = j_m$,

$$(x | h_i) = (x | h_{i_n}) = c_{i_n} = c_i = c_{j_m} = (y | h_{j_m}) = (y | h_i).$$

Since both x and y belong to the closed linear span of the family $(h_i)_{i \in I}$, this implies $x = y$. This argument also shows that the coefficients c_i in (3.7) are uniquely determined by x and given by $c_i = (x | h_i)$.

Example 3.20. The standard unit vectors $(0, \ldots, 0, 1, 0, \ldots)$, $n \geq 1$ (with the 1 on the nth place), form an orthonormal basis for ℓ^2.

In Section 3.5 we prove that the trigonometric system is an orthonormal basis for $L^2(0, 1)$ and the (suitably normalised) Hermite polynomials form an orthonormal basis for $L^2(\mathbb{R}, \gamma)$, where γ is the standard Gaussian measure on $\mathbb{R}$.

Theorem 3.21 (Orthonormal bases). *Let $(h_n)_{n \geq 1}$ be an orthonormal sequence in H. The following assertions are equivalent:*

(1) *$(h_n)_{n \geq 1}$ is a basis;*
(2) *$(h_n)_{n \geq 1}$ has dense linear span;*
(3) *if $x \in H$ satisfies $(x | h_n) = 0$ for all $n \geq 1$, then $x = 0$.*

Proof The equivalence (2)$\Leftrightarrow$(3) is immediate from the fact that a subspace is dense if and only if its orthogonal complement is trivial.

(1)$\Rightarrow$(2): This implication is trivial, because by assumption every $x \in H$ can be approximated by the partial sums of a series representation of the form $\sum_{n \geq 1} c_n h_n$, and these partial sums belong to the linear span of $(h_n)_{n \geq 1}$.

$(2) \Rightarrow (1)$: Suppose the linear span of $(h_n)_{n \geqslant 1}$ is dense in H and fix an arbitrary $x \in H$. We must prove that x admits a representation as a convergent sum $\sum_{n \geqslant 1} c_n h_n$.

For each $N \geqslant 1$ the mapping

$$P_N x := \sum_{n=1}^{N} (x|h_n) h_n$$

is a projection that maps H onto the span H_N of $(h_n)_{n=1}^{N}$. If $x \perp H_N$, then $(x|h_n) = 0$ for $n = 1, \ldots, N$ and therefore $P_N x = 0$. It follows that the projection P_N is orthogonal. This implies that P_N is contractive. Therefore, with $c_n := (x|h_n)$,

$$\sum_{n=1}^{N} |c_n|^2 = \|P_N x\|^2 \leqslant \|x\|^2.$$

This being true for all $N \geqslant 1$, it follows that $\sum_{n \geqslant 1} |c_n|^2 < \infty$ and therefore the sum $y := \sum_{n \geqslant 1} c_n h_n$ is convergent in H by Proposition 3.18. For all $n \geqslant 1$ we have $(y|h_n) = c_n = (x|h_n)$, and since the span of $(h_n)_{n \geqslant 1}$ is dense in H this implies $x = y = \sum_{n \geqslant 1} c_n h_n$. $\square$

When $(x_n)_{n \geqslant 1}$ is a (finite or infinite) linearly independent sequence in H, we may construct an orthonormal sequence $(h_n)_{n \geqslant 1}$ with the property that

$$\mathrm{span}\{x_1, \ldots, x_k\} = \mathrm{span}\{h_1, \ldots, h_k\}, \quad k \geqslant 1,$$

as follows. Set $h_1 := x_1 / \|x_1\|$. Suppose the orthonormal vectors $h_1, \ldots, h_k$ have been chosen subject to the condition that $H_j := \mathrm{span}\{x_1, \ldots, x_j\}$ equals $\mathrm{span}\{h_1, \ldots, h_j\}$ for all $j = 1, \ldots, k$. By linear independence, the subspace $H_{k+1} := \mathrm{span}\{x_1, \ldots, x_{k+1}\}$ has dimension $k + 1$. The orthogonal complement in H_{k+1} of the k-dimensional subspace H_k has dimension 1, and therefore we may select a norm one vector $h_{k+1} \in H_{k+1}$ orthogonal to H_k. Then $H_{k+1} = \mathrm{span}\{h_1, \ldots, h_{k+1}\}$ as desired. This procedure is called *Gram–Schmidt orthogonalisation*.

Theorem 3.22 (Orthonormal bases and separability). *A Hilbert space has an orthonormal basis if and only if it is separable.*

Proof 'If': By assumption we can find a (finite or infinite) sequence $(x_n)_{n \geqslant 1}$ with dense span in H. By passing to a subsequence, we may assume that the elements of the sequence are linearly independent. By Gram–Schmidt orthogonalisation we construct an orthonormal sequence $(h_n)_{n \geqslant 1}$ with the property that for all $k \geqslant 1$ the linear span of $\{h_1, \ldots, h_k\}$ equals the linear span of $\{x_1, \ldots, x_k\}$. Since the linear span of $(x_n)_{n \geqslant 1}$ is dense in H, the sequence $(h_n)_{n \geqslant 1}$ is an orthonormal basis of H by Theorem 3.21.

'Only if': If $(h_n)_{n \geqslant 1}$ is an orthonormal basis of H, its linear span is dense. $\square$

Corollary 3.23. *Any two infinite-dimensional separable Hilbert spaces are isometrically isomorphic.*

Proof   Suppose that the Hilbert spaces H_1 and H_2 are separable and pick orthonormal bases $(h_n^{(1)})_{n \geqslant 1}$ and $(h_n^{(2)})_{n \geqslant 1}$. The operator U sending $h_n^{(1)}$ to $h_n^{(2)}$ for each $n \geqslant 1$ is isometric by the Parseval identity and has dense range. In particular U is injective. By Proposition 1.21, U has also closed range and therefore U is surjective.   □

Definition 3.24 (Maximal orthonormal systems). A *maximal orthonormal system* is a family $(h_i)_{i \in I}$, where I is a nonempty set and:

(i) $(h_i | h_j) = \delta_{ij}$ for all $i, j \in I$;
(ii) if $h \perp h_i$ for all $i \in I$, then $h = 0$.

In a separable Hilbert space, every maximal orthonormal system is countable and can therefore be relabelled into an orthonormal basis.

Theorem 3.25 (Maximal orthonormal systems). *Every nonzero Hilbert space has a maximal orthonormal system.*

Proof   Partially order the set of all orthonormal systems in the nonzero Hilbert space H by set inclusion. By Zorn's lemma (Theorem A.3) this set has a maximal element, say $(h_i)_{i \in I}$, where I is some index set. It is clear that condition (i) in the above definition holds. If there were a nonzero $h \in H$ perpendicular to each h_i, after normalising h to unit length we obtain a new orthonormal system properly containing $(h_i)_{i \in I}$, contradicting the maximality of $(h_i)_{i \in I}$. Therefore (ii) also holds.   □

3.5 Examples

In this final section we present two nontrivial examples of orthonormal bases.

3.5.a The Trigonometric System

In this example $\mathbb{T}$ denotes the unit circle in the complex plane, parametrised by the interval $[-\pi, \pi]$ and equipped with the normalised Lebesgue measure $d\theta/2\pi$. We shall prove that the functions

$$e_n(\theta) := \exp(in\theta), \quad \theta \in [-\pi, \pi], \; n \in \mathbb{Z},$$

form an orthonormal basis for $L^2(\mathbb{T})$.

That $(e_n)_{n \in \mathbb{Z}}$ is an orthonormal sequence in $L^2(\mathbb{T})$ is evident from

$$(e_j | e_k) = \frac{1}{2\pi} \int_{-\pi}^{\pi} \exp(ij\theta)\overline{\exp(ik\theta)} \, d\theta = \frac{1}{2\pi} \int_{-\pi}^{\pi} \exp(i(j-k)\theta) \, d\theta = \delta_{jk}.$$

To prove that $(e_n)_{n \in \mathbb{Z}}$ is an orthonormal basis, by Theorem 3.21 it remains to be proved that the trigonometric polynomials, i.e., the functions of the form $\sum_{n=-N}^{N} c_n e_n$, are dense in $L^2(\mathbb{T})$. This can be deduced from the Stone–Weierstrass theorem (see Problem 3.11), but we prefer the following argument from Fourier Analysis which gives explicit approximants and some error bounds.

Definition 3.26 (Fourier coefficients). The *Fourier coefficients* of a function $f \in L^1(\mathbb{T})$ are defined as

$$\widehat{f}(n) := (f|e_n) = \frac{1}{2\pi} \int_{-\pi}^{\pi} f(\theta) \exp(-in\theta) \, d\theta, \quad n \in \mathbb{Z}.$$

Theorem 3.27. *For all $f \in C(\mathbb{T})$ we have*

$$\lim_{N \to \infty} \left\| f - \frac{1}{N} \sum_{n=0}^{N-1} \sum_{k=-n}^{n} \widehat{f}(k) e_k \right\|_{\infty} = 0.$$

Proof Fix $f \in C(\mathbb{T})$ with $\|f\|_{\infty} = 1$. We have

$$\widehat{f}(n) \exp(in\theta) = \frac{1}{2\pi} \int_{-\pi}^{\pi} \exp(in(\theta - \sigma)) f(\sigma) \, d\sigma = \frac{1}{2\pi} \int_{-\pi}^{\pi} \exp(in\sigma) f(\theta - \sigma) \, d\sigma$$

and therefore

$$f_N(\theta) := \frac{1}{N} \sum_{n=0}^{N-1} \sum_{k=-n}^{n} \widehat{f}(k) \exp(ik\theta) = \frac{1}{2\pi} \int_{-\pi}^{\pi} K_N(\sigma) f(\theta - \sigma) \, d\sigma,$$

where the *Fejér kernel K_N* is defined by

$$K_N(\theta) := \frac{1}{N} \sum_{n=0}^{N-1} \sum_{k=-n}^{n} \exp(ik\theta) = \frac{1}{N} \frac{\sin^2(\frac{1}{2} N\theta)}{\sin^2(\frac{1}{2}\theta)}; \tag{3.8}$$

the right-hand side identity is readily deduced from the geometric series. In view of the fact that $\frac{1}{2\pi} \int_{-\pi}^{\pi} K_N(\theta) \, d\theta = 1$, we have

$$\begin{aligned}
|f_N(\theta) - f(\theta)| &= \left| \frac{1}{2\pi} \int_{-\pi}^{\pi} K_N(s)(f(\theta - \sigma) - f(\theta)) \, d\sigma \right| \\
&\leqslant \frac{1}{2\pi} \int_{-\pi}^{\pi} K_N(s) |f(\theta - \sigma) - f(\theta)| \, d\sigma.
\end{aligned} \tag{3.9}$$

Fix $\varepsilon > 0$ and choose $0 < \delta < \pi$ so small that $\|f(\cdot - \sigma) - f\|_{\infty} < \varepsilon$ for all $|\sigma| < \delta$; this is possible since f is uniformly continuous. Then

$$\frac{1}{2\pi} \int_{-\delta}^{\delta} K_N(\sigma) |f(\theta - \sigma) - f(\theta)| \, d\sigma \leqslant \frac{\varepsilon}{2\pi} \int_{-\pi}^{\pi} K_N(\sigma) \, d\sigma = \varepsilon \tag{3.10}$$

and, by (3.8) and the normalisation $\|f\|_{\infty} = 1$,

$$\frac{1}{2\pi} \int_{\complement(-\delta,\delta)} K_N(\sigma) |f(\theta - \sigma) - f(\theta)| \, d\sigma \leqslant \frac{2}{N} \frac{1}{\sin^2(\frac{1}{2}\delta)}. \tag{3.11}$$

Combining (3.9) with (3.10) and (3.11) we obtain

$$\|f_N - f\|_\infty \leqslant \varepsilon + \frac{1}{N} \frac{2}{\sin^2(\frac{1}{2}\delta)},$$

so $\limsup_{N\to\infty} \|f_N - f\|_\infty \leqslant \varepsilon$. Since $\varepsilon > 0$ was arbitrary, this completes the proof. $\qquad\square$

This theorem implies that the trigonometric polynomials are dense in $C(\mathbb{T})$. Since this space is dense in $L^2(\mathbb{T})$ by Proposition 2.29, it follows that the trigonometric polynomials are dense in $L^2(\mathbb{T})$. Therefore, Theorem 3.21 implies:

Theorem 3.28. *The trigonometric polynomials form an orthonormal basis in $L^2(\mathbb{T})$.*

The theory of orthonormal bases can now be applied. It entails that every function $f \in L^2(\mathbb{T})$ has a unique series representation of the form $f = \sum_{n\in\mathbb{Z}} c_n e_n$, with convergence in $L^2(\mathbb{T})$ and coefficients given by $c_n = (f|e_n) = \widehat{f}(n)$. The resulting expansion

$$f = \sum_{n\in\mathbb{Z}} \widehat{f}(n) e_n$$

is called the *Fourier series* of f. By Parseval's identity we have

$$\frac{1}{2\pi} \int_{-\pi}^{\pi} |f(\theta)|^2 \, d\theta = \sum_{n\in\mathbb{Z}} |\widehat{f}(n)|^2,$$

and the mapping $f \mapsto (\widehat{f}(n))_{n\in\mathbb{Z}}$ is an isometry from $L^2(\mathbb{T})$ onto $\ell^2(\mathbb{Z})$.

By translation and scaling, the functions

$$\widetilde{e}_n(\theta) := \exp(2\pi i n\theta), \quad n \in \mathbb{Z},$$

form an orthonormal basis for $L^2(0,1)$. This can be used to prove:

Corollary 3.29 (Euler). $\displaystyle\sum_{n=1}^{\infty} \frac{1}{n^2} = \frac{\pi^2}{6}.$

Proof For the function $f(\theta) = \theta$ in $L^2(0,1)$ we have $(f|\widetilde{e}_0) = \frac{1}{2}$ and $(f|\widetilde{e}_n) = -\frac{1}{2\pi i n}$, so by Parseval's identity

$$\frac{1}{3} = \int_0^1 \theta^2 \, d\theta = \|f\|^2 = \sum_{n\in\mathbb{Z}} |(f|\widetilde{e}_n)|^2 = \left(\frac{1}{2}\right)^2 + 2\sum_{n=1}^{\infty} \left(\frac{1}{2\pi n}\right)^2 = \frac{1}{4} + \frac{1}{2\pi^2}\sum_{n=1}^{\infty} \frac{1}{n^2},$$

and the result follows. $\qquad\square$

Another proof will be given in Section 14.5.f.

The system of functions

$$1, \sqrt{2}\sin(2\pi n\theta), \sqrt{2}\cos(2\pi n\theta) : \quad n = 1, 2, \ldots \tag{3.12}$$

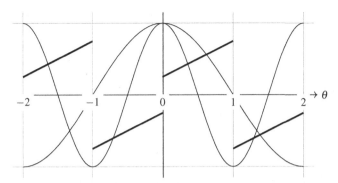

Figure 3.1 The function f, extended from $(0,1)$ to $(-2,2)$ by odd reflections (thick graph), and the functions $\cos(\pi n\theta/2)$ for $n = 1,2$. Notice that $\int_{-2}^{2} f(\theta)\cos(\pi nt/2)\,d\theta = 0$ for all $n \geqslant 1$, because $f(\theta)$ is odd about 0 and $\cos(\pi n\theta/2)$ is even about 0.

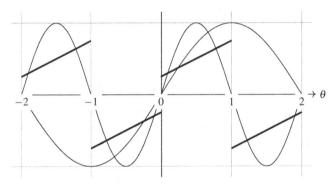

Figure 3.2 Idem, but now with the functions $\sin(\pi n\theta/2)$ for $n = 1,2$. This time we have $\int_{-2}^{2} f(\theta)\sin(\pi n\theta/2)\,d\theta = 0$ if $n \geqslant 1$ is odd, because f is odd about ± 1 and $\sin(\pi n\theta/2)$ is even about ± 1 for odd n.

is orthonormal in $L^2(0,1)$ and the trigonometric functions $\exp(2\pi in\theta)$, $n \in \mathbb{Z}$, are contained in their linear span. Hence this system forms an orthonormal basis by Theorem 3.21. Interestingly, from this we can deduce:

Theorem 3.30. *Each one of the two systems*

$$\sqrt{2}\sin(\pi n\theta): \quad n = 1,2,\ldots \tag{3.13}$$

and

$$\mathbf{1}, \sqrt{2}\cos(\pi n\theta): \quad n = 1,2,\ldots \tag{3.14}$$

forms an orthonormal basis for $L^2(0,1)$.

These bases arise naturally as the eigenvector bases of the Dirichlet and Neumann Laplacians in $L^2(0,1)$, respectively (see Example 12.23).

Proof   Given a function $f \in L^2(0,1)$, we extend it to an odd function in $L^2(-1,1)$. This function is extended to a function in $L^2(-2,2)$ whose restriction to $(0,2)$ is odd about the point 1 and whose restriction to $(-2,0)$ is odd about the point -1. If we expand the resulting function against the orthonormal basis for $L^2(-2,2)$ obtained by scaling the system (3.12), that is,

$$\frac{1}{2}\mathbf{1}, \frac{1}{\sqrt{2}}\sin(\pi n\theta/2), \frac{1}{\sqrt{2}}\cos(\pi n\theta/2): \quad n=1,2,\ldots,$$

then, due to the symmetries introduced by the odd reflections, only the coefficients corresponding to the system (3.13) with even indices can contribute, but not the ones with odd indices; nor do those of (3.14) contribute; see Figures 3.1 and 3.2. If we do the same with even extensions, only the coefficients corresponding to the system (3.14) with even indices can contribute. Restricting the resulting expansions in $L^2(-2,2)$ to $L^2(0,1)$, the desired expansions of $f \in L^2(0,1)$ in terms of the systems (3.13) and (3.14) are obtained.   □

3.5.b The Hermite Polynomials

In this section we prove that the (suitably normalised) Hermite polynomials form an orthonormal basis for $L^2(\mathbb{R}, \gamma)$, where γ is the standard Gaussian measure on $\mathbb{R}$. This is the Borel probability measure on $\mathbb{R}$ which is given, for Borel sets $B \subseteq \mathbb{R}$, by

$$\gamma(B) = \frac{1}{\sqrt{2\pi}} \int_B \exp\left(-\frac{1}{2}x^2\right) dx.$$

The Hermite polynomials will resurface in Chapters 9, 13, and 15 in connection with the spectral theorem, the Ornstein–Uhlenbeck semigroup, and second quantisation, respectively.

Definition 3.31. For $n \in \mathbb{N}$ the *Hermite polynomials* $H_n : \mathbb{R} \to \mathbb{R}$ are defined by the identity

$$H(t,x) := \exp\left(tx - \frac{1}{2}t^2\right) = \sum_{n=0}^{\infty} \frac{t^n}{n!} H_n(x), \quad t, x \in \mathbb{R}. \tag{3.15}$$

The first five Hermite polynomials are given by

$$H_0(x) = 1,$$
$$H_1(x) = x,$$
$$H_2(x) = x^2 - 1,$$
$$H_3(x) = x^3 - 3x,$$

$$H_4(x) = x^4 - 6x^2 + 3.$$

By induction one shows that

$$H_n(x) = \sum_{k=0}^{\lfloor n/2 \rfloor} \frac{(-1)^k}{2^k} \frac{n!}{k!(n-2k)!} x^{n-2k}, \quad n \in \mathbb{N}.$$

Proposition 3.32. *The Hermite polynomials have the following properties:*

(i) $H_n(-x) = (-1)^n H_n(x)$;

(ii) $H_{n+2}(x) = xH_{n+1}(x) - (n+1)H_n(x)$;

(iii) $H'_{n+1}(x) = (n+1)H_n(x)$;

(iv) H_n *is a monic polynomial of order* n.

Proof Property (i) follows from the identity $H(t,-x) = H(-t,x)$, (ii) from the identity $\frac{\partial H}{\partial t}(t,x) = (x-t)H(t,x)$, and (iii) from $\frac{\partial H}{\partial x}(t,x) = tH(t,x)$. Assertion (iv) follows from (ii) and the fact that $H_0(x) = 1$. $\square$

Theorem 3.33. *The sequence* $(\frac{1}{\sqrt{n!}} H_n)_{n \in \mathbb{N}}$ *forms an orthonormal basis for* $L^2(\mathbb{R}, \gamma)$.

Proof For all $s, t \in \mathbb{R}$ we have

$$\int_{-\infty}^{\infty} H(s,x) H(t,x) \, d\gamma(x) = \frac{1}{\sqrt{2\pi}} \int_{-\infty}^{\infty} \exp\left(-\frac{1}{2}(s^2 + t^2) + (s+t)x - \frac{1}{2}x^2\right) dx$$

$$= \frac{1}{\sqrt{2\pi}} \exp(st) \int_{-\infty}^{\infty} \exp\left(-\frac{1}{2}(s+t-x)^2\right) dx \qquad (3.16)$$

$$= \frac{1}{\sqrt{2\pi}} \exp(st) \int_{-\infty}^{\infty} \exp\left(-\frac{1}{2}y^2\right) dy = \exp(st).$$

Taking derivatives $\frac{\partial^{m+n}}{\partial s^m \partial t^n}$ at $s = t = 0$ on both sides of the identity in (3.16) gives

$$\int_{-\infty}^{\infty} H_m(x) H_n(x) \, d\gamma(x) = m! \delta_{mn}.$$

Since $m! \delta_{mn} = \sqrt{m!} \sqrt{n!} \delta_{mn}$, this shows that the sequence $(\frac{1}{\sqrt{n!}} H_n)_{n \in \mathbb{N}}$ is orthonormal in $L^2(\mathbb{R}, \gamma)$.

It remains to show that the span of the Hermite functions is dense in $L^2(\mathbb{R}, \gamma)$. A quick proof is obtained by making use of the injectivity of the Fourier transform (which is an immediate consequence of Theorem 5.20). If $f \in L^2(\mathbb{R}, \gamma)$ is orthogonal to every Hermite polynomial, then it is orthogonal to every polynomial. From this it follows that for all $z \in \mathbb{C}$,

$$F(z) := \int_{-\infty}^{\infty} f(x) e^{zx - \frac{1}{2}x^2} \, dx = \sum_{k=0}^{\infty} \sum_{m=0}^{\infty} \frac{z^k}{k!} \frac{(-1)^m}{2^m m!} \int_{-\infty}^{\infty} f(x) x^k x^{2m} \, dx = 0.$$

In particular we have $F(-it) = 0$. From this we infer that the Fourier transform of

$x \mapsto f(x)e^{-\frac{1}{2}x^2}$ vanishes identically. By the injectivity of the Fourier transform, this implies that $f(x)e^{-\frac{1}{2}x^2} = 0$ for almost all $x \in \mathbb{R}$, and therefore $f(x) = 0$ for almost all $x \in \mathbb{R}$.   □

The identity $H_{n+2}(x) = xH_{n+1}(x) - (n+1)H_n(x)$ of Proposition 3.32 is an example of a so-called *three point recurrence relation*. As we will see in Section 9.6, orthogonal polynomials in $L^2(\mathbb{R}, \mu)$, with μ a finite Borel measure on $\mathbb{R}$, always satisfy a three point recurrence relation, and conversely if a sequence of polynomials on $\mathbb{R}$ satisfies such a relation, then under a mild additional assumption these polynomials are orthogonal on $L^2(\mathbb{R}, \mu)$ for a suitable finite Borel measure μ on $\mathbb{R}$.

Theorem 3.33 admits an extension to higher dimensions; see Section 15.6.

3.5.c Tensor Bases

Let μ_j be a finite Borel measure on a compact metric space K_j for each $j = 1, \ldots, k$, and let $K = K_1 \times \cdots \times K_k$ and $\mu = \mu_1 \times \cdots \times \mu_k$ be their products. If $(f_n^{(j)})_{n \geq 1}$ is an orthonormal basis for $L^2(K_j, \mu_j)$ for each $j = 1, \ldots, k$, then the functions

$$f_n(x) := f_{n_1}^{(1)}(x_1) \cdots f_{n_k}^{(k)}(x_k), \quad n \in \{1, 2, \ldots\}^k,$$

form an orthonormal basis for $L^2(K, \mu)$. Orthonormality being clear, in view of Theorem 3.21 it remains to check that the span of the functions f_n is dense. This follows from the fact that $C(K)$ is dense in $L^2(K, \mu)$ by the observation in Remark 2.31 and the fact that functions of the form $g(x) := g^{(1)}(x_1) \cdots g^{(k)}(x_k)$ with $g^{(j)} \in C(K_j)$ for all $j = 1, \ldots, k$ are dense in $C(K)$ by Example 2.10. Since each of the functions $g^{(j)}$ can be approximated in $L^2(K_j, \mu_j)$ by linear combinations of the functions $f_n^{(j)}$, g can be approximated in $L^2(K, \mu)$ by linear combinations of the functions f_n.

This is a special case of a more general construction involving tensor products of Hilbert spaces (see Chapters 14 and 15, in particular (14.2)): if $(h_n^{(j)})_{n \geq 1}$ is an orthonormal basis for the Hilbert space H_j, $j = 1, \ldots, k$, then the vectors

$$h_n(x) := h_{n_1}^{(1)} \otimes \cdots \otimes h_{n_k}^{(k)}, \quad n \in \{1, 2, \ldots\}^k,$$

form an orthonormal basis for the Hilbert space tensor product $H = H_1 \otimes \cdots \otimes H_k$.

Problems

3.1   Show that equality $|(x|y)| = \|x\|\|y\|$ for the Cauchy–Schwarz inequality holds if and only if x and y are collinear (that is, both belong to some one-dimensional subspace).

3.2 Let $(x_n)_{n \geqslant 1}$ be a sequence in a Hilbert space H. Suppose that there exists $x \in H$ such that:

(i) $(x_n|y) \to (x|y)$ for all $y \in H$;

(ii) $\|x_n\| \to \|x\|$.

Show that $x_n \to x$ in H.

3.3 Provide the missing details in the proof of Proposition 3.9.

3.4 A Banach space X is called *strictly convex* if for all norm one vectors $x_0, x_1 \in X$ with $x_0 \neq x_1$ and $0 < \lambda < 1$ we have $\|(1-\lambda)x + \lambda y\| < 1$. Prove that every Hilbert space is strictly convex.

3.5 Give an example of a nonempty compact convex set C in a two-dimensional Banach space X along with a vector $x \in X$ such that the set

$$\left\{ c \in C : \|x - c\| = \min_{y \in C} \|x - y\| \right\}$$

consists of more than one element.

3.6 Let

$$C := \left\{ f \in C[0,1] : f \text{ is real-valued}, \ f(0) = 0, \ \int_0^1 f(t)\,dt = 0 \right\}.$$

(a) Check that C is a closed and convex subset of $C[0,1]$.

Let $g \in C[0,1]$ be the function defined by $g(t) := t$.

(b) Show that for any $f \in C$ we have $\|f - g\| > \frac{1}{2}$.

(c) Show that $\inf_{f \in C} \|f - g\| = \frac{1}{2}$.

This shows that C contains no point minimising the distance $d(g, C)$.

3.7 In this problem we determine some orthogonal complements.

(a) Let

$$Y := \{ f \in L^2(0,1) : f(t) = 0 \ \text{ for almost all } \ t \in (0, \tfrac{1}{2}) \}.$$

Show that Y is a closed subspace of $L^2(0,1)$ and find $Y^\perp$.

(b) Let

$$Y := \left\{ f \in L^2(0,1) : \int_0^1 f(t)\,dt = 0 \right\}.$$

Show that Y is a closed subspace of $L^2(0,1)$ and find $Y^\perp$.

3.8 Let $(h_n)_{n \geqslant 1}$ be a finite or infinite orthonormal sequence in a Hilbert space H. Prove that for all $x \in H$ we have *Bessel's inequality*

$$\sum_{n \geqslant 1} |(x|h_n)|^2 \leqslant \|x\|^2.$$

3.9   Let $(e_j)_{j \geqslant 1}$ be the sequence of standard unit vectors in ℓ^2, and let $F = \overline{\operatorname{span}}\{e_{2n-1} : n \geqslant 1\}$ and let $G = \overline{\operatorname{span}}\{e_{2n-1} + \frac{1}{n} e_{2n} : n \geqslant 1\}$.

   (a) Give explicit expressions for the orthogonal projections P_F and P_G.

   (b) Show that $F \cap G = \{0\}$.

   (c) Show that $F + G$ is dense, but not closed in ℓ^2.

3.10 Prove the identity (3.8).

3.11 Use the Stone–Weierstrass theorem to prove that the trigonometric polynomials are dense in $L^2(\mathbb{T})$.

3.12 Prove the following binomial identity for the Hermite polynomials: For all $n \in \mathbb{N}$ and $x, y \in \mathbb{R}$,

$$H_n(x+y) = \sum_{k=0}^{n} \binom{n}{k} x^{n-k} H_k(y).$$

3.13 Prove the following formula for the Hermite polynomials: For all $n \in \mathbb{N}$ and $x \in \mathbb{R}$,

$$H_n(x) = \frac{1}{\sqrt{2\pi}} \int_{-\infty}^{\infty} (x+iy)^n \exp\left(-\frac{1}{2} y^2\right) dy.$$

3.14 Define the polynomials L_n, $n \in \mathbb{N}$, by the generating function expansion

$$\frac{\exp(tx/(1+t))}{1+t} = \sum_{n \in \mathbb{N}} \frac{t^n}{n!} L_n(x).$$

These are the *Laguerre polynomials* normalised so as to become monic.

   (a) Compute the polynomials L_n for $n = 0, 1, 2, 3$.

   (b) Show that the polynomials L_n are monic, have degree n, and satisfy the recurrence relation

$$L_{n+2}(x) = (x - 2n + 3)L_{n+1}(x) - (n+1)^2 L_n(x), \quad n \in \mathbb{N}.$$

   (c) Prove that the sequence $(\frac{1}{n!} L_n)_{n \geqslant 0}$ is an orthonormal basis for $L^2(\mathbb{R}_+, e^{-x} dx)$.

3.15 The *Hardy space* $H^2(\mathbb{D})$ is the vector space of all holomorphic functions on $\mathbb{D}$ of the form $\sum_{n \in \mathbb{N}} c_n z^n$ with $\sum_{n \in \mathbb{N}} |c_n|^2 < \infty$.

   (a) Prove that $H^2(\mathbb{D})$ is a Hilbert space with respect to the norm

$$\|f\|_{H^2(\mathbb{D})} := \left(\sum_{n \in \mathbb{N}} |c_n|^2\right)^{1/2}.$$

Let the functions $e_n \in L^2(\mathbb{T})$ be defined by $e_n(\theta) := \exp(in\theta)$, $n \in \mathbb{N}$, $\theta \in [-\pi, \pi]$.

   (b) Show that for all $f = \sum_{n \in \mathbb{N}} c_n z^n \in H^2(\mathbb{D})$ the sum $f|_{\mathbb{T}} := \sum_{n \in \mathbb{N}} c_n e_n$ converges in $L^2(\mathbb{T})$ and that the mapping

$$f \mapsto f|_{\mathbb{T}}$$

sets up an isometric isomorphism from $H^2(\mathbb{D})$ onto the closed subspace of $L^2(\mathbb{T})$ of all functions whose negative Fourier coefficients vanish.

(c) For holomorphic functions $f : \mathbb{D} \to \mathbb{C}$ with power series expansion $f(z) = \sum_{n \in \mathbb{N}} c_n z^n$ and $0 < r < 1$ define

$$f_r := \sum_{n \in \mathbb{N}} c_n r^n e_n.$$

Show that for $f \in H^2(\mathbb{D})$ and $0 < r < 1$ we have $f_r \in L^2(\mathbb{T})$ and $\|f_r\|_{L^2(\mathbb{T})} \leqslant \|f|_{\mathbb{T}}\|_{L^2(\mathbb{T})}$, and show that

$$\lim_{r \uparrow 1} f_r = f|_{\mathbb{T}} \text{ in } L^2(\mathbb{T}).$$

(d) Show that a holomorphic function $f : \mathbb{D} \to \mathbb{C}$ belongs to $H^2(\mathbb{D})$ if and only if

$$\sup_{0 < r < 1} \|f_r\|_{L^2(\mathbb{T})} < \infty,$$

and that in this case we have

$$\|f\|_{H^2(\mathbb{D})} = \|f|_{\mathbb{T}}\|_{L^2(\mathbb{T})} = \sup_{0 < r < 1} \|f_r\|_{L^2(\mathbb{T})}.$$

(e) Show that all $f \in H^2(\mathbb{D})$, $0 < r < 1$, and $\theta \in [-\pi, \pi]$ we have

$$f(re^{i\theta}) = \frac{1}{2\pi} \int_{-\pi}^{\pi} f|_{\mathbb{T}}(\eta) P_r(\theta - \eta) \, d\eta,$$

where the *Poisson kernel* is given by

$$P_r(\theta) = \frac{1 - r^2}{1 - 2r\cos(\theta) + r^2}.$$

Hint: Begin by showing that $P_r(\theta) = \sum_{n \in \mathbb{Z}} r^{|n|} e^{in\theta}$.

3.16 We continue our study of the space $H^2(\mathbb{D})$.

(a) Show that for all $z_0 \in \mathbb{D}$ the function

$$k_{z_0}(z) := \frac{1}{1 - z\overline{z_0}}$$

belongs to $H^2(\mathbb{D})$ and

$$\|k_{z_0}\|_{H^2(\mathbb{D})} = \frac{1}{(1 - |z_0|^2)^{1/2}}.$$

(b) Show that for all $f \in H^2(\mathbb{D})$ and $z_0 \in \mathbb{D}$ we have

$$f(z_0) = (f|k_{z_0}).$$

(c) Use parts (a) and (b) to show that if $f_n \to f$ in $H^2(\mathbb{D})$, then $f_n \to f$ uniformly on every compact subset of $\mathbb{D}$.

3.17 The *disc algebra* $A(\mathbb{D})$ is the closed subspace of the Banach space $C(\overline{\mathbb{D}})$ consisting of those functions that are holomorphic on $\mathbb{D}$ (see Problem 2.38).

(a) Show that $A(\mathbb{D})$ is dense in $H^2(\mathbb{D})$ (see Problem 3.15 for its definition).
(b) Show that if $f \in C(\overline{\mathbb{D}})$, then we have $f|_{\mathbb{D}} \in H^2(\mathbb{D})$ if and only if $f \in A(\mathbb{D})$.
(c) Show that the restriction mapping $\rho : A(\mathbb{D}) \to C(\mathbb{T})$ given by

$$f \mapsto f|_{\mathbb{T}}$$

extends to an isometry from $H^2(\mathbb{D})$ onto $L^2(\mathbb{T})$. How does it relate to Problem 3.15?

3.18 Let $A^2(\mathbb{D})$ denote the subspace of $L^2(\mathbb{D})$ consisting of all square integrable holomorphic functions on $\mathbb{D}$. The goal of this problem is to show that $A^2(\mathbb{D})$ is a Hilbert space with orthonormal basis $(e_n)_{n \in \mathbb{N}}$ given by

$$e_n(z) := \left(\frac{n+1}{\pi}\right)^{1/2} z^n, \quad z \in \mathbb{D}, \ n \in \mathbb{N}.$$

(a) Show that $(e_n)_{n \in \mathbb{N}}$ is an orthonormal system in $L^2(\mathbb{D})$.
   Hint: Perform a computation in polar coordinates.
(b) Show that the closed linear span Y of $(e_n)_{n \in \mathbb{N}}$ in $L^2(\mathbb{D})$ is contained in $A^2(\mathbb{D})$.
   Hint: On the one hand, every $f \in Y$ can be written as a convergent sum $f = \sum_{n \in \mathbb{N}} c_n e_n$ with convergence in $L^2(\mathbb{D})$ (explain why). On the other hand, for each $z \in \mathbb{D}$ the sum $g(z) := \sum_{n \in \mathbb{N}} c_n (\frac{n+1}{\pi})^{1/2} z^n$ converges absolutely. Now use the fact that L^2-convergence implies pointwise almost everywhere convergence of a subsequence to show that $f(z) = g(z)$ for almost all $z \in \mathbb{D}$.
(c) Show that if a holomorphic function $f \in L^2(\mathbb{D})$ satisfies $(f|e_n) = 0$ for all $n \in \mathbb{N}$, then $f = 0$.
   Hint: Consider the Taylor expansion of f around 0.
(d) Combine the above to conclude that $A^2(\mathbb{D})$ is a Hilbert space with orthonormal basis $(e_n)_{n \in \mathbb{N}}$.
(e) How are the spaces $A^2(\mathbb{D})$ and $H^2(\mathbb{D})$ related?

3.19 On $\mathbb{C}$ we consider the measure

$$d\gamma_{\mathbb{C}}(z) = \frac{1}{\pi} e^{-|z|^2} dz,$$

where dz is the Lebesgue measure on $\mathbb{C}$.

(a) Show that $\gamma_{\mathbb{C}}$ is a probability measure satisfying $\int_{\mathbb{C}} |z|^2 \, d\gamma_{\mathbb{C}}(z) = 1$.

Let $A^2(\mathbb{C})$ denote the complex vector space of all entire functions $f : \mathbb{C} \to \mathbb{C}$ that are square integrable with respect to $\gamma_{\mathbb{C}}$,

$$\int_{\mathbb{C}} |f(z)|^2 \, d\gamma_{\mathbb{C}}(z) < \infty.$$

(b) Show that $A^2(\mathbb{C})$ is a Hilbert space with respect to the inner product

$$(f|g) = \int_{\mathbb{C}} f(z) \overline{g(z)} \, d\gamma_{\mathbb{C}}(z).$$

(c) Show that the functions

$$e_n(z) := \frac{z^n}{\sqrt{n!}}, \quad n \in \mathbb{N},$$

form an orthonormal basis for $A^2(\mathbb{C})$.

3.20 In this problem we continue our study of the Hilbert space $A^2(\mathbb{C})$.

(a) Using the mean value theorem, show that for all $w \in \mathbb{C}$ the mapping $f \mapsto f(w)$ is continuous from $A^2(\mathbb{C})$ to $\mathbb{C}$. Deduce that for all $w \in \mathbb{C}$ there exists a unique function $k_w \in A^2(\mathbb{C})$ such that

$$f(w) = \int_{\mathbb{C}} k_w(z) f(z) \, d\gamma_{\mathbb{C}}(z), \quad f \in A^2(\mathbb{C}), \ w \in \mathbb{C}.$$

(b) Show that this function is given by

$$k_w(z) = \exp(z\overline{w}), \quad w, z \in \mathbb{C}.$$

(c) Show that the orthogonal projection P in $L^2(\mathbb{C}, \gamma_{\mathbb{C}})$ onto $A^2(\mathbb{C})$ is given by

$$Pf(w) = \int_{\mathbb{C}} k_w(z) f(z) \, d\gamma_{\mathbb{C}}(z), \quad f \in L^2(\mathbb{C}, \gamma_{\mathbb{C}}), \ w \in \mathbb{C}.$$

3.21 This problem discusses the construction and elementary properties of conditional expectations. Let $(\Omega, \mathscr{F}, \mu)$ be a probability space and let $\mathscr{G}$ be a sub-σ-algebra of $\mathscr{F}$. Let $1 \leqslant p \leqslant \infty$ and denote by $L^p(\Omega, \mathscr{G})$ the subspace of $L^p(\Omega)$ consisting of all $f \in L^p(\Omega)$ having a $\mathscr{G}$-measurable pointwise defined representative.

(a) Show that $L^p(\Omega, \mathscr{G})$ is a closed subspace of $L^p(\Omega)$.

(b) Let $P_{\mathscr{G}}$ denote the orthogonal projection in $L^2(\Omega)$ onto $L^2(\Omega, \mathscr{G})$ and fix a function $f \in L^2(\Omega)$. Prove that $P_{\mathscr{G}} f$ is the unique element of $L^2(\Omega; \mathscr{G})$ such that the following identity holds for all $f \in L^2(\Omega)$ and $G \in \mathscr{G}$:

$$\int_G f \, d\mu = \int_G P_{\mathscr{G}} f \, d\mu.$$

Hint: $f - P_{\mathscr{G}} f \perp \mathbf{1}_G$ in $L^2(\Omega)$.

(c) Prove that if $f \in L^2(\Omega)$ satisfies $f \geqslant 0$ μ-almost everywhere, then $P_{\mathscr{G}} f \geqslant 0$ μ-almost everywhere.

(d) Prove that if $f \in L^2(\Omega)$ satisfies $0 \leqslant f \leqslant 1$ μ-almost everywhere, then $0 \leqslant P_{\mathscr{G}} f \leqslant 1$ μ-almost everywhere.

(e) Prove that $|P_{\mathscr{G}} f| \leqslant P_{\mathscr{G}} |f|$ μ-almost everywhere.

(f) Prove that $P_{\mathscr{G}}$ restricts to a contractive projection in $L^\infty(\Omega)$ onto $L^\infty(\Omega; \mathscr{G})$ and extends to a contractive projection in $L^1(\Omega)$ onto $L^1(\Omega; \mathscr{G})$, and that the properties described in parts (b)–(e) extend to functions in these spaces. Here, a *projection* is understood to be a bounded operator P satisfying $P^2 = P$.

(g) Discuss the relation of this problem with Problem 2.17.

(h) Give an explicit expression for $P_{\mathscr{G}}$ in each of the following two cases:

    (i) $\Omega = (0, 1)$, $\mathscr{F}$ the Borel σ-algebra, μ the Lebesgue measure, and $\mathscr{G} = \{\varnothing, \Omega\}$;

    (ii) $\Omega = (-\frac{1}{2}, \frac{1}{2})$, $\mathscr{F}$ the Borel σ-algebra, μ the Lebesgue measure, and $\mathscr{G} = \{B \in \mathscr{F} : B = -B\}$.

(i) Use the Radon–Nikodým theorem (Theorem 2.46) to give an alternative proof of the existence of the projection $P_{\mathscr{G}}$ in $L^1(\Omega)$ of part (f).

3.22 In this problem we outline alternative proof of the existence part of the Radon–Nikodým theorem (Theorem 2.46) based on Hilbert space methods. Let $(\Omega, \mathscr{F}, \mu)$ be a σ-finite measure space and let the $\mathbb{K}$-valued measure ν on $(\Omega, \mathscr{F})$ be absolutely continuous with respect to μ.

We first assume that ν is a finite nonnegative measure.

(a) Show that there exists a measurable function $w \in L^1(\Omega, \mu)$ such that $w(\omega) > 0$ for μ-almost all $\omega \in \Omega$.

(b) Show that the mapping $f \mapsto \int_\Omega f \, d\nu$ is bounded on $L^2(\Omega, \lambda)$, where λ is the finite measure on $(\Omega, \mathscr{F})$ given by $\lambda(F) := \nu(F) + \int_F w \, d\mu$. Conclude that there exists a unique $h \in L^2(\Omega, \lambda)$ such that

$$\int_\Omega f \, d\nu = \int_\Omega f h \, d\lambda, \quad f \in L^2(\Omega, \lambda).$$

(c) Show that $0 \leqslant h \leqslant 1$ for λ-almost all $\omega \in \Omega$.
Hint: Apply the identity in part (b) to $f = \mathbf{1}_F$ with $F \in \mathscr{F}$.

(d) Show that $\mu(B) = 0$, where $B = \{\omega \in \Omega : h(\omega) = 1\}$.
Hint: Apply the identity in part (b) to $f = \mathbf{1}_B$.

(e) Show that there exists a nonnegative function $g \in L^1(\Omega, \mu)$ such that

$$\nu(F) = \int_F g \, d\mu, \quad F \in \mathscr{F}.$$

Hint: Apply the identity in part (b) to the function $f = 1 + h + \cdots + h^n$ and use parts (c), (d), and the monotone convergence theorem to show that the limit $g := \lim_{n \to \infty} (1 + h + \cdots + h^n) hw$ exists μ-almost everywhere and belongs to $L^1(\Omega, \mu)$.

This proves the Radon–Nikodým theorem for finite nonnegative measures ν.

(f) Deduce from this the general case.

3.23 Using Zorn's lemma, we will construct two nonequivalent Hilbertian norms on ℓ^2.

An indexed set $(v_i)_{i \in I}$ (where I is some index set) of a vector space V is called an *algebraic basis* if every $v \in V$ admits a unique (up to permutation of the terms) expansion of the form $v = \sum_{k=1}^n c_k v_{i_k}$ with $n \geqslant 1$ an integer and $c_1, \ldots, c_n$ scalars in K. Thus, every v is expressed as a *finite* linear combination of the v_i. The uniqueness assumption implies that the v_i are linearly independent. By a straightforward application of Zorn's lemma (partially order the set of all linearly independent subsets of V by set inclusion) every vector space has an algebraic basis.

Select an algebraic basis $(h_i)_{i \in I}$ in ℓ^2 which contains the standard unit vectors. Now remove one of the vectors that have been added to the standard unit basis vectors, say h_{i_0}, and denote the resulting codimension one subspace by X.

(a) Prove that X is dense in ℓ^2.

Define a linear mapping $\phi : \ell^2 \to \mathbb{K}$ by $\ell \equiv 0$ on X and $\phi(h_{i_0}) := 1$.

(b) Prove that ϕ is not continuous.

On X we define a new norm $\|\cdot\|$ as follows. Let $b : I \setminus \{i_0\} \to I$ be a bijection (which exists since both sets are infinite; prove this). For $j_1, \ldots, j_n \in I \setminus \{i_0\}$ and scalars $c_1, \ldots, c_n \in \mathbb{K}$ define

$$\left\| \sum_{k=1}^n c_k h_{j_k} \right\| := \left\| \sum_{k=1}^n c_k h_{b(j_k)} \right\|.$$

This sets up an isometry $B : X \simeq \ell^2$. We extend this norm to ℓ^2 by defining

$$\|h + c h_{i_0}\|^2 := \|h\|^2 + |c|^2, \quad h \in X, \ c \in \mathbb{K}.$$

This defines a norm on ℓ^2.

(c) Show that ℓ^2 is a Hilbert space with respect to the norm $\|\cdot\|$ and that X is a closed subspace.

(d) Prove that X is not closed in ℓ^2 with respect to the norm $\|\cdot\|$ (thus there is no analogue of Corollary 1.36 for subspaces of finite codimension). Deduce that $\|\cdot\|$ and $\|\cdot\|$ are not equivalent.

(e) Does this result contradict the fact that $(\ell^2, \|\cdot\|)$ and $(\ell^2, \|\cdot\|)$ are isometrically isomorphic (both being separable Hilbert spaces)?

(f) Refine the construction so as to answer the question of Problem 1.10 to the negative.

3.24 This problem provides an example of a linear operator on ℓ^2 which fails to be bounded. We return to Problem 3.23 and use the notation introduced there. Define the mapping

$$\pi : \ell^2 \to \ell^2, \quad \sum_{i \in F} c_i x_i \mapsto \sum_{i \in F \setminus \{i_0\}} c_i x_i \quad \text{for finite sets } F \subseteq I.$$

(a) Prove that π is linear, satisfies $\pi^2 = \pi$, and has range X.

(b) Prove that π fails to be bounded.

3.25 This problem assumes familiarity with the language of probability theory. Let $(B_t)_{t \in [0,1]}$ be a standard Brownian motion on a probability space $(\Omega, \mathscr{F}, \mathbb{P})$. For each $t \in [0,1]$, let $\mathscr{F}_t$ denote the σ-algebra generated by the family $(B_s)_{s \in [0,t]}$.

(a) Show that if $0 \leqslant s < t \leqslant 1$, the increment $B_t - B_s$ is independent of $\mathscr{F}_s$.

Let $H_0^2((0,1) \times \Omega)$ denote the subspace of $L^2((0,1) \times \Omega)$ consisting of all stochastic processes $\xi = (\xi_t)_{t \in [0,1]}$ of the form

$$\xi_t(\omega) = \sum_{n=0}^{N-1} a_n(\omega) \mathbf{1}_{(t_n, t_{n+1}]}(t), \quad t \in [0,1], \ \omega \in \Omega,$$

where $0 = t_0 < t_1 < \cdots < t_N = 1$ and each $a_n \in L^2(\Omega)$ is $\mathscr{F}_{t_n}$-measurable. For such processes we define

$$\int_0^1 \xi_t \, dB_t := \sum_{n=0}^{N-1} a_n (B_{t_{n+1}} - B_{t_n}).$$

(b) Show that for all $\xi \in H_0^2((0,1) \times \Omega)$ we have $\int_0^1 \xi_t \, dB_t \in L^2(\Omega)$ and

$$\left\| \int_0^1 \xi_t \, dB_t \right\|_{L^2(\Omega)}^2 = \| \xi \|_{L^2((0,1) \times \Omega)}^2.$$

Let $H^2((0,1) \times \Omega)$ denote the closure of $H_0^2((0,1) \times \Omega)$ in $L^2((0,1) \times \Omega)$.

(c) Deduce that the mapping $\xi \mapsto \int_0^1 \xi_t \, dB_t$ admits a unique extension to an isometry from $H^2((0,1) \times \Omega)$ into $L^2(\Omega)$, the so-called *Itô isometry*.

The random variable $\int_0^1 \xi_t \, dB_t$ is called the *Itô stochastic integral* of ξ with respect to the Brownian motion $(B_t)_{t \in [0,1]}$.

4

Duality

The present chapter is devoted to the study of duality of Banach spaces. We begin by characterising the duals of various classical Banach spaces, and then proceed to proving the Hahn–Banach theorems. These theorems provide the existence of functionals with certain desirable properties. The remainder of the chapter is concerned with applications of these theorems.

4.1 Duals of the Classical Banach Spaces

Recall that the *dual* of a Banach space X is the Banach space $X^* := \mathscr{L}(X, \mathbb{K})$. For $x \in X$ and $x^* \in X^*$, the scalar $x^*(x) \in \mathbb{K}$ is denoted by $\langle x, x^* \rangle$, that is, we write

$$x^*(x) =: \langle x, x^* \rangle.$$

The Hahn–Banach theorems guarantee an abundance of nontrivial functionals in the dual of any Banach space. In many concrete situations, however, it is possible to completely describe the dual space. It will be our first task to do this for some classical Banach spaces discussed in Chapter 2.

4.1.a Finite-Dimensional Spaces

It is instructive to start with duality of finite-dimensional spaces. As we have seen, every finite-dimensional Banach space is isomorphic to $\mathbb{K}^d$ for some integer $d \geqslant 1$. The dual of $\mathbb{K}^d$ is determined as follows.

Every $\xi \in \mathbb{K}^d$ determines an element $\phi_\xi \in (\mathbb{K}^d)^*$ by the prescription

$$\phi_\xi(x) := x \cdot \xi = \sum_{n=1}^d x_n \xi_n, \quad x \in \mathbb{K}^d.$$

Indeed, the Cauchy–Schwarz inequality implies $|\phi_\xi(x)| \leq \|x\| \|\xi\|$, from which it follows that ϕ_ξ is bounded and $\|\phi_\xi\| \leq \|\xi\|$. Conversely, every $\phi \in (\mathbb{K}^d)^*$ is of this form. To see this, let $e_1, \ldots, e_d$ be the standard unit vectors of $\mathbb{K}^d$ and set $\xi_n := \phi(e_n)$. Then $\xi := (\xi_1, \ldots, \xi_d) \in \mathbb{K}^d$ and, for all $x = (x_1, \ldots, x_d) = \sum_{n=1}^{d} x_n e_n$,

$$\phi(x) = \phi\left(\sum_{n=1}^{d} x_n e_n\right) = \sum_{n=1}^{d} x_n \phi(e_n) = \sum_{n=1}^{d} x_n \xi_n = x \cdot \xi = \phi_\xi(x).$$

It follows that $\phi = \phi_\xi$. Moreover, $\|\xi\|^2 = \phi_\xi(\overline{\xi}) \leq \|\phi_\xi\| \|\xi\|$. Together with the inequality $\|\phi_\xi\| \leq \|\xi\|$ it follows that $\|\phi_\xi\| = \|\xi\|$.

In summary, the correspondence $\phi_\xi \leftrightarrow \xi$ establishes an isometric isomorphism

$$(\mathbb{K}^d)^* \simeq \mathbb{K}^d.$$

4.1.b Sequence Spaces

The above proof scheme can easily be extended to identify the duals of the infinite-dimensional sequence spaces c_0 and ℓ^p. We begin by proving that the dual of c_0 can be identified with ℓ^1. Every $\xi \in \ell^1$ determines an element $\phi_\xi \in (c_0)^*$ by the prescription

$$\phi_\xi(x) := \sum_{n \geq 1} x_n \xi_n, \quad x \in c_0.$$

Indeed,

$$|\phi_\xi(x)| \leq \left(\sup_{n \geq 1} |x_n|\right) \sum_{n \geq 1} |\xi_n| = \|x\|_\infty \|\xi\|_1,$$

so ϕ_ξ is bounded and $\|\phi_\xi\| \leq \|\xi\|_1$. Conversely, every $\phi \in (c_0)^*$ is of this form. To see this, let $(e_n)_{n \geq 1}$ be the sequence of standard unit vectors of c_0 and set $\xi_n := \phi(e_n)$. We claim that $\sum_{n \geq 1} |\xi_n| < \infty$. To see this, choose scalars $c_n \in \mathbb{K}$ of modulus one such that $c_n \xi_n = |\xi_n|$. The sequence $(c_1, \ldots, c_N, 0, 0, \ldots) = \sum_{n=1}^{N} c_n e_n$ belongs to c_0 and has norm one, and

$$\sum_{n=1}^{N} |\xi_n| = \sum_{n=1}^{N} c_n \xi_n = \phi\left(\sum_{n=1}^{N} c_n e_n\right) \leq \|\phi\|.$$

Since $N \geq 1$ was arbitrary, this establishes the claim, with bound $\|\xi\|_1 \leq \|\phi\|$. It follows that $\xi = (\xi_1, \xi_2, \ldots)$ belongs to ℓ^1 and for all $x \in c_0$ we have

$$\phi(x) = \phi\left(\sum_{n \geq 1} x_n e_n\right) = \sum_{n \geq 1} x_n \phi(e_n) = \sum_{n \geq 1} x_n \xi_n = \phi_\xi(x).$$

It follows that $\phi = \phi_\xi$, and the preceding bounds combine to the norm equality $\|\phi_\xi\| = \|\xi\|_1$. In summary, the correspondence $\phi_\xi \leftrightarrow \xi$ establishes an isometric isomorphism

$$(c_0)^* \simeq \ell^1.$$

In much the same way one proves that the dual of ℓ^p, $1 \leqslant p < \infty$, can be represented as ℓ^q, where $\frac{1}{p} + \frac{1}{q} = 1$. More precisely, every element of ℓ^q defines a bounded functional $\phi_\xi \in (\ell^p)^*$ of norm $\|\phi_\xi\| \leqslant \|\xi\|_p$ by the same formula as before, this time using Hölder's inequality

$$|\phi_\xi(x)| = \left| \sum_{n \geqslant 1} x_n \xi_n \right| \leqslant \|x\|_p \|\xi\|_q.$$

Conversely, every bounded functional is of this form. To see this let $(e_n)_{n \geqslant 1}$ be the sequence of standard unit vectors of ℓ^p and set $\xi_n := \phi(e_n)$. We claim that $(\xi_n)_{n \geqslant 1}$ belongs to ℓ^q. The case $p = 1$ and $q = \infty$ is trivial, for $|\xi_n| \leqslant \|\phi\| \|e_n\| = \|\phi\|$, $n \geqslant 1$, so $\|\xi\|_\infty \leqslant \|\phi\|$. Therefore we only consider the case $1 < p < \infty$, in which case also $1 < q < \infty$. To prove that $\sum_{n \geqslant 1} |\xi_n|^q < \infty$ it obviously suffices to show that

$$\sum_{n=1}^{N} |\xi_n|^q \leqslant \|\phi\|^q, \quad N \geqslant 1. \tag{4.1}$$

Fix $N \geqslant 1$ and put $x^{(N)} := (c_1 |\xi_1|^{q/p}, \ldots, c_N |\xi_N|^{q/p}, 0, 0, \ldots)$, where the scalars $c_n \in \mathbb{K}$ are chosen in such a way that $c_n \xi_n = |\xi_n|$. This sequence belongs to ℓ^p, with norm

$$\|x^{(N)}\|_p^p = \sum_{n=1}^{N} |\xi_n|^q.$$

Since $\frac{1}{p} + \frac{1}{q} = 1$ implies $\frac{q}{p} + 1 = q$,

$$\sum_{n=1}^{N} |\xi_n|^q = \left| \sum_{n=1}^{N} c_n |\xi_n|^{q/p} \xi_n \right| = |\phi(x^{(N)})| \leqslant \|x^{(N)}\|_p \|\phi\| = \left(\sum_{n=1}^{N} |\xi_n|^q \right)^{1/p} \|\phi\|$$

and therefore $(\sum_{n=1}^{N} |\xi_n|^q)^{1/q} \leqslant \|\phi\|$, using once more that $\frac{1}{p} + \frac{1}{q} = 1$. This proves (4.1). Since $N \geqslant 1$ was arbitrary it follows that $\xi = (\xi_1, \xi_2, \ldots)$ belongs to ℓ^q with norm $\|\xi\|_q \leqslant \|\phi\|$, and for all $x \in \ell^p$ we have

$$\phi(x) = \phi\left(\sum_{n \geqslant 1} x_n e_n \right) = \sum_{n \geqslant 1} x_n \phi(e_n) = \sum_{n \geqslant 1} x_n \xi_n = \phi_\xi(x).$$

It follows that $\phi = \phi_\xi$, and the preceding bounds combine to the norm equality $\|\phi_\xi\| = \|\xi\|_q$. In summary, the correspondence $\phi_\xi \leftrightarrow \xi$ establishes an isometric isomorphism

$$(\ell^p)^* \simeq \ell^q, \quad 1 \leqslant p < \infty, \quad \frac{1}{p} + \frac{1}{q} = 1.$$

At the end of Section 4.2 we show that this result does not extend to $p = \infty$.

4.1.c Spaces of Continuous Functions

Definition 4.1 (Locally compact spaces). A topological space X is called *locally compact* if every point $x \in X$ is contained in an open set with compact closure.

For example, the spaces $\mathbb{K}^d$ are locally compact.

When X is a locally compact topological space, we let $C_0(X)$ denote the space of continuous functions $f : X \to \mathbb{K}$ *vanishing at infinity*, that is, for every $\varepsilon > 0$ there exists a compact set $K \subseteq X$ such that $|f(x)| < \varepsilon$ for all $x \in \complement K$. With respect to the supremum norm, $C_0(X)$ is a Banach space; the proof is similar to that for c_0. Note that $C_0(X) = C(X)$ if X is compact.

In what follows we assume that X is a locally compact Hausdorff space and endow X with its Borel σ-algebra. The space $M(X)$ of $\mathbb{K}$-valued Borel measures on X has been introduced in Section 2.4. As was shown there, $M(X)$ is a Banach space with respect to the variation norm $\|\mu\| = |\mu|(X)$. Every $\mu \in M(X)$ determines a bounded functional $\phi_\mu \in (C_0(X))^*$ given by

$$\phi_\mu(f) := \int_X f \, \mathrm{d}\mu.$$

By Proposition 2.49 it satisfies

$$|\phi_\mu(f)| \leqslant \int_X |f| \, \mathrm{d}|\mu| \leqslant \|f\|_\infty \|\mu\|$$

and therefore

$$\|\phi_\mu\| \leqslant \|\mu\|.$$

A Borel measure $\mu \in M(X)$ is said to be *Radon* if its variation $|\mu|$ is Radon (see Definition E.20), that is, if for every Borel subset B of X and all $\varepsilon > 0$ there is a compact set $K \subseteq X$ and an open set $U \subseteq X$ such that $K \subseteq B \subseteq U$ and $|\mu|(U \setminus K) < \varepsilon$; these properties are referred to as *inner regularity with compact sets* and *outer regularity*, respectively. By $M_R(X)$ we denote the space of all Radon measures on X. It follows readily from the definitions that $M_R(X)$ is a closed subspace of $M(X)$, and therefore it is a Banach space with respect to the variation norm. The next theorem identifies this space as the dual of $C_0(X)$:

Theorem 4.2 (Riesz representation theorem). *Let X be a locally compact Hausdorff space. For every $\phi \in (C_0(X))^*$ there exists a unique Radon measure $\mu \in M_R(X)$ such that $\phi = \phi_\mu$, that is,*

$$\langle f, \phi \rangle = \int_X f \, \mathrm{d}\mu, \quad f \in C_0(X).$$

This measure satisfies $\|\mu\| = \|\phi\|$. The correspondence $\phi \leftrightarrow \mu$ establishes an isometric *isomorphism*

$$(C_0(X))^* \simeq M_{\mathbb{R}}(X).$$

The representing measure μ is nonnegative if and only if ϕ is positivity preserving.

In the special case of a compact metric space we have $M_R(X) = M(X)$ by Proposition E.21 and we obtain an isometric isomorphism

$$(C_0(X))^* \simeq M(X).$$

For the proof of Theorem 4.2 we need the following version of Urysohn's lemma.

Proposition 4.3. *Let X be a locally compact Hausdorff space. If $K \subseteq U \subseteq X$ with K compact and U open, then there exists a function $f \in C_c(X)$ with support contained in U such that $0 \leqslant g \leqslant 1$ pointwise on X and $g \equiv 1$ on K.*

Proof Cover K with finitely many open sets $U_1, \ldots, U_k$, each of which has compact closure. Then K is contained in the open set $(U_1 \cup \cdots \cup U_k) \cap U$ and this set has compact closure. Using this set instead of U, we may now appeal to Urysohn's lemma (Proposition C.11. $\qquad\square$

Proof of Theorem 4.2 Uniqueness is immediate from the norm equality $\|\phi\| = \|\mu\|$, which holds for any representing measure $\mu \in M_{\mathbb{R}}(X)$; this follows from the argument of Step 4 of the proof.

The existence proof will be given in four steps.

Step 1 – We begin with the case of positivity preserving functionals ϕ. In this step we prove the existence of a nonnegative representing measure $\mu \in M(X)$ for such functionals. The Radon property of μ is shown in Step 2.

Let $\mathscr{U}$ denote the collection of open subsets of X. For $U \in \mathscr{U}$ and $f \in C_c(X)$ we write

$$f \prec U$$

if $0 \leqslant f \leqslant 1$ and the support of f is a compact set contained in U. Define

$$\mu(U) := \sup\{\langle f, \phi \rangle : f \in C_c(X), \, f \prec U\}$$

with the convention that $\mu(\varnothing) := 0$. Note that

$$0 \leqslant \mu(U) \leqslant \mu(X) = \sup\{\langle f, \phi \rangle : 0 \leqslant f \leqslant 1, \, f \in C_c(X)\} \leqslant \|\phi\|.$$

Let us show that μ is countably subadditive on $\mathscr{U}$. To this end, suppose that $f \prec \bigcup_{j \geqslant 1} U_j$ with $U_j \in \mathscr{U}$ for all $j \geqslant 1$. Since $\operatorname{supp}(f)$ is compact, it is contained in some finite union $\bigcup_{j=1}^k U_j$. For every x in the compact support of f choose an open set V_x with compact closure such that $x \in V_x$. By compactness, $\operatorname{supp}(f)$ is contained in a finite union $V := \bigcup_{n=1}^N V_{x_n}$. For $j = 1, \ldots, k$ set $V_j := U_j \cap V$. Then $\operatorname{supp}(f)$ is contained in $\bigcup_{j=1}^k V_j$, and this union has compact closure. Hence we may use Theorem C.12 to select a partition of unity $(g_j)_{j=1}^k$ relative to the sets V_j, $j = 1, \ldots, k$, such that $\sum_{j=1}^k g_j \equiv 1$ on supp(f). Then $g_j \in C_c(X)$ and $g_j \prec V_j$ for $j = 1, \ldots, k$; from $V_j \subseteq U_j$ we infer that also $g_j \prec U_j$. Hence also $fg_j \prec U_j$, and therefore

$$\langle f, \phi \rangle = \left\langle f \sum_{j=1}^k g_j, \phi \right\rangle = \sum_{j=1}^k \langle fg_j, \phi \rangle \leqslant \sum_{j=1}^k \mu(U_j) \leqslant \sum_{j \geqslant 1} \mu(U_j).$$

This being true for all $0 \leqslant f \in C_c(X)$ satisfying $f \prec \bigcup_{j \geqslant 1} U_j$, it follows that

$$\mu\left(\bigcup_{j \geqslant 1} U_j\right) \leqslant \sum_{j \geqslant 1} \mu(U_j)$$

as claimed.

In what follows we freely use the notation and terminology introduced in Appendix E. Let $\mu^* : 2^X \to [0, \infty]$ be the outer measure associated with μ through (E.1), that is,

$$\mu^*(A) := \inf\left\{ \sum_{j \geqslant 1} \mu(U_j) : A \subseteq \bigcup_{j \geqslant 1} U_j, \text{ where } U_j \in \mathscr{U} \text{ for all } j \geqslant 1 \right\}$$

for $A \in 2^X$ (see Lemma E.6). By the definition of an outer measure and the countable subadditivity of μ we have, for any set $A \in 2^X$,

$$\mu^*(A) = \inf\left\{ \mu(U) : A \subseteq U, \text{ where } U \in \mathscr{U} \right\}. \tag{4.2}$$

Clearly, $\mu^*(A) \geqslant 0$. We also note that

$$\mu^*(U) = \mu(U) \text{ for all } U \in \mathscr{U},$$

that is, μ^* extends μ. This fact is used repeatedly below.

We claim that $\mathscr{U}$ is contained in the set

$$\mathscr{M}_{\mu^*} := \left\{ A \in 2^X : \mu^*(Q) = \mu^*(Q \cap A) + \mu^*(Q \cap \complement A), \quad Q \in 2^X \right\}.$$

To prove this, let $U \in \mathscr{U}$, that is, let U be an open subset of X. By the subadditivity of outer measures we have $\mu^*(Q) \leqslant \mu^*(Q \cap U) + \mu^*(Q \cap \complement U)$. The reverse inequality trivially holds if $\mu^*(Q) = \infty$, so it suffices to check the inequality for $Q \in 2^X$ satisfying $\mu^*(Q) < \infty$. Fix an arbitrary $\varepsilon > 0$. Choose an open set V such that $Q \subseteq V$ and $\mu(V) < \mu^*(Q) + \varepsilon$; this is possible by (4.2). Let $f, g \in C_c(X)$ satisfy

$$f \prec U \cap V, \quad \mu(U \cap V) < \langle f, \phi \rangle + \varepsilon,$$

respectively

$$g \prec V \cap \complement(\text{supp} f), \quad \mu(V \cap \complement(\text{supp} f)) < \langle g, \phi \rangle + \varepsilon.$$

Such functions f and g exist by the definition of μ. Then, using the linearity of ϕ along with the facts that $f + g \prec V$ (as f and g have disjoint supports both contained in V) and $Q \cap \complement U \subseteq V \cap \complement(\text{supp} f)$ (which follows from $Q \subseteq V$ and $\text{supp}(f) \subseteq U$),

$$\begin{aligned}
\mu^*(Q \cap U) + \mu^*(Q \cap \complement U) &\leqslant \mu^*(U \cap V) + \mu^*(V \cap \complement(\text{supp} f)) \\
&= \mu(U \cap V) + \mu(V \cap \complement(\text{supp} f)) \\
&\leqslant \langle f, \phi \rangle + \langle g, \phi \rangle + 2\varepsilon = \langle f + g, \phi \rangle + 2\varepsilon \\
&\leqslant \mu(V) + 2\varepsilon \leqslant \mu^*(Q) + 3\varepsilon.
\end{aligned}$$

Since $\varepsilon > 0$ was arbitrary, this gives the desired result.

By Theorem E.7, $\mathcal{M}_{\mu^*}$ is a σ-algebra and μ^* restricts to a measure on $\mathcal{M}_{\mu^*}$. Since $\mathcal{U}$ is contained in $\mathcal{M}_{\mu^*}$, so is $\sigma(\mathcal{U}) = \mathcal{B}(X)$, the Borel σ-algebra of X. Thus we find that the restriction of μ^* to $\mathcal{B}(X)$ is a measure. Since we have already seen that $\mu^*(U) = \mu(U)$ for all $U \in \mathcal{U}$, by slight abuse of notation we shall denote the measure on $\mathcal{B}(X)$ thus obtained by μ. The bound $\mu(X) \leqslant \|\phi\|$ shows that μ is a finite measure. Nonnegativity of μ follows from the nonnegativity of μ^*.

Next we check that μ represents the functional ϕ. To this end we first claim that if $A, B \in \mathcal{B}(X)$ and $g \in C_0(X)$ satisfy $\mathbf{1}_A \leqslant g \leqslant \mathbf{1}_B$, then

$$\mu(A) \leqslant \langle g, \phi \rangle \leqslant \mu(B). \tag{4.3}$$

Indeed, since A is contained in the set $\{g \geqslant 1\}$ and this set is contained in the open set $\{g > 1 - \delta\}$, we have, for any $0 < \delta < 1$,

$$\begin{aligned}
\mu(A) \leqslant \mu\{g > 1 - \delta\} &= \sup\left\{ \langle f, \phi \rangle : f \in C_c(X), \, f \prec \{g > 1 - \delta\} \right\} \\
&\leqslant \langle \frac{g}{1 - \delta}, \phi \rangle = \frac{1}{1 - \delta} \langle g, \phi \rangle
\end{aligned}$$

using the positivity of ϕ and the fact that on the set $\{g > 1 - \delta\}$ we have $f \leqslant 1 \leqslant g/(1 - \delta)$ pointwise. Since $0 < \delta < 1$ was arbitrary, it follows that $\mu(A) \leqslant \langle g, \phi \rangle$. In the same way, for all $\delta > 0$ we have

$$\mu(B) \geqslant \mu\{g > 0\} = \sup\left\{ \langle f, \phi \rangle : f \in C_c(X), \, f \prec \{g > 0\} \right\} \geqslant \langle (g - \delta)^+, \phi \rangle,$$

using that $(g - \delta)^+$ belongs to $C_c(X)$ and satisfies $(g - \delta)^+ \prec \{g > 0\}$. Since $(g - \delta)^+ \to g$ in $C_0(X)$ as $\delta \downarrow 0$, it follows that $\mu(B) \geqslant \langle g, \phi \rangle$. This proves (4.3).

Let $0 \leqslant f \in C_0(X)$, fix $\varepsilon > 0$, and for $\delta \geqslant 0$ write $f_\delta(\xi) := \min\{f(\xi), \delta\}$. Then

$$f = \sum_{k \geqslant 0} (f_{(k+1)\varepsilon} - f_{k\varepsilon}).$$

There is no convergence issue here since functions in $C_0(X)$ are bounded, so at most finitely many terms in this sum are nonzero. From the inequalities

$$\varepsilon \mathbf{1}_{\{f \geqslant (k+1)\varepsilon\}} \leqslant f_{(k+1)\varepsilon} - f_{k\varepsilon} \leqslant \varepsilon \mathbf{1}_{\{f \geqslant k\varepsilon\}},$$

on the one hand we obtain

$$\varepsilon\mu\{f \geqslant (k+1)\varepsilon\} \leqslant \int_X f_{(k+1)\varepsilon} - f_{k\varepsilon}\,d\mu \leqslant \varepsilon\mu\{f \geqslant k\varepsilon\},$$

while combining them with (4.3) gives

$$\varepsilon\mu\{f \geqslant (k+1)\varepsilon\} \leqslant \langle f_{(k+1)\varepsilon} - f_{k\varepsilon}, \phi \rangle \leqslant \varepsilon\mu\{f \geqslant k\varepsilon\}.$$

It follows that

$$\left| \int_X f_{(k+1)\varepsilon} - f_{k\varepsilon}\,d\mu - \langle f_{(k+1)\varepsilon} - f_{k\varepsilon}, \phi \rangle \right| \leqslant \varepsilon\mu\{k\varepsilon < f \leqslant (k+1)\varepsilon\}$$

and consequently

$$\left| \int_X f\,d\mu - \langle f, \phi \rangle \right| \leqslant \varepsilon \sum_{k \geqslant 0} \mu\{k\varepsilon < f \leqslant (k+1)\varepsilon\} \leqslant \varepsilon\mu(X).$$

Since $\varepsilon > 0$ was arbitrary, this proves that $\int_X f\,d\mu = \langle f, \phi \rangle$ as desired. By the linearity of both sides, this identity extends to arbitrary $f \in C_0(X)$.

Step 2 – We prove next that μ is a Radon measure. Outer regularity is clear from the constructions, and inner regularity with compact sets will be proved in two steps: (i) First we prove that if U is open in X, then for every $\varepsilon > 0$ there is a compact set $K \subseteq U$ such that $\mu(U \setminus K) < \varepsilon$; (ii) We then use this to deduce the analogous result for general Borel sets B in X.

(i): Let U be open in X. Pick $f \in C_c(X)$ such that $f \prec U$ and $\int_X f\,d\mu > \mu(U) - \varepsilon$ and let K be its support. Then $K \subseteq U$ and we have $\mu(K) \geqslant \int_X f\,d\mu > \mu(U) - \varepsilon$ since $0 \leqslant f \leqslant \mathbf{1}_K$. But then $\mu(U \setminus K) < \varepsilon$.

(ii): Suppose next that B is a Borel set in X. By outer regularity there is an open set $V \subseteq X$ such that $B \subseteq V$ and $\mu(V \setminus B) < \varepsilon$. By what we just proved there is a compact set $L \subseteq X$ such that $L \subseteq V$ and $\mu(V \setminus L) < \varepsilon$. Using outer regularity once more, choose an open set W such that $V \setminus B \subseteq W$ and $\mu(W) < \varepsilon$. Let $K := L \setminus W$. Then K is compact, contained in B, and

$$\mu(K) = \mu(L) - \mu(L \cap W) > (\mu(V) - \varepsilon) - \mu(W) > (\mu(B) - \varepsilon) - \mu(W) > \mu(B) - 2\varepsilon.$$

It follows that $\mu(B \setminus K) < 2\varepsilon$ and the claim is proved.

This completes the proof of the theorem for positivity preserving functionals, except for the norm identity $\|\mu\| = \|\phi\|$ which will be proved, for general functionals ϕ, in Step 4.

Step 3 – The *real part* of a functional $\phi \in (C_0(X))^*$ is defined, for $f = u + iv \in C_0(X)$ with u, v real-valued, by

$$\mathrm{Re}\,\phi(f) := \mathrm{Re}\langle u, \phi \rangle + i\,\mathrm{Re}\langle v, \phi \rangle.$$

It is clear that $\operatorname{Re}\phi$ is additive, and in combination with the identities

$$
\begin{aligned}
\operatorname{Re}\phi((a+bi)f) &= \operatorname{Re}\phi((au-bv)+i(bu+av)) \\
&= \operatorname{Re}\langle au-bv,\phi\rangle + i\operatorname{Re}\langle bu+av,\phi\rangle \\
&= (a+bi)(\operatorname{Re}\langle u,\phi\rangle + i\operatorname{Re}\langle v,\phi\rangle) = (a+bi)\operatorname{Re}\phi(f)
\end{aligned}
$$

we see that $\operatorname{Re}\phi$ is linear. Boundedness is clear, and therefore $\operatorname{Re}\phi \in (C_0(X))^*$. The functional $\operatorname{Im}\phi \in (C_0(X))^*$ is defined similarly. Both $\operatorname{Re}\phi$ and $\operatorname{Im}\phi$ are *real*, in the sense that they map real-valued functions to real numbers, and we have $\phi = \operatorname{Re}\phi + i\operatorname{Im}\phi$.

Suppose now that $\phi \in (C_0(X))^*$ is real. In analogy with the formulas for the positive and negative parts of a real measure we define, for functions $0 \leqslant f \in C_0(X)$,

$$
\phi^+(f) := \sup\{\langle g,\phi\rangle : g \in C_0(X), \ 0 \leqslant g \leqslant f\}. \tag{4.4}
$$

We claim that ϕ^+ is the restriction of a real-linear functional on $C_0(X;\mathbb{R})$ of norm at most $\|\phi\|$. This will follow from Theorem 4.5. This theorem implies that the dual of $C_0(X)$ is a Banach lattice and gives a general formula for the positive part of functionals in the dual of a Banach lattice of which (4.4) is a special case. For the reader's convenience, however, here we give a self-contained proof of the claim.

It is clear that $|\phi^+(f)| \leqslant \|f\|\|\phi\|$ and $0 = \phi^+(0) \leqslant \phi^+(f)$. It is also clear that $\phi^+(cf) = c\phi^+(f)$ for scalars $c \geqslant 0$. If $0 \leqslant g_1 \leqslant f_1$ and $0 \leqslant g_2 \leqslant f_2$, then $0 \leqslant g_1 + g_2 \leqslant f_1 + f_2$, so

$$
\phi^+(f_1+f_2) \geqslant \phi(g_1+g_2) = \phi(g_1)+\phi(g_2).
$$

Taking the supremum over all admissible g_1 and g_2 gives the inequality $\phi^+(f_1+f_2) \geqslant \phi^+(f_1) + \phi^+(f_2)$. To prove the converse inequality let $0 \leqslant g \leqslant f_1 + f_2$ with $f_1, f_2 \geqslant 0$, and set $g_1 := f_1 \wedge g$ and $g_2 := g - g_1$. Then $0 \leqslant g_1 \leqslant f_1$ and $0 \leqslant g_2 \leqslant f_2$, so

$$
\phi(g) = \phi(g_1)+\phi(g_2) \leqslant \phi^+(f_1)+\phi^+(f_2)
$$

and therefore $\phi^+(f_1+f_2) \leqslant \phi^+(f_1)+\phi^+(f_2)$. This proves the additivity of ϕ^+ on the cone of nonnegative functions in $C_0(X)$.

For functions $f \in C_0(X;\mathbb{R})$ we define $\phi^+(f) := \phi^+(f^+) - \phi^+(f^-)$. It is routine to check that ϕ^+ is real-linear on $C_0(X;\mathbb{R})$. Moreover,

$$
|\phi^+(f)| \leqslant \max\{\phi^+(f^+), \phi^+(f^-)\} \leqslant \|\phi\|\max\{\|f^+\|, \|f^-\|\} = \|\phi\|\|f\|.
$$

This completes the proof of the claim.

The functional $\phi^- = \phi^+ - \phi$ is real-linear and bounded on $C_0(X;\mathbb{R})$ and the definition of ϕ^+ implies that ϕ^- is positive. This gives the representation $\phi = \phi^+ - \phi^-$ with $\phi^\pm$ bounded, linear, and positive.

Since linear combinations of Radon measures are Radon, these reductions make it possible to apply Step 2 to obtain a Radon measure $\mu \in M_{\mathbb{R}}(X)$ representing ϕ.

Step 4 – The only thing left to be shown is that the norm equality $\|\mu\| = \|\phi\|$ holds for any representing Radon measure μ. This will be accomplished by invoking the Radon–Nikodým theorem (Theorem 2.46), or rather, the result of Example 2.48 which follows from it. It asserts that there exists a function $h \in L^1(\Omega, |\mu|)$ such that $|h| = 1$ $|\mu|$-almost everywhere and $\mu(B) = \int_B h \, d|\mu|$ for all Borel sets $B \subseteq X$. By the usual arguments, this implies

$$\int_X f \, d\mu = \int_X f h \, d|\mu|, \quad f \in C_c(X).$$

We claim that $C_c(X)$ is dense in $C_0(X)$. Indeed, for given $f \in C_0(X)$ and $\varepsilon > 0$, let K be a compact set such that $|f| < \varepsilon$ outside K, and apply Proposition 4.3 to obtain a function $g \in C_c(X)$ such that $0 \leqslant g \leqslant 1$ pointwise on X and $g \equiv 1$ on K. Then $fg \in C_c(X)$ and $\|f - fg\|_\infty \leqslant \varepsilon$. This proves the claim. Since $L^1(X, |\mu|)$ is isometrically contained in $M(X)$ it also follows that $C_c(X)$ is norming for $L^1(X, |\mu|)$.

Combining these observations, we obtain

$$\|\phi\| = \sup_{\substack{\|f\| \leqslant 1 \\ f \in C_c(X)}} |\langle f, \phi \rangle|$$

$$= \sup_{\substack{\|f\| \leqslant 1 \\ f \in C_c(X)}} \left| \int_X f \, d\mu \right| = \sup_{\substack{\|f\| \leqslant 1 \\ f \in C_c(X)}} \left| \int_X f h \, d|\mu| \right| = \|h\|_{L^1(X, |\mu|)} = |\mu|(X) = \|\mu\|$$

and the proof is complete. $\qquad\qquad\qquad\qquad\qquad\qquad\qquad\qquad\qquad\qquad\qquad\qquad\square$

The duality between spaces of continuous functions and spaces of Borel measures shows how elements of Measure Theory emerge naturally from considerations involving only linearity and topology (namely, from the problem of finding the continuous linear functionals of a space of continuous functions).

4.1.d Spaces of Integrable Functions

Let $(\Omega, \mathscr{F}, \mu)$ be a measure space and let $1 \leqslant p, q \leqslant \infty$ satisfy $\frac{1}{p} + \frac{1}{q} = 1$. By Hölder's inequality, every function $g \in L^q(\Omega)$ defines a functional $\phi_g \in (L^p(\Omega))^*$ by setting

$$\phi_g(f) := \int_\Omega f g \, d\mu, \quad f \in L^p(\Omega),$$

and we have $\|\phi_g\| \leqslant \|g\|_q$. If $1 \leqslant p < \infty$ and the measure space is σ-finite, every functional arises in this way:

Theorem 4.4 (Dual of $L^p(\Omega)$). *Let $(\Omega, \mathscr{F}, \mu)$ be a σ-finite measure space and let $1 \leqslant p < \infty$ and $\frac{1}{p} + \frac{1}{q} = 1$. For every $\phi \in (L^p(\Omega))^*$ there exists a unique $g \in L^q(\Omega)$ such that $\phi = \phi_g$, that is,*

$$\langle f, \phi \rangle = \int_\Omega f g \, d\mu, \quad f \in L^p(\Omega),$$

and it satisfies $\|g\|_q = \|\phi\|$. *The correspondence $\phi_g \leftrightarrow g$ establishes an isometric isomorphism*

$$(L^p(\Omega))^* \simeq L^q(\Omega), \quad 1 \leqslant p < \infty, \quad \frac{1}{p} + \frac{1}{q} = 1.$$

Proof Uniqueness is immediate from the norm identity $\|g\|_q = \|\phi\|$ which, by Hölder's inequality and Proposition 2.26, holds for any representing function $g \in L^q(\Omega)$.

The existence proof will be given in two steps.

Step 1 – In this step we prove the theorem for the special case $\mu(\Omega) < \infty$. Let $\phi \in (L^p(\Omega))^*$ be arbitrary and fixed. Then

$$\nu(A) := \langle \mathbf{1}_A, \phi \rangle, \quad A \in \mathscr{F},$$

defines a $\mathbb{K}$-valued measure. Indeed, if the sets $A_n \in \mathscr{F}$ are disjoint, then by dominated convergence $\lim_{n \to \infty} \sum_{j=1}^n \mathbf{1}_{A_j} = \mathbf{1}_{\bigcup_{j \geqslant 1} A_j}$ in $L^p(\Omega)$, and therefore

$$\nu\Big(\bigcup_{j \geqslant 1} A_j\Big) = \lim_{n \to \infty} \sum_{j=1}^n \langle \mathbf{1}_{A_j}, \phi \rangle = \lim_{n \to \infty} \sum_{j=1}^n \nu(A_j)$$

by the boundedness of ϕ. Clearly ν is absolutely continuous with respect to μ. By the Radon–Nikodým theorem (Theorem 2.46) we have $\nu = g\,d\mu$ for a unique $g \in L^1(\Omega)$. By the definition of ν and linearity, this means that

$$\langle f, \phi \rangle = \int_\Omega fg\,d\mu \quad \text{for all simple functions } f. \tag{4.5}$$

We wish to prove that $g \in L^q(\Omega)$ with $\|g\|_q \leqslant \|\phi\|$ and that the identity $\langle f, \phi \rangle = \int_\Omega fg\,d\mu$ holds for all $f \in L^p(\Omega)$. For $n = 1, 2, \ldots$ let $g_n := g\mathbf{1}_{\Omega_n}$ with $\Omega_n = \{|g| \leqslant n\}$. These functions are bounded and for all simple functions f we have, by (4.5),

$$\Big| \int_\Omega fg_n\,d\mu \Big| = \Big| \int_\Omega f\mathbf{1}_{\Omega_n} g\,d\mu \Big| = |\langle f\mathbf{1}_{\Omega_n}, \phi \rangle| \leqslant \|f\mathbf{1}_{\Omega_n}\|_p \|\phi\| \leqslant \|f\|_p \|\phi\|.$$

Since the simple functions are dense in $L^p(\Omega)$, Proposition 2.26 implies that $g_n \in L^q(\Omega)$ and $\|g_n\|_q \leqslant \|\phi\|$. This being true for all $n \geqslant 1$, Fatou's lemma implies that $g \in L^q(\Omega)$ and $\|g\|_q \leqslant \|\phi\|$.

Now that we know this, the density of the simple functions in $L^p(\Omega)$ and Hölder's inequality imply that (4.5) extends to arbitrary $f \in L^p(\Omega)$. This completes the proof of the theorem in the case $\mu(\Omega) < \infty$.

Step 2 – The general σ-finite case follows by an exhaustion argument as follows. Choose an increasing sequence $\Omega_1 \subseteq \Omega_2 \subseteq \ldots$ of sets of finite measure such that $\bigcup_{n \geqslant 1} \Omega_n = \Omega$. By restriction to functions supported on Ω_n, every $\phi \in (L^p(\Omega))^*$ restricts to a functional in $(L^p(\Omega_n))^*$, denoted by ϕ_n, of norm $\|\phi_n\| \leqslant \|\phi\|$. By the previous step, ϕ_n is represented by a unique function $g_n \in L^q(\Omega_n)$ of norm $\|g_n\|_q \leqslant \|\phi_n\| \leqslant \|\phi\|$. Moreover, by uniqueness we see that if $m \leqslant n$, then $g_n|_{\Omega_m} = g_m$, since both represent ϕ_m. We can thus define a measurable function $g : \Omega \to \mathbb{K}$ by setting $g := g_n$ on Ω_n for $n \geqslant 1$. This function satisfies $\|g\|_q = \sup_{n \geqslant 1} \|g_n\|_q \leqslant \|\phi\|$. Moreover, if $f \in L^p(\Omega)$, then by the continuity of ϕ and the dominated convergence theorem,

$$\langle f, \phi \rangle = \lim_{n \to \infty} \langle \mathbf{1}_{\Omega_n} f, \phi \rangle = \lim_{n \to \infty} \int_{\Omega_n} f g_n \, d\mu = \lim_{n \to \infty} \int_{\Omega_n} f g \, d\mu = \int_{\Omega} f g \, d\mu.$$

$\square$

For $1 < p < \infty$ the σ-finiteness assumption can be omitted; see Problem 4.3.

4.1.e Hilbert Spaces

The Riesz representation theorem (Theorem 3.15) states that every bounded functional on a Hilbert space H is of the form ψ_h for some unique $h \in H$, where

$$\psi_h(g) = (g|h), \quad g \in H.$$

Moreover, we have equality of norms $\|h\| = \|\psi_h\|$. The identification $\psi_h \leftrightarrow h$ therefore provides a bijective and norm-preserving correspondence

$$H^* \longleftrightarrow H.$$

It is important to observe that this correspondence is linear if $\mathbb{K} = \mathbb{R}$, but conjugate-linear if $\mathbb{K} = \mathbb{C}$. This is a consequence of the conjugate-linearity of inner products with respect to their second variable. Indeed, from $\psi_{ch}(x) = (x|ch) = \bar{c}(x|h) = \bar{c}\psi_h(x)$ it follows that

$$\psi_{ch} = \bar{c}\psi_h.$$

In contrast, the correspondence $\phi_x \leftrightarrow x$ in each of the Sections 4.1.a–4.1.d is linear both when $\mathbb{K} = \mathbb{R}$ and $\mathbb{K} = \mathbb{C}$.

4.1.f Banach Lattices

The objective of this section is to prove the following theorem.

Theorem 4.5. *With respect to the natural partial order given by*

$$x^* \leqslant y^* \Leftrightarrow \langle x, x^* \rangle \leqslant \langle x, y^* \rangle \ \text{for all} \ 0 \leqslant x \in X$$

the dual X^* of a Banach lattice X is a Banach lattice. Moreover, for all $0 \leqslant x \in X$ we have

$$\langle x, x^* \wedge y^* \rangle = \inf \{ \langle x - y, x^* \rangle + \langle y, y^* \rangle : 0 \leqslant y \leqslant x \},$$
$$\langle x, x^* \vee y^* \rangle = \sup \{ \langle x - y, x^* \rangle + \langle y, y^* \rangle : 0 \leqslant y \leqslant x \}.$$

A special case of the second formula has already been encountered in (4.4).

For the proof of the theorem we need the following lemma. If V is a vector lattice, then for $0 \leqslant v \in V$ we write $[0,v] := \{u \in V : 0 \leqslant u \leqslant v\}$.

Lemma 4.6 (Decomposition property). *If V is a vector lattice, then for all $0 \leqslant v, v' \in V$ we have*

$$[0,v] + [0,v'] = [0,v+v'].$$

Proof Let $w \in [0, v+v']$; we must show that there exist $u \in [0,v]$ and $u' \in [0,v']$ such that $u + u' = w$. We claim that $u := v \wedge w$ and $u' := w - u$ have the required properties. It is clear that $u \in [0,v]$, $u' \geqslant 0$, and $u + u' = w$, and by Proposition 2.52(2) we have

$$v' - u' = v' - w + u = (v' - w) + v \wedge w$$
$$= ((v' - w) + v) \wedge ((v' - w) + w) = ((v + v') - w) \wedge v' \geqslant 0;$$

this proves that $u' \in [0, v']$. $\qquad\square$

Proof of Theorem 4.5 First note that if $0 \leqslant y \leqslant x$, then also $0 \leqslant x - y \leqslant x$ and therefore $\|y\| \leqslant \|x\|$ and $\|x - y\| \leqslant \|x\|$. It follows that

$$|\langle x - y, x^* \rangle + \langle y, y^* \rangle| \leqslant \|x\|(\|x^*\| + \|y^*\|),$$

showing that the infima and suprema on the right-hand sides of the formulas in the statement of the theorem are finite. Lemma 4.6 implies that the right-hand sides are additive on the positive cone X^+ of X. To see this, let $x, x' \in X$. Then

$$\inf\{\langle x + x' - y, x^* \rangle + \langle y, y^* \rangle : y \in [0, x + x']\}$$
$$= \inf\{\langle x - u, x^* \rangle + \langle x' - u', x^* \rangle + \langle u, y^* \rangle + \langle u', y^* \rangle : u \in [0,x],\, u' \in [0,x']\}$$
$$= \inf\{\langle x - u, x^* \rangle + \langle u, y^* \rangle : u \in [0,x]\} + \inf\{\langle x' - u', x^* \rangle + \langle u', y^* \rangle : u' \in [0,x']\}.$$

The corresponding identity for the suprema is proved in the same way. Since the right-hand sides are also homogeneous with respect to scalar multiplication by nonnegative scalars, they uniquely extend to linear mappings from X to $\mathbb{R}$ and therefore define elements of X^*. Thus we may *define* functionals $x^* \wedge y^*$ and $x^* \vee y^*$ in X^* by the right-hand sides of the formulas in the statement of the theorem.

We begin by showing that the functionals $x^* \wedge y^*$ and $x^* \vee y^*$ thus defined are the greatest lower bound and the least upper bound for the pair $\{x^*, y^*\}$, respectively. We will present the argument for $x^* \wedge y^*$; the proof for $x^* \vee y^*$ is entirely similar.

It is clear that for all $x \geqslant 0$ we have $\langle x, x^* \wedge y^* \rangle \leqslant \langle x, x^* \rangle$. This means that $x^* \wedge y^* \leqslant x^*$, and in the same way we see that $x^* \wedge y^* \leqslant y^*$. This shows that $x^* \wedge y^*$ is a lower bound for the pair $\{x^*, y^*\}$. To prove that it is the greatest lower bound we must show that if $z^* \leqslant x^*$ and $z^* \leqslant y^*$, then $z^* \leqslant x^* \wedge y^*$. But this is easy: if $0 \leqslant y \leqslant x$, then

$$\langle x, z^* \rangle = \langle x - y, z^* \rangle + \langle y, z^* \rangle \leqslant \langle x - y, x^* \rangle + \langle y, y^* \rangle,$$

and therefore for all $x \geqslant 0$ we obtain

$$\langle x, z^* \rangle \leqslant \inf \{ \langle x - y, x^* \rangle + \langle y, y^* \rangle : 0 \leqslant y \leqslant x \} = \langle x, x^* \wedge y^* \rangle,$$

that is, $z^* \leqslant x^* \wedge y^*$.

This proves that the pair $(X^*, \leqslant)$ is a lattice. It is clear from the definition of the partial order $\leqslant$ on X^* that if $x^*, y^* \in X^*$ satisfy $x^* \leqslant y^*$, then $cx^* \leqslant cy^*$ for all $0 \leqslant c \in \mathbb{R}$ and $x^* + z^* \leqslant y^* + z^*$. It follows that $(X^*, \leqslant)$ is a vector lattice and that the identities in the statement of the theorem are satisfied.

Since X^* is complete, all that remains to be shown is that $|x^*| \leqslant |y^*|$ implies $\|x^*\| \leqslant \|y^*\|$. The assumption is equivalent to the statement that for all $x \geqslant 0$ we have

$$\sup \{ \langle x - y, x^* \rangle - \langle y, x^* \rangle : 0 \leqslant y \leqslant x \} \leqslant \sup \{ \langle x - y, y^* \rangle - \langle y, y^* \rangle : 0 \leqslant y \leqslant x \},$$

that is,

$$\sup \{ \langle z, x^* \rangle : -x \leqslant z \leqslant x \} \leqslant \sup \{ \langle z, y^* \rangle : -x \leqslant z \leqslant x \}.$$

This, combined with the identity

$$\|z^*\| = \sup_{\|x\| \leqslant 1} \langle x, z^* \rangle = \sup_{\|x\| \leqslant 1} \sup \{ \langle z, z^* \rangle : -x \leqslant z \leqslant x \}$$

(which follows from the fact that $-x \leqslant z \leqslant x$ implies $|x| \leqslant |z|$ and hence $\|z\| \leqslant \|x\|$), gives $\|x^*\| \leqslant \|y^*\|$ as desired. $\qquad \square$

4.2 The Hahn–Banach Extension Theorem

We now turn to one of the main pillars of Functional Analysis, the Hahn–Banach theorem. This is a collective name for a number of closely related results, all of which assert the existence of certain nontrivial functionals with desirable properties. These results come in two flavours: as extension theorems asserting the extendability of functionals that are given *a priori* on a subspace and as separation theorems asserting that certain disjoint subsets can be separated by means of functionals.

The present section is concerned with Hahn–Banach extension theorems. We begin with a version for *real* vector spaces whose proof exploits the order structure of the real line.

Theorem 4.7 (Hahn–Banach extension theorem for real vector spaces). *Let V be a real vector space and let $p : V \to \mathbb{R}$ be* sublinear, *that is, for all $v, v' \in V$ and $t \geqslant 0$ we have*

$$p(v + v') \leqslant p(v) + p(v'), \quad p(tv) = t p(v).$$

If $W \subseteq V$ is a subspace and $\phi : W \to \mathbb{R}$ is a linear mapping satisfying

$$\phi(w) \leqslant p(w), \quad w \in W,$$

then there exists a linear mapping $\Phi : V \to \mathbb{R}$ that extends ϕ, and satisfies

$$\Phi(v) \leqslant p(v), \quad v \in V.$$

Proof We may assume that W is a proper subspace of V. Fix $w, w' \in W$ and $v \in V \setminus W$. From $\phi(w) + \phi(w') = \phi(w + w') \leqslant p(w + w')$ and $p(w + w') \leqslant p(w - v) + p(v + w')$ we obtain $\phi(w) - p(w - v) \leqslant p(w' + v) - \phi(w')$. With $\alpha := \sup_{w \in W} (\phi(w) - p(w - v))$ we therefore have

$$\phi(w) - \alpha \leqslant p(w - v), \quad \phi(w') + \alpha \leqslant p(w' + v), \quad w, w' \in W.$$

Let W_1 denote the linear span of W and v. Then $\phi_1 : W_1 \to \mathbb{R}$, $\phi_1(w + tv) := \phi(w) + t\alpha$, is linear, extends ϕ and satisfies $\phi_1(w_1) \leqslant p(w_1)$ for all $w_1 \in W_1$. To see this, note that for $w \in W$ and $t > 0$,

$$\phi_1(w + tv) = \phi(w) + t\alpha = t(\phi(t^{-1}w) + \alpha) \leqslant t p(t^{-1}w + v) = p(w + tv)$$

and

$$\phi_1(w - tv) = \phi(w) - t\alpha = t(\phi(t^{-1}w) - \alpha) \leqslant t p(t^{-1}w - v) = p(w - tv)$$

while for $t = 0$ we have $\phi_1(w) = \phi(w)$. This proves that $\phi_1(w_1) \leqslant p(w_1)$ for all $w_1 \in W_1$. The proof can now be finished by an appeal to Zorn's lemma (Lemma A.3), applied to all linear extensions ϕ' of ϕ satisfying the inequality $\phi' \leqslant p$. $\qquad\square$

This result is used to give a second version of the theorem which is also valid over the complex scalars.

Theorem 4.8 (Hahn–Banach extension theorem for vector spaces). *Let V be a (real or complex) vector space and let $p : V \to [0, \infty)$ be a* seminorm, *that is, for all $v, v' \in V$ and $t \in \mathbb{K}$ we have*

$$p(v + v') \leqslant p(v) + p(v'), \quad p(tv) = |t| p(v).$$

If W is a subspace of V and $\phi : W \to \mathbb{K}$ is a linear mapping satisfying

$$|\phi(w)| \leqslant p(w), \quad w \in W,$$

then there exists a linear mapping $\Phi : V \to \mathbb{K}$ that extends ϕ and satisfies

$$|\Phi(v)| \leqslant p(v), \quad v \in V.$$

Proof First we consider the case $\mathbb{K} = \mathbb{R}$. The assumptions imply $\phi(w) \leqslant p(w)$ for all $w \in W$, and therefore by Theorem 4.7 the mapping $\phi : W \to \mathbb{R}$ admits a linear extension

$\Phi : V \to \mathbb{R}$ satisfying $\Phi(v) \leqslant p(v)$ for all $v \in V$. Also, $-\Phi(v) = \Phi(-v) \leqslant p(-v) = p(v)$, and therefore $|\Phi(v)| \leqslant p(v)$ for all $v \in V$.

Next we consider the case $\mathbb{K} = \mathbb{C}$. Let us write $\phi = \mathrm{Re}\,\phi + i\,\mathrm{Im}\,\phi$, where $\mathrm{Re}\,\phi$ and $\mathrm{Im}\,\phi$ are the real and imaginary parts of ϕ; these functions are real-linear. From

$$\phi(x) = -i\phi(ix) = -i(\mathrm{Re}\,\phi(ix) + i\,\mathrm{Im}\,\phi(ix)) = \mathrm{Im}\,\phi(ix) - i\,\mathrm{Re}\,\phi(ix),$$

with $\mathrm{Im}\,\phi(ix)$ and $\mathrm{Re}\,\phi(ix)$ real, we infer that $\mathrm{Im}\,\phi(x) = -\,\mathrm{Re}\,\phi(ix)$. Hence,

$$\phi(x) = \mathrm{Re}\,\phi(x) - i\,\mathrm{Re}\,\phi(ix).$$

The real-valued function $\psi := \mathrm{Re}\,\phi$ satisfies the assumptions of the previous theorem and thus extends to a real-linear mapping $\Psi : V \to \mathbb{R}$ satisfying $\Psi \leqslant p$. We now define

$$\Phi(v) := \Psi(v) - i\Psi(iv).$$

Then Φ extends ϕ, Φ is real-linear, and since also

$$\Phi(iv) = \Psi(iv) - i\Psi(-v) = \Psi(iv) + i\Psi(v) = i(\Psi(v) - i\Psi(iv)) = i\Phi(v),$$

Φ is actually complex-linear. Finally, for $t \in \mathbb{C}$ with $|t| = 1$ such that $t\Phi(v) = |\Phi(v)|$,

$$|\Phi(v)| = t\Phi(v) = \Phi(tv) \overset{(*)}{=} \Psi(tv) \leqslant p(tv) = |t|p(v) = p(v).$$

Here $(*)$ follows from the definition of Φ, noting that $\Phi(tv) = |\Phi(v)|$ is nonnegative and therefore $\Phi(tv) = \mathrm{Re}\,\Phi(tv)$, while at the same time $\mathrm{Re}\,\Phi(tv) = \Psi(tv)$ by the definition of Φ and the fact that Ψ is real-valued. $\qquad\square$

In the setting of normed spaces, from Theorem 4.8 we infer the following result.

Theorem 4.9 (Hahn–Banach extension theorem for normed spaces). *Let X be a normed space and let $Y \subseteq X$ be a subspace. Then every functional $y^* \in Y^*$ has an extension to a functional $x^* \in X^*$ that satisfies*

$$\|x^*\| = \|y^*\|.$$

Here, of course, $\|x^*\|$ is the norm of x^* as an element of X^* and $\|y^*\|$ is the norm of y^* as an element of Y^*. Such obvious conventions will be in place throughout the text.

Proof Given a functional $y^* \in Y^*$, we apply Theorem 4.8 to $V = X$, $W = Y$, $\phi(y) := \langle y, y^* \rangle$ for $y \in Y$, and $p(x) := \|x\|\|y^*\|$ for $x \in X$. $\qquad\square$

Remark 4.10. The proof of Theorem 4.7 depends on the Axiom of Choice through the use of Zorn's lemma. If X is separable and a countable dense sequence $(x_n)_{n \geqslant 1}$ is given, Theorem 4.9 can be proved without invoking Zorn's lemma as follows. Revisiting the proof of Theorem 4.7, starting from a functional $y^* \in Y^*$ one defines Y_n to be the span of Y and $\{x_1, \ldots, x_n\}$ and inductively extends y^* to Y_1, then to Y_2, and so forth. On the span of the spaces Y_n, $n \geqslant 1$, we thus obtain a well-defined functional of norm at most

$\|y^*\|$. Since this subspace is dense, by Proposition 1.18 this functional uniquely extends to a functional with the same norm on all of X.

Recall the identity

$$\|x^*\| = \sup_{\|x\| \leqslant 1} |\langle x, x^* \rangle|,$$

which is nothing but the definition of the operator norm of x^* as an element of $\mathscr{L}(X, \mathbb{K})$. As a consequence of the Hahn–Banach theorem we obtain the following dual expression for the norm of elements $x \in X$:

Corollary 4.11. *For all $x \in X$ we have*

$$\|x\| = \sup_{\|x^*\| \leqslant 1} |\langle x, x^* \rangle|.$$

In particular, if $\langle x, x^* \rangle = 0$ for all $x^* \in X^*$, then $x = 0$.

Proof Fix an arbitrary $x \in X$. If $x = 0$ the asserted identity trivially holds, so we may assume that $x \neq 0$. Let Y be the one-dimensional subspace of X spanned by x and define $y_0^* \in Y^*$ by $\langle tx, y_0^* \rangle := t\|x\|$. Then $\|y_0^*\| = 1$. Let $x_0^* \in X^*$ be a Hahn–Banach extension provided by Theorem 4.9, that is, $x_0^*|_Y = y_0^*$ and $\|x_0^*\| = \|y_0^*\| = 1$. Then

$$\|x\| = \langle x, y_0^* \rangle = \langle x, x_0^* \rangle \leqslant \sup_{\|x^*\| \leqslant 1} |\langle x, x^* \rangle|,$$

while trivially

$$\sup_{\|x^*\| \leqslant 1} |\langle x, x^* \rangle| \leqslant \sup_{\|x^*\| \leqslant 1} \|x\| \|x^*\| = \|x\|.$$

$\square$

The next application of Theorem 4.9 provides a condition for recognising proper closed subspaces.

Corollary 4.12. *If Y is a proper closed subspace of a Banach space X, then for every $x_0 \in X \setminus Y$ there exists an $x^* \in X^*$ such that*

$$\langle x_0, x^* \rangle \neq 0 \ \text{ and } \ \langle y, x^* \rangle = 0 \ \text{ for all } \ y \in Y.$$

Proof Fix an element $x_0 \in X \setminus Y$. Without loss of generality we may assume that $\|x_0\| = 1$. Let X_0 denote the span of Y and x_0. On X_0 we can uniquely define a linear scalar-valued mapping ϕ by declaring $\phi(y) := 0$ for all $y \in Y$ and $\phi(x_0) := 1$. The idea is to prove that $\phi : X_0 \to \mathbb{K}$ is bounded. Once this has been shown, the result follows from the Hahn–Banach extension theorem.

We claim that there is a constant $C > 0$ such that

$$\|x_0 + y\| \geqslant C\|x_0\|, \quad y \in Y.$$

If such a constant does not exist, for every $n \geqslant 1$ one can find a $y_n \in Y$ so that

$$\|x_0 + y_n\| < \frac{1}{n}\|x_0\|, \quad n = 1, 2, \ldots$$

Then $\lim_{n\to\infty} \|x_0 + y_n\| = 0$, and therefore $x_0 \in \overline{Y} = Y$. This contradiction proves the claim.

By the claim, for any scalar $a \in \mathbb{K}$ and $y \in Y$,

$$\|ax_0 + y\| = |a|\|x_0 + a^{-1}y\| \geqslant C|a|\|x_0\| = C\|ax_0\|.$$

Hence

$$|\phi(ax_0 + y)| = |a| = |a|\|x_0\| = \|ax_0\| \leqslant \frac{1}{C}\|ax_0 + y\|.$$

This proves that ϕ is bounded on X_0 and $\|\phi\|_{X_0^*} \leqslant 1/C$. $\qquad\square$

Definition 4.13 (Complemented subspaces). A closed linear subspace X_0 of a normed space X is said to be *complemented* if there exists a closed linear subspace X_1 of X such that

- $X_0 + X_1 = X$;
- $X_0 \cap X_1 = \{0\}$.

Here, $X_0 + X_1 := \{x_0 + x_1 : x_0 \in X_0, x_1 \in X_1\}$. In this situation we have

$$X = X_0 \oplus X_1$$

as a direct sum in the sense discussed in Section 1.1.b.

Definition 4.14 (Projections). A *projection* is an operator $P \in \mathscr{L}(X)$ satisfying $P^2 = P$.

Notice that boundedness of P is taken to be part of the definition. If P is a projection, then so is $I - P$ and the range $\mathsf{R}(P)$ of P equals the null space $\mathsf{N}(I - P)$. This implies that $\mathsf{R}(P)$ is closed and we have a direct sum decomposition

$$X = \mathsf{N}(P) \oplus \mathsf{R}(P).$$

Thus we have shown the following simple result:

Proposition 4.15. *If a closed subspace X_0 of a normed space X is the range of a projection in X, then X_0 is complemented.*

Conversely, if $X = X_0 \oplus X_1$ is a direct sum decomposition of a Banach space, then the natural projections associated with it are bounded; this will be proved in the next chapter (see Proposition 5.10).

We have seen in Corollary 1.36 that finite-dimensional subspaces are always closed. As an application of the Hahn–Banach theorem we prove next the stronger assertion that they are always complemented. For later use we also include an analogue for subspaces of finite codimension, which does not require the use of the Hahn–Banach theorem. A subspace X_0 of a Banach space X is said to have *finite codimension* if there exists a finite-dimensional subspace Y of X such that $X_0 \cap Y = \{0\}$ and $X_0 + Y = X$. In this situation we define the *codimension* of X_0 to be the dimension of Y and denote this number by $\mathrm{codim}\, X_0$. The following argument shows that this number is well defined. If Y_0 and Y_1 are subspaces with the said properties, then for every $y_0 \in Y_0$ there are unique $x_0 \in X$ and $y_1 \in Y_1$ such that $y_0 = x_0 + y_1$. The mapping $y_0 \mapsto y_1$ from Y_0 to Y_1 is easily seen to be linear. By the same procedure we obtain a well-defined mapping from Y_1 to Y_0, and it is clear that these mappings are each other's inverses. Hence they are isomorphisms of finite-dimensional vector spaces and therefore $\dim Y_0 = \dim Y_1$.

Subspaces of finite dimension are closed, but subspaces of finite codimension need not be closed, at least when we accept the Axiom of Choice: under this assumption, in Problem 3.23 a dense subspace of ℓ^2 with codimension one is constructed, and such subspaces cannot be closed. If X_0 is a *closed* subspace of finite codimension, it is easily checked that

$$\mathrm{codim}\, X_0 = \dim(X/X_0).$$

Proposition 4.16. *Let Y be a subspace of a normed space X. Then the following assertions hold:*

(1) *if $\dim(Y) < \infty$, then Y is closed and complemented;*
(2) *if $\mathrm{codim}(Y) < \infty$ and Y is closed, then Y is complemented.*

Proof (1): Let Y be a finite-dimensional subspace of X. By Corollary 1.36, Y is closed. To prove that Y is complemented we show that Y is the range of a projection in X.

Let $(y_n)_{n=1}^N$ be a basis for Y. Then every $y \in Y$ admits a unique representation $y = \sum_{n=1}^N c_n(y) y_n$ with coefficients $c_n(y) \in \mathbb{K}$. The mappings $c_n : y \mapsto c_n(y)$ are linear, and since linear mappings on finite-dimensional normed spaces are bounded, we have $c_n \in Y^*$. By the Hahn–Banach theorem we may extend each c_n to a functional $x_n^* \in X^*$. Consider the (bounded) linear operator P on X defined by

$$Px := \sum_{n=1}^N \langle x, x_n^* \rangle y_n.$$

It is clear that P maps X into Y and from $\langle y_m, x_n^* \rangle = \delta_{mn}$ we see that

$$Py_m = \sum_{n=1}^N \langle y_m, x_n^* \rangle y_n = y_m.$$

This shows that P maps X *onto* Y. The preceding identity, applied to the element $Px \in Y$, also shows that $P^2 x = P(Px) = Px$, so P is a projection.

(2): Since finite-dimensional subspaces are closed, this is immediate from the definitions. $\qquad\square$

We next identify the duals of closed subspaces, quotients, and direct sums. For this purpose we need the first part of the following definition. The second part is included for reasons of symmetry of presentation and will be needed later.

Definition 4.17 (Annihilators and pre-annihilators). Let X be a Banach space.

(i) The *annihilator* of a subset $A \subseteq X$ is the set

$$A^\perp := \{x^* \in X^* : \langle x, x^* \rangle = 0, \quad x \in A\}.$$

(ii) The *pre-annihilator* of a subset $B \subseteq X^*$ is the set

$$^\perp B := \{x \in X : \langle x, x^* \rangle = 0, \quad x^* \in B\}.$$

Proposition 4.18. *Let X be a Banach space and let Y be a closed subspace of X. Then $Y^\perp$ is a closed subspace of X^* and we have the following assertions:*

(1) *the mapping $i : X^*/Y^\perp \to Y^*$ defined by $i(x^* + Y^\perp) := x^*|_Y$ is well defined and induces an isometric isomorphism*

$$Y^* \simeq X^*/Y^\perp;$$

(2) *the mapping $j : Y^\perp \to (X/Y)^*$ defined by $jx^*(x+Y) := \langle x, x^* \rangle$ is well defined and induces an isometric isomorphism*

$$(X/Y)^* \simeq Y^\perp.$$

Proof The easy proof that $Y^\perp$ is a closed subspace of X^* is left as an exercise.

(1): Let $y^* \in Y^*$ be given, and let $x^* \in X^*$ be an extension with the same norm as provided by the Hahn–Banach theorem. If $\phi \in Y^\perp$, then x^* and $x^* + \phi$ both restrict to y^*. This means that we obtain a well-defined linear surjection from $X^*/Y^\perp$ to Y^*. This mapping is also injective, for if $x^* + Y^\perp$ is mapped to the zero element of Y^*, then $x^*|_Y = 0$ and therefore $x^* \in Y^\perp$, so $x^* + Y^\perp$ is the zero element of $X^*/Y^\perp$. We must show that the resulting bijection is an isometry. On the one hand,

$$\|x^* + Y^\perp\|_{X^*/Y^\perp} = \inf_{\phi \in Y^\perp} \|x^* + \phi\| \leqslant \|x^*\| = \|y^*\| = \|x^*|_Y\| = \|i(x^* + Y)\|.$$

On the other hand, for all $\phi \in Y^\perp$ we have

$$\|i(x^* + Y)\| = \|y^*\| = \sup_{\|y\| \leqslant 1} |\langle y, y^* \rangle| = \sup_{\|y\| \leqslant 1} |\langle y, x^* + \phi \rangle| \leqslant \|x^* + \phi\|$$

and therefore, taking the infimum over all $\phi \in Y^\perp$,

$$\|i(x^* + Y)\| \leqslant \inf_{\phi \in Y^\perp} \|x^* + \phi\| = \|x^* + Y^\perp\|_{X^*/Y^\perp}.$$

(2): It is clear that j is well defined. Fix an arbitrary $x^* \in Y^\perp$. Given $\varepsilon > 0$, choose

$x_0 \in X$ such that $\|x_0 + Y\|_{X/Y} = 1$ and $\|jx^*\| \leqslant |\langle x_0 + Y, jx^* \rangle| + \varepsilon$. Choose $y_0 \in Y$ such that $\|x_0 + y_0\| \leqslant 1 + \varepsilon$. Then,

$$|\langle x_0 + Y, jx^* \rangle| = |\langle x_0, x^* \rangle| = |\langle x_0 + y_0, x^* \rangle| \leqslant \|x_0 + y_0\| \|x^*\| \leqslant (1 + \varepsilon)\|x^*\|.$$

It follows that $\|jx^*\|_{(X/Y)^*} \leqslant (1 + \varepsilon)\|x^*\| + \varepsilon$. Since $\varepsilon > 0$ was arbitrary, we find that

$$\|jx^*\|_{(X/Y)^*} \leqslant \|x^*\|.$$

In the converse direction we have

$$\|jx^*\|_{(X/Y)^*} = \sup_{\|x+Y\|_{X/Y} \leqslant 1} |\langle x + Y, jx^* \rangle| = \sup_{\|x+Y\|_{X/Y} \leqslant 1} |\langle x, x^* \rangle| \geqslant \sup_{\|x\| \leqslant 1} |\langle x, x^* \rangle| = \|x^*\|,$$

where we used that $\|x\| \leqslant 1$ implies $\|x + Y\|_{X/Y} \leqslant 1$.

It follows that j is isometric. To see that it is also surjective, let $\phi \in (X/Y)^*$ be given. The linear mapping $x \mapsto \phi(x + Y)$ is bounded, noting that both $x \mapsto x + Y$ and ϕ are bounded. It thus defines an element $x_\phi^* \in X^*$, and this functional annihilates Y. From

$$\langle x + Y, jx_\phi^* \rangle = \langle x, x_\phi^* \rangle = \langle x + Y, \phi \rangle$$

it follows that $jx_\phi^* = \phi$. $\qquad\square$

The duality of direct sum decompositions is discussed in Proposition 4.27.

The Hahn–Banach theorem, through Corollary 4.11, offers a technique to reduce certain vector-valued questions to their scalar-valued counterparts. By way of example we demonstrate this technique by reproving some calculus rules for the vector-valued Riemann integral of Proposition 1.45. A second example is given in Problem 4.2 where the Cauchy integral formula is extended to vector-valued holomorphic functions.

Second proof of Proposition 1.45, parts (2) *and* (3). (2): If $f(t_0) \neq f(t_1)$ for certain $t_0 \neq t_1$ in I, Corollary 4.11 provides us with a functional $x^* \in X^*$ such that $\langle f(t_0), x^* \rangle \neq \langle f(t_1), x^* \rangle$. Consider the scalar-valued function $\langle f, x^* \rangle(t) := \langle f(t), x^* \rangle$ obtained by applying X^* pointwise. This function is continuous on $[0, 1]$ and continuously differentiable on $(0, 1)$ with $\langle f, x^* \rangle' = \langle f', x^* \rangle = 0$. Therefore $\langle f, x^* \rangle$ is constant by the scalar-valued version of the proposition. This contradicts the choice of x^*.

(3): By the scalar-valued version of the proposition,

$$\left\langle f(1) - f(0) - \int_0^1 f(t)\,dt, x^* \right\rangle = \langle f(1), x^* \rangle - \langle f(0), x^* \rangle - \int_0^1 \langle f(t), x^* \rangle\,dt = 0$$

for all $x^* \in X^*$. Corollary 4.11 implies that $f(1) - f(0) - \int_0^1 f(t)\,dt = 0$. $\qquad\square$

Using duality we can give the following version of the Pettis measurability theorem (Theorem 1.47):

Theorem 4.19 (Pettis measurability theorem, second version). *A function* $f : \Omega \to X$ *is strongly measurable if and only if f takes its values in a separable closed subspace of X and is weakly measurable, that is,* $\langle f, x^* \rangle : \Omega \to \mathbb{K}$ *is measurable for all* $x^* \in X^*$.

Proof The 'only if' part follows from the first version of the Pettis measurability theorem and the trivial fact that strong measurability implies weak measurability. For the 'if' part, choose a dense sequence $(x_k)_{k \geqslant 1}$ in a closed separable subspace X_0 of X where f takes its values. By the Hahn–Banach theorem, for every $k \geqslant 1$ there is a unit vector $x_k^* \in X^*$ such that $|\langle x_k, x_k^* \rangle| = \|x_k\|$. Then for all $k \geqslant 1$ we have $\sup_{n \geqslant 1} |\langle x_k, x_n^* \rangle| = \|x_k\|$, and by a simple approximation argument this implies that $\sup_{n \geqslant 1} |\langle x, x_n^* \rangle| = \|x\|$ for all $x \in X_0$. Then, for all $x_0 \in X_0$,

$$\omega \mapsto \|f(\omega) - x_0\| = \sup_{n \geqslant 1} |\langle f(\omega) - x_0, x_n^* \rangle|$$

is a measurable function. Now the result follows from Theorem 1.47. $\square$

Theorem 4.19 is accompanied by the following uniqueness result.

Proposition 4.20. *Let* $(\Omega, \mathscr{F}, \mu)$ *be a measure space. If* $f : \Omega \to X$ *is a strongly measurable function and for all* $x^* \in X^*$ *we have* $\langle f, x^* \rangle = 0$ μ*-almost everywhere, then* $f = 0$ μ*-almost everywhere.*

In the same way one proves that if $\langle f, x^* \rangle = 0$ pointwise for all $x^* \in X^*$, then $f = 0$ pointwise.

Proof Let $(x_n^*)_{n \geqslant 1}$ be a sequence in X^* separating the points of a closed subspace X_0 in which f takes its values; such a sequence exists by the argument in the proof of Theorem 4.19. Since $\langle f, x_n^* \rangle = 0$ outside a μ-null set N_n, we conclude that $f = 0$ on the complement of the μ-null set $\bigcup_{n \geqslant 1} N_n$. $\square$

Corollary 4.11 has the interesting consequence that every Banach space can be isometrically identified with a closed subspace of the bi-dual $X^{**} := (X^*)^*$ in a natural way. More specifically, given an element $x \in X$ we define a mapping $Jx : X^* \to \mathbb{K}$ by

$$Jx(x^*) := \langle x, x^* \rangle.$$

It is clear that this mapping is bounded and therefore it defines an element of the bi-dual X^{**}. By the corollary, its norm is given by

$$\|Jx\| = \sup_{\|x^*\| \leqslant 1} |\langle x^*, Jx \rangle| = \sup_{\|x^*\| \leqslant 1} |\langle x, x^* \rangle| = \|x\|,$$

and therefore the mapping $J : x \mapsto Jx$ is isometric. It is also linear, and therefore we have proved:

Proposition 4.21 (Isometric embedding into the bi-dual). *The operator J is an isometric embedding of X into* X^{**}.

The image of X under J is closed in X^{**}; this is immediate from the fact that J is isometric. It may happen that $J(X)$ is a proper subspace of X^{**}. For instance, the bidual of c_0 is ℓ^∞. Examples of Banach spaces for which we have $J(X) = X^{**}$ include all Hilbert spaces and the spaces ℓ^p and $L^p(\Omega)$ for $1 < p < \infty$. Spaces with this property are called *reflexive* and enjoy some pleasant properties, some of which will be discussed in Section 4.7.b.

We conclude this section by filling in a detail that was left open in our treatment of the duality of ℓ^p, namely, that the duality $(\ell^p)^* \simeq \ell^q$ with $\frac{1}{p} + \frac{1}{q} = 1$, which has been shown to hold for $1 \leqslant p < \infty$ in Section 4.1.b, does not hold for $p = \infty$.

Consider the closed subspace Y of ℓ^∞ consisting of all convergent sequences, and define $y^* \in Y^*$ as

$$\langle y, y^* \rangle := \lim_{n \to \infty} y_n$$

for $y = (y_n)_{n \geqslant 1} \in Y$. Let $x^* \in (\ell^\infty)^*$ be any Hahn–Banach extension of y^*. We claim that there exists no $z \in \ell^1$ such that $\langle z, x \rangle = \langle x, x^* \rangle$ for all $x \in \ell^\infty$. Indeed, let $z \in \ell^1$ be given. Given $0 < \varepsilon < 1$ we can choose $N \geqslant 1$ so large that $\sum_{n > N} |z_n| < \varepsilon$. Consider now the sequence $x^N := (0, 0, \ldots, 0, 1, 1, 1, \ldots) \in Y$, with N zeroes at the beginning. Then

$$\langle x^N, x^* \rangle = \langle x^N, y^* \rangle = \lim_{n \to \infty} x_n^N = 1$$

while on the other hand

$$|\langle z, x^N \rangle| = \left| \sum_{n \geqslant 1} z_n x_n^N \right| = \left| \sum_{n > N} z_n \right| \leqslant \sum_{n > N} |z_n| < \varepsilon < 1.$$

This shows that $\langle z, x^N \rangle \neq \langle x^N, x^* \rangle$.

4.3 Adjoint Operators

The Hahn–Banach theorem will now be used to show that when X and Y are Banach spaces and $T \in \mathscr{L}(X, Y)$ is a bounded operator, there exists a unique bounded operator $T^* \in \mathscr{L}(Y^*, X^*)$ of norm $\|T^*\| = \|T\|$ such that

$$\langle Tx, y^* \rangle = \langle x, T^* y^* \rangle$$

for all $x \in X$ and $y^* \in Y^*$. When X and Y are Hilbert spaces, the Riesz representation theorem will be used to prove the existence of a unique bounded operator $T^* \in \mathscr{L}(Y, X)$ of norm $\|T^*\| = \|T\|$ such that

$$(Tx | y) = (x | T^* y), \quad x, y \in X.$$

4.3.a The Banach Space Adjoint

Let X and Y be Banach spaces.

Proposition 4.22. *For every bounded operator $T \in \mathscr{L}(X,Y)$ there exists a unique bounded operator $T^* \in \mathscr{L}(Y^*,X^*)$ such that*

$$\langle Tx, y^* \rangle = \langle x, T^*y^* \rangle, \quad x \in X, \ y^* \in Y^*. \tag{4.6}$$

Furthermore,

$$\|T^*\| = \|T\|.$$

Proof The idea is to take the left-hand side of (4.6) as a definition for the operator defined by the right-hand side. More precisely, for any given $y^* \in Y^*$ we may define a linear mapping $T^*y^* : X \to \mathbb{K}$ by

$$(T^*y^*)x := \langle Tx, y^* \rangle.$$

This mapping is bounded, of norm $\|T^*y^*\| \leqslant \|T\|\|y^*\|$, since

$$|(T^*y^*)x| \leqslant \|Tx\|\|y^*\| \leqslant \|T\|\|x\|\|y^*\|.$$

Accordingly T^*y^* defines an element of X^*. The resulting mapping $T^* : Y^* \to X^*$ which maps $y^* \in Y^*$ to the element $T^*y^* \in X^*$ is linear. By the above estimate T^* is bounded, of norm $\|T^*\| \leqslant \|T\|$. It is clear from the definitions that $\langle Tx, y^* \rangle = \langle x, T^*y^* \rangle$ for all $x \in X$ and $y^* \in Y^*$.

Turning to uniqueness, if $S : Y^* \to X^*$ is an operator satisfying $\langle Tx, y^* \rangle = \langle x, Sy^* \rangle$ for all $x \in X$ and $y^* \in Y^*$, then $\langle x, T^*y^* \rangle = \langle x, Sy^* \rangle$ for all $x \in X$ and $y^* \in Y^*$, so $T^*y^* = Sy^*$ for all $y^* \in Y^*$, and so $T^* = S$.

Finally,

$$\|T^*\| = \sup_{\|y^*\| \leqslant 1} \|T^*y^*\| = \sup_{\|y^*\| \leqslant 1} \sup_{\|x\| \leqslant 1} |\langle x, T^*y^* \rangle|$$
$$= \sup_{\|x\| \leqslant 1} \sup_{\|y^*\| \leqslant 1} |\langle Tx, y^* \rangle| = \sup_{\|x\| \leqslant 1} \|Tx\| = \|T\|,$$

using Corollary 4.11 in the penultimate step. □

It is clear that $I_X^* = I_{X^*}$, where I_X and I_{X^*} are the identity operators on X and X^*, respectively. For all $T_1, T_2 \in \mathscr{L}(X,Y)$ and $c_1, c_2 \in \mathbb{K}$ we have

$$(c_1 T_1 + c_2 T_2)^* = c_1 T_1^* + c_2 T_2^*$$

and for all $T \in \mathscr{L}(X,Y)$ and $S \in \mathscr{L}(Y,Z)$ we have

$$(S \circ T)^* = T^* \circ S^*.$$

Definition 4.23 (Adjoint operator). The bounded operator T^* is called the *adjoint* of T.

Example 4.24. Let $X = \mathbb{K}^n$, $Y = \mathbb{K}^m$, and let $A \in \mathscr{L}(\mathbb{K}^n, \mathbb{K}^m)$. With respect to the standard unit bases we represent A as an $m \times n$ matrix with coefficients $a_{ij} \in \mathbb{K}$. With respect to the same basis, and using the identification $\xi \leftrightarrow \phi_\xi$ of Section 4.1.a, its adjoint A^* is represented as the $n \times m$ matrix with coefficients a_{ji}. Stated differently, the matrix associated with A^* is the transpose of the matrix associated with A.

Example 4.25. The adjoint of the kernel operator T_k on $L^2(0,1)$ given by

$$T_k f(t) := \int_0^1 k(t,s) f(s) \, ds,$$

where we assume that $k \in L^2((0,1) \times (0,1))$ (see Example 1.30) is the kernel operator T_{k^*} on $L^2(0,1)$ given by

$$T_{k^*} g(t) = \int_0^1 k^*(t,s) g(s) \, ds \quad \text{with} \quad k^*(t,s) = k(s,t).$$

Example 4.26. Let $1 \leqslant p < \infty$ and $\frac{1}{p} + \frac{1}{q} = 1$. The adjoint of the left (right) shift on ℓ^p is the right (left) shift on ℓ^q. The adjoint of the left (right) translation on $L^p(\mathbb{R})$ is the right (left) translation on $L^q(\mathbb{R})$.

As a first application we prove three simple duality results:

Proposition 4.27. *Let X be a Banach space. Then:*

(1) *if $i : X \to Y$ is an (isometric) isomorphism of X onto another Banach space Y, then the adjoint operator $i^* : Y^* \to X^*$ is an (isometric) isomorphism of their duals.*
(2) *if Y is a closed subspace of X and $i : X \to Y$ is the inclusion mapping, then the adjoint operator $i^* : X^* \to Y^*$ is the restriction mapping given by $i^* x^* = x^*|_Y$;*
(3) *If X admits a direct sum decomposition*

$$X = X_0 \oplus X_1$$

with associated projections π_0 and π_1, then X^* admits a direct sum decomposition

$$X^* = X_0^* \oplus X_1^*$$

with associated projections π_0^* and π_1^*. More precisely, we have the direct sum decomposition $X^* = \pi_0^* X^* \oplus \pi_1^* X^*$, and for $k = 0, 1$ the mapping $\pi_k^* x^* \mapsto i_k^* x^*$ defines an isometric isomorphisms from $\pi_k^* X^*$ onto X_k^*.

Proof The proofs are routine. We leave the proofs of (1) and (2) to the reader and write out a proof of (3). For $k \in \{0, 1\}$ we have, writing $x = x_0 + x_1$ along the decomposition $X = X_0 \oplus X_1$,

$$\|\pi_k^* x^*\| = \sup_{\|x\| \leqslant 1} |\langle \pi_k x, x^* \rangle| = \sup_{\|x_k\| \leqslant 1} |\langle x_k, x^* \rangle| = \sup_{\|x_k\| \leqslant 1} |\langle i_k x_k, x^* \rangle| = \|i_k^* x^*\|.$$

This establishes the isometric isomorphisms of the second part of (3). The other statements are immediate consequences. □

We conclude with a simple observation about the bi-adjoint operator $T^{**} := (T^*)^*$. Identifying X with a closed subspace of X^{**} by means of the natural isometric embedding $J : X \to X^{**}$ (see Proposition 4.21), the restriction of T^{**} to X equals T. Indeed, denoting by $J : X \to X^{**}$ the natural embedding, the claim follows from

$$\langle y^*, T^{**} Jx \rangle = \langle T^* y^*, Jx \rangle = \langle x, T^* y^* \rangle = \langle Tx, y^* \rangle = \langle y^*, JTx \rangle$$

so that $T^{**} Jx = JTx$ as claimed.

4.3.b The Hilbert Space Adjoint

Let H and K be Hilbert spaces. If $T \in \mathscr{L}(H,K)$ is a bounded operator, its adjoint $T^* \in \mathscr{L}(K^*, H^*)$ is a bounded operator acting in the reverse direction between their duals. By the Riesz representation theorem, the duals H^* and K^* can be canonically identified with H and K. Under these identifications, the adjoint of an operator $T \in \mathscr{L}(H,K)$ can be re-interpreted as an operator acting from K to H. Although the identifications are conjugate-linear, as an operator from K to H the adjoint of T is nevertheless linear. This is the content of the next proposition which, incidentally, admits a straightforward direct proof which does not call upon the Hahn–Banach theorem.

Proposition 4.28. *For every bounded operator $T \in \mathscr{L}(H,K)$ there exists a unique bounded operator $T^\star \in \mathscr{L}(K,H)$ such that*

$$(Tx|y) = (x|T^\star y), \quad x \in H, \ y \in K.$$

Furthermore,

$$\|T\| = \|T^\star\| = \|T^\star T\|^{1/2}.$$

Proof Let $y \in K$ be fixed and define a mapping $\phi = \phi_{y,T} : H \to \mathbb{K}$ by

$$\phi(x) := (Tx|y).$$

From $|\phi(x)| \leqslant \|Tx\|\|y\| \leqslant \|T\|\|x\|\|y\|$ we see that ϕ is bounded with norm at most $\|T\|\|y\|$. Hence by the Riesz representation theorem there is a unique element $T^\star y \in H$ with norm $\|T^\star y\| = \|\phi\|$ such that

$$\phi(x) = (x|T^\star y).$$

Combining the two identities we obtain $(Tx|y) = (x|T^\star y)$.

We must show that the mapping $T^\star : y \mapsto T^\star y$ is linear and bounded. Additivity is easy and homogeneity with respect to scalar multiplication follows from

$$(x|T^\star(cy)) = (Tx|cy) = \overline{c}(Tx|y) = \overline{c}(x|T^\star y) = (x|cT^\star y),$$

which implies $T^\star(cy) = cT^\star y$.

Next we show that $T^\star$ is bounded. This follows from what we already know. Indeed, we have

$$\|T^\star y\| = \|\phi\| \leqslant \|T\|\|y\|,$$

so $T^\star$ is bounded of norm $\|T^\star\| \leqslant \|T\|$. Writing $T^{\star\star} := (T^\star)^\star$, from

$$(T^{\star\star}x|y) = \overline{(y|T^{\star\star}x)} = \overline{(T^\star y|x)} = (x|T^\star y) = (Tx|y)$$

it follows that $T^{\star\star}x = Tx$ for all $x \in H$ and therefore $T^{\star\star} = T$. Hence, by what we just proved applied to $T^\star$, $\|T\| = \|T^{\star\star}\| \leqslant \|T^\star\|$. We conclude that equality holds: $\|T\| = \|T^\star\|$.

Next we prove the identity $\|T^\star T\|^{1/2} = \|T\|$. Clearly $\|T^\star T\| \leqslant \|T\|\|T^\star\| = \|T\|^2$, and in the converse direction we have

$$\begin{aligned}
\|T^\star T\| = \sup_{\|x\| \leqslant 1} \|T^\star Tx\| &= \sup_{\|x\| \leqslant 1} \sup_{\|y\| \leqslant 1} |(T^\star Tx|y)| \\
&= \sup_{\|y\| \leqslant 1} \sup_{\|x\| \leqslant 1} |(T^\star Tx|y)| \\
&= \sup_{\|y\| \leqslant 1} \sup_{\|x\| \leqslant 1} |(Tx|Ty)| \\
&\geqslant \sup_{\|x\| \leqslant 1} |(Tx|Tx)| = \sup_{\|x\| \leqslant 1} \|Tx\|^2 = \|T\|^2.
\end{aligned}$$

Finally we prove uniqueness. If $U \in \mathscr{L}(K,H)$ is bounded and $(x|Uy) = (Tx|y)$ for all $x \in H$ and $y \in K$, then $Uy = T^\star y$ for all $y \in K$, so $U = T^\star$. □

By definition we have

$$(Tx|y) = (x|T^\star y)$$

for all $x, y \in H$. Symmetrically, we also have

$$(T^\star x|y) = (x|Ty).$$

This can be seen by noting that $(T^\star x|y) = \overline{(y|T^\star x)} = \overline{(Ty|x)} = (x|Ty)$.

It is clear that $I^\star = I$ and for all $T, U \in \mathscr{L}(H)$ and $c \in \mathbb{K}$ we have

$$(T+U)^\star = T^\star + U^\star, \quad (cT)^\star = \overline{c}T^\star,$$

and, as we have seen in the proof of Proposition 4.28,

$$T^{\star\star} = T.$$

Definition 4.29 (Hilbert space adjoint). The bounded operator $T^\star$ is called the *Hilbert space adjoint* of T.

The relation between the Hilbert space adjoint $T^\star$ (which is a bounded operator on H) and the adjoint operator T^* (which is a bounded operator on the dual space H^*) is given by

$$T^* \psi_h = \psi_{T^\star h}$$

as elements of H^*, where ψ_h and $\psi_{T^\star h}$ are the functionals in H^* associated with h and $T^\star h$, respectively. Indeed, this follows from

$$\langle x, T^* \psi_h \rangle = \langle Tx, \psi_h \rangle = (Tx|h) = (x|T^\star h) = \langle x, \psi_{T^\star h} \rangle.$$

Here, the brackets $\langle \cdot, \cdot \rangle$ denote the duality between H and its dual H^*.

Example 4.30. Here are some examples of Hilbert space adjoints. They should be compared with Examples 4.24, 4.25, and 4.26, respectively.

(i) As in Example 4.24, let $A \in \mathscr{L}(\mathbb{K}^n, \mathbb{K}^m)$ be represented as an $m \times n$ matrix with coefficients $a_{ij} \in \mathbb{K}$. Viewing $\mathbb{K}^n$ and $\mathbb{K}^m$ as finite-dimensional Hilbert spaces, its Hilbert space adjoint $A^\star$ may be represented as the $n \times m$ matrix with coefficients $\overline{a_{ji}}$. Stated differently, the matrix associated with $A^\star$ is the Hermitian transpose of the matrix associated with A.

(ii) The Hilbert space adjoint of the kernel operator T_k on $L^2(0,1)$ of Example 4.25 is the kernel operator $T_{k^\star}$ on $L^2(0,1)$ given by

$$T_{k^\star} g(t) = \int_0^1 k^\star(t,s)\, g(s)\, \mathrm{d}s \quad \text{with} \quad k^\star(t,s) = \overline{k(s,t)}.$$

(iii) The adjoint of the left (right) shift in $\ell^2(\mathbb{Z})$ is the right (left) shift. Similarly, the adjoint of the left (right) translation in $L^2(\mathbb{R})$ is the right (left) translation.

For later reference we state a useful decomposition result. Versions for Banach spaces are given in Proposition 5.14 and Theorem 5.15.

Proposition 4.31. *If $T \in \mathscr{L}(H,K)$ is a bounded operator, then H and K admit orthogonal decompositions*

$$H = \mathsf{N}(T) \oplus \overline{\mathsf{R}(T^\star)}, \quad K = \mathsf{N}(T^\star) \oplus \overline{\mathsf{R}(T)}.$$

In particular,

(1) *T is injective if and only if $T^\star$ has dense range;*
(2) *T has dense range if and only if $T^\star$ is injective.*

Proof If $x \perp \overline{\mathsf{R}(T^\star)}$, then $(Tx|y) = (x|T^\star y) = 0$ for all $y \in K$ and therefore $Tx = 0$, so $x \in \mathsf{N}(T)$. Conversely, if $x \in \mathsf{N}(T)$, then $(x|T^\star y) = (Tx|y) = 0$ for all $y \in K$ implies that $x \perp \mathsf{R}(T^\star)$ and hence $x \perp \overline{\mathsf{R}(T^\star)}$. This proves the orthogonal decomposition for H. The decomposition for K follows from it applying it to $T^\star$ and using that $T^{\star\star} = T$. $\quad\square$

4.4 The Hahn–Banach Separation Theorem

In what follows, X is a normed space. Corollary 4.12 can be interpreted as a separation theorem, in that it guarantees the existence of a functional separating a closed subspace from a given element not contained in it. The following result provides a far-reaching generalisation:

Theorem 4.32 (Hahn–Banach separation theorem). *Let C and D be disjoint nonempty convex sets in X, with C open. Then there exists an $x^* \in X^*$ such that the sets $\langle C, x^* \rangle$ and $\langle D, x^* \rangle$ are disjoint.*

Proof We prove the theorem in three steps.

Step 1 – First we prove the theorem for the real scalar field and $D = \{x_0\}$. Replacing C and x_0 by $C - y_0$ and $x_0 - y_0$ for some fixed $y_0 \in C$, we may assume without loss of generality that $0 \in C$.

Define the *Minkowski functional* of C as the mapping $\lambda_C : X \to [0, \infty)$ given by

$$\lambda_C(x) := \inf\{t > 0 : t^{-1}x \in C\}.$$

Since C is convex, open, and contains 0, we have $\lambda_C(x) < 1$ if and only if $x \in C$. We claim that λ_C enjoys the following two properties:

  (i) $\lambda_C(x+y) \leqslant \lambda_C(x) + \lambda_C(y)$ for all $x, y \in X$;
  (ii) $\lambda_C(tx) = t\lambda_C(x)$ for all $t \geqslant 0$.

To prove (i), fix $\varepsilon > 0$ and let $s, t > 0$ be such that $s^{-1}x \in C$ and $t^{-1}y \in C$, with $s < \lambda_C(x) + \varepsilon$ and $t < \lambda_C(x) + \varepsilon$. Then

$$(s+t)^{-1}(x+y) = \frac{s}{s+t}s^{-1}x + \frac{t}{s+t}t^{-1}y$$

is a convex combination of the elements $s^{-1}x, t^{-1}y \in C$ and therefore belongs to C. It follows that

$$\lambda_C(x+y) \leqslant s+t \leqslant \lambda_C(x) + \lambda_C(y) + 2\varepsilon.$$

Since $\varepsilon > 0$ was arbitrary, this establishes (i). Assertion (ii) is obvious.

We now apply Theorem 4.7 to the linear span W of x_0 and the linear mapping $\phi : W \to \mathbb{R}$ given by $\phi(tx_0) := t$ for $t \in \mathbb{R}$. In view of $\lambda_C(x_0) \geqslant 1$, for all $t \geqslant 0$ it satisfies

$$\phi(tx_0) = t \leqslant t\lambda_C(x_0) = \lambda_C(tx_0).$$

Hence we may apply the theorem and obtain a linear mapping $x^* : X \to \mathbb{R}$ extending ϕ which satisfies $x^*(x) \leqslant \lambda_C(x)$ for all $x \in X$. For all $x \in C$ it satisfies $x^*(x) \leqslant \lambda_C(x) < 1$ and for all $x \in -C$ it satisfies $-x^*(x) = x^*(-x) \leqslant \lambda_C(-x) = \lambda_{-C}(x) < 1$. It follows that $|x^*(x)| < 1$ for all x in the open set $C \cap -C$ containing 0. This proves that $x^* \in X^*$.

Since $\langle x, x^* \rangle \leqslant \lambda_C(x) < 1$ for all $x \in C$ and $\langle x_0, x^* \rangle = \phi(x_0) = 1$, this functional has the required properties.

Step 2 – In the case of complex scalars and $D = \{x_0\}$, upon restricting scalar multiplication to the reals, Step 1 provides us with a real-linear mapping $x_{\mathbb{R}}^* : X \to \mathbb{R}$ such that $|x_{\mathbb{R}}^*(x)| < 1$ for all $x \in C \cap -C$ and $x_{\mathbb{R}}^*(x_0) \notin x_{\mathbb{R}}^*(C)$. Then, as in the proof of Theorem 4.8, $x^*(x) := x_{\mathbb{R}}^*(x) - ix_{\mathbb{R}}^*(ix)$ is complex-linear and bounded, and satisfies $x^*(x_0) \notin x^*(C)$ (by comparing real parts).

Step 3 – Now we prove the general case. The set $C - D$ is open and convex, and from $C \cap D = \varnothing$ it follows that $0 \notin C - D$. From Step 2 we obtain a functional $x^* \in X^*$ such that $0 \notin \langle C - D, x^* \rangle$, which is the same as saying that $\langle C, x^* \rangle \cap \langle D, x^* \rangle = \varnothing$. $\qquad \square$

Corollary 4.33. *Suppose $(x_n)_{n \geqslant 1}$ is a sequence in X and suppose that there exists an $x \in X$ such that*

$$\lim_{n \to \infty} \langle x_n, x^* \rangle = \langle x, x^* \rangle, \quad x^* \in X^*.$$

Then there exists a sequence $(y_n)_{n \geqslant 1}$ in the convex hull of $(x_n)_{n \geqslant 1}$ such that

$$\lim_{n \to \infty} y_n = x$$

with convergence in norm.

Proof Denote by D the closure of the convex hull of $(x_n)_{n \geqslant 1}$. Our task is to prove that $x \in D$. Suppose that this is not the case. Then Theorem 4.32 provides us with a functional $x^* \in X^*$ separating D from (a small enough open ball C around) x. This functional also separates $(x_n)_{n \geqslant 1}$ from x, in contradiction to the assumptions of the corollary. $\qquad \square$

As an application of the Hahn–Banach separation theorem we prove the following result.

Theorem 4.34. *For all $x^{**} \in X^{**}$, $x_1^*, \ldots, x_N^* \in X^*$, and $\varepsilon > 0$ there exists an $x \in X$ such that $\|x\| < \|x^{**}\| + \varepsilon$ and*

$$\langle x, x_n^* \rangle = \langle x_n^*, x^{**} \rangle, \quad n = 1, \ldots, N.$$

The proof uses an elementary version of the open mapping theorem (Theorem 5.8):

Lemma 4.35. *Let T be a bounded operator from a normed space X onto a finite-dimensional normed space Y. Then T maps open sets to open sets.*

Proof Let $(y_n)_{n=1}^d$ be a basis for Y and choose a sequence $(x_n)_{n=1}^d$ in X such that $Tx_n = y_n$ for $n = 1, \ldots, d$. Let X_0 denote the linear span of $(x_n)_{n=1}^d$. The restriction $T_0 := T|_{X_0} : X_0 \to Y$ is bounded and bijective. By Corollary 1.37, its inverse is bounded. This implies that T_0 maps open sets to open sets.

Now let U be open in X. For every $u \in U$ let $r_u > 0$ be such that $B_{X_0}(u; r_u) = u + r_u B_{X_0} \subseteq U$. Then $U = \bigcup_{u \in U}(u + r_u B_{X_0})$ and therefore

$$T(U) = \bigcup_{u \in U}(Tu + r_u T(B_{X_0}))$$

is open since $T(B_{X_0}) = T_0(B_{X_0})$ is open. $\qquad\square$

Proof of Theorem 4.34 We prove the theorem in two steps.

Step 1 – Let $x_1^*, \ldots, x_N^* \in X^*$ and $c_1, \ldots, c_N \in \mathbb{K}$ be given. In this step we prove that if there exists a constant $M \geqslant 0$ such that for all $\lambda_1, \ldots, \lambda_N \in \mathbb{K}$ we have

$$\left| \sum_{n=1}^N \lambda_n c_n \right| \leqslant M \left\| \sum_{n=1}^N \lambda_n x_n^* \right\|, \qquad (4.7)$$

then there exists an $x \in X$ such that $\|x\| < M + \varepsilon$ and

$$\langle x, x_n^* \rangle = c_n, \quad n = 1, \ldots, N.$$

Consider the mapping $T : x \mapsto (\langle x, x_n^* \rangle)_{n=1}^N$ from X into $\mathbb{K}^N$. We must prove that $T(B(0; M + \varepsilon))$, which by Lemma 4.35 is an open subset of the finite-dimensional space $\mathsf{R}(T)$, contains $(c_n)_{n=1}^N$. Suppose, for a contradiction, that this not true. Then $T(B(0; M + \varepsilon))$ is an open subset of $\mathbb{K}^N$ not containing $(c_n)_{n=1}^N$. The Hahn–Banach separation theorem provides us with a sequence $(\lambda_n)_{n=1}^N \in \mathbb{K}^N$ such that

$$\sum_{n=1}^N \lambda_n c_n \notin \left\{ \left\langle x, \sum_{n=1}^N \lambda_n x_n^* \right\rangle : \|x\| < M + \varepsilon \right\}.$$

Multiplying with an appropriate scalar of modulus one, it follows that also

$$\left| \sum_{n=1}^N \lambda_n c_n \right| \notin \left\{ \left\langle x, \sum_{n=1}^N \lambda_n x_n^* \right\rangle : \|x\| < M + \varepsilon \right\}.$$

By scaling, the right-hand side set contains the interval $[0, M\|\sum_{n=1}^N \lambda_n x_n^*\|]$. This contradicts the assumption (4.7).

Step 2 – Returning to the assumptions of the theorem, fix $x^{**} \in X^{**}$ and $x_1^*, \ldots, x_N^* \in X^*$, and set $c_n := \langle x_n^*, x^{**} \rangle$ for $n = 1, \ldots, N$. For all $\lambda_1, \ldots, \lambda_N \in \mathbb{K}$ we have

$$\left| \sum_{n=1}^N \lambda_n c_n \right| = \left| \left\langle \sum_{n=1}^N \lambda_n x_n^*, x^{**} \right\rangle \right| \leqslant \left\| \sum_{n=1}^N \lambda_n x_n^* \right\| \|x^{**}\|,$$

so the assumptions of Step 1 are satisfied with $M = \|x^{**}\|$. $\qquad\square$

4.5 The Krein–Milman Theorem

Extreme points play an important role in many applications of Functional Analysis. For instance, in Quantum Mechanics pure states are the extreme points of the convex set of all states (see Chapter 15).

Definition 4.36 (Extreme points). An *extreme point* of a convex subset C of a vector space is an element $v \in C$ such that if $v = (1 - \lambda)v_0 + \lambda v_1$ with $v_0, v_1 \in C$ and $0 < \lambda < 1$, then $v_0 = v_1 = v$.

Stated differently, extreme points are points of C which cannot be realised in a nontrivial way as a convex combination of other points of C.

Example 4.37. Let C denote the set of all probability measures on a given measure space $(\Omega, \mathscr{F}, \mu)$. Viewing C as a closed convex subset of $M(\Omega)$, the Banach space of $\mathbb{K}$-valued measures on $(\Omega, \mathscr{F})$, we claim that a probability measure μ is an extreme point of C if and only if μ is *atomic*, that is, whenever $A = A_0 \cup A_1$ with disjoint $A_0, A_1 \in \mathscr{F}$, then $\min\{\mu(A_0), \mu(A_1)\} = 0$.

To prove the claim, suppose first that $\mu \in C$ and $\mu = (1 - \lambda)\mu_0 + \lambda\mu_1$ with $\mu_0, \mu_1 \in C$ and $0 < \lambda < 1$. If $\mu_0 \neq \mu_1$, there is a set $A \in \mathscr{F}$ such that $\mu_0(A) \neq \mu_1(A)$. Interchanging μ_0 and μ_1 if necessary, we may assume that $0 \leqslant \mu_0(A) < \mu_1(A) \leqslant 1$. Then $\mu_1(A) > 0$ implies $\mu(A) > 0$, and $\mu_0(\complement A) > 0$ implies $\mu(\complement A) > 0$, and therefore μ is not atomic. This proves that every atomic measure is an extreme point of C.

Conversely, if $\mu \in C$ is not atomic, then there exists a set $A \in \mathscr{F}$ such that $\mu(A) > 0$ and $\mu(\complement A) > 0$. Consider the probability measures $\mu_0, \mu_1 \in C$ given by

$$\mu_0(B) := (\mu(A))^{-1}\mu(B \cap A), \quad \mu_1(B) := (\mu(\complement A))^{-1}\mu(B \cap \complement A), \quad B \in \mathscr{F}.$$

With $\lambda := \mu(\complement A)$ we have $0 < \lambda < 1$ and $\mu = (1 - \lambda)\mu_0 + \lambda\mu_1$, so μ is not extreme.

As a special case, if K is a compact Hausdorff space, the extreme points of the set of all Borel probability measures on K are the Dirac measures supported on K. To see this, suppose that μ is atomic and let S be its support, that is, S is the complement of the union of all open sets of μ-measure zero. If S is not a singleton, then it contains two distinct points, say x_0 and x_1. Since K is Hausdorff, they are contained in disjoint open sets U_0 and U_1. By the definition of support, $\mu(U_0) > 0$ and $\mu(U_1) > 0$, and μ is not atomic. This proves that S is a singleton, say $S = \{x\}$, and therefore $\mu = \delta_x$.

Closed convex sets need not have any extreme points:

Example 4.38. In $L^1(0,1)$, the closed convex set

$$C = \{f \in L^1(0,1) : f \geqslant 0, \|f\|_1 = 1\}$$

has no extreme points. Indeed, let $f \in C$. The mapping $\phi(\delta) := \|\mathbf{1}_{(0,\delta)}f\|_1$ is continuous from $[0,1]$ to $[0,1]$, and satisfies $\phi(0) = 0$ and $\phi(1) = 1$. Hence there exists $0 < \delta < 1$

such that $\phi(\delta) = \frac{1}{2}$. Then $f = \frac{1}{2}g + \frac{1}{2}h$, where $g, h \in C$ are given by $g = 21_{(0,\delta)}f$ and $h = 21_{(\delta,1)}f$, so f is not an extreme point of C.

As an application of the Hahn–Banach theorem we can prove the following result about existence of extreme points. Recall that the *(closed) convex hull* of a subset S of a Banach space is the smallest (closed) convex set containing S.

Theorem 4.39 (Krein–Milman). *Every compact convex subset of a Banach space is the closed convex hull of its extreme points.*

Proof Let K be a compact convex subset of the Banach space X. We first prove the existence of extreme points of K, and then prove that K is the closed convex hull of its extreme points.

A *face* in K is a nonempty closed convex subset F of K whose elements can only be realised as convex combinations of elements in F, that is, whenever $x \in F$ satisfies $x = (1 - \lambda)x_0 + \lambda x_1$ with $x_0, x_1 \in K$ and $0 < \lambda < 1$, then $x_0, x_1 \in F$. We make two useful observations:

  (i) $x \in K$ is an extreme point of K if and only if $\{x\}$ is a face of K;
 (ii) if F is a face of K and F' is a face of F, then F' is a face of K.

Claim (i) is evident. For (ii), if $x \in F'$ is given as $x = (1 - \lambda)x_0 + \lambda x_1$ with $x_0, x_1 \in K$ and $0 < \lambda < 1$, then $x_0, x_1 \in F$ since $x \in F$ and F is a face of K. Then $x_0, x_1 \in F'$ since F' is a face of F.

Step 1 – Let $\mathscr{K}$ denote the collection of faces in K. This collection is nonempty, for it contains K. We partially order $\mathscr{K}$ by declaring $K_1 \leqslant K_2$ if $K_2 \subseteq K_1$. By the finite intersection property (see Appendix C), any totally ordered subset $\mathscr{L}$ of $\mathscr{K}$ has nonempty intersection; the singletons consisting of elements in this intersection are upper bounds for $\mathscr{L}$. Hence we can apply Zorn's lemma and obtain that $\mathscr{K}$ has a maximal element, say F. We claim that F is a singleton, say $F = \{x\}$. By (i), this means that x is an extreme point of K.

To prove the claim, assume the contrary and let $x_0, x_1 \in F$ be two distinct points. By the Hahn–Banach theorem there exists an $x^* \in X^*$ such that $\mathrm{Re}\langle x_0, x^* \rangle \neq \mathrm{Re}\langle x_1, x^* \rangle$. Let

$$F_0 := \left\{ x \in F : \mathrm{Re}\langle x, x^* \rangle = \inf_{y \in F} \mathrm{Re}\langle y, x^* \rangle \right\}.$$

Then F_0 is a proper closed subset of F, which is nonempty since the compactness of F implies that the infimum is a minimum. If an element $x \in F_0$ can be represented as $x = (1 - \lambda)x' + \lambda x''$ with $x', x'' \in F$ and $0 < \lambda < 1$, then

$$(1 - \lambda)\mathrm{Re}\langle x', x^* \rangle + \lambda \mathrm{Re}\langle x'', x^* \rangle = \inf_{y \in F} \mathrm{Re}\langle y, x^* \rangle.$$

If $x' \notin F_0$, then $\mathrm{Re}\langle x', x^* \rangle > \inf_{y \in F} \mathrm{Re}\langle y, x^* \rangle$. Since also $\mathrm{Re}\langle x'', x^* \rangle \geqslant \inf_{y \in F} \mathrm{Re}\langle y, x^* \rangle$, the above inequality cannot hold. The same contradiction is reached if $x'' \notin F_0$. We conclude that $x', x'' \in F_0$ and F_0 is a face of F.

Now (ii) implies that F_0 is a face of K. Since $F_0 \gneq F$, this contradicts the maximality of F. This completes the proof of the claim that F is a singleton. It follows that K has an extreme point.

Step 2 – Let L denote the closed convex hull of all extreme points of K. We wish to show that $L = K$. Reasoning by contradiction, suppose that L is a proper subset of K and fix an element $x_0 \in K \setminus L$. By the Hahn–Banach separation theorem (Theorem 4.32) there exists an $x^* \in X^*$ such that $\langle x_0, x^* \rangle \notin \langle L, x^* \rangle$. Multiplying x^* with an appropriate scalar if necessary, we may assume that

$$\mathrm{Re}\langle x_0, x^* \rangle < \inf_{y \in L} \mathrm{Re}\langle y, x^* \rangle.$$

As in Step 1 we see that the set

$$F := \left\{ x \in K : \mathrm{Re}\langle x, x^* \rangle = \inf_{y \in K} \mathrm{Re}\langle y, x^* \rangle \right\}$$

is a nonempty face of K. Moreover, Step 1 applied to F shows that F has an extreme point x_1. By (i) and (ii), x_1 is also an extreme point of K. On the other hand from

$$\mathrm{Re}\langle x_1, x^* \rangle = \inf_{y \in K} \mathrm{Re}\langle y, x^* \rangle \leqslant \mathrm{Re}\langle x_0, x^* \rangle < \inf_{y \in L} \mathrm{Re}\langle y, x^* \rangle$$

we infer that $x_1 \notin L$. Since L contains all extreme points of K we have arrived at a contradiction. □

4.6 The Weak and Weak* Topologies

Some of the more advanced applications of the Hahn–Banach theorem can be conveniently formulated in terms of certain topologies generated by bounded functionals. The two most important ones are the weak topology of a Banach space and the weak* topology of its dual.

4.6.a Definition and Elementary Properties

Definition 4.40 (Weak topologies). Let V and W be vector spaces and let $\beta : V \times W \to \mathbb{K}$ be a bilinear mapping. The *weak topology of V generated by W* is the smallest topology τ on V with the property that the linear mapping $v \mapsto \beta(v, w)$ is continuous for all $w \in W$.

This topology is obtained as the intersection of all topologies in V for which all linear mappings $v \mapsto \beta(v, w)$, $w \in W$, are continuous. The family of topologies with this property is nonempty, for it always contains the power set topology of V.

By necessity, the weak topology τ must contain every set of the form

$$U_{v_0,w_0,\varepsilon} := \{v \in V : |\beta(v - v_0, w_0)| < \varepsilon\},$$

noting that this set is the inverse image under the continuous mapping $v \mapsto \beta(v, w_0)$ of the open ball $B(\beta(v_0, w_0); \varepsilon)$ in $\mathbb{K}$.

We claim that τ coincides with the topology τ' generated by the sets $U_{v_0,w_0,\varepsilon}$, where v_0, w_0, and ε range over V, W, and $(0, \infty)$ respectively. The observation just made implies that $\tau' \subseteq \tau$. In the opposite direction, for every $w \in W$ the inverse under $\beta(\cdot, w)$ of every open ball belongs to τ', so every $\beta(\cdot, w)$ is continuous with respect to τ'. Since τ is the smallest topology with this property we have $\tau \subseteq \tau'$. This establishes the claim.

It follows from the claim that a set $U \subseteq V$ belongs to τ if and only if it can be written as a union of finite intersections of sets of the form $U_{v_0,w_0,\varepsilon}$. Indeed, the collection τ'' of sets that can be written this way is a topology which contains every set $U_{v_0,w_0,\varepsilon}$, and therefore we have $\tau' \subseteq \tau''$. By the preceding observation, this means that $\tau \subseteq \tau''$. In the converse direction, the fact that topologies are closed under taking unions and finite intersections implies that every set in τ'' belongs to τ.

Proposition 4.41. *In the above setting, a sequence $(v_n)_{n \geqslant 1}$ converges to v with respect to τ if and only if*

$$\lim_{n \to \infty} \beta(v_n - v, w) = 0, \quad w \in W.$$

Proof The 'only if' part follows from the fact that if $v_n \to v$ with respect to τ, then for all $\varepsilon > 0$ and $w \in W$ we have $v_n \in U_{v,w,\varepsilon}$ for all large enough n. For the 'if' part we note that if $U \in \tau$ contains v, then the observation preceding the statement of the proposition allows us to find $U_{v^{(1)},w^{(1)},\varepsilon^{(1)}}, \ldots, U_{v^{(k)},w^{(k)},\varepsilon^{(k)}}$ such that

$$x \in \bigcap_{j=1}^{k} U_{v^{(j)},w^{(j)},\varepsilon^{(j)}} \subseteq U.$$

Since we assume that $\beta(v_n - v, w^{(j)}) \to 0$ for $j = 1, \ldots, k$, for large enough n we have $v_n \in \bigcap_{j=1}^{k} U_{v^{(j)},w^{(j)},\varepsilon^{(j)}}$ and hence $v_n \in U$. $\qquad\square$

The duality between a Banach space and its dual leads to two special cases of interest:

Definition 4.42 (The weak and weak* topologies). Let X be a Banach space.

(i) The *weak topology of X* is the topology induced by X^*.
(ii) The *weak* topology of X^** is the topology induced by X.

It is implicit that in (i) we use the bilinear mapping from $X \times X^*$ to $\mathbb{K}$ given by $(x, x^*) \mapsto \langle x, x^* \rangle$; in (ii) we use the bilinear mapping from $X^* \times X$ to $\mathbb{K}$ given by $(x^*, x) \mapsto \langle x, x^* \rangle$. For these topologies, Proposition 4.41 takes the following form:

Corollary 4.43. *Let X be a Banach space. The following assertions hold:*

(1) *a sequence $(x_n)_{n \geqslant 1}$ in X converges to $x \in X$ with respect to the weak topology of X if and only if $\lim_{n \to \infty} \langle x_n - x, x^* \rangle = 0$ for all $x^* \in X^*$;*
(2) *a sequence $(x_n^*)_{n \geqslant 1}$ in X^* converges to $x^* \in X^*$ with respect to the weak* topology of X^* if and only if $\lim_{n \to \infty} \langle x, x_n^* - x^* \rangle = 0$ for all $x \in X$.*

Convergence with respect to the weak and weak* topologies will be referred to as *weak convergence* and *weak* convergence*, respectively. The following result is immediate from the Hahn–Banach separation theorem:

Proposition 4.44 (Closed convex sets are weakly closed). *Every closed convex set in a Banach space is closed in the weak topology.*

The next result characterises functionals that are continuous with respect to the weak and weak* topologies.

Proposition 4.45. *Let X be a Banach space. The following assertions hold:*

(1) *a linear mapping $\phi : X \to \mathbb{K}$ is continuous with respect to the weak topology if and only if it belongs to X^*;*
(2) *a linear mapping $\phi : X^* \to \mathbb{K}$ is continuous with respect to the weak* topology if and only if it belongs to X, that is, there exists an $x \in X$ such that $\phi(x^*) = \langle x, x^* \rangle$ for all $x^* \in X^*$.*

Proof (1): We only need to prove the 'only if' part. If ϕ is weakly continuous, then $\phi^{-1}(B_{\mathbb{K}})$ contains a weakly open set containing the origin. Since weakly open sets are open, this set contains a ball $B(0; r)$ with $r > 0$. This means that $\phi \in X^*$ and $\|\phi\| \leqslant 1/r$.

(2): Again we only need to prove the 'only if' part. If ϕ is weakly continuous, then $\phi^{-1}(B_{\mathbb{K}})$ contains a weak* open set U containing the origin. Since weak* open sets are weakly open, part (1) shows that $\phi \in X^{**}$. The set U contains a set of the form

$$U' := \{x^* \in X^* : |\langle x_j, x^* \rangle| < \varepsilon, \quad j = 1, \ldots, k\}$$

for suitable $\varepsilon > 0$, $k \geqslant 1$, and $x_1, \ldots, x_k \in X$. Let X_0 denote the span of $x_1, \ldots, x_k$. This space is finite-dimensional and therefore closed, and by Proposition 4.16 it is complemented. The proof of this proposition shows that X_0 is the range of a projection π_0. Let $\pi_1 := I - \pi_0$ be the complementary projection and denote by X_1 its range.

Viewing π_0 as a bounded operator from X onto X_0, its second adjoint π_0^{**} is a bounded operator from X^{**} to X_0^{**}. The space X_0, being finite-dimensional, can be identified with its second dual, the identification being given by the natural inclusion mapping $J : X_0 \to X_0^{**}$ which is surjective in this case (the details are spelled out in Example 4.57). Thus we may identify $\pi_0^{**} \phi =: x$ with an element of X_0. We will show that $\phi(x^*) = \langle x, x^* \rangle$ for all $x^* \in X^*$.

Let $X^* = \mathrm{R}(\pi_0^*) \oplus \mathrm{R}(\pi_1^*)$ be the direct sum decomposition associated with the adjoint projections π_0^* and π_1^*. If $x_1^* \in \mathrm{R}(\pi_1^*)$, then $\langle x_j, x_1^* \rangle = 0$ for all $j = 1, \ldots, k$, so $\langle x, x_1^* \rangle = 0$. By the definition of U' this implies that $c x_1^* \in U' \subseteq \phi^{-1}(B_{\mathbb{K}})$ for all $c \in \mathbb{K}$, that is, $\phi(c x_1^*) \in B_{\mathbb{K}}$ for all $c \in \mathbb{K}$, and this is only possible if $\phi(x_1^*) = 0$. For all $x^* = x_0^* + x_1^* \in \mathrm{R}(\pi_0^*) \oplus \mathrm{R}(\pi_1^*) = X^*$ we thus obtain

$$\langle x, x^* \rangle = \langle x, x_0^* \rangle = \langle x_0^*, \pi_0^{**} \phi \rangle = \langle \pi_0^* x_0^*, \phi \rangle = \langle x_0^*, \phi \rangle = \langle x^*, \phi \rangle = \phi(x^*).$$

$\square$

We apply the second part of this proposition to prove a version of the Hahn–Banach separation theorem for the weak* topology.

Proposition 4.46. *If F is a weak* closed convex subset of X^* and $x_0^* \notin F$, then there exists an element $x_0 \in X$ such that $\langle x_0, x^* \rangle \notin \langle x_0, F \rangle$.*

Proof Suppose first that $\mathbb{K} = \mathbb{R}$. By definition of the weak* topology there exists a weak* open set U of the form

$$U = \{x^* \in X^* : |\langle x_j, x^* \rangle| < \varepsilon, \ j = 1, \ldots, k\}$$

for suitable $x_1, \ldots, x_k \in X$ and $\varepsilon > 0$, such that $(x_0^* + U) \cap F = \varnothing$. By the Hahn–Banach separation theorem there exists an element $x_0^{**} \in X^{**}$ separating $x_0^* + U$ from F. This forces x_0^{**} to be bounded on U, for otherwise the convexity and symmetry of U implies that $\langle U, x_0^{**} \rangle = \mathbb{R}$ and then the set $\langle x_0^* + U, x_0^{**} \rangle$ would contain the set $\langle F, x_0^* \rangle$, contradicting the choice of x_0^{**}.

Since x_0^{**} is bounded on U, x_0^{**} is continuous with respect to the weak* topology of X. Hence by Proposition 4.45 it can be identified with an element $x_0 \in X$. This element has the desired properties.

This concludes the proof in the case $\mathbb{K} = \mathbb{R}$. If $\mathbb{K} = \mathbb{C}$ we apply the result for real scalars to the real Banach space $X_{\mathbb{R}}$ obtained by restricting scalar multiplication to real scalars. $\square$

Using the notation introduced in Definition 4.17 we have the following characterisation of weak and weak* closures of subspaces.

Proposition 4.47. *Let X be a Banach space. The following assertions hold:*

(1) *for every subspace Y of X we have $^\perp(Y^\perp) = \overline{Y}^{\mathrm{weak}} = \overline{Y}$;*
(2) *for every subspace Y of X^* we have $(^\perp Y)^\perp = \overline{Y}^{\mathrm{weak}^*}$.*

Proof (1): The inclusion $^\perp(Y^\perp) \supseteq \overline{Y}^{\mathrm{weak}}$ follows from the easy observation that pre-annihilators are weakly closed, and the equality $\overline{Y}^{\mathrm{weak}} = \overline{Y}$ is a consequence of Proposition 4.44. To prove the inclusion $^\perp(Y^\perp) \subseteq \overline{Y}$, let $x \notin \overline{Y}$. By Corollary 4.12 we can find $x_0^* \in Y^\perp$ such that $\langle x, x_0^* \rangle \neq 0$. This means that x is not in the pre-annihilator of $Y^\perp$.

(2): This is proved in the same way, except that now we use Proposition 4.46.   □

4.7 The Banach–Alaoglu Theorem

We have seen in Chapter 1 that the closed unit ball of a Banach space is compact if and only if the space is finite-dimensional. In this section we prove that the closed unit ball of every dual Banach space is compact with respect to the weak* topology, and use this to prove that a Banach space is reflexive if and only if its closed unit ball is compact with respect to the weak topology.

4.7.a The Theorem

In preparation for the main result of this section, Theorem 4.50, we have the following simple result. It extends Proposition 3.16 to separable Banach spaces.

Proposition 4.48. *Let X be a separable Banach space and let $(x_n^*)_{n \geqslant 1}$ be a bounded sequence in the dual space X^*. Then there exists a subsequence $(x_{n_k}^*)_{k \geqslant 1}$ and an $x^* \in X^*$ such that*

$$\lim_{k \to \infty} \langle x, x_{n_k}^* \rangle = \langle x, x^* \rangle, \quad x \in X.$$

Proof Let $(x_j)_{j \geqslant 1}$ be a countable set with dense linear span X_0 in X. By a diagonal argument, there exists a subsequence $(x_{n_k}^*)_{k \geqslant 1}$ of $(x_n^*)_{n \geqslant 1}$ such that the limit $\phi(x_j) := \lim_{k \to \infty} \langle x_j, x_{n_k}^* \rangle$ exists for all $j \geqslant 1$. Then, by linearity, the limit $\phi(x) := \lim_{k \to \infty} \langle x, x_{n_k}^* \rangle$ exists for all $x \in X_0$. Clearly $x \mapsto \phi(x)$ is linear, and from $|\phi(x)| \leqslant \sup_{k \geqslant 1} \|x_{n_k}^*\|$ we see that ϕ is bounded as a mapping from X_0 to $\mathbb{K}$. Since X_0 is dense in X, Proposition 1.18 implies that ϕ has a unique bounded extension of the same norm to all of X. Denoting this extension also by ϕ, it follows from Proposition 1.19 that $\lim_{k \to \infty} \langle x, x_{n_k}^* \rangle = \phi(x)$ for all $x \in X$. Thus $x^* := \phi$ has the required properties.   □

In contrast to the Hilbert space case considered in Proposition 3.16, the separability assumption in Proposition 4.48 cannot be omitted:

Example 4.49. Consider the sequence $(\phi_n)_{n \geqslant 1}$ of coordinate functionals $x \mapsto x_n$ on $X = \ell^\infty$. Given a subsequence $(\phi_{n_k})_{k \geqslant 1}$, let $x \in \ell^\infty$ be any element such that $x_{n_k} = (-1)^k$ for all $k \geqslant 1$. Then $\langle x, \phi_{n_k} \rangle = (-1)^k$ fails to converge as $k \to \infty$.

Proposition 4.48 can be viewed as a sequential version of the following theorem.

Theorem 4.50 (Banach–Alaoglu). *The closed unit ball of every dual Banach space is compact with respect to the weak* topology.*

Proof Let $\overline{B}_{\mathbb{K}} = \{a \in \mathbb{K} : |a| \leqslant 1\}$ and $\overline{B}_{X^*} = \{x^* \in X^* : \|x^*\| \leqslant 1\}$, where X is a given Banach space. In the remainder of the proof we think of $\overline{B}_{X^*}$ as endowed with the weak* topology inherited from X^*.

By Tychonov's theorem (Theorem C.14), the product space

$$K := \prod_{x \in X} \|x\| \cdot \overline{B}_{\mathbb{K}}$$

is compact with respect to the product topology. Denoting elements of K as $k = (k_x)_{x \in X}$, consider the mapping from $\phi : \overline{B}_{X^*} \to K$ given by

$$\phi(x^*) := (\langle x, x^* \rangle)_{x \in X}. \tag{4.8}$$

Let R denote its range. We first prove that R is a closed subset of K. To this end let $r \in \overline{R}$. As a first step we show that r is linear in the sense that $a_0 r_{x_0} + a_1 r_{x_1} = r_{a_0 x_0 + a_1 x_1}$ for all $a_0, a_1 \in \mathbb{K}$ and $x_0, x_1 \in X$. Fix an arbitrary $\varepsilon > 0$. By the definitions of the weak* and product topologies, the set

$$U = \left\{ k \in K : |k_{x_0} - r_{x_0}| < \varepsilon, |k_{x_1} - r_{x_1}| < \varepsilon, |k_{a_0 x_0 + a_1 x_1} - r_{a_0 x_0 + a_1 x_1}| < \varepsilon \right\}$$

is open in K and contains r, and therefore U intersects R. This means that there is an $x_0^* \in \overline{B}_{X^*}$ such that $((\langle x, x_0^* \rangle))_{x \in X} \in U$, that is,

$$|\langle x_0, x_0^* \rangle - r_{x_0}| < \varepsilon, \ |\langle x_1, x_0^* \rangle - r_{x_1}| < \varepsilon, \ |\langle a_0 x_0 + a_1 x_1, x_0^* \rangle - r_{a_0 x_0 + a_1 x_1}| < \varepsilon.$$

Then,

$$\begin{aligned} |a_0 r_{x_0} + a_1 r_{x_1} - r_{a_0 x_0 + a_1 x_1}| &\leqslant |a_0 r_{x_0} - a_0 \langle x_0, x_0^* \rangle| + |a_1 r_{x_1} - a_1 \langle x_1, x_0^* \rangle| \\ &\quad + |a_1 \langle x_0, x_0^* \rangle - a_1 \langle x_1, x_0^* \rangle - r_{a_0 x_0 + a_1 x_1}| \\ &\leqslant |a_0| \varepsilon + |b_0| \varepsilon + \varepsilon. \end{aligned}$$

Since $\varepsilon > 0$ was arbitrary, this proves the linearity of r.

Next we prove that $r = (\langle x, x^* \rangle)_{x \in X}$ for some $x^* \in X^*$, which is the same as saying that $r \in R$. Since we already know that r is linear, we must prove that r is bounded in the sense that $|r_x| \leqslant C \|x\|$ for all $x \in X$. To this end let $x_0 \in X$ and $\varepsilon > 0$. Arguing as before, from $|\langle x_0, x_0^* \rangle - r_{x_0}| < \varepsilon$ we infer

$$|r_{x_0}| \leqslant |\langle x_0, x_0^* \rangle| + \varepsilon \leqslant \|x_0\| + \varepsilon.$$

Since $x_0 \in X$ and $\varepsilon > 0$ were arbitrary, it follows that r is bounded of norm at most 1 and thus defines an element $x^* \in \overline{B}_{X^*}$ in the sense that $r_x = \langle x, x^* \rangle$ for all $x \in X$. This completes the proof that R is closed. As a consequence, R is compact, it being a closed subset of the compact set K.

Since the mapping ϕ defined by (4.8) is injective, the inverse mapping $\phi^{-1} : R \to \overline{B}_{X^*}$ is well defined. We claim that this mapping is continuous. Since the sets $V_{x_0^*, x_0, \varepsilon} \cap \overline{B}_{X^*}$

generate the weak* topology of $\overline{B}_{X^*}$ it suffices to check that their images under ϕ are open in R. But these are the sets $\{k \in K : |k_{x_0} - \langle x_0, x_0^* \rangle| < \varepsilon\} \cap R$, which are open in R. The weak* compactness of $\overline{B}_{X^*} = \phi^{-1}(R)$ now follows from the compactness of R. $\qquad\square$

A topological space τ is said to be *metrisable* if the underlying set admits a metric whose open sets are precisely the sets of τ. The next result shows that if X is separable, then the weak* topology of the unit ball of X^* is metrisable. As a result, Proposition 4.48 can alternatively be deduced from Theorem 4.50 by using that compactness and sequential compactness are equivalent for metric spaces.

Proposition 4.51. *If X is a separable Banach space, then the weak* topology of the closed unit ball of X^* is metrisable.*

Proof Let $(x_n)_{n \geq 1}$ be a dense sequence in the closed unit ball $\overline{B}_X$ of X. Such a sequence exists since X is separable. It is easily checked that the formula

$$d(x^*, y^*) := \sum_{n \geq 1} \frac{1}{2^n} \frac{|\langle x_n, x^* - y^* \rangle|}{1 + |\langle x_n, x^* - y^* \rangle|}$$

defines a metric d on $\overline{B}_{X^*}$ and that the identity mapping $I_{X^*} : (\overline{B}_{X^*}, \text{weak}^*) \to (\overline{B}_{X^*}, d)$ is continuous. In particular I_{X^*} maps compact subsets of $(\overline{B}_{X^*}, \text{weak}^*)$ to compact subsets of $(\overline{B}_{X^*}, d)$. The Banach–Alaoglu theorem asserts that $(\overline{B}_{X^*}, \text{weak}^*)$ is compact. Since closed subsets of a compact space are compact and compact sets are closed, I_{X^*} maps closed subsets of $(\overline{B}_{X^*}, \text{weak}^*)$ to closed subsets of $(\overline{B}_{X^*}, d)$. Thus the continuous mapping I_{X^*} has continuous inverse and the result follows. $\qquad\square$

As an application of the Banach–Alaoglu theorem we have the following density result.

Theorem 4.52 (Goldstine). *Let X be a Banach space. The following assertions hold:*

(1) *X is weak* dense in X^{**};*
(2) *$\overline{B}_X$ is weak* dense in $\overline{B}_{X^{**}}$.*

Proof It suffices to prove the second assertion; the first follows from it by normalising elements $x \in X$ to unit length. Arguing by contradiction, suppose that $x_0^{**} \in \overline{B}_{X^{**}}$ is not contained in the weak* closure F of B_X. Then by Proposition 4.46 there exists an element $x_0^* \in X^*$ such that $\langle x_0^*, x_0^{**} \rangle \notin \langle x_0^*, F \rangle$. By multiplying with a scalar may assume that $\|x_0^*\| = 1$. Then $B_X \subseteq F \subseteq \overline{B}_{X^{**}}$ implies $B_{\mathbb{K}} \subseteq \langle x_0^*, F \rangle \subseteq \overline{B}_{\mathbb{K}}$. By the Banach–Alaoglu theorem $\overline{B}_{X^{**}}$ is weak* compact, hence so is F, and therefore $\langle x_0^*, F \rangle$ is a compact subset of $\mathbb{K}$. We conclude that $\langle x_0^*, F \rangle = \overline{B}_{\mathbb{K}}$. As a consequence we have $|\langle x_0^*, x_0^{**} \rangle| > 1$. But this contradicts the fact that $\|x_0^{**}\| \leq 1$ and $\|x_0^*\| = 1$. $\qquad\square$

4.7.b Reflexivity

Recall that if X is a Banach space, we can use the natural isometry $J : X \to X^{**}$ given by $\langle x^*, Jx \rangle := \langle x, x^* \rangle$ to identify X with a closed subspace of X^{**}.

Definition 4.53 (Reflexivity). A Banach space X is called *reflexive* if the mapping $J : X \to X^{**}$ is surjective.

The Banach–Alaoglu theorem implies the following characterisation of reflexivity.

Theorem 4.54 (Reflexivity and weak compactness of the unit ball). *A Banach space is reflexive if and only if its closed unit ball is weakly compact.*

Proof The 'only if' part follows from the Banach–Alaoglu theorem, noting that the canonical embedding $J : X \to X^{**}$ maps $\overline{B}_X$ onto $\overline{B}_{X^{**}}$ and that, under the identification of X and X^{**}, the weak topology of X equals the weak* topology of X^{**}.

The 'if' part follows from Goldstine's theorem (Theorem 4.52). Indeed, if $\overline{B}_X$ is weakly compact, then its image under the canonical embedding J is weak* compact in X^{**}: this follows from the observation that J is continuous as a mapping from (X, weak) to (X^{**}, weak^*), which in turn is a trivial consequence of the definitions of the weak and weak* topologies. It follows that $\overline{B}_X$ is weak* compact as a subset of the closed unit ball $\overline{B}_{X^{**}}$. On the other hand, Goldstine's theorem says that $\overline{B}_X$ is weak* dense as a subset of $\overline{B}_{X^{**}}$. Hence we must have $\overline{B}_X = \overline{B}_{X^{**}}$, and this implies $X = X^{**}$. $\quad\square$

Corollary 4.55. *Let X be a Banach space. The following assertions hold:*

(1) *if X is reflexive, then every closed subspace of X is reflexive;*
(2) *X is reflexive if and only if X^* is reflexive;*
(3) *if X is isomorphic to a Banach space Y, then X is reflexive if and only if Y is reflexive.*

Proof (1): If Y is a closed subspace of X, the closed unit ball $\overline{B}_Y$ is the intersection of the set $\overline{B}_X$, which is weakly compact by Theorem 4.54, and the set Y, which is weakly closed by Corollary 4.12. As a result, $\overline{B}_Y$ is weakly compact, and Y is reflexive by another application of Theorem 4.54.

(2): If X is reflexive, the weak* and weak topologies of X^* coincide. As a result, $\overline{B}_{X^*}$ is weakly compact by the Banach–Alaoglu theorem, and therefore X^* is reflexive by Theorem 4.54. If X^* is reflexive, then X^{**} is reflexive by what we just proved, and then X, viewed as a closed subspace of X^{**}, is reflexive by part (1).

(3): If $i : X \to Y$ is an isomorphism, then $i^{**} : X^{**} \to Y^{**}$ is an isomorphism by Proposition 4.27 applied twice. Denoting the natural isometries of X and Y into their second duals by J_X and J_Y, one easily checks that $i^{**} \circ J_X = J_Y \circ i^{**}$. This identity implies that J_X is surjective if an only if J_Y is surjective. $\quad\square$

Part (3) can be also be deduced from Theorem 4.54; we leave this as an easy exercise.

Corollary 4.56. *Every bounded sequence in a reflexive Banach space has a weakly convergent subsequence.*

Proof   Let $(x_n)_{n \geqslant 1}$ be a bounded sequence in the reflexive Banach space X and let Y be its closed span. Then Y is separable, and Y is reflexive by Corollary 4.55. Proceeding as in the proof of Proposition 4.51, the weakly compact set $\overline{B}_Y$ is metrisable and we may argue by sequential compactness of the resulting metric space.   □

Example 4.57. The following are examples of reflexive Banach spaces:

- finite-dimensional Banach spaces;
- the spaces ℓ^p with $1 < p < \infty$;
- the spaces $L^p(\Omega)$ with $1 < p < \infty$;
- Hilbert spaces.

To prove that finite-dimensional Banach spaces X are reflexive we use the fact that such spaces are isomorphic to $\mathbb{K}^d$, where $d = \dim(X)$. The isomorphism $\mathbb{K}^d \simeq (\mathbb{K}^d)^*$ dualises to an isomorphism $(\mathbb{K}^d)^* \simeq (\mathbb{K}^d)^{**}$, and one easily checks that the composition of these isomorphisms equals the canonical embedding $J : \mathbb{K}^d \to (\mathbb{K}^d)^{**}$. In particular, this embedding is surjective. It follows that $\mathbb{K}^d$ is reflexive, and therefore X is reflexive by Corollary 4.55.

In the same way the second and third examples follow from the isometric identifications $(\ell^p)^* = \ell^q$ and $(L^p(\Omega))^* = L^q(\Omega)$, $\frac{1}{p} + \frac{1}{q} = 1$. Strictly speaking we have shown the identification $(L^p(\Omega))^* = L^q(\Omega)$ only for σ-finite measure spaces, but for $1 < p < \infty$ the σ-finiteness assumption is redundant (see Problem 4.3).

That Hilbert spaces are reflexive is a consequence of the Riesz representation theorem (Theorem 3.15) which sets up a conjugate-linear identification of the dual H^* of a Hilbert space with the Hilbert space H itself. Applying the theorem twice and composing the identifications of H with H^* and H^* with H^{**}, and again the resulting identification H with H^{**} equals the natural embedding $J : H \to H^{**}$.

Example 4.58. The spaces c_0, ℓ^1 and ℓ^∞ are nonreflexive: for c_0 this follows from the fact that $c_0^{**} = \ell^\infty$ and for $\ell^1 = c_0^*$ and $\ell^\infty = (\ell^1)^*$ this follows from Corollary 4.55. The spaces $C(K)$ and $L^1(\Omega)$ are nonreflexive except when they are finite-dimensional. Indeed, it is easy to find closed subspaces isomorphic to c_0 or ℓ^1, respectively (take the closed linear span of any sequence of norm one vectors with disjoint supports), and nonreflexivity again follows from Corollary 4.55.

4.7.c Translation Invariant Operators on $L^1(\mathbb{R}^d)$

In this section and the next we give two nontrivial applications of the Banach–Alaoglu theorem or, more precisely, its sequential version contained in Proposition 4.48. The first application is concerned with characterising translation invariant operators on $L^1(\mathbb{R}^d)$ as convolutions with a finite Borel measure. It is complemented by Theorem 5.34 in the next chapter, which characterises the translation invariant operators on $L^2(\mathbb{R}^d)$ as the Fourier multiplier operators.

Lemma 4.59. *If $g \in L^1(\mathbb{R}^d)$ satisfies $\int_{\mathbb{R}^d} g(x)\phi(x)\,dx = 0$ for all $\phi \in C_c^\infty(D)$, then $g = 0$ almost everywhere on D.*

Proof This follows from the uniqueness part of Theorem 4.2, viewing $g\,dx$ as a finite Borel measure on $\mathbb{R}^d$ □

Theorem 4.60 (Translation invariant operators on $L^1(\mathbb{R}^d)$). *If T is a bounded operator on $L^1(\mathbb{R}^d)$ commuting with every translation, then T is the convolution with respect to a (necessarily unique) measure $\mu \in M(\mathbb{R}^d)$; more precisely, for all $f \in C_c(\mathbb{R}^d)$ and $x \in \mathbb{R}^d$ we have*

$$Tf(x) = \int_{\mathbb{R}^d} f(x-y)\,d\mu(y).$$

Moreover, $\|T\| \leqslant \|\mu\|$.

Proof Fix a function $\eta \in L^1(\mathbb{R}^d)$ satisfying $\int_{\mathbb{R}^d} \eta(x)\,dx = 1$. For $\varepsilon > 0$ the mollified functions $\eta_\varepsilon(x) := \varepsilon^{-d}\eta(\varepsilon^{-1}x)$ belong to $L^1(\mathbb{R}^d)$ and satisfy $\|\eta_\varepsilon\|_1 = \|\eta\|_1$. By viewing the functions $T\eta_\varepsilon \in L^1(\mathbb{R}^d)$ as densities of finite Borel measures on $\mathbb{R}^d$ we may identify them with finite Borel measures $\mu_\varepsilon \in M(\mathbb{R}^d)$. By Theorem 4.2 and Proposition E.16, $M(\mathbb{R}^d)$ can be identified with the dual of $C_0(\mathbb{R}^d)$. Hence by the sequential version of the Banach–Alaoglu theorem (Proposition 4.48), some subsequence $(T\eta_{\varepsilon_n})_{n\geqslant 1}$ converges weak* to a measure $\mu \in M(\mathbb{R}^d)$. Then, for all $g \in C_0(\mathbb{R}^d)$,

$$\lim_{n\to\infty} \int_{\mathbb{R}^d} g(y)T\eta_{\varepsilon_n}(y)\,dy = \int_{\mathbb{R}^d} g(y)\,d\mu(y).$$

Applying this to the functions $y \mapsto g(x+y) = (\tau_x g)(y)$, where τ_x is translation over x, upon letting $n \to \infty$ we obtain

$$\int_{\mathbb{R}^d} g(y)T\eta_{\varepsilon_n}(y-x)\,dy = \int_{\mathbb{R}^d} g(x+y)T\eta_{\varepsilon_n}(y)\,dy \to \int_{\mathbb{R}^d} g(x+y)\,d\mu(y).$$

By Fubini's theorem, a change of variables, the commutation assumption, and Proposition 2.34, for all $f \in C_c(\mathbb{R}^d)$ this implies

$$\int_{\mathbb{R}^d} g(x)\int_{\mathbb{R}^d} f(x-y)\,d\mu(y)\,dx = \int_{\mathbb{R}^d}\int_{\mathbb{R}^d} f(x-y)g(x)\,dx\,d\mu(y)$$

$$= \int_{\mathbb{R}^d} f(x)\int_{\mathbb{R}^d} g(x+y)\,d\mu(y)\,dx$$

$$= \lim_{n \to \infty} \int_{\mathbb{R}^d} f(x) \int_{\mathbb{R}^d} g(y) T \eta_{\varepsilon_n}(y-x) \, dy \, dx$$

$$= \lim_{n \to \infty} \int_{\mathbb{R}^d} f(x) \int_{\mathbb{R}^d} g(y) \tau_{-x} T \eta_{\varepsilon_n}(y) \, dy \, dx$$

$$= \lim_{n \to \infty} \int_{\mathbb{R}^d} f(x) \int_{\mathbb{R}^d} g(y) T \tau_{-x} \eta_{\varepsilon_n}(y) \, dy \, dx$$

$$\overset{(*)}{=} \lim_{n \to \infty} \int_{\mathbb{R}^d} T^* g(y) \int_{\mathbb{R}^d} f(x) \eta_{\varepsilon_n}(y-x) \, dx \, dy$$

$$= \int_{\mathbb{R}^d} T^* g(y) f(y) \, dy = \int_{\mathbb{R}^d} g(y) T f(y) \, dy.$$

In $(*)$ we identified $g \in C_0(\mathbb{R}^d)$ with a function in $L^\infty(\mathbb{R}^d) = (L^1(\mathbb{R}^d))^*$.

Since the above identities hold for all $g \in C_0(\mathbb{R}^d)$, it follows from Lemma 4.59 that $\int_{\mathbb{R}^d} f(x-y) \, d\mu(y) = T f(x)$ for almost all $x \in \mathbb{R}^d$, and hence by continuity for all $x \in \mathbb{R}^d$. This proves that T is of the asserted form, and the bound $\|T\| \leqslant \|\mu\|$ follows from the observation preceding the theorem.

It remains to establish uniqueness of the measure μ. Suppose that $\mu \in M(\mathbb{R}^d)$ satisfies

$$\int_{\mathbb{R}^d} \tau_x f(-y) \, d\mu(y) = \int_{\mathbb{R}^d} f(x-y) \, d\mu(y) = 0$$

for all $x \in \mathbb{R}^d$ and $f \in C_c(\mathbb{R}^d)$. Then

$$\int_{\mathbb{R}^d} \phi(y) \, d\mu(y) = 0, \quad \phi \in C_c(\mathbb{R}^d).$$

Since $C_c(\mathbb{R}^d)$ is dense in $C_0(\mathbb{R}^d)$, by Theorem 4.2 this implies $\mu = 0$. □

4.7.d Prokhorov's Theorem

The aim of this section is to prove a compactness result of fundamental importance in Probability Theory, known as Prokhorov's theorem. For its statement we need the following terminology.

Definition 4.61 (Uniform tightness). A collection P of Borel measures on a topological space X is called *uniformly tight* if for every $\varepsilon > 0$ there exists a compact set K in X such that $\mu(X \setminus K) < \varepsilon$ for all $\mu \in P$.

Definition 4.62 (Weak convergence). A sequence $(\mu_n)_{n \geqslant 1}$ of Borel probability measures on a topological space X is said to *converge weakly* to a Borel probability measure μ on X if

$$\lim_{n \to \infty} \int_X f \, d\mu_n = \int_X f \, d\mu, \quad f \in C_b(X),$$

where $C_b(X)$ denotes the Banach space of bounded continuous functions on X.

Viewing Borel probability measures on X as functionals in the dual space $(C_b(X))^*$, weak convergence in the sense of the above definition is precisely weak* convergence in the sense discussed in the present chapter. The terminology 'weak convergence' is firmly established in the Probability Theory literature, however.

Theorem 4.63 (Prokhorov's theorem). *For a metric space X, the following assertions hold:*

(1) *if a family P of Borel probability measures on X is uniformly tight, then every sequence in P contains a weakly convergent subsequence;*
(2) *if X is separable and complete, then every weakly convergent sequence of Borel probability measures on X is uniformly tight.*

For the proof of this theorem we need the following characterisation of weak convergence, known as the *Portmanteau theorem*.

Proposition 4.64. *Let μ_n, $n \geqslant 1$, and μ be Borel probability measures on a metric space X. The following assertions are equivalent:*

(1) $\lim_{n\to\infty} \mu_n = \mu$ *weakly;*
(2) *for all open subsets U of X we have $\mu(U) \leqslant \liminf_{n\to\infty} \mu_n(U)$;*
(3) *for all closed subsets F of X we have $\limsup_{n\to\infty} \mu_n(F) \leqslant \mu(F)$.*

Proof (1)$\Rightarrow$(2): Let $U \subseteq X$ be open. For each $k \geqslant 1$ let

$$U^{(k)} := \left\{ x \in U : d(x, \complement U) > \frac{1}{k} \right\}.$$

Since U is open we have $U = \bigcup_{k\geqslant 1} U^{(k)}$ and $\mu(U) = \lim_{k\to\infty} \mu(U^{(k)})$.

The functions $f_k(x) := \min\{1, k\, d(x, \complement U)\}$ belong to $C_b(X)$ and satisfy $0 \leqslant f_k \leqslant 1$, $f_k = 0$ on $\complement U$, and $f_k = 1$ on $U^{(k)}$.

For each $k \geqslant 1$,

$$\mu(U^{(k)}) \leqslant \int_X f_k \, d\mu = \lim_{n\to\infty} \int_X f_k \, d\mu_n = \liminf_{n\to\infty} \int_X f_k \, d\mu_n \leqslant \liminf_{n\to\infty} \mu_n(U).$$

Now we pass to the limit $k \to \infty$.

The equivalence (2)$\Leftrightarrow$(3) follows by taking complements.

It remains to prove that (2) and (3) together imply (1). Let $U_1, \dots, U_k$ be open sets in X and consider a function of the form $g = \sum_{j=1}^k c_j \mathbf{1}_{U_j}$. We have, using (2),

$$\int_X g \, d\mu = \sum_{j=1}^k c_j \mu(U_j) \leqslant \liminf_{n\to\infty} \sum_{j=1}^k c_j \mu_n(U_j) = \liminf_{n\to\infty} \int_X g \, d\mu_n.$$

Similarly, for $\overline{g} = \sum_{j=1}^{k} c_j \mathbf{1}_{\overline{U_j}}$ we have, using (3),

$$\limsup_{n \to \infty} \int_X \overline{g} \, d\mu_n \leqslant \sum_{j=1}^{k} c_j \limsup_{n \to \infty} \mu_n(\overline{U_j}) \leqslant \sum_{j=1}^{k} c_j \mu(\overline{U_j}) = \int_X \overline{g} \, d\mu.$$

Let $f \in C_b(X)$ be a real-valued function and choose $a, b \in \mathbb{R}$ such that $a < f(x) < b$ for all $x \in X$. There are at most countably many $r \in (a, b)$ such that the set $\{x \in X : f(x) = r\}$ has nonzero μ-measure. Let R denote the set of these numbers r. Fix $\varepsilon > 0$ and let $\pi = \{t_0, \ldots, t_k\}$ be a partition of $[a, b]$ with $\text{mesh}(\pi) < \varepsilon$ such that $t_0 = a$, $t_k = b$, and $t_j \notin R$ for all $j = 1, \ldots, k-1$. Put $U_j := \{x \in X : f(x) \in (t_{j-1}, t_j)\}$, $j = 1, \ldots, k$, and

$$g := \sum_{j=1}^{k} t_{j-1} \mathbf{1}_{U_j}, \quad h := \sum_{j=1}^{k} t_j \mathbf{1}_{U_j}.$$

With the above notation, $g(x) \leqslant f(x) \leqslant \varepsilon + g(x)$ and $f(x) \leqslant \overline{h}(x) \leqslant \varepsilon + f(x)$ whenever $f(x) \neq t_j$ for $j = 1, \ldots, k-1$. Since $\mu\{x \in X : f(x) = t_j\} = 0$ for $j = 1, \ldots, k-1$,

$$\limsup_{n \to \infty} \int_X f \, d\mu_n \leqslant \limsup_{n \to \infty} \int_X \overline{h} \, d\mu_n \leqslant \int_X \overline{h} \, d\mu \leqslant \varepsilon + \int_X f \, d\mu$$

$$\leqslant 2\varepsilon + \int_X g \, d\mu \leqslant 2\varepsilon + \liminf_{n \to \infty} \int_X g \, d\mu_n \leqslant 2\varepsilon + \liminf_{n \to \infty} \int_X f \, d\mu_n.$$

Since $\varepsilon > 0$ was arbitrary, this concludes the proof for real-valued functions f. In the case of complex scalars, the result for complex-valued f follows from it by considering real and imaginary parts separately. □

The proof of part (1) of Prokhorov's theorem relies on the following lemma.

Lemma 4.65. *Let* $(\Omega, \mathscr{F})$ *be a measurable space and let* $A_1 \subseteq A_2 \subseteq \ldots$ *be an increasing sequence of sets in* $\mathscr{F}$ *such that* $\bigcup_{j \geqslant 1} A_j = \Omega$. *For each* $j \geqslant 1$ *let* μ_j *be a measure on* A_j. *If the measures* μ_j *are increasing in the sense that* $\mu_j|_{A_i} \geqslant \mu_i$ *whenever* $j \geqslant i$, *then*

$$\mu(B) := \lim_{j \to \infty} \mu_j(B \cap A_j)$$

defines a measure on Ω.

Proof If $B \in \mathscr{F}$, then for $j \geqslant i$ we have $\mu_j(B \cap A_j) \geqslant \mu_j(B \cap A_i) = \mu_j|_{A_i}(B \cap A_i) \geqslant \mu_i(B \cap A_i)$. Therefore the limit defining $\mu(B)$ exists, and $\mu(B) = \sup_{j \geqslant 1} \mu_j(B \cap A_j)$.

It is clear that $\mu(\varnothing) = 0$. To prove that μ is countably additive, let $B = \bigcup_{n \geqslant 1} B_n$ with disjoint measurable sets B_n. On the one hand,

$$\mu(B) = \sup_{j \geqslant 1} \mu_j(B \cap A_j) = \sup_{j \geqslant 1} \sum_{n \geqslant 1} \mu_j(B_n \cap A_j)$$

$$\leqslant \sum_{n \geqslant 1} \sup_{j \geqslant 1} \mu_j(B_n \cap A_j) = \sum_{n \geqslant 1} \mu(B_n).$$

On the other hand, for each $j_0 \geq 1$ we have

$$\mu(B) = \sup_{j \geq 1} \mu_j(B \cap A_j) \geq \mu_{j_0}(B \cap A_{j_0}) = \sum_{n \geq 1} \mu_{j_0}(B_n \cap A_{j_0}).$$

Hence, by monotone convergence,

$$\mu(B) \geq \lim_{j_0 \to \infty} \sum_{n \geq 1} \mu_{j_0}(B_n \cap A_{j_0}) = \sum_{n \geq 1} \lim_{j_0 \to \infty} \mu_{j_0}(B_n \cap A_{j_0}) = \sum_{n \geq 1} \mu(B_n).$$

$\square$

Proof of Theorem 4.63   We begin with the proof of part (1). Assuming that P is uniformly tight, we must prove that there is a Borel probability measure μ on X such that $\mu_n \to \mu$ weakly.

Choose an increasing sequence of compact sets $K_j \subseteq X$ such that $\mu_n(K_j) \geq 1 - 1/2^j$ for all $j \geq 1$ and $n \geq 1$. Replacing X by $\overline{\bigcup_{j \geq 1} K_j}$, we may assume that X is separable.

Step 1 – Identifying each restriction $\mu_n|_{K_j}$ with an element of $(C(K_j))^*$, by a diagonal argument we find a subsequence $(\mu_{n_k})_{k \geq 1}$ such that for all $j \geq 1$ the sequence $(\mu_{n_k}|_{K_j})_{k \geq 1}$ is weak* convergent in $(C(K_j))^*$ to some Borel measure ν_j on K_j; this argument uses the sequential version of the Banach–Alaoglu theorem (Proposition 4.48) and the separability of the spaces $C(K_j)$ (Proposition 2.8). Hence by Proposition 4.64,

$$\nu_j(K_j) \geq \limsup_{k \to \infty} \mu_{n_k}(K_j) \geq 1 - 2^{-j}.$$

Step 2 – We claim that if $j \geq i$, then $\nu_j|_{K_i} \geq \nu_i$. To this end fix a number $\varepsilon > 0$ and a function $f \in C(K_i)$ satisfying $0 \leq f(x) \leq 1$ for all $x \in K_i$. Using Theorem C.13 we extend f to a function in $C(K_j)$ satisfying $0 \leq f(x) \leq 1$ for all $x \in K_j$, and let $f_m \in C(K_j)$ be defined by $f_m(x) := (1 - m \cdot d(x, K_i))^+ f(x)$, $x \in K_j$. Choosing m large enough, say $m \geq m_\varepsilon$, we may assume that

$$\int_{K_j \setminus K_i} f_m \, d\nu_j < \varepsilon.$$

Since $f_m = f$ on K_i we find

$$\int_{K_i} f \, d\nu_j - \int_{K_i} f \, d\nu_i \geq -\varepsilon + \int_{K_j} f_m \, d\nu_j - \int_{K_i} f_m \, d\nu_i$$

$$= -\varepsilon + \lim_{k \to \infty} \underbrace{\left(\int_{K_j} f_m \, d\mu_{n_k} - \int_{K_i} f_m \, d\mu_{n_k} \right)}_{\geq 0} \geq -\varepsilon.$$

Since $\varepsilon > 0$ was arbitrary, this proves that $\int_{K_i} f \, d\nu_j \geq \int_{K_i} f \, d\nu_i$.

Step 3 – We apply Lemma 4.65 to see that

$$\mu(B) := \lim_{j \to \infty} \nu_j(B \cap K_j) = \sup_{j \geq 1} \nu_j(B \cap K_j)$$

defines a Borel measure μ on $X_0 := \bigcup_{j \geqslant 1} K_j$. We may extend μ to a Borel measure on all of X by extending it identically 0 outside X_0. Clearly, $\mu(X) \leqslant 1$ and

$$\mu(X) \geqslant \mu(K_j) \geqslant \nu_j(K_j) \geqslant 1 - 2^{-j}.$$

This proves a couple of things at the same time, namely that μ is a probability measure and that μ is tight.

It remains to prove that $\lim_{n \to \infty} \mu_{n_k} = \mu$ weakly. For this, it suffices to prove that $\lim_{n \to \infty} \int_X f \, d\mu_{n_k} = \int_X f \, d\mu$ for all $f \in C_b(X)$ satisfying $0 \leqslant f \leqslant 1$. Fixing such a function, choose a sequence $(f_m)_{m \geqslant 1}$ of simple functions satisfying $0 \leqslant f_m \leqslant 1$ for all $m \geqslant 1$ and $f_m \to f$ uniformly as $m \to \infty$. Fix j_0 so large that $2^{-j_0} < \varepsilon$. Then, for $m \geqslant 1$ large enough and all $j \geqslant j_0$,

$$\limsup_{k \to \infty} \left| \int_X f \, d\mu - \int_X f \, d\mu_{n_k} \right| \leqslant 2^{-j+1} + \limsup_{k \to \infty} \left| \int_{K_j} f \, d\mu - \int_{K_j} f \, d\mu_{n_k} \right|$$

$$\leqslant 2\varepsilon + \left| \int_{K_j} f \, d\mu - \int_{K_j} f \, d\nu_j \right|$$

$$\leqslant 4\varepsilon + \left| \int_{K_{j_0}} f \, d\mu - \int_{K_{j_0}} f \, d\nu_j \right|$$

$$\leqslant 6\varepsilon + \left| \int_{K_{j_0}} f_m \, d\mu - \int_{K_{j_0}} f_m \, d\nu_j \right|.$$

Since $\lim_{j \to \infty} \nu_j(B \cap K_{j_0}) = \lim_{j \to \infty} \nu_j(B \cap K_{j_0} \cap K_j) = \mu(B \cap K_{j_0})$ for all Borel sets B in X and each function f_m is simple, upon letting $j \to \infty$ we obtain

$$\lim_{j \to \infty} \int_{K_{j_0}} f_m \, d\nu_j = \int_{K_{j_0}} f_m \, d\mu.$$

Consequently,

$$\limsup_{k \to \infty} \left| \int_X f \, d\mu - \int_X f \, d\mu_{n_k} \right| \leqslant 6\varepsilon.$$

Since $\varepsilon > 0$ was arbitrary this proves the weak convergence.

We now turn to the proof of part (2). Since X is separable, we may pick a dense sequence $(x_n)_{n \geqslant 1}$ in X. For every integer $k \geqslant 1$ the open balls $B(x_n; \frac{1}{k})$, $n \geqslant 1$, cover X. Fix $\varepsilon > 0$ and choose the integers $N_k \geqslant 1$ such that

$$\mu\left(\bigcup_{n=1}^{N_k} B(x_n; \tfrac{1}{k}) \right) > 1 - \frac{\varepsilon}{2^k}.$$

By Proposition 4.64, this implies that for all large enough j, say for $j \geqslant j_0$, we have

$$\mu_j\left(\bigcup_{n=1}^{N_k} B(x_n; \tfrac{1}{k}) \right) > 1 - \frac{\varepsilon}{2^k}.$$

The set

$$K = \bigcap_{k \geqslant 1} \bigcup_{n=1}^{N_k} \overline{B}(x_n; \tfrac{1}{k})$$

is closed and totally bounded. The completeness of X therefore implies that K is compact. Moreover, for $j \geqslant j_0$,

$$\mu_j(\complement K) \leqslant \sum_{k \geqslant 1} \mu_j\left(\complement \bigcup_{n=1}^{N_k} \overline{B}(x_n; \tfrac{1}{k})\right) \leqslant \sum_{k \geqslant 1} \mu_j\left(\complement \bigcup_{n=1}^{N_k} B(x_n; \tfrac{1}{k})\right) \leqslant \sum_{k \geqslant 1} \frac{\varepsilon}{2^k} = \varepsilon.$$

$\square$

Problems

4.1   A bounded operator is said to be of *finite rank* if its range is finite-dimensional. Show that every finite rank operator $T \in \mathscr{L}(X,Y)$ is of the form

$$Tx = \sum_{n=1}^{N} \langle x, x_n^* \rangle y_n, \quad x \in X,$$

for certain $y_n \in Y$ and $x_n^* \in X^*$.

4.2   Consider an open set $D \subseteq \mathbb{C}$ and a complex Banach space X. A function $f : D \to X$ is said to be *holomorphic* if for all $z_0 \in D$ the limit

$$\lim_{z \to z_0} \frac{f(z) - f(z_0)}{z - z_0}$$

exists in X. Use the Hahn–Banach theorem to prove that the Cauchy theorem and the Cauchy integral formula hold for holomorphic functions $f : D \to X$ defined on an open set D in $\mathbb{C}$:

$$\frac{1}{2\pi i} \int_{\{|z-z_0|=r\}} f(z)\, \mathrm{d}z = 0$$

and

$$\frac{1}{2\pi i} \int_{\{|z-z_0|=r\}} \frac{f(z)}{z - z_0}\, \mathrm{d}z = f(z_0).$$

Here it is assumed that $z_0 \in D$ and $r > 0$ is so small that $\{|z - z_0| = r\}$ is contained in D; this contour is oriented counterclockwise.

4.3   Let $1 \leqslant p, q \leqslant \infty$ satisfy $\frac{1}{p} + \frac{1}{q} = 1$.

(a) Prove that the identification $(L^p(\Omega))^* = L^q(\Omega)$, remains true in the non-σ-finite case if $1 < p < \infty$.

Hint: Given $\phi \in (L^p(\Omega))^*$, there is a sequence $(f_n)_{n \geqslant 1}$ in $L^p(\Omega)$ such that $\|\phi\| = \sup_{n \geqslant 1} |\langle f_n, \phi \rangle|$. The σ-algebra generated by this sequence is σ-finite.

(b) Show, by way of example, that part (a) does not extend to $p = 1$.

4.4 Prove that if X^* is separable, then X is separable.

Hint: There is a sequence $(x_n)_{n \geqslant 1}$ in X such that $\|x^*\| = \sup_{n \geqslant 1} |\langle x_n, x^* \rangle|$ for a countable dense set of functionals x^* in X^*.

4.5 Let X be a Banach space.

(a) Show that for all nonzero $x^* \in X^*$ we have an isomorphism of Banach spaces

$$X/N(x^*) \simeq \mathbb{K},$$

where $N(x^*) := \{x \in X : \langle x, x^* \rangle = 0\}$.

(b) Let $Q : X \to X/N(x^*)$ denote the quotient mapping. Prove that if $\|x^*\| = 1$, then for all $x \in X$ we have

$$\|Qx\| = |\langle x, x^* \rangle|.$$

Hint: For the inequality '$\leqslant$', begin by showing that for any $0 < \varepsilon < 1$ there must exist $c \in \mathbb{K}$ and $y \in X$ such that $\|cx + y\| = 1$ and $|\langle cx + y, x^* \rangle| \geqslant 1 - \varepsilon$.

4.6 Let Y be a proper closed subspace of X and let $x_0 \in X \setminus Y$. As in Corollary 4.12, on the span X_0 of Y and x_0 define $\phi(y) := 0$ for all $y \in Y$ and $\phi(x_0) := 1$. It was shown in the corollary that ϕ is bounded. Show that its norm is given by

$$\|\phi\|_{X_0^*} = \frac{1}{d(x_0, Y)}.$$

4.7 Let X be a Banach space. For a set $A \subseteq X$ and an element $x \in X$ we denote by $d(x, A) = \inf_{y \in A} \|x - y\|$ the distance from x to A.

(a) Let X be a Banach space, let $X_0 \subseteq X$ be a proper closed subspace, and let $x \in X \setminus X_0$. Prove that there exists an $x^* \in X^*$ with $\|x^*\| = 1$ such that $\langle x, x^* \rangle = d(x, X_0)$ and $x^*|_{X_0} = 0$.

Hint: Let $Y = \text{span}(X_0, x)$. Prove that the mapping $x_0^* : Y \to \mathbb{K}$ given by

$$x_0^*(x_0 + tx) := td(x, X_0), \quad x_0 \in X_0, \ t \in \mathbb{K},$$

is linear, belongs to Y^*, has norm $\|x_0^*\|_{Y^*} = 1$, and satisfies $x_0^*|_{X_0} = 0$. Apply the Hahn–Banach theorem to extend x_0^* to a functional on X.

(b) Using the result of part (a), show that there exists $x^* \in (L^\infty(0,1))^*$ such that

$$\langle f, x^* \rangle = \int_{[0,1]} f(t) \, dt, \quad f \in C[0,1],$$

but

$$\langle \mathbf{1}_{(0,\frac{1}{2})}, x^* \rangle \neq \int_{[0,1]} \mathbf{1}_{(0,\frac{1}{2})}(t) \, dt.$$

4.8 We take a look at Banach spaces containing – and contained in – ℓ^∞.

(a) Let X be a Banach space, Y be a closed subspace of X, and let $T_0 : Y \to \ell^\infty$ be a bounded operator. Show that there exists a bounded operator $T : X \to \ell^\infty$ such that $T|_Y = T_0$ and $\|T\| = \|T_0\|$.
Hint: Apply the Hahn–Banach theorem 'coordinatewise'.

(b) Using the result of part (a), prove that if a closed subspace Y of a Banach space X is isomorphic to ℓ^∞, then Y is complemented in X.

(c) Show that every separable Banach space X is isometrically isomorphic to a closed subspace of ℓ^∞. More precisely, show that if X is a separable Banach space, then there exists a closed subspace Y of ℓ^∞ and an isometric isomorphism T from X onto Y.
Hint: Use the Hahn–Banach theorem in combination with the separability to make a clever choice of a sequence of functionals $(x_n^*)_{n \geq 1}$ in X^*, and consider the mapping $T : x \mapsto (\langle x, x_n^* \rangle)_{n \geq 1}$.

4.9 Find an example of a two-dimensional Banach space X and a functional on one of its closed one-dimensional subspaces which has infinitely many extensions to a functional on X of the same norm.

4.10 Recall from Problem 3.4 that a Banach space X is called *strictly convex* if for all norm one vectors $x_0, x_1 \in X$ with $x_0 \neq x_1$ and $0 < \lambda < 1$ we have

$$\|(1-\lambda)x + \lambda y\| < 1.$$

This problem shows that if the dual X^* of a Banach space is strictly convex, then every functional on a closed subspace of X has a *unique* Hahn–Banach extension of the same norm.

Let Y be closed subspace of a Banach space X.

(a) Prove that for all $x^* \in X^*$ we have

$$d(x^*, Y^\perp) = \|x^*|_Y\|_{Y^*}.$$

The closed subspace Y is said to have the *Haar property* if for all $x \in X \setminus Y$ there exists a unique $y \in Y$ such that $d(x, Y) = \|x - y\|$.

(b) Prove that a functional $y^* \in Y^*$ has a unique extension to a functional in X^* of the same norm if and only if the annihilator $Y^\perp$ has the Haar property as a closed subspace of X^*.

(c) Prove that if X is strictly convex, then every closed subspace Y of X has the Haar property.

4.11 Prove that if $T : X \to Y$ is an isomorphism from X onto its range in Y, then the adjoint operator $T^* : Y^* \to X^*$ is surjective.

Provide proofs of parts (1) and (2) of Proposition 4.27.

4.12 Let H_1 and H_2 be Hilbert spaces and X be a Banach space. Show that for $T_1 \in \mathscr{L}(H_1, X)$ and $T_2 \in \mathscr{L}(H_2, X)$ the following assertions are equivalent:

(1) there is a constant $C \geqslant 0$ such that $\|T_1^* x^*\| \leqslant C \|T_2^* x^*\|$ for all $x^* \in X^*$;

(2) $T_1(H_1) \subseteq T_2(H_2)$.

Hint: First show that we may assume that T_1 and T_2 are injective.

4.13 Let $1 \leqslant p < \infty$.

(a) Let $T : L^p(0,1) \to \mathbb{K}$ be the linear mapping defined by

$$Tf := \int_0^1 f(s)\,\mathrm{d}s, \quad f \in L^p(0,1).$$

Show that T is bounded and find an expression for T^*.

(b) Let $T : L^p(0,1) \to L^p(0,1)$ be the linear operator defined by

$$(Tf)(t) := \int_0^t f(s)\,\mathrm{d}s, \quad t \in (0,1),\ f \in L^p(0,1).$$

Show that T is bounded and find an expression for T^*.

(c) Let T be the linear operator of part (b), now viewed as an operator from $L^p(0,1)$ into $C[0,1]$. Show that T is bounded and find an expression for T^*.

4.14 This problem presents a way to evaluate the norm of a $d \times d$ matrix A with complex coefficients, viewed as a bounded operator on $\mathbb{C}^d$.

(a) Prove that all eigenvalues of $A^\star A$ are nonnegative real numbers and that the eigenspaces corresponding to different eigenvalues are orthogonal.

(b) Conclude from Proposition 4.28 and part (a) that

$$\|A\|^2 = \max\{\lambda \geqslant 0 : \lambda \text{ is an eigenvalue of } A^\star A\}.$$

4.15 Let H_0 be a closed subspace of H and let $i : H_0 \to H$ be the inclusion mapping. Show that the adjoint $i^\star : H \to H_0$ is the orthogonal projection in H onto H_0, viewed as a mapping from H to H_0.

4.16 Let T be a bounded operator on H.

(a) Show that if T is a contraction (that is, $\|T\| \leqslant 1$), then for each $x \in H$ we have $Tx = x$ if and only if $T^\star x = x$. Conclude that H admits an orthogonal direct sum decomposition

$$H = \mathsf{N}(I - T) \oplus \overline{\mathsf{R}(I - T)}.$$

Hint: If $Tx = x$, show that $T^\star x - x \perp x$ and deduce that $T^\star x = x$.

(b) Let T be a contraction and define

$$S_n := \frac{1}{n} \sum_{k=0}^{n-1} T^k, \quad n \geqslant 1.$$

Using Proposition 4.31 and the result of part (a), prove that

$$\lim_{n \to \infty} S_n x = \begin{cases} x & \text{if } x \in N(I-T), \\ 0 & \text{if } x \perp N(I-T). \end{cases}$$

4.17 Let K be a compact Hausdorff space and let $\phi \in (C(K))^*$ be an element with the following two properties:

   (i) $\phi(1) = 1$;
   (ii) $\phi(fg) = \phi(f)\phi(g)$ for all $f, g \in C(K)$.

   Prove that $\phi(f) = f(x)$ for some $x \in K$.
   Hint: Show that ϕ, as an element of $M(K)$, is supported on a singleton.

4.18 Find the extreme points of the closed convex set

$$C = \{f \in L^2(0,1) : f \geqslant 0, \|f\|_2 \leqslant 1\}.$$

4.19 Let C be a closed convex subset of a separable Banach space X. Prove that there exists a sequence $(x_n^*)_{n \geqslant 1}$ of norm one elements in X^* and a sequence $(F_n)_{n \geqslant 1}$ of closed sets in $\mathbb{K}$ such that

$$C = \bigcap_{n \geqslant 1} \{x \in X : \langle x, x_n^* \rangle \in F_n\}.$$

   Hint: Separate C from the elements of a dense sequence in its complement using the Hahn–Banach separation theorem.

4.20 Let X and Y be Banach spaces. Prove the following assertions:

   (a) a linear operator $T : X \to Y$ is continuous with respect to the weak topologies of X and Y if and only if it is bounded;
   (b) a linear operator $S : Y^* \to X^*$ is continuous with respect to the weak* topologies of Y^* and X^* if and only if it is the adjoint of a bounded operator $T : X \to Y$.

4.21 Prove that the weak topology of a Banach space X coincides with the norm topology if and only if X is finite-dimensional.

4.22 Prove that c_0, $C[0,1]$, $C_b(D)$, and $L^1(\Omega)$ are norm closed and weak* dense in ℓ^∞, $L^\infty[0,1]$, $L^\infty(D)$, and $M(\Omega)$, respectively.

4.23 Prove that if X is a locally compact Hausdorff space, then the linear span of the Dirac measures δ_ξ, $\xi \in X$, is weak* dense in $M(X)$.

4.24 Find an example of a sequence $(x_n^*)_{n \geqslant 1}$ in a dual Banach space X^* with the following two properties:

(i) there exists an $x^* \in X^*$ such that $\lim_{n\to\infty}\langle x, x_n^*\rangle = \langle x, x^*\rangle$ for all $x \in X$;

(ii) no sequence contained in the convex hull of $(x_n^*)_{n\geqslant 1}$ converges to x^* with respect to the norm of X^*.

Compare with Corollary 4.33.

4.25 Prove the following converse to Proposition 4.51: If the weak* topology of the closed unit ball of a dual Banach space X^* is metrisable, then X is separable.

Hint: Complete the details of the following argument. Let d be a metric which induces the weak* topology of $\overline{B}_{X^*}$. Then $(\overline{B}_{X^*}, d)$ is a compact metric space and therefore the Banach space $C(\overline{B}_{X^*}, d)$ is separable by Proposition 2.8. Now observe that X is isometrically contained in $C(\overline{B}_{X^*}, d)$ in a natural way.

4.26 Let X be a Banach space.

(a) Let $x_1, \ldots, x_N \in X$ and $c_1, \ldots, c_N \in \mathbb{K}$ and a constant $M \geqslant 0$ be given. Prove that the following are equivalent:

(1) there exists $x^* \in X^*$ such that $\|x^*\| \leqslant M$ and

$$\langle x_n, x^*\rangle = c_n, \quad n = 1, \ldots, N.$$

(2) for all $\lambda_1, \ldots, \lambda_N \in \mathbb{K}$ we have

$$\left| \sum_{n=1}^{N} \lambda_n c_n \right| \leqslant M \left\| \sum_{n=1}^{N} \lambda_n x_n \right\|.$$

(b) Use the Banach–Alaoglu theorem to extend the result of part (a) to infinite sequences $(x_n)_{n\geqslant 1}$ and $(c_n)_{n\geqslant 1}$.

4.27 Show that the weak topology of a weakly compact subset of a separable Banach space is metrisable.

4.28 Using the result of the preceding problem, show that if K is a weakly compact subspace of a Banach space, then every sequence $(x_n)_{n\geqslant 1}$ contained in K has a weakly convergent subsequence.

4.29 Using the result of the preceding problem, show that $C[0, 1]$ and $L^1(0, 1)$ are non-reflexive by checking that their closed unit balls contain sequences that fail to converge weakly.

4.30 As an application of the Banach–Alaoglu theorem, prove that there exist functionals $\phi \in (\ell^\infty)^*$ such that for all $x = (x_n)_{n\geqslant 1} \in \ell^\infty$ we have:

(i) $\langle x, \phi\rangle \geqslant 0$ whenever $x \geqslant 0$;

(ii) $\langle x, \phi\rangle = \langle Sx, \phi\rangle$, where $S : (x_n)_{n\geqslant 1} \mapsto (x_{n+1})_{n\geqslant 1}$ is the left shift;

(iii) $\langle x, \phi\rangle = \lim_{n\to\infty} x_n$ whenever $\lim_{n\to\infty} x_n$ exists.

Functionals with these properties are called *Banach limits.*

Hint: Consider the functionals $\phi_n(x) := \frac{1}{n}\sum_{j=1}^{n} x_j$.

4.31 Let $(x_n)_{n \geqslant 1}$ be the sequence in ℓ^∞ defined by $x_n = (0, \ldots, 0, 1, 1, \ldots), n \geqslant 1$.
$$\underbrace{}_{n \text{ times}}$$

   (a) Show that this sequence has no weakly convergent subsequence.
       Hint: Use the result of Problem 4.30.
   (b) Why doesn't this contradict the Banach–Alaoglu theorem?

4.32 Let X be a separable Banach space and let X_0 be a closed subspace of X isomorphic to c_0. Our aim is to show that X_0 is complemented in X.

   (a) Use the Hahn–Banach theorem to show that there exists a *bounded* sequence
       $(x_n^*)_{n \geqslant 1}$ in X^* such that for all $y \in c_0$ and $n \geqslant 1$ we have $\langle jy, x_n^* \rangle = y_n$, where
       $j : c_0 \to X_0$ is the isomorphism mapping c_0 onto X_0.
       Hint: Consider the adjoint of the operator $j^{-1} : X_0 \to c_0$.
   (b) Suppose that $x^* \in X^*$ is such that $\lim_{n \to \infty} \langle x, x_{n_k}^* \rangle = \langle x, x^* \rangle$ for all $x \in X$ and
       some subsequence $(x_{n_k}^*)_{k \geqslant 1}$ of $(x_{n_k}^*)_{n \geqslant 1}$. Show that $\langle x_0, x^* \rangle = 0$ for all $x_0 \in X_0$.
   (c) Use the Banach–Alaoglu theorem to deduce that $\lim_{n \to \infty} d(x_n^*, X_0^\perp) = 0$.
   (d) Suppose that $\|x_n^*\| \leqslant R$ for all $n \geqslant 1$. Use Proposition 4.51 and part (c) to conclude that there exists a sequence $(y_n^*)_{n \geqslant 1}$ in $\overline{B}(0; R)$ such that $\lim_{n \to \infty} \langle x, x_n^* - y_n^* \rangle = 0$ for all $x \in X$ and $\langle x_0, y_n^* \rangle = 0$ for all $x_0 \in X_0$ and all $n \geqslant 1$.
   (e) Show that the mapping $P : x \mapsto (\langle x, x_n^* - y_n^* \rangle)_{n \geqslant 1}$ is well defined and bounded
       from X into c_0 and that $j \circ P$ is a projection in X whose range equals X_0.

4.33 Show that ℓ^1 has the *Schur property*: If $\lim_{n \to \infty} x_n = x$ weakly in ℓ^1, then
      $\lim_{n \to \infty} x_n = x$ strongly in ℓ^1.

4.34 Let $(\Omega, \mathscr{F}, \mu)$ be a probability space. Let $(f_n)_{n \geqslant 1}$ be a bounded sequence in $L^1(\Omega)$
      which is *uniformly integrable*, that is,
$$\lim_{r \to \infty} \sup_{n \geqslant 1} \|\mathbf{1}_{\{|f_n| > r\}} f\|_1 = 0.$$

      Show that $(f_n)_{n \geqslant 1}$ contains a weakly convergent subsequence by completing the
      details of the following argument.

   (a) For $k = 1, 2, \ldots$ the sequence defined by $f_n^{(k)} := \mathbf{1}_{\{|f_n| \leqslant k\}} f_n$ contains a subsequence that is weakly convergent in $L^2(\Omega)$, and hence weakly convergent in
       $L^1(\Omega)$. Denote by $f^{(k)}$ their weak limits in $L^1(\Omega)$.
   (b) Show that $\|f^{(k)} - f^{(\ell)}\|_1 \leqslant \liminf_{n \to \infty} \|f_n^{(k)} - f_n^{(\ell)}\|_1$ and the latter tends to 0
       by uniform integrability.
   (c) Conclude that the limit $\lim_{k \to \infty} f^{(k)} = f$ exists in $L^1(\Omega)$ and that $\lim_{n \to \infty} f_n = f$ weakly in $L^1(\Omega)$.

4.35 Deduce Theorem 4.52 from Theorem 4.34.

4.36 Prove the various identifications made in the discussion following Example 4.57.

4.37 Prove that if Y is a closed subspace of a reflexive Banach space X, then the quotient space X/Y is reflexive.

4.38 Show that if X is a Banach lattice, then an element $x \in X$ satisfies $x \geqslant 0$ if and only if $\langle x, x^* \rangle \geqslant 0$ for all $x^* \in X^*$ satisfying $x^* \geqslant 0$.

4.39 Show that if X is a Banach lattice and $x^* \in X^*$ satisfies $x^* \geqslant 0$, then for all $x \in X$ we have the following assertions:

(a) $\langle x^+, x^* \rangle = \sup\{\langle x, y^* \rangle : 0 \leqslant y^* \leqslant x^*\}$;

(b) $\langle |x|, x^* \rangle = \sup\{\langle x, y^* \rangle : |y^*| \leqslant x^*\}$.

4.40 Under the assumptions of Theorem 4.2, let $\mu \in M_R(X)$ represent the functional $\phi \in (C_0(X))^*$. Show that the measures representing ϕ^+, ϕ^-, and $|\phi|$, are μ^+, μ^-, and $|\mu|$, respectively.

4.41 For $1 \leqslant p < \infty$ consider the space $\ell^p[0, 1]$ introduced in Problem 2.34.

(a) Show that there is a natural isometric isomorphism $(\ell^p[0, 1])^* \simeq \ell^q[0, 1]$, where $\frac{1}{p} + \frac{1}{q} = 1$.

(b) Show that the function $F : [0, 1] \to \ell^p[0, 1]$ given by

$$(F(t))(s) = \begin{cases} 1, & s = t, \\ 0, & s \neq t, \end{cases}$$

has the following properties:

(i) $t \mapsto \langle F(t), g \rangle$ is measurable for all $g \in \ell^q[0, 1]$;

(ii) $t \mapsto F(t)$ fails to be strongly measurable.

4.42 Let $(A_n)_{n \geqslant 1}$ be a sequence of disjoint intervals of positive measure $|A_n|$ in the interval $[0, 1]$ and define $f : [0, 1] \to c_0$ by

$$f(t) = \sum_{n \geqslant 1} \frac{1}{|A_n|} \mathbf{1}_{A_n}(t) u_n,$$

where $(u_n)_{n \geqslant 1}$ is the sequence of standard unit vectors of c_0.

(a) Show that f is strongly measurable.

(b) Show that for all $x^* \in c_0^* = \ell^1$ the integral $\int_0^1 \langle f(t), x^* \rangle \, dt$ is well defined.

(c) Show that f fails to be Bochner integrable.

4.43 Consider the mapping $f : (0, 1) \to L^\infty(0, 1)$ given by $f(t) := \mathbf{1}_{(0, t)}$.

(a) Show that $\langle f, x^* \rangle$ is measurable for all $x^* \in (L^\infty(0, 1))^*$.

Hint: Monotone scalar-valued functions are measurable.

(b) Show that f fails to be Bochner integrable.

5

Bounded Operators

In the first chapter, bounded operators have been introduced and some of their basic properties were established. This chapter treats some of their deeper properties. In Sections 5.1–5.3 we begin with three results, each of which expresses a certain robustness property of the class of bounded operators: the uniform boundedness theorem (Theorem 5.2), the open mapping theorem (Theorem 5.8), and the closed graph theorem (Theorem 5.12). Completeness plays a critical role through their dependence on the Baire category theorem. In Section 5.4 we present the fourth main result of this chapter, the closed range theorem (Theorem 5.15).

As simple as the definition of a bounded operator may seem, in practice it can be quite hard to establish boundedness of a given linear operator. This applies in particular to some of the most important operators in Analysis, such as the Fourier–Plancherel transform and the Hilbert transform. Their properties are studied in fair detail in Sections 5.5 and 5.6. The final Section 5.7 discusses the Riesz–Thorin interpolation theorem (Theorem 5.38) and its companion, the Marcinkiewicz interpolation theorem (Theorem 5.46).

5.1 The Uniform Boundedness Theorem

The proof of the uniform boundedness theorem, as well as the proofs of some other results in this chapter, depend on the Baire category theorem.

5.1.a The Baire Category Theorem

Theorem 5.1 (Baire category theorem). *Let X be a nonempty complete metric space. Let $F_1, F_2, \ldots$ be closed subsets of X such that*

$$X = \bigcup_{n \geqslant 1} F_n.$$

Then at least one of the sets F_n has nonempty interior.

Proof   Assuming that all sets F_n have empty interior, we prove the existence of an $x \in X$ not contained in any one of the F_n's.

Pick an $x_1 \in \complement F_1$. This is possible, for otherwise we have $F_1 = X$ and F_1 contains open balls. Since F_1 is closed, $\complement F_1$ is open and therefore contains an open ball $B(x_1; r_1)$. By shrinking the radius a bit, we may even assume that the closed ball $\overline{B}(x_1; r_1)$ is contained in $\complement F_1$ and, moreover, that $0 < r_1 \leqslant 1$. The ball $B(x_1; r_1)$ is not contained in F_2 and consequently the open set $B(x_1; r_1) \setminus F_2$ is nonempty. By the same reasoning as before, this set contains a closed ball $\overline{B}(x_2; r_2)$ with radius $0 < r_2 \leqslant \frac{1}{2}$. Continuing in this way we obtain a decreasing sequence of closed balls $\overline{B}(x_1; r_1) \supseteq \overline{B}(x_2; r_2) \supseteq \dots$ with $0 < r_n \leqslant \frac{1}{n}$. The sequence $(x_n)_{n \geqslant 1}$ is a Cauchy sequence, and therefore has a limit x, by the completeness of X. It is clear that $x \in \bigcap_{n \geqslant 1} \overline{B}(x_n; r_n)$, and therefore $x \notin \bigcup_{n \geqslant 1} F_n$. $\square$

5.1.b The Uniform Boundedness Theorem

The uniform boundedness theorem infers uniform boundedness of a family of bounded operators from their pointwise boundedness.

Theorem 5.2 (Uniform boundedness theorem). *Let $(T_i)_{i \in I}$ be a family of bounded operators from a Banach space X into a normed space Y. If*

$$\sup_{i \in I} \|T_i x\| < \infty, \quad x \in X,$$

then

$$\sup_{i \in I} \|T_i\| < \infty.$$

Proof   For each $i \in I$ the sets $\{x \in X : \|T_i x\| \leqslant n\}$ are closed by the continuity of the operator T_i. Since the intersection of closed sets is closed, the sets

$$F_n := \left\{x \in X : \sup_{i \in I} \|T_i x\| \leqslant n\right\} = \bigcap_{i \in I} \{x \in X : \|T_i x\| \leqslant n\}$$

are closed. Moreover, their union equals X. By the Baire category theorem, at least one of them, say F_{n_0}, has nonempty interior. Accordingly there exist $x_0 \in X$ and $r_0 > 0$ such that $B(x_0; r_0) \subseteq F_{n_0}$.

Fix an index $i \in I$. For any $x \in X$ with norm $\|x\| < r_0$ we write $x = x_0 - (x_0 - x)$ and note that both x_0 and $x_0 - x$ belong to $B(x_0; r_0)$. As a consequence,

$$\|T_i x\| \leqslant \|T_i x_0\| + \|T_i(x_0 - x)\| \leqslant n_0 + n_0 = 2n_0.$$

Hence, for all $x \in X$ with norm $\|x\| < 1$,

$$\|T_i x\| \leqslant 2n_0 / r_0.$$

This implies that $\|T_i\| \leqslant 2n_0/r_0$. This being true for all $i \in I$, we have shown that $\sup_{i \in I} \|T_i\| \leqslant 2n_0/r_0$. □

We continue with some typical applications.

Proposition 5.3. *Let X be a Banach space, Y be a normed space, and suppose that $T_n : X \to Y$, $n \geqslant 1$, are bounded operators such that $\lim_{n \to \infty} T_n x =: Tx$ exists for all $x \in X$. Then the operators T_n, $n \geqslant 1$, are uniformly bounded, the mapping $x \mapsto Tx$ is linear and bounded, and*

$$\|T\| \leqslant \liminf_{n \to \infty} \|T_n\|.$$

Proof The uniform boundedness theorem implies $\sup_n \|T_n\| < \infty$. The remaining assertions are proved by the argument of Proposition 1.19. □

Proposition 5.4 (Boundedness of bilinear mappings). *Let X, Y, Z be normed spaces and suppose that at least one of the spaces X and Y is a Banach space. Let $B : X \times Y \to Z$ be linear and bounded in both variables separately. Then there exists a constant $C \geqslant 0$ such that*

$$\|B(x, y)\| \leqslant C \|x\| \|y\|, \quad x \in X, \ y \in Y.$$

In particular, B is jointly continuous.

Proof Assume that X is a Banach space (if Y is a Banach space we interchange the roles of X and Y). For each $y \in Y$, $T_y x := B(x, y)$ defines an element of $\mathscr{L}(X, Z)$ since B is bounded in its first variable. Also, for each $x \in X$ we have $\sup_{\|y\| \leqslant 1} \|T_y x\| < \infty$ since B is bounded in its second variable. Since X is a Banach space, the uniform boundedness theorem shows that $\{T_y : \|y\| \leqslant 1\}$ is uniformly bounded in $\mathscr{L}(X, Z)$. With $M := \sup_{\|y\| \leqslant 1} \|T_y\|$ we then obtain, for all $y \in Y$ with $\|y\| \leqslant 1$,

$$\|B(x, y)\| = \|T_y x\| \leqslant M \|x\|.$$

By a scaling argument for the second variable, this implies the claim as stated. □

The same proof works if we assume that B is linear in the first variable, conjugate-linear in the second variable, and bounded in both variables separately; this observation will be useful in the context of Hilbert spaces.

The following proposition and its corollary give an application of the uniform boundedness theorem to duality.

Proposition 5.5 (Weakly bounded sets are bounded). *A subset S of a Banach space X is bounded if and only if it is weakly bounded, that is, the set $\langle S, x^* \rangle := \{\langle x, x^* \rangle : x \in S\}$ is bounded for all $x^* \in X^*$.*

Proof Only the 'if' part needs proof. Suppose that S is weakly bounded. Its image $J(S)$ in X^{**} under the natural isometric embedding $J : X \to X^{**}$ (see Section 4.2) is a family of bounded operators from X^* to $\mathbb{K}$, and by assumption these operators are pointwise bounded. By the uniform boundedness theorem, these operators are uniformly bounded in operator norm. This is the same as saying that $J(S)$ is a bounded subset of X^{**}. Since J is isometric, it follows that S is bounded. □

Corollary 5.6. *Let X be a Banach space. The following assertions hold:*

(1) *if $\lim_{n\to\infty}\langle x,x_n^*\rangle = \langle x,x^*\rangle$ for all $x \in X$, then $(x_n^*)_{n\geqslant 1}$ is bounded;*
(2) *if $\lim_{n\to\infty}\langle x_n,x^*\rangle = \langle x,x^*\rangle$ for all $x^* \in X^*$, then $(x_n)_{n\geqslant 1}$ is bounded.*

Proof Part (1) follows directly from the uniform boundedness theorem and part (2) is a special case of Proposition 5.5. □

5.2 The Open Mapping Theorem

The next main theorem is the open mapping theorem. Among other things it implies that a bijective bounded operator between Banach spaces has a bounded inverse (and hence is an isomorphism). Its proof relies on the following lemma, in which we use subscripts to tell apart open balls in X and Y.

Lemma 5.7. *Let X be a Banach space, Y a normed space, and let $T \in \mathscr{L}(X,Y)$ be a bounded operator. If $0 < r, R < \infty$ are such that*

$$B_Y(0;r) \subseteq \overline{T(B_X(0;R))},$$

then

$$B_Y(0;r) \subseteq T(B_X(0;2R)).$$

As is apparent from the proof, the constant 2 may be replaced by $1 + \varepsilon$ for any fixed $\varepsilon > 0$.

Proof Fix an arbitrary $y_0 \in B_Y(0;r)$. Then $y_0 \in \overline{T(B_X(0;R))}$, so we can write

$$y_0 = Tx_1 + y_1 \quad \text{with } x_1 \in B_X(0;R) \text{ and } \|y_1\| < \tfrac{1}{2}r.$$

Then $2y_1 \in B_Y(0;r)$, so

$$2y_1 = Tx_2 + y_2 \quad \text{with } x_2 \in B_X(0;R) \text{ and } \|y_2\| < \tfrac{1}{2}r.$$

Then $2y_2 \in B_Y(0;r)$, so

$$2y_2 = Tx_3 + y_3 \quad \text{with } x_3 \in B_X(0;R) \text{ and } \|y_3\| < \tfrac{1}{2}r.$$

Continuing this way, for all $N \in \mathbb{N}$ we obtain

$$\begin{aligned}
y_0 &= Tx_1 + y_1 \\
&= Tx_1 + \tfrac{1}{2}Tx_2 + \tfrac{1}{2}y_2 \\
&= Tx_1 + \tfrac{1}{2}Tx_2 + \tfrac{1}{4}Tx_3 + \tfrac{1}{4}y_3 \\
&= \dots \\
&= Tx_1 + \tfrac{1}{2}Tx_2 + \tfrac{1}{4}Tx_3 + \dots + \tfrac{1}{2^N}Tx_{N+1} + \tfrac{1}{2^N}y_{N+1}.
\end{aligned}$$

Clearly, $\lim_{N \to \infty} \frac{1}{2^N} y_{N+1} = 0$ and

$$\sum_{k=0}^{\infty} \frac{1}{2^k} \|x_{k+1}\| < \sum_{k=0}^{\infty} \frac{R}{2^k} = 2R.$$

This implies that the sum $\sum_{k=0}^{\infty} \frac{1}{2^k} x_{k+1}$ converges in X, by the completeness of X. The boundedness of T implies that the sum $\sum_{k=0}^{\infty} \frac{1}{2^k} Tx_{k+1}$ converges in Y to $T \sum_{k=0}^{\infty} \frac{1}{2^k} x_{k+1}$, and therefore

$$y_0 = \lim_{N \to \infty} \left(\left(\sum_{k=0}^{N} \frac{1}{2^k} Tx_{k+1} \right) + \frac{1}{2^N} y_{N+1} \right) = T \sum_{k=0}^{\infty} \frac{1}{2^k} x_{k+1} \in T(B_X(0;2R)).$$

$\square$

Theorem 5.8 (Open mapping theorem). *Let X and Y be Banach spaces. If $T \in \mathscr{L}(X,Y)$ is bounded and surjective, then T maps open sets to open sets.*

Proof Set $F_n := \overline{T(B_X(0;n))}$. The surjectivity of T implies that $Y = \bigcup_{n \geqslant 1} F_n$. Therefore, by the Baire category theorem (Theorem 5.1), some F_{n_0} has nonempty interior. This means that there exist $y_0 \in Y$ and $r_0 > 0$ such that

$$B_Y(y_0;r_0) \subseteq \overline{T(B_X(0;n_0))}.$$

In view of $T(-x) = -Tx$, we then also have

$$B_Y(-y_0;r_0) \subseteq \overline{T(B_X(0;n_0))}.$$

Writing $y = \frac{1}{2}(y_0 + y) + \frac{1}{2}(-y_0 + y)$, it follows that

$$B_Y(0;r_0) \subseteq \tfrac{1}{2} \cdot \overline{T(B_X(0;n_0))} + \tfrac{1}{2} \cdot \overline{T(B_X(0;n_0))} = \overline{T(B_X(0;n_0))}.$$

Now we can invoke the lemma and find

$$B_Y(0;r_0) \subseteq T(B_X(0;2n_0)).$$

Let U be an open set in X; we wish to prove that $T(U)$ is open. To this end let $Tx \in T(U)$ be given, with $x \in U$; we wish to prove that $T(U)$ contains the open ball $B_Y(Tx;\rho)$ for some $\rho > 0$.

Since U is open, there is an $\varepsilon > 0$ such that $B_X(x;\varepsilon) \subseteq U$. Let $\delta := \varepsilon/(2n_0)$. Then

$T(U)$ contains $Tx + T(B_X(0;\varepsilon)) = Tx + \delta T(B_X(0;2n_0))$, and the latter contains the open ball $Tx + \delta B_Y(0;r_0) = B_Y(Tx;\delta r_0)$. $\qquad\square$

Corollary 5.9. *Let X and Y be Banach spaces. If $T \in \mathscr{L}(X,Y)$ is bijective, then T is an isomorphism.*

Proof The fact that T maps open sets to open sets can now be reformulated as saying that $(T^{-1})^{-1}(U)$ is open for every open set U, so T^{-1} is continuous. $\qquad\square$

It was noted in Proposition 4.15 that if a subspace X_0 of a normed space X is the range of a projection in X, then X_0 is complemented in X. As an application of the closed graph theorem we prove the following converse:

Proposition 5.10. *A closed subspace of a Banach space X is complemented if and only if it is the range of a projection in X.*

Proof It remains to prove the 'only if' part. If $X = X_0 \oplus X_1$ is a direct sum decomposition, then $\|\|x\|\| := \|x_0\| + \|x_1\|$, with $x = x_0 + x_1$ along the decomposition $X = X_0 \oplus X_1$, defines a complete norm on X, and the mapping $x \mapsto x$ is bounded (in fact, contractive) from $(X, \|\| \cdot \|\|)$ to X by the triangle inequality. By Corollary 5.9, its inverse is bounded as well. The boundedness of the projections π_0 and π_1 from X to X_0 and X_1 immediately follows from this, noting that they are bounded (in fact, contractive) from $(X, \|\| \cdot \|\|)$ to X_0 and X_1. $\qquad\square$

5.3 The Closed Graph Theorem

Let X and Y be Banach spaces. The *graph* of a mapping $T : X \to Y$ is the set

$$\mathsf{G}(T) := \{(x, Tx) : x \in X\}$$

in $X \times Y$. If T is linear, $\mathsf{G}(T)$ is a linear subspace of $X \times Y$. Endowing $X \times Y$ with the norm

$$\|(x,y)\|_1 := \|x\| + \|y\| \tag{5.1}$$

turns this space into a Banach space and it is easy to check that if T is bounded, then $\mathsf{G}(T)$ is closed $X \times Y$. Since all product norms on $X \times Y$ are equivalent (see Example 1.33), the particular choice of product norm made in (5.1) is immaterial.

Definition 5.11 (Closed operators). A linear operator $T : X \to Y$ is *closed* if its graph is closed in $X \times Y$.

Every bounded linear operator is closed. In the converse direction we have the following result.

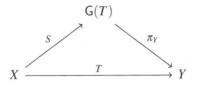

Figure 5.1 Proof of the closed graph theorem

Theorem 5.12 (Closed graph theorem). *Let X and Y be Banach spaces. If a linear operator $T : X \to Y$ is closed, then T is bounded.*

Proof Let $\pi_X : X \times Y \to X$ and $\pi_Y : X \times Y \to Y$ be given by $\pi_X(x,y) := x$ and $\pi_Y(x,y) := y$. Both mappings are bounded and of norm one. By assumption $Z := G(T)$ is a closed subspace of $X \times Y$, hence a Banach space with respect to the inherited norm. Consider the linear operator $S : X \to Z$ given by $Sx := (x, Tx)$. This operator is a bijection whose inverse S^{-1} is the bounded operator π_X. By Corollary 5.9 the inverse S of S^{-1} is bounded. Hence also $T = \pi_Y \circ S$ is bounded. $\qquad\square$

As an application of the closed graph theorem we prove the following variation of Proposition 2.26.

Proposition 5.13. *Let $1 \leqslant p,q \leqslant \infty$ satisfy $\frac{1}{p} + \frac{1}{q} = 1$. Let (Ω, μ) be a measure space, which is assumed to be σ-finite if $p = \infty$. A measurable function f belongs to $L^p(\Omega)$ if and only if $fg \in L^1(\Omega)$ for all $g \in L^q(\Omega)$. In that case we have*

$$\|f\|_p = \sup_{\|g\|_q \leqslant 1} \int_{\Omega} |fg| \, d\mu.$$

Proof The 'only if' part is immediate from Hölder's inequality. To prove the 'if' part we may assume that f is not identically 0. Using Corollary 2.21, the mapping $g \mapsto fg$ is easily seen to be closed as a mapping from $L^q(\Omega)$ to $L^1(\Omega)$: if $g_n \to g$ in $L^q(\Omega)$ and $fg_n \to h$ in $L^1(\Omega)$, we may pass to a subsequence such that $g_{n_k} \to g$ and $fg_{n_k} \to h$ μ-almost everywhere, and therefore $fg = \lim_{n\to\infty} fg_n = h$ μ-almost everywhere. By the closed graph theorem, the operator $g \mapsto fg$ is bounded. It follows that the assumptions of Proposition 2.26 are satisfied, with M the norm of the operator $g \mapsto fg$. This gives that $g \in L^q(\Omega)$ with bound

$$\|f\|_p \leqslant \sup_{\|g\|_q \leqslant 1} \int_{\Omega} |fg| \, d\mu.$$

Hölder's inequality gives the opposite bound. $\qquad\square$

As a further illustration of the use of the closed graph theorem, let us deduce Proposition 5.10 from it. Let $X = X_0 \oplus X_1$ be a direct sum decomposition, and consider the linear mapping $\pi_0 : (x_0, x_1) \mapsto x_0$. In what follows we suggestively write (x_0, x_1) for the element $x_0 + x_1$ of X. To prove that π_0 is bounded, we shall prove that its graph is closed. Suppose that $(x_0^n, x_1^n) \to (x_0, x_1)$ and $(x_0^n, 0) \to (y_0, y_1)$ in X. Then also

$$(0, x_1^n) = (x_0^n, x_1^n) - (x_0^n, 0) \to (x_0, x_1) - (y_0, y_1) = (x_0 - y_0, x_1 - y_1)$$

in X. The closedness of X_0 and X_1 in X implies that $(y_0, y_1) = \lim_{n \to \infty}(x_0^n, 0)$ belongs to X_0 and $(x_0 - y_0, x_1 - y_1) = \lim_{n \to \infty}(0, x_1^n)$ belongs to X_1. This forces $y_0 = x_0$ and $y_1 = 0$. It follows that $(y_0, y_1) = (x_0, 0) = \pi_0(x_0, x_1)$. This proves that π_0 is closed. Therefore, by the closed graph theorem, π_0 is bounded.

5.4 The Closed Range Theorem

As a warm-up for the main result of this section we begin with a simple application of the Hahn–Banach theorem. Recall that the *annihilator* of a subset A of X is the set

$$A^\perp := \{x^* \in X^* : \langle x, x^* \rangle = 0 \text{ for all } x \in A\}$$

and the *pre-annihilator* of a subset B of X^* is the set

$${}^\perp B := \{x \in X : \langle x, x^* \rangle = 0 \text{ for all } x^* \in B\}.$$

Proposition 5.14. *For any operator $T \in \mathscr{L}(X, Y)$, where X and Y are Banach spaces, we have*

$$\overline{R(T)} = {}^\perp(N(T^*)).$$

In particular, T has dense range if and only if T^* is injective.

Proof $\subseteq$: If $y = Tx \in R(T)$, then for all $y^* \in N(T^*)$ we have $\langle y, y^* \rangle = \langle x, T^* y^* \rangle = 0$, and therefore $y \in {}^\perp(N(T^*))$. This proves $R(T) \subseteq {}^\perp(N(T^*))$. The result now follows from the fact that ${}^\perp(N(T^*))$ is closed.

$\supseteq$: If $x_0 \notin \overline{R(T)} =: Y_0$, by Corollary 4.12 there exists a $y^* \in Y^*$ such that $\langle x_0, y^* \rangle \neq 0$ and $y^*|_{Y_0} \equiv 0$. For all $x \in X$, $\langle x, T^* y^* \rangle = \langle Tx, y^* \rangle = 0$ and therefore $y^* \in N(T^*)$. Since $\langle x_0, y^* \rangle \neq 0$ we have $x_0 \notin {}^\perp(N(T^*))$. $\qquad\square$

In Sections 7.2 and 7.3 we will encounter interesting classes of operators whose ranges are closed. For such operators, the closed range theorem provides a 'dual' variant of Proposition 5.14 which is considerably harder to prove. We need this theorem in our discussion of duality of Fredholm operators in Section 7.3.

Theorem 5.15 (Closed range theorem). *Let the operator $T \in \mathscr{L}(X, Y)$ be given, where X and Y are Banach spaces. If $R(T)$ is closed, then*

$$R(T^*) = (N(T))^\perp.$$

As a consequence, $R(T^*)$ is weak closed.*

Proof Once we have proved the identity $R(T^*) = (N(T))^{\perp}$, the weak* closedness of $R(T^*)$ follows from the general observation that annihilators are weak* closed.

$\subseteq$: If $x^* = T^*y^* \in R(T^*)$, then for all $x \in N(T)$ and we have $\langle x, x^* \rangle = \langle Tx, y^* \rangle = 0$. This shows that $x^* \in (N(T))^{\perp}$.

$\supseteq$: Suppose that $x^* \in (N(T))^{\perp}$. For elements $y = Tx \in R(T)$ we define

$$\phi(y) := \langle x, x^* \rangle.$$

To see that this is well defined, suppose that we also have $y = Tx'$. Then $T(x - x') = y - y = 0$ implies $x - x' \in N(T)$ and therefore $\langle x - x', x^* \rangle = 0$. This gives the well-definedness as claimed.

For all $z \in N(T)$ we have $\phi(y) = \langle x - z, x^* \rangle$ and therefore

$$|\phi(y)| \leqslant \|x - z\| \|x^*\|.$$

By taking the infimum over all $z \in N(T)$ we obtain

$$|\phi(y)| \leqslant d(x, N(T)) \|x^*\|.$$

We claim that the closedness of the range of T implies the existence of a constant $C \geqslant 0$ such that

$$d(x, N(T)) \leqslant C\|Tx\|. \tag{5.2}$$

Taking this for granted for the moment, we first show how to complete the proof. From (5.2) we obtain the estimate

$$|\phi(y)| \leqslant C\|Tx\| \|x^*\| = C\|y\| \|x^*\|,$$

proving that ϕ is bounded as a functional defined on $R(T)$. By the Hahn–Banach theorem, we obtain an element $y^* \in Y^*$ extending ϕ. For all $x \in X$ we obtain

$$\langle x, T^*y^* \rangle = \langle Tx, y^* \rangle = \phi(Tx) = \langle x, x^* \rangle,$$

so $x^* = T^*y^* \in R(T^*)$.

It remains to prove (5.2). For this we note that

$$d(x, N(T)) = \|x + N(T)\|_{X/N(T)}. \tag{5.3}$$

The operator T induces a well-defined and bounded quotient operator $T_/$ from $X/N(T)$ onto $R(T)$ by setting $T_/(x + N(T)) := Tx$. It is clear that $T_/$ is injective and surjective, and since $R(T)$ is closed we may apply the open mapping theorem and obtain that $T_/$ maps $X/N(T)$ isomorphically onto $R(T)$. Denoting by C the norm of its inverse we obtain the desired estimate from (5.3) and

$$\|x + N(T)\|_{X/N(T)} \leqslant C\|T_/(x + N(T))\| = C\|Tx\|.$$

This completes the proof of (5.2).   □

5.5 The Fourier Transform

In the present section and the next we study two nontrivial examples of bounded operators: the Fourier–Plancherel transform and the Hilbert transform. It is not an exaggeration to state that, at least from the point of view of the theory of partial differential equations, these rank among the most important bounded operators in all of Analysis.

5.5.a Definition and General Properties

Definition 5.16 (Fourier transform). The *Fourier transform* of a function $f \in L^1(\mathbb{R}^d)$ is the function $\widehat{f} : \mathbb{R}^d \to \mathbb{C}$ defined by

$$\widehat{f}(\xi) := \frac{1}{(2\pi)^{d/2}} \int_{\mathbb{R}^d} f(x) \exp(-ix \cdot \xi) \, dx, \quad \xi \in \mathbb{R}^d, \tag{5.4}$$

where $x \cdot \xi := \sum_{j=1}^{d} x_j \xi_j$.

It is evident that $\widehat{f} \in L^\infty(\mathbb{R}^d)$ and $\|\widehat{f}\|_\infty \leqslant (2\pi)^{-d/2}\|f\|_1$. This shows that the operator $\mathscr{F} : f \mapsto \widehat{f}$, which will be referred to as the *Fourier transform*, defines a bounded operator from $L^1(\mathbb{R}^d)$ to $L^\infty(\mathbb{R}^d)$.

Remark 5.17. In certain situations it is useful to absorb the constant $(2\pi)^{-d/2}$ into the measure. Denoting the resulting *normalised Lebesgue measure* by

$$dm(x) = (2\pi)^{-d/2} \, dx,$$

one may interpret the Fourier transform as the operator from $L^1(\mathbb{R}^d, m) \to L^\infty(\mathbb{R}^d, m)$ given by

$$\widehat{f}(\xi) := \int_{\mathbb{R}^d} f(x) \exp(-ix \cdot \xi) \, dm(x), \quad \xi \in \mathbb{R}^d.$$

The advantage of this point of view is that this operator is contractive. In many applications, however, working with the normalised Lebesgue measure is somewhat artificial, and for this reason we stick with (5.4) most of the time.

The dominated convergence theorem implies that for all $f \in L^1(\mathbb{R}^d)$ the function $\widehat{f}$ is sequentially continuous, hence continuous. More is true: the following lemma shows that $\widehat{f}$ belongs to $C_0(\mathbb{R}^d)$, the space of continuous functions vanishing at infinity.

Theorem 5.18 (Riemann–Lebesgue lemma). *For all $f \in L^1(\mathbb{R}^d)$ we have $\widehat{f} \in C_0(\mathbb{R}^d)$.*

Proof By separation of variables one sees that $\lim_{|\xi|\to\infty} |\widehat{f}(\xi)| = 0$ for step functions $f = \sum_{i=1}^{n} c_i \mathbf{1}_{Q_i}$ where $n \geqslant 1$, $c_i \in \mathbb{C}$ and Q_i cubes with sides parallel to the coordinate axes $(1 \leqslant i \leqslant n)$. Indeed, if $Q = \prod_{j=1}^{d} [a_j, b_j]$ is such a cube, then

$$\widehat{f}(\xi) = \frac{1}{(2\pi)^{d/2}} \int_{\mathbb{R}^d} \prod_{j=1}^{d} \mathbf{1}_{[a_j,b_j]} \exp(-ix_j\xi_j)\, dx$$

$$= \frac{1}{(2\pi)^{d/2}i} \prod_{j=1}^{d} \frac{1}{\xi_j} (\exp(-ia_j\xi_j) - \exp(-ib_j\xi_j))$$

$$= \frac{1}{(2\pi)^{d/2}i} \prod_{j=1}^{d} \exp(-ib_j\xi_j) \frac{\exp(i(b_j - a_j)\xi_j) - 1}{\xi_j}.$$

If $|\xi| \geqslant r$, then at least one coordinate satisfies $|\xi_{j_0}| \geqslant r/\sqrt{d}$ and then

$$|\widehat{f}(\xi)| \leqslant \frac{1}{(2\pi)^{d/2}} \frac{2\sqrt{d}}{r} \prod_{j \neq j_0} M_j,$$

where the constants

$$M_j = \sup_{y \in \mathbb{R} \setminus \{0\}} \left| \frac{\exp(i(b_j - a_j)y) - 1}{y} \right|$$

are finite. This proves that $\widehat{f} \in C_0(\mathbb{R}^d)$ as claimed.

Since the functions of the form considered above are dense in $L^1(\mathbb{R}^d)$ by (the proof of) Proposition 2.29, and since $C_0(\mathbb{R}^d)$ is a closed subspace of $L^\infty(\mathbb{R}^d)$, by Proposition 1.18 the Fourier transform extends uniquely to a bounded operator from $L^1(\mathbb{R}^d)$ to $C_0(\mathbb{R}^d)$. Identifying $C_0(\mathbb{R}^d)$ with a closed subspace of $L^\infty(\mathbb{R}^d)$, this extension agrees with the Fourier transform. $\qquad\square$

We continue with an inversion theorem for the Fourier transform. Its proof is based on a simple lemma.

Lemma 5.19. *For $\lambda > 0$ the Fourier transform of the function*

$$g^{(\lambda)}(x) = \frac{1}{(2\pi)^{d/2}} \exp\left(-\frac{1}{2}\lambda|x|^2\right)$$

equals

$$\widehat{g^{(\lambda)}}(\xi) = \frac{1}{(2\pi\lambda)^{d/2}} \exp\left(-\frac{1}{2}|\xi|^2/\lambda\right).$$

Proof First let $d = 1$. Completing squares and using Cauchy's theorem to shift the path of integration, we find

$$\frac{1}{(2\pi)^{1/2}} \int_{\mathbb{R}} g^{(\lambda)}(x) \exp(-ix \cdot \xi)(x)\, dx$$

$$= \frac{1}{2\pi} \int_{\mathbb{R}} \exp\left(-\frac{1}{2}\lambda\left[(x+i\xi/\lambda)^2 + \xi^2/\lambda^2\right]\right) dx$$

$$= \frac{1}{2\pi} \exp\left(-\frac{1}{2}\xi^2/\lambda\right) \int_{\mathbb{R}+i\xi/\lambda} \exp\left(-\frac{1}{2}\lambda z^2\right) dz$$

$$= \frac{1}{2\pi} \exp\left(-\frac{1}{2}\xi^2/\lambda\right) \int_{\mathbb{R}} \exp\left(-\frac{1}{2}\lambda z^2\right) dz = \frac{1}{(2\pi\lambda)^{1/2}} \exp\left(-\frac{1}{2}\xi^2/\lambda\right).$$

The general case follows from this by separation of variables. $\qquad\square$

A different proof based on the Picard–Lindelöf theorem is outlined in Problem 5.19.

Theorem 5.20 (Fourier inversion theorem). *If $f \in L^1(\mathbb{R}^d)$ satisfies $\widehat{f} \in L^1(\mathbb{R}^d)$, then for almost all $x \in \mathbb{R}^d$ we have the identity*

$$f(x) = \frac{1}{(2\pi)^{d/2}} \int_{\mathbb{R}^d} \widehat{f}(\xi) \exp(ix\cdot\xi) \, d\xi.$$

In particular this result implies that the Fourier transform is injective as a mapping from $L^1(\mathbb{R}^d)$ to $L^\infty(\mathbb{R}^d)$. A more general injectivity result will be proved in Theorem 5.30.

Proof   By Lemma 5.19, the function $g(x) = (2\pi)^{-d/2} \exp\left(-\frac{1}{2}|x|^2\right)$ satisfies

$$g(x) = \widehat{g}(x) = \frac{1}{(2\pi)^{d/2}} \int_{\mathbb{R}^d} g(\xi) \exp(-ix\cdot\xi) \, d\xi = \frac{1}{(2\pi)^{d/2}} \int_{\mathbb{R}^d} g(\xi) \exp(ix\cdot\xi) \, d\xi,$$

where the last identity uses that g is real-valued, so that taking complex conjugates leaves the expression unchanged. Substituting x/λ for x we obtain

$$g_\lambda(x) := \lambda^{-d} g(\lambda^{-1}x) = \frac{1}{(2\pi)^{d/2}} \int_{\mathbb{R}^d} g(\lambda\xi) \exp(ix\cdot\xi) \, d\xi.$$

By Proposition 2.34 and Corollary 2.21, after passing to an appropriate subsequence $\lambda_j \downarrow 0$ we have $g_{\lambda_j} * f(x) \to f(x)$ for almost all $x \in \mathbb{R}^d$ as $j \to \infty$. Using the above, it follows that for almost all $x \in \mathbb{R}^d$ we have

$$f(x) = \lim_{j\to\infty} \int_{\mathbb{R}^d} g_{\lambda_j}(y) f(x-y) \, dy$$

$$= \lim_{j\to\infty} \int_{\mathbb{R}^d} \left(\frac{1}{(2\pi)^{d/2}} \int_{\mathbb{R}^d} g(\lambda_j\xi) \exp(iy\cdot\xi) \, d\xi\right) f(x-y) \, dy$$

$$= \lim_{j\to\infty} \int_{\mathbb{R}^d} \exp(ix\cdot\xi) \left(\frac{1}{(2\pi)^{d/2}} \int_{\mathbb{R}^d} f(x-y) \exp(-i(x-y)\cdot\xi) \, dy\right) g(\lambda_j\xi) \, d\xi$$

$$= \lim_{j\to\infty} \frac{1}{(2\pi)^{d/2}} \int_{\mathbb{R}^d} g(\lambda_j\xi) \exp(ix\cdot\xi) \widehat{f}(\xi) \, d\xi$$

$$= \frac{1}{(2\pi)^{d/2}} \int_{\mathbb{R}^d} \widehat{f}(\xi) \exp(ix\cdot\xi) \, d\xi,$$

where the last step is justified by dominated convergence, which can be used here since $\widehat{f}$ is integrable, g is bounded, and $g(\lambda_j \xi) \to g(0) = 1$ pointwise as $j \to \infty$. $\qquad\square$

The Fourier transform of the translate $\tau_h f$ of a function $f \in L^1(\mathbb{R})$ is given by

$$\widehat{\tau_h f}(\xi) = \frac{1}{\sqrt{2\pi}} \int_{-\infty}^{\infty} f(x+h) \exp(-ix\xi)\, dx = \exp(ih\xi)\widehat{f}(\xi).$$

It follows that if $\widehat{f}(\xi_0) = 0$ for some $\xi_0 \in \mathbb{R}$, then $\widehat{\tau_h f}(\xi_0) = 0$ for all $h \in \mathbb{R}$. Therefore the linear span of the set of translates of f is contained in $\{g \in L^1(\mathbb{R}) : \widehat{g}(\xi_0) = 0\}$, which is a proper closed subspace of $L^1(\mathbb{R})$. Thus, a necessary condition in order that the linear span of the set of translates of a function $f \in L^1(\mathbb{R})$ be dense in $L^1(\mathbb{R})$ is that $\widehat{f}$ be zero-free. Strikingly, this necessary condition is also sufficient. This is the content of the next theorem, which will be proved by operator theoretic methods in Section 13.1.b.

Norbert Wiener, 1894–1964

Theorem 5.21 (Wiener's Tauberian theorem). *If the Fourier transform of a function $f \in L^1(\mathbb{R})$ is zero-free, then the span of the set of all translates of f is dense in $L^1(\mathbb{R})$.*

5.5.b The Plancherel Theorem

The Fourier transform enjoys an important L^2 boundedness property.

Theorem 5.22 (Plancherel, preliminary version). *If $f \in L^1(\mathbb{R}^d) \cap L^2(\mathbb{R}^d)$, then $\widehat{f} \in L^2(\mathbb{R}^d)$ and*

$$\|\widehat{f}\|_2 = \|f\|_2.$$

Proof Since $f \in L^1(\mathbb{R}^d)$, $\widehat{f}$ is bounded and $\xi \mapsto \exp\left(-\frac{1}{2}\lambda|\xi|^2\right)|\widehat{f}(\xi)|^2$ is integrable for all $\lambda > 0$, and

$$\int_{\mathbb{R}^d} \exp\left(-\frac{1}{2}\lambda|\xi|^2\right)|\widehat{f}(\xi)|^2\, d\xi$$

$$= \int_{\mathbb{R}^d} \exp\left(-\frac{1}{2}\lambda|\xi|^2\right)\widehat{f}(\xi)\overline{\widehat{f}(\xi)}\, d\xi$$

$$= \frac{1}{(2\pi)^{d/2}} \int_{\mathbb{R}^d} g^{(\lambda)}(\xi) \int_{\mathbb{R}^d} f(x)\exp(-ix\cdot\xi)\, dx \int_{\mathbb{R}^d} \overline{f(y)}\exp(iy\cdot\xi)\, dy\, d\xi$$

$$= \int_{\mathbb{R}^d} \int_{\mathbb{R}^d} \left(\frac{1}{(2\pi)^{d/2}} \int_{\mathbb{R}^d} g^{(\lambda)}(\xi)\exp(-i(x-y)\xi)\, d\xi\right) f(x)\overline{f(y)}\, dx\, dy$$

$$= \int_{\mathbb{R}^d} \int_{\mathbb{R}^d} \frac{1}{(2\pi\lambda)^{d/2}} \exp\left(-\frac{1}{2}|x-y|^2/\lambda\right) f(x)\overline{f(y)} \, dx \, dy$$

$$= \int_{\mathbb{R}^d} f * \phi_{\sqrt{\lambda}}(y)\overline{f(y)} \, dy,$$

where $\phi_\mu(x) := \mu^{-d}\phi(\mu^{-1}x)$ with $\phi(y) = (2\pi)^{-d/2}\exp\left(-\frac{1}{2}|y|^2\right)$; the change of order of integration is justified by the absolute integrability of the integrand. Applying Proposition 2.34 we find that $\lim_{\lambda\downarrow 0} f * \phi_{\sqrt{\lambda}} = f$ in $L^2(\mathbb{R}^d)$. Then,

$$\lim_{\lambda\downarrow 0} \int_{\mathbb{R}^d} f * \phi_{\sqrt{\lambda}}(y)\overline{f(y)} \, dy = \int_{\mathbb{R}^d} f(y)\overline{f(y)} \, dy = \|f\|_2^2.$$

On the other hand,

$$\lim_{\lambda\downarrow 0} \int_{\mathbb{R}^d} \exp\left(-\frac{1}{2}\lambda|\xi|^2\right)|\widehat{f}(\xi)|^2 \, d\xi = \int_{\mathbb{R}^d} |\widehat{f}(\xi)|^2 \, d\xi = \|\widehat{f}\|_2^2$$

by dominated convergence. This completes the proof. $\qquad\square$

Consider the vector space

$$\mathscr{F}^2(\mathbb{R}^d) := \{f \in L^1(\mathbb{R}^d) \cap L^2(\mathbb{R}^d) : \widehat{f} \in L^1(\mathbb{R}^d) \cap L^2(\mathbb{R}^d)\}.$$

There is some redundancy in this definition, for if $f \in L^1(\mathbb{R}^d) \cap L^2(\mathbb{R}^d)$, then $\widehat{f} \in L^2(\mathbb{R}^d)$ by the Plancherel theorem. The advantage of the above format is that it brings out the symmetry between f and $\widehat{f}$ explicitly. The interest of this space is explained by the following two observations.

Lemma 5.23. *The Fourier transform maps $\mathscr{F}^2(\mathbb{R}^d)$ bijectively into itself.*

Proof Injectivity of $f \mapsto \widehat{f}$ follows from the Plancherel theorem and surjectivity from the Fourier inversion theorem, which implies if $f \in \mathscr{F}^2(\mathbb{R}^d)$, then f is the Fourier transform of the function $\xi \mapsto \widehat{f}(-\xi)$ in $\mathscr{F}^2(\mathbb{R}^d)$. $\qquad\square$

Lemma 5.24. *$\mathscr{F}^2(\mathbb{R}^d)$ is dense in $L^2(\mathbb{R}^d)$.*

Proof Since $C_c^\infty(\mathbb{R}^d)$ is dense in $L^2(\mathbb{R}^d)$ by Proposition 2.29, it suffices to show that every $f \in C_c^\infty(\mathbb{R}^d)$ belongs to $\mathscr{F}^2(\mathbb{R}^d)$. Integrating by parts, for all $f \in C_c^\infty(\mathbb{R}^d)$ and multi-indices $\alpha \in \mathbb{N}^d$ we have

$$\widehat{\partial^\alpha f}(\xi) = i^{|\alpha|}\xi^\alpha \widehat{f}(\xi),$$

where $\partial^\alpha := \partial_1^{\alpha_1} \circ \cdots \circ \partial_d^{\alpha_d}$ with ∂_j the partial derivative in the jth direction, $\xi^\alpha := \xi_1^{\alpha_1} \cdots \xi_d^{\alpha_d}$, and $|\alpha| := \alpha_1 + \cdots + \alpha_d$. Since Fourier transforms of integrable functions are bounded, this implies that $\xi \mapsto (1 + |\xi|^k)\widehat{f}(\xi)$ is bounded for every integer $k \geq 1$. The desired result follows from this. $\qquad\square$

Combining these lemmas with Proposition 1.18, we obtain the following improved version of Theorem 5.22.

Theorem 5.25 (Plancherel). *The restriction of the Fourier transform to $L^1(\mathbb{R}^d) \cap L^2(\mathbb{R}^d)$ has a unique extension to an isometry from $L^2(\mathbb{R}^d)$ onto itself.*

Definition 5.26 (Fourier–Plancherel transform). This isometry of $L^2(\mathbb{R}^d)$ is called the *Fourier–Plancherel transform*.

With slight abuse of notation we denote the Fourier–Plancherel transform again by $\mathscr{F}: f \mapsto \widehat{f}$. It is important to realise that $\widehat{f}$ is no longer given by the pointwise formula (5.4). In fact, for functions $f \in L^2(\mathbb{R}^d)$ the integrand in (5.4) is not even integrable unless $f \in L^1(\mathbb{R}^d) \cap L^2(\mathbb{R}^d)$.

Remark 5.27. Theorem 5.25 also holds with respect to the normalised Lebesgue measure $dm(x) = (2\pi)^{-d/2} dx$: the restriction to $L^1(\mathbb{R}^d, m) \cap L^2(\mathbb{R}^d, m)$ of the Fourier transform as defined in Remark 5.17 extends to an isometry from $L^2(\mathbb{R}^d, m)$ onto itself.

For later use we record two further properties of the Fourier–Plancherel transform.

Proposition 5.28. *For all $f, g \in L^2(\mathbb{R}^d)$ we have*

$$\int_{\mathbb{R}^d} f(x)\widehat{g}(x)\,dx = \int_{\mathbb{R}^d} \widehat{f}(x)g(x)\,dx.$$

Proof For $f, g \in \mathscr{F}^2(\mathbb{R}^d)$ the identity follows by writing out the Fourier transforms and using Fubini's theorem:

$$\int_{\mathbb{R}^d} f(x)\widehat{g}(x)\,dx = \int_{\mathbb{R}^d} \frac{1}{(2\pi)^{d/2}} \int_{\mathbb{R}^d} f(x)g(\xi)\exp(-ix \cdot \xi)\,d\xi\,dx$$

$$= \int_{\mathbb{R}^d} \frac{1}{(2\pi)^{d/2}} \int_{\mathbb{R}^d} f(x)g(\xi)\exp(-ix \cdot \xi)\,dx\,d\xi = \int_{\mathbb{R}^d} \widehat{f}(\xi)g(\xi)\,d\xi.$$

The general case follows by approximation, using that $\mathscr{F}^2(\mathbb{R}^d)$ is dense in $L^2(\mathbb{R}^d)$ by Lemma 5.24. $\qquad\square$

The appearance of the factor $(2\pi)^{d/2}$ in the next proposition is an artefact of our convention to normalise the Fourier transform with this factor, but not the convolution.

Proposition 5.29. *Let $f \in L^1(\mathbb{R}^d)$ and let $g \in L^1(\mathbb{R}^d)$ or $g \in L^2(\mathbb{R}^d)$. For almost all $\xi \in \mathbb{R}^d$ we have*

$$\widehat{f * g}(\xi) = (2\pi)^{d/2}\widehat{f}(\xi)\widehat{g}(\xi).$$

Proof If $g \in C_c(\mathbb{R}^d)$, then $f * g \in L^1(\mathbb{R}^d) \cap L^2(\mathbb{R}^d)$ by Young's inequality, and by Fubini's theorem and a change of variables we obtain

$$\widehat{f * g}(\xi) = \frac{1}{(2\pi)^{d/2}} \int_{\mathbb{R}^d} \left(\int_{\mathbb{R}^d} f(x-y)g(y)\,dy \right) \exp(-ix \cdot \xi)\,dx$$

$$= \int_{\mathbb{R}^d} \left(\frac{1}{(2\pi)^{d/2}} \int_{\mathbb{R}^d} f(x-y)\exp(-i(x-y) \cdot \xi)\,dx \right) \exp(-iy \cdot \xi)g(y)\,dy$$

$$= \int_{\mathbb{R}^d} \left(\frac{1}{(2\pi)^{d/2}} \int_{\mathbb{R}^d} f(u) \exp(-iu \cdot \xi) \, du \right) g(y) \exp(-iy \cdot \xi) \, dy$$
$$= (2\pi)^{d/2} \widehat{f}(\xi) \widehat{g}(\xi).$$

This proves the identity for $g \in C_c(\mathbb{R}^d)$. For $p \in \{1, 2\}$ and $\frac{1}{p} + \frac{1}{q} = 1$, Young's inequality implies that the L^q-function $\widehat{f * g}$ depends continuously on the L^p-norm of g. Since $C_c(\mathbb{R}^d)$ is dense in $L^p(\mathbb{R}^d)$ by Proposition 2.29, it follows that the identity extends to arbitrary functions $g \in L^p(\mathbb{R}^d)$. $\square$

From Proposition 2.43 we know that if μ is a real or complex measure, its variation $|\mu|$ is a finite measure. Accordingly, the Lebesgue integrals of bounded Borel functions with respect to μ are well defined. In particular we can define the *Fourier transform* of a real or complex Borel measure μ on $\mathbb{R}^d$ by

$$\widehat{\mu}(\xi) := \frac{1}{(2\pi)^{d/2}} \int_{\mathbb{R}^d} \exp(-ix \cdot \xi) \, d\mu(x), \quad \xi \in \mathbb{R}^d.$$

From

$$|\widehat{\mu}(\xi)| \leqslant \frac{1}{(2\pi)^{d/2}} \int_{\mathbb{R}^d} d|\mu| = \frac{1}{(2\pi)^{d/2}} \|\mu\|,$$

where $\|\mu\| = |\mu|(\mathbb{R}^d)$ is the variation norm of μ, we see that $\widehat{\mu}$ is a bounded function, and it is continuous by dominated convergence. Thus we have $\widehat{\mu} \in C_b(\mathbb{R}^d)$ and $\|\widehat{\mu}\|_\infty \leqslant (2\pi)^{-d/2} \|\mu\|$. The Riemann–Lebesgue lemma does not extend to the present setting, as is demonstrated by the identity $\widehat{\delta_0} = \mathbf{1}$.

The Fourier inversion theorem (Theorem 5.20) implies that the Fourier transform is injective as an operator from $L^1(\mathbb{R}^d)$ to $L^\infty(\mathbb{R}^d)$. More generally we have the following result.

Theorem 5.30 (Injectivity of the Fourier transform). *If μ is a real or complex Borel measure on $\mathbb{R}^d$ satisfying $\widehat{\mu}(\xi) = 0$ for all $\xi \in \mathbb{R}^d$, then $\mu = 0$.*

Proof To prove that $\mu = 0$, by the uniqueness part of the Riesz Representation theorem (Theorem 4.2) it suffices to show that

$$\int_{\mathbb{R}^d} f \, d\mu = 0, \quad f \in C_c(\mathbb{R}^d). \tag{5.5}$$

Fix $0 < \varepsilon < 1$ and $f \in C_c(\mathbb{R}^d)$. We may assume that $\|f\|_\infty \leqslant 1$. Let $r > 0$ be so large that the support of f is contained in a cube $[-r, r]^d$ satisfying $|\mu|(\complement[-r, r]^d) \leqslant \varepsilon$. By the Stone–Weierstrass theorem (Theorem 2.5) there exists a linear combination $p : \mathbb{R}^d \to \mathbb{C}$ of the functions of the form $x \mapsto \exp(2\pi i kx \cdot \xi / 2r)$ with $\xi \in \mathbb{R}^d$ and $k \in \mathbb{Z}$ (that is, a 'trigonometric polynomial of period $2r$') such that $\sup_{x \in [-r,r]^d} |f(x) - p(x)| \leqslant \varepsilon$. Then, noting that $\|p\|_\infty \leqslant 1 + \varepsilon$,

$$\left| \int_{\mathbb{R}^d} f \, d\mu \right| = \left| \int_{[-r,r]^d} f \, d\mu \right| \leqslant \int_{[-r,r]^d} |f - p| \, d|\mu| + \left| \int_{[-r,r]^d} p \, d\mu \right|$$

$$\leqslant \varepsilon \|\mu\| + \left| \int_{\mathbb{R}^d} p \, d\mu - \int_{\complement[-r,r]^d} p \, d\mu \right|$$

$$\leqslant \varepsilon \|\mu\| + \underbrace{\left| \int_{\mathbb{R}^d} p \, d\mu \right|}_{=0} + \varepsilon(1 + \varepsilon) = \varepsilon \|\mu\| + \varepsilon(1 + \varepsilon).$$

The equality in the last step follows from the assumption that $\widehat{\mu}$ vanishes, as it implies $\int_{\mathbb{R}^d} p \, d\mu = 0$. Since $\varepsilon > 0$ was arbitrary, this proves (5.5). $\qquad\square$

For later reference we mention that this theorem also admits a discrete version, the proof of which is an even more direct application of the Stone–Weierstrass theorem (see Problem 5.17). The *Fourier coefficients* of a real or complex Borel measure μ on the unit circle $\mathbb{T}$ are defined by

$$\widehat{\mu}(n) := \frac{1}{2\pi} \int_{-\pi}^{\pi} \exp(-in\theta) \, d\mu(\theta), \quad n \in \mathbb{Z}.$$

Theorem 5.31 (Injectivity of the Fourier transform on the circle). *If μ is a real or complex Borel measure on $\mathbb{T}$ satisfying $\widehat{\mu}(n) = 0$ for all $n \in \mathbb{Z}$, then $\mu = 0$.*

If μ is real-valued, it suffices to have $\widehat{\mu}(n) = 0$ for all $n \in \mathbb{N}$, for then $\widehat{\mu}(-n) = \overline{\widehat{\mu}(n)}$ implies that $\widehat{\mu}(n) = 0$ for all $n \in \mathbb{Z}$.

5.5.c Fourier Multiplier Operators

The Plancherel theorem provides a method for constructing nontrivial bounded operators on $L^2(\mathbb{R}^d)$ as follows. Given $m \in L^\infty(\mathbb{R}^d)$ and $f \in L^2(\mathbb{R}^d)$, the function

$$m\widehat{f} : \xi \mapsto m(\xi)\widehat{f}(\xi)$$

belongs to $L^2(\mathbb{R}^d)$ and therefore the same is true for its inverse Fourier–Plancherel transform $(m\widehat{f})^{\vee}$.

Definition 5.32 (Fourier multiplier operators). For functions $m \in L^\infty(\mathbb{R}^d)$, the bounded operator on $L^2(\mathbb{R}^d)$ defined by

$$T_m : f \mapsto (m\widehat{f})^{\vee}$$

is called the *Fourier multiplier operator* with *symbol m*.

The operator T_m is bounded of norm $\|T_m\| = \|g \mapsto mg\|_{\mathscr{L}(L^2(\mathbb{R}^d))} = \|m\|_\infty$ (cf. Example 1.29). We have the elementary properties

$$T_{m_1+m_2} = T_{m_1} + T_{m_2}, \quad T_{m_1 m_2} = T_{m_1} \circ T_{m_2}.$$

Fourier multipliers can be characterised by commutation properties. This fact depends on the following lemma.

Lemma 5.33. *Let $(\Omega, \mathscr{F}, \mu)$ be a σ-finite measure space and let $1 \leqslant p \leqslant \infty$. For a bounded operator T on $L^p(\Omega)$ the following assertions are equivalent:*

(1) *T is a pointwise multiplier, that is, there exists a function $m \in L^\infty(\Omega)$ such that $Tf = mf$ for all $f \in L^p(\Omega)$;*
(2) *T commutes with all pointwise multipliers, that is, $TM_\phi = M_\phi T$ for all $\phi \in L^\infty(\Omega)$, where $M_\phi f = \phi f$.*

If $\Omega = \mathbb{R}^d$ with Lebesgue measure and $1 \leqslant p < \infty$, then (1) and (2) are equivalent to

(3) *$TM_{e_\xi} = M_{e_\xi} T$ for all $\xi \in \mathbb{R}^d$, where $e_\xi(x) = \exp(ix \cdot \xi)$ for $x \in \mathbb{R}^d$.*

Proof It is trivial that (1) implies (2). If, conversely, T commutes with every pointwise multiplier, then for all $f \in L^p(\Omega) \cap L^\infty(\Omega)$ we have

$$Tf = T(f\mathbf{1}) = TM_f\mathbf{1} = M_fT\mathbf{1} = fT\mathbf{1} = M_{T\mathbf{1}}f.$$

Hence for all $f \in L^p(\Omega) \cap L^\infty(\Omega)$ we have

$$\|M_{T\mathbf{1}}f\|_p = \|Tf\|_p \leqslant \|T\|\|f\|_p.$$

Since $L^p(\Omega) \cap L^\infty(\Omega)$ is dense in $L^p(\Omega)$, this implies that pointwise multiplication by $T\mathbf{1}$ extends to a bounded operator on $L^p(\Omega)$. This forces $T\mathbf{1} \in L^\infty(\Omega)$ (see the observation at the end of Section 2.3.a; this is where the σ-finiteness assumption is used). Setting $m := T\mathbf{1}$, we obtain $T = M_m$. This proves that (2) implies (1).

Now suppose that $\Omega = \mathbb{R}^d$ with Lebesgue measure. It is trivial that (2) implies (3). Suppose now that (3) holds, with $1 \leqslant p < \infty$. Fix $\varepsilon > 0$ and $f \in L^p(\mathbb{R}^d)$, and choose $r > 0$ so large that $\|f_r - f\|_p < \varepsilon$, where $f_r := f|_{[-r;r]^d}$. If p is a linear combination of $2r$-periodic trigonometric exponentials, (3) implies

$$T(pf_r) = pTf_r.$$

By the Stone–Weierstrass theorem, every $\phi \in C([-r,r]^d)$ can be uniformly approximated by linear combinations p_n of such trigonometric exponentials. Applying the preceding identity to p_n and taking limits in $L^p(\mathbb{R}^d)$, we find that

$$T(\phi f_r) = \phi Tf_r, \quad \phi \in C([-r,r]^d).$$

If $\phi \in C_b(\mathbb{R}^d)$, this implies that $T(\phi_r f_r) = \phi_r T f_r$. As $r \to \infty$ we have $\phi_r f_r \to \phi f$ and $\phi_r T f_r \to \phi T f$ in $L^p(\mathbb{R}^d)$, and therefore

$$\phi T f = T(\phi f), \quad \phi \in C_b(\mathbb{R}^d).$$

Finally, if $\phi \in L^\infty(\mathbb{R}^d)$, then by the regularity of the Lebesgue measure and the use of Urysohn functions we can find a sequence of functions $\phi_n \in C_b(\mathbb{R}^d)$ converging to ϕ pointwise almost everywhere and such that $\sup_{n \geqslant 1} \|\phi_n\|_\infty < \infty$. Since $\phi_n T f \to \phi T f$ and $\phi_n f \to \phi f$ in $L^p(\mathbb{R}^d)$, we conclude that

$$T(\phi f) = \phi T f, \quad \phi \in L^\infty(\mathbb{R}^d).$$

This proves that (2) holds. $\qquad\square$

As an application we have the following characterisation of translation invariant operators on $L^2(\mathbb{R}^d)$.

Theorem 5.34 (Translation invariant operators on $L^2(\mathbb{R}^d)$). *If T is a bounded operator on $L^2(\mathbb{R}^d)$ commuting with every translation, then T is a Fourier multiplier operator, that is, there exists a (necessarily unique) function $m \in L^\infty(\mathbb{R}^d)$ such that $Tf = \mathscr{F}^{-1}(m\mathscr{F}f)$ for all $f \in L^2(\mathbb{R}^d)$.*

Proof Using the notation of Lemma 5.33 and letting $\tau_y f(x) := f(x+y)$, easy calculations show that $M_{e_\xi}\mathscr{F} = \mathscr{F}\tau_\xi$ and $\tau_\xi \mathscr{F}^{-1} = \mathscr{F}^{-1}M_{e_\xi}$ for all $\xi \in \mathbb{R}^d$. These identities imply that the operator $\widetilde{T} = \mathscr{F}T\mathscr{F}^{-1}$ has the property $M_{e_\xi}\widetilde{T} = \widetilde{T}M_{e_\xi}$ for all $\xi \in \mathbb{R}^d$, and the lemma implies that $\widetilde{T}$ is a pointwise multiplier. This means that T is a Fourier multiplier. $\qquad\square$

5.6 The Hilbert Transform

A case of special interest concerns the multiplier

$$m(\xi) = -i\,\mathrm{sign}(\xi), \quad \xi \in \mathbb{R}.$$

In order to obtain an explicit representation for the Fourier multiplier operator T_m we observe that $m = -i(\mathbf{1}_{\mathbb{R}_+} - \mathbf{1}_{\mathbb{R}_-})$ and consider the functions

$$n_a^\pm(\xi) := \exp(-a|\xi|)\mathbf{1}_{\mathbb{R}_\pm}(\xi), \quad a > 0.$$

Then

$$\widecheck{n_a^+}(x) = \frac{1}{\sqrt{2\pi}}\int_0^\infty \exp(-a\xi)\exp(ix\xi)\,\mathrm{d}\xi = \frac{1}{\sqrt{2\pi}}\frac{1}{a-ix}$$

and

$$\widecheck{n_a^-}(x) = \frac{1}{\sqrt{2\pi}} \int_{-\infty}^0 \exp(a\xi)\exp(ix\xi)\,d\xi = \frac{1}{\sqrt{2\pi}}\frac{1}{a+ix}.$$

Formally letting $a \downarrow 0$, in view of Proposition 5.29 one expects that

$$\begin{aligned}
T_{-i\,\mathrm{sign}}f &= -i\lim_{a\downarrow 0}(T_{n_a^+}f - T_{n_a^-}f) \\
&= -\frac{i}{\sqrt{2\pi}}\lim_{a\downarrow 0}(\widecheck{n_a^+}*f - \widecheck{n_a^-}*f) \\
&= -\frac{i}{2\pi}\lim_{a\downarrow 0}\left(\frac{1}{a-i(\cdot)} - \frac{1}{a+i(\cdot)}\right)*f = \frac{1}{\pi(\cdot)}*f.
\end{aligned}$$

This suggests the formula

$$T_{-i\,\mathrm{sign}}f(x) = \frac{1}{\pi}\int_{-\infty}^{\infty} \frac{f(x-y)}{y}\,dy.$$

The above argument is nonrigorous and the convolution with the nonintegrable function $1/x$ is not even well defined as an operator acting on $L^2(\mathbb{R})$. The next theorem turns the above formal derivation into a rigorous argument.

Theorem 5.35 (Hilbert transform as a Fourier multiplier operator). *The Fourier multiplier operator* $H := T_{-i\,\mathrm{sign}}$ *is given by*

$$Hf(\cdot) = \lim_{\varepsilon\downarrow 0}\frac{1}{\pi}\int_{\{|y|>\varepsilon\}} \frac{f(\cdot - y)}{y}\,dy, \quad f \in L^2(\mathbb{R}),$$

the convergence being in the sense of $L^2(\mathbb{R})$.

Proof Setting

$$H_\varepsilon f := \frac{1}{\pi}\int_{\{|y|>\varepsilon\}} \frac{f(\cdot - y)}{y}\,dy,$$

we see that H_ε is the operator of convolution with the integrable function $\phi_\varepsilon(x) = \frac{1}{\pi x}\mathbf{1}_{\{|x|>\varepsilon\}}$. The Fourier transform of this function can be rewritten as

$$\begin{aligned}
\widehat{\phi_\varepsilon}(\xi) &= \frac{1}{\pi\sqrt{2\pi}}\int_{\{|x|>\varepsilon\}}\frac{1}{x}\exp(-ix\xi)\,dx \\
&= -\frac{1}{\pi\sqrt{2\pi}}\mathrm{sign}(\xi)\lim_{R\to\infty}\int_{i[-R,-\varepsilon]\cup i[\varepsilon,R]}\exp(z|\xi|)\frac{dz}{z} \\
&= -\frac{i}{\pi\sqrt{2\pi}}\mathrm{sign}(\xi)\lim_{R\to\infty}\int_{\frac{1}{2}\pi}^{\frac{3}{2}\pi}\exp\left(|\xi|Re^{i\theta}\right) - \exp\left(|\xi|\varepsilon e^{i\theta}\right)\,d\theta,
\end{aligned}$$

where the last step used Cauchy's theorem applied to the boundary of the part of the annulus $\{\varepsilon < |z| < R\}$ in the left half-plane. Now,

$$\left| \int_{\frac{1}{2}\pi}^{\frac{3}{2}\pi} \exp\left(|\xi| R e^{i\theta} \right) d\theta \right| \leq \int_{\frac{1}{2}\pi}^{\frac{3}{2}\pi} \exp(|\xi| R \cos\theta) d\theta$$

$$= 2 \int_0^{\frac{1}{2}\pi} \exp(-|\xi| R \sin\theta) d\theta$$

$$\leq 2 \int_0^{\frac{1}{2}\pi} \exp(-(2/\pi)|\xi| R\theta) d\theta = \frac{1 - \exp(-|\xi| R)}{|\xi| R/\pi},$$

where we used that $\sin\theta \geq (2/\pi)\theta$ for $\theta \in [0, \frac{1}{2}\pi]$. As the expression on the right-hand side tends to 0 as $R \to \infty$, we find

$$\widehat{\phi_\varepsilon}(\xi) = -\frac{i}{\pi\sqrt{2\pi}} \text{sign}(\xi) \int_{\frac{1}{2}\pi}^{\frac{3}{2}\pi} \exp\left(|\xi| \varepsilon e^{i\theta} \right) d\theta.$$

As $\varepsilon \downarrow 0$, the integral on the right-hand side tends to π for every $\xi \in \mathbb{R}$. Hence by dominated convergence

$$\lim_{\varepsilon\downarrow 0} \left\| \widehat{\phi_\varepsilon} - \left(-\frac{i}{\sqrt{2\pi}} \text{sign} \right) \right\|_2^2 = \int_{\mathbb{R}} \left| \frac{1}{\pi} \int_{\frac{1}{2}\pi}^{\frac{3}{2}\pi} \exp\left(|\xi| \varepsilon e^{i\theta} \right) d\theta - 1 \right|^2 d\xi = 0.$$

As a result, by using Proposition 5.29 we obtain

$$\lim_{\varepsilon\downarrow 0} \widehat{H_\varepsilon f} = \sqrt{2\pi} \lim_{\varepsilon\downarrow 0} \widehat{\phi_\varepsilon} \widehat{f} = -i \, \text{sign} \cdot \widehat{f}$$

with convergence in $L^2(\mathbb{R})$. Therefore, by the Plancherel theorem and the definition of H,

$$\lim_{\varepsilon\downarrow 0} H_\varepsilon f = (-i \, \text{sign} \cdot \widehat{f})^\vee = Hf$$

with convergence in $L^2(\mathbb{R})$. $\qquad\square$

Definition 5.36 (Hilbert transform). The operator $H = T_{-i\,\text{sign}}$ is called the *Hilbert transform*.

The Hilbert transform has deep applications in Harmonic Analysis and the theory of partial differential equations. It will resurface in our treatment of the Poisson semigroup in Chapter 13. Its connection with harmonic functions is pointed out in Problem 5.22.

The final theorem of this section gives a characterisation of the Hilbert transform in terms of commutation properties. A *dilation* on $L^2(\mathbb{R}^d)$ is an operator of the form

$$D_\delta f(x) := f(\delta x), \quad x \in \mathbb{R}^d,$$

where $\delta > 0$.

Theorem 5.37 (Characterisation of the Hilbert transform). *If T is a bounded operator on $L^2(\mathbb{R})$ commuting with every translation and dilation, then T is a linear combination of the identity operator and the Hilbert transform.*

Proof Theorem 5.34 tells us that T is a Fourier multiplier operator, say $T = T_m$ with $m \in L^\infty(\mathbb{R})$. For $\delta > 0$, simple calculations give

$$T_m D_\delta f(x) = T_{D_\delta m} f(\delta x)$$

and

$$D_\delta T_m f(x) = T_m f(\delta x)$$

for almost all $x \in \mathbb{R}$. It follows that $T_m D_\delta = D_\delta T_m$ if and only if $T_{D_\delta m} = T_m$, that is, if and only if $m(\delta \xi) = m(\xi)$ for almost all $\xi \in \mathbb{R}$. This is true for all $\delta > 0$ if and only if m is constant almost everywhere on both $\mathbb{R}_+$ and $\mathbb{R}_-$. Hence $m = a\mathbf{1} + b\,\text{sign}$ for suitable $a, b \in \mathbb{C}$. The result now follows from Theorem 5.35. $\qquad\qquad\square$

5.7 Interpolation

In general it can be difficult to establish L^p-boundedness of operators acting in spaces of measurable functions. In such situations, interpolation theorems may be helpful. They serve to establish L^p-boundedness in situations where suitable boundedness properties can be established for 'endpoint' exponents p_0 and p_1 satisfying $p_0 \leqslant p \leqslant p_1$. In typical applications one takes $p_0 \in \{1, 2\}$ and $p_1 \in \{2, \infty\}$ (cf. Sections 5.7.b and 5.7.c).

5.7.a The Riesz–Thorin Interpolation Theorem

Theorem 5.38 (Riesz–Thorin interpolation theorem). *Let $(\Omega, \mathscr{F}, \mu)$ and $(\Omega', \mathscr{F}', \mu')$ be measure spaces and let $1 \leqslant p_0, p_1, q_0, q_1 \leqslant \infty$. Let*

$$T_0 : L^{p_0}(\Omega) \to L^{q_0}(\Omega'), \quad T_1 : L^{p_1}(\Omega) \to L^{q_1}(\Omega')$$

be bounded operators which are consistent in the sense that $Tf := T_0 f = T_1 f$ μ'-almost surely for all $f \in L^{p_0}(\Omega) \cap L^{p_1}(\Omega)$. Assume furthermore that

$$\|T_0 f\|_{L^{q_0}(\Omega')} \leqslant A_0 \|f\|_{L^{p_0}(\Omega)}, \quad f \in L^{p_0}(\Omega),$$

$$\|T_1 f\|_{L^{q_1}(\Omega')} \leqslant A_1 \|f\|_{L^{p_1}(\Omega)}, \quad f \in L^{p_1}(\Omega).$$

Let $0 < \theta < 1$ and set

$$\frac{1}{p_\theta} := \frac{1-\theta}{p_0} + \frac{\theta}{p_1}, \quad \frac{1}{q_\theta} := \frac{1-\theta}{q_0} + \frac{\theta}{q_1}.$$

Then the common restriction T of T_0 and T_1 to $L^{p_0}(\Omega) \cap L^{p_1}(\Omega)$ maps this space into $L^{q_\theta}(\Omega')$ and extends uniquely to a bounded operator

$$T : L^{p_\theta}(\Omega) \to L^{q_\theta}(\Omega')$$

satisfying

$$\|Tf\|_{L^{q_\theta}(\Omega')} \leqslant A_0^{1-\theta} A_1^\theta \|f\|_{L^{p_\theta}(\Omega)}, \quad f \in L^{p_\theta}(\Omega).$$

We begin with a simple lemma, which corresponds to the special case where the interpolated operator is the identity operator.

Lemma 5.39. *Let $(\Omega, \mathscr{F}, \mu)$ be a measure space, let $1 \leqslant r_0 \leqslant r_1 \leqslant \infty$ and $0 < \theta < 1$, and set $\frac{1}{r_\theta} := \frac{1-\theta}{r_0} + \frac{\theta}{r_1}$. Then for all $f \in L^{r_0}(\Omega) \cap L^{r_1}(\Omega)$ we have $f \in L^{r_\theta}(\Omega)$*

$$\|f\|_{r_\theta} \leqslant \|f\|_{r_0}^{1-\theta} \|f\|_{r_1}^\theta.$$

Proof Write $|f|^{r_\theta} = |f|^{(1-\theta)r_\theta} |f|^{\theta r_\theta}$ and apply Hölder's inequality. $\qquad\square$

Consider the open strip

$$\mathbb{S} := \{z \in \mathbb{C} : 0 < \operatorname{Re} z < 1\}.$$

Lemma 5.40 (Three lines lemma). *Suppose that $F : \overline{\mathbb{S}} \to \mathbb{C}$ is a bounded continuous function, holomorphic on $\mathbb{S}$, and satisfying*

$$\sup_{v \in \mathbb{R}} |F(iv)| \leqslant A_0, \quad \sup_{v \in \mathbb{R}} |F(1+iv)| \leqslant A_1.$$

Then for all $0 < \theta < 1$ we have

$$\sup_{v \in \mathbb{R}} |F(\theta + iv)| \leqslant A_0^{1-\theta} A_1^\theta.$$

Proof Let F satisfy the assumptions of the lemma with constants A_0 and A_1. For each $\varepsilon > 0$ the function $F_\varepsilon(z) := A_0^{z-1} A_1^{-z} \exp(\varepsilon z(z-1)) F(z)$ satisfies the assumptions of the lemma with constants $A_{0,\varepsilon} = A_{1,\varepsilon} = 1$. Moreover, $\lim_{v \to \infty} |F_\varepsilon(u+iv)| = 0$ uniformly with respect to $u \in [0,1]$. Hence for large enough R we have $|F_\varepsilon| \leqslant 1$ on the boundary of the rectangle $\operatorname{Re} z \in [0,1], |\operatorname{Im} z| \leqslant R$. The maximum modulus principle implies that $|F_\varepsilon| \leqslant 1$ on this rectangle, and by letting $R \to \infty$ we find that $|F_\varepsilon| \leqslant 1$ on $\overline{\mathbb{S}}$. The lemma now follows by letting $\varepsilon \downarrow 0$. $\qquad\square$

Inspection of the proof reveals that the boundedness assumption on F can be relaxed. It cannot be entirely dispensed with, however: the function $F(z) = \exp(\exp(\pi(z-1)))$ is bounded on the lines $\operatorname{Re} z = 0$ and $\operatorname{Re} z = 1$, but unbounded on the line $\operatorname{Re} z = \frac{1}{2}$.

Proof of Theorem 5.38 There is no loss of generality in assuming that $A_0 > 0$ and

$A_1 > 0$. If $p_0 = p_1 = \infty$, then for all $f \in L^\infty(\Omega)$ we have $Tf \in L^{q_0}(\Omega') \cap L^{q_1}(\Omega')$ and therefore, by Lemma 5.39,

$$\|Tf\|_{q_\theta} \leqslant \|Tf\|_{q_0}^{1-\theta} \|Tf\|_{q_1}^\theta \leqslant A_0^{1-\theta} A_1^\theta \|f\|_\infty^{1-\theta} \|f\|_\infty^\theta = A_0^{1-\theta} A_1^\theta \|f\|_\infty.$$

This settles that case $p_0 = p_1 = \infty$. In the rest of the proof we may therefore assume that $\min\{p_0, p_1\} < \infty$. This assumption implies $p_\theta < \infty$.

For $z \in \overline{\mathbb{S}}$ define $p_z, q_z \in \mathbb{C}$ by the relations

$$\frac{1}{p_z} = \frac{1-z}{p_0} + \frac{z}{p_1}, \quad \frac{1}{q_z'} = \frac{1-z}{q_0'} + \frac{z}{q_1'},$$

where q_0' and q_1' are the conjugate exponents of q_0 and q_1, respectively. Let $a : \Omega \to \mathbb{C}$ and $b : \Omega' \to \mathbb{C}$ be μ- and μ'-simple functions and define, for each $z \in \overline{\mathbb{S}}$, the μ- and μ'-simple functions $f_z \in L^{p_0}(\Omega) \cap L^{p_1}(\Omega)$ and $g_z \in L^{q_0'}(\Omega') \cap L^{q_1'}(\Omega')$ by

$$f_z(\omega) = \mathbf{1}_{\{a \neq 0\}} |a(\omega)|^{p_\theta/p_z} \frac{a(\omega)}{|a(\omega)|}, \quad g_z(\omega') = \mathbf{1}_{\{b \neq 0\}} |b(\omega')|^{q_\theta'/q_z'} \frac{b(\omega')}{|b(\omega')|}. \tag{5.6}$$

Then $Tf_z \in L^{q_0}(\Omega') \cap L^{q_1}(\Omega')$, and the function $F : \overline{\mathbb{S}} \to \mathbb{C}$ defined by

$$F(z) := \int_{\Omega'} (Tf_z) \cdot g_z \, d\mu'$$

is easily checked to be bounded and continuous on $\overline{\mathbb{S}}$, holomorphic on $\mathbb{S}$, and for all $v \in \mathbb{R}$ we have

$$|F(iv)| \leqslant A_0 \|f_{iv}\|_{p_0} \|g_{iv}\|_{q_0'} \leqslant A_0 \|a\|_{p_\theta}^{p_\theta/p_0} \|b\|_{q_\theta'}^{q_\theta'/q_0'}$$

and similarly

$$|F(1+iv)| \leqslant A_1 \|a\|_{p_\theta}^{p_\theta/p_1} \|b\|_{q_\theta'}^{q_\theta'/q_1'}.$$

Hence, by (5.6) and the three lines lemma (Lemma 5.40),

$$\left| \int_{\Omega'} (Ta) \cdot b \, d\mu' \right| = |F(\theta)| \leqslant A_0^{1-\theta} A_1^\theta \|a\|_{p_\theta}^{(1-\theta)p_\theta/p_0} \|b\|_{q_\theta'}^{(1-\theta)q_\theta'/q_0'} \|a\|_{p_\theta}^{\theta p_\theta/p_1} \|b\|_{q_\theta'}^{\theta q_\theta'/q_1'}$$

$$= A_0^{1-\theta} A_1^\theta \|a\|_{p_\theta} \|b\|_{q_\theta'}.$$

Taking the supremum over all μ'-simple functions $b \in L^{q_\theta'}(\Omega')$ of norm at most one, by Hölder's inequality and Propositions 2.26 and 2.28 (here we use the assumption $p_\theta < \infty$) we obtain

$$\|Ta\|_{q_\theta} \leqslant A_0^{1-\theta} A_1^\theta \|a\|_{p_\theta}.$$

Since the μ-simple functions are dense in $L^{p_\theta}(\Omega)$, this proves that the restriction of T to the μ-simple functions has a unique extension to a bounded operator $\widetilde{T}$ from $L^{p_\theta}(\Omega)$ into $L^{q_\theta}(\Omega')$ of norm at most $A_0^{1-\theta} A_1^\theta$.

It remains to be checked that $\widetilde{T} f = T f$ for all $f \in L^{p_\theta}(\Omega)$. To this end, we may assume $p_0 \leqslant p_1$. Selecting $y > 0$ such that $\mu\{|f| = y\} = 0$ and replacing f by $y^{-1} f$, we may assume that $\mu\{|f| = 1\} = 0$. Write $f = \mathbf{1}_{\{|f| > 1\}} f + \mathbf{1}_{\{|f| \leqslant 1\}} f =: f^0 + f^1$ and observe that $f^j \in L^{p_j}(\Omega)$ $(j = 0, 1)$. If $f_n \to f$ in $L^{p_\theta}(\Omega)$ with each f_n μ-simple, then, with obvious notation, $f_n^j \to f^j$ in $L^{p_j}(\Omega)$ and therefore $\widetilde{T} f^j = \lim_{n\to\infty} \widetilde{T} f_n^j = \lim_{n\to\infty} T f_n^j = T f^j$ in $L^{q_j}(\Omega')$. As a consequence $\widetilde{T} f = \widetilde{T} f^0 + \widetilde{T} f^1 = T f^0 + T f^1 = T f$. $\qquad\square$

Up to this point we have implicitly assumed that the scalar field is complex. Suppose now that the scalar field is real. We may extend a bounded operator $T : L^p(\Omega) \to L^q(\Omega')$ to a bounded operator $T_{\mathbb{C}} : L^p(\Omega; \mathbb{C}) \to L^q(\Omega'; \mathbb{C})$ by setting

$$T_{\mathbb{C}}(u + iv) := Tu + iTv$$

for real-valued $u, v \in L^p(\Omega)$. The triangle inequality implies the trivial bounds

$$\|T\| \leqslant \|T_{\mathbb{C}}\| \leqslant 2\|T\|.$$

If T is a positivity preserving operator (that is, $f \geqslant 0$ implies $Tf \geqslant 0$), then the identity

$$|a + ib| = \sup_{\theta \in [0, 2\pi]} |a \cos\theta + b \sin\theta|$$

(for a proof, rotate the point $(a, b) \in \mathbb{R}^2$ to the positive x-axis) together with the inequality $|Tg| \leqslant T|g|$ for real-valued $g \in L^p(\Omega)$ (which follows from (2.20)) implies the pointwise bound

$$|T_{\mathbb{C}} f| = |Tu + iTv| = \sup_{\theta \in [0, 2\pi]} |(Tu) \cos\theta + (Tv) \sin\theta|$$

$$\leqslant \sup_{\theta \in [0, 2\pi]} T|u \cos\theta + Tv \sin\theta| \leqslant T|u + iv| = T|f|.$$

This implies the norm bound $\|T_{\mathbb{C}}\| \leqslant \|T\|$ and hence equality

$$\|T_{\mathbb{C}}\| = \|T\|. \tag{5.7}$$

With some additional work, the equality (5.7) can be extended to arbitrary bounded operators T and exponents $1 \leqslant p \leqslant q < \infty$. The proof is based on the observation that for all $z = a + bi \in \mathbb{C}$ and $1 \leqslant q < \infty$ we have

$$|z| = \frac{1}{\|\gamma\|_q} (\widetilde{\mathbb{E}} |a\gamma_1 + b\gamma_2|^q)^{1/q}, \tag{5.8}$$

where $\gamma, \gamma_1, \gamma_2$ are real-valued standard Gaussian random variables defined on some probability space $\widetilde{\Omega}$, with γ_1 and γ_2 independent, and $\widetilde{\mathbb{E}}$ denotes the expectation. Indeed, if $|z| = 1$, then $a\gamma_1 + b\gamma_2$ is another standard Gaussian random variable and

$$(\widetilde{\mathbb{E}} |a\gamma_1 + b\gamma_2|^q)^{1/q} = \|\gamma_1\|_q.$$

The general case follows from this by scaling.

Assuming $1 \leqslant p \leqslant q < \infty$, from (5.8) in combination with Fubini's theorem we obtain

$$\|\gamma\|_q^q \|T_\mathbb{C}(u+iv)\|_{L^q(\Omega';\mathbb{C})}^q = \|\gamma\|_q^q \int_{\Omega'} |Tu+iTv|^q \, d\mu'$$

$$= \int_{\Omega'} \widetilde{\mathbb{E}} |\gamma_1 Tu + \gamma_2 Tv|^q \, d\mu' = \|\gamma_1 Tu + \gamma_2 Tv\|_{L^q(\Omega';L^q(\tilde{\Omega}))}^q$$

$$= \|T(\gamma_1 u + \gamma_2 v)\|_{L^q(\tilde{\Omega};L^q(\Omega'))}^q \leqslant \|T\|^q \|\gamma_1 u + \gamma_2 v\|_{L^q(\tilde{\Omega};L^p(\Omega))}^q$$

$$\leqslant \|T\|^q \|\gamma_1 u + \gamma_2 v\|_{L^p(\Omega;L^q(\tilde{\Omega}))}^q = \|\gamma\|_q^q \|T\|^q \|u + iv\|_{L^p(\Omega;\mathbb{C})}^q.$$

In the penultimate step we used the continuous version of Hölder's inequality (Problem 2.29). This proves (5.7) for $1 \leqslant p \leqslant q < \infty$. In the range $q < p$, (5.7) is generally false, as is shown by a classical counterexample due to M. Riesz.

The Riesz–Thorin theorem may now be extended to the case of real scalars and exponents $1 \leqslant p \leqslant q < \infty$ as follows. Suppose that the assumptions of the theorem are satisfied, except that all spaces are real. Apply the Riesz–Thorin theorem to the complexified operators $S_0 := (T_0)_\mathbb{C}$ and $S_1 := (T_1)_\mathbb{C}$, we obtain bounded operators S_θ from $L^{p_\theta}(\Omega;\mathbb{C})$ to $L^{q_\theta}(\Omega';\mathbb{C})$ of norm at most $A_0^{1-\theta} A_1^\theta$. This operator maps functions in $L^{p_0}(\Omega) \cap L^{p_1}(\Omega)$ to functions in $L^{q_\theta}(\Omega')$. Since $L^{p_0}(\Omega) \cap L^{p_1}(\Omega)$ is dense in $L^{p_\theta}(\Omega)$, by approximation it follows that S_θ maps

Felix Hausdorff, 1868–1942

real-valued functions in $L^{p_\theta}(\Omega)$ to real-valued functions in $L^{q_\theta}(\Omega')$. Stated differently, S_θ restricts to a bounded operator, denoted T_θ, from $L^{p_\theta}(\Omega)$ to $L^{q_\theta}(\Omega')$ of norm at most $A_0^{1-\theta} A_1^\theta$. On $L^{p_0}(\Omega) \cap L^{p_1}(\Omega)$, T_θ coincides with the common restriction of T_0 and T_1.

Informally stated, this discussion shows that the Riesz–Thorin theorem extends, with the same constant, to the case of real scalars if we assume T to be positivity preserving or the exponents satisfy $1 \leqslant p_j \leqslant q_j < \infty$ for $j = 0, 1$.

5.7.b The Hausdorff–Young Theorem

This brief section and the next are devoted to some applications of the Riesz–Thorin theorem.

As we have seen in Section 5.5, the Fourier transform is bounded from $L^1(\mathbb{R}^d)$ to $L^\infty(\mathbb{R}^d)$ and its restriction to $L^1(\mathbb{R}^d) \cap L^2(\mathbb{R}^d)$ extends to an isometry from $L^2(\mathbb{R}^d)$ onto itself. The Fourier transform with respect to the normalised Lebesgue measure $dm(x) = (2\pi)^{-d/2} \, dx$ defined in Remark 5.17 is contractive from $L^1(\mathbb{R}^d, m)$ to $L^\infty(\mathbb{R}^d, m)$, and its restriction to $L^1(\mathbb{R}^d, m) \cap L^2(\mathbb{R}^d, m)$ extends to an isometry from $L^2(\mathbb{R}^d, m)$ onto itself by Remark 5.27. Accordingly, the Riesz–Thorin theorem implies:

Theorem 5.41 (Hausdorff–Young). *Let $1 \leqslant p \leqslant 2$ and $\frac{1}{p} + \frac{1}{q} = 1$. The restriction to $L^1(\mathbb{R}^d) \cap L^2(\mathbb{R}^d)$ of the Fourier transform has a unique extension to a bounded operator from $L^p(\mathbb{R}^d)$ to $L^q(\mathbb{R}^d)$. With respect to the normalised Lebesgue measure, the Fourier transform has a unique extension to a contraction from $L^p(\mathbb{R}^d, m)$ to $L^q(\mathbb{R}^d, m)$.*

A similar result holds for the Fourier transform on the circle (see Problem 5.27).

5.7.c L^p-Boundedness of the Hilbert Transform

A second application of the Riesz–Thorin theorem is the following theorem due to M. Riesz about L^p-boundedness of the Hilbert transform.

Theorem 5.42 (Riesz). *For all $1 < p < \infty$ the restriction of the Hilbert transform to $L^2(\mathbb{R}) \cap L^p(\mathbb{R})$ has a unique extension to a bounded operator on $L^p(\mathbb{R})$.*

The proof of Theorem 5.42 is based on a couple of lemmas.

Lemma 5.43. *If $f \in C_c^1(\mathbb{R})$, then $Hf \in L^p(\mathbb{R})$ for all $2 \leqslant p \leqslant \infty$.*

Proof Let I be a bounded interval containing the support of f. The pointwise identity

$$H_\varepsilon f(x) = \frac{1}{\pi} \int_\varepsilon^\infty \frac{f(x-y) - f(x+y)}{y} \, dy$$
$$= \frac{1}{\pi} \int_\varepsilon^\infty \mathbf{1}_{(-I+x) \cup (I-x)}(y) \frac{f(x-y) - f(x+y)}{y} \, dy, \quad x \in \mathbb{R},$$

implies the bound

$$|H_\varepsilon f(x)| \leqslant \frac{1}{\pi} \cdot 2|I| \cdot 2\|f'\|_\infty, \quad x \in \mathbb{R}. \tag{5.9}$$

As $\varepsilon \downarrow 0$, we have $H_\varepsilon f \to Hf$ in $L^2(\mathbb{R})$ by Theorem 5.35 and, upon passing to an almost everywhere convergent subsequence, (5.9) implies that $Hf \in L^\infty(\mathbb{R})$. This gives the result. $\qquad\square$

Lemma 5.44. *The Hilbert transform of a real-valued function $u \in L^2(\mathbb{R})$ is the unique real-valued function $v \in L^2(\mathbb{R})$ such that the Fourier–Plancherel transform of $u + iv$ vanishes on $\mathbb{R}_-$.*

Proof That the Fourier transform of $u + iHu$ vanishes on $\mathbb{R}_-$ is immediate from Theorem 5.35. In the converse direction, let $u, v \in L^2(\mathbb{R})$ be real-valued such that the Fourier–Plancherel transform of $u + iv$ vanishes on $\mathbb{R}_-$. Then for almost all $\xi > 0$ we have

$$0 = \widehat{u}(-\xi) + i\widehat{v}(-\xi) = \overline{\widehat{u}(\xi)} + i\overline{\widehat{v}(\xi)} = \overline{\widehat{u}(\xi)} - i\overline{\widehat{v}(\xi)},$$

so $\widehat{v}(-\xi) = i\widehat{u}(-\xi)$ and $\widehat{v}(\xi) = -i\widehat{u}(\xi)$. Hence, for almost all $\xi \in \mathbb{R}$,

$$\widehat{v}(\xi) = -i\,\mathrm{sign}(\xi)\widehat{u}(\xi).$$

By Theorem 5.35 this proves that $v = Hu$. □

In the next lemma we use that if $\phi \in C_c^2(\mathbb{R})$, then $\widehat{\phi''}(\xi) = -|\xi|^2\widehat{\phi}(\xi)$ is bounded, and therefore $\widehat{\phi}$ is integrable.

Lemma 5.45 (Cotlar). *Let H be the Hilbert transform on $L^2(\mathbb{R})$. For all real-valued $u \in C_c^2(\mathbb{R})$ we have*

$$(Hu)^2 = u^2 + 2H(uHu).$$

Proof Let $u, v \in C_c^2(\mathbb{R})$ be real-valued functions. By Theorem 5.35 the Fourier transforms of $u + iHu$ and $v + iHv$ are integrable and vanish on $\mathbb{R}_-$, and by Proposition 5.29 the same is true for the Fourier transform of

$$(u \cdot Hv + Hu \cdot v) + i(Hu \cdot Hv - u \cdot v) = -i(u + iHu)(v + iHv).$$

By Lemma 5.44, this implies

$$Hu \cdot Hv - u \cdot v = H(u \cdot Hv + Hu \cdot v).$$

Cotlar's identity follows by taking $u = v$. □

Proof of Theorem 5.42 The proof consists of three steps. First we prove the theorem for exponents $p = 2^n$ with $n \in \mathbb{N}$ by Cotlar's identity, then for $2 < p < \infty$ by interpolation, and finally for $1 < p < 2$ by duality.

Step 1 – In this step we show that if H is L^p-bounded for some $2 \leqslant p < \infty$, then H is L^{2p}-bounded. The proof also gives a bound for $\|H\|_{2p}$ in terms of $\|H\|_p$. In what follows we set $\|H\|_p =: c_p$.

Let $u \in C_c^2(\mathbb{R})$ be real-valued. Then $Hu \in L^{2p}(\mathbb{R})$ by Lemma 5.43. By Cotlar's identity and Hölder's inequality,

$$\|(Hu)^2\|_p \leqslant \|u^2\|_p + 2\|H\|_p\|uHu\|_p \leqslant \|u^2\|_p + 2c_p\|u\|_{2p}\|Hu\|_{2p}.$$

Using the identity $\|v^2\|_p = \|v\|_{2p}^2$, this gives

$$\|Hu\|_{2p}^2 \leqslant \|u\|_{2p}^2 + 2c_p\|u\|_{2p}\|Hu\|_{2p},$$

or equivalently,

$$(\|Hu\|_{2p} - c_p\|u\|_{2p})^2 \leqslant (1 + c_p^2)\|u\|_{2p}^2.$$

It follows that

$$\|Hu\|_{2p} - c_p\|u\|_{2p} \leqslant \sqrt{1 + c_p^2}\|u\|_{2p}$$

and hence

$$\|Hu\|_{2p} \leqslant \left(c_p + \sqrt{1 + c_p^2}\right)\|u\|_{2p}.$$

By considering real and imaginary parts separately, at the expense of an additional constant 2 this inequality extends to arbitrary $u \in C_c^2(\mathbb{R})$. Since $C_c^2(\mathbb{R})$ is dense in $L^{2p}(\mathbb{R})$, it follows that the restriction of H to $C_c^2(\mathbb{R})$ uniquely extends to a bounded operator on $L^{2p}(\mathbb{R})$. Obviously, this operator extends the restriction of H to $L^2(\mathbb{R}) \cap L^{2p}(\mathbb{R})$.

Step 2 – Since H is L^2-bounded, Step 1 implies that H is L^{2^n}-bounded for all $n \in \mathbb{N}$. The Riesz–Thorin theorem then implies that H is L^p-bounded for all $2 \leqslant p < \infty$.

Step 3 – Finally suppose that $1 < p < 2$ and let $\frac{1}{p} + \frac{1}{q} = 2$. For $f, g \in C_c^2(\mathbb{R})$ one easily checks that

$$\int_{\mathbb{R}} Hf \cdot \overline{g} \, dx = -\int_{\mathbb{R}} f \cdot \overline{Hg} \, dx$$

and consequently

$$\left|\int_{\mathbb{R}} Hf \cdot \overline{g} \, dx\right| \leqslant \|f\|_p \|Hg\|_q \leqslant c_q \|f\|_p \|g\|_q,$$

where $c_q = \|H\|_q$. Since $C_c^2(\mathbb{R})$ is dense in $L^q(\mathbb{R})$ by Proposition 2.29, Proposition 2.26 implies that $Hf \in L^p(\mathbb{R})$ and $\|Hf\|_p \leqslant c_q \|f\|_p$. This proves that H is L^p-bounded, with $\|H\|_p \leqslant c_q$ (in fact we have equality here, since we can also apply this argument in the opposite direction). $\qquad\square$

5.7.d The Marcinkiewicz Interpolation Theorem

In this final section we prove another L^p-interpolation theorem, the Marcinkiewicz interpolation theorem. It has the virtue of requiring less stringent conditions at the endpoints and the operator to be interpolated does not even need to be linear. On the downside, the constant obtained from the proof is rather poor. The theorem elaborates on the observation, made after the proof of the Hardy–Littlewood maximal theorem, that the proof of the L^p-bound essentially only depended on the weak L^1-bound.

By $(L^{p_0} + L^{p_1})(\mathbb{R}^d)$ we denote the vector space of all $f \in L_{\text{loc}}^1(\mathbb{R}^d)$ that admit a decomposition $f = f_0 + f_1$ with $f_0 \in L^{p_0}(\mathbb{R}^d)$ and $f_1 \in L^{p_1}(\mathbb{R}^d)$.

Theorem 5.46 (Marcinkiewicz interpolation theorem). *Let $1 \leqslant p_0 < p < p_1 \leqslant \infty$ and suppose that $T : (L^{p_0} + L^{p_1})(\mathbb{R}^d) \to (L^{p_0} + L^{p_1})(\mathbb{R}^d)$ is a* subadditive *mapping in the sense that for all $f \in L^{p_0}$ and $g \in L^{p_1}(\mathbb{R}^d)$ we have*

$$|T(f + g)| \leqslant |T(f)| + |T(g)| \quad \text{almost everywhere.}$$

Suppose furthermore that for $j = 1, 2$ there are constants $C_{d,p_j} \geqslant 0$, depending only on *d* and p_j, *such that*

$$|\{T(f) > t\}| \leqslant \left(\frac{C_{d,p_j}}{t}\right)^{p_j} \|f\|_{p_j}^{p_j}, \quad f \in L^{p_j}(\mathbb{R}^d),$$

if $1 < p_1 < \infty$; *if* $p_1 = \infty$ *we replace the assumption regarding* p_1 *by*

$$\|T(f)\|_\infty \leqslant C_{d,\infty}\|f\|_\infty, \quad f \in L^\infty(\mathbb{R}^d).$$

Then T maps $L^p(\mathbb{R}^d)$ *into* $L^p(\mathbb{R}^d)$ *and*

$$\|T(f)\|_p \leqslant C_{d,p}\|f\|_p, \quad f \in L^p(\mathbb{R}^d),$$

where $C_{d,p}$ *is a constant independent of f.*

A weak L^q-bound holds if *T* is L^q-bounded in the sense that $\|T(f)\|_q \leqslant C_{d,q}\|f\|_q$ for all $f \in L^q(\mathbb{R}^d)$.

Proof We give the proof for $1 < p_1 < \infty$; the case $p_1 = \infty$ proceeds along the lines of Theorem 2.38, requiring small changes that are left to the reader. Fixing $t > 0$, we split $f = f_0 + f_1$ with $f_0 \in L^{p_0}(\mathbb{R}^d)$ and $f_1 \in L^{p_1}(\mathbb{R}^d)$ by taking

$$f_0 = \mathbf{1}_{\{|f| \geqslant t/2\}} f, \quad f_1 = \mathbf{1}_{\{|f| < t/2\}} f.$$

From the subadditivity of *T* we obtain

$$\{|T(f)| > t\} \subseteq \{|T(f_0)| > t/2\} + \{|T(f_1)| > t/2\}$$

and therefore

$$|\{|T(f)| > t\}| \leqslant |\{T(f_0)| > t/2\}| + |\{T(f_1)| > t/2\}|.$$

Combining the assumptions with Fubini's theorem and proceeding as in the proof of Theorem 2.38, after some computations we arrive at

$$\begin{aligned}
\int_{\mathbb{R}^d} |T(f)(x)|^p \, dx &= p \int_0^\infty t^{p-1} |\{|T(f)| > t\}| \, dt \\
&\leqslant p \int_0^\infty t^{p-1} \left(\left(\frac{C_{d,p_0}}{t/2}\right)^{p_0} \int_{\{|f| \geqslant t/2\}} |f(x)|^{p_0} \, dx\right) dt \\
&\quad + p \int_0^\infty t^{p-1} \left(\left(\frac{C_{d,p_1}}{t/2}\right)^{p_1} \int_{\{|f| < t/2\}} |f(x)|^{p_1} \, dx\right) dt \\
&\leqslant C_{d,p}^p \int_{\mathbb{R}^d} |f(x)|^p \, dx,
\end{aligned}$$

where $C_{d,p}^p = 2^p p \left(\dfrac{C_{d,p_0}^{p_0}}{p - p_0} + \dfrac{C_{d,p_1}^{p_1}}{p_1 - p}\right).$ $\qquad\square$

Problems

5.1 Let $(x_n)_{n\geqslant 1}$ be a sequence with dense linear span in a Banach space X. Using Baire's theorem, prove that this linear span equals X if and only if $\dim X < \infty$.

5.2 Using the Baire category theorem, prove that there exists no norm on $L^1_{\text{loc}}(\mathbb{R}^d)$ that turns this space into a Banach lattice.

Hint: Use Theorem 2.57.

5.3 This problem outlines a proof of the uniform boundedness theorem that does not appeal to the Baire category theorem.

Suppose that $(T_i)_{i\in I}$ is an operator family such that:

(i) $\sup_{i\in I}\|T_i x\| =: C_x < \infty$ for all $x \in X$;
(ii) $\sup_{i\in I}\|T_i\| = \infty$.

For $n = 1, 2, \ldots$ choose indices $i_n \in I$ and vectors $x_n \in X$ such that

$$\frac{1}{4\cdot 3^n}\|T_{i_n}\| \geqslant \sum_{m=1}^{n-1} C_{x_m} + n, \quad \|x_n\| \leqslant \frac{1}{3^n}, \quad \|T_{i_n}x_n\| \geqslant \frac{3}{4\cdot 3^n}\|T_{i_n}\|.$$

Let $x := \sum_{n\geqslant 1} x_n$. By writing

$$T_{i_n}x = \sum_{m=1}^{n-1} T_{i_n}x_m + T_{i_n}x_n + \sum_{m=n+1}^{\infty} T_{i_n}x_m$$

and estimating these terms, prove that $\|T_{i_n}x\| \geqslant n$ for all $n \geqslant 1$. This contradiction proves the result.

5.4 Let X be the linear span of the standard basis vectors of ℓ^2 and let $P_n : \ell^2 \to \mathbb{K}$ denote the orthogonal projection in ℓ^2 onto the nth coordinate. Show that $nP_n x \to 0$ for all $x \in X$ and $\|nP_n\| = n \to \infty$. Conclude that the completeness assumption cannot be omitted from the uniform boundedness theorem.

5.5 The aim of this problem is to prove that a weakly holomorphic function is holomorphic. Let us start with the definitions of these notions. We fix an open set $D \subseteq \mathbb{C}$ and a complex Banach space X. A function $f : D \to X$ is said to be:

- *holomorphic*, if for all $z_0 \in D$ the limit

$$\lim_{z\to z_0} \frac{f(z) - f(z_0)}{z - z_0}$$

exists in X (see Problem 4.2);
- *weakly holomorphic*, if for all $z_0 \in D$ and $x^* \in X^*$ the limit

$$\lim_{z\to z_0} \left\langle \frac{f(z) - f(z_0)}{z - z_0}, x^* \right\rangle$$

exists in $\mathbb{C}$.

Obviously every holomorphic function $f : D \to X$ is weakly holomorphic.

Now let $f : D \to X$ be weakly holomorphic, fix $z_0 \in D$, and let $r > 0$ be so small that the closed disc $\{z \in \mathbb{C} : |z - z_0| \leqslant r\}$ is contained in D.

(a) Applying Proposition 5.5 to the set

$$U = \left\{ \frac{1}{h - g} \left(\frac{f(z_0 + h) - f(z_0)}{h} - \frac{f(z_0 + g) - f(z_0)}{g} \right) : |g|, |h| < \frac{r}{2} \right\}$$

and using the Cauchy integral formula for X-valued holomorphic functions (see Problem 4.2), prove that there is a constant $M \geqslant 0$ such that for all $|g|, |h| < r/2$ we have

$$\left\| \frac{f(z_0 + h) - f(z_0)}{h} - \frac{f(z_0 + g) - f(z_0)}{g} \right\| \leqslant M |h - g|.$$

(b) Deduce that every weakly holomorphic function $f : D \to X$ is holomorphic.

5.6 State and prove an analogue of Proposition 5.5 for the weak* topology.

5.7 Using the open mapping theorem, show that there exists no complete norm $\||| \cdot \|||$ on $C[0, 1]$ with the property that

$$\||| f_n - f \||| \to 0 \iff f_n \to f \text{ pointwise.}$$

5.8 A sequence $(x_n)_{n \geqslant 1}$ in X is called a *Schauder basis* if for every $x \in X$ admits a unique representation as a convergent sum $x = \sum_{n \geqslant 1} c_n x_n$ with $c_n \in \mathbb{K}$ for all $n \geqslant 1$.

Let $(x_n)_{n \geqslant 1}$ be Schauder basis in X, and let Y be the vector space of all scalar sequences $c = (c_n)_{n \geqslant 1}$ such that the sum $\sum_{n \geqslant 1} c_n x_n$ converges in X.

(a) Show that

$$\|c\|_Y := \sup_{N \geqslant 1} \left\| \sum_{n=1}^{N} c_n x_n \right\|$$

defines a norm on Y and that Y is a Banach space with respect to this norm.

(b) Show that $c \mapsto \sum_{n \geqslant 1} c_n x_n$ is an isomorphism from Y onto X.

(c) Conclude that the *coordinate projections*

$$P_k : \sum_{n \geqslant 1} c_n x_n \mapsto c_k$$

are bounded and that $\sup_{k \geqslant 1} \|P_k\| < \infty$.

5.9 For $f \in L^1(-\pi, \pi)$ and $N \in \mathbb{N}$ define the functions

$$s_N f(t) := \sum_{n=-N}^{N} \widehat{f}(n) \exp(int), \quad t \in [-\pi, \pi],$$

where $\widehat{f}(n)$ is the nth Fourier coefficient of f, that is,

$$\widehat{f}(n) := \frac{1}{2\pi}\int_{-\pi}^{\pi} f(s)\exp(-ins)\,ds, \quad n \in \mathbb{Z}.$$

By the results of Section 3.5.a, for all $f \in L^2(-\pi,\pi)$ we have

$$f = \lim_{N\to\infty} s_N f = \sum_{n\in\mathbb{Z}} \widehat{f}(n)\exp(in(\cdot))$$

with convergence in $L^2(-\pi,\pi)$; the series on the right-hand side is the Fourier series of f. One might express the hope that if $f \in C[-\pi,\pi]$ is periodic, then its Fourier series converges to f with respect to the norm of $C[-\pi,\pi]$. The aim of this problem is to prove that this is wrong in a strong sense: there exists a function $f \in C[-\pi,\pi]$ that is periodic in the sense that $f(-\pi) = f(\pi)$ and whose Fourier series diverges at $t = 0$.

(a) Show that $s_N f(t) = \frac{1}{2\pi}\int_{-\pi}^{\pi} f(s)D_N(t-s)\,ds$, where the *Dirichlet kernel* is given by

$$D_N(t) := \sum_{n=-N}^{N} \exp(int) = \frac{\sin\left(N+\frac{1}{2}\right)t}{\sin\left(\frac{1}{2}t\right)}.$$

(b) Show that $\|\Lambda_N\| = \|D_N\|_1$, where the linear map $\Lambda_N : C_{\text{per}}[-\pi,\pi] \to \mathbb{C}$ is given by $\Lambda_N f := s_N f(0)$.
    Hint: To prove the inequality $\|\Lambda_N\| \geqslant \|D_N\|_1$, approximate $\text{sign}(D_N)$ pointwise almost everywhere by a sequence of continuous periodic functions f_n of norm $\leqslant 1$, set $g_n(t) = f_n(-t)$, and use dominated convergence to obtain

$$\lim_{n\to\infty} \Lambda_N(g_n) = \lim_{n\to\infty} \frac{1}{2\pi}\int_{-\pi}^{\pi} f_n(t)D_N(t)\,dt = \frac{1}{2\pi}\int_{-\pi}^{\pi} |D_N(t)|\,dt.$$

    Fill in the missing details.

(c) Show that $\lim_{N\to\infty}\|D_N\|_1 = \infty$.
    Hint: Use that $|\sin(x)| \leqslant |x|$ for all $x \in \mathbb{R}$, and then perform some careful estimates on the resulting integral.

(d) Apply the uniform boundedness theorem to prove that $s_N f(0) \not\to f(0)$ as $N \to \infty$ for some $f \in C_{\text{per}}[-\pi,\pi]$.

5.10 Let $(a_n)_{n\geqslant 1}$ be a scalar sequence with the property that the sum $\sum_{n\geqslant 1} a_n b_n$ converges for all scalar sequences $(b_n)_{n\geqslant 1}$ satisfying $\sum_{n\geqslant 1}|b_n|^2 < \infty$.

(a) Show that $\sum_{n\geqslant 1} a_n b_n$ converges absolutely for all scalar sequences $(b_n)_{n\geqslant 1}$ satisfying $\sum_{n\geqslant 1}|b_n|^2 < \infty$.

(b) Show that $\sum_{n\geqslant 1}|a_n|^2 < \infty$.
    Hint: Apply the closed graph theorem to the mapping $T : \ell^2 \to \ell^1$ defined by $T : (b_n)_{n\geqslant 1} \mapsto (a_n b_n)_{n\geqslant 1}$. Conclude that $(a_n)_{n\geqslant 1}$ defines a bounded functional on ℓ^2.

5.11 Consider a Banach space with a direct sum decomposition $X = X_0 \oplus X_1$. Prove that the projections onto the summands define isomorphisms of Banach spaces

$$X/X_0 \simeq X_1, \quad X/X_1 \simeq X_0.$$

5.12 Let X be a Banach space with direct sum decomposition $X = X_0 \oplus X_1$. Show that if $T_0 : X_0 \to Y$ and $T_1 : X_1 \to Y$ are bounded operators, then the operator $T := T_0 \oplus T_1$ from X to Y defined by

$$T(x_0 + x_1) := T_0 x_0 + T_1 x_1$$

is bounded. What can be said about the norm of T?

Hint: First show that $\|\|x_0 + x_1\|\| := \|x_0\| + \|x_1\|$ is an equivalent norm on X.

5.13 The aim of this problem is to prove that the set of surjective operators from a Banach space X into a Banach space Y is open in $\mathscr{L}(X,Y)$.

(a) Let $T \in \mathscr{L}(X,Y)$ be a surjective operator. Show that there is a constant $A \geqslant 0$ such that for all $y \in Y$ there exists an $x \in X$ such that $\|x\| \leqslant A\|y\|$ and $Tx = y$.

(b) Let $T \in \mathscr{L}(X,Y)$ be a bounded operator. Suppose there exist constants $A \geqslant 0$ and $0 \leqslant B < 1$ such that for all $y \in Y$ with $\|y\| \leqslant 1$ there exists an $x \in X$ such that $\|x\| \leqslant A$ and $\|Tx - y\| \leqslant B$. Show that T is surjective.

Hint: Look into the proof of Lemma 5.7.

(c) Show that if $T \in \mathscr{L}(X,Y)$ is surjective and $S \in \mathscr{L}(X,Y)$ is a bounded operator satisfying $\|S\| < 1/A$, where A is the constant of part (a), then $T + S$ is surjective.

Hint: Apply the first part with $B = A\|S\|$.

5.14 Let X be a vector space.

(a) Suppose that $\| \cdot \|$ and $\| \cdot \|'$ are two norms on X, each of which turns X into a Banach space. Show that if there exists a constant $C \geqslant 0$ such that $\|x\| \leqslant C\|x\|'$ for all $x \in X$, then the two norms are equivalent.

(b) Find the mistake in the following "proof" that *every two norms* $\| \cdot \|$ and $\| \cdot \|'$ turning X into a Banach space are equivalent. Define

$$\|\|x\|\| := \|x\| + \|x\|', \quad x \in X.$$

This is a norm which turns X into a Banach space and we have $\|x\| \leqslant \|\|x\|\|$ and $\|x\|' \leqslant \|\|x\|\|$. Hence part (b) implies that $\|x\|$ and $\|\|x\|\|$ are equivalent and that $\|x\|'$ and $\|\|x\|\|$ are equivalent. It follows that $\| \cdot \|$ and $\| \cdot \|'$ are equivalent.

5.15 Let $(\Omega, \mathscr{F}, \mu)$ be a measure space, let $1 \leqslant p \leqslant \infty$, and suppose that $f : \Omega \to X$ is a function which has the property that the scalar-valued function $\omega \mapsto \langle f(\omega), x^* \rangle$ belongs to $L^p(\Omega)$ for every $x^* \in X^*$.

(a) Show that the mapping $T : X^* \to L^p(\Omega)$ defined by $x^* \mapsto \langle f(\cdot), x^* \rangle$ is closed.

(b) Deduce that there exists a constant $C \geqslant 0$ such that

$$\|\langle f(\cdot), x^* \rangle\|_{L^p(\Omega)} \leqslant C\|x^*\|, \quad x^* \in X^*.$$

5.16 Let $(\Omega, \mathscr{F}, \mu)$ be a finite measure space and let X be a Banach space.

(a) Let $f : \Omega \to X$ be a strongly measurable function. Show that if there is an exponent $1 < p \leqslant \infty$ such that $\langle f(\cdot), x^* \rangle \in L^p(\Omega)$ for all $x^* \in X^*$, then there exists a unique element $x_f \in X$, the *Pettis integral* of f with respect to μ, such that

$$\langle x_f, x^* \rangle = \int_\Omega \langle f(\omega), x^* \rangle \, d\mu(\omega), \quad x^* \in X^*.$$

Hint: The integrals $\int_\Omega \mathbf{1}_{\{\|f\| \leqslant n\}} f \, d\mu$ are well defined as Bochner integrals.

(b) Show that the result of part (a) fails for $p = 1$.

Hint: Let $(A_n)_{n \geqslant 1}$ be a sequence of disjoint intervals of positive measure $|A_n|$ in the interval $(0, 1)$ and consider the function $f : (0, 1) \to c_0$ defined by

$$f(t) := \sum_{n \geqslant 1} \frac{1}{|A_n|} \mathbf{1}_{A_n}(t) e_n, \quad t \in (0, 1),$$

where $(e_n)_{n \geqslant 1}$ is the sequence of standard unit vectors in c_0.

5.17 Write out a proof of Theorem 5.31.

5.18 For $n \geqslant 1$ and $f \in L^2(\mathbb{R}^d)$ let

$$\mathscr{F}_n f(\xi) := \frac{1}{(2\pi)^{d/2}} \int_{[-n,n]^d} \exp(-ix \cdot \xi) f(x) \, dx.$$

Show that $\mathscr{F}_n$ maps $L^2(\mathbb{R}^d)$ into itself and defines a bounded operator on $L^2(\mathbb{R}^d)$, and show that for all $f \in L^2(\mathbb{R}^d)$ we have the following identity for the Fourier–Plancherel transform:

$$\mathscr{F} f = \lim_{n \to \infty} \mathscr{F}_n f,$$

where the limit is taken in $L^2(\mathbb{R}^d)$.

5.19 It was shown in Lemma 5.19 that for $\lambda > 0$ the Fourier transform of $g(x) = (2\pi)^{-d/2} \exp\left(-\frac{1}{2}\lambda|x|^2\right)$ is given as

$$\widehat{g}(\xi) = (2\pi\lambda)^{-d/2} \exp\left(-\frac{1}{2}|\xi|^2/\lambda\right).$$

Give an alternative proof of this identity by completing the following steps:

(a) it suffices to prove the identity for $\lambda = 1$ and in dimension $d = 1$;

(b) the function $u(x) := (2\pi)^{-d/2} \exp\left(-\frac{1}{2}x^2\right)$ solves the differential equation

$$u'(x) + xu(x) = 0;$$

(c) the Fourier transform of u also satisfies the differential equation;

(d) apply the Picard–Lindelöf theorem (Theorem 2.12).

5.20 Consider the Fourier–Plancherel transform $\mathscr{F} : f \mapsto \hat{f}$ on $L^2(\mathbb{R}^d)$.

(a) Show that $\mathscr{F}^2 = R$, where $Rf(x) := f(-x)$ is the reflection operator on $L^2(\mathbb{R}^d)$.

(b) Deduce that $\mathscr{F}^4 = I$.

5.21 Prove that the Hilbert transform H on $L^2(\mathbb{R})$ satisfies $H^2 = -I$.

5.22 This problem establishes a connection between the Hilbert transform and the theory of harmonic functions.

For real-valued functions $f \in L^2(\mathbb{R})$ we define $u_f : \mathbb{R} \times (0, \infty) \to \mathbb{R}$

$$u_f(x, y) := p_y * f(x), \quad x \in \mathbb{R}, \ y > 0,$$

where

$$p_y(x) := \frac{1}{\pi} \frac{y}{x^2 + y^2}, \quad x \in \mathbb{R}, \ y > 0,$$

is the *Poisson kernel*.

(a) Show that u_f is *harmonic*, that is, $u \in C^2(\mathbb{R} \times (0, \infty))$ and $\Delta u \equiv 0$.

(b) Show that $u_f + i u_{Hf}$ is holomorphic.

5.23 Let $f \in L^1(\mathbb{R})$ satisfy $\hat{f}(-\xi) = 0$ for almost all $\xi \geq 0$.

(a) Show that for all $y > 0$ the function $p_y * f$, where p_y is the Poisson kernel introduced in the preceding problem, is integrable and its Fourier transform belongs to $L^1(\mathbb{R}) \cap L^2(\mathbb{R})$.

Hint: Compute the Fourier transform of p_y.

(b) Using Fourier inversion, prove that the function

$$g(x + iy) := p_y * f(x)$$

is holomorphic on the open half-plane $\{\operatorname{Im} z = x + iy > 0\}$.

(c) Using Proposition 2.34, show that

$$\lim_{y \downarrow 0} \|g(\cdot + iy) - f(\cdot)\|_{L^1(\mathbb{R})} = 0.$$

Somewhat informally, part (c) states that every L^1-function whose Fourier transform vanishes on the negative half-line is the boundary value (in the L^1-sense) of a holomorphic function on the upper half-plane.

(d) State and prove a version of part (c) for the disc.

5.24 We use the notation introduced in Problem 2.27. Let $\mathscr{F}$ be the Fourier–Plancherel transform on $L^2(\mathbb{R}^d)$ and let X be a Banach space. On the space $L^2(\mathbb{R}^d) \otimes X$ we define the linear operator $\mathscr{F} \otimes I$ by

$$(\mathscr{F} \otimes I)(f \otimes x) := (\mathscr{F} f) \otimes x, \quad f \in L^2(\mathbb{R}^d), \ x \in X.$$

(a) Show that this operator is well defined.

(b) Show that if $X = \ell^p$ with $1 \leqslant p \leqslant \infty$, then $\mathscr{F} \otimes I$ extends to a bounded operator on $L^2(\mathbb{R}; \ell^p)$ if and only if $p = 2$.

Hint: For $1 \leqslant p < 2$ consider the functions

$$f_N := \sum_{n=0}^{N} f(\cdot + 2\pi n) \otimes e_{n+1},$$

where $(e_n)_{n \geqslant 1}$ is the sequence of standard unit vectors in ℓ^p and $0 \neq f \in C_c(\mathbb{R})$ has support in the interval $(-\pi, \pi)$; for $2 < p \leqslant \infty$ use the functions

$$f_N := \sum_{n=0}^{N} e^{-in(\cdot)} f \otimes e_{n+1}.$$

5.25 Let $1 \leqslant p \leqslant \infty$. Young's inequality implies that the convolution of a function $f \in L^1(\mathbb{R}^d)$ with a function $g \in L^p(\mathbb{R}^d)$ belongs to $L^p(\mathbb{R}^d)$ and $\|f * g\|_p \leqslant \|f\|_1 \|g\|_p$.

(a) Write out the proof of this result obtained by taking $r = 1$ in the proof of Proposition 2.33.

The special case of Young's inequality just stated can be reformulated as saying that for every $g \in L^p(\mathbb{R}^d)$ the convolution operator $C_g : f \mapsto f * g$ is bounded from $L^1(\mathbb{R}^d)$ to $L^p(\mathbb{R}^d)$ with norm

$$\|C_g\|_{\mathscr{L}(L^1(\mathbb{R}^d), L^p(\mathbb{R}^d))} \leqslant \|g\|_p.$$

(b) Let $\frac{1}{p} + \frac{1}{q} = 1$. Using Hölder's inequality, show that the restriction of C_g to $L^1(\mathbb{R}^d) \cap L^q(\mathbb{R}^d)$ extends uniquely to a bounded operator from $L^q(\mathbb{R}^d)$ to $L^\infty(\mathbb{R}^d)$ of norm

$$\|C_g\|_{\mathscr{L}(L^q(\mathbb{R}^d), L^\infty(\mathbb{R}^d))} \leqslant \|g\|_p.$$

(c) Use the Riesz–Thorin interpolation theorem to obtain an alternative proof of the general form of Young's inequality.

5.26 Let $(\Omega, \mathscr{F}, \mu)$ be a measure space and let $1 \leqslant p, q < \infty$ satisfy $\frac{1}{p} + \frac{1}{q} = 1$. The aim of this problem is to use the Riesz–Thorin interpolation theorem to derive the *Clarkson inequalities*:

(1) if $1 \leqslant p \leqslant 2$, then for all $f, g \in L^p(\Omega)$ we have

$$(\|f + g\|_p^p + \|f - g\|_p^p)^{1/p} \leqslant 2^{1/p}(\|f\|_p^p + \|g\|_p^p)^{1/p}$$

and

$$(\|f + g\|_p^q + \|f - g\|_p^q)^{1/q} \leqslant 2^{1/q}(\|f\|_p^p + \|g\|_p^p)^{1/p};$$

(2) if $2 \leqslant p < \infty$, then for all $f, g \in L^p(\Omega)$ we have

$$(\|f+g\|_p^p + \|f-g\|_p^p)^{1/p} \leqslant 2^{1/q}(\|f\|_p^p + \|g\|_p^p)^{1/p}$$

and

$$(\|f+g\|_p^p + \|f-g\|_p^p)^{1/p} \leqslant 2^{1/p}(\|f\|_p^q + \|g\|_p^q)^{1/q}.$$

Let $1 \leqslant r, s \leqslant \infty$. On the cartesian product $L^r(\Omega) \times L^r(\Omega)$ we consider the norm

$$\|(f,g)\|_{r,s} := (\|f\|_r^s + \|g\|_r^s)^{1/s}.$$

(a) Show that the resulting normed space $X_{r,s}(\Omega)$ is complete.

On $X_{r,s}(\Omega)$ we consider the operator

$$T : (f,g) \mapsto (f+g, f-g).$$

Its norm will be denoted by $\|T\|_{r,s}$.

(b) Show that $\|T\|_{1,1} = 2$ and $\|T\|_{2,2} = \sqrt{2}$.
(c) Deduce the first inequality in (1).
(d) Show that for all $f, g \in L^1(\Omega)$ we have $\|T(f,g)\|_{1,\infty} \leqslant \|(f,g)\|_{1,1}$.
(e) Deduce the second inequality in (1).
(f) Prove the inequalities in (2).
(g) Prove that $L^p(\Omega)$ is strictly convex, that is, $\|f\|_p = \|g\|_p = 1$ with $f \neq g$ implies $\|\frac{1}{2}(f+g)\|_p < 1$.

5.27 State and prove an analogue of the Hausdorff–Young theorem for the circle.
5.28 Write out the details of the proof of the Marcinkiewicz interpolation theorem for the case $p_1 = \infty$.

6

Spectral Theory

Spectral theory is the branch of operator theory that seeks to extend the theory of eigenvalues to an infinite-dimensional setting. Much of its power derives from the observation that, away from the spectrum of a bounded operator T, the operator-valued function $\lambda \mapsto (\lambda - T)^{-1}$ is holomorphic. This makes it possible to import results from the theory of functions into operator theory. For instance, the fact that bounded operators on nonzero Banach spaces have nonempty spectra is deduced from Liouville's theorem, and the Cauchy integral formula can be used to introduce a functional calculus for functions holomorphic in an open set containing the spectrum of T.

6.1 Spectrum and Resolvent

In Linear Algebra, a complex number λ is said to be an *eigenvalue* of an $n \times n$ matrix A with complex coefficients if there exists a nonzero vector $x \in \mathbb{C}^n$ such that $Ax = \lambda x$. The number λ is an eigenvalue if and only if $\lambda I - A$ fails to be invertible, or equivalently, if and only if $\det(\lambda I - A) = 0$. Writing out the determinant we obtain the so-called characteristic polynomial in the variable λ, which has n zeroes (counting multiplicities) by the main theorem of Algebra. Our first task will be to investigate to what extent these results generalise to bounded operators acting on a Banach space.

Throughout the chapter, T denotes a bounded operator acting on a complex space X. We work over the complex scalars; this convention will remain force throughout the rest of this work.

Definition 6.1 (Resolvent and spectrum). The *resolvent set* of an operator $T \in \mathscr{L}(X)$ is the set $\rho(T)$ consisting of all $\lambda \in \mathbb{C}$ for which the operator $\lambda I - T$ is *boundedly invertible*, by which we mean that there exists a bounded operator U on X such that

$$(\lambda I - T)U = U(\lambda I - T) = I.$$

The *spectrum* of T is the complement of the resolvent set of T:

$$\sigma(T) := \mathbb{C} \setminus \rho(T).$$

From now on we shall write $\lambda - T$ instead of $\lambda I - T$. It is customary to write

$$R(\lambda, T) := (\lambda - T)^{-1}$$

for the *resolvent operator* of T at the point $\lambda \in \rho(T)$. By the open mapping theorem (Theorem 5.8), a complex number λ belongs to $\rho(T)$ if and only if $\lambda - T$ is a bijection on X.

Example 6.2. The spectrum of an $n \times n$ matrix with complex coefficients, viewed as a bounded operator acting on $\mathbb{C}^n$, equals its set of eigenvalues.

In the present setting, a complex number λ is said to be an *eigenvalue* of the bounded operator $T \in \mathscr{L}(X)$ if $Tx = \lambda x$ for some nonzero vector $x \in X$; such a vector is then said to be an *eigenvector*. The set $\sigma_{\mathrm{p}}(T)$ of all eigenvalues of T is called the *point spectrum* of T. If λ is an eigenvalue of T, then $\lambda - T$ is not injective and therefore not invertible. As a result, eigenvalues belong to the spectrum. In contrast to the finite-dimensional situation, however, points in the spectrum need not be eigenvalues:

Example 6.3. The right shift T on ℓ^2, given by the right shift

$$T : (c_1, c_2, \dots) \mapsto (0, c_1, c_2, \dots)$$

has no eigenvalues. Indeed, the identity $Tc = \lambda c$ can be written out as

$$(0, c_1, c_2, \dots) = (\lambda c_1, \lambda c_2, \dots).$$

If $\lambda \neq 0$, comparison of the entries of these sequences inductively gives $c_n = 0$ for all $n \geqslant 1$. If $\lambda = 0$, the identity reads $(0, c_1, c_2, \dots) = (0, 0, \dots)$ and again we obtain $c_n = 0$ for all $n \geqslant 1$. In both cases we find that $Tc = \lambda c$ only admits the zero solution $c = 0$.

The spectrum of T equals the closed unit disc: $\sigma(T) = \overline{\mathbb{D}}$. This can be proved directly (see Problem 6.4) or by the following argument based on results proved below. By Proposition 6.18 we have $\sigma(T) = \sigma(T^*)$. The adjoint operator T^* is readily identified as the left shift $(c_1, c_2, \dots) \mapsto (c_2, c_3, \dots)$. For each $\lambda \in \mathbb{C}$ with $|\lambda| < 1$, the element $(1, \lambda, \lambda^2, \dots) \in \ell^2$ is an eigenvector for this operator with eigenvalue λ. It follows that $\mathbb{D} \subseteq \sigma(T)$. Since by Lemma 6.7 the spectrum of a bounded operator is closed, this forces $\overline{\mathbb{D}} \subseteq \sigma(T)$. On the other hand, by Lemma 6.6, the fact that T is a contraction implies that $\sigma(T) \subseteq \overline{\mathbb{D}}$.

Remark 6.4. In contrast to the case of matrices in finite dimensions, the existence of a left inverse does not imply the existence of a right inverse and vice versa. For example, the right shift in ℓ^2 has a left inverse, namely by the left shift, but not a right inverse; similarly the left shift in ℓ^2 has a right inverse, namely the right shift, but not a left inverse. This is the reason for insisting on the existence of a two-sided inverse in Definition 6.1. If both a left inverse U_l and a right inverse U_r exist, then necessarily $U_l = U_r$ and this operator is a two-sided inverse.

If the bounded operators T and U are boundedly invertible, then so is TU and

$$(TU)^{-1} = U^{-1}T^{-1}.$$

Invertibility of TU by itself does not imply invertibility of T or U; a counterexample is obtained by taking T and U be the left and right shift on ℓ^2. However we do have the following result.

Lemma 6.5. *The product of two commuting bounded operators is invertible if and only if each of the operators is invertible.*

Proof The 'if' part is clear. For the 'only if' part, suppose that $TU = UT$ is invertible. The invertibility of TU implies that T is surjective and U is injective, and likewise the invertibility of UT implies that U is surjective and T is injective. It follows that both T and U are bijective, and by the open mapping theorem their inverses are bounded. □

Our first main result, Theorem 6.11, asserts that the spectrum of a bounded operator acting on a nonzero Banach space is always a *nonempty compact* subset of the complex plane. This will be deduced from a series of lemmas.

Lemma 6.6 (Neumann series). *If $\|T\| < 1$, then $I - T$ is boundedly invertible and its inverse is given by the absolutely convergent series*

$$(I - T)^{-1} = \sum_{n=0}^{\infty} T^n.$$

As a consequence, the spectrum of a bounded operator T is contained in the closed disc $\{z \in \mathbb{C} : |z| \leqslant \|T\|\}$.

Proof The absolute convergence in $\mathscr{L}(X)$ of the series follows from $\sum_{n=0}^{\infty} \|T^n\| \leqslant \sum_{n=0}^{\infty} \|T\|^n < \infty$. By the completeness of $\mathscr{L}(X)$, the series $\sum_{n=0}^{\infty} T^n$ converges in $\mathscr{L}(X)$. The identity in the statement of the lemma is a consequence of the identities

$$(I - T) \sum_{n=0}^{N} T^n = \sum_{n=0}^{N} T^n (I - T) = I - T^{N+1},$$

valid for all $N \geqslant 1$. Upon letting $N \to \infty$ they give

$$(I - T) \sum_{n=0}^{\infty} T^n = \sum_{n=0}^{\infty} T^n (I - T) = I,$$

which means that the bounded operator $\sum_{n=0}^{\infty} T^n$ is a two-sided inverse for $I - T$.

To prove the second assertion, let $T \in \mathscr{L}(X)$ be an arbitrary bounded operator. By the first assertion, for all $\lambda \in \mathbb{C}$ with $\lambda > \|T\|$ the operator $\lambda - T = \lambda(I - T/\lambda)$ is boundedly invertible. □

As an application we prove that the spectrum is always a closed subset of $\mathbb{C}$.

Lemma 6.7. *The spectrum $\sigma(T)$ is a closed subset of $\mathbb{C}$. More precisely, if $\lambda \in \rho(T)$, then $\rho(T)$ contains the open ball with centre λ and radius $r = 1/\|R(\lambda, T)\|$. If $|\lambda - \mu| \leqslant \delta r$ with $0 \leqslant \delta < 1$, then*

$$\|R(\mu, T)\| \leqslant \frac{1}{1 - \delta}\|R(\lambda, T)\|.$$

Proof If $S \in \mathscr{L}(X)$ is boundedly invertible and $U \in \mathscr{L}(X)$ has norm $\|U\| \leqslant \delta r$ with $r := 1/\|S^{-1}\|$, then $\|S^{-1}U\| \leqslant \delta < 1$ and therefore, by Lemma 6.6, $S - U = S(I - S^{-1}U)$ is boundedly invertible and

$$\|(S - U)^{-1}\| \leqslant \|S^{-1}\|\|(I - S^{-1}U)^{-1}\| = \|S^{-1}\|\left\|\sum_{n=0}^{\infty}(S^{-1}U)^n\right\| \leqslant \frac{1}{r}\sum_{n=0}^{\infty}\delta^n = \frac{1}{r}\frac{1}{1 - \delta}.$$

□

Lemma 6.8 (Resolvent identity). *For all $\lambda, \mu \in \rho(T)$ we have*

$$R(\lambda, T) - R(\mu, T) = (\mu - \lambda)R(\lambda, T)R(\mu, T).$$

Proof Multiply both sides with the invertible operator $(\mu - T)(\lambda - T)$. □

Definition 6.9 (Holomorphy). Let Ω be an open subset of $\mathbb{C}$. A function $f : \Omega \to X$ is *holomorphic* if for all $z_0 \in \Omega$ the limit

$$\lim_{z \to z_0}\frac{f(z) - f(z_0)}{z - z_0}$$

exists in X.

Some properties of Banach space-valued functions have already been explored in Problems 4.2 and 5.5.

Lemma 6.10. *The function $\lambda \mapsto R(\lambda, T)$ is holomorphic on $\rho(T)$ and satisfies*

$$\lim_{|\lambda| \to \infty}\|R(\lambda, T)\| = 0.$$

Proof Continuity of the mapping $\lambda \mapsto R(\lambda, T)$ follows from the resolvent identity and the bound in Lemma 6.7. To prove the holomorphy of $\lambda \mapsto R(\lambda, T)$ on $\rho(T)$ we use the resolvent identity and the continuity of $\lambda \mapsto R(\lambda, T)$ to obtain

$$\lim_{\mu \to \lambda}\frac{R(\lambda, T) - R(\mu, T)}{\lambda - \mu} = -\lim_{\mu \to \lambda}R(\lambda, T)R(\mu, T) = -(R(\lambda, T))^2.$$

For $|\lambda| > 2\|T\|$ the Neumann series gives

$$\|R(\lambda,T)\| = |\lambda|^{-1}\|(I-\lambda^{-1}T)^{-1}\| \leqslant |\lambda|^{-1}\sum_{n=0}^{\infty}\|\lambda^{-1}T\|^n \leqslant 2|\lambda|^{-1}.$$

This proves the second assertion. □

We are ready for the first main result of this chapter:

Theorem 6.11 (Nonemptiness of the spectrum). *If T is a bounded operator on a non-zero Banach space, then $\sigma(T)$ is a nonempty compact subset of the closed disc $\{\lambda \in \mathbb{C} : |\lambda| \leqslant \|T\|\}$.*

Proof Containment in $\{\lambda \in \mathbb{C} : |\lambda| \leqslant \|T\|\}$ and closedness of the spectrum have already been proved in Lemmas 6.6 and 6.7, respectively. Since bounded closed subsets of $\mathbb{C}$ are compact, this gives the compactness of $\sigma(T)$.

Suppose, for a contradiction, that $\sigma(T) = \varnothing$. Then the function $\lambda \mapsto R(\lambda,T)$ is holomorphic on $\mathbb{C}$. By Lemma 6.10 it is also bounded. Now we are in a position to apply Liouville's theorem: for all $x \in X$ and $x^* \in X^*$ we find that $\lambda \mapsto \langle R(\lambda,T)x, x^*\rangle$ is constant. Its limit for $|\lambda| \to \infty$ is zero, and therefore $\langle R(\lambda,T)x, x^*\rangle = 0$ for all $\lambda \in \rho(T)$ and all $x \in X$ and $x^* \in X^*$. By the Hahn–Banach theorem, $R(\lambda,T)x = 0$ for all $\lambda \in \rho(T)$ and all $x \in X$. It follows that $R(\lambda,T) = 0$ for all $\lambda \in \rho(T)$. This implies $X = \{0\}$. □

Instead of using duality to reduce matters to scalar-valued functions one may note that the proof of Liouville's theorem generalises *mutatis mutandis* to holomorphic functions with value in a Banach space.

We have seen in Lemma 6.10 that the resolvent $\lambda \mapsto R(\lambda,T)$ is holomorphic on $\rho(T)$. The next result shows that the topological boundary $\partial\rho(T) := \overline{\rho(T)} \setminus \rho(T)$ is a natural barrier for holomorphy for this function.

Proposition 6.12. *If $\lambda_n \to \lambda$ in $\mathbb{C}$, with each $\lambda_n \in \rho(T)$ and $d(\lambda_n, \sigma(T)) \to 0$, then*

$$\lim_{n\to\infty}\|R(\lambda_n,T)\| = \infty.$$

Proof By Lemma 6.7, if $\mu \in \rho(T)$, then the open ball $B(\mu; \|R(\mu,T)\|^{-1})$ is contained in $\rho(T)$. This implies the more precise assertion that for all $\mu \in \rho(T)$ we have $d(\mu, \sigma(T)) \geqslant \|R(\mu,T)\|^{-1}$, that is, $\|R(\mu,T)\| \geqslant 1/d(\mu, \sigma(T))$. □

An immediate application is the following analytic continuation result.

Corollary 6.13. *If $D \subseteq \mathbb{C}$ is a connected open set intersecting the resolvent set of a bounded operator T, and if $\lambda \mapsto R(\lambda,T)$ extends holomorphically to D, then $D \subseteq \rho(T)$.*

A more substantial application of Proposition 6.12 is the following result about the spectra of isometries. Recall that an *isometry* is an operator $T \in \mathscr{L}(X,Y)$ such that $\|Tx\| = \|x\|$ for all $x \in X$.

Corollary 6.14. *The spectrum of an isometry is either contained in the unit circle* $\mathbb{T}$ *or else equals the closed unit disc* $\overline{\mathbb{D}}$.

Proof First note that an isometry T has norm $\|T\| = 1$, so that $\sigma(T) \subseteq \overline{\mathbb{D}}$ by the second assertion of Lemma 6.6.

If $\mu \in \rho(T)$ with $0 < |\mu| < 1$, then, using that T is an isometry, for all $x \in X$ we find

$$\|x\| = \|(\mu - T)R(\mu,T)x\| \geqslant \|TR(\mu,T)x\| - \|\mu R(\mu,T)x\| = (1 - |\mu|)\|R(\mu,T)x\|$$

and therefore $\|R(\mu,T)\| \leqslant 1/(1 - |\mu|)$. In view of Proposition 6.12, this implies that for all $0 < r < 1$ the disc $\{\mu \in \mathbb{C} : |\mu| \leqslant r\}$ does not contain boundary points of $\sigma(T)$. This being true for all $0 < r < 1$ it follows that either the open unit disc $\mathbb{D}$ is contained in $\sigma(T)$ or else $\mathbb{D}$ contains no points of $\sigma(T)$. In the former case we have $\sigma(T) = \overline{\mathbb{D}}$, as $\sigma(T)$ is closed and contained in $\overline{\mathbb{D}}$; in the latter case we have $\sigma(T) \subseteq \mathbb{T}$. $\qquad\square$

This result is the best possible in the following sense: both the closed unit disc and any nonempty closed subset of the unit circle can be realised as the spectrum of suitable isometries. For instance, the left and right shift on ℓ^2 provide examples of the former, and if $K \subseteq \mathbb{T}$ is a nonempty closed set, then the bounded operator on $C(K)$ given by $(Tf)(z) = zf(z)$ is easily verified to have spectrum equal to K.

We have the following continuity result for spectra:

Proposition 6.15 (Lower semicontinuity of the spectrum). *Let* Ω *be an open set in the complex plane containing* $\sigma(T)$. *Then there exists a* $\delta > 0$ *such that if the bounded operator* T' *satisfies* $\|T - T'\| < \delta$, *then* $\sigma(T') \subseteq \Omega$.

Proof In the proof of Lemma 6.10 we have seen that $\lim_{|\lambda| \to \infty} \|R(\lambda,T)\| = 0$. In particular, $\sup_{|\lambda| > 2\|T\|} \|R(\lambda,T)\| < \infty$. By the continuity of $\lambda \mapsto R(\lambda,T)$ we also have $\sup_{\{|\lambda| \leqslant 2\|T\|\} \cap \{\lambda \notin \Omega\}} \|R(\lambda,T)\| < \infty$. Combining these, we find that

$$\sup_{\lambda \notin \Omega} \|R(\lambda,T)\| < \infty.$$

Denote this supremum by M. If $\|T - T'\| < 1/M$ and $\lambda \notin \Omega$, then from

$$\lambda - T' = (\lambda - T)[I + R(\lambda,T)(T - T')]$$

we infer that $\lambda - T'$ is invertible, noting that $I + R(\lambda,T)(T - T')$ is invertible since $\|R(\lambda,T)(T - T')\| < M \cdot 1/M = 1$. $\qquad\square$

It has already been noted that eigenvalues belong to the spectrum, and that a bounded operator need not have any eigenvalues (see Example 6.3). We now prove a useful result that makes up for this to some extent.

Definition 6.16 (Approximate eigenspectrum). The number $\lambda \in \mathbb{C}$ is called an *approximate eigenvalue* of the operator T if there exists a sequence $(x_n)_{n \geqslant 1}$ in X with the following two properties:

(1) $\|x_n\| = 1$ for all $n \geqslant 1$;

(2) $\|Tx_n - \lambda x_n\| \to 0$ as $n \to \infty$.

In this context the sequence $(x_n)_{n \geqslant 1}$ is called an *approximate eigensequence* for λ. The set of all approximate eigenvalues is called the *approximate point spectrum*.

Every eigenvalue is an approximate eigenvalue. Approximate eigenvalues belong to the spectrum, for if λ were an approximate eigenvalue belonging to the resolvent set of T, we would arrive at the contradiction

$$1 = \|x_n\| = \|R(\lambda, T)(Tx_n - \lambda x_n)\| \leqslant \|R(\lambda, T)\| \|Tx_n - \lambda x_n\| \to 0 \quad \text{as } n \to \infty.$$

Proposition 6.17. *The boundary of* $\sigma(T)$ *consists of approximate eigenvalues.*

Proof If $\lambda \in \partial \sigma(T)$, then there exists a sequence $(\lambda_n)_{n \geqslant 1}$ in $\rho(T)$ converging to λ. Using Proposition 6.12 and the uniform boundedness principle, we find a vector $x \in X$ such that $\|R(\lambda_n, T)x\| \to \infty$. The vectors

$$x_n := \frac{R(\lambda_n, T)x}{\|R(\lambda_n, T)x\|}$$

then define an approximate eigensequence: this follows from

$$\|Tx_n - \lambda x_n\| = \frac{\|[(T - \lambda_n) + (\lambda_n - \lambda)]R(\lambda_n, T)x\|}{\|R(\lambda_n, T)x\|} \leqslant \frac{\|x\|}{\|R(\lambda_n, T)x\|} + |\lambda_n - \lambda| \to 0.$$

$\square$

We have the following duality result:

Proposition 6.18. *The spectrum of the adjoint of a bounded operator* T *equals*

$$\sigma(T^*) = \sigma(T).$$

Proof If $\lambda \in \rho(T)$, then

$$(\lambda - T^*)[R(\lambda, T)]^* = [R(\lambda, T)(\lambda - T)]^* = I_X^* = I_{X^*},$$

and similarly $[R(\lambda, T)]^*(\lambda - T^*) = I_{X^*}$, from which it follows that $\lambda \in \rho(T^*)$ and $R(\lambda, T^*) = [R(\lambda, T)]^*$. This proves the inclusion $\sigma(T^*) \subseteq \sigma(T)$.

To complete the proof we show that $\rho(T^*) \subseteq \rho(T)$. Applying what we just proved to T^* we obtain $\rho(T^*) \subseteq \rho(T^{**})$. Fix $\lambda \in \rho(T^*)$. Identifying X with its natural isometric image in X^{**} (see Proposition 4.21), we wish to prove that the restriction of $R(\lambda, T^{**})$ to X maps X into itself. This restriction will be a two-sided inverse for $\lambda - T$, proving that $\lambda \in \rho(T)$.

By Proposition 5.14 the range $\mathsf{R}(\lambda - T)$ is dense in X. Moreover, for all $x \in X$ we have

$$x = R(\lambda, T^{**})(\lambda - T^{**})x = R(\lambda, T^{**})(\lambda - T)x.$$

This shows that $R(\lambda, T^{**})$ maps the dense subspace $\mathsf{R}(\lambda - T)$ of X into X, and therefore it maps all of X into X, by the boundedness of $R(\lambda, T^{**})$ and the closedness of X in X^{**}. The restriction of $R(\lambda, T^{**})$ to X is therefore a bounded operator on X, which is a left inverse to $\lambda - T$. It is also a right inverse, since in X^{**} we have the identities

$$(\lambda - T)R(\lambda, T^{**})x = (\lambda - T^{**})R(\lambda, T^{**})x = x.$$

This proves that $\lambda - T$ is invertible, with inverse $R(\lambda, T^{**})|_X$. Hence $\rho(T^*) \subseteq \rho(T)$, or equivalently $\sigma(T) \subseteq \sigma(T^*)$. $\square$

We continue with an observation about relative spectra. Let $\mathscr{A} \subseteq \mathscr{L}(X)$ be a *unital closed subalgebra*, that is, $\mathscr{A}$ is a closed subspace of $\mathscr{L}(X)$ closed under taking compositions and containing the identity operator I. For an operator $T \in \mathscr{A}$ we define $\rho_{\mathscr{A}}(T)$ as the set of all $\lambda \in \mathbb{C}$ for which the operator $\lambda - T$ boundedly invertible in $\mathscr{A}$, that is, there exists an operator $U \in \mathscr{A}$ such that $(\lambda - T)U = U(\lambda - T) = I$. We further set $\sigma_{\mathscr{A}}(T) := \mathbb{C} \setminus \rho_{\mathscr{A}}(T)$. By redoing the proofs of Lemmas 6.6 and 6.7, $\rho_{\mathscr{A}}(T)$ is an open set and $\sigma_{\mathscr{A}}(T)$ is a closed set contained in the closed disc of radius $\|T\|$, and therefore $\sigma_{\mathscr{A}}(T)$ is compact.

It is evident that $\rho_{\mathscr{A}}(T) \subseteq \rho(T)$ and therefore

$$\sigma(T) \subseteq \sigma_{\mathscr{A}}(T).$$

The next result provides a partial converse.

Proposition 6.19. *Let $\mathscr{A} \subseteq \mathscr{L}(X)$ be a unital closed subalgebra and let $T \in \mathscr{A}$. Then*

$$\partial \sigma_{\mathscr{A}}(T) \subseteq \sigma(T).$$

Proof Let $\lambda \in \partial \sigma_{\mathscr{A}}(T)$ and let $\lambda_n \to \lambda$ in $\mathbb{C}$ with each λ_n in $\rho_{\mathscr{A}}(T)$. By redoing the proof of Proposition 6.12 we obtain that $\|R(\lambda_n, T)\| \to \infty$ as $n \to \infty$. Since $\rho(T)$ is open and the resolvent $\lambda \mapsto R(\lambda, T)$ continuous with respect to the operator norm, this implies that $\lambda \in \sigma(T)$. $\square$

The example of the left shift T on $X = \ell^2(\mathbb{Z})$ and the unital closed subalgebra $\mathscr{A}$ generated by the identity and the right shift, shows that this result is the best possible: here one has $\partial \sigma_{\mathscr{A}}(T) = \sigma(T) = \mathbb{T}$ and $\sigma_{\mathscr{A}}(T) = \overline{\mathbb{D}}$.

6.2 The Holomorphic Functional Calculus

If $f : \mathbb{C} \to \mathbb{C}$ is an entire function and T is a bounded operator on X, we may define a bounded operator $f(T)$ on X as follows. Writing f as a convergent power series about $z = 0$,

$$f(z) = \sum_{n=0}^{\infty} \frac{f^{(n)}(0)}{n!} z^n,$$

we define

$$f(T) := \sum_{n=0}^{\infty} \frac{f^{(n)}(0)}{n!} T^n.$$

This series converges absolutely in $\mathscr{L}(X)$ since the same is true for the power series of $f(z)$ for every $z \in \mathbb{C}$. The mapping $f \mapsto f(T)$ is called the *entire functional calculus* of T and has the following properties, the easy proofs of which we leave to the reader:

(i) if $f(z) = z^n$ with $n \in \mathbb{N}$, then $f(T) = T^n$;
(ii) $f(T)g(T) = (fg)(T)$;
(iii) $g(f(T)) = (g \circ f)(T)$.

This calculus may be used to define operators such as $\exp(T)$, $\sin(T)$, $\cos(T)$, and so forth. There is a beautiful way to extend the entire functional calculus to a larger class of holomorphic functions, namely by replacing power series expansions by the Cauchy integral formula

$$f(z_0) = \frac{1}{2\pi i} \int_{\Gamma} \frac{f(\lambda)}{\lambda - z_0} \, d\lambda. \tag{6.1}$$

Here, f is assumed to be a holomorphic function on an open set Ω in the complex plane containing z_0, and Γ is a suitable contour winding about z_0 in Ω counterclockwise once. Formally substituting T for z_0 and interpreting $1/(\lambda - T)$ as $R(\lambda, T)$, one is led to conjecture that a holomorphic functional calculus may be defined by the formula

$$f(T) := \frac{1}{2\pi i} \int_{\Gamma} f(\lambda) R(\lambda, T) \, d\lambda. \tag{6.2}$$

Since $\lambda \mapsto R(\lambda, T)$ is continuous on $\rho(T)$ with respect to the operator norm of $\mathscr{L}(X)$, after parametrising Γ the integral in (6.2) is well defined as a Riemann integral with values in $\mathscr{L}(X)$ (see Section 1.1).

In order to flesh out a set of conditions on Ω, Γ, and T to make this idea work we first take a closer look at the precise assumptions in the Cauchy integral formula (6.1). These are that Ω is an open set in the complex plane containing z_0 and Γ is a piecewise continuously differentiable closed contour in $\Omega \setminus \{z_0\}$ with the following two properties:

(i) the winding number of Γ about the point z_0 equals 1;
(ii) the winding number of Γ about every point in $\complement\Omega$ equals 0.

Here, the *winding number* of Γ about a point z is the (integer) number

$$w(\Gamma; z_0) := \frac{1}{2\pi i} \int_{\Gamma} \frac{1}{\lambda - z_0} \, d\lambda.$$

More generally we can admit finite unions of such contours, as long as their union satisfies (i) and (ii). If $\Gamma = \Gamma_1 \cup \cdots \cup \Gamma_k$ is such a union, we define

$$w(\Gamma; z_0) := \frac{1}{2\pi i} \sum_{j=1}^{k} \int_{\Gamma_j} \frac{1}{\lambda - z_0} \, d\lambda.$$

Condition (ii) is satisfied if $\Omega = \Omega_1 \cup \cdots \cup \Omega_k$ is a finite union of disjoint convex sets and Γ_j is a piecewise continuously differentiable closed contour in $\Omega_j \setminus \{z_0\}$.

Turning to the discussion of (6.2), we need to fix a similar set of assumptions regarding Ω and Γ while letting the operator T take the role of z_0. We require that Ω is an open set in the complex plane containing $\sigma(T)$ and Γ is a piecewise continuously differentiable closed contour in $\Omega \setminus \sigma(T)$ with the following two properties:

   (i) the winding number of Γ about every point $z_0 \in \sigma(T)$ equals 1;
   (ii) the winding number of Γ about every point in $\complement\Omega$ equals 0.

It is easy to see that such contours always exist. If the conditions (i) and (ii) are met we say that Γ is an *admissible contour* for $\sigma(T)$ in Ω. As before we admit the possibility that Γ is a finite union of such contours; for such contours we interpret (6.2) as

$$f(T) := \frac{1}{2\pi i} \sum_{j=1}^{k} \int_{\Gamma_j} f(\lambda) R(\lambda, T) \, d\lambda.$$

For example, if $\sigma(T) = K_1 \cup K_2$ is the union of two disjoint compact sets with $K_j \subseteq \Omega_j$, where $\Omega := \Omega_1 \cup \Omega_2$ is a disjoint union of open convex sets, we may select contours Γ_j with winding number 1 about every point in K_j and consider $\Gamma = \Gamma_1 \cup \Gamma_2$ as an admissible contour for $\sigma(T)$ in Ω. This example is relevant for defining spectral projections, where one uses the holomorphic functions $f = \mathbf{1}_{\Omega_1}$ and $f = \mathbf{1}_{\Omega_2}$ (see Proposition 6.23).

For an open set Ω in the complex plane we denote by $H(\Omega)$ the vector space of all holomorphic functions on Ω.

Theorem 6.20 (Holomorphic functional calculus). *Let $\Omega \subseteq \mathbb{C}$ be an open set containing $\sigma(T)$. For functions $f \in H(\Omega)$ we define*

$$f(T) := \frac{1}{2\pi i} \int_{\Gamma} f(\lambda) R(\lambda, T) \, d\lambda,$$

where Γ is an admissible contour for $\sigma(T)$ in Ω. The resulting operators $f(T)$ are well defined and have the following properties:

   (i) *the operator $f(T)$ is independent of the admissible contour Γ;*
   (ii) *for entire functions the holomorphic and entire functional calculi agree;*
   (iii) *for $f_\mu(\lambda) = 1/(\mu - \lambda)$ with $\mu \in \rho(T)$ we have $f_\mu(T) = R(\mu, T)$;*
   (iv) *for all $f, g \in H(\Omega)$ we have $f(T)g(T) = (fg)(T)$;*
   (v) *for all $f \in H(\Omega)$ we have $f(T^*) = (f(T))^*$.*

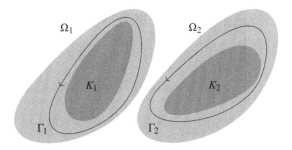

Figure 6.1 The contour $\Gamma = \Gamma_1 \cup \Gamma_2$ around $\sigma(T) = K_1 \cup \Gamma_2$ in $\Omega = \Omega_1 \cup \Omega_2$

Proof (i): This follows by applying the Cauchy theorem to the scalar-valued integrals

$$\frac{1}{2\pi i} \int_\Gamma f(\lambda)\langle R(\lambda,T)x, x^*\rangle \, d\lambda$$

and then using the Hahn–Banach theorem. Alternatively one may extend *mutatis mutandis* the proof of the Cauchy theorem to functions with values in a Banach space.

(ii): For $f_n(z) = z^n$ with $n \in \mathbb{N}$ we have, with Γ a circle of radius $r > \|T\|$ oriented counterclockwise,

$$\begin{aligned}
f_n(T) &= \frac{1}{2\pi i} \int_\Gamma \lambda^n R(\lambda,T) \, d\lambda \\
&= \frac{1}{2\pi i} \int_\Gamma \lambda^{n-1} \sum_{k=0}^\infty \lambda^{-k} T^k \, d\lambda = \sum_{k=0}^\infty \left(\frac{1}{2\pi i} \int_\Gamma \lambda^{n-1-k} \, d\lambda\right) T^k = T^n.
\end{aligned}$$

Here we first used the Neumann series for $R(\lambda,T) = \lambda^{-1}(I - \lambda^{-1}T)^{-1}$ (noting that $|\lambda| > \|T\|$ for $\lambda \in \Gamma$), then we interchanged integration and summation (which is justified by the absolute convergence of the series, uniformly in $\lambda \in \Gamma$), and finally we used that

$$\frac{1}{2\pi i} \int_\Gamma \lambda^j \, d\lambda = \begin{cases} 1, & j = -1; \\ 0, & j \in \mathbb{Z} \setminus \{-1\}. \end{cases}$$

By linearity, this proves that the holomorphic calculus agrees with the entire functional calculus for polynomials. The general case follows by approximating an entire function by its power series and noting that this approximation is uniform on bounded sets.

(iv): Let Γ and Γ' be admissible contours for $\sigma(T)$ in Ω, with Γ' to the interior of Γ (more precisely, the outer contour Γ should have winding number one with respect to every point on the inner contour Γ'). By the resolvent identity and Fubini's theorem,

$$
\begin{aligned}
f(T)g(T) &= \frac{1}{(2\pi i)^2} \int_\Gamma \int_{\Gamma'} f(\lambda)g(\mu)R(\lambda,T)R(\mu,T)\,d\mu\,d\lambda \\
&= \frac{1}{(2\pi i)^2} \int_\Gamma \int_{\Gamma'} f(\lambda)g(\mu)(\lambda-\mu)^{-1}[R(\mu,T)-R(\lambda,T)]\,d\mu\,d\lambda \\
&= \frac{1}{(2\pi i)^2} \int_\Gamma \int_{\Gamma'} f(\lambda)g(\mu)(\lambda-\mu)^{-1}R(\mu,T)\,d\mu\,d\lambda \\
&= \frac{1}{(2\pi i)^2} \int_{\Gamma'} \int_{\Gamma} f(\lambda)g(\mu)(\lambda-\mu)^{-1}R(\mu,T)\,d\lambda\,d\mu \\
&= \frac{1}{2\pi i} \int_{\Gamma'} f(\mu)g(\mu)R(\mu,T)\,d\mu = (fg)(T).
\end{aligned}
$$

Here we used that

$$
\frac{1}{2\pi i} \int_{\Gamma'} g(\mu)(\lambda-\mu)^{-1}\,d\mu = 0 \quad\text{and}\quad \frac{1}{2\pi i} \int_{\Gamma} f(\lambda)(\lambda-\mu)^{-1}\,d\lambda = f(\mu).
$$

(iii): The identity $(\mu-\lambda)\cdot f_\mu(\lambda) = f_\mu(\lambda)\cdot(\mu-\lambda) = 1$ gives, via (ii) and (iv), that $(\mu-T)f_\mu(T) = f_\mu(T)(\mu-T) = I$. It follows that $f_\mu(T) = R(\mu,T)$.

(v): This follows from Proposition 6.18 and the continuity of the mapping $S \mapsto S^*$, which allows one to 'take adjoint under the integral sign'. □

Theorem 6.21 (Spectral mapping theorem). *Let $\Omega \subseteq \mathbb{C}$ be an open set containing $\sigma(T)$. For all $f \in H(\Omega)$ we have*

$$
\sigma(f(T)) = f(\sigma(T)).
$$

Proof We begin with the proof of the inclusion $\sigma(f(T)) \subseteq f(\sigma(T))$. Fix $\lambda \notin f(\sigma(T))$; our aim is to show that $\lambda \notin \sigma(f(T))$. Let U be an open set containing the (compact) set $f(\sigma(T))$ but not λ. Then $\Omega' := \Omega \cap f^{-1}(U)$ is an open subset of Ω containing $\sigma(T)$ and $\lambda \notin f(\Omega')$.

The function f is holomorphic on Ω', and so is $g_\lambda(z) = (\lambda - f(z))^{-1}$, by the choice of Ω'. By the multiplicativity of the holomorphic functional calculus applied to $H(\Omega')$,

$$
(\lambda - f(T))g_\lambda(T) = g_\lambda(T)(\lambda - f(T)) = I,
$$

from which we infer that $\lambda \notin \sigma(f(T))$.

Turning to the converse inclusion $f(\sigma(T)) \subseteq \sigma(f(T))$, let $\lambda \in \Omega$. By the theory of functions in one complex variable we have $f(\lambda) - f(z) = (\lambda - z)h(z)$ for some $h \in H(\Omega)$, so $f(\lambda) - f(T) = (\lambda - T)h(T)$. If $\lambda \in \sigma(T)$, then $\lambda - T$ is noninvertible. Since $\lambda - T$ and $h_\lambda(T)$ commute, Lemma 6.5 implies that $f(\lambda) - f(T)$ is noninvertible and therefore $f(\lambda) \in \sigma(f(T))$. □

Theorem 6.22 (Composition). *Let $\Omega \subseteq \mathbb{C}$ be an open set containing $\sigma(T)$, let $f \in$*

$H(\Omega)$, and let $\Omega' \subseteq \mathbb{C}$ be an open set containing $\sigma(f(T))$. Then for all $g \in H(\Omega')$ we have

$$g(f(T)) = (g \circ f)(T).$$

Proof Let Γ' be an admissible contour for $\sigma(f(T))$ in Ω', and let $\widetilde{\Omega}$ be an open subset of Ω containing $\sigma(T)$, chosen in such a way that Γ' has winding number 1 about every point of $f(\widetilde{\Omega})$. Let Γ be an admissible contour for $\sigma(T)$ in $\widetilde{\Omega}$.

If μ is a point on Γ', then $h_\mu(\lambda) := (\mu - f(\lambda))^{-1}$ defines a function in $H(\widetilde{\Omega})$ and

$$h_\lambda(T)(\mu - f(T)) = (\mu - f(T))h_\lambda(T) = I.$$

It follows that $\mu \in \rho(f(T))$ and

$$R(\mu, f(T)) = h_\lambda(T) = \frac{1}{2\pi i} \int_\Gamma (\mu - f(\lambda))^{-1} R(\lambda, T) \, d\lambda.$$

Hence, using Fubini's theorem to justify the change of order of integration,

$$
\begin{aligned}
g(f(T)) &= \frac{1}{2\pi i} \int_{\Gamma'} g(\mu) R(\mu, f(T)) \, d\mu \\
&= \left(\frac{1}{2\pi i}\right)^2 \int_{\Gamma'} \int_\Gamma g(\mu)(\mu - f(\lambda))^{-1} R(\lambda, T) \, d\lambda \, d\mu \\
&= \left(\frac{1}{2\pi i}\right)^2 \int_\Gamma \int_{\Gamma'} g(\mu)(\mu - f(\lambda))^{-1} R(\lambda, T) \, d\mu \, d\lambda \\
&= \frac{1}{2\pi i} \int_\Gamma g(f(\lambda)) R(\lambda, T) \, d\lambda = (g \circ f)(T).
\end{aligned}
$$

In the penultimate identity we used that $\frac{1}{2\pi i} \int_{\Gamma'} g(\mu)(\mu - f(\lambda))^{-1} \, d\mu = g(f(\lambda))$ by the Cauchy integral formula. $\qquad\square$

An interesting application of the holomorphic calculus arises when the spectrum is the disjoint union of two nonempty disjoint compact sets.

Theorem 6.23 (Spectral projections). *Suppose that $\sigma(T)$ is the union of two nonempty disjoint compact sets K_1 and K_2, Ω_1 and Ω_2 are disjoint open sets containing K_1 and K_2, and let Γ_1 and Γ_2 be admissible contours for K_1 and K_2 in Ω_1 and Ω_2, respectively. The operators*

$$P_j := \frac{1}{2\pi i} \int_{\Gamma_j} R(\lambda, T) \, d\lambda, \quad j \in \{1, 2\},$$

are projections, their ranges X_1 and X_2 are invariant under T, and we have a direct sum decomposition $X = X_1 \oplus X_2$. Moreover,

$$\sigma(T|_{X_j}) = K_j, \quad j \in \{1, 2\}.$$

Proof   To see that P_j is a projection we just note that $P_j = f_j(T)$, where $f_j : \Omega_1 \cup \Omega_2 \to \mathbb{C}$ is the holomorphic function which is 1 on Ω_j and 0 elsewhere. By the multiplicativity of the holomorphic calculus, P_j is bounded and

$$P_j^2 = (f_j(T))^2 = f_j^2(T) = f_j(T) = P_j.$$

Also, $f_1 + f_2 \equiv 1$ implies that $P_1 + P_2 = I$. We further have

$$P_1 P_2 = f_1(T) f_2(T) = (f_1 f_2)(T) = 0$$

since $f_1 f_2 = 0$, and similarly $P_2 P_1 = 0$. Consequently, $R(P_1) \cap R(P_2) = \{0\}$, for if $x = P_1 x_1 = P_2 x_2$, then $P_1 x = P_1 P_2 x_2 = 0$ and $P_2 x = P_2 P_1 x_1 = 0$ and therefore $x = (P_1 + P_2)x = 0 + 0 = 0$. It follows that we have a direct sum decomposition $X = R(P_1) \oplus R(P_2)$. Since T obviously commutes with P_j, it follows that T maps $X_j = R(P_j)$ into itself. It follows that T restricts to a bounded operator $T_j := T|_{X_j}$ on X_j.

Let $\mu \in \mathbb{C} \setminus K_j$. We show that $\mu \in \rho(T_j)$. Define

$$S_{\mu,j} := \frac{1}{2\pi i} \int_{\Gamma_j} (\mu - \lambda)^{-1} R(\lambda, T) \, d\lambda,$$

where Γ_j is an admissible contour for K_j in Ω_j with μ on the exterior. Writing $\mu - T_j = (\mu - \lambda) + (\lambda - T_j)$ and using that $T P_j x = T_j P_j x$ for all $x \in X$, we find

$$(\mu - T_j) S_{\mu,j} P_j x = S_{\mu,j}(\mu - T_j) P_j x = \frac{1}{2\pi i} \int_{\Gamma_j} (\mu - \lambda)^{-1} R(\lambda, T)(\mu - T_j) P_j x \, d\lambda$$

$$= \frac{1}{2\pi i} \int_{\Gamma_j} R(\lambda, T) P_j x + (\mu - \lambda)^{-1} P_j x \, d\lambda$$

$$= \frac{1}{2\pi i} \int_{\Gamma_j} R(\lambda, T) P_j x \, d\lambda = P_j^2 x = P_j x,$$

which shows that $(\mu - T_j) S_{\mu,j} = S_{\mu,j}(\mu - T_j) = I$ on $R(P_j)$. This proves that $\mu \in \rho(T_j)$. We have shown that $\sigma(T_j) \subseteq K_j$. In particular this implies that $\sigma(T_0) \cap \sigma(T_1) = \varnothing$.

Next we claim that $\sigma(T) \subseteq \sigma(T_1) \cup \sigma(T_2)$; this concludes the proof since it gives

$$K_1 \cup K_2 = \sigma(T) \subseteq \sigma(T_1) \cup \sigma(T_2) \subseteq K_1 \cup K_2$$

and therefore equality holds at all steps. To prove the claim it suffices to note that if $\mu \notin \sigma(T_1) \cup \sigma(T_2)$, then $\mu \in \rho(T_1) \cap \rho(T_2)$ and $R(\mu, T_1) \oplus R(\mu, T_2)$ is a two-sided inverse to $\mu - T = (\mu - T_1) \oplus (\mu - T_2)$. $\square$

As a further application of the holomorphic calculus we prove the following exact formula for the *spectral radius*

$$r(T) := \sup\{|\lambda| : \lambda \in \sigma(T)\}$$

of a bounded operator $T \in \mathscr{L}(X)$ (with the convention $\sup \varnothing = 0$ to deal with the trivial case $X = \{0\}$). Since the spectrum $\sigma(T)$ is contained in the closed disc with radius $\|T\|$ we have $r(T) \leqslant \|T\|$. More generally, by the spectral mapping theorem, $r(T)^n = r(T^n) \leqslant \|T^n\|$, so $r(T) \leqslant \|T^n\|^{1/n}$. We actually have equality:

Theorem 6.24 (Gelfand). *For every bounded operator $T \in \mathscr{L}(X)$ we have*

$$r(T) = \inf_{n \geqslant 1} \|T^n\|^{1/n} = \lim_{n \to \infty} \|T^n\|^{1/n}.$$

Existence of the right-hand side limit is part of the assertion.

Proof We have already observed that

$$r(T) \leqslant \inf_{n \geqslant 1} \|T^n\|^{1/n}.$$

The theorem will be proved once we show that

$$\limsup_{n \to \infty} \|T^n\|^{1/n} \leqslant r(T).$$

Israel Gelfand, 1913–2009

Fix $\varepsilon > 0$ arbitrary and let Γ denote the circular contour about the origin with radius $R = r(T) + \varepsilon$, oriented counterclockwise. Then

$$T^n = \frac{1}{2\pi i} \int_\Gamma \lambda^n R(\lambda, T) \, d\lambda$$

implies

$$\|T^n\| \leqslant \frac{1}{2\pi} \cdot 2\pi R \cdot R^n \sup_{|\lambda|=R} \|R(\lambda, T)\|,$$

the supremum being finite in view of the continuity of $\lambda \mapsto R(\lambda, T)$. Taking nth roots and passing to the limit superior for $n \to \infty$, this implies $\limsup_{n \to \infty} \|T^n\|^{1/n} \leqslant R = r(T) + \varepsilon$. Since $\varepsilon > 0$ was arbitrary, this completes the proof. $\qquad\square$

As an application we have the following stability result.

Theorem 6.25 (Lyapunov's stability theorem). *If $A \in \mathscr{L}(X)$ is a bounded operator with $\sigma(A) \subseteq \{z \in \mathbb{C} : \mathrm{Re}\, z < 0\}$, then*

$$\lim_{t \to \infty} \|e^{tA}\| = 0.$$

Proof Since $\sigma(A)$ is compact, there is a $\delta > 0$ such that $\sigma(A) \subseteq \{z \in \mathbb{C} : \mathrm{Re}\, z \leqslant -\delta\}$. By the spectral mapping theorem (Theorem 6.21) this implies that $\sigma(e^A) \subseteq \{z \in \mathbb{C} : |z| \leqslant e^{-\delta}\}$. Stated differently, we have $r(e^A) \leqslant e^{-\delta} < 1$. By Gelfand's theorem (Theorem 6.24) there exists an integer $n_0 \geqslant 1$ such that $M_0 := \|e^{n_0 A}\| < 1$.

By estimating the power series defining e^{sA}, $\|e^{sA}\| \leqslant e^{s\|A\|}$ for all $s \geqslant 0$. Let now $t \geqslant 0$ and write $t = kn_0 + r$ with $k \in \mathbb{N}$ and $r \in [0, n_0)$. Then

$$\|e^{tA}\| = \|e^{kn_0 A} e^{rA}\| \leqslant \|e^{n_0 A}\|^k e^{r\|A\|} \leqslant M_0^k e^{n_0 \|A\|}.$$

Since $M_0 < 1$ this gives the result (with exponential rate), for $t \to \infty$ implies $k \to \infty$. $\square$

The interpretation of this theorem is as follows. Consider the initial value problem

$$\begin{cases} u'(t) = Au(t), & t \geqslant 0, \\ u(0) = u_0, \end{cases}$$

where $A \in \mathscr{L}(X)$ is bounded and $u_0 \in X$ is given. By differentiating the power series defining e^{tA} we see that this problem is solved by the function $u(t) := e^{tA} u_0$. Lyapunov's theorem now gives the following sufficient spectral criterion for stability of this solution: if the spectrum of A is contained in the open left-half plane, then $\lim_{t \to \infty} \|u(t)\| = 0$.

Problems

6.1   Prove the following improvement to Lemma 6.6: if $T \in \mathscr{L}(X)$ is such that the sum $\sum_{n=0}^{\infty} T^n x$ converges for all $x \in X$, then $I - T$ is invertible. What is its inverse?

6.2   Show that if $\lambda, \mu \in \rho(T)$ satisfy $|\lambda - \mu| \leqslant \delta \|R(\lambda, T)\|^{-1}$ with $0 \leqslant \delta < 1$, then

$$\|R(\mu, T) - R(\lambda, T)\| \leqslant \frac{\delta}{1 - \delta} \|R(\lambda, T)\|.$$

6.3   Prove in an elementary way, by multiplying power series, that $e^{wT} e^{zT} = e^{(w+z)T}$ for all complex numbers $w, z \in \mathbb{C}$.

6.4   For $1 \leqslant p \leqslant \infty$ consider the right shift operator $T_r \in \mathscr{L}(\ell^p)$:

$$T : (a_1, a_2, a_3, \ldots) \mapsto (0, a_1, a_2, \ldots).$$

Give a direct proof of the fact (see Example 6.3) that $\sigma(T) = \{\lambda \in \mathbb{C} : |\lambda| \leqslant 1\}$.

6.5   We compute the spectra of some multiplier operators.

(a) Let $m \in C[0, 1]$ and define $T_m \in \mathscr{L}(C[0, 1])$ by

$$(T_m f)(t) = m(t) f(t), \quad f \in C[0, 1], \ t \in [0, 1].$$

Show that $\sigma(T_m)$ coincides with the range of m.

(b) Let $m \in L^\infty(0, 1)$, let $1 \leqslant p \leqslant \infty$, and define $T_m \in \mathscr{L}(L^p(0, 1))$ by

$$(T_m f)(t) = m(t) f(t), \quad f \in L^p(0, 1), \ t \in (0, 1).$$

Show that $\sigma(T_m)$ coincides with the *essential range* of m, that is, the set of all $\lambda \in \mathbb{K}$ such that for any open set $U \subseteq \mathbb{K}$ containing λ the set $\{t \in (0,1) : m(t) \in U\}$ has positive measure.

Hint: First show that λ is not contained in the essential range of m if and only if $\frac{1}{m-\lambda}$ is well defined almost everywhere and essentially bounded.

6.6 Let T_m be the Fourier multiplier on $L^2(\mathbb{R}^d)$ with symbol $m \in L^\infty(\mathbb{R}^d)$.

    (a) Show that $\sigma(T_m)$ equals the essential range of m.
       Hint: Use the result of the preceding problem.

    (b) Show that if f is a holomorphic function defined on an open set containing $\sigma(T_m)$, then $f \circ m \in L^\infty(\mathbb{R}^d)$ and $T_{f \circ m} = f(T_m)$, the latter being defined by the holomorphic calculus.

6.7 Show that every nonempty compact subset of $\mathbb{C}$ can be realised as the spectrum of some bounded operator.

6.8 Show that if P, Q are projections on a Banach space X satisfying $\|P - Q\| < 1$, then $\dim(R(P)) = \dim(R(Q))$ (admitting the possibility $\infty = \infty$).

Hint: The invertibility of $I - (P - Q)$ implies $PQ = P(I - P + Q) = P$.

6.9 Let $\mathscr{F}$ be the Fourier–Plancherel transform on $L^2(\mathbb{R}^d)$.

    (a) Recalling that $\mathscr{F}^4 = I$ (see Problem 5.20), apply the spectral mapping theorem to see that $\sigma(\mathscr{F}) \subseteq \{\pm 1, \pm i\}$.

    (b) Show that $(\mathscr{F} - iI)(\mathscr{F} + I)(\mathscr{F} + iI)(\mathscr{F} - I) = 0$ and $(\mathscr{F} + I)(\mathscr{F} + iI)(\mathscr{F} - I) \neq 0$, and deduce that $i \in \sigma(\mathscr{F})$.

    (c) Prove that $\sigma(\mathscr{F}) = \{\pm 1, \pm i\}$.

6.10 Determine the spectrum of the Hilbert transform H on $L^2(\mathbb{R}^d)$.

Hint: Use the result of Problem 5.21.

6.11 Suppose that we have a direct sum decomposition $X = X_0 \oplus X_1$. Prove that if $T \in \mathscr{L}(X)$ leaves both X_0 and X_1 invariant, then

$$\sigma(T) = \sigma(T|_{X_0}) \cup \sigma(T|_{X_1}),$$

viewing $T|_{X_0}$ and $T|_{X_1}$ as bounded operators on the Banach spaces X_0 and X_1.

6.12 Let S and T be bounded operators on a Banach space X. Prove that

$$\sigma(ST) \setminus \{0\} = \sigma(TS) \setminus \{0\}.$$

Hint: Use the Neumann series to relate the resolvents of ST and TS.

6.13 Prove the claims about the example below Proposition 6.19.

6.14 The aim of this problem is to prove *Chernoff's theorem*: If T is a bounded operator on a Banach space X satisfying $\sup_{n \in \mathbb{N}} \|T^n\| =: M < \infty$, then for all $n \in \mathbb{N}$ we have

$$\|\exp(n(T - I))x - T^n x\| \leqslant \sqrt{n} M \|Tx - x\|.$$

(a) Show that

$$\|\exp(n(T-I))x - T^n x\| \leqslant e^{-n} \sum_{k=0}^{\infty} \frac{n^k}{k!} \|T^k x - T^n x\|$$

$$\leqslant e^{-n} M \|Tx - x\| \sum_{k=0}^{\infty} \left(\frac{n^k}{k!}\right)^{1/2} \left(\frac{n^k}{k!}\right)^{1/2} |n-k|.$$

(b) Show that

$$\sum_{k=0}^{\infty} \frac{n^k}{k!}(n-k)^2 = ne^n.$$

(c) Combine (a), (b), and the Cauchy–Schwarz inequality to complete the proof.

6.15 Let T be a *power bounded* operator on a Banach space X, that is, T is invertible and $\sup_{k \in \mathbb{Z}} \|T^k\| < \infty$, with the property that $\sigma(T) = \{1\}$.

    (a) Using the holomorphic calculus and its properties, explain that we can define the bounded operators $S := -i \log T$ and $\sin(nS)$, $n \in \mathbb{N}$.

    (b) Show that $\sin(nS) = \frac{1}{2i}(T^n - T^{-n})$, $n \in \mathbb{N}$.

    (c) Using the spectral mapping theorem, show that $\sigma(nS) = \sigma(\sin(nS)) = \{0\}$.

We now use that if $\sum_{k=0}^{\infty} c_k z^k$ denotes the Taylor series of the principal branch of $\arcsin z$ at $z = 0$, then $c_k \geqslant 0$ for all $k \in \mathbb{N}$ and $\sum_{k=0}^{\infty} c_k = \arcsin(1) = \frac{\pi}{2}$.

    (d) Show that $nS = \arcsin(\sin(nS))$, where the latter is again defined by means of the holomorphic calculus, and deduce that

$$\|nS\| \leqslant \frac{\pi}{2} \sup_{k \in \mathbb{Z}} \|T^k\|.$$

Conclude that $S = 0$ and $T = e^{iS} = I$.

6.16 Let $T \in \mathscr{L}(X)$ and let f be a nonzero holomorphic function on a connected open set Ω containing $\sigma(T)$. Show that if $f(T) = 0$, then $\sigma(T)$ is a finite set.

7

Compact Operators

This chapter studies the class of compact operators. By definition, these are the operators that map bounded sets to relatively compact sets. Examples include integral operators on various Banach spaces of functions over a compact domain. Because of this, compact operators have important applications in the theory of partial differential equations and Mathematical Physics. After establishing some generalities we prove the Riesz–Schauder theorem, which asserts that the nonzero part of the spectrum of a compact operator is discrete and consists of eigenvalues.

The final section of this chapter presents an introduction to the theory of Fredholm operators. These are the operators that are invertible modulo a compact operator, and their degree of noninvertibility is quantified by the so-called Fredholm index. As an example we prove the Gohberg–Krein–Noether theorem, which states that a Toeplitz operator with continuous zero-free symbol is Fredholm and its index equals the negative winding number of their symbol.

7.1 Compact Operators

Let X and Y be Banach spaces.

Definition 7.1 (Compact operators). An operator $T \in \mathscr{L}(X,Y)$ is *compact* if it maps bounded sets to relatively compact sets.

Since every bounded set in X is contained in a multiple of the unit ball $B_X = \{x \in X : \|x\| < 1\}$, a bounded operator is compact if and only if TB_X is relatively compact. Furthermore, using that a subset of a Banach space is relatively compact if and only if it is relatively sequentially compact, a linear operator T is compact if and only if $(Tx_n)_{n \geqslant 1}$ has a convergent subsequence for every bounded sequence $(x_n)_{n \geqslant 1}$ in X.

The set of all compact operators is a linear subspace of $\mathscr{L}(X,Y)$. It is clear that cT is compact if T is compact, for any scalar $c \in \mathbb{K}$, and if S and T are compact, then also

$S+T$ is compact: for if SB_X and TB_X are contained in the compact sets K and L, then $(S+T)B_X$ is contained in the compact set $K+L$ (this set is the image of the compact set $K \times L$ under the continuous image $(x_1,x_2) \mapsto x_1 + x_2$ from $X \times X$ to X). It is also a two-sided ideal in $\mathscr{L}(X,Y)$, in the sense that if $T \in \mathscr{L}(X,Y)$ is compact and $S \in \mathscr{L}(X',X)$ and $U \in \mathscr{L}(Y,Y')$ are bounded, then $UTS \in \mathscr{L}(X',Y')$ is compact. Indeed, if C is a bounded set in X', then $S(C)$ is bounded in X, so $TS(C)$ is contained in a compact set K of Y, and then $UTS(C)$ is contained in the compact set $U(K)$.

Example 7.2. As an immediate corollary to Theorem 1.38, the identity operator on a Banach space X is compact if and only if X is finite-dimensional.

Example 7.3. A bounded operator is said to be of *finite rank* if its range is finite-dimensional. Since bounded sets in finite-dimensional spaces are relatively compact, every finite rank operator is compact.

Example 7.4 (Integral operators on $C(K)$). Let μ be a finite Borel measure on compact metric space K and let $k : K \times K \to \mathbb{K}$ be continuous. Then the operator $T : C(K) \to C(K)$,

$$Tf(x) := \int_K k(x,y)f(y)\,d\mu(y), \quad f \in C(K),\ x \in K,$$

is well defined and bounded by Example 1.30. Let us show that T is compact. Let $(f_n)_{n \geq 1}$ be a bounded sequence in $C(K)$. We claim that the bounded sequence $(Tf_n)_{n \geq 1}$ is equicontinuous. Once we have shown this, the Arzelà–Ascoli theorem (Theorem 2.11) implies that this sequence is relatively compact, hence has a convergent subsequence. This implies that T is compact.

To check the equicontinuity we first note that $K \times K$ is compact and hence k is uniformly continuous, in the sense that given any $\varepsilon > 0$ we can find $\delta > 0$ in K such that $d(x,x') + d(y,y') < \delta$ implies $|k(x,y) - k(x',y')| < \varepsilon$. Then, if $d(x,x') < \delta$,

$$|Tf_n(x) - Tf_n(x')| \leq \int_K |k(x,y) - k(x',y)||f_n(y)|\,d\mu(y) \leq \varepsilon M \mu(K),$$

where $M = \sup_{n \geq 1} \|f_n\|_\infty$. The equicontinuity follows immediately from this.

The next proposition shows that the compact operators form a closed subspace in $\mathscr{L}(X,Y)$. This subspace will be denoted by $\mathscr{K}(X,Y)$ and we write $\mathscr{K}(X) := \mathscr{K}(X,X)$.

Proposition 7.5. *If $\lim_{n \to \infty} \|T_n - T\| = 0$ with each T_n compact, then T is compact. In particular, uniform limits of finite rank operators are compact.*

Proof For any $\varepsilon > 0$ we can choose an index $n_\varepsilon \geq 1$ such that $\|T_{n_\varepsilon} - T\| < \varepsilon$. Since $T_{n_\varepsilon}B_X$ is relatively compact and $TB_X \subseteq T_{n_\varepsilon}B_X + B(0;\varepsilon)$, the relative compactness of TB_X follows from Proposition 1.40. $\square$

The following converse of the second assertion of Proposition 7.5 holds for Hilbert space operators:

Proposition 7.6. *Let H and K be Hilbert spaces. An operator $T \in \mathscr{L}(H,K)$ is compact if and only if it is the uniform limit of finite rank operators.*

Proof It remains to prove the 'only if' part. Let T be compact, say $TB_H \subseteq C$ with $C \subseteq K$ compact, where B_H is the open unit ball of H. Fix an arbitrary $\varepsilon > 0$ and let $B(y_1;\varepsilon),\ldots,B(y_N;\varepsilon)$ be an open cover of C. Let Y denote the linear span of $\{y_1,\ldots,y_N\}$ and let P be the orthogonal projection in K onto Y. This projection is of finite rank, and therefore PT is of finite rank. For any $x \in H$ of norm $\|x\| < 1$ we have $Tx \in C$, so $\|Tx - y_n\| < \varepsilon$ for some $1 \leqslant n \leqslant N$. Then, noting that $Py_n = y_n$ and using that $\|P\| \leqslant 1$,

$$\|Tx - PTx\| \leqslant \|Tx - y_n\| + \|y_n - PTx\| < \varepsilon + \|P(y_n - Tx)\| \leqslant \varepsilon + \|y_n - Tx\| < 2\varepsilon.$$

Taking the supremum over all $x \in H$ with $\|x\| < 1$ we obtain $\|T - PT\| \leqslant 2\varepsilon$. $\qquad\square$

Example 7.7 (Integral operators on $L^2(K,\mu)$). Let μ be a finite Borel measure on a compact metric space K and let $k : K \times K \to \mathbb{K}$ be square integrable. Then the operator $T : L^2(K) \to L^2(K)$,

$$Tf(x) := \int_K k(x,y)f(y)\,d\mu(y), \quad f \in L^2(K), \ x \in K,$$

is well defined and bounded by Example 1.30. Let us prove that T is compact.

Fix $\varepsilon > 0$. Since $C(K \times K)$ is dense in $L^2(K \times K, \mu \times \mu)$ (see Remark 2.31), we may choose $\kappa \in C(K \times K)$ such that $\|\kappa - k\|_2 < \varepsilon$. Since κ is uniformly continuous we can find $\delta > 0$ such that $|\kappa(x,y) - \kappa(x',y')| < \varepsilon$ whenever $d(x,x') + d(y,y') < \delta$. Starting from a finite cover of K by open balls with diameter at most $\frac{1}{2}\delta$, we can write $K = B_1 \cup \cdots \cup B_n$ with disjoint Borel sets B_j of diameter at most $\frac{1}{2}\delta$. Set

$$\widetilde{k} := \sum_{j,k=1}^{n} \kappa(x_j,y_k)\mathbf{1}_{B_j \times B_k},$$

where $x_j \in B_j$ and $y_k \in B_k$ are chosen arbitrarily. For this function we have $\|\kappa - \widetilde{k}\|_\infty \leqslant \varepsilon$. Then, $\|\kappa - \widetilde{k}\|_2 \leqslant \varepsilon(\mu \times \mu)(K \times K)^{1/2} = \varepsilon\mu(K)$ and hence

$$\|\widetilde{k} - k\|_2 \leqslant \|\widetilde{k} - \kappa\|_2 + \|\kappa - k\|_2 < \varepsilon(1 + \mu(K)). \tag{7.1}$$

The integral operator with kernel $\widetilde{k}$, which we denote by $\widetilde{T}$, is given explicitly as

$$\widetilde{T} = \sum_{j=1}^{n} \left(\sum_{k=1}^{n} \kappa(x_j,y_k)\mu(B_k) \right) \mathbf{1}_{B_j}.$$

This shows that $\widetilde{T}$ is of finite rank and therefore compact. By (1.4) (with k replaced by

$\widetilde{k} - k$) and (7.1) we have

$$\|\widetilde{T} - T\| \leqslant \|\widetilde{k} - k\|_2 < \varepsilon(1 + \mu(K)).$$

Since $\varepsilon > 0$ was arbitrary this proves that T can be approximated in the operator norm by compact operators.

We conclude this section with a duality result for compact operators.

Proposition 7.8. *An operator $T \in \mathscr{L}(X,Y)$ is compact if and only if its adjoint $T^* \in \mathscr{L}(Y^*,X^*)$ is compact.*

Proof First we prove the 'only if' part. Let T be compact and let K denote the closure of TB_X. By assumption, K is a compact subset of Y. By restriction, every $y^* \in Y^*$ determines a function in $C(K)$ given by $y^*(y) := \langle y, y^* \rangle$ for $y \in K$. Moreover, if $(y_n^*)_{n \geqslant 1}$ is a bounded sequence in Y^*, the corresponding functions are uniformly bounded and equicontinuous; the latter follows from $|\langle x - x', y_n^* \rangle| \leqslant M\|x - x'\|$ with $M := \sup_{n \geqslant 1} \|y_n^*\|$. Hence, by the Arzelà–Ascoli theorem, there is a subsequence $(y_{n_j}^*)_{j \geqslant 1}$ such that

$$\lim_{j,k \to \infty} \|y_{n_k}^* - y_{n_j}^*\|_{C(K)} = 0.$$

Then,

$$\lim_{j,k \to \infty} \|T^* y_{n_k}^* - T^* y_{n_j}^*\| = \lim_{j,k \to \infty} \sup_{\|x\| \leqslant 1} |\langle x, T^* y_{n_k}^* - T^* y_{n_j}^* \rangle|$$

$$= \lim_{j,k \to \infty} \sup_{\|x\| \leqslant 1} |\langle Tx, y_{n_k}^* - y_{n_j}^* \rangle| \leqslant \lim_{j,k \to \infty} \|T\| \|y_{n_k}^* - y_{n_j}^*\|_\infty = 0.$$

Thus we have shown that $(T^* y_n^*)_{j \geqslant 1}$ has a convergent subsequence. It follows that T^* is compact.

To prove the 'if' part suppose that T^* is compact. Then, by what we just proved, T^{**} is compact as an operator from X^{**} to Y^{**}. Identifying X with a closed subspace of X^{**} in the natural way, the restriction of T^{**} to X maps X to Y and equals T. Since the restriction of a compact operator is compact, the compactness of T^{**} implies the compactness of T. $\qquad\square$

7.2 The Riesz–Schauder Theorem

As we have seen in earlier examples, the spectrum of a bounded operator need not contain eigenvalues. This is in sharp contrast to the situation in finite dimensions, where the spectra of matrices consist of eigenvalues. The aim of the present section is to show that compact operators spectrally resemble matrices to some degree. The main result is Theorem 7.11, which shows that the nonzero part of the spectrum of a compact operator is discrete and consists of eigenvalues.

Lemma 7.9. *If $T \in \mathscr{L}(X)$ is compact, then:*

(1) $N(I - T)$ *is finite-dimensional;*
(2) $R(I - T)$ *is closed.*

Proof (1): Let B_Y denote the unit ball of $Y := N(I - T)$. We have $Ty = y$ for all $y \in Y$, so the compactness of T implies that $B_Y = TB_Y$ is relatively compact. By Theorem 1.38, this implies that Y is finite-dimensional.

(2): Let $Y := N(I - T)$ and consider the linear mapping $S : X/Y \to X$ defined by

$$S(x + Y) := (I - T)x, \quad x \in X.$$

Then S is well defined and bounded as a quotient operator. We claim that there exists a constant $c > 0$ such that

$$\|S(x + Y)\| \geqslant c\|x + Y\|, \quad x \in X. \tag{7.2}$$

If this were false we would be able to find elements $x_n \in X$ satisfying $\|x_n + Y\| = 1$ and $\|S(x_n + Y)\| < 1/n$. Then $(I - T)x_n = S(x_n + Y) \to 0$. By the compactness of T we can find a subsequence such that $Tx_{n_k} \to x_0$ for some $x_0 \in X$. Then,

$$\lim_{k \to \infty} x_{n_k} = \lim_{k \to \infty} [(I - T)x_{n_k} + Tx_{n_k}] = 0 + x_0 = x_0.$$

By the boundedness of $I - T$,

$$(I - T)x_0 = \lim_{k \to \infty} (I - T)x_{n_k} = \lim_{k \to \infty} S(x_{n_k} + Y) = 0,$$

and therefore $x_0 \in N(I - T) = Y$. The contradiction $0 = \|x_0 + Y\| = \lim_{k \to \infty} \|x_{n_k} + Y\| = 1$ concludes the proof of (7.2).

By Proposition 1.21, (7.2) implies that S is injective and the range $R(S)$ of S is closed. Finally, $R(S) = R(I - T)$ and therefore the range of $I - T$ is closed. □

In order to describe the spectra of compact operators we need the following lemma.

Lemma 7.10. *Let $T \in \mathscr{L}(X)$ be a compact operator. Then $I - T$ is injective if and only if $I - T$ is surjective.*

Proof We begin with the proof of the 'only if' part. We assume that $I - T$ is injective but not surjective and deduce a contradiction.

By assumption, $X_1 := R(I - T)$ is a proper subspace of X and by Lemma 7.9 this subspace is closed. It is clear that $TX_1 \subseteq X_1$. Let T_1 denote the restriction of T to X_1. This operator is compact.

The operator $I - T_1 : X_1 \to X_1$ is not surjective, since $I - T : X \to X_1$ is bijective and X_1 is a proper subspace of X. The same argument shows that $X_2 := R(I - T_1)$ is a proper closed subspace of X_1. It is clear that $TX_2 \subseteq X_2$. Let T_2 denote the restriction of T to

X_2. Continuing as above we obtain a strictly decreasing sequence of closed subspaces $X_1 \supsetneq X_2 \supsetneq \ldots$ each of which is T-invariant, such that

$$X_{n+1} = (I - T)X_n, \quad n = 1, 2, \ldots$$

By Lemma 1.39 we can select vectors $x_n \in X_n \setminus X_{n+1}$ of norm one such that

$$\inf_{y \in X_{n+1}} \|x_n - y\| \geqslant \frac{1}{2}. \tag{7.3}$$

Since T is compact, $(Tx_n)_{n \geqslant 1}$ has a convergent subsequence $(Tx_{n_k})_{k \geqslant 1}$. Then, for $\ell > k$,

$$\|Tx_{n_k} - Tx_{n_\ell}\| = \|x_{n_k} + (T - I)x_{n_k} - Tx_{n_\ell}\| \geqslant \frac{1}{2},$$

where we use (7.3) along with $(T - I)x_{n_k} \in X_{n_k+1}$ and $Tx_{n_\ell} \in X_{n_\ell} \subseteq X_{n_k+1}$. This contradicts the convergence of $(Tx_{n_k})_{k \geqslant 1}$.

Turning to the 'if' part, assume that $I - T$ is surjective. If $T^*x^* = x^*$ for some $x^* \in X^*$, then by writing an arbitrary $x \in X$ as $(I - T)y$, we find $\langle x, x^* \rangle = \langle y, (I - T^*)x^* \rangle = 0$ for all $x \in X$, so $x^* = 0$. This shows that $I - T^*$ is injective. Applying the preceding step to the operator T^*, which is compact by Proposition 7.8, it follows that $I - T^*$ is surjective as well. Then from the preceding argument, applied to T^*, it follows that $I - T^{**}$ is injective, and hence $I - T$ is injective. □

Theorem 7.11 (Riesz–Schauder). *Let $T \in \mathscr{L}(X)$ be a compact operator. Then:*

(1) *every nonzero $\lambda \in \sigma(T)$ is an eigenvalue of T and the eigenspace*

$$E_\lambda := \{x \in X : Tx = \lambda x\}$$

is finite-dimensional;

(2) *for every $r > 0$, the number of eigenvalues satisfying $|\lambda| \geqslant r$ is finite;*

(3) *if $\dim(X) = \infty$, then $0 \in \sigma(T)$.*

Proof (1): Let $0 \neq \lambda \in \sigma(T)$ and suppose that λ is not an eigenvalue of T. Then $I - \lambda^{-1}T$ is injective and hence, by the preceding lemma, surjective. It follows that $I - \lambda^{-1}T$ is invertible, and this implies that $\lambda \in \rho(T)$.

Since T acts as a multiple of the identity on the subspace E_λ and T is compact, the identity operator on E_λ is compact. By Theorem 1.38, this implies that E_λ is finite-dimensional.

(2): Suppose there is a sequence of distinct eigenvalues λ_n, all of which satisfy $|\lambda_n| \geqslant r$. Let $x_n \in X$ be eigenvectors for λ_n of norm one.

Let Y_n denote the linear span of $\{x_1, \ldots, x_n\}$, $n \geqslant 1$, and set $Y_0 := \{0\}$. By the distinctness of the eigenvalues, the vectors x_n are linearly independent (this is easily proved with induction on n). Therefore $\dim(Y_n) = n$. In particular, Y_n is a proper subspace of Y_{n+1}. For $y \in Y_n$, say $y = \sum_{j=1}^{n} c_j x_j$, we have

$$Ty = \sum_{j=1}^{n} c_j \lambda_j x_j \in Y_n$$

and

$$(\lambda_n - T)y = \sum_{j=1}^{n} c_j (\lambda_n - \lambda_j) x_j = \sum_{j=1}^{n-1} c_j (\lambda_n - \lambda_j) x_j \in Y_{n-1}.$$

Lemma 1.39 shows that for every $n \geqslant 1$ it is possible to find a vector $y_n \in Y_n$ of norm one such that $\|y - y_n\| \geqslant \frac{1}{2}$ for all $y \in Y_{n-1}$. Arguing as in the proof of Lemma 7.10, for $n > m$ these vectors satisfy the lower bound

$$\|Ty_n - Ty_m\| = \|\lambda_n y_n + (T - \lambda_n)y_n - Ty_m\|$$
$$= |\lambda_n| \left\| y_n - \frac{Ty_m - (T - \lambda_n)y_n}{\lambda_n} \right\| \geqslant \frac{1}{2} |\lambda_n| \geqslant \frac{1}{2} r,$$

using that $Ty_m \in Y_m \subseteq Y_{n-1}$ and $(T - \lambda_n)y_n \in Y_{n-1}$, and therefore $(Ty_n)_{n \geqslant 1}$ cannot have a convergent subsequence. This contradicts the compactness of T.

(3): If T is invertible, then $T^{-1} B_X$ is bounded and $B_X = T(T^{-1} B_X)$ is relatively compact. Therefore X must be finite-dimensional. $\square$

For compact normal operators on a Hilbert space, a more direct proof is sketched in Problem 9.2.

Definition 7.12. The number $\dim(E_\lambda)$ is called the *geometric multiplicity* of λ.

Suppose now that $T \in \mathscr{L}(X)$ is compact and that $0 \neq \lambda \in \sigma(T)$. Then λ is an isolated point of $\sigma(T)$, that is, for small enough $r > 0$ we have $B(\lambda; r) \cap \sigma(T) = \{\lambda\}$. With $0 < r' < r$ and $\Gamma = \partial B(\lambda, r')$ oriented counterclockwise, the spectral projection corresponding to λ (see Theorem 6.23) is given by

Juliusz Schauder, 1899–1943

$$P_\lambda := \frac{1}{2\pi i} \int_\Gamma R(\mu, T) \, d\mu.$$

By Theorem 6.23 the range $X_\lambda := P_\lambda X$ is invariant under T and $\sigma(T|_{X_\lambda}) = \{\lambda\}$. In particular, $T|_{X_\lambda}$ is invertible. This operator is also compact as an operator on X_λ. This is only possible if $\dim(X_\lambda)$ is finite. Thus we have proved:

Corollary 7.13. *Let T be a compact operator on a Banach space X. For all nonzero $\lambda \in \sigma(T)$ the range X_λ of the spectral projection P_λ is finite-dimensional.*

This argument can be used to give the following alternative proof that nonzero elements λ in the spectrum of a compact operator T are eigenvalues. Since X_λ is finite-dimensional, upon choosing a basis we may identify X_λ with a space $\mathbb{C}^d$ and represent $T|_{X_\lambda}$ as a $d \times d$ matrix. Since $\sigma(T|_{X_\lambda}) = \{\lambda\}$, it follows that λ is an eigenvalue of this matrix, and hence of T.

Definition 7.14. The number $\dim(X_\lambda)$ is called the *algebraic multiplicity* of λ.

Proposition 7.15. *Let $T \in \mathscr{L}(X)$ be a compact operator. Then the geometric multiplicity of every nonzero $\lambda \in \sigma(T)$ is less than or equal to its algebraic multiplicity.*

Proof If $Tx = \lambda x$ and Γ is as before, then $P_\lambda x = (\frac{1}{2\pi i} \int_\Gamma (\mu - \lambda)^{-1} \, d\mu)x = x$. This shows that the eigenspace E_λ is contained in the range of the projection P_λ. $\qquad\square$

Example 7.16 (Jordan normal form). Consider a $k \times k$ Jordan block

$$
J_\lambda = \begin{pmatrix}
\lambda & 1 & 0 & \cdots & 0 \\
0 & \lambda & 1 & \ddots & \vdots \\
\vdots & \ddots & \ddots & \ddots & 0 \\
\vdots & & \ddots & \ddots & 1 \\
0 & \cdots & & 0 & \lambda
\end{pmatrix}
$$

and its resolvent

$$
R(\mu, J_\lambda) = \begin{pmatrix}
(\lambda - \mu)^{-1} & (\lambda - \mu)^{-2} & & \cdots & (\lambda - \mu)^{-k} \\
0 & (\lambda - \mu)^{-1} & (\lambda - \mu)^{-2} & & \vdots \\
\vdots & & \ddots & \ddots & \ddots & \vdots \\
\vdots & & & \ddots & \ddots & (\lambda - \mu)^{-2} \\
0 & & \cdots & & 0 & (\lambda - \mu)^{-1}
\end{pmatrix}.
$$

If A is a matrix with $\lambda \in \sigma(A)$ and if J_λ is the corresponding block in the Jordan normal form of A, i It follows that the spectral projection corresponding to λ is given by

$$
P_\lambda = \frac{1}{2\pi i} \int_\Gamma R(\mu, A) \, d\lambda = I_\lambda,
$$

where I_λ is the diagonal matrix with 1's on the diagonal entries corresponding to the Jordan block J_λ and 0's elsewhere. It follows that the algebraic multiplicity ν_λ of λ equals the sum of dimension of all Jordan blocks with λ on the diagonal.

7.3 Fredholm Theory

Throughout this section we let X and Y be Banach spaces.

7.3.a The Fredholm Alternative

From the results in the preceding section we know that if T is a compact operator on a Banach space X, then every nonzero $\lambda \in \sigma(T)$ is an eigenvalue and the corresponding eigenspace is finite-dimensional. The next theorem (when applied to the compact operator $\lambda^{-1}T$) asserts that the dimension of the eigenspace is equal to the codimension of the range of $\lambda - T$. This generalises the elementary result in Linear Algebra that for a $d \times d$ matrix A we have $\dim N(A) + \dim R(A) = d$.

Theorem 7.17 (Fredholm alternative). *If $T \in \mathscr{L}(X)$ is compact, then*

$$\dim N(I - T) = \operatorname{codim} R(I - T).$$

This theorem contains Lemma 7.10 as a special case. The proof of the theorem is based on the following geometric lemma. Recall that if $Y \subseteq X^*$, then

$$^{\perp}Y = \{x \in X : \langle x, x^* \rangle = 0 \text{ for all } x^* \in Y\}.$$

Lemma 7.18. *If Y is a finite-dimensional subspace of X^*, then $^{\perp}Y$ has finite codimension in X and $\operatorname{codim}(^{\perp}Y) = \dim Y$.*

Proof Let $x_1^*, \ldots, x_d^*$ be a basis of Y and consider the mapping from X to $\mathbb{K}^d$,

Ivar Fredholm, 1866–1927

$$\psi : x \mapsto (\langle x, x_1^* \rangle, \ldots, \langle x, x_d^* \rangle).$$

We claim that this mapping is surjective. Indeed, if $\xi \in \mathbb{K}^d$ is such that $\psi(x) \cdot \xi = 0$ for all $x \in X$, that is, $\sum_{j=1}^d \langle x, x_j^* \rangle \xi_j = 0$ for all $x \in X$, then $\sum_{j=1}^d \xi_j x_j^* = 0$ and therefore $\xi = 0$ by linear independence. This proves the claim. As a consequence there exist $x_j \in X$ such that $\psi(x_j) = e_j$, the jth unit vector of $\mathbb{K}^d$. The resulting sequence $x_1, \ldots, x_d$ has the property that

$$\langle x_i, x_j^* \rangle = \delta_{ij}, \quad i, j = 1, \ldots, d.$$

The vectors x_j are linearly independent, for if $\sum_{j=1}^d c_j x_j = 0$, then $c_k = \langle \sum_{j=1}^d c_j x_j, x_k^* \rangle = 0$ for all $k = 1, \ldots, d$.

Now suppose that an arbitrary $x \in X$ is given and set $\widetilde{x} := x - \sum_{j=1}^d c_j x_j$, where $c_j := \langle x, x_j^* \rangle$. Then $\langle \widetilde{x}, x_k^* \rangle = \langle x, x_k^* \rangle - c_k = 0$ for all $k = 1, \ldots, d$, so $\widetilde{x} \in {}^{\perp}Y$. This shows that

$X = (^\perp Y) + X_0$, where X_0 is the linear span of $x_1, \ldots, x_d$. If $x \in (^\perp Y) \cap X_0$, then we can write $x = \sum_{j=1}^{d} c_j x_j$ since $x \in X_0$, and we have $c_k = \sum_{j=1}^{d} c_j \langle x_j, x_k^* \rangle = 0$ for all $k = 1, \ldots, d$ since $x \in {}^\perp Y$. It follows that $x = 0$. Thus we obtain the direct sum decomposition $X = (^\perp Y) \oplus X_0$, and therefore $\operatorname{codim}(^\perp Y) = \dim X_0 = d = \dim(Y)$. $\qquad\square$

Proof of Theorem 7.17 We begin by recalling that, by Lemma 7.9, $\mathsf{N}(I - T)$ is finite-dimensional and $\mathsf{R}(I - T)$ is closed. Hence Proposition 5.14, applied to $I - T$, implies that $\mathsf{R}(I - T) = {}^\perp(\mathsf{N}(I - T^*))$ and therefore, by Lemma 7.18 (which can be applied because Lemma 7.9 applied to the compact operator T^* gives $\dim \mathsf{N}(I - T^*) < \infty$),

$$\operatorname{codim} \mathsf{R}(I - T) = \operatorname{codim}(^\perp(\mathsf{N}(I - T^*))) = \dim \mathsf{N}(I - T^*).$$

Thus it remains to prove that $d := \dim \mathsf{N}(I - T) = \dim \mathsf{N}(I - T^*) =: d^*$.

Step 1 – We first prove that $d^* \leqslant d$. Reasoning by contradiction, suppose that $d^* > d$. Since $\mathsf{N}(I - T)$ is finite-dimensional, by Proposition 4.16(1) we have a direct sum decomposition

$$X = \mathsf{N}(I - T) \oplus Y$$

for some closed subspace Y of X. Also, since $\mathsf{R}(I - T)$ is closed and has finite codimension d^*, by Proposition 4.16(2) we have a direct sum decomposition

$$X = \mathsf{R}(I - T) \oplus Z \tag{7.4}$$

for some closed subspace Z of X of dimension d^*. Since $d < d^*$ there is an injective linear map $L : \mathsf{N}(I - T) \to Z$ that is not surjective. Set $S := T + L \circ \pi$, where π is the projection in X onto $\mathsf{N}(I - T)$ along Y. Since L is a finite rank operator, it is compact and hence also $L \circ \pi$ is compact.

We claim that $\mathsf{N}(I - S) = \{0\}$. Indeed, if $Sx = x$, then

$$0 = Sx - x = \underbrace{Tx - x}_{\in \mathsf{R}(I-T)} + \underbrace{L\pi x}_{\in Z}$$

and therefore (7.4) implies $Tx - x = 0$ and $L\pi x = 0$. The first of these identities means that $x \in \mathsf{N}(I - T)$, so $\pi x = x$, and then the second of these identities takes the form $Lx = 0$. The injectivity of L then implies that $x = 0$. This proves the claim.

By Lemma 7.10, $\mathsf{R}(I - S) = X$. To arrive at a contradiction, let $z \in Z \setminus \mathsf{R}(L)$ and choose $x \in X$ such that $x - Sx = z$. Then

$$\underbrace{x - Tx}_{\in \mathsf{R}(I-T)} - \underbrace{L\pi x}_{\in Z} = \underbrace{z}_{\in Z}$$

and by (7.4) this implies $x - Tx = 0$ and $z = L\pi x$. The second of these identities contradicts our assumption that $z \in Z \setminus \mathsf{R}(L)$.

Step 2 – Having established that $d^* \leqslant d$, we now prove the opposite inequality $d \leqslant d^*$

by a duality argument. Setting $d^{**} := \dim \mathsf{N}(I - T^{**})$, applying Step 1 to the compact operator T^* gives $d^{**} \leqslant d^*$. Identifying X with a closed subspace of X, T^{**} is an extension of T and therefore $d \leqslant d^*$. □

7.3.b Application to Integral Equations

As an application of the foregoing theory we turn to the problem of finding a function $u \in C[0,1]$ solving inhomogeneous integral equations of the form

$$\lambda u(s) = f(s) + \int_0^1 k(s,t)u(t)\,dt, \quad s \in [0,1]. \tag{H_f}$$

Here $f \in C[0,1]$ is given, $k : [0,1] \times [0,1] \to \mathbb{K}$ is continuous, and λ is a nonzero scalar. Under a *solution* of this equation we understand a function $u \in C[0,1]$ satisfying (H_f) for all $s \in [0,1]$. In order to study existence of solutions it is useful to also consider the homogeneous equation corresponding to $f = 0$,

$$\lambda u(s) = \int_0^1 k(s,t)u(t)\,dt, \quad s \in [0,1], \tag{H_0}$$

as well as the 'dual' homogeneous problem

$$\lambda v(s) = \int_0^1 k(t,s)v(t)\,dt, \quad s \in [0,1]. \tag{H_0^*}$$

Solutions to these problems are defined in the same way.

Theorem 7.19 (Fredholm alternative for integral equations). *Let* $k : [0,1] \times [0,1] \to \mathbb{K}$ *be continuous and let* $\lambda \neq 0$ *be fixed. Then:*

(1) *if the homogeneous problem* (H_0) *has no nonzero solution, then for all* $f \in C[0,1]$ *the inhomogeneous problem* (H_f) *has a unique solution* u *in* $C[0,1]$;
(2) *if the homogeneous problem* (H_0) *has a nonzero solution, then it has at most finitely many linearly independent nonzero solutions, and for a given* $f \in C[0,1]$ *the inhomogeneous problem* (H_f) *has a solution if and only if*

$$\int_0^1 f(t)v(t)\,dt = 0$$

for all $v \in L^1(0,1)$ *satisfying the dual homogeneous problem* (H_0^*).

Proof By the result of Example 7.4 the operator $T : C[0,1] \to C[0,1]$,

$$Tu(s) := \int_0^1 k(s,t)u(t)\,dt, \quad s \in [0,1],$$

is compact. Using this operator, the problem (H_f) can be abstractly formulated as

$$(\lambda - T)u = f.$$

If the homogeneous problem $(\lambda - T)u = 0$ has no nonzero solution, then λ is not an eigenvalue. Since we are assuming that $\lambda \neq 0$ it follows that $\lambda \notin \sigma(T)$ by Theorem 7.11. Therefore, $\lambda - T$ is invertible and the inhomogeneous problem $(\lambda - T)u = f$ is uniquely solved by $u = (\lambda - T)^{-1}f$.

If the homogeneous problem $(\lambda - T)u = 0$ has a nonzero solution, then λ is an eigenvalue, in which case the space of solutions u equals the eigenspace corresponding to λ, which is finite-dimensional. In that case, the inhomogeneous problem $(\lambda - T)u = f$ has a solution $u \in C[0,1]$ if and only if $f \in \mathsf{R}(\lambda - T) = {}^{\perp}\mathsf{N}(\lambda - T^*)$; here we use Proposition 5.14 along with the fact that $\lambda - T$ has closed range. Stated differently, problem $(\lambda - T)u = f$ has a solution $u \in C[0,1]$ if and only if

$$\langle f, x^* \rangle = 0, \quad x^* \in \mathsf{N}(\lambda - T^*).$$

To make this condition more explicit we recall from Section 4.1.c that the dual of $C[0,1]$ is the space of complex Borel measures on $[0,1]$, the duality between functions ϕ and measures μ being given by $\langle f, \mu \rangle = \int_0^1 f \, d\mu$. For such measures μ we compute

$$\langle g, T^* \mu \rangle = \langle Tg, \mu \rangle = \int_0^1 \int_0^1 k(s,t) g(t) \, dt \, d\mu(s)$$
$$= \int_0^1 g(t) \int_0^1 k(s,t) \, d\mu(s) \, dt = \langle g, \nu \rangle,$$

where the $\mathbb{K}$-valued measure ν is given by

$$\nu(B) = \langle \mathbf{1}_B, \nu \rangle = \int_B \int_0^1 k(s,t) \, d\mu(s) \, dt$$

for Borel sets $B \subseteq [0,1]$. Now $\mu \in \mathsf{N}(\lambda - T^*)$ if and only if $\lambda \mu(B) = \nu(B)$ for all Borel sets $B \subseteq [0,1]$, that is, if and only if

$$\int_B \lambda \, d\mu(t) = \int_B \int_0^1 k(s,t) \, d\mu(s) \, dt$$

for all Borel sets $B \subseteq [0,1]$. In this case μ is absolutely continuous with respect to the Lebesgue measure dt. Then, by the Radon–Nikodým theorem, $d\mu = v \, dt$, where $v \in L^1(0,1)$ satisfies

$$\lambda v(t) = \int_0^1 k(s,t) \, d\mu(s) = \int_0^1 k(s,t) v(s) \, ds$$

for almost all $t \in (0,1)$. Since both sides are continuous functions of t, the equality holds for all $t \in [0,1]$. This means that v solves (H_0^*). $\qquad\qquad \square$

7.3.c Fredholm Operators

Let X and Y be Banach spaces. The following definition is suggested by the Fredholm alternative (Theorem 7.17):

Definition 7.20 (Fredholm operators). A bounded operator $T \in \mathscr{L}(X,Y)$ is called a *Fredholm operator* if it has the following properties:

   (i) $\dim \mathsf{N}(T) < \infty$;
   (ii) $\operatorname{codim} \mathsf{R}(T) < \infty$.

The *index* of such an operator is defined as

$$\operatorname{ind}(T) := \dim \mathsf{N}(T) - \operatorname{codim} \mathsf{R}(T).$$

Example 7.21. Here are some examples of Fredholm operators:

   (i) If T is a compact operator, then $I - T$ is Fredholm with index $\operatorname{ind}(T) = 0$. This is a restatement of the Fredholm alternative.
   (ii) The left and right shift on ℓ^p, $1 \leqslant p \leqslant \infty$, are Fredholm with indices 1 and -1, respectively.
   (iii) For every zero-free $\phi \in C(\mathbb{T})$ the Toeplitz operator T_ϕ on the Hardy space $H^2(\mathbb{D})$ is Fredholm with index $\operatorname{ind}(T_\phi) = -w(\phi)$, where $w(\phi)$ is the winding number of ϕ. This is the content of Noether–Gohberg–Krein theorem in Section 7.3.d, where the relevant definitions can be found.

We begin our analysis of Fredholm operators with the observation that such operators have closed range. As a result, $\operatorname{codim} \mathsf{R}(T)$ equals the dimension of the quotient Banach space $Y/\mathsf{R}(T)$.

Proposition 7.22. *If the range of a bounded operator $T \in \mathscr{L}(X,Y)$ has finite codimension, then it is closed.*

Proof Let Y_0 be a finite-dimensional subspace of Y such that $\mathsf{R}(T) \cap Y_0 = \{0\}$ and $\mathsf{R}(T) + Y_0 = Y$. Then Y_0 is closed and the bounded operator $S : X \times Y_0 \to Y$ defined by $S(x,y_0) := Tx + y_0$ is surjective. By the open mapping theorem, S is open. In particular, $S(X \times (Y_0 \setminus \{0\}))$ is open. Clearly this set is the complement of $S(X \times \{0\}) = \mathsf{R}(T)$ and therefore $\mathsf{R}(T)$ is closed. □

Theorem 7.23 (Atkinson). *For a bounded operator $T \in \mathscr{L}(X,Y)$ the following assertions are equivalent:*

(1) *T is Fredholm;*
(2) *there exist a bounded operator $S \in \mathscr{L}(Y,X)$ and compact operators $K \in \mathscr{L}(X)$ and $L \in \mathscr{L}(Y)$ such that*

$$ST = I - K, \quad TS = I - L.$$

If these equivalent conditions hold, the operator S is Fredholm with index

$$\operatorname{ind}(S) = -\operatorname{ind}(T).$$

*Moreover, S can be chosen in such a way that $K = I - ST$ and $L = I - TS$ are finite rank
projections satisfying* $\dim(R(K)) = \dim(N(T))$ *and* $\dim(R(L)) = \text{codim}(R(T))$.

Proof (1)$\Rightarrow$(2): By Propositions 4.16 and 7.22 there exist closed subspaces $X_0 \subseteq X$
and $Y_0 \subseteq Y$ such that $\text{codim}(X_0) < \infty$, $\dim(Y_0) < \infty$, and

$$X = N(T) \oplus X_0, \quad Y = R(T) \oplus Y_0.$$

Let P and Q denote the corresponding projections in X and Y onto X_0 and $R(T)$, re-
spectively. The restriction $T_0 := T|_{X_0}$ is a bijection from X_0 onto $R(T)$, injectivity and
surjectivity both being clear. Since $R(T)$ is closed, the open mapping theorem implies
that the inverse mapping $S_0 := T_0^{-1}$ is bounded as an operator from $R(T)$ onto X_0. Define
$S \in \mathscr{L}(Y, X)$ by $S := S_0 \circ Q$. Then for all $x \in X$ and $y \in Y$ we have

$$STx = S_0 QTx = S_0 Tx = S_0 TPx = Px = x - Kx \quad \text{with} \ K = I - P$$

and

$$TSy = TS_0 Qy = Qy = y - Ly \quad \text{with} \ L = I - Q.$$

Since $I - P$ and $I - Q$ are the projections onto the finite-dimensional subspaces $N(T)$
and Y_0, these projections are of finite rank and hence compact. It also follows that
$\dim(R(K)) = \dim(N(T))$ and $\dim(R(L)) = \text{codim}(R(T))$.

(2)$\Rightarrow$(1): We have $N(T) \subseteq N(ST)$ and hence

$$\dim(N(T)) \leqslant \dim(N(ST)) = \dim(N(I - K)) < \infty.$$

Likewise $R(T) \supseteq R(TS)$ and hence

$$\text{codim}(R(T)) \leqslant \text{codim}(R(TS)) = \text{codim}(R(I - L)) < \infty,$$

the finiteness of the codimension being a consequence of Theorem 7.17. This completes
the proof of the equivalence (1)$\Leftrightarrow$(2).

It remains to prove the identity $\text{ind}(S) = -\text{ind}(T)$. Using the notation introduced
before we have $N(S) = N(S_0 Q) = N(Q) = Y_0$, so

$$\dim(N(S)) = \dim(Y_0) = \text{codim}(R(T))$$

and likewise $R(S) = R(S_0 Q) = X_0$, so

$$\text{codim}(R(S)) = \text{codim}(X_0) = \dim(N(T)).$$

As a result,

$$\text{ind}(S) = \dim(N(S)) - \text{codim}(R(S)) = \text{codim}(R(T)) - \dim(N(T)) = -\text{ind}(T).$$

$\square$

A concise way of stating the equivalence (1)⇔(2) is by introducing the *Calkin alge-bra*

$$\mathscr{L}(X,Y)/\mathscr{K}(X,Y).$$

Theorem 7.23 states that an operator $T \in \mathscr{L}(X,Y)$ is Fredholm if and only if its equiv-alence class in $\mathscr{L}(X,Y)/\mathscr{K}(X,Y)$ is invertible in the sense that there exists an operator $S \in \mathscr{L}(Y,X)$ such that $ST = I$ mod $\mathscr{K}(X)$ and $TS = I$ mod $\mathscr{K}(Y)$.

Proposition 7.24. *If $T_1 \in \mathscr{L}(X,Y)$ and $T_2 \in \mathscr{L}(Y,Z)$ are Fredholm, where Z is another Banach space, then $T_2 T_1 \in \mathscr{L}(X,Z)$ is Fredholm and*

$$\mathrm{ind}(T_2 T_1) = \mathrm{ind}(T_1) + \mathrm{ind}(T_2).$$

Proof Let $S_1 \in \mathscr{L}(Y,X)$ and $S_2 \in \mathscr{L}(Z,Y)$ be such that

$$S_1 T_1 = I - K_1, \quad T_1 S_1 = I - L_1, \quad S_2 T_2 = I - K_2, \quad T_2 S_2 = I - L_2,$$

with K_1, K_2, L_1, L_2 compact. Then

$$(S_1 S_2)(T_2 T_1) = S_1(I - K_2)T_1 = I - K_1 - S_1 K_2 T_1 =: I - K_3,$$

where $K_3 = K_1 + S_1 K_2 T_1$ is compact. Likewise

$$(T_2 T_1)(S_1 S_2) = T_2(I - L_1)S_2 = I - L_2 - T_2 L_1 S_2 =: I - L_3,$$

where $L_3 = L_2 + T_2 L_1 S_2$ is compact. Hence Atkinson's theorem implies that $T_2 T_1$ is Fredholm. To compute its index, let X_1, Y_1, Y_2, Z_2 be finite dimensional subspaces such that

$$X = \mathrm{N}(T_1) \oplus X_1, \quad Y = \mathrm{R}(T_1) \oplus Y_1 = \mathrm{N}(T_2) \oplus Y_2, \quad Z = \mathrm{R}(T_2) \oplus Z_2.$$

We have $x \in \mathrm{N}(T_2 T_1)$ if and only if

$$x \in \mathrm{N}(T_1) \oplus \{x_1 \in X_1 : T_1 x_1 \in \mathrm{N}(T_2)\} = \mathrm{N}(T_1) \oplus (T_1|_{X_1})^{-1}(\mathrm{R}(T_1) \cap \mathrm{N}(T_2)).$$

Since T_1 acts as an isomorphism from X_1 onto $\mathrm{R}(T_1)$, we have

$$\dim(\mathrm{N}(T_2 T_1)) = \dim(\mathrm{N}(T_1)) + \dim(\mathrm{R}(T_1) \cap \mathrm{N}(T_2)).$$

Furthermore we have

$$\dim(\mathrm{N}(T_2)) = \dim(\mathrm{R}(T_1) \cap \mathrm{N}(T_2)) + \dim(Y_1 \cap \mathrm{N}(T_2)).$$

Combined with the previous identity this gives

$$\dim(\mathrm{N}(T_2 T_1)) = \dim(\mathrm{N}(T_1)) + \dim(\mathrm{N}(T_2)) - \dim(Y_1 \cap \mathrm{N}(T_2)). \tag{7.5}$$

Next, we have

$$Z = \mathrm{R}(T_2) \oplus Z_2 = T_2(\mathrm{R}(T_1) \oplus Y_1) \oplus Z_2$$

and therefore

$$\text{codim}(\mathsf{R}(T_2 T_1)) = \text{codim}(\mathsf{R}(T_2)) + \text{codim}(\mathsf{R}(T_1)) - \dim(Y_1 \cap \mathsf{N}(T_2)). \qquad (7.6)$$

It follows from (7.5) and (7.6) that

$$\begin{aligned}
\text{ind}(T_2 T_1) &= \dim(\mathsf{N}(T_2 T_1)) - \text{codim}(\mathsf{R}(T_2 T_1)) \\
&= \dim(\mathsf{N}(T_1)) + \dim(\mathsf{N}(T_2)) - \text{codim}(\mathsf{R}(T_2)) - \text{codim}(\mathsf{R}(T_1)) \\
&= \text{ind}(T_1) + \text{ind}(T_2).
\end{aligned}$$

$\square$

Another proof is proposed in Problem 14.17.

The next three propositions show that Fredholmness is preserved under various operations.

Proposition 7.25. *If $T \in \mathscr{L}(X,Y)$ is Fredholm and $K \in \mathscr{K}(X,Y)$ is compact, then $T + K$ is Fredholm and*

$$\text{ind}(T + K) = \text{ind}(T).$$

Proof   If $S \in \mathscr{L}(Y,X)$ and compact operators $L_1 \in \mathscr{L}(X)$ and $L_2 \in \mathscr{L}(Y)$ are such that $ST = I - L_1$ and $TS = I - L_2$, then $S(T + K) = I - L_1 + SK = I - M_1$ with $M_1 = L_1 - SK$ compact, and $(T + K)S = I - L_2 + KS = I - M_2$ with $M_2 = L_2 - KS$ compact. Hence $T + K$ is Fredholm by Atkinson's theorem. Moreover, by Proposition 7.24,

$$0 = \text{ind}(I - M_1) = \text{ind}(S(T + K)) = \text{ind}(S) + \text{ind}(T + K),$$

so $\text{ind}(T + K) = -\text{ind}(S) = \text{ind}(T)$ by the identity for indices in Atkinson's theorem.

$\square$

The set of Fredholm operators is open in $\mathscr{L}(X,Y)$:

Proposition 7.26 (Dieudonné). *For any Fredholm operator $T \in \mathscr{L}(X,Y)$ there exists a number $\delta > 0$ such that for all $U \in \mathscr{L}(X,Y)$ with $\|U\| < \delta$ the operator $T + U$ is Fredholm and*

$$\text{ind}(T + U) = \text{ind}(T).$$

Proof   The proof is a variation on the proof of the openness of the set of invertible bounded operators. Let $S \in \mathscr{L}(Y,X)$ be such that $ST = I - K$ and $TS = I - L$ with $K \in \mathscr{L}(X)$ and $L \in \mathscr{L}(Y)$ compact. Then

$$S(T + U) = I - K + SU, \quad (T + U)S = I - L + US.$$

If $\|U\| < \delta := \|S\|^{-1}$, then $I + SU$ and $I + US$ are boundedly invertible and

$$(I + SU)^{-1} S(T + U) = I - (I + SU)^{-1} K,$$

$$(T+U)S(I+US)^{-1} = I - L(I+US)^{-1},$$

where $M := (I+SU)^{-1}K$ and $N := L(I+US)^{-1}$ are compact. Noting that

$$(I+SU)^{-1}S = \sum_{n=0}^{\infty} (SU)^n S = \sum_{n=0}^{\infty} S(US)^n = S(I+SU)^{-1}$$

by the Neumann series, Atkinson's theorem implies that $T+U$ is Fredholm.

Next, by Proposition 7.24 and Theorem 7.17,

$$\text{ind}(I+SU)^{-1} + \text{ind}(S) + \text{ind}(T+U) = \text{ind}(I-M) = 0$$

and $\text{ind}(I+SU)^{-1} = 0$ by invertibility. By the identity for indices in Atkinson's theorem it follows that

$$\text{ind}(T+U) = -\text{ind}(S) = \text{ind}(T).$$

$\square$

Proposition 7.27. *If $T \in \mathcal{L}(X,Y)$ is Fredholm, then $T^* \in \mathcal{L}(Y^*,X^*)$ is Fredholm and*

$$\text{ind}(T^*) = -\text{ind}(T).$$

Proof If $S \in \mathcal{L}(Y,X)$ is bounded and $K \in \mathcal{L}(X)$ and $L \in \mathcal{L}(Y)$ are compact operators such that $ST = I - K$ and $TS = I - L$, then $S^*T^* = I - L^*$ and $T^*S^* = I - K^*$ with K^* and L^* compact. Hence T^* is Fredholm by Atkinson's theorem.

We claim that

$$\dim(\mathsf{N}(T^*)) = \text{codim}(\mathsf{R}(T)) \tag{7.7}$$

and

$$\dim(\mathsf{N}(T)) = \text{codim}(\mathsf{R}(T^*)). \tag{7.8}$$

Together, these identities imply that $\text{ind}(T^*) = -\text{ind}(T)$. We give a detailed proof of (7.8) and indicate the changes that need to be made to prove (7.7).

Since T^* is Fredholm we have a direct sum decomposition

$$X^* = \mathsf{R}(T^*) \oplus W \tag{7.9}$$

with $W \subseteq X^*$ a finite-dimensional subspace.

If $x_1, \ldots, x_k$ is a basis for $\mathsf{N}(T)$, by the Hahn–Banach extension theorem we obtain $x_1^*, \ldots, x_k^* \in X^*$ such that

$$\langle x_i, x_j^* \rangle = \delta_{ij}, \quad 1 \leqslant i, j \leqslant k. \tag{7.10}$$

Let Z denote the span of $x_1^*, \ldots, x_k^*$ in X^*. We claim that

$$\mathsf{R}(T^*) \cap Z = \{0\}.$$

Indeed, if $x^* \in Z$, say $x^* = \sum_{j=1}^{k} c_j x_j^*$, then

$$\langle x_i, x^* \rangle = c_i, \quad 1 \leqslant i \leqslant k.$$

If we also have $x^* \in \mathsf{R}(T^*)$, say $x^* = T^* y^*$, then from $x_i \in \mathsf{N}(T)$ we obtain

$$c_i = \langle x_i, x^* \rangle = \langle T x_i, y^* \rangle = 0, \quad 1 \leqslant i \leqslant k.$$

This implies $x^* = 0$ and proves the claim.

Now, for any fixed $x^* \in X^*$, set

$$\xi^* := x^* - \sum_{j=1}^{k} \langle x_j, x^* \rangle x_j^*.$$

Then, for $i = 1, \ldots, k$,

$$\langle x_i, \xi^* \rangle = \langle x_i, x^* \rangle - \sum_{j=1}^{k} \langle x_j, x^* \rangle \langle x_i, x_j^* \rangle = \langle x_i, x^* \rangle - \langle x_i, x^* \rangle = 0.$$

This means that $\xi^* \in \mathsf{N}(T)^{\perp}$. By Theorem 5.15, this implies that $\xi^* \in \mathsf{R}(T^*)$. Since $x^* - \xi^* \in Z$ it follows that $\mathsf{R}(T^*) + Z = X^*$. Together with $Z \cap \mathsf{R}(T^*) = \{0\}$ it follows that we have a direct sum decomposition

$$X^* = \mathsf{R}(T^*) \oplus Z. \tag{7.11}$$

From (7.9) and (7.11) it follows that $\dim(W) = \dim(Z)$, and $\dim(Z) = \dim(\mathsf{N}(T))$ and $\dim(W) = \operatorname{codim}(\mathsf{R}(T^*))$. This completes the proof of (7.8).

The proof of (7.7) proceeds along the same lines, interchanging the roles of T and T^*. We now consider a basis $x_1^*, \ldots, x_k^*$ for $\mathsf{N}(T^*)$ and use the Hahn–Banach theorem to obtain $x_1^{**}, \ldots, x_k^{**} \in X^{**}$ such that

$$\langle x_i^*, x_j^{**} \rangle = \delta_{ij}, \quad 1 \leqslant i, j \leqslant k.$$

At this point we invoke Theorem 4.34 to obtain $x_1, \ldots, x_k \in X$ such that

$$\langle x_j, x_i^* \rangle = \delta_{ij}, \quad 1 \leqslant i, j \leqslant k.$$

With this analogue of (7.10) at hand the proof can be completed as before. $\qquad \square$

7.3.d The Noether–Gohberg–Krein Theorem

Let $\mathbb{D}$ and $\mathbb{T}$ denote the open unit disc and unit circle in the complex plane, respectively. We think of $\mathbb{T}$ as parametrised by $\theta \in [-\pi, \pi]$ and equipped with the normalised Lebesgue measure $\mathrm{d}\theta / 2\pi$. The *Hardy space* $H^2(\mathbb{D})$ is the Hilbert space of all holomorphic functions on $\mathbb{D}$ of the form $\sum_{n \in \mathbb{N}} c_n z^n$ with

$$\sum_{n \in \mathbb{N}} |c_n|^2 < \infty.$$

Since every square summable sequence $(c_n)_{n\in\mathbb{N}}$ defines a convergent power series $\sum_{n\in\mathbb{N}} c_n z^n$ holomorphic on $\mathbb{D}$, the correspondence between a power series and its coefficient sequence sets up an isometric isomorphism between $H^2(\mathbb{D})$ and $\ell^2(\mathbb{N})$. With respect to the norm

$$\|f\|_{H^2(\mathbb{D})} := \left(\sum_{n\in\mathbb{N}} |c_n|^2 \right)^{1/2},$$

$H^2(\mathbb{D})$ is a Hilbert space. For $n \in \mathbb{N}$ consider the functions $e_n \in L^2(\mathbb{T})$ defined by

$$e_n(\theta) := \exp(in\theta).$$

Since $(e_n)_{n\in\mathbb{N}}$ is an orthonormal sequence in $L^2(\mathbb{T})$, every square summable sequence $(c_n)_{n\in\mathbb{N}}$ defines a convergent sum $\sum_{n\in\mathbb{N}} c_n e_n$ in $L^2(\mathbb{T})$. Denoting this sum by f, its Fourier coefficients are given by

$$\widehat{f}(n) = \begin{cases} c_n, & n \geqslant 0, \\ 0, & n \leqslant -1. \end{cases}$$

Conversely, if all negative Fourier coefficients of a function $f \in L^2(\mathbb{T})$ vanish, then $f = \sum_{n\in\mathbb{N}} \widehat{f}(n) e_n$ as a convergent sum in $L^2(\mathbb{T})$. Since the Fourier coefficients of functions in $L^2(\mathbb{T})$ are square summable, we obtain an isometric isomorphism between $H^2(\mathbb{D})$ and the closed subspace of $L^2(\mathbb{T})$ consisting of all functions whose negative Fourier coefficients vanish. In what follows we identity $H^2(\mathbb{D})$ with this closed subspace of $L^2(\mathbb{T})$. As such, $H^2(\mathbb{D})$ is the range of the *Riesz projection*

$$P : \sum_{n\in\mathbb{Z}} \widehat{f}(n) e_n \mapsto \sum_{n\in\mathbb{N}} \widehat{f}(n) e_n$$

in $L^2(\mathbb{T})$ which discards the terms in the Fourier series with negative indices.

Given a function $\phi \in L^\infty(\mathbb{T})$ we define the bounded operator M_ϕ on $L^2(\mathbb{T})$ by pointwise multiplication,

$$M_\phi f := \phi f, \quad f \in L^2(\mathbb{T}).$$

When M_ϕ is applied to a function $f \in H^2(\mathbb{D})$, the resulting function ϕf generally does not belong to $H^2(\mathbb{D})$, but the Riesz projection will take us back to $H^2(\mathbb{D})$. This motivates the following definition.

Definition 7.28 (Toeplitz operators). Given a function $\phi \in L^\infty(\mathbb{T})$, the *Toeplitz operator with symbol* ϕ is the operator T_ϕ on $H^2(\mathbb{D})$ given by

$$T_\phi f := P(\phi f), \quad f \in H^2(\mathbb{D}),$$

where P is the Riesz projection.

Every Toeplitz operator T_ϕ is bounded of norm

$$\|T_\phi\| \leqslant \|P\| \|M_\phi\| \leqslant \|\phi\|_\infty. \tag{7.12}$$

Its Hilbert space adjoint is given by $T_\phi^\star = T_{\overline{\phi}}$; this follows from

$$(T_\phi f | g)_{H^2(\mathbb{D})} = (\phi f | g)_{L^2(\mathbb{T})} = (f | \overline{\phi} g)_{L^2(\mathbb{T})} = (f | T_{\overline{\phi}} g)_{H^2(\mathbb{D})}.$$

The following theorem shows that a Toeplitz operator with continuous and zero-free symbol $\phi \in C(\mathbb{T})$ is Fredholm, and its index equals the negative of the winding number of the closed contour in $\mathbb{C} \setminus \{0\}$ parametrised by ϕ. As we have seen in Section 6.2, for piecewise C^1 functions ϕ, the winding number is given analytically by the contour integral

$$w(\phi) = \frac{1}{2\pi i} \int_\phi \frac{dz}{z} = \frac{1}{2\pi i} \int_{-\pi}^\pi \frac{\phi'(t)}{\phi(t)} \, dt.$$

For functions ϕ that are merely continuous, the winding number can be defined as follows. It is an elementary theorem in Algebraic Topology that there exists a unique integer $n \in \mathbb{Z}$ such that ϕ is homotopic to the curve $\theta \mapsto e_n(\theta)$. By definition, this means that there exists a *homotopy* from ϕ to e_n in $\mathbb{C} \setminus \{0\}$, that is, a continuous function

$$h : [0,1] \times [-\pi, \pi] \to \mathbb{C} \setminus \{0\}$$

such that for all $\theta \in [-\pi, \pi]$ we have

$$h(0, \theta) = \phi(\theta), \quad h(1, \theta) = e_n(\theta).$$

Setting $h_t(\theta) := h(t, \theta)$, we think of the curves $h_t : [-\pi, \pi] \to \mathbb{C} \setminus \{0\}$ as continuously deforming $\phi = h_0$ to $e_n = h_1$. The *winding number* $w(\phi)$ of ϕ is defined to be this integer:

$$w(\phi) := n.$$

In particular, the winding number of e_n equals n. It is an easy consequence of Cauchy's theorem that this definition agrees with the analytic definition given earlier if ϕ is piecewise C^1.

Theorem 7.29 (Noether–Gohberg–Krein). *If the function $\phi \in C(\mathbb{T})$ is zero-free, then the Toeplitz operator T_ϕ is Fredholm on $H^2(\mathbb{D})$ and*

$$\operatorname{ind}(T_\phi) = -w(\phi).$$

This theorem is remarkable, as it computes an analytic quantity (the index) in terms of a topological one (the winding number). The main ingredient in the proof is the following lemma, which implies that the mapping $\phi \mapsto T_\phi$ from $C(\mathbb{T})$ to $\mathscr{L}(H^2(\mathbb{D}))$ is multiplicative up to a compact operator.

Lemma 7.30. *For all $\phi, \psi \in C(\mathbb{T})$ the operator $T_\phi T_\psi - T_{\phi\psi}$ is compact on $H^2(\mathbb{D})$.*

Proof By the estimate (7.12), the Weierstrass approximation theorem (Theorem 2.3), and the fact that uniform limits of compact operators are compact (Proposition 7.5), it suffices to prove the lemma for trigonometric polynomials ϕ and ψ.

Let $\phi = \sum_{m=-M}^{M} c_m e_m$ and $\psi = \sum_{n=-N}^{N} d_n e_n$ be trigonometric polynomials. For $j \geqslant N$ we have

$$
(T_\phi T_\psi - T_{\phi\psi})e_j
$$
$$
= \sum_{m=-M}^{M} \sum_{n=-N}^{N} c_m d_n P(e_m P(e_n e_j)) - \sum_{m=-M}^{M} \sum_{n=-N}^{N} c_m d_n P(e_m e_n e_j) = 0
$$

since $n + j \geqslant 0$ and hence $e_m P(e_n e_j) = e_m P(e_{n+j}) = e_m e_{n+j} = e_m e_n e_j$ in each summand. By linearity and density, this shows that $(T_\phi T_\psi - T_{\phi\psi})f = 0$ for all f in the closed linear span of $\{e_j : j \geqslant N\}$. This implies that $T_\psi T_\phi - T_{\phi\psi}$ is a finite rank operator (of rank at most N) and hence compact. $\qquad\square$

In particular this lemma implies that the commutator $T_\phi T_\psi - T_\psi T_\phi$ is compact. For functions $\phi, \psi \in C^2(\mathbb{T})$ a more precise result will be proved in Section 14.5.d.

Proof of Theorem 7.29 It follows from the lemma that the mapping

$$
J : C(\mathbb{T}) \to \mathscr{L}(H^2(\mathbb{D}))/\mathscr{K}(H^2(\mathbb{D}))
$$

given by

$$
J(\phi) := T_\phi + \mathscr{K}(H^2(\mathbb{D}))
$$

is multiplicative:

$$
J(\phi)J(\psi) = J(\phi\psi), \quad \phi, \psi \in C(\mathbb{T}).
$$

If ϕ is zero-free, then $1/\phi$ defines an element of $C(\mathbb{T})$ and

$$
J(1/\phi)J(\phi) = J(\phi)J(1/\phi) = J(\mathbf{1}) = I + \mathscr{K}(H^2(\mathbb{D})).
$$

Stated differently, there exist compact operators $K, L \in \mathscr{K}(H^2(\mathbb{D}))$ such that with $S := T_{1/\phi}$ we have

$$
ST_\phi = I - K, \quad T_\phi S = I - L.
$$

By Atkinson's theorem this implies that T_ϕ is Fredholm.

It remains to compute the index. To this end let $w(\phi) = n$ be the winding number of ϕ and let $h : [0,1] \times [-\pi, \pi] \to \mathbb{C} \setminus \{0\}$ be a homotopy from ϕ to e_n. By the continuity of the mapping $\phi \mapsto T_\phi$ and Dieudonné's theorem (Theorem 7.26), the mapping

$$
t \mapsto \mathrm{ind}(T_{h_t}), \quad t \in [0,1],
$$

is locally constant, and hence constant, where $h_t(s) := h(t,s)$. Since $h_0 = \phi$ and $h_1 = e_n$, in particular we obtain

$$\mathrm{ind}(T_\phi) = \mathrm{ind}(T_{e_n}).$$

Moreover, from

$$T_{e_n} \sum_{j \in \mathbb{N}} c_j e_j = P\Big(\sum_{j \in \mathbb{N}} c_j e_{j+n} \Big) = \sum_{j=(-n)\wedge 0}^{\infty} c_j e_{j+n}$$

we see that

$$\dim \mathsf{N}(T_{e_n}) = \begin{cases} 0, & n \geqslant 0, \\ -n, & n \leqslant -1, \end{cases} \qquad \mathrm{codim}\, \mathsf{R}(T_{e_n}) = \begin{cases} n, & n \geqslant 0, \\ 0, & n \leqslant -1, \end{cases}$$

so that $\mathrm{ind}(T_{e_n}) = -n$. $\square$

The following result clarifies why the symbol was assumed to be zero-free.

Theorem 7.31 (Hartman–Wintner). *If $\phi \in C(\mathbb{T})$ is such that the Toeplitz operator T_ϕ is Fredholm on $H^2(\mathbb{D})$, then ϕ is zero-free.*

Proof   Since $\mathsf{N}(T_\phi)$ is finite-dimensional and hence complemented, we have a direct sum decomposition $H^2(\mathbb{D}) = X_0 \oplus \mathsf{N}(T_\phi)$. Denote by π the projection onto $\mathsf{N}(T_\phi)$ along X_0. The operator T_ϕ restricts to an injective bounded operator from X_0 onto $\mathsf{R}(T_\phi)$, and the latter is a closed subspace of $H^2(\mathbb{D})$ by Proposition 7.22. Hence by the open mapping theorem there exists a constant $C > 0$ such that

$$\|T_\phi f_0\| \geqslant C\|f_0\|, \quad f_0 \in X_0.$$

For $f \in H^2(\mathbb{D})$ write $f = f_0 + g$ along the above decomposition. Then, since $\|\|f\|\| := C\|f_0\| + \|g\|$ is an equivalent norm on $H^2(\mathbb{D})$,

$$\|T_\phi f\| + \|\pi f\| = \|T_\phi f_0\| + \|g\| \geqslant C\|f_0\| + \|g\| \geqslant C'\|f\|,$$

where $C' > 0$ is a constant independent of f. For all $g \in L^2(\mathbb{T})$ we thus obtain

$$\|T_\phi Pg\| + \|\pi Pg\| \geqslant C'\|Pg\| \geqslant C'(\|g\| - \|(I-P)g\|),$$

where P is the Riesz projection. Let $U \in \mathscr{L}(L^2(\mathbb{T}))$ be the bounded operator given by $Ug(\theta) := \exp(i\theta)g(\theta)$. Applying the preceding estimate to $U^n g$ in place of g and using that U^n and U^{-n} are isometric, for all $g \in L^2(\mathbb{T})$ we obtain

$$\|U^{-n}T_\phi PU^n g\| + \|U^{-n}\pi PU^n g\| + C'\|U^{-n}(I-P)U^n g\| \geqslant C'\|g\|. \tag{7.13}$$

For every trigonometric polynomial g we have $U^{-n}PU^n g \to g$ in $L^2(\mathbb{T})$. Since these polynomials are dense in $L^2(\mathbb{T})$ and the operators $U^\pm$ all have norm one, it follows that

$$U^{-n}PU^n g \to g, \quad g \in L^2(\mathbb{T}).$$

This implies that $U^{-n}(I - P)U^n g \to 0$ for all $g \in L^2(\mathbb{T})$ and, using that U commutes with M_ϕ,

$$U^{-n}T_\phi PU^n g = U^{-n}PM_\phi PU^n g = (U^{-n}PU^n)M_\phi(U^{-n}PU^n)g \to M_\phi g$$

for all $g \in L^2(\mathbb{T})$. Also, $(U^n g|h) \to 0$ for all $g, h \in L^2(\mathbb{T})$ and therefore, since π is of finite rank,

$$\pi PU^n g \to 0, \quad g \in L^2(\mathbb{T}).$$

Passing to the limit in (7.13), we obtain

$$\|M_\phi g\| \geqslant C'\|g\|, \quad g \in L^2(\mathbb{T}).$$

Since $T_\phi^* = T_{\bar\phi}$ is Fredholm, we also obtain

$$\|M_\phi^* g\| = \|M_\phi g\| \geqslant C'\|g\|, \quad g \in L^2(\mathbb{T}).$$

It follows that M_ϕ is invertible (indeed, the inequality for M_ϕ gives injectivity and closed range, and the inequality for M_ϕ^* gives that M_ϕ has dense range). This is only possible if ϕ is zero-free (the inverse is then given by $M_{1/\phi}$). $\qquad\square$

Corollary 7.32. *For all $\phi \in C(\mathbb{T})$, the norm of the Toeplitz operator T_ϕ is given by*

$$\|T_\phi\| = \|\phi\|_\infty.$$

Proof Denote by M_ϕ the pointwise multiplication operator $f \mapsto \phi f$ on $L^2(\mathbb{T})$. We have $\sigma(M_\phi) = \{\phi(\theta) : \theta \in \mathbb{T}\}$. If $\lambda - T_\phi = T_{\lambda-\phi}$ is invertible, then $T_{\lambda-\phi}$ is Fredholm with index zero, and therefore $\lambda - \phi$ is zero-free by Theorem 7.31. But then $M_{\lambda-\phi} = \lambda - M_\phi$ is invertible. This argument shows that $\sigma(M_\phi) \subseteq \sigma(T_\phi)$. Now the corollary follows from the inequalities

$$\|\phi\|_\infty = \sup\{|\phi(\theta)| : \theta \in \mathbb{T}\} = \sup\{|\lambda| : \lambda \in \sigma(M_\phi)\}$$
$$\leqslant \sup\{|\lambda| : \lambda \in \sigma(T_\phi)\} \leqslant \|T_\phi\| \leqslant \|\phi\|_\infty. \tag{7.14}$$

$\qquad\square$

In the next theorem we denote by T_z the Toeplitz operator with symbol $\phi(z) = z$. Its adjoint is the Toeplitz operator $T_z^* = T_{\bar z}$ with symbol $\bar\phi(z) = \bar z$. Identifying $H^2(\mathbb{D})$ with $\ell^2(\mathbb{N})$ by identifying the function $z \mapsto z^n$ with the nth unit vector e_n, the operators T_z and $T_{\bar z}$ correspond to the right and left shift on $\ell^2(\mathbb{N})$, respectively. With these identifications in mind, the following theorem can be interpreted as giving a precise description of the closed subalgebra of $\ell^2(\mathbb{N})$ generated by the right and left shift.

Theorem 7.33 (Coburn). *Let $\mathscr{T}$ denote the closed subalgebra in $\mathscr{L}(H^2(\mathbb{D}))$ generated by T_z and T_z^*. Denoting by $\mathscr{K}$ the space of compact operators on $H^2(\mathbb{D})$, we have:*

(1) $\mathscr{T} = \{T_\phi + K : \phi \in C(\mathbb{T}), K \in \mathscr{K}\}$;

(2) *the mapping $\pi : T_\phi + K \mapsto \phi$ induces a multiplicative isometric isomorphism*

$$\mathscr{T}/\mathscr{K} \approx C(\mathbb{T}).$$

As a consequence, the representation of elements in $\mathscr{T}$ as the sum of a Toeplitz operator and a compact operator is unique, and we have the short exact sequence

$$0 \longrightarrow \mathscr{K} \longrightarrow \mathscr{T} \xrightarrow{\;\pi\;} C(\mathbb{T}) \longrightarrow 0.$$

In the final statement we used standard terminology from Algebraic Topology: a sequence of mappings is *exact* if the range of every operator in the sequence equals the null space of the next operator in the sequence.

In the proof, as well as in later chapters, we need the following notation. For elements g, h of a Hilbert space H, we denote by $g \bar{\otimes} h$ the rank one operator on H defined by

$$(g \bar{\otimes} h)x := (x|h)g, \quad x \in H.$$

The bar in this notation serves to emphasise the fact that $\bar{\otimes}$ is not a tensor product, but rather its sesquilinear counterpart in the sense that for all $c \in \mathbb{K}$ and $x \in H$ we have

$$(cg) \bar{\otimes} h = c(g \bar{\otimes} h), \quad g \bar{\otimes} (ch) = \bar{c}(g \bar{\otimes} h).$$

For norm one vectors $h \in H$, the operator $h \bar{\otimes} h$ is the orthogonal projection onto the one-dimensional subspace of H spanned by h.

Proof The crucial observation is that for all $\phi \in C(\mathbb{T})$ and $K \in \mathscr{K}$ we have

$$\|T_\phi + K\| \geqslant \|T_\phi\|. \tag{7.15}$$

In order to prove this it suffices to show that $\sigma(T_\phi) \subseteq \sigma(T_\phi + K)$, for this implies the claim via

$$\|T_\phi + K\| \geqslant r(T_\phi + K) \geqslant r(T_\phi) = \|T_\phi\|,$$

the last identity being a consequence of the proof of (7.14). To prove the spectral inclusion we argue as follows. Suppose $\lambda \in \mathbb{C}$ is such that $\lambda - (T_\phi + K) = T_{\lambda - \phi} - K$ is invertible. Then this operator is Fredholm with index 0. By Dieudonné's theorem, this implies that $T_{\lambda - \phi}$ is Fredholm with index 0. It remains to prove that this implies the invertibility of $T_{\lambda - \phi}$.

Suppose now that $\psi \in C(\mathbb{T})$ is such that T_ψ has index 0 but is not invertible. Then T_ψ has a nontrivial null space. By Proposition 7.27, $T_\psi^\star = T_{\overline{\psi}}$ has index 0 and fails to be invertible, hence also this operator has a nontrivial null space. This means that there are nonzero $g, h \in H^2(\mathbb{D})$ such that $P(\psi g) = P(\overline{\psi} h) = 0$, that is, ψg and $\overline{\psi} h$ have only negative Fourier coefficients. Invoking some standard results from the theory of Hardy spaces (see the Notes), this can be shown to imply $\psi = 0$.

Applying the preceding argument to $\psi := \lambda - \phi$ it follows that if $T_{\lambda-\phi}$ were non-invertible, then $\phi \equiv \lambda$. But then $T_{\lambda-\phi} - K = -K$ is compact and hence noninvertible, contradicting our assumption. This concludes the proof of (7.15).

(1): The inclusion '$\subseteq$' is a consequence of Lemma 7.30. To prove the inclusion '$\supseteq$' we must prove that $\mathscr{T}$ contains all Toeplitz operators with continuous symbol and all compact operators. Given a function $\phi \in C(\mathbb{T})$, we use the Stone–Weierstrass theorem to find a sequence of trigonometric polynomials $p_n \to \phi$ in $C(\mathbb{T})$. Then $T_{p_n} \to T_\phi$ in operator norm by (7.12). Since $T_{p_n} = p_n(T_z)$ we have $T_{p_n} \in \mathscr{T}$, and since $\mathscr{T}$ is closed this implies $T_\phi \in \mathscr{T}$.

We prove next that $\mathscr{T}$ contains every compact operator. Let $S := T_z$ for brevity, where z is shorthand for the function $z \mapsto z$. We have $S^\star S = I$ and $I - SS^\star = P_0$, the orthogonal projection onto the constant functions. These identities show that both I and P_0 belong to $\mathscr{T}$. Clearly,

$$\mathscr{I} := \mathscr{T} \cap \mathscr{K}$$

is a closed ideal in $\mathscr{T}$ which is closed under taking adjoints, and since P_0 is compact we have $P_0 \in \mathscr{I}$. We will show that $\mathscr{I} = \mathscr{K}$.

Fix an arbitrary $f \in H^2(\mathbb{D})$ and let $\varepsilon > 0$. There is a polynomial p such that $\|p(S)\mathbf{1} - f\| < \varepsilon$. Then $P_0(p(S))^\star \in \mathscr{I}$ and, since $P_0 h = (h|\mathbf{1})\mathbf{1}$,

$$\|P_0(p(S))^\star - (\mathbf{1} \bar\otimes f)\| = \sup_{\|g\|=\|h\|=1} |(P_0(p(S))^\star g - (\mathbf{1} \bar\otimes f)g|h)|$$

$$= \sup_{\|g\|=\|h\|=1} |(g|p(S)\mathbf{1}) - (g|f))||(\mathbf{1}|h)|$$

$$= \|p(S)\mathbf{1} - f\| < \varepsilon.$$

In the same way, for any $g \in H^2(\mathbb{D})$ and $\varepsilon > 0$ there is a polynomial q such that $\|q(S)\mathbf{1} - g\| < \varepsilon$. Then,

$$\|q(S)(\mathbf{1} \bar\otimes f) - (g \bar\otimes f)\| = \sup_{\|h\|=\|h'\|=1} |(q(S)(\mathbf{1} \bar\otimes f)h|h') - ((g \bar\otimes f)h|h')|$$

$$= \sup_{\|h\|=\|h'\|=1} |(q(S)\mathbf{1}|h') - (g|h')||(h|f)|$$

$$= \|q(S)\mathbf{1} - g\|\|f\| < \varepsilon\|f\|.$$

Since $\varepsilon > 0$ was arbitrary and $\mathscr{I}$ is closed, it follows that every rank one operator $g \bar\otimes f$ is contained in $\mathscr{I}$. By linearity, the same is true for every finite rank operator. Since the finite rank operators are dense in $\mathscr{K}$ by Proposition 7.6, it follows that $\mathscr{K} \subseteq \mathscr{I}$.

(2): From (7.15) it follows that

$$\|T_\phi + \mathscr{K}\| = \inf_{K \in \mathscr{K}} \|T_\phi + K\| \geqslant \|T_\phi\|.$$

Compact Operators

Together with the trivial inequality $\|T_\phi\| \geqslant \inf_{K \in \mathscr{K}} \|T_\phi + K\|$ we conclude that

$$\|T_\phi + \mathscr{K}\| = \|T_\phi\| = \|\phi\|_\infty.$$

This shows that the mapping $T_\phi + K \mapsto \phi$ is well defined and isometric. Clearly it is surjective, and therefore it is an isometric isomorphism. Its multiplicativity follows from Lemma 7.30. □

Using some elementary facts from the theory of C^*-algebras, a more transparent alternative proof of Theorem 7.31 can be given as a corollary to Theorem 7.33. This proof is sketched in the Notes to this chapter.

Remark 7.34. Identifying $H^2(\mathbb{D})$ with $\ell^2(\mathbb{N})$ as indicated above, the short exact sequence of the theorem induces a short exact sequence

$$0 \longrightarrow \mathscr{K}(\ell^2(\mathbb{N})) \longrightarrow \mathscr{T}(\ell^2(\mathbb{N})) \xrightarrow{\ \overline{\pi}\ } C(\mathbb{T}) \longrightarrow 0,$$

where $\mathscr{K}(\ell^2(\mathbb{N}))$ and $\mathscr{T}(\ell^2(\mathbb{N}))$ denote, respectively, the compact operators acting on $\ell^2(\mathbb{N})$ and the closed algebra generated by the left and right shift in $\ell^2(\mathbb{N})$, and $\overline{\pi}$ is the operator induced by π under the identifications made.

Problems

7.1   Give an alternative proof of Proposition 7.5 by using the equivalence of compactness and sequential compactness.

7.2   Let X and Y be Banach spaces. Prove that if X is infinite-dimensional and $T \in \mathscr{L}(X,Y)$ is compact, then there exists a sequence $(x_n)_{n \geqslant 1}$ of norm one vectors in X such that $\lim_{n \to \infty} T x_n = 0$ in Y.

7.3   Let $(m_n)_{n \geqslant 1}$ be a bounded scalar sequence.

   (a) For $1 \leqslant p < \infty$, show that the multiplication operator $(c_n)_{n \geqslant 1} \mapsto (m_n c_n)_{n \geqslant 1}$ is compact on ℓ^p if and only if $\lim_{n \to \infty} m_n = 0$.

   (b) Does the same result hold for ℓ^∞? And for c_0?
   Hint: Compare with Problem 2.32.

7.4   Let $1 \leqslant p < q \leqslant \infty$.

   (a) Prove that the inclusion mapping $\ell^p \subseteq \ell^q$ is not compact.

   (b) Prove that the inclusion mapping $L^q(0,1) \subseteq L^p(0,1)$ is not compact.
   Hint: Look up Khintchine's inequality for the Rademacher functions $f_n(\theta) = \text{sign}(\sin(2\pi 2^n \theta))$.

7.5   For $1 \leqslant p \leqslant \infty$ consider the bounded operator $T_p : L^p(0,1) \to C[0,1]$,

$$T_p f(t) := \int_0^t f(s)\, ds, \quad t \in [0,1].$$

(a) Show that if $1 < p \leqslant \infty$, then T_p is compact.

(b) Is T_1 compact?

7.6 For any fixed $k \geqslant 1$, find a bounded operator T acting on a Hilbert space such that T^{k+1} is compact but T^k is not.

7.7 Let $g \in C[0,1]$ be given. Show that the multiplication operator $T_g : f \mapsto fg$ on $C[0,1]$ is compact if and only if $g = 0$.

7.8 Let X and Y be Banach spaces. Show that an operator $T \in \mathscr{L}(X,Y)$ is compact if and only if there is a sequence $(x_n^*)_{n \geqslant 1}$ in X^* such that $\lim_{n \to \infty} x_n^* = 0$ in X^* and

$$\|Tx\| \leqslant \sup_{n \geqslant 1} |\langle x, x_n^* \rangle|, \quad x \in X.$$

Hint: For the 'if' part consider T as the composition of mappings

$$x \mapsto (\langle x, x_n^* \rangle)_{n \geqslant 1} \mapsto Tx$$

and argue as in Problem 7.3; for the 'only if' part use the result of Problem 1.23 and the compactness of T^*.

7.9 The aim of this problem is to show how part (1) of Theorem 7.11 can be deduced from part (2). Let X be a Banach space and let $T \in \mathscr{L}(X)$ be compact.

(a) Using Proposition 6.17, show that every nonzero $\lambda \in \partial \sigma(T)$ is an eigenvalue.

(b) Using part (a) and part (2) of Theorem 7.11, deduce that every nonzero $\lambda \in \sigma(T)$ is an eigenvalue.

7.10 Let X be a Banach space. Show that for a bounded operator $T \in \mathscr{L}(X)$ the following assertions are equivalent:

(1) T is compact;

(2) $\exp(T) - I$ is compact.

Hint: To prove the implication (2)$\Rightarrow$(1) show that for large enough integers k we have $T = (\exp(kT) - I)f_k(T)$, where $f_k(z) = z/(e^{kz} - 1)$ is holomorphic in a neighbourhood of $\sigma(T)$.

7.11 Let T be a compact operator on a Banach space X, let $0 \neq \lambda \in \sigma(T)$, and let ν be its algebraic multiplicity. Let $X_\lambda := P_\lambda X$ be the range of the spectral projection associated with the point λ. Prove the following assertions:

(a) a vector $x \in X$ belongs to X_λ if and only if $(\lambda - T)^k x = 0$ for some $k \geqslant 1$;

(b) for all $x \in X_\lambda$ we have $(\lambda - T)^\nu x = 0$;

(c) deduce that $X_\lambda = \mathsf{N}(\lambda - T)^\nu$.

7.12 Let X be a Banach space. In this problem we write $[T]$ for the element $T + \mathscr{K}(X)$

of the Calkin algebra $\mathscr{L}(X)/\mathscr{K}(X)$. Show that the multiplication $[S] \circ [T] := [ST]$ is well defined on $\mathscr{L}(X)/\mathscr{K}(X)$ and satisfies

$$\|[S] \circ [T]\|_{\mathscr{L}(X)/\mathscr{K}(X)} \leqslant \|[S]\|_{\mathscr{L}(X)/\mathscr{K}(X)} \|[T]\|_{\mathscr{L}(X)/\mathscr{K}(X)}.$$

In the terminology introduced in the Notes to this chapter, this shows that the Calkin algebra $\mathscr{L}(X)/\mathscr{K}(X)$ is a (unital) Banach algebra.

7.13 Let X be a Banach space and $T \in \mathscr{L}(X)$ be a bounded operator. Show that if T^k is compact for some integer $k \geqslant 1$, then $I + T$ is Fredholm. What is its index?

7.14 Prove that the two definitions of the winding number of a piecewise C^1 curve discussed in Section 7.3 agree.

7.15 Let $\phi \in L^\infty(\mathbb{T})$. Prove the following assertions:

(a) the operator M_ϕ on $L^2(\mathbb{T})$ defined by

$$M_\phi f := \phi f, \quad f \in L^2(\mathbb{T}),$$

maps $H^2(\mathbb{D})$ into itself if and only if $\phi \in H^\infty(\mathbb{D})$, that is, upon identifying ϕ with a function in $L^2(\mathbb{T})$ whose negative Fourier coefficient vanish we have $\phi \in L^\infty(\mathbb{T})$;

(b) if $\phi \in L^\infty(\mathbb{T})$ and $\psi \in H^\infty(\mathbb{T})$, then the associated Toeplitz operators satisfy

$$T_\phi T_\psi f = T_{\phi\psi} f, \quad f \in H^2(\mathbb{D});$$

(c) if $\phi, \psi \in H^\infty(\mathbb{T})$, then the associated Toeplitz operators satisfy

$$T_\psi T_\phi f = T_{\phi\psi} f, \quad f \in H^2(\mathbb{D}).$$

7.16 Using notation introduced in Section 7.3.d, prove the following uniqueness result: If $T \in \mathscr{T}$ satisfies

$$T = T_\phi + K = T_\psi + L$$

with $\phi, \psi \in C(\mathbb{T})$ and $K, L \in \mathscr{K}(H^2(\mathbb{D}))$, then $\phi = \psi$ and $K = L$.

7.17 Show that $T \in \mathscr{T}$ is Fredholm if and only if $T = T_\phi + K$, where $\phi \in C(\mathbb{T})$ is zero-free and $K \in \mathscr{K}(H^2(\mathbb{D}))$.

8

Bounded Operators on Hilbert Spaces

The identification of a Hilbert space H with its dual via the Riesz representation theorem makes it possible to consider a bounded operator T and its adjoint simultaneously on H. This leads to the important classes of selfadjoint, unitary, and normal operators. Their spectral theory is particularly rich. Its full power comes to bear only in the next chapter, where we prove the spectral theorem for bounded normal operators. The present chapter discusses the elementary theory and, for normal operators T, establishes a generalisation of the holomorphic calculus to a calculus for continuous functions on the spectrum $\sigma(T)$. Using this calculus, we prove a number of nontrivial results such as the existence of a unique positive square root of a positive operator and a polar decomposition for general bounded operators. In the last section we establish the celebrated Sz.-Nagy theorem on the existence of unitary dilations for Hilbert space contractions.

8.1 Selfadjoint, Unitary, and Normal Operators

Throughout this chapter, H is a complex Hilbert space. The following proposition is key to several proofs in this chapter. The example of rotation over $\frac{1}{2}\pi$ in $\mathbb{R}^2$ shows that its counterpart for real Hilbert spaces fails.

Proposition 8.1. *If $T \in \mathscr{L}(H)$ satisfies $(Tx|x) = 0$ for all $x \in H$, then $T = 0$.*

Proof For all $x, y \in H$, from $(T(x+y)|x+y) = 0$ we obtain

$$(Tx|y) + (Ty|x) = 0. \tag{8.1}$$

Replacing y by iy we obtain $-i(Tx|y) + i(Ty|x) = 0$. Multiplying both sides with i gives

$$(Tx|y) - (Ty|x) = 0. \tag{8.2}$$

Adding (8.1) and (8.2) gives $(Tx|y) = 0$ for all $x, y \in H$. This implies the result. $\square$

The trick used in the proof is called *polarisation*.

Definition 8.2 (Normal, unitary, selfadjoint, and positive operators). An operator $T \in \mathcal{L}(H)$ is called:

- *positive*, if $(Tx|x) \geqslant 0$ for all $x \in H$;
- *selfadjoint*, if $T = T^\star$;
- *unitary*, if $TT^\star = T^\star T = I$;
- *normal*, if $TT^\star = T^\star T$.

Here, $T^\star$ is the Hilbert space adjoint of T (cf. Section 4.3.b). Every positive operator is selfadjoint: for if T is positive, then $(Tx|x)$ is positive and therefore $(T^\star x|x) = \overline{(x|T^\star x)} = \overline{(Tx|x)} = (Tx|x)$, that is, $((T - T^\star)x|x) = 0$ for all $x \in X$. Proposition 8.1 now implies that $T - T^\star = 0$. Selfadjoint operators and unitary operators are normal.

The classes of positive, selfadjoint, unitary, and normal operators can be viewed as operator analogues of the positive real numbers, the real numbers, the complex numbers of modulus one, and the complex numbers, respectively. A number of results support this view:

- every selfadjoint operator is the difference of two positive operators;
- every invertible operator is the composition of an invertible positive operator and a unitary operator;
- a bounded operator is unitary if and only if it is the complex exponential of a selfadjoint operator.

The first result follows from the spectral theorem in the next chapter (see Problem 9.12), the second and the 'if' part of the third will be proved in the present chapter, and the 'only if' part of the third is again a consequence of the spectral theorem (see Problem 9.13).

In terms of spectra we have the following characterisations:

- a normal operator is unitary if and only if its spectrum is contained in the unit circle;
- a normal operator is selfadjoint if and only if its spectrum is contained in the real line;
- a normal operator is positive if and only if its spectrum is contained in $[0, \infty)$;
- a normal operator is an orthogonal projection if and only if its spectrum is contained in $\{0, 1\}$.

The last of these results indicated that orthogonal projections can be viewed as the analogous to the 'Boolean' set $\{0, 1\}$. It also implies that normal projections are orthogonal.

Let us begin by proving an operator analogue of the decomposition of a complex number into real and imaginary parts.

Proposition 8.3. *For every operator $T \in \mathcal{L}(H)$ there exist unique selfadjoint operators $A, B \in \mathcal{L}(H)$ such that $T = A + iB$.*

Proof The operators $A := \frac{1}{2}(T + T^\star)$ and $B := \frac{1}{2i}(T - T^\star)$ are selfadjoint and $T = A + iB$. Suppose we also have $T = A' + iB'$ with A' and B' selfadjoint. Put $U := B - B'$. Then $(iU)^\star = -iU^\star = -iU$ and also $iU = i(B - B') = (T - A) - (T - A') = A' - A$, so $(iU)^\star = (A' - A)^\star = A' - A = iU$. It follows that $iU = -iU$ and therefore $B = B'$. This in turn implies $A = A'$. $\qquad\square$

A complex number satisfies $|z| = 1$ if and only if there is a real number x such that $z = e^{ix}$. The operator analogue of the 'if' part is contained in the next proposition.

Proposition 8.4. *If $T \in \mathscr{L}(H)$ is selfadjoint, then e^{iT} is unitary.*

Proof From the expansion $e^{iT} = \sum_{n=0}^{\infty} \frac{i^n}{n!} T^n$ we see that $(e^{iT})^\star = \sum_{n=0}^{\infty} \frac{(-i)^n}{n!} T^n = e^{-iT}$. It is elementary to check that $e^{iT} e^{-iT} = e^{-iT} e^{iT} = I$ by writing out the defining power series and multiplying them. Alternatively, this identity follows from the multiplicativity of the entire calculus of T applied with $f(z) = \exp(iz)$ and $g(z) = \exp(-iz)$. $\qquad\square$

We have the following simple characterisation of unitary operators:

Proposition 8.5. *For an operator $U \in \mathscr{L}(H)$ the following assertions are equivalent:*

(1) *U is unitary;*
(2) *U is surjective and $\|Ux\| = \|x\|$ for all $x \in H$;*
(3) *U is surjective and $(Ux|Uy) = (x|y)$ for all $x, y \in H$.*

Proof (1)$\Rightarrow$(3): If U is unitary, then U is invertible (with inverse $U^{-1} = U^\star$) and therefore U is surjective. Moreover, $(Ux|Uy) = (x|U^\star Uy) = (x|y)$.

(3)$\Rightarrow$(2): Take $x = y$.

(2)$\Rightarrow$(1): We have $(U^\star Ux|x) = (Ux|Ux) = \|Ux\|^2 = \|x\|^2 = (x|x)$ for all $x \in H$ and therefore $U^\star U = I$. It follows that $U^\star$ is a left inverse to U. The assumptions further imply that U is surjective and injective, hence invertible. The inverse must be equal to the left inverse, which is therefore $U^\star$. It follows that also $UU^\star = I$. $\qquad\square$

The right shift on ℓ^2 shows that the surjectivity assumption cannot be omitted from (2) and (3).

Example 8.6. The left and right shifts on $\ell^2(\mathbb{Z})$ are unitary. Indeed, the adjoint of the left (right) shift is the right (left) shift, so in either case the adjoint equals the inverse. Similarly, left and right translations on $L^2(\mathbb{R})$ are unitary.

Example 8.7. The Fourier–Plancherel transform on $L^2(\mathbb{R}^d)$ is unitary; this follows from Theorem 5.25 and Proposition 8.5.

Projections in Hilbert spaces are orthogonal if and only if they are selfadjoint:

Proposition 8.8. *For a projection $P \in \mathscr{L}(H)$ the following assertions are equivalent:*

(1) P is orthogonal, *that is, its null space and range are orthogonal;*
(2) P is selfadjoint.

Proof (1)$\Rightarrow$(2): If P is orthogonal, then $x - Px \perp Py$ for all $x, y \in H$, noting that $x - Px \in N(P)$ (since $P(x - Px) = Px - P^2x = Px - Px = 0$) and $Py \in R(P)$. Therefore,

$$(x|Py) = (Px|Py) + (x - Px|Py) = (Px|Py)$$

and similarly

$$(Px|y) = (Px|Py) + (Px|y - Py) = (Px|Py),$$

so $(Px|y) = (x|Py)$ and P is selfadjoint.

(2)$\Rightarrow$(1): If P is a selfadjoint projection, then

$$(x - Px|Py) = (P^*(x - Px)|y) = (P(x - Px)|y) = 0$$

since $P = P^2$. Since every element in $N(P)$ is of the form $x - Px$, this shows that $N(P) \perp R(P)$, that is, the projection P is orthogonal. $\qquad\square$

We now turn to the study of some spectral properties of Hilbert space operators. From Proposition 6.18 we recall that for every bounded operator T on a Banach space we have

$$\sigma(T^*) = \sigma(T).$$

A similar result holds for the spectrum of the Hilbert space adjoint $T^\star$:

Proposition 8.9. *For all $T \in \mathscr{L}(H)$ we have*

$$\sigma(T^\star) = \overline{\sigma(T)},$$

where the bar denotes complex conjugation.

Proof The proof follows the lines of Proposition 6.18 but is simpler because of the Riesz representation theorem. The idea is to prove that $\lambda \in \rho(T)$ if and only if $\overline{\lambda} \in \rho(T^\star)$, and that in this case $(R(\lambda, T))^\star = R(\overline{\lambda}, T^\star)$.

First suppose that $\lambda \in \rho(T)$. Then

$$(\overline{\lambda} - T^\star)(R(\lambda, T))^\star = (\lambda - T)^\star(R(\lambda, T))^\star = (R(\lambda, T)(\lambda - T))^\star = I^\star = I.$$

In the same way it is shown that $(R(\lambda, T))^\star(\overline{\lambda} - T^\star) = I$. It follows that $\overline{\lambda} \in \rho(T^\star)$ and $R(\overline{\lambda}, T^\star) = (R(\lambda, T))^\star$.

If $\lambda \in \rho(T^\star)$, applying what we just proved to $T^\star$ gives $\overline{\lambda} \in \rho(T^{\star\star}) = \rho(T)$ and $R(\overline{\lambda}, T) = R(\overline{\lambda}, T^{\star\star}) = (R(\lambda, T^\star))^\star.$ $\qquad\square$

For unitary operators we have the following simple result.

Proposition 8.10. *If $U \in \mathscr{L}(H)$ is unitary, then $\sigma(U)$ is contained in the unit circle.*

Proof Since U is an invertible isometry, this follows from Corollary 6.14. □

Alternatively we can apply the spectral mapping theorem of the holomorphic calculus, which gives $\sigma(U^\star) = \sigma(U^{-1}) = (\sigma(U))^{-1}$. Since both $\sigma(U)$ and $\sigma(U^\star)$ are contained in the closed unit disc, $\sigma(U)$ must therefore be contained in the unit circle.

In the converse direction, a normal operator whose spectrum is contained in the unit circle is unitary. The proof of this fact is harder and will be given in Corollary 9.18.

The next result describes the spectrum of selfadjoint operators.

Theorem 8.11 (Spectrum of selfadjoint operators). *An operator $T \in \mathscr{L}(H)$ is selfadjoint if and only if $(Tx|x) \in \mathbb{R}$ for all $x \in H$. If T is selfadjoint on H, then*

$$\|T\| = \sup_{\|x\|\leqslant 1} |(Tx|x)| = \max\{|m|, |M|\}$$

and

$$\{m, M\} \subseteq \sigma(T) \subseteq [m, M],$$

where $m := \inf_{\|x\|=1}(Tx|x)$ and $M := \sup_{\|x\|=1}(Tx|x)$.

Proof If $(Tx|x) \in \mathbb{R}$, then $(T^\star x|x) = \overline{(x|T^\star x)} = \overline{(Tx|x)} = (Tx|x)$. Hence if $(Tx|x) \in \mathbb{R}$ for all $x \in H$, then $T = T^\star$ by Proposition 8.1 applied to $T - T^\star$. Conversely if $T = T^\star$, then $(Tx|x) = (x|Tx) = \overline{(Tx|x)}$ and therefore $(Tx|x) \in \mathbb{R}$.

Next we prove that $\sigma(T) \subseteq \mathbb{R}$. To this end let $\lambda = \alpha + i\beta$ with $\alpha, \beta \in \mathbb{R}$ and $\beta \neq 0$; we wish to prove that $\lambda \in \rho(T)$. For all $x \in H$ we have

$$\|(\lambda - T)x\|\|x\| \geqslant |((\lambda - T)x|x)| = |\alpha(x|x) - (Tx|x) + i\beta(x|x)| \geqslant |\beta|\|x\|^2,$$

using that $(Tx|x) \in \mathbb{R}$ in the last step. By Proposition 1.21 this implies that $\lambda - T$ is injective and has closed range. Replacing λ by $\overline{\lambda}$, we also conclude that $\overline{\lambda} - T$ is injective and has closed range. By Proposition 4.31, this implies that $\lambda - T = (\overline{\lambda} - T)^\star$ has dense range. We conclude that $\lambda - T$ is both injective and surjective, hence invertible, and therefore $\lambda \in \rho(T)$.

By now we have shown that $\sigma(T) \subseteq \mathbb{R}$. Next we show that $\sigma(T) \subseteq [m, M]$. Let $\lambda = M + \delta$ with $\delta > 0$. Then, by the definition of M, for all $x \in H$ with $\|x\| = 1$ we have

$$\|(\lambda - T)x\|\|x\| \geqslant ((\lambda - T)x|x) = M(x|x) - (Tx|x) + \delta(x|x) \geqslant \delta(x|x) = \delta\|x\|^2.$$

The same argument as before shows that $\lambda - T$ is both injective and surjective, hence invertible, and therefore $\lambda \in \rho(T)$. This proves that $(M, \infty) \subseteq \rho(T)$. Applying this result to $-T$ (and replacing $[m, M]$ with $[-M, -m]$) we also obtain $(-\infty, m) \subseteq \rho(T)$. This completes the proof that $\sigma(T) \subseteq [m, M]$.

We prove next that $\|T\| = \max\{|m|, |M|\}$; this implies $\|T\| = \sup_{\|x\|\leqslant 1} |(Tx|x)|$. Replacing T by $-T$ if necessary, we may assume that $|m| \leqslant |M|$. Clearly we then have

$|M| = \sup_{\|x\|=1} |(Tx|x)| \leqslant \|T\|$. To prove the converse inequality $\|T\| \leqslant |M|$, note that for all $x \in H$ with $\|x\| = 1$ and all $\mu > 0$ we have

$$
\begin{aligned}
4\|Tx\|^2 &= (T(\mu x + \mu^{-1}Tx)|\mu x + \mu^{-1}Tx) - (T(\mu x - \mu^{-1}Tx)|\mu x - \mu^{-1}Tx) \\
&\leqslant |M|\,\|\mu x + \mu^{-1}Tx\|^2 + |m|\,\|\mu x - \mu^{-1}Tx\|^2 \\
&\leqslant |M|\,\|\mu x + \mu^{-1}Tx\|^2 + |M|\,\|\mu x - \mu^{-1}Tx\|^2 \\
&= 2|M|\left(\mu^2\|x\|^2 + \frac{1}{\mu^2}\|Tx\|^2\right) = 2|M|\left(\mu^2 + \frac{1}{\mu^2}\|Tx\|^2\right),
\end{aligned}
$$

where the first inequality follows from the definitions of m and M and the next equality uses the parallelogram identity. Taking $\mu^2 = \|Tx\|$ we obtain, for all $x \in H$ with $\|x\| = 1$,

$$
4\|Tx\|^2 \leqslant 2|M|(\|Tx\| + \|Tx\|) = 4|M|\,\|Tx\|.
$$

It follows that $\|Tx\| \leqslant |M|$ for all $x \in H$ with $\|x\| = 1$, so $\|T\| \leqslant |M|$.

The last thing to prove is that $m, M \in \sigma(T)$. We prove this for M; the result for m follows by considering $-T$. Replacing T by $T - m$ we may assume that $0 = m \leqslant M$. Then $(Tx|x) \geqslant 0$ for all $x \in H$ and therefore

$$
M = \sup_{\|x\|=1} (Tx|x) = \sup_{\|x\|=1} |(Tx|x)| = \|T\|.
$$

Choose a sequence $(x_n)_{n\geqslant 1}$ of norm one vectors such that $\lim_{n\to\infty}(Tx_n|x_n) = M$. Then,

$$
\begin{aligned}
\|(M-T)x_n\|^2 &= ((M-T)x_n|(M-T)x_n) \\
&= M^2\|x_n\|^2 - 2M(Tx_n|x_n) + \|Tx_n\|^2 \\
&\leqslant M^2 - 2M(Tx_n|x_n) + \|T\|^2 = M^2 - 2M(Tx_n|x_n) + M^2,
\end{aligned}
$$

which tends to $M^2 - 2M^2 + M^2 = 0$ as $n \to \infty$. This implies that M is an approximate eigenvalue of T. $\qquad\square$

In the converse direction, a normal operator whose spectrum is contained in the real line is selfadjoint. This will be proved in Corollary 9.18.

It is an immediate consequence of Theorem 8.11 that the norm of a selfadjoint operator equals its spectral radius. More generally this is true for normal operators; see Proposition 8.13. This equality of norm and spectral radius can sometimes be used to determine the norm of an operator. We illustrate this by determining the norm of the Volterra operator.

Example 8.12 (Volterra operator). From Example 1.31 we recall that the *Volterra operator* is the operator $T \in \mathscr{L}(L^2(0,1))$ given by the indefinite integral

$$
(Tf)(s) := \int_0^s f(t)\,dt, \quad f \in L^2(0,1),\ s \in (0,1).
$$

The operator T fails to be selfadjoint (it even fails to be normal, see Problem 8.12), but we have $\|T\| = \|S\|$, where $S \in \mathscr{L}(L^2(0,1))$ is defined by

$$(Sf)(s) := \int_0^{1-s} f(t)\,dt, \quad f \in L^2(0,1),\ s \in (0,1),$$

as is immediate from the identity $(Sf)(s) = (Tf)(1-s)$. This identity also implies

$$
\begin{aligned}
(f|S^\star g) = (Sf|g) &= \int_0^1 (Tf)(1-s)\overline{g}(s)\,ds = \int_0^1 (Tf)(s)\overline{g}(1-s)\,ds \\
&= \int_0^1 \int_0^s f(t)\overline{g}(1-s)\,dt\,ds = \int_0^1 \int_t^1 f(t)\overline{g}(1-s)\,ds\,dt \\
&= \int_0^1 \int_0^{1-t} f(t)\overline{g}(s)\,ds\,dt = \int_0^1 f(t)\overline{Tg(1-t)}\,dt = (f|Sg),
\end{aligned}
$$

which shows that S is selfadjoint. By Example 7.4 S is compact, and therefore by Theorem 7.11 every nonzero $\lambda \in \sigma(S)$ is an eigenvalue. To compute the spectral radius $r(S)$ we therefore have to determine the set of nonzero eigenvalues of S.

Suppose that $\lambda \neq 0$ is an eigenvalue of S and let f be an eigenfunction. Then

$$f(s) = \frac{1}{\lambda} \int_0^{1-s} f(t)\,dt \tag{8.3}$$

for almost all $s \in (0,1)$, and the right-hand side is a continuous function of s. It follows that $f \in C[0,1]$ and that (8.3) holds for all $s \in [0,1]$. Then the same argument shows that in fact $f \in C^1[0,1]$, and applying the same argument once more gives $f \in C^2[0,1]$. Differentiating (8.3) twice gives

$$\lambda^2 f''(s) = -f(s), \quad s \in [0,1],$$

subject to the initial conditions $f(1) = 0$ and $f'(0) = 0$. The reader may check that this problem admit a solution if and only if $\frac{1}{\lambda} = \frac{\pi}{2} + \pi n$ for some $n \in \mathbb{Z}$, and that the solutions $f_n(s) = \cos\left(\left(\frac{\pi}{2} + \pi n\right)s\right)$ are indeed eigenfunctions of S. The largest eigenvalue of S therefore equals $\frac{2}{\pi}$. We conclude that

$$\|T\| = \|S\| = r(S) = \frac{2}{\pi}.$$

The remainder of this section is devoted to studying some spectral properties of normal operators. Our first aim is to prove that the spectral radius of a normal operator equals the operator norm. For selfadjoint operators this has already been observed as a consequence of Theorem 8.11, and for unitary operators this is an immediate consequence of Proposition 8.10.

Proposition 8.13. *A operator $T \in \mathscr{L}(H)$ is normal if and only if*

$$\|Tx\| = \|T^\star x\|, \quad x \in H.$$

If T is normal, then $\|T\|^n = \|T^n\|$ for all $n \in \mathbb{N}$, and therefore $r(T) = \|T\|$.

Proof If T is normal, then

$$\|Tx\|^2 = (Tx|Tx) = (x|T^\star Tx) = (x|TT^\star x) = (T^\star x|T^\star x) = \|T^\star x\|^2.$$

In the converse direction, the equality implies $((T^\star T - TT^\star)x|x) = 0$ for all $x \in H$ and therefore $T^\star T - TT^\star = 0$ by Proposition 8.1. This proves the first assertion.

If T is normal, then for all norm one vectors $x \in H$ we have

$$\|T^\star Tx\|^2 = ((T^\star T)^2 x|x) = ((T^\star)^2 T^2 x|x) = \|T^2 x\|^2$$

and therefore, since $\|T^\star T\| = \|T\|^2$ by Proposition 4.28,

$$\|T\|^2 = \|T^\star T\| = \|T^2\|. \tag{8.4}$$

Suppose the identity $\|T^n\| = \|T\|^n$ has been proved for $n = 2, \ldots, k$. Then, for all norm one vectors $x \in H$,

$$\|Tx\|^{2k} = \|T^k x\|^2 = (T^\star T^k x|T^{k-1} x)$$
$$\leqslant \|T^\star T^k x\| \|T^{k-1} x\| = \|T^{k+1} x\| \|T^{k-1} x\| \leqslant \|T^{k+1}\| \|T^{k-1}\| = \|T^{k+1}\| \|T\|^{k-1},$$

using (8.4) and the inductive assumption. This results in the identity $\|T\|^{k+1} \leqslant \|T^{k+1}\|$. Since the reverse inequality holds trivially, we conclude that $\|T^{k+1}\| = \|T\|^{k+1}$.

The final assertion follows from the spectral radius formula (Theorem 6.24). $\qquad \square$

Recall that $\lambda \in \mathbb{C}$ is called an *approximate eigenvalue* of T if there exists a sequence $(x_n)_{n \geqslant 1}$ in X such that $\|x_n\| = 1$ for all $n \geqslant 1$ and $\lim_{n \to \infty} \|Tx_n - \lambda x_n\| \to 0$. By Proposition 6.17 the boundary spectrum of any bounded operator on a Banach space consists of approximate eigenvalues. For normal operators on Hilbert spaces more is true:

Proposition 8.14. *Every point in the spectrum of a normal operator $T \in \mathscr{L}(H)$ is an approximate eigenvalue.*

Proof Suppose $\lambda \in \mathbb{C}$ is not an approximate eigenvalue. Then $\lambda - T$ is injective (otherwise λ would be an eigenvalue), and $\|\lambda x - Tx\| = \|\bar{\lambda} x - T^\star x\|$ implies that also $(\lambda - T)^\star$ is injective, that is, $\lambda - T$ has dense range. Let us prove that $\lambda - T$ has closed range.

Let $(x_n)_{n \geqslant 1}$ be a sequence in H such that $\lim_{n \to \infty} (\lambda - T)x_n = y$ in H. Then

$$\lim_{N \to \infty} \sup_{n,m \geqslant N} \|(\lambda - T)(x_n - x_m)\| = 0.$$

Unless we have $\lim_{m,n \to \infty} \|x_n - x_m\| = 0$, normalisation allows us to construct an approximate eigensequence to arrive at a contradiction. Thus $\lim_{m,n \to \infty} \|x_n - x_m\| = 0$, which means that $(x_n)_{n \geqslant 1}$ is Cauchy and therefore converges to a limit x. Then $y = (\lambda - T)x$.

We have shown that $\lambda - T$ is surjective. Since this operator is also injective, it follows that $\lambda \in \rho(T)$. $\qquad \square$

Theorem 7.11 implies that every nonzero element of the spectrum of a compact operator is both an isolated point and an eigenvalue. The final result of this section states that for normal operators on a Hilbert space, all isolated points in the spectrum are eigenvalues; no compactness assumption is needed. Normality cannot be omitted: the Volterra operator has spectrum $\{0\}$, but 0 is not an eigenvalue (see Problem 8.12).

If T is a bounded operator on H, for $\lambda \in \mathbb{C}$ we set $E_\lambda := \{x \in X : Tx = \lambda x\}$. Thus λ is an eigenvalue for T if and only if E_λ is nonzero. We recall that spectral projections have been defined in Theorem 6.23.

Theorem 8.15 (Isolated points are eigenvalues). *Let $T \in \mathscr{L}(H)$ be a normal operator and let λ be an isolated point in $\sigma(T)$. Then λ is an eigenvalue for T and the spectral projection $P^{\{\lambda\}}$ corresponding to $\{\lambda\}$ equals the orthogonal projection P_λ onto E_λ.*

The proof uses the following simple observation.

Lemma 8.16. *Let $T \in \mathscr{L}(H)$. If Y be a closed subspace of H, then:*

(1) *if Y is invariant under T, then $Y^\perp$ is invariant under $T^\star$;*
(2) *if Y is invariant under T and $T^\star$, then $Y^\perp$ is invariant under T and $T^\star$ and*

$$(T|_Y)^\star = T^\star|_Y \ \text{ and } \ (T|_{Y^\perp})^\star = T^\star|_{Y^\perp}$$

as operators in $\mathscr{L}(Y)$ and $\mathscr{L}(Y^\perp)$, respectively.

In particular, if T is selfadjoint (respectively, normal) and Y is invariant under T and $T^\star$, then $T|_Y$ and $T|_{Y^\perp}$ are selfadjoint (respectively, normal).

Proof If Y is invariant under T, then for all $y \in Y$ and $y^\perp \in Y^\perp$ we have $(y|T^\star y^\perp) = (Ty|y^\perp) = 0$. This proves (1). The first assertion of (2) follows as well, and if Y is invariant under T and $T^\star$, then for all $y, y' \in Y$ we have

$$(y|(T|_Y)^\star y') = (T|_Y y|y') = (Ty|y') = (y|T^\star y') = (y|(T^\star)|_Y y').$$

This proves the first identity of (2). The second is proved in the same way. The final assertion is an immediate consequence of (2). $\qquad\square$

Proof of Theorem 8.15 Replacing T by $T - \lambda$ we may assume that $\lambda = 0$. Let $P^{\{0\}}$ denote the spectral projection corresponding to $\{0\}$ and denote its range by $E^{\{0\}}$. We wish to prove that 0 is an eigenvalue for T, that $E_0 = E^{\{0\}}$, and that $P^{\{0\}}$ is an orthogonal projection. Once these facts have been proved, it follows that $P^{\{0\}}$ and P_0 are orthogonal projections onto the same closed subspace of H and therefore are equal.

As we have seen in Theorem 6.23, T maps $E^{\{0\}}$ into itself, and if we denote by $T^{\{0\}} := T|_{E^{\{0\}}}$ the restriction of T to $E^{\{0\}}$, then $\sigma(T^{\{0\}}) = \{0\}$. In particular this implies that $E^{\{0\}} \neq \{0\}$ (trivially, every operator on $\{0\}$ has empty spectrum).

Since T is normal, the formula for the spectral projection of Theorem 6.23 implies

$$T^\star P^{\{0\}}x = \frac{1}{2\pi i}\int_\Gamma T^\star R(\lambda,T)x\,\mathrm{d}\lambda = \frac{1}{2\pi i}\int_\Gamma R(\lambda,T)T^\star x\,\mathrm{d}\lambda = P^{\{0\}}T^\star x,$$

where Γ is a circular contour of small enough radius surrounding 0. This shows that $T^\star$ leaves $E^{\{0\}}$ invariant. Hence by Lemma 8.16, the restricted operator $T^{\{0\}}$ is normal as an operator on $E^{\{0\}}$. By Proposition 8.13, $\|T^{\{0\}}\| = r(T^{\{0\}}) = 0$ and therefore $T^{\{0\}} = 0$. This means that $Tx_0 = T^{\{0\}}x_0 = 0$ for all $x_0 \in E^{\{0\}}$, so $E^{\{0\}} \subseteq E_0$. Moreover, since $E^{\{0\}}$ is nontrivial, 0 is an eigenvalue of T.

For all $y \in E_0$ we have $Ty = 0$ and therefore

$$P^{\{0\}}y = \frac{1}{2\pi i}\int_\Gamma R(\lambda,T)y\,\mathrm{d}\lambda = \frac{1}{2\pi i}\int_\Gamma R(\lambda,T)(I - \lambda^{-1}T)y\,\mathrm{d}\lambda = \frac{1}{2\pi i}\int_\Gamma \lambda^{-1}y\,\mathrm{d}\lambda = y$$

with Γ as before. It follows that $y \in \mathsf{R}(P^{\{0\}}) = E^{\{0\}}$. This proves the inclusion $E_0 \subseteq E^{\{0\}}$.

It remains to be shown that $P^{\{0\}}$ equals the orthogonal projection onto E_0. Since $P^{\{0\}}$ is normal (this follows from the integral formula for $P^{\{0\}}$), this is a consequence of Corollary 9.18. The reader may check that no circularity is introduced; the proof of this corollary does not depend on the present result. $\qquad\square$

Corollary 8.17. *The geometric and algebraic multiplicity of every nonzero element in the spectrum of a compact normal operator coincide.*

This justifies the terminology *multiplicity* to denote the geometric and algebraic multiplicity of such a point.

We have the following commutation theorem for normal operators.

Theorem 8.18 (Fuglede–Putnam–Rosenblum)**.** *If $T \in \mathcal{L}(H)$ is normal and $S \in \mathcal{L}(H)$ is bounded and satisfies*

$$ST = TS,$$

then

$$ST^\star = T^\star S.$$

Proof We prove the more general result that if T_1, T_2 are normal and S is bounded such that $ST_1 = T_2S$, then $ST_1^\star = T_2^\star S$.

Step 1 – Let $V \in \mathcal{L}(H)$ be an arbitrary bounded operator. Expanding the exponential as a power series and taking adjoints termwise, we obtain

$$[\exp(V^\star - V)]^\star = \exp(V - V^\star) = [\exp(V^\star - V)]^{-1}$$

and therefore $\exp(V - V^\star)$ is unitary.

Step 2 – By induction, the assumption $ST_1 = T_2S$ implies $ST_1^n = T_2^n S$ for all $n \in$

$\mathbb{N}$ and therefore $S \exp(T_1) = \exp(T_2)S$. Since the normality of an operator T implies $\exp(T^\star - T) = \exp(T^\star)\exp(-T)$, this identity and the result of Step 1 imply

$$\exp(T_2^\star)S\exp(-T_1^\star) = \exp(T_2^\star - T_2)\exp(T_2)S\exp(-T_1^\star)$$
$$= \exp(T_2^\star - T_2)S\exp(T_1)\exp(-T_1^\star)$$
$$= \exp(T_2^\star - T_2)S\exp(T_1 - T_1^\star).$$

Since the two exponentials on the right-hand side are unitary, this gives

$$\| \exp(T_2^\star)S\exp(-T_1^\star)\| = \|S\|.$$

Applying this inequality to the normal operators $\bar{z}T_1$ and $\bar{z}T_2$ it follows that

$$\| \exp(zT_2^\star)S\exp(-zT_1^\star)\| = \|S\|,$$

so the entire function $f(z) = \exp(zT_2^\star)S\exp(-zT_1^\star)$ is bounded. By Liouville's theorem it is constant, so in particular $\exp(T_2^\star)S\exp(-T_1^\star) = f(1) = f(0) = S$, that is, $S\exp(T_1^\star) = \exp(T_2^\star)S$. Expanding the exponentials as power series and comparing terms we obtain $ST_1^\star = T_2^\star S$. $\qquad\square$

We finish with an observation about relative spectra. Let $\mathscr{A} \subseteq \mathscr{L}(H)$ be a unital closed $\star$-*subalgebra*, that is, $\mathscr{A}$ is a unital subalgebra of $\mathscr{L}(H)$ closed under taking adjoints. For such subalgebras we have the following improvement to Proposition 6.19:

Proposition 8.19. *Let $\mathscr{A} \subseteq \mathscr{L}(H)$ be a unital closed $\star$-subalgebra and let $T \in \mathscr{A}$. Then*

$$\sigma_{\mathscr{A}}(T) = \sigma(T).$$

Proof By Proposition 6.19 and the observation preceding it, for all $T \in \mathscr{A}$ we have

$$\partial\sigma_{\mathscr{A}}(T) \subseteq \sigma(T) \subseteq \sigma_{\mathscr{A}}(T).$$

First let $S \in \mathscr{A}$ be a selfadjoint operator. We claim that $\sigma_{\mathscr{A}}(S) \subseteq \mathbb{R}$. Indeed, if we had $\lambda \in \sigma_{\mathscr{A}}(S)$ with $\lambda \notin \mathbb{R}$, then $\sigma_{\mathscr{A}}(S)$ would have boundary points not belonging to $\mathbb{R}$. But $\partial\sigma_{\mathscr{A}}(S) \subseteq \sigma(S) \subseteq \mathbb{R}$ since S is selfadjoint. This proves the claim. It implies that $\partial\sigma_{\mathscr{A}}(S) = \sigma_{\mathscr{A}}(S)$, and since also $\partial\sigma_{\mathscr{A}}(S) \subseteq \sigma(S) \subseteq \sigma_{\mathscr{A}}(S)$ we obtain $\sigma_{\mathscr{A}}(S) = \sigma(S)$.

Suppose next that $T \in \mathscr{A}$ is invertible in $\mathscr{L}(H)$. Then also $T^\star$ is invertible in $\mathscr{L}(H)$, and hence so is $S = T^\star T$. Moreover, since $\mathscr{A}$ is closed under taking adjoints and compositions, the operator $S := T^\star T$ belongs to $\mathscr{A}$. By what we just proved, $\sigma_{\mathscr{A}}(S) = \sigma(S)$, so $T^\star T$ is invertible in $\mathscr{A}$. But then $T^{-1} = (T^\star T)^{-1}T^\star$ belongs to $\mathscr{A}$ as well.

We have shown that if $T \in \mathscr{L}(H)$ and $0 \in \rho(T)$, then $0 \in \rho_{\mathscr{A}}(T)$. Applying this result to $\lambda - T$ gives the inclusion $\rho(T) \subseteq \rho_{\mathscr{A}}(T)$, that is, $\sigma_{\mathscr{A}}(T) \subseteq \sigma(T)$. $\qquad\square$

8.2 The Continuous Functional Calculus

In Chapter 6 we have seen how to associate a bounded operator $f(T)$ with a bounded operator T when f is holomorphic in an open neighbourhood of $\sigma(T)$. Here we will prove that for normal operators T acting on a Hilbert space, the functional calculus $f \mapsto f(T)$ can be extended to continuous functions on $\sigma(T)$.

8.2.a The Continuous Functional Calculus for Selfadjoint Operators

We begin with the case of selfadjoint operators.

Theorem 8.20 (Continuous functional calculus for selfadjoint operators). *Let $T \in \mathscr{L}(H)$ be a selfadjoint operator. Then there exists a unique continuous linear mapping $f \mapsto f(T)$ from $C(\sigma(T))$ to $\mathscr{L}(H)$ with the following properties:*

  (i) *if $f(z) = z^n$ with $n \in \mathbb{N}$, then $f(T) = T^n$;*
 (ii) *for all $f, g \in C(\sigma(T))$ we have $(fg)(T) = f(T)g(T)$;*
(iii) *for all $f \in C(\sigma(T))$ we have $\overline{f}(T) = (f(T))^\star$;*
(iv) *for all $f \in C(\sigma(T))$ we have $\|f(T)\| = \|f\|_\infty$.*

The operators $f(T)$ are normal, and $f(T)$ is selfadjoint if and only if f is real-valued.

Proof   For polynomials $p(z) = \sum_{n=0}^N c_n z^n$ we define $p(T) := \sum_{n=0}^N c_n T^n$. These operators are normal and satisfy (i), (ii), and (iii). Moreover, by the spectral mapping theorem for the holomorphic calculus,

$$\|p(T)\| = \sup\{|\lambda| : \lambda \in \sigma(p(T))\} = \sup\{|\lambda| : \lambda \in p(\sigma(T))\} = \|p\|_\infty$$

and therefore (iv) holds.

By the Weierstrass approximation theorem, the polynomials are dense in $C(\sigma(T))$. Therefore, by (iv) and an approximation argument, the mapping $p \mapsto p(T)$ has a unique extension to an isometry from $C(\sigma(T))$ into $\mathscr{L}(H)$, and (ii)–(iv) again hold.

Since normality is inherited in passing to operator norm limits, the operators $f(T)$ are normal. Property (iii) implies that if f is real-valued, then $f(T)$ is selfadjoint. Conversely, if $f(T)$ is selfadjoint, then $f(T) = \overline{f}(T)$ by property (iii) and therefore $\|f - \overline{f}\|_{C(\sigma(T))} = 0$ by property (iv), so $f = \overline{f}$ is real-valued. $\square$

8.2.b The Continuous Functional Calculus for Normal Operators

Every polynomial in the real variables x and y can be written as a polynomial in the variables z and $\overline{z}$ by substituting $z = x + iy$, $\overline{z} = x - iy$. For example, $x^2 + y^2 = z\overline{z}$. For polynomials $p(z, \bar{z}) = \sum_{i,j=0}^{k} c_{ij} z^i \bar{z}^j$ and normal operators $T \in \mathcal{L}(H)$, we define

$$p(T, T^\star) := \sum_{i,j=0}^{k} c_{ij} T^i \overline{T}^{\star j}.$$

The crucial result that enables us to extend the continuous functional calculus to normal operators is the following spectral mapping theorem.

Proposition 8.21. *If $T \in \mathcal{L}(H)$ is normal and p is a polynomial in z and $\bar{z}$, then*

$$\sigma(p(T, T^\star)) = \{p(\lambda, \bar{\lambda}) : \lambda \in \sigma(T)\}.$$

Proof By Proposition 8.14, every $\lambda \in \sigma(T)$ is an approximate eigenvalue of T, that is, there exists a sequence $(x_n)_{n \geqslant 1}$ of norm one vectors such that $\lim_{n \to \infty} T x_n - \lambda x_n = 0$. Then $\lim_{n \to \infty} T^\star x_n - \bar{\lambda} x_n = 0$ by Proposition 8.13. This implies $\lim_{n \to \infty} p(T, T^\star) x_n - p(\lambda, \bar{\lambda}) x_n = 0$, so $p(\lambda, \bar{\lambda})$ is an approximate eigenvalue for $p(T, T^\star)$. In particular, $p(\lambda, \bar{\lambda}) \in \sigma(p(T, T^\star))$. This proves the inclusion '$\supseteq$'.

For the inclusion '$\subseteq$', fix an arbitrary $\mu \in \sigma(p(T, T^\star))$. We wish to prove the existence of a $\lambda \in \sigma(T)$ such that $p(\lambda, \bar{\lambda}) = \mu$.

Step 1 – Fixing $\varepsilon > 0$, we claim that there is a nonzero closed subspace Y of H, invariant under both T and $T^\star$, such that

$$\|(p(T, T^\star) - \mu I)|_Y\| < \varepsilon. \tag{8.5}$$

To prove the claim, let $S := p(T, T^\star) - \mu I$. This operator is normal and we have $0 \in \sigma(S)$. Let $R := S^\star S$. Arguing as above, we find that $0 \in \sigma(R)$. Consider the continuous function $f : [0, \infty) \to [0, 1]$ given by

$$f(t) := \begin{cases} 1, & 0 \leqslant t \leqslant \varepsilon/2; \\ 2(1 - t/\varepsilon), & \varepsilon/2 \leqslant t \leqslant \varepsilon; \\ 0, & t \geqslant \varepsilon, \end{cases}$$

and let $f(R)$ be the selfadjoint operator obtained from the continuous functional calculus for selfadjoint operators (Theorem 8.20). We will show that

$$Y := \{x \in H : f(R)x = x\}$$

has the desired properties.

Since T commutes with R, it commutes with $f(R)$, and therefore Y is invariant under T. By the same reasoning, Y is invariant under $T^\star$. Moreover, by the properties of continuous functional calculus, for all $y \in Y$ we have

$$\|Ry\| = \|Rf(R)y\| \leqslant \|t \mapsto tf(t)\|_{C(\sigma(R))} \|y\| = \|t \mapsto tf(t)\|_{C[0,\varepsilon]} \|y\| \leqslant \varepsilon \|y\|.$$

This implies

$$\|Sy\|^2 = (Ry|y) \leqslant \|Ry\|\|y\| \leqslant \varepsilon\|y\|^2.$$

This gives (8.5). The claim will be proved once we have checked that Y is nonzero. If $f(2t) \neq 0$ for some $t \geqslant 0$, then $2t \leqslant \varepsilon$ and therefore $f(t) = 1$. By multiplicativity,

$$\|(I - f(R))f(2R)\| = \big\|t \mapsto (1 - f(t))f(2t)\big\|_{C(\sigma(R))} = 0,$$

where $f(2R) := g(R)$ with $g(t) := f(2t)$. It follows that $\mathrm{R}(f(2R)) \subseteq Y$. But $\mathrm{R}(f(2R))$ is nonzero since

$$\|f(2R)\| = \big\|t \mapsto |f(2t)|\big\|_{C(\sigma(R))} \geqslant |f(0)| = 1.$$

Step 2 – Given $\varepsilon > 0$, let Y be the closed subspace of Step 1. Since Y is nonzero we have $\sigma(T|_Y) \neq \varnothing$ by Theorem 6.11. Pick an arbitrary $\lambda \in \sigma(T|_Y)$. Since Y is invariant under both T and $T^\star$, the restricted operator $T|_Y$ is normal as an operator in $\mathscr{L}(Y)$ by Lemma 8.16, and therefore λ is an approximate eigenvalue of $T|_Y$ and hence of T. In particular, we can find a norm one vector $y \in Y$ such that $\|Ty - \lambda y\| < \varepsilon$.

Step 3 – Up to this point, $\varepsilon > 0$ was fixed. Applying Step 2 to a sequence $\varepsilon_n \downarrow 0$, we obtain nonzero closed subspaces Y_n of H, norm one vectors $y_n \in Y_n$, and points $\lambda_n \in \sigma(T)$ such that $Ty_n - \lambda_n y_n \to 0$ as $n \to \infty$. Passing to a subsequence, we may assume that $\lambda_n \to \lambda$, and then $Ty_n - \lambda y_n \to 0$ as $n \to \infty$. It follows that λ is an approximate eigenvalue of T, with approximate eigensequence $(y_n)_{n \geqslant 1}$. By the argument of the first part of the proof,

$$\lim_{n \to \infty} p(T, T^\star)y_n - p(\lambda, \overline{\lambda})y_n = 0.$$

On the other hand, by the inequality of Step 1 applied to ε_n, we also have

$$\|p(T, T^\star)y_n - \mu y_n\| < \varepsilon_n$$

for every $n \geqslant 1$, and therefore we must have $p(\lambda, \overline{\lambda}) = \mu$.   □

With this theorem at hand we can extend the continuous functional calculus to normal operators. Repeating the proof of Theorem 8.20 we obtain the following result.

Theorem 8.22 (Continuous functional calculus for normal operators). *Let $T \in \mathscr{L}(H)$ be a normal operator. Then there exists a unique continuous linear mapping $f \mapsto f(T)$ from $C(\sigma(T))$ to $\mathscr{L}(H)$ with the following properties:*

(i) *if $f(z) = z^m \overline{z}^n$ with $m, n \in \mathbb{N}$, then $f(T) = T^m T^{\star n}$;*
(ii) *for all $f, g \in C(\sigma(T))$ we have $(fg)(T) = f(T)g(T)$;*
(iii) *for all $f \in C(\sigma(T))$ we have $\overline{f}(T) = (f(T))^\star$;*
(iv) *for all $f \in C(\sigma(T))$ we have $\|f(T)\| = \|f\|_\infty$.*

The operators $f(T)$ are normal, and $f(T)$ is selfadjoint if and only if f is real-valued.

The next theorem extends Proposition 8.21 to continuous functions defined on $\sigma(T)$.

Theorem 8.23 (Spectral mapping theorem). *If $T \in \mathscr{L}(H)$ is normal, then for all $f \in C(\sigma(T))$ we have*

$$\sigma(f(T)) = f(\sigma(T)).$$

Proof The proof of the inclusion $\sigma(f(T)) \subseteq f(\sigma(T))$ follows the lines of Theorem 6.21. Indeed, if $\lambda \notin f(\sigma(T))$, the function $g_\lambda = 1/f_\lambda$ with $f_\lambda(z) := \lambda - f(z)$ is continuous on $\sigma(T)$, and by multiplicativity we obtain

$$g_\lambda(T)(\lambda - f(T)) = (\lambda - f(T))g_\lambda(T) = (f_\lambda g_\lambda)(T) = \mathbf{1}(T) = I,$$

so $\lambda \in \rho(f(T))$ and $R(\lambda, f(T)) = g_\lambda(T)$. This gives the stated inclusion.

For the converse inclusion, let $\lambda \in \sigma(T)$ be arbitrary and fixed and let $\mu = f(\lambda)$. Using the Stone–Weierstrass theorem, choose polynomials p_n such that

$$\lim_{n \to \infty} \sup_{z \in \sigma(T)} |p_n(z, \bar{z}) - f(z)| = 0.$$

We may assume that $p_n(\lambda, \bar{\lambda}) = \mu$. Then, by Proposition 8.21, $\mu \in \sigma(p_n(T, T^\star))$. Also, by property (iv) of the functional calculus, $\lim_{n \to \infty} \|p_n(T, T^\star) - f(T)\| = 0$. By lower semicontinuity (Proposition 6.15), this implies $\mu \in \sigma(f(T))$. $\qquad\square$

Corollary 8.24. *Let $T \in \mathscr{L}(H)$ be a normal operator and let $f \in C(\sigma(T))$. Then $f(T)$ is positive if and only if f is nonnegative.*

Proof If f is nonnegative, the spectral mapping theorem gives $\sigma(f(T)) = f(\sigma(T)) \subseteq [0, \infty)$, and therefore the selfadjoint operator $f(T)$ is positive by Theorem 8.11. Conversely, if $f(T)$ is positive, then $\sigma(f(T)) \subseteq [0, \infty)$ by Theorem 8.11, and therefore $f(\sigma(T)) \subseteq [0, \infty)$ by the spectral mapping theorem. $\qquad\square$

The next theorem extends Proposition 6.22 to continuous functions defined on $\sigma(T)$.

Theorem 8.25 (Composition). *Let $T \in \mathscr{L}(H)$ be normal. For all $f \in C(\sigma(T))$ and $g \in C(f(\sigma(T))) = C(\sigma(f(T)))$ we have $g \circ f \in C(\sigma(T))$ and*

$$g(f(T)) = (g \circ f)(T).$$

Proof First let $p(z) = z^m \bar{z}^n$ with $m, n \in \mathbb{N}$. Then, by the properties of the continuous calculus,

$$p(f(T)) = (f(T))^m (\overline{f}(T))^n = (f^m \overline{f}^n)(T) = (p \circ f)(T).$$

By linearity, the this identity extends to polynomials p. If $g \in C(\sigma(f(T)))$ is an arbitrary continuous function, the identity follows by approximating g uniformly by polynomials p_n via the Stone–Weierstrass theorem to obtain

$$\|p_n(f(T)) - g(f(T))\| = \|p_n - g\|_{C(\sigma(f(T)))} \to 0 \quad \text{as } n \to \infty$$

and

$$\|(p_n \circ f)(T) - (g \circ f)(T)\| = \|p_n \circ f - g \circ f\|_{C(\sigma(T))} \to 0 \quad \text{as } n \to \infty.$$

$\square$

We finally check consistency with the holomorphic calculus.

Theorem 8.26. *Let $T \in \mathscr{L}(H)$ be normal. If $f \in H(\Omega)$, where Ω is an open set containing $\sigma(T)$, then $f(T)$ agrees with the operator defined through the holomorphic calculus.*

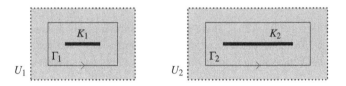

Figure 8.1 Proof of Theorem 8.26: $\sigma(T) = K_1 \cup K_2$, $\Gamma = \Gamma_1 \cup \Gamma_2$

Proof The set Ω is the union of at most countably many disjoint connected open sets, and by compactness the set $\sigma(T)$ is contained in finitely many of them. Hence there is no loss of generality in assuming that $\sigma(T) = \bigcup_{j=1}^{k} K_j \subseteq \bigcup_{j=1}^{k} \Omega_j = \Omega$, with the sets K_j compact and contained in the connected open sets Ω_j, which are disjoint. Choose bounded open sets U_j such that $K_j \subseteq U_j \subseteq \overline{U}_j \subseteq \Omega_j$, $k = 1, \ldots, k$, in such a way that $\mathbb{C} \setminus \bigcup_{j=1}^{k} \overline{U}_j$ is the union of at most finitely many disjoint connected open sets. Finally, let $\Gamma = \bigcup_{j=1}^{k} \Gamma_j$ be an admissible contour for $\sigma(T)$ in the sense described in Section 6.2, with Γ_j contained in U_j.

By Runge's theorem there exists a sequence of rational functions r_n such that $r_n \to f$ uniformly in $\overline{U}$, where $U = \bigcup_{j=1}^{k} U_j$. The operators $r_n(T)$ agree in the continuous calculus and the holomorphic calculus because this equality holds in the case of polynomials and for resolvents, and equality for rational functions follows from this by the multiplicativity of both calculi. Denoting by $f^{(c)}(T)$ and $f^{(h)}(T)$ the operators defined by the continuous calculus and the holomorphic calculus, respectively, it follows that

$$f^{(c)}(T) = \lim_{n \to \infty} r_n(T) = f^{(h)}(T)$$

with convergence in operator norm; the first equality is a consequence of property (iv) and the second follows from the estimate

$$\|r_n(T) - f^{(h)}(T)\| \leqslant \frac{1}{2\pi} \int_{\Gamma} |r_n(z) - f(z)| \|R(z,T)\| \, |dz|$$
$$\leqslant \frac{|\Gamma|}{2\pi} \|r_n - f\|_{C(\overline{U})} \sup_{z \in \Gamma} \|R(z,T)\|,$$

with $|\Gamma|$ the length of Γ. □

8.2.c Applications of the continuous functional calculus

We now turn to some applications of the continuous functional calculus.

Proposition 8.27 (Square roots). *If $T \in \mathscr{L}(H)$ is positive, there exists a unique positive operator $S \in \mathscr{L}(H)$ such that $S^2 = T$.*

Henceforth, this operator S will be denoted by $T^{1/2}$.

Proof Since T is positive, T is selfadjoint and its spectrum is contained in $[0, \infty)$ by Theorem 8.11. Hence, $f(t) = \sqrt{t}$ is a well-defined continuous function on $\sigma(T)$. The operator $S := f(T)$ is positive by Corollary 8.24, and it satisfies $S^2 = T$ by the properties of the continuous functional calculus. It remains to prove uniqueness. Suppose $\widetilde{S}$ is another positive operator with the property that $\widetilde{S}^2 = T$. With $f(t) = \sqrt{t}$, $g(t) = t^2$, and $h(t) = t$ we have, by the properties of the continuous functional calculus and Theorem 8.25,

$$f(\widetilde{S}^2) = f(g(\widetilde{S})) = (f \circ g)(\widetilde{S}) = h(\widetilde{S}) = \widetilde{S}.$$

It follows that $S = f(S^2) = f(T) = f(\widetilde{S}^2) = \widetilde{S}$. This completes the uniqueness proof. □

Definition 8.28 (Modulus of an operator). The *modulus* of an operator $T \in \mathscr{L}(H)$ is the positive operator $|T| := (T^\star T)^{1/2}$.

Corollary 8.29. *If $T \in \mathscr{L}(H)$ is normal operator, then $|T| = f(T)$, where $f(z) := |z|$.*

Proof Let $g(z) := \bar{z}z$. Then $f^2 = g$ and therefore $|T|^2 = T^\star T = g(T) = f^2(T) = (f(T))^2$ by the multiplicativity of the continuous functional calculus. Since by Corollary 8.24 the operator $f(T)$ is positive, the result follows by taking square roots. □

We continue with a polar decomposition result. In view of future applications we phrase it for bounded operators $T \in \mathscr{L}(H, K)$, where H and K are Hilbert spaces. The *modulus* of such an operator is the positive operator $|T| := (T^\star T)^{1/2}$ on H. An operator $U \in \mathscr{L}(H, K)$ will be called *unitary* if it is isometric and surjective. A *partial isometry* is a bounded operator $V \in \mathscr{L}(H, K)$ for which there exists orthogonal direct sum decomposition $H = H_0 \oplus H_0^\perp$ such that V is isometric from H_0 into K and zero on $H_0^\perp$.

Theorem 8.30 (Polar decomposition). *Let $T \in \mathscr{L}(H, K)$. The following assertions hold:*

(1) *T admits a representation $T = U|T|$, with U a partial isometry from H to K which is isometric from $\overline{R(|T|)}$ onto $\overline{R(T)}$;*

(2) *if T is invertible, then T admits a unique representation $T = U|T|$ with U unitary from H onto K.*

Proof (1): From

$$\|Tx\|^2 = (T^\star Tx|x) = \big(|T|x\big||T|x\big) = \big\|\,|T|x\big\|^2$$

it follows that the mapping $U_0 : |T|x \mapsto Tx$, viewed as a linear operator from $\mathsf{R}(|T|)$ onto $\mathsf{R}(T)$, is well defined and isometric, and by density it extends to an isometry from $\overline{\mathsf{R}(|T|)}$ onto $\overline{\mathsf{R}(T)}$. Moreover, $T = U_0|T|$. Along the orthogonal decomposition $H = \overline{\mathsf{R}(|T|)} \oplus (\overline{\mathsf{R}(|T|)})^\perp$ we extend U_0 identically zero on $(\mathsf{R}(|T|))^\perp$ to obtain the desired partial isometry U.

(2): The operator $T^\star T$ is positive and invertible. Hence $|T| = (T^\star T)^{1/2}$ is invertible as well, by the spectral mapping theorem. Set $U := T|T|^{-1}$. From

$$U^\star U = |T|^{-1}T^\star T|T|^{-1} = |T|^{-1}|T|^2|T|^{-1} = I$$

and the fact that U is invertible it follows that $U^\star = U^{-1}$ and U is unitary. To prove that U is unique suppose that $T = U|T| = \widetilde{U}|T|$ with both U and $\widetilde{U}$ unitary. Then $|T| = U^\star U|T| = U^\star \widetilde{U}|T|$, and since $|T|$ is invertible this implies $U^\star \widetilde{U} = I$. Multiplying both sides with U gives $\widetilde{U} = U$. $\qquad\square$

8.3 The Sz.-Nagy Dilation Theorem

The last section of this chapter is devoted to a proof of the celebrated Sz.-Nagy dilation theorem, which asserts that every Hilbert space contraction has a unitary dilation. Since it poses no additional difficulties, we take a rather general approach starting from an arbitrary group G with unit element e; the Sz.-Nagy dilation theorem is obtained by considering $G = \mathbb{Z}$.

Definition 8.31 (Positive definiteness). A mapping $T : G \to \mathscr{L}(H)$ is called *positive definite* if for all finite choices of $g_1, \dots, g_N \in G$ and $h_1, \dots, h_N \in H$ we have

$$\sum_{m,n=1}^{N} (T(g_m^{-1}g_n)h_n|h_m) \geqslant 0.$$

Definition 8.32 (Representations, unitary representations). A mapping $U : G \to \mathscr{L}(H)$ is called a *representation of G on H* if $U(e) = I$ and $U(g_1)U(g_2) = U(g_1g_2)$ for all $g_1, g_2 \in G$. A *unitary representation* is a representation whose constituting operators are unitaries.

The following result connects these two notions.

Proposition 8.33. *Let $U : G \to \mathscr{L}(\widetilde{H})$ be a unitary representation of G on a Hilbert* space $\widetilde{H}$. Let $J : H \to \widetilde{H}$ be an isometric embedding of another Hilbert space H into $\widetilde{H}$. Then the function $T : G \to \mathscr{L}(H)$ given by

$$T(g) := J^\star U(g)J,$$

is positive definite and satisfies $T(e) = I$ and $T^\star(g) = T(g^{-1})$ for all $g \in G$.

Proof The identity $T(e) = I$ follows from $U(e) = I$ and $J^\star J = I$, and the identity $U(g^{-1}) = (U(g))^\star$ implies $T^\star(g) = J^\star U^\star(g)J = J^\star(U(g))^{-1}J = J^\star U(g^{-1})J = T(g^{-1})$. To prove positive definiteness, let $g_1, \dots, g_N \in G$ and $h_1, \dots, h_N \in H$. Then

$$\sum_{m,n=1}^{N} (T(g_m^{-1} g_n)h_n | h_m) = \sum_{m,n=1}^{N} (U(g_m^{-1})U(g_n')Jh_n | Jh_m) = \left\| \sum_{n=1}^{N} U(g_n)Jh_n \right\|^2 \geqslant 0.$$

$\square$

The next theorem establishes that, conversely, every positive definite function $T : G \to \mathscr{L}(H)$ satisfying $T(e) = I$ and $T^\star(g) = T(g^{-1})$ for all $g \in G$ arises in this way.

Theorem 8.34 (Unitary dilations). *Let $T : G \to \mathscr{L}(H)$ be a positive definite function satisfying $T(e) = I$ and $T^\star(g) = T(g^{-1})$ for all $g \in G$. There exist a Hilbert space $\widetilde{H}$, an isometric embedding $J : H \to \widetilde{H}$, and a unitary representation $U : G \to \mathscr{L}(\widetilde{H})$ such that*

$$T(g)h = J^\star U(g)Jh, \quad h \in H.$$

Proof Let V be the vector space of all functions $f : G \to H$ that vanish outside a finite set. We claim that

$$(f_1 | f_2) := \sum_{g,g' \in G} (T(g^{-1}g')f_1(g') | f_2(g))$$

defines a sesquilinear mapping from $V \times V$ to $\mathbb{C}$. Writing $f_m = \sum_{j=1}^{k} \mathbf{1}_{\{g_j\}} \otimes h_j^{(m)}$, $m = 1, 2$ (allowing the possibility that some of the $h_j^{(i)}$ are zero), we have

$$(f_1 | f_2) = \sum_{g,g' \in G} \sum_{i,j=1}^{k} \mathbf{1}_{\{g_i\}}(g')\mathbf{1}_{\{g_j\}}(g)(T(g^{-1}g')h_i^{(1)} | h_j^{(2)}) = \sum_{i,j=1}^{k} (T(g_j^{-1}g_i)h_i^{(1)} | h_j^{(2)}). \tag{8.6}$$

We claim that $(f_1 | f_2) = \overline{(f_2 | f_1)}$ for all $f_1, f_2 \in V$ and $(f | f) \geqslant 0$ for all $f \in V$. A similar computation as above gives

$$\overline{(f_2 | f_1)} = \sum_{i,j=1}^{k} (h_i^{(1)} | T(g_i^{-1}g_j)h_j^{(2)}). \tag{8.7}$$

Since $T^\star(g) = T(g^{-1})$ for all $g \in G$, the right-hand sides of (8.6) and (8.7) are equal, thus proving the identity $(f_1 | f_2) = \overline{(f_2 | f_1)}$. Positive definiteness implies $(f | f) \geqslant 0$.

The properties established in the claim suffice for the validity of the Cauchy–Schwarz inequality. It may happen, however, that $(f|f) = 0$ for certain nonzero functions f in V, so this sesquilinear form may fail to be an inner product. For this reason we consider the vector space quotient V/N, where

$$N = \{f \in V : (f|f) = 0\}.$$

Let us prove that N is indeed a subspace of V. It is clear that $cf \in N$ for all $c \in \mathbb{C}$ and $f \in N$. Furthermore, if $f, f' \in N$, then $(f|f') = 0$ by the Cauchy–Schwarz inequality, and from this it follows that $f + f' \in N$.

On the quotient space V/N, the sesquilinear mapping $(\cdot|\cdot)$ induces the inner product

$$(f + N|g + N) := (f|g).$$

Define $\widetilde{H}$ to be the Hilbert space completion of V/N with respect to this inner product. To realise H as a closed subspace of $\widetilde{H}$ we identify elements $h \in H$ with the class modulo N of the functions $f_h : G \to H$ given by $f_h = \mathbf{1}_{\{e\}} \otimes h$. Then

$$(f_{h_1}|f_{h_2}) = \sum_{g,g' \in G} (T(g^{-1}g')f_{h_1}(g')|f_{h_2}(g)) = (T(e)h_1|h_2) = (h_1|h_2)$$

since $T(e) = I$. This implies that the mapping $J : h \mapsto f_h + N$ is isometric from H into $\widetilde{H}$.

The linear mapping $U : V \to V$ given by

$$(U(g)f)(g') := f(g^{-1}g'), \quad f \in V, \; g, g' \in G,$$

is well defined and preserves inner products; in particular it maps N into itself. Indeed, if $f \in N$, then by a change of variables we have

$$
\begin{aligned}
(U(g)f_1|U(g)f_2) &= \sum_{g',g'' \in G} (T(g'^{-1}g'')f_1(g^{-1}g'')|f_2(g^{-1}g')) \\
&= \sum_{g',g'' \in G} (T((g^{-1}g')^{-1}g^{-1}g'')f_1(g^{-1}g'')|f_2(g^{-1}g')) \\
&= \sum_{g'',g''' \in G} (T(g''^{-1}g''')f(g''')|f(g'')) = (f_1|f_2),
\end{aligned}
$$

and the asserted properties follow. Moreover,

$$U(g_1)U(g_2)f(g) = f(g_2^{-1}g_1^{-1}g) = f((g_1g_2)^{-1}g) = U(g_1g_2)f(g). \tag{8.8}$$

Upon passing to the quotient, we obtain a well-defined linear mapping, denoted by $\widetilde{U}(g)$, on V/N which preserves inner products. Therefore $\widetilde{U}(g)$ extends to an isometry from $\widetilde{H}$ into itself, which we once again denote by $\widetilde{U}(g)$, and by passing to the quotient in (8.8) we see that $\widetilde{U}(g_1)\widetilde{U}(g_2) = \widetilde{U}(g_1g_2)$, that is, the resulting mapping $\widetilde{U} : G \to \mathscr{L}(\widetilde{H})$ is a homomorphism. Since each operator $\widetilde{U}(g)$ preserves inner products, in order to prove that $\widetilde{U}(g)$ is unitary it suffices to prove that it is surjective, and for this it suffices to prove that each $U(g)$ is surjective. However, the latter is immediate from the definition, which implies that every finitely supported H-valued function on G is in the range of $U(g)$. It follows that $\widetilde{U}$ is a unitary representation of G on $\widetilde{H}$.

This representation has the desired properties: for all $g, g' \in G$ and $h, h' \in H$ we have

$$U(g)f_h(g') = (U(g)(\mathbf{1}_{\{e\}} \otimes h))(g') = \mathbf{1}_{\{e\}}(g^{-1}g')h = (\mathbf{1}_{\{g\}} \otimes h)(g')$$

and consequently, by (8.6),

$$(J^{\star}\widetilde{U}(g)Jh|h') = (\widetilde{U}(g)f_h|f_{h'}) = (\mathbf{1}_{\{g\}} \otimes h|\mathbf{1}_{\{e\}} \otimes h') = (T(g)h|h').$$

□

The theorem will be applied in the following situation:

Lemma 8.35. *If $T \in \mathscr{L}(H)$ is a contraction, the mapping $S : \mathbb{Z} \to \mathscr{L}(H)$ defined by*

$$S(n) := \begin{cases} T^n, & n \geqslant 1, \\ I, & n = 0, \\ (T^{\star})^{-n}, & n \leqslant -1, \end{cases}$$

is positive definite and satisfies $S^{\star}(n) = S(-n)$ for all $n \in \mathbb{Z}$.

Proof Since T is a contraction, from $((I - T^{\star}T)x|x) = \|x\|^2 - \|Tx\|^2 \geqslant 0$ it follows that $I - T^{\star}T$ is a positive operator. As consequence, by Proposition 8.27 the *defect operator*

$$D_T := (I - T^{\star}T)^{1/2}$$

is well defined and positive, and we have

$$\|D_Tx\|^2 = ((I - T^{\star}T)^{1/2}x|(I - T^{\star}T)^{1/2}x) = ((I - T^{\star}T)x|x) = \|x\|^2 - \|Tx\|^2.$$

Define

$$\ell^2(H) := \left\{ \mathbf{h} = (h_n)_{n \geqslant 1} \subseteq H : \sum_{n \geqslant 1} \|h_n\|^2 < \infty \right\}.$$

With respect to the inner product $(\mathbf{g}|\mathbf{h}) := \sum_{n \geqslant 1}(g_n|h_n)$, $\ell^2(H)$ is a Hilbert space; completeness is proved in the same way as for ℓ^2. We define the operator $\widetilde{T}$ on $\ell^2(H)$ by

$$\widetilde{T} : \mathbf{h} \mapsto (Th_1, D_Th_1, h_2, h_3, \dots).$$

Clearly

$$\|\widetilde{T}\mathbf{h}\|^2_{\ell^2(H)} = \|Th_1\|^2 + \|D_Th_1\|^2 + \sum_{n \geqslant 2} \|h_n\|^2 = \|\mathbf{h}\|^2,$$

so $\widetilde{T}$ is isometric. Define $\widetilde{S} : \mathbb{Z} \to \mathscr{L}(H)$ by

$$\widetilde{S}(n) := \begin{cases} \widetilde{T}^n, & n \geqslant 1; \\ I, & n = 0; \\ (\widetilde{T}^\star)^{-n}, & n \leqslant -1, \end{cases}$$

where $\widetilde{T}^\star$ is the Hilbert space adjoint of $\widetilde{T}$. We make the trivial but crucial observation that

$$(S(n-m)h|h') = (\widetilde{S}(n-m)Jh|Jh'), \quad h, h' \in H, \ m, n \geqslant 1,$$

where $J : H \to \ell^2(H)$ is defined by $Jh := (h, 0, 0, \dots)$. It follows that for all choices of $h_1, \dots, h_N \in H$ we have (with the convention $\widetilde{T}^0 = I$)

$$
\begin{aligned}
\sum_{m,n=1}^{N} (S(n-m)h_n|h_m) &= \sum_{m,n=1}^{N} (\widetilde{S}(n-m)Jh_n|Jh_m) \\
&= \sum_{1 \leqslant m \leqslant n \leqslant N} (\widetilde{T}^{n-m}Jh_n|Jh_m) + \sum_{1 \leqslant n < m \leqslant N} ((\widetilde{T}^\star)^{m-n}Jh_n|Jh_m) \\
&= \sum_{1 \leqslant m \leqslant n \leqslant N} (\widetilde{T}^{n-m}Jh_n|Jh_m) + \sum_{1 \leqslant n < m \leqslant N} (Jh_n|\widetilde{T}^{m-n}Jh_m) \\
&\overset{(*)}{=} \sum_{1 \leqslant m \leqslant n \leqslant N} (\widetilde{T}^n Jh_n|\widetilde{T}^m Jh_m) + \sum_{1 \leqslant n < m \leqslant N} (\widetilde{T}^n Jh_n|\widetilde{T}^m Jh_m) \\
&= \left\| \sum_{k=1}^{N} \widetilde{T}^k Jh_k \right\|^2 \geqslant 0,
\end{aligned}
$$

where $(*)$ uses that $\widetilde{T}$ is an isometry and consequently $(\widetilde{T}g|\widetilde{T}h) = (g|h)$. This proves that S is positive definite.

The identity $S^\star(n) = S(-n)$ for $n \in \mathbb{Z}$ is clear from the definition. $\qquad\square$

Combining the lemma with the theorem, we arrive at the following result.

Theorem 8.36 (Sz.-Nagy dilation theorem). *If $T \in \mathscr{L}(H)$ is a contraction, then there exist a Hilbert space $\widetilde{H}$, an isometric embedding $J : H \to \widetilde{H}$, and a unitary operator $U \in \mathscr{L}(\widetilde{H})$ such that*

$$T^n = J^\star U^n J, \quad n \in \mathbb{N}.$$

In this context the operator U is said to be a *unitary dilation* of T. As a simple example, the left (right) shift on $\ell^2(\mathbb{Z})$ is a unitary dilation of the left (right) shift on ℓ^2.

Problems

8.1 Prove the *Hellinger–Toeplitz theorem*: If $T : H \to H$ is a linear mapping satisfying

$$(Tx|y) = (x|Ty), \quad x, y \in H,$$

then T is bounded.

Hint: Apply the uniform boundedness theorem to the operators $T_x : H \to \mathbb{K}$ given by $T_x y := (Tx|y)$.

8.2 Give a direct proof of Proposition 8.10 based on a Neumann series argument for U and U^{-1}.

8.3 Let $T \in \mathscr{L}(H)$ be selfadjoint. The aim of this problem is to deduce the inclusion $\sigma(T) \subseteq \mathbb{R}$ from Proposition 8.10 (or the preceding problem). Fix $\lambda \in \sigma(T)$.

(a) Show that

$$e^{i\lambda} - e^{iT} = e^{i\lambda}(\lambda - T)\Big(\sum_{n=1}^{\infty} \frac{i^n}{n!}(T - \lambda)^{n-1}\Big)$$

and conclude that $e^{i\lambda} - e^{iT}$ fails to be invertible.

(b) Combine this with Proposition 8.10 to conclude that $\lambda \in \mathbb{R}$.

8.4 Show that a projection $P \in \mathscr{L}(H)$ is an orthogonal projection if and only if

$$\|Ph\| \leqslant \|h\|, \quad h \in H.$$

Hint: The latter condition implies that $\|P(g+ch)\|^2 \leqslant \|g+ch\|^2$ for all $c \in \mathbb{K}$ and $g, h \in H$. Now consider $g \in \mathsf{R}(P)$ and $h \in \mathsf{N}(P)$, and vary c.

8.5 Let H_0 be a closed subspace of H and let $T \in \mathscr{L}(H)$. Prove that if both T and $T^\star$ leave H_0 invariant, then $\sigma(T|_{H_0}) \subseteq \sigma(T)$.

8.6 Show that the norm of an operator $T \in \mathscr{L}(H)$ is given by

$$\|T\|^2 = \inf\{\lambda > 0 : T^\star T \leqslant \lambda I\},$$

where $T^\star T \leqslant \lambda I$ means that $\lambda - T^\star T$ is positive.

8.7 Let $T \in \mathscr{L}(H)$ be selfadjoint and let $\lambda \in \mathbb{R}$.

(a) Show that if $\sigma(T) = \{\lambda\}$, then $T = \lambda I$.

(b) Show that if $\sigma(PTP) = \{\lambda\}$, where P is an orthogonal projection with $\mathsf{N}(P)$ finite-dimensional, then $\lambda - T$ is compact; if, in addition, $P \neq I$, then $\lambda = 0$ and T is compact.

8.8 The *numerical range* of an operator $T \in \mathscr{L}(H)$ is the set

$$W(T) := \{(Tx|x) : \|x\| = 1\}.$$

The *numerical radius* of T is defined by

$$w(T) := \sup\{|\lambda| : \lambda \in W(T)\}.$$

Prove the following assertions:

(a) $\frac{1}{2}\|T\| \leqslant w(T) \leqslant \|T\|$.

Hint: To prove the first inequality use the identity

$$4(Tx|y) = (T(x+y)|x+y) - (T(x-y)|x-y)$$
$$+ i(T(x+iy)|x+iy) - i(T(x-iy)|x-iy).$$

(b) T is selfadjoint if and only if $W(T) \subseteq \mathbb{R}$.

Hint: Consider the operator $i(T - T^\star)$.

(c) If $W(T) = \{\lambda\}$ for some $\lambda \in \mathbb{C}$, then $T = \lambda I$.

8.9 In this problem we prove the *Toeplitz–Hausdorff theorem* which asserts that the numerical range of any operator $T \in \mathscr{L}(H)$ is a convex subset of $\mathbb{C}$.

(a) Show that for all $\lambda, \mu \in \mathbb{C}$ we have $W(\lambda T + \mu I) = \lambda W(T) + \mu$. Conclude that in order to prove the Toeplitz–Hausdorff theorem it suffices to establish that $\{0, 1\} \subseteq W(T)$ implies $[0, 1] \subseteq W(T)$.

In what follows we fix an operator $T \in \mathscr{L}(H)$ such that $\{0, 1\} \subseteq W(T)$. We prove that $[0, 1] \subseteq W(T)$.

Choose norm one vectors $x, y \in H$ such that $(Tx|x) = 0$ and $(Ty|y) = 1$.

(b) Define $g : [0, 2\pi] \to \mathbb{C}$ by

$$g(t) := e^{-it}(Tx|y) + e^{it}(Ty|x).$$

Using that $g(t + \pi) = -g(t)$ for $t \in [0, \pi]$, show that either there exists $t_0 \in [0, \pi]$ such that $g(t_0) = 0$ or else there exists $t_0 \in [0, 2\pi]$ such that $g(t_0) > 0$.

Set $\widetilde{y} := e^{it_0}y$.

(c) Show that x and $\widetilde{y}$ are linearly independent.

Define $z : [0, 1] \to H$ and $f : [0, 1] \to \mathbb{C}$ by

$$z(t) := \frac{(1-t)x + t\widetilde{y}}{\|(1-t)x + t\widetilde{y}\|}, \quad f(t) := (Tz(t)|z(t)).$$

These functions are well defined by part (c).

(d) Show that f is continuous, real-valued, and satisfies $f(0) = 0$ and $f(1) = 1$. Deduce that $[0, 1] \subseteq W(T)$.

8.10 Prove that for all $T \in \mathscr{L}(H)$ we have $\sigma(T) \subseteq \overline{W(T)}$.

Hint: First prove that approximate eigenvalues belong to $\overline{W(T)}$. Then apply the Toeplitz–Hausdorff theorem in combination with Proposition 6.17.

8.11 Show that if $S, T \in \mathscr{L}(H)$ are positive operators, then $\sigma(ST) \subseteq [0, \infty)$.

Hint: Apply the result of Problem 6.12 to the operators $S^{1/2}$ and $S^{1/2}T$.

8.12 Consider the Volterra operator T on $L^2(0,1)$ of Example 1.31:

$$(Tf)(t) = \int_0^t f(s)\,ds, \quad f \in L^2(0,1),\, t \in [0,1].$$

We sketch two proofs that $\sigma(T) = \{0\}$.

(a) Show that for all $0 \neq \lambda \in \mathbb{C}$ and $g \in L^2(0,1)$ the equation $(\lambda - T)f = g$ has a unique solution in $L^2(0,1)$. Deduce that $\sigma(T) = \{0\}$.

A second proof is obtained by estimating the norm of T^n:

(b) Show that $\|T^n\| \leq \frac{1}{n!}$ for all $n = 1,2,\ldots$
   Hint: First show that

$$(T^n f)(t) = \int_0^t \int_0^{t_{n-1}} \cdots \int_0^{t_1} f(s)\,ds\,dt_1 \cdots dt_{n-1}$$
$$= \int_0^t f(s) \int_s^t \int_s^{t_{n-1}} \cdots \int_s^{t_2} dt_1 \cdots dt_{n-2}\,dt_{n-1}\,ds.$$

(c) Using Theorem 6.24, conclude that $\sigma(T) = \{0\}$.

(d) Show that 0 is not an eigenvalue of T.

(e) Deduce from (a) or (c) that T is not normal.

8.13 Let T_m be a Fourier multiplier operator on $L^2(\mathbb{R}^d)$ with symbol $m \in L^\infty(\mathbb{R}^d)$.

(a) Show that T_m is normal.

(b) Show that if f is a continuous function on the essential range of m (see Problem 6.5), then $f(T_m)$ is well defined through the continuous functional calculus and equal to the Fourier multiplier $T_{f \circ m}$ with symbol $f \circ m \in L^\infty(\mathbb{R}^d)$.

(c) Compare this result with Problem 6.6.

8.14 Show that the rotation operator on $L^2(\mathbb{T})$ defined by $R_\theta f(e^{ix}) := f(e^{i(x-\theta)})$ is unitary and find its spectrum.
   Hint: Distinguish the cases $\theta/2\pi \in \mathbb{Q}$ and $\theta/2\pi \notin \mathbb{Q}$.

8.15 Prove that if $T \in \mathscr{L}(H)$ is an isometry, then there exist Hilbert spaces G and K such that we have an isometric isomorphism of Hilbert spaces

$$H \simeq \ell^2(G) \oplus K,$$

where $\ell^2(G)$ is the Hilbert space of all square summable sequences $g = (g_n)_{n \geq 1}$ in G with norm $\|g\|_{\ell^2(G)}^2 = \sum_{n \geq 1} \|g_n\|^2$ and that along this decomposition we have

$$T \simeq S \oplus U,$$

where S is the right shift on $\ell^2(G)$, that is, S maps the sequence $g_1, g_2, \ldots$ to $0, g_1, g_2, \ldots$ and U is a unitary operator on K. This decomposition is known as the *Wold decomposition*.
   Hint: For $n \in \mathbb{N}$ let $H_n := \mathsf{R}(T^n)$, and for $n \geq 1$ let G_n denote the orthonormal

complement of H_n in H_{n-1}. Show that the spaces G_n are all isometric as Hilbert spaces and set $K := \bigcap_{n \in \mathbb{N}} H_n$.

8.16 This problem sketches an alternative proof of the Sz.-Nagy dilation theorem. Let $T \in \mathscr{L}(H)$ be a contraction and $D_T = (I - T^\star T)^{1/2}$ the associated defect operator.

A *dilation* of a bounded operator T on H is a bounded operator $\widetilde{T}$ on a Hilbert space $\widetilde{H}$ containing H isometrically as a closed subspace such that

$$T^n = P\widetilde{T}^n J, \quad n \in \mathbb{N},$$

where J is the inclusion mapping from H into $\widetilde{H}$ and $P = J^\star$ is the orthogonal projection of $\widetilde{H}$ onto H, viewed as a mapping from $\widetilde{H}$ onto H.

(a) Show that $TD_T = D_{T^\star}T$.

On the Hilbert space

$$\ell^2(H) := \left\{ h = (h_n)_{n \geqslant 1} \subseteq H : \sum_{n \geqslant 1} \|h_n\|^2 < \infty \right\}$$

we consider the operator $S : h \mapsto (Th_1, D_T h_1, h_2, h_3, \dots)$.

(b) Show that S is an isometry, that is, $\|Sh\| = \|h\|$ for all $h \in \ell^2(H)$.

(c) Show that $T^n = J^\star S^n J$ for all $n \in \mathbb{N}$, where $J : H \to \ell^2(H)$ is given by $h \mapsto (h, 0, 0, \dots)$. Conclude that S is a dilation of T.

(d) Show that a dilation of a dilation is a dilation.

To complete the proof of the theorem it suffices to show that every isometry has a unitary dilation. Accordingly, in the rest of the problem we consider an isometry S on a Hilbert space G.

(e) Show that under these assumptions we have $D_S = 0$.

On the Hilbert space direct sum $G \oplus G$ define the operator

$$U := \begin{pmatrix} S & D_{S^\star} \\ D_S & -S^\star \end{pmatrix} = \begin{pmatrix} S & D_{S^\star} \\ 0 & -S^\star \end{pmatrix}.$$

(f) Show that

$$U^\star := \begin{pmatrix} S^\star & 0 \\ D_{S^\star} & -S \end{pmatrix}.$$

(g) Show that $S^\star D_{S^\star} = D_S S^\star = 0$ and use this to prove that U is unitary.

(h) Show that U is a dilation of S.

   Hint: First compute U^2 and use this for finding U^{2k} and U^{2k+1}.

9

The Spectral Theorem for Bounded Normal Operators

In this chapter we show that normal operators admit a spectral representation as sums or integrals of orthogonal projections. We begin by showing that every compact normal operator T admits the spectral decomposition

$$T = \sum_{n \geqslant 1} \lambda_n P_n,$$

where $(\lambda_n)_{n \geqslant 1}$ is the sequence of eigenvalues of T and $(P_n)_{n \geqslant 1}$ is the sequence of orthogonal projections onto the corresponding eigenspaces. For arbitrary bounded normal operators T, the main result of this chapter, the spectral theorem for bounded normal operators, provides an analogous representation as an integral

$$T = \int_{\sigma(T)} \lambda \, dP(\lambda).$$

9.1 The Spectral Theorem for Compact Normal Operators

Throughout this chapter we let H be a Hilbert space.

From Linear Algebra we know that normal matrices can be orthogonally diagonalised. This result admits the following extension to compact normal operators on H:

Theorem 9.1 (Spectral theorem for compact normal operators). *Let $T \in \mathscr{L}(H)$ be a compact normal operator and let $(\lambda_n)_{n \geqslant 1}$ be the (finite or infinite) sequence of its distinct eigenvalues. Let $(E_n)_{n \geqslant 1}$ be the corresponding sequence of eigenspaces, and let $(P_n)_{n \geqslant 1}$ be the associated sequence of orthogonal projections. Then:*

(1) *the spaces E_n are pairwise orthogonal and have dense linear span;*
(2) *we have*

$$T = \sum_{n \geqslant 1} \lambda_n P_n$$

281

with convergence in the operator norm of $\mathscr{L}(H)$.

Proof   The proof of the theorem uses the properties spectra of compact operators on Banach spaces established in Theorem 7.11. In the present situation, where the compact operator acts on a Hilbert space, the proof of this theorem can be considerably shortened; see Problem 9.2.

(1): From Proposition 8.13 (applied to $T - \lambda$) we see that $Tx - \lambda x = 0$ if and only if $T^\star x - \overline{\lambda}x = 0$, so λ is an eigenvalue for T if and only if $\overline{\lambda}$ is an eigenvalue for $T^\star$ and the eigenspaces coincide.

If $y_m \in E_m$ and $y_n \in E_n$ are nonzero vectors, then

$$\lambda_m(y_m|y_n) = (Ty_m|y_n) = (y_m|T^\star y_n) = (y_m|\overline{\lambda_n}y_n) = \lambda_n(y_m|y_n).$$

If $\lambda_m \neq \lambda_n$, then this is possible only if $(y_m|y_n) = 0$. This gives $E_n \perp E_m$ for $n \neq m$.

Let $E := \bigoplus_{n \geq 1} E_n$ denote the closed linear span of the spaces E_n, $n \geq 1$. We wish to prove that $E = H$. Suppose the contrary. Then $E^\perp$ is a nonzero closed subspace of H. Since $TE_n \subseteq E_n$ for all $n \geq 1$ we have $TE \subseteq E$. Furthermore, if $x \in E_n$, then $T^\star x = \overline{\lambda}_n x \in E_n$, so $T^\star E_n \subseteq E_n$. This being true for all $n \geq 1$, it follows that $T^\star E \subseteq E$. Hence, by Lemma 8.16 we have $TE^\perp \subseteq E^\perp$ and the restriction $T^\perp$ of T to $E^\perp$ is normal. Moreover, by the very construction of E, $T^\perp$ has no eigenvalues. By Theorem 7.11, this implies that $\sigma(T^\perp) \subseteq \{0\}$, and since $\sigma(T^\perp) \neq \varnothing$ it follows that $\sigma(T^\perp) = \{0\}$. Now Proposition 8.13 implies that $T^\perp = 0$. This means that every element of $E^\perp$ is an eigenvector of $T^\perp$ with eigenvalue 0. This contradicts the observation just made that that $T^\perp$ has no eigenvalues and completes the proof that $E = H$.

(2): Let P_n denote the orthogonal projection onto E_n. Then for all $x \in H$ we have $x = \sum_{n \geq 1} P_n x$ with convergence in H. This is clear for every $x \in E_n$, and since the span of the spaces E_n is dense in H and the operators $\sum_{n=1}^N P_n$ are orthogonal projections and hence have norm one, the convergence extends to all $x \in H$ by Proposition 1.19. It follows that the sum $\sum_{n \geq 1} TP_n x$ converges as well, with sum Tx.

Fix $\varepsilon > 0$. The set $\Lambda_\varepsilon := \{n \geq 1 : |\lambda_n| > \varepsilon\}$ is finite by Theorem 7.11. Let $N \geq 1$ be so large that $\Lambda_\varepsilon \subseteq \{1, 2, \ldots, N\}$. Fixing $x \in H$ and writing $x_n := P_n x$, by orthogonality we have

$$\left\| Tx - \sum_{n=1}^N \lambda_n P_n x \right\|^2 = \left\| \sum_{n \geq 1} Tx_n - \sum_{n=1}^N \lambda_n x_n \right\|^2 = \left\| \sum_{n \geq N+1} \lambda_n x_n \right\|^2$$

$$= \sum_{n \geq N+1} |\lambda_n|^2 \|x_n\|^2 \leqslant \varepsilon^2 \sum_{n \geq N+1} \|x_n\|^2 \leqslant \varepsilon^2 \|x\|^2.$$

Taking the supremum over all $x \in H$ with $\|x\| \leqslant 1$ we obtain

$$\left\| T - \sum_{n=1}^N \lambda_n P_n \right\|^2 \leqslant \varepsilon^2.$$

This completes the proof. □

Let H and K be Hilbert spaces. For $h \in H$ and $k \in K$ we denote by $k \bar{\otimes} h$ the operator in $\mathscr{L}(H, K)$ defined by

$$(k \bar{\otimes} h)x := (x|h)k, \quad x \in H.$$

If $H = K$ and $h \in H$ has norm one, then $h \bar{\otimes} h$ is the orthogonal projection onto the subspace spanned by h. If $T \in \mathscr{L}(H)$ is a compact normal operator, the eigenspaces corresponding to nonzero eigenvalues are finite-dimensional. Choosing orthonormal bases for each of them, from Theorem 9.1 we obtain a representation

$$T = \sum_{n \geqslant 1} \lambda_n h_n \bar{\otimes} h_n$$

with convergence in the operator norm, where now $(\lambda_n)_{n \geqslant 1}$ is the sequence of nonzero eigenvalues of T *repeated according to multiplicities* and $(h_n)_{n \geqslant 1}$ is an associated orthonormal sequence of eigenvectors. Strictly speaking, the spectral theorem gives convergence of sum for 'blockwise' summation 'per eigenspace', but the proof of the theorem may be repeated to obtain the convergence as stated. The geometric and algebraic multiplicities of the eigenvalues coincide by Corollary 8.17, so we can unambiguously speak about their *multiplicity*.

Theorem 9.1 allows us to deduce the following general representation theorem for compact operators acting on a Hilbert space. It strengthens Proposition 7.6 which asserted that such operators can be approximated in operator norm by finite rank operators.

Theorem 9.2 (Singular value decomposition). *Let $T \in \mathscr{L}(H, K)$ be a compact operator, where K is another Hilbert space. Then T admits a decomposition*

$$T = \sum_{n \geqslant 1} \lambda_n k_n \bar{\otimes} h_n$$

with convergence in the operator norm, where $(\lambda_n)_{n \geqslant 1}$ is the sequence of nonzero eigenvalues of the compact operator $(T^\star T)^{1/2}$ repeated according to multiplicities, and $(h_n)_{n \geqslant 1}$ and $(k_n)_{n \geqslant 1}$ are orthonormal sequences in H and K respectively, the former consisting of eigenvectors of $(T^\star T)^{1/2}$.

The proof depends on the following lemma.

Lemma 9.3. *If $S \in \mathscr{L}(H)$ is a positive compact operator, then its square root $S^{1/2}$ is compact.*

Proof By Theorem 9.1 we have $S = \sum_{n \geqslant 1} \nu_n Q_n$, with $(\nu_n)_{n \geqslant 1}$ the (nonnegative) sequence of distinct nonzero eigenvalues of S and $(Q_n)_{n \geqslant 1}$ the sequence of orthogonal projections onto the corresponding eigenspaces. Fix $\varepsilon > 0$ and let $N \geqslant 1$ be so large that

$v_n \leqslant \varepsilon$ for all $n \geqslant N$. Then for $N' \geqslant N$ and $x \in H$ we have, by orthogonality,

$$\left\| \sum_{n=N}^{N'} v_n^{1/2} Q_n x \right\|^2 = \sum_{n=N}^{N'} v_n \|Q_n x\|^2 \leqslant \varepsilon \sum_{n=N}^{N'} \|Q_n x\|^2 \leqslant \varepsilon \|x\|^2.$$

This implies that the sum $R := \sum_{n \geqslant 1} v_n^{1/2} Q_n$ converges in operator norm. We have $R \geqslant 0$ and $R^2 = \sum_{n \geqslant 1} v_n Q_n = S$, so $R = S^{1/2}$. This operator is the limit in operator norm of the finite rank operators $\sum_{n=1}^{N} v_n^{1/2} Q_n$, $N \geqslant 1$, and therefore it is compact. □

Proof of Theorem 9.2   By Theorem 9.1, applied to $|T| := (T^\star T)^{1/2}$, which is compact by Lemma 9.3, we arrive at a representation

$$|T| = \sum_{n \geqslant 1} \lambda_n h_n \bar{\otimes} h_n,$$

with convergence in the operator norm, where $(\lambda_n)_{n \geqslant 1}$ is the sequence of eigenvalues of $|T|$ repeated according to multiplicities and the orthonormal sequence $(h_n)_{n \geqslant 1}$ consists of eigenvectors of $|T|$. Let $T = U|T|$ with U an isometry from $\overline{R(|T|)}$ onto $\overline{R(T)}$ as in Theorem 8.30. The sequence $(k_n)_{n \geqslant 1}$ defined by $k_n := U h_n$ is orthonormal in K and

$$T = \sum_{n \geqslant 1} \lambda_n k_n \bar{\otimes} h_n$$

with convergence in the operator norm. □

As a second application of Theorem 9.1 we record the following formulas for the eigenvalues of a compact positive operator.

Theorem 9.4 (Min-max theorem). *Let $T \in \mathscr{L}(H)$ be compact and positive, and let $\lambda_1 \geqslant \lambda_2 \geqslant \cdots \geqslant 0$ be the sequence of its nonzero eigenvalues repeated according to multiplicities. Then for all $n \geqslant 1$ we have*

$$\lambda_n = \inf_{\substack{Y \subseteq H \\ \dim(Y) = n-1}} \sup_{\substack{\|y\|=1 \\ y \perp Y}} (Ty|y) = \inf_{\substack{Y \subseteq H \\ \dim(Y) = n-1}} \sup_{\substack{\|y\|=1 \\ y \perp Y}} \|Ty\|$$

where the infima are taken over all subspaces Y of H of dimension $n - 1$.

Proof   For $n = 1$ the only subspace Y to be considered is $\{0\}$. In this case both suprema are taken over all norm one vectors $y \in H$ and are equal to $\|T\| = \sup_{\|y\| \leqslant 1} \|Ty\| = \sup_{\|y\| \leqslant 1} (Ty|y) = \lambda_1$ by Theorem 8.11; here we use that T is positive. In the remainder of the proof we may therefore assume that $n \geqslant 2$.

Using Theorem 9.1 we select an orthonormal basis $(h_j)_{j \geqslant 1}$ for H such that $Th_j = \lambda_j h_j$ for all $j \geqslant 1$. Let $Y \subseteq H$ be any subspace of dimension $n - 1$ and let H_n denote the linear span of the vectors $h_1, \ldots, h_n$. Then $Y^\perp \cap H_n$ is a nonzero subspace of H, so it contains a norm one vector y. Writing $y = \sum_{j=1}^{n} c_j h_j$ with $\sum_{j=1}^{n} |c_j|^2 = 1$, we have

$$(Ty|y) = \sum_{j=1}^{n} \lambda_j |c_j|^2 \geqslant \lambda_n \sum_{j=1}^{n} |c_j|^2 = \lambda_n.$$

This proves the inequality

$$\lambda_n \leqslant \inf_{\substack{Y \subseteq H \\ \dim(Y)=n-1}} \sup_{\substack{\|y\|=1 \\ y \perp Y}} (Ty|y).$$

The inequality

$$\inf_{\substack{Y \subseteq H \\ \dim(Y)=n-1}} \sup_{\substack{\|y\|=1 \\ y \perp Y}} (Ty|y) \leqslant \inf_{\substack{Y \subseteq H \\ \dim(Y)=n-1}} \sup_{\substack{\|y\|=1 \\ y \perp Y}} \|Ty\|$$

holds trivially. To prove the inequality

$$\inf_{\substack{Y \subseteq H \\ \dim(Y)=n-1}} \sup_{\substack{\|y\|=1 \\ y \perp Y}} \|Ty\| \leqslant \lambda_n,$$

let $y \perp H_{n-1}$ have norm one. Then $y = \sum_{j \geqslant n} (y|h_j) h_j$ and $\sum_{j \geqslant n} |(y|h_j)|^2 = 1$. Hence,

$$\|Ty\|^2 = \left\| \sum_{j \geqslant n} \lambda_j (y|h_j) h_j \right\|^2 = \sum_{j \geqslant n} \lambda_j^2 |(y|h_j)|^2 \leqslant \lambda_n^2 \sum_{j \geqslant n} |(y|h_j)|^2 = \lambda_n^2,$$

and the result follows. $\qquad \square$

Corollary 9.5. *If $S, T \in \mathscr{L}(H)$ are compact operators satisfying $0 \leqslant S \leqslant T$, and if $\lambda_1 \geqslant \lambda_2 \geqslant \cdots > 0$ and $\mu_1 \geqslant \mu_2 \geqslant \cdots > 0$ are their sequences of nonzero eigenvalues, both repeated according to multiplicities, then for all $n \geqslant 1$ we have $\lambda_n \leqslant \mu_n$.*

9.2 Projection-Valued Measures

This section and the next deal with the preliminaries needed to state and prove the spectral theorem for bounded normal operators.

Let $(\Omega, \mathscr{F})$ be a measurable space.

Definition 9.6 (Projection-valued measures). A *projection-valued measure* on a measurable space $(\Omega, \mathscr{F})$ is a mapping $P : \mathscr{F} \to \mathscr{L}(H)$ that assigns to every set $F \in \mathscr{F}$ an orthogonal projection $P_F := P(F) \in \mathscr{L}(H)$ such that the following conditions are satisfied:

(i) $P_\Omega = I$;

(ii) for all $x \in H$ the mapping

$$F \mapsto (P_F x | x), \quad F \in \mathscr{F},$$

defines a measure on $(\Omega, \mathscr{F})$.

For $x \in H$ the measure defined by (ii) is denoted by P_x. Thus, for all $F \in \mathscr{F}$,

$$(P_F x | x) = P_x(F) = \int_\Omega 1_F \, dP_x.$$

From

$$P_x(\Omega) = (P_\Omega x | x) = (x | x) = \|x\|^2$$

we see that P_x is a finite measure.

We make some easy observations:

- $P_\varnothing = 0$.

Indeed, the additivity of P_x, applied to $\Omega = \Omega \cup \varnothing$ implies

$$(x | x) = P_x(\Omega) = P_x(\Omega \cup \varnothing) = P_x(\Omega) + P_x(\varnothing) = (x | x) + P_x(\varnothing)$$

and therefore $(P_\varnothing x | x) = P_x(\varnothing) = 0$ for all $x \in H$.

- *If $F_1, F_2 \in \mathscr{F}$ are disjoint, then the ranges of P_{F_1} and P_{F_2} are orthogonal.*

Indeed, by additivity we have

$$(P_{F_1 \cup F_2} x | x) = P_x(F_1 \cup F_2) = P_x(F_1) + P_x(F_2) = (P_{F_1} x | x) + (P_{F_2} x | x)$$

for all $x \in H$, so $P_{F_1 \cup F_2} = P_{F_1} + P_{F_2}$ by Proposition 8.1. Therefore,

$$P_{F_1} + P_{F_2} = P_{F_1 \cup F_2} = P_{F_1 \cup F_2}^2 = (P_{F_1} + P_{F_2})^2 = P_{F_1} + P_{F_1} P_{F_2} + P_{F_2} P_{F_1} + P_{F_2}.$$

It follows that

$$P_{F_1} P_{F_2} + P_{F_2} P_{F_1} = 0.$$

Then,

$$(P_{F_1} P_{F_2})^2 = P_{F_1} (P_{F_2} P_{F_1}) P_{F_2} = -P_{F_1} (P_{F_1} P_{F_2}) P_{F_2} = -P_{F_1} P_{F_2}$$

and similarly

$$(P_{F_2} P_{F_1})^2 = -P_{F_2} P_{F_1}.$$

Adding these identities gives

$$(P_{F_1} P_{F_2})^2 + (P_{F_2} P_{F_1})^2 = 0.$$

But

$$((P_{F_1} P_{F_2})^2 x | x) = (P_{F_1} P_{F_2} x | P_{F_2} P_{F_1} x) = -(P_{F_1} P_{F_2} x | P_{F_1} P_{F_2} x) = -\|P_{F_1} P_{F_2} x\|^2$$

and similarly

$$((P_{F_2}P_{F_1})^2 x | x) = -\|P_{F_2}P_{F_1}x\|^2.$$

Adding, we find

$$\|P_{F_1}P_{F_2}x\|^2 + \|P_{F_2}P_{F_1}x\|^2 = 0.$$

This is only possible if

$$\|P_{F_1}P_{F_2}x\|^2 = \|P_{F_2}P_{F_1}x\|^2 = 0.$$

Since this holds for all $x \in H$, we conclude that $P_{F_2}P_{F_1} = P_{F_2}P_{F_1} = 0$. In particular, P_{F_1} and P_{F_2} have orthogonal ranges.

- *For all $F_1, F_2 \in \mathscr{F}$ we have $P_{F_1 \cap F_2} = P_{F_1}P_{F_2} = P_{F_2}P_{F_1}$.*

In the special case of disjoint sets this has just been proved, with all three expressions equal to 0. From this special case it follows that

$$\begin{aligned} P_{F_1}P_{F_2} &= (P_{F_1 \setminus F_2} + P_{F_1 \cap F_2})(P_{F_2 \setminus F_1} + P_{F_1 \cap F_2}) \\ &= P_{F_1 \setminus F_2}P_{F_2 \setminus F_1} + P_{F_1 \cap F_2}P_{F_2 \setminus F_1} + P_{F_1 \setminus F_2}P_{F_1 \cap F_2} + P_{F_1 \cap F_2}^2 \\ &= 0 + 0 + 0 + P_{F_1 \cap F_2}. \end{aligned}$$

Reversing the roles of F_1 and F_2 gives the other identity.

Example 9.7. If $T \in \mathscr{L}(H)$ is a compact normal operator, Theorem 9.1 implies that the mapping $\{\lambda\} \mapsto P_\lambda$, where P_λ is the orthogonal projection onto the eigenspace of λ, extends to a projection-valued measure on $\sigma(T)$.

9.3 The Bounded Functional Calculus

Let $(\Omega, \mathscr{F})$ be a measurable space. The Banach space of all bounded measurable functions $f : \Omega \to \mathbb{C}$, endowed with the supremum norm $\|f\|_\infty = \sup_{\omega \in \Omega} |f(\omega)|$, is denoted by $B_b(\Omega)$.

Theorem 9.8 (Bounded functional calculus). *Let $P : \mathscr{F} \to \mathscr{L}(H)$ be a projection-valued measure. There exists a unique linear mapping $\Phi : B_b(\Omega) \to \mathscr{L}(H)$ with the following properties:*

(i) *for all $F \in \mathscr{F}$ we have $\Phi(1_F) = P_F$;*
(ii) *for all $f, g \in B_b(\Omega)$ we have $\Phi(fg) = \Phi(f)\Phi(g)$;*
(iii) *for all $f \in B_b(\Omega)$ we have $\Phi(\overline{f}) = (\Phi(f))^\star$;*
(iv) *for all $f \in B_b(\Omega)$ we have $\|\Phi(f)\| \leqslant \|f\|_\infty$;*
(v) *for all $f_n, f \in B_b(\Omega)$, if $\sup_{n \geqslant 1} \|f_n\|_\infty < \infty$ and $f_n \to f$ pointwise on Ω, then for all $x \in H$ we have $\Phi(f_n)x \to \Phi(f)x$.*

Moreover, for all $x \in H$ and $f \in B_b(\Omega)$ we have

$$(\Phi(f)x|x) = \int_\Omega f \, dP_x \tag{9.1}$$

and

$$\|\Phi(f)x\|^2 = \int_\Omega |f|^2 \, dP_x. \tag{9.2}$$

The operators $\Phi(f)$ are normal, and if f is real-valued (respectively, takes values in $[0, \infty)$) they are selfadjoint (respectively, positive).

Proof For $F \in \mathscr{F}$ we set $\Phi(\mathbf{1}_F) := P_F$, which is (i), and extend this definition by linearity to simple functions f. It is routine to verify that $\Phi(f)$ is well defined for such functions and that (ii) and (iii) hold.

If $f = \sum_{j=1}^k c_j \mathbf{1}_{F_j}$ is a simple function with disjoint supporting sets $F_j \in \mathscr{F}$, the orthogonality of the vectors $P_{F_j} x$ gives

$$\|\Phi(f)x\|^2 = \sum_{j=1}^k |c_j|^2 \|P_{F_j}x\|^2 \leqslant \max_{1 \leqslant j \leqslant k} |c_j|^2 \|x\|^2 = \|f\|_\infty^2 \|x\|^2.$$

It follows that $\|\Phi(f)\| \leqslant \|f\|_\infty$. For general $f \in B_b(\Omega)$ we can find a sequence of simple functions f_n converging to f uniformly on Ω and satisfying $\|f_n\|_\infty \leqslant \|f\|_\infty$. From $\|\Phi(f_n) - \Phi(f_m)\| \leqslant \|f_n - f_n\|_\infty$ we infer that the operators $\Phi(f_n)$ form a Cauchy sequence in $\mathscr{L}(H)$. It follows that the limit

$$\Phi(f) := \lim_{n \to \infty} \Phi(f_n)$$

exists, with convergence in the operator norm, and it is routine to check that the limit is independent of the choice of approximating sequence. Moreover,

$$\|\Phi(f)\| \leqslant \limsup_{n \to \infty} \|f_n\|_\infty \leqslant \|f\|_\infty,$$

which gives (iv). The general case of (ii) and (iii) now follows by approximation.

To prove (v) we first establish (9.1) and (9.2). If $f = \sum_{j=1}^k c_j \mathbf{1}_{F_j}$ is a simple function with disjoint supporting sets $F_j \in \mathscr{F}$, then

$$(\Phi(f)x|x) = \sum_{j=1}^k c_j(P_{F_j}x|x) = \sum_{j=1}^k c_j P_x(F_j) = \int_\Omega f \, dP_x.$$

This gives (9.1) for simple functions. The general case follows by approximation and dominated convergence. Similarly,

$$\|\Phi(f)x\|^2 = \sum_{j=1}^k |c_j|^2 \|P_{F_j}x\|^2 = \sum_{j=1}^k |c_j|^2 (P_{F_j}x|x) = \int_\Omega |f|^2 \, dP_x.$$

This gives (9.2) for simple functions. Again the general case follows by approximation and dominated convergence. Property (v) follows from (9.2), applied to the functions $f_n - f$, and dominated convergence.

To prove normality of $\Phi(f)$, note that (i) and (iii) imply

$$(\Phi(f))^\star \Phi(f) = \Phi(\overline{f}) \Phi(f) = \Phi(|f|^2) = \Phi(f) \Phi(\overline{f}) = \Phi(f)(\Phi(f))^\star.$$

Selfadjointness (respectively, positivity) for real-valued (respectively, nonnegative) f is immediate from (iii) (respectively, (9.1)). □

In what follows we shall write

$$\Phi(f) = \int_\Omega f \, dP = \int_\Omega f(\lambda) \, dP(\lambda)$$

for functions $f \in B_b(\Omega)$; the rigorous interpretation of these integrals is through (9.1).

Proposition 9.9 (Substitution). *Let $(\Omega, \mathscr{F})$ and $(\Omega', \mathscr{F}')$ be measurable spaces and let $f : \Omega \to \Omega'$ be a measurable mapping. If $P : \mathscr{F} \to \mathscr{L}(H)$ is a projection-valued measure, then the mapping $Q : \mathscr{F}' \to \mathscr{L}(H)$ defined by*

$$Q_{F'} := P_{f^{-1}(F')}, \quad F' \in \mathscr{F}',$$

is a projection-valued measure. Denoting by Φ and Ψ the bounded functional calculi of P and Q, for all $g \in B_b(\Omega')$ we have

$$\Phi(g \circ f) = \Psi(g).$$

Proof The elementary verification that Q is a projection-valued measure is left as an exercise. For all $F' \in \mathscr{F}'$ and $x \in H$,

$$\int_\Omega \mathbf{1}_{F'} \circ f \, dP_x = \int_\Omega \mathbf{1}_{f^{-1}(F')} \, dP_x = (P_{f^{-1}(F')} x | x) = (Q_{F'} x | x) = \int_{\Omega'} \mathbf{1}_{F'} \, dQ_x.$$

By linearity and monotone convergence, it follows that for nonnegative functions $g \in B_b(\Omega')$ and $x \in H$ we have

$$\int_\Omega g \circ f \, dP_x = \int_{\Omega'} g \, dQ_x,$$

that is, $(\Phi(g \circ f) x | x) = (\Psi(g) x | x)$. For nonnegative $g \in B_b(\Omega')$ the result now follows from Proposition 8.1. For general $g \in B_b(\Omega')$ the result follows by splitting into real and imaginary parts and considering their positive and negative parts. □

Due to the absence of a reference measure μ on Ω we had to work with the Banach space $B_b(\Omega)$ rather than with a Lebesgue space $L^\infty(\Omega, \mu)$. However, the projection-valued measure P can be used to define a Lebesgue-type space $L^\infty(\Omega, P)$ as follows.

Definition 9.10 (*P-Essential boundedness*). A measurable function $f : \Omega \to \mathbb{C}$ is said to be *P-essentially bounded* if $P(\{|f| > r\}) = 0$ for some $r \geqslant 0$. We define $L^\infty(\Omega, P)$ to be the space of all equivalence classes of P-essentially bounded measurable functions, identifying the functions f and g when $P(\{f \neq g\}) = 0$.

With respect to the norm

$$\|f\|_{L^\infty(\Omega,P)} := \inf\{r \geqslant 0 : P(\{|f| > r\}) = 0\}$$

the space $L^\infty(\Omega, P)$ is easily checked to be a Banach space.

For functions $f \in L^\infty(\Omega, P)$ we obtain a well-defined bounded operator $\Phi(f)$, the properties (i) and (ii) holds again, and (iv) improves to *equality*:

Proposition 9.11. *If $P : \mathscr{F} \to \mathscr{L}(H)$ is a projection-valued measure, then for all $f \in L^\infty(\Omega, P)$ we have*

$$\|\Phi(f)\| = \|f\|_{L^\infty(\Omega,P)}.$$

Proof The upper bound '$\leqslant$' is an immediate consequence of part (ii) of Theorem 9.8. The lower bound '$\geqslant$' is proved by observing that the definition of the P-essential supremum implies that for all $\varepsilon > 0$ the projection P_{F_ε} is nonzero, where

$$F_\varepsilon := \{|f| > (1 - \varepsilon)\|f\|_{L^\infty(\Omega,P)}\}.$$

Then, for all $x \in \mathsf{R}(P_{F_\varepsilon})$,

$$\|\Phi(f)x\|^2 = \int_\Omega |f|^2 \, dP_x \geqslant (1 - \varepsilon)^2 \|f\|^2_{L^\infty(\Omega,P)} \int_\Omega \mathbf{1}_{F_\varepsilon} \, dP_x$$

$$= (1 - \varepsilon)^2 \|f\|^2_{L^\infty(\Omega,P)} \|P_{F_\varepsilon} x\|^2 = (1 - \varepsilon)^2 \|f\|^2_{L^\infty(\Omega,P)} \|x\|^2.$$

This shows that $\|\Phi(f)\| \geqslant (1 - \varepsilon)\|f\|_{L^\infty(\Omega,P)}$. Since $\varepsilon > 0$ was arbitrary, the result follows from this. $\qquad\square$

We now turn to the special case of projection-valued measures defined on the Borel σ-algebra $\mathscr{B}(K)$ of a compact subset K of the complex plane. In that case we can consider the function

$$\mathrm{id}(\lambda) := \lambda.$$

The properties of the operator $\Phi(\mathrm{id})$ are summarised in the next proposition.

Proposition 9.12. *Let $K \subseteq \mathbb{C}$ be compact and let $P : \mathscr{B}(K) \to \mathscr{L}(H)$ be a projection-valued measure. Define the bounded operator $T_P \in \mathscr{L}(H)$ by $T_P := \Phi(\mathrm{id})$, that is,*

$$T_P := \Phi(\mathrm{id}) = \int_K \lambda \, dP(\lambda).$$

Then:

(1) *T_P is normal;*

(2) *the spectrum of T_P is contained in K;*

(3) *the support of P equals $\sigma(T_P)$ in the following sense:*

    (i) *$P_{K \cap U} \neq 0$ for all open sets $U \subseteq \mathbb{C}$ such that $\sigma(T_P) \cap U \neq \varnothing$;*

    (ii) *$P_B = 0$ for all Borel sets $B \subseteq K$ such that $\sigma(T_P) \cap B = \varnothing$.*

The operator T_P is selfadjoint if $K \subseteq \mathbb{R}$, positive if $K \subseteq [0, \infty)$, unitary if $K \subseteq \mathbb{T}$, and an orthogonal projection if $K \subseteq \{0, 1\}$.

Proof Let $\Phi : B_{\mathrm{b}}(K) \to \mathscr{L}(H)$ be the bounded functional calculus associated with P and write $T_P =: T$ for brevity.

Part (1) is immediate from the properties of the bounded calculus.

If $\lambda_0 \in \complement K$, the functions $\lambda \mapsto \lambda_0 - \lambda$ and $\lambda \mapsto (\lambda_0 - \lambda)^{-1}$ are bounded on K and multiply to $\mathbf{1}$. In view of $\Phi(\mathbf{1}) = P_K = I$ we have $\lambda_0 - T = \Phi(\lambda_0 - \mathrm{id})$, and property (iii) of the bounded calculus shows that $\lambda_0 - T$ is a two-sided inverse of $\Phi((\lambda_0 - \mathrm{id})^{-1})$. It follows that $\lambda_0 \in \rho(T)$ and $R(\lambda_0, T) = \Phi((\lambda_0 - \mathrm{id})^{-1})$. This proves (2).

Next we show that $P_B = 0$ for all Borel sets $B \subseteq K$ such that $\sigma(T) \cap B = \varnothing$.

Step 1 – Suppose first, for a contradiction, that there is a Borel set $B \subseteq K$ such that $\overline{B} \cap \sigma(T) = \varnothing$ and $P_B = \mathbf{1}_B(T) \neq 0$. By the additivity of P, there exists a half-open rectangle R_1 of sufficiently small diameter $\rho > 0$ such that $\overline{R_1} \cap \sigma(T) = \varnothing$ and $P_{B_1} = \mathbf{1}_{B_1}(T) \neq 0$, where $B_1 = B \cap R_1$. Proceeding inductively we obtain a sequence of nested half-open rectangles $R_1 \supseteq R_2 \supseteq \ldots$ such that $\mathrm{diam}(R_n) \leqslant 2^{-n+1}\rho$, $\overline{R_n} \cap \sigma(T) = \varnothing$, and $P_{B_n} = \mathbf{1}_{B_n}(T) \neq 0$ for $B_n := B \cap R_n$. Let $\bigcap_{n \geqslant 1} \overline{R_n} =: \{\lambda_0\}$ and note that $\lambda_0 \in \overline{B}$.

Let $x_n \in \mathrm{R}(\mathbf{1}_{B_n}(T))$ have norm one. Since $\mathbf{1}_{B_n}(T)$ is a projection we have $x_n = \mathbf{1}_{B_n}(T)x_n$, and for all $y \in H$ we obtain, using the multiplicativity of Φ,

$$\|Tx_n - \lambda_0 x_n\|^2 = ((T - \lambda_0)^\star(T - \lambda_0)\mathbf{1}_{B_n}(T)x_n | x_n) = \int_K |\lambda - \lambda_0|^2 \mathbf{1}_{B_n}(\lambda) \, \mathrm{d}P_{x_n}(\lambda)$$

and therefore

$$\|Tx_n - \lambda_0 x_n\|^2 \leqslant \sup_{\lambda \in K} |\lambda - \lambda_0|^2 (P_{B_n} x_n | x_n) \leqslant \mathrm{diam}^2(B_n) \leqslant 2^{-2n+2}.$$

This means that λ_0 is an approximate eigenvalue for T, so $\lambda_0 \in \sigma(T)$. We also have $\lambda_0 \in \overline{B}$ and $\overline{B} \cap \sigma(T) = \varnothing$. This contradiction proves that $P_B = 0$ for all Borel sets $B \subseteq K$ whose closure is disjoint of $\sigma(T)$.

Step 2 – Now consider a general Borel set $B \subseteq K$ disjoint with $\sigma(T)$. The Borel set

$$B^{(n)} := B \cap \left\{ \lambda \in K : d(\lambda, \sigma(T)) \geqslant \frac{1}{n} \right\}$$

has closure disjoint from $\sigma(T)$ and consequently $P_{B^{(n)}} = 0$ for all $n \geqslant 1$ by what we already proved. In particular we have $P_x(B^{(n)}) = 0$ for all $x \in H$, and by monotone convergence it follows that $P_x(B) = (P_B x | x) = 0$ for all $x \in X$. This implies $P_B = 0$.

This completes the proof of the support property (i). It implies that there is no loss of generality in assuming that $K = \sigma(T)$. Assuming this in the rest of the proof, we now turn to the proof of the support property (ii).

Let $U \subseteq \mathbb{C}$ be an open set such that $\sigma(T) \cap U \neq \varnothing$ and suppose, for a contradiction, that $P_{\sigma(T) \cap U} = 0$. Then $P_B = 0$ for all Borel sets $B \subseteq \sigma(T) \cap U$. This implies that $\int_{\sigma(T)} f \, dP = 0$ for all simple functions f supported on such sets, and by approximation the same is true for all bounded Borel functions supported in $\sigma(T) \cap U$. In particular,

$$\int_{\sigma(T)} \mathbf{1}_{\sigma(T) \cap U} \lambda \, dP(\lambda) = 0.$$

Let $\widetilde{P} : \sigma(T) \setminus U \to \mathscr{L}(H)$ be the restriction of P to $\sigma(T) \setminus U$. Since

$$\widetilde{P}_{\sigma(T) \setminus U} = P_{\sigma(T) \setminus U} = P_{\sigma(T)} = I,$$

$\widetilde{P}$ is a projection-valued measure. Denoting $\widetilde{T} := \widetilde{\Phi}(\mathrm{id})$ the associated operator, we have

$$\widetilde{T} = \int_{\sigma(T) \setminus U} \lambda \, d\widetilde{P}(\lambda) = \int_{\sigma(T)} \mathbf{1}_{\sigma(T) \setminus U} \lambda \, dP(\lambda) = \int_{\sigma(T)} \lambda \, dP(\lambda) = T$$

and therefore $\sigma(T) = \sigma(\widetilde{T}) \subseteq \sigma(T) \setminus U$ by (2), which is absurd.

If $K \subseteq \mathbb{R}$ (respectively $K \subseteq [0, \infty)$), then $(Tx|x) \in \mathbb{R}$ (respectively $(Tx|x) \in [0, \infty)$) for all $x \in H$ and therefore T is selfadjoint (respectively positive). If $K \subseteq \mathbb{T}$, then T is invertible by part (2) and

$$T^\star = \int_K \overline{\lambda} \, dP(\lambda) = \int_K \lambda^{-1} \, dP(\lambda) = T^{-1}$$

and therefore T is unitary. If $K \subseteq \{0, 1\}$, then $T = 0$ if $K = \{0\}$ and $T = \int_K \lambda \, dP(\lambda) = P_{\{1\}}$ if $1 \in K$. In both cases we see that T is an orthogonal projection. $\qquad\square$

It follows from the proposition that P restricts to a projection-valued measure on $\sigma(T_P)$ in a natural way. Accordingly we have

$$T_P = \int_{\sigma(T_P)} \lambda \, dP(\lambda).$$

The spectral theorem for bounded normal operators, which will be proved in Section 9.4, asserts that conversely for every normal operator $T \in \mathscr{L}(H)$ there exists a unique projection-valued measure P on $\sigma(T)$ such that $T = T_P$, that is,

$$T = \int_{\sigma(T)} \lambda \, dP(\lambda).$$

This will allow us to prove converses to the four implications in the final assertion in the proposition (see Corollary 9.18).

We have the following uniqueness result:

Proposition 9.13 (Uniqueness). *Let P and $\widetilde{P}$ be projection-valued measures on a compact set $K \subseteq \mathbb{C}$ and define the operators T_P and $T_{\widetilde{P}}$ as before. If $T_P = T_{\widetilde{P}}$, then $P = \widetilde{P}$.*

Proof Let us write $T := T_P = T_{\widetilde{P}}$. Then $T = \Phi(\mathrm{id}) = \widetilde{\Phi}(\mathrm{id})$ and $T^* = \Phi(\overline{\mathrm{id}}) = \widetilde{\Phi}(\overline{\mathrm{id}})$, where Φ and $\widetilde{\Phi}$ are the bounded calculi associated with P and $\widetilde{P}$, respectively, and $\mathrm{id}(\lambda) = \lambda$. By the multiplicativity of the calculi,

$$T^m T^{*n} = \Phi(\mathrm{id}^m \overline{\mathrm{id}}^n) = \Phi(\mathrm{id})^m \Phi(\overline{\mathrm{id}})^n = \widetilde{\Phi}(\mathrm{id})^m \widetilde{\Phi}(\overline{\mathrm{id}})^n = \widetilde{\Phi}(\mathrm{id}^m \overline{\mathrm{id}}^n).$$

It follows that $p(T) = \Phi(p) = \widetilde{\Phi}(p)$ for all functions $p(z) = q(z,\bar{z})$ with q a polynomial in two variables, and then $f(T) = \Phi(f) = \widetilde{\Phi}(f)$ for all $f \in C(\sigma(T))$ by approximation using the Stone–Weierstrass theorem (Theorem 2.5). This means that

$$\int_{\sigma(T)} f \, \mathrm{d}P = \int_{\sigma(T)} f \, \mathrm{d}\widetilde{P}, \quad f \in C(\sigma(T)).$$

If R is an open rectangle in $\mathbb{C}$, and if $0 \leqslant f_n \uparrow \mathbf{1}_R$ pointwise on K with each f_n continuous on K, we find that

$$P_x(\sigma(T) \cap R) = \int_{\sigma(T)} \mathbf{1}_R \, \mathrm{d}P_x = \lim_{n \to \infty} \int_{\sigma(T)} f_n \, \mathrm{d}P_x$$

$$= \lim_{n \to \infty} \int_{\sigma(T)} f_n \, \mathrm{d}\widetilde{P}_x = \int_{\sigma(T)} \mathbf{1}_R \, \mathrm{d}\widetilde{P}_x = \widetilde{P}_x(\sigma(T) \cap R).$$

This means that $P_x(\sigma(T) \cap R) = \widetilde{P}_x(\sigma(T) \cap R)$ for all open rectangles R. By Dynkin's lemma (Lemma E.4), this implies that $P_x(B) = \widetilde{P}_x(B)$ and therefore

$$(P_B x | x) = P_x(B) = \widetilde{P}_x(B) = (\widetilde{P}_B x | x)$$

for all $x \in H$ and Borel subsets B of $\sigma(T)$. It follows that $P_B = \widetilde{P}_B$ for all Borel subsets B of $\sigma(T)$. Since P and $\widetilde{P}$ are supported on $\sigma(T)$ this completes the proof. $\qquad\square$

9.4 The Spectral Theorem for Bounded Normal Operators

We are now ready to state and prove the spectral theorem for bounded normal operators.

Theorem 9.14 (Spectral theorem for bounded normal operators). *Let $T \in \mathcal{L}(H)$ be a normal operator. There exists a unique projection-valued measure P on $\sigma(T)$ such that*

$$T = \int_{\sigma(T)} \lambda \, \mathrm{d}P(\lambda).$$

For the proof of Theorem 9.14 we need the following elementary consequence of the Riesz representation theorem.

Proposition 9.15. *Let* $\mathfrak{a} : H \times H \to \mathbb{C}$ *be a sesquilinear mapping with the property that there exists a constant* $C \geqslant 0$ *such that*

$$|\mathfrak{a}(x,y)| \leqslant C \|x\| \|y\|, \quad x,y \in H.$$

Then there exists a unique operator $A \in \mathscr{L}(H)$ *such that*

$$\mathfrak{a}(x,y) = (Ax|y), \quad x,y \in H.$$

Moreover, $\|A\| \leqslant C$, *where* C *is the boundedness constant of* $\mathfrak{a}$.

Proof For all $y \in H$, the mapping $x \mapsto \mathfrak{a}(x,y)$ is a bounded functional on H, and the Riesz representation theorem gives a unique $w = w(y) \in H$ satisfying

$$\mathfrak{a}(x,y) = (x|w(y)), \quad x \in H.$$

Let $B : y \mapsto w = w(y)$ be the resulting mapping. Then,

$$\mathfrak{a}(x,y) = (x|By), \quad x,y \in H.$$

We claim that $B : H \to H$ is a bounded operator. Indeed if $c_1, c_2 \in \mathbb{K}$ and $y_1, y_2 \in H$, for all $x \in H$ we have

$$(x|B(c_1 y_1 + c_2 y_2)) = \mathfrak{a}(x, c_1 y_1 + c_2 y_2) = \overline{c}_1 \mathfrak{a}(x,y_1) + \overline{c}_2 \mathfrak{a}(x,y_2)$$
$$= \overline{c}_1 (x|By_1) + \overline{c}_2 (x|By_2) = (x|c_1 By_1 + c_2 By_2).$$

Since this equality holds for all $x \in H$ it follows that B is linear. Furthermore,

$$\|Bx\|^2 = (Bx|Bx) = |(Bx|Bx)| = |\mathfrak{a}(Bx,x)| \leqslant C\|Bx\| \|x\|.$$

Consequently $\|Bx\| \leqslant C\|x\|$ for all $x \in H$, so B is bounded with $\|B\| \leqslant C$. The operator $A := B^\star$ has the required properties. $\qquad\square$

Proof of Theorem 9.14 We begin with existence. For $x, y \in H$ consider the linear mapping $\phi_{x,y} : C(\sigma(T)) \to \mathbb{C}$,

$$\phi_{x,y}(f) := (f(T)x|y),$$

where $f(T)$ is given by the continuous functional calculus of T. The bound $\|f(T)\| \leqslant \|f\|_\infty$ implies that $\phi_{x,y}$ is bounded and

$$\|\phi_{x,y}\| \leqslant \|x\| \|y\|.$$

By the Riesz representation theorem (Theorem 4.2) there exists a unique complex Borel measure $P_{x,y}$ on $\sigma(T)$ such that

$$(f(T)x|y) = \int_{\sigma(T)} f \, dP_{x,y}, \quad f \in C(\sigma(T)). \tag{9.3}$$

Note that

$$\|P_{x,y}\| = \sup_{\|f\|_\infty \leq 1} |(f(T)x|y)| \leq \|x\| \|y\|$$

since $\|f(T)\| \leq \|f\|_\infty$. Hence by Proposition 9.15, for all $f \in B_b(\sigma(T))$ there exists a unique bounded operator on H, which we denote by $f(T)$, such that

$$(f(T)x|y) = \int_{\sigma(T)} f \, dP_{x,y} \tag{9.4}$$

for all $x, y \in H$. The bound on the norm of $P_{x,y}$ implies

$$\|f(T)\| \leq \|f\|_\infty, \quad f \in B_b(\sigma(T)).$$

For $f \in C(\sigma(T))$, (9.4) is consistent with (9.3). Taking $f(\lambda) = \lambda$ and $x = y$, and noting that $P_x := P_{x,x}$ is a finite Borel measure on $\sigma(T)$, we obtain the identity in the statement of the theorem,

$$(Tx|x) = \int_{\sigma(T)} \lambda \, dP_x(\lambda),$$

except that it remains to be proved that the measures P_x come from a projection-valued measure. The remainder of the proof is devoted to showing that this is indeed the case.

For Borel sets $B \subseteq \sigma(T)$ we define the bounded operator $P_B \in \mathscr{L}(H)$ by

$$P_B := \mathbf{1}_B(T).$$

This operator is positive since $\mathbf{1}_B$ is nonnegative, contractive, and for all $x \in H$ we have

$$(P_B x|x) = \int_{\sigma(T)} \mathbf{1}_B \, dP_x = P_x(B).$$

Proving that P_B is a projection takes a bit of effort since we must proceed from first principles; once we have established the existence of a projection-valued measure P generating the measures P_x, various steps in the argument appear as special cases of properties of the bounded functional calculus for P already established in Theorem 9.8.

We first assume that U is open as a subset of $\sigma(T)$ (that is, we assume that $U = U' \cap \sigma(T)$ for some open set $U' \subseteq \mathbb{C}$). Choose a sequence of real-valued continuous functions $f_n \in C(\sigma(T))$ satisfying $0 \leq f_n \uparrow \mathbf{1}_U$ pointwise. Then, for all $x \in H$,

$$(P_U x|x) = \int_{\sigma(T)} \mathbf{1}_U \, dP_x = \lim_{n \to \infty} \int_{\sigma(T)} f_n \, dP_x = \lim_{n \to \infty} (f_n(T)x|x) \tag{9.5}$$

by the monotone convergence theorem. This implies $(P_U x|y) = \lim_{n \to \infty} (f_n(T)x|y)$ for all $x, y \in H$ by polarisation. Then,

$$(P_U^2 x|y) = \lim_{n \to \infty} (f_n(T)x|P_U y) = \lim_{n \to \infty} \lim_{m \to \infty} (f_n(T)x|f_m(T)y)$$

$$\overset{(i)}{=} \lim_{n \to \infty} \lim_{m \to \infty} (f_m(T)f_n(T)x|y) \overset{(ii)}{=} \lim_{n \to \infty} \lim_{m \to \infty} ((f_m f_n)(T)x|y) \overset{(iii)}{=} (P_U x|y),$$

where (i) uses that f_m is real-valued, implying that $f_m(T)$ is selfadjoint, (ii) follows from the multiplicativity of the continuous calculus, and (iii) follows by repeating the steps of (9.5). This proves that P_U is a projection.

For general Borel sets $B \subseteq \sigma(T)$ we use the outer regularity of the measure P_x (see Proposition E.16) to find, for every $n \geqslant 1$, an open set U_n (depending on x) containing B and such that $U_n \setminus B$ has P_x-measure less than $1/n$. Then

$$0 \leqslant (P_{U_n \setminus B} x | x) = P_x(U_n \setminus B) < \frac{1}{n}.$$

It follows that $P_{U_n \setminus B}$ is a positive operator whose positive square root satisfies

$$\|P_{U_n \setminus B}^{1/2} x\|^2 = (P_{U_n \setminus B} x | x) < \frac{1}{n}.$$

Fix $x \in H$. By Theorem 8.11, $0 \leqslant P_{U_n \setminus B} \leqslant I$ implies $\sigma(P_{U_n \setminus B}) \subseteq [0, 1]$, and the spectral mapping theorem (Theorem 8.23) implies that $\sigma(P_{U_n \setminus B}^{1/2}) \subseteq [0, 1]$. Another application of Theorem 8.11 gives $0 \leqslant P_{U_n \setminus B}^{1/2} \leqslant I$ and therefore $\|P_{U_n \setminus B}^{1/2}\| \leqslant 1$. As a consequence,

$$\|P_{U_n \setminus B} x\| \leqslant \|P_{U_n \setminus B}^{1/2}\| \|P_{U_n \setminus B}^{1/2} x\| \leqslant \|P_{U_n \setminus B}^{1/2} x\| < \frac{1}{\sqrt{n}}.$$

From $P_{U_n} - P_B = P_{U_n \setminus B}$ we obtain $P_B P_{U_n} - P_{U_n} P_B = P_{U_n} P_{U_n \setminus B} - P_{U_n \setminus B} P_{U_n}$, and therefore

$$|((P_B P_{U_n} - P_{U_n} P_B) x | x)| = |(P_B P_{U_n} x | x) - (P_B x | P_{U_n} x)| \leqslant \frac{2}{\sqrt{n}} \|x\|.$$

Since $P_B^2 - P_{U_n}^2 = (P_B + P_{U_n})(P_B - P_{U_n}) + (P_B P_{U_n} - P_{U_n} P_B)$, it follows that

$$|((P_B^2 - P_B) x | x)| \leqslant |((P_B^2 - P_{U_n}^2) x | x)| + \underbrace{|((P_{U_n}^2 - P_{U_n}) x | x)|}_{=0} + |((P_{U_n} - P_B) x | x)|$$

$$\leqslant \left(\frac{1}{\sqrt{n}} \|(P_B + P_{U_n}) x\| + \frac{2}{\sqrt{n}} \|x\| \right) + 0 + \frac{1}{\sqrt{n}} \|x\| \leqslant \frac{5}{\sqrt{n}} \|x\|.$$

This being true for all $n \geqslant 1$, we conclude that $P_B^2 x = P_B x$. Since $x \in H$ was arbitrary, this proves that P_B is a projection, which is orthogonal by selfadjointness.

Uniqueness follows from Proposition 9.13.   □

Example 9.16. On $L^2(0, 1)$, the *position operator* X is defined by

$$X f(x) := x f(x), \quad x \in (0, 1), \ f \in L^2(0, 1).$$

We have $\sigma(X) = [0, 1]$, and the projection-valued measure of X is given by

$$P_B f = \mathbf{1}_B f$$

for all $f \in L^2(0, 1)$ and Borel subsets B of $[0, 1]$ (see Problem 9.9). In particular, $\mathbf{1}_B(X) = 0$ for any Borel null set B of $[0, 1]$. As a consequence, for all $\phi \in L^\infty(0, 1)$ the operator

$\phi(X)$ is well defined as a bounded operator on $L^2(0,1)$. In fact we have $\phi(X) = T_\phi$, where $T_\phi f(x) := \phi(x)f(x)$. The operators $\phi(X)$ arise quite naturally as follows:

Proposition 9.17. *An operator $S \in \mathscr{L}(L^2(0,1))$ commutes with the operator X (that is, $SX = XS$) if and only if there exists $\phi \in L^\infty(0,1)$ such that $S = \phi(X)$.*

Proof The 'if' part being easy, we concentrate on the 'only if' part. Let $S \in \mathscr{L}(L^2(0,1))$ be such that $SX = XS$. To show that there exists $\phi \in L^\infty(0,1)$ such that $S = \phi(X)$, a natural candidate for ϕ is the function $S\mathbf{1}$ (which *a priori* is an element of $L^2(0,1)$). For $f_n(x) := x^n$ with $n \in \mathbb{N}$ we have $f_n = X^n f_0 = X^n \mathbf{1}$ and therefore

$$S f_n = SX^n \mathbf{1} = X^n S\mathbf{1} = X^n \phi = [x \mapsto f_n(x)\phi(x)].$$

By the Weierstrass approximation theorem, the polynomials are dense in $C[0,1]$ and hence in $L^2(0,1)$ (as $C[0,1]$ is dense in $L^2(0,1)$). By a limiting argument we conclude that $Sf = [x \mapsto f(x)\phi(x)]$. The boundedness of S implies the boundedness of the multiplier $f \mapsto \phi f$, which in turn implies that $\phi \in L^\infty(0,1)$ (cf. Remark 2.27). □

We are now in a position to prove the assertions made in Section 8.1.

Corollary 9.18. *Let $T \in \mathscr{L}(H)$ be a normal operator.*

(1) *T is selfadjoint if and only if $\sigma(T) \subseteq \mathbb{R}$;*
(2) *T is positive if and only if $\sigma(T) \subseteq [0,\infty)$;*
(3) *T is unitary if and only if $\sigma(T) \subseteq \mathbb{T}$;*
(4) *T is an orthogonal projection if and only if $\sigma(T) \subseteq \{0,1\}$.*

Furthermore, if T is a projection, then it is an orthogonal projection.

Proof The 'only if' statements have already been proved in Chapter 8. The 'if' statements follow from Theorem 9.14, either by combining it with the final assertion of Proposition 9.12 or by the following direct reasoning.

Write $T = \int_{\sigma(T)} \lambda \, dP(\lambda)$ as in Theorem 9.14. If $\sigma(T)$ is contained in the real line, then, by Theorem 9.8,

$$T^\star = \int_{\sigma(T)} \overline{\lambda} \, dP(\lambda) = \int_{\sigma(T)} \lambda \, dP(\lambda) = T.$$

If $\sigma(T)$ is contained in the nonnegative half-line, then for all $x \in H$ we have

$$(Tx|x) = \int_{\sigma(T)} \lambda \, dP_x(\lambda) \geqslant 0.$$

If $\sigma(T)$ is contained in the unit circle, then T is invertible and

$$T^\star = \int_{\sigma(T)} \overline{\lambda} \, dP(\lambda) = \int_{\sigma(T)} \lambda^{-1} \, dP(\lambda) = T^{-1}.$$

If T is a projection, then $\sigma(P) \subseteq \{0, 1\}$ and

$$T = \int_{\{0,1\}} \lambda \, dP(\lambda) = 0 \cdot P_{\{0\}} + 1 \cdot P_{\{1\}} = P_{\{1\}},$$

which is an orthogonal projection.   □

As the example of the Volterra operator V shows (see Example 8.12 and Problem 8.12), normality cannot be omitted in parts (1), (2), and (4). The operator $I + V$ shows that normality cannot be omitted in part (3).

For normal operators $T \in \mathscr{L}(H)$ and functions $f \in B_b(\sigma(T))$, the operator $\Phi(f) \in \mathscr{L}(H)$ defined in terms of the projection-valued measure of T by is denoted by $f(T)$,

$$f(T) := \Phi(f) = \int_{\sigma(T)} f \, dP.$$

The properties of the bounded calculus for Φ translate into corresponding properties for the mapping $f \mapsto f(T)$:

Theorem 9.19 (Bounded functional calculus for normal operators). *Let $T \in \mathscr{L}(H)$ be normal. Then:*

  (i) *for $f(z) = z^m \bar{z}^n$ we have $f(T) = T^m T^{*n}$;*
  (ii) *for all $f, g \in B_b(\sigma(T))$ we have $(fg)(T) = f(T)g(T)$;*
  (iii) *for all $f \in B_b(\sigma(T))$ we have $\bar{f}(T) = (f(T))^\star$;*
  (iv) *for all $f \in B_b(\sigma(T))$ we have $\|f(T)\| \leqslant \|f\|_\infty$;*
  (v) *for all $f_n, f \in B_b(\sigma(T))$, if $\sup_{n \geqslant 1} \|f_n\|_\infty < \infty$ and $f_n \to f$ pointwise on $\sigma(T)$, then for all $x \in H$ we have $f_n(T)x \to f(T)x$.*

Moreover, for all $x \in H$ and $f \in B_b(\sigma(T))$ we have

$$(f(T)x|x) = \int_\Omega f \, dP_x$$

and

$$\|f(T)x\|^2 = \int_\Omega |f|^2 \, dP_x.$$

The operators $f(T)$ are normal, and if f is real-valued (respectively, nonnegative) they are selfadjoint (respectively, positive).

Proof Everything but (i) follows from Theorem 9.8; (i) follows from (ii) and (iii).   □

Theorem 9.20 (Composition). *Let $T \in \mathscr{L}(H)$ be a normal operator, let $f \in C(\sigma(T))$, and put $K := \sigma(f(T)) = f(\sigma(T))$. Then:*

  (i) *the projection-valued measure $P^{(f)}$ of $f(T)$ is given by*

$$P^{(f)}(B) = P_{f^{-1}(B)}, \quad B \in \mathscr{B}(K).$$

(ii) *for all $g \in B_b(K)$ we have*

$$(g \circ f)(T) = g(f(T)).$$

Proof Once the first assertion has been proved, the second is merely a restatement of the substitution rule of Proposition 9.9.

Let P and $P^{(f)}$ denote the projection-valued measures on $\sigma(T)$ and K of the normal operators T and $f(T)$, respectively, and define the projection-valued measure Q on K by $Q_B := P_{f^{-1}(B)}$ for Borel sets $B \subseteq K$ as in Proposition 9.9. By this proposition, applied to the function $g(\lambda) = \lambda$ on K, for all $x \in H$ we have

$$(f(T)x|x) = \int_{\sigma(T)} g \circ f \, dP_x = \int_{f(\sigma(T))} g \, dQ_x = \int_{\sigma(f(T))} \lambda \, dQ_x(\lambda).$$

This implies that both $P^{(f)}$ and Q represent $f(T)$. By the uniqueness result of Theorem 9.13, $P^{(f)} = Q$. This gives part (i); part (ii) now follows from Proposition 9.9. □

It is of some interest to revisit the case of a compact normal operator T. The spectrum of T is then a finite or infinite sequence $(\lambda_n)_{n \geqslant 1}$ with 0 as its only possible limit point. In Theorem 8.15 we have already shown that for any nonzero $\lambda \in \sigma(T)$, the orthogonal projection P_λ onto the corresponding eigenspace equals the spectral projection $P^{\{\lambda\}}$ of Theorem 6.23.

Proposition 9.21. *Let $T \in \mathscr{L}(H)$ be a compact normal operator and let P be its projection-valued measure. Then for all nonzero $\lambda \in \sigma(T)$,*

$$P_{\{\lambda\}} = P^{\{\lambda\}} = P_\lambda.$$

Proof Putting $\widetilde{P}_{\{\lambda\}} := P_\lambda$ and extending this definition by putting $\widetilde{P}_{\{0\}} := 0$ if 0 is not an eigenvalue, the spectral theorem for compact normal operators implies that $\widetilde{P}$ defines a projection-valued measure and

$$(Tx|x) = \sum_{\lambda \in \sigma(T)} \lambda (P_\lambda x|x) = \sum_{\lambda \in \sigma(T) \setminus \{0\}} \lambda (P_\lambda x|x)$$

$$= \int_{\sigma(T) \setminus \{0\}} \lambda \, d\widetilde{P}_x(\lambda) = \int_{\sigma(T)} \lambda \, d\widetilde{P}_x(\lambda).$$

Hence, by the uniqueness theorem for projection-valued measures, $\widetilde{P} = P$. □

Further results along this line are given in Problem 9.7 and Theorem 10.58.

We conclude this section with two famous results due to von Neumann.

Theorem 9.22 (von Neumann). *If $T \in \mathscr{L}(H)$ is a contraction, then for all polynomials p in one complex variable we have*

$$\|p(T)\| \leqslant \sup_{|z|=1} |p(z)|.$$

Proof First suppose that U is a unitary operator. Then $\sigma(U)$ is contained in the unit circle. Since unitaries are normal, by Theorem 9.14 we have $U = \int_{\sigma(U)} \lambda \, dP(\lambda)$, where P is the projection-valued measure of U. By Theorem 8.22 and the fact that $\sigma(U) \subseteq \mathbb{T}$,

$$\|p(U)\| = \sup_{z \in \sigma(U)} |p(z)| \leqslant \sup_{|z|=1} |p(z)|. \tag{9.6}$$

Next let T be a contraction. By the Sz.-Nagy dilation theorem (Theorem 8.36), T has a unitary dilation U, so that $T^n = J^* U^n J$ for some isometric operator J and all $n \in \mathbb{N}$. Then $p(T) = J^* p(U) J$, so $\|p(T)\| \leqslant \|p(U)\|$ and the result follows from (9.6). $\qquad\square$

As an application of the spectral theorem for normal operators we prove von Neumann's theorem on pairs of commuting selfadjoint operators. Two projection-valued measures $P : \mathscr{F} \to \mathscr{P}(H)$ and $P' : \mathscr{F} \to \mathscr{P}(H)$ are said to *commute* if P_F and $P'_{F'}$ commute for all $F, F' \in \mathscr{F}$.

Lemma 9.23. *Let $P^1, \ldots, P^k$ be commuting projection-valued measures on compact Hausdorff spaces $K_1, \ldots, K_k$, respectively. There exists a unique projection-valued measure P on $K = K_1 \times \cdots \times K_k$ such that*

$$P_{B_1 \times \cdots \times B_k} = P^1_{B_1} \circ \cdots \circ P^k_{B_k}$$

for all Borel sets $B_j \subseteq K_j$, $j = 1, \ldots, k$.

Proof For Borel sets $B_j \subseteq K_j$, $j = 1, \ldots, k$, and $f := \mathbf{1}_{B_1 \times \cdots \times B_k}$ define

$$\Phi(f) := P^1_{B_1} \circ \cdots \circ P^k_{B_k}.$$

We extend this definition by linearity to functions f on K which are linear combinations of Borel rectangles. Using the commutativity assumptions it is easily checked that this is well defined and that for such f we have

$$\|\Phi(f)\| \leqslant \|f\|_\infty.$$

We may use this to define, for functions $f \in C(K)$, a well-defined bounded operator $\Phi(f)$, by uniform approximation by simple functions of the above form. In the same way as in the proof of Theorem 9.14, there exists a projection-valued measure P on K such that (9.3) holds for all $f \in C(K)$ and $x \in H$, that is,

$$(\Phi(f)x|x) = \int_K f \, dP_x, \quad f \in C(K), \ x \in H.$$

This projection-valued measure has the desired properties. Its uniqueness can be proved using the method of Proposition 9.13. $\qquad\square$

Theorem 9.24 (von Neumann). *Two selfadjoint operators $T_1, T_2 \in \mathscr{L}(H)$ commute if* *and only if there exist a normal operator $S \in \mathscr{L}(H)$ and continuous functions $f_1, f_2 : \sigma(S) \to \mathbb{R}$ such that*

$$T_1 = f_1(S), \quad T_2 = f_2(S).$$

Proof The 'if' part follows from the multiplicativity of the Borel calculus of S. The point is to prove the 'only if' part. To this end let P^1 and P^2 denote the projection-valued measures of T_1 and T_2 on $\sigma(T_1)$ and $\sigma(T_2)$ respectively, and let P be the projection-valued measure on $K := \sigma(T_1) \times \sigma(T_2) \subseteq \mathbb{R}^2$ as in Lemma 9.23. Let $L = \{z \in \mathbb{C} : \mathrm{Re}\, z \in \sigma(T_1), \mathrm{Im}\, z \in \sigma(T_2)\} \subseteq \mathbb{C}$, that is, we identify K with a rectangle L in the complex plane. Under this identification, P induces a projection-valued measure on L, denoted by Q. The operator

$$S := T_Q = \int_L \lambda \, dQ(\lambda)$$

is normal. We will prove that $T_1 = f_1(S)$ and $T_2 = f_2(S)$ with $f_1(z) = \mathrm{Re}\, z$ and $f_2(z) = \mathrm{Im}\, z$. For all $x \in H$ we have, using that Q_x is supported on $\sigma(S)$,

$$(f_1(S)x|x) = \int_{\sigma(S)} f_1(\lambda) \, dQ_x(\lambda).$$

We claim that the image measure of Q_x under f_1 equals P_x^1. Indeed, for all Borel sets $B_1 \subseteq \sigma(T_1)$ we have

$$f_1(Q_x)(B_1) = Q_x(B_1 + i\sigma(T_2)) = (Q_{B_1 + i\sigma(T_2)}x|x)$$
$$= (P_{B_1 \times \sigma(T_2)}x|x) = (P_{B_1}^1 P_{\sigma(T_2)}^2 x|x) = (P_{B_1}^1 x|x) = P_x^1(B_1),$$

where we used that $P^2(\sigma(T_2)) = I$. So $f_1(Q_x) = P_x^1$ as claimed. It now follows that

$$\int_L f_1(\lambda) \, dQ_x(\lambda) = \int_{\sigma(T_1)} \mu \, dP_x^1(\mu) = (T_1 x|x).$$

Combining things we find that $(f_1(S)x|x) = (T_1 x|x)$. This being true for all $x \in H$, we conclude that $f_1(S) = T_1$. The identity $f_2(S) = T_2$ is proved similarly. $\qquad\square$

9.5 The Von Neumann Bicommutant Theorem

In this section we prove a result of fundamental importance in the theory of operator algebras, von Neumann's celebrated bicommutant theorem. We also start the proof, to be completed in the next chapter, of the identification of the bicommutant of a single self-adjoint operator on a separable Hilbert space as being precisely the bounded functional calculus of this operator.

We begin by introducing the relevant terminology.

Definition 9.25 (Commutant). The *commutant* of a subset $\mathscr{T} \subseteq \mathscr{L}(H)$ is the set

$$\mathscr{T}' := \{S \in \mathscr{L}(H) : ST = TS \text{ for all } T \in \mathscr{T}\}.$$

The *bicommutant* of a subset $\mathscr{T} \subseteq \mathscr{L}(H)$ is the set $\mathscr{T}'' := (\mathscr{T}')'$.

Definition 9.26 (Strong and weak operator topologies). The *strong operator topology on* $\mathscr{L}(H)$ is the smallest topology τ on $\mathscr{L}(H)$ with the property that the linear mappings $T \mapsto Tx$ are continuous for all $x \in H$. The *weak opera-*

John von Neumann, 1903–1957

tor topology on $\mathscr{L}(H)$ is the smallest topology τ on $\mathscr{L}(H)$ with the property that linear mappings $T \mapsto (Tx|y)$ are continuous for all $x, y \in H$.

Definition 9.26 has natural counterparts for $\mathscr{L}(X)$ with X a Banach space, but these will not be needed.

In the same way as was explained in Section 4.6 for the weak and weak* topologies, the strong operator topology is generated by the sets of the form

$$\{T \in \mathscr{L}(H) : \|(T - T_0)x\| < \varepsilon\}$$

with $\varepsilon > 0, x \in H$, and $T_0 \in \mathscr{L}(H)$, and likewise the weak operator topology is generated by the sets of the form

$$\{T \in \mathscr{L}(H) : |((T - T_0)x|y)| < \varepsilon\}$$

with $\varepsilon > 0$, $x, y \in H$, and $T_0 \in \mathscr{L}(H)$.

For every set $\mathscr{T} \subseteq \mathscr{L}(H)$, the commutant $\mathscr{T}'$ is closed in the weak operator topology. To see this, suppose that $T_0 \notin \mathscr{T}'$. Then there exist an operator $S \in \mathscr{T}$, vectors $x, y \in H$, and a number $\delta > 0$ such that $|(T_0Sx|y) - (ST_0x|y)| = \delta$. The set

$$U := \{T \in \mathscr{L}(H) : |((T_0 - T)Sx|y)| < \delta/2, |((T_0 - T)x|S^\star y)| < \delta/2\}$$

is open in the weak operator topology, contains T_0, and every $T \in U$ fails to commute with S. It follows that $U \cap \mathscr{T}' = \varnothing$.

Recall that a *subalgebra* of $\mathscr{L}(H)$ is a subspace of $\mathscr{L}(H)$ closed under taking compositions. A *⋆-subalgebra* of $\mathscr{L}(H)$ is a subalgebra of $\mathscr{L}(H)$ closed under taking Hilbert space adjoints. A subalgebra is said to be *unital* if it contains the identity operator.

Theorem 9.27 (von Neumann bicommutant theorem). *For a unital ⋆-subalgebra $\mathscr{A}$ of $\mathscr{L}(H)$ the following assertions are equivalent:*

(1) $\mathscr{A} = \mathscr{A}''$;
(2) $\mathscr{A}$ is weakly closed;
(3) $\mathscr{A}$ is strongly closed.

A $\star$-subalgebra $\mathscr{A}$ of $\mathscr{L}(H)$ which is closed with respect to the operator norm is called a $C^\star$-*algebra*. This is not the commonly used definition (the standard definition is mentioned in the Notes to Chapter 7), but one of the main theorems on the structure of $C^\star$-algebras establishes that this definition is equivalent to the standard one. A unital $\star$-subalgebra $\mathscr{A}$ of $\mathscr{L}(H)$ satisfying the equivalent conditions of Theorem 9.27 is called a *von Neumann algebra*.

Proof The implications (1)$\Rightarrow$(2)$\Rightarrow$(3) are clear.

(3)$\Rightarrow$(1): We proceed in two steps.

Step 1 – Fix $x_0 \in H$ and let P denote the orthogonal projection in H onto the closure Y of the subspace $\{Tx_0 : T \in \mathscr{A}\}$. Since $I \in \mathscr{A}$ we have $Px_0 = x_0$.

We claim that Y is invariant under all $T \in \mathscr{A}$: for if $T \in \mathscr{A}$ and $y \in Y$, say $y = \lim_{n\to\infty} T_n x_0$ with $T_n \in \mathscr{A}$ for all $n \geqslant 1$, then $Ty = \lim_{n\to\infty} TT_n x_0$ with $TT_n \in \mathscr{A}$ for all $n \geqslant 1$, and therefore $Ty \in Y$. Similarly $Y^\perp$ is invariant under all $T \in \mathscr{A}$: for if $T \in \mathscr{A}$ and $x \in Y^\perp$, then for all $y \in Y$ we have $(Tx|y) = (x|T^\star y) = 0$ since $T^\star \in \mathscr{A}$ and therefore $T^\star y \in Y$.

By the claim, for all $T \in \mathscr{A}$ and $x \in H$ we have $TPx \in Y$ and $T(I-P)x \in Y^\perp$, and therefore
$$TPx = PTPx = PT(Px + (I-P)x) = PTx, \quad x \in H.$$
We conclude that $TP = PT$ for all $T \in \mathscr{A}$, that is, $P \in \mathscr{A}'$.

Let $T_0 \in \mathscr{A}''$ be fixed. Then $PT_0 = T_0 P$ since $P \in \mathscr{A}'$, and this implies that $T_0 x_0 = T_0 P x_0 = P T_0 x_0 \in Y$. By the definition of Y this means that for all $\varepsilon > 0$ there exists an element $T \in \mathscr{A}$ such that
$$\|T_0 x_0 - Tx_0\| < \varepsilon.$$

Step 2 – To show that every strongly open set containing the operator $T_0 \in \mathscr{A}''$ intersects $\mathscr{A}$ it suffices to show that, for any choice of $x_1, \ldots, x_k \in H$ and $\varepsilon > 0$, there exists $T \in \mathscr{A}$ such that
$$\|(T_0 - T)x_j\| < \varepsilon, \quad j = 1, \ldots, k. \tag{9.7}$$
In what follows we set $\mathbf{x_0} := (x_1, \ldots, x_k)$.

For $S \in \mathscr{L}(H)$ let $\rho(S) \in \mathscr{L}(H^k)$ be given by
$$\rho(S)(h_1, \ldots, h_k) := (Sh_1, \ldots, Sh_k).$$
We claim that
$$\rho(T_0) \in (\rho(\mathscr{A}))''.$$

Indeed, suppose that $\mathbf{S} = (S_{ij})_{i,j=1}^k \in (\rho(\mathscr{A}))'$. This means that $\mathbf{S}\rho(T)\mathbf{y} = \rho(T)\mathbf{S}\mathbf{y}$ for all $T \in \mathscr{A}$ and $\mathbf{y} = (y_1, \ldots, y_k) \in H^k$, that is,

$$\sum_{j=1}^k S_{ij}Ty_j = \sum_{j=1}^k TS_{ij}y_j, \quad i = 1, \ldots, k, \ y_1, \ldots, y_k \in H,$$

which implies that for all $1 \leqslant i, j \leqslant k$ we have $S_{ij} \in \{T\}'$ for all $T \in \mathscr{A}$, so $S_{ij} \in \mathscr{A}'$. But this clearly implies that $\rho(T_0)$ commutes with $\mathbf{S}$.

We now apply Step 1, with H, $\mathscr{A}$, and T_0 replaced by H^k, $\rho(\mathscr{A})$, and $\rho(T_0)$ respectively. This gives an operator $T \in \mathscr{A}$ such that $\|(\rho(T_0) - \rho(T))\mathbf{x_0}\| < \varepsilon$, that is,

$$\sum_{j=1}^k \|(T_0 - T)x_j\|^2 < \varepsilon^2.$$

In particular, (9.7) follows from this.

We have shown that every strongly open set containing an element from $\mathscr{A}''$ intersects $\mathscr{A}$. This means that $\mathscr{A}$ is strongly dense in $\mathscr{A}''$. Since $\mathscr{A}$ was assumed to be strongly closed, it follows that $\mathscr{A} = \mathscr{A}''$. $\qquad\square$

The next theorem establishes a beautiful connection between bicommutants and the bounded functional calculus.

Theorem 9.28 (von Neumann, bicommutant of a selfadjoint operator). *Let H be separable and let $T \in \mathscr{L}(H)$ be selfadjoint. Then*

$$\{T\}'' = \{f(T) : f \in B_{\mathrm{b}}(\sigma(T))\}.$$

Proof The inclusion '$\supseteq$' holds for normal operators T and arbitrary Hilbert spaces H, and is proved as follows. The Fuglede–Putnam–Rosenblum theorem (Theorem 8.18) implies that $T^* \in \{T\}''$. It follows that every operator of the form $p(T, T^*)$, with p a polynomial in z and $\bar{z}$, is contained in $\{T\}''$. By the Stone–Weierstrass theorem, the same is true for every function $f \in C(\sigma(T))$. By pointwise approximation from below, the result extends to indicator functions $f = \mathbf{1}_U$ with $U \subseteq \sigma(T)$ relatively open. For all $x \in H$, the outer regularity of P_x implies that $P_B x = \lim_{n\to\infty} P_{U_n} x$ whenever the relatively open sets $U_1 \supseteq U_2 \supseteq \cdots \supseteq B$ satisfy $\lim_{n\to\infty} P_x(U_n \setminus B) = 0$. Applying this to Sx and x with $S \in \{T\}'$, as a consequence we obtain

$$P_B S x = \lim_{n\to\infty} P_{U_n} S x = \lim_{n\to\infty} S P_{U_n} x = S P_B x, \quad x \in X,$$

which shows that $P_B \in \{S\}'$ for all $S \in \{T\}'$, that is, $P_B \in \{T\}''$. This, in turn, implies that if $f \in B_{\mathrm{b}}(\sigma(T))$ and $S \in \{T\}'$, then, upon approximating f with simple functions,

$$Sf(T) = S\left(\int_{\sigma(T)} f \, \mathrm{d}P\right) = \left(\int_{\sigma(T)} f \, \mathrm{d}P\right)S = f(T)S$$

and therefore $f(T) \in \{T\}''$ for all $f \in B_{\mathrm{b}}(\sigma(T))$.

The proof of the inclusion '$\subseteq$' will be given in the next chapter, as it uses the spectral theory for unbounded selfadjoint operators developed in that chapter. $\qquad\square$

9.6 Application to Orthogonal Polynomials

In this final section we present an interesting application of the spectral theorem to orthogonal polynomials. Let μ be a Borel measure on the real line satisfying the condition

$$\int_{-\infty}^{\infty} |x|^n \, d\mu(x) < \infty, \quad n \in \mathbb{N}. \tag{9.8}$$

We further assume that the support of μ is not a finite set. Suppose that $(p_n)_{n \in \mathbb{N}}$ is a sequence of polynomials with real coefficients satisfying the following two assumptions:

(i) for all $n \in \mathbb{N}$ we have $\deg(p_n) = n$;
(ii) for all $m, n \in \mathbb{N}$ with $m \neq n$ we have the orthogonality relation

$$\int_{-\infty}^{\infty} p_m(x) p_n(x) \, d\mu(x) = 0. \tag{9.9}$$

For $n = 0$, (i) is understood to mean that $p_0 \neq 0$. By linearity, (i) and (ii) imply

$$\int_{-\infty}^{\infty} x^m p_n(x) \, d\mu(x) = 0 \quad \text{whenever } m < n.$$

Proposition 9.29. *Let μ be a Borel measure on the real line with the properties stated above. For any sequence of polynomials $(p_n)_{n \in \mathbb{N}}$ with real coefficients satisfying the conditions* (i) *and* (ii)*, there exist real numbers A_n, B_n, C_n ($n \in \mathbb{N}$) satisfying $C_0 = 0$ and $A_{n-1}C_n > 0$ ($n \geqslant 1$) such that, with $p_{-1} \equiv 0$,*

$$x p_n = A_n p_{n+1} + B_n p_n + C_n p_{n-1}, \quad n \in \mathbb{N}.$$

Proof Since $x p_n$ is a polynomial of degree $n + 1$, it is of the form $x p_n = \sum_{j=0}^{n+1} c_{j,n} p_j$ with all coefficients $c_{j,n}$ real-valued. Setting $c_{j,n} := 0$ for $j \geqslant n + 2$, (9.9) implies

$$\int_{-\infty}^{\infty} x p_n(x) p_m(x) \, d\mu(x) = c_{m,n} N_m, \quad \text{where } N_m = \int_{-\infty}^{\infty} p_m^2(x) \, d\mu(x).$$

Since the support of μ is not finite we have $N_m \neq 0$ for all $m \in \mathbb{N}$. For $n \in \mathbb{N}$ the polynomial $p_n(x)$ is orthogonal to $x p_m(x)$ for all $m = 0, 1, \ldots, n - 2$. This forces $c_{m,n} = 0$ for all $m = 0, 1, \ldots, n - 2$. This, in turn, implies

$$x p_n = c_{n+1,n} p_{n+1} + c_{n,n} p_n + c_{n-1,n} p_{n-1}, \quad n \in \mathbb{N}.$$

This gives the three point recurrence relation with $A_n = c_{n+1,n}$, $B_n = c_{n,n}$, and $C_n =$

$c_{n-1,n}$, with convention $C_0 = c_{-1,0} = 0$, say. Since the degree of xp_n is $n+1$ we have $A_n \neq 0$. Also, for $n \geqslant 1$,

$$
\begin{aligned}
A_{n-1}N_n = c_{n,n-1}N_n &= \int_{-\infty}^{\infty} xp_n(x)p_{n-1}(x)\,\mathrm{d}\mu(x) \\
&= \int_{-\infty}^{\infty} xp_{n-1}(x)p_n(x)\,\mathrm{d}\mu(x) = c_{n-1,n}N_{n-1} = C_nN_{n-1},
\end{aligned}
\tag{9.10}
$$

and therefore $A_{n-1}C_n > 0$ for $n \geqslant 1$. $\qquad\square$

The polynomial p_n has norm one in $L^2(\mathbb{R}, \mu)$ if and only if $N_n = 1$. Hence if the p_n are orthonormal, (9.10) gives $0 \neq A_{n-1} = C_n$ for all $n \geqslant 1$. As an application of the spectral theorem we show that, conversely, for every sequence of polynomials satisfying the three point recurrence relation subject to the conditions $0 \neq A_{n-1} = C_n$ for all $n \geqslant 1$, and satisfying the additional boundedness assumption

$$
\sup_{n\in\mathbb{N}} \max\{|A_n|, |B_n|, |C_n|\} < \infty,
$$

there exists a finite Borel measure μ on the real line with respect to which the polynomials are orthonormal.

Theorem 9.30 (Three-point recurrence). *For every sequence $(p_n)_{n\in\mathbb{N}}$ of polynomials satisfying the three point recurrence relation*

$$
xp_n = A_np_{n+1} + B_np_n + C_np_{n-1}, \quad n \in \mathbb{N},
$$

with $p_{-1} \equiv 0$, subject to the conditions $0 \neq A_{n-1} = C_n$ for all $n \geqslant 1$ and

$$
\sup_{n\in\mathbb{N}} \max\{|A_n|, |B_n|, |C_n|\} < \infty,
$$

there exists a finite Borel measure μ on the real line which satisfies (9.8) and such that the sequence $(p_n)_{n\in\mathbb{N}}$ is orthonormal in $L^2(\mathbb{R}, \mu)$.

Proof   Without loss of generality we may assume $p_0 \equiv 1$.
On the Hilbert space $\ell^2(\mathbb{N})$ we consider the bounded operator

$$
Te_n := A_ne_{n+1} + B_ne_n + C_ne_{n-1}, \quad n \in \mathbb{N},
$$

with the understanding that $Te_0 := A_0e_1 + B_0e_0$; boundedness of T is a consequence of the boundedness assumption on A_n, B_n, and C_n. Since $A_{n-1} = C_n$, from

$$
\begin{aligned}
(e_n|T^\star e_m) &= A_n\delta_{m,n+1} + B_n\delta_{m,n} + A_{n-1}\delta_{m,n-1} \\
&= A_{m-1}\delta_{m-1,n} + B_m\delta_{m,n} + A_m\delta_{m+1,n} = (e_n|Te_m)
\end{aligned}
$$

(which is checked by hand also to hold if $n = 0$ or $m = 0$) we see that T is selfadjoint. Let P be its projection-valued measure and define $\mu := P_{e_0}$. Then μ is a finite measure supported on $\sigma(T)$, which is a compact set since T is bounded.

Define a linear operator from $\ell_{00}^2(\mathbb{N})$, the span of the vectors e_n in $\ell^2(\mathbb{N})$, into $L^2(\mathbb{R}, \mu)$ by setting $Ue_n := p_n$ for $n \in \mathbb{N}$. We further define the bounded operator $M : L^2(\mathbb{R}, \mu) \to L^2(\mathbb{R}, \mu)$ by $Mf(x) := xf(x)$; the boundedness of M follows from the fact that μ is supported in a bounded interval I. From

$$UTe_n = A_n p_{n+1} + B_n p_n + A_{n-1} p_{n-1} = x p_n = MUe_n$$

we see that $UT = MU$ as linear operators from $\ell_{00}^2(\mathbb{N})$ to $L^2(\mathbb{R}, \mu)$. By a simple induction argument, $UT^n = M^n U$ for all $n \in \mathbb{N}$.

We claim that U extends to a unitary operator from $\ell^2(\mathbb{N})$ to $L^2(\mathbb{R}, \mu)$. First we check that U has dense range. By the Stone–Weierstrass theorem, the functions $\varepsilon_k(x) := e^{2\pi ikx/|I|}$, $k \in \mathbb{Z}$, can be uniformly approximated by polynomials, and the injectivity of the Fourier transform of finite Borel measures (Theorem 5.31) implies that the span of the functions ε_k, $k \in \mathbb{Z}$, is dense in $L^2(I, \mu)$, hence in $L^2(\mathbb{R}, \mu)$. These observations imply that U has dense range. Next, from $UT^n e_0 = M^n p_0 = x^n$ and $\mu = P_{e_0}$ we obtain

$$(UT^m e_0 | UT^n e_0) = (x^m | x^n) = \int_I x^{m+n} \, dP_{e_0}(x) = (T^{m+n} e_0 | e_0) = (T^m e_0 | T^n e_0)$$

using the functional calculus of T. The span of the vectors $T^n e_0$, $n \in \mathbb{N}$, being dense in $\ell^2(\mathbb{N})$, this concludes the proof that U extends to a unitary operator. It now follows from

$$(p_m | p_n) = (Ue_m | Ue_n) = (e_m | e_n) = \delta_{mn}$$

that the polynomials p_n are orthonormal in $L^2(\mathbb{R}, \mu)$. $\qquad\square$

Example 9.31. We have already encountered two examples of orthogonal polynomials.

(i) The Hermite polynomials H_n, $n \in \mathbb{N}$, have been introduced in Section 3.5.b. They are orthogonal with respect to the Gaussian measure $\frac{1}{\sqrt{2\pi}} \exp\left(-\frac{1}{2}x^2\right) dx$ on the real line and satisfy the three point recurrence relation $H_0(x) = 1$, $H_1(x) = x$, and

$$H_{n+2}(x) = xH_{n+1}(x) - (n+1)H_n(x), \quad n \in \mathbb{N}.$$

(ii) The monic Laguerre polynomials L_n, $n \in \mathbb{N}$, have been introduced in Problem 3.14 as a scaled version of the Laguerre polynomials. They are orthogonal with respect to the measure $\mathbf{1}_{\mathbb{R}_+}(x) \exp(-x)$ on the real line and satisfy the three point recurrence relation $L_0(x) = 1$, $L_1(x) = x - 1$, and

$$L_{n+2}(x) = (x - 2n + 3)L_{n+1}(x) - (n+1)^2 L_n(x), \quad n \in \mathbb{N}.$$

Problems

9.1 Let $T \in \mathscr{L}(H)$ be a compact normal operator with spectral decomposition

$$T = \sum_{n \geqslant 1} \lambda_n P_n,$$

where $(\lambda_n)_{n \geqslant 1}$ is the (finite or infinite) sequence of eigenvalues of T. Prove that if $f : \sigma(T) \to \mathbb{C}$ is a bounded function, then for all $x \in H$ we have

$$f(T)x = \sum_{n \geqslant 1} f(\lambda_n) P_n x$$

with convergence in H. Does the sum $\sum_{n \geqslant 1} f(\lambda_n) P_n$ converge to $f(T)$ in $\mathscr{L}(H)$?

9.2 Let $T \in \mathscr{L}(H)$ be a compact normal operator. Give a direct proof (that is, without invoking Theorem 7.11) of the following two statements:

(a) If λ is a nonzero element of $\sigma(T)$, then λ is an eigenvalue.
  Hint: Use the fact that $\mathsf{R}(\lambda - T)$ is closed (Lemma 7.9) to establish the equivalences $\mathsf{N}(\lambda - T) = \{0\} \Leftrightarrow \mathsf{N}(\overline{\lambda} - T^*) = \{0\} \Leftrightarrow \mathsf{R}(\lambda - T) = H$.

(b) If T has infinitely many distinct eigenvalues λ_n, then $\lim_{n \to \infty} \lambda_n = 0$.
  Hint: Choose eigenvectors $Th_n = \lambda_n h_n$ and define $H_0 := \{0\}$ and $H_n :=$ span$\{h_1, \ldots, h_n\}$ for $n = 1, 2, \ldots$ Show that $H_{n-1} \subsetneq H_n$ and choose norm one vectors $x_n \in H_n \cap H_{n-1}^{\perp}$. Show that if $n > m$, then $\|T x_m - T x_n\| \geqslant |\lambda_n|$.

9.3 Let $T \in \mathscr{L}(H)$ be a selfadjoint operator with projection-valued measure P. Prove the following results.

(a) If $t \in \mathbb{R}$, then for all $x \in H$ we have

$$\lim_{\varepsilon \to 0} i\varepsilon R(t + i\varepsilon, T)x = P_{\{t\}}x.$$

(b) If $a, b \in \mathbb{R}$ with $a < b$, then for all $x \in H$ we have *Stone's formula*

$$\lim_{\varepsilon \downarrow 0} \frac{1}{2\pi i} \int_a^b R(t - i\varepsilon, T)x - R(t + i\varepsilon, T)x \, dt = \frac{1}{2}(P_{[a,b]}x + P_{(a,b)}x).$$

  Hint: Show first that

$$\lim_{\varepsilon \downarrow 0} \frac{1}{2\pi i} \int_a^b \frac{1}{\lambda - i\varepsilon - t} - \frac{1}{\lambda + i\varepsilon - t} \, dt = \begin{cases} 0, & t \notin [a,b], \\ \frac{1}{2}, & t \in \{a,b\}, \\ 1, & t \in (a,b). \end{cases}$$

9.4 Let $T_1, T_2 \in \mathscr{L}(H)$ be selfadjoint operators with projection-valued measures $P^{(1)}$ and $P^{(2)}$, respectively. Prove that the following assertions are equivalent:

(1) the projection-valued measures $P^{(1)}$ and $P^{(2)}$ commute, that is, for all Borel sets B_1, B_2 in $\mathbb{R}$ we have

$$P^{(1)}_{B_1} P^{(2)}_{B_2} = P^{(2)}_{B_2} P^{(1)}_{B_1};$$

(2) the resolvents of T_1 and T_2 commute, that is, for all $\lambda_1 \in \rho(T_1)$ and $\lambda_2 \in \rho(T_2)$ we have

$$R(\lambda_1, T_1)R(\lambda_2, T_2) = R(\lambda_2, T_2)R(\lambda_1, T_1);$$

(3) for all $t_1, t_2 \in \mathbb{R}$ we have

$$\exp(it_1 T_1)\exp(it_2 T_2) = \exp(it_2 T_2)\exp(it_1 T_1).$$

Hint: For the implication (2)$\Rightarrow$(1) use the results of the preceding problem; for implication (3)$\Rightarrow$(1) write $\exp(it_1 T_1)$ and $\exp(it_2 T_2)$ as spectral integrals with respect to $P^{(1)}$ and $P^{(2)}$ and use the properties of the Fourier–Plancherel transform to deduce that for all $f, g \in \mathscr{F}^2(\mathbb{R})$ we have $\hat{f}(T_1)\hat{g}(T_2) = \hat{g}(T_2)\hat{f}(T_1)$ and hence, for all $f, g \in \mathscr{F}^2(\mathbb{R})$,

$$f(T_1)g(T_2) = g(T_2)f(T_1).$$

By approximation with functions in $\mathscr{F}^2(\mathbb{R})$, deduce that

$$P^{(1)}_{(a_1,b_1)} P^{(2)}_{(a_2,b_2)} = P^{(2)}_{(a_2,b_2)} P^{(1)}_{(a_1,b_1)}$$

for all $a_1, a_2, b_1, b_2 \in \mathbb{R}$ with $a_1 < b_1$ and $a_2 < b_2$.

9.5 Let $T \in \mathscr{L}(H)$ be a normal operator with projection-valued measure P. Let B be a Borel subset of $\sigma(T)$.

(a) Show that T leaves the range of P_B invariant.
(b) Show that $\sigma(T|_{R(P_B)}) \subseteq \overline{B}$.

9.6 Prove that if $T \in \mathscr{L}(H)$ is normal, then

$$\|T\| = \sup\{|(Tx|x)| : \|x\| = 1\}.$$

Hint: Fix $\lambda_0 \in \sigma(T)$ with $|\lambda_0| = \|T\|$ (why does such λ_0 exist?) and $\varepsilon > 0$, and consider the projection $P_{B(\lambda_0;\varepsilon)}$, where P is the projection-valued measure of T. Show that if $x \in B(\lambda_0;\varepsilon)$ has norm one, then $|(Tx|x)| \geqslant \|T\| - \varepsilon$.

9.7 Let $T \in \mathscr{L}(H)$ be a normal operator with projection-valued measure P.

(a) Show that $N(T) = R(P_{\{0\}})$.
Hint: For the inclusion $\subseteq$, write $\sigma(T) \setminus \{0\}$ as a countable union of Borel sets B_n, each of which has the property that $\inf\{|\lambda| : \lambda \in B_n\} > 0$, and consider

$$f_n(z) = \begin{cases} 1/f(z), & z \in B_n, \\ 0, & z \in \sigma(T) \setminus B_n. \end{cases}$$

(b) Conclude that if $\lambda \in \sigma(T)$ is an eigenvalue, then $P_{\{\lambda\}}$ equals the orthogonal projection P_λ onto the eigenspace E_λ.

(c) Conclude that if $\lambda \in \sigma(T)$ is an isolated point, then λ is eigenvalue and $P_{\{\lambda\}} = P_\lambda$ equals the spectral projection associated with $\{\lambda\}$.

9.8 Show that the space $L^\infty(\Omega, P)$ introduced in Definition 9.10 is a Banach space.

9.9 Prove the following two claims made in Example 9.16:

(a) The position operator X on $H = L^2(0,1)$ defined by

$$Xf(x) := xf(x), \quad x \in (0,1), \ f \in L^2(0,1),$$

has spectrum $\sigma(X) = [0,1]$.

(b) The projection-valued measure of X is given by $P_B f = \mathbf{1}_B f$ for all Borel subsets B of $[0,1]$ and $f \in L^2(0,1)$.

9.10 Find the projection-valued measures of the following unitary operators:

(a) the right shift on $\ell^2(\mathbb{Z})$;

(b) translation over t on $L^2(\mathbb{R})$;

(c) rotation over θ on $L^2(\mathbb{T})$;

(d) the Fourier transform on $L^2(\mathbb{R}^d)$.

Hint: For parts (c) and (d) revisit Problems 8.14 and 6.9, respectively.

9.11 Let $k \in L^2((0,1) \times (0,1))$ satisfy $k(s,t) = \overline{k(t,s)}$ almost everywhere. Find the projection-valued measure of the selfadjoint integral operator T on $L^2(0,1)$,

$$Tf(t) := \int_0^1 k(t,s)f(s)\,\mathrm{d}s, \quad f \in L^2(0,1).$$

9.12 Let $T \in \mathcal{L}(H)$ be selfadjoint.

(a) Show that there exist positive operators T^+ and T^- such that $T = T^+ - T^-$. Hint: Consider the functions $f^+(t) := t^+$ and $f^-(t) := t^-$.

(b) Show that these operators are unique if we also ask that $T^+ T^- = T^- T^+ = 0$.

9.13 Prove that if $U \in \mathcal{L}(H)$ is unitary, there exists a selfadjoint operator T such that $U = e^{iT}$ and $\sigma(T) \subseteq [-\pi, \pi]$. Is this operator T unique?

Hint: Write $U = \int_{\sigma(T)} \lambda \, \mathrm{d}P(\lambda)$ as in Theorem 9.14. Find a projection-valued measure Q on $[0,1]$ whose image under $t \mapsto e^{it}$ is P.

9.14 This problem outlines another proof of the spectral theorem for normal operators.

(a) Explain how the proof of the spectral theorem for normal operators simplifies for selfadjoint operators.

(b) Deduce the spectral theorem for normal operators T from the selfadjoint case by considering the selfadjoint operators $\frac{1}{2}(T + T^\star)$ and $\frac{1}{2i}(T - T^\star)$, and applying Lemma 9.23 to their projection-valued measures.

10

The Spectral Theorem for Unbounded Normal Operators

Up to this point we have been dealing exclusively with bounded operators. In order to make the functional analytic apparatus applicable to the study of partial differential equations we need to accommodate differential operators into the theory. This leads to the notion of an unbounded operator as a linear operator defined only on a suitable subspace, the domain of the operator. Of special interest are unbounded selfadjoint and normal operators, and the main goal of this chapter is to extend the spectral theorems of the preceding chapter to these classes of operators.

10.1 Unbounded Operators

Throughout this chapter, X and Y are Banach spaces.

Definition 10.1 (Linear operators). A *linear operator* from X to Y is a pair $(A, D(A))$, where $D(A)$ is a subspace of X and $A : D(A) \to Y$ is a linear operator. The subspace $D(A)$ is called the *domain* of A. A linear operator is *densely defined* when $D(A)$ is a dense subspace of X.

When no confusion is likely to arrive we omit the domain from the notation and write A instead of $(A, D(A))$.

It is perfectly allowable that $D(A) = X$, so in particular every bounded operator $A : X \to Y$ is a linear operator in the sense of the above definition. More generally it may happen that there exists a constant $C \geqslant 0$ such that $\|Ax\| \leqslant C\|x\|$ for all $x \in D(A)$. In this situation, A admits a unique extension to a bounded operator (of norm at most C) defined on the closure of $D(A)$. The interest in the above definition arises from the fact that many interesting examples of *unbounded* linear operators exist, that is, linear operators for which such a constant C does not exist. Typical examples, treated in more detail below, include differential operators and multiplication operators with unbounded multipliers.

The terms 'linear operators' and 'unbounded operators' are often used interchangeably. With the latter terminology, however, it becomes somewhat ambiguous whether bounded operators are to be considered as special cases of unbounded operators. To avoid such trivial issues we generally prefer the terminology 'linear operator', which is neutral in this respect.

10.1.a Closed Operators

A linear operator from X to Y is bounded if and only if its graph is closed in $X \times Y$, the 'if' part being the content of the closed graph theorem. This motivates the following definition.

Definition 10.2 (Closed operators). A linear operator A from X to Y is called *closed* when its *graph*

$$G(A) := \{(x, Ax) : x \in D(A)\}$$

is closed in $X \times Y$.

Every bounded operator from X to Y is closed, and by the closed graph theorem a closed operator with domain $D(A) = X$ is bounded.

If A is a linear operator with domain $D(A)$, then A is bounded (in fact, contractive) as an operator defined on the normed space $D(A)$ endowed with the *graph norm*

$$\|x\|_{D(A)} := \|x\| + \|Ax\|, \quad x \in D(A).$$

This follows from the trivial inequality

$$\|Ax\| \leqslant \|x\| + \|Ax\| = \|x\|_{D(A)}.$$

The following proposition gives a necessary and sufficient condition for closedness in terms of the graph norm.

Proposition 10.3. *A linear operator is closed if and only if its domain is a Banach space with respect to its graph norm.*

Proof 'If': Suppose that the domain $D(A)$ of the linear operator A is complete with respect to its graph norm. To prove that A is closed we must show that its graph is closed, or equivalently, sequentially closed, in $X \times Y$. Let $(x_n, Ax_n)_{n \geqslant 1}$ be a sequence converging to some limit (x, y) in $X \times Y$. We must check that (x, y) belongs to the graph of A. By the properties of product norms, we have $x_n \to x$ in X and $Ax_n \to y$ in Y. In particular, the sequences $(x_n)_{n \geqslant 1}$ and $(Ax_n)_{n \geqslant 1}$ are Cauchy in X and Y respectively. Then the sequence $(x_n)_{n \geqslant 1}$ is Cauchy in $D(A)$ since

$$\|x_n - x_m\|_{D(A)} = \|x_n - x_m\| + \|Ax_n - Ax_m\| \to 0 \quad \text{as } m, n \to \infty.$$

By the completeness of $D(A)$, the sequence $(x_n)_{n\geqslant 1}$ converges in $D(A)$, say $x_n \to x'$ in $D(A)$. This means that $x_n \to x'$ in X and $Ax_n \to Ax'$ in Y. Comparing limits we find that $x' = x$ and $Ax' = y$. This means that $(x,y) = (x',Ax')$ belongs to $D(A)$.

'Only if': Assume now that A is closed and that $(x_n)_{n\geqslant 1}$ is a Cauchy sequence in $D(A)$, so $(x_n)_{n\geqslant 1}$ is a Cauchy sequence in X and $(Ax_n)_{n\geqslant 1}$ is a Cauchy sequence in Y. Then $((x_n, Ax_n))_{n\geqslant 1}$ is a Cauchy sequence in $X \times Y$. This Cauchy sequence is contained in the graph of A. This graph is closed by our assumption, and since closed subspaces of Banach spaces are Banach spaces, this Cauchy sequence converges in $X \times Y$ to a limit contained in the graph of A, say $(x_n, Ax_n) \to (x, Ax)$ in $X \times Y$. This implies that $x_n \to x$ in X and $Ax_n \to Ax$ in Y, which is the same as saying that $x_n \to x$ in $D(A)$ with respect to the graph norm. We have thus shown that every Cauchy sequence in $D(A)$ is convergent. $\qquad\square$

The following proposition gives a convenient sequential criterion for a linear operator to be closed. The proof is already implicitly contained in the proof of Proposition 10.3 and the reader is invited to make this explicit.

Proposition 10.4. *A linear operator A with domain $D(A)$ is closed if and only if the following holds: whenever $x_n \to x$ in X, with $x_n \in D(A)$ for all n, and $Ax_n \to y$ in Y, then $x \in D(A)$ and $Ax = y$.*

This criterion is used to prove closedness in the next two examples.

Example 10.5. The derivative operator, as a linear operator in $C[0,1]$ with domain $C^1[0,1]$, is densely defined and closed. The density of $C^1[0,1]$ in $C[0,1]$ is clear (by the Weierstrass approximation theorem we can even approximate continuous functions with polynomials). To prove closedness, suppose that $f_n \to f$ in $C[0,1]$, with $f_n \in C^1[0,1]$ for all n, and $f_n' \to g$ in $C[0,1]$. We must prove that $f \in C^1[0,1]$ and $f' = g$. For all $x \in [0,1]$ we have

$$f(x) - f(0) = \lim_{n\to\infty} f_n(x) - f_n(0) = \lim_{n\to\infty} \int_0^x f_n'(y)\,\mathrm{d}y = \int_0^x g(y)\,\mathrm{d}y,$$

using the uniform convergence of f_n' to g in the last step. The right-hand side is a continuously differentiable function, with derivative g. This proves that $f \in C^1[0,1]$ and $f' = g$.

An analogous result holds for weak derivatives in $L^p(D)$, where D is an open subset of $\mathbb{R}^d$; see Section 11.1.a.

Example 10.6. Let $(\Omega, \mathscr{F}, \mu)$ be a measure space and let $1 \leqslant p \leqslant \infty$. Given a measurable function $m : \Omega \to \mathbb{K}$ we may define

$$D(A_m) := \{f \in L^p(\Omega) : mf \in L^p(\Omega)\},$$
$$A_m f := mf, \quad f \in D(A_m).$$

We claim that the linear operator A_m is closed. Moreover, if $1 \leqslant p < \infty$, then A_m is densely defined.

To prove closedness, let $f_n \to f$ in $L^p(\Omega)$ with each f_n in $\mathsf{D}(A_m)$ and $A_m f_n \to g$ in $L^p(\Omega)$. We must show that $f \in \mathsf{D}(A_m)$ and $A_m f = g$. By passing to a subsequence we may assume that both convergences also hold pointwise μ-almost everywhere. Then, for μ-almost all $\omega \in \Omega$,

$$g(\omega) = \lim_{n \to \infty} A_m f_n(\omega) = \lim_{n \to \infty} m(\omega) f_n(\omega) = m(\omega) f(\omega).$$

This proves that $mf \in L^p(\Omega)$ and $mf = g$ μ-almost everywhere, hence as elements of $L^p(\Omega)$. Equivalently, this says that $f \in \mathsf{D}(A_m)$ and $A_m f = g$.

Now let $1 \leqslant p < \infty$. By dominated convergence, $\lim_{N \to \infty} \mathbf{1}_{\{|m| \leqslant N\}} f = f$ for all $f \in L^p(\Omega)$, with convergence in the norm of $L^p(\Omega)$. Since $\mathbf{1}_{\{|m| \leqslant N\}} f \in \mathsf{D}(A_m)$, this shows that $\mathsf{D}(A_m)$ is dense in $L^p(\Omega)$.

Example 10.7. If A is a closed operator and B is bounded, then the operator $A + B$ with domain $\mathsf{D}(A + B) := \mathsf{D}(A)$ defined by $(A + B)x := Ax + Bx$ for $x \in \mathsf{D}(A)$ is closed. The easy proof is left as an exercise.

Example 10.8. If A is an injective closed operator (in particular, if A is an injective bounded operator), its inverse A^{-1}, with domain $\mathsf{D}(A^{-1}) = \mathsf{R}(A)$, is closed. This is immediate by noting that the graph of A^{-1} equals

$$\{(y, A^{-1}y) : y \in \mathsf{D}(A^{-1})\} = \{(Ax, x) : x \in \mathsf{D}(A)\}$$

and that the latter is closed in $Y \times X$ since $\{(x, Ax) : x \in \mathsf{D}(A)\}$ is closed in $X \times Y$.

Further examples will be given later on. We highlight two of them:

Example 10.9. The adjoint A^* of a densely defined linear operator A acting between Banach spaces is closed by Proposition 10.18. Likewise, by Proposition 10.22, the Hilbertian adjoint $A^\star$ of a densely defined linear operator A acting between Hilbert spaces is closed.

Example 10.10. Generators of C_0-semigroups are closed by Proposition 13.4.

It frequently happens that linear operators are initially defined on a 'too small' domain to be closed, but can be extended to a closed operator on a larger domain. Typical examples of this situation arise in connection with differential operators, which initially can be defined on compactly supported smooth functions only.

When A and B are linear operators satisfying $\mathsf{D}(A) \subseteq \mathsf{D}(B)$ and $Ax = Bx$ for all $x \in \mathsf{D}(A)$, we call B an *extension* of A, notation:

$$A \subseteq B.$$

Definition 10.11 (Closability and closure). A linear operator is said to be *closable* if it has a closed extension, or equivalently, if the closure of its graph is the graph of a linear operator. The unique linear operator $\overline{A}$ whose graph is the closure of the graph of a closable operator A is called the *closure* of A; it is the smallest closed extension of A.

We have the following analogue of Proposition 10.4:

Proposition 10.12. *A linear operator A with domain $D(A)$ is closable if and only if the following holds: whenever $x_n \to 0$ in X and $Ax_n \to y$ in Y, with all x_n in $D(A)$, then $y = 0$.*

Proof We only need to prove the 'if' part, the 'only if' part being trivial. Denote the closure of $G(A)$ by $\overline{G}$. We must prove that, under the stated condition, $\overline{G}$ is the graph of a linear operator B. This is the case if and only if $(x, y_1) \in \overline{G}$, $(x, y_2) \in \overline{G}$ implies $y_1 = y_2$, for in that case we may define $D(B)$ to be the set of all $x \in X$ such that $(x, y) \in \overline{G}$; for $x \in B$ we may then define $Bx := y$, where $y \in Y$ is the unique element such that $(x, y) \in \overline{G}$. By a limiting argument, the linearity of A implies that the operator B thus defined is linear. Clearly it extends A and its graph $G(B) = \overline{G}$ is closed.

Suppose, therefore, that $(x, y_1) \in \overline{G}$ and $(x, y_2) \in \overline{G}$. Then $(0, y_1 - y_2) \in \overline{G}$ since $\overline{G}$ is a linear subspace of $X \times Y$, and this means that there exists a sequence $(x_n, Ax_n) \to (0, y_1 - y_2)$ in $X \times Y$. But then $x_n \to 0$ in X and $Ax_n \to y_1 - y_2$ in Y. By our assumption, this forces $y_1 - y_2 = 0$. $\qquad\square$

Example 10.13. In the setting of Examples 10.5 and 10.6, a closable operator is obtained by replacing the domain $D(A)$ of the operator A by any smaller subspace Y. The closure of the operator thus obtained equals A if and only if Y is dense in $D(A)$ with respect to the graph norm.

Example 10.14. It is shown in Proposition 10.36 in the next section that every densely defined symmetric operator acting in a Hilbert space is closable.

Example 10.15. Let D be an nonempty open subset of $\mathbb{R}^d$, let $1 \leqslant p \leqslant \infty$, and let $\alpha \in \mathbb{N}^d$ be a multi-index. In $L^p(D)$ we consider the linear operator A with domain $C_c^\infty(D)$ defined by

$$Af := \partial^\alpha f, \quad f \in C_c^\infty(D),$$

where $\partial^\alpha = \partial_1^{\alpha_1} \circ \cdots \circ \partial_d^{\alpha_d}$, with $\partial_j = \partial/\partial x_j$ the jth directional derivative. We claim that A is closable. Indeed, suppose the functions $f_n \in C_c^\infty(D)$ satisfy $f_n \to 0$ and $Af_n \to g$ in $L^p(D)$. Integrating by parts, for all $\phi \in C_c^\infty(D)$ we obtain

$$\int_D g\phi \, dx = \lim_{n \to \infty} \int_D (\partial^\alpha f_n)\phi \, dx = (-1)^{|\alpha|} \lim_{n \to \infty} \int_D f_n \partial^\alpha \phi \, dx = 0, \qquad (10.1)$$

where $|\alpha| = \alpha_1 + \cdots + \alpha_d$; the last step follows by Hölder's inequality (cf. Corollary

2.25). It is shown in Proposition 11.5 in the next chapter that (10.1) implies $g = 0$ almost everywhere.

Example 10.16. Let D be an nonempty open subset of $\mathbb{R}^d$ and let $1 \leqslant p \leqslant \infty$. In $L^p(D)$ we consider the linear operator A with domain $C_c^\infty(D)$ defined by

$$Af := \Delta f, \quad f \in C_c^\infty(D),$$

where $\Delta f = \partial_1^2 f + \cdots + \partial_d^2 f$ is the Laplacian of f. In the same way as in the previous example one shows that this operator is closable. Various explicit descriptions of its closure can be given, some of which are discussed in Chapters 11–13; see in particular Sections 11.1.e, 12.3, and 13.6.c.

10.1.b The Adjoint Operator

When A is a densely defined linear operator from X to Y, we may uniquely define a linear operator A^* from Y^* to X^* by defining its domain $\mathsf{D}(A^*)$ to be the set of all $y^* \in Y^*$ with the property that there exists an element $x^* \in X^*$ such that

$$\langle x, x^* \rangle = \langle Ax, y^* \rangle, \quad x \in \mathsf{D}(A).$$

Since $\mathsf{D}(A)$ is dense in X, the element $x^* \in X^*$ (if it exists) is unique and we can set

$$A^* y^* := x^*, \quad y^* \in \mathsf{D}(A^*).$$

Thus, by definition, we have the identity

$$\langle Ax, y^* \rangle = \langle x, A^* y^* \rangle, \quad x \in \mathsf{D}(A), \, y^* \in \mathsf{D}(A^*).$$

Definition 10.17 (Adjoint operator). The operator A^* is called the *adjoint* of A.

The adjoint of a closable densely defined operator A equals the adjoint of the closure $\overline{A}$, for if $x^* \in X^*$ and $y^* \in Y^*$ are such that $\langle x, x^* \rangle = \langle Ax, y^* \rangle$ for all $x \in \mathsf{D}(A)$, by continuity this identity extends to all $x \in \mathsf{D}(\overline{A})$.

Proposition 10.18. *If A is a densely defined linear operator from X to Y, then A^* is weak* closed in the sense that its graph is weak* closed in $Y^* \times X^*$ and we have*

$$\mathsf{G}(A^*) = (J(\mathsf{G}(A)))^\perp,$$

where $J : X \times Y \to Y \times X$ is defined by $J(x,y) = (-y,x)$. If A is densely defined and closed, then A^* is weak* densely defined in the sense that its domain is weak* dense.

Proof The pairing

$$\langle (y,x), (y^*,x^*) \rangle := \langle y, y^* \rangle + \langle x, x^* \rangle$$

allows us to identify $Y^* \times X^*$ with the dual of $Y \times X$. By the definition of the adjoint operator we have $(y^*, x^*) \in G(A^*)$ if and only if

$$\langle (-Ax, x), (y^*, x^*) \rangle = 0, \quad x \in D(A).$$

This proves the identity $G(A^*) = (J(G(A)))^\perp$. Since annihilators are weak* closed, this proves that A^* is weak* closed.

If $D(A^*)$ is not weak* dense, then by Proposition 4.46 there exists a nonzero element $y_0 \in Y$ such that $\langle y_0, y^* \rangle = 0$ for all $y^* \in D(A^*)$. By assumption $G(A)$ is closed in $X \times Y$ and $(0, y_0) \notin G(A)$. It follows that $J(G(A))$ is a closed subspace of $Y \times X$ not containing $J(0, y_0) = (-y_0, 0)$, so by the Hahn–Banach theorem there exists an element $(y_0^*, x_0^*) \in Y^* \times X^*$ annihilating $J(G(A))$ but not $(-y_0, 0)$. In other words,

$$\langle x, x_0^* \rangle = \langle Ax, y_0^* \rangle, \quad x \in D(A),$$

and

$$\langle y_0, y_0^* \rangle \neq 0.$$

The first equality implies that $y_0^* \in D(A^*)$, so the second one implies that y_0 does not vanish against every element of $D(A^*)$, contradicting the choice of y_0. □

If the linear operators A and B act from X to Y, we define the operator $A + B$ acting from X to Y by

$$D(A + B) := D(A) \cap D(B),$$
$$(A + B)x := Ax + Bx, \quad x \in D(A + B).$$

If A acts from X to Y and B acts from Y to another Banach space Z, we define the operator BA acting from X to Z by

$$D(BA) := \{x \in D(A) : Ax \in D(B)\},$$
$$BAx := B(Ax) \quad x \in D(BA).$$

There is of course *a priori* no guarantee that $D(A + B)$ and $D(BA)$ contain any nontrivial elements even when both A and B are densely defined.

Proposition 10.19. *Let A and B be densely defined operators acting in the ways indicated above. Then:*

(1) *if $A \subseteq B$, then $B^* \supseteq A^*$;*
(2) *if $A + B$ is densely defined, then $A^* + B^* \subseteq (A + B)^*$, with equality if B is bounded;*
(3) *if BA is densely defined, then $A^* B^* \subseteq (BA)^*$, with equality if B is bounded.*

Proof Part (1) is immediate from the definitions.

If $y^* \in D(A^* + B^*) = D(A^*) \cap D(B^*)$, then for all $x \in D(A + B) = D(A) \cap D(B)$ we have $\langle (A + B)x, y^* \rangle = \langle Ax, y^* \rangle + \langle Bx, y^* \rangle = \langle x, A^* y^* + B^* y^* \rangle$, so $y^* \in D((A + B)^*)$ and

$(A+B)^*y^* = A^*y^* + B^*y^*$. If B is bounded and $y^* \in D((A+B)^*)$, then for all $x \in D(A+B) = D(A)$ we have $\langle Ax, y^* \rangle = \langle (A+B)x, y^* \rangle - \langle Bx, y^* \rangle = \langle x, (A+B)^*y^* \rangle - \langle x, B^*y^* \rangle$, so that $y^* \in D(A^*) = D(A^* + B^*)$ and $A^*y^* = (A+B)^*y^* - B^*y^*$. This gives (2).

If $z^* \in D(A^*B^*)$, then $z^* \in D(B^*)$ and $B^*z^* \in D(A^*)$, and for all $x \in D(BA)$ we have $\langle (BA)x, z^* \rangle = \langle Ax, B^*z^* \rangle = \langle x, A^*B^*z^* \rangle$, so that $z^* \in D((BA)^*)$ and $(BA)^*z^* = A^*B^*z^*$. If B is bounded and $z^* \in D((BA)^*)$, then for all $x \in D(BA) = D(A)$ we have $\langle Ax, B^*z^* \rangle = \langle BAx, z^* \rangle = \langle x, (BA)^*z^* \rangle$, so $B^*z^* \in D(A^*)$ and $z^* \in D(A^*B^*)$ and $A^*B^*z^* = (BA)^*z^*$. This gives (3). $\square$

We have the following useful duality criterion to decide whether an element belongs to the domain of an operator.

Proposition 10.20. *Let A be a densely defined closed operator from X to Y. If $x \in X$ and $y \in Y$ are such that $\langle y, y^* \rangle = \langle x, A^*y^* \rangle$ for all $y^* \in D(A^*)$, then $x \in D(A)$ and $Ax = y$.*

Proof By the Hahn–Banach theorem, the result follows once we have checked that $\langle (x,y), (x^*,y^*) \rangle = 0$ for all $(x^*,y^*) \in (G(A))^\perp$. Indeed, this gives

$$(x,y) \in {}^\perp((G(A))^\perp) = \overline{G(A)}^{\text{weak}} = G(A),$$

where the first identity follows from Proposition 4.47 and the second from the fact that closed subspaces are weakly closed by the Hahn–Banach theorem (see Proposition 4.44).

Fix an arbitrary $(x^*,y^*) \in (G(A))^\perp$. For all $x \in D(A)$ we have $(x,Ax) \in G(A)$ and therefore

$$0 = \langle (x,Ax), (x^*,y^*) \rangle = \langle x, x^* \rangle + \langle Ax, y^* \rangle.$$

This means that $y^* \in D(A^*)$ and $A^*y^* = -x^*$. Hence,

$$\langle (x,y), (x^*,y^*) \rangle = \langle x, -A^*y^* \rangle + \langle y, y^* \rangle = 0.$$

$\square$

In what follows we let H and K be Hilbert spaces. When A is a densely defined operator acting from H to K, the Riesz representation theorem may be used to identify the adjoint A^*, which acts from K^* to H^*, with a linear operator $A^\star$ acting from K to H. Thus, by definition, an element $k \in K$ belongs to $D(A^\star)$ if there exists a (necessarily unique) element $h \in H$ such that

$$(x|h) = (Ax|k), \quad x \in D(A),$$

and in that case $A^\star k = h$. Thus we have the identity

$$(x|A^\star k) = (Ax|k), \quad x \in D(A), \, k \in D(A^\star).$$

Definition 10.21 (Hilbert space adjoint). The operator A^* is called the *Hilbert space adjoint* of A.

Proposition 10.18 admits the following Hilbertian version. We denote by $H \oplus K$ the Hilbert space obtained by endowing the cartesian product $H \times K$ with the inner product

$$((h,k)|(h',k')) := (h|h') + (k|k').$$

Proposition 10.22. *If A is a densely defined linear operator from H to K, then A^* is closed and we have*

$$\mathsf{G}(A^\star) = (J(\mathsf{G}(A)))^\perp$$

in the sense of orthogonal complements, where $J : H \oplus K \to K \oplus H$ is defined by $J(x,y) := (-y,x)$. If A is densely defined and closed, then A^* is densely defined.

Proof In Hilbert spaces the weak topology and the weak* topology agree, and a subspace is weakly dense (respectively, weakly closed) if and only if it is dense (respectively, closed). Hence everything follows from Proposition 10.18, except the statement that $\mathsf{G}(A^\star)$ is the orthogonal complement of $J(\mathsf{G}(A))$ in $K \oplus H$. This follows from

$$(k,h) \in \mathsf{G}(A^\star) \Leftrightarrow (Ax|k) = (x|h) \text{ for all } x \in \mathsf{D}(A)$$
$$\Leftrightarrow (k,h) \perp (-Ax,x) \text{ for all } x \in \mathsf{D}(A) \Leftrightarrow (k,h) \perp J(\mathsf{G}(A)).$$

$\square$

Mutatis mutandis, Proposition 10.19 admits a Hilbertian version as well; we leave this as an exercise to the reader.

Proposition 10.23. *If A is a densely defined closed operator acting from H to K, then $A = A^{\star\star}$ with equality of domains.*

Proof If Z is a subspace of $H \oplus K$, then

$$(k,h) \perp J(Z) \Leftrightarrow J(h,k) \in Z^\perp \Leftrightarrow (k,h) \in J(Z^\perp),$$

which shows that $(J(Z))^\perp = J(Z^\perp)$. Using Proposition 10.22 it follows that $\mathsf{G}(A^\star) = (J(\mathsf{G}(A)))^\perp = J((\mathsf{G}(A))^\perp)$ and

$$\mathsf{G}(A^{\star\star}) = (J(\mathsf{G}(A^\star)))^\perp = (JJ((\mathsf{G}(A))^\perp))^\perp = (\mathsf{G}(A))^{\perp\perp} = \mathsf{G}(A).$$

$\square$

For operators acting in Hilbert spaces we have the following extension of Proposition 4.31, the proof of which is almost *verbatim* the same:

Proposition 10.24. *If A is a densely defined closed operator from H into another Hilbert space K, then H and K admit orthogonal decompositions*

$$H = \mathsf{N}(A) \oplus \overline{\mathsf{R}(A^\star)}, \quad K = \mathsf{N}(A^\star) \oplus \overline{\mathsf{R}(A)}.$$

In particular,

(1) *A is injective if and only if $A^\star$ has dense range;*
(2) *A has dense range if and only if $A^\star$ is injective.*

10.1.c The Spectrum

The spectrum of a linear operator is defined in the same way as for bounded operators, except that explicit attention has to be paid to domains.

Definition 10.25 (Resolvent and spectrum). The *resolvent set* of a linear operator A acting in X is the set $\rho(A)$ consisting of all $\lambda \in \mathbb{C}$ for which the operator $\lambda I - A$ has a two-sided inverse, that is, there exists a bounded operator U on X such that:

(i) for all $x \in \mathsf{D}(A)$ we have $U(\lambda I - A)x = x$;
(ii) for all $x \in X$ we have $Ux \in \mathsf{D}(A)$ and $(\lambda I - A)Ux = x$.

The *spectrum* of A is defined as the complement of the resolvent set of A:

$$\sigma(A) := \mathbb{C} \setminus \rho(A).$$

We emphasise that, although A is allowed to be unbounded, the two-sided inverse $U = (\lambda I - A)^{-1}$ is required to be bounded. It is customary to write

$$R(\lambda, A) := (\lambda - A)^{-1}$$

for $\lambda \in \rho(A)$. As in the bounded case the *resolvent identity* holds: if $\lambda, \mu \in \rho(A)$, then

$$R(\lambda, A) - R(\mu, A) = (\mu - \lambda) R(\lambda, A) R(\mu, A). \tag{10.2}$$

By the observation in Example 10.8, a linear operator with nonempty resolvent set is closed. The proofs of Lemmas 6.7, the holomorphy of the resolvent (contained as part of Lemma 6.10), and Propositions 6.12 and 6.17 carry over *verbatim*, and Proposition 1.21 carries over with an obvious adaptation of the proof. For the reader's convenience we state the results here:

Proposition 10.26. *If A is closed and satisfies $\|Ax\| \geqslant C\|x\|$ for some $C > 0$ and all $x \in \mathsf{D}(A)$, then A is injective and has closed range.*

Proposition 10.27. *The spectrum $\sigma(A)$ is a closed subset of $\mathbb{C}$. More precisely, if $\lambda \in \rho(A)$, then $B(\lambda;r) \subseteq \rho(A)$ with $r = 1/\|R(\lambda,A)\|$. Moreover, if $|\lambda - \mu| \leqslant \delta r$ with $0 \leqslant \delta < 1$, then*

$$\|R(\mu,A)\| \leqslant \frac{\delta}{1-\delta}\|R(\lambda,A)\|.$$

Proposition 10.28. *The function $\lambda \mapsto R(\lambda,A)$ is holomorphic on $\rho(A)$, and its complex derivative is given by $-R(\lambda,A)^2$.*

Proposition 10.29. *If $\lambda_n \to \lambda$ in $\mathbb{C}$, with each $\lambda_n \in \rho(A)$ and with $\lambda \in \partial\rho(A)$, then*

$$\lim_{n\to\infty}\|R(\lambda_n,A)\| = \infty.$$

By the result of Examples 10.7 and 10.8 a linear operator which is not closed has full spectrum $\sigma(A) = \mathbb{C}$. For if we had $\lambda \in \rho(A)$, then the boundedness of $R(\lambda,A) = (\lambda - A)^{-1}$ exhibits $\lambda - A$ as the inverse of a bounded (hence closed) operator, and hence A is closed.

The following proposition gives a simple but powerful uniqueness result:

Proposition 10.30. *Let A and B be linear operators on a Banach space X. If $\rho(A) \cap \rho(B) \neq \varnothing$ and B is an extension of A, then $A = B$ with equality of domains.*

Proof Fix an arbitrary $\lambda \in \rho(A) \cap \rho(B)$. Then for all $x \in X$ we have $R(\lambda,A)x \in D(A) \subseteq D(B)$ and

$$(\lambda - B)R(\lambda,A)x = (\lambda - A)R(\lambda,A)x = x.$$

Multiplying both sides from the left with $R(\lambda,B)$ gives $R(\lambda,A)x = R(\lambda,B)x$. Since $x \in X$ was arbitrary, we conclude that $R(\lambda,A) = R(\lambda,B)$ and therefore $D(A) = D(B)$. □

The following result is proved in the same way as Propositions 6.18 and 8.9.

Proposition 10.31. *If A as a densely defined operator in a Banach space X, then*

$$\sigma(A^*) = \sigma(A).$$

If A as a densely defined operator in a Hilbert space H, then

$$\sigma(A^*) = \overline{\sigma(A)}.$$

For later use we compute the spectrum of a simple diagonal operator.

Proposition 10.32. *Let A be a densely defined closed operator in a separable Hilbert space H with an orthonormal basis $(h_n)_{n\geqslant 1}$ of eigenvectors. If $\rho(A) \neq \varnothing$ and the corresponding eigenvalue sequence $(\lambda_n)_{n\geqslant 1}$ satisfies $\lim_{n\to\infty}|\lambda_n| = \infty$, then*

$$\sigma(A) = \{\lambda_n : n \geqslant 1\}.$$

Proof   Let $\mu \notin \{\lambda_n : n \geqslant 1\}$. The assumption $|\lambda_n| \to \infty$ implies that $\inf_{n\geqslant 1} |\mu - \lambda_n| =:$ $\delta > 0$, and therefore the mapping

$$R_\mu : h_n \mapsto \frac{1}{\mu - \lambda_n} h_n$$

has a unique extension to a bounded operator on H of norm at most $1/\delta$. It is clear that this operator is injective, so its inverse R_μ^{-1} is closed. Hence the operator $B := \mu - R_\mu^{-1}$ with domain $D(B) := R(R_\mu)$ is closed. Clearly $\mu \in \rho(B)$ and $R(\mu, B) = R_\mu$. Moreover,

$$Bh_n = \mu h_n - (\mu - \lambda_n)h_n = \lambda_n h_n = A h_n, \quad n \geqslant 1.$$

We claim that the linear span Y of the vectors h_n, $n \geqslant 1$, is dense in $D(B)$ with respect to the graph norm. Indeed, let $g \in D(B)$. Then $g \in R(R_\mu)$, say $g = R_\mu h$ with $h = \sum_{j\geqslant 1} c_j h_j \in H$. Let P_k denote the orthogonal projection onto the span of $h_1, \dots, h_k$. Then $P_k g \to g$ in H as $k \to \infty$. Also, $P_k R_\mu = R_\mu P_k$ implies $P_k g \in R(R_\mu) = D(B)$ and

$$(\mu - B)P_k g = (\mu - B)P_k R_\mu h$$

$$= (\mu - B) \sum_{j=1}^{k} \frac{c_j}{\mu - \lambda_j} h_j = \sum_{j=1}^{k} c_j h_j \to h = R_\mu^{-1} g = (\mu - B)g$$

as $k \to \infty$. This implies $BP_k g \to Bg$. It follows that $P_k g \to g$ in $D(B)$ as claimed. Since Y is contained in $D(A)$ and A is closed, it follows that $B \subseteq A$.

Now let $\mu_0 \in \rho(A)$. Then $\mu_0 \notin \{\lambda_n : n \geqslant 1\}$ since every λ_n is an eigenvalue for A, and therefore $\mu_0 \in \rho(B)$ by what we just proved. Proposition 10.30 now implies $A = B$. But then, by what we already proved for B, every $\mu \notin \{\lambda_n : n \geqslant 1\}$ belongs to $\rho(A)$.   □

We conclude this section with two useful elaborations on Proposition 10.27. In the first, we write $\Sigma_\varphi := \{z \in \mathbb{C} \setminus \{0\} : |\arg(z)| < \varphi\}$, the argument being taken in $(-\pi, \pi)$.

Lemma 10.33. *Let A be a linear operator acting in X. If the open half-line $(0, \infty)$ is contained in $\rho(A)$ and*

$$\sup_{\lambda > 0} \|\lambda R(\lambda, A)\| =: M < \infty,$$

then $M \geqslant 1$, and for all $\varphi \in (0, \frac{1}{2}\pi)$ with $\sin \varphi < 1/M$ we have $\Sigma_\varphi \subseteq \rho(A)$ and

$$\sup_{\lambda \in \Sigma_\varphi} \|\lambda R(\lambda, A)\| \leqslant \frac{M}{1 - M \sin \varphi}.$$

Proof   For $x \in D(A)$ we have $\lambda R(\lambda, A)x = x + R(\lambda, A)Ax \to x$ as $\lambda \to \infty$, from which it follows that $M \geqslant 1$.

Proposition 10.27 implies that for every $\mu > 0$ the open ball with centre μ and radius $1/\|R(\mu, A)\|$ is contained in $\rho(A)$. The union of these balls is a sector; we shall now verify that the sine of its angle equals at least $1/M$.

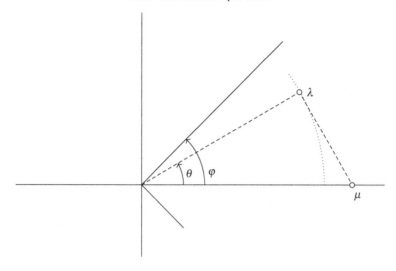

Figure 10.1 The proof of Lemma 10.33

Let $\varphi \in (0, \frac{1}{2}\pi)$ satisfy $\sin\varphi < 1/M$. Fix $\lambda \in \overline{\Sigma}_\varphi$ and let $\mu > 0$ be determined by the requirement that the triangle spanned by 0, λ, μ has a right angle at λ (thus, by Pythagoras, $|\lambda - \mu|^2 + |\lambda|^2 = |\mu|^2$, so $\mu = -|\lambda|^2/|\operatorname{Re}\lambda|$). Let θ denote the angle of λ with the positive real line. See Figure 10.1. Then $|\lambda - \mu|/|\mu| = \sin\theta < \sin\varphi < 1/M$, so $|\lambda - \mu| < |\mu|/M \leqslant 1/\|R(\mu,A)\|$. Hence $\lambda \in \rho(A)$, and estimating for the Neumann series gives

$$\|R(\lambda,A)\| \leqslant \|R(\mu,A)\| \sum_{n=0}^{\infty} \frac{|\lambda - \mu|^n}{|\mu|^n} \|\mu R(\mu,A)\|^n$$

$$\leqslant \frac{M}{|\mu|} \sum_{n=0}^{\infty} (\sin\varphi)^n M^n \leqslant \frac{M}{1 - M\sin\varphi} \cdot \frac{1}{|\lambda|}.$$

$\square$

A typical application of this lemma is the second part of the next corollary, which extends a uniform bound on $\lambda R(\lambda,A)$ on a half-plane to a larger sector. For reasons of completeness we also include its counterpart for uniform bounds on $R(\lambda,A)$.

Lemma 10.34. *Suppose that the half-plane* $\mathbb{C}_+ = \{\operatorname{Re}\lambda > 0\} = \{|\arg(\lambda)| < \frac{1}{2}\pi\}$ *is contained in the resolvent set* $\rho(A)$ *of the linear operator A acting in X. Then:*

(1) *if* $\sup_{\lambda \in \mathbb{C}_+} \|R(\lambda,A)\| < \infty$, *then there exists a* $\delta > 0$ *such that* $\{\operatorname{Re}\lambda > -\delta\} \subseteq \rho(A)$

and

$$\sup_{\operatorname{Re}\lambda > -\delta} \|R(\lambda, A)\| < \infty;$$

(2) *if* $\sup_{\lambda \in \mathbb{C}_+} \|\lambda R(\lambda, A)\| < \infty$, *then there exists a* $\delta > 0$ *such that* $\{|\arg \lambda| < \frac{1}{2}\pi + \delta\} \subseteq \rho(A)$ *and*

$$\sup_{|\arg\lambda| < \frac{1}{2}\pi + \delta} \|\lambda R(\lambda, A)\| < \infty.$$

Proof Proposition 10.29 implies that in case (1) we have $i\mathbb{R} \subseteq \rho(A)$, and that in case (2) we have $i\mathbb{R} \setminus \{0\} \subseteq \rho(A)$. The result now follows from Proposition 10.27 applied to the points $\lambda \in i\mathbb{R}$ (in case (1)) and Lemma 10.33 applied to the operators $\pm iA$ (in case (2)).   □

10.2 Unbounded Selfadjoint Operators

In what follows we let H be a Hilbert space.

Definition 10.35 (Symmetric and positive operators)**.** A linear operator A acting in H is called:

- *symmetric*, if for all $x, y \in \mathsf{D}(A)$ we have $(Ax|y) = (x|Ay)$.
- *positive*, if for all $x \in \mathsf{D}(A)$ we have $(Ax|x) \geqslant 0$.

Over the complex scalars, positive operators are symmetric. Indeed, if A is positive, then for all $x \in \mathsf{D}(A)$ we have $(Ax|x) = \overline{(Ax|x)} = (x|Ax)$. By polarisation (as in the proof of Proposition 8.1, this requires working over the complex scalars) this implies $(Ax|y) = (x|Ay)$ for all $x, y \in \mathsf{D}(A)$.

It is an immediate consequence of Definition 10.35 and the definition of $A^\star$ that if A is densely defined and symmetric, then $\mathsf{D}(A) \subseteq \mathsf{D}(A^\star)$ and $Ax = A^\star x$ for all $x \in \mathsf{D}(A)$, that is, $A^\star$ is an extension of A. Since $A^\star$ is closed, we have shown:

Proposition 10.36. *If A is a densely defined symmetric operator, then A is closable and $A^\star$ is a closed extension of A.*

In general, $\mathsf{D}(A)$ may be strictly smaller than $\mathsf{D}(A^\star)$. A simple example is the Laplace operator Δ on $L^2(\mathbb{R}^d)$ with domain $C_c^\infty(\mathbb{R}^d)$: this operator is densely defined and symmetric but not closed, and therefore $\Delta^\star$ is a proper extension of Δ. This motivates the following definition.

Definition 10.37 (Selfadjoint operators)**.** A densely defined operator A in H is called *selfadjoint* if $A = A^\star$, that is, if $\mathsf{D}(A) = \mathsf{D}(A^\star)$ and $Ah = A^\star h$ for all $h \in \mathsf{D}(A) = \mathsf{D}(A^\star)$.

The operator A is called *essentially selfadjoint* if it is closable and its closure $\overline{A}$ is self-adjoint.

By Propositions 10.23 and 10.36, a densely defined symmetric operator A is selfadjoint if and only if A^* is symmetric.

Example 10.38 (Multipliers). Let $(\Omega, \mathcal{F}, \mu)$ be a measure space and let $m : \Omega \to \mathbb{C}$ be a measurable function. It has been shown in Example 10.6 that the linear operator M_m in $L^2(\Omega)$ defined by

$$D(M_m) := \{f \in L^2(\Omega) : mf \in L^2(\Omega)\},$$
$$M_m f := mf, \quad f \in D(M_m),$$

is densely defined and closed. It is immediate from the definition of the Hilbert space adjoint that $M_m^* = M_{\overline{m}}$ with equality of domains $D(M_{\overline{m}}) = D(M_m)$. As a consequence, M_m is selfadjoint in $L^2(\Omega)$ if and only if m is real-valued μ-almost everywhere.

Example 10.39 (Fourier multipliers). Let $m : \mathbb{R}^d \to \mathbb{R}$ be real-valued and measurable, and let A_m denote the (possibly unbounded) Fourier multiplier in $L^2(\mathbb{R}^d)$ defined by

$$D(A_m) := \{f \in L^2(\mathbb{R}^d) : m\widehat{f} \in L^2(\mathbb{R}^d)\},$$
$$A_m f := (m\widehat{f})^{\vee}, \quad f \in D(A_m). \tag{10.3}$$

Let us prove that A_m is selfadjoint in $L^2(\mathbb{R}^d)$. The symmetry of the multiplier M_m considered in the previous example implies that A_m is symmetric: for all $f, g \in D(A_m)$ we have $\widehat{f}, \widehat{g} \in D(M_m)$ and, by the Plancherel theorem,

$$(A_m f | g) = (M_m \widehat{f} | \widehat{g}) = (\widehat{f} | M_m \widehat{g}) = (f | A_m g).$$

By Proposition 10.36 this implies $D(A_m) \subseteq D(A_m^*)$. Conversely, if $g \in D(A_m^*)$ and $A_m^* g = h$, then for $f \in D(A_m)$ we have, since $\widehat{f} \in D(M_m)$,

$$(\widehat{f} | \widehat{h}) = (f | h) = (A_m f | g) = (\widehat{A_m f} | \widehat{g}) = (M_m \widehat{f} | \widehat{g}).$$

This means that $\widehat{g} \in D(M_m^*)$. Hence, by the previous example, $\widehat{g} \in D(M_m)$. This means that $m\widehat{g} \in L^2(\mathbb{R}^d)$ and therefore $g \in D(A_m)$.

Two special cases are of special interest:

Example 10.40 (The momentum operator). The multiplier $m(\xi) = \xi$ gives rise to the operator $\frac{1}{i}\frac{d}{dx}$ on $L^2(\mathbb{R})$. With the domain given by (10.3) this operator is selfadjoint. With the notation and techniques developed in Section 11.1.e this domain is seen to be the Sobolev space $H^1(\mathbb{R}) = W^{1,2}(\mathbb{R})$.

Example 10.41 (The Laplacian). The multiplier $m(\xi) = -|\xi|^2$ gives rise to the Laplace operator $\Delta = \sum_{j=1}^{d} \frac{\partial^2}{\partial x_j^2}$ on $L^2(\mathbb{R}^d)$. With the domain given by (10.3) this operator is selfadjoint. With the notation and techniques developed in Section 11.1.e this domain is seen to be the Sobolev space $H^2(\mathbb{R}^d) = W^{2,2}(\mathbb{R}^d)$.

The following version of Theorem 8.11 holds.

Proposition 10.42. *If A is selfadjoint, then $\sigma(A) \subseteq \mathbb{R}$. If, in addition, A is positive, then $\sigma(A) \subseteq [0,\infty)$.*

Proof This may be established by repeating parts of the proof of Theorem 8.11, using Propositions 10.26 and 10.24 instead of Proposition 1.21 and 4.31, respectively.   □

The following proposition provides a sufficient condition for selfadjointness.

Proposition 10.43. *If A is densely defined and symmetric and $\rho(A) \cap \mathbb{R} \neq \varnothing$, then A is selfadjoint.*

Proof The nonemptiness of the resolvent set implies that A is closed (cf. Example 10.8). The operator A^* is closed as well. The symmetry of A implies $D(A) \subseteq D(A^*)$, and if $\lambda \in \rho(A) \cap \mathbb{R}$, then $\lambda \in \rho(A^*)$ in view of Proposition 10.31. The identity $A = A^*$ with equality of domains therefore follows from Proposition 10.30.   □

An efficient proof of the next proposition is obtained by noting that Proposition 10.22 implies the following criterion for selfadjointness: *a densely defined operator A in H is selfadjoint if and only if*

$$(J(\mathsf{G}(A)))^\perp = \mathsf{G}(A),$$

where $J(x,y) = (-y,x)$ for $x,y \in H$.

Proposition 10.44. *If the linear operator A is selfadjoint, injective, and has dense range, then its inverse A^{-1} with domain $D(A^{-1}) = R(A)$ is selfadjoint.*

Proof From

$$(x,y) \in \mathsf{G}(A^{-1}) \;\Leftrightarrow\; (y,x) \in \mathsf{G}(A) \;\Leftrightarrow\; J(x,-y) \in \mathsf{G}(A)$$
$$\Leftrightarrow\; (x,-y) = J(\mathsf{G}(A)) \;\Leftrightarrow\; (x,y) = J(\mathsf{G}(-A))$$

we see that $\mathsf{G}(A^{-1}) = J(\mathsf{G}(-A))$. Applying J to both sides gives $J(\mathsf{G}(A^{-1})) = \mathsf{G}(-A)$. Hence, since $-A$ is selfadjoint, by the above criterion

$$\mathsf{G}(A^{-1}) = J(\mathsf{G}(-A)) = (\mathsf{G}(-A))^\perp = (J(\mathsf{G}(A^{-1})))^\perp.$$

Applying the criterion once more, this proves that A^{-1} is selfadjoint.   □

As a simple application of Proposition 10.44 we record the following result.

Corollary 10.45. *Let A be a densely defined closed positive operator in H. If $I + A$ has dense range, then A is selfadjoint.*

Proof From $\|(I+A)x\|\|x\| \geqslant ((I+A)x|x) \geqslant \|x\|^2$ we see that $\|(I+A)x\| \geqslant \|x\|$ for all $x \in D(A)$. Since A (and hence $I+A$) is closed, by Proposition 10.26 this implies that $I+A$ is injective and has closed range. Since $I+A$ also has dense range, $I+A$ is surjective and the inverse $(I+A)^{-1}$ is well defined as a linear operator. The bound $\|(I+A)x\| \geqslant \|x\|$ implies that $(I+A)^{-1}$ is bounded (and in fact contractive). At the same time, this bounded operator is positive and therefore selfadjoint. Proposition 10.44 therefore implies that $I+A$, hence also A, is selfadjoint. $\qquad\square$

As an application of Corollary 10.45 we have the following sufficient condition for selfadjointness.

Theorem 10.46 (Selfadjointness of $A^\star A$). *If A is a densely defined closed operator from H into another Hilbert space K, then:*

(1) *the operator $A^\star A$ is selfadjoint and positive;*
(2) $D(A^\star A)$ *is dense in* $D(A)$ *with respect to the graph norm.*

This operator will be revisited in Proposition 12.18 in connection with the theory of forms.

Proof (1): We check that the operator $A^\star A$, which is obviously positive, satisfies the assumptions of Corollary 10.45.

By Proposition 10.22 we have the orthogonal decomposition

$$H \oplus K = G(A^\star) \oplus J(G(A)).$$

Hence for any $u \in K$ we can find $x \in D(A)$ and $y \in D(A^\star)$ such that

$$(0,u) = (y,A^\star y) + J(x,Ax) = (y-Ax, A^\star y+x).$$

It follows that $y = Ax$, which implies $x \in D(A^\star A)$, and $u = A^\star y + x = (I+A^\star A)x$. This proves that $I+A^\star A$ is surjective.

(2): To prove density of $D(A^\star A)$ in $D(A)$ with respect to the graph norm, suppose that $x \in D(A)$ is such that $(x|y)_{D(A)} = 0$ for all $y \in D(A^\star A)$, where $(x|y)_{D(A)} := (x|y) + (Ax|Ay)$ is the inner product of $D(A)$, viewed as a Hilbert space with respect to this inner product (completeness being a consequence of the closedness of A; see Proposition 10.3). Then

$$0 = (x|y) + (x|A^\star Ay) = (x|(I+A^\star A)y)$$

for all $y \in D(A^\star A)$. Since $I+A^\star A$ is surjective, this means that $(x|z) = 0$ for all $z \in D(A)$, so $x = 0$. $\qquad\square$

We finish this section with another useful criterion for selfadjointness.

Theorem 10.47. *For a densely defined symmetric operator A in H the following assertions are equivalent:*

(1) *A is selfadjoint;*
(2) *A is closed and* $N(A^* + i) = N(A^* - i) = \{0\}$*;*
(3) $R(A + i) = R(A - i) = H$.

Proof (1)⇒(2): If A is selfadjoint, then $A = A^*$ is closed by Proposition 10.23. If $x \in D(A^*)$ satisfies $(A^* + i)x = 0$, then $x \in D(A)$ and $Ax = A^*x = -ix$, and

$$-i(x|x) = (Ax|x) = (x|A^*x) = i(x|x)$$

implies $x = 0$. In the same way $(A^* - i)x = 0$ implies $x = 0$.

(2)⇒(3): By the same argument as in the proof of Proposition 4.31, the injectivity of $(A \pm i)^* = A^* \mp i$ implies that $A \pm i$ has dense range (and conversely; this will be used in the proof of the next implication). By the same argument as in the proof of Theorem 8.11, the symmetry of A implies $\|(A \pm i)x\| \geqslant \|x\|$ for all $x \in D(A)$, and since A is closed, Proposition 10.26 implies that the ranges of $A \pm i$ are closed. We conclude that both ranges equal H.

(3)⇒(1): Fix an arbitrary $h \in D(A^*)$. The assumption $R(A - i) = H$ implies that there exists an $h' \in D(A)$ such that $(A - i)h' = (A^* - i)h$. Since A^* extends A, we have $h' \in D(A^*)$ and $A^*h' = Ah'$. It follows that $(A^* - i)h' = (A^* - i)h$. As was noted in the proof of the previous implication, the assumption $R(A + i) = H$ implies that $A^* - i$ is injective. It follows that $h = h'$. Since $h' \in D(A)$, this implies that $h \in D(A)$ and $Ah = A^*h$, the latter since A^* extends A.

This shows that A extends A^*. Since A^* extends A, these operators are equal.   □

The theory of selfadjoint operators is taken up again in Section 12.2 in connection with the theory of forms.

10.3 Unbounded Normal Operators

Having dealt with unbounded selfadjoint operators, we now turn to unbounded normal operators.

10.3.a Definition and General Properties

Definition 10.48 (Normal operators). A linear operator A in H is said to be *normal* if it is closed, densely defined, and satisfies

$$A^*A = AA^*.$$

The equality $A^*A = AA^*$ is shorthand for equality of the domains

$$D(A^*A) := \{x \in D(A) : Ax \in D(A^*)\},$$

$$D(AA^\star) := \{x \in D(A^\star) : A^\star x \in D(A)\},$$

along with equality

$$A^\star Ax = AA^\star x$$

for all x in this common domain.

Since $A^{\star\star} = A$ by Proposition 10.23, a densely defined closed operator A is normal if and only if its adjoint $A^\star$ is normal.

Proposition 10.49. *If A is a normal operator, then:*

(1) $D(A) = D(A^\star)$;
(2) $\|Ax\| = \|A^\star x\|$ *for all $x \in D(A) = D(A^\star)$;*
(3) *if $A \subseteq B$ with B normal, then $A = B$.*

Proof (1) and (2): The normality of A implies that if $x \in D(A^\star A) = D(AA^\star)$, then $x \in D(A), x \in D(A^\star)$, and

$$\|Ax\|^2 = (A^\star Ax|x) = (AA^\star x|x) = \|A^\star x\|^2.$$

By Theorem 10.46, $D(A^\star A)$ is dense in $D(A)$, so for any $x \in D(A)$ we may choose a sequence $x_n \to x$ with each $x_n \in D(A^\star A) = D(AA^\star)$ and with convergence in the graph norm of $D(A)$. Then $x_m, x_n \in D(A^\star)$, and from

$$\lim_{n,m\to\infty} \|A^\star x_n - A^\star x_m\| = \lim_{n,m\to\infty} \|Ax_n - Ax_m\| = 0$$

we infer that $A^\star x_n \to y$ for some $y \in H$. From the closedness of $A^\star$ (Proposition 10.23) we infer that $x \in D(A^\star)$ and $A^\star x = y$. This argument shows that $D(A) \subseteq D(A^\star)$.

Since $A^\star$ is normal, what we just proved can be applied to $A^\star$. This, together with Proposition 10.23, gives the reverse inclusion $D(A^\star) \subseteq D(A^{\star\star}) = D(A)$.

(3): If $A \subseteq B$ with A and B normal, then by (1), Proposition 10.19(1), and another application of (1),

$$D(B) = D(B^\star) \subseteq D(A^\star) = D(A).$$

Together with the assumption $D(A) \subseteq D(B)$ this implies $D(A) = D(B)$. $\qquad\square$

10.3.b The Measurable Functional Calculus

Projection-valued measures give rise to normal operators:

Theorem 10.50 (Measurable functional calculus). *Let $(\Omega, \mathscr{F})$ be a measurable space, let $P : \mathscr{F} \to \mathscr{L}(H)$ be a projection-valued measure, and let $f : \Omega \to \mathbb{C}$ be a measurable function. There exists a unique normal operator $\Phi(f)$ in H satisfying*

$$D(\Phi(f)) = \left\{x \in H : \int_\Omega |f|^2 \,dP_x < \infty\right\},$$

$$(\Phi(f)x|x) = \int_\Omega f \, dP_x, \quad x \in D(\Phi(f)).$$

For all $x \in D(\Phi(f))$ we have

$$\|\Phi(f)x\|^2 = \int_\Omega |f|^2 \, dP_x. \tag{10.4}$$

Furthermore, if $f_n, f, g : \Omega \to \mathbb{C}$ are measurable functions, then:

(1) $\Phi(f)\Phi(g) \subseteq \Phi(fg)$ with $D(\Phi(f)\Phi(g)) = D(\Phi(fg)) \cap D(\Phi(g))$;
(2) $\Phi(f)^\star = \Phi(\overline{f})$;
(3) *if $0 \leqslant |f_n| \leqslant |f|$ and $\lim_{n\to\infty} f_n = f$ pointwise on Ω, then $D(\Phi(f)) \subseteq D(\Phi(f_n))$ and*

$$\lim_{n\to\infty} \Phi(f_n)x = \Phi(f)x, \quad x \in D(\Phi(f)).$$

The operator $\Phi(f)$ is selfadjoint if and only if f is real-valued P_x-almost everywhere for all $x \in H$.

It follows from (1) that

$$\Phi(f)\Phi(g) = \Phi(fg) \iff D(\Phi(fg)) \subseteq D(\Phi(g)). \tag{10.5}$$

This is trivially the case if g is bounded, for then $D(\Phi(g)) = H$. In that case $\Phi(g)$ is bounded and equals the operator given by the bounded calculus of Theorem 9.8. This fact, and the properties of the bounded calculus, will be frequently used in the proof below.

Example 10.51. Under the above assumptions, it follows from (10.5) that

$$\Phi(f^n) = (\Phi(f))^n, \quad n = 1, 2, \dots$$

To prove this, proceeding by induction it suffices to check that $\Phi(f^{k+1}) = \Phi(f^k)\Phi(f)$ for all $k = 1, 2, \dots$ By (10.5), this operator identity holds if and only if $D(\Phi(f^{k+1})) \subseteq D(\Phi(f))$. If $x \in D(\Phi(f^{k+1}))$, then $\int_\Omega |f|^{2k+2} \, dP_x < \infty$. Since P_x is a finite measure, this implies $\int_\Omega |f|^2 \, dP_x < \infty$, that is, $x \in D(\Phi(f))$.

Proof of Theorem 10.50   For the moment define D_f to be the set $\{x \in H : \int_\Omega |f|^2 \, dP_x < \infty\}$.

If $x, y \in D_f$ and $g = \sum_{j=1}^k c_j \mathbf{1}_{B_j}$ is a simple function satisfying $0 \leqslant g \leqslant |f|^2$, with $c_j \geqslant 0$ and disjoint sets $B_j \in \mathscr{F}$, then by the Cauchy–Schwarz inequality for the sesquilinear forms $(x, y) \mapsto (P_{B_j}x|y)$,

$$\int_\Omega g \, dP_{x+y} = \sum_{j=1}^k c_j (P_{B_j}(x+y)|x+y)$$

$$\leqslant \sum_{j=1}^k c_j \left((P_{B_j}x|x) + 2(P_{B_j}x|x)^{1/2}(P_{B_j}y|y)^{1/2} + (P_{B_j}y|y) \right)$$

$$\leqslant 2 \sum_{j=1}^{k} c_j \big((P_{B_j} x | x) + (P_{B_j} y | y) \big)$$

$$= 2 \int_{\Omega} g \, dP_x + 2 \int_{\Omega} g \, dP_y \leqslant 2 \int_{\Omega} |f|^2 \, dP_x + 2 \int_{\Omega} |f|^2 \, dP_y.$$

Taking the supremum over all such simple functions g, we obtain

$$\int_{\Omega} |f|^2 \, dP_{x+y} \leqslant 2 \int_{\Omega} |f|^2 \, dP_x + 2 \int_{\Omega} |f|^2 \, dP_y.$$

This shows that D_f is closed under addition. The identity

$$\int_{\Omega} |f|^2 \, dP_{cx} = |c|^2 \int_{\Omega} |f|^2 \, dP_x$$

is evident for simple functions f, follows for general functions f by approximation, and shows that D_f is also closed under scalar multiplication. It follows that D_f is a linear subspace of H.

To prove that D_f is dense in H fix an arbitrary $x \in H$ and for $n = 1, 2, \ldots$ let $B_n := \{|f| \leqslant n\}$. Then for all $x \in R(P_{B_n})$ and $B \in \mathscr{F}$,

$$\int_{\Omega} \mathbf{1}_B \, dP_x = (P_B x | x) = (P_B P_{B_n} x | x) = (P_{B \cap B_n} x | x) = \int_{B_n} \mathbf{1}_B \, dP_x,$$

so by linearity and monotone convergence,

$$\int_{\Omega} |f|^2 \, dP_x = \int_{B_n} |f|^2 \, dP_x \leqslant n^2 P_x(B_n) \leqslant n^2 P_x(\Omega) = n^2 \|x\|^2.$$

This implies that $R(P_{B_n})$ is contained in D_f. Since $\Omega = \bigcup_{n \geqslant 1} B_n$, monotone convergence implies that $\|P_{B_n} x\|^2 = (P_{B_n} x | x) = \int_{\Omega} \mathbf{1}_{B_n} \, dP_x \to \int_{\Omega} \mathbf{1} \, dP_x = \|x\|^2$ and therefore

$$\|x - P_{B_n} x\| = \|x\|^2 - 2(P_{B_n} x | x) + \|P_{B_n} x\|^2 \to 0$$

as $n \to \infty$. This proves that x belongs to the closure of D_f.

For simple functions $g = \sum_{j=1}^{k} c_j \mathbf{1}_{B_j}$ we set

$$\Phi(g) := \sum_{j=1}^{k} c_j P_{B_j}.$$

It is routine to check that this is well defined and that (10.4) holds for g. If $x \in D_f$, then $f \in L^2(\Omega, P_x)$. If $g_n \to f$ in $L^2(\Omega, P_x)$ with each g_n simple, then

$$\|\Phi(g_n) x - \Phi(g_m) x\|^2 = \int_{\Omega} |g_n - g_m|^2 \, dP_x \to 0 \quad \text{as } n, m \to \infty.$$

Consequently for $x \in D_f$ we may define

$$\Phi(f) x := \lim_{n \to \infty} \Phi(g_n) x.$$

This is well defined, and the validity of (10.4) for g_n implies the validity of (10.4) for f.

In this way we obtain a well-defined linear operator $\Phi(f) : D_f \to H$. In what follows we view $\Phi(f)$ as a linear operator in H with domain $D(\Phi(f)) = D_f$. The closedness of $\Phi(f)$ follows from (2) applied to $\overline{f}$ and Proposition 10.18. Normality of $\Phi(f)$ is an easy consequence of (1) applied with $g = \overline{f}$, noting that $D(\Phi(|f|^2)) \subseteq D(\Phi(\overline{f}))$ follows from Hölder's inequality or noting that $\int_\Omega |f|^2 \, dP_x \leqslant \int_\Omega 1 + |f|^4 \, dP_x$.

By (2), $\Phi(f)$ is selfadjoint if and only if $\Phi(f)^\star = \Phi(f)$, and by (10.4) applied to $\overline{f} - f$, this holds if and only if f is real-valued P_x-almost everywhere for all $x \in H$.

(3): By (10.4) applied to $f - f_n$, for $x \in D(\Phi(f))$ we have $x \in D(\Phi(f - f_n))$ and

$$\lim_{n \to \infty} \|\Phi(f)x - \Phi(f_n)x\| = \lim_{n \to \infty} \int_\Omega |f - f_n|^2 \, dP_x = 0$$

by dominated convergence, so $\lim_{n \to \infty} \Phi(f_n)x = \Phi(f)x$.

In what follows, for $n = 1, 2, \ldots$ let

$$f_n := \mathbf{1}_{\{|f| \leqslant n\}} f.$$

(1): First let f be bounded and measurable and g be measurable. For all $x \in D(\Phi(g))$ we have $x \in D(\Phi(fg))$, and using (3), the boundedness of $\Phi(f)$, and the multiplicativity of the Borel calculus for bounded normal operators,

$$\Phi(f)\Phi(g)x = \lim_{n \to \infty} \Phi(f)\Phi(g_n)x = \lim_{n \to \infty} \Phi(fg_n)x = \Phi(fg)x.$$

Hence by (10.4),

$$\int_\Omega |f|^2 \, dP_{\Phi(g)x} = \int_\Omega |fg|^2 \, dP_x.$$

This being true for all bounded measurable functions f, by monotone convergence it is true for all measurable functions f. Hence, if f and g are measurable, we infer that for elements $x \in D(\Phi(g))$ we have $\Phi(g)x \in D(\Phi(f))$ if and only if $x \in D(\Phi(fg))$. This is the same as saying that (1) holds.

(2): Let $x \in D(\Phi(f))$ and $y \in D(\Phi(\overline{f})) = D(\Phi(f))$. Then, by (3) and the properties of the Borel calculus for bounded normal operators,

$$(\Phi(f)x|y) = \lim_{n \to \infty} (\Phi(f_n)x|y) = \lim_{n \to \infty} (x|\Phi(\overline{f_n})y) = (x|\Phi(\overline{f})y).$$

This shows that $y \in D(\Phi(f)^\star)$ and $\Phi(f)^\star y = \Phi(\overline{f})y$. We have thus proved the inclusion $\Phi(\overline{f}) \subseteq \Phi(f)^\star$. For the converse inclusion let $y \in D(\Phi(f)^\star)$. We wish to prove that $y \in D(\Phi(f)) = D(\Phi(\overline{f}))$, that is, that $\int_\Omega |f|^2 \, dP_y < \infty$.

Let $z := \Phi(f)^\star y$. We claim that

$$\Phi(\mathbf{1}_{\{|f| \leqslant n\}})z = \Phi(\overline{f_n})y. \tag{10.6}$$

It follows from (1), applied with $g = \mathbf{1}_{\{|f| \leqslant n\}}$, that for all $x \in H$ we have

$$\Phi(\mathbf{1}_{\{|f| \leqslant n\}})x \in D(\Phi(f)) \quad \text{and} \quad \Phi(f)\Phi(\mathbf{1}_{\{|f| \leqslant n\}})x = \Phi(f_n)x.$$

Then, for all $x \in H$,

$$(x|\Phi(1_{\{|f|\leqslant n\}})z) = (\Phi(1_{\{|f|\leqslant n\}})x|\Phi(f)^*y)$$
$$= (\Phi(f)\Phi(1_{\{|f|\leqslant n\}})x|y) = (\Phi(f_n)x|y) = (x|\Phi(\overline{f_n})y),$$

using the conjugation property of the Borel calculus for bounded normal operators in the last step. This proves the claim (10.6).

By (10.4) and (10.6),

$$\int_\Omega |f_n|^2 \, dP_y = \|\Phi(\overline{f_n})y\|^2 = \|\Phi(1_{\{|f|\leqslant n\}})z\|^2 = \int_\Omega 1_{\{|f|\leqslant n\}} \, dP_z$$

and therefore

$$\int_\Omega |f|^2 \, dP_y = \lim_{n\to\infty} \int_\Omega |f_n|^2 \, dP_y = \lim_{n\to\infty} \int_\Omega 1_{\{|f|\leqslant n\}} \, dP_z \leqslant \int_\Omega 1 \, dP_z = \|z\|^2,$$

so that $y \in D(\Phi(f))$. This completes the proof of the identity $\Phi(f) = \Phi(\overline{f})^*$. $\qquad\square$

The following substitution rule extends Proposition 9.9.

Proposition 10.52. *Let* $(\Omega, \mathscr{F})$ *and* $(\Omega', \mathscr{F}')$ *be measurable spaces and let* $f : \Omega \to \Omega'$ *be a measurable mapping. If* $P : \mathscr{F} \to \mathscr{L}(H)$ *is a projection-valued measure, then the mapping* $Q : \mathscr{F}' \to \mathscr{L}(H)$ *defined by*

$$Q_B := P_{f^{-1}(B)}, \quad B \in \mathscr{F}',$$

is a projection-valued measure. Denoting by Φ *and* Ψ *the measurable functional calculi of* P *and* Q, *for all measurable functions* $g : \Omega' \to \mathbb{C}$ *we have*

$$\Phi(g \circ f) = \Psi(g)$$

with equality of domains.

Proof The proof is the same as that of Proposition 9.9, except that some domain issues have to be taken care of. Following this proof, for all nonnegative measurable functions g on Ω' and $x \in X$ we obtain

$$\int_\Omega g \circ f \, dP_x = \int_{\Omega'} g \, dQ_x, \tag{10.7}$$

the finiteness of one of these integrals implying the finiteness of the other. Applying this with g replaced by $|g|^2$, it follows that $x \in D(\Phi(g \circ f))$ if and only if $x \in D(\Psi(g))$. For all x in this common domain, (10.7) can be rewritten as $(\Phi(g \circ f)x|x) = (\Psi(g)x|x)$, and by polarisation this implies that for all x, y in this common domain we have $(\Phi(g \circ f)x|y) = (\Psi(g)x|y)$. The operator $\Psi(g)$, being normal, is densely defined and therefore this identity holds for all $y \in H$. The result now follows. $\qquad\square$

10.4 The Spectral Theorem for Unbounded Normal Operators

The proof of the spectral theorem for unbounded normal operators proceeds by a reduction to the bounded case. The basic idea is to exploit the fact that the mapping

$$\zeta : z \mapsto \frac{z}{(1+|z|^2)^{1/2}}$$

maps the complex plane bijectively onto the open unit disc $\mathbb{D}$. This suggests that if A is a normal operator, then

$$Z_A := A(I + A^\star A)^{-1/2}$$

is a normal contraction on H. This is indeed the case, as will be proved in Proposition 10.55. It follows that $\sigma(Z_A) \subseteq \overline{\mathbb{D}}$. By the spectral theorem for bounded normal operators, there exists a projection-valued measure Q on $\overline{\mathbb{D}}$ such that

$$Z_A = \int_{\overline{\mathbb{D}}} \lambda \, \mathrm{d}Q(\lambda).$$

We now define a projection-valued measure P on $\mathbb{C}$ by setting $P_B := Q_{\zeta(B)}$ for Borel sets $B \subseteq \mathbb{C}$, and use Proposition 10.52 to show that

$$A = \int_{\mathbb{C}} \lambda \, \mathrm{d}P(\lambda).$$

In the same way, the uniqueness of P for representing A is reduced to the uniqueness of Q for representing Z_A.

Some technical details need to be addressed to turn this simple idea into a rigorous proof: one has to deal with subtle domain issues and with the fact that ζ maps $\mathbb{C}$ onto the open unit disc, whereas Q is supported on the closed unit disc.

We start with the proof that Z_A is well defined as a contractive normal operator on H. This is accomplished in Proposition 10.55, for which we need two lemmas.

Lemma 10.53. *Let A be a closed operator in H, let $T \in \mathscr{L}(H)$, and let $f \in C(\sigma(T))$. Then:*

(1) *if T is selfadjoint and $TA \subseteq AT$, then $f(T)A \subseteq Af(T)$;*
(2) *if T is normal and $TA \subseteq AT$ and $T^\star A \subseteq AT^\star$, then $f(T)A \subseteq Af(T)$.*

Proof We only prove the second assertion, the first being an immediate consequence.

We have $T^2 A = T(TA) \subseteq T(AT) = (TA)T \subseteq (AT)T = AT^2$. Continuing by induction we see that $T^k A \subseteq AT^k$ for all $k \in \mathbb{N}$. In the same way it is seen that $(T^\star)^k A \subseteq AT(^\star)^k$ for all $k \in \mathbb{N}$. These inclusions imply that

$$p(T, T^\star)A \subseteq Ap(T, T^\star)$$

for all polynomials p in the variables z and $\bar{z}$; notation is as in Section 8.2.b. By the

Stone–Weierstrass theorem there exist polynomials p_n in the variables z and $\bar{z}$ such that $p_n(z,\bar{z}) \to f(z)$ uniformly with respect to $z \in \sigma(T)$. Then, by the properties of the continuous functional calculus for normal operators (Theorem 8.22),

$$\|p_n(T,T^\star) - f(T)\| = \sup_{z \in \sigma(T)} |p_n(z,\bar{z}) - f(z)| \to 0 \text{ as } n \to \infty.$$

The inclusions $p_n(T)A \subseteq Ap_n(T)$ imply that if $x \in D(A)$, then $p_n(T)x \in D(A)$ and

$$\lim_{n \to \infty} Ap_n(T)x = \lim_{n \to \infty} p_n(T)Ax = f(T)Ax.$$

Since also $\lim_{n \to \infty} p_n(T)x = f(T)x$, the closedness of A implies that $f(T)x \in D(A)$ and $Af(T)x = f(T)Ax$. This gives the result. □

We have seen in Theorem 10.46 that if A is a densely defined closed operator in H, then $A^\star A$ is selfadjoint, and by Proposition 10.42 we have $\sigma(A^\star A) \subseteq [0,\infty)$. This allows us to define

$$T_A := (I + A^\star A)^{-1}.$$

This operator is bounded and positive, and if A is normal we have $T_A = T_{A^\star}$.

Lemma 10.54. *If A is normal, then for all $x \in D(A)$ we have $T_A x \in D(A)$ and*

$$AT_A x = T_A Ax.$$

Proof Let $x \in D(A)$. Then $y := T_A x \in D(A^\star A) \subseteq D(A)$, $Ay = AT_A x \in D(A^\star)$, and $A^\star Ay = x - T_A x \in D(A)$, so $Ay \in D(AA^\star) = D(A^\star A)$. Combining this with $(I + AA^\star)A = A + (AA^\star)A = A + A(A^\star A) = A(I + A^\star A)$, it follows that

$$AT_A x = [T_A(I + AA^\star)]AT_A x = T_A A[(I + A^\star A)T_A]x = T_A Ax.$$

□

Proposition 10.55. *If A is a normal operator, then:*

(1) *the range of $T_A^{1/2}$ is densely contained in $D(A)$;*

(2) *the operator $Z_A := AT_A^{1/2}$ is contractive and we have $T_A = I - Z_A^\star Z_A$;*

(3) *Z_A is normal and $Z_A^\star = Z_{A^\star}$.*

Proof (1) and (2): We have $R(T_A) \subseteq D(A^\star A) \subseteq D(A)$ and therefore the operator AT_A is well defined on all of H. As a composition of a bounded operator and a closed operator, it is closed and therefore bounded by the closed graph theorem. The operator T_A is bounded and positive, and the injectivity of T_A implies the injectivity of its square root $T_A^{1/2}$. By selfadjointness and Proposition 4.31, this square root has dense range. For

$y = T_A^{1/2}x$ in this range we have $T_A^{1/2}y = T_Ax \in D(A^\star A) \subseteq D(A)$, so $y \in D(Z_A) := \{h \in H : T_A^{1/2}h \in D(A)\}$ and

$$\|Z_Ay\|^2 = \|AT_Ax\|^2 = (A^\star AT_Ax|T_Ax) = (x|T_Ax) - (T_Ax|T_Ax) \leqslant (x|T_Ax) = (y|y) = \|y\|^2.$$

Since the range of $T_A^{1/2}$ is dense, so is $D(Z_A)$ and therefore, with respect to the norm of H, Z_A is contractive from its dense domain into H. The operator Z_A is also closed, for if $y_n \to y$ in H with $y_n \in D(Z_A)$ and $Z_Ay_n = AT_A^{1/2}y_n \to y'$ in H, the closedness of A implies $T_A^{1/2}y \in D(A)$ and $AT_A^{1/2}y = y'$; but then $y \in D(Z_A)$ and $Z_Ay = y'$. Thus Z_A is closed, densely defined, and contractive with respect to the norm of H. This forces $D(Z_A) = H$.

We have already shown that the range of $T_A^{1/2}$ is contained in $D(A)$. To see that this inclusion is dense with respect to the graph norm it suffices to note that $R(T_A) = D(A^\star A)$ is dense in $D(A)$ with respect to the graph norm of $D(A)$ by Theorem 10.46. The inclusions $R(T_A) \subseteq R(T_A^{1/2}) \subseteq D(A)$ therefore imply that the inclusion $R(T_A^{1/2}) \subseteq D(A)$ is dense with respect to the graph norm of $D(A)$.

By Lemma 10.54, for $x \in D(A)$ we have $T_Ax \in D(A)$ and $AT_Ax = T_AAx$, so $T_AA \subseteq AT_A$, and then Lemma 10.53 implies

$$T_A^{1/2}A \subseteq AT_A^{1/2}. \tag{10.8}$$

Also, for $x \in D(A)$,

$$(Z_A^\star Z_Ax|x) = (AT_A^{1/2}x|AT_A^{1/2}x) = (T_A^{1/2}Ax|T_A^{1/2}Ax)$$
$$= (T_AAx|Ax) = (AT_Ax|Ax) = (A^\star AT_Ax|x) = ((I - T_A)x|x).$$

Since both T_A and Z_A are bounded, the identity $(Z_A^\star Z_Ax|x) = ((I - T_A)x|x)$ extends to arbitrary $x \in H$. This implies the operator identity $T_A = I - Z_A^\star Z_A$.

(3): Since A is normal we have $T_{A^\star} = T_A$. Since this operator is selfadjoint and $A^\star$ is normal, it follows that

$$Z_{A^\star} = A^\star T_{A^\star}^{1/2} = A^\star T_A^{1/2} = A^\star(T_A^{1/2})^\star \overset{\text{(i)}}{=} (T_A^{1/2}A)^\star \overset{\text{(ii)}}{\supseteq} (AT_A^{1/2})^\star = Z_A^\star,$$

where both (i) and (ii) follow from Proposition 10.19 and (10.8). Both Z_A and $Z_{A^\star}$ are bounded (the latter by Proposition 10.55 applied to $A^\star$), and therefore

$$Z_{A^\star} = Z_A^\star.$$

Normality of Z_A now follows from

$$(Z_A^\star Z_Ax|x) = (Z_Ax|Z_Ax) = (AT_A^{1/2}x|AT_A^{1/2}x) = (A^\star AT_A^{1/2}x|T_A^{1/2}x)$$

and

$$(Z_AZ_A^\star x|x) = (Z_A^\star x|Z_A^\star x) = (Z_{A^\star}x|Z_{A^\star}x) = (AA^\star T_{A^\star}^{1/2}x|T_{A^\star}^{1/2}x),$$

observing that the two right-hand sides are equal since A is normal. $\qquad\square$

Now we are ready for stating and proving the main result of this section.

Theorem 10.56 (Spectral theorem for normal operators). *For every normal operator A there exists a unique projection-valued measure P on $\sigma(A)$ such that*

$$A = \int_{\sigma(A)} \lambda \, dP(\lambda).$$

Proof Consider the mapping

$$\zeta : z \mapsto \frac{z}{(1+|z|^2)^{1/2}} \tag{10.9}$$

which maps the complex plane bijectively onto the open unit disc $\mathbb{D}$, with inverse

$$\zeta^{-1} : w \mapsto \frac{w}{(1-|w|^2)^{1/2}}.$$

Define the projection-valued measure P on $\mathbb{C}$ by

$$P_B := Q_{\zeta(B)}, \quad B \in \mathscr{B}(\mathbb{C}),$$

where Q is the projection-valued measure of the normal contraction Z_A, which is supported on $\sigma(Z_A)$. Since Z_A is contractive, $\sigma(Z_A)$ is contained in $\overline{\mathbb{D}}$. It will be convenient to think of Q as supported on $\overline{\mathbb{D}}$. The proof that P has the desired properties and is unique is carried out in several steps.

In what follows we let Φ and Ψ denote the measurable functional calculi of P and Q.

Step 1 – Let $\mathrm{id}(\lambda) := \lambda$. We begin by proving the inclusion

$$\mathrm{R}(T_A^{1/2}) \subseteq \mathrm{D}(\Phi(\mathrm{id})),$$

where $\Phi(\mathrm{id}) = \int_{\mathbb{C}} \mathrm{id} \, dP$.

Let $\rho \in C_c(\mathbb{C})$ satisfy $0 \leqslant \rho \leqslant 1$ pointwise. Using Proposition 10.52, the fact that $\rho \circ \zeta^{-1}$ has compact support in $\mathbb{D}$, and the fact that Q is supported on $\sigma(Z_A) \subseteq \overline{\mathbb{D}}$,

$$\int_{\mathbb{C}} \rho(z)|z|^2 \, dP_x(z) = \int_{\mathbb{D}} \rho(\zeta^{-1}(\lambda))|\zeta^{-1}(\lambda)|^2 \, dQ_x(\lambda)$$

$$= \int_{\overline{\mathbb{D}}} \rho(\zeta^{-1}(\lambda))|\zeta^{-1}(\lambda)|^2 \, dQ_x(\lambda)$$

$$= \int_{\sigma(Z_A)} \rho(\zeta^{-1}(\lambda))|\zeta^{-1}(\lambda)|^2 \, dQ_x(\lambda) = (\phi(Z_A)x|x),$$

where $\phi \in C(\sigma(Z_A))$ is the function

$$\phi(\lambda) = \rho(\zeta^{-1}(\lambda))|\zeta^{-1}(\lambda)|^2 = \rho(\zeta^{-1}(\lambda))|\lambda|^2(1-|\lambda|^2)^{-1}.$$

Suppose now that $x \in \mathsf{R}(T_A^{1/2})$, say

$$x = T_A^{1/2} y = (I - |Z_A|^2)^{1/2} y \tag{10.10}$$

for some $y \in H$; the second identity follows from $T_A = I - Z_A^\star Z_A = I - |Z_A|^2$. Using the multiplicativity of the continuous calculus of Z_A twice, with $\psi(\lambda) := \rho(\zeta^{-1}(\lambda))|\lambda|^2$ we have

$$(\phi(Z_A)x|x) = (\phi(Z_A)(I - |Z_A|^2)y|y) = (\psi(Z_A)y|y) = (\rho(\zeta^{-1}(Z_A))|Z_A|^2 y|y)$$

and therefore

$$\int_{\mathbb{C}} \rho(z)|z|^2 \, dP_x(z) = (\phi(Z_A)x|x) = (\rho(\zeta^{-1}(Z_A))|Z_A|^2 y|y) = \|\rho^{1/2}(\zeta^{-1}(Z_A))Z_A y\|^2$$

$$\leqslant \|\lambda \mapsto \rho^{1/2}(\zeta^{-1}(\lambda))\|_\infty \|Z_A y\|^2 \leqslant \|Z_A y\|^2 = \|Ax\|^2,$$

keeping in mind that $x \in \mathsf{R}(T_A^{1/2}) \subseteq \mathsf{D}(A)$. Applying this to a sequence $\rho_n \in C_c(\mathbb{C})$ satisfying $0 \leqslant \rho_n \uparrow 1$ pointwise as $n \to \infty$, by monotone convergence we obtain

$$\int_{\mathbb{C}} |\mathrm{id}|^2 \, dP_x = \int_{\mathbb{C}} |z|^2 \, dP_x(z) \leqslant \|Ax\|^2 < \infty.$$

This proves that $x \in \mathsf{D}(\Phi(\mathrm{id}))$.

Step 2 – We now prove that for $x = T_A^{1/2} y \in \mathsf{R}(T_A^{1/2})$ we have

$$\int_{\mathbb{C}} \lambda \, dP_x(\lambda) = (Ax|x).$$

Repeating the reasoning in Step 1 with $\rho \in C_c(\mathbb{C})$ as before, with

$$\widetilde{\phi}(\lambda) := \rho(\zeta^{-1}(\lambda))\zeta^{-1}(\lambda) = \rho(\zeta^{-1}(\lambda))\lambda(1 - |\lambda|^2)^{-1/2}$$

we obtain

$$\int_{\mathbb{C}} \rho(z)z \, dP_x(z) = (\widetilde{\phi}(Z_A)x|x) = (\rho(\zeta^{-1}(Z_A))Z_A(I - |Z_A|^2)^{1/2} y|y).$$

Applying this to a sequence $\rho_n \in C_c(\mathbb{C})$ satisfying $0 \leqslant \rho_n \uparrow 1$ pointwise as $n \to \infty$, by dominated convergence, the convergence property of the bounded functional calculus, and (10.10), we obtain

$$\int_{\mathbb{C}} z \, dP_x(z) = \lim_{n \to \infty} \int_{\mathbb{C}} \rho_n(z)z \, dP_x(z) = \lim_{n \to \infty} (\rho_n(\zeta^{-1}(Z_A))Z_A(I - Z_A^\star Z_A)^{1/2} y|y)$$

$$= (Z_A(I - Z_A^\star Z_A)^{1/2} y|y) = (Z_A T_A^{1/2} y|y) = (Ax|x).$$

Step 3 – Since both A and $\Phi(\mathrm{id})$ are closed, and since $\mathsf{R}(T_A^{1/2})$ is dense in $\mathsf{D}(A)$ by Proposition 10.55, the result of Step 2 implies that $A \subseteq \Phi(\mathrm{id})$. Since both operators are normal, the identity $A = \Phi(\mathrm{id}) = \int_{\mathbb{C}} \lambda \, dP(\lambda)$ follows from Proposition 10.49.

Step 4 – It remains to prove uniqueness of P. We will do so by reducing matters to the uniqueness of Q.

Suppose that

$$A = \int_{\sigma(A)} \lambda \, d\widetilde{P}(\lambda)$$

for a projection-valued measure $\widetilde{P}$ on $\sigma(A)$. Let $\widetilde{\Phi}$ denote the measurable calculus associated with $\widetilde{P}$. We have $\widetilde{\Phi}(\mathrm{id}) = A$ and $\widetilde{\Phi}(\overline{\mathrm{id}}) = A^\star$ by Theorem 10.50(2). By the multiplicativity,

$$\widetilde{\Phi}(\mathbf{1}_{\{|\mathrm{id}| \leqslant n\}} |\mathrm{id}|^2) = \widetilde{\Phi}(\mathbf{1}_{\{|\mathrm{id}| \leqslant n\}} \overline{\mathrm{id}}) \widetilde{\Phi}(\mathbf{1}_{\{|\mathrm{id}| \leqslant n\}} \mathrm{id}).$$

Taking limits $n \to \infty$ using Theorem 10.50(3), for $x \in D(A^\star A)$ we have $x \in D(\widetilde{\Phi}(|\mathrm{id}|^2))$ and

$$\widetilde{\Phi}(|\mathrm{id}|^2)x = \widetilde{\Phi}(\overline{\mathrm{id}})\widetilde{\Phi}(\mathrm{id})x = A^\star A x.$$

Similar arguments show that

$$\widetilde{\Phi}((1 + |\mathrm{id}|^2)^{-1})x = (\widetilde{\Phi}(1 + |\mathrm{id}|^2))^{-1}x = (I + A^\star A)^{-1}x = T_A x.$$

Since $D(A^\star A)$ is dense, this identity extends to arbitrary $x \in H$. Then, as in Step 1 of the proof of Theorem 10.56, the multiplicativity for the measurable calculus for bounded selfadjoint operators and the uniqueness of positive square roots gives

$$\widetilde{\Phi}((1 + |\mathrm{id}|^2)^{-1/2}) = T_A^{1/2}.$$

Hence,

$$\widetilde{\Phi}(\zeta)x = \widetilde{\Phi}(\mathrm{id}(1 + |\mathrm{id}|^2)^{-1/2})x = A T_A^{1/2} x = Z_A x, \tag{10.11}$$

where $\zeta : \mathbb{C} \to \mathbb{D}$ is the bijection of (10.9). Since $D(A^\star A)$ is dense in H, this identity extends to arbitrary $x \in H$.

Consider the projection-valued measure $\widetilde{Q}$ on $\mathbb{D}$ given by $\widetilde{Q}_B := \widetilde{P}_{\zeta^{-1}(B)}$. In what follows we view ζ as a measurable mapping from $\mathbb{C}$ to $\overline{\mathbb{D}}$. With $\rho \in C_c(\mathbb{C})$ as before, by (10.11) and Proposition 10.52 we have

$$(\widetilde{\Phi}(\rho)Z_A x | x) = (\widetilde{\Phi}(\rho)\widetilde{\Phi}(\zeta)x | x) = (\widetilde{\Phi}(\rho\zeta)x | x)$$
$$= \lim_{n \to \infty} \int_{\mathbb{C}} \mathbf{1}_{\{|\zeta| \leqslant n\}} \rho \zeta \, d\widetilde{P}_x = \lim_{n \to \infty} \int_{\mathbb{D}} \mathbf{1}_{\{|\mu| \leqslant n\}} \rho(\zeta^{-1}(\mu)) \mu \, d\widetilde{Q}_x(\mu)$$
$$= \int_{\mathbb{D}} \rho(\zeta^{-1}(\mu)) \mu \, d\widetilde{Q}_x(\mu) = \int_{\overline{\mathbb{D}}} \rho(\zeta^{-1}(\mu)) \mu \, d\widetilde{Q}_x(\mu).$$

Applying this to a sequence $\rho_n \in C_c(\mathbb{C})$ satisfying $0 \leqslant \rho_n \uparrow \mathbf{1}$ pointwise as $n \to \infty$, by the

convergence property of the bounded functional calculus and dominated convergence
we obtain

$$(Z_A x|x) = \lim_{n\to\infty}(\widetilde{\Phi}(\rho_n)Z_A x|x) = \lim_{n\to\infty}\int_{\mathbb{D}}\rho_n(\zeta^{-1}(\mu))\mu\,d\widetilde{Q}_x(\mu) = \int_{\mathbb{D}}\mu\,d\widetilde{Q}_x(\mu).$$

By Proposition 9.12, the support of $\widetilde{Q}$ is contained in $\sigma(Z_A)$, and therefore we have

$$Z_A = \int_{\sigma(Z_A)}\mu\,d\widetilde{Q}(\mu).$$

This shows that $\widetilde{Q}$ is the projection-valued measure of Z_A. Now Proposition 9.13 implies
that $\widetilde{Q} = Q$ and hence $\widetilde{P} = P$. □

For normal operators A and measurable functions $f : \sigma(A) \to \mathbb{C}$, the operator $\Phi(f)$
defined in terms of the projection-valued measure P of A by the calculus of Theorem
10.50 will be denoted by $f(A)$:

$$f(A) := \Phi(f) = \int_{\sigma(A)}f\,dP.$$

In the same way as for bounded normal operators in Theorem 9.19, the properties of
the bounded calculus Φ translate into corresponding properties for the mapping $f \mapsto
f(A)$. The result of Example 10.51 says that for any normal operator A and measurable
function $f : \sigma(A) \to \mathbb{C}$,

$$(f^n)(A) = (f(A))^n, \quad n = 1, 2, \ldots$$

If P is a projection-valued measure on a measurable space $(\Omega, \mathscr{F})$ and $f : \Omega \to \mathbb{C}$ is
measurable, the *P-essential range* of f is the set $\mathsf{R}_P(f)$ of all $z \in \mathbb{C}$ such that $P_{E_{z,r}} \neq 0$
for all $r > 0$, where

$$E_{z,r} := \{\omega \in \Omega : |f(\omega) - z| < r\}.$$

It is easy to see that $\mathsf{R}_P(f)$ is a closed set contained in $\overline{f(\Omega)}$.

Theorem 10.57 (Spectral mapping theorem). *Let A be normal with projection-valued
measure P, and let $f : \sigma(A) \to \mathbb{C}$ be measurable. Then*

$$\sigma(f(A)) = \mathsf{R}_P(f) \subseteq \overline{f(\sigma(A))}.$$

If f is continuous, then

$$\sigma(f(A)) = \overline{f(\sigma(A))}.$$

Proof   Let $z \in \complement\mathsf{R}_P(f)$. Since $\mathsf{R}_P(f)$ is closed, the function $g_z : \lambda \mapsto (z - f(\lambda))^{-1}$ is
well defined P-almost everywhere and bounded on $\sigma(A)$, and therefore the operator
$g_z(A)$ is bounded. Moreover, $(z - f(A))g_z(A) = g_z(A)(z - f(A)) = A$ by Theorem 10.50
and the boundedness of g_z. It follows that $g_z(A)$ is a two-sided inverse for $z - f(A)$. This
proves the inclusion $\sigma(f(A)) \subseteq \mathsf{R}_P(f)$.

Suppose next that $z \in R_P(f)$. Since P is supported on $\sigma(A)$, for $n = 1, 2, \ldots$ the orthogonal projections $P_{E_{z,1/n}}$ are nonzero, where $E_{z,1/n} = \{\lambda \in \sigma(A) : |f(\lambda) - z| < \frac{1}{n}\}$. In particular, $E_{z,1/n} \neq \varnothing$, and this implies that $d(z, f(\sigma(A))) < \frac{1}{n}$. This being true for all $n \geqslant 1$, it follows that $z \in \overline{f(\sigma(A))}$.

If f is continuous and $\mu \in \mathbb{C}$ is such that $f(\mu) \notin \sigma(f(A)) = R_P(f)$, then for $\varepsilon > 0$ small enough the relatively open set

$$N := \{\lambda \in \sigma(A) : |f(\lambda) - f(\mu)| < \varepsilon\}$$

satisfies $P_N = 0$. The continuity of f implies that $B(\mu; \delta) \cap \sigma(A) \subseteq N$ for some small enough $\delta > 0$. But then $\mu \notin R_P(\mathrm{id}) = \sigma(\mathrm{id}(A)) = \sigma(A)$. This proves the inclusion $f(\sigma(A)) \subseteq \sigma(f(A))$, which self-improves to $\overline{f(\sigma(A))} \subseteq \sigma(f(A))$ since $\sigma(f(A))$ is closed. □

The following result gives necessary and sufficient condition for the presence of eigenvalues for the operators $f(A)$.

Theorem 10.58 (Eigenvalues). *Let A be a normal operator with projection-valued measure P, and let $f : \sigma(A) \to \mathbb{C}$ be measurable. For $\mu \in \mathbb{C}$ let $N_f(\mu) := \{\lambda \in \sigma(A) : f(\lambda) = \mu\}$. The following assertions are equivalent:*

(1) *μ is an eigenvalue of $f(A)$;*
(2) *$P_{N_f(\mu)} \neq 0$.*

In this situation, $P_{N_f(\mu)}$ is the orthogonal projection onto the corresponding eigenspace, and for a vector $x \in H$ the following assertions are equivalent:

(3) *$x \in D(f(A))$ and $f(A)x = \mu x$;*
(4) *$P_{N_f(\mu)}x = x$.*

Proof Upon replacing f by $f - \mu$ we may assume that $\mu = 0$. Set $N_f := N_f(0)$ for brevity.

If $x \in D(f(A))$ satisfies $f(A)x = 0$, then $f(\lambda) = 0$ for P_x-almost all $\lambda \in \sigma(A)$ by (10.4). This is equivalent to saying that P_x-almost every point of $\sigma(A)$ is contained in N_f, that is, $P_x(\sigma(A) \setminus N_f) = 0$. This, in turn, is equivalent to saying that $P_{\sigma(A) \setminus N_f}x = 0$, that is, $x - P_{N_f}x = 0$. Conversely, if $P_{N_f}x = x$, or equivalently, if $f = 0$ P_x-almost everywhere on $\sigma(A)$, then $x \in D(\Phi(f)) = D(f(A))$ by the definition of $D(\Phi(f))$ in Theorem 10.50.

This proves the equivalence (3)⇔(4). This equivalence also establishes that P_{N_f} is the orthogonal projection onto the eigenspace $\{x \in D(f(A)) : f(A)x = 0\}$. It further shows that if 0 is an eigenvalue of $f(A)$, with eigenvector $x \in D(f(A))$, then

$$P_{N_f}x = P_{\sigma(A)}x - P_{\sigma(A) \setminus N_f}x = P_{\sigma(A)}x = \|x\|^2 \neq 0$$

since $x \neq 0$. This proves the implication (1)⇒(2). If (2) holds, there exists a nonzero $x \in H$ with $P_{N_f(\mu)}x = x$, and (1) follows from the implication (4)⇒(3). □

Corollary 10.59. *Let A be a normal operator and let $f : \sigma(A) \to \mathbb{C}$ be measurable. If $x \in D(A)$ satisfies $Ax = \lambda x$, then $x \in D(f(A))$ and $f(A)x = f(\lambda)x$.*

As in the bounded case, we can use the functional calculus to define square roots:

Proposition 10.60. *If A is a positive selfadjoint operator, then A admits a unique positive selfadjoint square root $A^{1/2}$.*

Proof The operator $f(A)$ with $f(\lambda) = \lambda^{1/2}$ is selfadjoint and positive, and squares to A by the result of Example 10.51. This proves existence.

To prove uniqueness, suppose that B is a positive selfadjoint operator satisfying $B^2 = A$. Let P and Q be the projection-valued measures of A and B; both are supported on $[0, \infty)$ since A and B are selfadjoint and positive. Let R be the projection-valued measure on $[0, \infty)$ defined by $R_C = Q_{C^2}$ for Borel sets $C \subseteq [0, \infty)$. By Theorem 10.52 and the result of Example 10.51,

$$\int_{[0,\infty)} \lambda \, dR = \int_{[0,\infty)} \lambda^2 \, dQ = B^2 = A = \int_{[0,\infty)} \lambda \, dP.$$

It follows that both R and P are projection-valued measures representing A. By the uniqueness part of Theorem 10.56 we therefore have $R = P$. But then

$$A^{1/2} = \int_{[0,\infty)} \lambda^{1/2} \, dP = \int_{[0,\infty)} \lambda^{1/2} \, dR = \int_{[0,\infty)} \lambda \, dQ = B.$$

$\square$

If A is normal, we may use the measurable calculus to define $|A| := f(A)$, where $f(\lambda) = |\lambda|$. Furthermore, $A^\star A$ is selfadjoint and positive, so it has a unique selfadjoint and positive square root $(A^\star A)^{1/2}$ by Proposition 10.60. The next corollary extends Corollary 8.29 to unbounded normal operators:

Corollary 10.61. *For every normal operator A we have $D(|A|) = D(A)$ and*

$$(A^\star A)^{1/2} = |A|.$$

Proof With $\mathrm{id}(\lambda) = \lambda$ we have $D(|A|) = D(\Phi(|\mathrm{id}|)) = D(\Phi(\mathrm{id})) = D(A)$, the middle identity being immediate from the definition of these domains. Applying the result of Example 10.51 twice we obtain, with $f(\lambda) = |\lambda|$,

$$|A|^2 = f(A)f(A) = f^2(A) = (\overline{\mathrm{id}} \circ \mathrm{id})(A) = \overline{\mathrm{id}}(A)\mathrm{id}(A) = A^\star A,$$

where the penultimate identity is proved as in Example 4.9. The identity $|A| = (|A|^2)^{1/2}$ now follows by taking positive square roots using Proposition 10.60. $\square$

We proceed with some examples.

Example 10.62 (Multiplication operators). Let $(\Omega, \mathscr{F}, \mu)$ be a measure space and let $m : \Omega \to \mathbb{C}$ be a measurable function. The linear operator A_m defined by

$$D(A_m) := \{f \in L^2(\Omega, \mu) : mf \in L^2(\Omega, \mu)\},$$
$$A_m f := mf, \quad f \in D(A_m),$$

is normal, its spectrum $\sigma(A_m)$ equals the μ-essential range of m, and its projection-valued measure is given by

$$P_B f = \mathbf{1}_{m^{-1}(B)} f$$

for $B \in \mathscr{B}(\sigma(A_m))$ and $f \in L^2(\Omega, \mu)$.

Example 10.63 (Fourier multiplication operators). Let $m : \mathbb{R}^d \to \mathbb{C}$ be a measurable function. The linear operator T_m defined by

$$D(T_m) := \{f \in L^2(\mathbb{R}^d) : \mathscr{F}f \in D(A_m)\},$$
$$T_m f := \mathscr{F}^{-1} A_m \mathscr{F} f, \quad f \in D(T_m),$$

where A_m is the operator of the preceding example, is normal, its spectrum equals $\sigma(T_m) = \sigma(A_m)$, and its projection-valued measure is given by

$$P_B f = \mathscr{F}^{-1}(\mathbf{1}_{m^{-1}(B)} \mathscr{F}f) = T_{\mathbf{1}_{m^{-1}(B)}}$$

for $B \in \mathscr{B}(\sigma(T_m))$ and $f \in L^2(\mathbb{R}^d)$. Thus, the projections in the range of the projection-valued measure of the Fourier multiplier operator T_m are Fourier multiplier operators themselves.

To conclude we complete the proof of Theorem 9.28.

Proof of Theorem 9.28, the inclusion '$\subseteq$' We split the proof into four steps.

Step 1 – Let $(h_n)_{n \geqslant 1}$ be a dense sequence in H and set

$$g_n := h_n - \sum_{k=1}^{n-1} \pi_k h_n, \quad n \geqslant 1,$$

where π_k is the orthogonal projection onto the closed linear span H_k of the sequence $(T^j g_k)_{j \in \mathbb{N}}$. We further set $\pi_0 := 0$.

We claim that $T\pi_k = \pi_k T$ for all $k \in \mathbb{N}$. For $k = 0$ this is trivial and for $k \geqslant 1$ this follows from the following argument. If $h \in R(\pi_k) = H_k$, then $Th \in H_k = R(\pi_k)$ and therefore $T\pi_k h = Th = \pi_k Th$. If $h \perp R(\pi_k)$, then $h \perp T^j g_k$ for all $j \in \mathbb{N}$ and then $(Th | T^j g_k) = (h | T^{j+1} g_k) = 0$ by selfadjointness, so $Th \perp T^j g_k$ for all $j \in \mathbb{N}$ and consequently $Th \perp R(\pi_k)$; this implies $T\pi_k h = 0 = \pi_k Th$.

We claim that

$$\pi_j \pi_k = \pi_k \pi_j = 0, \quad j, k \in \mathbb{N}, \ j \neq k. \tag{10.12}$$

It is clear that (10.12) holds for $j = 0$ and $k = 1$. Suppose (10.12) has been proved for all indices $0 \leqslant j \neq k \leqslant n - 1$. If $j < n$, then

$$\pi_j g_n = \pi_j h_n - \pi_j \sum_{k=1}^{n-1} \pi_k h_n = \pi_j h_n - \pi_j h_n = 0$$

by the induction hypothesis. In particular,

$$\pi_j T^k g_n = T^k \pi_j g_n = 0, \quad k \in \mathbb{N},$$

from which it follows that $H_n = \mathsf{R}(\pi_n)$ is contained in $\mathsf{N}(\pi_j)$, so $H_n \perp H_j$. This implies that $\pi_j \pi_n = \pi_n \pi_j = 0$, completing the induction step.

Next we claim that $\pi h := \sum_{j \geqslant 1} \pi_j h = h$ for all $h \in H$. By density it suffices to prove this for the vectors $h = h_n$ from the dense sequence. For these vectors,

$$\pi h_n = \pi g_n + \pi \sum_{k=1}^{n-1} \pi_k h_n = g_n + \sum_{k=1}^{n-1} \pi_k h_n = h_n,$$

where we used that $g_n \in H_n$ (by the definition of H_n) along with the orthogonality of the projections π_n to see that $\pi g_n = \pi_n g_n = g_n$ and $\pi \pi_k = \pi_k$.

Choose a sequence of numbers $c_n > 0$ such that the sum $\sum_{n \geqslant 1} c_n g_n$ converges; for instance, one could choose $c_n = 2^{-n}(\|g_n\| + 1)^{-n}$. Denote its sums by x. Let H_x denote the closed linear span of the sequence $(T^k x)_{k \in \mathbb{N}}$ and let π_x denote the orthogonal projection onto H_x. By the same argument as before the operators π_x and T commute.

Step 2 – Let $S \in \{T\}''$ be arbitrary and fixed. Since $x \in H_x$ and $\pi_x \in \{T\}'$ it follows that $S\pi_x = \pi_x S$. As a result, $Sx \in H_x$. Since H_x is the closed linear span of the vectors $T^j x$, $j \in \mathbb{N}$, it follows that there exist polynomials p_n such that $p_n(T)x \to Sx$. Then,

$$\|p_n(T)x - p_m(T)x\|^2 = \int_{\sigma(T)} |p_n - p_m|^2 \, dP_x,$$

where P is the projection-valued measure of T. Since the left-hand side tends to 0 as $m, n \to \infty$, the sequence $(p_n)_{n \geqslant 1}$ is Cauchy in $L^2(\sigma(T), P_x)$. Let $f \in L^2(\sigma(T), P_x)$ be its limit and pick any pointwise defined measurable function representing it. With slight abuse of notation, this representative is denoted by f again.

Step 3 – Relative to the projection-valued measure P, by the measurable calculus of Theorem 10.50 the operator $f(T) := \Phi(f)$ with domain

$$\mathsf{D}(f(T)) = \left\{ h \in H : \int_\Omega |f|^2 \, dP_h < \infty \right\}$$

is well defined and normal. To complete the proof we show that $\mathsf{D}(f(T)) = H$, $f(T)$ is bounded (hence *a posteriori* f can be taken to be bounded), and $S = f(T)$.

We begin by noting that $x \in \mathsf{D}(f(T))$ and

$$Sx = \int_{\sigma(T)} f \, dP_x = \lim_{n \to \infty} \int_{\sigma(T)} p_n \, dP_x = \lim_{n \to \infty} p_n(T)x = f(T)x.$$

If $V \in \mathscr{L}(H)$ commutes with T, then $V \in \{T\}'$, so V commutes with S. We claim that $Vx \in D(f(T))$ and

$$f(T)Vx = Vf(T)x = VSx.$$

Indeed, first we note that

$$\int_{\sigma(T)} |p_n - p_m|^2 \, dP_{Vx} = \|(p_n(T) - p_m(T))Vx\|^2$$
$$= \|V(p_n(T) - p_m(T))x\|^2 \to \|V(S-S)x\|^2 = 0.$$

Since $p_n \to f$ in $L^2(\sigma(T), P_x)$, after passing to a subsequence if necessary, we may assume that $p_n \to f$ pointwise outside a Borel set N satisfying $P_x(N) = 0$. But, since V commutes with P_N,

$$|P_{Vx}(N)| = \left| \int_{\sigma(T)} 1_N P_{Vx} \right| = |(P_N Vx|Vx)| = |(P_N x|V^\star Vx)|$$
$$\leqslant \|P_N x\| \|V^\star Vx\| = (P_N x|x)^{\frac{1}{2}} \|V^\star Vx\| = 0.$$

This implies that also $p_n \to f$ pointwise P_{Vx}-almost everywhere. This allows us to invoke the dominated convergence theorem to conclude that for every $N \geqslant 1$,

$$\int_{\sigma(T)} |f|^2 \wedge N \, dP_{Vx} = \lim_{n \to \infty} \int_{\sigma(T)} |p_n|^2 \wedge N \, dP_{Vx}$$
$$\leqslant \limsup_{n \to \infty} \int_{\sigma(T)} |p_n|^2 \, dP_{Vx}$$
$$= \limsup_{n \to \infty} \|p_n(T)Vx\|^2$$
$$= \limsup_{n \to \infty} \|Vp_n(T)x\|^2$$
$$\leqslant \|V\|^2 \limsup_{n \to \infty} \|p_n(T)x\|^2 = \|V\|^2 \int_{\sigma(T)} |f|^2 \, dP_x.$$

Letting $N \to \infty$ and applying the monotone convergence theorem, this proves that $Vx \in D(f(T))$. In the same way, for all $m \in \mathbb{N}$ we have

$$\|(f - p_n(T))Vx\|^2 = \int_{\sigma(T)} |f - p_m|^2 \, dP_{Vx} = \limsup_{n \to \infty} \|V(p_n - p_m)(T)x\|^2.$$

Choose $N \geqslant 1$ so large that $\|p_n(T)x - p_m(T)x\| < \varepsilon$ for all $n, m \geqslant N$. For such indices we have, using that $p_m(T)V = Vp_m(T)$,

$$\|f(T)Vx - Vf(T)x\| \leqslant \|f(T)Vx - p_m(T)Vx\| + \|Vp_m(T)x - Vf(T)x\|^2$$
$$= \limsup_{n \to \infty} \|V(p_n - p_m)(T)x\|^2 + \lim_{n \to \infty} \|V(p_m - p_n)(T)x\|^2$$
$$\leqslant 2\|V\|^2 \varepsilon^2.$$

Since $\varepsilon > 0$ was arbitrary, then proves that $f(T)Vx = Vf(T)x$. The required identity now follows from

$$f(T)Vx = Vf(T)x = VSx = SVx.$$

Step 4 – We apply the result of Step 3 to the bounded operator $V_{kmn} := c_m^{-1} P_n T^k \pi_m$, where $P_n := P_{\{|f| \leqslant n\}}$. In view of $\pi_m x = \sum_{n \geqslant 1} c_n \pi_m g_n = c_m g_m$ we have

$$f(T)P_n T^k g_m = c_m^{-1} f(T) P_n T^k \pi_m x = f(T) V_{kmn} x$$
$$= SV_{kmn}x = c_m^{-1} SP_n T^k \pi_m x = SP_n T^k g_m.$$

It follows that $f(T)P_n h = SP_n h$ for all h in the linear span of the sequence $(T^k g_m)_{k \in \mathbb{N}}$, which is dense in H_m. Since the spaces H_m, $m \geqslant 1$, span a dense subspace of H this implies that there is a dense subspace Y of H such that $f(T)P_n h = SP_n h$ for all $n \geqslant 1$ and $h \in Y$. Since $f(T)P_n$ is bounded, this identity extends to arbitrary $h \in H$.

Fix an arbitrary $h \in D(f(T))$. Then $P_n h \to h$ and $f(T)P_n h = SP_n h \to Sh$, so the closedness of $f(T)$ implies that $f(T)h = Sh$. Since S is bounded and $f(T)$ is closed, this forces $D(f(T)) = H$ and $f(T) = S$.   □

Problems

10.1 Show that $\lambda \mapsto R(\lambda, A)$ is holomorphic as an $\mathscr{L}(X, D(A))$-valued mapping.

10.2 Let A be a densely defined operator in a Banach space X which is bounded with respect to the norm of X, that is, there is a constant $C \geqslant 0$ such that $\|Ax\| \leqslant C\|x\|$ for all $x \in D(A)$. Prove that A is closable, $D(\overline{A}) = X$, and $\overline{A}$ is bounded with $\|\overline{A}x\| \leqslant C\|x\|$ for all $x \in X$.

10.3 Let A be a densely defined closed operator in a Banach space X and suppose there is a subspace Y, contained in $D(A)$ and dense in X, such that $Ay = 0$ for all $y \in Y$. Does it follow that $Ax = 0$ for all $x \in D(A)$? What happens if the closedness assumption is dropped?

10.4 Show that if A and B are linear operators in a complex Hilbert space such that $D(A) = D(B)$ and $(Ax|x) = (Bx|x)$ for all $x \in D(A) = D(B)$, then $A = B$.

10.5 Define the linear operator A in $L^2(0,1)$ by $D(A) := C[0,1]$ and

$$Af := f(0)\mathbf{1}, \quad f \in D(A).$$

Show that A is densely defined but nonclosable.

10.6 Let A be any nonclosable operator in a Hilbert space H. Show that the operator B on the Hilbert space direct sum $H \oplus H$ defined by $D(B) := D(A) \oplus \{0\}$ and

$$B(x,0) := (0,Ax), \quad (x,0) \in D(B),$$

is symmetric and nonclosable. This example shows that the densely definedness assumption cannot be omitted from Proposition 10.36.

10.7 Provide the details to Example 10.7.

10.8 Let M be an arbitrary nonempty closed subset of $\mathbb{C}$. In the Banach space $C_b(M)$ of bounded continuous functions on M consider the linear operator A given by

$$D(A) := \{f \in C_b(M) : z \mapsto zf(z) \in C_b(M)\},$$
$$Af(z) := zf(z), \quad f \in D(A), \, z \in M.$$

(a) Show that A is a closed operator.

(b) Show that $\sigma(A) = M$.

10.9 In $C[0,1]$ consider the linear operators $A_1 f := f'$ and $A_2 f := f'$ with domains $D(A_1) := C^\infty[0,1]$ and $D(A_2) := C_c^\infty(0,1)$.

(a) Show that A_1 is closable and find the domain of its closure.

(b) Show that A_2 is closable and find the domain of its closure.

10.10 Let A be a densely defined closed linear operator from a Banach space X to a Banach space Y. Prove or disprove:

(a) for all $T \in \mathscr{L}(X)$, the operator AT with domain $D(AT) = \{x \in X : Tx \in D(A)\}$ is closed;

(b) for all $T \in \mathscr{L}(Y)$ the operator TA with domain $D(TA) = D(A)$ is closed.

10.11 Let A be a densely defined closed linear operator from a Banach space X to a Banach space Y. Prove or disprove:

(a) for all $T \in \mathscr{L}(X)$ we have $D((AT)^*) = D(T^*A^*)$;

(b) for all $T \in \mathscr{L}(Y)$ we have $D((TA)^*) = D(A^*T^*)$.

10.12 Give a direct proof of Proposition 10.44.

10.13 Give a proof of Proposition 10.24.

10.14 Let $(\Omega, \mathscr{F}, \mu)$ be a measure space, X be a Banach space, and suppose that $f : \Omega \to X$ is Bochner integrable with respect to μ.

(a) Prove *Hille's theorem*: If A is a closed linear operator in a Banach space X, f takes its values in $D(A)$ μ-almost everywhere, and the μ-almost everywhere defined function $Af : \Omega \to X$ is μ-Bochner integrable, then $\int_\Omega f \, d\mu \in D(A)$ and

$$A \int_\Omega f \, d\mu = \int_\Omega Af \, d\mu.$$

Hint: Show that $\omega \mapsto (f(\omega), Af(\omega))$ is Bochner integrable as a function with values in $X \times X$ and hence, by the result of Problem 1.26, as a function with values in the graph $G(A)$.

(b) Justify the identity

$$\frac{\partial}{\partial t} \int_0^1 f(t,s)\,\mathrm{d}s = \int_0^1 \frac{\partial}{\partial t} f(t,s)\,\mathrm{d}s$$

by providing conditions on f so that the result of part (a) can be applied.

10.15 Extend the results of Problems 9.3 and 9.4 to unbounded selfadjoint operators.

10.16 Prove the claims in Examples 10.62 and 10.63.

10.17 Combining Examples 10.41 and 10.63, find the projection-valued measure of the Laplace operator Δ on $L^2(\mathbb{R}^d)$, viewed as a selfadjoint operator in this space with domain $\mathsf{D}(\Delta) = H^2(\mathbb{R}^d)$.

10.18 Let A be a normal operator with projection-valued measure P, and let $B \subseteq \sigma(A)$ be a bounded Borel subset. Show that $P_B x \in \mathsf{D}(A)$ for all $x \in H$.

10.19 Let A be a selfadjoint operator with projection-valued measure P. Show that for all $\mu \in \mathbb{C} \setminus \mathbb{R}$ we have the following formula for the resolvent of A:

$$R(\mu,A) = \int_{\sigma(A)} \frac{1}{\lambda - \mu}\,\mathrm{d}P(\lambda).$$

10.20 Let A be a normal operator with projection-valued measure P, and let $f,g \in B_b(\sigma(A))$. Show that if $f = g$ P-almost everywhere (in the sense that there is a Borel set N such that $P_N = 0$ and $f = g$ on $\complement N$), then $f(A) = g(A)$.

10.21 Let A be a normal operator.

(a) Show that A is bounded if and only if $\sigma(A)$ is bounded.

(b) Find necessary and sufficient conditions on a given Borel function f on $\sigma(A)$ in order that $f(A)$ be bounded.

10.22 Let A be a normal operator and let f be a Borel function on $\sigma(A)$. Show that $f(A)$ is injective with dense range if and only if $f \neq 0$ P-almost everywhere, where P is the projection-valued measure of A, and that in this case we have $(f(A))^{-1} = (1/f)(A)$.

Hint: Explain how $(1/f)(A)$ can be defined through the measurable functional calculus. Then use Theorem 10.50 to check that $\mathsf{D}((1/f)(A)(f(A)) = \mathsf{D}(f(A))$. Conclude that $(f(A))^{-1} \subseteq (1/f)(A)$. To get the reverse inclusion apply the preceding to f^{-1} instead of f.

11

Boundary Value Problems

Having developed some of the core results of Functional Analysis, we now turn to applications to partial differential equations. This chapter is concerned with boundary value problems.

11.1 Sobolev Spaces

We begin by developing some elements of the theory of Sobolev spaces. Our aims are relatively modest, in that we only discuss those aspects of the theory that are needed for the purposes of the present chapter.

Throughout this chapter we assume that $d \geqslant 1$ is an integer and D is a nonempty open subset of $\mathbb{R}^d$.

Multi-Index Notation A d-tuple $\alpha = (\alpha_1, \ldots, \alpha_d) \in \mathbb{N}^d$ is called a *multi-index* of dimension d. Its *order* is the nonnegative integer

$$|\alpha| := \alpha_1 + \cdots + \alpha_d.$$

We also define

$$\alpha! := \alpha_1! \cdots \alpha_d!.$$

We write $\alpha \leqslant \beta$ if $\alpha_j \leqslant \beta_j$ for all $j = 1, \ldots, d$, and in such cases we define $\alpha - \beta := (\alpha_1 - \beta_1, \ldots, \alpha_d - \beta_d)$ and

$$\binom{\alpha}{\beta} := \frac{\alpha!}{\beta!(\alpha - \beta)!}.$$

For $x \in \mathbb{R}^d$ and $\alpha \in \mathbb{N}^d$ we set

$$x^\alpha := x_1^{\alpha_1} \cdots x_d^{\alpha_d}.$$

349

Similarly we define

$$\partial^\alpha := \partial_1^{\alpha_1} \circ \cdots \circ \partial_d^{\alpha_d},$$

where ∂_j is the partial derivative in the jth direction. By a standard Calculus result, for $C^{|\alpha|}$-functions the order in which the derivatives are taken is unimportant.

Test Functions By $C^\infty(D)$ we denote the space of functions $f : D \to \mathbb{K}$ having continuous derivatives $\partial^\alpha f$ on D of all orders $\alpha \in \mathbb{N}^d$, and by $C_c^\infty(D)$ its subspace consisting of all functions compactly supported in D; recall that this means that the closure of the set $\{x \in D : f(x) \neq 0\}$ is a compact set contained in D. Elements of $C_c^\infty(D)$ are referred to as *test functions* on D. The existence of test functions with various additional properties is established in Problem 2.9.

In the same way one defines the space $C^k(D)$, $k \in \mathbb{N}$, as the space of functions $f : D \to \mathbb{K}$ having continuous derivatives $\partial^\alpha f$ on D of all orders $\alpha \in \mathbb{N}^d$ satisfying $|\alpha| \leqslant k$ (with the convention that $C^0(D) = C(D)$), and $C_c^k(D)$ as its subspace of compactly supported functions.

By $C^\infty(\overline{D})$ we denote the space of all functions in $C^\infty(D)$ and with the property that $\partial^\alpha f$ has a continuous extension to $\overline{D}$ for all $\alpha \in \mathbb{N}^d$. The spaces $C^k(\overline{D})$ are defined similarly, by considering only the multi-indices satisfying $|\alpha| \leqslant k$.

A measurable function $f : D \to \mathbb{K}$ is called *locally integrable* if its restriction to every open set U with compact closure contained in D is integrable. The space of all locally integrable functions $f : D \to \mathbb{K}$ is denoted by $L^1_{\text{loc}}(D)$; as always we identify functions that are equal almost everywhere. In our study of weak derivatives we need the following result on convolutions.

Proposition 11.1. *Let k be a nonnegative integer. If $f \in L^1_{\text{loc}}(\mathbb{R}^d)$ and $g \in C_c^k(\mathbb{R}^d)$, then the convolution $f * g$ is pointwise well defined and belongs to $C^k(\mathbb{R}^d)$, and we have*

$$\partial^\alpha(f * g) = f * (\partial^\alpha g)$$

for all multi-indices $\alpha \in \mathbb{N}^d$ satisfying $|\alpha| \leqslant k$.

Proof First note that the convolution integrals defining $f * g$ and $f * (\partial^\alpha g)$ are pointwise well defined as Lebesgue integrals.

Step 1 – We begin with the case $k = 0$. Let $f \in L^1_{\text{loc}}(\mathbb{R}^d)$ and $g \in C_c(\mathbb{R}^d)$ be given. Choose $r > 0$ such that the support of g is contained in the ball $B(0; r)$. By uniform continuity, given $\varepsilon > 0$ there exists $\delta > 0$ such that for all $u, u' \in \mathbb{R}^d$ with $|u - u'| < \delta$ we have $|g(u) - g(u')| < \varepsilon$. Hence, for all $x, x' \in \mathbb{R}^d$ with $|x - x'| < \delta$,

$$|(f * g)(x) - (f * g)(x')| \leqslant \int_{\mathbb{R}^d} |f(y)| |g(x - y) - g(x' - y)| \, dy$$

$$= \int_{B(x; r + \delta)} |f(y)| |g(x - y) - g(x' - y)| \, dy$$

$$\leqslant \varepsilon \int_{B(x;r+\delta)} |f(y)|\,dy,$$

noting that $g(x-y) = g(x'-y) = 0$ for $y \notin B(x;r+\delta)$. This proves the continuity of $f * g$ at the point $x \in \mathbb{R}^d$.

Step 2 – Next consider the case $k = 1$. Fix $1 \leqslant j \leqslant d$ and let $e_j \in \mathbb{R}^d$ denote the unit vector along the jth coordinate axis. Let $r > 0$ be such that the support of g is contained in the ball $B(0;r)$. Given $\varepsilon > 0$, choose $\delta > 0$ such that $|u - u'| < \delta$ implies $|\partial_j g(u) - \partial_j g(u')| < \varepsilon$. If $0 < h < \delta$, then for all $y \in \mathbb{R}^d$ we obtain

$$\left| \frac{1}{h}\big(g(y+he_j) - g(y)\big) - \partial_j g(y) \right| = \left| \frac{1}{h}\int_0^h \partial_j g(y+te_j) - \partial_j g(y)\,dt \right|$$

$$\leqslant \frac{1}{h}\int_0^h |\partial_j g(y+te_j) - \partial_j g(y)|\,dt \leqslant \varepsilon.$$

Taking the supremum over y, this shows that

$$\lim_{h \downarrow 0} \left\| \frac{1}{h}\big(g(\cdot+he_j) - g(\cdot)\big) - \partial_j g(\cdot) \right\|_\infty = 0.$$

As a consequence, for all $x \in \mathbb{R}^d$ we have

$$\lim_{h \to 0} \frac{1}{h}\big((f*g)(x+he_j) - (f*g)(x)\big)$$

$$= \lim_{h \to 0} \frac{1}{h}\int_{\mathbb{R}^d} f(y)\big(g(x+he_j-y) - g(x-y)\big)\,dy$$

$$= \lim_{h \to 0} \frac{1}{h}\int_{B(0;r+\delta)} f(x-y)\big(g(y+he_j) - g(y)\big)\,dy$$

$$= \int_{B(0;r+\delta)} f(x-y)\partial_j g(y)\,dy = \int_{\mathbb{R}^d} f(x-y)\partial_j g(y)\,dy,$$

where the penultimate step is justified by the uniform convergence of the difference quotient and the fact that f is integrable on bounded sets. This proves the differentiability of $f * g$ in the jth direction, with $\partial_j(f*g) = f*(\partial_j g)$, and the derivative is continuous by Step 1 applied to the function $\partial_j g \in C_c(\mathbb{R}^d)$.

Step 3 – The result for $k \geqslant 2$ follows by repeating the argument of Step 2 inductively. $\qquad\square$

The following version of Theorem C.12 will be useful.

Proposition 11.2 (Smooth partition of unity). *Let*

$$F \subseteq U_1 \cup \cdots \cup U_k,$$

where $F \subseteq \mathbb{R}^d$ is compact and the sets $U_j \subseteq \mathbb{R}^d$ are open for all $j = 1, \ldots, k$. Then there exist nonnegative functions $f_j \in C_c^\infty(U_j)$, $j = 1, \ldots, k$, such that

$$f_1 + \cdots + f_k \equiv 1 \text{ on } F.$$

Here, we think of the functions f_j as elements of $C_c^\infty(\mathbb{R}^d)$ with support in U_j.

Proof Taking intersections with an open ball containing F, there is no loss of generality in assuming that the sets U_j are bounded. Since F is compact and the sets U_j are open, there exists a $\delta > 0$ such that $F_\delta \subseteq U_1^\delta \cup \cdots \cup U_k^\delta$, where $F_\delta := \{x \in \mathbb{R}^d : d(x, F) \leqslant \delta\}$ and $U_j^\delta = \{x \in U_j : d(x, \complement U_j) > \delta\}$. Theorem C.12 provides us with nonnegative continuous functions $g_j : \mathbb{R}^d \to [0, 1]$ supported in U_j^δ such that

$$g_1 + \cdots + g_k \equiv 1 \text{ on } F_\delta.$$

Choose a nonnegative test function $\phi \in C_c^\infty(\mathbb{R}^d)$ with compact support in the open ball $B(0; \delta)$ and satisfying $\int_{\mathbb{R}^d} \phi \, dx = 1$. The functions $f_j := g_j * \phi$ are smooth by Proposition 11.1 and have the desired properties. $\qquad \square$

11.1.a Weak Derivatives

In order to make the body of theorems in Functional Analysis applicable to the theory of partial differential equations it is desirable to be able to discuss derivatives of functions in $L^p(D)$. The difficulty is that for such functions, the classical pointwise definition of differentiability through limits of difference quotients does not make sense since their values are well defined only almost everywhere. This necessitates an approach that is insensitive to redefining functions on sets of measure zero. Such an approach is provided by the notion of a *weak derivative*. With the help of weak derivatives we then introduce the class of Sobolev spaces, which provides the L^p-analogues of the classical spaces of continuously differentiable functions.

If $f \in C^k(D)$, then for all test functions $\phi \in C_c^\infty(D)$ and multi-indices $\alpha \in \mathbb{N}^d$ with $|\alpha| \leqslant k$ we have the integration by parts formula

$$\int_D f(x) \partial^\alpha \phi(x) \, dx = (-1)^{|\alpha|} \int_D g(x) \phi(x) \, dx, \qquad (11.1)$$

where $g = \partial^\alpha f$. Using a smooth partition of unity (Proposition 11.2), the proof of this identity can be reduced to the situation where the support of ϕ is contained in an open rectangle contained within D; for such ϕ, the formula follows by separation of variables and integration by parts on intervals in dimension one.

This motivates the following definition.

Definition 11.3 (Weak derivatives). Let $f \in L_{\text{loc}}^1(D)$. A function $g \in L_{\text{loc}}^1(D)$ is said to be a *weak derivative of order* $\alpha \in \mathbb{N}^d$ of f if for all $\phi \in C_c^\infty(D)$ we have

$$\int_D f(x) \partial^\alpha \phi(x) \, dx = (-1)^{|\alpha|} \int_D g(x) \phi(x) \, dx. \qquad (11.2)$$

A function $f \in L_{\text{loc}}^1(D)$ is said to be *weakly differentiable of order* k if it has weak derivatives $\partial^\alpha f \in L_{\text{loc}}^1(D)$ for all multi-indices satisfying $|\alpha| \leqslant k$.

Remark 11.4. The definition of a weak derivative of order α can equivalently be stated by using functions $\phi \in C_c^k(D)$ for any integer $k \geqslant |\alpha|$. To see this, suppose that $g \in L^1_{\text{loc}}(D)$ is a weak derivative of order α for the function $f \in L^1_{\text{loc}}(D)$. We wish to prove that the integration by parts formula (11.1) holds for functions $\phi \in C_c^k(D)$. To this end we claim that there exist functions $\phi_n \in C_c^\infty(D)$ such that $\phi_n \to \phi$ and $\partial^\alpha \phi_n \to \partial^\alpha \phi$ uniformly. Once this has been shown, (11.2) follows from

$$\int_D f(x) \partial^\alpha \phi(x)\, dx = \lim_{n \to \infty} \int_D f(x) \partial^\alpha \phi_n(x)\, dx$$
$$= (-1)^{|\alpha|} \lim_{n \to \infty} \int_D g(x) \phi_n(x)\, dx = (-1)^{|\alpha|} \int_D g(x) \phi(x)\, dx.$$

To prove the claim, let $\eta \in C_c^\infty(\mathbb{R}^d)$ be supported in the unit ball $B(0; 1)$ of $\mathbb{R}^d$ and satisfy $\int_{\mathbb{R}^d} \eta\, dx = 1$. For $n \geqslant 1$ let $\eta^{(n)}(x) = n^d \eta(nx)$. We extend ϕ identically zero outside D and define, for $y \in D$,

$$\phi_n(y) := \eta^{(n)} * \phi(y) = \int_{\mathbb{R}^d} \eta^{(n)}(y - x) \phi(x)\, dx.$$

Since $\eta^{(n)}$ is supported in $B(0; \frac{1}{n})$, for sufficiently large n the functions ϕ_n are compactly supported in D. They are also smooth and hence belong to $C_c^\infty(D)$ by Proposition 11.1, and the desired convergence properties follow by elementary calculus arguments.

The following proposition implies that weak derivatives, if they exist, are necessarily uniquely defined up to a null set. This allows us to speak of *the* weak derivative of order α of a function f, and denote it by $\partial^\alpha f$. The proposition could be proved along the lines of Lemma 4.59, but it will be instructive to present a proof based on mollification.

In what follows we write

$$U \Subset D$$

to express that the closure of U is compact and contained in D.

Proposition 11.5. *If a function $g \in L^1_{\text{loc}}(D)$ satisfies*

$$\int_D g(x) \phi(x)\, dx = 0$$

for all $\phi \in C_c^\infty(D)$, then $g = 0$ almost everywhere on D.

This proposition may be proved in exactly the same way as Lemma 4.59, but it is instructive to give a proof by mollification here.

Proof Let B be any open ball such that $B \Subset D$ and let $\psi \in C_c^\infty(D)$ satisfy $\psi \equiv 1$ on B. Pick a mollifier $\eta \in C_c^\infty(\mathbb{R}^d)$ satisfying $\int_{\mathbb{R}^d} \eta\, dx = 1$. For $y \in \mathbb{R}^d$ set $\eta_{\varepsilon,y}(x) := \eta_\varepsilon(y - x)$, where $\eta_\varepsilon(x) = \varepsilon^{-d}\eta(\varepsilon^{-1}x)$ for $x \in \mathbb{R}^d$. Extending ψ and g identically zero outside D and noting that $\eta_{\varepsilon,y}\psi \in C_c^\infty(D)$, the assumption implies that for all $y \in \mathbb{R}^d$ we have

$$\eta_\varepsilon * (\psi g)(y) = \int_{\mathbb{R}^d} \eta_\varepsilon(y-x)\psi(x)g(x)\,dx = \int_D \eta_{\varepsilon,y}(x)\psi(x)g(x)\,dx = 0.$$

By Proposition 2.34 we have $\eta_\varepsilon * (\psi g) \to \psi g$ in $L^1(\mathbb{R}^d)$ as $\varepsilon \downarrow 0$. It follows that $\psi g = 0$ almost everywhere on $\mathbb{R}^d$ and therefore $g = 0$ almost everywhere on B. Since this is true for every open ball $B \Subset D$, the result follows. $\square$

The following simple observation will be used repeatedly without further comment.

Lemma 11.6. *If $f \in L_{\mathrm{loc}}^1(D)$ and $\alpha \in \mathbb{N}^d$ is a multi-index, then:*

(1) *if f has a weak derivative of order α and D' is a nonempty open subset of D, then $f|_{D'}$ has a weak derivative of order α given by*

$$\partial^\alpha(f|_{D'}) = (\partial^\alpha f)|_{D'};$$

(2) *if f has a weak derivative g of order α and g has weak derivative h of order β, then f has a weak derivative of order $\alpha + \beta$ given by h, that is,*

$$\partial^\beta(\partial^\alpha f) = \partial^{\alpha+\beta} f.$$

Proof (1): We consider only test functions $\phi \in C_c^\infty(D')$ in (11.2). Extending them identically 0 to test functions defined on all of D, we obtain

$$\int_{D'} f(x)\partial^\alpha\phi(x)\,dx = \int_D f(x)\partial^\alpha\phi(x)\,dx$$
$$= (-1)^{|\alpha|}\int_D g(x)\phi(x)\,dx = (-1)^{|\alpha|}\int_{D'} g(x)\phi(x)\,dx.$$

(2): For all $\phi, \psi \in C_c^\infty(D)$ we have

$$\int_D f(x)\partial^\alpha\phi(x)\,dx = (-1)^{|\alpha|}\int_D g(x)\phi(x)\,dx$$

and

$$\int_D g(x)\partial^\beta\psi(x)\,dx = (-1)^{|\beta|}\int_D h(x)\psi(x)\,dx,$$

and the result follows by applying the first identity with $\phi = \partial^\beta\psi$. $\square$

Example 11.7. The classical integration by parts formula (11.1) says that functions in $C^k(D)$ are weakly differentiable of order k, with weak derivatives given by their classical derivatives.

Example 11.8. The function $f(x) = |x|$ has a weak derivative on $\mathbb{R}$, given by

$$\text{sign}(x) = \begin{cases} 1, & x > 0, \\ -1, & x < 0. \end{cases}$$

This follows from

$$\int_{-\infty}^{\infty} |x| \phi'(x) \, dx = \int_{0}^{\infty} x \phi'(x) \, dx - \int_{-\infty}^{0} x \phi'(x) \, dx$$

$$= -\int_{0}^{\infty} \phi(x) \, dx + \int_{-\infty}^{0} \phi(x) \, dx = -\int_{-\infty}^{\infty} \text{sign}(x) \phi(x) \, dx.$$

A far-reaching generalisation of this example is given in Theorem 11.23.

Example 11.9. We claim that the function $f(x) = \text{sign}(x)$ has no weak derivative on $\mathbb{R}$. Suppose, for a contradiction, that $g \in L^1_{\text{loc}}(\mathbb{R})$ is a weak derivative of f. The restrictions of g to $\mathbb{R}_+$ and $\mathbb{R}_-$ are weak derivatives of the corresponding restrictions of f. But these restrictions, being constant functions, have classical derivatives both equal to 0. Since classical derivatives are weak derivatives, it follows that $g \equiv 0$ on both $\mathbb{R}_+$ and $\mathbb{R}_-$ almost everywhere and therefore $g \equiv 0$ on $\mathbb{R}$ almost everywhere. We then arrive at the contradiction that, for all test functions $\phi \in C_c^{\infty}(\mathbb{R})$,

$$0 = -\int_{-\infty}^{\infty} g(x) \phi(x) = \int_{-\infty}^{\infty} \text{sign}(x) \phi'(x) \, dx$$

$$= \int_{0}^{\infty} \phi'(x) \, dx - \int_{-\infty}^{0} \phi'(x) \, dx = (0 - \phi(0)) - (\phi(0) - 0) = -2\phi(0).$$

This proves the claim.

We have the following version of Proposition 11.1. In its statement, we think of f as being defined on $\mathbb{R}^d$ by zero extension.

Proposition 11.10. *Let $f \in L^1_{\text{loc}}(D)$ have a weak derivative of order α on D. Suppose that $\eta \in C_c^{\infty}(\mathbb{R}^d)$ has support in $B(0; r)$ for some $r > 0$, and let the open set $U \Subset D$ satisfy $d(U, \partial D) > r$. Then the function $\eta * f$ has weak and classical derivatives of order α on U, and both are given by*

$$\partial^{\alpha}(\eta * f) = (\partial^{\alpha} \eta) * f = \eta * (\partial^{\alpha} f). \tag{11.3}$$

Proof Proposition 11.1 shows that $\eta * f \in C^k(\mathbb{R}^d)$ and the first equality in (11.3) holds. For all $\phi \in C_c^{\infty}(U)$ we have, using Fubini's theorem twice,

$$\int_U (\eta * f)(x) \partial^{\alpha} \phi(x) \, dx = \int_{\mathbb{R}^d} (\eta * f)(x) \partial^{\alpha} \phi(x) \, dx$$

$$= \int_{\mathbb{R}^d} \left(\int_{\mathbb{R}^d} \eta(y) f(x - y) \, dy \right) \partial^{\alpha} \phi(x) \, dx$$

$$= \int_{\mathbb{R}^d} \eta(y) \left(\int_{\mathbb{R}^d} f(x-y) \partial^\alpha \phi(x) \, dx \right) dy$$

$$\overset{(*)}{=} \int_{B(0;r)} \eta(y) \left(\int_D f(x) \partial^\alpha \phi(x+y) \, dx \right) dy$$

$$\overset{(**)}{=} (-1)^{|\alpha|} \int_{B(0;r)} \eta(y) \left(\int_D \partial^\alpha f(x) \phi(x+y) \, dx \right) dy$$

$$= (-1)^{|\alpha|} \int_{\mathbb{R}^d} \eta(y) \left(\int_{\mathbb{R}^d} (\partial^\alpha f)(x-y) \phi(x) \, dx \right) dy$$

$$= (-1)^{|\alpha|} \int_{\mathbb{R}^d} \phi(x) \left(\int_{\mathbb{R}^d} \eta(y) (\partial^\alpha f)(x-y) \, dy \right) dx$$

$$= (-1)^{|\alpha|} \int_{\mathbb{R}^d} (\eta * \partial^\alpha f)(x) \phi(x) \, dx$$

$$= (-1)^{|\alpha|} \int_U (\eta * \partial^\alpha f)(x) \phi(x) \, dx.$$

The identities $(*)$ and $(**)$ are justified by the assumptions that ϕ is supported in U, η is supported in $B(0;r)$, and $d(U, \partial D) > r$ (and therefore $\phi(\cdot + y) \in C_c(D)$ for all $y \in B(0;r)$) and f has a weak derivative of order α on D. This proves that $\eta * f$ has a weak derivative of order α on U given by (11.3). □

In the proof of the next proposition we will use the fact that for all $1 \leqslant p \leqslant \infty$, the operator

$$f \mapsto \partial^\alpha f$$

is closed as a linear operator in $L^p(D)$ with domain

$$D(\partial^\alpha) := \{ f \in L^p(D) : f \text{ has a weak derivative of order } \alpha \text{ in } L^p(D) \}.$$

This domain of course depends on p, but we suppress this from the notation. To prove that ∂^α is a closed operator, suppose that $f_n \to f$ in $L^p(D)$, with $f_n \in D(\partial^\alpha)$ for all n, and $\partial^\alpha f_n \to g$ in $L^p(D)$. We must prove that $f \in D(\partial^\alpha)$ and $\partial^\alpha f = g$. For all $\phi \in C_c^\infty(D)$ we have

$$\int_D f_n \partial^\alpha \phi \, dx = (-1)^{|\alpha|} \int_D \partial^\alpha f_n \phi \, dx.$$

Passing to the limit $n \to \infty$ in this formula (which is possible by Hölder's inequality, thanks to the fact that test functions belong to $L^q(D)$ with $\frac{1}{p} + \frac{1}{q} = 1$) we obtain

$$\int_D f \partial^\alpha \phi \, dx = (-1)^{|\alpha|} \int_D g \phi \, dx.$$

This means that the function $g \in L^p(D)$ is a weak derivative of f of order α.

By $L_{\text{loc}}^p(D)$ we denote the space of measurable functions whose restrictions to all sets $U \Subset D$ belong to $L^p(U)$, identifying functions that are equal almost everywhere on D.

Proposition 11.11. *Let* $1 \leqslant p < \infty$. *A function* $f \in L^p_{\text{loc}}(D)$ *admits a weak derivative of order* α *in* $L^p_{\text{loc}}(D)$ *if and only if there exist a sequence of functions* $f_n \in C^\infty_c(D)$ *and a function* $g \in L^p_{\text{loc}}(D)$ *such that for all open sets* $U \Subset D$ *we have:*

(i) $f_n \to f$ *in* $L^p(U)$;
(ii) $\partial^\alpha f_n \to g$ *in* $L^p(U)$.

In this situation we have $\partial^\alpha f = g$.

Proof 'If': Let $U \Subset D$. Since ∂^α is closed as an operator in $L^p(U)$, (i) and (ii) imply that $f \in \mathsf{D}(\partial^\alpha)$ and $\partial^\alpha f = g$. If $U_1 \Subset D$ and $U_2 \Subset D$ are open sets with nonempty intersection, the resulting weak derivatives $g_1 \in L^p(U_1)$ and $g_2 \in L^p(U_2)$ agree on $U_1 \cap U_2$ by Proposition 11.5. Hence by piecing together these weak derivatives we obtain a well-defined function $g \in L^p_{\text{loc}}(D)$. Since every test function is supported in one of the sets U under consideration, g is seen to be a weak derivative of order α for f.

'Only if': For $r > 0$ let $D_r = \{x \in D : d(x, \partial D) > r\}$. For every $\varepsilon > 0$, choose $\psi_\varepsilon \in C^\infty_c(\mathbb{R}^d)$ such that $0 \leqslant \psi_\varepsilon \leqslant 1$ pointwise, $\psi_\varepsilon \equiv 1$ on $D_{2\varepsilon}$, and $\psi_\varepsilon \equiv 0$ on $\complement D_\varepsilon$. Let $\eta \in C^\infty_c(\mathbb{R}^d)$ have support in $B(0;1)$ and satisfy $\int_{\mathbb{R}^d} \eta \, dx = 1$, and define $\eta_\varepsilon := \varepsilon^{-d} \eta(\varepsilon^{-1} x)$. For $n \geqslant 1$ define

$$f_n := \psi_{1/n} \cdot (f * \eta_{1/n}),$$

where we think of f as a function on $\mathbb{R}^d$ by zero extension. We will prove that the functions f_n have the required properties, with $g = \partial^\alpha f$.

We have $f_n \in C^\infty_c(D)$ and, by Proposition 11.10 and the classical product rule,

$$\partial^\alpha f_n = \psi_{1/n} \cdot ((\partial^\alpha f) * \eta_{1/n}) + \sum_{\substack{0 \leqslant \beta \leqslant \alpha \\ \beta \neq 0}} \binom{\alpha}{\beta} (\partial^\beta \psi_{1/n}) \cdot \partial^{\alpha - \beta}(f * \eta_{1/n}).$$

Given an open set $U \Subset D$, let $N \geqslant 1$ be so large that $\overline{U} \subseteq D_{2/N}$. For all $n \geqslant N$ we have $\psi_{1/n} \equiv 1$ on U and

$$\begin{aligned}
\mathbf{1}_U(x) \cdot (f * \eta_{1/n})(x) &= \mathbf{1}_U(x) \int_{\mathbb{R}^d} f(x-y)\eta_{1/n}(y) \, dy \\
&= \mathbf{1}_U(x) \int_{\mathbb{R}^d} \mathbf{1}_{U+B(0;1/N)}(x-y)f(x-y)\eta_{1/n}(y) \, dy \\
&= \mathbf{1}_U(x) \cdot ((\mathbf{1}_{U+B(0;1/N)}f) * \eta_{1/n})(x), \quad x \in D.
\end{aligned}$$

with $U + B(0; 1/N) \Subset D$. Since $\mathbf{1}_{U+B(0;1/N)}f \in L^p(D)$, by Proposition 2.34 we obtain

$$\mathbf{1}_U f_n = \mathbf{1}_U \psi_{1/n} \cdot (f * \eta_{1/n}) = \mathbf{1}_U \cdot ((\mathbf{1}_{U+B(0;1/N)}f) * \eta_{1/n}) \to \mathbf{1}_U \mathbf{1}_{U+B(0;1/N)}f = \mathbf{1}_U f$$

as $n \to \infty$, with convergence in $L^p(D)$. Also, for all $n \geqslant N$ we have $\partial^\alpha(f * \eta_{1/n}) =$

$(\partial^\alpha f) * \eta_{1/n}$ on U, as well as $\partial^\beta \psi_{1/n} \equiv 0$ on U for all $0 \leqslant \beta \leqslant \alpha$ with $\beta \neq 0$. Therefore, by the same reasoning,

$$\mathbf{1}_U \partial^\alpha f_n = \mathbf{1}_U \cdot ((\mathbf{1}_{U+B(0;1/N)} \partial^\alpha f) * \eta_{1/n}) \to \mathbf{1}_U \partial^\alpha f$$

as $n \to \infty$, with convergence in $L^p(D)$. $\qquad\square$

As an application of this proposition we prove the following result on existence of lower-order weak derivatives.

Theorem 11.12 (Existence of lower order weak derivatives). *If a function $f \in L^1_{\mathrm{loc}}(D)$ admits a weak derivative $\partial^\alpha f$ for some $\alpha \in \mathbb{N}^d$, then it admits weak derivatives $\partial^\beta f$ for all $\beta \in \mathbb{N}^d$ satisfying $0 \leqslant \beta \leqslant \alpha$. If both f and $\partial^\alpha f$ belong to $L^p(D)$, then so do the weak derivatives $\partial^\beta f$.*

For notational simplicity we give the proof only in dimension $d = 1$; the argument carries over without difficulty to higher dimensions. The crucial step is contained in the next lemma.

Lemma 11.13. *For all $k \geqslant 1$ there is a constant $C_k \geqslant 0$ such that for all $f \in C^k_c(\mathbb{R})$ and $1 \leqslant p < \infty$ we have*

$$\|f^{(k-1)}\|_p \leqslant C_k(\|f\|_p + \|f^{(k)}\|_p).$$

Proof Let $\zeta \in C^\infty_c(\mathbb{R})$ satisfy $\zeta(0) = 1$ and $\zeta'(0) = \cdots = \zeta^{(k)}(0) = 0$. Combining the identity

$$\frac{\mathrm{d}}{\mathrm{d}t}(\zeta'(t)f(x+t)) = \zeta'(t)f'(x+t) + \zeta''(t)f(x+t),$$

which follows from $\frac{\mathrm{d}}{\mathrm{d}t}f(x+t) = f'(x+t)$, with the identity

$$\frac{\mathrm{d}}{\mathrm{d}t}(\zeta(t)f'(x+t)) = \zeta(t)f''(x+t) + \zeta'(t)f'(x+t),$$

which follows from $\frac{\mathrm{d}}{\mathrm{d}t}f'(x+t) = f''(x+t)$, we arrive at

$$\frac{\mathrm{d}}{\mathrm{d}t}(\zeta(t)f'(x+t)) = \zeta(t)f''(x+t) + \frac{\mathrm{d}}{\mathrm{d}t}(\zeta'(t)f(x+t)) - \zeta''(t)f(x+t).$$

Integrating and using that ζ is compactly supported and satisfies $\zeta(0) = 1$ and $\zeta'(0) = 0$,

$$f'(x) = -\int_0^\infty \zeta(t)f''(x+t)\,\mathrm{d}t + \int_0^\infty \zeta''(t)f(x+t)\,\mathrm{d}t.$$

If $f \in C^3_c(\mathbb{R})$ we can apply this identity with f replaced by f'. Integrating by parts and using that $\zeta''(0) = 0$, we obtain

$$f''(x) = -\int_0^\infty \zeta(t)f'''(x+t)\,\mathrm{d}t + \int_0^\infty \zeta''(t)f'(x+t)\,\mathrm{d}t$$

$$= -\int_0^\infty \zeta(t) f'''(x+t)\,dt - \int_0^\infty \zeta'''(t) f(x+t)\,dt.$$

Continuing inductively, for $f \in C_c^k(\mathbb{R})$ with $k \geqslant 2$ we arrive at the identity

$$f^{(k-1)}(x) = -\int_0^\infty \zeta(t) f^{(k)}(x+t)\,dt + (-1)^k \int_0^\infty \zeta^{(k)}(t) f(x+t)\,dt.$$

Taking L^p-norms and moving norms inside the integral using Proposition 1.44, we obtain

$$\|f^{(k-1)}\|_p \leqslant \int_0^\infty |\zeta(t)| \|f(\cdot + t)\|_p + |\zeta^{(k)}(t)| \|f^{(k)}(\cdot + t)\|_p \,dt$$

$$\leqslant L \max\{\|\zeta\|_\infty, \|\zeta^{(k)}\|_\infty\}(\|f\|_p + \|f^{(k)}\|_p),$$

where L is the length of an bounded interval containing the compact support of ζ. $\square$

Proof of Theorem 11.12 Again we limit ourselves to dimension $d = 1$ for notational convenience, and set $k := \alpha$. The cases $k = 0, 1$ being trivial, we assume that $k \geqslant 2$.

Let $f \in L^1_{\text{loc}}(D)$ have weak derivative $\partial^k f$. We will prove that f has a weak derivative $\partial^{k-1} f$; once we know that, lower order weak derivative are obtained by repeatedly applying this result.

Let $(f_n)_{n \geqslant 1}$ be a sequence in $C_c^\infty(D)$ as stated in Proposition 11.11, that is, for all open sets $U \Subset D$ we have:

(i) $f_n \to f$ in $L^1(U)$;
(ii) $\partial^k f_n^{(k)} \to g$ in $L^1(U)$.

By the lemma, for all $1 \leqslant p < \infty$ we have

$$\|\partial^{k-1} f_n - \partial^{k-1} f_m\|_1 \leqslant C_k(\|f_n - f_m\|_1 + \|\partial^k f_n - \partial^k f_m\|_p).$$

Since the right-hand side tends to 0 as $m, n \to \infty$, by completeness there exists a function $g_U \in L^1(U)$ such that $\lim_{n \to \infty} \partial^{k-1} f_n = g$ in $L^1(U)$. As in the proof of Proposition 11.11 these functions may be glued together to a well-defined function $g \in L^1_{\text{loc}}(D)$, and this function is easily checked to be a weak derivative of order $k - 1$ of f.

If f and $\partial^k f$ belong to $L^p(D)$, and $(f_n)_{n \geqslant 1}$ is a sequence in $C_c^\infty(D)$ such that (i) and (ii) hold with $L^1(U)$ replaced by $L^p(U)$, then by the estimate of the lemma, the functions g_U belong to $L^p(U)$ with norm $\|g_U\|_{L^p(U)} \leqslant C_k \|f\|_{L^p(U)} \leqslant \|f\|_{L^p(D)}$. This implies that also $\|g\|_{L^p(D)} \leqslant C_k \|f\|_{L^p(D)}$. $\square$

As an application we prove the following product rule.

Proposition 11.14 (Product rule). *If $f \in L^1_{\text{loc}}(D)$ admits a weak derivative of order α, then so does the pointwise product ψf for every $\psi \in C^\infty(D)$, and we have the Leibniz* *formula*

$$\partial^\alpha (fg) = \sum_{0 \leqslant \beta \leqslant \alpha} \binom{\alpha}{\beta} (\partial^\beta f)(\partial^{\alpha - \beta} g).$$

The lower-order weak derivatives $\partial^\beta f$ exists thanks to Theorem 11.12. In most applications, however, the function f belongs to suitable Sobolev spaces and the existence of all lower order weak derivatives is part of the assumptions. The proposition admits variations in terms of the assumptions on f and ψ (see, for instance, Problem 11.12).

Proof   We only need to prove this for multi-indices of order one, since the general case then follows by an induction argument. Fix $1 \leqslant j \leqslant d$. Since for all $\phi \in C_c^\infty(D)$ we have $\phi \psi \in C_c^\infty(D)$, the classical product rule for $\psi \phi$ gives

$$\int_D \big(\psi(x)\partial_j \phi(x) + \partial_j \psi(x)\phi(x)\big) f(x)\,\mathrm{d}x = -\int_D \psi(x)\phi(x)\partial_j f(x)\,\mathrm{d}x.$$

After rearranging, this says that the weak derivative $\partial_j(\psi f)$ exists and is given by $(\partial_j \psi)f + \psi \partial_j f$.   □

For the proof of the next proposition we need a simple lemma on extensions.

Lemma 11.15. *Let $f \in L_{\mathrm{loc}}^1(D)$ be supported in an open set $U \Subset D$ and let $\widetilde{D}$ be an open set containing D. If f admits a weak derivative $\partial^\alpha f$ on D, then the zero extension $\widetilde{f}$ of f to $\widetilde{D}$ admits a weak derivative $\partial^\alpha \widetilde{f}$ on $\widetilde{D}$, given by the zero extension of $\partial^\alpha f$.*

Proof   Fix an arbitrary test function $\eta \in C^\infty(\widetilde{D})$ with support in D and such that $\eta \equiv 1$ on U. For all $\phi \in C_c^\infty(\widetilde{D})$, the (classical) derivatives $\partial^\alpha \phi$ and $\partial^\alpha(\eta \phi)$ agree on U, and therefore

$$\int_{\widetilde{D}} \widetilde{f}\partial^\alpha \phi \,\mathrm{d}x = \int_D f \partial^\alpha(\eta \phi)\,\mathrm{d}x = (-1)^{|\alpha|}\int_D (\partial^\alpha f)\eta \phi \,\mathrm{d}x = (-1)^{|\alpha|}\int_{\widetilde{D}} (\widetilde{\partial^\alpha f})\phi \,\mathrm{d}x,$$

since $\eta \phi \in C_c^\infty(D)$.   □

The *gradient* of a weakly differentiable function f is the function $\nabla f \in L_{\mathrm{loc}}^1(D; \mathbb{K}^d)$ defined by

$$\nabla f := (\partial_1 f, \ldots, \partial_d f).$$

An open subset U of $\mathbb{R}^d$ is said to be *(pathwise) connected* if any two points $x, y \in U$ can be joined by a continuous curve in U, that is, there exists a continuous mapping $\varphi : [0,1] \to U$ such that $\varphi(0) = x$ and $\varphi(1) = y$.

Proposition 11.16. *Let $f \in L_{\mathrm{loc}}^1(D)$ be weakly differentiable. If $\nabla f = 0$ almost everywhere on an open connected subset U of D, then f is almost everywhere equal to a constant on U.*

Proof Let B be an open ball whose closure is contained in U and choose a test function $\psi \in C_c^\infty(D)$ such that $\psi \equiv 1$ on B. By Proposition 11.14, ψf is weakly differentiable on D and $\nabla(\psi f) = f\nabla\psi + \psi\nabla f$. In particular, $\nabla(\psi f) \equiv 0$ on B.

Extending f and ψ identically zero to all of $\mathbb{R}^d$, by Lemma 11.15 the function ψf is weakly differentiable as a function on $\mathbb{R}^d$, with a weak gradient in $L_{\text{loc}}^1(\mathbb{R}^d)$ that equals the zero extension of the weak gradient of ψf on D. By slight abuse of notation, both will be denoted by $\nabla(\psi f)$.

Pick mollifier $\eta \in C_c^\infty(\mathbb{R}^d)$ satisfying $\int_{\mathbb{R}^d} \eta \, dx = 1$, and set $\eta_\varepsilon(x) := \varepsilon^{-d}\eta(\varepsilon^{-1}x)$, $x \in \mathbb{R}^d$ for $\varepsilon > 0$. By Proposition 11.10 we have $\eta_\varepsilon * (\psi f) \in C^\infty(\mathbb{R}^d)$ and

$$\nabla(\eta_\varepsilon * (\psi f)) = \eta_\varepsilon * \nabla(\psi f) \quad \text{on } \mathbb{R}^d.$$

In particular, $\nabla(\eta_\varepsilon * (\psi f)) \equiv 0$ on B. Since the function $\eta_\varepsilon * (\psi f)$ is C^∞, classical calculus arguments imply that this function is constant on B. Taking the $L^1(\mathbb{R}^d)$-limit as $\varepsilon \downarrow 0$ using Proposition 2.34, and passing to an almost everywhere convergent subsequence, it is seen that ψf is constant almost everywhere on B. Indeed, this shows that f is the almost everywhere limit of a sequence of functions each of which is constant on B. Since $\psi \equiv 1$ on B it follows that f is constant almost everywhere on B. This being true for every open ball B contained in U, this gives the result. $\square$

11.1.b The Sobolev Spaces $W^{k,p}(D)$

By Hölder's inequality, every $f \in L^p(D)$ with $1 \leqslant p \leqslant \infty$ belongs to $L_{\text{loc}}^1(D)$. This suggests the following definition.

Definition 11.17 (Sobolev spaces). For $k \in \mathbb{N}$ and $1 \leqslant p \leqslant \infty$ the *Sobolev space* $W^{k,p}(D)$ is the space of all $f \in L^p(D)$ whose weak derivatives of all orders $|\alpha| \leqslant k$ exist and belong to $L^p(D)$.

Endowed with the norm

$$\|f\|_{W^{k,p}(D)} := \sum_{|\alpha| \leqslant k} \|\partial^\alpha f\|_{L^p(D)},$$

$W^{k,p}(D)$ is a Banach space. Indeed, if $(f_n)_{n \geqslant 1}$ is a Cauchy sequence in $W^{k,p}(D)$, then for all $|\alpha| \leqslant k$ the sequence of weak derivatives $(\partial^\alpha f_n)_{n \geqslant 1}$ is a Cauchy sequence in $L^p(D)$ and hence convergent to some $f^{(\alpha)} \in L^p(D)$. Set $f := f^{(0)}$. Using Hölder's inequality as in Corollary 2.25, we may pass to the limit $n \to \infty$ in the identity

$$\int_D f_n(x)\partial^\alpha \phi(x) \, dx = (-1)^{|\alpha|} \int_D \partial^\alpha f_n(x)\phi(x) \, dx, \quad \phi \in C_c^\infty(D).$$

It follows that f has weak derivatives of all orders $|\alpha| \leqslant k$ given by $\partial^\alpha f = f^{(\alpha)}$. Having observed this, it is clear that $f \in W^{k,p}(D)$ and $f_n \to f$ in $W^{k,p}(D)$.

For $1 \leqslant p < \infty$, the linear operators ∂^α are closable as operators in $L^p(D)$ with initial domain $C_c^\infty(D)$; this follows by an argument similar to the one just given. By Proposition 2.29, these operators are densely defined. This prompts the question which functions in $W^{k,p}(D)$ can be approximated, in the norm of this space, by test functions. A first result into this direction is the following lemma. We return to this question in Section 11.1.c.

Proposition 11.18 (Local approximation by test functions). *For all $f \in W^{1,p}(D)$ with $1 \leqslant p < \infty$ there exists a sequence of test functions $f_n \in C_c^\infty(D)$ such that for every open set $U \Subset D$ we have*

$$\lim_{n \to \infty} f_n|_U = f|_U \quad in \ W^{1,p}(U).$$

Proof   The functions f_n constructed in the proof of Proposition 11.11 have the required properties.   □

Proposition 11.19 (Chain rule). *Let $\rho : \mathbb{K} \to \mathbb{K}$ be a C^1-function with bounded derivative and satisfying $\rho(0) = 0$, and let $1 \leqslant p \leqslant \infty$. For all $f \in W^{1,p}(D)$ we have $\rho \circ f \in W^{1,p}(D)$ and*

$$\partial_j(\rho \circ f) = (\rho' \circ f)\partial_j f, \quad j = 1, \dots, d.$$

Proof   The condition $\rho(0) = 0$ and the boundedness of ρ' imply that $|\rho(t)| \leqslant M|t|$ with $M := \sup_{t \in \mathbb{K}} |\rho'(t)|$ and therefore $f \in L^p(D)$ implies that $\rho \circ f \in L^p(D)$. The boundedness of ρ' implies that $\rho' \circ f$ is bounded and therefore $\partial_j f \in L^p(D)$ implies that $(\rho' \circ f)\partial_j f \in L^p(D)$. To conclude the proof it therefore suffices to check that $(\rho' \circ f)\partial_j f$ is a weak derivative in the jth direction of $\rho \circ f$.

Fix $\phi \in C_c^\infty(D)$ and let $U \Subset D$ be an open set containing the compact support of ϕ. By Proposition 11.18 we can find functions $f_n \in C_c^\infty(D)$ such that $f_n|_U \to f|_U$ in $W^{1,p}(U)$. Since $|\rho \circ f_n - \rho \circ f| \leqslant M|f_n - f|$ and $f_n \to f$ in $L^p(U)$, Hölder's inequality and the classical chain rule imply that

$$\int_D (\rho \circ f)(\partial_j)\phi \, dx = \lim_{n \to \infty} \int_D (\rho \circ f_n)(\partial_j \phi) \, dx = -\lim_{n \to \infty} \int_D (\rho' \circ f_n)(\partial_j f_n)\phi \, dx.$$

Upon passing to a subsequence if necessary, by Corollary 2.21 we may assume that $f_n \to f$ and $\partial_j f_n \to \partial_j f$ almost everywhere on D and that there exists a function $0 \leqslant g \in L^p(U)$ such that $|\partial_j f_n| \leqslant g$ almost everywhere on U for every n. Then $|(\rho' \circ f_n)(\partial_j f_n)\phi| \leqslant M|g||\phi|$ almost everywhere on U for every n, and since $|g||\phi| \in L^1(U)$ we may use dominated convergence to obtain

$$\lim_{n \to \infty} \int_D (\rho' \circ f_n)(\partial_j f_n)\phi \, dx = \int_D (\rho' \circ f)(\partial_j f)\phi \, dx,$$

noting that on both sides the integrands vanish outside U. This completes the proof.   □

Proposition 11.20 (Substitution rule). *Let D and D' be nonempty open subsets of $\mathbb{R}^d$ and suppose that $\rho : D \to D'$ is a C^k-diffeomorphism with $k \geqslant 1$. Let $1 \leqslant p < \infty$. A* *function $f \in L^1_{\text{loc}}(D)$ is weakly differentiable if and only if $f \circ \rho^{-1} \in L^1_{\text{loc}}(D')$ is weakly differentiable. Denoting the space variables of D and D' by x and y, respectively, in that case we have*

$$\partial_i f = \sum_{j=1}^{d} \frac{\partial \rho_j}{\partial x_i} \partial_j (f \circ \rho^{-1}) \circ \rho.$$

Proof For smooth functions f this is the substitution rule from Calculus, and the general case follows by local approximation with smooth functions. □

11.1.c The Sobolev Spaces $W_0^{1,p}(D)$

We now introduce a class of spaces which play an important role in the theory of partial differential equations, where they provide the L^p-setting for studying boundary value problems subject to Dirichlet boundary conditions.

Definition 11.21 (The spaces $W_0^{1,p}(D)$). For $1 \leqslant p < \infty$ we define $W_0^{1,p}(D)$ to be the closure of $C_c^\infty(D)$ in $W^{1,p}(D)$.

The following result gives a simple sufficient condition for membership of $W_0^{1,p}(D)$:

Proposition 11.22. *Let $U \Subset D$ be open and let $1 \leqslant p < \infty$. If $f \in W^{1,p}(D)$ vanishes outside U, then $f \in W_0^{1,p}(D)$.*

Proof Since $\overline{U}$ is compact and does not intersect ∂D we have $\delta := d(\overline{U}, \partial D) > 0$. Fix a function $\eta \in C_c^\infty(\mathbb{R}^d)$ compactly supported in the ball $B(0;1)$ and satisfying $\int_{\mathbb{R}^d} \eta \, dx = 1$, and set $\eta_\varepsilon(x) := \varepsilon^{-d} \eta(\varepsilon^{-1} x)$ for $0 < \varepsilon < \delta$. Each η_ε has compact support in $B(0;\varepsilon)$. Extending f identically zero outside D, the condition $0 < \varepsilon < \delta$ implies that the convolution $\eta_\varepsilon * f$ is compactly supported in D.

As $\varepsilon \downarrow 0$, by Proposition 2.34 we have

$$\eta_\varepsilon * f \to f \text{ in } L^p(\mathbb{R}^d), \text{ and hence in } L^p(D). \tag{11.4}$$

By Proposition 11.10 the weak derivatives of $\eta_\varepsilon * f$ are given by

$$\partial_j(\eta_\varepsilon * f) = \eta_\varepsilon * \partial_j f, \quad j = 1, \ldots, d.$$

These weak derivatives belong to $L^p(D)$, so $\eta_\varepsilon * f$ restricts to an element of $W^{1,p}(D)$.
By Proposition 2.34,

$$\partial_j(\eta_\varepsilon * f) = \eta_\varepsilon * \partial_j f \to \partial_j f \text{ in } L^p(\mathbb{R}^d), \text{ and hence in } L^p(D). \tag{11.5}$$

By (11.4) and (11.5) we have $\eta_\varepsilon * f \to f$ in $W^{1,p}(D)$. Finally we observe that each $\eta_\varepsilon * f$ is C^∞ (by Proposition 11.1) and compactly supported in D. □

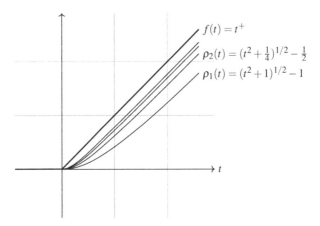

$$f(t) = t^+$$
$$\rho_2(t) = (t^2 + \tfrac{1}{4})^{1/2} - \tfrac{1}{2}$$
$$\rho_1(t) = (t^2 + 1)^{1/2} - 1$$

Figure 11.1 The functions $\rho_n(t) = (t^2 + \tfrac{1}{n^2})^{1/2} - \tfrac{1}{n}$

It is trivial that if $f \in W^{1,p}(D)$, then $\mathrm{Re}\,f, \mathrm{Im}\,f \in W^{1,p}(D)$, and similarly if $f \in W_0^{1,p}(D)$, then $\mathrm{Re}\,f, \mathrm{Im}\,f \in W_0^{1,p}(D)$. The next proposition asserts that the positive part f^+ of a real-valued function f in $W^{1,p}(D)$ belongs to $W^{1,p}(D)$ and provides an explicit expression for its weak derivatives; moreover, if $f \in W_0^{1,p}(D)$, then $f^+ \in W_0^{1,p}(D)$. The analogous results then also hold for the negative part $f^- = f^+ - f$ and the absolute value $|f| = f^+ + f^-$.

Theorem 11.23 (Positive parts). *Let $1 \leqslant p < \infty$. Then:*

(1) *for all real-valued functions $f \in W^{1,p}(D)$ we have $f^+ \in W^{1,p}(D)$ and*

$$\partial_j f^+ = \mathbf{1}_{\{f>0\}} \partial_j f, \quad j = 1, \dots, d;$$

and if $f \in W_0^{1,p}(D)$, then $f^+ \in W_0^{1,p}(D)$;
(2) *the mapping $f \mapsto f^+$ is continuous with respect to the norm of $W^{1,p}(D)$.*

Proof   The idea of the proof is to approximate the function $\rho : t \mapsto t^+ = \max\{t, 0\}$ with C^1-functions ρ_n in such a way that for all $t \in \mathbb{R}$ we have

(i) $0 \leqslant \rho_n(t) \uparrow t^+$;
(ii) $0 \leqslant \rho_n'(t) \uparrow \mathbf{1}_{(0,\infty)}(t)$.

For instance, the choice $\rho_n(t) = (t^2 + \tfrac{1}{n^2})^{1/2} - \tfrac{1}{n}$ for $t > 0$ and $\rho_n(t) = 0$ for $t \leqslant 0$ will do; see Figure 11.1.

Step 1 – Let f be a real-valued function in $W^{1,p}(D)$. By Proposition 11.19 we have

$\rho_n \circ f \in W^{1,p}(D)$ and, for test functions $\phi \in C_c^\infty(D)$,

$$-\int_D (\rho_n \circ f)(x)\partial_j \phi(x)\,dx = \int_D (\rho'_n \circ f)(x)\partial_j f(x)\phi(x)\,dx.$$

By (i), (ii), and dominated convergence we obtain

$$-\int_D f^+(x)\partial_j \phi(x)\,dx = \int_D \mathbf{1}_{\{f>0\}}(x)\partial_j f(x)\phi(x)\,dx.$$

This proves the first part of (1).

Step 2 – To prove (2), let $f_n \to f$ in $W^{1,p}(D)$ with all functions real-valued. We must show that $f_n^+ \to f^+$ in $W^{1,p}(D)$. It is clear that $f_n^+ \to f^+$ in $L^p(D)$, so it remains to show that $\partial_j f_n^+ \to \partial_j f^+$ for all $1 \leqslant j \leqslant d$. By Step 1 this is equivalent to showing that $\mathbf{1}_{\{f_n>0\}}\partial_j f_n \to \mathbf{1}_{\{f>0\}}\partial_j f$ in $L^p(D)$ for all $1 \leqslant j \leqslant d$.

By Corollary 2.21 we can choose a subsequence such that $f_{n_k} \to f$ and $|f_{n_k}| \leqslant g$ almost everywhere, where $0 \leqslant g \in L^p(D)$, and $\partial_j f_{n_k} \to \partial_j f$ and $|\partial_j f_{n_k}| \leqslant h_j$ almost everywhere, where $0 \leqslant h_j \in L^p(D)$, for all $1 \leqslant j \leqslant d$. Then $\mathbf{1}_{\{f_{n_k}>0\}} \to \mathbf{1}_{\{f>0\}}$ almost everywhere, so dominated convergence implies that $\mathbf{1}_{\{f_{n_k}>0\}}\partial_j f_{n_k} \to \mathbf{1}_{\{f>0\}}\partial_j f$ in $L^p(D)$ for all $1 \leqslant j \leqslant d$.

Applying the above argument to subsequences of $(f_n)_{n\geqslant 1}$, we thus find that every subsequence $(f_{n_m})_{m\geqslant 1}$ of $(f_n)_{n\geqslant 1}$ contains a further subsequence $(f_{n_{m_k}})_{k\geqslant 1}$ such that $f_{n_{m_k}}^+ \to f^+$ in $W^{1,p}(D)$. This implies that $f_n^+ \to f^+$ in $W^{1,p}(D)$.

Step 3 – It remains to prove the second part of (1). Suppose that $f \in W_0^{1,p}(D)$ is real-valued; we must prove that $f^+ \in W_0^{1,p}(D)$. Choose functions $f_n \in C_c^\infty(D)$ such that $f_n \to f$ in $W^{1,p}(D)$. Then $f_n^+ \to f^+$ in $W^{1,p}(D)$ by Step 2. Thus it suffices to show that every f_n^+ can be approximated by functions in $C_c^\infty(D)$. This is accomplished by a mollifier argument.

Fix $n \geqslant 1$. Since f_n is compactly supported in D, its support has positive distance δ_n to the boundary of D. If $\eta \in C_c^\infty(\mathbb{R}^d)$ has compact support in $B(0;1)$ and satisfies $\int_{\mathbb{R}^d} \eta\,dx = 1$, then for $0 < \varepsilon < \delta_n$ the function $\eta_\varepsilon * f_n^+$ (where $\eta_\varepsilon(x) = \varepsilon^{-d}\eta(\varepsilon^{-1}x)$) is smooth by Proposition 11.1, has compact support in D, and by Proposition 2.34 we have $\eta_\varepsilon * f_n^+ \to f_n^+$ in $L^p(D)$ as $\varepsilon \downarrow 0$. Likewise, by (11.3) applied to f_n^+,

$$\partial_j(\eta_\varepsilon * f_n^+) = \eta_\varepsilon * (\partial_j f_n^+) \to \partial_j f_n^+$$

with convergence in $L^p(D)$. □

The following proposition connects the space $W_0^{1,p}(D)$ with Dirichlet boundary conditions.

Theorem 11.24. *Let D be bounded and let $1 \leqslant p < \infty$. For all $f \in W^{1,p}(D) \cap C(\overline{D})$ the following assertions hold:*

(1) *if $f|_{\partial D} \equiv 0$, then $f \in W_0^{1,p}(D)$;*

(2) *if ∂D is C^1 and $f \in W_0^{1,p}(D)$, then $f|_{\partial D} \equiv 0$.*

Proof (1): By considering real and imaginary parts separately we may assume that f is real-valued, and by Theorem 11.23 we may even assume that f is nonnegative. Since $f - \frac{1}{k} \to f$ in $W^{1,p}(D)$ as $k \to \infty$ and taking positive parts is continuous in $W^{1,p}(D)$ by Theorem 11.23, we see that $f_k := (f - \frac{1}{k})^+ \to f^+ = f$ in $W^{1,p}(D)$ as $k \to \infty$. Hence to prove that $f \in W_0^{1,p}(D)$ it suffices to prove that $f_k \in W_0^{1,p}(D)$ for all $n \geqslant 1$.

By continuity, each f_k vanishes in a neighbourhood of the compact set ∂D. Then Proposition 11.22 shows that f_k can be approximated in $W^{1,p}(D)$ by functions $f_{k,n} \in C_c^\infty(D)$.

(2): By the definition of a C^1-boundary and a partition of unity argument, it suffices to prove that for all $f \in W_0^{1,p}(\mathbb{R}_+^d) \cap C(\overline{\mathbb{R}_+^d})$ we have $f|_{\partial \mathbb{R}_+^d} \equiv 0$; here,

$$\mathbb{R}_+^d := \{x \in \mathbb{R}^d : x_d > 0\}.$$

Let $\phi_n \to f$ in $W_0^{1,p}(\mathbb{R}_+^d)$ with $\phi_n \in C_c(\mathbb{R}_+^d)$ for all $n \geqslant 1$. For all $x = (x', x_d) \in \mathbb{R}_+^d = \mathbb{R}^{d-1} \times (0, \infty)$,

$$|\phi_n(x', x_d)| \leqslant \int_0^{x_d} |\partial_d \phi_n(x', y)| \, \mathrm{d}y.$$

Integrating over $|x'| < 1$ and averaging over $0 < x_d < \varepsilon$ with $0 < \varepsilon < 1$, we obtain

$$\frac{1}{\varepsilon} \int_0^\varepsilon \int_{\{|x'|<1\}} |\phi_n(x', x_d)| \, \mathrm{d}x' \, \mathrm{d}x_d \leqslant \frac{1}{\varepsilon} \int_0^\varepsilon \int_{\{|x'|<1\}} \int_0^{x_d} |\partial_d \phi_n(x', y)| \, \mathrm{d}y \, \mathrm{d}x' \, \mathrm{d}x_d$$

$$= \int_0^\varepsilon \int_{\{|x'|<1\}} \frac{1}{\varepsilon} \int_{x_d}^\varepsilon |\partial_d \phi_n(x', y)| \, \mathrm{d}x_d \, \mathrm{d}x' \, \mathrm{d}y$$

$$\leqslant \int_0^\varepsilon \int_{\{|x'|<1\}} |\partial_d \phi_n(x', y)| \, \mathrm{d}x' \, \mathrm{d}y.$$

Letting $n \to \infty$, we obtain

$$\frac{1}{\varepsilon} \int_0^\varepsilon \int_{\{|x'|<1\}} |f(x', x_d)| \, \mathrm{d}x' \, \mathrm{d}x_d \leqslant \int_0^\varepsilon \int_{\{|x'|<1\}} |\partial_d f(x', y)| \, \mathrm{d}x' \, \mathrm{d}y.$$

Letting $\varepsilon \downarrow 0$, we obtain

$$\int_{\{|x'|<1\}} |f(x', x_d)| \, \mathrm{d}x' = 0.$$

This implies $f(x', 0) = 0$ for almost all $x' \in \mathbb{R}^{d-1}$ and, since f is continuous on $\overline{\mathbb{R}_+^d}$, $f(x', 0) = 0$ for all $x' \in \mathbb{R}^{d-1}$.   □

Theorem 11.25. *For all $1 \leqslant p < \infty$ we have*

$$W_0^{1,p}(\mathbb{R}^d) = W^{1,p}(\mathbb{R}^d).$$

Proof Clearly, $W_0^{1,p}(\mathbb{R}^d) \subseteq W^{1,p}(\mathbb{R}^d)$. To prove the reverse inclusion we must show that every $f \in W^{1,p}(\mathbb{R}^d)$ can be approximated in $W^{1,p}(\mathbb{R}^d)$ by functions in $C_c^\infty(\mathbb{R}^d)$.

Let $\psi_n \in C_c^\infty(\mathbb{R}^d)$ satisfy $0 \leqslant \psi_n \leqslant 1$ pointwise on $\mathbb{R}^d$, $\psi_n(x) = 1$ for all $|x| \leqslant n$, and $|\nabla \psi_n| \leqslant 1$ on $\mathbb{R}^d$. Proposition 11.14 implies that the functions $\psi_n f$ belong to $W^{1,p}(\mathbb{R}^d)$, and by dominated convergence we have $\psi_n f \to f$ and $\partial_j(\psi_n f) = \psi_n \partial_j f + f \partial_j \psi_n \to \partial_j f$ in $L^p(\mathbb{R}^d)$ as $n \to \infty$, using that $\partial_j \psi_n \to 0$ and $\psi_n \to 1$ pointwise on $\mathbb{R}^d$ with uniform bounds $|\partial_j \psi_n| \leqslant 1$ and $|\psi_n| \leqslant 1$ to justify dominated convergence. It follows that $\psi_n f \to f$ in $W^{1,p}(\mathbb{R}^d)$.

Accordingly it suffices to prove that every compactly supported function in $W^{1,p}(\mathbb{R}^d)$ can be approximated, in the norm of $W^{1,p}(\mathbb{R}^d)$, by test functions. This was accomplished in Proposition 11.22. ∎

With the same method of proof one obtains the following density result.

Theorem 11.26. *For all $k \in \mathbb{N}$ and $1 \leqslant p < \infty$ the space $C_c^\infty(\mathbb{R}^d)$ is dense in $W^{k,p}(\mathbb{R}^d)$.*

11.1.d Extension Operators

Let $k \in \mathbb{N}$ be an integer that is kept fixed throughout this section. We say that D has a C^k-*boundary*, if for every $x_0 \in \partial D$ there exist open sets $U, V \subseteq \mathbb{R}^d$, with $x_0 \in U$, and a C^k-diffeomorphism $\rho : U \to V$ with the following properties:

(i) $\rho(D \cap U) = \{x \in V : x_d > 0\}$;
(ii) $\rho(\partial D \cap U) = \{x \in V : x_d = 0\}$;
(iii) there exists a constant $C > 0$ such that

$$C^{-1} \leqslant |\det(D\rho(x))| \leqslant C, \quad x \in U,$$

where D is the total derivative of ρ.

See Figure 11.2.

In this situation, by Proposition 11.20 applied inductively, a function $u \in L_{\text{loc}}^1(U)$ belongs to $W^{k,p}(U)$ if and only if the function $v := u \circ \rho^{-1} \in L_{\text{loc}}^1(V)$ belongs to $W^{k,p}(V)$, and in this case there exists a constant $C > 0$ such that

$$C^{-1}\|v\|_{W^{k,p}(V)} \leqslant \|u\|_{W^{k,p}(U)} \leqslant C\|v\|_{W^{k,p}(V)}. \tag{11.6}$$

Theorem 11.27 (Density of $C^\infty(\overline{D})$ in $W^{k,p}(D)$). *If D is bounded with C^k-boundary, then for all $1 \leqslant p < \infty$ the space $C^\infty(\overline{D})$ is dense in $W^{k,p}(D)$.*

We actually prove the stronger result that for any $f \in W^{k,p}(D)$ there exists a sequence of functions $f_n \in C^\infty(\mathbb{R}^d)$ whose restrictions to D satisfy $\lim_{n \to \infty} \|f_n - f\|_{W^{k,p}(D)} = 0$.

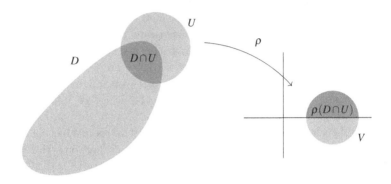

Figure 11.2 The definition of a C^k-boundary

Proof The proof proceeds in three steps. The first step deals with the case where D is (a bounded open subset of) $\mathbb{R}^d_+ := \{x \in \mathbb{R}^d : x_d > 0\}$. The second step, commonly referred to as 'straightening the boundary', uses the definition of a C^k-domain to carry over the result of Step 1 to the open sets U in the definition. The third step patches together the local results of Step 2 by means of a partition of unity argument.

Step 1 – In this first step we prove that if $f \in W^{k,p}(\mathbb{R}^d_+)$, then there exists a sequence of functions $f_n \in C^\infty_c(\mathbb{R}^d)$ whose restrictions to $\mathbb{R}^d_+$ satisfy $f_n \to f$ in $W^{k,p}(\mathbb{R}^d_+)$ as $n \to \infty$. Let $\psi \in C^\infty_c(\mathbb{R}^d)$ satisfy $\mathbf{1}_{B(0;1)} \leqslant \psi \leqslant \mathbf{1}_{B(0;2)}$ and let $M := \sup_{|\alpha| \leqslant k} \|\partial^\alpha \psi\|_\infty$. Then for all $n \geqslant 1$ the function $\psi_n(x) := \psi(x/n)$ satisfies $\mathbf{1}_{B(0;n)} \leqslant \psi_n \leqslant \mathbf{1}_{B(0;2n)}$ and $\|\partial^\alpha \psi_n\|_\infty \leqslant n^{-|\alpha|}\|\partial^\alpha \psi\|_\infty \leqslant M$. By the product rule and dominated convergence, this implies that $\psi_n f \to f$ in $W^{k,p}(\mathbb{R}^d_+)$. It follows that we may assume that f has bounded support.

For $t > 0$ define the functions
$$f_t(x) := f(x+te_d), \quad x \in \mathbb{R}^d_+,$$
where e_d is the dth unit vector of $\mathbb{R}^d$. Clearly we have $f_t \in W^{k,p}(\mathbb{R}^d_+)$, with weak derivatives
$$\partial^\alpha f_t(x) = (\partial^\alpha f)(x+te_d), \quad x \in \mathbb{R}^d_+. \tag{11.7}$$
The L^p-continuity of translations (Proposition 2.32) therefore implies that $f_t \to f$ in $W^{k,p}(\mathbb{R}^d_+)$. As a result, it suffices to approximate each f_t in $W^{k,p}(\mathbb{R}^d_+)$ with functions in $\overline{C^\infty_c(\mathbb{R}^d_+)}$.

In the remainder of this step we fix $t > 0$. Let $h \in W^{k,p}(\mathbb{R}^d)$ be a function whose restriction to $\{x \in \mathbb{R}^d : x_d > t\}$ agrees with the restriction of f on that set. Such a function may be found by multiplying f with a test function $\psi \in C^\infty_c(\mathbb{R}^d_+)$ satisfying $\psi \equiv$

1 on $\text{supp}(f) \cap \{x \in \mathbb{R}_+^d : x_d > t\}$; this results in a function with the desired properties by Proposition 11.14 and Lemma 11.15. Letting $h_t(x) := h(x + te_d)$ for $x \in \mathbb{R}^d$, it follows that for almost all $x \in \mathbb{R}_+^d$ we have

$$(\partial^\alpha h_t)(x) = \partial^\alpha h(x + te_d) = \partial^\alpha f(x + te_d) = \partial^\alpha f_t(x).$$

Choose a mollifier $\eta \in C_c^\infty(\mathbb{R}^d)$ satisfying $\int_{\mathbb{R}^d} \eta \, dx = 1$, and let $\eta_\varepsilon(x) := \varepsilon^{-d} \eta(\varepsilon^{-1}x)$ for $\varepsilon > 0$ and $x \in \mathbb{R}^d$. By Proposition 11.10, the functions $h_{t,\varepsilon} := \eta_\varepsilon * h_t$ belong to $C^\infty(\mathbb{R}^d)$ and for all multi-indices $|\alpha| \leqslant k$ we have

$$\partial^\alpha h_{t,\varepsilon} = \eta_\varepsilon * (\partial^\alpha h_t). \tag{11.8}$$

By (11.7), (11.8), and Proposition 2.34, for $\varepsilon \downarrow 0$ we then obtain

$$\|\partial^\alpha h_{t,\varepsilon} - \partial^\alpha f_t\|_{L^p(\mathbb{R}_+^d)} = \|\eta_\varepsilon * \partial^\alpha h_t - \partial^\alpha h_t\|_{L^p(\mathbb{R}_+^d)} \to 0.$$

This gives de desired approximation.

Step 2 – Let $x_0 \in \partial D$ be fixed. Choose open sets $U, V \subseteq \mathbb{R}^d$, with $x_0 \in U$, and a C^k-diffeomorphism $\rho : U \to V$ with the properties (i)–(iii) in the definition of a C^k-domain. In this step we assume that $f \in W^{k,p}(D)$ has its support in an open set $\widetilde{U} \Subset U$.

Let $g : V \cap \mathbb{R}_+^d \to \mathbb{R}$ be defined by $g := f \circ \rho^{-1}$. Then $g \in W^{k,p}(V \cap \mathbb{R}_+^d)$ by (11.6). Since $\rho(\widetilde{U}) \Subset V$, the same argument as in the proof of Lemma 11.15 shows that the zero extension $\widetilde{g}$ of g to all of $\mathbb{R}_+^d$ belongs to $W^{k,p}(\mathbb{R}_+^d)$.

By Step 1 we may choose functions $g_n \in C_c^\infty(\mathbb{R}^d)$ such that $g_n \to \widetilde{g}$ in $W^{k,p}(\mathbb{R}_+^d)$. Fix a test function $\zeta \in C_c^\infty(V)$ such that $\zeta \equiv 1$ on $\rho(\widetilde{U})$. Then also $\zeta g_n \to \zeta \widetilde{g}$ in $W^{k,p}(\mathbb{R}_+^d)$, and on $V \cap \mathbb{R}_+^d$ we have $\zeta \widetilde{g} = g$. Replacing g_n by ζg_n if necessary, we may therefore assume without loss of generality that $g_n \in C_c^\infty(\widetilde{V})$, where $\widetilde{V} \Subset V$ contains the support of ζ. Let $f_n := g_n \circ \rho$. Then $f_n \in C_c^\infty(U)$. It follows that $f_n \in C^\infty(\mathbb{R}^d)$ by zero extension, and $f_n \to f$ in $W^{k,p}(D)$ by (11.6).

Step 3 – Now let $f \in W^{k,p}(D)$ be arbitrary. Since ∂D is compact, as in Step 1 we can find open sets U_m and V_m and C^k-diffeomorphisms $\rho_m : U_m \to V_m$, $m = 1, \ldots, M$, as well as open sets $\widetilde{U}_m \Subset U_m$, $m = 1, \ldots, M$, in such a way that $\partial D \subseteq \bigcup_{m=1}^M \widetilde{U}_m$. By adding one open set $\widetilde{U}_0 \Subset D$ we may arrange that $\overline{D} \subseteq \bigcup_{m=0}^M \widetilde{U}_m$. Let $(\eta_m)_{m=0}^M$ be a smooth partition of unity for $\overline{D}$ subordinate to this cover, that is, $\eta_m \in C_c^\infty(\widetilde{U}_m)$ for $m = 0, 1, \ldots, M$ and

$$\sum_{m=0}^M \eta_m \equiv 1 \text{ on } \overline{D}, \quad 0 \leqslant \eta_m \leqslant 1, \quad m = 0, 1, \ldots, M.$$

Let $f^{(m)} := \eta_m f$. By Lemma 11.15 the zero extension of $f^{(0)}$ belongs to $W^{k,p}(\mathbb{R}^d)$, and therefore by Theorem 11.25 and we can find $f_n^{(0)} \in C^\infty(\mathbb{R}^d)$ such that $f_n^{(0)} \to f^{(0)}$ in $W^{k,p}(D)$. For $1 \leqslant m \leqslant M$ the function $f^{(m)}$ is supported in $\widetilde{U}_m$ and by Step 2 we can find $f_n^{(m)} \in C^\infty(\mathbb{R}^d)$ such that $f_n^{(m)} \to f^{(m)}$ in $W^{k,p}(D)$.

Finally let $f_n := \sum_{m=0}^{M} f_n^{(m)}$. Then $f_n \in C^\infty(\mathbb{R}^d)$ and, as $n \to \infty$,

$$\|f_n - f\|_{W^{k,p}(D)} \leqslant \sum_{m=0}^{M} \|f_n^{(m)} - f^{(m)}\|_{W^{k,p}(D)} \to 0.$$

The restrictions of f_n to $\overline{D}$ have the required properties. $\square$

The boundary of a C^k-domain is locally the graph of a C^k-function in $d-1$ 'horizontal' coordinates, and in fact this could be taken as an alternative definition of a C^k-domain. By translating f in the remaining 'vertical' direction, the use of the C^k-diffeomorphism ρ and the substitution rule can be avoided and all constructions can be carried out directly in U. This is the reason we have been rather brief in explaining the fine details of the substitution rule and its use in the present proof. Nevertheless we prefer the approach presented here, as it brings out clearly the idea that constructions involving the boundary can locally be reduced to hyperplanes $\{x \in \mathbb{R}^d : x_d = 0\}$ using C^k-diffeomorphisms. The advantage of this becomes even more clear in the proof of the next theorem.

Theorem 11.28 (Extension operator). *Let D be bounded and have a C^k-boundary. Then there exists a linear mapping $E : L^1_{loc}(D) \to L^1_{loc}(\mathbb{R}^d)$ with the following properties:*

(i) *for all $f \in L^1_{loc}(D)$ we have $(Ef)|_D = f$;*
(ii) *for all $1 \leqslant p < \infty$ there exists a constant $C \geqslant 0$ such that for all $\ell = 0, 1, \ldots, k$, and $f \in W^{\ell,p}(D)$ we have $Ef \in W^{\ell,p}(\mathbb{R}^d)$ and*

$$\|Ef\|_{W^{\ell,p}(\mathbb{R}^d)} \leqslant C\|f\|_{W^{\ell,p}(D)}, \quad f \in W^{\ell,p}(\mathbb{R}^d).$$

Proof   We proceed in three steps.

Step 1 – Let the integer $0 \leqslant \ell \leqslant k$ be fixed. For $f \in L^1_{loc}(\mathbb{R}^d_+)$ and multi-indices $|\alpha| \leqslant \ell$ define $E_\alpha f \in L^1_{loc}(\mathbb{R}^d)$ by

$$E_\alpha f(x) := \begin{cases} f(x), & x \in \mathbb{R}^d_+, \\ \sum_{j=1}^{\ell+1} (-j)^{\alpha_d} c_j f(x_1, \ldots, x_{d-1}, -jx_d), & x \notin \mathbb{R}^d_+, \end{cases}$$

where the scalars c_j are chosen in such a way that

$$\sum_{j=1}^{\ell+1} (-j)^k c_j = 1, \quad k = 0, 1, \ldots, \ell.$$

By the theory of Vandermonde determinants this system of $\ell+1$ equations is uniquely solvable for $c_1, \ldots, c_{\ell+1}$.

It is easily verified that if $f \in C^\ell(\overline{\mathbb{R}_+^d})$, then $E_0 f \in C^\ell(\mathbb{R}^d)$ (due to the choice of the coefficients c_j) and $\partial^\alpha E_0 f = E_\alpha \partial^\alpha f$. Thus

$$\|\partial^\alpha E_0 f\|_{L^p(\mathbb{R}^d)}^p = \int_{\mathbb{R}_+^d} |\partial^\alpha f|^p \, dx + \int_{\complement \mathbb{R}_+^d} \left| \sum_{j=1}^{\ell+1} (-j)^{\alpha_d} c_j f(x_1, \ldots, x_{d-1}, -jx_d) \right|^p dx$$

$$\leqslant C_{\alpha,\ell,p}^p \|\partial^\alpha f\|_{L^p(\mathbb{R}_+^d)}^p.$$

This gives the bound

$$\|E_0 f\|_{W^{\ell,p}(\mathbb{R}^d)} \leqslant C_{k,\ell,p} \|f\|_{W^{\ell,p}(\mathbb{R}_+^d)}.$$

By Theorem 11.27 this bound extends to functions $f \in W^{\ell,p}(\mathbb{R}_+^d)$.

Step 2 – Now let D be bounded and open. By Theorem 11.27 it suffices to prove the existence of a linear mapping $E : L_{\mathrm{loc}}^1(D) \to L_{\mathrm{loc}}^1(\mathbb{R}^d)$ such that for all $f \in L_{\mathrm{loc}}^1(D)$ we have $(Ef)|_D = f$ and for all $1 \leqslant p < \infty$, $\ell = 0, 1, \ldots, k$, and $f \in C_c^\infty(\overline{D})$ we have

$$\|Ef\|_{W^{\ell,p}(\mathbb{R}^d)} \leqslant C \|f\|_{W^{\ell,p}(D)}$$

with a constant $C \geqslant 0$ depending only on ℓ, p, and D.

Let $f \in L_{\mathrm{loc}}^1(D)$ and $0 \leqslant \ell \leqslant k$ be given. Using the notation of the proof of Theorem 11.27, set $f_m := \eta_m f$, $m = 0, 1, \ldots, M$. Let h_0 be the zero extension of f_0 to $\mathbb{R}^d$ and, for $m = 1, \ldots, M$, let g_m denote the zero extension to $\mathbb{R}_+^d$ of the function $f_m \circ \rho_m^{-1} \in L^1(V_m \cap \mathbb{R}_+^d)$. Let E_0 be the extension operator of Step 1 and define $Ef := \sum_{m=0}^M h_m$, where for $m = 1, \ldots, M$ we set

$$h_m := \begin{cases} (E_0 g_m) \circ \rho_m & \text{on } U_m, \\ 0 & \text{on } \complement U_m. \end{cases}$$

Then $Ef \equiv f$ on $\overline{D}$. If we now assume that $f \in C^\infty(\overline{D})$ and fix $1 \leqslant p < \infty$, then $f_0 = \eta_0 f \in C^\infty(\overline{D})$ and $\|h_0\|_{W^{\ell,p}(\mathbb{R}^d)} \leqslant C_{\ell,p,D} \|f\|_{W^{\ell,p}(D)}$. For $m = 1, \ldots, M$ we have $h_m \in C^\infty(U_m)$ with support in $\widetilde{U}_m$. Therefore $h_m \in C^\infty(\mathbb{R}^d)$ and, using (11.6) twice,

$$\|h_m\|_{W^{\ell,p}(D)} = \|h_m\|_{W^{\ell,p}(U_m)} \leqslant C_1 \|E_0 g_m\|_{W^{\ell,p}(V_m)} \leqslant C_1 \|E_0 g_m\|_{W^{\ell,p}(\mathbb{R}^d)}$$

$$\leqslant C_2 \|g_m\|_{W^{\ell,p}(\mathbb{R}_+^d)} = C_2 \|g_m\|_{W^{\ell,p}(V_m \cap \mathbb{R}_+^d)}$$

$$\leqslant C_3 \|f_m\|_{W^{\ell,p}(U_m \cap D)} \leqslant C_4 \|f\|_{W^{\ell,p}(D)}.$$

It follows that

$$\|Ef\|_{W^{\ell,p}(\mathbb{R}^d)} \leqslant \sum_{m=0}^M \|h_m\|_{W^{\ell,p}(\mathbb{R}^d)} \leqslant C_5 \|f\|_{W^{\ell,p}(D)},$$

with all constants only depending on ℓ, p, D. $\qquad\square$

11.1.e Bessel Potential Spaces

A function $f \in L^p(D)$, with $1 \leqslant p \leqslant \infty$, is said to admit a *weak L^p-Laplacian* if there exists a function $g \in L^p(D)$ such that for all test functions $\phi \in C_c^\infty(D)$ we have

$$\int_D f\Delta\phi \, dx = \int_D g\phi \, dx.$$

This defines a linear operator $\Delta_{p,D}$ in $L^p(D)$, the *weak L^p-Laplacian on D*, with domain

$$\mathsf{D}(\Delta_{p,D}) := \{f \in L^p(D) : f \text{ admits a weak } L^p\text{-Laplacian on } D\}.$$

By the argument of Example 10.16, this operator is closed. In what follows we will denote weak Laplacians simply by Δ.

Among other things, as an application of the Plancherel theorem the next theorem establishes that the domain of the weak Laplacian in $L^2(\mathbb{R}^d)$ equals $W^{2,2}(\mathbb{R}^d)$.

Theorem 11.29 (Fourier analytic characterisation, weak Laplacian). *Let $k \geqslant 0$ be a nonnegative integer. For a function $f \in L^2(\mathbb{R}^d)$ the following assertions are equivalent:*

(1) *f belongs to $W^{k,2}(\mathbb{R}^d)$;*
(2) *$\xi \mapsto (1+|\xi|^2)^{k/2}\widehat{f}(\xi)$ belongs to $L^2(\mathbb{R}^d)$.*

Moreover,

$$f \mapsto \left\| \xi \mapsto (1+|\xi|^2)^{k/2}\widehat{f}(\xi) \right\|_2$$

defines an equivalent norm on $W^{k,2}(\mathbb{R}^d)$. For $k = 2$, (1) and (2) are equivalent to:

(3) *f admits a weak Laplacian in $L^2(\mathbb{R}^d)$.*

Moreover, $\mathsf{D}(\Delta) = W^{2,2}(\mathbb{R}^d)$ and $f \mapsto \|f\|_2 + \|\Delta f\|_2$ defines an equivalent norm on $W^{2,2}(\mathbb{R}^d)$.

Definition 11.30 (The Bessel potential spaces $H^s(\mathbb{R}^d)$). For real numbers $s \geqslant 0$ the subspace of all $f \in L^2(\mathbb{R}^d)$ such that

$$\xi \mapsto (1+|\xi|^2)^{s/2}\widehat{f}(\xi)$$

belongs to $L^2(\mathbb{R}^d)$ is denoted by $H^s(\mathbb{R}^d)$ and is called the *Bessel potential space* with smoothness exponent s. With respect to the norm

$$\|f\|_{H^s(\mathbb{R}^d)} := \left\| \xi \mapsto (1+|\xi|^2)^{s/2}\widehat{f}(\xi) \right\|_{L^2(\mathbb{R}^d)} \tag{11.9}$$

this space is a Hilbert space. The easy verification is left as an exercise.

For noninteger values of s, the spaces $H^s(\mathbb{R}^d)$ play an important role in the regularity theory for partial differential equations.

The equivalence of (1) and (2) of Theorem 11.29 asserts:

Theorem 11.31 ($W^{k,2}(\mathbb{R}^d) = H^k(\mathbb{R}^d)$)**.** *If $k \geqslant 0$ is a nonnegative integer, then*

$$W^{k,2}(\mathbb{R}^d) = H^k(\mathbb{R}^d)$$

with equivalence of norms.

The proof of Theorem 11.29 relies on the following lemma, where for vectors $\xi \in \mathbb{R}^d$ and multi-indices $\alpha \in \mathbb{N}^d$ we use the short-hand notation

$$\xi^\alpha := \xi_1^{\alpha_1} \cdots \xi_d^{\alpha_d}.$$

Lemma 11.32. *A function $f \in L^2(\mathbb{R}^d)$ admits a weak derivative $\partial^\alpha f$ belonging to $L^2(\mathbb{R}^d)$ if and only if $\xi \mapsto \xi^\alpha \widehat{f}(\xi)$ belongs to $L^2(\mathbb{R}^d)$, and in that case we have*

$$\widehat{\partial^\alpha f}(\xi) = i^{|\alpha|} \xi^\alpha \widehat{f}(\xi)$$

for almost all $\xi \in \mathbb{R}^d$.

Proof For test functions $\phi \in C_c^\infty(\mathbb{R}^d)$ and $\xi \in \mathbb{R}^d$, an integration by parts gives

$$\begin{aligned}
\widehat{\partial^\alpha \phi}(\xi) &= \frac{1}{(2\pi)^{d/2}} \int_{\mathbb{R}^d} \exp(-ix \cdot \xi) \partial^\alpha \phi(x) \, dx \\
&= \frac{i^{|\alpha|} \xi^\alpha}{(2\pi)^{d/2}} \int_{\mathbb{R}^d} \exp(-ix \cdot \xi) \phi(x) \, dx = i^{|\alpha|} \xi^\alpha \widehat{\phi}(\xi).
\end{aligned} \tag{11.10}$$

Also, by differentiating under the integral, we see that $\widehat{\phi}$ is smooth and

$$\partial^\alpha \widehat{\phi}(\xi) = (-i)^{|\alpha|} (x^\alpha \phi(x))^\widehat{}(\xi). \tag{11.11}$$

'If': Suppose that $\xi \mapsto \xi^\alpha \widehat{f}(\xi)$ belongs to $L^2(\mathbb{R}^d)$ and denote its inverse Fourier–Plancherel transform by g_α. Using the Plancherel isometry and (11.10), for real-valued $\phi \in C_c^\infty(\mathbb{R}^d)$ we obtain

$$\begin{aligned}
(-1)^{|\alpha|} \int_{\mathbb{R}^d} f(x) \partial^\alpha \phi(x) \, dx &= (-1)^{|\alpha|} (f | \partial^\alpha \phi) = (-1)^{|\alpha|} (\widehat{f} | \widehat{\partial^\alpha \phi}) \\
&= i^{|\alpha|} \int_{\mathbb{R}^d} \xi^\alpha \widehat{f}(\xi) \widehat{\phi}(\xi) \, d\xi = i^{|\alpha|} \int_{\mathbb{R}^d} g_\alpha(x) \phi(x) \, dx.
\end{aligned}$$

Considering real and imaginary parts separately, the identity extends to complex-valued $\phi \in C_c^\infty(\mathbb{R}^d)$. This shows that f has weak derivative $\partial^\alpha f = i^{|\alpha|} g_\alpha$ in $L^2(\mathbb{R}^d)$.

'Only if': Fix a test function $\phi \in C_c^\infty(\mathbb{R}^d)$. Using Proposition 5.28 and (11.11) we obtain

$$\begin{aligned}
\int_{\mathbb{R}^d} \widehat{\partial^\alpha f}(\xi) \phi(\xi) \, d\xi &= \int_{\mathbb{R}^d} (\partial^\alpha f)(\xi) \widehat{\phi}(\xi) \, d\xi = (-1)^{|\alpha|} \int_{\mathbb{R}^d} f(\xi) \partial^\alpha \widehat{\phi}(\xi) \, d\xi \\
&= i^{|\alpha|} \int_{\mathbb{R}^d} f(\xi) (x^\alpha \phi(x))^\widehat{}(\xi) \, d\xi = i^{|\alpha|} \int_{\mathbb{R}^d} \xi^\alpha \widehat{f}(\xi) \phi(\xi) \, d\xi.
\end{aligned}$$

By Proposition 11.5 this implies $\widehat{\partial^\alpha f}(\xi) = i^{|\alpha|} \xi^\alpha \widehat{f}(\xi)$ for almost all $\xi \in \mathbb{R}^d$. Since by assumption $\partial^\alpha f \in L^2(\mathbb{R}^d)$, the Plancherel theorem implies that $\widehat{\partial^\alpha f} \in L^2(\mathbb{R}^d)$. This shows that $\xi \mapsto \xi^\alpha \widehat{f}(\xi)$ belongs to $L^2(\mathbb{R}^d)$.   □

Proof of Theorem 11.29   We begin with the proof of the equivalence (1)⇔(2) for general integers $k \geqslant 0$. The heart of the matter is contained in the two-sided estimate

$$(1 + |\xi|^2)^{k/2} \approx_{d,k} \left(\sum_{|\alpha| \leqslant k} |\xi^\alpha|^2 \right)^{1/2}, \tag{11.12}$$

which follows from the binomial identity. Here, the notation $A \approx_{d,k} B$ is short-hand for the existence of constants $c_{d,k}, c'_{d,k} \geqslant 0$, both depending only on d and k, such that $A \leqslant c_{d,k} B$ and $B \leqslant c'_{d,k} A$.

(1)⇒(2): First let $f \in W^{k,2}(\mathbb{R}^d)$. By (11.12), Lemma 11.32, and Plancherel's theorem,

$$\int_{\mathbb{R}^d} (1 + |\xi|^2)^k |\widehat{f}(\xi)|^2 \, d\xi \approx_{d,k} \sum_{|\alpha| \leqslant k} \int_{\mathbb{R}^d} |\xi^\alpha \widehat{f}(\xi)|^2 \, d\xi$$

$$= \sum_{|\alpha| \leqslant k} \int_{\mathbb{R}^d} |\widehat{\partial^\alpha f}(\xi)|^2 \, d\xi$$

$$= \sum_{|\alpha| \leqslant k} \int_{\mathbb{R}^d} |\partial^\alpha f(x)|^2 \, dx \approx_{d,k} \|f\|^2_{W^{k,2}(\mathbb{R}^d)}.$$

This shows that $\xi \mapsto (1 + |\xi|^2)^{k/2} \widehat{f}(\xi)$ belongs to $L^2(\mathbb{R}^d)$, and that its L^2-norm is equivalent to the norm of f in $W^{k,2}(\mathbb{R}^d)$.

(2)⇒(1): Suppose that $f \in L^2(\mathbb{R}^d)$ is a function with the property that $\xi \mapsto (1 + |\xi|^2)^{k/2} \widehat{f}(\xi)$ belongs to $L^2(\mathbb{R}^d)$. Then $\xi \mapsto \xi^\alpha \widehat{f}(\xi)$ belongs to $L^2(\mathbb{R}^d)$ for all multi-indices $\alpha \in \mathbb{N}^d$ satisfying $|\alpha| \leqslant k$, and therefore f has weak derivatives $\partial^\alpha f$ for all $|\alpha| \leqslant k$ by Lemma 11.32. This shows that $f \in W^{k,2}(\mathbb{R}^d)$.

For $k = 2$, (1) obviously implies (3). Conversely, if (3) holds (with $\Delta f = g$), then by the same argument as in the 'only if' part of Lemma 11.32 we find that $\xi \mapsto |\xi|^2 \widehat{f}(\xi)$ belongs to $L^2(\mathbb{R}^d)$ (and equals $\widehat{g}(\xi)$ for almost all $\xi \in \mathbb{R}^d$), and therefore (2) holds. The equivalence of norms again follows from the Plancherel theorem and (11.12).   □

11.2 The Poisson Problem $-\Delta u = f$

The results developed above are now applied to study the Poisson problem.

11.2.a Dirichlet Boundary Conditions

We recall from Theorem 11.31 that $H^k(\mathbb{R}^d) = W^{k,2}(\mathbb{R}^d)$ with equivalent norms. For nonempty open subsets D of $\mathbb{R}^d$ it is customary to *define*

$$H^k(D) := W^{k,2}(D), \quad H_0^1(D) := W_0^{1,2}(D),$$

where $W_0^{1,2}(D)$ is the closure of $C_c^\infty(D)$ in $W^{1,2}(D)$. We endow $H^k(D)$ with the norm

$$\|f\|_{H^k(D)}^2 := \sum_{|\alpha| \leqslant k} \|\partial^\alpha f\|_2^2,$$

where the summation extends over all multi-indices $\alpha \in \mathbb{N}^d$ of order at most k. This norm is associated with the inner product

$$(f|g)_{H^k(D)} = \sum_{|\alpha| \leqslant k} (\partial^\alpha f | \partial^\alpha g)_2$$

and it turns $H^k(D)$ into a Hilbert space. As a closed subspace of $H^1(D)$, the space $H_0^1(D)$ is a Hilbert space as well.

Let us now take a look at the *Poisson problem with Dirichlet boundary conditions:*

$$\begin{cases} -\Delta u &= f \text{ on } D, \\ u|_{\partial D} &= 0, \end{cases} \tag{11.13}$$

where $f \in L^2(D)$ is a given function and $\Delta = \sum_{i=1}^d \partial_i^2$ is the Laplace operator. Multiplying both sides of the equation (11.13) with a test function $\phi \in C_c^\infty(D)$ and integrating, we obtain the following integrated version of the problem:

$$\int_D (\Delta u)\phi \, dx = -\int_D f\phi \, dx,$$

which, after a formal integration by parts (which can be rigorously justified if $u \in C^2(D)$), can be rewritten as

$$\int_D \nabla u \cdot \nabla \phi \, dx = \int_D f\phi \, dx. \tag{11.14}$$

This formal derivation justifies the following definition.

Definition 11.33 (Weak solutions). A function $u \in H_0^1(D)$ is called a *weak solution* of the Poisson problem with Dirichlet boundary conditions (11.13) if

$$\int_D \nabla u \cdot \nabla \phi \, dx = \int_D f\phi \, dx, \quad \phi \in C_c^\infty(D).$$

The notion of weak solution makes sense for inhomogeneities $f \in L^2(D)$. In the special case $f \in C(\overline{D})$, a *classical solution* may be defined as a function $u \in C^2(D) \cap C(\overline{D})$ satisfying the equations of the Poisson problem, $-\Delta u = f$ on D and $u|_{\partial D} = 0$ pointwise.

Classical solutions may not exist, however, even when $f \in C_c(D)$ (see Problem 11.26). The advantage of working with weak solutions is that integrated equation in (11.14) involves only the first derivative and both integrals in (11.14) can be interpreted as inner products. This makes the problem of finding weak solutions amenable to Hilbert space methods. The requirement that u be in $H_0^1(D)$ implements the boundary condition $u|_{\partial D} = 0$, as evidenced by Theorem 11.24.

Having admitted membership of $H_0^1(D)$ as a valid way of implementing Dirichlet boundary conditions, one may still be tempted to look for solutions in $H_0^1(D) \cap H^2(D)$. It can be shown that this works (in the sense that a unique weak solution in the sense of Definition 11.33 belonging to this space exists) if D is assumed to be bounded with C^2-boundary (see Remark 11.38). Without additional assumptions on D, however, such solutions need not exist.

Before attacking the problem (11.13) using Hilbert space methods, we pause to emphasise that in some special cases this problem is simple enough to admit a direct elementary solution. For instance, for $d = 1$ and $D = (a, b)$ we may integrate the equation twice and determine the two integration constants by substituting the boundary conditions, which take the form $u(a) = u(b) = 0$. After some computations one arrives at

$$u(x) = \int_a^b k(x, y) f(y) \, dy, \quad x \in (a, b),$$

where

$$k(x, y) = \frac{1}{b - a} \begin{cases} (b - y)(x - a), & x \leqslant y, \\ (b - x)(y - a), & y \leqslant x, \end{cases}$$

Henri Poincaré, 1854–1912

is the so-called *Green function* for the Poisson problem on (a, b) with Dirichlet boundary conditions. The reader may check (see Problem 11.25) that the function u thus defined belongs to $H_0^1(a, b)$ and is indeed a weak solution of (11.13).

It is unclear, however, how to extend this elementary method to higher dimensions. In contrast, the Hilbert space method adopted here works in arbitrary dimensions. Our main tool is the following inequality which is of interest in its own right.

Theorem 11.34 (Poincaré inequality). *Let D be contained in $R := (0, r) \times \mathbb{R}^{d-1}$ and let $1 \leqslant p < \infty$. Then the following estimate holds:*

$$\|f\|_p \leqslant r p^{-1/p} \left\| \frac{\partial f}{\partial x_1} \right\|_p^p, \quad f \in W_0^{1,p}(D).$$

As a consequence, $\|f\|_{W_0^{1,p}(D)} := \|\nabla f\|_{L^p(D; \mathbb{K}^d)}$ defines an equivalent norm on $W_0^{1,p}(D)$.

Proof   First assume that $f \in C_c^\infty(D)$. By extending f identically zero on $R \setminus D$ we may view f as a function in $C_c^\infty(R)$. For $1 \leqslant p < \infty$ and $x_1 \in [0,r]$ we have, using Hölder's inequality and the fact that $f(0,x_2,\ldots,x_d) = 0$,

$$|f(x_1,x_2,\ldots,x_d)|^p = \left| \int_0^{x_1} \frac{\partial f}{\partial x_1}(t,x_2,\ldots,x_d)\,dt \right|^p \leqslant x_1^{p/q} \int_0^{x_1} \left| \frac{\partial f}{\partial x_1}(t,x_2,\ldots,x_d) \right|^p dt,$$

where $\frac{1}{p} + \frac{1}{q} = 1$. Integrating both sides over R, we obtain

$$\begin{aligned}
\|f\|_p^p &= \int_R |f(x_1,x_2,\ldots,x_d)|^p\,dx \\
&\leqslant \int_R \left(x_1^{p/q} \int_0^r \left| \frac{\partial f}{\partial x_1}(t,x_2,\ldots,x_d) \right|^p dt \right) dx \\
&= \int_0^r x_1^{p/q}\,dx_1 \int_{\mathbb{R}^{d-1}} \int_0^r \left| \frac{\partial f}{\partial x_1}(t,x_2,\ldots,x_d) \right|^p dt\,dx_2 \cdots dx_d \\
&= \frac{r^p}{p} \left\| \frac{\partial f}{\partial x_1} \right\|_p^p.
\end{aligned}$$

Since $C_c^\infty(D)$ is dense in $W_0^{1,p}(D)$, the estimate for a general $f \in W_0^{1,p}(D)$ follows by approximation.

The equivalence of the norms follows from

$$\|\nabla f\|_p \leqslant \|f\|_p + \|\nabla f\|_p \leqslant (rp^{-1/p} + 1)\|\nabla f\|_p,$$

where we used the trivial estimate $\left\| \frac{\partial f}{\partial x_1} \right\|_p \leqslant \|\nabla f\|_p$. $\square$

We now take $p = 2$ and recall the notation $H^1(D) = W^{1,2}(D)$ and $H_0^1(D) = W_0^{1,2}(D)$. Poincaré's inequality then states that

$$\|f\|_{H_0^1(D)} := \|\nabla f\|_2, \quad f \in H_0^1(D),$$

defines an equivalent norm on $H_0^1(D)$. With respect to this norm, $H_0^1(D)$ is again a Hilbert space, this time with respect to the inner product

$$((f_1|f_2))_{H_0^1(D)} := (\nabla f_1|\nabla f_2)_2.$$

Theorem 11.35 (Poisson problem, Dirichlet boundary conditions). *If D is bounded, then for every $f \in L^2(D)$ the Poisson problem (11.13) admits a unique weak solution $u \in H_0^1(D)$. Moreover, there exists a constant $C \geqslant 0$, independent of f, such that*

$$\|u\|_{H_0^1(D)} \leqslant C\|f\|_2.$$

Proof   By the Cauchy–Schwarz inequality and Poincaré's inequality, the linear mapping $L : g \mapsto \int_D g\overline{f}\,dx$ is bounded from $H_0^1(D)$ to $\mathbb{K}$:

$$|L(g)| \leqslant \|g\|_2\|f\|_2 \leqslant C\|\nabla g\|_2\|f\|_2 = C\|g\|_{H_0^1(D)}\|f\|_2.$$

Therefore L defines a bounded functional on $H_0^1(D)$. Hence, by the Riesz representation theorem, there exists a unique $u \in H_0^1(D)$ such that

$$L(g) = ((g|u))_{H_0^1(D)}, \quad g \in H_0^1(D), \tag{11.15}$$

and it satisfies $\|u\|_{H_0^1(D)} = \|L\| \leqslant C\|f\|_2$. Writing out the identity (11.15), it takes the form

$$\int_D \nabla g \cdot \overline{\nabla u}\, dx = \int_D g\overline{f}\, dx, \quad g \in H_0^1(D). \tag{11.16}$$

In particular, (11.16) holds for all $g \in C_c^\infty(D)$, since $C_c^\infty(D)$ is contained in $H_0^1(D)$. Replacing g by $\overline{g}$ and taking conjugates on both sides, we see that u is a weak solution of (11.13).

If $v \in H_0^1(D)$ is another weak solution, then

$$\int_D (\nabla u - \nabla v) \cdot \nabla \phi\, dx = 0, \quad \phi \in C_c^\infty(D).$$

Since $C_c^\infty(D)$ is dense in $H_0^1(D)$, it follows that

$$\int_D (\nabla u - \nabla v) \cdot \nabla g\, dx = 0, \quad g \in H_0^1(D),$$

and applying this to $\overline{g}$ gives $((u - v|g))_{H_0^1(D)} = 0$ for all $g \in H_0^1(D)$. This implies $u - v = 0$ in $H_0^1(D)$. $\square$

We prove next that the weak solution actually belongs to $H_0^1(D) \cap H_{\mathrm{loc}}^2(D)$, where $H_{\mathrm{loc}}^2(D)$ is the space of all $f \in L_{\mathrm{loc}}^1(D)$ with the property that $f|_U \in H^2(U)$ for all open sets $U \Subset D$. Defining the space $L_{\mathrm{loc}}^2(D)$ similarly, this will follow from the following lemma.

Lemma 11.36. *If $f \in H^1(D)$ admits a weak Laplacian in $L_{\mathrm{loc}}^2(D)$, then $f \in H_{\mathrm{loc}}^2(D)$.*

Proof Let U, U' be bounded open sets such that $U \Subset U' \Subset D$, and let $\psi \in C_c^\infty(U')$ satisfy $\psi \equiv 1$ on U. It is routine to check that if we view ψf as an element of $L^2(\mathbb{R}^d)$, it admits a weak Laplacian belonging to $L^2(\mathbb{R}^d)$ given by the Leibniz formula

$$\Delta(\psi f) = (\Delta \psi)f + \psi h + \sum_{j=1}^d (\partial_j \psi)g_j,$$

where $g_j := \partial_j f \in L^2(D)$ and $h := \Delta f \in L_{\mathrm{loc}}^2(D)$ are the weak directional derivatives and the weak Laplacian of f on D, respectively; we view all terms as functions defined on all of $\mathbb{R}^d$ by zero extension.

Theorem 11.29 then implies that $\psi f \in H^2(\mathbb{R}^d)$. Since $(\psi f)|_U = f|_U$, it follows that $f|_U$ belongs to $H^2(U)$. $\square$

Theorem 11.37. *Let D be bounded. The weak solution u of the Poisson problem* $-\Delta u = f$ *with* $f \in L^2(D)$, *subject to Dirichlet boundary conditions, belongs to* $H^1_0(D) \cap H^2_{loc}(D)$.

Proof The very definition of a weak solution implies that u admits a weak Laplacian belonging to $L^2(D)$, given by $\Delta u = -f$. The result now follows from the lemma. □

Remark 11.38. If D is bounded and has a C^2-boundary, the weak solution belongs to $H^2(D)$. This follows from Theorem 11.28 and Lemma 11.36.

The next proposition shows that a function in $H^1_0(D)$ solves the Poisson problem with Dirichlet boundary conditions if and only if it minimises a certain energy functional.

Theorem 11.39 (Variational characterisation). *Let D be bounded and let* $f \in L^2(D)$. *For a function* $u_0 \in H^1_0(D)$ *the following assertions are equivalent:*

(1) u_0 *is the weak solution of the Poisson problem* $-\Delta u = f$ *on D subject to Dirichlet boundary conditions;*
(2) u_0 *minimises the energy functional* $E : H^1_0(D) \to \mathbb{R}$ *defined by*

$$E(u) := \frac{1}{2} \int_D |\nabla u|^2 \, dx - \operatorname{Re} \int_D u \overline{f} \, dx.$$

Proof We use the notation

$$\mathfrak{a}(u,v) := \int_D \nabla u \cdot \overline{\nabla v} \, dx, \quad L(u) := \int_D u \overline{f} \, dx.$$

With this notation, for all $t \in \mathbb{R}$ and $u_0, u \in H^1_0(D)$ we have

$$E(u_0 + tu) = E(u_0) + t \operatorname{Re}(\mathfrak{a}(u,u_0) - Lu) + \frac{1}{2}t^2 \mathfrak{a}(u,u). \tag{11.17}$$

(1)$\Rightarrow$(2): Suppose that u_0 is a weak solution, that is, $u \in H^1_0(D)$ and $\mathfrak{a}(\phi, u_0) = L(\phi)$ for all $\phi \in C^\infty_c(D)$. By density, this identity extends to arbitrary $\phi \in H^1_0(D)$. Applying the identity with $\phi = u$ and taking $t = 1$ in (11.17), for all nonzero $u \in H^1_0(D)$ we obtain

$$E(u_0 + u) = E(u_0) + \frac{1}{2}\mathfrak{a}(u,u) \geqslant E(u_0), \tag{11.18}$$

and, by Poincaré's inequality, the inequality is in fact strict. It follows that u_0 is a minimiser of E in $H^1_0(D)$.

(2)$\Rightarrow$(1): Suppose conversely that u_0 minimises E in $H^1_0(D)$. The identity (11.17) implies that for all $u \in H^1_0(D)$ the function $t \mapsto E(u_0 + tu)$ is differentiable in t and

$$0 = \frac{d}{dt}\Big|_{t=0} E(u_0 + tu) = \operatorname{Re}(\mathfrak{a}(u,u_0) - Lu).$$

Over the real scalar field this implies that $\mathfrak{a}(u,u_0) - Lu = 0$. Over the complex scalar field we apply the preceding identity with u replaced by iu to find that also $\operatorname{Im}(\mathfrak{a}(u,u_0) -$

Lu) $= 0$, and again it follows that $\mathfrak{a}(u, u_0) - Lu = 0$. In both cases we conclude that u_0 is a weak solution.   □

The existence and uniqueness of a weak solution implies that for each $f \in L^2(D)$ the energy functional

$$E(u) := \frac{1}{2} \int_D |\nabla u|^2 \, dx - \mathrm{Re} \int_D u\overline{f} \, dx$$

has a unique minimiser in $H_0^1(D)$. In the above proof, uniqueness was reflected by the strictness of the inequality in (11.18).

Theorem 11.39 is a special case of a more general result on the existence and uniqueness of minimisers for suitable nonlinear functionals defined on Hilbert spaces; see Problem 11.31.

11.2.b Neumann Boundary Conditions

Having dealt with Dirichlet boundary conditions, we shall now take a look at the *Poisson problem with Neumann boundary conditions:*

$$\begin{cases} -\Delta u & = f \ \text{ on } D, \\ \frac{\partial u}{\partial \nu} & = 0 \ \text{ on } \partial D, \end{cases} \tag{11.19}$$

where $\frac{\partial u}{\partial \nu} := \nabla u \cdot \nu$ is the partial derivative in the direction of the outward normal ν along ∂D and $f \in L^2(D)$ is a given function. The treatment of this example in higher dimensions requires some familiarity with standard techniques from partial differential equations, as the notion of an outward normal is meaningful only under some regularity assumptions on the boundary ∂D. For the present treatment C^1-regularity of the boundary suffices.

To motivate the notion of weak solutions we need Green's theorem: *If D is bounded with C^1-boundary, then for all $u \in C^2(\overline{D})$ and $v \in C^1(\overline{D})$ we have*

$$\int_D \nabla u \cdot \nabla v \, dx = -\int_D v \Delta u \, dx + \int_{\partial D} v \frac{\partial u}{\partial \nu} \, dS,$$

where dS is the normalised surface measure on ∂D. We forget the boundary condition for a moment and ask ourselves which information is conveyed by the integrated equation

$$\int_D \nabla u \cdot \nabla \phi \, dx = \int_D f\phi \, dx \tag{11.20}$$

if it is to hold for all $\phi \in C^\infty(\overline{D})$ (and not just for all $C_c^\infty(D)$, since that would ignore what happens at the boundary). By Green's theorem,

$$\int_D \phi \Delta u \, dx - \int_{\partial D} \phi \frac{\partial u}{\partial \nu} \, dS = -\int_D f\phi \, dx. \tag{11.21}$$

Since we wish to solve $-\Delta u = f$, we substitute this relation into (11.21) and find that

$$\int_{\partial D} \phi \frac{\partial u}{\partial v} \, dS = 0.$$

This can only hold for all $\phi \in C^\infty(\overline{D})$ if

$$\frac{\partial u}{\partial v} = 0 \ \text{ on } \partial D,$$

that is, *if Neumann boundary conditions hold.*

By considering $\phi \equiv 1$ we see that the identity (11.20) can only hold for all $\phi \in C^\infty(\overline{D})$ if f satisfies the compatibility condition

$$\int_D f \, dx = 0.$$

More generally the integral of f against every locally constant function should vanish. Likewise, solutions of (11.20) cannot be unique: if u is a solution (in whatever weak sense), then also $u + C$ is a solution, for any locally constant function C. In order to simplify matters we henceforth assume that D is connected, so that the only locally constant functions are the constant functions. Under this assumption we get rid of integration constants by imposing the constraint

$$\int_D u \, dx = 0,$$

that is, the average of u over D should vanish. Accordingly we define

$$H^1_{\mathrm{av}}(D) := \left\{ u \in H^1(D) : \int_D u \, dx = 0 \right\}.$$

This discussion leads to the following weak formulation of problem (11.19).

Definition 11.40 (Weak solutions). Let D be bounded and connected and let $f \in L^2(D)$ satisfy $\int_D f \, dx = 0$. A function $u \in H^1_{\mathrm{av}}(D)$ is called a *weak solution* of the Poisson problem (11.19) if

$$\int_D \nabla u \cdot \nabla \phi \, dx = \int_D f \phi \, dx, \quad \phi \in C^\infty(\overline{D}).$$

Our treatment of the Poisson problem with Dirichlet boundary conditions crucially depended on the Poincaré inequality for $H^1_0(D)$. The treatment of Neumann boundary conditions proceeds analogously, the role of $H^1_0(D)$ being taken over by $H^1_{\mathrm{av}}(D)$. We will prove a version of the Poincaré inequality for this space in Theorem 11.42. Its proof depends on the following compactness result.

Theorem 11.41 (Rellich–Kondrachov compactness theorem). *If D is bounded and $1 \leqslant p < \infty$, then:*

(1) *the inclusion mapping from $W^{1,p}_0(D)$ into $L^p(D)$ is compact;*

(2) *if D has C^1-boundary, the inclusion mapping from $W^{1,p}(D)$ into $L^p(D)$ is compact.*

Proof We first prove (1) and deduce (2) from it by means of the extension theorem.

(1): We must show that the unit ball of $W_0^{1,p}(D)$ is relatively compact in $L^p(D)$. By extending the elements of $W_0^{1,p}(D)$ identically zero outside D we may view this unit ball as a bounded subset, which we denote by B, of $L^p(\mathbb{R}^d)$, and it suffices to prove that this set is relatively compact in $L^p(\mathbb{R}^d)$. For this purpose we use the Fréchet–Kolmogorov theorem, or rather, its corollary for bounded domains (Corollary 2.36). According to this corollary we must check that

$$\lim_{|h| \to 0} \sup_{f \in B} \| \tau_h f - f \|_p = 0. \tag{11.22}$$

Here, $\tau_h f$ is the translate of f over h, that is, $\tau_h f(x) = f(x+h)$. To prove this, first let $f \in B \cap C_c^\infty(D)$ and extend f to identically zero to all of $\mathbb{R}^d$. For $r > 0$ let $D_r := \{x \in \mathbb{R}^d : d(x,D) < r\}$. If $|h| < \frac{1}{2}r$, then by Hölder's inequality, Fubini's theorem, and a change of variables,

$$\int_{\mathbb{R}^d} |\tau_h f - f|^p \, dx = \int_{D_{\frac{1}{2}r}} |f(x+h) - f(x)|^p \, dx$$

$$\leqslant \int_{D_{\frac{1}{2}r}} \left(\int_0^1 \left| \frac{d}{dt} f(x+th) \right| dt \right)^p dx$$

$$\leqslant \int_{D_{\frac{1}{2}r}} \int_0^1 \left| \frac{d}{dt} f(x+th) \right|^p dt \, dx$$

$$\leqslant |h|^p \int_{D_{\frac{1}{2}r}} \int_0^1 |\nabla f(x+th)|^p \, dt \, dx$$

$$= |h|^p \int_0^1 \int_{D_{\frac{1}{2}r}} |\nabla f(x+th)|^p \, dx \, dt$$

$$\leqslant |h|^p \int_0^1 \int_{D_r} |\nabla f(y)|^p \, dy \, dt$$

$$= |h|^p \int_D |\nabla f(y)|^p \, dy \leqslant |h|^p \|f\|_{W_0^1(D)}^p \leqslant |h|^p,$$

keeping in mind that $\|f\|_{W_0^1(D)} \leqslant 1$ since $f \in B$. The above estimate holds for any $f \in B \cap C_c^\infty(D)$. Since $C_c^\infty(D)$ is dense in $W_0^{1,p}(D)$ this estimate extends to arbitrary $f \in B$. This proves that if $|h| < \frac{1}{2}r$, then

$$\sup_{f \in B} \| \tau_h f - f \|_p \leqslant |h|$$

and (11.22) follows.

(2): Let $D \Subset D'$, where $D' \subseteq \mathbb{R}^d$ is some larger bounded domain. Let $\psi \in C_c^\infty(\mathbb{R}^d)$ be compactly supported in D' and satisfy $\psi \equiv 1$ on D. As in Theorem 11.28 let $E_D :$ $W^{1,p}(D) \to W^{1,p}(\mathbb{R}^d)$ be a bounded extension operator, let $M_\psi : W^{1,p}(\mathbb{R}^d) \to W_0^{1,p}(D')$ be the multiplier given by $f \mapsto \psi f$ (we use Proposition 11.22 to see that this operator maps into $W_0^{1,p}(D')$ as claimed), let $i_{D'} : W_0^{1,p}(D') \to L^p(D')$ be the inclusion mapping (which is compact by Step 1), and let $R_{D',D} : L^p(D') \to L^p(D)$ the restriction operator $f \mapsto f|_{D'}$. Then the inclusion mapping $j_D : W^{1,p}(D) \to L^p(D)$ factors as

$$j_D = R_{D',D} \circ i_{D'} \circ M_\psi \circ E_D$$

and is therefore compact. $\qquad\square$

As an application of the Rellich–Kondrachov theorem we have the following variant of Poincaré's inequality. Define

$$W_{av}^{1,p}(D) := \left\{ u \in W^{1,p}(D) : \int_D u\,dx = 0 \right\}.$$

Theorem 11.42 (Poincaré–Wirtinger inequality). *Let D be bounded and connected with C^1-boundary. Let $1 \leqslant p < \infty$. Then there exists a constant $C = C_{p,D}$ such that for all $f \in W_{av}^{1,p}(D)$ the following estimate holds:*

$$\|f\|_p \leqslant C\|\nabla f\|_p.$$

In particular, $\|\|f\|\|_{W_{av}^{1,p}(D)} := \|\nabla f\|_p$ *defines an equivalent norm on* $W_{av}^{1,p}(D)$.

Proof We argue by contradiction. If the theorem were false we could find a sequence $(f_n)_{n \geqslant 1}$ in $W_{av}^{1,p}(D)$ such that $\|f_n\|_p \geqslant n\|\nabla f_n\|_p$ for $n = 1, 2, \ldots$ By scaling we may assume that $\|f_n\|_p = 1$, so that $\|\nabla f_n\|_p \leqslant \frac{1}{n}$.

Since $(f_n)_{n \geqslant 1}$ is bounded in $W_{av}^{1,p}(D)$ we may use the Rellich–Kondrachov theorem to extract a subsequence $(f_{n_k})_{k \geqslant 1}$ of $(f_n)_{n \geqslant 1}$ that converges, with respect to the norm of $L^p(D)$, to some $f \in L^p(D)$. Also $\|\nabla f_{n_k}\|_p \leqslant \frac{1}{n_k} \to 0$ as $n \to \infty$, and therefore the closedness of ∇ as an operator from $L^p(D)$ to $L^p(D;\mathbb{K}^d)$ implies that $f \in \mathsf{D}(\nabla) = W^{1,p}(D)$ and $\nabla f = 0$. In view of

$$\int_D f\,dx = \lim_{k \to \infty} \int_D f_{n_k}\,dx = 0 \qquad (11.23)$$

we even have $f \in W_{av}^{1,p}(D)$. But $\nabla f = 0$ implies, via Proposition 11.16, that f is a constant almost everywhere. In view of (11.23) this is only possible if $f = 0$ almost everywhere. We thus arrive at the contradiction $0 = \|f\|_p = \lim_{k \to \infty} \|f_{n_k}\|_p = 1$. $\qquad\square$

We are now in a position to solve the Poisson problem with Neumann boundary conditions.

Theorem 11.43 (Poisson problem, Neumann boundary conditions). *Let D be bounded and connected with C^1-boundary. For every $f \in L^2(D)$ satisfying*

$$\int_D f \, dx = 0$$

the Poisson problem (11.19) *admits a unique weak solution $u \in H^1_{av}(D)$. Moreover, there exists a constant $C \geqslant 0$, independent of f, such that*

$$\|u\|_{H^1_{av}(D)} \leqslant C\|f\|_2.$$

Proof The argument follows the proof of Theorem 11.35 with minor adjustments. By Theorem 11.42,

$$\|g\|_{H^1_{av}(D)} := \|\nabla g\|_2$$

defines an equivalent norm on $H^1_{av}(D)$. This norm arises from the inner product

$$((g|h))_{H^1_{av}(D)} := (\nabla g|\nabla h)_2.$$

In the rest of the proof we shall consider $H^1_{av}(D)$ with this norm.

By the Cauchy–Schwarz inequality and the Poincaré–Wirtinger inequality, the linear mapping $L : g \mapsto \int_D g\overline{f} \, dx$ is bounded from $H^1_{av}(D)$ to $\mathbb{K}$:

$$|L(g)| \leqslant \|g\|_2\|f\|_2 \leqslant C\|\nabla g\|_2\|f\|_2 = C\|g\|_{H^1_{av}(D)}\|f\|_2.$$

Therefore L defines a bounded functional on $H^1_{av}(D)$. Hence, by the Riesz representation theorem there exists a unique $u \in H^1_{av}(D)$ such that

$$L(g) = ((g|u))_{H^1_{av}(D)}, \quad g \in H^1_{av}(D),$$

and it satisfies $\|u\|_{H^1_{av}(D)} = \|L\| \leqslant C\|f\|_2$. Writing out this identity, it takes the form

$$\int_D \nabla g \cdot \overline{\nabla u} \, dx = \int_D g\overline{f} \, dx, \quad g \in H^1_{av}(D). \tag{11.24}$$

For an arbitrary $g \in H^1(D)$ we may write $g = m + (g - m)$ with $m := \int_D g \, dx$. Then $g - m \in H^1_{av}(D)$. Since (11.24) also holds with g replaced by the constant function m, it follows that (11.24) holds for all $g \in H^1(D)$. In particular it holds for all $g \in C^\infty(\overline{D})$, since such functions belong to $H^1(D)$. Taking conjugates on both sides we see that u is a weak solution of (11.19).

If v is another weak solution, then

$$\int_D (\nabla u - \nabla v) \cdot \nabla \phi \, dx = 0, \quad \phi \in C^\infty(\overline{D}).$$

Since $C^\infty(\overline{D})$ is dense in $H^1(D)$ by Theorem 11.27, $((u - v|g))_{H^1_{av}(D)} = 0$ for all $g \in H^1_{av}(D)$. This implies that $u - v = 0$ in $H^1_{av}(D)$. $\qquad\square$

The analogue of Theorem 11.37 holds, with the same proof:

Theorem 11.44. *Let D be bounded and connected with C^1-boundary and let $f \in L^2(D)$ satisfy $\int_D f \, dx = 0$. The weak solution of the Poisson problem $-\Delta u = f$, subject to Neumann boundary conditions, belongs to $H^1(D) \cap H^2_{loc}(D)$.*

If D is bounded and has a C^2-boundary, the weak solution u can again be shown to belong to $H^2(D)$.

A variational characterisation of weak solutions in the spirit of Theorem 11.39 can be given:

Theorem 11.45 (Variational characterisation of the solution). *Let D be bounded and connected with C^1-boundary, and let $f \in L^2(D)$ satisfy $\int_D f \, dx = 0$. For a function $u_0 \in H^1_{av}(D)$ the following assertions are equivalent:*

(1) *u_0 is the weak solution of the Poisson problem $-\Delta u = f$ on D subject to Neumann boundary conditions;*
(2) *u_0 minimises the energy functional $E : H^1(D) \to \mathbb{R}$ defined by*

$$E(u) := \frac{1}{2} \int_D |\nabla u|^2 \, dx - \text{Re} \int_D u\overline{f} \, dx.$$

Proof The proof is very similar to that in the case of Dirichlet boundary conditions. We use the notation

$$\mathfrak{a}(u,v) := \int_D \nabla u \cdot \overline{\nabla v} \, dx, \quad L(u) := \int_D u\overline{f} \, dx.$$

With this notation, for all $t \in \mathbb{R}$ and $u_0, u \in H^1(D)$ we have

$$E(u_0 + tu) = E(u_0) + t \, \text{Re}(\mathfrak{a}(u,u_0) - Lu) + \frac{1}{2}t^2\mathfrak{a}(u,u). \tag{11.25}$$

(1)⇒(2): Suppose that u_0 is a weak solution, that is, $u_0 \in H^1_{av}(D)$ and $\mathfrak{a}(\phi,u_0) = L(\phi)$ for all $\phi \in C^\infty(\overline{D})$. By density, this identity extends to arbitrary $\phi \in H^1(D)$ by approximation. Applying the identity with $\phi = u$ and $t = 1$ in (11.25), for all nonzero $u \in H^1(D)$ we obtain

$$E(u_0 + u) = E(u_0) + \frac{1}{2}\mathfrak{a}(u,u) \geqslant E(u_0),$$

and, by the Poincaré–Wirtinger inequality, the inequality is strict if $u \in H^1_{av}(D)$. It follows that u_0 is a minimiser of E in $H^1(D)$.

(2)⇒(1): Suppose conversely that $u_0 \in H^1_{av}(D)$ minimises E in $H^1(D)$. The identity (11.25) implies that for all $u \in H^1_{av}(D)$ the function $t \mapsto E(u_0 + tu)$ is differentiable in t and

$$0 = \frac{d}{dt}\bigg|_{t=0} E(u_0 + tu) = \text{Re}(\mathfrak{a}(u,u_0) - Lu).$$

Over the real scalar field this implies that $\mathfrak{a}(u, u_0) - Lu = 0$. Over the complex scalar field we apply the preceding identity with u replaced by iu to find that also $\mathrm{Im}(\mathfrak{a}(u, u_0) - Lu) = 0$, and again it follows that $\mathfrak{a}(u, u_0) - Lu = 0$. In both cases we conclude that u_0 is a weak solution. $\qquad\qquad\qquad\qquad\qquad\qquad\qquad\qquad\qquad\qquad\qquad\qquad\qquad\qquad\square$

11.2.c The Elliptic Problem $\lambda u - \Delta u = f$

The results of the preceding two sections admit straightforward modifications for the *elliptic problem*

$$\lambda u - \Delta u = f$$

for $\mathrm{Re}\,\lambda > 0$ and $f \in L^2(D)$, subject to Dirichlet or Neumann boundary conditions. As always we assume that D is open and bounded in $\mathbb{R}^d$, and in the case of Neumann boundary conditions we furthermore assume that D is connected and has C^1-boundary.

To treat the case of Dirichlet boundary conditions we define a *weak solution* to be a function $u \in H_0^1(D)$ such that

$$\int_D \lambda u\phi + \nabla u \cdot \nabla \phi \, \mathrm{d}x = \int_D f\phi \, \mathrm{d}x, \quad \phi \in C_c^\infty(D).$$

Repeating the steps of the proofs of Theorem 11.35 and 11.39 one obtains:

Theorem 11.46 (Elliptic problem, Dirichlet boundary conditions). *If D is bounded, then for all $\mathrm{Re}\,\lambda > 0$ and $f \in L^2(D)$ the elliptic problem $\lambda u - \Delta u = f$ subject to Dirichlet boundary conditions admits a unique weak solution. For $\lambda > 0$ this weak solution is the unique minimiser of the energy functional $E : H_0^1(D) \to \mathbb{R}$ defined by*

$$E(u) := \frac{1}{2} \int_D |\nabla u|^2 + \lambda |u|^2 \, \mathrm{d}x - \mathrm{Re} \int_D u\overline{f} \, \mathrm{d}x.$$

In the case of Neumann boundary conditions, the presence of additional term λu has the effect of simplifying the heuristic reasoning motivating the definition of a weak solution in Section 11.2.b, in that the averaging conditions are no longer needed. Repeating the argument, it is found that a *weak solution* should now be defined to be an element $u \in H^1(D)$ such that

$$\lambda \int_D u\phi + \nabla u \cdot \nabla \phi \, \mathrm{d}x = \int_D f\phi \, \mathrm{d}x, \quad \phi \in C^\infty(\overline{D}).$$

Repeating the steps of the proofs of Theorem 11.43 and 11.45 one obtains:

Theorem 11.47 (Elliptic problem, Neumann boundary conditions). *If D is bounded and connected with C^1-boundary, then for all $\mathrm{Re}\,\lambda > 0$ and $f \in L^2(D)$ the elliptic problem $\lambda u - \Delta u = f$ subject to Neumann boundary conditions admits a unique weak solution. For $\lambda > 0$ this weak solution is the unique minimiser of the energy functional*

$E : H_0^1(D) \to \mathbb{R}$ *defined by*

$$E(u) := \frac{1}{2} \int_D |\nabla u|^2 + \lambda |u|^2 \, dx - \text{Re} \int_D u \overline{f} \, dx.$$

We limit the present treatment of the elliptic problem to the above two theorems. In the next two chapters we will develop powerful techniques that allow us to give precise L^2-estimates for the solutions u in terms of the data f and to extend these estimates to L^p for $1 \leqslant p < \infty$.

11.3 The Lax–Milgram Theorem

The considerations of the previous section depended crucially on the use of the Riesz representation theorem as an abstract tool to prove the existence and uniqueness of solutions. This technique can be generalised to more general classes of boundary value problems by using a more flexible version of Riesz representation theorem, the so-called Lax–Milgram theorem.

11.3.a The Theorem

In what follows, V is a Hilbert space. The reason for using the letter V is that in applications, typical choices are $V = H_0^1(D)$ and $V = H^1(D)$, where D is an open subset of $\mathbb{R}^d$. In such settings the letter H will be reserved for the space $L^2(D)$. In order to prevent possible confusion, the inner product and norm of V will be denoted by $(\cdot|\cdot)_V$ and $\|\cdot\|_V$, respectively.

Definition 11.48 (Forms). A *form on V* is a sesquilinear mapping $\mathfrak{a} : V \times V \to \mathbb{K}$. A form $\mathfrak{a}$ on V is called *bounded* if there exists a constant $C \geqslant 0$ such that

$$|\mathfrak{a}(u,v)| \leqslant C\|u\|_V \|v\|_V, \quad u, v \in V.$$

In the language of forms, Proposition 9.15 asserts that if $\mathfrak{a} : V \times V \to \mathbb{K}$ is a bounded form, then there exists a unique bounded operator A on V such that

$$\mathfrak{a}(v, v') = (Av|v')_V \quad \text{for all } v, v' \in V.$$

Moreover, $\|A\|_V \leqslant C$, where C is the boundedness constant of $\mathfrak{a}$.

Definition 11.49 (Accretive and coercive forms). A form $\mathfrak{a}$ on V is called *accretive* if

$$\text{Re} \, \mathfrak{a}(v, v) \geqslant 0, \quad v \in V,$$

and *coercive* if there exists a constant $\alpha > 0$ such that

$$\text{Re} \, \mathfrak{a}(v, v) \geqslant \alpha \|v\|_V^2, \quad v \in V.$$

For bounded coercive forms we have the following version of Proposition 9.15.

Theorem 11.50 (Lax–Milgram). *If $\mathfrak{a}$ is a bounded coercive form on V, then:*

(1) *the bounded operator A associated with $\mathfrak{a}$ is boundedly invertible and $\|A^{-1}\|_V \leqslant \alpha^{-1}$, where α is the coercivity constant of $\mathfrak{a}$;*

(2) *for every bounded functional $L : V \to \mathbb{K}$ there exists a unique $v' \in V$ such that*

$$L(v) = \mathfrak{a}(v, v'), \quad v \in V.$$

Moreover, $\|v'\|_V \leqslant \|A^{-1}\|_V \|L\|$.

Proof We proceed in two steps.

Step 1 – Let A be the bounded operator provided by Proposition 9.15. The estimate

$$\alpha \|v\|^2 \leqslant \operatorname{Re} \mathfrak{a}(v, v) \leqslant |\mathfrak{a}(v, v)| = |(Av|v)_V| \leqslant \|Av\|_V \|v\|_V$$

implies that $\alpha \|v\|_V \leqslant \|Av\|_V$ for all $v \in V$. From this we infer that A is one-to-one and has closed range $\mathsf{R}(A)$ in V (see Proposition 1.21). The operator A is also surjective, for otherwise there exists a nonzero element $v^\perp \in (\mathsf{R}(A))^\perp$ and we arrive at the contradiction

$$0 < \alpha \|v^\perp\|_V^2 \leqslant \operatorname{Re} \mathfrak{a}(v^\perp, v^\perp) = \operatorname{Re}(Av^\perp|v^\perp)_V = 0.$$

By the open mapping theorem, A has bounded inverse. The estimate $\alpha \|v\|_V \leqslant \|Av\|_V$ now implies that $\|A^{-1}\|_V \leqslant \alpha^{-1}$.

Step 2 – Given a bounded functional $L : V \to \mathbb{K}$, by the Riesz representation theorem there exists a unique $v_0 \in V$ such that $L(v) = (v|v_0)_V$ for all $v \in V$. Moreover, it satisfies $\|v_0\|_V = \|L\|$. Since A, and hence $A^\star$, is boundedly invertible, there exists a unique $v' \in V$ satisfying $A^\star v' = v_0$. Then

$$L(v) = (v|v_0)_V = (v|A^\star v')_V = (Av|v')_V = \mathfrak{a}(v, v'), \quad v \in V,$$

and

$$\|v'\|_V \leqslant \|(A^\star)^{-1}\|_V \|v_0\|_V = \|A^{-1}\|_V \|L\|.$$

This proves the existence part as well as the estimate for the norm of v'. To prove uniqueness, suppose that also $L(v) = \mathfrak{a}(v, v'')$ for some $v'' \in V$ and all $v \in V$. Then $\mathfrak{a}(v, v' - v'') = 0$ for all $v \in V$. Taking $v = v' - v''$, coercivity gives $0 \leqslant \alpha \|v' - v''\|_V^2 \leqslant \operatorname{Re} \mathfrak{a}(v' - v'', v' - v'') = 0$. This implies $v' = v''$. $\qquad\square$

Part (2) of the theorem provides a generalisation of the Riesz representation theorem with the inner product replaced by a bounded coercive form $\mathfrak{a}$. If $\mathfrak{a}$ is *symmetric*, that is, $\mathfrak{a}(v, v') = \overline{\mathfrak{a}(v', v)}$ for all $v, v' \in V$ (some authors refer to this as $\mathfrak{a}$ being *Hermitian*), then $\mathfrak{a}(v, v')$ defines an inner product on V generating an equivalent norm. In this situation the Lax–Milgram theorem is an immediate consequence of the Riesz representation theorem. The principal interest in the theorem lies in the nonsymmetric case.

11.3.b The Sturm–Liouville Problem

As an application of the Lax–Milgram theorem, generalising the results on the Poisson problem we shall consider the *Sturm–Liouville problem with Dirichlet boundary conditions* on a nonempty bounded open subset D of $\mathbb{R}^d$:

$$\begin{cases} -\operatorname{div}(a\nabla u) + qu &= f \ \text{ on } D, \\ u|_{\partial D} &= 0, \end{cases} \tag{11.26}$$

where we make the following assumptions:

- the function f belongs to $L^2(D)$;
- the matrix-valued function $a : D \to M_d(\mathbb{K})$ has bounded measurable coefficients and is *coercive* in the sense that there is a constant $\alpha > 0$ such that for almost all $x \in D$ we have

$$\operatorname{Re} \sum_{i,j=1}^{d} a_{ij}(x)\xi_i\overline{\xi}_j \geqslant \alpha|\xi|^2, \quad \xi \in \mathbb{K}^d;$$

- the function $q : D \to \mathbb{K}$ is bounded and measurable and satisfies $\operatorname{Re} q(x) \geqslant 0$ for almost all $x \in D$.

A function $u \in H_0^1(D)$ is called a *weak solution* of (11.26) if for all $\phi \in C_c^\infty(D)$ we have

$$\int_D a\nabla u \cdot \nabla\phi \, dx + \int_D qu\phi \, dx = \int_D f\phi \, dx.$$

Theorem 11.51 (Sturm–Liouville problem). *Under the above assumptions on D, a, q, and f, (11.26) admits a unique weak solution u in $H_0^1(D)$. Moreover, there exists a constant $C \geqslant 0$ independent of f such that*

$$\|u\|_{H_0^1(D)} \leqslant C\|f\|_2.$$

Proof The proof is a straightforward adaptation of the proof of existence and uniqueness of the Poisson problem with Dirichlet boundary conditions. This time we apply the Lax–Milgram theorem to the form $\mathfrak{a} : H_0^1(D) \times H_0^1(D) \to \mathbb{K}$,

$$a(u,v) := \int_D a^\star\nabla u \cdot \overline{\nabla v}\, dx + \int_D \overline{q} u\overline{v}\, dx,$$

where $a_{ij}^\star = \overline{a_{ji}}$. This form is bounded and coercive: boundedness follows from

$$\begin{aligned} |\mathfrak{a}(u,v)| &\leqslant \|a\|_\infty\|\nabla u\|_2\|\nabla v\|_2 + \|q\|_\infty\|u\|_2\|v\|_2 \\ &\leqslant 2\max\{\|a\|_\infty, \|q\|_\infty\}\|u\|_{H_0^1(D)}\|v\|_{H_0^1(D)}, \end{aligned}$$

and coercivity from

$$\operatorname{Re}\mathfrak{a}(v,v) = \operatorname{Re}\int_D a^\star\nabla v \cdot \overline{\nabla v}\, dx + \operatorname{Re}\int_D \overline{q}|v|^2\, dx$$

$$\geqslant \int_D \operatorname{Re} a^\star \nabla v \cdot \overline{\nabla v} \, dx = \int_D \operatorname{Re} a \nabla v \cdot \overline{\nabla v} \, dx \geqslant \alpha \|\nabla v\|_2^2 = \alpha \|\|\nabla v\|\|_{H_0^1(D)}^2,$$

where $\|\|v\|\|_{H_0^1(D)} := \|\nabla v\|_2$ is the equivalent norm on $H_0^1(D)$ considered before. $\qquad\square$

The case of Neumann boundary conditions can be handled similarly and is left to the reader (see Problem 11.30).

Problems

11.1 Show that for all $f \in L^p(0,1)$ with $1 \leqslant p \leqslant \infty$ the function

$$I_f(x) := \int_0^x f(y) \, dy, \quad x \in (0,1),$$

belongs to $W^{1,p}(0,1)$ and its weak derivative is given by $\partial I_f = f$. Moreover, the mapping $f \mapsto I_f$ from $L^p(0,1)$ to $W^{1,p}(0,1)$ is bounded.

11.2 Let $f \in W^{1,p}(0,1)$ with $1 \leqslant p \leqslant \infty$.

(a) Show that f is equal almost everywhere to a (unique) continuous function $\tilde{f} \in C[0,1]$.
Hint: the function $f - \int_0^\cdot f(y) \, dy$ has weak derivative 0.

(b) Show that the resulting mapping $f \mapsto \tilde{f}$ from $W^{1,p}(0,1)$ to $C[0,1]$ is bounded.

(c) Show that a function $f \in W^{1,p}(0,1)$ with $1 \leqslant p < \infty$ belongs to $W_0^{1,p}(0,1)$ if and only if its continuous version $\tilde{f}$ satisfies $\tilde{f}(0) = \tilde{f}(1) = 0$.

11.3 Give a direct proof that $C^\infty[0,1]$ is dense in $W^{1,p}(0,1)$ for all $1 \leqslant p < \infty$.

11.4 Fix $1 < p < \infty$ and $f \in W^{1,p}(0,1)$, and let $\tilde{f} \in C[0,1]$ be its continuous version (see Problem 11.2).

(a) Suppose that $\tilde{f}(0) = 0$. Show that $x \mapsto \frac{f(x)}{x}$ belongs to $L^p(0,1)$ with

$$\left\| x \mapsto \frac{f(x)}{x} \right\|_p \leqslant \frac{p}{p-1} \|f'\|_p.$$

Hint: Use Young's inequality for $(\mathbb{R}_+, \frac{dx}{x})$ from Problem 2.25 with

$$f(x) = \begin{cases} x^{1/p} f'(x), & x \in [0,1], \\ 0, & x \in (1,\infty). \end{cases}$$

(b) Suppose that $x \mapsto \frac{f(x)}{x}$ belongs to $L^p(0,1)$. Show that $\tilde{f}(0) = 0$.
Hint: Argue by contradiction.

(c) Define

$$f(x) = \frac{1}{1 - \log x}, \quad x \in (0,1).$$

Show that $f \in W^{1,1}(0,1)$ and $\tilde{f}(0) = 0$, but $\frac{f(x)}{x} \notin L^1(0,1)$.

11.5 Determine whether the function $f \in L^1((-1,1) \times (-1,1))$ given by

$$f(x,y) := |xy|, \quad (x,y) \in (-1,1) \times (-1,1)$$

has weak derivatives of order one. If 'no', provide a proof; if 'yes', compute the weak derivatives $\partial_1 f$ and $\partial_2 f$.

11.6 Let D be bounded and fix $1 \leqslant p < \infty$. Let $r \in \mathbb{R}$ satisfy $r > 1 - \frac{d}{p}$.

(a) Show that the function $f(x) := |x|^r$ belongs to $W^{1,p}(D)$, and compute its weak partial derivatives.

(b) Let $\{x_n : n \geqslant 1\}$ be a countable dense set in D. Show that the function

$$g(x) := \sum_{n \geqslant 1} \frac{1}{2^n}|x - x_n|^r$$

belongs to $W^{1,p}(D)$.

(c) Show that if $d \geqslant 2$ and $1 - \frac{d}{p} < r < 0$, then g is unbounded on every open subset of D.

11.7 For $1 \leqslant p < \infty$ we consider the weak derivative ∂ as a linear operator in $L^p(0,1)$ with domain $\mathsf{D}(\partial) := C_c^\infty(0,1)$.

(a) Show that ∂ is closable.

(b) Show that the domain of the closure of ∂ equals $W_0^{1,p}(D)$.

(c) Show that this closure has a proper closed extension, given by the weak derivative with domain $W^{1,p}(0,1)$.

(d) Why doesn't this contradict the result of Proposition 10.30?

11.8 Show that if a function $f \in L^2_{\mathrm{loc}}(D)$ admits a weak Laplacian in $L^2_{\mathrm{loc}}(D)$, then it belongs $f \in H^2_{\mathrm{loc}}(D)$.
Hint: First prove that $\psi f \in H^1(D)$ for every test function $\psi \in C_c^\infty(D)$. Then use Lemma 11.36.

11.9 Show that if $f \in W^{1,p}(D)$ with $1 < p < \infty$, then $\nabla f = 0$ almost everywhere on the set $\{x \in \mathbb{R}^d : f(x) = 0\}$.
Hint: In the real-valued case, $\nabla f = \nabla(f^+) - \nabla(f^-)$.

11.10 Is $H^1(D)$ a Banach lattice?

11.11 Show that if a real-valued function $f \in L^1_{\mathrm{loc}}(D)$ admits weak derivatives $\partial_j f$ and $\rho : \mathbb{R} \to \mathbb{K}$ is a C^1-function with bounded derivative, then $\rho \circ f$ admits weak derivatives given by

$$\partial_j(\rho \circ f) = (\rho' \circ f)\partial_j f.$$

11.12 Let $1 \leqslant p, q, r \leqslant \infty$ satisfy $\frac{1}{p} + \frac{1}{q} = \frac{1}{r}$. Prove that if $f \in W^{k,p}(D)$ and $g \in W^{k,q}(D)$, then $fg \in W^{k,r}(D)$, and for all multi-indices α with $|\alpha| \leqslant k$ we have the Leibniz formula

$$\partial^{\alpha}(fg) = \sum_{0 \leqslant \beta \leqslant \alpha} \binom{\alpha}{\beta} (\partial^{\beta} f)(\partial^{\alpha-\beta} g).$$

11.13 Does Theorem 11.41(2) extend to the case $p = \infty$?

11.14 Let $1 \leqslant p \leqslant \infty$ and consider the inclusion mapping $f \mapsto \tilde{f}$ from $W^{1,p}(0,1)$ to $C[0,1]$ of Problem 11.2.

  (a) Show that for $< p \leqslant \infty$ the inclusion $W^{1,p}(0,1) \subseteq C[0,1]$ is compact. *Hint:* Use the Arzelà–Ascoli theorem.

  (b) Show that the inclusion $W^{1,1}(0,1) \subseteq C[0,1]$ fails to be compact. *Hint:* Approximate $\mathbf{1}_{(\frac{1}{2},1)}$ pointwise by a sequence of piecewise linear functions that is bounded in $W^{1,1}(0,1)$.

11.15 Show that the inclusion mapping $W^{1,2}(\mathbb{R}) \subseteq L^2(\mathbb{R})$ fails to be compact.

11.16 Let $f \in W^{1,\infty}(D)$.

  (a) Suppose that $\eta \in L^1(\mathbb{R}^d)$ has support in the unit ball $B(0;1)$ of $\mathbb{R}^d$ and satisfies $\int_{\mathbb{R}^d} \eta(x)\,dx = 1$. For $\varepsilon > 0$ denote $\eta_\varepsilon(x) := \varepsilon^{-d}\eta(\varepsilon^{-1}x)$. Show that the convolution $f_\varepsilon := \eta_\varepsilon * f$ satisfies the pointwise bound

$$|\nabla f_\varepsilon(x)| \leqslant \|\nabla f\|_\infty, \quad x \in D_\varepsilon,$$

  where $D_\varepsilon := \{x \in D : d(x, \partial D) > \varepsilon\}$ and ∇f is the weak gradient of f.

  (b) Show that if D is convex, then

$$|f_\varepsilon(x) - f_\varepsilon(y)| \leqslant \|\nabla f\|_\infty |x - y|, \quad x,y \in D_\varepsilon.$$

  (c) Deduce that if D is convex, then for every $f \in W^{1,\infty}(D)$ there exists a Lipschitz continuous function $g : D \to \mathbb{K}$ such that $f = g$ almost everywhere, with Lipschitz constant $L_g \leqslant \|\nabla f\|_\infty$.

  (d) Show that the result of part (c) fails for the nonconvex open set in $\mathbb{R}^2$ obtained by removing the nonnegative part of the x-axis from $B(0;1)$.

11.17 Show that if $f \in W_0^{1,p}(D)$ with $1 \leqslant p < \infty$ and D' is an open set containing D, then the function $\tilde{f} : D' \to \mathbb{K}$ defined by

$$\tilde{f}(x) := \begin{cases} f(x), & x \in D, \\ 0, & x \in D' \setminus D, \end{cases}$$

belongs to $W_0^{1,p}(D')$.

11.18 Show that there exists a constant $C \geqslant 0$ such that for all $f \in C_c^\infty(\mathbb{R}^d)$ we have

$$\|\nabla f\|_2 \leqslant C \|f\|_2^{1/2} \|\Delta f\|^{1/2}.$$

Show that this inequality extends to functions $f \in W^{2,2}(\mathbb{R}^d)$.

Hint: Start by showing that

$$\|\nabla f\|_2 \leqslant C(\|f\|_2 + \|\Delta f\|)$$

for some (possibly different) constant $C \geqslant 0$. Then apply this inequality with $f(cx)$ in place of $f(x)$ and optimise over $c > 0$.

11.19 For $h \in \mathbb{R}^d$ let

$$D_j^t f(x) := \frac{1}{t}(f(x + te_j) - f(x)), \quad 1 \leqslant j \leqslant d,\ t \in \mathbb{R} \setminus \{0\},$$

where e_j is the jth standard unit vector of $\mathbb{R}^d$.

(a) Prove that if $f \in W^{1,p}(\mathbb{R}^d)$ with $1 \leqslant p \leqslant \infty$, then

$$\|D_j^t f\|_p \leqslant \|\partial_j f\|_p, \quad 1 \leqslant j \leqslant d,\ t \in \mathbb{R} \setminus \{0\},$$

where $\partial_j f$ denotes the jth partial derivative of f.

(b) Prove that if $1 < p < \infty$ and there exists a constant $C \geqslant 0$ such that for all $f \in L^p(\mathbb{R}^d)$ we have

$$\|D_j^t f\|_p \leqslant C, \quad 1 \leqslant j \leqslant d,\ t \in \mathbb{R} \setminus \{0\},$$

then $f \in W^{1,p}(\mathbb{R}^d)$ and $\|\partial_j f\|_p \leqslant C$ for all $1 \leqslant j \leqslant d$.

11.20 Let $1 \leqslant p < \infty$. The aim of this problem is to show that for all $f \in W^{1,p}(\mathbb{R})$ we have $f' = \lim_{h \to 0} D_h f$ in $L^p(\mathbb{R})$, where

$$D_h f(x) := \frac{f(x+h) - f(x)}{h}, \quad x \in \mathbb{R},\ h \neq 0.$$

(a) Let $h \neq 0$. Show that

$$T_h f(x) := \frac{1}{h} \int_x^{x+h} f(t)\, dt, \quad x \in \mathbb{R},$$

defines a bounded operator on $L^p(\mathbb{R})$ of norm $\|T_h\| \leqslant 1$.

Hint: Show that $T_h f = \frac{1}{h} \mathbf{1}_{[0,1]}(-\frac{1}{h} \cdot) * f$, where $*$ denotes the convolution product, and use Young's inequality.

(b) Show that for all $f \in C_c^1(\mathbb{R})$ we have $f' = \lim_{h \to 0} D_h f$ in $L^p(\mathbb{R})$.

(c) Deduce that for all $f \in W^{1,p}(\mathbb{R})$ we have $f' = \lim_{h \to 0} D_h f$ in $L^p(\mathbb{R})$.

11.21 Show that for all $s \geqslant 0$ the norm given by (11.9) turns $H^s(\mathbb{R}^d)$ into a Hilbert space.

11.22 Using Fourier analytic methods, prove the following special case of the Sobolev embedding theorem: If $k > d/2$, then every $f \in H^k(\mathbb{R}^d)$ is equal almost everywhere to a function belonging to $C_0(\mathbb{R}^d)$. Moreover, the inclusion mapping $H^k(\mathbb{R}^d) \subseteq C_0(\mathbb{R}^d)$ is continuous.

11.23 The aim of this problem is to prove another special case of the Sobolev embedding theorem. By completing the following steps, show that if $d < p < \infty$, then every $f \in W^{1,p}(D)$ is equal almost everywhere to a continuous function on D.

(a) First let $D = \mathbb{R}^d$. Show that if $f \in C_c^1(\mathbb{R}^d)$ and the function $\eta \in C_c^1(-1,1)$ satisfies $\eta(0) = 1$, then

$$f(0) = \int_0^\infty \int_{\partial B(0;1)} \eta(r) \sum_{j=1}^d y_j \partial_j f(ry) + f(ry)\eta'(y) \, dS(y) \, dr,$$

where S is the surface measure, and hence

$$|f(0)| \leqslant C \int_{B(0;1)} \frac{1}{|x|^d} (|\nabla f(x)| + |f(x)|) \, dx$$

for some constant $C \geqslant 0$ independent of f.

(b) Apply Hölder's inequality to obtain the bound

$$|f(0)| \leqslant C' \|f\|_{W^{1,p}(\mathbb{R}^d)}$$

for some constant $C' \geqslant 0$ independent of f.

(c) By translation, conclude that

$$\|f\|_\infty \leqslant C' \|f\|_{W^{1,p}(\mathbb{R}^d)}.$$

Use a density argument to prove that if $f \in W^{1,p}(\mathbb{R}^d)$, then it is equal almost everywhere to a bounded continuous function.

(d) For general domains use a localisation argument.

11.24 This problem is a continuation of the preceding one.

(a) Show that if $1 \leqslant p, q < \infty$ satisfy $d(\frac{1}{p} - \frac{1}{q}) < 1$, then every $f \in W^{1,p}(\mathbb{R}^d)$ belongs to $L^q(\mathbb{R}^d)$ and

$$\|f\|_{L^q(\mathbb{R}^d)} \leqslant C \|f\|_{W^{1,p}(\mathbb{R}^d)}$$

for some constant $C \geqslant 0$ independent of f.

Hint: Starting from the formulas of the preceding problem, use a translation argument in combination with Young's inequality (see the hint of Problem 11.20).

(b) By repeatedly applying the inequality of part (a), deduce an embedding result for functions in $W^{k,p}(\mathbb{R}^d)$ into the space of bounded continuous functions.

11.25 Consider the Green function on the unit interval $[0,1]$ (see Section 11.2.a):

$$k(x,y) := \begin{cases} (1-x)y, & 0 \leqslant y \leqslant x, \\ (1-y)x, & x \leqslant y \leqslant 1. \end{cases}$$

(a) Show that the associated integral operator

$$T_k f(x) := \int_0^1 k(x,y) f(y) \, dy$$

on $L^2(0,1)$ is compact and has eigenvalues $1/(n\pi)^2$, $n = 1,2,3\ldots$ with corresponding eigenfunctions $x \mapsto \sin(n\pi x)$.

Let now $f \in L^2(0,1)$ be given and define $u \in L^2(0,1)$ by

$$u(x) := T_k f(x), \quad x \in (0,1).$$

(b) Show that $u \in H_0^1(0,1)$.

(c) Show that u is a weak solution of the Poisson problem with Dirichlet boundary conditions

$$\begin{cases} -u'' = f \text{ on } (0,1), \\ u(0) = u(1) = 0. \end{cases}$$

11.26 In this problem we consider the Poisson problem with Dirichlet boundary conditions on the unit disc $\mathbb{D} = B(0;1)$ in $\mathbb{R}^2$:

$$\begin{cases} -\Delta u = f \text{ on } \mathbb{D}, \\ u|_{\partial\mathbb{D}} = 0. \end{cases} \tag{11.27}$$

If $f \in L^2(\mathbb{D})$, we know from Theorem 11.35 that (11.27) has a unique weak solution $u \in H_0^1(\mathbb{D})$. The aim of this problem is to show that, even for functions $f \in C_c(\mathbb{D})$, (11.27) may not admit a classical solution $u \in C^2(\mathbb{D}) \cap C(\overline{\mathbb{D}})$.

Define $v: B(0;\frac{1}{2}) \setminus \{0\} \to \mathbb{R}$ by

$$v(x,y) := (x^2 - y^2) \log \left| \log \left((x^2 + y^2)^{1/2} \right) \right|, \quad (x,y) \in B(0;\tfrac{1}{2}).$$

(a) Show that $v \in C^2(B(0;\frac{1}{2}) \setminus \{0\})$ and compute v_x, v_y, v_{xx}, and v_{yy}.

(b) Show that

$$\lim_{(x,y)\to(0,0)} v(x,y) = \lim_{(x,y)\to(0,0)} v_x(x,y) = \lim_{(x,y)\to(0,0)} v_y(x,y) = 0.$$

Conclude v can be extended to a function in $C^1\left(\overline{B(0;\frac{1}{2})}\right)$.

Hint: For $\varepsilon > 0$ one has $|\log s| \leqslant s^{-\varepsilon}$ for s small and $|\log s| \leqslant s^\varepsilon$ for s large.

(c) Show that $\Delta v\colon B(0;\frac{1}{2})\setminus\{0\}\to\mathbb{R}$ has a continuous extension to $\overline{B(0;\frac{1}{2})}$, which we denote by $g\colon\overline{B(0;\frac{1}{2})}\to\mathbb{R}$. Moreover, show that

$$\int_{B(0;\frac{1}{2})} v\Delta\phi\,dx = \int_{B(0;\frac{1}{2})} g\phi\,dx, \quad \phi\in C_c^\infty(B(0;\tfrac{1}{2})).$$

Hint: Use Green's theorem on $B(0;\frac{1}{2})\setminus\overline{B(0;\varepsilon)}$ with $\varepsilon>0$ and let $\varepsilon\downarrow 0$.

(d) Let $\eta\in C_c^\infty(\mathbb{D})$ be compactly supported in $B(0;\frac{1}{2})$ with $\eta=1$ on $B(0;\frac{1}{4})$. Show that $u:=-\eta v$ belongs to $H_0^1(\mathbb{D})$ and is the weak solution of (11.27) with $f:=\eta g+2\nabla\eta\cdot\nabla v+(\Delta\eta)v$.

(e) Show that $\lim_{x\to 0} v_{xx}(x,x)=\infty$ and deduce from this that $u\notin C^2(\mathbb{D})$. Conclude that (11.27) does not admit a classical solution u.
Hint: Use Theorem 11.35.

11.27 Prove Theorems 11.46 and 11.47.

11.28 The aim of this problem is to solve the Poisson problem with *inhomogeneous boundary conditions*

$$\begin{cases} -\Delta u = f & \text{on } D, \\ u|_{\partial D} = g, \end{cases} \tag{11.28}$$

where $D\subseteq\mathbb{R}^d$ is bounded and $f\in L^2(D)$ and $g\in C(\partial D)$ are given functions. We assume the function g *admits an $H^1(D)$-extension*, by which we mean that there exists a function $\widetilde{g}\in H^1(D)\cap C(\overline{D})$ such that $\widetilde{g}|_{\partial D}=g$. Under these assumptions, a function $u\in H^1(D)$ is called a *weak solution* of the Poisson problem with Dirichlet boundary conditions (11.28) if

$$\int_D \nabla u\cdot\nabla\phi\,dx = \int_D f\phi\,dx, \quad \phi\in C_c^\infty(D),$$

and $u-\widetilde{g}\in H_0^1(D)$. The condition $u-\widetilde{g}\in H_0^1(D)$ is the rigorous way to express the boundary condition $u|_{\partial D}=g$.

(a) The condition $u-\widetilde{g}\in H_0^1(D)$ explicitly refers to the extension $\widetilde{g}$. By using Theorem 11.24, show that the fulfilment of this condition does not depend on the particular choice of the extension.

(b) Prove that for every $f\in L^2(D)$ the Poisson problem (11.28) has a unique weak solution $u\in H^1(D)$.
Hint: For $u\in H_0^1(D)$, show that

$$L(u) := \int_D u\overline{f}\,dx - \int_D \nabla u\cdot\overline{\nabla\widetilde{g}}\,dx$$

defines a bounded functional on $H_0^1(D)$ and apply the Riesz representation theorem.

11.29 Let D be a bounded and let $g \in C(\partial D)$ be arbitrary. Let u be a classical solution g of the *Dirichlet problem*, that is, the problem (11.28) with $f \equiv 0$. Prove that the following assertions are equivalent:

(1) u has *finite energy*, that is, $\int_D |\nabla u|^2 \, dx < \infty$;
(2) u is a weak solution;
(3) g has an H^1-extension.

Hint: For the proof of (3)⇒(2), let $D' \Subset D$. The function $g' := u|_{D'}$ has an $H^1(D')$-extension, given by $u|_{D'}$. Hence by the result of the preceding problem, the problem

$$\begin{cases} \Delta v = 0 \text{ on } D', \\ v|_{\partial D} = g', \end{cases}$$

has a unique weak solution $\tilde{u} \in H^1(D')$. Prove that $\tilde{u} = u$ almost everywhere on D' and that the restriction of $\tilde{u} - u$ to D' belongs to $H_0^1(D')$.

11.30 Discuss the Sturm–Liouville problem with Neumann boundary conditions.

11.31 Let H be a Hilbert space and let $h \in H$ be a given element. Show that the nonlinear functional $E : H \to \mathbb{R}$ defined by

$$E(u) := \frac{1}{2}\|u\|^2 - \operatorname{Re}(u|h)$$

has a unique minimiser by completing the following steps.

(a) Show that E is continuous and bounded from below, that is,

$$m := \inf_{u \in H} E(u) > -\infty.$$

(b) Using the parallelogram identity, show that for all $u, v \in H$ we have

$$\frac{1}{4}\|u - v\|^2 \leqslant (E(u) - \alpha) + (E(v) - \alpha).$$

(c) Deduce that if $(u_n)_{n \geqslant 1}$ is a sequence in H such that $\lim_{n \to \infty} E(u_n) = m$, then this sequence is Cauchy.

(d) Prove that $u := \lim_{n \to \infty} u_n$ is the unique element of H minimising E.

11.32 Let V be a Hilbert space and consider a bounded coercive form $\mathfrak{a} : V \times V \to \mathbb{K}$. Let $L : V \to \mathbb{K}$ be a bounded functional. By the Lax–Milgram theorem there is a unique $u_V \in V$ satisfying $\mathfrak{a}(v, u_V) = L(v)$ for all $v \in V$.

Suppose now that W is a closed subspace of V. By the Lax–Milgram theorem, applied to the restriction of $\mathfrak{a}$ to $W \times W$, there is a unique $u_W \in W$ satisfying $\mathfrak{a}(w, u_W) = L(w)$ for all $w \in W$.

(a) Show that $\mathfrak{a}(u_V - u_W, w) = 0$ for all $w \in W$.

(b) Show that

$$\|u_V - u_W\| \leqslant C\alpha^{-1} \inf_{w \in W} \|u_V - w\|, \tag{11.29}$$

with $C \geqslant 0$ and $\alpha > 0$ the boundedness and coercivity constants of $\mathfrak{a}$.

(c) Show that if $\mathfrak{a}$ is symmetric, that is, $\mathfrak{a}(v_1, v_2) = \overline{\mathfrak{a}(v_2, v_1)}$ for all $v_1, v_2 \in V$, then

$$\|u_V - u_W\| \leqslant \sqrt{C\alpha^{-1}} \inf_{w \in W} \|u_V - w\|. \tag{11.30}$$

The *quasi-optimality estimates* in (11.29) and (11.30) are known as *Céa's lemma*.

11.33 In this problem we outline an application to the so-called *finite element method* for the Poisson problem (11.13) on the unit interval $(0, 1)$ with datum $f \in L^2(0, 1)$:

$$\begin{cases} -\Delta u = f & \text{on } (0, 1), \\ u(0) = u(1) = 0. \end{cases} \tag{11.31}$$

In what follows we endow $V := H_0^1(0, 1)$ with the norm $\|v\|_{H_0^1(0,1)} := \|v'\|_2$. By the Poincaré inequality, this norm is equivalent to the Sobolev norm $\|v\|_2 + \|v'\|_2$.

Consider a partition $\pi = \{x_0, \ldots, x_N\}$ of the interval $[0, 1]$, that is, we assume that $0 = x_0 < x_1 < \cdots < x_{N-1} < x_N = 1$. Let V_π denote the closed subspace consisting of all $v \in V$ that are linear on each of the intervals $[x_{n-1}, x_n]$.

(a) Show that there exist unique elements $u \in H_0^1(0, 1)$ and $u_\pi \in V_\pi$ such that

$$\begin{aligned} \mathfrak{a}(v, u) &= \int_0^1 v(x)\overline{f(x)}\, dx, \quad v \in V, \\ \mathfrak{a}(v, u_\pi) &= \int_0^1 v(x)\overline{f(x)}\, dx, \quad v \in V_\pi. \end{aligned} \tag{11.32}$$

Using the results of Problem 11.32, prove the quasi-optimality estimate

$$\|u - u_\pi\|_{H_0^1(0,1)} \leqslant \inf_{v \in V_\pi} \|u - v\|_{H_0^1(0,1)}.$$

(b) Show that for all $v \in H_0^1(0, 1) \cap H^2(0, 1)$ we have $\pi v \in H^1(0, 1)$ and

$$\|v - \pi v\|_{H_0^1(0,1)} \leqslant h_\pi \|v''\|_{L^2(0,1)},$$

where $\pi v \in V_\pi$ is obtained by piecewise linear interpolation of the values $v(x_n)$, $0 \leqslant n \leqslant N$ and $h_\pi := \max_{1 \leqslant n \leqslant N} |x_n - x_{n-1}|$ is the mesh of π.

Hint: Fix $x \in [0, 1] \setminus \pi$ and choose $1 \leqslant n \leqslant N$ such that $x_{n-1} < x < x_n$. Then

$$(\pi v)'(x) = \frac{v(x_n) - v(x_{n-1})}{x_n - x_{n-1}}$$

and, since $v \in H^2(0,1)$,

$$v(x_n) = v(x_{n-1}) + (x_n - x_{n-1})v'(x_{n-1}) + \int_{x_{n-1}}^{x_n} (x_n - y)v''(y)\,\mathrm{d}y.$$

Rewriting the latter as

$$v'(x_{n-1}) - \frac{v(x_n) - v(x_{n-1})}{x_n - x_{n-1}} = -\int_{x_{n-1}}^{x_n} \frac{x_n - y}{x_n - x_{n-1}} v''(y)\,\mathrm{d}y$$

and using that $\left|\frac{y - x_{n-1}}{x_n - x_{n-1}}\right| \leqslant 1$ and $\left|\frac{x_n - y}{x_n - x_{n-1}}\right| \leqslant 1$, show that

$$|v'(x) - (\pi v)'(x)| \leqslant \int_{x_{n-1}}^{x_n} |v''(y)|\,\mathrm{d}y.$$

(c) Let the assumptions of Theorem 11.35 be satisfied with $d = 1$ and $D = (0,1)$, and let $u \in H_0^1(0,1) \cap H^2(0,1)$ be the weak solution of the Poisson problem (11.31) (see Theorem 11.37). Prove that

$$\|u - u_\pi\|_{H_0^1(0,1)} \leqslant h_\pi \|u''\|_{L^2(0,1)}.$$

Since the norm $\|v\|_{H_0^1(0,1)}$ is equivalent to the Sobolev norm $\|v\|_2 + \|v'\|_2$, the result of part (c) shows that u_π and its weak derivative u'_π provide good approximations of u and its weak derivative u' in the $L^2(0,1)$-norm if h_π is small.

The approximate solution u_π can be constructed explicitly as follows. Every $u \in V_\pi$ can be written uniquely as a finite linear combination $u = \sum_{n=1}^{N-1} c_n \psi_n$, where $\psi_n \in V_\pi$ is the piece-wise linear function given by the requirements

$$\psi_n(x_m) = \begin{cases} 1, & n = m, \\ 0, & n \neq m, \end{cases}$$

since these functions form a basis for V_π. By definition, u_π is the unique element of V_π solving (11.32) which, for our boundary value problem, takes the form

$$\int_0^1 v'u'_\pi\,\mathrm{d}x = \int_0^1 vf\,\mathrm{d}x, \quad v \in V_\pi.$$

Since the functions ψ_n form a basis for V_π, this holds if and only if

$$\int_0^1 \psi'_n u'_\pi\,\mathrm{d}x = \int_0^1 \psi_n f\,\mathrm{d}x, \quad n = 1, \ldots, N-1.$$

Writing $u_\pi = \sum_{m=1}^{N-1} c_m \psi_m$, our task is reduced to determining the coefficients $c_1, \ldots, c_{N-1}$ from the equation system of $N - 1$ linear equations

$$\sum_{m=1}^{N-1} c_m \int_0^1 \psi'_m \psi'_n\,\mathrm{d}x = \int_0^1 \psi_n f\,\mathrm{d}x, \quad n = 1, \ldots, N-1.$$

The functions ψ'_n take nonzero constant values on the intervals (x_{n-1}, x_n) and

$(x_n, x_{n,n+1})$ and vanish on the remaining sub-intervals. It follows from this that $\int_0^1 \psi_m' \psi_n' \, dx = 0$ unless $m - n \in \{-1, 0, 1\}$. Therefore the computation of the co-efficients c_m reduces to a matrix problem of the form $Sc = d$, where the so-called *stiffness matrix* S is the $(N-1) \times (N-1)$ matrix whose coefficients

$$s_{nm} = \int_0^1 \psi_n' \psi_m' \, dx$$

vanish off the diagonal and the two neighbouring off-diagonals, and

$$d_n = \int_0^1 \psi_n f \, dx.$$

This problem is easy to solve with numerical methods from Linear Algebra.

12

Forms

This chapter develops elements of the theory of sesquilinear forms and uses it to define and study certain bounded and unbounded operators, including second order differential operators such as the Laplace operator subject to Dirichlet and Neumann boundary conditions.

12.1 Forms

In the previous chapter we proved existence and uniqueness of weak solutions of the Poisson problem $-\Delta u = f$ on a nonempty bounded open subset $D \subseteq \mathbb{R}^d$ for functions $f \in H = L^2(D)$ by exploiting the properties of the sesquilinear mapping $\mathfrak{a} : V \times V \to \mathbb{K}$,

$$\mathfrak{a}(u,v) \mapsto \int_D \nabla u \cdot \overline{\nabla v} \, dx, \qquad (12.1)$$

where $V = H^1(D)$ or a suitable closed subspace thereof. If the matrix-valued function $a : D \to M_d(\mathbb{K})$ is coercive, the sesquilinear mapping

$$\mathfrak{a}(u,v) \mapsto \int_D a\nabla u \cdot \overline{\nabla v} \, dx \qquad (12.2)$$

played the same role in solving the Sturm–Liouville problem. In each of these cases, the key ingredient was the Poincaré inequality, which can be phrased in terms of $\mathfrak{a}$ as

$$\operatorname{Re} \mathfrak{a}(v,v) \geqslant \alpha \|v\|^2, \quad v \in V,$$

where $\alpha > 0$ is a positive constant and the norm is taken in H.

In order to study these matters from an abstract point of view it will be useful to interpret a form $\mathfrak{a}$ defined on a subspace V of a Hilbert space H as one in H with domain $\mathsf{D}(\mathfrak{a}) = V$, in the same way as the notion of a bounded operator has been generalised to that of a linear (possibly unbounded) operator A defined on a domain $\mathsf{D}(A)$.

Definition 12.1 (Forms, accretivity and coercivity). A *form* in a Hilbert space H is a pair $(\mathfrak{a}, D(\mathfrak{a}))$, where $D(\mathfrak{a})$ is a subspace of H, the *domain* of $\mathfrak{a}$, and $\mathfrak{a} : D(\mathfrak{a}) \times D(\mathfrak{a}) \to \mathbb{K}$ is a sesquilinear mapping. A form $(\mathfrak{a}, D(\mathfrak{a}))$ is called *accretive* if

$$\operatorname{Re} \mathfrak{a}(x, x) \geqslant 0, \quad x \in D(\mathfrak{a}),$$

and *coercive* if there exists a constant $\alpha > 0$ such that

$$\operatorname{Re} \mathfrak{a}(x, x) \geqslant \alpha \|x\|^2, \quad x \in D(\mathfrak{a}).$$

In what follows, H always denotes a Hilbert space. Definitions 11.48 and 11.49 are recovered in the special case $D(\mathfrak{a}) = H$. When no confusion is likely to arise, we omit $D(\mathfrak{a})$ from the notation and denote the form by $\mathfrak{a}$.

Example 12.2. The forms in $H = L^2(D)$ defined by (12.1) and (12.2) are accretive and continuous on the domain $D(\mathfrak{a}) = H^1(D)$, and coercive on the domains $D(\mathfrak{a}) = H_0^1(D)$ and $D(\mathfrak{a}) = H_{\mathrm{av}}^1(D)$.

If $\mathfrak{a}$ is an accretive form in H, then

$$(x|y)_{\mathfrak{a}} := \operatorname{Re} \mathfrak{a}(x, y) + (x|y), \quad x, y \in D(\mathfrak{a}), \tag{12.3}$$

defines an inner product on $D(\mathfrak{a})$; here, $(x|y)$ is the inner product of x and y in H and

$$\operatorname{Re} \mathfrak{a} := \frac{1}{2}(\mathfrak{a} + \mathfrak{a}^\star)$$

is the *symmetric part* of $\mathfrak{a}$, given by $\mathfrak{a}^\star(x, y) := \overline{\mathfrak{a}(y, x)}$. The inner product (12.3) induces a norm on $D(\mathfrak{a})$ given by

$$\|x\|_{\mathfrak{a}} = (x|x)_{\mathfrak{a}}^{1/2}.$$

Warning:

$$\operatorname{Re} \mathfrak{a}(x, y) = \frac{1}{2}(\mathfrak{a}(x, y) + \overline{\mathfrak{a}(y, x)})$$

should not be confused with

$$\operatorname{Re}(\mathfrak{a}(x, y)) = \frac{1}{2}(\mathfrak{a}(x, y) + \overline{\mathfrak{a}(x, y)}).$$

The former defines a sesquilinear form, the *real part* of $\mathfrak{a}$, but the latter generally does not. It is true, however, that $\operatorname{Re} \mathfrak{a}(x, x) = \operatorname{Re}(\mathfrak{a}(x, x))$ for all $x \in D(\mathfrak{a})$.

From Definition 11.48 we recall that a form $\mathfrak{a}$ on V is said to be *bounded* if there exists a constant $C \geqslant 0$ such that

$$|\mathfrak{a}(u, v)| \leqslant C \|u\|_V \|v\|_V, \quad u, v \in V.$$

This definition can be extended to forms in H as follows.

Definition 12.3 (Continuous forms). An accretive form $\mathfrak{a}$ in H is called *continuous* if there exists a constant $C \geqslant 0$ such that

$$|\mathfrak{a}(x,y)| \leqslant C\|x\|_\mathfrak{a}\|y\|_\mathfrak{a}, \quad u,y \in D(\mathfrak{a}).$$

A sufficient condition for continuity will be given in Proposition 13.40.

12.1.a Closed Forms

The following definition is motivated by the simple fact, observed in Proposition 10.3, that a linear operator A is closed if and only if its domain $D(A)$ is a Banach space with respect to the graph norm.

Definition 12.4 (Closed forms). An accretive form $\mathfrak{a}$ in H is called *closed* if $D(\mathfrak{a})$ is a Hilbert space with respect to the norm $\|\cdot\|_\mathfrak{a}$.

The following two propositions express some robustness properties of closed forms. Among other things, the first proposition clarifies the relation between Definition 12.1, where accretivity of forms in H was defined through the condition

$$\mathrm{Re}\,\mathfrak{a}(x,x) \geqslant \alpha\|x\|^2, \quad x \in D(\mathfrak{a}),$$

and Definition 11.49, where accretivity of a form on a Hilbert space V was defined through the condition

$$\mathrm{Re}\,\mathfrak{a}(x,x) \geqslant \alpha\|x\|_V^2, \quad x \in V.$$

In the former case, one could view $\mathfrak{a}$ as a form on $V = D(\mathfrak{a})$ and ask why norms are taken in H rather than in V. As it turns out, except for the numerical value of the constant, this leads to the same definition.

Proposition 12.5. *A closed form $\mathfrak{a}$ in H is accretive (respectively coercive, continuous) if and only if $\mathfrak{a}$, as a form on the Hilbert space $V = (D(\mathfrak{a}), \|\cdot\|_\mathfrak{a})$, is accretive (respectively coercive, bounded).*

Proof Only the assertion concerning coercivity needs proof. We must prove that there exists a constant $\alpha > 0$ such that

$$\mathrm{Re}\,\mathfrak{a}(x,x) \geqslant \alpha\|x\|^2, \quad x \in D(\mathfrak{a}), \tag{12.4}$$

if and only if there exists a constant $\beta > 0$ such that

$$\mathrm{Re}\,\mathfrak{a}(x,x) \geqslant \beta\|x\|_\mathfrak{a}^2, \quad x \in D(\mathfrak{a}). \tag{12.5}$$

If (12.4) holds, then for all $x \in D(\mathfrak{a})$ we have $(1+\alpha)\,\mathrm{Re}\,\mathfrak{a}(x,x) \geqslant \alpha\,\mathrm{Re}\,\mathfrak{a}(x,x) + \alpha\|x\|^2 = \alpha\|x\|_\mathfrak{a}^2$ and therefore (12.5) holds with $\beta = \frac{\alpha}{1+\alpha}$.

Conversely, if (12.5) holds, then for all $x \in D(\mathfrak{a})$ we have $\mathrm{Re}\,\mathfrak{a}(x,x) \geqslant \beta\|x\|_\mathfrak{a}^2 = \beta(\mathrm{Re}\,\mathfrak{a}(x,x) + \|x\|^2)$. This forces $0 < \beta < 1$ and (12.4) holds with $\alpha = \frac{\beta}{1-\beta}$. $\qquad\square$

Proposition 12.6. *Let* $\mathfrak{a}$ *be a closed accretive form in* H. *If* $V := \mathsf{D}(\mathfrak{a})$ *admits an inner product* $(\cdot|\cdot)_V$ *turning* V *into a Hilbert space such that the inclusion mapping from* V *into* H *is bounded, then the associated norm* $\|\cdot\|_V$ *is equivalent to the norm* $\|\cdot\|_{\mathfrak{a}}$.

Proof Define a norm $\|\|\cdot\|\|$ on V by

$$\|\|v\|\| := \|v\|_V + \|v\|_{\mathfrak{a}}, \quad v \in V.$$

We claim that V is complete with respect to $\|\|\cdot\|\|$. Indeed, if $(v_n)_{n \geqslant 1}$ is a Cauchy sequence with respect to $\|\|\cdot\|\|$, then it is Cauchy with respect to both $\|\cdot\|_V$ and $\|\cdot\|_{\mathfrak{a}}$. By completeness there exist $v', v'' \in V$ such that $\lim_{n\to\infty} \|v_n - v'\|_V = \lim_{n\to\infty} \|v_n - v''\|_{\mathfrak{a}} = 0$. Since the inclusion mapping from V into H is bounded with respect to both norms, we also have $\lim_{n\to\infty} \|v_n - v'\| = \lim_{n\to\infty} \|v_n - v''\| = 0$ in H. It follows that $v' = v''$ as elements in H, hence also as elements of V. Setting $v := v' = v''$, we then have $\lim_{n\to\infty} \|\|v_n - v\|\| = 0$, proving the completeness of V with respect to $\|\|\cdot\|\|$. Since $\|u\|_V \leqslant \|\|u\|\|$ and $\|u\|_{\mathfrak{a}} \leqslant \|\|u\|\|$ for all $u \in V$, the open mapping theorem can be applied to find that both $\|\cdot\|_V$ and $\|\cdot\|_{\mathfrak{a}}$ are equivalent to $\|\|\cdot\|\|$. $\qquad\square$

12.1.b Gelfand Triples

Motivated by Proposition 12.6 we shall now consider the abstract setting where we are given a Hilbert space V which is *continuously embedded* into another Hilbert space H, meaning that there exists a bounded injective operator $i : V \to H$. We shall write

$$(\cdot|\cdot) \text{ and } \|\cdot\|,$$

respectively

$$(\cdot|\cdot)_V \text{ and } \|\cdot\|_V,$$

for the inner products and norms of H and V. Identifying elements of V with their images in H, without loss of generality we may (and will) assume that, as a set, V is a subspace of H and i is the inclusion mapping. We write

$$V \hookrightarrow H$$

to summarise this state of affairs.

Definition 12.7 (Gelfand triples). A *Gelfand triple* is a triple (i, V, H), where H and V are Hilbert spaces and $i : V \hookrightarrow H$ is a continuous and dense embedding.

Example 12.8 (Gelfand triples from closed forms). If $\mathfrak{a}$ is a densely defined closed accretive form in H, then $(i, \mathsf{D}(\mathfrak{a}), H)$, with i the inclusion mapping from $\mathsf{D}(\mathfrak{a})$ into H, is a Gelfand triple.

The concrete examples covered by Example 12.2 will be discussed in Section 12.3, where the connection with weak solutions to boundary value problems will be made. This connection will be made more explicit in operator theoretic terms in Section 12.4.

Our main aim is to connect Gelfand triples with the theory of closed operators. We will prove that if (i, V, H) is a Gelfand triple and $\mathfrak{a}$ is a bounded accretive form on V, then it is possible to associate a densely defined closed linear operator A with $\mathfrak{a}$ such that $D(A) \subseteq V$ and

$$(Au|v) = \mathfrak{a}(u, v), \quad u \in D(A), \ v \in V.$$

Moreover, suitable bounds on the resolvent of A can be given.

We start with some preparations.

Definition 12.9 (Conjugate dual). The *conjugate dual* of a Hilbert space V is the vector space V' of all mappings $\phi : V \to \mathbb{K}$ that are conjugate-linear in the sense that

$$\phi(u + v) = \phi(u) + \phi(v), \quad \phi(cv) = \bar{c}\phi(v), \quad u, v \in V, \ c \in \mathbb{K},$$

and *bounded* in the sense that

$$|\phi(v)| \leqslant C\|v\|_V, \quad v \in V,$$

where $C \geqslant 0$ is a constant independent of v.

It is routine to check that the space V' is a Banach space in a natural way with norm

$$\|\phi\|_{V'} := \sup_{\|v\|_V \leqslant 1} |\phi(v)|.$$

In the presence of a continuous embedding $i : V \hookrightarrow H$, every element $h \in H$ defines an element $\phi_h \in V'$ in a natural way by defining

$$\phi_h(v) := (h|i(v)), \quad v \in V,$$

and we have

$$\|\phi_h\|_{V'} \leqslant \sup_{\|v\|_V \leqslant 1} \|h\|\|i(v)\| \leqslant \|i\|\|h\|. \tag{12.6}$$

As a mapping from H to V', the mapping $\phi : h \mapsto \phi_h$ is linear. Additivity is clear, and for the scalar multiplication we have

$$\phi_{ch}(v) = (ch|i(v)) = c(h|i(v)) = c\phi_h(v),$$

so $\phi_{ch} = c\phi_h$. The estimate (12.6) shows that this mapping is bounded with norm $\|\phi\| \leqslant \|i\|$. We claim that if the inclusion mapping i has dense range, then ϕ is injective. Indeed, if $\phi_h = 0$, then for all $v \in V$ we have $(h|i(v)) = \phi_h(v) = 0$, and since i has dense range this is only possible if $h = 0$.

Composing i and ϕ, every $v \in V$ defines an element $j(v) := (\phi \circ i)v$ in V', and we have

$$j(v)(u) = \phi_{iv}(u) = (i(v)|i(u)), \quad u, v \in V.$$

The mapping $j : V \to V'$ thus obtained is linear.

Proposition 12.10. *If $i : V \hookrightarrow H$ has dense range, then the mapping $\phi : H \to V'$ is injective and has dense range.*

Proof Injectivity has already been observed, so it remains to prove the dense range property. The Riesz representation theorem sets up a norm-preserving conjugate-linear bijection $\rho : V \to V^*$, and a norm-preserving conjugate-linear bijection $\sigma : V^* \to V'$ is obtained by mapping a functional $v^* \in V^*$ to the conjugate-linear mapping $v' \in V'$ given by $v'(v) := \overline{\langle v, v^* \rangle}$. Combining these identifications, we obtain a norm-preserving linear bijection $\sigma \circ \rho : V \to V'$. By Proposition 4.31 the injectivity of i implies that its adjoint i^* has dense range in V, and $\sigma \circ \rho$ maps this range to a dense subspace of V'. The claim follows from this by observing that $\phi = \sigma \circ \rho \circ i^*$, since for all $h \in H$ and $v \in V$ we have

$$((\sigma \circ \rho \circ i^*)h)(v) = \overline{\langle v, (\rho \circ i^*)h \rangle} = \overline{(v|i^*h)_V} = (i^*h|v)_V = (h|i(v)) = \phi_h(v).$$

$\square$

From now on we assume that V is densely embedded in H, omit the mappings i, j, ϕ, and think of V as a dense subspace of H and H as a dense subspace of V'.

Definition 12.11 (The linear operator associated with a form). *The operator A associated with a densely defined form $\mathfrak{a}$ in H is defined by*

$$u \in D(A) \text{ and } Au = h \iff u \in D(\mathfrak{a}) \text{ and } (h|v) = \mathfrak{a}(u,v) \text{ for all } v \in D(\mathfrak{a}).$$

Since $D(\mathfrak{a})$ is dense in H, the element $h \in H$ is uniquely defined and thus A is well defined as a linear operator in H, linearity being clear from the definition.

Without imposing further properties on $\mathfrak{a}$ this definition is not very useful. Under appropriate additional assumptions on $\mathfrak{a}$, the next theorem provides some interesting properties of the associated operator.

Theorem 12.12 (Resolvent estimate – bounded coercive forms in V). *Let (i, V, H) be a Gelfand triple and let A be the linear operator in H associated with a bounded coercive form $\mathfrak{a}$ on V. Then A is densely defined and closed, and for all $\lambda \in \mathbb{C}$ with $\mathrm{Re}\,\lambda > 0$ we have $-\lambda \in \rho(A)$ and*

$$\|(\lambda + A)^{-1}\| \leqslant \frac{1}{\mathrm{Re}\,\lambda}, \quad \|(\lambda + A)^{-1}\| \leqslant \left(1 + \frac{C}{\alpha}\right)\frac{1}{|\lambda|},$$

where C and α are the boundedness and coercivity constants of $\mathfrak{a}$.

Proof Fix $\lambda \in \mathbb{C}$ with $\mathrm{Re}\,\lambda > 0$. As a form on V,

$$\mathfrak{a}_\lambda(u,v) := \mathfrak{a}(u,v) + \lambda(u|v), \quad u,v \in V,$$

is bounded and coercive: this follows from

$$|\mathfrak{a}_\lambda(u,v)| \leqslant |\mathfrak{a}(u,v)| + |\lambda|\,\|u\|\,\|v\| \leqslant C\|u\|_V\|v\|_V + |\lambda|\,\|i\|^2\|u\|_V\|v\|_V$$

and

$$\mathrm{Re}\,\mathfrak{a}_\lambda(v,v) = \mathrm{Re}\,\mathfrak{a}(v,v) + \mathrm{Re}\,\lambda(v|v) \geqslant \mathrm{Re}\,\mathfrak{a}(v,v) \geqslant \alpha\|v\|_V^2. \tag{12.7}$$

Denote by A_V the bounded operator on V associated with $\mathfrak{a}$ through Proposition 9.15, so that $(A_V u|v)_V = \mathfrak{a}(u,v)$ for all $u,v \in V$. The bounded operator on V associated with $\mathfrak{a}_\lambda$ equals $A_{V,\lambda} := A_V + \lambda i^* i$. By the Lax–Milgram theorem applied to the form $\mathfrak{a}_\lambda$, $A_{V,\lambda}$ is boundedly invertible with $\|A_{V,\lambda}^{-1}\|_{\mathscr{L}(V)} \leqslant \alpha^{-1}$. Composing $A_{V,\lambda}$ with the isometric isomorphism $\sigma \circ \rho$ from V onto V' from the proof of Proposition 12.10, we may identify $A_{V,\lambda}$ with a bounded operator $A'_{V,\lambda}$ from V to V' which is boundedly invertible and satisfies $\|A'^{-1}_{V,\lambda}\|_{\mathscr{L}(V',V)} \leqslant \alpha^{-1}$.

Let R_λ denote the restriction of $A'^{-1}_{V,\lambda}$ to H, viewed as a bounded operator from H to H. As such it is bounded and injective. Define the closed operator $(B_\lambda, \mathrm{D}(B_\lambda))$ in H by $\mathrm{D}(B_\lambda) := \mathrm{R}(R_\lambda)$ and $B_\lambda := R_\lambda^{-1} - \lambda$. To see that B_λ is densely defined in H, note that

$$\mathrm{D}(B_\lambda) = \mathrm{R}(R_\lambda) = \mathrm{R}(A'^{-1}_{V,\lambda}|_H) = \{A'^{-1}_{V,\lambda}h : h \in H\}$$

is dense in V (and hence in H) since $A'^{-1}_{V,\lambda} : V' \to V$ is an isomorphism, V is dense in H, and H is dense in V'. For all $u, f \in H$,

$$
\begin{aligned}
u \in \mathrm{D}(B_\lambda) \text{ and } B_\lambda u = f &\Leftrightarrow u \in \mathrm{R}(R_\lambda) \text{ and } R_\lambda^{-1} u = \lambda u + f \\
&\Leftrightarrow u \in V \text{ and } A_{V,\lambda} u = \lambda u + f \\
&\Leftrightarrow u \in V \text{ and } (f|v) = \mathfrak{a}(u,v) \text{ for all } v \in V \\
&\Leftrightarrow u \in \mathrm{D}(A) \text{ and } Au = f.
\end{aligned}
$$

It follows that $A = B_\lambda$, so A is densely defined and closed, and $\lambda + A = \lambda + B_\lambda = R_\lambda^{-1}$. This, in turn, implies that $\lambda + A$ is injective and surjective (the latter since R_λ is defined on all of H) and hence boundedly invertible.

For $v \in \mathrm{D}(A)$, the accretivity of $\mathfrak{a}$ gives

$$
\begin{aligned}
\|(\lambda + A)v\|\,\|v\| &\geqslant |((\lambda + A)v|v)| \\
&\geqslant \mathrm{Re}((\lambda + A)v|v) = \mathrm{Re}\,\lambda\|v\|^2 + \mathrm{Re}\,\mathfrak{a}(v,v) \geqslant \mathrm{Re}\,\lambda\|v\|^2.
\end{aligned}
$$

This gives the first resolvent estimate.

Fix an arbitrary $h \in H$. Defining $u := (\lambda + A)^{-1}h = R_\lambda h \in V$ and using that $h = (\lambda + A)u = R_\lambda^{-1}u = A'_{V,\lambda}u = A_{V,\lambda}u$, we have

$$\mathfrak{a}_\lambda(u,v) = \mathfrak{a}(u,v) + \lambda(u|v) = (A_{V,\lambda}u|v) = (h|v), \quad v \in V. \tag{12.8}$$

Taking $v := u$ in (12.8), by (12.7) we obtain

$$\alpha\|u\|_V^2 \leqslant \operatorname{Re}\mathfrak{a}_\lambda(u,u) = \operatorname{Re}(h|u) \leqslant \|h\|\|u\|. \tag{12.9}$$

By (12.8) and (12.9),

$$|\lambda|\|u\|^2 \leqslant |(h|u)| + |\mathfrak{a}(u,u)| \leqslant \|h\|\|u\| + C\|u\|_V^2 \leqslant \left(1 + \frac{C}{\alpha}\right)\|h\|\|u\|,$$

where C is the boundedness constant of $\mathfrak{a}$. Substituting back the definition of u we obtain the bound

$$|\lambda|\|(\lambda + A)^{-1}h\| \leqslant \left(1 + \frac{C}{\alpha}\right)\|h\|, \quad h \in H.$$

This gives the second resolvent estimate. □

In applications, V often arises as the domain of a densely defined closed form $\mathfrak{a}$ in H (cf. Example 12.8). In this setting, Theorem 12.12 implies the following result.

Corollary 12.13. *Let A be the linear operator in H associated with a densely defined closed continuous accretive form $\mathfrak{a}$ in H. Then A is densely defined and closed, for all $\lambda \in \mathbb{C}$ with $\operatorname{Re}\lambda > 0$ we have $-\lambda \in \rho(A)$ and*

$$\|(\lambda + A)^{-1}\| \leqslant \frac{1}{\operatorname{Re}\lambda}, \quad \operatorname{Re}\lambda > 0,$$

and for all $\delta > 0$ there exists a constant $C_\delta \geqslant 0$ such that

$$\|(\lambda + A)^{-1}\| \leqslant \frac{C_\delta}{|\lambda|}, \quad \operatorname{Re}\lambda \geqslant \delta.$$

Proof Consider the Hilbert space $V = (\mathrm{D}(\mathfrak{a}), \|\cdot\|_\mathfrak{a})$ and let $i : V \hookrightarrow H$ be the inclusion mapping. By Proposition 12.5 and its proof, for all $\delta > 0$ the form

$$\mathfrak{a}^\delta(u,v) := \mathfrak{a}(u,v) + \delta(u|v) = \mathfrak{a}(u,v) + \delta(i^*iu|v)_V, \quad u,v \in V,$$

is bounded and coercive as a form on V, with boundedness constant $C + \delta\|i^*i\|$ and coercivity constant $\delta/(1 + \delta)$. The operator associated with $\mathfrak{a}^\delta$ is $A + \delta$. By Theorem 12.12 this operator (and hence A itself) is densely defined and closed, and for all $\operatorname{Re}\lambda > 0$ the operator $\lambda + \delta + A$ is boundedly invertible and satisfies the resolvent bounds

$$\|(\lambda + \delta + A)^{-1}\| \leqslant \frac{1}{\operatorname{Re}\lambda}, \quad \|(\lambda + \delta + A)^{-1}\| \leqslant \left(1 + \frac{C + \delta\|i^*i\|}{\delta/(1+\delta)}\right)\frac{1}{|\lambda|}.$$

Since $\delta > 0$ was arbitrary, the corollary follows from this. □

Further properties of the operators A in the theorem and its corollary will be obtained in the next chapter (see Theorem 13.35). Without the continuity assumption it is still possible to prove a version of the first resolvent estimate (see Theorem 13.34).

An elegant application of the corollary is the following duality result. Recall that if $\mathfrak{a}$ is a form in H, we define $\mathfrak{a}^\star(x,y) := \overline{\mathfrak{a}(y,x)}$ for $x,y \in D(\mathfrak{a})$.

Corollary 12.14 ($A^\star$ is associated with $\mathfrak{a}^\star$). *Let A be the densely defined closed operator in H, and suppose that one of the following two conditions is satisfied:*

(1) *A is the operator associated with a closed continuous accretive form $\mathfrak{a}$ in H;*
(2) *A is the operator associated with a bounded coercive form $\mathfrak{a}$ on V, where (i,V,H) is a Gelfand triple.*

Then $A^\star$ is the densely closed defined operator associated with the form $\mathfrak{a}^\star$.

Proof (1): Since A is densely defined, $D(\mathfrak{a})$ is dense. Since $D(\mathfrak{a}^\star) = D(\mathfrak{a})$ by definition, it follows that $\mathfrak{a}^\star$ is densely defined. From

$$(v|v)_{\mathfrak{a}^\star} = \mathrm{Re}\,\overline{\mathfrak{a}(v,v)} + (v|v) = \mathrm{Re}\,\mathfrak{a}(v,v) + (v|v) = (v|v)_{\mathfrak{a}}, \quad v \in V,$$

it follows that $\mathfrak{a}^\star$ is continuous and accretive. Let B denote the densely defined closed operator associated with $\mathfrak{a}^\star$. If $x \in D(B)$, then for all $y \in D(A)$ we have

$$(y|Bx) = \overline{(Bx|y)} = \overline{\mathfrak{a}^\star(x,y)} = \mathfrak{a}(y,x) = (Ay|x).$$

It follows that $x \in D(A^\star)$ and $A^\star x = Bx$. This shows that $B \subseteq A^\star$.

Next let $x \in D(A^\star)$. By Corollary 12.13 applied to $\mathfrak{a}^\star$, the operator $I + B$ is invertible, so there exists $y \in D(B)$ such that $(I + A^\star)x = (I + B)y$. Since $B \subseteq A^\star$, we have $y \in D(A^\star)$ and $(I + A^\star)x = (I + A^\star)y$. By Corollary 12.13 applied to $\mathfrak{a}$, the operator $I + A$ is invertible and therefore so is its adjoint $I + A^\star$. It follows that $x = y \in D(B)$. This shows that $A^\star \subseteq B$.

(2): This is proved in the same way, this time using Theorem 12.12. $\qquad\square$

12.1.c Closable Forms

We return to the setting of forms in a Hilbert space H considered at the beginning of Section 12.1.

Definition 12.15 (Closable forms). An accretive form $\mathfrak{a}$ in H is called *closable* if there exists a closed accretive form $\widetilde{\mathfrak{a}}$ in H extending $\mathfrak{a}$, that is, $\widetilde{\mathfrak{a}}$ is closed and accretive, $D(\mathfrak{a}) \subseteq D(\widetilde{\mathfrak{a}})$, and $\widetilde{\mathfrak{a}}(u,v) = \mathfrak{a}(u,v)$ for all $u,v \in D(\mathfrak{a})$.

The following proposition, in which we view $D(\mathfrak{a})$ as a (not necessarily complete) normed space with norm $\|\cdot\|_{\mathfrak{a}}$, gives a useful necessary and sufficient condition for a form $\mathfrak{a}$ in H to be closable. It should be compared with Proposition 10.12.

Proposition 12.16. *For a continuous accretive form* $\mathfrak{a}$ *in* H *the following assertions are equivalent:*

(1) $\mathfrak{a}$ *is closable;*
(2) *every Cauchy sequence in* $D(\mathfrak{a})$ *converging to* 0 *in* H *converges to* 0 *in* $D(\mathfrak{a})$.

The hard implication is (2)$\Rightarrow$(1). It is tempting to try to prove it as follows. By continuity, $\mathfrak{a}$ extends to an accretive form $\widetilde{\mathfrak{a}}$ on the completion of $D(\mathfrak{a})$ with respect to the norm $\|\cdot\|_{\mathfrak{a}}$. It is not clear, however, whether the inclusion mapping of $D(\mathfrak{a})$ into H extends to an embedding of its completion into H. This difficulty explains why we have to proceed more carefully.

Proof Set $V := D(\mathfrak{a})$ with norm $\|\cdot\|_V := \|\cdot\|_{\mathfrak{a}}$. We note that assertion (2) can be equivalently stated as follows:

(2') Whenever a sequence $(v_n)_{n\geqslant 1}$ in V satisfies $\lim_{n\to\infty} v_n = 0$ in H and

$$\lim_{m,n\to\infty} \operatorname{Re}\mathfrak{a}(v_m - v_n, v_m - v_n) = 0,$$

then $\lim_{n\to\infty} \operatorname{Re}\mathfrak{a}(v_n, v_n) = 0$.

(1)$\Rightarrow$(2): Suppose that $\mathfrak{a}$ has a closed extension $\widetilde{\mathfrak{a}}$ whose domain $D(\mathfrak{a}) =: \widetilde{V}$ is complete with respect to $\|\cdot\|_{\widetilde{\mathfrak{a}}}$. If $(u_n)_{n\geqslant 1}$ is a sequence in V such that $\lim_{n\to\infty} u_n = 0$ in H and $\lim_{m,n\to\infty} \operatorname{Re}\mathfrak{a}(u_m - u_n, u_m - u_n) = 0$, then the sequence $(u_n)_{n\geqslant 1}$ is Cauchy with respect to $\|\cdot\|_{\mathfrak{a}}$ and hence, since $\widetilde{\mathfrak{a}}$ extends $\mathfrak{a}$, with respect to $\|\cdot\|_{\widetilde{\mathfrak{a}}}$. Since $\widetilde{V}$ is complete with respect to the norm $\|\cdot\|_{\mathfrak{a}}$, the sequence is convergent in $\widetilde{V}$, say to $\widetilde{u} \in \widetilde{V}$. The sequence $(u_n)_{n\geqslant 1}$ is Cauchy in H as well, and since $\widetilde{V}$ embeds in H we have $u_n \to \widetilde{u}$ in H. Since we assumed that $u_n \to 0$ in H it follows that $\widetilde{u} = 0$. Hence, $\lim_{n\to\infty} u_n = 0$ with respect to $\|\cdot\|_{\mathfrak{a}}$, and this in turn implies that $\lim_{n\to\infty} \operatorname{Re}\mathfrak{a}(u_n, u_n) = 0$.

(2)$\Rightarrow$(1): The proof proceeds in three steps.

Step 1 – Define $\overline{V}$ to be the set of all $\overline{v} \in H$ for which there exists a Cauchy sequence $(v_n)_{n\geqslant 1}$ in V such that $\lim_{n\to\infty} v_n = \overline{v}$ in H. In what follows we refer to a sequence with these properties as an *approximating sequence* for $\overline{v}$.

We begin by showing that the limit

$$\overline{\mathfrak{a}}(\overline{u}, \overline{v}) := \lim_{n\to\infty} \mathfrak{a}(u_n, v_n)$$

exists whenever $(u_n)_{n\geqslant 1}$ and $(v_n)_{n\geqslant 1}$ are approximating sequences for $\overline{u}, \overline{v} \in \overline{V}$, and that this limit is independent of the choice of approximating sequences.

To begin with the existence of the limit, we note that for all $m, n \geqslant 1$

$$
\begin{aligned}
|\mathfrak{a}(u_m, v_m) - \mathfrak{a}(u_n, v_n)| &= |\mathfrak{a}(u_m - u_n, v_m) + \mathfrak{a}(u_n, v_m - v_n)| \\
&\leqslant C\|u_m - u_n\|_{\mathfrak{a}} \sup_{m\geqslant 1} \|v_m\|_{\mathfrak{a}} + C\|v_m - v_n\|_{\mathfrak{a}} \sup_{n\geqslant 1} \|u_n\|_{\mathfrak{a}}
\end{aligned}
\tag{12.10}
$$

by the continuity of $\mathfrak{a}$. Since $(u_n)_{n\geqslant 1}$ and $(v_n)_{n\geqslant 1}$ are Cauchy in V, they are bounded and we conclude from (12.10) that $(\mathfrak{a}(u_n,v_n))_{n\geqslant 1}$ is a Cauchy sequence, hence convergent.

As to the well-definedness of the limit, suppose that $\overline{u}$ and $\overline{v}$ are approximated by the sequences $(u'_n)_{n\geqslant 1}$ and $(v'_n)_{n\geqslant 1}$ with the properties as stated. Then, by a similar estimate,

$$|\mathfrak{a}(u_n,v_n) - \mathfrak{a}(u'_n,v'_n)| \leqslant C\|u_n - u'_n\|_{\mathfrak{a}} \sup_{m\geqslant 1} \|v_n\|_{\mathfrak{a}} + C\|v_n - v'_n\|_{\mathfrak{a}} \sup_{n\geqslant 1} \|u'_n\|_{\mathfrak{a}}.$$

Now

$$\|u_n - u'_n\|_{\mathfrak{a}}^2 = \|u_n - u'_n\|^2 + \mathrm{Re}\,\mathfrak{a}(u_n - u'_n, u_n - u'_n).$$

The first term on the right-hand side tends to 0 as $n \to \infty$ since $u_n \to u$ and $u'_n \to u$. The second term tends to 0 because $(u_n - u'_n)_{n\geqslant 1}$ is an approximating sequence for 0, so $(2')$ can be applied with v_n replaced by $u_n - u'_n$; to see this, note that

$$\mathrm{Re}\,\mathfrak{a}((u_m - u'_m) - (u_n - u'_n),(u_m - u'_m) - (u_n - u'_n))$$
$$= \mathrm{Re}\,\mathfrak{a}((u_m - u_n) - (u'_m - u'_n),(u_m - u_n) - (u'_m - u'_n))$$
$$\leqslant C\|u_m - u_n\|_{\mathfrak{a}}^2 + 2C\|u_m - u_n\|_{\mathfrak{a}}\|u'_m - u'_n\|_{\mathfrak{a}} + C\|u'_m - u'_n\|_{\mathfrak{a}}^2$$

by the continuity of $\mathfrak{a}$, and all three terms on the right-hand side tend to 0 and $m,n \to \infty$ since both $(u_n)_{n\geqslant 1}$ and $(u'_n)_{n\geqslant 1}$ are approximating sequences for $\overline{u}$, and therefore Cauchy with respect to $\|\cdot\|_{\mathfrak{a}}$. In the same way we obtain $\|v_n - v'_n\|_{\mathfrak{a}}^2 \to 0$ and the proof can be completed as before.

It is clear that $V \subseteq \overline{V}$ and that the resulting mapping $\overline{\mathfrak{a}} : \overline{V} \times \overline{V} \to \mathbb{C}$ is sesquilinear, so it defines a form, is continuous and accretive, and extends $\mathfrak{a}$.

Step 2 – We show that V is dense in $\overline{V}$ with respect to the norm $\|\cdot\|_{\overline{\mathfrak{a}}}$. To this end let $\overline{v} \in \overline{V}$ and let $(v_n)_{n\geqslant 1}$ be an approximating sequence. We claim that $\lim_{n\to\infty} \|v_n - \overline{v}\|_{\overline{\mathfrak{a}}} = 0$. Since we already know that $\lim_{n\to\infty} v_n = \overline{v}$, it suffices to prove that $\lim_{n\to\infty} \mathrm{Re}\,\overline{\mathfrak{a}}(v_n - \overline{v}, v_n - \overline{v}) = 0$. This follows from

$$\lim_{n\to\infty} \mathrm{Re}\,\overline{\mathfrak{a}}(v_n - \overline{v}, v_n - \overline{v}) = \lim_{n\to\infty} \lim_{m\to\infty} \mathrm{Re}\,\mathfrak{a}(v_n - v_m, v_n - v_m) = 0,$$

the first of these identities being a consequence of the definition of $\overline{\mathfrak{a}}$ along with the fact that $v_n - v_m \to v_n - \overline{v}$ in H and $\mathrm{Re}\,\mathfrak{a}((v_n - v_m) - (v_n - v_\ell),(v_n - v_m) - (v_n - v_\ell)) \to 0$ as $\ell,m \to \infty$ by the continuity of $\mathfrak{a}$ as in the previous step.

Step 3 – To prove that $\overline{\mathfrak{a}}$ is closed, suppose first that $(v_n)_{n\geqslant 1}$ is a sequence in V which is Cauchy with respect to $\|\cdot\|_{\overline{\mathfrak{a}}}$. This means that $(v_n)_{n\geqslant 1}$ is Cauchy in H and

$$\lim_{m,n\to\infty} \mathrm{Re}\,\overline{\mathfrak{a}}(v_m - v_n, v_m - v_n) = 0.$$

Let $\lim_{n\to\infty} v_n =: \overline{v}$, the convergence being in H. Since $\overline{\mathfrak{a}}$ extends $\mathfrak{a}$ we have

$$\lim_{m,n\to\infty} \mathrm{Re}\,\mathfrak{a}(v_m - v_n, v_m - v_n) = 0.$$

The very definition of $\overline{V}$ implies that $\overline{v} \in \overline{V}$, and as in Step 2 we have

$$\lim_{n\to\infty} \|v_n - \overline{v}\|_{\overline{a}}^2 = \lim_{n\to\infty} \operatorname{Re} \overline{a}(v_n - \overline{v}, v_n - \overline{v}) + \|v_n - \overline{v}\|^2$$

$$= \lim_{n\to\infty} \lim_{m\to\infty} \operatorname{Re} a(v_n - v_m, v_n - v_m) + \|v_n - \overline{v}\|^2 = 0.$$

Suppose next that $(\overline{v}_n)_{n\geqslant 1}$ is a sequence in $\overline{V}$ which is Cauchy with respect to $\|\cdot\|_{\overline{a}}$. Since V is dense in $\overline{V}$ by Step 2, we may choose elements $v_n \in V$ such that $\|v_n - \overline{v}_n\| < 1/n$. Then $(v_n)_{n\geqslant 1}$ is Cauchy in $\overline{V}$, and by what we just proved it has a limit $\overline{v}$ in V. Then $\overline{v}$ is also a limit for $(\overline{v}_n)_{n\geqslant 1}$. □

The form $\overline{a}$ constructed in the above proof is called the *closure* of a. Further properties of $\overline{a}$ are discussed in Problem 12.5.

12.2 The Friedrichs Extension Theorem

It has been shown in Corollary 10.45 that if A is a densely defined operator which is positive in the sense that $(Ax|x) \geqslant 0$ for all $x \in D(A)$ and has the property that $I + A$ has dense range, then A is selfadjoint. The next theorem states that if we give up the dense range condition, selfadjoint extensions still exist.

Theorem 12.17 (Friedrichs extension). *Let A be a densely defined positive operator acting in a complex Hilbert space H. Then:*

(1) *the form a in H given by* $D(a) := D(A)$ *and*

$$a(x,y) := (Ax|y), \quad x, y \in D(A),$$

   is densely defined, positive, continuous, and closable;

(2) *the operator associated with the closure of a is a positive selfadjoint extension of A.*

Proof (1): It is clear that a is densely defined and positive, and continuity of a follows from the Cauchy–Schwarz inequality (Proposition 3.3):

$$|a(x,y)|^2 = |(Ax|y)|^2 \leqslant |(Ax|x)||(Ay|y)| = a(x,x)a(y,y) \leqslant \|x\|_a^2 \|y\|_a^2.$$

Here, the positivity of A was used to see that $a(x,x) = (Ax|x) \geqslant 0$ and hence $a(x,x) = \operatorname{Re} a(x,x) \leqslant \|x\|_a^2$. To prove that a is closable we check the criterion of Proposition 12.16. Keeping in mind that $a(x,x) \geqslant 0$ for all $x \in D(A)$, pick a sequence $(v_n)_{n\geqslant 1}$ in V such that $\lim_{n\to\infty} v_n = 0$ in H and $\lim_{m,n\to\infty} a(v_m - v_n, v_m - v_n) = 0$. We must show that $\lim_{n\to\infty} a(v_n, v_n) = 0$.

Given $\varepsilon > 0$, for large enough m, n we have

$$0 \leqslant a(v_m - v_n, v_m - v_n) = (Av_m - Av_n|v_m - v_n)$$

$$= (Av_m|v_m) + (Av_n|v_n) - 2\operatorname{Re}(Av_m|v_n) < \varepsilon.$$

Fixing m, upon letting $n \to \infty$ and using that $v_n \to 0$, we obtain

$$0 \leqslant (Av_m|v_m) + \limsup_{n\to\infty}(Av_n|v_n) \leqslant \varepsilon.$$

Since A is positive, this can only happen if $\limsup_{n\to\infty}(Av_n|v_n) \leqslant \varepsilon$, and since $\varepsilon > 0$ was arbitrary this forces $\lim_{n\to\infty} \mathfrak{a}(v_n, v_n) = 0$.

(2): By (1) the form $\mathfrak{a}$ is densely defined, continuous, closable, and satisfies $\mathfrak{a}(v,v) \geqslant 0$ for all $v \in D(\mathfrak{a})$. Its closure $\overline{\mathfrak{a}}$ enjoys the same properties, and therefore Corollary 12.13 allows us to associate a positive operator B with $\sigma(B) \subseteq \{\operatorname{Re}\lambda \geqslant 0\}$. By Proposition 10.43 (which applies since positive operators are symmetric; here we use the assumption that the scalar field is complex, cf. the remark after Definition 10.35), this implies that B is selfadjoint. Alternatively one may observe that the positivity of A implies that $\mathfrak{a}^\star = \mathfrak{a}$ and hence $\overline{\mathfrak{a}}^\star = \overline{\mathfrak{a}}$, and therefore $B = B^\star$ by Corollary 12.14. $\qquad\square$

If A is a densely defined closed operator from H to another Hilbert space K, then by Theorem 10.46 the operator $A^\star A$ with domain $D(A^\star A) = \{x \in D(A) : Ax \in D(A^\star)\}$ is positive and selfadjoint. The next result relates this operator with the theory of forms.

Proposition 12.18. *Let A be a densely defined closed operator from H to another Hilbert space K. The form $\mathfrak{a}$ given by $D(\mathfrak{a}) := D(A)$ and*

$$\mathfrak{a}(u,v) := (Au|Av), \quad u,v \in D(A),$$

is closed, continuous, and accretive, and $A^\star A$ coincides with the operator associated with $\mathfrak{a}$.

Proof Densely definedness, continuity, and accretivity are clear. For $v \in D(\mathfrak{a})$ we have

$$\|v\|_{\mathfrak{a}}^2 = \|v\|^2 + \mathfrak{a}(v,v) = \|v\|^2 + \|Av\|^2,$$

from which we deduce that $\|\cdot\|_{\mathfrak{a}}$ is equivalent to the graph norm of A. Since A is closed, $D(\mathfrak{a}) = D(A)$ is complete with respect to $\|\cdot\|_{\mathfrak{a}}$ and the closedness of $\mathfrak{a}$ follows.

Let B be the operator associated with $\mathfrak{a}$. Then $(Bu|u) = \mathfrak{a}(u,u) = \|Au\|^2 \geqslant 0$ for all $u \in D(B)$, so B is positive. By the definition of the domain of an operator associated with a form we have

$$u \in D(B) \iff u \in V \text{ and } \exists f \in D(\mathfrak{a}) : (f|v) = \mathfrak{a}(u,v) \text{ for all } v \in D(\mathfrak{a})$$
$$\iff u \in V \text{ and } \exists f \in D(A) : (f|v) = (Au|Av) \text{ for all } v \in D(A)$$
$$\iff u \in D(A), \ Au \in D(A^\star), \text{ and } Bu = A^\star(Au).$$

This shows that $B = A^\star A$. $\qquad\square$

12.3 The Dirichlet and Neumann Laplacians

We now turn to some examples that connect the theory developed in the preceding sections to the boundary value problems studied in the previous chapter.

12.3.a The Laplace Operator

Let $V := H^1(\mathbb{R}^d) = W^{1,2}(\mathbb{R}^d)$ and consider the sesquilinear form $\mathfrak{a}$ on V defined by

$$\mathfrak{a}(u,v) := \int_{\mathbb{R}^d} \nabla u \cdot \overline{\nabla v} \, dx, \quad u, v \in V.$$

This form is bounded and positive on V; the easy proof is left to the reader. We claim that the densely defined closed operator A in $L^2(\mathbb{R}^d)$ associated with $\mathfrak{a}$ equals $-\Delta$, where Δ is the weak Laplacian in $L^2(\mathbb{R}^d)$ with domain $\mathsf{D}(-\Delta) = H^2(\mathbb{R}^d)$ (cf. Theorem 11.29).

To prove the claim we begin by noting that if $u \in H^2(\mathbb{R}^d)$, then $\partial_j u \in H^1(\mathbb{R}^d) = W^{1,2}(\mathbb{R}^d)$ by Theorems 11.29 and 11.31, and therefore

$$\mathfrak{a}(u,v) = -\sum_{j=1}^{d} \int_{\mathbb{R}^d} \partial_j^2 u(x) \overline{v(x)} \, dx = -(\Delta u | v) \tag{12.11}$$

for all $v \in C_c^\infty(\mathbb{R}^d)$. By approximation this identity extends to all $v \in H^1(\mathbb{R}^d)$. This means that $u \in \mathsf{D}(A)$ and $Au = -\Delta u$.

Conversely, if $u \in \mathsf{D}(A)$, then $u \in H^1(\mathbb{R}^d)$ and for all $v \in C_c^\infty(\mathbb{R}^d)$ we have

$$\int_{\mathbb{R}^d} Au(x) \overline{v(x)} \, dx = (Au | v) = \mathfrak{a}(u,v) = \int_{\mathbb{R}^d} \nabla u \cdot \overline{\nabla v} \, dx = -\int_{\mathbb{R}^d} u(x) \overline{\Delta v(x)} \, dx$$

by the definition of weak derivatives. This shows that u admits a weak Laplacian given by $\Delta u = -Au$ in the sense of Theorem 11.29, and therefore $u \in H^2(\mathbb{R}^d)$ by this theorem.

Another description of the operator Δ can be given on the basis of Theorem 10.46 and Proposition 12.18. These results identify the operator associated with the form $\mathfrak{a}$ defined by (12.11) to be $-\nabla^*\nabla$ with domain $\mathsf{D}(\nabla^*\nabla) = \{f \in \mathsf{D}(\nabla) : \nabla f \in \mathsf{D}(\nabla)\}$, where $\mathsf{D}(\nabla) = H^1(\mathbb{R}^d)$.

Summarising this discussion, we have proved:

Theorem 12.19. *The following operators in $L^2(\mathbb{R}^d)$ are equal, with equal domains:*

(1) *the weak Laplacian Δ with domain*

$$\mathsf{D}(\Delta) = \{f \in L^2(D) : f \text{ admits a weak Laplacian in } L^2(\mathbb{R}^d)\};$$

(2) *the operator $-A$, where A is the operator in $L^2(\mathbb{R}^d)$ associated with the form $\mathfrak{a}$ on $H^1(\mathbb{R}^d)$ given by*

$$\mathfrak{a}(u,v) := \int_{\mathbb{R}^d} \nabla u \cdot \overline{\nabla v} \, dx;$$

(3) *the operator* $-\nabla^\star\nabla$ *with domain*

$$D(\nabla^\star\nabla) = \{f \in D(\nabla) : \nabla f \in D(\nabla^\star)\},$$

where ∇ *is the weak gradient, viewed as a densely defined closed operator from* $L^2(\mathbb{R}^d)$ *to* $L^2(\mathbb{R}^d, \mathbb{C}^d)$ *with domain* $D(\nabla) = H^1(\mathbb{R}^d)$.

A fourth description of Δ will be added to this list in Section 13.6.c, namely, as the generator of the heat semigroup on $L^2(\mathbb{R}^d)$.

12.3.b The Dirichlet Laplace Operator

Let D be a nonempty bounded open subset of $\mathbb{R}^d$. As before we write

$$H_0^1(D) := W_0^{1,2}(D).$$

Let $V := H_0^1(D)$, viewed as a dense subspace of $L^2(D)$, and consider the form $\mathfrak{a}_{\mathrm{Dir}}$ on V given by

$$\mathfrak{a}_{\mathrm{Dir}}(u,v) := \int_D \nabla u \cdot \overline{\nabla v}\,\mathrm{d}x, \quad u,v \in V. \tag{12.12}$$

This form is bounded, positive, and coercive (by the Poincaré inequality) as a form on V. The densely defined closed operator in $L^2(D)$ associated with it is denoted by $-\Delta_{\mathrm{Dir}}$. The operator Δ_{Dir} is called the *Dirichlet Laplacian* on $L^2(D)$.

To substantiate the claim that Δ_{Dir} correctly models the Dirichlet boundary condition, consider a function $u \in C^2(\overline{D})$ which satisfies $u|_{\partial D} = 0$. If $v \in C_c^2(D)$, an integration by parts gives

$$(\Delta u|v) = \int_D (\Delta u)\bar{v}\,\mathrm{d}x = -\int_D \nabla u \cdot \overline{\nabla v}\,\mathrm{d}x = -\mathfrak{a}_{\mathrm{Dir}}(u,v),$$

where the last identity is justified by the fact that u belongs to $H_0^1(D)$ by Theorem 11.24. Since $C_c^2(D)$ is dense in $H_0^1(D)$ it follows that $u \in D(\Delta_{\mathrm{Dir}})$ and $\Delta_{\mathrm{Dir}}u = \Delta u$.

Using Theorem 11.37 it follows that

$$D(\Delta_{\mathrm{Dir}}) = \{u \in H_0^1(D) \cap H_{\mathrm{loc}}^2(D) : \Delta u \in L^2(D)\}, \tag{12.13}$$

To prove this we must show that a function $u \in H_0^1(D)$ belongs to $H_{\mathrm{loc}}^2(D)$ if and only if there exists $f \in L^2(D)$ such that

$$\int_D f\bar{v}\,\mathrm{d}x = \int_D \nabla u \cdot \overline{\nabla v}\,\mathrm{d}x, \quad v \in H_0^1(D). \tag{12.14}$$

If such a function f exists, then u is the weak solution of the Poisson problem $-\Delta u = f$ and Theorem 11.37 implies that $u \in H_{\mathrm{loc}}^2(D)$. In the converse direction, suppose that

$u \in H_0^1(D)$ belongs to $H_{\text{loc}}^2(D)$. If $\phi \in C_c^\infty(D)$ is a given test function, select an open set $U \Subset D$ containing the support of ϕ and use the fact that $u \in H^2(U)$ to see that

$$\int_D \nabla u \cdot \nabla \phi \, dx = \int_U \nabla u \cdot \nabla \phi \, dx = -\int_U (\Delta u)\phi \, dx = -\int_D (\Delta u)\phi \, dx.$$

Since $\Delta u \in L^2(D)$, both sides depend continuously on ϕ with respect to the norm of $H_0^1(D)$. Since $\phi \in C_c^\infty(D)$ is dense in $H_0^1(D)$, it follows that this identity extends to arbitrary $\phi \in H_0^1(D)$. This proves that u satisfies (12.14) with $f := \Delta u \in L^2(D)$.

The result of Remark 11.38 also implies, by the same reasoning, that if D has a C^2-boundary, this domain characterisation improves to

$$\mathsf{D}(\Delta_{\text{Dir}}) = H_0^1(D) \cap H^2(D).$$

12.3.c The Neumann Laplace Operator

As before we let D be nonempty bounded open subset of $\mathbb{R}^d$. As a variation of the preceding example, we may take $V := H^1(D) = W^{1,2}(D)$, viewed as a dense subspace of $H := L^2(D)$, and consider the form $\mathfrak{a}_{\text{Neum}}$ on V given by

$$\mathfrak{a}_{\text{Neum}}(u,v) := \int_D \nabla u \cdot \overline{\nabla v} \, dx, \quad u, v \in V.$$

The only difference with (12.12) is the different choice of the space V. This form is bounded and positive as a form on V. The densely defined closed operator in $L^2(D)$ associated with it is denoted by $-\Delta_{\text{Neum}}$. The operator Δ_{Neum} is called the *Neumann Laplacian* on $L^2(D)$.

To substantiate the claim that Δ_{Neum} correctly models the Neumann boundary condition, let us assume for the moment that D has a C^1-boundary. Consider a function $u \in C^2(\overline{D})$ which satisfies $\frac{\partial f}{\partial \nu}\big|_{\partial D} = 0$, where ν is the outward normal vector on ∂D. If $v \in C^2(\overline{D})$, then Green's formula asserts that

$$(\Delta u | v) = \int_D (\Delta u)\overline{v} \, dx = \int_{\partial D} \frac{\partial u}{\partial \nu} \overline{v} \, dS - \int_D \nabla u \cdot \overline{\nabla v} \, dx$$
$$= -\int_D \nabla u \cdot \overline{\nabla v} \, dx = -\mathfrak{a}_{\text{Neum}}(u,v),$$

where S is the normalised surface measure on ∂D. Since $C^2(\overline{D})$ is dense in $H^1(D)$ by Theorem 11.27, it follows that $u \in \mathsf{D}(\Delta_{\text{Neum}})$ and $\Delta_{\text{Neum}} u = \Delta u$.

As for the Dirichlet Laplacian, Theorem 11.44 implies that

$$\mathsf{D}(\Delta_{\text{Neum}}) = \{u \in H^1(D) \cap H_{\text{loc}}^2(D) : \Delta u \in L^2(D)\}. \tag{12.15}$$

If D has a C^2-boundary, this improves to

$$\mathsf{D}(\Delta_{\text{Neum}}) = H^2(D).$$

12.3.d Selfadjointness

The following result is an immediate consequence of Theorem 12.17 (noting that the forms involved are closed):

Theorem 12.20 (Selfadjointness of the Laplacian). *Let Δ denote the Laplacian on $L^2(\mathbb{R}^d)$ or the Dirichlet or Neumann Laplacian on $L^2(D)$ with $D \subseteq \mathbb{R}^d$ nonempty, bounded, and open. Then $-\Delta$ is positive and selfadjoint.*

These three operators also fall into the setting of Theorem 10.46. Indeed, by Proposition 12.18, all three Laplacians are of the form $\nabla^\star \nabla$, where ∇ is the gradient viewed as a densely defined closed operator from H to K, where $H = L^2(U)$ and $K = L^2(U; \mathbb{R}^d)$ with $U \in \{\mathbb{R}^d, D\}$. This gives an alternative proof of their selfadjointness.

12.3.e Operators in Divergence Form

Let D be a nonempty bounded open subset of $\mathbb{R}^d$ and consider a matrix-valued function $a : D \to M_d(\mathbb{K})$ satisfying the following conditions:

(i) the coefficients $a_{ij} : D \to \mathbb{K}$ are measurable and bounded;

(ii) for all $x \in D$ and $\xi \in \mathbb{K}^d$ we have $\operatorname{Re} \sum_{i,j=1}^d a_{ij}(x)\xi_i \overline{\xi}_j \geqslant 0$.

Condition (ii) is an accretivity condition and is more general than its coercive counterpart used in our treatment of the Sturm–Liouville problem in the preceding chapter.

Under the assumptions (i) and (ii), a bounded accretive form $\mathfrak{a}_a$ on both $V := H_0^1(D)$ (in the case of Dirichlet boundary conditions) and $V := H^1(D)$ (in the case of Neumann boundary conditions) can be defined by

$$\mathfrak{a}_a(u,v) := \int_D a\nabla u \cdot \overline{\nabla v}\, dx, \quad u,v \in V.$$

The operator on $L^2(D)$ associated with $\mathfrak{a}_a$ is usually denoted by

$$-\operatorname{div}(a\nabla)$$

in recognition of the fact that (at least formally) the Hilbert space adjoint of ∇ equals $-\operatorname{div}$. The operator $\operatorname{div}(a\nabla)$ is often referred to as a second order differential operator in *divergence form*. This operator is selfadjoint if the coefficients satisfy the symmetry condition $a_{ij} = \overline{a_{ji}}$.

12.4 The Poisson Problem Revisited

We now revisit the Poisson problem $-\Delta u = f$ by viewing it as a special instance of the abstract problem

$$Au = x,$$

where A is assumed to be a closed operator acting in a Banach space X, $x \in X$ is a given element, and $u \in X$ is the unknown. One could define a *strong solution* as an element $u \in D(A)$ such that $Au = x$, but this is not what we did in Section 11.2. Instead, we considered *weak solutions* defined in terms of the sesquilinear form with which A is associated. A third option is to use duality to define a *scalar solution* to be an element $u \in X$ with the property that

$$\langle u, A^* x^* \rangle = \langle x, x^* \rangle, \quad x^* \in D(A^*).$$

In line with standard functional analytic terminology it would be more appropriate to call *this* a weak solution, but the usage of the term 'weak solution' in connection with integration by parts using test functions is well established.

Proposition 12.21. *Let A be a densely defined closed linear operator on a Banach space X and let $x \in X$ be a given element. For an element $u \in X$ the following assertions are equivalent:*

(1) *u is a strong solution of $Au = x$, that is, $u \in D(A)$ and $Au = x$;*
(2) *u is a scalar solution of $Au = x$, that is, $\langle u, A^* x^* \rangle = \langle x, x^* \rangle$ for all $x^* \in D(A^*)$.*

If $X = H$ is a Hilbert space and A is the operator associated with a densely defined form $\mathfrak{a}$ in H, then (1) and (2) are equivalent to:

(3) *u is a weak solution of $Au = x$, that is, $u \in D(\mathfrak{a})$ and $\mathfrak{a}(u, v) = (x|v)$ for all $v \in V$.*

Proof The implications (1)$\Rightarrow$(2) and (1)$\Rightarrow$(3) are trivial. The implication (2)$\Rightarrow$(1) is an immediate consequence of Proposition 10.20. Finally, if (3) holds, then by the definition of the associated operator we have $u \in D(A)$ and $Au = x$, so (1) holds. $\square$

In the special case where $A = \Delta$, with Δ the Dirichlet or Neumann Laplacian, Proposition 12.21 implies that every weak solution of the Poisson problem $-\Delta u = f$ with $f \in L^2(D)$ is in fact a strong solution. In view of the domain identifications (12.13) and (12.15), this recovers the maximal regularity results of Theorems 11.37 and 11.44.

For the sake of completeness we also mention a maximal regularity result for the Poisson problem on $-\Delta u = f$ on the full space $\mathbb{R}^d$.

Theorem 12.22 (Maximal regularity for $\mathbb{R}^d$). *Let $f \in L^2(\mathbb{R}^d)$. If u is a weak solution of the Poisson problem $-\Delta u = f$ on $\mathbb{R}^d$, then $u \in H^2(\mathbb{R}^d)$.*

Proof If u is a weak solution, an integration by parts gives that u admits a weak Laplacian. The result now follows from Theorem 11.29. □

12.5 Weyl's Theorem

This section is a digression from the main line of development and is dedicated to a proof of Weyl's celebrated asymptotic formula for the number of eigenvalues of Dirichlet Laplacian.

12.5.a Spectrum of the Dirichlet and Neumann Laplacians

As a warm-up we compute the spectrum of Δ_{Dir} and Δ_{Neum} in $L^2(0,1)$.

Example 12.23. Both $-\Delta_{\mathrm{Dir}}$ and $-\Delta_{\mathrm{Neum}}$ are positive and selfadjoint as operators in $L^2(0,1)$ (by Theorem 12.20) and their spectrum is contained in $[0,\infty)$ (by Theorem 12.17). We will use the fact that every $u \in C^2[0,1]$ is included in their domains, that the Laplacians of such a function u are given by taking classical second derivatives pointwise, and that a function $u \in C^2[0,1]$ belongs to $H_0^1(0,1)$ if and only if $u(0) = u(1) = 0$; we leave the elementary proof to the reader (see Problem 11.2 for a more precise result).

The functions $u_n(\theta) = \sin(\pi n\theta)$, $n \geqslant 1$, satisfy $-\Delta_{\mathrm{Dir}}u_n = -u_n'' = \pi^2 n^2 u_n$ and obey Dirichlet boundary conditions. Moreover, by Theorem 3.30, these functions form an orthonormal basis for $L^2(0,1)$. By Proposition 10.32, this implies

$$\sigma(-\Delta_{\mathrm{Dir}}) = \left\{ \pi^2 n^2 : n = 1, 2, \ldots \right\}.$$

with eigenfunctions $u_n(\theta) = \sin(\pi n\theta)$. Likewise, the functions $v_n(\theta) = \cos(\pi n\theta)$, $n \geqslant 0$, satisfy $-\Delta_{\mathrm{Neum}}u_n = -u_n'' = \pi^2 n^2 u_n$ and obey Neumann boundary conditions. Again by Theorem 3.30, and form an orthonormal basis for $L^2(0,1)$. This implies

$$\sigma(-\Delta_{\mathrm{Neum}}) = \left\{ \pi^2 n^2 : n = 0, 1, 2, \ldots \right\}.$$

Turning to higher dimensions, begin with a simple observation.

Proposition 12.24. *Let D be a nonempty bounded open subset of $\mathbb{R}^d$.*

(1) Δ_{Dir} *is both injective and surjective, and hence invertible;*
(2) *if, in addition, D is connected and has C^1-boundary, the null space of Δ_{Neum} consists of the constant functions and its range is the orthogonal complement of the constant functions. In particular, its range is closed and*

$$\dim \mathsf{N}(\Delta_{\mathrm{Neum}}) = \operatorname{codim} \mathsf{R}(\Delta_{\mathrm{Neum}}) = 1.$$

Extending the corresponding definition for bounded operators, a *Fredholm operator* is a closed operator whose null space is finite-dimensional and whose range has finite codimension. With the same proof as in the bounded case, the second condition implies that the range is closed. The *index* of such an operator A is defined as

$$\mathrm{ind}(A) := \dim \mathrm{N}(A) - \mathrm{codim}\, \mathrm{R}(A).$$

Proposition 12.24 implies that both Δ_{Dir} and Δ_{Neum} (the latter under the stated more restrictive assumptions on D) are Fredholm operators with index 0.

Proof (1): If $\Delta_{\mathrm{Dir}} u = 0$ for some $u \in \mathrm{D}(\Delta_{\mathrm{Dir}})$, then $u \in H_0^1(D)$ and u is a strong solution, and hence a weak solution, of the Dirichlet Poisson problem with $f = 0$. By the uniqueness of weak solutions it follows that $u = 0$. Likewise surjectivity follows from the existence of weak solutions for any $f \in L^2(D)$ combined with Proposition 12.21, according to which weak solutions are strong solutions.

(2): This is proved in the same way, using that the problem $-\Delta u = f$ with Neumann boundary conditions has a weak solution for a given $f \in L^2(D)$ if and only if $\int_D f \, dx = 0$, and that uniqueness of weak solutions holds in $H_{\mathrm{av}}^1(D) = \{u \in H^1(D) : \int_D u \, dx = 0\}$. $\square$

For the proof of the next theorem we isolate a lemma that will also be useful in the next chapter.

Lemma 12.25. *Let A be a closed operator on a Banach space X with nonempty resolvent set and suppose that $R(\lambda_0, A)$ is compact for some $\lambda_0 \in \rho(A)$. Then:*

(1) *for all $\lambda \in \rho(A)$ the resolvent operator $R(\lambda, A)$ is compact;*
(2) *for all $\lambda \in \rho(A)$ the following spectral mapping theorem holds:*

$$\sigma(R(\lambda, A)) \setminus \{0\} = \left\{ \frac{1}{\lambda - \mu} : \mu \in \sigma(A) \right\};$$

(3) *every $\mu \in \sigma(A)$ is an eigenvalue with finite multiplicity;*
(4) *for all $\mu \in \sigma(A)$, the eigenspace of the eigenvalue μ for A and the eigenspace of the eigenvalue $\frac{1}{\lambda - \mu}$ for $R(\lambda, A)$ coincide;*
(5) *$\sigma(A)$ is either finite, or it is a sequence diverging to ∞ in absolute value.*

Proof Once (1) and (2) have been established, (3)–(5) are easy consequences of the Riesz–Schauder theorem for compact operators (Theorem 7.11).

(1): This is immediate from the resolvent identity.

(2): Fix $\lambda \in \rho(A)$.

'$\subseteq$': Let $\nu \in \sigma(R(\lambda, A)) \setminus \{0\}$. By the Riesz–Schauder theorem, ν is an eigenvalue for $R(\lambda, A)$. If $R(\lambda, A)x = \nu x$, then $x \in \mathrm{D}(A)$ and $x = \nu(\lambda - A)x$, so $Ax = (\lambda - \frac{1}{\nu})x = \mu x$ with $\mu := \lambda - \frac{1}{\nu}$. It follows that μ is an eigenvalue of A and $\nu = \frac{1}{\lambda - \mu}$.

'⊇': If $\frac{1}{\lambda - \mu} \in \rho(R(\lambda, A))$, then by elementary computation it follows that $\mu \in \rho(A)$ and $R(\mu, A) = \frac{1}{\lambda - \mu} R(\lambda, A) R(\frac{1}{\lambda - \mu}, R(\lambda, A))$.

(3) and (4): For all $x \in X$ and $\mu \in \sigma(A)$ we have $x \in D(A)$ and $Ax = \mu x$ if and only if $R(\lambda, A)x = \frac{1}{\lambda - \mu} x$. This gives (4). By (1) and the Riesz–Schauder theorem, $\sigma(R(\lambda, A)) \setminus \{0\}$ consists of eigenvalues of finite multiplicity. If $\mu \in \sigma(A)$, then $\frac{1}{\lambda - \mu}$ is an eigenvalue for $R(\lambda, A)$ of finite multiplicity by (2), and then (2) and (4) show that μ is an eigenvalue for A of the same finite multiplicity.

(5): If $\sigma(A)$ is an infinite set, then so is $\sigma(R(\lambda, A))$. By the Riesz–Schauder theorem, $\sigma(R(\lambda, A))$ can only accumulate at 0, so $\sigma(A)$ can only accumulate at infinity.   □

Theorem 12.26. *Let D be a nonempty bounded open subset of* $\mathbb{R}^d$. *Then:*

(1) *the spectrum of* $-\Delta_{\mathrm{Dir}}$ *is of the form*

$$\sigma(-\Delta_{\mathrm{Dir}}) = \{\lambda_1, \lambda_2, \ldots\} \quad \text{with} \quad 0 < \lambda_1 < \lambda_2 < \cdots \to \infty;$$

(2) *if, in addition, D is connected and has* C^1*-boundary, the spectrum of* $-\Delta_{\mathrm{Neum}}$ *is of the form*

$$\sigma(-\Delta_{\mathrm{Neum}}) = \{\lambda_1, \lambda_2, \ldots\} \quad \text{with} \quad 0 = \lambda_1 < \lambda_2 < \cdots \to \infty.$$

In either case, each λ_j *is an eigenvalue with finite-dimensional eigenspace.*

Proof Let $A := -\Delta_{\mathrm{Dir}}$ (in the case of Dirichlet boundary conditions) or $A := -\Delta_{\mathrm{Neum}}$ (in the case of Neumann boundary conditions). We claim that, under the respective assumptions on D, the resolvent operators $R(\lambda, A)$ are compact for all $\lambda \in \rho(A)$. To prove the claim we recall that $D(A)$ is contained in $V := H_0^1(D)$ (in the case of Dirichlet boundary conditions), respectively in $V := H^1(D)$ (in the case of Neumann boundary conditions). By the Rellich–Kondrachov theorem (Theorem 11.41), in either case the inclusion mapping from V into $L^2(D)$ is compact. The compactness of $R(\lambda, A)$ now follows by viewing it as the composition of three bounded operators, one of which is compact: (i) $R(\lambda, A)$, viewed as a bounded operator from $L^2(D)$ to $D(A)$, (ii) the inclusion mapping from $D(A)$ into V, which is bounded by the closed graph theorem, the closedness of A, and the boundedness of the inclusion mappings from both $D(A)$ and V into $L^2(D)$, and (iii) the compact inclusion mapping from V into $L^2(D)$.

Since A is positive and selfadjoint (by Theorem 12.20) we have $\sigma(A) \subseteq [0, \infty)$ (by Proposition 10.42). By Proposition 12.24 we have $0 \in \rho(-\Delta_{\mathrm{Dir}})$ and $0 \in \sigma(-\Delta_{\mathrm{Neum}})$. The result now follows from Lemma 12.25.   □

As a variation on the min-max theorem for compact positive Hilbert space operators (Theorem 9.4), we prove an explicit formula for the Dirichlet and Neumann eigenvalues of the Laplace operator on a nonempty bounded open set $D \subseteq \mathbb{R}^d$; in the case of Neumann boundary conditions we make the additional assumption that D is connected and has C^1-boundary. We denote by $0 < \lambda_1 \leqslant \lambda_2 \leqslant \ldots$ and $0 = \mu_1 \leqslant \mu_2 \leqslant \ldots$ the sequences of eigenvalues of $-\Delta_{\text{Dir}}$ and $-\Delta_{\text{Neum}}$, respectively, taking multiplicities into account.

Theorem 12.27 (Courant–Fischer). *With the notation just introduced,*

(1) *for all $n \geqslant 1$ we have*

$$\lambda_n = \inf_{\substack{Y \subseteq H_0^1(D) \\ \dim(Y)=n}} \sup_{\substack{y \in Y \\ y \neq 0}} \frac{\|\nabla y\|_{L^2(D)}^2}{\|y\|_{L^2(D)}^2}, \tag{12.16}$$

where the infima are taken over all subspaces Y of dimension n;

(2) *if, in addition, D is connected and has a C^1-boundary, then for all $n \geqslant 1$ we have*

$$\mu_n = \inf_{\substack{Y \subseteq H^1(D) \\ \dim(Y)=n}} \sup_{\substack{y \in Y \\ y \neq 0}} \frac{\|\nabla y\|_{L^2(D)}^2}{\|y\|_{L^2(D)}^2},$$

where the infima are taken over all subspaces Y of dimension n.

Proof We present the case of Dirichlet eigenvalues, the proof for Neumann eigenvalues being entirely similar (the zero eigenvalue μ_1 does not create difficulties since it has multiplicity 1; here we use the connectedness assumption).

We write $\Delta := \Delta_{\text{Dir}}$ and choose an orthonormal basis $(h_j)_{j \geqslant 1}$ in $L^2(D)$ such that $-\Delta h_j = \lambda_j h_j$ for all $j \geqslant 1$. As was shown in the proof of Theorem 12.26, such a sequence exists by the spectral theorem applied to the compact positive operator Δ^{-1}; this theorem also implies that the span of this sequence is dense in $L^2(D)$.

Set $H_0 := H_0^1(D)$ and, for $n \geqslant 1$,

$$H_n := \left\{ f \in H_0^1(D) : (f|h_j) = 0, \ j = 1, \ldots, n \right\}.$$

Step 1 – Fix $f \in L^2(D)$ and set $f_n := \sum_{j=1}^n c_j h_j$ with $c_j := (f|h_j)$. Since $(h_j)_{j \geqslant 1}$ is an orthonormal bases for $L^2(D)$ we have $f_n \to f$ in $L^2(D)$ as $n \to \infty$.

Clearly, $f - f_n \perp f_n$ in $L^2(D)$. We claim that if $f \in H_0^1(D)$, then also $f - f_n \perp f_n$ in $H_0^1(D)$. In view of

$$(g|g')_{H_0^1(D)} = (g|g') + (\nabla g|\nabla g'), \quad g, g' \in H_0^1(D),$$

this amounts to showing that

$$(\nabla(f - f_n)|\nabla f_n) = 0.$$

For all $j, k \geqslant 1$ we have

$$(\nabla h_j|\nabla h_k) = -(\Delta h_j|h_k) = \lambda_j(h_j|h_k) = \lambda_j \delta_{jk}.$$

and therefore

$$(\nabla f_n | \nabla f_n) = \sum_{j=1}^{n} |c_j|^2 \lambda_j.$$

Also, for $j \geqslant 1$ we have $\lambda_j \geqslant 0$ and

$$(\nabla f | \nabla h_j) = -(f | \Delta h_j) = \lambda_j (f | h_j) = c_j \lambda_j$$

and therefore

$$(\nabla f | \nabla f_n) = \sum_{j=1}^{n} \overline{c_j} (\nabla f | \nabla h_j) = \sum_{j=1}^{n} \overline{c_j} \cdot c_j \lambda_j = \sum_{j=1}^{n} |c_j|^2 \lambda_j. \qquad (12.17)$$

It follows that

$$(\nabla (f - f_n) | \nabla f_n) = \sum_{j=1}^{n} |c_j|^2 \lambda_j - \sum_{j=1}^{n} |c_j|^2 \lambda_j = 0.$$

This proves the claim.

By what we just proved,

$$\|\nabla f\|^2 = \|\nabla (f - f_n)\|^2 + \|\nabla f_n\|^2 \geqslant \|\nabla f_n\|^2.$$

This shows that the sequence $(f_n)_{n \geqslant 1}$ is bounded in $H_0^1(D)$. By Proposition 3.16, some subsequence $(f_{n_k})_{k \geqslant 1}$ converges weakly to a limit $\overline{f}$ in $H_0^1(D)$. Since also $f_n \to f$ in $L^2(D)$ we must have $\overline{f} = f$. Thus $f_{n_k} \to f$ weakly in $H_0^1(D)$. Since bounded operators are weakly continuous and ∇ is bounded from $H_0^1(D)$ to $L^2(R; \mathbb{C}^d)$, this implies $\nabla f_{n_k} \to \nabla f$ weakly in $L^2(D; \mathbb{C}^d)$. By (12.17) it then follows that

$$\|\nabla f\|^2 = \lim_{k \to \infty} (\nabla f | \nabla f_{n_k}) = \lim_{k \to \infty} \sum_{j=1}^{n_k} |c_j|^2 \lambda_j = \sum_{j \geqslant 1} |c_j|^2 \lambda_j. \qquad (12.18)$$

Step 2 – Let $Y \subseteq H_0^1(D)$ be any subspace of dimension n. Since H_{n-1} has codimension $n - 1$ in $H_0^1(D)$, the intersection $H_{n-1} \cap Y$ is a nonzero subspace of $H_0^1(D)$ and hence contains a nonzero element f. Applying the results of Step 1 to f and noting that $(f | h_j) = c_j = 0$ for $j = 1, \ldots, n - 1$, by (12.18) we have

$$\|\nabla f\|^2 = \sum_{j \geqslant n} \lambda_j |c_j|^2 \geqslant \lambda_n \sum_{j \geqslant n} |c_j|^2 = \lambda_n \|f\|^2.$$

This proves the inequality '$\leqslant$' in (12.16).

Step 3 – If f belongs to the span of $\{h_1, \ldots, h_n\}$, then

$$\|\nabla f\|^2 = \sum_{j=1}^{n} |c_j|^2 \lambda_j \leqslant \lambda_n \sum_{j \geqslant 1} |c_j|^2 = \lambda_n \|f\|^2.$$

This proves the inequality '$\geqslant$' in (12.16). $\qquad \square$

Corollary 12.28. *For all $n \geqslant 1$ we have $\mu_n \leqslant \lambda_n$.*

12.5.b Weyl's theorem

The following celebrated theorem of Weyl gives an asymptotic expression for the number of Dirichlet eigenvalues in the interval $[0, r]$ as $r \to \infty$.

Theorem 12.29 (Weyl). *Let D be a nonempty bounded open subset of $\mathbb{R}^d$ satisfying $|\partial D| = 0$, let $0 < \lambda_1 < \lambda_2 < \dots$ the sequence of eigenvalues of $-\Delta_{\mathrm{Dir}}$ on $L^2(D)$, taking multiplicities into account, and for $r > 0$ let*

$$N_D(r) := \max\{n \geqslant 1 : \lambda_n \leqslant r\}.$$

Then

$$\lim_{r \to \infty} \frac{N_D(r)}{r^{d/2}} = \frac{\omega_d}{(2\pi)^d} |D|,$$

where $\omega_d = \pi^{d/2}/\Gamma(1 + \frac{1}{2}d)$ is the volume of the unit ball in $\mathbb{R}^d$.

The condition $|\partial D| = 0$ is satisfied if the boundary is a rectifiable curve.

Before turning to the proof it is instructive to revisit Example 12.23. For the Dirichlet Laplacian in $L^2(0, 1)$ we obtain

$$N_D(r) = \max\{n \geqslant 1 : \pi^2 n^2 \leqslant r\}.$$

On the other hand, $\omega_1 = |(-1, 1)| = 2$ and $|D| = |(0, 1)| = 1$. It follows that

$$\lim_{r \to \infty} \frac{N_D(r)}{r^{1/2}} = \frac{1}{\pi}, \quad \frac{\omega_1}{2\pi} |D| = \frac{2}{2\pi} \cdot 1 = \frac{1}{\pi}.$$

The main lemma needed for the proof of Weyl's theorem is a monotonicity result.

Lemma 12.30. *Let D_1 and D_2 be nonempty bounded open subsets of $\mathbb{R}^d$ with $D_1 \subseteq D_2$. Then the corresponding Dirichlet eigenvalues, taking multiplicities into account, satisfy*

$$\lambda_{n, D_1} \geqslant \lambda_{n, D_2}, \quad n \geqslant 1.$$

As a consequence, $N_{D_1}(r) \leqslant N_{D_2}(r)$ for all $r > 0$.

Hermann Weyl, 1885–1955

Proof This follows from the Courant–Fischer theorem, observing that zero extensions of functions in $H_0^1(D_1)$ belong to $H_0^1(D_2)$.   □

The analogue of this lemma fails for Neumann boundary conditions. It is for this reason that we only present Weyl's theorem for Dirichlet eigenvalues. The case of Neumann boundary conditions is discussed in the Notes to this chapter.

Proof of Theorem 12.29 For an open subset U of $\mathbb{R}^d$ we denote the Dirichlet Laplacian in $L^2(U)$ by Δ_U.

Step 1 – The theorem is true if $D = \prod_{j=1}^d (a_j, b_j)$ is an open rectangle. To prove this there is no loss of generality in assuming that $a_j = 0$ for all $j = 1, \ldots, d$. By the results of Example 12.23 and Section 3.5.c, the eigenfunctions for $-\Delta_D$ are the functions

$$u_n(x) = \prod_{j=1}^d \sin(n_j \pi x_j / b_j), \quad x = (x_1, \ldots, x_d) \in D, \tag{12.19}$$

where $n = (n_1, \ldots, n_d)$ with each n_j in $\mathbb{N}_1 := \{n \in \mathbb{N} : n \geqslant 1\}$. The corresponding eigenvalues are the positive real numbers $\lambda_n = \pi^2 \sum_{j=1}^d n_j^2 / b_j^2$. Hence,

$$N_D(r) = \#\left\{ n \in \mathbb{N}_1^d : \sum_{j=1}^d \frac{n_j^2}{b_j^2} \leqslant \frac{r}{\pi^2} \right\}.$$

As $r \to \infty$, this is asymptotic to $2^{-d} \pi^{-d} r^{d/2} \omega_d \prod_{j=1}^d b_j$, namely, a fraction $1/2^d$ (the 'positive quadrant') of the volume enclosed by the ellipse $\sum_{j=1}^d x_j^2 / b_j^2 = r/\pi^2$. Thus,

$$\lim_{r \to \infty} \frac{N_D(r)}{\dfrac{r^{d/2} \omega_d}{(2\pi)^d} \displaystyle\prod_{j=1}^d b_j} = 1.$$

Since $\prod_{j=1}^d b_j$ equals $|D|$, this is precisely what we wanted to prove.

Step 2 – If U_1 and U_2 are disjoint open subsets of $\mathbb{R}^d$ and $U = U_1 \cup U_2$ is their union, then for all $r > 0$ we have

$$N_U(r) = N_{U_1}(r) + N_{U_2}(r). \tag{12.20}$$

Indeed, $f \in L^2(U)$ is an eigenfunction for $-\Delta_U$, then for $j = 1, 2$ the restriction $f|_{U_j}$ is either an eigenfunction for $-\Delta_{U_j}$ with the same eigenvalue or else it vanishes almost everywhere on U_j. If the first alternative happens for both $j = 1, 2$, then f contributes 2 to $N_R(r)$ (since we count multiplicities) and $f|_{U_j}$ contributes 1 to $N_{R_j}(r)$ for $j = 1, 2$. If the second alternative happens for either $j = 1$ or $j = 2$ (both cannot happen, for this would imply that f vanishes almost everywhere on U), then f contributes 1 to $N_R(r)$ and one of the $f|_{U_j}$ contributes 1 to $N_{R_j}(r)$ and the other one contributes 0.

In the converse direction, if $f_j \in L^2(R_j)$ is an eigenfunction for the Dirichlet Laplacian on R_j, then its zero extension to R is an eigenfunction for the Dirichlet Laplacian on R. This contributes 1 to $N_R(r)$ and 1 to $N_{R_j}(r)$.

Step 3 – By Steps 1 and 2 (applied inductively), the theorem holds for finite unions of open rectangles. Hence, if $R_1, \ldots, R_k$ are disjoint open rectangles contained in D and

R denotes their union, then, by Lemma 12.30,

$$\frac{\omega_d}{(2\pi)^d}|R| = \lim_{r\to\infty}\frac{N_R(r)}{r^{d/2}} \leqslant \liminf_{r\to\infty}\frac{N_D(r)}{r^{d/2}}.$$

Approximating D from within by such finite unions, we obtain

$$\frac{\omega_d}{(2\pi)^d}|D| \leqslant \liminf_{r\to\infty}\frac{N_D(r)}{r^{d/2}}. \tag{12.21}$$

Let $\rho > 0$ be so large that $\overline{D} \subseteq (-\rho,\rho)^d$. By the same reasoning,

$$\frac{\omega_d}{(2\pi)^d}|(-\rho,\rho)^d \setminus \overline{D}| \leqslant \liminf_{r\to\infty}\frac{N_{(-\rho,\rho)^d\setminus\overline{D}}(r)}{r^{d/2}}. \tag{12.22}$$

Using Step 1, (12.20), (12.21) and (12.22), and the assumption $|\partial D| = 0$, we obtain

$$\begin{aligned}
\frac{\omega_d}{(2\pi)^d}(2\rho)^d &= \lim_{r\to\infty}\frac{N_{(-\rho,\rho)^d}(r)}{r^{d/2}} \geqslant \liminf_{r\to\infty}\frac{N_D(r)}{r^{d/2}} + \liminf_{r\to\infty}\frac{N_{(-\rho,\rho)^d\setminus\overline{D}}(r)}{r^{d/2}} \\
&\geqslant \frac{\omega_d}{(2\pi)^d}(|D| + |(-\rho,\rho)^d \setminus \overline{D}|) = \frac{\omega_d(2\rho)^d}{(2\pi)^d}.
\end{aligned} \tag{12.23}$$

It follows that we have equality throughout. Hence, using (12.20) and Step 1 once more,

$$\begin{aligned}
\frac{\omega_d(2\rho)^d}{(2\pi)^d} &= \liminf_{r\to\infty}\frac{N_D(r)}{r^{d/2}} + \liminf_{r\to\infty}\frac{N_{(-\rho,\rho)^d\setminus\overline{D}}(r)}{r^{d/2}} \\
&= \liminf_{r\to\infty}\frac{N_D(r)}{r^{d/2}} + \frac{\omega_d}{(2\pi)^d}(2\rho)^d - \limsup_{r\to\infty}\frac{N_D(r)}{r^{d/2}}
\end{aligned}$$

and therefore the limit inferior and the limit superior are equal. It follows that the limit $\lim_{r\to\infty}\frac{N_D(r)}{r^{d/2}}$ exists. Finally, if we had strict inequality (12.21), that is, if

$$\frac{\omega_d}{(2\pi)^d}|D| < \liminf_{r\to\infty}\frac{N_D(r)}{r^{d/2}},$$

then in (12.23) the second inequality would be strict. $\square$

If one imagines a bounded open set D in $\mathbb{R}^d$ as a 'drum', the eigenvalues of the negative Dirichlet Laplacian on $L^2(D)$ can be interpreted as the 'frequencies' of the drum. This prompted the famous question of Mark Kac: "Can one hear the shape of a drum?". In its mathematical formulation, the question is whether the shape of D, up to an isometry of $\mathbb{R}^d$, is determined by its sequence of frequencies. Without further assumptions on D, in general the answer is negative. Nevertheless, Weyl's theorem implies that the volume $|D|$ of D can be recovered from the spectrum.

Problems

12.1 Let (i,V,H) be a Gelfand triple and let A be the linear operator in H associated with a bounded accretive form $\mathfrak{a}$ on V. Prove that the inclusion mapping from $D(A)$ into V is bounded.

12.2 Let (i,V,H) be a Gelfand triple. A form $\mathfrak{a}$ on V is said to be *elliptic* if there exist $\lambda > 0$ and $\alpha > 0$ such that

$$\operatorname{Re}\mathfrak{a}(v,v) + \lambda \|v\|^2 \geqslant \alpha \|v\|_V^2, \quad v \in V.$$

(a) Show that the form $\mathfrak{a}$ on V is elliptic if and only if the form

$$\mathfrak{a}_\lambda(u,v) := \mathfrak{a}(u,v) + \lambda(u|v), \quad u,v \in V,$$

is coercive on V.

(b) Prove a version of Corollary 12.13 for operators A associated with an elliptic form $\mathfrak{a}$ on V.

12.3 Let (i,V,H) be a Gelfand triple, let $\mathfrak{a}$ be a coercive form on V, and let $B \in \mathscr{L}(V,H)$ be bounded. Show that the form $\mathfrak{a}_B$ on V defined by

$$\mathfrak{a}_B(u,v) := \mathfrak{a}(u,v) + (Bu|v), \quad u,v \in V,$$

is bounded and elliptic.

12.4 Revisiting the conditions imposed in the treatment of the Sturm–Liouville problem in Section 11.3.b, let $D \subseteq \mathbb{R}^d$ be open and bounded and let $a : D \to M_d(\mathbb{C})$ be a function with bounded measurable coefficients such that

$$\operatorname{Re} \sum_{i,j=1}^d a_{ij}(x)\xi_i\overline{\xi}_j \geqslant \alpha|\xi|^2, \quad \xi \in \mathbb{C}^d,$$

for some $\alpha > 0$ and almost all $x \in D$. Let $b : D \to \mathbb{K}^d$ have bounded measurable coordinate functions and let $c : D \to \mathbb{K}$ be bounded and measurable. Show that the form

$$\mathfrak{a}(u,v) := \int_D a\nabla u \cdot \overline{\nabla v}\,dx + \int_D b \cdot \nabla u\overline{v}\,dx + \int_D cu\overline{v}\,dx, \quad u,v \in H_0^1(D),$$

is elliptic.

12.5 Prove that the form $\overline{\mathfrak{a}}$ constructed in the proof of Proposition 12.16 has the following minimality property: If $\widetilde{\mathfrak{a}} : \widetilde{V} \times \widetilde{V} \to \mathbb{C}$ is a closed form extending the closable continuous accretive form $\mathfrak{a}$, then $\widetilde{\mathfrak{a}}$ extends $\overline{\mathfrak{a}}$.

12.6 Prove the following facts for $d = 1$ and $D = (a,b)$:

(a) $D(\Delta_{\mathrm{Dir}}) = \{f \in H^2(D) : f(a) = f(b) = 0\}$;

(b) $D(\Delta_{\mathrm{Neum}}) = \{f \in H^2(D) : f'(a) = f'(b) = 0\}$.

12.7 Let V be a Hilbert space, let $\mathfrak{a} : V \times V \to \mathbb{K}$ be a bounded coercive form and let $L : V \to \mathbb{K}$ be a bounded functional.

(a) Show that the *energy functional*

$$E(x) := \frac{1}{2} \operatorname{Re} \mathfrak{a}(x,x) - \operatorname{Re} L(x)$$

is bounded from below.

Fix a nonempty closed convex subset C of V.

(b) Let $(x_n)_{n \geqslant 1}$ be a sequence in C such that

$$\lim_{n \to \infty} E(x_n) = \inf_{x \in C} E(x) =: E > -\infty.$$

Prove that this sequence is Cauchy in V.

Hint: The convexity of C implies that $\frac{1}{2}(x_n + x_m) \in C$. Then use the identity

$$E(\frac{1}{2}(x_n + x_m)) = \frac{1}{2}E(x_n) + \frac{1}{2}E(x_m) - \frac{1}{8} \operatorname{Re} \mathfrak{a}(x_n - x_m, x_n - x_m).$$

(c) Prove that $x := \lim_{n \to \infty} x_n$ is the unique element of C minimising E.

(d) Compare this result with Problem 11.31.

12.8 We take a look at Example 12.23 from a Calculus perspective.

(a) For which $\lambda \in \mathbb{R}$ does the problem

$$\begin{cases} -u'' = \lambda u & \text{on } (0,1), \\ u(0) = u(1) = 0, \end{cases}$$

admit a C^2-solutions? For these values of λ, find all C^2-solutions.

(b) Do the same for Neumann boundary conditions $u'(0) = u'(1) = 0$.

(c) Explain why this is not enough to determine the spectra of the Dirichlet and Neumann Laplacians in $L^2(0,1)$.

12.9 Provide the details of the proof that all Dirichlet eigenfunctions on a cube are given by (12.19).

13
Semigroups of Linear Operators

In this chapter we set up a functional analytic framework for the study of linear and non-linear initial value problems. This includes the treatment of parabolic problems such as the heat equation and hyperbolic problems such as the wave equation. From the operator-theoretic perspective the main challenge is to arrive at a thorough understanding of linear equations. This is achieved through the theory of C_0-semigroups developed in the present chapter. Once this is done, nonlinear equations are handled by perturbation techniques.

13.1 C_0-Semigroups

Equations of mathematical physics describing systems involving time evolution can often be cast in the abstract form

$$\begin{cases} u'(t) = Au(t) + f(t, u(t)), & t \in [0, T], \\ u(0) = u_0, \end{cases}$$

where the unknown is a function u from the time interval $[0, T]$ into a Banach space X, the operator A is a linear, usually unbounded, operator acting in X, $f : [0, T] \times X \to X$ is a given function, and the initial value u_0 is assumed to be an element of X. This initial value problem is referred to as the *abstract Cauchy problem* associated with A and f. In applications, typically X is a Banach space of functions suited for the particular problem and A is a partial differential operator. For instance, for the heat equation on a bounded open subset D of $\mathbb{R}^d$ subject to Dirichlet boundary conditions one could choose $X = L^2(D)$ and take A to be the Dirichlet Laplacian studied in the previous chapter.

If A is a *bounded* operator, the unique solution u of the linear abstract Cauchy problem

$$\begin{cases} u'(t) = Au(t), & t \in [0, T], \\ u(0) = u_0, \end{cases} \tag{ACP}$$

429

is given by

$$u(t) = e^{tA}u_0 = \sum_{n=0}^{\infty} \frac{t^n}{n!} A^n u_0, \quad t \in [0,T].$$

The operators e^{tA} may be thought of as 'solution operators' mapping the initial value u_0 to the solution $e^{tA}u_0$ at time t. For unbounded operators A this simple strategy does not work since we run into convergence and domain issues. In the case of selfadjoint operators A and, more generally, normal operators A acting in a Hilbert space, one could instead use the functional calculus of Chapter 10 to define the exponentials e^{tA}. This would still limit the scope and applicability of the theory considerably. In order to set up a more general and flexible framework we take a more abstract approach which is motivated by the properties of the exponentials e^{tA} for bounded operators A: they satisfy $e^{0A} = I$ and $e^{tA}e^{sA} = e^{(t+s)A}$, and the mapping $t \mapsto e^{tA}$ is continuous with respect to the operator norm.

13.1.a Definition and General Properties

Throughout this chapter, X is a Banach space and H is a Hilbert space.

The preceding discussion suggests the following definition.

Definition 13.1 (C_0-Semigroups). A family $S = (S(t))_{t \geq 0}$ of bounded operators acting on X is called a C_0-*semigroup* if the following three properties are satisfied:

(S1) $S(0) = I$;
(S2) (semigroup property) $S(t)S(s) = S(t+s)$ for all $t, s \geq 0$;
(S3) (strong continuity) $\lim_{t \downarrow 0} \|S(t)x - x\| = 0$ for all $x \in X$.

Its *infinitesimal generator*, or briefly the *generator*, is the linear operator A defined by

$$D(A) = \left\{ x \in X : \lim_{t \downarrow 0} \frac{1}{t}(S(t)x - x) \text{ exists in } X \right\},$$

$$Ax = \lim_{t \downarrow 0} \frac{1}{t}(S(t)x - x), \quad x \in D(A).$$

The idea is to interpret the orbit

$$u(t) := S(t)u_0$$

as the 'solution' of the linear problem (ACP). To find a precise way to make this idea rigorous, and to subsequently cover also nonlinear initial value problems, is among the main objectives of this chapter.

Remark 13.2 (Strong convergence versus uniform convergence). The properties of e^{tA} suggest replacing (S3) by the stronger condition $\lim_{t \downarrow 0} \|S(t) - I\| = 0$. As it turns out, however, this condition forces the generator A to be bounded (see Problem 13.2). This renders the theory useless, as it would fail to cover equations in which A is a differential operator acting in Banach space X of functions. In a sense the strong convergence imposed in (S3) is also more natural, as it gives the continuity with respect to the norm of X of the 'solution' $u(t) = S(t)u_0$ (see Proposition 13.4).

The next two propositions collect some elementary properties of C_0-semigroups and their generators.

Proposition 13.3. *Let S be a C_0-semigroup on X. There exist $M \geqslant 1$ and $\omega \in \mathbb{R}$ such that $\|S(t)\| \leqslant Me^{\omega t}$ for all $t \geqslant 0$.*

Proof There exists a number $\delta > 0$ such that $\sup_{t \in [0,\delta]} \|S(t)\| =: \sigma < \infty$. Indeed, otherwise we could find a sequence $t_n \downarrow 0$ such that $\lim_{n \to \infty} \|S(t_n)\| = \infty$. By the uniform boundedness theorem, this implies the existence of an $x \in X$ such that $\sup_{n \geqslant 1} \|S(t_n)x\| = \infty$, contradicting the strong continuity assumption (S3).

By the semigroup property (S2), for $t \in [(k-1)\delta, k\delta]$ it follows that $\|S(t)\| \leqslant \sigma^k \leqslant \sigma^{1+t/\delta}$, where the second inequality uses that $\sigma \geqslant 1$ by (S1). This proves the proposition, with $M = \sigma$ and $\omega = \frac{1}{\delta}\log\sigma$. $\qquad\square$

We will frequently use the trivial observation that if A generates the C_0-semigroup $(S(t))_{t \geqslant 0}$, then for all scalars μ the linear operator $A - \mu$ generates the C_0-semigroup $(e^{-\mu t}S(t))_{t \geqslant 0}$. For $\mu > \omega$, with ω as in Proposition 13.3, this rescaled semigroup has exponential decay in operator norm.

Proposition 13.4. *Let S be a C_0-semigroup on X with generator A. The following assertions hold:*

(1) *for all $x \in X$ the orbit $t \mapsto S(t)x$ is continuous for $t \geqslant 0$;*
(2) *for all $x \in D(A)$ the orbit $t \mapsto S(t)x$ is continuously differentiable for $t \geqslant 0$, we have $S(t)x \in D(A)$, and*

$$\frac{\mathrm{d}}{\mathrm{d}t}S(t)x = AS(t)x = S(t)Ax, \quad t \geqslant 0;$$

(3) *for all $x \in X$ and $t \geqslant 0$ we have $\int_0^t S(s)x\,\mathrm{d}s \in D(A)$ and*

$$A\int_0^t S(s)x\,\mathrm{d}s = S(t)x - x,$$

and if $x \in D(A)$, then both sides are equal to $\int_0^t S(s)Ax\,\mathrm{d}s$;
(4) *the generator A is a densely defined closed operator.*

Proof The proof uses the calculus rules for Banach space-valued Riemann integrals (Proposition 1.45).

(1): Right continuity of $t \mapsto S(t)x$ follows from the right continuity at $t = 0$ (S3) and the semigroup property (S2). For left continuity, observe that by the semigroup property, for $0 < h < t$ we have

$$\|S(t)x - S(t-h)x\| \leqslant \|S(t-h)\|\,\|S(h)x - x\| \leqslant \sup_{s \in [0,t]} \|S(s)\|\,\|S(h)x - x\|,$$

where the supremum is finite by Proposition 13.3.

(2): Fix $x \in D(A)$ and $t \geqslant 0$. By the semigroup property we have

$$\lim_{h \downarrow 0} \frac{1}{h}(S(t+h)x - S(t)x) = S(t) \lim_{h \downarrow 0} \frac{1}{h}(S(h)x - x) = S(t)Ax.$$

This proves all assertions except left differentiability. For $t > 0$ we note that

$$\lim_{h \downarrow 0} \frac{1}{h}(S(t)x - S(t-h)x) = \lim_{h \downarrow 0} S(t-h)\left(\frac{1}{h}(S(h)x - x)\right) = S(t)Ax,$$

where we used that $x \in D(A)$ and the fact that the convergence $\lim_{h \downarrow 0} S(t-h)y = S(t)y$ for all $y \in X$ implies convergence uniformly on compact sets by Proposition 1.42.

(3): The first identity follows from

$$\lim_{h \downarrow 0} \frac{1}{h}(S(h) - I)\int_0^t S(s)x\,ds = \lim_{h \downarrow 0} \frac{1}{h}\left(\int_0^t S(s+h)x\,ds - \int_0^t S(s)x\,ds\right)$$

$$= \lim_{h \downarrow 0} \frac{1}{h}\left(\int_t^{t+h} S(s)x\,ds - \int_0^h S(s)x\,ds\right)$$

$$= S(t)x - x,$$

where we first did a substitution and then used the continuity of $t \mapsto S(t)x$. The identity for $x \in D(A)$ follows by integrating the identity of part (2), or by noting that

$$\lim_{h \downarrow 0} \frac{1}{h}(S(h) - I)\int_0^t S(s)x\,ds = \lim_{h \downarrow 0} \frac{1}{h}\int_0^t S(s)\left(\frac{1}{h}(S(h)x - x)\right)ds = \int_0^t S(s)Ax\,ds,$$

where the convergence under the integral is justified by the fact that the convergence of the difference quotients $\frac{1}{h}(S(h)x - x)$ to Ax implies uniform convergence of the integrands on $[0,t]$.

(4): Denseness of $D(A)$ follows from (1) and the first part of (3): for any $x \in X$, the latter implies that $\int_0^t S(s)x\,ds \in D(A)$ for all $t > 0$, while the former implies that $\lim_{t \downarrow 0} \frac{1}{t}\int_0^t S(s)x\,ds = x$.

To prove that A is closed we must check that the graph $G(A) = \{(x, Ax) : x \in D(A)\}$ is closed in $X \times X$. Suppose that $(x_n)_{n \geqslant 1}$ is a sequence in $D(A)$ such that $\lim_{n \to \infty} x_n = x$ and $\lim_{n \to \infty} Ax_n = y$ in X. Then, by the second part of (3),

$$\frac{1}{h}(S(h)x - x) = \lim_{n \to \infty} \frac{1}{h}(S(h)x_n - x_n) = \lim_{n \to \infty} \frac{1}{h}\int_0^h S(s)Ax_n\,ds = \frac{1}{h}\int_0^h S(s)y\,ds.$$

Passing to the limit for $h \downarrow 0$, this gives $x \in D(A)$ and $Ax = y$. $\qquad\square$

We have just seen that the generator of a C_0-semigroup is always densely defined and closed. As a consequence of the latter, $D(A)$ is a Banach space with respect to its graph norm. In various applications it is of interest to know when a subspace Y, which is dense in X and contained in $D(A)$, is dense as a subspace of $D(A)$. If this is the case, Y is called a *core* for A. The next result gives a simple sufficient condition.

Proposition 13.5. *Let S be a C_0-semigroup with generator A on X. If Y is a subspace of $D(A)$ which is dense in X and invariant under each operator $S(t)$, $t \geq 0$, then Y is dense in $D(A)$.*

Proof The operator $A - \lambda$ is the generator of the C_0-semigroup $(e^{-\lambda t} S(t))_{t \geq 0}$. Hence, by the exponential boundedness of S, replacing A by $A - \lambda$ for sufficiently large $\lambda > 0$ we may assume that $\lim_{t \to \infty} \|S(t)\| = 0$.

Fix $x \in D(A)$ and choose a sequence $(y_n)_{n \geq 1}$ in Y such that $\lim_{n \to \infty} y_n = Ax$ in X. Fix $t > 0$. Then

$$\lim_{n \to \infty} \int_0^t S(s) y_n \, ds = \int_0^t S(s) Ax \, ds = S(t)x - x$$

in X and

$$\lim_{n \to \infty} A \int_0^t S(s) y_n \, ds = \lim_{n \to \infty} S(t) y_n - y_n = S(t) Ax - Ax.$$

It follows that

$$\lim_{n \to \infty} \int_0^t S(s) y_n \, ds = S(t)x - x \ \text{ in } \ D(A).$$

The identity

$$\|S(t)x - x\|_{D(A)} = \|S(t)x - x\| + \|S(t)Ax - Ax\|$$

implies that the restriction of S to $D(A)$ is strongly continuous with respect to the graph norm of $D(A)$, and for this reason we may approximate the integrals $\int_0^t S(s) y_n \, ds$ by Riemann sums in the norm of $D(A)$. By the invariance of Y under S, these Riemann sums belong to Y. It follows that for each $t > 0$ and $\varepsilon > 0$ there is a $y_{t,\varepsilon} \in Y$ such that

$$\|(S(t)x - x) - y_{t,\varepsilon}\|_{D(A)} < \varepsilon.$$

As $t \to \infty$, $\|S(t)x\|_{D(A)} = \|S(t)x\| + \|S(t)Ax\| \to 0$, and therefore, for large enough $t > 0$,

$$\|y_{t,\varepsilon} - x\|_{D(A)} \leq \varepsilon + \|S(t)x\|_{D(A)} < 2\varepsilon.$$

This shows that x can be approximated in $D(A)$ by elements of Y. $\qquad\square$

This proposition is often helpful in determining the domain of the generator explicitly when the semigroup is given; see for instance Section 13.6.b.

The proof of the next proposition uses the following version of the product rule. It is proved in the same way as the product rule in calculus; uniform convergence on compact sets follows from Proposition 1.42.

Lemma 13.6. *Let $I \subseteq \mathbb{R}$ be an open interval and let $S : I \to \mathscr{L}(X)$ and $T : I \to \mathscr{L}(X)$ be strongly continuous functions. Let $t_0 \in I$ and $x \in X$ be fixed. If*

(i) *$t \mapsto S(t)x$ is differentiable at t_0, with derivative*

$$\frac{\mathrm{d}}{\mathrm{d}t}\Big|_{t=t_0} S(t)x =: S'(t_0)x,$$

(ii) *$t \mapsto T(t)S(t_0)x$ is differentiable at t_0, with derivative*

$$\frac{\mathrm{d}}{\mathrm{d}t}\Big|_{t=t_0} T(t)S(t_0)x =: T'(t_0)S(t_0)x,$$

then $t \mapsto T(t)S(t)x$ is differentiable at t_0, with derivative

$$\frac{\mathrm{d}}{\mathrm{d}t}\Big|_{t=t_0} T(t)S(t)x = T'(t_0)S(t_0)x + T(t_0)S'(t_0)x.$$

Proof For $t \in I \setminus \{t_0\}$ we have

$$\frac{T(t)S(t)x - T(t_0)S(t_0)x}{t - t_0}$$

$$= \frac{T(t)S(t)x - T(t)S(t_0)x}{t - t_0} + \frac{T(t)S(t_0)x - T(t_0)S(t_0)x}{t - t_0}$$

$$=: (I) + (II).$$

By assumption, (II) tends to $T'(t_0)S(t_0)x$ as $t \to t_0$. Concerning (I), fix $\delta > 0$ small enough so that $(t_0 - \delta, t_0 + \delta)$ is contained in I, set $I_{t_0,\delta} := (t_0 - \delta, t_0) \cup (t_0, t_0 + \delta)$, and consider the relatively compact set

$$C_{t_0,\delta} := \left\{ \frac{S(t)x - S(t_0)x}{t - t_0} - S'(t_0)x : t \in I_{t_0,\delta} \right\}.$$

For $t \in I_{t_0,\delta}$ we have

$$\left\| \frac{T(t)S(t)x - T(t)S(t_0)x}{t - t_0} - T(t_0)S'(t_0)x \right\|$$

$$\leqslant \left\| T(t)\left(\frac{S(t)x - S(t_0)x}{t - t_0} - S'(t_0)x \right) \right\| + \left\| (T(t) - T(t_0))S'(t_0)x \right\|$$

$$\leqslant \sup_{y \in C_{t_0,\delta}} \| T(t)y - T(t_0)y \| + \left\| T(t_0)\left(\frac{S(t)x - S(t_0)x}{t - t_0} - S'(t_0)x \right) \right\|$$

$$+ \| (T(t) - T(t_0))S'(t_0)x \|.$$

The strong continuity of T and the fact that strong convergence implies strong convergence on compact sets imply that the first and third terms on the right-hand side tend to 0 as $t \to t_0$. The second term tends to 0 by the assumptions on S. $\square$

A C_0-semigroup is uniquely determined by its generator:

Proposition 13.7. *If A is the generator of the C_0-semigroups S and T, then $S(t) = T(t)$ for all $t \geqslant 0$.*

Proof By Lemma 13.6, for all $t > 0$ and $x \in D(A)$ the function $\phi_t(s) := S(t-s)T(s)x$ is continuously differentiable on $[0,t]$ with derivative $\phi'_t(s) = -AS(t-s)T(s)x + S(t-s)AT(s)x = 0$, and therefore ϕ_t is constant by Proposition 1.45. Hence, $S(t)x = \phi_t(0) = \phi_t(t) = T(t)x$. This being true for all x in the dense subspace $D(A)$ of X, it follows that $S(t) = T(t)$. □

The next proposition identifies the resolvent of the generator as the Laplace transform of the semigroup.

Proposition 13.8. *Let A be the generator of a C_0-semigroup S on X, and fix constants $M \geqslant 1$ and $\omega \in \mathbb{R}$ such that $\|S(t)\| \leqslant Me^{\omega t}$ for all $t \geqslant 0$. Then $\{\lambda \in \mathbb{C} : \operatorname{Re}\lambda > \omega\} \subseteq \rho(A)$, and on this set the resolvent of A is given by*

$$R(\lambda,A)x = \int_0^\infty e^{-\lambda t} S(t)x \, dt, \quad x \in X.$$

As a consequence, for $\operatorname{Re}\lambda > \omega$ we have

$$\|R(\lambda,A)\| \leqslant \frac{M}{\operatorname{Re}\lambda - \omega}.$$

Proof Fix $x \in X$ and define $R_\lambda x := \int_0^\infty e^{-\lambda t} S(t)x \, dt$. Using the semigroup property (S2) and a substitution, we obtain the identity

$$\lim_{h \downarrow 0} \frac{1}{h}(S(h) - I)R_\lambda x = \lim_{h \downarrow 0} \frac{1}{h}\left(e^{ht}\int_h^\infty e^{-\lambda t} S(t)x \, dt - \int_0^\infty e^{-\lambda t} S(t)x \, dt\right)$$

$$= \lim_{h \downarrow 0} \frac{e^{ht} - 1}{h}\int_h^\infty e^{-\lambda t} S(t)x \, dt - \lim_{h \downarrow 0} \frac{1}{h}\int_0^h e^{-\lambda t} S(t)x \, dt$$

$$= \lambda R_\lambda x - x,$$

from which it follows that $R_\lambda x \in D(A)$ and $AR_\lambda x = \lambda R_\lambda x - x$. This shows that the bounded operator R_λ is a right inverse for $\lambda - A$.

Integrating by parts and using that $\frac{d}{dt}S(t)x = S(t)Ax$ for $x \in D(A)$ we obtain, for $x \in D(A)$,

$$\lambda \int_0^T e^{-\lambda t} S(t)x \, dt = -e^{-\lambda T} S(T)x + x + \int_0^T e^{-\lambda t} S(t)Ax \, dt.$$

Since $\operatorname{Re}\lambda > \omega$, sending $T \to \infty$ gives $\lambda R_\lambda x = x + R_\lambda Ax$. This shows that R_λ is also a left inverse.

The estimate for the resolvent follows from

$$\left\|\int_0^\infty e^{-\lambda t} S(t)x \, dt\right\| \leqslant \int_0^\infty e^{-\operatorname{Re}\lambda t}\|S(t)x\| \, dt \leqslant M\|x\|\int_0^\infty e^{(\omega - \operatorname{Re}\lambda)t} \, dt = \frac{M}{\operatorname{Re}\lambda - \omega}\|x\|.$$

□

Combining this result with Proposition 10.30 we obtain the result that a C_0-semigroup is determined by its generator:

Proposition 13.9. *If A and B generate C_0-semigroups on X, and if B is an extension of A, then $A = B$.*

Proof Proposition 13.8 implies that the resolvent sets of A and B share a common half-plane. The equality $A = B$ then follows from Proposition 10.30. □

For operators satisfying the resolvent estimate of Proposition 13.8 for real λ, we have the following convergence result.

Proposition 13.10. *Let A be a densely defined closed operator acting in X, and suppose that for some $\omega \in \mathbb{R}$ we have $\{\lambda > \omega\} \subseteq \rho(A)$ and*

$$\|R(\lambda, A)\| \leqslant \frac{M}{\lambda - \omega}, \quad \lambda > \omega.$$

Then for all $x \in X$ we have

$$\lim_{\lambda \to \infty} \lambda R(\lambda, A)x = x.$$

Proof First let $x \in D(A)$ and fix an arbitrary $\mu \in \rho(A)$. Then $x = R(\mu, A)y$ for $y := (\mu - A)x$. By the resolvent identity and the above estimate on the resolvent we obtain

$$\lim_{\lambda \to \infty} \lambda R(\lambda, A)x = \lim_{\lambda \to \infty} \lambda R(\lambda, A)R(\mu, A)y$$

$$= \lim_{\lambda \to \infty} \frac{\lambda}{\lambda - \mu}(R(\mu, A) - R(\lambda, A))y = R(\mu, A)y = x.$$

For general $x \in X$ the claim then follows by approximation with elements from $D(A)$, using the uniform boundedness of the resolvent for $\lambda \geqslant \omega + 1$. □

The final result of this section gives a useful sufficient condition for a semigroup of operators to be strongly continuous. We need the following terminology. A family of bounded operators $S = (S(t))_{t \geqslant 0}$ on X is said to be a *weakly continuous semigroup* if conditions (S1) and (S2) in Definition 13.1 hold and (S3) is replaced by the condition that for all $x \in X$ and $x^* \in X^*$ one has $\lim_{t \downarrow 0} \langle S(t)x, x^* \rangle = \langle x, x^* \rangle$.

Theorem 13.11 (Phillips). *Every weakly continuous semigroup is strongly continuous.*

Proof Let

$$X_0 := \Big\{ x \in X : \lim_{t \downarrow 0} \|S(t)x - x\| = 0 \Big\}.$$

It is evident that X_0 is a linear subspace of X. We wish to show that $X_0 = X$.

Arguing as in the proof of Proposition 13.3 we see that the family $\{S(t) : 0 \leqslant t \leqslant 1\}$ is uniformly bounded. A first consequence is that X_0 is a *closed* subspace of X. Next we note that the weak continuity of $t \mapsto S(t)x$ along with the fact that closed subspaces are weakly closed (Proposition 4.44) implies that each orbit $t \mapsto S(t)x$ is contained in a separable closed subspace of X. It follows that we can apply the Pettis measurability theorem (Theorem 4.19) and conclude that every orbit $t \mapsto S(t)x$ is strongly measurable. It follows from these considerations that the Bochner integrals $x_t := \frac{1}{t}\int_0^t S(s)x\,ds$ are well defined.

Fix $x \in X$ and $0 < t < \frac{1}{2}$. For $0 < s < t$,

$$\|S(s)x_t - x_t\| = \frac{1}{t}\left\| \int_0^t S(s+r)x\,dr - \int_0^t S(r)x\,dr \right\|$$
$$= \frac{1}{t}\left\| \int_t^{t+s} S(r)x\,dr - \int_0^s S(r)x\,dr \right\| \leqslant 2s \cdot \frac{1}{t}\left(\sup_{0 \leqslant r \leqslant 1} \|S(r)\| \right)\|x\|.$$

This shows that $x_t^* \in X_0$.

Suppose now, for a contradiction, that $X_0 \neq X$. Then there exists an $x \in X \setminus X_0$ and by the Hahn–Banach theorem we can find an $x^* \in X^*$ which vanishes on X_0 but not on x. Then, with $x_t = \frac{1}{t}\int_0^t S(s)x\,ds$ as before,

$$0 = \lim_{t \downarrow 0} \langle x_t, x^* \rangle = \lim_{t \downarrow 0} \frac{1}{t}\int_0^t \langle S(s)x, x^* \rangle\,ds = \langle x, x^* \rangle \neq 0,$$

a contradiction. $\qquad\square$

13.1.b C_0-Groups

Instead of considering only forward time we could also include backward time. This leads to the notion of a C_0-group.

Definition 13.12 (C_0-groups). A C_0-*group* is a family $S = (S(t))_{t \in \mathbb{R}}$ of bounded operators acting on X with the following properties:

(G1) $S(0) = I$;
(G2) $S(t)S(s) = S(t+s)$ for all $t, s \in \mathbb{R}$;
(G3) $\lim_{t \to 0} \|S(t)x - x\| = 0$ for all $x \in X$.

Its *infinitesimal generator*, or briefly its *generator*, is the linear operator A defined by

$$D(A) := \left\{ x \in X : \lim_{t \to 0} \frac{1}{t}(S(t)x - x) \text{ exists} \right\},$$
$$Ax := \lim_{t \to 0} \frac{1}{t}(S(t)x - x), \quad x \in D(A).$$

It is evident from the definition that if A generates a C_0-group $(S(t))_{t \in \mathbb{R}}$, then both $(S(t))_{t \geqslant 0}$ and $(S(-t))_{t \geqslant 0}$ are C_0-semigroups. Denoting their generators by A_+ and A_-, it is evident that $D(A) \subseteq D(A_+) \cap D(A_-)$ and that for all $x \in D(A)$ we have $Ax = A_+x = -A_-x$. In fact, more is true:

Proposition 13.13. *A linear operator A in X generates a C_0-group $(S(t))_{t \in \mathbb{R}}$ if and only if both A and $-A$ generate C_0-semigroups. These semigroups are $(S(t))_{t \geq 0}$ and $(S(-t))_{t \geq 0}$, respectively.*

Proof If A generates a C_0-group $(S(t))_{t \in \mathbb{R}}$ and $x \in D(A_+)$, then

$$\lim_{t \uparrow 0} \left\| \frac{1}{t}(S(t)x - x) - A_+ x \right\| = \lim_{t \uparrow 0} \left\| -\frac{1}{t}S(-1) \int_{1+t}^{1} S(s) A_+ x \, ds - A_+ x \right\| = 0.$$

Since also $\lim_{t \downarrow 0} \frac{1}{t}(S(t)x - x) = A_+ x$ it follows that $x \in D(A)$ and $Ax = A_+ x$. In combination with the inclusion $D(A) \subseteq D(A_+)$ it follows that $D(A) = D(A_+)$ and therefore $A = A_+$. In the same way one proves that $D(A) = D(A_-)$ and $A = -A_-$.

For the converse, suppose that A and $-A$ generate C_0-semigroups $(S_+(t))_{t \geq 0}$ and $(S_-(t))_{t \geq 0}$ respectively. By Lemma 13.6, for $x \in D(A) = D(-A)$ the function $t \mapsto S_-(t)S_+(t)x$ is continuously differentiable and

$$\frac{d}{dt} S_-(t)S_+(t)x = -AS_-(t)S_+(t)x + S_-(t)AS_+(t)x = 0,$$

where we used that $S(t)$ commutes with A. It follows from Proposition 1.45 that the function $t \mapsto S_-(t)S_+(t)x$ is constant, and evaluation at $t = 0$ shows that $S_-(t)S_+(t)x = x$ for all $t \geq 0$. Since $D(A)$ is dense this identity extends to arbitrary $x \in X$. This proves that $S_-(t)$ is a left inverse for $S_+(t)$. Interchanging the roles of $S_-(t)$ and $S_+(t)$ we find that $S_-(t)$ is also a right inverse for $S_+(t)$. As a result, $S_+(t)$ is invertible and $(S_+(t))^{-1} = S_-(t)$ for all $t \geq 0$. For $t \in \mathbb{R}$ define

$$S(t) := \begin{cases} S_+(t), & t \geq 0, \\ S_-(t), & t < 0. \end{cases}$$

With what we have proved it is trivial to verify that $(S(t))_{t \geq 0}$ is a C_0-group and that A is its generator. $\qquad\square$

Proposition 13.8, applied to the semigroups generated by $\pm A$, implies:

Corollary 13.14. *If A generates a uniformly bounded C_0-group on X, then $\sigma(A) \subseteq i\mathbb{R}$.*

The spectrum of the generator of a C_0-semigroup may be empty (an example is given in Problem 13.4). This is contrasted by the second part of the following result. For a uniformly bounded C_0-group S on X and $f \in L^1(\mathbb{R})$ we define $S(f) \in \mathcal{L}(X)$ by

$$S(f)x := \int_{-\infty}^{\infty} f(t)S(t)x \, dt, \quad x \in X. \tag{13.1}$$

Theorem 13.15. *If A generates a uniformly bounded C_0-group S on X, then:*

(1) *if the Fourier transform of a function $f \in L^1(\mathbb{R})$ is compactly supported and vanishes in a neighbourhood of $i\sigma(A)$, then $S(f) = 0$;*

(2) *if $X \neq \{0\}$, then $\sigma(A) \neq \varnothing$.*

Proof (1): For all $\delta > 0$ and $s \in \mathbb{R}$ we have $\pm\delta - is \in \rho(A)$, and for all $x \in X$ we have the identities

$$R(\delta - is, A)x = \int_0^\infty e^{-(\delta - is)t} S(t)x \, dt$$

and

$$R(-\delta - is, A)x = -R(\delta + is, -A) = -\int_0^\infty e^{-(\delta + is)t} S(-t)x \, dt.$$

Hence by dominated convergence, Fourier inversion (Theorem 5.20), Fubini's theorem, and Propositions 13.8 and 13.13,

$$
\begin{aligned}
S(f)x &= \lim_{\delta \downarrow 0} \int_{-\infty}^\infty e^{-\delta|t|} f(t) S(t)x \, dt \\
&= \frac{1}{\sqrt{2\pi}} \lim_{\delta \downarrow 0} \int_{-\infty}^\infty e^{-\delta|t|} \left(\int_{-\infty}^\infty e^{ist} \widehat{f}(s) \, ds \right) S(t)x \, dt \\
&= \frac{1}{\sqrt{2\pi}} \lim_{\delta \downarrow 0} \int_{-\infty}^\infty \widehat{f}(s) \left(\int_{-\infty}^\infty e^{-\delta|t|} e^{ist} S(t)x \, dt \right) ds \\
&= \frac{1}{\sqrt{2\pi}} \lim_{\delta \downarrow 0} \int_{-\infty}^\infty \widehat{f}(s) (R(\delta - is, A) - R(-\delta - is, A))x \, ds.
\end{aligned}
$$

By dominated convergence, this identity immediately implies (1).

(2): Suppose that $\sigma(A) = \varnothing$. The result of part (1) implies that $S(f) = 0$ for all $f \in L^1(\mathbb{R})$ whose Fourier transform has compact support. We claim that such functions are dense in $L^1(\mathbb{R})$. To see this, fix an arbitrary nonzero function $\phi \in C_c^\infty(\mathbb{R})$. Its inverse Fourier transform $\psi := \check{\phi}$ belongs to $L^1(\mathbb{R})$ (since $\phi^{(k)} \in L^1(\mathbb{R})$ implies that $|x|^k \psi(x)$ is bounded for all $k \in \mathbb{N}$). Since ψ is nonzero by the injectivity of the (inverse) Fourier transform, after multiplying with an appropriate scalar we may assume that $\int_\mathbb{R} \psi \, dx = 1$. By Proposition 2.34 we then have $\lim_{\varepsilon \downarrow 0} \psi_\varepsilon * f = f$ in $L^1(\mathbb{R})$, where $\psi_\varepsilon(x) := \varepsilon^{-d} \phi(\varepsilon^{-1}x)$, and the Fourier transforms $\widehat{\psi_\varepsilon * f} = \sqrt{2\pi} \widehat{\psi_\varepsilon} \widehat{f}$ are compactly supported. This proves the claim.

By approximation we obtain that $S(f) = 0$ for all $f \in L^1(\mathbb{R})$. In particular, taking $f_0(t) := e^{-t}$ for $t \geqslant 0$ and $f_0(t) := 0$ for $t < 0$, Proposition 13.8 implies that $R(1, A) = S(f_0) = 0$. Since $\mathsf{R}(R(1, A)) = \mathsf{D}(A)$ is dense in X, this implies that $X = \{0\}$. $\quad\square$

We will use this theorem to give a proof of Wiener's Tauberian theorem (Theorem 5.21). Recall that this theorem asserts that if the Fourier transform of a function $f \in L^1(\mathbb{R})$ is zero-free, then the span of the set of all translates of f is dense in $L^1(\mathbb{R})$.

We start with some preparations. If S is a uniformly bounded C_0-group on a Banach space X, we define

$$I_S := \{f \in L^1(\mathbb{R}) : S(f) = 0\},$$

where $S(f)$ is given by (13.1). The *Arveson spectrum* of S is the set

$$\mathrm{Sp}(S) := \{\omega \in \mathbb{R} : \widehat{f}(\omega) = 0 \text{ for all } f \in I_S\}.$$

The key to proving Wiener's Tauberian theorem is the following result, which is of independent interest.

Theorem 13.16. *Let S be a uniformly bounded C_0-group S with generator A on X. Then* $\mathrm{Sp}(S) = i\sigma(A)$.

Proof First let $\omega \in \mathbb{R}$ satisfy $\omega \notin i\sigma(A)$. Noting that $\sigma(A) \subseteq i\mathbb{R}$, we choose a function $f \in L^1(\mathbb{R})$ whose Fourier transform is compactly supported and vanishes in a neighbourhood of $i\sigma(A)$ but not on ω. By Theorem 13.15, $S(f) = 0$, so $f \in I_S$. But then $\widehat{f}(\omega) \neq 0$ implies that $\omega \notin \mathrm{Sp}(S)$.

Conversely, let $\omega \in i\sigma(A)$. Since $\sigma(A) \subseteq i\mathbb{R}$ and since the topological boundary of $\sigma(A)$ is always contained in the approximate point spectrum (see Section 10.1.c, where it was observed that the corresponding result for bounded operators, Proposition 6.17, extends to unbounded operators), $-i\omega$ is contained in the approximate point spectrum of A. Hence we may choose a sequence $(x_n)_{n \geqslant 1}$ of norm one vectors in X, with $x_n \in D(A)$ for all $n \geqslant 1$, such that $\lim_{n \to \infty} \|Ax_n + i\omega x_n\| \to 0$. In view of

$$S(t)x_n - e^{-i\omega t}x_n = \int_0^t e^{i\omega s} S(s)(A + i\omega)x_n \, ds \to 0 \text{ as } n \to \infty,$$

$(x_n)_{n \geqslant 1}$ is an approximate eigensequence of $S(t)$ with approximate eigenvalue $e^{-i\omega t}$.

Let $f \in L^1(\mathbb{R})$. By dominated convergence,

$$\lim_{n \to \infty} \int_{-\infty}^{\infty} f(t)(S(t)x_n - e^{-i\omega t}x_n) \, dt = 0.$$

Thus, using that $\|x_n\| = 1$,

$$\|S(f)\| \geqslant \lim_{n \to \infty} \|S(f)x_n\| = \lim_{n \to \infty} \left\| \int_{-\infty}^{\infty} f(t)S(t)x_n \, dt \right\| = \left| \int_{-\infty}^{\infty} e^{-i\omega t} f(t) \, dt \right| = |\widehat{f}(\omega)|.$$

This inequality implies that $\widehat{f}(\omega) = 0$ for all $f \in I_S$. Therefore, $\omega \in \mathrm{Sp}(S)$. $\qquad\square$

The *right translation group* is the C_0-group $U = (U(t))_{t \in \mathbb{R}}$ on $L^1(\mathbb{R})$ defined by

$$U(t)f(s) := f(s - t), \qquad s, t \in \mathbb{R}.$$

Note that $U(f)g = f * g$ for all $f, g \in L^1(\mathbb{R})$, where $*$ denotes convolution.

We are now ready for the proof of Wiener's Tauberian theorem.

Proof of Theorem 5.21 Let $f \in L^1(\mathbb{R})$ be a function whose Fourier transform is zero-free and let $X := \overline{\mathrm{span}}\{U(t)f : t \in \mathbb{R}\}$. We wish to prove that $X = L^1(\mathbb{R})$. Consider the quotient space $Y := L^1(\mathbb{R})/X$ and let $U_Y = (U_Y(t))_{t \in \mathbb{R}}$ denote the associated quotient translation group on Y. This group is strongly continuous and bounded. For all

$g \in L^1(\mathbb{R})$ we have $U(f)g = f * g = g * f = U(g)f$. By the translation invariance of X, $U(g)f \in X$. Hence $U(f)g \in X$, so $U_Y(f)(g + X) = 0$ for all $g \in L^1(\mathbb{R})$. It follows that $U_Y(f) = 0$. On the other hand, by assumption we have $\widehat{f}(\omega) \neq 0$ for all $\omega \in \mathbb{R}$. Therefore, $\mathrm{Sp}(U_Y) = \varnothing$. We conclude that $Y = \{0\}$ and $X = L^1(\mathbb{R})$. $\qquad\square$

13.2 The Hille–Yosida Theorem

The main theorem on generation of C_0-semigroups is the *Hille–Yosida theorem*, which gives necessary and sufficient conditions in terms of resolvent growth. We only need the version for contraction semigroups, which is somewhat easier to state and prove. Its extension to general semigroups can be done via the same reductions that will be used in the proof of Theorem 13.18 (see Problem 13.1).

Theorem 13.17 (Hille–Yosida). *For a densely defined closed linear operator A in X the following assertions are equivalent:*

(1) *A generates a C_0-semigroup of contractions on X;*
(2) *$\{\lambda \in \mathbb{C} : \mathrm{Re}\,\lambda > 0\} \subseteq \rho(A)$ and*

$$\|R(\lambda, A)\| \leqslant \frac{1}{\mathrm{Re}\,\lambda}, \quad \mathrm{Re}\,\lambda > 0;$$

(3) *$\{\lambda \in \mathbb{R} : \lambda > 0\} \subseteq \rho(A)$ and*

$$\|R(\lambda, A)\| \leqslant \frac{1}{\lambda}, \quad \lambda > 0.$$

Proof The implication (1)$\Rightarrow$(2) follows from Propositions 13.4 and 13.8, and the implication (2)$\Rightarrow$(3) is trivial.

Assume now that (3) holds. For the bounded operators $A_n := nAR(n, A) = n^2 R(n, A) - nI$, $n \geqslant 1$, Proposition 13.10 implies $\lim_{n \to \infty} A_n x = Ax$ for all $x \in D(A)$. Also,

$$\|e^{tA_n}\| \leqslant e^{n^2 \|R(n,A)\| t} e^{-nt} \leqslant e^{nt} e^{-nt} = 1. \tag{13.2}$$

Fix $x \in D(A)$ and $t \geqslant 0$. The identity

$$e^{tA_n} x - e^{tA_m} x = \int_0^t \frac{\mathrm{d}}{\mathrm{d}s} [e^{(t-s)A_m} e^{sA_n} x] \, \mathrm{d}s = \int_0^t e^{(t-s)A_m} e^{sA_n} (A_n x - A_m x) \, \mathrm{d}s$$

and the contractivity estimate (13.2) imply

$$\|e^{tA_n} x - e^{tA_m} x\| \leqslant t \|A_n x - A_m x\|,$$

and therefore $(e^{tA_n} x)_{n \geqslant 1}$ is Cauchy in X for all $x \in D(A)$. Hence, the limit $S(t)x := \lim_{n \to \infty} e^{tA_n} x$ exists for all $x \in D(A)$. By the uniform boundedness of the operators e^{tA_n} guaranteed by (13.2), this limit in fact exists for all $x \in X$. Moreover, for each $t \geqslant 0$

the resulting mapping $x \mapsto S(t)x$ is linear and contractive. It remains to verify that the contractions $S(t)$, $t \geqslant 0$, form a C_0-semigroup on X and that A is its generator.

It is clear that $S(0) = I$. The semigroup property follows from

$$S(t)S(s)x = \lim_{n\to\infty} e^{tA_n}e^{sA_n}x = \lim_{n\to\infty} e^{(t+s)A_n}x = S(t+s)x,$$

using the uniform boundedness of the sequence $(e^{tA_n})_{n\geqslant 1}$ in the first equality and the properties of the power series of the exponential function in the second.

Next we prove the strong continuity. For $x \in D(A)$ we have

$$S(t)x - x = \lim_{n\to\infty} e^{tA_n}x - x = \lim_{n\to\infty} \int_0^t e^{sA_n}A_n x \, ds = \int_0^t S(s)Ax \, ds,$$

where we used that

$$\|e^{sA_n}A_n x - S(s)Ax\| \leqslant \|e^{sA_n}(A_n x - Ax)\| + \|(e^{sA_n} - S(s))Ax\|$$

$$\leqslant \|(A_n x - Ax)\| + \|(e^{sA_n} - S(s))Ax\| \to 0.$$

Therefore, for $x \in D(A)$,

$$\lim_{t\downarrow 0} S(t)x - x = \lim_{t\downarrow 0} \int_0^t S(s)Ax \, ds = 0.$$

Once again the strong continuity for general $x \in X$ follows from this by approximation.

It remains to check that A equals the generator of S, which we denote by B. By what we have already proved, for $x \in D(A)$ we have

$$\lim_{t\downarrow 0} \frac{1}{t}(S(t)x - x) = \lim_{t\downarrow 0} \frac{1}{t} \int_0^t S(s)Ax \, ds = Ax,$$

so $x \in D(B)$ and $Bx = Ax$. Since both A and B are closed and share a half-line in their resolvent sets, Proposition 10.30 implies that $A = B$. $\qquad\square$

As an application we have the following perturbation result.

Theorem 13.18 (Perturbation). *Let A be the generator of a C_0-semigroup S on X and let B be a bounded operator on X, then $A + B$ generates a C_0-semigroup on X.*

Here it is understood that $D(A + B) = D(A)$ and $(A + B)x = Ax + Bx$ for $x \in D(A)$. The proof of the theorem shows that if $\|S(t)\| \leqslant Me^{\omega t}$, then $\|S_B(t)\| \leqslant Me^{(\omega + \|B\|)t}$.

Proof We prove the theorem in three steps. We begin with two reductions.

Step 1 – Choose $M \geqslant 1$ and $\omega \in \mathbb{R}$ such that $\|S(t)\| \leqslant Me^{\omega t}$ for all $t \geqslant 0$. The operator $A - \omega$ is the generator of the C_0-semigroup $(e^{-\omega t}S(t))_{t\geqslant 0}$, and this semigroup satisfies $\|e^{-\omega t}S(t)\| \leqslant M$. Since

$$A + B = (A - \omega) + (B + \omega)$$

and $B + \omega$ is bounded, this argument shows that it is enough to prove the theorem for *uniformly bounded* semigroups.

Step 2 – We now assume that the semigroup generated by A is uniformly bounded, say by a constant $M \geqslant 1$. From

$$\|x\| \leqslant \sup_{t \geqslant 0} \|S(t)x\| \leqslant M\|x\|$$

it follows that

$$\|\|x\|\| := \sup_{t \geqslant 0} \|S(t)x\|$$

defines an equivalent norm on X. With respect to this norm, for all $x \in X$ we have

$$\|\|S(s)x\|\| = \sup_{t \geqslant 0} \|S(s+t)x\| \leqslant \sup_{r \geqslant 0} \|S(r)x\| = \|\|x\|\|.$$

This argument shows that we may assume that our semigroup is a *semigroup of contractions*.

Step 3 – By the previous two steps it suffices to prove the theorem for generators of contraction semigroups.

Fix $\lambda \in \mathbb{C}$ with $\mathrm{Re}\,\lambda > 0$. Then $\lambda \in \rho(A)$ and $\|R(\lambda, A)\| \leqslant (\mathrm{Re}\,\lambda)^{-1}$. Because

$$(\lambda - (A+B)) = (I - BR(\lambda, A))(\lambda - A)$$

and $\|BR(\lambda, A)\| \leqslant (\mathrm{Re}\,\lambda)^{-1}\|B\|$, for $\mathrm{Re}\,\lambda > \|B\|$ the operator $I - BR(\lambda, A)$ is invertible, and the Neumann series for its inverse gives

$$\|R(\lambda, A)\|\|(I - BR(\lambda, A))^{-1}\| \leqslant (\mathrm{Re}\,\lambda)^{-1}(I - (\mathrm{Re}\,\lambda)^{-1}\|B\|)^{-1} = (\mathrm{Re}\,\lambda - \|B\|)^{-1}.$$

Hence, for $\mathrm{Re}\,\lambda > \|B\|$, the operator $\lambda - (A+B)$ is invertible and its inverse satisfies

$$\|R(\lambda, A+B)\| \leqslant (\mathrm{Re}\,\lambda - \|B\|)^{-1}.$$

The operator $A + B - \|B\|$ is then invertible for $\mathrm{Re}\,\lambda > 0$ and satisfies

$$\|(\lambda - (A+B - \|B\|))\| \leqslant (\mathrm{Re}\,\lambda)^{-1}.$$

By the Hille–Yosida theorem this operator generates a C_0-semigroup T of contractions. Then $A + B$ generates the C_0-semigroup given by $S_B(t) := e^{t\|B\|}T(t)$.

Clearly, $\|S_B(t)\| \leqslant e^{t\|B\|}$. Remembering that we made two reductions, reversing them gives the estimate given after the statement of the theorem. $\square$

In general there is no closed-form expression for $S_B(t)$, but we do have the so-called *variation of constants identity*

$$S_B(t)x = S(t)x + \int_0^t S(t-s)BS_B(s)x\,\mathrm{d}s.$$

The proof of this identity is simple: for $x \in D(A) = D(A+B)$, using Lemma 13.6 we differentiate the function $\phi(s) = S(t-s)S_B(s)x$ using the product rule and get

$$\phi'(s) = -AS(t-s)S_B(s)x + S(t-s)(A+B)S_B(s) = S(t-s)BS_B(s)x.$$

Integrating this identity over the interval $[0,t]$ gives the required result.

As a consequence of this identity we see that the norm of the difference is of the order

$$\|S(t) - S_B(t)\| = O(t) \quad \text{as } t \downarrow 0.$$

We continue with a useful approximation formula, by means of which it is possible to deduce information about the semigroup from information about the properties of the resolvent along the positive real line. It will be used later on to prove the positivity of the heat semigroup under Dirichlet and Neumann boundary conditions.

To motivate the result we recall Euler's formula for the exponential, which entails that for all $a \in \mathbb{R}$ and $t \geqslant 0$,

$$e^{ta} = \lim_{n\to\infty} \left(1 - \frac{t}{n}a\right)^{-n} = \lim_{n\to\infty} \left(\frac{n}{t}\left(\frac{n}{t} - a\right)^{-1}\right)^n.$$

Theorem 13.19 (Euler's formula). *Let A be the generator of a C_0-semigroup S on X. Then for all $x \in X$ and $t > 0$ we have*

$$S(t)x = \lim_{n\to\infty} \left(\frac{n}{t}R(\frac{n}{t},A)\right)^n x.$$

Proof By Proposition 10.28 the resolvent of A is holomorphic with complex derivative given by $\frac{d}{d\lambda}R(\lambda,A) = -R(\lambda,A)^2$. By induction this implies

$$\frac{d^n}{d\lambda^n}R(\lambda,A) = (-1)^n n! R(\lambda,A)^{n+1}. \tag{13.3}$$

On the other hand, repeated differentiation under the integral in the Laplace transform representation $R(\lambda,A)x = \int_0^\infty e^{-\lambda s}S(s)x\,ds$ gives

$$\frac{d^n}{d\lambda^n}R(\lambda,A)x = (-1)^n \int_0^\infty s^n e^{-\lambda s}S(s)x\,ds.$$

Substituting $s = rt$ and specialising to $\lambda = n/t$ we obtain

$$\frac{d^n}{d\lambda^n}R(\lambda,A)x\bigg|_{\lambda=n/t} = (-1)^n t^{n+1}\int_0^\infty (re^{-r})^n S(rt)x\,dr. \tag{13.4}$$

Combining (13.3) and (13.4), and using the identity

$$\frac{n^{n+1}}{n!}\int_0^\infty (re^{-r})^n \, dt = 1, \tag{13.5}$$

we arrive at

$$\left(\frac{n}{t}R(\frac{n}{t},A)\right)^{n+1}x - S(t)x = \frac{n^{n+1}}{n!}\int_0^\infty (re^{-r})^n S(rt)x\,dr - S(t)x$$

$$= \frac{n^{n+1}}{n!} \int_0^\infty (re^{-r})^n (S(rt)x - S(t)x)\, dr.$$

Fixing $x \in X$ and $\varepsilon > 0$, by strong continuity we may choose $0 < a < 1 < b < \infty$ in such a way that

$$\sup_{a \leqslant r \leqslant b} \|S(rt)x - S(t)x\| < \varepsilon.$$

We split the integral into three parts I_1, I_2, and I_3 corresponding to $[0,a]$, $[a,b]$, and $[b,\infty)$ and estimate each part separately, using that $u \mapsto ue^{-u}$ is increasing on $[0,1]$ and decreasing on $[1,\infty)$. For the first integral, using the elementary bound

$$\frac{n^n}{n!} \leqslant \frac{e^n}{\sqrt{2\pi n}} \tag{13.6}$$

we obtain

$$\|I_1\| \leqslant \frac{n^{n+1}}{n!}(ae^{-a})^n \int_0^a \|S(rt)x - S(t)x\|\, dr \leqslant \frac{1}{\sqrt{2\pi}} n^{1/2} e^n (ae^{-a})^n \cdot 2 \sup_{0 \leqslant s \leqslant t} \|S(s)\| \|x\|,$$

which tends to 0 as $n \to \infty$ since $ae^{-a} < e^{-1}$. Next, using (13.5),

$$\|I_2\| \leqslant \frac{n^{n+1}}{n!} \int_a^b (re^{-r})^n \varepsilon\, dr \leqslant \varepsilon \frac{n^{n+1}}{n!} \int_0^\infty (re^{-r})^n\, dr = \varepsilon.$$

To estimate I_3 we choose $M \geqslant 1$ and $\omega \in \mathbb{R}$ such that $\|S(t)\| \leqslant Me^{\omega t}$. Choose $0 < \delta < 1$ so small that $be^{1-b(1-\delta)} < 1$; this is possible since $be^{-b} < e^{-1}$. For all $n \geqslant \delta^{-1}(1 + |\omega| \max\{t,1\})$, by (13.6) we have

$$\|I_3\| \leqslant \frac{n^{n+1}}{n!} \int_b^\infty r^n e^{-r(1-\delta)n} e^{-r(1+|\omega|\max\{t,1\})} M(e^{\omega rt} + e^{\omega r}) \|x\|\, dr$$

$$\leqslant \frac{n^{n+1}}{n!} \frac{1}{(1-\delta)^n} \cdot 2M\|x\| \int_b^\infty \left(r(1-\delta)\right) e^{-r(1-\delta)})^n e^{-r}\, dr$$

$$\leqslant \frac{e^n}{\sqrt{2\pi}} n^{1/2} (be^{-b(1-\delta)})^n \cdot 2M\|x\| \int_b^\infty e^{-r}\, dr$$

$$\leqslant \frac{1}{\sqrt{2\pi}} n^{1/2} (be^{1-b(1-\delta)})^n \cdot 2M\|x\|,$$

which tends to 0 as $n \to \infty$ by the choice of δ; we used the monotonicity of $u \mapsto ue^{-u}$ on $[1,\infty)$ to bound $(r(1-\delta))e^{-r(1-\delta)}$ by $(b(1-\delta))e^{-b(1-\delta)}$.

Collecting the estimates, we have shown that

$$\lim_{n \to \infty} \left(\frac{n}{t} R(\frac{n}{t}, A)\right)^{n+1} x = S(t)x.$$

This is almost the result we want, except for the power $n+1$ instead of n. To correct for this we argue as follows. By Proposition 13.8,

$$\left\| \left(\frac{n}{t}R(\frac{n}{t},A)\right)^{n+1} x - \left(\frac{n}{t}R(\frac{n}{t},A)\right)^{n} x \right\| \leqslant \left\| \left(\frac{n}{t}R(\frac{n}{t},A)\right)^{n} \left(\frac{n}{t}R(\frac{n}{t},A)x - x\right) \right\|$$

$$\leqslant M\left(1 - \frac{t}{n}|\omega|\right)^{-n} \left\| \frac{n}{t}R(\frac{n}{t},A)x - x \right\|.$$

As $n \to \infty$, by Euler's formula and Proposition 13.10 we have

$$\left(1 - \frac{t}{n}|\omega|\right)^{-n} \to e^{|\omega|t}, \quad \left\| \frac{n}{t}R(\frac{n}{t},A)x - x \right\| \to 0,$$

and therefore

$$\lim_{n\to\infty} \left(\frac{n}{t}R(\frac{n}{t},A)\right)^{n} x = \lim_{n\to\infty} \left(\frac{n}{t}R(\frac{n}{t},A)\right)^{n+1} x = S(t)x.$$

□

We continue with a simple result about compact semigroups.

Proposition 13.20 (Compact semigroups)**.** *Let A be the generator of a C_0-semigroup S on X. If $S(t)$ is a compact operator for every $t > 0$, then:*

(1) *the semigroup is uniformly continuous for $t > 0$;*
(2) *the resolvent operator $R(\lambda,A)$ is compact for every $\lambda \in \rho(A)$;*
(3) *the spectrum of A is finite or countable and consists of isolated eigenvalues, and the corresponding eigenspaces are finite-dimensional;*
(4) *for all $t > 0$ we have the* spectral mapping formula

$$\sigma(S(t)) \setminus \{0\} = \exp(t\sigma(A)).$$

Moreover, the eigenspaces corresponding to $\lambda \in \sigma(A)$ and $e^{\lambda t} \in \sigma(S(t))$ coincide.

Proof (1): Fix $s > 0$. For $t > s/2$ we have

$$\|S(t) - S(s)\| = \sup_{\|x\|\leqslant 1} \|(S(t - s/2) - S(s/2))S(s/2)x\|.$$

Since $S(s/2)\overline{B}_X$ is relatively compact, by Proposition 1.42 this implies $\lim_{t\to s} \|S(t) - S(s)\| = 0$.

(2): Choose $M \geqslant 1$ and $\omega \in \mathbb{R}$ such that $\|S(t)\| \leqslant Me^{\omega t}$ for all $t \geqslant 0$. We first claim that the compactness of the semigroup operators $S(t)$ for $t > 0$ implies that $R(\mu,A)$ is compact for all $\operatorname{Re}\mu > \omega$. For all $x \in X$ and $t > 0$ we have

$$S(t)R(\mu,A)x - R(\mu,A)x = \int_0^t S(s)AR(\mu,A)x,$$

and therefore

$$\|S(t)R(\mu,A) - R(\mu,A)\| \leqslant t \sup_{s\in[0,t]} \|S(s)\|\|AR(\mu,A)\|,$$

where $AR(\mu,A) = \mu R(\mu,A) - I$ is a bounded operator. Since $S(t)R(\mu,A)$ is compact for every $t > 0$, from Proposition 7.5 we obtain that $R(\mu,A)$ is compact. This proves the claim. The compactness of $R(\lambda,A)$ for arbitrary $\lambda \in \rho(A)$ now follows from the resolvent identity (10.2).

(3): This follows from Lemma 12.25.

(4): This follows from (3) and the next proposition. $\qquad\qquad\qquad\qquad\qquad\square$

In the next proposition we denote by $\sigma_p(B)$ the *point spectrum* of a bounded or unbounded operator B, that is, the set of its eigenvalues.

Proposition 13.21 (Spectral mapping theorem for the point spectrum). *Let A be the generator of a C_0-semigroup S on X. Then*

$$\sigma_p(S(t))\backslash\{0\} = \exp\big(t\sigma_p(A)\big), \quad t \geqslant 0.$$

Moreover, the eigenspaces corresponding to $\lambda \in \sigma_p(A)$ and $e^{\lambda t} \in \sigma_p(S(t))$ coincide.

Proof If $x \in D(A)$ is an eigenvector of A corresponding to the eigenvalue λ, the identity

$$\int_0^t e^{\lambda(t-s)}S(s)(\lambda - A)x\,\mathrm{d}s = (e^{\lambda t} - S(t))x$$

shows that $S(t)x = e^{\lambda t}x$, that is, $e^{\lambda t}$ is an eigenvalue of $S(t)$ with eigenvector x. This proves the inclusion $\sigma_p(S(t))\backslash\{0\} \supseteq \exp\big(t\sigma_p(A)\big)$.

The inclusion $\sigma_p(S(t))\backslash\{0\} \subseteq \exp\big(t\sigma_p(A)\big)$ is proved as follows. Fix $t > 0$ and suppose that $x \in X$ is an eigenvector of $S(t)$ corresponding to a nonzero eigenvalue μ. Then $\mu = e^{\lambda t}$ for some $\lambda \in \mathbb{C}$. The identity $S(t)x = e^{\lambda t}x$ implies that the map $s \mapsto e^{-\lambda s}S(s)x$ is periodic with period t. Since this map is not identically zero, the uniqueness theorem for the Fourier transform implies that (after scaling the interval $[0,t]$ to $[0,2\pi]$) at least one of its Fourier coefficients is nonzero. Thus, there exists an integer $k \in \mathbb{Z}$ such that with $\lambda_k := \lambda + 2\pi ik/t$ we have

$$x_k := \frac{1}{t}\int_0^t e^{-\lambda_k s}S(s)x\,\mathrm{d}s \neq 0.$$

We will show that λ_k is an eigenvalue of A with eigenvector x_k.

Choose $M \geqslant 1$ and $\omega \in \mathbb{R}$ such that $\|S(t)\| \leqslant Me^{\omega t}$. By the t-periodicity of $s \mapsto e^{-\lambda s}S(s)x$, for all $\mathrm{Re}\,\nu > \omega$ we have

$$R(\nu,A)x = \int_0^\infty e^{-\nu s}S(s)x\,\mathrm{d}s = \sum_{n=0}^\infty \int_{nt}^{(n+1)t} e^{-\nu s}S(s)x\,\mathrm{d}s$$

$$= \sum_{n=0}^\infty \int_0^t e^{-\nu s}S(s)(e^{-\nu nt}S(nt)x)\,\mathrm{d}s = \sum_{n=0}^\infty e^{(\lambda-\nu)nt}\int_0^t e^{-\nu s}S(s)x\,\mathrm{d}s \quad (13.7)$$

$$= \frac{1}{1-e^{(\lambda-\nu)t}}\int_0^t e^{-\nu s}S(s)x\,\mathrm{d}s = \frac{1}{1-e^{(\lambda_k-\nu)t}}\int_0^t e^{-\nu s}S(s)x\,\mathrm{d}s.$$

Since the integral on the right-hand side is an entire function of the variable v, this shows that the map $v \mapsto R(v,A)x$ can be holomorphically extended to $\mathbb{C}\backslash\{\lambda + 2\pi in/t : n \in \mathbb{Z}\}$. Denoting this extension by F, by (13.7) and the definition of x_k we have

$$\lim_{v \to \lambda_k} (v - \lambda_k)F(v) = x_k.$$

Also, by (13.7) and the t-periodicity of $s \mapsto e^{-\lambda s}S(s)x$,

$$\lim_{v \to \lambda_k} (\lambda_k - A)(v - \lambda_k)F(v)$$

$$= \lim_{v \to \lambda_k} \frac{v - \lambda_k}{1 - e^{(\lambda_k - v)t}} \left((I - e^{-vt}S(t))x + (\lambda_k - v) \int_0^t e^{-vs}S(s)x\,ds \right) = \frac{1}{t}(0+0) = 0.$$

From the closedness of A it follows that $x_k \in D(A)$ and $(\lambda_k - A)x_k = 0$.

It remains to prove the final statement on coincidence of eigenspaces. Let us denote the eigenspaces corresponding to $\lambda \in \sigma_p(A)$ and $e^{\lambda t} \in \sigma_p(S(t))$ by E_λ and $E_{t,\lambda}$, respectively. The first part of the proof shows that $E_\lambda \subseteq E_{t,\lambda}$. Denote by F_λ and $F_{t,\lambda}$ the closed linear spans of $\{S(t)x : x \in E_\lambda\}$ and $\{S(t)x : x \in E_{t,\lambda}\}$. Then $E_\lambda = F_\lambda \subseteq F_{t,\lambda}$, and the second part of the proof shows that $F_{t,\lambda} \subseteq E_\lambda$ (because the vector x_k belongs to $F_{t,\lambda}$). Putting these inclusions together, we obtain

$$E_\lambda = F_\lambda \subseteq F_{t,\lambda} \subseteq E_\lambda \quad \text{and} \quad E_\lambda \subseteq E_{t,\lambda} \subseteq F_{t,\lambda} \subseteq E_\lambda,$$

and therefore all these subspaces coincide.   □

13.3 The Abstract Cauchy Problem

Having set up the general theory of C_0-semigroups, it is time to put them to use in solving abstract Cauchy problems.

13.3.a The Inhomogeneous Cauchy Problem

If A is the generator of a C_0-semigroup S on X, then by Proposition 13.4 for initial values $u_0 \in D(A)$ the function

$$u(t) := S(t)u_0, \quad t \geqslant 0, \tag{13.8}$$

solves the initial value problem (ACP),

$$\begin{cases} u'(t) = Au(t), & t \in [0,T], \\ u(0) = u_0, \end{cases}$$

in the sense that u is continuously differentiable, takes values in $D(A)$, and satisfies the equation pointwise in time. A function u with these properties is called a *classical solution.* However, the definition (13.8) makes sense for arbitrary $u_0 \in X$, not just for $u_0 \in D(A)$, and for all $u_0 \in X$ the function $u(t) = S(t)u_0$ solves the following integrated version of (ACP):

$$u(t) = u_0 + A \int_0^t u(s)\,\mathrm{d}s, \quad t \in [0, T]. \tag{13.9}$$

Indeed, by Proposition 13.4(3), for arbitrary $u_0 \in X$ we have $\int_0^t S(s)u_0\,\mathrm{d}s \in D(A)$ and $A \int_0^t S(s)u_0\,\mathrm{d}s = S(t)u_0 - u_0$, confirming that (13.9) holds for $u(t) = S(t)u_0$. This observation leads to the notion of *strong solution* which we develop next in the more general context of the inhomogeneous Cauchy problem

$$\begin{cases} u'(t) = Au(t) + f(t), & t \in [0, T], \\ u(0) = u_0, \end{cases} \tag{IACP}$$

with initial value $u_0 \in X$. We assume that f belongs to $L^1(0, T; X)$, the space of all strongly measurable functions $f : (0, T) \to X$ such that

$$\|f\|_1 := \int_0^T \|f(t)\|\,\mathrm{d}t < \infty,$$

identifying functions that are equal almost everywhere. In same way as in the scalar-valued case one shows that $L^1(0, T; X)$ is Banach space.

Definition 13.22 (Strong solutions). A *strong solution* of (IACP) is a continuous function $u : [0, T] \to X$ such that for all $t \in [0, T]$ we have $\int_0^t u(s)\,\mathrm{d}s \in D(A)$ and

$$u(t) = u_0 + A \int_0^t u(s)\,\mathrm{d}s + \int_0^t f(s)\,\mathrm{d}s.$$

We proceed with an existence and uniqueness result for strong solutions of (IACP). It is based on the following lemma.

Lemma 13.23. *Let $f \in L^1(0, T; X)$. Then:*

(1) *for all $t \in [0, T]$ the function $s \mapsto S(t - s)f(s)$ has a strongly measurable representative and is integrable on $[0, t]$;*

(2) *the function $t \mapsto \int_0^t S(t - s)f(s)\,\mathrm{d}s$ is continuous on $[0, T]$.*

Proof (1): Choose a strongly measurable representative for f, which we denote again by f, as well as a sequence of simple functions f_n converging to f pointwise. Each function $s \mapsto S(t - s)f_n(s)$ is strongly measurable, since it is a linear combination of functions of the form $s \mapsto \mathbf{1}_B(s)S(t - s)x$ with $B \subseteq [0, T]$ a Borel subset, and such functions are strongly measurable because continuous functions on an interval are strongly measurable and if g is strongly measurable and B is a Borel set, then $\mathbf{1}_B g$ is strongly measurable. By Proposition 1.48, the pointwise limit $s \mapsto S(t-s)f(s)$ is strongly measurable. Integrability follows from the estimate $\|S(t-s)f(s)\| \leqslant M\|f(s)\|$, where $M = \sup_{t \in [0,T]} \|S(t)\|$, and the integrability of f.

(2): Let $0 \leqslant t \leqslant t' \leqslant T$. Then

$$\left\| \int_0^{t'} S(t'-s)f(s)\,ds - \int_0^t S(t-s)f(s)\,ds \right\|$$

$$\leqslant \left\| \int_t^{t'} S(t'-s)f(s)\,ds \right\| + \left\| \int_0^t S(t'-s)f(s)\,ds - \int_0^t S(t-s)f(s)\,ds \right\|.$$

The first term on the right-hand side can be bounded above by $M \int_t^{t'} \|f(s)\|\,ds$ which tends to 0 by dominated convergence as $t' - t \to 0$. The second term tends to 0 by dominated convergence as well: for simple functions f this follows from the strong continuity and local boundedness of the semigroup, and for general $f \in L^1(0,T;X)$ this follows by approximation by simple functions in the $L^1(0,T;X)$-norm. $\qquad\square$

Theorem 13.24 (Existence and uniqueness). *For all $u_0 \in X$ and $f \in L^1(0,T;X)$ the problem* (IACP) *admits a unique strong solution u. It is given by the convolution formula*

$$u(t) = S(t)u_0 + \int_0^t S(t-s)f(s)\,ds. \tag{13.10}$$

If $f \in L^p(0,T;X)$ with $1 \leqslant p < \infty$, then $u \in L^p(0,T;X)$.

The function $t \mapsto S(t)u_0 + \int_0^t S(t-s)f(s)\,ds$ is usually referred to as the *mild solution* of (IACP).

Proof For the existence part we will show that the right-hand side of (13.10) defines a strong solution. By Lemma 13.23, this function is continuous.

We begin by showing that $\int_0^t u(s)\,ds \in D(A)$. We have $\int_0^t S(s)u_0\,ds \in D(A)$ by Proposition 13.4. To prove that $\int_0^t \int_0^s S(s-r)f(r)\,dr\,ds \in D(A)$ we apply the definition of A. As $h \downarrow 0$ we have, using Fubini's theorem,

$$\frac{1}{h}(S(h)-I) \int_0^t \int_0^s S(s-r)f(r)\,dr\,ds = \int_0^t \frac{1}{h}(S(h)-I) \int_r^t S(s-r)f(r)\,ds\,dr$$

$$= \int_0^t \frac{1}{h}(S(h)-I) \int_0^{t-r} S(s)f(r)\,ds\,dr$$

$$\to \int_0^t A \int_0^{t-r} S(s)f(r)\,ds\,dr$$

$$= \int_0^t S(t-r)f(r) - f(r)\,dr$$

$$= u(t) - S(t)u_0 - \int_0^t f(r)\,dr.$$

The convergence is justified by the dominated convergence theorem since for all $0 < h < 1$ we have the following pointwise bound with respect to the variable r:

$$\left\| \frac{1}{h}(S(h) - I) \int_0^{t-r} S(s)f(r)\,ds \right\| = \frac{1}{h} \left\| \int_{t-r}^{t-r+h} S(s)f(r)\,ds - \int_0^h S(s)f(r)\,ds \right\|$$

$$\leqslant \frac{1}{h} \int_{t-r}^{t-r+h} \|S(s)f(r)\|\,ds + \frac{1}{h} \int_{t-r}^{t-r+h} \|S(s)f(r)\|\,ds$$

$$\leqslant 2M_t \|f(r)\|,$$

where $M_t := \sup_{0 \leqslant \tau \leqslant t+1} \|S(\tau)\|$.

The above computation shows that $\int_0^t u(s)\,ds \in D(A)$ and

$$A \int_0^t u(s)\,ds = A \int_0^t S(s)u_0\,ds + A \int_0^t \int_0^s S(s - r)f(r)\,dr\,ds$$

$$= (S(t)u_0 - u_0) + \left(u(t) - S(t)u_0 - \int_0^t f(r)\,dr \right).$$

This shows that the function u given by (13.10) is a strong solution.

To prove uniqueness, suppose that u and $\widetilde{u}$ are strong solutions of (IACP). It follows from the definition that u and $\widetilde{u}$ are continuous. Set $v := u - \widetilde{u}$. Then v is continuous, $\int_0^t v(s)\,ds \in D(A)$, and $v(t) = A \int_0^t v(s)\,ds$ for all $t \in [0,T]$.

Fix $0 < t \leqslant T$ and define $w : [0,t] \to X$ by $w(s) = S(t - s) \int_0^s v(r)\,dr$. This function is differentiable with derivative

$$w'(s) = S(t - s)v(s) - S(t - s)A \int_0^s v(r)\,dr = S(t - s)v(s) - S(t - s)v(s) = 0.$$

It follows that w is constant. Hence

$$\int_0^t v(r)\,dr = w(t) = w(0) = 0.$$

Since this is true for all $0 < t \leqslant T$ and v is continuous, it follows that $v(r) = 0$ for all $r \in [0,T]$.

The final assertion is a consequence of Young's inequality. $\qquad\square$

The solution u depends continuously on u_0 in the norm of $C([0,T];X)$, the Banach space of all continuous functions from $[0,T]$ to X. Indeed, if $\widetilde{u}_0$ is another initial value and the corresponding unique strong solution is denoted by $\widetilde{u}$, then

$$\|u(t) - \widetilde{u}(t)\| \leqslant \|S(t)\| \|u_0 - \widetilde{u}_0\| \leqslant M \|u_0 - \widetilde{u}_0\|,$$

where $M := \sup_{t \in [0,T]} \|S(t)\|$, and therefore

$$\|u - \widetilde{u}\|_\infty \leqslant M \|u_0 - \widetilde{u}_0\|.$$

Unique solvability plus continuous dependence on the initial value is usually summarised as *well-posedness*. Thus, the inhomogeneous problem (IACP) is well posed for strong solutions.

13.3.b The Semilinear Cauchy Problem

In this section we study a class of nonlinear evolution equations of the form

$$\begin{cases} u'(t) = Au(t) + f(t, u(t)), & t \in [0, T], \\ u(0) = u_0. \end{cases} \tag{SCP}$$

Equations of this form are referred to as *semilinear* equations. We assume that A generates a C_0-semigroup S on X and that the initial value u_0 lies in X. We make the following assumptions on the function $f : [0, T] \times X \to X$:

(i) (Strong measurability) for all $x \in X$ the function $t \mapsto f(t, x)$ is strongly measurable on $[0, T]$;

(ii) (Linear growth) there exists a constant $C \geqslant 0$ such that

$$\|f(t, x)\| \leqslant C(1 + \|x\|), \quad t \in [0, T], \ x \in X;$$

(iii) (Lipschitz continuity) there exists a constant $L \geqslant 0$ such that

$$\|f(t, x) - f(t, x')\| \leqslant L\|x - x'\|, \quad t \in [0, T], \ x, x' \in X.$$

Under these assumptions, in force throughout this section, we will prove existence, uniqueness, and continuous dependence on the initial conditions of mild solutions. Thus (SCP) is well posed for mild solutions.

Definition 13.25 (Mild solutions). A function $u : [0, T] \to X$ is called a *mild solution* of (SCP) if it is continuous and satisfies

$$u(t) = S(t)u_0 + \int_0^t S(t - s)f(s, u(s)) \, ds, \quad t \in [0, T].$$

To see that this is well defined we must check that the integral converges as a Bochner integral in X. Lemma 13.26 takes care of this. Taking the lemma for granted for the moment, let us first motivate the definition.

First of all, it generalises the formula of Theorem 13.24 for the strong solution of the inhomogeneous problem. Perhaps more importantly, we shall prove that every *classical solution is a mild solution*. To prepare for this, suppose that $u : [0, T] \to X$ is not just continuous but continuously differentiable and takes values in $D(A)$. Then it makes sense to ask whether u satisfies (SCP) in a pointwise sense. If it does, we call u a *classical solution*. Let us assume that this is the case. Multiplying the equation for $s \in [0, t]$ on both sides with $S(t - s)$ and integrating, we obtain

$$\int_0^t S(t - s)u'(s) \, ds = \int_0^t S(t - s)Au(s) \, ds + \int_0^t S(t - s)f(s, u(s)) \, ds.$$

On the other hand, an integration by parts, using that $u(0) = u_0$ and $S'(t)x = S(t)Ax$ for $x \in D(A)$, gives the identity

$$\int_0^t S(t-s)u'(s)\,ds = u(t) - S(t)u_0 + \int_0^t S(t-s)Au(s)\,ds.$$

Substituting this identity into the preceding one, the identity defining a mild solution is obtained.

In general there is no reason to expect existence of classical solutions, but, under the standing assumptions (i)–(iii) formulated above, a unique mild solution always exists. In the definition of a mild solution, no differentiability or $D(A)$-valuedness is imposed, and this is precisely what makes things work.

As promised we now check that the integral in Definition 13.25 is well defined as a Bochner integral in X. The next result extends Lemma 13.23 to the present situation.

Lemma 13.26. *Let* $f : [0,T] \times X \to X$ *satisfy the conditions* (i)–(iii) *and suppose that* $u : [0,T] \to X$ *is continuous. Then:*

(1) *the functions* $s \mapsto f(s,u(s))$ *and* $s \mapsto S(t-s)f(s,u(s))$ *have strongly measurable representatives and are integrable;*
(2) *the function* $t \mapsto \int_0^t S(t-s)f(s,u(s))\,ds$ *is continuous on* $[0,T]$.

Proof (1): First let $v = \sum_{j=1}^k \mathbf{1}_{I_j} \otimes x_j$ be an X-valued step function, where the intervals $I_j \subseteq [0,T]$ are disjoint. If $s \in I_j$, then $f(s,v(s)) = f(s,x_j)$ and therefore $s \mapsto v(s,f(s))$ belongs to $L^1(0,T;X)$, with

$$\int_0^T \|f(s,v(s))\|\,ds = \sum_{j=1}^k \int_{I_j} \|f(s,x_j)\|\,ds \leqslant C \sum_{j=1}^k |I_j|(1 + \|x_j\|)$$

using the linear growth assumption. If v' is another X-valued step function, from the Lipschitz continuity assumption (iii) we obtain the estimate

$$\int_0^T \|f(s,v(s)) - f(s,v'(s))\|\,ds \leqslant LT\|v - v'\|_\infty.$$

Since $u : [0,T] \to X$ is continuous, we can find step functions $u_n : [0,T] \to X$ such that $\|u - u_n\|_\infty \leqslant 1/n$. Then, for $m,n \geqslant N$,

$$\|u_n - u_m\|_\infty \leqslant \|u_n - u\|_\infty + \|u - u_m\|_\infty \leqslant \frac{1}{n} + \frac{1}{m} \leqslant \frac{2}{N},$$

so the functions $s \mapsto f(s,u_n(s))$ form a Cauchy sequence in $L^1(0,T;X)$. By the completeness of $L^1(0,T;X)$ they tend to a limit, say g. Moreover, after passing to a subsequence, we may assume that $\lim_{n\to\infty} f(s,u_n(s)) = g(s)$ for almost all $s \in [0,T]$. By modifying the functions on a common Borel null set as in the proof of Lemma 13.23

we may even assume that the convergence holds pointwise. On the other hand, for all $s \in [0,T]$ we have

$$\|f(s,u_n(s)) - f(s,u(s))\| \leqslant \frac{L}{n}.$$

It follows that $g(s) = f(s,u(s))$ for almost all $s \in [0,T]$. In particular, this proves that $s \mapsto f(s,u(s))$ has a strongly measurable representative and belongs to $L^1(0,T;X)$.

(2): This follows by applying Lemma 13.23 to the function $s \mapsto f(s,u(s))$. $\qquad \square$

We are now ready to state and prove our main result:

Theorem 13.27 (Well-posedness of the semilinear problem). *Under the assumptions* (i)–(iii) *formulated at the beginning of the section, the semilinear problem* (SCP) *admits a unique mild solution* $u \in C([0,T];X)$. *This solution depends continuously, in the norm of* $C([0,T];X)$, *on the initial condition* $u_0 \in X$.

Proof To obtain existence and uniqueness we define a nonlinear mapping Φ from $C([0,T];X)$ to itself by

$$(\Phi(v))(t) := S(t)u_0 + \int_0^t S(t-s)f(s,v(s))\,ds, \quad t \in [0,T], \ v \in C([0,T];X).$$

We have already observed in Lemma 13.26 that the integrand is integrable, and the continuity of $\Phi(v)$ follows from the strong continuity of the semigroup and Lemma 13.23. It follows that Φ is well defined as a mapping of $C([0,T];X)$ into itself. We now reuse the idea in the proof of Lemma 2.14 and set, for a parameter $\lambda > 0$ to be chosen in a moment,

$$\|g\|_\lambda := \sup_{t \in [0,T]} e^{-\lambda t}\|g(t)\|.$$

This defines an equivalent norm on $C([0,T];X)$. By the Lipschitz continuity assumption, for all $v, w \in C([0,T];X)$ and $t \in [0,T]$ we have

$$\|(\Phi(v))(t) - (\Phi(w))(t)\| \leqslant \int_0^t \|S(t-s)(f(s,v(s)) - f(s,w(s)))\|\,ds$$

$$\leqslant LM \int_0^t e^{\lambda s}e^{-\lambda s}\|v(s) - w(s)\|\,ds$$

$$\leqslant LM \int_0^t e^{\lambda s}\|v - w\|_\lambda \,ds = \frac{LM}{\lambda}(e^{\lambda t} - 1)\|v - w\|_\lambda,$$

where $M = \sup_{t \in [0,T]} \|S(t)\|$. It follows that

$$\|\Phi(v) - \Phi(w)\|_\lambda \leqslant \frac{LM}{\lambda}(1 - e^{-\lambda t})\|v - w\|_\lambda \leqslant \frac{LM}{\lambda}\|v - w\|_\lambda.$$

If we choose $\lambda > LM$, the mapping Φ is a uniform contraction on $C([0,T];X)$ and therefore has a unique fixed point $u \in C([0,T];X)$ by the Banach fixed point theorem (Theorem 2.13). Then,

$$u(t) = (\Phi(u))(t) = S(t)u_0 + \int_0^t S(t-s)f(s,u(s))\,\mathrm{d}s, \quad t \in [0,T],$$

so u is a mild solution. Conversely, any mild solution is a fixed point of Φ, and since Φ has a unique fixed point the mild solution u is unique.

To complete the proof we check the continuous dependence of the mild solution on the initial value u_0. If $\widetilde{u}_0$ is another initial value and the corresponding unique mild solution is denoted by $\widetilde{u}$, estimating as before we obtain

$$\|u(t) - \widetilde{u}(t)\| \leqslant \|S(t)\|\|u_0 - \widetilde{u}_0\| + \int_0^t \|S(t-s)(f(s,u(s)) - f(s,\widetilde{u}(s)))\|\,\mathrm{d}s$$

$$\leqslant M\|u_0 - \widetilde{u}_0\| + \frac{LM}{\lambda}(e^{\lambda t} - 1)\|u - \widetilde{u}\|_\lambda$$

and therefore

$$\|u - \widetilde{u}\|_\lambda \leqslant M\|u_0 - \widetilde{u}_0\| + \frac{LM}{\lambda}\|u - \widetilde{u}\|_\lambda.$$

Choosing $\lambda = 2LM$ gives

$$\frac{1}{2}\|u - \widetilde{u}\|_{2LM} \leqslant M\|u_0 - \widetilde{u}_0\|$$

and the desired continuity follows, keeping in mind that $\|\cdot\|_{2LM}$ is an equivalent norm on $C([0,T];X)$. $\qquad\square$

For this this to be useful, one must have ways to 'translate' nonlinearities occurring in concrete partial differential equations into our abstract framework. We demonstrate how this works by means of an example. Consider the following semilinear heat equation on a nonempty bounded open subset D of $\mathbb{R}^d$:

$$\begin{cases} \dfrac{\partial u}{\partial t}(t,\xi) = \Delta u(t,\xi) + b(u(t,\xi)), & \xi \in D, \quad t \in [0,T], \\[2mm] u(t,\xi) = 0, & \xi \in \partial D, \quad t \in [0,T], \\[2mm] u(0,\xi) = u_0(\xi), & \xi \in D. \end{cases}$$

We assume that $b : \mathbb{R} \to \mathbb{R}$ is Lipschitz continuous, with Lipschitz constant L:

$$|b(\xi) - b(\xi')| \leqslant L|\xi - \xi'|, \quad \xi, \xi' \in \mathbb{R}.$$

The assumption that b only depends on the solution is made for simplicity; the more general case where b also depends on time can be treated in the same way.

To cast this problem into a semilinear abstract Cauchy problem we assume that the initial value u_0 belongs to $L^2(D)$. The above problem may then be written in the form

$$\begin{cases} u'(t) = Au(t) + B(u(t)), \\ u(0) = u_0, \end{cases}$$

where A is the Dirichlet Laplacian on $L^2(D)$, which generates an analytic C_0-contraction semigroup on this space (see Proposition 13.47), and $B : L^2(D) \to L^2(D)$ is the *Nemytskii mapping* associated with b:

$$(B(x))(\xi) := b(x(\xi)), \quad x \in L^2(D).$$

The next proposition checks that this mapping is well defined, of linear growth, and Lipschitz continuous on $L^2(D)$ (and, with the same proof, on $L^p(D)$ with $1 \leqslant p < \infty$).

Proposition 13.28. *Under the above assumptions on b, the Nemytskii mapping $B : L^2(D) \to L^2(D)$ is well defined, of linear growth, and Lipschitz continuous (in the sense that $f(t,x) := B(x)$ satisfies conditions* (ii) *and* (iii) *at the beginning of this section).*

Proof Let us first check that $B(x) \in L^2(D)$ for all $x \in L^2(D)$. Using the triangle inequality in $L^2(D)$, for all $x \in L^2(D)$ we have

$$\begin{aligned} \|B(x)\|_{L^2(D)} &= \left(\int_D |b(x(\xi))|^2 \, d\xi \right)^{1/2} \\ &\leqslant \left(\int_D |b(x(\xi)) - b(0)|^2 \, d\xi \right)^{1/2} + \left(\int_D |b(0)|^2 \, d\xi \right)^{1/2} \\ &\leqslant L \left(\int_D |x(\xi) - 0|^2 \, d\xi \right)^{1/2} + |b(0)| \left(\int_D 1 \, d\xi \right)^{1/2} = L\|x\|_2 + |D|^{1/2} |b(0)|, \end{aligned}$$

where $|D|$ stands for the Lebesgue measure of D. This proves that B is well defined and of linear growth.

Lipschitz continuity follows by a similar estimate. For all $x, y \in L^2(D)$,

$$\begin{aligned} \|B(x) - B(y)\|_{L^2(D)} &= \left(\int_D |b(x(\xi)) - b(y(\xi))|^2 \, d\xi \right)^{1/2} \\ &\leqslant L \left(\int_D |x(\xi) - y(\xi)|^2 \, d\xi \right)^{1/2} = \|x - y\|_{L^2(D)}. \end{aligned}$$

$\square$

We have thus shown that all assumptions of Theorem 13.27 are fulfilled. Accordingly we obtain unique solvability of the semilinear heat equation, in the sense that the corresponding abstract Cauchy problem admits a unique mild solution.

13.4 Analytic Semigroups

Analytic semigroups provide an abstract framework for discussing a class of initial value problems, referred to in the partial differential equations literature as *parabolic*. An important characteristic of this class of problems is that solutions are smooth.

13.4.a The Main Result

For $\omega \in (0, \pi)$ consider the open sector

$$\Sigma_\omega := \{z \in \mathbb{C} \setminus \{0\} : |\arg(z)| < \omega\},$$

where the argument is taken in $(-\pi, \pi)$.

Definition 13.29 (Analytic C_0-semigroups). A C_0-semigroup S on X is called *analytic on Σ_ω* if for all $x \in X$ the function $t \mapsto S(t)x$ extends holomorphically to Σ_ω and satisfies

$$\lim_{z \in \Sigma_\omega, z \to 0} S(z)x = x.$$

We call S an *analytic C_0-semigroup* if it is an analytic C_0-semigroup on Σ_ω for some $\omega \in (0, \pi)$.

If S is an analytic C_0-semigroup on Σ_ω, then for all $z_1, z_2 \in \Sigma_\omega$ we have

$$S(z_1)S(z_2) = S(z_1 + z_2).$$

Indeed, for each $x \in X$ the functions $z_1 \mapsto S(z_1)S(t)x$ and $S(z_1 + t)x$ are holomorphic extensions of $s \mapsto S(s + t)x$ and are therefore equal. Repeating this argument, the functions $z_2 \mapsto S(z_1)S(z_2)x$ and $S(z_1 + z_2)x$ are holomorphic extensions of $t \mapsto S(z_1 + t)x$ and are therefore equal.

As in the proof of Proposition 13.3, the uniform boundedness theorem implies that if S is an analytic C_0-semigroup on Σ_ω, then the operators $S(z)$ is uniformly bounded on $\Sigma_{\omega'} \cap B(0; r)$ for every $0 < \omega' < \omega$ and $r \geqslant 0$. The same argument as in Proposition 13.3 then gives exponential boundedness on $\Sigma_{\omega'}$ for all $0 < \omega' < \omega$, in the sense that there are constants $M' \geqslant 1$ and $c' = c_{\omega'} \in \mathbb{R}$ such that

$$\|S(z)\| \leqslant M' e^{c'|z|}, \quad z \in \Sigma_{\omega'}.$$

We say that S is a *bounded analytic C_0-semigroup on Σ_ω* if S is an analytic C_0-semigroup on Σ_ω and the operators $S(z)$ are uniformly bounded on Σ_ω. *Analytic C_0-contraction semigroups on Σ_ω* are defined similarly. There is a rather subtle point here: Boundedness and contractivity are imposed on a sector, not just on the positive real line. That this makes a difference is shown by simple example of the rotation group on $\mathbb{C}^2$, given by

$$S(t) = \begin{pmatrix} \cos t & -\sin t \\ \sin t & \cos t \end{pmatrix}.$$

For each $t \in \mathbb{R}$ we have $\|S(t)\| = 1$. Upon replacing t by a complex parameter z the group extends holomorphically to the entire complex plane, but it is unbounded on every sector Σ_ω with $0 < \omega < \pi$. It may even happen that a bounded analytic C_0-semigroup is contractive on the positive real line, yet fails to be an analytic C_0-contraction semigroup; an example of such a semigroup on $\mathbb{C}^2$ is discussed in Problem 13.9.

Theorem 13.30 (Bounded analytic semigroups, complex characterisation). *For a densely defined closed operator A in X the following assertions are equivalent:*

(1) *A generates a bounded analytic C_0-semigroup on Σ_η for some $\eta \in (0, \frac{1}{2}\pi)$;*
(2) *there exists $\theta \in (\frac{1}{2}\pi, \pi)$ such that $\Sigma_\theta \subseteq \rho(A)$ and*

$$\sup_{\lambda \in \Sigma_\theta} \|\lambda R(\lambda, A)\| < \infty.$$

Denoting the suprema of all admissible η and θ in (1) and (2) by $\omega_{\mathrm{holo}}(A)$ and $\omega_{\mathrm{res}}(A)$ respectively, we have

$$\omega_{\mathrm{res}}(A) = \frac{1}{2}\pi + \omega_{\mathrm{holo}}(A).$$

Under the equivalent conditions (1) and (2) we have the inverse Laplace transform representation

$$S(t)x = \frac{1}{2\pi i} \int_\Gamma e^{\lambda t} R(\lambda, A) x \, d\lambda, \quad t > 0, \ x \in X, \tag{13.11}$$

where $\Gamma = \Gamma_{\theta', B}$ is the upwards oriented boundary of $\Sigma_{\theta'} \setminus B$, for any $\theta' \in (\frac{1}{2}\pi, \theta)$ and any closed ball B centred at the origin.

Note that (2) implies $\sigma(A) \cap i\mathbb{R} \subseteq \{0\}$.

Proof By Cauchy's theorem, if the integral representation holds for some choice of $\theta' \in (\frac{1}{2}\pi, \theta)$ and a closed ball B centred at the origin, then it holds for any such θ' and B.

(1)$\Rightarrow$(2): We start with the preliminary observation that if a linear operator $\widetilde{A}$ generates a uniformly bounded C_0-semigroup $\widetilde{S}$ on X, then, by Proposition 13.8, the open right half-plane $\mathbb{C}_+ = \{\mathrm{Re}\,\lambda > 0\}$ is contained in the resolvent set of A and we have the bound $\|R(\lambda, \widetilde{A})\| \leqslant M/\mathrm{Re}\,\lambda$ for all $\lambda \in \mathbb{C}_+$. Moreover, for all $\theta \in (0, \frac{1}{2}\pi)$ and $\lambda \in \Sigma_\theta$ we have $\mathrm{Re}\,\lambda \geqslant |\lambda| \cos(\theta)$ and therefore

$$\sup_{\lambda \in \Sigma_\theta} \|\lambda R(\lambda, \widetilde{A})\| \leqslant \frac{M}{\cos \theta}.$$

Now if A generates a C_0-semigroup which is bounded on a sector Σ_η with $\eta \in (0, \frac{1}{2}\pi)$, say by a constant M, we can apply the above reasoning to the bounded semigroups

$S(e^{i\eta'}t)$, with $\eta' \in (0,\eta)$, and obtain (2). Optimising the various choices of angles we obtain the inequality

$$\omega_{\text{res}}(A) \geqslant \frac{1}{2}\pi + \omega_{\text{holo}}(A).$$

(2)$\Rightarrow$(1): The idea is to define the semigroup operators by the integral representation given in the statement of the theorem, and prove that they define a bounded C_0-semigroup which has the properties stated in part (1).

Once we have this, it is fairly straightforward to deduce (1) with $\eta = \theta - \frac{1}{2}\pi$; this is done in the second step and gives the inequality

$$\omega_{\text{holo}}(A) \geqslant \omega_{\text{res}}(A) - \frac{1}{2}\pi.$$

Step 1 – Let $\eta := \theta - \frac{1}{2}\pi$. For any $\zeta \in \Sigma_\eta$ let

$$S(\zeta)x := \frac{1}{2\pi i}\int_\Gamma e^{\lambda\zeta}R(\lambda,A)x\,\mathrm{d}\lambda, \quad x \in X,$$

where Γ is the boundary of $\Sigma_{\theta'} \setminus B$ with $\theta' \in (\frac{1}{2}\pi,\theta)$ any number such that $\frac{1}{2}\pi + |\arg(\zeta)| < \theta' < \theta$ and B is any closed ball centred at the origin; see Figure 13.1. This integral converges absolutely, defines a bounded operator $S(\zeta)$ on X, and the function $\zeta \mapsto S(\zeta)x$ is holomorphic on Σ_η.

The proof of the semigroup property proceeds much in the same way as the proof of the multiplicativity of the holomorphic calculus. Fix $\zeta, \zeta' \in \Sigma_\eta$ and choose contours Γ and Γ' as above, with Γ to the left of Γ'. Then, by the resolvent identity (10.2), Cauchy's theorem, Fubini's theorem, and the Cauchy integral formula,

$$\begin{aligned}
S(\zeta')S(\zeta)x &= \frac{1}{(2\pi i)^2}\int_\Gamma\int_{\Gamma'} e^{\lambda\zeta+\mu\zeta'}R(\lambda,A)R(\mu,A)x\,\mathrm{d}\lambda\,\mathrm{d}\mu \\
&= \frac{1}{(2\pi i)^2}\int_\Gamma\int_{\Gamma'} e^{\lambda\zeta+\mu\zeta'}\frac{R(\lambda,A)x - R(\mu,A)x}{\mu-\lambda}\,\mathrm{d}\lambda\,\mathrm{d}\mu \\
&= \frac{1}{(2\pi i)^2}\int_\Gamma\int_{\Gamma'} e^{\lambda\zeta+\mu\zeta'}\frac{R(\lambda,A)x}{\mu-\lambda}\,\mathrm{d}\mu\,\mathrm{d}\lambda \\
&= \frac{1}{2\pi i}\int_\Gamma e^{\lambda\zeta+\lambda\zeta'}R(\lambda,A)x\,\mathrm{d}\lambda = S(\zeta+\zeta')x.
\end{aligned}$$

Put $M := \sup_{\lambda\in\Sigma_\theta}\|\lambda R(\lambda,A)\|$ and fix $\zeta \in \Sigma_\eta$. To estimate the norm of $S(\zeta)x$, by Cauchy's theorem we may take $\Gamma = \Gamma_{\theta',B_r}$ with $B_r = B(0;r)$ the ball of radius r and centre 0, where we take $\frac{1}{2}\pi + |\arg(\zeta)| < \theta' < \theta$ as before; the choice of $r > 0$ will be made shortly. The arc $\{|z| = r,\ |\arg(z)| \leqslant \theta'\}$ contributes at most

$$\frac{1}{2\pi}\cdot 2\theta' r\cdot\exp(r|\zeta|)\frac{M}{r} = \frac{\theta'M}{\pi}\exp(r|\zeta|),$$

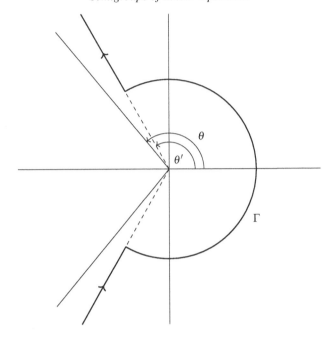

Figure 13.1 The contour Γ

while each of the rays $\{|z| \geqslant r, \arg(z) = \pm\theta'\}$ contributes at most

$$\frac{1}{2\pi} \cdot \frac{M}{r} \int_r^\infty \exp(-\rho|\zeta||\cos(\theta')|)\,d\rho \leqslant \frac{1}{2\pi} \cdot \frac{M}{r|\zeta||\cos(\theta')|}.$$

It follows that

$$\|S(\zeta)\| \leqslant \frac{M}{\pi}\left(\theta' \exp(r|\zeta|) + \frac{1}{r|\zeta||\cos(\theta')|}\right).$$

Taking $r = 1/|\zeta|$ and letting $\theta' \uparrow \theta$ we obtain the uniform bound

$$\|S(\zeta)\| \leqslant \frac{M}{\pi}\left(\theta e + \frac{1}{|\cos(\theta)|}\right), \quad \zeta \in \Sigma_\eta, \ \eta = \theta - \frac{1}{2}\pi.$$

It remains to prove strong continuity on each sector $\Sigma_{\eta'}$ with $0 < \eta' < \eta$. Let $x \in D(A)$. Fix $\zeta \in \Sigma_{\eta'}$ and write $x = R(\mu,A)y$ with $\mu \in \Sigma_\theta \setminus \Sigma_{\theta'}$, where $\frac{1}{2}\pi + |\arg(\zeta)| < \theta' < \theta$ as before. Inserting this in the integral expression for $S(\zeta)x$, using the resolvent identity to rewrite $R(\lambda,A)R(\mu,A)y = (R(\lambda,A) - R(\mu,A))/(\mu - \lambda)$, and arguing as above, we find that the integral corresponding to the term with $R(\mu,A)$ vanishes by

Cauchy's theorem and the choice of μ and obtain

$$S(\zeta)x = \frac{1}{2\pi i} \int_\Gamma e^{\lambda\zeta}(\mu - \lambda)^{-1} R(\lambda, A) y \, d\lambda.$$

Letting $\zeta \to 0$ inside $\Sigma_{\eta'}$ we see that $S(\zeta)x$ converges to

$$\frac{1}{2\pi i} \int_\Gamma (\mu - \lambda)^{-1} R(\lambda, A) y \, d\lambda = R(\mu, A) y = x$$

by dominated convergence.

This proves the convergence $S(\zeta)x \to x$ for $x \in D(A)$ as $\zeta \to 0$ inside $\Sigma_{\eta'}$. In view of the uniform boundedness of $S(\zeta)$ on $\Sigma_{\eta'}$, the convergence for general $x \in X$ follows from this. □

The following result characterises analytic C_0-semigroups directly in terms of the semigroup and its generator, without reference to the resolvent. Its importance lies in the smoothing property revealed by (2): the semigroup operators $S(t)$ map every $x \in X$ into the smaller subspace $D(A)$ for all $t > 0$.

Theorem 13.31 (Bounded analytic semigroups, real characterisation). *Let A be the generator of a C_0-semigroup S on X. The following assertions are equivalent:*

(1) *S is bounded analytic;*
(2) *$S(t)x \in D(A)$ for all $x \in X$ and $t > 0$, and*

$$\sup_{t>0} t \|AS(t)\| < \infty.$$

Remark 13.32. By writing $S(t) = [S(\frac{t}{n})]^n$, part (2) self-improves as follows: for all $x \in X$ and $t > 0$ we have $S(t)x \in D(A^n)$ for all $x \in X$ and $t > 0$, and

$$\sup_{t>0} t^n \|A^n S(t)\| =: C_n < \infty.$$

This will be used in the proof below.

Proof of Theorem 13.31 (1)$\Rightarrow$(2): Fix $t > 0$ and $x \in X$. Arguing as in the proof of Proposition 13.8, from the integral representation (13.11) with $\Gamma = \partial(\Sigma_{\theta'} \setminus B)$ we deduce that $S(t)x \in D(A)$ and

$$AS(t)x = \frac{1}{2\pi i} \int_\Gamma e^{\lambda t} R(\lambda, A) Ax \, d\lambda.$$

The integral on the right-hand side converges absolutely since

$$\sup_{\lambda \in \Gamma} \|AR(\lambda, A)\| = \sup_{\lambda \in \Gamma} \|\lambda R(\lambda, A) - I\| < \infty.$$

By estimating this integral and letting the radius of the ball B in the definition of Γ tend to 0, it follows moreover that

$$t\|AS(t)x\| \leqslant \frac{M}{\pi}\|x\| \int_0^\infty t e^{\rho t \cos\theta'} \, d\rho = \frac{M}{\pi|\cos\theta'|}\|x\|,$$

where M is the supremum in the preceding line.

(2)$\Rightarrow$(1): For all $x \in D(A^n)$ the mapping $t \mapsto S(t)x$ is n times continuously differentiable and $S^{(n)}(t)x = A^n S(t)x = (AS(t/n))^n x$. Since $D(A^n)$ is dense in X, the boundedness of $AS(t/n)$ and closedness of the nth derivative in $C([0,T];X)$ together imply that the same conclusion holds for $x \in X$. Moreover,

$$\|S^{(n)}(t)x\| \leqslant \frac{C^n n^n}{t^n}\|x\|,$$

where C is the supremum in (2). From the inequality $n! \geqslant n^n/e^n$ (which follows from Stirling's inequality) we obtain that for each $t > 0$ the series

$$S(z)x := \sum_{n=0}^\infty \frac{1}{n!}(z-t)^n S^{(n)}(t)x$$

converges absolutely on every ball $B(t; rt/eC)$ with $0 < r < 1$ and defines a holomorphic function there. The union of all these balls is the sector Σ_η with $\sin\eta = 1/eC$ (cf. the argument in the proof of the next lemma). We shall complete the proof by showing that $S(z)$ is uniformly bounded and satisfies $\lim_{z\to 0} S(z)x = x$ in $\Sigma_{\eta'}$ for each $0 < \eta' < \eta$. To this end we fix $0 < r < 1$ so that the union of the balls $B(t; rt/eC)$ equals $\Sigma_{\eta'}$. For $z \in B(t; rt/eC)$ we have

$$\|S(z)x\| \leqslant \sum_{n=0}^\infty \frac{1}{n!} r^n (t/eC)^n \frac{C^n n^n}{t^n}\|x\| \leqslant \sum_{n=0}^\infty r^n \|x\| = \frac{\|x\|}{1-r}.$$

This proves uniform boundedness on the sectors $\Sigma_{\eta'}$. To prove strong continuity it then suffices to consider $x \in D(A)$, for which it follows from estimating the identity

$$S(z)x - x = e^{i\theta} \int_0^r S(se^{i\theta})Ax \, ds$$

where $z = re^{i\theta}$. $\square$

13.4.b The Lumer–Phillips Theorem

The main result of this section is the Lumer–Phillips theorem, which gives a characterisation of analytic C_0-semigroups of contractions in Hilbert spaces. We begin with a useful lemma about extending resolvent bounds from a half-line to a sector.

In Hilbert spaces with have the following characterisation of contractive analytic C_0-semigroups (for an extension to Banach spaces see Problem 13.18).

Theorem 13.33 (Lumer–Phillips, analytic contraction semigroups). *Let A be a densely defined closed operator in a Hilbert space H and let* $0 < \eta < \frac{1}{2}\pi$. *The following assertions are equivalent:*

(1) *A generates a contractive analytic C_0-semigroup on H on the sector* Σ_η;
(2) $\mu - A$ *has dense range for some* $\mu > 0$ *and* $-(Ax|x) \in \overline{\Sigma_{\frac{1}{2}\pi-\eta}}$ *for all* $x \in D(A)$.

Proof By a multiplicative scaling of A we may assume that $\mu = 1$.

(1)⇒(2): Let $|\eta'| < \eta$, $x \in D(A)$, and consider the function $f(t) := \mathrm{Re}(S(te^{i\eta'})x|x)$. Observe that $f(0) = \|x\|^2$. Furthermore, for all $t \geqslant 0$ we have

$$|f(t)| = |(S(te^{i\eta'})x|x)| \leqslant \|S(te^{i\eta'})x\|\|x\| \leqslant \|x\|^2,$$

where we used that S is contractive on Σ_η. From these two observations we infer that $f'(0) \leqslant 0$. On the other hand, differentiating f gives

$$f'(t) = \mathrm{Re}(e^{i\eta'}S(te^{i\eta'})Ax|x).$$

Evaluating at $t = 0$ gives

$$\mathrm{Re}(e^{i\eta'}(Ax|x)) \leqslant 0.$$

This can only be true for all $|\eta'| < \eta$ if $(Ax|x) \in -\overline{\Sigma_{\frac{1}{2}\pi-\eta}}$.

(2)⇒(1): Set $\lambda := re^{i\eta'}$ with $r > 0$ and $|\eta'| < \eta$. We want to show that for all $x \in D(A)$ we have $\|(\lambda - A)x\| \geqslant r\|x\| = |\lambda|\|x\|$. For this we may assume that $\|x\| = 1$.

From $\lambda \in \Sigma_\eta$ and $-(Ax|x) \in \overline{\Sigma_{\frac{1}{2}\pi-\eta}}$ it is easy to see that $|\lambda - (Ax|x)| \geqslant |\lambda|$. See Figure 13.2. As a consequence,

$$\|(\lambda - A)x\| \geqslant |((\lambda - A)x|x)| = |\lambda - (Ax|x)| \geqslant |\lambda| = |\lambda|\|x\|. \tag{13.12}$$

From this inequality and Proposition 10.26 we infer that $\lambda - A$ has closed range. Therefore, to show that this operator is invertible, it suffices to show that it has dense range. This will be deduced from the assumption that $I - A$ has dense range. Since $I - A$ has also closed range, we have in fact $1 \in \rho(A)$. Now suppose, for a contradiction, that some $\lambda_1 \in \Sigma_\eta$ belongs to $\sigma(A)$. Set $\lambda_t := (1 - t) + t\lambda_1$. Then $\lambda_t \in \Sigma_\eta$ for all $t \in [0,1]$. Let $t_0 := \inf\{t \in [0,1] : \lambda_t \in \sigma(A)\}$. Then $t_0 \in (0,1]$ and $\lim_{t\uparrow t_0}\|R(\lambda_t,A)\| = \infty$ since resolvent norms diverge as we approach the boundary of the spectrum by Proposition 10.29. This clearly contradicts (13.12), which tells us that $\|R(\lambda_t,A)\| \leqslant |\lambda_t|^{-1} \leqslant m^{-1}$, where $m = \min_{0 \leqslant t \leqslant 1} |(1 - t) + t\lambda_1|$.

We have now shown that $\Sigma_\eta \subseteq \rho(A)$ and $\|R(\lambda,A)\| \leqslant |\lambda|^{-1}$ on this sector. A similar argument shows that for all $0 < \theta < \frac{1}{2}\pi + \eta$ we have $\Sigma_\theta \subseteq \rho(A)$ and $\|R(\lambda,A)\| \leqslant M|\lambda|^{-1}$ for some constant $M \geqslant 0$ depending on θ. Theorem 13.30 then implies that the semigroup generated by A is bounded analytic on every sector $\Sigma_{\eta'}$ with $0 < \eta' < \eta$. Its contractivity on these sectors is obtained by applying the Hille–Yosida theorem to the operators $e^{i\eta'}A$. □

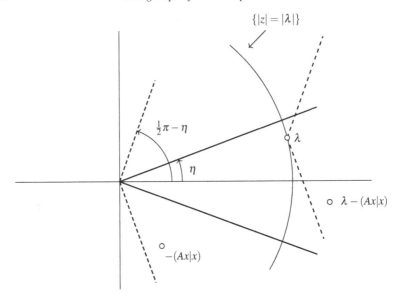

Figure 13.2 $|\lambda - (Ax|x)| \geqslant |\lambda|$

The conditions of the theorem are satisfied if $-A$ is a positive selfadjoint operator on H. In that case, the semigroup S generated by A is given by

$$S(t) = \int_{\sigma(-A)} e^{-t\lambda} \, dP(\lambda),$$

where P is the projection-valued measure associated with $-A$. More generally, this formula can be used to associate a C_0-semigroup of contractions with every normal operator A; see Theorem 13.61.

With the same proofs, both Theorem 13.33 and its corollary extend to $\eta = 0$, provided we interpret Σ_0 as the positive real line and replace 'analytic C_0-semigroup of contractions' by 'C_0-semigroup of contractions'. The theorem then takes the following form:

Theorem 13.34. *Let A be a densely defined operator on a Hilbert space H. The following assertions are equivalent:*

(1) *A generates a C_0-semigroup of contractions on H;*
(2) *$\mu - A$ has dense range for some $\mu > 0$ and $-\mathrm{Re}(Ax|x) \geqslant 0$ for all $x \in D(A)$.*

The condition '$-\mathrm{Re}(Ax|x) \geqslant 0$ for all $x \in D(A)$' says that $-A$ is *accretive*. Since the open half-line $(0, \infty)$ is contained in the resolvent set of any operator generating a

C_0-semigroup of contractions, in Theorems 13.33 and 13.34, the condition '$\mu - A$ has dense range for some $\mu > 0$' may be replaced by '$1 \in \rho(A)$'. An accretive operator $-A$ satisfying $1 \in \rho(A)$ is also called a *maximal accretive* operator, or briefly, an *m-accretive operator*.

13.4.c Semigroups Associated with Forms

The first estimate of Theorem 12.12 shows that the assumptions of the Hille–Yosida theorem are fulfilled. The second estimate, combined with Corollary 12.13 and Lemma 10.34 applied to $A - \delta$ and A, implies that the conditions of Theorem 13.30 are fulfilled. Thus we obtain the following result.

Theorem 13.35 (Bounded analytic semigroups via forms). *Let A be a linear operator in a Hilbert space H. Then:*

(1) *if $-A$ is the operator associated with a densely defined closed continuous accretive form in H, then A generates a C_0-contraction semigroup on H, and for all $\delta > 0$ the operator $A - \delta$ generates a bounded analytic C_0-semigroup on H;*

(2) *if $-A$ is the operator associated with a bounded coercive form on a Hilbert space V densely embedded in H, then then A generates a C_0-contraction semigroup on H that extends to a bounded analytic C_0-semigroup on H.*

Example 13.36 (Operators in Divergence Form I). Let D be a nonempty bounded open subset of $\mathbb{R}^d$. In $H = L^2(D)$ we consider the divergence form operators

$$A_a = \operatorname{div}(a\nabla)$$

of Section 12.3.e subject to Dirichlet conditions. As in that section we assume that the matrix-valued function $a : D \to M_d(\mathbb{C})$ satisfies

(i) the functions $a_{ij} : D \to \mathbb{C}$ are measurable and bounded;

(ii) there exists a constant $\alpha > 0$ such that

$$\operatorname{Re} \sum_{i,j=1}^{d} a_{ij}(x)\xi_i \overline{\xi}_j \geqslant 0, \quad \xi \in \mathbb{C}^d.$$

The operator $-A_a$ is rigorously defined as the densely defined closed operator associated with the form

$$\mathfrak{a}_a(u,v) = \int_D a\nabla u \cdot \overline{\nabla v}\, dx$$

on $V = H_0^1(D)$. The form $\mathfrak{a}_a$ is satisfies the assumptions of the first part of Theorem 13.35.

If the accretivity assumption

$$\text{Re} \sum_{i,j=1}^{d} a_{ij}(x)\xi_i\overline{\xi}_j \geqslant 0$$

of Section 12.3.e is replaced by the coercivity condition

$$\text{Re} \sum_{i,j=1}^{d} a_{ij}(x)\xi_i\overline{\xi}_j \geqslant \alpha|\xi|^2$$

of Section 11.3.b, with $\alpha > 0$, then the second part of Theorem 13.35 can be applied.

We show next that generators of analytic C_0-contraction semigroups are obtained if the range of $\mathfrak{a}$ is contained in the closure of a subsector strictly contained in the open right-half plane.

Definition 13.37 (Sectorial forms). Let $0 < \omega \leqslant \frac{1}{2}\pi$. A form $\mathfrak{a}$ on H is called ω-*sectorial* if

$$\mathfrak{a}(v) := \mathfrak{a}(v,v) \in \overline{\Sigma_\omega}, \quad v \in D(\mathfrak{a}).$$

Theorem 13.38 (Analytic contraction semigroups via forms). *Let H be a Hilbert space and let A be the densely defined closed operator in H associated with a densely defined closed form $\mathfrak{a}$ that is ω-sectorial for some $0 < \omega < \frac{1}{2}\pi$. Then $-A$ generates an analytic C_0-semigroup of contractions on the sector $\Sigma_{\frac{1}{2}\pi-\omega}$.*

Proof This is an immediate consequence of Theorem 13.33. □

Example 13.39 (Operators in divergence form II). Consider again the divergence form operator

$$A_a := \text{div}(a\nabla)$$

in $L^2(D)$, subject to Dirichlet conditions. As before we assume that D is a nonempty bounded open subset of $\mathbb{R}^d$. We now assume that the matrix-valued function $a : D \to M_d(\mathbb{C})$ satisfies

(i) the functions $a_{ij} : D \to \mathbb{C}$ are measurable and bounded;
(ii) there exists a constant $\alpha > 0$ such that

$$\sum_{i,j=1}^{d} a_{ij}(x)\xi_i\overline{\xi}_j \geqslant \alpha|\xi|^2, \quad \xi \in \mathbb{C}^d.$$

The *uniform ellipticity* condition (ii) is stronger than the corresponding condition of Example 13.36, in that no real parts are taken. It implies that the form $\mathfrak{a}_a$ of Example 13.36 takes values in $[0,\infty)$, so it is ω-sectorial for all $\omega \in (0, \frac{1}{2}\pi)$. Accordingly, $-A_a$ generates an analytic C_0-semigroup of contractions on every sector Σ_θ with $\theta \in (0, \frac{1}{2}\pi)$.

Sectorial forms of angle less than $\frac{1}{2}\pi$ are continuous and accretive; this clarifies the relationship between Theorems 13.35 and 13.38. Accretivity is clear, and continuity follows from the following proposition.

Proposition 13.40. *Let* $\mathfrak{a}$ *be an* ω-*sectorial form on* H *with* $0 < \omega < \frac{1}{2}\pi$. *Then* $\mathfrak{a}$ *is continuous and for all* $u, v \in D(A)$ *we have*

$$|\mathfrak{a}(u,v)| \leqslant (1+\tan\omega)(\operatorname{Re}\mathfrak{a}(u))^{1/2}(\operatorname{Re}\mathfrak{a}(v))^{1/2},$$

where $\operatorname{Re}\mathfrak{a} = \frac{1}{2}(\mathfrak{a} + \mathfrak{a}^\star)$ *with* $\mathfrak{a}^\star(u,v) := \overline{\mathfrak{a}(v,u)}$.

Proof By the Cauchy–Schwarz inequality applied to the (symmetric) form $\operatorname{Re}\mathfrak{a}$,

$$|\operatorname{Re}\mathfrak{a}(u,v)| \leqslant (\operatorname{Re}\mathfrak{a}(u))^{1/2}(\operatorname{Re}\mathfrak{a}(v))^{1/2}.$$

If $\operatorname{Re}\mathfrak{a}(u) = 0$, the desired inequality follows from this. In the rest of the proof we mat therefore assume that $\operatorname{Re}\mathfrak{a}(u) > 0$.

Consider the form $\operatorname{Im}\mathfrak{a} = \frac{1}{2i}(\mathfrak{a} - \mathfrak{a}^\star)$. Fix $u, v \in D(\mathfrak{a})$. Replacing v by $e^{i\theta}v$ if necessary we may assume that $\operatorname{Im}\mathfrak{a}(u,v) \in \mathbb{R}$. Then $\operatorname{Im}\mathfrak{a}(u,v) = \operatorname{Im}\mathfrak{a}(v,u)$ and therefore, by ω-sectoriality,

$$
\begin{aligned}
|\operatorname{Im}\mathfrak{a}(u,v)| &= \frac{1}{4}\left|\operatorname{Im}\mathfrak{a}(u+v,u+v) - \operatorname{Im}\mathfrak{a}(u-v,u-v)\right| \\
&\leqslant \frac{1}{4}\tan\omega\big(\operatorname{Re}\mathfrak{a}(u+v,u+v) + \operatorname{Re}\mathfrak{a}(u-v,u-v)\big) \\
&= \frac{1}{2}\tan\omega\big(\operatorname{Re}\mathfrak{a}(u) + \operatorname{Re}\mathfrak{a}(v)\big).
\end{aligned}
$$

Replacing u and v with $\sqrt{\varepsilon}u$ and $v/\sqrt{\varepsilon}$ gives

$$|\operatorname{Im}\mathfrak{a}(u,v)| \leqslant \frac{1}{2}\tan\omega\big(\sqrt{\varepsilon}\operatorname{Re}\mathfrak{a}(u) + \frac{1}{\sqrt{\varepsilon}}\operatorname{Re}\mathfrak{a}(v)\big).$$

Taking $\varepsilon := \operatorname{Re}\mathfrak{a}(v)/\operatorname{Re}\mathfrak{a}(u)$, we obtain

$$|\operatorname{Im}\mathfrak{a}(u,v)| \leqslant \tan\omega\big((\operatorname{Re}\mathfrak{a}(u))^{1/2}(\operatorname{Re}\mathfrak{a}(v))^{1/2}\big).$$

Together with the estimate for $|\operatorname{Re}\mathfrak{a}(u,v)|$, this gives the result. $\qquad\square$

13.4.d Maximal Regularity

In Section 13.3 we have seen that the mild solution u of the inhomogeneous problem $u' = Au + f$ with initial condition $u(0) = u_0$, which is given by

$$u(t) = S(t)u_0 + \int_0^t S(t-s)f(s)\,\mathrm{d}s, \quad t \geqslant 0,$$

is also a strong solution, that is, for all $t \geqslant 0$ we have $\int_0^t u(s)\,\mathrm{d}s \in D(A)$ and

$$u(t) = u_0 + A \int_0^t u(s)\,\mathrm{d}s + \int_0^t f(s)\,\mathrm{d}s.$$

In general it cannot be asserted, however, that

$$u(t) = u_0 + \int_0^t Au(s)\,\mathrm{d}s + \int_0^t f(s)\,\mathrm{d}s,$$

the problem being that u need not take values in $D(A)$ almost everywhere, and even if it does so we cannot be certain that $s \mapsto Au(s)$ is integrable on intervals $(0,t)$. The aim of the present section is to prove that these things do hold if A generates a bounded analytic C_0-semigroup on a Hilbert space.

Theorem 13.41 (Maximal regularity). *Let A be the generator of a bounded analytic C_0-semigroup S on a Hilbert space H. Then for all $f \in L^2(\mathbb{R}_+;H)$ the mild solution $u = u_f$ of the inhomogeneous problem $u' = Au + f$ with initial condition $u(0) = 0$ enjoys the following properties:*

(1) *u belongs to $D(A)$ almost everywhere, Au belongs to $L^2(\mathbb{R}_+;H)$, and for almost all $t \geqslant 0$ we have*

$$u(t) = \int_0^t Au(s)\,\mathrm{d}s + \int_0^t f(s)\,\mathrm{d}s;$$

(2) *we have*

$$\|Au\|_2 \leqslant C\|f\|_2,$$

where $C = \sup_{\xi \in \mathbb{R}\setminus\{0\}} \|AR(i\xi,A)\|$.

By the Lebesgue differentiation theorem (Theorem 2.39, or rather its the vector-valued version which is proved in exactly the same way), the identity in (1) implies that u is differentiable almost everywhere and that the pointwise identity

$$u'(t) = Au(t) + f(t)$$

holds for almost all $t \geqslant 0$. This, in combination with (2), implies that also u' belongs to $L^2(\mathbb{R}_+;H)$ (with estimate $\|u'\|_2 \leqslant (C+1)\|f\|_2$). This explains the name 'maximal regularity' attached to the theorem.

We begin with a reduction to a class of "nice" functions f. To this end, for subspaces F and Y of $L^2(\mathbb{R}_+)$ and H respectively, we introduce the notation $F \otimes Y$ for the vector space of all linear combinations of functions $f : \mathbb{R}_+ \to H$ of the form $f = \phi \otimes y$ with $\phi \in F$ and $y \in Y$, where

$$(\phi \otimes y)(t) := \phi(t)y, \quad t \geqslant 0.$$

If F and Y are dense in $L^2(\mathbb{R}_+)$ and H respectively, then $F \otimes Y$ is a dense subspace of

$L^2(\mathbb{R}_+;H)$. This is because the dt-simple functions are dense in $L^2(\mathbb{R}_+;H)$ and every such function is a linear combination of functions of the form $\mathbf{1}_B \otimes h$ with B a Borel set of finite measure and h an element of H; we now approximate $\mathbf{1}_B$ with functions in F (with respect to the norm of $L^2(\mathbb{R}_+)$) and h with elements in Y (with respect to the norm of H).

In what follows we consider the dense subspaces $F = C_c^1(\mathbb{R}_+)$ and $Y = D(A)$. For functions $f \in C_c^1(\mathbb{R}_+) \otimes D(A)$ the mild solution of the problem $u' = Au + f$ with initial condition $u(0) = 0$, given by

$$u(t) = \int_0^t S(t-s)f(s)\,ds, \quad t \geqslant 0,$$

is continuously differentiable in H, takes values in $D(A)$, and satisfies $u'(t) = Au(t) + f(t)$ for every $t \geqslant 0$.

Lemma 13.42. *Let A be the generator of a bounded analytic C_0-semigroup S on a Hilbert space H. If there exists a constant $C \geqslant 0$ such that for all $f \in C_c^1(\mathbb{R}_+) \otimes D(A)$ the mild solution u associated with f satisfies $Au \in L^2(\mathbb{R}_+;H)$ and*

$$\|Au\|_2 \leqslant C\|f\|_2,$$

where $C \geqslant 0$ is a constant independent of f, then the assertions (1) *and* (2) *of Theorem 13.41 hold for all $f \in L^2(\mathbb{R}_+;H)$, with the same constant C.*

Proof Since $C_c^1(\mathbb{R}_+) \otimes D(A)$ is dense in $L^2(\mathbb{R}_+;H)$, for any $f \in L^2(\mathbb{R}_+;H)$ we may choose functions $f_n \in C_c^1(\mathbb{R}_+) \otimes D(A)$ converging to f in $L^2(\mathbb{R}_+;H)$. Writing u and u_n for the mild solutions corresponding to f and f_n respectively, the assumptions imply that the functions Au_n form a Cauchy sequence in $L^2(\mathbb{R}_+;H)$ and therefore converge to a limit v in $L^2(\mathbb{R}_+;H)$.

Using the Cauchy–Schwarz inequality for $L^2(0,T;H)$ and taking the supremum over $t \in [0,T]$, we obtain

$$\|u_n - u_m\|_{C([0,T];H)} \leqslant T^{1/2}(\|Au_n - Au_m\|_{L^2(0,T;H)} + \|f_n - f_m\|_{L^2(0,T;H)}).$$

It follows that the functions u_n converge uniformly on every interval $[0,T]$ to a function u. As a result, for all $t \geqslant 0$ we obtain

$$u(t) = \int_0^t v(s)\,ds + \int_0^t f(s)\,ds.$$

Since A is closed, a standard subsequence argument furthermore gives that v takes values in $D(A)$ almost surely and $v = Au$ in $L^2(\mathbb{R}_+;H)$. $\qquad\square$

The proof of Theorem 13.41 relies on the observation that the Fourier–Plancherel transform $\mathscr{F}$ on $L^2(\mathbb{R})$ extends to an isometry from $L^2(\mathbb{R};H)$ onto itself, defining

$$\mathscr{F}(\mathbf{1}_B \otimes h) := (\mathscr{F}\mathbf{1}_B) \otimes h$$

for Borel sets $B \subseteq \mathbb{R}$ of finite measure and elements $h \in H$, and extending this definition by linearity. That this extension enjoys the stated properties can be proved in exactly the same way as in the scalar-valued case, repeating the proof given for that case word by word with the obvious adjustments.

Proof of Theorem 13.41   For functions $f \in C_c^1(\mathbb{R}_+) \otimes D(A)$, the mild solution u associated with f takes values in $D(A)$ and satisfies $Au = Vf$ in $L^2(\mathbb{R}_+;H)$, where

$$Vf(t) := \int_0^t AS(t-s)f(s)\,\mathrm{d}s, \quad t \geqslant 0.$$

Thus the assumptions of Lemma 13.42 are satisfied if we can show that $Vf \in L^2(\mathbb{R}_+;H)$ for all $f \in C_c^1(\mathbb{R}_+) \otimes D(A)$ and

$$\|Vf\|_2 \leqslant C\|f\|_2, \quad f \in C_c^1(\mathbb{R}_+) \otimes D(A),$$

where C is the constant from the statement of the theorem. In order to set the stage for the Fourier transform we translate this into a statement about functions defined on the full real line. Let $K(t) := AS(t)$ for $t > 0$ and $K(t) := 0$ for $t \leqslant 0$, and define

$$\overline{V}f(t) := \int_{-\infty}^{\infty} K(t-s)f(s)\,\mathrm{d}s, \quad f \in C_c^1(\mathbb{R}_+) \otimes D(A), \, t \in \mathbb{R}.$$

Then $\overline{V}f = Vf$ for functions $f \in C_c^1(\mathbb{R}_+) \otimes D(A)$ (where on the left-hand side we think of f as being extended identically zero to all of $\mathbb{R}$), so it suffices to prove that $\overline{V}$ maps $C_c^1(\mathbb{R}_+) \otimes D(A)$ into $L^2(\mathbb{R};H)$ with bound

$$\|\overline{V}f\|_2 \leqslant C\|f\|_2, \quad f \in C_c^1(\mathbb{R}_+) \otimes D(A). \tag{13.13}$$

This will be achieved by showing that

$$\overline{V}f = T_m f, \quad f \in C_c^1(\mathbb{R}_+) \otimes D(A), \tag{13.14}$$

where T_m is the (operator-valued) Fourier multiplier operator on $L^2(\mathbb{R};H)$ with

$$m(\xi) := AR(i\xi, A) = i\xi R(i\xi, A) - I, \quad \xi \in \mathbb{R} \setminus \{0\},$$

that is,

$$T_m f = \mathscr{F}^{-1}(m\mathscr{F}f), \quad f \in L^2(\mathbb{R};H).$$

To see that the operator T_m is well defined and bounded, we note that since A generates a bounded analytic C_0-semigroup the function

$$m(\xi) := AR(i\xi, A) = i\xi R(i\xi, A) - I, \quad \xi \in \mathbb{R} \setminus \{0\},$$

is uniformly bounded. Moreover, by holomorphy, this function is continuous from $\mathbb{R} \setminus$

$\{0\}$ into $\mathcal{L}(H)$. As a consequence, the mapping $g \mapsto mg$ given almost everywhere by applying $m(\xi)$ to $g(\xi)$ is well defined and bounded on $L^2(\mathbb{R};H)$ as required, with norm

$$\|T_m\| = \sup_{\eta \in \mathbb{R}\setminus\{0\}} \|AR(i\eta, A)\|.$$

This gives (13.14) as well as (13.13) with the correct value for C.

In order to prove the identity (13.14) we must show that

$$\widehat{\overline{V}f} = m\widehat{f}, \quad f \in C_c^1(\mathbb{R}_+) \otimes D(A).$$

At least formally, the operator $\overline{V}$ has the form of a convolution with K, so in view of Proposition 5.29 one is led to believe that the identity $\mathscr{F}\overline{V}f = \mathscr{F}(K * f) = \sqrt{2\pi}\widehat{K}\widehat{f} = m\widehat{f}$ should hold since, at least formally,

$$\sqrt{2\pi}\widehat{K}(\xi) = \int_{-\infty}^{\infty} e^{-it\xi} K(t)\,dt$$

$$= \int_0^{\infty} e^{-it\xi} AS(t)\,dt = A \int_0^{\infty} e^{-it\xi} S(t)\,dt = AR(i\xi, A) = m(\xi),$$

and this would give the desired result. None of the steps in this formal argument is rigorous, however, and the remainder of the proof is devoted to presenting a rigorous version of it.

We "mollify" both $\overline{V}$ and m by defining, for $r > 0$, the regularising operator

$$\begin{aligned} B(r) &= -(1 - rA)^{-1}A(r - A)^{-1} \\ &= -r^{-1}(r^{-1} - A)^{-1}[r - (r - A)](r - A)^{-1} \\ &= -(r^{-1} - A)^{-1}(r - A)^{-1} + r^{-1}(r^{-1} - A)^{-1}. \end{aligned} \tag{13.15}$$

If $y \in R(A)$, say $y = Ax$, then

$$\begin{aligned} -(r^{-1} - A)^{-1}(r - A)^{-1}y &= (r^{-1} - A)^{-1}(r - A)^{-1}[(r - A) - r]x \\ &= (r^{-1} - A)^{-1}x - r(r - A)^{-1}(r^{-1} - A)^{-1}x. \end{aligned}$$

As $r \downarrow 0$ we have $r^{-1}(r^{-1} - A)^{-1}y \to y$ by Propositions 13.10 and 13.8, and the latter one implies $\|r(r - A)^{-1}\| \leqslant M$ and $\|(r^{-1} - A)^{-1}\| \leqslant M/r^{-1}$, where M is as in the proposition. Combining these observations, we find that

$$\lim_{r \downarrow 0} B(r)Ax = Ax, \quad x \in D(A). \tag{13.16}$$

Moreover, (13.15) implies that $\|B(r)\| \leqslant (M/r^{-1})(M/r) + M = M^2 + M$.

Define $m_r(\xi) = m(\xi)B(r)$ for $\xi \neq 0$, and

$$\overline{V}_r f(t) := \int_0^t AS(t - s)B(r)f(s)\,ds = \int_{-\infty}^{\infty} K_r(t - s)f(s)\,ds, \quad f \in C_c^1(\mathbb{R}_+) \otimes D(A),$$

where $K_r(t) = AS(t)B(r)$ for $t > 0$ and $K_r(t) = 0$ otherwise.

By Theorem 13.31 and the uniform boundedness of $S(t)$ for $t > 0$,

$$\|K_r(t)\| = \|A^2 S(t)\| \, \|(1 - rA)^{-1}\| \, \|(r - A)^{-1}\| \leqslant C_r / t^2$$

and

$$\|K_r(t)\| \leqslant \|S(t)\| \, \|A(1 - rA)^{-1}\| \, \|A(r - A)^{-1}\| \leqslant C_r,$$

where C_r is independent of $t > 0$. It follows that $K_r \in L^1(\mathbb{R}; \mathscr{L}(H))$ and thus, by dominated convergence and Proposition 13.8,

$$\sqrt{2\pi} \widehat{K}_r(\xi) = \lim_{\eta \downarrow 0} \int_0^\infty e^{-(\eta + i\xi)t} K_r(t) \, dt = \lim_{\eta \downarrow 0} A(\eta + i\xi - A)^{-1} B(r) = m_r(\xi).$$

Therefore, $\overline{V}_r = T_{m_r}$ on $C_c^1(\mathbb{R}_+) \otimes D(A)$.

We now let $r \downarrow 0$. By (13.16) and dominated convergence, for $f = g \otimes x \in C_c^1(\mathbb{R}_+) \otimes D(A)$ and $t \in \mathbb{R}$ we have

$$\overline{V}_r f(t) = \int_0^t g(s) S(t - s) B(r) Ax \, ds \to \int_{-\infty}^\infty g(s) S(t - s) Ax \, ds = \overline{V} f(t)$$

for every $t \in \mathbb{R}$. Similarly, by (13.16) and the uniform boundedness of the operators $B(r)$,

$$m_r(\xi) \widehat{f}(\xi) = \widehat{g}(\xi)(i\xi - A)^{-1} B(r) Ax \to \widehat{g}(\xi)(i\xi - A)^{-1} Ax = m(\xi) \widehat{f}(\xi),$$

with convergence in $L^2(\mathbb{R}; H)$. Therefore, $T_{m_r} f \to T_m f$ in $L^2(\mathbb{R}; H)$, and along an appropriate subsequence we also have almost everywhere convergence. This shows that $\overline{V} f = T_m f$, and by linearity this implies $\overline{V} f = T_m f$ for all $f \in C_c^1(\mathbb{R}_+) \otimes D(A)$. $\square$

We demonstrate the usefulness of maximal regularity by proving local existence for the time-dependent inhomogeneous Cauchy problem

$$\begin{cases} u'(t) = A(t)u(t) + f(t), & t \in [0, T], \\ u(0) = 0, \end{cases} \tag{13.17}$$

where $(A(t))_{t \in [0,T]}$ is a family of densely defined closed operators in a Hilbert space H. We make the following assumptions:

- each domain $D(A(t))$ is isomorphic to a fixed Banach space D which is continuously and densely embedded in H;
- the mapping $t \mapsto A(t) \in \mathscr{L}(D, H)$ is continuous on $[0, T]$;
- the operator $A(0)$ is invertible and generates a bounded analytic C_0-semigroup on H.

The idea is to rewrite the problem in the form

$$u'(t) = A(0)u(t) + g_u(t) \quad \text{with} \quad g_u(t) := (A(t) - A(0))u(t) + f(t).$$

Now let $0 < a \leqslant T$ and consider a fixed function $u \in L^2(0,a;D)$. Referring to Theorem 13.24, denote by $K_a(u)$ the mild solution of the inhomogeneous problem

$$\begin{cases} u'(t) = A(0)u(t) + g_u(t), & t \in [0,a], \\ u(0) = 0. \end{cases} \tag{13.18}$$

Then, at least formally, the solutions of (13.17) are the fixed points of K_a. The maximal regularity of $A(0)$ will now be used to show that K_a is a uniform contraction (that is, its norm is strictly smaller than one) in $L^2(0,a;D)$ provided $0 < a \leqslant T$ is small enough. The Banach fixed point theorem then gives the existence of a unique fixed point for K_a in $L^2(0,a;D)$. This fixed point will be called the *solution* on $(0,a)$.

Indeed, if $u_1, u_2 \in L^2(0,a;D)$, then $K_a(u_1) - K_a(u_2)$ equals the solution u of

$$u'(t) = A(0)u(t) + g_{u_1}(t) - g_{u_2}(t), \quad u(0) = 0.$$

Since D is isomorphic to $D(A(0))$, which is a Banach space with respect to the norm $x \mapsto \|A(0)x\|$ since $0 \in \rho(A(0))$, we obtain with the maximal regularity inequality for the problem (13.18) on the interval $(0,a)$ that

$$\begin{aligned} \|K_a(u_1) - K_a(u_2)\|_{L^2(0,a;D)} &= \|A(0)u\|_{L^2(0,a;H)} \\ &\leqslant C\|g_{u_1} - g_{u_2}\|_{L^2(0,a;H)} \\ &= C\|[A(\cdot) - A(0)](u_1 - u_2)\|_{L^2(0,a;H)} \\ &\leqslant C \sup_{t \in [0,a]} \|A(t) - A(0)\|_{\mathscr{L}(D,H)} \|u_1 - u_2\|_{L^2(0,a;D)}. \end{aligned}$$

To justify the first inequality we extend the inhomogeneity $g_{u_1} - g_{u_2}$ identically 0 on $[a,\infty)$ and observe that the mild solution for the resulting inhomogeneous problem on $\mathbb{R}_+$ restricts to u on the interval $(0,a)$.

If the constant $a > 0$ is small enough, then

$$\sup\{\|A(t) - A(0)\| : t \leqslant a\} < 1/C$$

and $\|K_a\|_{\mathscr{L}(L^2(0,a;D))} < 1$, and the Banach fixed point theorem provides a unique solution for (13.17) on $(0,a)$.

13.5 Stone's Theorem

If A is a positive selfadjoint operator in a Hilbert space H, then $-A$ satisfies the conditions of Theorem 13.33. Denote by S the analytic C_0-semigroup of contractions generated by $-A$. It can be shown (see Problem 13.12) that for all $t \in \mathbb{R}$ and $x \in H$ the limit

$$U(t)x := \lim_{s \downarrow 0} S(s - it)x$$

exists and that the family $(U(t))_{t\in\mathbb{R}}$ is a C_0-group of unitary operators with generator iA. The main result of this section is Stone's theorem, which asserts that for any selfadjoint operator A it is true that the operator iA generate C_0-groups of unitary operators. For the proof of this theorem we need the following auxiliary result. A more precise version for bounded selfadjoint operators has been proved in Theorem 8.11.

Proposition 13.43. *If A is a selfadjoint operator in H, then $\sigma(A) \subseteq \mathbb{R}$ and*

$$\|R(\lambda,A)\| \leq \frac{1}{|\operatorname{Im}\lambda|}, \quad \lambda \in \mathbb{C}\setminus\mathbb{R}.$$

Proof Let $\lambda = \alpha + i\beta$ with $\beta \neq 0$. For all $x \in D(A)$ we have $(Ax|x) = (x|Ax) = \overline{(Ax|x)}$ and therefore $(Ax|x) \in \mathbb{R}$. Then,

$$\|(\lambda - A)x\|\|x\| \geq |((\lambda - A)x|x)| = |\alpha(x|x) - (Ax|x) + i\beta(x|x)| \geq |\beta|\,\|x\|^2$$

and therefore $\|(\lambda - A)x\| \geq \beta\,\|x\|$. This implies that $\lambda - A$ is injective and by Proposition 10.26 it has closed range. The same argument can be applied to $\overline{\lambda}$ and allows us to conclude that $\overline{\lambda} - A$ is injective and has closed range. Moreover, using that $((\overline{\lambda} - A)x|y) = (x|(\lambda - A)y)$, the injectivity of $\overline{\lambda} - A$ implies that $\lambda - A$ has dense range.

We conclude that $\lambda - A$ is bijective, hence invertible, and from the inequality $\|(\lambda - A)x\| \geq |\beta|\,\|x\|$ we see that $\|R(\lambda,A)\| \leq 1/|\beta|$. □

Theorem 13.44 (Stone). *For a densely defined operator A in H, the following assertions are equivalent:*

(1) *A is selfadjoint;*
(2) *iA is the generator of a C_0-group of unitary operators.*

Proof (1)$\Rightarrow$(2): By Proposition 13.43, $\sigma(A)$ is contained in the real line and for all $\lambda \in \mathbb{C}\setminus\mathbb{R}$ we have $\|R(\lambda,A)\| \leq 1/|\operatorname{Im}\lambda|$. Hence, by the Hille–Yosida theorem (Theorem 13.17), the operators $\pm iA$ generate C_0-contraction semigroups $S_\pm$. Hence, by Proposition 13.13, iA generates a C_0-group of contractions given by

$$U(t) := \begin{cases} S_+(t), & t \geq 0, \\ S_-(t), & t \leq 0. \end{cases}$$

Also, since $(iA)^\star = -iA^\star = -iA$, we have $S_-(t) = S_+^\star(t)$ and vice versa, from which it follows that the operators $U(\pm t)$ are unitary.

(2)$\Rightarrow$(1): Suppose that iA generates the unitary group $(U(t))_{t\in\mathbb{R}}$. From $U(-t) = (U(t))^{-1} = U^\star(t)$ we see that $(U^\star(t))_{t\in\mathbb{R}}$ is a C_0-group as well. To determine its generator, which we call B for the moment, suppose that $x \in D(A)$ and $h \in D(B)$. Then

$$(x|Bh) = \lim_{t\to 0}\frac{1}{t}(x|U^\star(t)h - h) = \lim_{t\to 0}\frac{1}{t}(U(t)x - x|h) = (iAx|h).$$

This shows that $h \in D(A^*)$ and $-iA^* = (iA)^*h = Bh$. In the converse direction, if $h \in D(A^*)$, then for all $x \in D(A)$ we have

$$(x| -iA^*h) = (iAx|h) = \lim_{t \to 0} \frac{1}{t}(U(t)x - x|h) = \lim_{t \to 0} \frac{1}{t}(x|U^*(t)h - h) = (x|Bh).$$

This shows that $h \in D(B)$ and $Bh = -iA^*h$. We conclude that $B = -iA^*$ with equal domains. The identity

$$\frac{1}{t}(U(-t)x - x) = \frac{1}{t}(U^*(t)x - x)$$

then shows that $x \in D(A)$ if and only if $x \in D(A^*)$ and $-iAx = Bx = -iA^*x$. $\qquad\square$

Some applications of this theorem will be given in the next section (see Sections 13.6.g and 13.6.h).

13.6 Examples

In this section we collect some important examples of C_0-semigroups and C_0-groups.

13.6.a Multiplication Semigroups

Let $(\Omega, \mathscr{F}, \mu)$ be a measure space and let $m : \Omega \to \mathbb{K}$ be measurable with real part bounded from below:

$$\inf_{\omega \in \Omega} \operatorname{Re} m(\omega) =: M > -\infty.$$

The operators

$$S(t)f := e^{-tm}f, \quad t \geqslant 0,$$

are bounded on $L^p(\Omega)$, $1 \leqslant p \leqslant \infty$, with norm $\|S(t)\| \leqslant e^{-tM}$. We will prove that S is a C_0-semigroup on $L^p(\Omega)$ for $1 \leqslant p < \infty$, with generator A given by

$$\begin{aligned} D(A) &:= \{f \in L^p(\Omega) : mf \in L^p(\Omega)\}, \\ Af &:= -mf, \quad f \in D(A). \end{aligned} \tag{13.19}$$

Fix $1 \leqslant p < \infty$. The semigroup properties (S1) and (S2) are clear and (S3) follows by dominated convergence. To prove (13.19) let $f \in L^p(\Omega)$ be such that $mf \in L^p(\Omega)$. For μ-almost all $\omega \in \Omega$ we have

$$S(t)f(\omega) - f(\omega) = e^{-tm(\omega)}f(\omega) - f(\omega) = -m(\omega)f(\omega)\int_0^t e^{-sm(\omega)}\,\mathrm{d}s.$$

Also, by Proposition 13.4, $S(t)f - f = A\int_0^t S(s)f\,\mathrm{d}s$. It follows that

$$A\int_0^t S(s)f\,\mathrm{d}s = -mf\int_0^t e^{-sm}\,\mathrm{d}s.$$

Next we note that

$$\lim_{t \downarrow 0} \frac{1}{t} \int_0^t S(s) f \, ds = f$$

and

$$\lim_{t \downarrow 0} A \frac{1}{t} \int_0^t S(s) f \, ds = -\lim_{t \downarrow 0} m f \frac{1}{t} \int_0^t e^{-sm} \, ds = -mf$$

in $L^p(\Omega)$. Since A is closed this implies $f \in D(A)$ and $Af = -mf$.
Conversely, if $f \in D(A)$, then the limit

$$\lim_{t \downarrow 0} \frac{1}{t} (e^{-tm} f - f)$$

exists in $L^p(\Omega)$ and equals Af. Since convergence in $L^p(\Omega)$ implies μ-almost every-where convergence along a subsequence, there is a sequence $t_n \downarrow 0$ such that

$$Af(\omega) = \lim_{n \to \infty} \frac{1}{t_n} (e^{-t_n m(\omega)} f(\omega) - f(\omega))$$

for μ-almost all $\omega \in \Omega$. Clearly,

$$\lim_{n \to \infty} \frac{1}{t_n} (e^{-t_n m(\omega)} f(\omega) - f(\omega)) = -m(\omega) f(\omega).$$

It follows that $mf \in L^p(\Omega)$ and $Af = -mf$.

13.6.b The Translation Group

On the space $L^p(\mathbb{R})$, $1 \leqslant p < \infty$, the formula

$$(S(t)f)(x) := f(x+t), \quad x \in \mathbb{R}, \ t \in \mathbb{R},$$

defines a C_0-group S. Its generator A is given by

$$D(A) = W^{1,p}(\mathbb{R}),$$
$$Af = f', \quad f \in D(A).$$

The group properties (G1) and (G2) are clear and (G3) follows from Proposition 2.32.
To prove that $D(A) = W^{1,p}(\mathbb{R})$ and $Af = f'$, we first note that for $f \in C_c^1(\mathbb{R})$ we have

$$S(t)f(x) - f(x) = f(x+t) - f(x) = \int_0^t f'(x+s) \, ds = \int_0^t S(s) f'(x) \, ds.$$

It follows that $S(t)f - f = \int_0^t S(s) f' \, ds$. Also, $S(t)f - f = A \int_0^t S(s) f \, ds$. It follows that

$$A \int_0^t S(s) f \, ds = \int_0^t S(s) f' \, ds.$$

Next we note that

$$\lim_{t \to 0} \frac{1}{t} \int_0^t S(s) f \, ds = f$$

and

$$\lim_{t \to 0} A \frac{1}{t} \int_0^t S(s) f \, ds = \lim_{t \to 0} \frac{1}{t} \int_0^t S(s) f' \, ds = f'$$

in $L^p(\mathbb{R})$. Since A is closed this implies $f \in D(A)$ and $Af = f'$.

Since $C_c^1(\mathbb{R})$ is dense in $L^p(\mathbb{R})$ and invariant under translations, from Proposition 13.5 we infer that $C_c^1(\mathbb{R})$ is dense in $D(A)$. Since $C_c^1(\mathbb{R})$ is also dense in $W^{1,p}(\mathbb{R})$ and $\|f\|_{D(A)} = \|f\| + \|Af\| = \|f\| + \|f'\| = \|f\|_{W^{1,p}(\mathbb{R})}$ for all $f \in C_c^1(\mathbb{R})$, it follows that $D(A) = W^{1,p}(\mathbb{R})$ and $Af = f'$ for all $f \in D(A) = W^{1,p}(\mathbb{R})$.

13.6.c The Heat Semigroup

The Heat Semigroup on $\mathbb{R}^d$ For $1 \leqslant p < \infty$ and $t > 0$ we define a linear operator $H(t)$ on $L^p(\mathbb{R}^d)$ by

$$H(t) f(x) = K_t * f(x), \quad f \in C_c(\mathbb{R}^d), x \in \mathbb{R}^d, \tag{13.20}$$

where

$$K_t(x) := (4\pi t)^{-d/2} e^{-|x|^2/4t}$$

is the *heat kernel*. Since $K_t \in L^1(\mathbb{R}^d)$ with $\|K_t\|_1 = 1$, it follows from Young's inequality (Proposition 2.33) that for all $1 \leqslant p < \infty$ the operators $H(t)$ are well defined and bounded on $L^p(\mathbb{R}^d)$ and satisfy

$$\|H(t) f\|_{L^p(\mathbb{R}^d)} \leqslant \|f\|_{L^p(\mathbb{R}^d)}.$$

We furthermore set $H(0) := I$, the identity operator on $L^p(\mathbb{R}^d)$. We will prove that the family $H = (H(t))_{t \geqslant 0}$ is a C_0-semigroup of contractions, the so-called *heat semigroup*, on $L^p(\mathbb{R}^d)$ and that its generator A is the weak L^p-Laplacian Δ. Thus the heat semigroup solves the linear heat equation

$$\begin{cases} \dfrac{\partial u}{\partial t}(t,x) = \Delta u(t,x), & t \geqslant 0, x \in \mathbb{R}^d, \\ u(0,x) = f(x), & x \in \mathbb{R}^d, \end{cases}$$

in the sense that its orbits satisfy $\frac{d}{dt} H(t) f = \Delta H(t) f$ and $H(0) = f$.

Step 1 – Fix $1 \leqslant p < \infty$. First we prove that H is a C_0-semigroup on $L^p(\mathbb{R}^d)$. For all $t > 0$, by Lemma 5.19 and a change of variables the Fourier transform of K_t is given by

$$\widehat{K_t}(\xi) := \frac{1}{(2\pi)^{d/2}} \int_{\mathbb{R}^d} K_t(x) e^{-ix\xi} \, dx = \frac{1}{(2\pi)^{d/2}} e^{-t|\xi|^2}, \quad \xi \in \mathbb{R}^d.$$

It follows that $(2\pi)^{d/2}\widehat{K_t}\widehat{K_s} = \widehat{K_{t+s}}$ for each $t, s > 0$, and by Proposition 5.29 this implies $H(t+s)f = H(t)H(s)f$ for all $f \in L^2(\mathbb{R}^d)$. In particular this identity holds for functions $f \in L^p(\mathbb{R}^d) \cap L^2(\mathbb{R}^d)$, and since we have already seen that the operators $H(t)$ are contractive on $L^p(\mathbb{R}^d)$ the identity $H(t+s)f = H(t)H(s)f$ extends to general functions $f \in L^p(\mathbb{R}^d)$, by the density of $L^p(\mathbb{R}^d) \cap L^2(\mathbb{R}^d)$ in $L^p(\mathbb{R}^d)$.

Strong continuity of the semigroup is an immediate consequence of Proposition 2.34.

Step 2 – We now prove that $A = \Delta$, the weak L^p-Laplacian, with equal domains. We begin by proving the inclusion $\mathsf{D}(\Delta) \subseteq \mathsf{D}(A)$ along with the fact that

$$Af = \Delta f, \quad f \in \mathsf{D}(A).$$

First let $f \in C_c^\infty(\mathbb{R}^d)$. For all $t > 0$ we have the pointwise identities

$$\frac{\partial}{\partial t}K_t = \Delta K_t, \quad \frac{\partial}{\partial t}K_t * f = \Delta K_t * f,$$

and therefore

$$H(t)f - f = K_t * f - f = \int_0^t \Delta K_s * f \, ds = \int_0^t \Delta H(s)f \, ds. \tag{13.21}$$

Since we are assuming that $f \in C_c^\infty(\mathbb{R}^d)$, all identities can be rigorously justified by elementary calculus arguments. By mollification and smooth cut-off, $C_c^\infty(\mathbb{R}^d)$ is dense in $\mathsf{D}(\Delta)$. Since all terms in the above identity depend continuously on the graph norm of $\mathsf{D}(\Delta)$, the identity extends to arbitrary functions $f \in \mathsf{D}(\Delta)$. Dividing both sides by t and passing to the limit $t \downarrow 0$, by the continuity of $t \mapsto \Delta H(t)f$ as an $L^p(\mathbb{R}^d)$-valued function we obtain that $f \in \mathsf{D}(A)$ and $Af = \Delta f$ as claimed. This completes the proof.

To prove the converse inclusion $\mathsf{D}(A) \subseteq \mathsf{D}(\Delta)$ we must show that every $f \in \mathsf{D}(A)$ admits a weak L^p-Laplacian. To this end we multiply both sides of (13.21) with a test function ϕ and integrate by parts. This results in the identity

$$\int_{\mathbb{R}^d}(H(t)f(x) - f(x))\phi(x)\,dx = \int_0^t \int_{\mathbb{R}^d} H(s)f(x)\Delta\phi(x)\,dx\,ds.$$

Dividing by t and passing to the limit $t \downarrow 0$, and using the assumption $f \in \mathsf{D}(A)$, we obtain the identity

$$\int_{\mathbb{R}^d} Af(x)\phi(x)\,dx = \int_{\mathbb{R}^d} f(x)\Delta\phi(x)\,dx.$$

This identity precisely expresses that f admits a weak Laplacian, given by the function $Af \in L^p(\mathbb{R}^d)$.

By Theorem 11.29, for $p = 2$ we have

$$\mathsf{D}(A) = \mathsf{D}(\Delta) = W^{2,2}(\mathbb{R}^d). \tag{13.22}$$

Remark 13.45. For $1 < p < \infty$ one has the analogous equality

$$\mathsf{D}(A) = \mathsf{D}(\Delta) = W^{2,p}(\mathbb{R}^d),$$

but this is highly nontrivial and depends on the L^p-boundedness of the Riesz transforms (see the Notes to Chapter 5). For $d = 1$ there is the following more elementary argument that also works for $p = 1$. The inclusion $W^{2,p}(\mathbb{R}^d) \subseteq D(\Delta)$ being clear for any dimension d, the point is to prove the inclusion $D(\Delta) \subseteq W^{2,p}(\mathbb{R})$. If $f \in L^p(\mathbb{R})$ admits a weak L^p-Laplacian $\Delta f = f''$, Theorem 11.12 implies that f admits a weak derivative f' belonging to $L^p(\mathbb{R})$. This shows that f belongs to $W^{2,p}(\mathbb{R})$.

Remark 13.46. There is a slightly different route to the identification $A = \Delta$ for $p = 2$ which depends on the fact that each of the operators $H(t)$ is a Fourier multiplication operator associated with the multiplier $m_t(\xi) = \exp(-t|\xi|^2)$. Defining $\widetilde{H}(t)g := m_t g$ we obtain a multiplication semigroup $\widetilde{H}$ on $L^2(\mathbb{R}^d)$ which is strongly continuous and whose generator $\widetilde{A}$ is given by

$$D(\widetilde{A}) = \{g \in L^2(\mathbb{R}^d) : \xi \mapsto |\xi|^2 g(\xi) \in L^2(\mathbb{R}^d)\} = H^2(\mathbb{R}^d),$$
$$\widetilde{A}g(\xi) = -|\xi|^2 g(\xi), \quad g \in D(\widetilde{A}), \ \xi \in \mathbb{R}^d;$$

this follows from the results proved in Section 13.6.a. This semigroup is related to the heat semigroup through the identity

$$H(t) = \mathscr{F}^{-1} \circ \widetilde{H}(t) \circ \mathscr{F}, \quad t \geq 0,$$

from which it follows that a function $f \in L^2(\mathbb{R}^d)$ belongs to the domain of the generator A of H if and only if $\mathscr{F}f = \widehat{f}$ belongs to the domain of the generator $\widetilde{A}$ of $\widetilde{H}$, in which case the identity

$$Af = \mathscr{F}^{-1} \circ \widetilde{A} \circ \mathscr{F}f$$

holds. As we have seen, this is the case if and only if $f \in H^2(\mathbb{R}^d)$. Since $H^2(\mathbb{R}^d) = W^{2,2}(\mathbb{R}^d)$ up to equivalence of norm, this implies (13.22).

The Heat Semigroup on Bounded Domains Let D be a nonempty bounded open set in $\mathbb{R}^d$.

Proposition 13.47. *The Dirichlet and Neumann Laplacians on $L^2(D)$ generate analytic C_0-semigroups of selfadjoint contractions on every sector of angle less than $\frac{1}{2}\pi$.*

Proof Everything follows from the Lumer–Phillips theorem (Theorem 13.33), except the selfadjointness of the semigroup operators which follows from Euler's theorem (Theorem 13.19) after noting that for $\lambda > 0$ the resolvent operators are selfadjoint.

To check the conditions of Lumer–Phillips theorem, let us denote the Dirichlet and Neumann Laplacians by Δ. The operator $-\Delta$ is positive and selfadjoint by Theorem 12.20), and therefore $I - \Delta$ is injective. Dualising and using selfadjointness, this in turn implies that $I - \Delta$ has dense range. This verifies the first condition of Lumer–Phillips theorem; the second follows immediately from the positivity of $-\Delta$. $\square$

Alternatively, Proposition 13.47 can be deduced from the spectral theorem for self-adjoint operators (as such the result is a special case of Theorem 10.56, where further details are provided). This gives the representation

$$S(z) = \int_{[0,\infty)} e^{-zt} \, dP(t), \quad \mathrm{Re}\, z > 0,$$

where P is the projection-valued measure associated with the Laplacian under consideration (see Example 13.62).

The Dirichlet and Neumann heat semigroups on $L^2(D)$ are positivity preserving, that is, they map nonnegative functions to nonnegative functions. From the physics point of view it is natural to expect that heat semigroups should have this property, as they are meant to describe the time evolution of heat distributions. Positivity of the heat semigroup on $L^2(\mathbb{R}^d)$ is evident from the explicit representation though convolution with the heat kernel.

Theorem 13.48 (Positivity). *Let D be bounded. Then the C_0-semigroups on $L^2(D)$ generated by Δ_{Dir} and Δ_{Neum} are positivity preserving.*

Proof Let A denote the Dirichlet or Neumann Laplacian on $L^2(D)$ and let S be the C_0-contraction semigroup generated by A on $L^2(D)$. We must prove that $S(t)f \geqslant 0$ for all $t \geqslant 0$ whenever $f \in L^2(D)$ satisfies $f \geqslant 0$. In what follows we fix such a function f.

Step 1 – We first prove that for all $\lambda > 0$ we have $g := R(\lambda, A)f \geqslant 0$. Since the positive and negative parts g^+ and g^- of g have disjoint supports, they are orthogonal in $L^2(D)$ and therefore $(g^{\pm}|g) = \pm\|g^{\pm}\|^2$. Furthermore, Theorem 11.23 implies that if $g \in H^1(D)$, then $g^{\pm} \in H^1(D)$ and $\partial_j g^{\pm} = \pm\mathbf{1}_{\{\pm g > 0\}}\partial_j g$, and this in turn implies that $\int_D \nabla g \cdot \nabla g^{\pm} \, dx = \pm \int_D \|\nabla g^{\pm}\|^2 \, dx$. Combination of these facts gives

$$0 \leqslant \lambda\|g^-\|^2 = \lambda(g^-|g^-) = -\lambda(g|g^-) = -(f|g^-) - (Ag|g^-)$$
$$\leqslant -(Ag|g^-) = \int_D \nabla g \cdot \nabla g^- \, dx = -\int_D \|\nabla g^-\|^2 \, dx \leqslant 0,$$

the middle inequality being a consequence of the fact that $f \geqslant 0$, and the equality following it being a consequence of the definition of $-A$ as the operator associated with the form on the right-hand side of the equality. This proves that $g^- = 0$ in $L^2(D)$, so $R(\lambda, A)f = g = g^+ \geqslant 0$.

Step 2 – The positivity of the operators $S(t)$ follows from the result of Step 1 via the Euler formula (Theorem 13.19)

$$S(t)f = \lim_{n\to\infty} \left(\frac{n}{t}R\left(\frac{n}{t}, A\right)\right)^n f, \quad f \in L^2(D).$$

$\square$

Above we have seen that the Laplace operator Δ generates a C_0-semigroup of contractions on $L^p(\mathbb{R}^d)$ for all $1 \leqslant p < \infty$. For bounded open subsets D of $\mathbb{R}^d$, up to this point we have only considered the analogues of this semigroup on the space $L^2(D)$. We prove next that the Dirichlet and Neumann Laplacians also generate C_0-semigroups of contractions on the space $L^p(D)$ for $1 \leqslant p < \infty$. This will be derived from an abstract result on L^p-boundedness of submarkovian operators which we discuss first.

Let $(\Omega, \mathscr{F}, \mu)$ be a finite measure space. A bounded operator T on $L^2(\Omega)$ is called *doubly submarkovian* if it has the following properties:

(i) $Tf \geqslant 0$ for all $f \geqslant 0$;
(ii) $T\mathbf{1} \leqslant \mathbf{1}$ and $T^*\mathbf{1} \leqslant \mathbf{1}$.

Such operators enjoy the following extension property.

Theorem 13.49 (L^p-Boundedness of doubly submarkovian operators). *Let $(\Omega, \mathscr{F}, \mu)$ be a finite measure space and let T be a doubly submarkovian operator on $L^2(\Omega)$. Then for all $1 \leqslant p \leqslant \infty$, the restriction of T to $L^2(\Omega) \cap L^p(\Omega)$ has a unique extension to a contraction on $L^p(\Omega)$.*

Proof For all $f \in L^2(\Omega)$,

$$\|Tf\|_1 = \| \,|Tf|\, \|_1 \leqslant \| T|f| \, \|_1 = \langle T|f|, \mathbf{1} \rangle$$
$$= \big(T|f| \,\big|\, \mathbf{1}\big) = \big(|f| \,\big|\, T^*\mathbf{1}\big) \leqslant \big(|f| \,\big|\, \mathbf{1}\big) = \langle |f|, \mathbf{1} \rangle = \| \,|f|\, \|_1 = \|f\|_1,$$

where $\langle \cdot, \cdot \rangle$ denotes the L^1-L^∞ duality and $(\cdot | \cdot)$ the L^2-inner product. It follows that T has a unique extension to a contraction on $L^1(\Omega)$. By similar reasoning, for all $f \in L^2(\Omega)$ and $g \in L^\infty(\Omega)$ we have

$$|\langle f, Tg \rangle| \leqslant \langle |f|, T|g| \rangle \leqslant \langle |f|, T\mathbf{1} \rangle \|g\|_\infty \leqslant \langle |f|, \mathbf{1} \rangle \|g\|_\infty = \|f\|_1 \|g\|_\infty.$$

Since $L^2(\Omega)$ is dense in $L^1(\Omega)$, it follows that $\|Tg\|_\infty \leqslant \|g\|_\infty$. It follows that T restricts to a contraction on $L^\infty(\Omega)$.

Boundedness and contractivity for $1 < p < \infty$ now follow from the Riesz–Thorin interpolation theorem. $\qquad\square$

Turning to the L^p-boundedness of the heat semigroup, we begin with the case of Neumann boundary conditions.

Theorem 13.50 (L^p-Boundedness, Neumann boundary conditions). *Let D be bounded and let S_{Neum} denote the C_0-semigroup generated by Δ_{Neum} to $L^2(D)$. For all $1 \leqslant p < \infty$, the restriction of $S_{\mathrm{Neum}}(t)$ to $L^2(D) \cap L^p(D)$ uniquely extends to a C_0-semigroup of positivity preserving contractions on $L^p(D)$.*

Proof By Proposition 13.47 and Theorem 13.48, for all $t \geqslant 0$ the operator $S_{\mathrm{Neum}}(t)$ is selfadjoint and positivity preserving. From $\Delta_{\mathrm{Neum}}\mathbf{1} = 0$ it follows that $S_{\mathrm{Neum}}^*(t)\mathbf{1} =$

$S_{\text{Neum}}(t)\mathbf{1} = \mathbf{1}$ and therefore the operators $S_{\text{Neum}}(t)$ are doubly submarkovian. Applying Theorem 13.49 we obtain that for all $1 \leqslant p < \infty$ and $t \geqslant 0$ the restriction of $S_{\text{Neum}}(t)$ to $L^2(D) \cap L^p(D)$ has a unique extension, also denoted by $S_{\text{Neum}}(t)$, to a contraction on $L^p(D)$.

By Hölder's inequality, the strong continuity of S_{Neum} on $L^2(D)$ implies that for all $1 \leqslant p < 2$ and $f \in L^2(D)$ we have

$$\|S_{\text{Neum}}(t)f - f\|_p \leqslant |D|^{1/r} \|S_{\text{Neum}}(t)f - f\|_2 \to 0$$

as $t \downarrow 0$, where $\frac{1}{2} + \frac{1}{r} = \frac{1}{p}$. Since $\|S_{\text{Neum}}(t)\|_p \leqslant 1$ for all $t \geqslant 0$, the density of $L^2(D)$ in $L^p(D)$ implies that the strong continuity extends all $f \in L^p(D)$. For $2 < p < \infty$ we use selfadjointness to see that for all $f \in L^2(D) \cap L^p(D)$ and $g \in L^2(D) \cap L^q(D)$ with $\frac{1}{p} + \frac{1}{q} = 1$ we have

$$\begin{aligned}
|\langle S_{\text{Neum}}(t)f - f, g\rangle| &= |(S_{\text{Neum}}(t)f - f|\bar{g})| = |(f|S_{\text{Neum}}(t)\bar{g} - \bar{g})| \\
&= \langle f, S_{\text{Neum}}(t)g - g\rangle| \leqslant \|f\|_p \|S_{\text{Neum}}(t)g - g\|_q \to 0
\end{aligned} \tag{13.23}$$

applying in the last step what we already proved to the exponent $1 < q < 2$. Using the contractivity of the operators $S_{\text{Neum}}(t)$ on $L^p(D)$ and $L^q(D)$, by density (13.23) extends to arbitrary $f \in L^p(D)$ and $g \in L^q(D)$. It follows that the semigroup S_{Neum} is weakly continuous on $L^p(D)$. By Theorem 13.11, this implies its strong continuity. Positivity on $L^p(D)$ follows from the positivity of $L^2(D)$ by a density argument. $\qquad\square$

Our next aim is to prove the following analogue of Theorem 13.50 for the Dirichlet heat semigroup.

Theorem 13.51 (L^p-Boundedness, Dirichlet boundary conditions). *Let D be a bounded open subset of $\mathbb{R}^d$ and let S_{Dir} denote the C_0-semigroup generated by Δ_{Dir} to $L^2(D)$. For all $1 \leqslant p < \infty$, the restriction of S_{Dir} to $L^2(D) \cap L^p(D)$ extends to a C_0-semigroup of positivity preserving contractions on $L^p(D)$.*

The heart of the matter is to prove the following resolvent inequality.

Lemma 13.52. *For all $0 \leqslant f \in L^2(D)$ and $\lambda > 0$ we have*

$$0 \leqslant R(\lambda, \Delta_{\text{Dir}})f \leqslant R(\lambda, \Delta_{\text{Neum}})f.$$

Proof Fix $\lambda > 0$ and $0 \leqslant f \in L^2(D)$. Then $0 \leqslant u := R(\lambda, \Delta_{\text{Dir}})f \in \mathsf{D}(\Delta_{\text{Dir}})$ and $0 \leqslant v := R(\lambda, \Delta_{\text{Neum}})f \in \mathsf{D}(\Delta_{\text{Neum}})$ and

$$\lambda u - \Delta_{\text{Dir}}u = f = \lambda v - \Delta_{\text{Neum}}v. \tag{13.24}$$

The theorem will be proved by showing that this implies $u \leqslant v$.

Fix a nonnegative test function $\phi \in C_c^\infty(D)$. Multiplying (13.24) on both sides with ϕ

and integrating by parts, we arrive at

$$\lambda \int_D u\phi \, dx + \int_D \nabla u \nabla \phi \, dx = \lambda \int_D v\phi \, dx + \int_D \nabla v \nabla \phi \, dx. \tag{13.25}$$

We claim that this equality extends to all nonnegative functions $\phi \in H_0^1(D)$. Indeed, if $\phi_n \to \phi$ in $H_0^1(D)$ with $\phi_n \in C_c^\infty(D)$ for all $n \geqslant 1$, then $\phi_n^+ \to \phi$ in $H_0^1(D)$ by Theorem 11.23. Since each ϕ_n^+ has compact support, mollification with a nonnegative compactly supported smooth mollifier allows us to approximate ϕ_n^+ by nonnegative test functions in $H_0^1(D)$ as in Proposition 11.22. Applying (13.25) to these test functions and taking limits, the claim is obtained.

Next we claim that $(u-v)^+$ belongs to $H_0^1(D)$, the point here being that $u \in H_0^1(D)$ but $v \in H^1(D)$. To prove the claim let $u_k \to u$ in $H_0^1(D)$ with $u_k \in C_c^\infty(D)$. Since $v \geqslant 0$, for each $k \geqslant 1$ the function $(u_k - v)^+$ is supported in the compact support of u_k and therefore it belongs to $H_0^1(D)$ by Proposition 11.22. By another application of Theorem 11.23 it then follows that $(u-v)^+ = \lim_{k\to\infty}(u_k - v)^+$ belongs to $H_0^1(D)$ as claimed.

By (13.25), which we may now apply to $\phi = (u-v)^+$,

$$\lambda \int_D u(u-v)^+ \, dx + \int_D \nabla u \nabla (u-v)^+ \, dx = \lambda \int_D v(u-v)^+ \, dx + \int_D \nabla v \nabla (u-v)^+ \, dx.$$

As a consequence,

$$\lambda \int_D (u-v)^{+2} \, dx = \lambda \int_D (u-v)(u-v)^+ \, dx$$

$$= \int_D \nabla(v-u)\nabla(u-v)^+ \, dx = -\int_D |\nabla(u-v)^+|^2 \, dx \leqslant 0,$$

arguing as in the proof of Theorem 13.48 in the last step. This implies that $(u-v)^+ \leqslant 0$, that is, $u \leqslant v$. $\qquad\square$

Proof of Theorem 13.51 By the lemma and Euler's formula (Theorem 13.19),

$$S_{\mathrm{Dir}}(t)f = \lim_{n\to\infty} \left(\frac{n}{t}R(\frac{n}{t},\Delta_{\mathrm{Dir}})\right)^n f \leqslant \lim_{n\to\infty} \left(\frac{n}{t}R(\frac{n}{t},\Delta_{\mathrm{Neum}})\right)^n f = S_{\mathrm{Neum}}(t)f.$$

In particular this implies $S_{\mathrm{Dir}}(t)\mathbf{1} \leqslant S_{\mathrm{Neum}}(t)\mathbf{1} = \mathbf{1}$. Together with positivity and self-adjointness, this implies that the operators $S_{\mathrm{Dir}}(t)$ are doubly submarkovian. The proof can now be finished along the lines of that of Theorem 13.50. $\qquad\square$

13.6.d The Poisson Semigroup

Let $1 \leqslant p < \infty$. For $t > 0$ we define the operator $P(t)$ on $L^p(\mathbb{R}^d)$ by convolution with the *Poisson kernel*

$$p_t(x) = \frac{c_d t}{(t^2 + |x|^2)^{\frac{1}{2}(d+1)}}, \quad t > 0, \; x \in \mathbb{R}^d, \tag{13.26}$$

where $c_d = \Gamma(\frac{1}{2}(d+1))/\pi^{\frac{1}{2}(d+1)}$ with $\Gamma(t) = \int_0^\infty x^{t-1} e^{-x} dx$ the Euler Gamma function. The change of variables $y = x/t$ gives the norm estimate

$$\|p_t\|_{L^1(\mathbb{R}^d)} = c_d \int_{\mathbb{R}^d} \frac{1}{(1+|y|^2)^{\frac{1}{2}(d+1)}} \, dy$$

$$= c_d \sigma_{d-1} \int_0^\infty \frac{1}{(1+r^2)^{\frac{1}{2}(d+1)}} r^{d-1} \, dr = c_d \sigma_{d-1} \cdot \frac{1}{2}\sqrt{\pi} \, \frac{\Gamma(\frac{1}{2}d)}{\Gamma(\frac{1}{2}(d+1))} = 1,$$

where $\sigma_{d-1} = 2\pi^{d/2}/\Gamma(\frac{1}{2}d)$ is the surface area of the unit sphere in $\mathbb{R}^d$. Young's inequality guarantees that the operators $P(t)$ are well defined and contractive on $L^p(\mathbb{R}^d)$. For $d = 1$, the formula (13.26) takes the simpler form

$$p_t(x) = \frac{1}{\pi} \frac{t}{t^2 + x^2}, \quad t > 0, \, x \in \mathbb{R}.$$

Define

$$P(t)f := p_t * f$$

for $t > 0$ and $f \in L^p(\mathbb{R}^d)$ and supplement this definition by setting $P(0) := I$. To see that the operators $P(t)$ satisfy the semigroup property we will show that the Fourier transform of p_t is given by

$$\widehat{p_t}(\xi) = \frac{1}{(2\pi)^{d/2}} \exp(-t|\xi|). \tag{13.27}$$

To prove this identity we compute the inverse Fourier transform of the exponential on the right-hand side. By a standard contour integration argument, for $\gamma > 0$ we have

$$e^{-\gamma} = \frac{1}{\pi} \int_{\mathbb{R}} \frac{e^{i\gamma y}}{1+y^2} \, dy.$$

Writing $\frac{1}{1+y^2} = \int_0^\infty e^{-(1+y^2)u} \, du$, using Fubini's theorem to interchange the order of integration, and Lemma 5.19 and a change of variables to evaluate the inner integral, we obtain

$$e^{-\gamma} = \frac{1}{\pi} \int_{\mathbb{R}} \int_0^\infty e^{i\gamma y} e^{-(1+y^2)u} \, du \, dy$$

$$= \frac{1}{\pi} \int_0^\infty e^{-u} \left(\int_{\mathbb{R}} e^{i\gamma y} e^{-y^2 u} \, dy \right) du = \frac{1}{\sqrt{\pi}} \int_0^\infty \frac{e^{-u}}{\sqrt{u}} e^{-\gamma^2/4u} \, du.$$

We apply this with $\gamma = t|\xi|$. Using Fubini's theorem, Lemma 5.19, and another substitution, we obtain

$$\frac{1}{(2\pi)^{d/2}} \int_{\mathbb{R}^d} \frac{1}{(2\pi)^{d/2}} \exp(-t|\xi|) \exp(ix \cdot \xi) \, d\xi$$

$$= \frac{1}{(2\pi)^d} \int_{\mathbb{R}^d} \frac{1}{\sqrt{\pi}} \int_0^\infty \frac{e^{-u}}{\sqrt{u}} e^{-t^2|\xi|^2/4u} \exp(ix \cdot \xi) \, du \, d\xi$$

$$= \frac{1}{\sqrt{\pi}} \int_0^\infty \frac{e^{-u}}{\sqrt{u}} \frac{1}{(2\pi)^d} \int_{\mathbb{R}^d} e^{-t^2|\xi|^2/4u} \exp(ix \cdot \xi) \, d\xi \, du$$

$$= \frac{1}{\sqrt{\pi}} \int_0^\infty \frac{e^{-u}}{\sqrt{u}} \left(\frac{u}{\pi t^2}\right)^{d/2} e^{-|x|^2 u/t^2} \, du$$

$$= \frac{1}{\pi^{\frac{1}{2}(d+1)}} \frac{1}{t^d} \int_0^\infty e^{-(t^2+|x|^2)u/t^2} u^{\frac{1}{2}(d-1)} \, du$$

$$= \frac{1}{\pi^{\frac{1}{2}(d+1)}} \frac{t}{(t^2+|x|^2)^{\frac{1}{2}(d+1)}} \int_0^\infty e^{-v} v^{\frac{1}{2}(d-1)} \, dv$$

$$= \frac{1}{\pi^{\frac{1}{2}(d+1)}} \frac{t}{(t^2+|x|^2)^{\frac{1}{2}(d+1)}} \Gamma(\tfrac{1}{2}(d+1)) = p_t(x).$$

This completes the proof of (13.27). Thanks to this identity, for $t, s > 0$ we obtain

$$\widehat{p_t * p_s} = (2\pi)^{d/2} \widehat{p_t} \widehat{p_s} = (2\pi)^{-d/2} \exp(-t|\xi|) \exp(-s|\xi|)$$
$$= (2\pi)^{-d/2} \exp(-(t+s)|\xi|) = \widehat{p_{t+s}}$$

and therefore $p_t * p_s = p_{t+s}$. It follows that for, say, $f \in C_c(\mathbb{R}^d)$,

$$P(t)P(s)f = p_t * (p_s * f) = (p_t * p_s) * f = p_{t+s} * f = P(t+s)f.$$

Since $C_c(\mathbb{R}^d)$ is dense in $L^p(\mathbb{R}^d)$ for $1 \leqslant p < \infty$ this proves that $P(t)P(s) = P(t+s)$ for all $t, s > 0$. This identity of course trivially extends to $t, s \geqslant 0$. Strong continuity of the family P on $L^p(\mathbb{R}^d)$ for $1 \leqslant p < \infty$ is an immediate consequence of Proposition 2.34.

We will determine the generator A of this semigroup for $p = 2$. It will turn out that

$$A = -(-\Delta)^{1/2}, \quad D(A) = H^1(\mathbb{R}^d).$$

Here, the square root first is defined by the functional calculus of the selfadjoint operator $-\Delta$ (see Proposition 10.60) or by the following more direct argument. Recall the definition of $H^1(\mathbb{R}^d)$ as the space of all $f \in L^2(\mathbb{R}^d)$ for which

$$\xi \mapsto (1+|\xi|^2)^{1/2} \widehat{f}(\xi)$$

belongs to $L^2(\mathbb{R}^d)$. In view of the trivial inequality $|\xi| \leqslant (1+|\xi|^2)^{1/2}$, for all $f \in H^1(\mathbb{R}^d)$ the function $\xi \mapsto |\xi| \widehat{f}(\xi)$ belongs to $L^2(\mathbb{R}^d)$. Thus we can define an operator $(B, D(B))$ by

$$Bf := (\xi \mapsto |\xi| \widehat{f}(\xi))^\vee, \quad D(B) = H^1(\mathbb{R}^d).$$

Note the formal analogy with the definition of Fourier multiplier operators; the only difference here is that the multiplier function $m(\xi) = |\xi|$ does not belong to $L^\infty(\mathbb{R}^d)$ and is therefore not covered by the definition of these operators. For $f \in H^2(\mathbb{R}^d)$ we similarly have

$$-\Delta f = (\xi \mapsto |\xi|^2 \widehat{f}(\xi))^\vee,$$

so that $Bf \in D(-\Delta)$ and $B^2 f := B(Bf) = -\Delta f$. This justifies the notation

$$(-\Delta)^{1/2} := B, \quad D((-\Delta)^{1/2}) = H^1(\mathbb{R}^d). \tag{13.28}$$

From

$$(Bf|f) = (|\cdot|\widehat{f}(\cdot)|\widehat{f}(\cdot)) = \int_{\mathbb{R}^d} |\xi||\widehat{f}(\xi)|^2 \, \mathrm{d}\xi \geqslant 0$$

we see that B is positive. The uniqueness part of Proposition 10.60 implies that B co-incides with the positive square root of $-\Delta$ obtained in the corollary by means of the functional calculus of $-\Delta$.

In dimension $d = 1$ the identity $|\xi| = -i\,\mathrm{sign}(\xi) \cdot i\xi$ implies that

$$(-\mathrm{d}^2/\mathrm{d}x^2)^{1/2} = H \circ \frac{\mathrm{d}}{\mathrm{d}x},$$

where H is the Hilbert transform (which, as we recall from Section 5.6, is the Fourier multiplier operator corresponding to the multiplier $\xi \mapsto -i\,\mathrm{sign}(\xi)$).

Theorem 13.53 (Poisson semigroup). *The Poisson semigroup on $L^2(\mathbb{R}^d)$ is generated by the selfadjoint operator $-(-\Delta)^{1/2}$.*

Selfadjointness follows from Example 10.39.

Proof We start by noting that a function $f \in L^2(\mathbb{R}^d)$ belongs to $H^1(\mathbb{R}^d)$ if and only if $\xi \mapsto |\xi|\widehat{f}(\xi)$ belongs to $L^2(\mathbb{R}^d)$. The 'only if' part has already been noted, and the 'if' part follows from the inequality $(1 + |\xi|^2)^{1/2} \leqslant 1 + |\xi|$ in the same way.

On $L^2(\mathbb{R}^d)$ we now define the multiplication semigroup Q by

$$Q(t)g(\xi) := e^{-t|\xi|}g(\xi)$$

for $t \geqslant 0$ and $g \in L^2(\mathbb{R}^d)$. As we have shown in Section 13.6.a, this is a C_0-semigroup whose generator $(C, D(C))$ is given by

$$Cg(\xi) = -|\xi|g(\xi)$$

for $g \in D(C) = \{g \in L^2(\mathbb{R}^d) : \xi \mapsto |\xi|g(\xi) \in L^2(\mathbb{R}^d)\}$. Evidently,

$$P(t) = \mathscr{F}^{-1} \circ Q(t) \circ \mathscr{F}, \quad t \geqslant 0,$$

from which it follows that a function $f \in L^2(\mathbb{R}^d)$ belongs to the domain of the generator A of P if and only if $\mathscr{F}f = \widehat{f}$ belongs to the domain of the generator C of Q, in which case the identity

$$Af = \mathscr{F}^{-1} \circ C \circ \mathscr{F}f$$

holds. By the observation at the beginning of the proof and (13.28), $\mathscr{F}f$ belongs to $D(C)$ if and only if $f \in H^1(\mathbb{R}^d) = D((-\Delta)^{1/2})$, and in that case we have

$$-(-\Delta)^{1/2}f = (\xi \mapsto -|\xi|\widehat{f})^{\vee} = \mathscr{F}^{-1} \circ C \circ \mathscr{F}f.$$

These considerations prove that $A = (-\Delta)^{1/2}$ with equality of their domains. □

The operator $(-\Delta)^{1/2}$ has an interesting connection with the wave equation which will be elaborated in Section 13.6.h.

13.6.e The Ornstein–Uhlenbeck Semigroup

In this section we assume some elementary knowledge about Gaussian random variables. Let

$$d\gamma(x) = \frac{1}{(2\pi)^{d/2}} \exp\left(-\frac{1}{2}|x|^2\right) dx$$

denote the standard Gaussian measure on $\mathbb{R}^d$. For $t \geqslant 0$ and $f \in C_c(\mathbb{R}^d)$, define the operator $OU(t)$ on $L^2(\mathbb{R}^d, \gamma)$ by

$$OU(t)f(x) := \int_{\mathbb{R}^d} f(e^{-t}x + \sqrt{1 - e^{-2t}}y) \, d\gamma(y), \quad x \in \mathbb{R}^d.$$

We will prove that OU is a C_0-semigroup on $L^p(\mathbb{R}^d, \gamma)$ for all $1 \leqslant p < \infty$. This semigroup is known as the *Ornstein–Uhlenbeck semigroup*. It plays a central role in the so-called Malliavin calculus, an infinite-dimensional Gaussian version of calculus which finds applications in, for example, the theory of stochastic (partial) differential equations and mathematical finance. Interestingly, this semigroup also makes its appearance in Quantum Field Theory, where its negative generator takes the role of the so-called bosonic number operator. This point of view will be taken up in Section 15.6.

Let us first show that each operator $OU(t)$ extends to a bounded operator on $L^p(\mathbb{R}^d, \gamma)$ of norm at most 1. By Hölder's inequality, for all $f \in C_c(\mathbb{R}^d)$ we have

$$\begin{aligned}
\|OU(t)f\|_{L^p(\mathbb{R}^d, \gamma)}^p &= \int_{\mathbb{R}^d} \left| \int_{\mathbb{R}^d} f(e^{-t}x + \sqrt{1 - e^{-2t}}y) \, d\gamma(y) \right|^p d\gamma(x) \\
&\leqslant \int_{\mathbb{R}^d} \int_{\mathbb{R}^d} \left| f(e^{-t}x + \sqrt{1 - e^{-2t}}y) \right|^p d\gamma(x) \, d\gamma(y) \\
&= \mathbb{E} \left| f(e^{-t}X + \sqrt{1 - e^{-2t}}Y) \right|^p,
\end{aligned}$$

where X and Y are independent standard Gaussian random variables defined on some probability space and $\mathbb{E}$ is the expectation with respect to the probability measure. Since $(e^{-t})^2 + (\sqrt{1 - e^{-2t}})^2 = 1$, the random variable $e^{-t}X + \sqrt{1 - e^{-2t}}Y$ is standard Gaussian again, so it is equal in distribution to X. Hence

$$\mathbb{E} \left| f(e^{-t}X + \sqrt{1 - e^{-2t}}Y) \right|^p = \mathbb{E}|f(X)|^p = \int_{\mathbb{R}^d} |f(x)|^p \, d\gamma(x) = \|f\|_{L^p(\mathbb{R}^d, \gamma)}^p.$$

This proves that $\|OU(t)f\|_p \leqslant \|f\|_p$.

Next we claim that $C_c^\infty(\mathbb{R}^d)$ is dense in $L^p(\mathbb{R}^d, \gamma)$. To this end we first approximate f by a function of the form ψf, where $0 \leqslant \psi \leqslant 1$ and $\psi \equiv 1$ on a large enough open ball $B(0;r)$ in $\mathbb{R}^d$. On each ball $B(0;r)$ the Gaussian density is bounded from below, and therefore convergence in the $L^2(B(0;r),\gamma)$-norm is equivalent to convergence in the $L^p(B(0;r))$-norm. Since $C_c^\infty(B(0;r))$ is dense in $L^p(B(0;r))$ the desired result follows.

Combining the above two steps it follows that the operators $OU(t)$ extend uniquely to contractions on $L^p(\mathbb{R}^d,\gamma)$. By a limiting argument involving the extraction of an almost everywhere convergent subsequence and dominated convergence, the defining formula for $OU(t)$ extends to arbitrary functions $f \in L^p(\mathbb{R}^d,\gamma)$, in the sense that for every $f \in L^p(\mathbb{R}^d,\gamma)$ the formula holds for almost all $x \in \mathbb{R}^d$.

Next we prove that OU is a C_0-semigroup on $L^p(\mathbb{R}^d,\gamma)$. It is clear that (S1) holds. To prove the semigroup property (S2) let us first fix a function $f \in C_c(\mathbb{R}^d)$. Then $OU(t)f \in C_b(\mathbb{R}^d)$ and

$$
\begin{aligned}
OU(t)OU(s)f(x) &= \int_{\mathbb{R}^d} OU(s)f(e^t x + \sqrt{1-e^{-2t}}y)\,\mathrm{d}\gamma(y) \\
&= \int_{\mathbb{R}^d}\int_{\mathbb{R}^d} f(e^{-s}(e^{-t}x + \sqrt{1-e^{-2t}}y) + \sqrt{1-e^{-2s}}z)\,\mathrm{d}\gamma(z)\,\mathrm{d}\gamma(y) \\
&= \mathbb{E}f(e^{-(t+s)}x + e^{-s}\sqrt{1-e^{-2t}}Y + \sqrt{1-e^{-2s}}Z),
\end{aligned}
$$

where Y and Z are independent standard Gaussians. In view of the identity

$$
(e^{-s}\sqrt{1-e^{-2t}})^2 + (\sqrt{1-e^{-2s}})^2 = 1 - e^{-2(t+s)},
$$

the random variable $e^{-s}\sqrt{1-e^{-2t}}Y + \sqrt{1-e^{-2s}}Z$ is equal in distribution to a Gaussian random variable with variance $1 - e^{-2(t+s)}$. Therefore

$$
\begin{aligned}
\mathbb{E}f(e^{-(t+s)}x &+ e^{-s}\sqrt{1-e^{-2t}}Y + \sqrt{1-e^{-2s}}Z) \\
&= \mathbb{E}f(e^{-(t+s)}x + \sqrt{1-e^{-2(t+s)}}Y) \\
&= \int_{\mathbb{R}^d} f(e^{-(t+s)}x + \sqrt{1-e^{-2(t+s)}}y)\,\mathrm{d}\gamma(y) = OU(t+s)f(x).
\end{aligned}
$$

This proves the identity $OU(t)OU(s)f = OU(t+s)$ for $f \in C_c(\mathbb{R}^d)$. Using the denseness of these functions in $L^p(\mathbb{R}^d,\gamma)$, the identity extends to general $f \in L^p(\mathbb{R}^d,\gamma)$.

To prove the strong continuity property (S3) we again first consider a function $f \in C_c(\mathbb{R}^d)$. By Hölder's inequality,

$$
\|OU(t)f - f\|_{L^p(\mathbb{R}^d,\gamma)}^p \leqslant \int_{\mathbb{R}^d}\int_{\mathbb{R}^d} |f(e^{-t}x + \sqrt{1-e^{-2t}}y) - f(x)|^p\,\mathrm{d}\gamma(x)\,\mathrm{d}\gamma(y).
$$

The right-hand side tends to 0 as $t \downarrow 0$ by dominated convergence. This gives strong continuity for functions $f \in C_c(\mathbb{R}^d)$. The general case follows again by approximation, keeping in mind that the operators $OU(t)$ are all contractive.

The generator of $(OU(t))_{t\geqslant 0}$ is traditionally denoted as L. We show next that it is given, for functions $f \in C_c^\infty(\mathbb{R}^d)$, by the *Ornstein–Uhlenbeck operator*

$$
Lf(x) = \Delta f(x) - x \cdot \nabla f(x). \tag{13.29}
$$

Before turning to the proof, we wish to point out an interesting feature of this formula. Multiplying both sides with a test function ϕ, integrating with respect to γ, and integrating by parts after having written out the Gaussian density, we obtain

$$\int_{\mathbb{R}^d} Lf(x)\overline{\phi(x)}\,d\gamma(x) = \frac{1}{(2\pi)^{d/2}} \int_{\mathbb{R}^d} (\Delta f(x) - x\cdot\nabla f(x))\overline{\phi(x)}\exp\left(-\frac{1}{2}|x|^2\right)dx$$

$$= -\frac{1}{(2\pi)^{d/2}} \int_{\mathbb{R}^d} \nabla f\cdot\overline{\nabla\phi}\exp\left(-\frac{1}{2}|x|^2\right)dx \qquad (13.30)$$

$$= -\int_{\mathbb{R}^d} \nabla f\cdot\overline{\nabla\phi}\,d\gamma(x).$$

Since $C_c^\infty(\mathbb{R}^d)$ is dense in $D(\nabla) = W^{1,2}(\mathbb{R}^d,\gamma)$, the Gaussian Sobolev space of all $f \in L^2(\mathbb{R}^d,\gamma)$ whose weak first order derivatives exist and belong to $L^2(\mathbb{R}^d,\gamma)$, for $p = 2$ the identity (13.30) identifies $-L$ as the operator associated with the closed, densely defined, accretive form $\mathfrak{a}_{OU}$ with domain $W^{1,2}(\mathbb{R}^d,\gamma)$ defined by

$$\mathfrak{a}_{OU}(f,g) = \int_{\mathbb{R}^d} \nabla f\cdot\overline{\nabla g}\,d\gamma(x).$$

Let us now turn to a proof of (13.29). Substituting $\sqrt{1-e^{-2t}}\,y = u$ and $e^{-t}x+u = v$ and writing out the Gaussian density, we arrive at

$$OU(t)f(x) = \frac{1}{(2\pi)^{d/2}}\left(\frac{1}{1-e^{-2t}}\right)^{d/2} \int_{\mathbb{R}^d} f(e^{-t}x+u)\exp\left(-\frac{1}{2}\frac{|u|^2}{1-e^{-2t}}\right)du$$

$$= \frac{1}{(2\pi)^{d/2}}\left(\frac{1}{1-e^{-2t}}\right)^{d/2} \int_{\mathbb{R}^d} f(v)\exp\left(-\frac{1}{2}\frac{|e^{-t}x-v|^2}{1-e^{-2t}}\right)dv \qquad (13.31)$$

$$= \int_{\mathbb{R}^d} M_t(x,v)f(v)\,dv.$$

This represents $OU(t)$ as an integral operator with kernel

$$M_t(x,y) = \frac{1}{(2\pi)^{d/2}}\left(\frac{1}{1-e^{-2t}}\right)^{d/2} \exp\left(-\frac{1}{2}\frac{|e^{-t}x-y|^2}{(1-e^{-2t})}\right).$$

The function M_t is called the *Mehler kernel* at time t. We can express it in terms of the heat kernel as $M_t(x,y) = K_{\frac{1}{2}(1-e^{-2t})}(e^{-t}x-y)$, and therefore we have the pointwise identity

$$OU(t)f(x) = H(\tfrac{1}{2}(1-e^{-2t}))f(e^{-t}x),$$

where H is the heat semigroup on $L^2(\mathbb{R}^d)$.

Let $f \in C_c^\infty(\mathbb{R}^d)$. Then $f \in W^{2,p}(\mathbb{R}^d)$ and therefore, by the results of Section 13.6.c, $f \in D(\Delta)$. Hence, we may differentiate the above identity at $t = 0$ and obtain

$$\lim_{t\downarrow 0}\frac{1}{t}(OU(t)f - f)(\cdot)$$

$$= \left[e^{-2t}\Delta H(\tfrac{1}{2}(1-e^{-2t})f(e^{-t}\cdot) - e^{-t}(\cdot)\cdot H(\tfrac{1}{2}(1-e^{-2t})\nabla f(e^{-t}\cdot)\right]_{t=0}$$

$$= \Delta f(\cdot) - (\cdot)\cdot\nabla f(\cdot).$$

In the middle expression, $H(s)\nabla g$ is short-hand for $\sum_{j=1}^{d} H(s)\partial_j g$. The combined use of the product rule and chain rule can be rigorously justified by going through the steps of the standard proof of the corresponding scalar analogue, which we leave as an exercise to the reader. It is important to point out that the limit is taken with respect to the norm of $L^p(\mathbb{R}^d)$, as we were dealing with the heat semigroup in $L^p(\mathbb{R}^d)$. However, since convergence in $L^p(\mathbb{R}^d)$ implies convergence in $L^p(\mathbb{R}^d, \gamma)$ it follows that the above differentiation can retrospectively be interpreted with respect to the norm of $L^p(\mathbb{R}^d, \gamma)$. This proves that $f \in D(L)$ and that the asserted formula for Lf holds.

Let us show next that the analogue of Theorem 12.19 holds for L. We have already identified $-L$ as the operator associated with the form $\mathfrak{a}_{OU}$. This, in combination with Theorem 10.46 and Proposition 12.18, implies that

$$-L = \nabla^* \nabla,$$

where the adjoint refers to the inner product of $L^2(\mathbb{R}^d, \gamma)$. Finally, *mutatis mutandis* the argument for the heat semigroup can be repeated to prove that in $L^2(\mathbb{R}^d, \gamma)$ the generator domain $D(L)$ equals the domain of the *weak L^2-Ornstein–Uhlenbeck operator* which is defined in the obvious way.

Remark 13.54. For $1 < p < \infty$ it can be shown that

$$D(L) = W^{2,p}(\mathbb{R}^d, \gamma),$$

the Gaussian Sobolev space of all $f \in L^p(\mathbb{R}^d, \gamma)$ admitting weak derivatives up to order 2, all of which belong to $L^p(\mathbb{R}^d, \gamma)$. The proof of this fact is beyond the scope of this work, even for the case $p = 2$.

For $d = 1$ one has the following representation for the Ornstein–Uhlenbeck semigroup in terms of the Hermite polynomials H_n discussed in Section 3.5.b:

$$OU(t)H_n = e^{-nt}H_n, \quad n \in \mathbb{N}. \tag{13.32}$$

This is a simple exercise based on the representation (13.29) and the recurrence relation for the Hermite polynomials discussed in Section 3.5.b, and it implies

$$\sigma(-L) = \mathbb{N}$$

(see Proposition 10.32). The formula (13.32) generalises to arbitrary dimension d once one has found an analogue of the Hermite basis for $L^2(\mathbb{R}^d, \gamma)$. This is accomplished in Section 15.6.a (see Theorems 15.57 and 15.58). An immediate consequence is that the Ornstein–Uhlenbeck semigroup extends holomorphically and contractively to the open right-half plane $\{z \in \mathbb{C} : \mathrm{Re}\, z > 0)$ and is strongly continuous on every sector Σ_ω with

$\omega < \frac{1}{2}\pi$. Another way to see this is to note that L is selfadjoint (by Proposition 10.43, noting that L is symmetric by the selfadjointness of the operators $OU(t)$ and satisfies $(0,\infty) \subseteq \rho(-L)$ by Proposition 13.8); the holomorphic extension to the right-half plane may now be defined by

$$OU(z) = \exp(-zL), \quad \mathrm{Re}\, z > 0.$$

Following this approach, strong continuity on the sectors Σ_ω with $\omega < \frac{1}{2}\pi$ will follow from Theorem 13.61. This discussion is summarised in the following theorem.

Theorem 13.55. *The operator $-L$ generates an analytic C_0-semigroup of contractions on every sector Σ_ω with $\omega < \frac{1}{2}\pi$.*

13.6.f The Hermite Semigroup

Let $0 \leqslant V \in L^1_{\mathrm{loc}}(\mathbb{R}^d)$ be given; in applications we think of V as a *potential*. Consider the forms $\mathfrak{a}_\Delta, \mathfrak{a}_V : C^\infty_c(\mathbb{R}^d) \times C^\infty_c(\mathbb{R}^d)$ defined by

$$\mathfrak{a}_\Delta(u,v) := \int_{\mathbb{R}^d} \nabla u(x) \cdot \overline{\nabla v(x)}\, dx, \quad \mathfrak{a}_V(u,v) := \int_{\mathbb{R}^d} V(x) u(x) \overline{v(x)}\, dx.$$

The form $\mathfrak{a}_\Delta + \mathfrak{a}_V$ is densely defined, closable, positive, and continuous. The operator A associated with its closure $\mathfrak{a}$ is densely defined, positive, and selfadjoint by Theorem 12.17. We denote this operator somewhat suggestively by $-\Delta + V$.

An interesting special case arises if we take $V(x) = |x|^2$. This results in the selfadjoint operator $-\Delta + |x|^2$ on $L^2(\mathbb{R}^d)$, the so-called *Hermite operator*. In Quantum Mechanics, the operator

$$H := -\frac{1}{2}\Delta + \frac{1}{2}|x|^2$$

is called the *quantum harmonic oscillator*. Let us here take a closer look at the C_0-contraction semigroup generated by $-H$.

Theorem 13.56 (The Hermite semigroup). *The C_0-semigroup S on $L^2(\mathbb{R}^d)$ generated by $-H + \frac{d}{2}I$ is unitarily equivalent to the Ornstein–Uhlenbeck semigroup OU on $L^2(\mathbb{R}^d; \gamma)$. More precisely, we have*

$$U^{-1}S(t)U = OU(t), \quad t \geqslant 0,$$

where $U : L^2(\mathbb{R}^d, \gamma) \to L^2(\mathbb{R}^d)$ is the unitary operator given by $U = D \circ E$, with $D : L^2(\mathbb{R}^d) \to L^2(\mathbb{R}^d)$ and $D : L^2(\mathbb{R}^d, \gamma) \to L^2(\mathbb{R}^d)$ given by

$$Df(x) = (\sqrt{2})^{d/2} f(\sqrt{2}x),$$

$$Ef(x) = \frac{1}{(2\pi)^{d/4}} \exp(-|x|^2/4) f(x).$$

Proof Fix $f \in C_c^\infty(\mathbb{R}^d)$. Recalling from (13.29) that

$$Lf = \Delta f - x \cdot \nabla f,$$

a somewhat tedious but straightforward computation gives the identity

$$Lf = U^{-1}\left(-H + \frac{d}{2}\right)Uf.$$

Passing to the resolvents, applying Theorem 13.19, and using the density of $C_c^\infty(\mathbb{R}^d)$ in $L^2(\mathbb{R}^d, \gamma)$, the identity for the semigroups follows from this. □

As an immediate consequence of this result and Theorem 13.55 we see that the Hermite semigroup extends to an analytic C_0-semigroup of contractions on every sector Σ_ω with $\omega < \frac{1}{2}\pi$.

It will follow from Theorem 15.58 that $\sigma(-L)$ equals $\mathbb{N} = \{0, 1, 2, \ldots\}$ and consists of eigenvalues (see Corollary 15.59). From this we see that the spectrum of the quantum Harmonic oscillator equals

$$\sigma(H) = \mathbb{N} + \frac{d}{2}$$

and consists of eigenvalues. The lowest eigenvalue $\frac{1}{2}d$ is the *ground state energy* of H.

13.6.g   The Schrödinger Group

We again consider the setting of Section 13.6.f and consider, for nonnegative potentials $V \in L^1_{loc}(\mathbb{R}^d)$ we consider the positive selfadjoint operator $A := -\Delta + V$ associated with the closure of the densely defined, closable, positive, and continuous form $\mathfrak{a} := \mathfrak{a}_\Delta + \mathfrak{a}_V$, where

$$\mathfrak{a}_\Delta(u, v) := \int_{\mathbb{R}^d} \nabla u(x) \cdot \overline{\nabla v(x)}\, dx, \quad \mathfrak{a}_V(u, v) := \int_{\mathbb{R}^d} V(x)u(x)\overline{v(x)}\, dx,$$

for $u, v \in C_c^\infty(\mathbb{R}^d)$. By Stone's theorem, the operator iA generates a unitary C_0-group on $L^2(\mathbb{R}^d)$, the so-called *Schrödinger group* with potential V. It solves the *Schrödinger equation* with potential V,

$$\frac{1}{i}\frac{\partial}{\partial t}u(t, x) = -\Delta u(t, x) + V(x)u(t, x), \quad t \geqslant 0, \; x \in \mathbb{R}^d. \tag{13.33}$$

The special case $V \equiv 0$ is of special interest:

Example 13.57 (Free Schrödinger group). The C_0-group $(S(t))_{t \in \mathbb{R}}$ on $L^2(\mathbb{R}^d)$ generated by the operator $A = i\Delta$ with domain $\mathsf{D}(A) = H^2(\mathbb{R}^d)$ is called the *free Schrödinger group*. For functions $f \in L^1(\mathbb{R}^d) \cap L^2(\mathbb{R}^d)$ and $t \neq 0$ it is given explicitly be the formula

$$S(t)f(x) = \frac{1}{(4\pi it)^{d/2}} \int_{\mathbb{R}^d} \exp\left(i\frac{|x - y|^2}{4t}\right)f(y)\, dy, \tag{13.34}$$

valid for almost all $x \in \mathbb{R}^d$. Note that, on a formal level, we have $S(t) = H(it)$, where $(H(z))_{\mathrm{Re}\,z>0}$ is the (holomorphic extension to the open right-half plane of the) heat semigroup generated by Δ given by (13.20). We refer to Problem 13.12 for a proof that the limit $H(it)f := \lim_{s\downarrow 0} S(s + it)f$ indeed exists for all $f \in L^2(\mathbb{R}^d)$; the point we are making here is that this limit is still represented by the explicit formula (13.20) for the heat semigroup evaluated at it. Taking the result of Problem 13.12 for granted, (13.34) follows from (13.20), with t replaced by $s + it$, by dominated convergence.

An alternative derivation can be given on the basis of the spectral theorem for self-adjoint operators; see Problem 13.23. This idea will be explored more systematically in Example 13.62.

The representation 13.34 implies that $S(t)f \in L^\infty(\mathbb{R}^d)$ for all $f \in L^1(\mathbb{R}^d) \cap L^2(\mathbb{R}^d)$ and $t \in \mathbb{R} \setminus \{0\}$, with bounds

$$\|S(t)f\|_\infty \leqslant \frac{1}{4\pi|t|}\|f\|_1, \quad \|S(t)f\|_2 \leqslant \|f\|_2,$$

the former by a direct estimate and the latter by Plancherel's theorem. By the Riesz–Thorin interpolation theorem, for all $t \in \mathbb{R} \setminus \{0\}$ the operators $S(t)$, when restricted to $L^1(\mathbb{R}^d) \cap L^2(\mathbb{R}^d)$, extend to bounded operators from $L^p(\mathbb{R}^d)$ to $L^q(\mathbb{R}^d)$ for all $1 \leqslant p \leqslant 2$ and $\frac{1}{p} + \frac{1}{q} = 1$, with bound

$$\|S(t)\|_{\mathscr{L}(L^p(\mathbb{R}^d), L^q(\mathbb{R}^d))} \leqslant \frac{1}{(4\pi|t|)^{\frac{1}{p}-\frac{1}{2}}} \tag{13.35}$$

for $t \neq 0$.

13.6.h The Wave Group

The Wave Group on Bounded Domains Let D be a nonempty bounded open set in $\mathbb{R}^d$. The space $H := H_0^1(D) \times L^2(D)$ is a Hilbert space with respect to the norm given by

$$\|(u,v)\|^2 = \|u\|_{H_0^1(D)}^2 + \|v\|_2^2,$$

where we consider the norm on $H_0^1(D)$ given by

$$\|u\|_{H_0^1(D)}^2 := \int_D |\nabla u|^2 \, \mathrm{d}x, \quad u \in H_0^1(D). \tag{13.36}$$

This norm is equivalent to the usual Sobolev norm on $H_0^1(D)$ by Poincaré's inequality. In H we define the operator A defined by

$$A := \begin{pmatrix} 0 & I \\ \Delta_{\mathrm{Dir}} & 0 \end{pmatrix}, \quad \mathsf{D}(A) := \mathsf{D}(\Delta_{\mathrm{Dir}}) \times H_0^1(D),$$

where Δ_{Dir} is the Dirichlet Laplacian on $L^2(D)$. We will prove that A is the generator of a unitary C_0-group W on H. This group solves the linear wave equation

$$
\begin{cases}
\dfrac{\partial^2 u}{\partial t^2}(t,x) = \Delta u(t,x), & t \in \mathbb{R},\ x \in D, \\[2mm]
u(0,x) = u_0(x), & x \in D, \\[2mm]
\dfrac{\partial v}{\partial t}(0,x) = v_0(x), & x \in D,
\end{cases}
$$

written as a system of first-order ODEs $u' = v$, $v' = \Delta u$, with initial value $u(0) = u_0$, $v(0) = v_0$, subject to Dirichlet boundary conditions.

The operator A is densely defined and an integration by parts gives, for $u = \begin{pmatrix} u_1 \\ u_2 \end{pmatrix} \in$ $D(A)$ and $v = \begin{pmatrix} v_1 \\ v_2 \end{pmatrix} \in D(A)$,

$$
\begin{aligned}
(Au|v) = \left(\begin{pmatrix} u_2 \\ \Delta u_1 \end{pmatrix} \middle| \begin{pmatrix} v_1 \\ v_2 \end{pmatrix} \right) &= \int_D \nabla u_2 \cdot \overline{\nabla v_1} + (\Delta u_1)\overline{v_2}\, dx \\
&= \int_D \nabla u_2 \cdot \overline{\nabla v_1} - \nabla u_1 \overline{\nabla v_2}\, dx \\
&= - \left(\begin{pmatrix} u_1 \\ u_2 \end{pmatrix} \middle| \begin{pmatrix} v_2 \\ \Delta v_1 \end{pmatrix} \right) = -(u|Av).
\end{aligned}
$$

This implies that $-iA$ is symmetric.

We next observe that $0 \in \rho(\Delta_{\mathrm{Dir}})$ by Theorem 12.26. This allows us to consider the bounded operator

$$
R := \begin{pmatrix} 0 & \Delta_{\mathrm{Dir}}^{-1} \\ I & 0 \end{pmatrix}
$$

on H. For $(u,v) \in H$ we have

$$
R \begin{pmatrix} u \\ v \end{pmatrix} = \begin{pmatrix} \Delta_{\mathrm{Dir}}^{-1} v \\ u \end{pmatrix} \in D(\Delta_{\mathrm{Dir}}) \times H_0^1(D) = D(A)
$$

and it is immediate to check that $AR = I$ and $RAh = h$ for $h \in D(A)$. This proves that A is boundedly invertible, that is, $0 \in \rho(A)$. An application of Proposition 10.43 now gives that $-iA$ is selfadjoint. Therefore, by Stone's theorem (Theorem 13.44) we obtain:

Theorem 13.58 (Wave group on bounded domains). *The operator A generates a unitary C_0-group $(W(t))_{t\in\mathbb{R}}$ on $H_0^1(D) \times L^2(D)$, provided $H_0^1(D)$ is endowed with the equivalent norm given by* (13.36).

The Wave Group on $\mathbb{R}^d$ Let us next consider the case $D = \mathbb{R}^d$, which is not covered by the above considerations since it was assumed that D be bounded. We have $H_0^1(\mathbb{R}^d) = H^1(\mathbb{R}^d)$ by Theorem 11.25 and Theorem 11.31 and $\Delta_{\mathrm{Dir}} = \Delta$ with $D(\Delta) = H^2(\mathbb{R}^d)$. This suggests considering in $H^1(\mathbb{R}^d) \times L^2(\mathbb{R}^d)$ the operator

$$A := \begin{pmatrix} 0 & I \\ \Delta & 0 \end{pmatrix}, \quad D(A) := H^2(\mathbb{R}^d) \times H^1(\mathbb{R}^d).$$

We will use the theory of Fourier multipliers to prove that A generates a C_0-group on $H^1(\mathbb{R}^d) \times L^2(\mathbb{R}^d)$ and give an explicit expression for it.

To motivate the upcoming expressions we first consider the matrix

$$A_a := \begin{pmatrix} 0 & 1 \\ -a & 0 \end{pmatrix},$$

where $a \geq 0$ is a nonnegative scalar, and compute its exponentials e^{tA_a}. The powers of A_a are given by

$$A_a^{2k} = \begin{pmatrix} (-a)^k & 0 \\ 0 & (-a)^k \end{pmatrix}, \quad A_a^{2k+1} = \begin{pmatrix} 0 & (-a)^k \\ (-a)^{k+1} & 0 \end{pmatrix}, \quad k \in \mathbb{N},$$

so that with $b := a^{1/2}$,

$$\begin{aligned}
e^{tA_a} &= \sum_{k=0}^{\infty} \begin{pmatrix} \dfrac{t^{2k}}{(2k)!}(-a)^k & \dfrac{t^{2k+1}}{(2k+1)!}(-a)^k \\[3mm] \dfrac{t^{2k+1}}{(2k+1)!}(-a)^{k+1} & \dfrac{t^k}{(2k)!}a^k \end{pmatrix} \\[3mm]
&= \sum_{k=0}^{\infty} \begin{pmatrix} (-1)^k \dfrac{t^{2k}}{(2k)!}b^{2k} & (-1)^k \dfrac{t^{2k+1}}{(2k+1)!}b^{2k} \\[3mm] -(-1)^k \dfrac{t^{2k+1}}{(2k+1)!}b^{2k+2} & (-1)^k \dfrac{t^{2k}}{(2k)!}b^{2k} \end{pmatrix} \\[3mm]
&= \begin{pmatrix} \cos tb & b^{-1}\sin tb \\ -b\sin tb & \cos tb \end{pmatrix}.
\end{aligned}$$

Substituting $-\Delta$ for a and $(-\Delta)^{1/2}$ for b, we arrive at the following guess for the expression for the wave group:

$$W(t) = \begin{pmatrix} \cos(t(-\Delta)^{1/2}) & (-\Delta)^{-1/2}\sin(t(-\Delta)^{1/2}) \\ -(-\Delta)^{1/2}\sin(t(-\Delta)^{1/2}) & \cos(t(-\Delta)^{1/2}) \end{pmatrix}, \quad t \geq 0.$$

We need to give a meaning to the operators occurring in this matrix, which can be accomplished by the functional calculus of $-\Delta$, or by interpreting them as Fourier multiplier operators as follows. The operators on the diagonal can be interpreted as Fourier multiplier operators on $L^2(\mathbb{R}^d)$ associated with the multipliers

$$m_{1,1;t}(\xi) = m_{2,2;t}(\xi) = \cos(t|\xi|);$$

this function belongs to $L^\infty(\mathbb{R}^d)$ with norm 1 for every $t \in \mathbb{R}$. Recalling the characterisation of $H^1(\mathbb{R}^d)$ as those functions f in $L^2(\mathbb{R}^d)$ for which $\xi \mapsto (1+|\xi|^2)^{1/2}\widehat{f}(\xi)$ is in $L^2(\mathbb{R}^d)$, we see moreover that $\cos\big(t(-\Delta)^{1/2}\big)$ maps $W^{1,2}(\mathbb{R}^d)$ into itself. Recalling the norm of $H^1(\mathbb{R}^d)$ given by (11.9), this argument also gives the estimates

$$\left\| \cos\big(t(-\Delta)^{1/2}\big) \right\|_{\mathscr{L}(L^2(\mathbb{R}^d))} \leqslant 1, \quad \left\| \cos\big(t(-\Delta)^{1/2}\big) \right\|_{\mathscr{L}(H^1(\mathbb{R}^d))} \leqslant 1.$$

Similarly we can associate a bounded operator from $H^1(\mathbb{R}^d)$ to $L^2(\mathbb{R}^d)$ with the multiplier

$$m_{2,1;t}(\xi) = -|\xi|\sin(t|\xi|).$$

Indeed, if $f \in H^1(\mathbb{R}^d)$, then

$$\left\| m_{2,1;t}(\xi)\widehat{f}(\xi) \right\|_{L^2(\mathbb{R}^d)} \leqslant \left(\sup_{\xi \in \mathbb{R}^d} \frac{|\xi|}{(1+|\xi|^2)^{1/2}} |\sin(t|\xi|)| \right) \|f\|_{H^1(\mathbb{R}^d)} \leqslant \|f\|_{H^1(\mathbb{R}^d)}$$

and therefore

$$\left\| -(-\Delta)^{1/2}\sin\big(t(-\Delta)^{1/2}\big)f \right\|_{\mathscr{L}(H^1(\mathbb{R}^d),L^2(\mathbb{R}^d))} \leqslant 1$$

for all $t \in \mathbb{R}$. In the same way the operators $(-\Delta)^{-1/2}\sin\big(t(-\Delta)^{1/2}\big)$ are interpreted as bounded operators from $L^2(\mathbb{R}^d)$ into $H^1(\mathbb{R}^d)$ given by the multipliers

$$m_{1,2;t}(\xi) = \frac{\sin(t|\xi|)}{|\xi|},$$

which satisfy (distinguish the cases $|\xi| \leqslant 1$ and $|\xi| > 1$)

$$\left\| m_{1,2;t}(\xi)\widehat{f}(\xi) \right\|_{H^1(\mathbb{R}^d)} \leqslant \left(\sup_{\xi \in \mathbb{R}^d} \frac{|\sin(t|\xi|)|}{|\xi|} (1+|\xi|^2)^{1/2} \right) \|\widehat{f}\|_{L^2(\mathbb{R}^d)}$$

$$\leqslant C(1+|t|)\|\widehat{f}\|_{L^2(\mathbb{R}^d)} = C(1+|t|)\|f\|_{L^2(\mathbb{R}^d)},$$

where C is a universal constant. Therefore

$$\left\| (-\Delta)^{-1/2}\sin\big(t(-\Delta)^{1/2}\big) \right\|_{\mathscr{L}(L^2(\mathbb{R}^d),H^1(\mathbb{R}^d))} \leqslant C(1+|t|)$$

for all $t \in \mathbb{R}$.

Theorem 13.59 (Wave group on $\mathbb{R}^d$). *The operator A generates a C_0-group $(W(t))_{t \in \mathbb{R}}$ on $H^1(\mathbb{R}^d) \times L^2(\mathbb{R}^d)$ which is given by*

$$W(t) = \begin{pmatrix} \cos\big(t(-\Delta)^{1/2}\big) & (-\Delta)^{-1/2}\sin\big(t(-\Delta)^{1/2}\big) \\ -(-\Delta)^{1/2}\sin\big(t(-\Delta)^{1/2}\big) & \cos\big(t(-\Delta)^{1/2}\big) \end{pmatrix}, \quad t \in \mathbb{R}.$$

Moreover, there is a constant $C \geqslant 0$ such that

$$\|W(t)\| \leqslant C(1 + |t|), \quad t \in \mathbb{R}.$$

Proof The group property follows from formal matrix multiplication, which can be made rigorous by noting that in the Fourier domain we are just multiplying matrices of scalar-valued multipliers, much like what we did in the treatment of the heat and Poisson semigroups. Once we have proved strong continuity, differentiation of the entries at $t = 0$ identifies $\begin{pmatrix} 0 & I \\ \Delta & 0 \end{pmatrix}$ as the generator by the same reasoning.

To prove strong continuity we note that $\|m_{1,1;t}\|_\infty \leqslant 1$ and $\lim_{t \to 0} m_{1,1;t}(\xi) = 1$ pointwise, implying that $\lim_{t \to 0} m_{1,1;t}\widehat{f} = \widehat{f}$ in $L^2(\mathbb{R}^d)$ by dominated convergence. Hence $\lim_{t \to 0} \cos\big(t(-\Delta)^{1/2}\big)f = f$ in $L^2(\mathbb{R}^d)$ for all $f \in L^2(\mathbb{R}^d)$ by Plancherel's theorem. The strong convergence of the other three terms is proved similarly. □

Remark 13.60. Notwithstanding the linear growth bound for $W(t)$, the *energy functional*

$$E(t) := \|\partial_t W(t)f\|_2^2 + \|\nabla W(t)f\|_2^2$$

is constant in time. For functions $f \in D(A)$ one has $E'(t) = 0$ by direct calculation, and the general case then follows by density.

We conclude with some informal remarks establishing a connection with the Poisson semigroup of Section 13.6.d. Since $(-\Delta)^{1/2}$ with domain $H^1(\mathbb{R}^d)$ is selfadjoint, $i(-\Delta)^{1/2}$ is the generator of a C_0-group $(U(t))_{t \in \mathbb{R}}$ of unitary operators on $L^2(\mathbb{R}^d)$ by Stone's theorem. For $f \in H^2(\mathbb{R}^d)$ we have $(-\Delta)^{1/2}f \in H^1(\mathbb{R}^d)$ and

$$\frac{d^2}{dt^2}U(t)f = [i(-\Delta)^{1/2}]^2 U(t)f = \Delta U(t)f$$

In this sense, $t \mapsto u(t,x) = U(t)f(x)$ satisfies the wave equation with initial condition $u(0,x) = f(x)$. We are neglecting the initial condition for the first derivative, however, and in fact we could run the same argument for $-i(-\Delta)^{1/2}$, which is the generator of the C_0-group $(U(-t))_{t \in \mathbb{R}}$ to find that its orbits also solve the wave equation with initial condition $u(0,x) = f(x)$. Interpreting $U(t)$ and $U(-t)$ as Fourier multipliers one sees that

$$\frac{1}{2}(U(t) + U(-t)) = \cos\big(t(-\Delta)^{1/2}\big),$$

which is the first entry in the matrix representation for the wave group. These operators solve the wave equation with initial conditions $u(0,x) = f(x)$ and $\frac{\partial u}{\partial t}(0,x) = 0$, the latter because of the cancellation of the derivatives of $U(t)f$ and $U(-t)f$. This argument is admittedly somewhat sketchy; the reader is invited to provide the rigorous details.

13.7 Semigroups Generated by Normal Operators

We begin with a general observation about semigroup generation by normal operators. Recall from Section 13.4 the notation

$$\Sigma_\omega := \{z \in \mathbb{C} \setminus \{0\} : |\arg(z)| < \omega\}$$

for the open sector of angle $\omega \in (0, \pi)$, arguments being taken in $(-\pi, \pi)$.

Theorem 13.61 (Semigroups generated by normal operators). *Let N be a normal operator in a Hilbert space H with associated projection-valued measure P. Then:*

(1) *if $\sigma(N)$ is contained in the closed right half-plane, then $-N$ is the generator of a C_0-semigroup of contractions S, given by*

$$S(t) = \int_{\sigma(N)} e^{-\lambda t} \, dP(\lambda), \quad t \geqslant 0;$$

(2) *if $\sigma(N)$ is contained in a closed sector of angle $0 < \theta < \frac{1}{2}\pi$, then $-N$ is the generator of an analytic C_0-semigroup of contractions $(S(z))_{z \in \Sigma_{\frac{1}{2}\pi - \theta}}$, given by*

$$S(z) = \int_{\sigma(N)} e^{-\lambda z} \, dP(\lambda), \quad z \in \Sigma_{\frac{1}{2}\pi - \theta}.$$

Proof We give a detailed proof of (1); the proof of (2) is entirely similar.

First of all, the operators $S(t)$ are well defined and contractive by Theorem 9.8(ii). We next check that the semigroup is strongly continuous. For all $x \in H$, dominated convergence gives

$$\lim_{t\downarrow 0}(S(t)x|x) = \lim_{t\downarrow 0} \int_{\sigma(N)} e^{-\lambda t} \, dP_x(\lambda) = \int_{\sigma(N)} dP_x = (x|x).$$

By a polarisation argument, this gives the weak continuity of the semigroup. By Theorem 13.11, this implies its strong continuity. It remains to be shown that N is its generator. If $x \in D(N)$, that is, if $\int_{\sigma(N)} |\lambda|^2 \, dP_x(\lambda) < \infty$, then by dominated convergence

$$\lim_{t\downarrow 0} \frac{1}{t}(S(t)x - x|x) = \lim_{t\downarrow 0} \int_{\sigma(N)} \frac{e^{-\lambda t} - 1}{t} \, dP_x(\lambda) = -\int_{\sigma(N)} \lambda \, dP_x(\lambda) = -(Nx|x).$$

$$(13.37)$$

The same argument proves that

$$\lim_{t\downarrow 0}\frac{1}{t^2}\|S(t)x-x\|^2 = \lim_{t\downarrow 0}\frac{1}{t^2}(S^\star(t)-I)S(t)x-x|x) = \int_{\sigma(N)}|\lambda|^2\,dP_x(\lambda) = \|Nx\|^2,$$
(13.38)

using the final identity in the statement of Theorem 10.50 in the last step.

By polarisation, (13.37) implies that

$$\lim_{t\downarrow 0}\frac{1}{t}(S(t)x-x|y) = (Nx|y), \quad x\in D(N),\ y\in D(N),$$

and hence, using that $\limsup_{t\downarrow 0}\frac{1}{t}\|S(t)x-x\| < \infty$ by (13.38), by approximation we obtain

$$\lim_{t\downarrow 0}\frac{1}{t}(S(t)x-x|y) = -(Nx|y), \quad x\in D(N),\ y\in H.$$

Denoting the generator of the semigroup by A, for $y\in D(A^\star)$ it follows that

$$-(Nx|y) = \lim_{t\downarrow 0}\frac{1}{t}(S(t)x-x|y) = \lim_{t\downarrow 0}\frac{1}{t}(x|S^\star(t)y-y) = (x|A^\star y).$$

This implies that $Nx\in D(A^{\star\star}) = D(A)$, referring to Proposition 10.23 for the equality of these domains.

We have thus proved that $-N\subseteq A$. Since $(0,\infty)$ is contained in the resolvent sets of both $-N$ (by assumption) and A (since it generates a C_0-contraction semigroup), Proposition 10.30 implies $A = -N$. $\qquad\square$

Some of the semigroup examples of the previous section can be constructed rather easily using the spectral theorem.

Example 13.62 (Heat semigroup, Poisson semigroup, free Schrödinger group, wave group revisited). Let P be the projection-valued measure on $\mathbb{R}$ associated with the negative Laplace operator $-\Delta$, viewed as a selfadjoint operator on $L^2(\mathbb{R}^d)$ (see Problem 10.17). The heat semigroup H is then given by

$$H(t) = \int_{\mathbb{R}}e^{-\lambda t}\,dP(\lambda), \quad t\geqslant 0,$$

and the free Schrödinger group by

$$S(t) = \int_{\mathbb{R}}e^{-i\lambda t}\,dP(\lambda), \quad t\geqslant 0.$$

The positive square root $(-\Delta)^{1/2}$ defined through Proposition 10.60 coincides with the unbounded Fourier multiplier operator corresponding to the multiplier $m(\xi) = |\xi|$. Using Proposition 10.52 to switch between the projection-valued measure Q of $-\Delta$ and

R of $(-\Delta)^{1/2}$, we see that the Poisson semigroup generated by the latter is given by

$$P(t) = \int_{[0,\infty)} e^{-\lambda t} \, dR(\lambda) = \int_{[0,\infty)} e^{-\lambda^{1/2}t} \, dQ(\lambda), \quad t \geqslant 0.$$

In the same way, the operators $\cos\big(t(-\Delta)^{1/2}\big)$ and $\sin\big(t(-\Delta)^{1/2}\big)$ featuring in the wave group are given by

$$\cos\Big(t(-\Delta)^{1/2}\Big) = \int_{[0,\infty)} \cos(t\lambda) \, dR(\lambda) = \int_{[0,\infty)} \cos\Big(t\lambda^{1/2}\Big) \, dQ(\lambda),$$

$$\sin\Big(t(-\Delta)^{1/2}\Big) = \int_{[0,\infty)} \sin(t\lambda) \, dR(\lambda) = \int_{[0,\infty)} \sin\Big(t\lambda^{1/2}\Big) \, dQ(\lambda).$$

Example 13.63 (Stone's theorem revisited). Let P be the projection-valued measure on $\mathbb{R}$ associated with a selfadjoint operator A on a Hilbert space H. Then the unitary C_0-group $(U(t))_{t\in\mathbb{R}}$ generated by iA is given by

$$U(t) = \int_{\mathbb{R}} e^{i\lambda t} \, dP(\lambda), \quad t \geqslant 0.$$

Problems

13.1 Let S be a C_0-semigroup on X with generator A, and suppose that $\|S(t)\| \leqslant Me^{\mu t}$ for all $t \geqslant 0$. Prove that

$$\|(\lambda - A)^{-k}\| \leqslant M/(\operatorname{Re}\lambda - \mu)^k, \quad \operatorname{Re}\lambda > \mu, \ k = 1, 2, \dots$$

Hint: By considering $A - \mu$ instead of A we may assume that $\mu = 0$. Under this assumption, observe that $\||R(\lambda, A)\|| \leqslant 1/\operatorname{Re}\lambda$, where $\||x\|| := \sup_{t\geqslant 0} \|S(t)x\|$ defines an equivalent norm on X.

Remark: The converse holds as well: If A is a densely defined operator on X satisfying the above inequalities, then A generates a C_0-semigroup on X satisfying $\|S(t)\| \leqslant Me^{\mu t}$ for all $t \geqslant 0$. This is the version of the Hille–Yosida theorem for arbitrary C_0-semigroups. The ambitious reader may try to prove this.

13.2 The aim of this problem is to prove that if A generates a C_0-semigroup S on X, then A is bounded if and only if

$$\lim_{t\downarrow 0} \|S(t) - I\| = 0.$$

(a) Show that if A is bounded, then $S(t) = e^{tA}$ and $\lim_{t\downarrow 0} \|S(t) - I\| = 0$.

(b) Use a Neumann series argument to prove that if $\lim_{t\downarrow 0} \|S(t) - I\| = 0$, then for small enough $t > 0$ the operators $T_t \in \mathscr{L}(X)$ defined by

$$T_t := \int_0^t S(s) \, ds$$

are invertible, and show that for such $t > 0$ we have

$$A = T_t^{-1}(S(t) - I).$$

13.3 Let A be the generator of a C_0-semigroup S on X. This problem gives a rigorous interpretation to the "formula" "$S(t) = e^{tA}$".

For each $h > 0$ consider the bounded operator $A(h)x := \frac{1}{h}(S(h)x - x)$.

(a) Choosing $M \geqslant 1$ and $\omega \geqslant 0$ so that $\|S(t)\| \leqslant Me^{\omega t}$ for all $t \geqslant 0$, show that

$$\|e^{tA(h)}\| \leqslant M\exp\left(\frac{t}{h}(e^{\omega h} - 1)\right).$$

Deduce that for all $0 < h \leqslant 1$ we have $\|e^{tA(h)}\| \leqslant Me^{t(e^\omega - 1)}$.

(b) Using the identity

$$S(t)x - e^{tA(h)}x = \int_0^t \frac{d}{ds}[e^{(t-s)A(h)}S(s)x]\,ds$$

deduce from part (a) that for all $x \in D(A)$ and $0 < h < 1$ we have

$$\|S(t)x - e^{tA(h)}x\| \leqslant tM^2 e^{t(\omega + e^\omega - 1)}\|Ax - A(h)x\|.$$

(c) Prove that for all $x \in X$ and $t \geqslant 0$ we have

$$\lim_{h\downarrow 0} e^{tA(h)}x = S(t)x.$$

(d) For $n \in \mathbb{N}$ with $n > \omega$, let $A_n := nAR(n,A)$ as in the proof of the Hille–Yosida theorem. Prove that for all $x \in X$ and $t \geqslant 0$ we have

$$\lim_{n\to\infty} e^{tA_n}x = S(t)x.$$

13.4 Let A be the generator of the C_0-semigroup S on $L^2(0,1)$ of left translations, inserting zeroes from the right. Show that $\sigma(A) = \varnothing$.

13.5 Let A be the generator of a C_0-semigroup S on X. The *adjoint semigroup* on X^* is the family $S^* = (S^*(t))_{t\geqslant 0}$, where $S^*(t) = (S(t))^*$ for $t \geqslant 0$.

(a) Show that the adjoint semigroup has the semigroup properties (S1) and (S2) but may fail (S3).

Set

$$X^\odot := \{x^* \in X^* : \lim_{t\downarrow 0}\|S^*(t)x^* - x^*\| = 0\}.$$

(b) Show that $X^\odot$ is a closed subspace of X^*.

(c) Show that the adjoint semigroup maps $X^\odot$ into itself and that its restriction to $X^\odot$ is a C_0-semigroup.

(d) Show that for all $x^* \in X^*$ and $t > 0$ there exists a unique element $\phi_{t,x^*} \in X^*$ satisfying

$$\langle x, \phi_{t,x^*} \rangle = \int_0^t \langle x, S^*(s)x^* \rangle \, ds.$$

(e) Show that for all $x^* \in X^*$ and $t > 0$ we have $\phi_{t,x^*} \in D(A^*)$ and

$$A^* \phi_{t,x^*} = S^*(t)x^* - x^*.$$

(f) Show that $X^\odot = \overline{D(A^*)}$ and deduce that $X^\odot$ is weak* dense in X^*.

(g) Show that

$$D(A^*) = \left\{ x^* \in X^* : \limsup_{t \downarrow 0} \frac{1}{t} \|S^*(t)x^* - x^*\| < \infty \right\}.$$

(h) Show that if X is reflexive, then S^* is a C_0-semigroup on X^* and A^* is its generator.

Hint: For the strong continuity apply Phillips's theorem (Theorem 13.11); for the identification of the generator use Proposition 10.30.

13.6 In this problem we prove a continuous analogue of the Sz.-Nagy dilation theorem (Theorem 8.36). Let S be a C_0-semigroup of contractions on a Hilbert space H.

(a) Show that the mapping $T : \mathbb{R} \to \mathscr{L}(H)$ defined by

$$T(t) := \begin{cases} S(t), & t > 0, \\ I, & t = 0, \\ (S(-t))^\star, & t < 0, \end{cases}$$

is positive definite.

Hint: For $t_1, \ldots, t_N \in \mathbb{Q}$ use Lemma 8.35 to show that for all $h_1, \ldots, h_N \in H$ we have $\sum_{m,n=1}^N (T(t_n - t_m)h_m | h_n) \geqslant 0$.

(b) Show that there exist a Hilbert space $\widetilde{H}$ containing H as a closed subspace and a C_0-group $(U(t))_{t \in \mathbb{R}}$ on $\widetilde{H}$ such that

$$T(t)h = PU(t)h, \quad t \geqslant 0, \ h \in H,$$

where P is the orthogonal projection of $\widetilde{H}$ onto H.

Hint: Combine the result of part (a) with Theorem 8.34 to obtain the dilation and use Theorem 13.11 to prove its strong continuity.

13.7 Let A be the generator of a C_0-semigroup on X. Prove the following *spectral inclusion formula*: for all $t \geqslant 0$ we have

$$\sigma(S(t)) \subseteq \exp(t\sigma(A)).$$

Hint: Suppose $e^{\lambda t} \in \rho(S(t))$ for some $\lambda \in \mathbb{C}$ and denote the inverse of $e^{\lambda t}I - S(t)$ by $Q_{\lambda,t}$. Show that the bounded operator B_λ defined by

$$B_\lambda x := \int_0^t e^{\lambda(t-s)} S(s) Q_{\lambda,t} x \, ds$$

is a two-sided inverse for $\lambda - A$.

13.8 Let A be the generator of a C_0-semigroup on X, and let $f \in L^1(0,T;X)$ be given and fixed. A function $u \in L^1(0,T;X)$ is said to be a *weak solution* of the inhomogeneous abstract Cauchy problem

$$\begin{cases} u'(t) = Au(t) + f(t), & t \in [0,T], \\ u(0) = u_0, \end{cases}$$

if for all $t \in [0,T]$ and $x^* \in D(A^*)$ we have

$$\langle u(t), x^* \rangle = \langle u_0, x^* \rangle + \int_0^t \langle u(s), A^* x^* \rangle \, ds + \int_0^t \langle f(s), x^* \rangle \, ds.$$

Prove that the inhomogeneous abstract Cauchy problem has a unique weak solution, and that it equals the unique strong solution.

13.9 This problem gives a two-dimensional example of a bounded analytic C_0-semigroup which is uniformly exponentially stable, contractive on $\mathbb{R}_+$, and fails to be contractive on any open sector containing $\mathbb{R}_+$.

On $\mathbb{C}^2$ consider the norm $\|\cdot\|_Q$ induced by the inner product $(x|y)_Q := (Qx|y)$, where

$$Q = \begin{pmatrix} 1 & 2 \\ 1 & 1 \end{pmatrix}.$$

(a) Show that $(\cdot|\cdot)_Q$ does indeed define an inner product on $\mathbb{C}^2$.

On $(\mathbb{C}^2, \|\cdot\|_Q)$ we consider the C_0-semigroup S,

$$S(t) = e^{-t/2} \begin{pmatrix} 1 & t \\ 0 & 1 \end{pmatrix}.$$

(b) Show that $\|S(t)\|_Q^2 = \frac{1}{2}e^{-t}(t^2 + 2 + t\sqrt{t^2+4})$ and conclude that $S(t)$ is contractive for all $t \geq 0$.

Hint: Use the fact that $\|S(t)\|_Q^2$ equals the largest eigenvalue of $S(t)S^\star(t)$ (see Problem 4.14), where the adjoint refers to the inner product $(\cdot|\cdot)_Q$.

(c) Show that S extends to an entire C_0-semigroup which is uniformly bounded on the open sector Σ_η for all $0 < \eta < \frac{1}{2}\pi$.

(d) Show that S fails to be contractive on any open sector Σ_η.

13.10 Let A be the generator of an analytic C_0-semigroup on X. Show that if B is a bounded operator on X, then $A + B$ generates an analytic C_0-semigroup on X. Also show that if A generates a bounded analytic C_0-semigroup, then so does $A + B - \|B\| I$.

Hint: First prove the second assertion.

13.11 Let A be the generator of an analytic C_0-contraction semigroup on a Hilbert space H. Show that the form $\mathfrak{a}$ on H with domain $\mathsf{D}(\mathfrak{a}) := \mathsf{D}(A)$ defined by $\mathfrak{a}(x,y) := -(Ax|y)$ is accretive, continuous, and closable.

13.12 Let A be the generator of a C_0-semigroup C_0-semigroup S on X which is bounded analytic on the open right-half plane. Show that for all $t \in \mathbb{R}$ and $x \in X$ the limit

$$T(t)x := \lim_{s\downarrow 0} S(s + it)x$$

exists and that the family $(T(t))_{t\in\mathbb{R}}$ is a uniformly bounded C_0-group of operators with generator iA.

Hint: Begin by observing that if $0 < s' < s < \infty$ and $-\infty < t' < t < \infty$, then

$$\|S(s + it)x - S(s' + it)x\| \leqslant M\|S(s - s')x - x\|,$$

where $M = \sup_{\operatorname{Re} z > 0} \|S(z)\|$. Deduce from this the existence of the limits. Deduce the semigroup properties and strong continuity in a similar manner. Finally show that if $x \in \mathsf{D}(A)$, then

$$\int_0^t T(s) iAx \, \mathrm{d}s = T(t)x - x$$

to deduce that $x \in \mathsf{D}(B)$, where B is the generator of $(U(t))_{t\in\mathbb{R}}$, and use this to conclude that $B = iA$.

13.13 Let A be the generator of an analytic C_0-semigroup S on X. Prove that if $\sigma(A) \subseteq \{z \in \mathbb{C} : \operatorname{Re} z < 0\}$, then S is *uniformly exponentially stable*, that is, there exists an exponent $\omega > 0$ such that $\sup_{t\geqslant 0} e^{\omega t} \|S(t)\| < \infty$.

Hint: Verify the assumptions of Theorem 13.30 for $\omega + A$ for small enough $\omega > 0$.

13.14 Let $(S(t))_{t\geqslant 0}$ be a C_0-semigroup on X and let $1 \leqslant p < \infty$. Prove the *Datko–Pazy theorem*: The semigroup $(S(t))_{t\geqslant 0}$ is uniformly exponentially stable (see Problem 13.13) if and only if the orbit $t \mapsto S(t)x$ belongs to $L^p(\mathbb{R}_+; X)$ for all $x \in X$.

Hint: Apply the uniform boundedness theorem and reason by contradiction.

13.15 Let S be a C_0-semigroup on X.

(a) Show that S is uniformly exponentially stable if and only if for some (equivalently, for all) $1 \leqslant p < \infty$ one has $S * f \in L^p(\mathbb{R}_+; X)$ for all $p \in L^1(\mathbb{R}_+; X)$, where

$$(S * f)(t) = \int_0^t S(t - s) f(s) \, \mathrm{d}s, \quad t \in \mathbb{R}_+.$$

Hint: For the 'if' part consider the functions $f(t) = e^{-\mu t} S(t)x$ to reduce matters to the Datko–Pazy theorem of the preceding problem.

(b) Does the analogous result hold for $p = \infty$?

13.16 Let A be the generator of a C_0-semigroup $(S(t))_{t \geqslant 0}$ on a Hilbert space H. Prove the *Gearhart–Prüss theorem*: The semigroup $(S(t))_{t \geqslant 0}$ is uniformly exponentially stable (see Problem 13.13) if and only if $\{\lambda \in \mathbb{C} : \operatorname{Re} \lambda > 0\} \subseteq \rho(A)$ and

$$\sup_{\operatorname{Re} \lambda > 0} \|R(\lambda, A)\| < \infty.$$

Hint: Suppose that $\|S(t)\| \leqslant Me^{\omega t}$. Complete the following steps:

(a) Extend the Fourier–Plancherel theorem to $L^2(\mathbb{R}^d; H)$.

(b) Prove that $t \mapsto R(s + it, A)x$ belongs to $L^2(\mathbb{R}; H)$ for all $x \in H$ and $s > \omega$.

By Lemma 10.34 there exists a $\delta > 0$ such that $\{\lambda \in \mathbb{C} : \operatorname{Re} \lambda > -\delta\} \subseteq \rho(A)$ and

$$\sup_{\operatorname{Re} \lambda > -\delta} \|R(\lambda, A)\| < \infty.$$

(c) Use the resolvent identity to prove that the function $t \mapsto R(it, A)x$ belongs to $L^2(\mathbb{R}; H)$ for all $x \in H$.

(d) Conclude that $t \mapsto S(t)x$ belongs to $L^2(\mathbb{R}; H)$ for all $x \in H$.

13.17 This problem shows that the Gearhart–Prüss theorem of Problem 13.16 does not extend to general Banach spaces.

Let $1 \leqslant p < q < \infty$ and let $X := L^p(1, \infty) \cap L^q(1, \infty)$. This space is a Banach space under the norm $\|f\| := \max\{\|f\|_p, \|f\|_q\}$ (see Problem 2.21). On X define the operators $S(t), t \geqslant 0$, by

$$(S(t)f)(x) := f(xe^t), \quad x > 1.$$

(a) Show that S is a C_0-semigroup on X with generator A given by

$$D(A) := \{f \in X : x \mapsto xf'(x) \in X\},$$
$$(Af)(x) := xf'(x), \quad x > 1, \ f \in D(A).$$

(b) Show that $\{\operatorname{Re} \lambda > -1/q\} \subseteq \rho(A)$ and that for all $\omega' > -1/q$ we have

$$\sup_{\operatorname{Re} \lambda > \omega'} \|R(\lambda, A)\| < \infty.$$

(c) Show that for all $\omega < -1/p$ we have $\lim_{t \to \infty} e^{-\omega t} \|S(t)\| = \infty$.

13.18 For $x \in X$ define the *subdifferential* of x. by

$$\partial(x) := \{x^* \in X^* : \|x^*\| = \|x\|, \ \langle x, x^* \rangle = \|x\|\|x^*\|\}.$$

(a) Show that $\partial(x) \neq \varnothing$.

(b) Show that if X is a Hilbert space, then for all $x \in X$ we have

$$\partial(x) = \{x\}.$$

(c) Let $1 < p < \infty$ and $\frac{1}{p} + \frac{1}{q} = 1$. Show that for $X = L^p(\Omega)$ and $f \in L^p(\Omega)$ we have

$$\partial(f) = \{g_{f,p}\},$$

where $g_{f,p} \in L^q(\Omega)$ is defined by writing $f(\omega) = e^{i\theta(\omega)}|f(\omega)|$ and setting $g_{f,p}(\omega) = e^{-i\theta(\omega)}|f(\omega)|^{p-1}$.

(d) Is the subdifferential always a singleton?

(e) Using subdifferentials, extend the Lumer–Phillips theorem (Theorem 13.33) to Banach spaces.

13.19 Discuss Examples 13.36 and 13.39 for Neumann boundary conditions.

13.20 Prove the identity (13.32).

13.21 Consider the wave groups W on a bounded open set $D \subseteq \mathbb{R}^d$ (as in Theorem 13.58) or on the full space $D = \mathbb{R}^d$ (as in Theorem 13.58) and denote their generators by A. Prove that if $f = (u, v) \in D(A)$, then the solution of the wave equation in the sense of semigroup theory, that is, the mapping $t \mapsto W(t)f$, belongs to $C^2(\mathbb{R}; L^2(D)) \cap C(\mathbb{R}; H^2(D))$.

Hint: Let $\mathcal{H} = H_0^1(D) \times L^2(D)$ be the Hilbert space on which the wave group acts. Start from the general observation that if $f \in D(A)$, then $t \mapsto W(t)f$ belongs to $C^1(\mathbb{R}; \mathcal{H}) \cap C(\mathbb{R}; D(A))$; this follows from general semigroup considerations. Then use the special structure of the wave operator A.

13.22 This problem gives some perspective on the bound $\|W(t)\| \leqslant C(1 + t)$ for the wave group over the domain $\mathbb{R}^d$ (Theorem 13.59). Let A be its generator.

(a) Is the operator $-iA$ selfadjoint? (Compare with Theorem 13.58.)

(b) Show that $A - I$ satisfies the conditions of the Lumer–Phillips theorem if we endow $H^1(\mathbb{R}^d)$ with the equivalent norm

$$\|u\|_{1,2}^2 := \int_D |u|^2 + |\nabla u|^2 \, dx, \quad u \in H^1(\mathbb{R}^d).$$

(Compare with (13.36).) Conclude that with respect to the resulting equivalent norm $\||\cdot\||$ on $H^1(\mathbb{R}^d) \times L^2(\mathbb{R}^d)$ we have $\||W(t)\|| \leqslant e^t$ for all $t \geqslant 0$. How does the norm $\||\cdot\||$ compare to the norm used in Theorem 13.59?

(c) Elaborating on the idea of part (b), show that for all $\varepsilon > 0$ the space $H^1(\mathbb{R}^d) \times L^2(\mathbb{R}^d)$ admits an equivalent norm $\||\cdot\||_\varepsilon$ such that $\||W(t)\||_\varepsilon \leqslant e^{\varepsilon t}$, $t \geqslant 0$.

13.23 Derive the formulas (13.20) and (13.34) for the heat semigroup and the free Schrödinger group from their representations in terms of the projection-valued measure associated with the Laplace operator (Example 13.62).

14

Trace Class Operators

This chapter is devoted to the study of trace class operators and the related class of Hilbert–Schmidt operators. In a sense that will be explained in the next chapter, we can think of positive trace class operators and the trace as noncommutative analogues of finite measures and the expectation. After proving some general properties of trace class operators, we compute traces in a number of interesting examples.

14.1 Hilbert–Schmidt Operators

Throughout this chapter we assume that H is a *separable* complex Hilbert space.

Definition 14.1 (Hilbert–Schmidt operators). A bounded operator $T \in \mathscr{L}(H)$ is called a *Hilbert–Schmidt operator* if

$$\sum_{n \geqslant 1} \|T h_n\|^2 < \infty$$

for some (equivalently, for every) orthonormal basis $(h_n)_{n \geqslant 1}$ of H.

To see that this definition is independent of the orthonormal basis $(h_n)_{n \geqslant 1}$, let $(h'_n)_{n \geqslant 1}$ be another orthonormal basis of H. If T_1 and T_2 are Hilbert–Schmidt, then

$$
\begin{aligned}
\sum_{n \geqslant 1} (T_1 h_n | T_2 h_n) &= \sum_{n \geqslant 1} \sum_{k \geqslant 1} (T_1 h_n | h'_k)(h'_k | T_2 h_n) \\
&= \sum_{k \geqslant 1} \sum_{n \geqslant 1} \overline{(T_1^\star h'_k | h_n)(h_n | T_2^\star h'_k)} = \sum_{k \geqslant 1} \overline{(T_1^\star h'_k | T_2^\star h'_k)}.
\end{aligned}
\tag{14.1}
$$

Using this identity with h_n replaced by h'_n,

$$\sum_{k \geqslant 1} \overline{(T_1^\star h'_k | T_2^\star h'_k)} = \sum_{n \geqslant 1} (T_1 h'_n | T_2 h'_n).$$

Taking $T_1 = T_2 = T$ we infer that for a Hilbert–Schmidt operator T, the quantity

$$\|T\|_{\mathscr{L}_2(H)} := \left(\sum_{n \geq 1} \|T h_n\|^2 \right)^{1/2}$$

is independent of the orthonormal basis $(h_n)_{n \geq 1}$ of H. It is clear that

$$\|T\| \leq \|T\|_{\mathscr{L}_2(H)}.$$

For $g, h \in H$ we recall the notation $g \bar{\otimes} h$ for the operator on H defined by

$$(g \bar{\otimes} h)x := (x|h)g, \quad x \in H.$$

Example 14.2 (Finite rank operators). Every finite rank operator is a Hilbert–Schmidt operator. Indeed, by a Gram–Schmidt argument we may represent T as

$$T = \sum_{j=1}^{k} g_j \bar{\otimes} h_j$$

with $g_1, \ldots, g_k \in H$ orthonormal in H and $h_1, \ldots, h_k \in H$. Completing to an orthonormal basis $(g_j)_{j \geq 1}$, we have

$$\sum_{n \geq 1} \|T g_n\|^2 = \sum_{n \geq 1} \sum_{j=1}^{k} |(g_n|h_j)|^2 = \sum_{j=1}^{k} \sum_{n \geq 1} |(g_n|h_j)|^2 = \sum_{j=1}^{k} \|h_j\|^2.$$

Example 14.3 (Integral operator with square integrable kernel). Let (Ω, μ) be a σ-finite measure space and let $k \in L^2(\Omega \times \Omega, \mu \times \mu)$ be given. Then

$$T f(s) := \int_{\Omega} k(s,t) f(t) \, d\mu(t), \quad s \in \Omega,$$

defines a Hilbert–Schmidt T operator on $L^2(\Omega, \mu)$, for if $(h_n)_{n \geq 1}$ is an orthonormal basis of $L^2(\Omega, \mu)$, then

$$\begin{aligned}
\|T\|_{\mathscr{L}_2(H)}^2 &= \sum_{n \geq 1} \int_{\Omega} \left| \int_{\Omega} k(s,t) h_n(t) \, d\mu(t) \right|^2 d\mu(s) \\
&= \int_{\Omega} \sum_{n \geq 1} \left| \int_{\Omega} k(s,t) h_n(t) \, d\mu(t) \right|^2 d\mu(s) \\
&= \int_{\Omega} \|k(s,\cdot)\|_{L^2(\Omega,\mu)}^2 \, d\mu(s) = \|k\|_{L^2(\Omega \times \Omega, \mu \times \mu)}^2.
\end{aligned}$$

As a special case, any $d \times d$ matrix $A = (a_{jk})_{1 \leq j,k \leq d}$ is a Hilbert–Schmidt operator as a linear operator on $\mathbb{C}^d$, and

$$\|A\|_{\mathscr{L}_2(\mathbb{C}^n)}^2 = \sum_{1 \leq j,k \leq d} |a_{jk}|^2.$$

A converse to this example will be stated at the end of this section.

Proposition 14.4. *The space $\mathscr{L}_2(H)$ of all Hilbert–Schmidt operators on H is a Hilbert space with respect to the inner product*

$$(T_1|T_2) := \sum_{n \geqslant 1} (T_1 h_n | T_2 h_n),$$

where $(h_n)_{n \geqslant 1}$ is any orthonormal basis of H.

Proof It is elementary to check that $(T_1|T_2) := \sum_{n \geqslant 1}(T_1 h_n|T_2 h_n)$ defines an inner product. Its independence of the choice of the basis follows from (14.1).

The triangle inequality in ℓ^2 implies that $\mathscr{L}_2(H)$ is a normed space. To prove completeness, suppose that $(T_n)_{n \geqslant 1}$ is a Cauchy sequence in $\mathscr{L}_2(H)$. Then $(T_n)_{n \geqslant 1}$ is a Cauchy sequence in $\mathscr{L}(H)$. Let $T \in \mathscr{L}(H)$ be its limit. If $(h_j)_{j \geqslant 1}$ is an orthonormal basis for H, then for all $n \geqslant 1$ we have

$$\sum_{j=1}^{n} \|Th_j\|^2 = \lim_{k \to \infty} \sum_{j=1}^{n} \|T_k h_j\|^2 \leqslant \lim_{k \to \infty} \|T_k\|_{\mathscr{L}_2(H)}^2 < \infty.$$

Upon letting $n \to \infty$, it follows that T is a Hilbert–Schmidt operator and

$$\|T\|_{\mathscr{L}_2(H)} \leqslant \lim_{k \to \infty} \|T_k\|_{\mathscr{L}_2(H)}.$$

Also,

$$\sum_{j=1}^{n} \|(T_k - T)h_j\|^2 = \lim_{m \to \infty} \sum_{j=1}^{n} \|(T_k - T_m)h_j\|^2 \leqslant \limsup_{m \to \infty} \|T_k - T_m\|_{\mathscr{L}_2(H)}^2.$$

It follows that $\lim_{k \to \infty} \|T_k - T\|_{\mathscr{L}_2(H)} \leqslant \lim_{k \to \infty} \limsup_{m \to \infty} \|T_k - T_m\|_{\mathscr{L}_2(H)}$. Since the latter tends to 0 as $k \to \infty$ it follows that $\lim_{k \to \infty} T_k = T$ in $\mathscr{L}_2(H)$. This proves completeness. $\square$

Proposition 14.5. *Every Hilbert–Schmidt operator is compact and can be approximated, in the Hilbert–Schmidt norm, by finite rank operators.*

Proof Let T be a Hilbert–Schmidt operator on H, let $(h_n)_{n \geqslant 1}$ be an orthonormal basis for H, and denote by P_N the orthogonal projection onto the span of $\{h_1, \ldots, h_N\}$. Then $P_N T$ is a finite rank operator and hence Hilbert–Schmidt, and we have

$$\limsup_{N \to \infty} \|P_N T - T\|^2 \leqslant \limsup_{N \to \infty} \|P_N T - T\|_{\mathscr{L}_2(H)}^2 = \limsup_{N \to \infty} \sum_{n \geqslant N+1} \|T h_n\|^2 = 0.$$

Each $P_N T$ is a finite rank operator, hence compact. Since uniform limits of compact operators are compact, it follows that T is compact. $\square$

Proposition 14.6. *A bounded operator $T \in \mathscr{L}(H)$ is a Hilbert–Schmidt operator if and only if $T^\star$ is a Hilbert–Schmidt operator, and in this case we have $\|T\|_{\mathscr{L}_2(H)} = \|T^\star\|_{\mathscr{L}_2(H)}$.*

Proof This is immediate from (14.1).   □

Hilbert–Schmidt operators have the following ideal property:

Proposition 14.7. *If T is a Hilbert–Schmidt operator and S and U are bounded, then STU is a Hilbert–Schmidt operator and*

$$\|STU\|_{\mathscr{L}_2(H)} \leqslant \|S\| \|T\|_{\mathscr{L}_2(H)} \|U\|.$$

Proof It is clear that ST is a Hilbert–Schmidt operator and

$$\|ST\|_{\mathscr{L}_2(H)} \leqslant \|S\| \|T\|_{\mathscr{L}_2(H)}.$$

Applying this to $U^\star$ and $T^\star$ using Proposition 14.6, it follows that $U^\star T^\star$ is a Hilbert–Schmidt operator and

$$\|U^\star T^\star\|_{\mathscr{L}_2(H)} \leqslant \|U^\star\| \|T^\star\|_{\mathscr{L}_2(H)} = \|U\| \|T\|_{\mathscr{L}_2(H)}.$$

Then $TU = (U^\star T^\star)^\star$ is a Hilbert–Schmidt operator and

$$\|TU\|_{\mathscr{L}_2(H)} = \|U^\star T^\star\|_{\mathscr{L}_2(H)} \leqslant \|U\| \|T\|_{\mathscr{L}_2(H)}.$$

Using the first step once more, this implies that STU is a Hilbert–Schmidt operator and satisfies the estimate in the statement of the proposition.   □

Theorem 14.8. *Let $(\Omega, \mathscr{F}, \mu)$ be a* separable *measure space, that is, the Hilbert space $L^2(\Omega, \mu)$ is separable. If $T \in \mathscr{L}_2(L^2(\Omega, \mu))$, there exists a unique $k \in L^2(\Omega \times \Omega, \mu \times \mu)$ such that for all $f \in L^2(\Omega, \mu)$ we have*

$$Tf(\omega) = \int_\Omega k(\omega, \omega') f(\omega') \, d\mu(\omega')$$

for μ-almost all $\omega \in \Omega$.

The proof of this theorem will be given in the next section.

14.2 Trace Class Operators

14.2.a The Singular Value Decomposition

Recall that a bounded operator $T \in \mathscr{L}(H)$ is called *positive* if $(Th|h) \geqslant 0$ for all $h \in H$. Every bounded positive operator is selfadjoint.

Definition 14.9 (Trace, of a positive operator). The *trace* of a positive operator $T \in \mathscr{L}(H)$ is the nonnegative extended-real number defined by

$$\operatorname{tr}(T) := \sum_{n \geqslant 1} (Th_n|h_n),$$

where $(h_n)_{n \geqslant 1}$ is any orthonormal basis of H.

To see that tr(T) is well defined, suppose that $(h_n)_{n \geq 1}$ and $(h'_n)_{n \geq 1}$ are orthonormal bases of H. Then, by the result already proved for Hilbert–Schmidt operators,

$$\sum_{n \geq 1} (T h'_n | h'_n) = \sum_{n \geq 1}^{N} \|T^{1/2} h'_n\|^2 = \sum_{n \geq 1}^{N} \|T^{1/2} h_n\|^2 = \sum_{n=1}^{N} (T h_n | h_n).$$

Definition 14.10 (Trace class operators). A bounded operator $T \in \mathscr{L}(H)$ is called a *trace class operator* if its modulus $|T| := (T^\star T)^{1/2}$ has finite trace.

Proposition 14.11. *If $T \in \mathscr{L}(H)$ is a trace class operator, then* $\|T\| \leq \text{tr}(|T|)$.

Proof This follows from

$$\|T\| = \|T^\star T\|^{1/2} = \sup_{\|h\|=1} \left(|T| h \,\big|\, |T| h \right)^{1/2}$$

$$= \sup_{\|h\|=1} \big\| |T| h \big\| = \big\| |T| \big\| = \sup_{\|h\| \leq 1} \left(|T| h \,\big|\, h \right) \leq \text{tr}(|T|),$$

using the identities of Proposition 4.28 and Theorem 8.11. □

Example 14.12 (Finite rank operators). Every finite rank operator is a trace class operator. Indeed, Theorem 9.2 gives a representation

$$T = \sum_{n \geq 1} \lambda_n g_n \bar{\otimes} h_n$$

with orthonormal sequences $(g_n)_{n \geq 1}$ and $(h_n)_{n \geq 1}$ in H and $\lambda_n \geq 0$ the singular values of T; the sum converges in the norm of $\mathscr{L}(H)$. We claim that the singular value sequence of T is finite. To prove this, by the orthogonality of the sequence $(g_n)_{n=1}^{N}$ we obtain

$$T^\star T = \left(\sum_{m \geq 1} \lambda_m h_m \bar{\otimes} g_m \right) \left(\sum_{n \geq 1} \lambda_n g_n \bar{\otimes} h_n \right) = \sum_{n \geq 1} \lambda_n^2 h_n \bar{\otimes} h_n.$$

Inspecting the proof of Theorem 9.2, we see that every h_n is an eigenvector of $T^\star T$. Since $T^\star T$ is of finite rank, the sum on the right-hand side must by a finite sum, say $T^\star T = \sum_{n=1}^{N} \lambda_n^2 h_n \bar{\otimes} h_n$. This proves the claim. This implies that $\text{tr}(|T|) = \sum_{n=1}^{N} \lambda_n$ is finite.

Proposition 14.13. *Every trace class operator is compact.*

Proof First assume that T is a positive trace class operator. Let $(h_n)_{n \geq 1}$ be an orthonormal basis of T. Then from

$$\sum_{n \geq 1} \|T^{1/2} h_n\|^2 = \sum_{n \geq 1} (T h_n | h_n) = \text{tr}(T) < \infty$$

we see that $T^{1/2}$ is a Hilbert–Schmidt operator and therefore compact. Hence also $T = (T^{1/2})^2$ is compact.

In the general case let $T = U|T|$ be the polar decomposition of T, with U an isometry from $\overline{R(|T|)}$ onto $\overline{R(T)}$ (see Theorem 8.30). Since the positive operator $|T|$ is a trace class operator, $|T|$ is compact, hence so is T. □

Let T be a compact operator, with polar decomposition $T = U|T|$. Viewing U as an isometry from $\overline{R(|T|)}$ onto $\overline{R(T)}$, its adjoint U^* is an isometry from $\overline{R(T)}$ onto $\overline{R(|T|)}$ satisfying $U^*U = I$, and consequently $|T| = U^*T$. It follows that $|T|$ is compact.

Definition 14.14 (Singular values). The *singular values* of a compact operator T on H are the nonzero eigenvalues of the compact operator $|T|$.

Since $|T|$ is positive, every singular value is a strictly positive real number, and since $|T|$ is compact, the set of singular values is finite or countable with 0 as its only possible accumulation point. We may therefore think of the set of singular values as a nonincreasing (finite or infinite) sequence $(\lambda_n)_{n \geqslant 1}$. This sequence, where each λ_n is repeated according to its multiplicity, is called the *singular value sequence*.

According to the singular value decomposition for compact operators (Theorem 9.2), every compact operator $T \in \mathscr{L}(H)$ admits a decomposition

$$T = \sum_{n \geqslant 1} \lambda_n g_n \bar{\otimes} h_n$$

with convergence in the operator norm, where $(\lambda_n)_{n \geqslant 1}$ is the singular value sequence of T and $(g_n)_{n \geqslant 1}$ and $(h_n)_{n \geqslant 1}$ are orthonormal sequences in H. The following theorem characterises trace class and Hilbert–Schmidt operators in terms of the sequence $(\lambda_n)_{n \geqslant 1}$. In order to state the two cases symmetrically, we use the notation $\|T\|_{\mathscr{L}_1(H)} := \mathrm{tr}(|T|)$. In the next section we prove that the set $\mathscr{L}_1(H)$ of all trace class operators on H is a Banach space.

Theorem 14.15 (Singular value decomposition). *Let $T \in \mathscr{L}(H)$ be compact, and let $(\lambda_n)_{n \geqslant 1}$ be its singular value sequence. Then:*

(1) *T is a trace class operator if and only if $\sum_{n \geqslant 1} \lambda_n < \infty$. In this case we have*

$$\|T\|_{\mathscr{L}_1(H)} = \sum_{n \geqslant 1} \lambda_n.$$

(2) *T is a Hilbert–Schmidt operator if and only if $\sum_{n \geqslant 1} \lambda_n^2 < \infty$. In this case we have*

$$\|T\|_{\mathscr{L}_2(H)}^2 = \sum_{n \geqslant 1} \lambda_n^2.$$

In either case we have

$$T = \sum_{n \geqslant 1} \lambda_n g_n \bar{\otimes} h_n$$

where $(g_n)_{n \geqslant 1}$ and $(h_n)_{n \geqslant 1}$ are orthonormal sequences in H, with convergence in the norm of $\mathscr{L}_1(H)$ in case (1) and convergence in the norm of $\mathscr{L}_2(H)$ in case (2). If T is positive we may take $(g_n)_{n\geqslant 1} = (h_n)_{n\geqslant 1}$.

Proof (1): Let $\mu_1 > \mu_2 > \dots$ be the sequence of *distinct* nonzero eigenvalues of $|T|$. Since $|T|$ is selfadjoint, the eigenspaces Y_k corresponding to μ_k are pairwise orthogonal, and since $|T|$ is compact they are finite-dimensional, say $\dim(Y_k) =: d_k$. By the spectral theorem for compact selfadjoint operators (Theorem 9.1) we have

$$|T| = \sum_{k\geqslant 1} \mu_k P_k$$

with convergence in the operator norm of $\mathscr{L}(H)$. Choosing orthonormal bases $(h_j^k)_{j=1}^{d_k}$ for Y_k we may write $P_k = \sum_{j=1}^{d_k} h_j^k \otimes h_j^k$ and

$$|T| = \sum_{k\geqslant 1} \mu_k \sum_{j=1}^{d_k} h_j^k \otimes h_j^k,$$

again with convergence in the operator norm of $\mathscr{L}(H)$. Fixing an orthonormal basis $(h_n')_{n\geqslant 1}$ of H, it follows that

$$\mathrm{tr}(|T|) = \sum_{n\geqslant 1} (|T|h_n'|h_n') = \sum_{n\geqslant 1}\sum_{k\geqslant 1} \mu_k \sum_{j=1}^{d_k} (h_n'|h_j^k)(h_j^k|h_n')$$

$$= \sum_{k\geqslant 1} \mu_k \sum_{j=1}^{d_k}\sum_{n\geqslant 1} |(h_n'|h_j^k)|^2 = \sum_{k\geqslant 1} d_k \mu_k = \sum_{n\geqslant 1} \lambda_n.$$

This gives the first assertion. Now let T be represented as $\sum_{n\geqslant 1}\lambda_n g_n \otimes h_n$, as in the discussion preceding the theorem. To prove convergence in the norm of $\mathscr{L}_1(H)$ of this sum, we note that

$$\left\| T - \sum_{n=1}^{N} \lambda_n g_n \otimes h_n \right\|_{\mathscr{L}_1(H)} = \left\| \sum_{n\geqslant N+1} \lambda_n g_n \otimes h_n \right\|_{\mathscr{L}_1(H)} = \sum_{n\geqslant N+1} \lambda_n,$$

the last identity being a consequence of the fact that both $(g_n)_{n\geqslant 1}$ and $(h_n)_{n\geqslant 1}$ are orthogonal sequences (cf. Example 14.18). As $N \to \infty$, the right-hand side tends to 0 and the required convergence follows.

(2): With the notation of (1), the doubly indexed sequence $(h_j^k)_{k\geqslant 1, 1\leqslant j\leqslant d_k}$ is an orthonormal basis for $\bigoplus_{k\geqslant 1} Y_k$ and we have

$$\sum_{k\geqslant 1}\sum_{j=1}^{d_k} \|T h_j^k\|^2 = \sum_{k\geqslant 1}\sum_{j=1}^{d_k} (|T|^2 h_j^k | h_j^k) = \sum_{k\geqslant 1}\sum_{j=1}^{d_k} \mu_k^2 = \sum_{n\geqslant 1} \lambda_n^2.$$

Since $|T|h = 0$ for $h \in Y_0 := \mathsf{N}(|T|)$, this gives the first assertion. Convergence in the norm of $\mathscr{L}_2(H)$ is proved by testing against an orthonormal basis $(h'_m)_{m\geqslant 1}$ containing $(h_n)_{n\geqslant 1}$ as a subsequence, which gives

$$\left\| T - \sum_{n=1}^{N} \lambda_n g_n \bar{\otimes} h_n \right\|^2_{\mathscr{L}_2(H)} = \left\| \sum_{n\geqslant N+1} \lambda_n g_n \bar{\otimes} h_n \right\|^2_{\mathscr{L}_2(H)} \leqslant \sum_{n\geqslant N+1} \lambda_n^2.$$

The right-hand side tends to 0 as $N \to \infty$. $\square$

At this point we briefly pause to insert a proof of Theorem 14.8.

Proof of Theorem 14.8   Let $T \in \mathscr{L}_2(L^2(\Omega,\mu))$ be given, with $L^2(\Omega,\mu)$ separable. By Theorem 14.15 we have

$$T = \sum_{n\geqslant 1} \lambda_n g_n \bar{\otimes} h_n,$$

where $(g_n)_{n\geqslant 1}$ and $(h_n)_{n\geqslant 1}$ are orthonormal sequences of $L^2(\Omega,\mu)$ and the nonnegative real numbers $\lambda_n \geqslant 0$ satisfy $\sum_{n\geqslant 1} \lambda_n^2 < \infty$. Then, for $f \in L^2(\Omega,\mu)$ and μ-almost all $\omega \in \Omega$,

$$Tf(\omega) = \sum_{n\geqslant 1} \lambda_n (f|h_n) g_n(\omega)$$

$$= \sum_{n\geqslant 1} \lambda_n \int_\Omega f(\omega')\overline{h_n(\omega')} g_n(\omega) \, \mathrm{d}\mu(\omega') = \int_\Omega k(\omega,\omega') f(\omega') \, \mathrm{d}\mu(\omega'),$$

where

$$k(\omega,\omega') := \sum_{n\geqslant 1} \lambda_n g_n(\omega)\overline{h_n(\omega')}$$

is square integrable since

$$\int_\Omega \int_\Omega |k(\omega,\omega')|^2 \, \mathrm{d}\mu(\omega) \, \mathrm{d}\mu(\omega')$$

$$= \int_\Omega \int_\Omega \sum_{m\geqslant 1}\sum_{n\geqslant 1} \lambda_m \lambda_n g_m(\omega)\overline{g_n(\omega)}\, \overline{h_m(\omega')} h_n(\omega') \, \mathrm{d}\mu(\omega) \, \mathrm{d}\mu(\omega')$$

$$= \sum_{m\geqslant 1}\sum_{n\geqslant 1} \lambda_m \lambda_n (g_m|g_n)(h_n|h_m) = \sum_{n\geqslant 1} \lambda_n^2 < \infty.$$

$\square$

Returning to the main line of development, Theorem 14.15 permits us to extend the trace to arbitrary trace class operators.

Theorem 14.16 (Trace).   *If $T \in \mathscr{L}(H)$ is a trace class operator, for any orthonormal basis $(h_n)_{n\geqslant 1}$ of H the sum*

$$\mathrm{tr}(T) := \sum_{n\geqslant 1} (Th_n|h_n)$$

converges absolutely, the sum is independent of the choice of the basis, and

$$|\mathrm{tr}(T)| \leqslant \mathrm{tr}(|T|).$$

Proof Let $T = U|T|$ be the polar decomposition of T, with U a partial isometry from $\overline{\mathsf{R}(|T|)}$ onto $\overline{\mathsf{R}(T)}$. First consider the special case where $(h_n)_{n \geqslant 1}$ is an orthonormal basis containing the sequences $(h_j^k)_{j=1}^{d_k}$, $k \geqslant 1$, from the proof of Theorem 14.15. Since T vanishes on $\mathsf{R}(|T|)^\perp = \mathsf{N}(|T|)$, upon taking $\lambda_n = 0$ if $h_n \in \mathsf{N}(|T|)$ we have $|T|h = \sum_{n \geqslant 1} \lambda_n (h|h_n) h_n$ for all $h \in H$. Then,

$$\sum_{n \geqslant 1} |(Th_n|h_n)| = \sum_{n \geqslant 1} |(U|T|h_n|h_n)| = \sum_{n \geqslant 1} \lambda_n |(Uh_n|h_n)| \leqslant \sum_{n \geqslant 1} \lambda_n = \mathrm{tr}(|T|).$$

It follows that $\mathrm{tr}(T) = \sum_{n \geqslant 1} (Th_n|h_n)$ with absolute convergence and $|\mathrm{tr}(T)| \leqslant \mathrm{tr}(|T|)$.

Now let $(h'_n)_{n \geqslant 1}$ be an arbitrary orthonormal basis. Then, with the notation as above,

$$\sum_{n \geqslant 1} |(Th'_n|h'_n)| = \sum_{n \geqslant 1} \left| \sum_{k \geqslant 1} (Th_k|h'_n)(h'_n|h_k) \right| \leqslant \sum_{n \geqslant 1} \sum_{k \geqslant 1} |(Th_k|h'_n)(h'_n|h_k)|$$

$$= \sum_{k \geqslant 1} \sum_{n \geqslant 1} |(|T|Uh_k|h'_n)(h'_n|h_k)| = \sum_{k \geqslant 1} \lambda_k \sum_{n \geqslant 1} |(Uh_k|h'_n)(h'_n|h_k)|$$

$$\leqslant \sum_{k \geqslant 1} \lambda_k \left(\sum_{n \geqslant 1} |(Uh_k|h'_n)|^2 \right)^{1/2} \left(\sum_{n \geqslant 1} |(h'_n|h_k)|^2 \right)^{1/2}$$

$$= \sum_{k \geqslant 1} \lambda_k \|Uh_k\| \|h_k\| \leqslant \sum_{k \geqslant 1} \lambda_k.$$

This shows that $\sum_{n \geqslant 1} (Th'_n|h'_n)$ is absolutely summable. Moreover,

$$\sum_{n \geqslant 1} (Th'_n|h'_n) = \sum_{n \geqslant 1} \sum_{k \geqslant 1} (Th'_n|h_k)(h_k|h'_n) = \sum_{k \geqslant 1} \sum_{n \geqslant 1} (Th'_n|h_k)(h_k|h'_n)$$

$$= \sum_{k \geqslant 1} \sum_{n \geqslant 1} \overline{(T^\star h_k|h'_n)(h'_n|h_k)} = \sum_{k \geqslant 1} \overline{(T^\star h_k|h_k)} = \sum_{k \geqslant 1} (Th_k|h_k),$$

where the change of summation order is justified by the previous estimates, which imply the absolute summability of the double summations. □

Definition 14.17 (Trace, of a trace class operator). The *trace* of a trace class operator $T \in \mathscr{L}(H)$ is defined by

$$\mathrm{tr}(T) := \sum_{n \geqslant 1} (Th_n|h_n),$$

where $(h_n)_{n \geqslant 1}$ is any orthonormal basis of H.

By Theorem 14.16, the trace is well defined.

Example 14.18 (Finite rank operators, continued). The trace of any finite rank operator $T = \sum_{n=1}^{N} g_n \bar{\otimes} h_n$ is given by

$$\text{tr}(T) = \sum_{n=1}^{N} \text{tr}(g_n \bar{\otimes} h_n) = \sum_{n=1}^{N} (g_n | h_n).$$

Here we use the fact that the trace of a rank one operator $g \bar{\otimes} h$ may be evaluated in terms of an orthonormal basis $(h'_n)_{n \geqslant 1}$ chosen such that $h'_1 = h/\|h\|$ to give

$$\text{tr}(g \bar{\otimes} h) = \sum_{n \geqslant 1} (g | h'_n)(h'_n | h) = (g | h).$$

If P is a (not necessarily orthogonal) projection onto an N-dimensional subspace, then

$$\text{tr}(P) = N.$$

To see this we write $P = \sum_{n=1}^{N} g_n \bar{\otimes} h_n$ with $g_1, \ldots, g_N$ orthonormal. From $Ph_n = \|h_n\|^2 g_n$ and $P^2 h_n = \|h_n\|^2 \sum_{m=1}^{N} (g_n | h_m) g_m$ we deduce that $(g_n | h_m) = \delta_{mn}$ and the result follows from the first part of the example.

More interesting examples will be given in Section 14.5.

We prove next that the set $\mathscr{L}_1(H)$ of all trace class operators on H is a vector space and, endowed with norm

$$\|T\|_{\mathscr{L}_1(H)} := \text{tr}(|T|),$$

a Banach space. We begin with the proof that $\mathscr{L}_1(H)$ is a vector space. It is evident that if T is a trace class operator, then so is cT for all $c \in \mathbb{C}$ and $\text{tr}(|cT|) = |c|\text{tr}(|T|)$. Additivity is less trivial and is based on a characterisation of trace class operators which we prove first. The crucial ingredient is the following lemma.

Lemma 14.19. *Let $T \in \mathscr{L}(H)$ be compact and let $(\lambda_n)_{n \geqslant 1}$ be its singular value sequence. Then for all $n \geqslant 1$ we have*

$$\sum_{j \geqslant 1} \lambda_j = \sup_{g,h} \left| \sum_j (Tg_j | h_j) \right| = \sup_{g,h} \sum_j |(Tg_j | h_j)|,$$

where the suprema are taken over all integers $k \geqslant 1$ and all finite orthonormal sequences $g = (g_j)_{j=1}^{k}$ and $h = (h_j)_{j=1}^{k}$ of length k in H.

Here we allow the possibility that all three expressions are infinite.

Proof   Without loss of generality we assume that $T \neq 0$. Let e_j be a normalised eigenvector for $|T|$ with strictly positive eigenvalue λ_j. Consider a polar decomposition $T = U|T|$, with U a partial isometry which is isometric from $\overline{R(|T|)}$ onto $\overline{R(T)}$. Then,

$$\sum_{j=1}^{n} \lambda_j = \sum_{j=1}^{n} (|T|e_j | e_j) = \sum_{j=1}^{n} (Te_j | Ue_j) = \left| \sum_{j=1}^{n} (Te_j | Ue_j) \right|,$$

using that $(x|y) = (Ux|Uy)$ for all $x, y \in \overline{R(|T|)}$ and that all λ_j are positive. For the same reason, $(Ue_j)_{j=1}^n$ is an orthonormal sequence. This gives the two inequalities '$\leqslant$'.

In the converse direction, let two orthonormal sequences $(g_j)_{j=1}^n$ and $(h_j)_{j=1}^n$ in H be given. Then, with the above notation, repetition of the second part of the proof of Theorem 14.16 gives

$$
\sum_{j=1}^n |(Tg_j|h_j)| = \sum_{j=1}^n \sum_{k\geqslant 1} |(U|T|e_k|h_j)(g_j|e_k)|
$$

$$
= \sum_{k\geqslant 1} \lambda_k \sum_{j=1}^n |(Ue_k|h_j)(g_j|e_k)|
$$

$$
\leqslant \sum_{k\geqslant 1} \lambda_k \Big(\sum_{j=1}^n |(Ue_k|h_j)|^2 \Big)^{1/2} \Big(\sum_{j=1}^n |(g_j|e_k)|^2 \Big)^{1/2}
$$

$$
\leqslant \sum_{k\geqslant 1} \lambda_k \|Ue_k\| \|e_k\| = \sum_{k\geqslant 1} \lambda_k.
$$

This concludes the proof of the equalities. □

Theorem 14.20 (Trace class operators). *For a bounded operator $T \in \mathscr{L}(H)$ the following assertions are equivalent:*

(1) *T is a trace class operator;*
(2) *we have*

$$
\sup_{g,h} \Big| \sum_{j\geqslant 1} (Tg_j|h_j) \Big| < \infty,
$$

the supremum being taken over all orthonormal sequences $g = (g_j)_{j\geqslant 1}$ and $h = (h_j)_{j\geqslant 1}$ of H;
(3) *we have*

$$
\sup_{g,h} \sum_{j\geqslant 1} |(Tg_j|h_j)| < \infty,
$$

the supremum being taken over all orthonormal sequences $g = (g_j)_{j\geqslant 1}$ and $h = (h_j)_{j\geqslant 1}$ of H.

In this situation, the suprema in (2) and (3) are in fact maxima, and we have

$$
\|T\|_{\mathscr{L}_1(H)} = \sup_{g,h} \Big| \sum_{j\geqslant 1} (Tg_j|h_j) \Big| = \sup_{g,h} \sum_{j\geqslant 1} |(Tg_j|h_j)|.
$$

Proof The equivalences follow from Lemma 14.19, which also gives the equalities in the final assertion of the theorem. To see that the suprema are in fact maxima, consider the sequences given by the singular value decomposition of Theorem 14.15. □

The trace class condition is stated in terms of summability of its singular value sequence. As a first application of Theorem 14.20 we show that if an operator is of trace class, then its eigenvalue sequence is absolutely summable:

Proposition 14.21. *For any trace class operator* $T \in \mathscr{L}(H)$, *with eigenvalue sequence* $(\lambda_n)_{n \geqslant 1}$ *repeated according to algebraic multiplicity, we have*

$$\sum_{n \geqslant 1} |\lambda_n| \leqslant \|T\|_{\mathscr{L}_1(H)}.$$

Proof We prove the proposition in two steps.

Step 1 – In this step we let T be any linear operator acting on a d-dimensional Hilbert space H, with eigenvalue sequence $(\lambda_j)_{j=1}^{d}$ repeated according to algebraic multiplicities. Our aim is to prove that there exists an orthonormal basis $(h_j)_{j=1}^{d}$ in H such that $(Th_j|h_j) = \lambda_j$ for all $j = 1, \ldots, d$.

As a first step we prove that there exists an orthonormal basis $(h_j)_{j=1}^{d}$ in H such that the matrix representation $(t_{ij})_{i,j=1}^{d}$ of T with respect to this bases is lower triangular, that is, it satisfies $t_{ij} = 0$ whenever $i < j$. We prove this by induction on the dimension d. For $d = 1$ there is nothing to be proved. Assume now that the claim has been proved for all dimensions less than d. Assuming now that the dimension equals d, let λ be an eigenvalue of T. Then $\lambda - T$ maps H into some $(d-1)$-dimensional subspace G of H. Since G is invariant under T, the induction hypothesis implies that there exists an orthonormal basis $(h_j)_{j=1}^{d-1}$ in G relative to which the matrix representation of $T|_G$ is lower triangular. Then,

$$((\lambda - T|_G)h_i|h_j) = \lambda(h_i|h_j) - t_{ij} = \lambda \cdot 0 - 0 = 0, \quad 1 \leqslant i < j \leqslant d-1.$$

Choose a norm one vector x_d in H orthogonal to G. Relative to the orthonormal basis $(h_j)_{j=1}^{d}$ the matrix representation of T is lower triangular and we have

$$((\lambda - T)h_i|h_j) = 0, \quad 1 \leqslant i < j \leqslant d.$$

This proves the claim.

With the orthonormal basis $(h_j)_{j=1}^{d}$ as in the claim, we have

$$\det(\lambda - T) = (\lambda - (Th_1|h_1)) \ldots (\lambda - (Th_d|h_d)).$$

But this implies that $((Th_j|h_j))_{j=1}^{d}$ is the eigenvalue sequence of T repeated according to algebraic multiplicities.

Step 2 – Let now H and T be as in the statement of the proposition, and let $(\lambda_n)_{n \geqslant 1}$ be the sequence of eigenvalues of T repeated according to algebraic multiplicities. Let $(x_n)_{n \geqslant 1}$ be a corresponding sequence of eigenvectors, that is, $x_n \neq 0$ and $Tx_n = \lambda_n x_n$ for all $n \geqslant 1$. For each $N \geqslant 1$ let H_N be the linear span of $x_1, \ldots, x_N$. Applying the result just mentioned to the restriction of T to H_N, we obtain an orthonormal basis $(h_n)_{n=1}^N$ for H_N such that

$$\sum_{n=1}^{N} |\lambda_n| = \sum_{n=1}^{N} |(T h_n | h_n)|.$$

By Theorem 14.20,

$$\sum_{n=1}^{N} |(T h_n | h_n)| \leqslant \|T\|_{\mathscr{L}_1(H)}.$$

This completes the proof. $\qquad\qquad\square$

We continue with some results about the structure of the set of trace class operators.

Theorem 14.22 (Sums). *If T_1 and T_2 are trace class operators on H, then so is their sum $T_1 + T_2$, and we have* $\mathrm{tr}(T_1 + T_2) = \mathrm{tr}(T_1) + \mathrm{tr}(T_2)$ *and*

$$\mathrm{tr}(|T_1 + T_2|) \leqslant \mathrm{tr}(|T_1|) + \mathrm{tr}(|T_2|).$$

Proof Let $(\lambda_n)_{n \geqslant 1}$, $(\mu_n)_{n \geqslant 1}$, and $(\nu_n)_{n \geqslant 1}$ denote the singular value sequences of T_1, T_2, and $T_1 + T_2$, respectively. Applying Theorem 14.20 first to the compact operator $T_1 + T_2$ and then to T_1 and T_2 separately, we obtain

$$\sum_{n \geqslant 1} \nu_n = \sup_{g,h} \left| \sum_{n \geqslant 1} ((T_1 + T_2) g_n | h_n) \right|$$

$$\leqslant \sup_{g,h} \left| \sum_{n \geqslant 1} (T_1 g_n | h_n) \right| + \sup_{g,h} \left| \sum_{n \geqslant 1} (T_2 g_n | h_n) \right| = \sum_{n \geqslant 1} \lambda_n + \sum_{n \geqslant 1} \mu_n,$$

where the suprema are taken over all orthonormal sequences $g = (g_n)_{n \geqslant 1}$ and $h = (h_n)_{n \geqslant 1}$ in H. $\qquad\qquad\square$

Theorem 14.23 (Completeness). *The normed space $\mathscr{L}_1(H)$ is a Banach space, and the finite rank operators are dense in this space.*

Proof Suppose $(T_n)_{n \geqslant 1}$ is a Cauchy sequence in $\mathscr{L}_1(H)$. Then $(T_n)_{k \geqslant 1}$ is a Cauchy sequence in $\mathscr{L}(H)$. Let $T \in \mathscr{L}(H)$ be its limit. Since each T_n is compact, so is T. Moreover, for all orthonormal sequences $(g_j)_{j \geqslant 1}$ and $(h_j)_{j \geqslant 1}$ in H and all $k \geqslant 1$,

$$\sum_{j=1}^{k} |(T g_j | h_j)| = \lim_{n \to \infty} \sum_{j=1}^{k} |(T_n g_j | h_j)| \leqslant \sup_{n \geqslant 1} \|T_n\|_{\mathscr{L}_1(H)} < \infty.$$

Therefore Theorem 14.20 implies that T is a trace class operator. Also,

$$\sum_{j=1}^{k} |((T_n - T) g_j | h_j)| = \lim_{m \to \infty} \sum_{j=1}^{k} |((T_n - T_m) g_j | h_j)| \leqslant \limsup_{m \to \infty} \|T_n - T_m\|_{\mathscr{L}_1(H)}.$$

Again by Theorem 14.20, this implies that

$$\|T_n - T\|_{\mathscr{L}_1(H)} \leqslant \limsup_{m \to \infty} \|T_n - T_m\|_{\mathscr{L}_1(H)}.$$

Since the latter tends to 0 as $n \to \infty$, it follows that $\lim_{n \to \infty} T_n = T$ in $\mathscr{L}_1(H)$. This proves that $\mathscr{L}_1(H)$ is complete.

To prove that the finite rank operators are dense it suffices to show that if T is a trace class operator, then the convergence in part (2) of Theorem 14.20 also takes place with respect to the norm of $\mathscr{L}_1(H)$. The partial sums in (2) are Cauchy in the norm of $\mathscr{L}_1(H)$ thanks to the absolute summability of the sequence $(\lambda_n)_{n \geqslant 1}$ and the fact that $\|g \bar{\otimes} h\|_{\mathscr{L}_1(H)} = \|g\|\|h\|$. Therefore, by the completeness of $\mathscr{L}_1(H)$, the partial sums converge to some operator $\widetilde{T} \in \mathscr{L}_1(H)$; but since we know already that the sum converges to T in $\mathscr{L}(H)$, we must have $\widetilde{T} = T$. $\quad\square$

Trace class operators have the following ideal property:

Theorem 14.24 (Ideal property). *If T is a trace class operator and if S and U are bounded, then STU is a trace class operator and*

$$\|STU\|_{\mathscr{L}_1(H)} \leqslant \|S\|\|T\|_{\mathscr{L}_1(H)}\|U\|.$$

For the proof we need a lemma which is of some independent interest:

Lemma 14.25. *Every contraction is a convex combination of four unitaries.*

Proof Let $T \in \mathscr{L}(H)$ be a contraction. The two operators $S_1 := \frac{1}{2}(T + T^\star)$ and $S_2 := \frac{1}{2i}(T - T^\star)$ are selfadjoint and satisfy $T = S_1 + iS_2$. The four operators $U_j^\pm := S_j \pm i(I - S_j^2)^{1/2}$ are unitary and satisfy $S_j = \frac{1}{2}(U_j^+ + U_j^-)$. $\quad\square$

Proof of Theorem 14.24 If U is a unitary operator, the operator TU is a trace class operator and $\|TU\|_{\mathscr{L}_1(H)} \leqslant \|T\|_{\mathscr{L}_1(H)}$ by Theorem 14.20. For contractions U, the same conclusion follows by combining the unitary case with Lemma 14.25 and the general case follows by scaling. The proof can now be finished as in Proposition 14.7. $\quad\square$

Proposition 14.26. *A bounded operator T is a trace class operator if and only if $T^\star$ is a trace class operator, and in this case we have $\mathrm{tr}(T^\star) = \overline{\mathrm{tr}(T)}$ and $\|T^\star\|_{\mathscr{L}_1(H)} = \|T\|_{\mathscr{L}_1(H)}$.*

Proof This is immediate from Theorem 14.20. $\quad\square$

Proposition 14.27. *If T is a trace class operator and S is bounded, then ST and TS are trace class operators and*

$$\mathrm{tr}(ST) = \mathrm{tr}(TS).$$

Proof That ST and TS are trace class operators follows from Theorem 14.24.

To prove the identity $\operatorname{tr}(ST) = \operatorname{tr}(TS)$ we first assume that S is unitary. If $(h_n)_{n \geqslant 1}$ is an orthonormal basis for H, then so is $(Sh_n)_{n \geqslant 1}$. Hence, since the trace is independent of the choice of basis,

$$\operatorname{tr}(ST) = \sum_{n \geqslant 1} (STh_n | h_n) = \sum_{n \geqslant 1} (STSh_n | Sh_n) = \sum_{n \geqslant 1} (TSh_n | h_n) = \operatorname{tr}(TS),$$

where we used that $S^*S = I$. The general case follows as in the previous proof by writing a contraction S as a convex combination of four unitaries. $\square$

We conclude with a proposition describing the relationship between trace class operators and Hilbert–Schmidt operators. As a preliminary observation note that the inner product of $\mathscr{L}_2(H)$ can be reinterpreted in terms of the trace: we have the *trace duality*

$$(T_1 | T_2) = \operatorname{tr}(T_2^\star T_1) = \operatorname{tr}(T_1 T_2^\star).$$

Proposition 14.28. *A bounded operator on H is a trace class operator if and only if it is the product of two Hilbert–Schmidt operators. If $T = S_2 S_1$ is such a decomposition, then*

$$\|T\|_{\mathscr{L}_1(H)} \leqslant \|S_1\|_{\mathscr{L}_2(H)} \|S_2\|_{\mathscr{L}_2(H)}.$$

Proof 'If': If S_1 and S_2 are Hilbert–Schmidt and $T := S_2 S_1$, then for all orthonormal sequences $(g_j)_{j=1}^n$ and $(h_j)_{j=1}^n$ in H and all $n \geqslant 1$,

$$\sum_{j=1}^n |(Tg_j | h_j)| \leqslant \sum_{j=1}^n \|S_1 g_j\| \|S_2^\star h_j\| \leqslant \left(\sum_{j=1}^n \|S_1 g_j\|^2 \right)^{1/2} \left(\sum_{j=1}^n \|S_2^\star h_j\|^2 \right)^{1/2}$$

$$\leqslant \|S_1\|_{\mathscr{L}_2(H)} \|S_2^\star\|_{\mathscr{L}_2(H)} = \|S_1\|_{\mathscr{L}_2(H)} \|S_2\|_{\mathscr{L}_2(H)}.$$

Letting $n \to \infty$, by Theorem 14.20, this implies that T is a trace class operator and satisfies the inequality $\|T\|_{\mathscr{L}_1(H)} \leqslant \|S_1\|_{\mathscr{L}_2(H)} \|S_2\|_{\mathscr{L}_2(H)}$.

'Only if': Using a polar decomposition $T = U|T|$ with U a partial isometry, take $S_1 = |T|^{1/2}$ and $S_2 = U|T|^{1/2}$. Since $|T|$ is a trace class operator, $|T|^{1/2}$ is a Hilbert–Schmidt operator, and hence so are S_1 and S_2. $\square$

14.3 Trace Duality

We have already noted that the space $\mathscr{L}_2(H)$ of Hilbert–Schmidt operators on H is a Hilbert space with respect to the inner product given by trace duality,

$$(T_1 | T_2) = \operatorname{tr}(T_1 T_2^\star), \quad T_1, T_2 \in \mathscr{L}_2(H).$$

The next theorem establishes that, by the same formula, $\mathscr{L}_1(H)$ can be identified isometrically as the dual of $\mathscr{K}(H)$, the closed subspace of $\mathscr{L}(H)$ consisting of all compact operators on H, and $\mathscr{L}(H)$ as the dual of $\mathscr{L}_1(H)$.

Theorem 14.29 (Trace duality). *By trace duality we have isometric isomorphisms*

$$(\mathscr{K}(H))^* \simeq \mathscr{L}_1(H) \quad and \quad (\mathscr{L}_1(H))^* \simeq \mathscr{L}(H).$$

More precisely, the following results hold:

(1) *for every* $T \in \mathscr{L}_1(H)$ *the mapping* $\phi_T : \mathscr{K}(H) \to \mathbb{C}$ *given by* $\phi_T(S) := \mathrm{tr}(ST)$ *is linear and bounded and satisfies*

$$\|\phi_T\|_{(\mathscr{K}(H))^*} = \|T\|_{\mathscr{L}_1(H)},$$

and, conversely, for every $\phi \in (\mathscr{K}(H))^*$ *there exists a unique* $T \in \mathscr{L}_1(H)$ *such that* $\phi = \phi_T$;

(2) *for every* $T \in \mathscr{L}(H)$ *the mapping* $\psi_T : \mathscr{L}_1(H) \to \mathbb{C}$ *given by* $\psi_T(S) := \mathrm{tr}(ST)$ *is linear and bounded and satisfies*

$$\|\psi_T\|_{(\mathscr{L}_1(H))^*} = \|T\|_{\mathscr{L}(H)},$$

and, conversely, for every $\psi \in (\mathscr{L}_1(H))^*$ *there exists a unique* $T \in \mathscr{L}(H)$ *such that* $\psi = \psi_T$.

The point of working with $\mathrm{tr}(ST)$ rather than $\mathrm{tr}(ST^\star)$ is that this makes the identifications of the duals into a linear correspondence rather than a conjugate-linear one.

Proof (1): Linearity of ϕ_T is clear and boundedness follows from Theorem 14.24, which also gives the upper bound

$$\|\phi_T\|_{(\mathscr{K}(H))^*} \leqslant \|T\|_{\mathscr{L}_1(H)}.$$

To conclude the proof of (1) it remains to show that every $\phi \in (\mathscr{K}(H))^*$ is of the form ϕ_T for some $T \in \mathscr{L}_1(T)$ and that the converse inequality $\|T\|_{\mathscr{L}_1(H)} \leqslant \|\phi_T\|_{(\mathscr{K}(H))^*}$ holds; this also gives uniqueness.

Let $\phi \in (\mathscr{K}(H))^*$ be given. The inclusion mapping $i : \mathscr{L}_2(H) \to \mathscr{K}(H)$ is continuous, so the same is true for the functional $\widetilde{\phi} := \phi \circ i : \mathscr{L}_2(H) \to \mathbb{C}$. By the Riesz representation theorem there exists a unique $T \in \mathscr{L}_2(H)$ such that

$$\phi(S) = \widetilde{\phi}(S) = (S|T)_{\mathscr{L}_2(H)}, \quad S \in \mathscr{L}_2(H).$$

Let $T = U|T|$ be its polar decomposition and let $(g_n)_{n \geqslant 1}$ be an orthonormal basis for $\mathsf{R}(|T|)$. Denoting by $P_N = \sum_{n=1}^{N} g_n \bar{\otimes} g_n$ the orthogonal projection onto the span of $g_1, \ldots, g_N$,

$$\sum_{n \geqslant 1} (|T|g_n|g_n) = \sum_{n \geqslant 1} (Tg_n|Ug_n) = \sum_{n \geqslant 1} (Ug_n|Tg_n)$$

$$= \lim_{N \to \infty} (P_N U P_N | T)_{\mathscr{L}_2(H)} = \lim_{n \to \infty} \widetilde{\phi}(P_N U P_N),$$

where the nonnegativity of the first expression justifies the second equality. Because ϕ is continuous it follows that

$$|\widetilde{\phi}(P_N U P_N)| = |\phi(i P_N U P_N)| \leqslant \|i\| \|P_N U P_N\| \|\phi\| \leqslant \|\phi\|,$$

which proves that T is a trace class operator with $\|T\|_{\mathscr{L}_1(H)} = \mathrm{tr}(|T|) \leqslant \|\phi\|$.
We now show that $\phi = \phi_{T^\star}$. For all $g, h \in H$ we have

$$\phi_{T^\star}(g \bar\otimes h) = \mathrm{tr}((g \bar\otimes h) \circ T^\star) = (g|Th).$$

On the other hand, if $(h_n)_{n \geqslant 1}$ is an orthonormal basis such that $h_1 = h$,

$$\phi(g \bar\otimes h) = (g \bar\otimes h | T) = \sum_{n \geqslant 1} ((g \bar\otimes h) h_n | T h_n) = (g|Th).$$

By linearity, this proves the identity $\phi_{T^\star}(S) = \phi(S)$ for all finite rank operators S. Since these are dense in $\mathscr{K}(H)$ by Proposition 7.6, it follows that $\phi = \phi_{T^\star}$ as claimed.

(2): Again linearity is clear and boundedness follows from Theorem 14.24, which also gives the upper bound $\|\psi_T\|_{(\mathscr{L}_1(H))^*} \leqslant \|T\|_{\mathscr{L}(H)}$. The converse inequality follows from

$$\|T\|_{\mathscr{L}(H)} = \sup_{\|x\|,\|y\| \leqslant 1} |(Tx|y)| = \sup_{\|x\|,\|y\| \leqslant 1} |\mathrm{tr}(T \circ (x \bar\otimes y))|$$

$$= \sup_{\|x\|,\|y\| \leqslant 1} |\psi_T(x \bar\otimes y)| \leqslant \|\psi_T\| \sup_{\|x\|,\|y\| \leqslant 1} \|x \bar\otimes y\|_{\mathscr{L}_1(H)} = \|\psi_T\|,$$

which also gives uniqueness.

To conclude the proof of (2) it remains to show that every $\psi \in (\mathscr{L}_1(H))^*$ is of the form ψ_T for some (necessarily unique) $T \in \mathscr{L}(H)$. By the Riesz representation theorem, for any $h \in H$ there is a unique element $Th \in H$ such that

$$(g|Th) = \psi(g \bar\otimes h), \quad g \in H,$$

and the mapping $h \mapsto Th$ is linear. From the identity it is immediate that T is bounded, with $\|T\| \leqslant \|\psi\|$. As in the proof of (1), for all finite rank operators S we have $\psi(S) = \psi_{T^\star}(S)$. By Theorem 14.23 the finite rank operators are dense in $\mathscr{L}_1(H)$. Therefore, $\psi = \psi_{T^\star}$. $\qquad \square$

14.4 The Partial Trace

If we think of the trace as the noncommutative analogue of the expectation, the partial trace of a trace class operator is then the noncommutative analogue of the conditional expectation of a random variable.

Using the notation of Appendix B we introduce the following definition.

Definition 14.30 (Hilbert space tensor product). The *Hilbert space tensor product* of the Hilbert spaces $H_1, \ldots, H_N$ is the completion of the algebraic tensor product $H_1 \otimes \cdots \otimes H_N$ with respect to the norm obtained from the inner product

$$\left(\sum_{i=1}^{k} g_1^{(i)} \otimes \cdots \otimes g_N^{(i)} \,\Big|\, \sum_{j=1}^{\ell} h_1^{(j)} \otimes \cdots \otimes h_N^{(j)} \right) := \sum_{i=1}^{k} \sum_{j=1}^{\ell} \prod_{n=1}^{N} \big(g_n^{(i)} \,\big|\, h_n^{(j)} \big).$$

With slight abuse of notation the Hilbert space tensor product of $H_1, \ldots, H_N$ is denoted again by $H_1 \otimes \cdots \otimes H_N$. We leave it to the reader to check that if $H_1, \ldots, H_N$ are separable, with an orthonormal basis $(h_j^{(n)})_{j \geq 1}$ for each H_n, then the tensors $h_{j_1}^{(1)} \otimes \cdots \otimes h_{j_N}^{(N)}$ form an orthonormal basis for $H_1 \otimes \cdots \otimes H_N$.

If $(\Omega_1, \mu_1), \ldots, (\Omega_N, \mu_N)$ are σ-finite measure spaces, then the linear mapping from $L^2(\Omega_1, \mu_1) \otimes \cdots \otimes L^2(\Omega_N, \mu_N)$ into $L^2(\Omega_1 \times \cdots \times \Omega_N, \mu_1 \times \cdots \times \mu_N)$ defined by

$$f_1 \otimes \cdots \otimes f_N \mapsto \left[(\omega_1, \ldots, \omega_N) \mapsto \prod_{n=1}^{N} f_n(\omega_n) \right]$$

extends uniquely to an isometric isomorphism

$$L^2(\Omega_1, \mu_1) \otimes \cdots \otimes L^2(\Omega_N, \mu_N) \simeq L^2(\Omega_1 \times \cdots \times \Omega_N, \mu_1 \times \cdots \times \mu_N). \tag{14.2}$$

Now let H and K be Hilbert spaces. If $S \in \mathscr{L}(H)$, then the operator $S \otimes I$, defined on the algebraic tensor product of H and K by

$$(S \otimes I)(h \otimes k) := Sh \otimes k$$

and extended by linearity, extends to a bounded operator on the Hilbert space tensor product $H \otimes K$ and

$$\|S \otimes I\| = \|S\|.$$

We leave the proof of this simple fact as an exercise to the reader; a more general version of this result will be proved in Section 15.6.c (see Proposition 15.65).

For $k \in K$ let $U_k : H \to H \otimes K$ be given by

$$U_k h := h \otimes k.$$

Its Hilbert space adjoint equals $U_k^{\star}(h \otimes k') = (k|k')h$.

Theorem 14.31 (Partial trace). *Let H and K be separable Hilbert spaces and let $T \in \mathscr{L}_1(H \otimes K)$. There exists a unique operator $\mathrm{tr}_K(T) \in \mathscr{L}_1(H)$ such that for all $S \in \mathscr{L}(H)$ we have*

$$\mathrm{tr}(\mathrm{tr}_K(T)S) = \mathrm{tr}(T(S \otimes I)). \tag{14.3}$$

The mapping $T \mapsto \mathrm{tr}_K(T)$ is called the *partial trace* with respect to K and is obtained by *tracing out K*.

Proof We claim that if $(k_n)_{n \geq 1}$ is an orthonormal basis of K, the sum

$$\mathrm{tr}_K(T) := \sum_{n \geq 1} U_{k_n}^{\star} T U_{k_n} \tag{14.4}$$

converges in $\mathscr{L}(H)$ and its sum has the required properties.

By Theorem 14.24, each operator $U_{k_n}^{\star} T U_{k_n}$ is a trace class operator. Hence by Theorem 14.20, for each $n \geq 1$ there exist orthonormal sequences $(g_j^{(n)})_{j \geq 1}$ and $(h_j^{(n)})_{j \geq 1}$ in H such that

$$\|U_{k_n}^{\star} T U_{k_n}\|_{\mathscr{L}_1(H)} = \sum_{j \geq 1} |(U_{k_n}^{\star} T U_{k_n} g_j^{(n)} | h_j^{(n)})|.$$

It follows that

$$\sum_{n \geq 1} \|U_{k_n}^{\star} T U_{k_n}\|_{\mathscr{L}_1(H)} = \sum_{n \geq 1} \sum_{j \geq 1} |(U_{k_n}^{\star} T U_{k_n} g_j^{(n)} | h_j^{(n)})|$$

$$= \sum_{n \geq 1} \sum_{j \geq 1} |(T(g_j^{(n)} \otimes k_n) | h_j^{(n)} \otimes k_n)| < \infty,$$

where the last step uses that T is a trace class operator and the sequences $(g_j^{(n)} \otimes k_n)_{j,n \geq 1}$ and $(h_j^{(n)} \otimes k_n)_{j,n \geq 1}$ are orthonormal in $H \otimes K$.

Next we check the required identity. If $(h_m)_{m \geq 1}$ is an orthonormal basis for H, then

$$\mathrm{tr}(\mathrm{tr}_K(T)S) = \sum_{n \geq 1} \mathrm{tr}(U_{k_n}^{\star} T U_{k_n} S) = \sum_{n \geq 1} \sum_{m \geq 1} (T U_{k_n} S h_m | U_{k_n} h_m)$$

$$= \sum_{n \geq 1} \sum_{m \geq 1} (T(S \otimes I)(h_m \otimes k_n) | h_m \otimes k_n) = \mathrm{tr}(T(S \otimes I)).$$

It remains to prove uniqueness. If A is a trace class operator on H such that

$$\mathrm{tr}(AS) = \mathrm{tr}(\mathrm{tr}_K(T)S)$$

for all $S \in \mathscr{L}(H)$, then Theorem 14.29 implies that $A = \mathrm{tr}_K(T)$. $\qquad\square$

Example 14.32 (Partial trace of a rank one projection). If $h \in H$ and $k \in K$ have norm one and $T = (h \otimes k) \bar{\otimes} (h \otimes k)$ is the corresponding rank one projection in $H \otimes K$, then $\mathrm{tr}_K(T)$ is the rank one projection $h \bar{\otimes} h$ in H:

$$\mathrm{tr}_K(T) = h \bar{\otimes} h.$$

Indeed, for all $S \in \mathscr{L}(H)$ we have

$$\mathrm{tr}(\mathrm{tr}_K(T)S) = \mathrm{tr}(T(S \otimes I)) = ((S \otimes I)(h \otimes k) | h \otimes k)$$

$$= (Sh \otimes k | h \otimes k) = (Sh | h)(k | k) = (Sh | h) = \mathrm{tr}((h \bar{\otimes} h)S).$$

The result now follows from the uniqueness part of Theorem 14.31.

In the terminology of the next chapter, the following proposition states that the partial trace of a state is again a state.

Proposition 14.33. *Let H and K be separable Hilbert spaces and let $T \in \mathscr{L}_1(H \otimes K)$. Then:*

(1) *if T has unit trace, then so has $\mathrm{tr}_K(T)$;*
(2) *if T is positive, then so is $\mathrm{tr}_K(T)$.*

Proof Both properties are immediate consequences of the formulas (14.3) and (14.4) for the partial trace. Indeed, the first implies that if $\mathrm{tr}(T) = 1$, then for orthonormal bases $(h_n)_{n \geqslant 1}$ and $(k_n)_{n \geqslant 1}$ of H and K,

$$
\begin{aligned}
\mathrm{tr}(\mathrm{tr}_K(T)) &= \sum_{n \geqslant 1} (\mathrm{tr}_K(T)h_n|h_n) \\
&= \sum_{n \geqslant 1} \mathrm{tr}(\mathrm{tr}_K(T)(h_n\bar{\otimes}h_n)) \\
&= \sum_{n \geqslant 1} \mathrm{tr}(T((h_n\bar{\otimes}h_n) \otimes I)) \\
&= \sum_{n \geqslant 1}\sum_{i,j \geqslant 1} (T((h_n\bar{\otimes}h_n) \otimes I)(h_i \otimes k_j)|h_i \otimes k_j) \\
&= \sum_{n \geqslant 1}\sum_{i,j \geqslant 1} (T((h_n\bar{\otimes}h_n)h_i \otimes k_j)|h_i \otimes k_j) \\
&= \sum_{n \geqslant 1}\sum_{j \geqslant 1} (Th_n \otimes k_j|h_n \otimes k_j) = \mathrm{tr}(T) = 1.
\end{aligned}
$$

This proves (1). Assertion (2) follows from

$$
(\mathrm{tr}_K(T)h|h) = \sum_{n \geqslant 1} (U_{k_n}^\star T U_{k_n} h|h) = \sum_{n \geqslant 1} (T U_{k_n} h|U_{k_n} h) \geqslant 0.
$$

$\square$

14.5 Trace Formulas

In this final section we illustrate the preceding theory by computing traces in a number of interesting situations.

14.5.a Lidskii's Theorem

If T is a linear operator acting on $\mathbb{C}^d$ whose matrix representation is in Jordan normal form, then

$$\operatorname{tr}(T) = \sum_{n=1}^{d} \lambda_n,$$

where $\lambda_1, \ldots, \lambda_d$ are the eigenvalues of T repeated according to their algebraic multiplicities; see Example 7.16. By the result of Step 1 of the proof of Proposition 14.21, this identity extends to arbitrary linear operators T acting on $\mathbb{C}^d$.

If T is a normal trace class operator on a Hilbert space H, the spectral theorem for compact normal operators allows us to select an orthonormal basis for H consisting of eigenvectors in the following way. For each of the eigenspaces corresponding to the eigenvalues of T we select an orthonormal basis. These eigenspaces are mutually orthogonal and the union of these bases, after a relabelling, is an orthonormal basis $(h_n)_{n \geq 1}$ for H. For this basis we have

$$\operatorname{tr}(T) = \sum_{n \geq 1} (T h_n | h_n) = \sum_{n \geq 1} \lambda_n,$$

where $(\lambda_n)_{n \geq 1}$ is the sequence of nonzero eigenvalues of T repeated according to their multiplicities; in the last sum we left out the indices corresponding to eigenvalue $\lambda_n = 0$ and did another relabelling.

The following deep result asserts that these formulas for the trace extend to general trace class operators:

Theorem 14.34 (Lidskii). *For every trace class operator T we have*

$$\operatorname{tr}(T) = \sum_{n \geq 1} \lambda_n,$$

where $(\lambda_n)_{n \geq 1}$ is the sequence of nonzero eigenvalues of T repeated according to their algebraic multiplicities.

Here we use the convention $\operatorname{tr}(T) = 0$ in case there are no nonzero eigenvalues. The absolute summability of the eigenvalue sequence has already been established in Proposition 14.21.

We present the beautiful proof of this theorem due to Simon, which is based on the theory of Fredholm determinants. In order to introduce these, we need some notation from multi-linear algebra. We refer to Appendix B for the definitions. The n-fold exterior product of a vector space V is denoted by $\Lambda^n V$. If T is a linear operator on V, then

$$\Lambda^n(T)(v_1 \wedge \cdots \wedge v_n) := T v_1 \wedge \cdots \wedge T v_n$$

defines a linear operator $\Lambda^n(T)$ on $\Lambda^n(V)$. It is the restriction to $\Lambda^n(V)$ of the n-fold tensor product $T^{\otimes n}$ acting on $V^{\otimes n}$. If S is another linear operator on V, then $\Lambda^n(ST) = \Lambda^n(S)\Lambda^n(T)$.

If H is a Hilbert space and T is bounded on H, then $\Lambda^n(T)$ is bounded on $\Lambda^n(H)$, which is a Hilbert space in a natural way, and its adjoint equals $(\Lambda^n(T))^\star = \Lambda^n(T^\star)$. From this we infer that $|(\Lambda^n(T))| = \Lambda^n(|T|)$. Thus if $(\mu_j)_{j \geqslant 1}$ is the singular value sequence of T, the singular values of $\Lambda^n(T)$ are $\mu_{j_1} \cdots \mu_{j_n}$ with $j_1 < \cdots < j_n$. It follows that $\Lambda^n(T)$ is a trace class operator and

$$\|\Lambda^n(T)\|_{\mathscr{L}_1(\Lambda^n H)} = \sum_{j_1 < \cdots < j_n} \mu_{j_n} \cdots \mu_{j_n} = \frac{1}{n!} \sum_{j_1,\dots,j_n \geqslant 1} \mu_{j_1} \cdots \mu_{j_n} = \frac{1}{n!}\|T\|_{\mathscr{L}_1(H)}^n.$$

(14.5)

For $n \times n$ matrices A we have the following identity relating the determinant to traces and exterior products, known as *MacMahon's formula*:

$$\det(1+A) = \sum_{k=0}^n \mathrm{tr}(\Lambda^k A),$$

with the convention that $\Lambda^0(T) = I$. A proof is sketched in Problem 14.13. Observing that $\Lambda^k(V) = \{0\}$ when $k > \dim(V)$, MacMahon's formula suggests the following definition.

Definition 14.35 (Fredholm determinant). Let $T \in \mathscr{L}(H)$ be a trace class operator. The *Fredholm determinant* of $I + T$ is defined as

$$\det(I + T) := \sum_{n \in \mathbb{N}} \mathrm{tr}(\Lambda^n(T)).$$

The sum on the right-hand side is absolutely convergent since

$$\sum_{n \in \mathbb{N}} |\mathrm{tr}(\Lambda^n(T))| \leqslant \sum_{n \geqslant 1} \|(\Lambda^n(T))\|_{\mathscr{L}_1(\Lambda^n H)} \leqslant \sum_{n \in \mathbb{N}} \frac{1}{n!}\|T\|_{\mathscr{L}_1(H)}^n = \exp\big(\|T\|_{\mathscr{L}_1(H)}\big).$$

(14.6)

The crucial step in the proof of Lidskii's theorem is to establish validity of the following identity for all trace class operators T and all $\mu \in \mathbb{C}$:

$$\det(I + \mu T) = \prod_{n \geqslant 1}(1 + \mu\lambda_n).$$

Here $(\lambda_n)_{n \geqslant 1}$ is the sequence of eigenvalues of T repeated according to algebraic multiplicities. Notice that Proposition 14.21 guarantees the convergence of the infinite product. Once this formula has been obtained, Lidskii's theorem is immediate by comparing the linear term of this product with the linear term in the definition $\det(I + \mu T) = \sum_{n \in \mathbb{N}} \mu^n \mathrm{tr}(\Lambda^n(T))$.

The remainder of this section is devoted to proving Lidskii's theorem. We fix a separable Hilbert space H and start with some preliminary results.

Lemma 14.36. *Let* $T \in \mathscr{L}(H)$ *be a trace class operator and let* $(\mu_n)_{n \geqslant 1}$ *be its singular value sequence, repeated according to multiplicities. Then*

$$|\det(I+T)| \leqslant \prod_{n \geqslant 1}(1+\mu_n).$$

Proof It follows from (14.5) that

$$|\det(I+T)| \leqslant \sum_{n \in \mathbb{N}} |\mathrm{tr}(\Lambda^n(T))| \leqslant \sum_{n \in \mathbb{N}} \|\Lambda^n(T)\|_{\mathscr{L}_1(\Lambda^n H)}$$

$$\leqslant \sum_{n \in \mathbb{N}} \frac{1}{n!} \sum_{j_1,\ldots,j_n \geqslant 1} \mu_{j_1} \cdots \mu_{j_n} \leqslant \prod_{n \geqslant 1}(1+\mu_n).$$

$\square$

Lemma 14.37. *Let* $T \in \mathscr{L}(H)$ *be a trace class operator. For all* $\varepsilon > 0$ *there exists a constant* $C_\varepsilon \geqslant 0$ *such that for all* $\lambda \in \mathbb{C}$ *we have*

$$|\det(I+\lambda T)| \leqslant C_\varepsilon \exp(\varepsilon|\lambda|).$$

Proof Using the inequality $|1+t| \leqslant \exp(|t|)$, Lemma 14.36 implies, for any $N \geqslant 1$,

$$|\det(I+\lambda T)| \leqslant \prod_{n \geqslant 1}(1+|\lambda||\mu_n|) \leqslant \prod_{n=1}^{N}(1+|\lambda||\mu_n|)\exp\Big(\sum_{n \geqslant N+1}|\lambda||\mu_n|\Big).$$

Fix $\varepsilon > 0$. If we choose $N \geqslant 1$ so large that $\sum_{n \geqslant N+1}\mu_n < \frac{1}{2}\varepsilon$ the desired estimate is obtained, with

$$C_\varepsilon := \sup_{\lambda \in \mathbb{C}} \prod_{n=1}^{N}(1+|\lambda||\mu_n|)\exp\Big(-\frac{1}{2}\varepsilon|\lambda|\Big).$$

$\square$

Lemma 14.38. *The map* $T \mapsto \det(I+T)$ *is continuous from* $\mathscr{L}_1(H)$ *to* $\mathbb{C}$.

Proof Suppose $T_j \to T$ in $\mathscr{L}_1(H)$ as $j \to \infty$. Fix $\varepsilon > 0$ and choose $N \geqslant 0$ so large that $\sum_{n \geqslant N+1} C^n/n! < \frac{1}{3}\varepsilon$, where $C := \sup_j \|T_j\|_{\mathscr{L}_1(H)}$. Then, by (14.6),

$$|\det(I+T_j) - \det(I+T)| \leqslant \frac{2}{3}\varepsilon + \sum_{n=1}^{N} \mathrm{tr}(|\Lambda^n(T_j)-\Lambda^n(T)|).$$

Denoting by P_n the orthogonal projection in $H^{\otimes n}$ onto $\Lambda^n(H)$, we have

$$\mathrm{tr}(|\Lambda^n(T_j)-\Lambda^n(T)|) = \mathrm{tr}(|P_n(T_j^{\otimes n}-T^{\otimes n})P_n|)$$

$$\leqslant \mathrm{tr}(|T_j^{\otimes n}-T^{\otimes n}|) \leqslant nC^{n-1}\|T_j-T\|_{\mathscr{L}_1(H)}.$$

If we choose $N \geqslant 1$ so large that also $\|T_j-T\|_{\mathscr{L}_1(H)} < \frac{1}{3}\varepsilon(\sum_{n=1}^{N} nC^{n-1})^{-1}$ for $j \geqslant N$, then $|\det(I+T_j)-\det(I+T)| < \varepsilon$ for $j \geqslant N$. $\square$

Lemma 14.39. *Let* T *be a bounded operator on* H *such that* $T = PTP$ *for some orthogonal projection* P *on* H *of finite rank* m. *Viewing* $P(I+T)P$ *as an operator on the* m-*dimensional Hilbert space* $\mathsf{R}(P)$, *we have*

$$\det(I+T) = \det(P(I+T)P).$$

Proof The identity $T = PTP$ implies that T is of rank at most m, and therefore we have $\Lambda^n(T) = 0$ for $n > m$. For $0 \leqslant n \leqslant m$ we have $\operatorname{tr}(\Lambda^n(T)) = \operatorname{tr}(\Lambda^n(PTP))$. Applying Definition 14.35 twice,

$$\det(I+T) = \sum_{n=0}^{m} \operatorname{tr}(\Lambda^n(T)) = \sum_{n=0}^{m} \operatorname{tr}(\Lambda^n(PTP))$$

$$= \sum_{n=0}^{\nu} \operatorname{tr}(\Lambda^n(PTP)|_{\mathsf{R}(P)}) = \det(I_{\mathsf{R}(P)} + PTP) = \det(P(I+T)P).$$

$\square$

Lemma 14.40. *If* $S, T \in \mathscr{L}(H)$ *are trace class operators, then*

$$\det(I+T)\det(I+S) = \det((I+T)(I+S)).$$

Proof First assume that T and S are both of finite rank. Let P be a finite rank projection in H whose range contains the ranges of T, $T^\star$, S, and $S^\star$. With m being the rank of P, from Lemma 14.39 along with the identity

$$\det(P(I+T)P) = \sum_{k=0}^{m} \operatorname{tr}(\Lambda^k(PTP)) = \operatorname{tr}(\Lambda^m(P(I+T)P))$$

and similarly for S, we obtain

$$\det(I+T)\det(I+S) = \operatorname{tr}(\Lambda^m(P(I+T)P))\operatorname{tr}(\Lambda^m(P(I+S)P))$$
$$= \operatorname{tr}(\Lambda^m(P(I+T)P)\Lambda^m(P(I+S)P))$$
$$= \operatorname{tr}(\Lambda^m(P(I+T)(I+S)P)) = \det((I+T)(I+S)).$$

Here we used that $\Lambda^m(\mathsf{R}(P))$ is one-dimensional, so that the trace is multiplicative on this space. This proves the lemma for finite rank operators T and S. By Lemma 14.38, the general case now follows by approximation. $\square$

Proposition 14.41. *If* $T \in \mathscr{L}(H)$ *is a trace class operator, then* $I+T$ *is invertible if and only if* $\det(I+T) \neq 0$.

Proof Suppose first that $I+T$ is invertible and let $S := -T(I+T)^{-1}$. Then S is a trace class operator and an easy computation gives $(I+T)(I+S) = I$. It follows from Lemma 14.40 that $\det(I+T)\det(I+S) = \det(I) = 1$, so $\det(I+T) \neq 0$.

If $I+T$ is not invertible, then -1 is an eigenvalue of T. Denoting the corresponding spectral projection by P, then from Lemma 14.40 and the commutation relation $TP = PT$ we obtain

$$\det(I + TP)\det(I + T(I - P)) = \det(I + TP + T(I - P) + TPT(I - P)) = \det(I + T).$$

Denote by v the algebraic multiplicity of -1. By Lemma 14.39 applied to TP, $\det(I + TP)$ is the determinant of a finite-dimensional noninvertible operator and therefore it equals 0. This proves that $\det(I + T) = \det(I + TP) = 0$. □

Proposition 14.42. *If $T \in \mathscr{L}(H)$ is a trace class operator with nonzero eigenvalue $-1/\mu_0$ of algebraic multiplicity v, then $F(\mu) = \det(I + \mu T)$ has a zero at μ_0 of multiplicity v.*

Proof Denoting by P the spectral projection associated with $-1/\mu_0$, we have

$$\det(I + \mu T) = \det(I + \mu TP)\det(I + \mu T(I - P))$$

and $\det(I + \mu T(I - P)) \neq 0$ by Proposition 14.41. The operator TP vanishes on the range of $I - P$ and its restriction to the range of P has spectrum $\{-1/\mu_0\}$. Thus, for $0 \leqslant n \leqslant v$,

$$\mathrm{tr}(\Lambda^n(\mu TP)) = \sum_{1 \leqslant j_1 < \cdots < j_n \leqslant v} \left(-\frac{\mu}{\mu_0}\right)^n = \binom{v}{n}\left(-\frac{\mu}{\mu_0}\right)^n$$

and consequently

$$\det(I + \mu TP) = \sum_{n=0}^{v} \binom{v}{n}\left(-\frac{\mu}{\mu_0}\right)^n = \left(1 - \frac{\mu}{\mu_0}\right)^v.$$

□

The next lemma from complex function theory is stated without proof.

Lemma 14.43. *Let F be an entire function whose zeroes $z_1, z_2, \ldots$ (counting multiplicities) satisfy $\sum_{n \geqslant 1} 1/|z_n| < \infty$. Assume furthermore that $F(0) = 1$ and that for all $\varepsilon > 0$ there exists a constant $C_\varepsilon \geqslant 0$ such that $|F(z)| \leqslant C_\varepsilon \exp(\varepsilon|z|)$. Then*

$$F(z) = \prod_{n \geqslant 1}\left(1 - \frac{z}{z_n}\right), \quad z \in \mathbb{C}.$$

Theorem 14.44. *If $T \in \mathscr{L}(H)$ is a trace class operator, with eigenvalue sequence $(\lambda_n)_{n \geqslant 1}$ repeated according to algebraic multiplicities, then for all $\mu \in \mathbb{C}$ we have*

$$\det(I + \mu T) = \prod_{n \geqslant 1}(1 + \mu \lambda_n).$$

Proof By Propositions 14.41 and 14.42, the zeroes of $F(\mu) := \det(I + \mu T)$, counting multiplicities, are precisely the points $-1/\lambda_n$. Proposition 14.21 and Lemma 14.37 show that the assumptions of Lemma 14.43 hold for this function. The result now follows from the lemma. □

Proof of Theorem 14.34 The linear term in the Taylor expansion of $\det(I + \mu T) = \sum_{n \in \mathbb{N}} \mathrm{tr}(\Lambda^n(T))$ equals $\mathrm{tr}(\Lambda^1(T)) = \mathrm{tr}(T)$. On the other hand, by Theorem 14.44, this term equals $\sum_{n \geqslant 1} \lambda_n$. $\qquad\square$

14.5.b Trace Formula for Integral Operators

The trace of an integral operator with continuous kernel can be computed as follows.

Theorem 14.45 (Mercer). *Let μ be a finite Borel measure on a compact metric space K. Let T be an integral operator on $L^2(K, \mu)$ of the form*

$$Tf(s) = \int_K k(s,t) f(t) \, d\mu(t)$$

with continuous kernel $k \in C(K \times K)$. Then:

(1) *if T is a trace class operator, then its trace is given by*

$$\mathrm{tr}(T) = \int_K k(t,t) \, d\mu(t);$$

(2) *if T is positive, that is, if $(Tf|f) \geqslant 0$ for all $f \in L^2(K, \mu)$, then T is a trace class operator.*

By an argument similar to that employed in the proof below, one sees that T is positive if and only if the kernel k is *positive definite* in the sense that for all integers $N \geqslant 1$ and all $t_1, \ldots, t_N \in S$ and $z_1, \ldots, z_N \in \mathbb{C}$ we have

$$\sum_{n,m=1}^{N} k(t_m, t_n) z_m \bar{z}_n \geqslant 0.$$

Proof It has been observed in Remark 2.31 that $L^2(K, \mu)$ is separable.

(1): Suppose that the integral operator T is a trace class operator. By Proposition 14.28 we have $T = S_2 S_1$ with S_1, S_2 Hilbert–Schmidt on $L^2(K, \mu)$. Accordingly, by Theorem 14.8 there exist $k_1, k_2 \in L^2(K \times K, \mu \times \mu)$ such that for μ-almost all $s \in K$ we have

$$Tf(s) = \int_K \int_K k_2(s,t) k_1(t,u) f(u) \, d\mu(u) \, d\mu(t).$$

As a result, for $\mu \times \mu$-almost all $(s,t) \in K \times K$ we have

$$k(s,t) = \int_K k_2(s,t) k_1(t,u) \, d\mu(u).$$

Then,

$$\mathrm{tr}(T) = \mathrm{tr}(S_2 S_1) = (S_1 | S_2^{\star})_{\mathscr{L}_2(L^2(K,\mu))}$$

$$\overset{(*)}{=} (k_1|\overline{k_2})_{L^2(K \times K, \mu \times \mu)}$$

$$= \int_K \int_K k_1(s,t)k_2(t,s) \, d\mu(s) \, d\mu(t) = \int_K k(s,s) \, d\mu(s),$$

where $(*)$ follows from the fact, which follows from Example 14.2 and Theorem 14.8, that the correspondence between Hilbert–Schmidt operators and their square integrable kernels is unitary.

(2): By the result of Example 7.7, T is compact, and the positivity of T implies that its singular value sequence equals its sequence of nonzero eigenvalues $(\lambda_n)_{n \geq 1}$, taking into account multiplicities. The rest of the proof is accomplished in two steps.

Step 1 – Let $(h_n)_{n \geq 1}$ be an orthonormal sequence of eigenvectors in $L^2(K, \mu)$ corresponding to the sequence $(\lambda_n)_{n \geq 1}$. The uniform continuity of k implies that T maps $L^2(K)$ into $C(K)$ and therefore $Th_n = \lambda_n h_n$ implies $h_n \in C(K)$ for all $n \geq 1$. As a consequence, for each $n \geq 1$ the kernel

$$k_n(s,t) := k(s,t) - \sum_{j=1}^n \lambda_j h_j(s)\overline{h_j(t)}, \quad s,t \in K,$$

is continuous.

Let $f \in L^2(K, \mu)$. Since T vanishes on the orthogonal complement of the closed linear span of $(h_n)_{n \geq 1}$, we have

$$(Tf|f) = \sum_{n \geq 1} \sum_{m \geq 1} \left(\lambda_n(f|h_n)h_n \Big| (f|h_m)h_m \right) = \sum_{n \geq 1} \lambda_n |(f|h_n)|^2$$

and therefore

$$\int_K \int_K k_n(s,t)f(t)\overline{f(s)} \, d\mu(t) \, d\mu(s)$$

$$= (Tf|f) - \sum_{j=1}^n \lambda_j \int_K \int_K f(t)\overline{h_j(t)} \, \overline{f(s)}h_j(s) \, d\mu(t) \, d\mu(s)$$

$$= \sum_{n \geq 1} \lambda_n |(f|h_n)|^2 - \sum_{j=1}^n \lambda_j |(f|h_j)|^2 \geq 0.$$

In particular, for any Borel sets B of positive μ-measure,

$$\frac{1}{(\mu(B))^2} \int_K \int_K k_n(s,t)\mathbf{1}_B(t)\mathbf{1}_B(s) \, d\mu(t) \, d\mu(s) \geq 0. \tag{14.7}$$

By a limiting argument (applying (14.7) to a sequence of balls $B(t;r_n)$ centred at a given point $t \in \mathrm{supp}(\mu)$ with radii $r_n \downarrow 0$), from this inequality and the continuity of k_n we obtain $k_n(t,t) \geq 0$ for μ-almost all $t \in K$ for all $n \geq 1$ and $t \in K$. Then,

$$0 \leq \int_K k_n(t,t) \, d\mu(t)$$

$$= \int_K k(t,t)\,\mathrm{d}\mu(t) - \sum_{j=1}^{n} \lambda_j \int_K |h_j(t)|^2 \,\mathrm{d}\mu(t) = \int_K k(t,t)\,\mathrm{d}\mu(t) - \sum_{j=1}^{n} \lambda_j.$$

Letting $n \to \infty$ we obtain that $T \in \mathscr{L}_1(H)$ and

$$\|T\|_{\mathscr{L}_1(H)} = \mathrm{tr}(T) = \sum_{j \geqslant 1} \lambda_j \leqslant \int_K k(t,t)\,\mathrm{d}\mu(t).$$

$\square$

In the positive case, the trace formula can be alternatively proved by the following more elementary argument. For $m = 1, 2, \ldots$ let $(K_n^{(m)})_{n=1}^{N_m}$ be a partition of K of mesh less than $1/m$. For $1 \leqslant n \leqslant N_m$ let $h_n^{(m)} := \mathbf{1}_{K_n^{(m)}} / \sqrt{\mu(K_n^{(m)})}$ (here, and in what follows, we discard those indices for which $\mu(K_n^{(m)}) = 0$ without expressing this in our notation in order not to overburden it). This sequence is orthonormal in $L^2(K,\mu)$. Using the uniform continuity of k we obtain

$$\lim_{m \to \infty} \sum_{n=1}^{N_m} (Th_n^{(m)} | h_n^{(m)}) = \lim_{m \to \infty} \sum_{n=1}^{N_m} \frac{1}{\mu(K_n^{(m)})} \int_{K_n^{(m)}} \int_{K_n^{(m)}} k(x,y)\,\mathrm{d}\mu(x)\,\mathrm{d}\mu(y)$$

$$= \lim_{m \to \infty} \sum_{n=1}^{N_m} \frac{1}{\mu(K_n^{(m)})} \int_{K_n^{(m)}} \int_{K_n^{(m)}} k(y,y)\,\mathrm{d}\mu(x)\,\mathrm{d}\mu(y)$$

$$= \lim_{m \to \infty} \sum_{n=1}^{N_m} \int_{K_n^{(m)}} k(y,y)\,\mathrm{d}\mu(y) = \int_K k(y,y)\,\mathrm{d}\mu(y).$$

Hence, by Theorem 14.20,

$$\int_K k(y,y)\,\mathrm{d}\mu(y) = \lim_{m \to \infty} \sum_{n=1}^{N_m} (Th_n^{(m)} | h_n^{(m)}) \leqslant \mathrm{tr}(T).$$

14.5.c Trace Formula for Fredholm Operators

The following theorem gives a formula for the index of a Fredholm operator in terms of traces.

Theorem 14.46 (Fedosov). *Let $T \in \mathscr{L}(H)$ be a Fredholm operator and let $S \in \mathscr{L}(H)$ be an operator such that both $I - ST$ and $I - TS$ are finite rank operators. Then the commutator $[T,S] = TS - ST$ is a trace class operator and*

$$\mathrm{tr}([T,S]) = \mathrm{ind}(T).$$

By Atkinson's theorem (Theorem 7.23), operators S with the stated properties always exist.

Proof The operator $[T,S] = (I - ST) - (I - TS)$ is of finite rank and hence a trace class operator. If $S' \in \mathscr{L}(H)$ is another operator such that $I - S'T$ and $I - TS'$ are of finite rank, then $R := S' - S$ is of finite rank. Indeed, $RT = (I - ST) - (I - S'T)$ is of finite rank and the range of T, being a Fredholm operator, has finite codimension; these fact are compatible only if R itself is of finite rank. As a consequence,

$$\mathrm{tr}(TS' - S'T) = \mathrm{tr}(TS - ST + TR - RT) = \mathrm{tr}(TS - ST) + \mathrm{tr}(TR - RT) = \mathrm{tr}(TS - ST),$$

using that R, being of finite rank, is a trace class operator and therefore $\mathrm{tr}(TR) = \mathrm{tr}(RT)$ by Proposition 14.27.

To prove the theorem it therefore suffices to prove it for the bounded operator $S \in \mathscr{L}(H)$ constructed in the proof of Theorem 7.23. This operator enjoys the following properties: (i) $I - ST$ and $I - TS$ are finite rank projections, and (ii) $\dim \mathsf{N}(T) = \dim \mathsf{R}(I - ST)$ and $\mathrm{codim}\, \mathsf{R}(T) = \dim \mathsf{R}(I - TS)$. Since the rank of a finite rank projection is equal to its trace (by Example 14.18), we have

$$\mathrm{ind}(T) = \dim \mathsf{N}(T) - \mathrm{codim}\, \mathsf{R}(T) = \mathrm{tr}(I - ST) - \mathrm{tr}(I - TS) = \mathrm{tr}(TS - ST).$$

$\square$

14.5.d Trace Formula for Commutators of Toeplitz Operators

From Section 7.3.d we recall that $H^2(\mathbb{D})$ is the vector space of all holomorphic functions on $\mathbb{D}$ of the form $\sum_{n \in \mathbb{N}} c_n z^n$ with $\sum_{n \in \mathbb{N}} |c_n|^2 < \infty$. Identifying it with the closed subspace of $L^2(\mathbb{T})$ consisting of all functions whose negative Fourier coefficients vanish, $H^2(\mathbb{D})$ is the range of the Riesz projection

$$P : \sum_{n \in \mathbb{Z}} \widehat{f}(n) e_n \mapsto \sum_{n \in \mathbb{N}} \widehat{f}(n) e_n$$

in $L^2(\mathbb{T})$, where $e_n(\theta) = e^{in\theta}$. This projection discards the terms in the Fourier series $(\widehat{f}(n))_{n \in \mathbb{Z}}$ of f corresponding to the negative indices $n = -1, -2, \ldots$

Given a function $\phi \in L^\infty(\mathbb{T})$, the *Toeplitz operator* with symbol ϕ has been defined as the bounded operator T_ϕ on $H^2(\mathbb{D})$ given by

$$T_\phi f := P(\phi f), \quad f \in H^2(\mathbb{D}).$$

It follows from Lemma 7.30 that for all $\phi, \psi \in C(\mathbb{T})$ the commutator

$$[T_\phi, T_\psi] = T_\phi T_\psi - T_\psi T_\phi$$

is compact. For functions $\phi, \psi \in C^2(\mathbb{T})$ we have the following stronger result.

Theorem 14.47 (Helton–Howe). *For all $\phi, \psi \in C^2(\mathbb{T})$ the commutator $[T_\phi, T_\psi]$ is a trace class operator and*

$$\operatorname{tr}([T_\phi, T_\psi]) = \frac{1}{2\pi i} \int_{-\pi}^{\pi} \phi(\theta) \psi'(\theta) \, d\theta. \tag{14.8}$$

Proof The proof is a matter of computation. First, for $n, m \in \mathbb{Z}$, $[T_{e_n}, T_{e_m}]$ is a finite rank operator of rank at most $\min\{|m|, |n|\}$, and therefore by Example 14.18 with

$$\|[T_{e_n}, T_{e_m}]\|_{\mathscr{L}_1(H^2(\mathbb{D}))} \leqslant \min\{|m|, |n|\}. \tag{14.9}$$

Second, for all $n, m \in \mathbb{Z}$ and $j \in \mathbb{N}$ we have $T_{e_n} T_{e_m} e_j = \lambda_j^{nm} e_{n+m+j}$ with $\lambda_j^{nm} \in \{0, 1\}$, so that

$$\operatorname{tr}([T_{e_n}, T_{e_m}]) = \sum_{j \geqslant 0} (\lambda_j^{nm} - \lambda_j^{mn})(e_{n+m+j}|e_j). \tag{14.10}$$

Case 1: $n + m \neq 0$. In that case (14.10) gives

$$\operatorname{tr}([T_{e_n}, T_{e_m}]) = 0 = \frac{1}{2\pi i} \int_{-\pi}^{\pi} e_n(\theta) e_m'(\theta) \, d\theta.$$

Case 2: $n + m = 0$ with $n \geqslant 0$. In that case $\lambda_j^{n,-n} = 0$ if $j < n$ and $\lambda_j^{n,-n} = 1$ if $j \geqslant n$, while always $\lambda_j^{-n,n} = 1$, and (14.10) gives

$$\operatorname{tr}([T_{e_n}, T_{e_m}]) = -n = -\frac{n}{2\pi} \int_{-\pi}^{\pi} e_n(\theta) e_{-n}(\theta) \, d\theta = \frac{1}{2\pi i} \int_{-\pi}^{\pi} e_n(\theta) e_m'(\theta) \, d\theta.$$

The case $n + m = 0$ with $n < 0$ is entirely similar.

This completes the proof of (14.8) for $\phi = e_n$ and $\psi = e_m$. Since both the left- and right-hand side of (14.8) are linear in both ϕ and ψ, for $\phi = \sum_{n \in \mathbb{N}} a_n e_n$ and $\psi = \sum_{n \in \mathbb{N}} b_n e_n$ we have

$$[T_\phi, T_\psi] = \sum_{m,n \in \mathbb{N}} a_n b_m [T_{e_n}, T_{e_m}] \tag{14.11}$$

and hence, taking traces,

$$\operatorname{tr}([T_\phi, T_\psi]) = \sum_{m,n \in \mathbb{N}} a_n b_m \operatorname{tr}([T_{e_n}, T_{e_m}])$$

$$= \sum_{m,n \in \mathbb{N}} a_n b_m \frac{1}{2\pi} \int_{-\pi}^{\pi} e_n(\theta) e_m'(\theta) \, d\theta = \frac{1}{2\pi i} \int_{-\pi}^{\pi} \phi(\theta) \psi'(\theta) \, d\theta,$$

provided the sum in (14.11) converges in $\mathscr{L}_1(H^2(\mathbb{D}))$. Keeping in mind (14.9), this can be guaranteed if we assume that ϕ and ψ are C^2, for then $|a_n|$ and $|b_n|$ are of order $O(\frac{1}{n^2})$ as $n \to \infty$ and

$$\sum_{m,n \in \mathbb{Z}} \frac{\min\{|m|, |n|\}}{(1+m^2)(1+n^2)} = \sum_{m \in \mathbb{Z}} \sum_{\substack{n \in \mathbb{Z} \\ |n| \leqslant |m|}} \frac{|n|}{(1+m^2)(1+n^2)} + \sum_{n \in \mathbb{Z}} \sum_{\substack{m \in \mathbb{Z} \\ |m| < |n|}} \frac{|m|}{(1+m^2)(1+n^2)}$$

$$\lesssim \sum_{k \in \mathbb{Z}} \frac{\log(1+|k|)}{1+k^2} < \infty.$$

$\square$

14.5.e Trace Formula for the Dirichlet Heat Semigroup

Let D be a bounded open subset of $\mathbb{R}^d$ satisfying $|\partial D| = 0$ and let S_{Dir} be the C_0-semigroup on $L^2(D)$ generated by the Dirichlet Laplacian Δ_{Dir} associated with D.

Theorem 14.48 (Trace formula for the Dirichlet heat semigroup). *For all $t > 0$ the operator $S_{\mathrm{Dir}}(t)$ is a trace class operator on $L^2(D)$ with*

$$\lim_{t \downarrow 0} t^{d/2} \mathrm{tr}(S_{\mathrm{Dir}}(t)) = \frac{|D|}{(4\pi)^{d/2}}.$$

For the proof of this formula we need the following lemma.

Lemma 14.49. *Let μ be a Borel measure on $[0, \infty)$ whose Laplace transform satisfies*

$$\mathscr{L}\mu(t) := \int_0^\infty e^{-tx} \, d\mu(x) < \infty$$

for all $t > 0$. If for some $r \geqslant 0$ and $a \in \mathbb{R}$ we have

$$\lim_{x \to \infty} x^{-r} \mu([0,x]) = a,$$

then

$$\lim_{t \downarrow 0} t^r \mathscr{L}\mu(t) = a\Gamma(1+r),$$

where $\Gamma(s) = \int_0^\infty x^{s-1} e^{-x} \, dx$, $s > 0$, is the Euler Gamma function.

Proof Integrating by parts and setting $v(x) := \mu([0,x])$, we have

$$t^r \mathscr{L}\mu(t) = t^r \int_0^\infty e^{-tx} \, d\mu(x) = t^{r+1} \int_0^\infty e^{-tx} v(x) \, dx$$

$$= t^r \int_0^\infty e^{-y} v\left(\frac{y}{t}\right) dy = \int_0^\infty e^{-y} (t+y)^r \left(1 + \frac{y}{t}\right)^{-r} v\left(\frac{y}{t}\right) dy.$$

By assumption we have $\lim_{x \to \infty} x^{-r} v(x) = a$, and therefore, for all $y > 0$,

$$\lim_{t \downarrow 0} \left(1 + \frac{y}{t}\right)^{-r} v\left(\frac{y}{t}\right) = a.$$

In particular we have $C := \sup_{x>0} (1+x)^{-r} v(x) < \infty$ and therefore

$$e^{-y}(t+y)^r \left(1 + \frac{y}{t}\right)^{-r} v\left(\frac{y}{t}\right) \leqslant C e^{-y}(t+y)^r,$$

and for $0 \leqslant t \leqslant 1$ we can bound the right-hand side by $Ce^{-y}(1+y)^r$. It follows that the dominated convergence theorem can be applied to obtain

$$\lim_{t \downarrow 0} t^r \mathscr{L}\mu(t) = a \int_0^\infty e^{-y} y^r \, dy = a\Gamma(1+r).$$

$\square$

Proof of Theorem 14.48   As was observed in the course of the proof of Theorem 12.26, the resolvent operators $R(\lambda, \Delta_{\text{Dir}})$ are compact, and this implies the compactness of the inclusion mapping of $D(\Delta_{\text{Dir}})$ into $L^2(D)$. By analyticity, for each $t > 0$ the operator $S_{\text{Dir}}(t)$ maps $L^2(D)$ into $D(\Delta_{\text{Dir}})$ boundedly, and therefore $S_{\text{Dir}}(t)$ is compact as a bounded operator on $L^2(D)$. We can now apply the spectral mapping formula Proposition 13.20. Evaluating the trace against an orthonormal basis consisting of eigenvectors we conclude that $S_{\text{Dir}}(t)$ is a trace class operator and

$$\text{tr}(S_{\text{Dir}}(t)) = \sum_{n \geqslant 1} e^{-\lambda_n t} < \infty,$$

where $0 < \lambda_1 < \lambda_2 < \cdots \to \infty$ is the enumeration, counting multiplicities, of the eigenvalues of Δ_{Dir}; the finiteness of this sum is a consequence of Weyl's theorem (Theorem 12.29). We now apply Lemma 14.49 to the Borel measure $\mu = \sum_{n \geqslant 1} \delta_{\{\lambda_n\}}$. Setting $N(x) := \max\{n \geqslant 1 : \lambda_n \leqslant x\}$, by Weyl's theorem we have

$$\lim_{x \to \infty} x^{-d/2} \mu([0,x]) = \lim_{x \to \infty} x^{-d/2} N(x) = \frac{\omega_d}{(2\pi)^d} |D|,$$

where $\omega_d = \pi^{d/2}/\Gamma(1 + \frac{1}{2}d)$ is the volume of the unit ball in $\mathbb{R}^d$. Lemma 14.49 allows us to conclude that

$$\lim_{t \downarrow 0} t^{d/2} \text{tr}(S_{\text{Dir}}(t)) = \lim_{t \downarrow 0} t^{d/2} \sum_{n \geqslant 1} e^{-\lambda_n t} = \lim_{t \downarrow 0} t^{d/2} \widehat{\mu}(t) = \frac{\omega_d}{(2\pi)^d} |D| \Gamma\left(1 + \frac{d}{2}\right) = \frac{|D|}{(4\pi)^{d/2}}.$$

$\square$

14.5.f Euler's Identity Revisited

Consider the Dirichlet Laplacian Δ_{Dir} on $L^2(0,1)$. As shown in Example 12.23, the spectrum of this operator equals

$$\sigma(\Delta_{\text{Dir}}) = \{-\pi^2 n^2 : n = 1, 2, \dots\}$$

and consists of the eigenvalues corresponding to the eigenfunctions $f_n(t) = \sin(n\pi t)$; as a consequence of Lemma 12.25, the spectrum of the inverse operator Δ_{Dir}^{-1} is given by

$$\sigma(\Delta_{\text{Dir}}^{-1}) = \left\{-\frac{1}{\pi^2 n^2} : n = 1, 2, 3, \dots\right\}$$

and it again consists of the eigenvalues. Since $-\Delta_{\text{Dir}}^{-1}$ is positive, it follows from Mercer's theorem that

$$\sum_{n=1}^{\infty} \frac{1}{\pi^2 n^2} = \operatorname{tr}(-\Delta_{\text{Dir}}^{-1}) = \int_0^1 k(t,t)\,dt,$$

where k is Green's function for the Poisson problem on the unit interval with Dirichlet boundary conditions. From Section 11.2.a we recall that it is given by

$$k(s,t) = \begin{cases} (1-t)s, & s \leqslant t, \\ (1-s)t, & t \leqslant s. \end{cases}$$

In view of

$$\int_0^1 k(t,t)\,dt = \int_0^1 (1-t)t\,dt = \frac{1}{6}$$

we recover Euler's identity

$$\sum_{n=1}^{\infty} \frac{1}{n^2} = \frac{\pi^2}{6}.$$

Problems

14.1 Show that if $T \in \mathscr{L}(H)$ is a trace class operator and $U \in \mathscr{L}(H)$ is unitary, then $UTU^\star$ is a trace class operator and $\operatorname{tr}(T) = \operatorname{tr}(UTU^\star)$.

14.2 Show that if $A \in \mathscr{L}(H)$ is a positive trace class operator, then $A \leqslant \operatorname{tr}(A)I$, that is, $\operatorname{tr}(A)I - A$ is positive.

14.3 Prove the following properties of the partial trace.

  (a) If $T \in \mathscr{L}_1(H \otimes K)$, then

$$\operatorname{tr}(\operatorname{tr}_K(T)) = \operatorname{tr}(T).$$

  (b) If $T \in \mathscr{L}_1(H \otimes K)$ and $A_1, A_2 \in \mathscr{L}(H)$, then $(A_1 \otimes I)T(A_2 \otimes I) \in \mathscr{L}_1(H \otimes K)$ and

$$\operatorname{tr}_K((A_1 \otimes I)T(A_2 \otimes I)) = A_1 \operatorname{tr}_K(T)A_2.$$

  (c) If $A \in \mathscr{L}_1(H)$ and $B \in \mathscr{L}_1(K)$, then $T := A \otimes B \in \mathscr{L}_1(H \otimes K)$ and

$$\operatorname{tr}_K(T) = \operatorname{tr}(B)A.$$

  (d) The adjoint of the mapping $T \mapsto \operatorname{tr}_K T$ from $\mathscr{L}_1(H \otimes K) \to \mathscr{L}_1(H)$ is the mapping $S \mapsto S \otimes I$ from $\mathscr{L}(H)$ to $\mathscr{L}(H \otimes K)$.

14.4 Show that if $S = x \bar{\otimes} x$ and $T = y \bar{\otimes} y$ with $\|x\| = \|y\| = 1$ are two rank one orthogonal projections, then

$$\|S - T\|^2 = 1 - |(x|y)|^2 = 1 - \mathrm{tr}(ST).$$

14.5 Consider a bounded operator $T \in \mathscr{L}(H)$. Show that the following assertions are equivalent:

(1) T is a trace class operator, respectively Hilbert–Schmidt;
(2) $\exp(T) - I$ is a trace class operator, respectively Hilbert–Schmidt.

Hint: Compare with Problem 7.10.

14.6 Prove the two assertions made after Definition 14.30.

14.7 Let $T : L^2(0,1) \to L^\infty(0,1)$ be a bounded operator, and let $(h_n)_{n \geq 1}$ be an orthonormal basis for $L^2(0,1)$.

(a) Show that for every $k \geq 1$ there exists a null set $N_k \subseteq (0,1)$ such that for all $c \in \mathbb{K}^k$ we have

$$\left| \sum_{j=1}^{k} c_j T h_j(t) \right| \leq \|T\|, \quad t \in (0,1) \setminus N_k.$$

(b) Deduce from part (a) that

$$\sum_{j=1}^{k} |T h_j(t)|^2 \leq \|T\|^2, \quad t \in (0,1) \setminus N_k.$$

Let $i : L^\infty(0,1) \to L^2(0,1)$ be the inclusion mapping.

(c) Show that $i \circ T$ is Hilbert–Schmidt on $L^2(0,1)$ and $\|T\|_{\mathscr{L}_2(L^2(0,1))} \leq \|T\|$.

14.8 Prove that if $T \in \mathscr{L}(H)$ is selfadjoint and $S \in \mathscr{L}(H)$ is compact, and if the commutator $[T, S]$ is a trace class operator, then $\mathrm{tr}[T, S] = 0$.
Hint: Compute the traces of $[T, S \pm S^\star]$ relative to orthonormal bases which diagonalise $S \pm S^\star$.

14.9 Let $S, T \in \mathscr{L}(H)$ be selfadjoint trace class operators. Let $f : \mathbb{R} \to \mathbb{R}$ be a convex C^1-function.

(a) Show that for all norm one vectors $h \in H$ we have

$$(f(T)h|h) \geq f((Th|h)) \geq f((Sh|h)) + f'((Sh|h))((T - S)h|h).$$

Hint: For the first inequality expand h against an orthonormal basis of eigenvectors of T.

(b) Deduce that

$$\mathrm{tr}(f(T)) \geq \mathrm{tr}(f(S)) + \mathrm{tr}(f'(S)(T - S)).$$

Hint: Show that if h is an eigenvector of S, then the right-hand side in the identity of part (a) equals $((f(S) + f'(S)(T - S))h|h)$.

14.10 Prove the following analogue of Proposition 14.21: If $T \in \mathscr{L}(H)$ is a Hilbert–Schmidt operator, with eigenvalue sequence $(\lambda_n)_{n \geqslant 1}$ repeated according to algebraic multiplicity, then

$$\sum_{n \geqslant 1} |\lambda_n|^2 \leqslant \|T\|_{\mathscr{L}_2(H)}^2.$$

14.11 Let T be a $d \times d$ matrix and let $(\lambda_n)_{n=1}^d$ and $(\mu_n)_{n=1}^d$ be the sequences of eigenvalues and singular values of T, respectively, repeated according to algebraic multiplicities. Show that

$$\prod_{n=1}^d |\lambda_n| = \prod_{n=1}^d \mu_n.$$

Hint: Use the result from Step 1 in the proof of Proposition 14.21 to see that $\det(T) = \prod_{n=1}^d \lambda_n$. Apply this to $|T|$.

14.12 Let $T \in \mathscr{L}(H)$ be compact and let $\mu_1 \geqslant \mu_2 \geqslant \cdots \geqslant 0$ be the sequence of its nonzero singular values, repeated according to multiplicities. Show that for all $n \geqslant 1$ we have

$$\mu_n = \inf_{\substack{Y \subseteq H \\ \dim(Y) = n-1}} \sup_{\substack{\|y\|=1 \\ y \perp Y}} \|Ty\|,$$

where the infima are taken over all subspaces Y of H of dimension $n - 1$.

Hint: Use Theorem 9.4.

14.13 Complete the following outline of a proof of MacMahon's formula $\det(1 + A) = \sum_{k=0}^n \mathrm{tr}(\Lambda^k(A))$ for complex $n \times n$ matrices A.

(a) Prove the formula for the special case when A is diagonalisable, by showing that in this case the formula reduces to the identity

$$\prod_{k=0}^n (1 + \lambda_k) = \sum_{k=0}^n \sum_{1 \leqslant i_1 < \cdots < i_k \leqslant n} \lambda_{i_1} \cdots \lambda_{i_k}.$$

(b) Complete the proof by showing that the diagonalisable matrices are dense in $M_n(\mathbb{C})$.

14.14 Prove the following symmetric analogue of MacMahon's formula: for complex $n \times n$ matrices A one has

$$\frac{1}{\det(I - A)} = \sum_{j=0}^n \mathrm{tr}(\Sigma^j(A)).$$

14.15 Let ϕ, ψ be smooth functions on the unit circle and let $f, g : \overline{\mathbb{D}} \to \mathbb{R}$ denote their harmonic extensions. Applying Green's theorem to $(f\frac{\partial g}{\partial x}, f\frac{\partial g}{\partial y})$, show that Theorem 14.47 implies the identity

$$\text{tr}([T_\phi, T_\psi]) = \frac{1}{2\pi i} \int_{\mathbb{D}} \frac{\partial f}{\partial x}\frac{\partial g}{\partial y} - \frac{\partial g}{\partial x}\frac{\partial f}{\partial y}\, dx\,dy.$$

14.16 Using Fedosov's theorem, prove that if T is a Fredholm operator on H and $T = U|T|$ is its polar decomposition, then

$$\text{ind}(T) = \text{tr}(UU^\star - U^\star U).$$

Hint: Show that $I - U^\star U$ and $I - UU^\star$ are the projections onto the null spaces of T and $T^\star$ respectively, and that $\text{codim}\,R(T) = \dim N(T^\star)$.

14.17 Use Fedosov's theorem to give an alternative proof of the identity

$$\text{ind}(T_1 T_2) = \text{ind}(T_1) + \text{ind}(T_2)$$

for Fredholm operators T_1 and T_2 acting on H.

14.18 Let A and B be bounded positive operators on H. Show that if AB is of trace class, then $\text{tr}(AB) \geqslant 0$.
Hint: Use Problem 6.12 to infer that $\sigma(AB) \setminus \{0\} = \sigma(A^{1/2}BA^{1/2}) \setminus \{0\}$ is contained in the interval $(0, \infty)$. Then apply Lidskii's theorem.

15

States and Observables

In this final chapter we apply some of the ideas developed in the preceding chapters to set up a functional analytic framework for Quantum Mechanics. More specifically, we will show how the replacement of Borel sets in classical mechanics by orthogonal projections in a Hilbert space leads, in a natural way, to the quantum mechanical formalism for states and observables.

15.1 States and Observables in Classical Mechanics

We start by taking a brief look at the notions of state and observable in classical mechanics from a rather abstract measure theoretic point of view.

15.1.a States

In classical mechanics, the *state space* of a physical system is a measurable space $(X, \mathscr{X})$, typically a manifold with its Borel σ-algebra. For example, the state space of an ensemble of N free moving point particles in $\mathbb{R}^3$ is $\mathbb{R}^{3N} \times \mathbb{R}^{3N}$ (three position coordinates x_j and three momentum coordinates p_j for each particle) and that of the harmonic oscillator (with physical constants normalised to unity) is the submanifold of $\mathbb{R} \times \mathbb{R}$ given by $x^2 + p^2 = 1$.

Definition 15.1 (States, pure states). Let $(X, \mathscr{X})$ be a measurable space.

  (i) A *state* is a probability measure ν on $(X, \mathscr{X})$.

  (ii) A *pure state* is an extreme point of the set of probability measures on $(X, \mathscr{X})$.

For a measurable set $B \in \mathscr{X}$, the number $\nu(B)$ is thought of as "the probability that the state is described by a point in B".

Thus we identify the "state" of a system with the ensemble of truth probabilities of certain questions about the system. For example, the exact positions and momenta of all particles in a gas container at a given time cannot be known with complete precision, but one might ask about the probability of finding a certain portion of the gas in a certain subset of the container.

Recall that a measure ν on $(X, \mathscr{X})$ is said to be *atomic* if, whenever we have $\nu(B) > 0$ and $B = B_0 \cup B_1$ with disjoint $B_0, B_1 \in \mathscr{X}$, it follows that either $\nu(B_0) = 0$ or $\nu(B_1) = 0$.

Proposition 15.2. *The pure states are precisely the atomic probability measures.*

Proof This was shown in Example 4.37. □

15.1.b Observables

Definition 15.3 (Observables). Let $(\Omega, \mathscr{F})$ be a measurable space. An Ω-*valued observable* is a measurable function $f : X \to \Omega$. An *elementary observable* is a $\{0, 1\}$-valued observable.

For example, the three position coordinates x_j and momentum coordinates p_j of a free moving particle in $\mathbb{R}^3$ are real-valued observables on the state space $X = \mathbb{R}^3 \times \mathbb{R}^3$, and so are the kinetic energy $|p|^2/2m$ (where the mass m is treated as a constant) and potentials $V(x)$.

If ν is a state on $(X, \mathscr{X})$ and $f : X \to \Omega$ is an observable, then for $F \in \mathscr{F}$ the number

$$\nu(f^{-1}(F)) = \nu(\{x \in X : f(x) \in F\})$$

belongs to the interval $[0, 1]$ and is interpreted as "the probability that measuring f results in a value in F when the system is in state ν."

15.1.c From Classical to Quantum

An elementary observable is of the form $\mathbf{1}_B$ with $B \in \mathscr{X}$. Its range equals $\{0, 1\}$ unless $B = \varnothing$ or $B = \mathscr{X}$, in which case one has $\mathbf{1}_\varnothing \equiv 0$ and $\mathbf{1}_X \equiv 1$. Orthogonal projections in a complex Hilbert space enjoy similar properties *spectrally*: if P is an orthogonal projection in a Hilbert space H, its spectrum equals $\sigma(P) = \{0, 1\}$ unless $P = 0$ or $P = I$; in these cases one has $\sigma(0) = \{0\}$ and $\sigma(I) = \{1\}$.

The basic idea that underlies Quantum Mechanics is to *replace elementary observables by orthogonal projections*. The set of all orthogonal projections in a complex Hilbert space H is denoted by $\mathscr{P}(H)$. This set is partially ordered in a natural way by declaring $P_1 \leqslant P_2$ to mean that the range of P_1 is contained in the range of P_2; this is equivalent to the statement that the operator $P_2 - P_1$ is positive. With respect to this partial ordering, $\mathscr{P}(H)$ is a lattice in the sense of Definition 2.50; for P_1 and P_2 in $\mathscr{P}(H)$

the greatest lower bound

$$P_1 \wedge P_2$$

in $\mathscr{P}(H)$ is the orthogonal projection onto $R(P_1) \cap R(P_2)$, and the least upper bound

$$P_1 \vee P_2$$

in $\mathscr{P}(H)$ is the orthogonal projection onto the closed subspace spanned by $R(P_1)$ and $R(P_2)$. In addition to these operators, the *negation* of an orthogonal projection $P \in \mathscr{P}(H)$ is the orthogonal projection

$$\neg P = I - P$$

onto the orthogonal complement of $R(P)$. One has the associative laws

$$(P_1 \wedge P_2) \wedge P_3 = P_1 \wedge (P_2 \wedge P_3), \quad (P_1 \vee P_2) \vee P_3 = P_1 \vee (P_2 \vee P_3)$$

and the identities

$$\neg(P_1 \wedge P_2) = \neg P_1 \vee \neg P_2, \quad \neg(P_1 \vee P_2) = \neg P_1 \wedge \neg P_2.$$

The important difference with the classical setting is that the distributive laws

$$P_1 \wedge (P_2 \vee P_3) = (P_1 \wedge P_2) \vee (P_1 \wedge P_3)$$
$$P_1 \vee (P_2 \wedge P_3) = (P_1 \vee P_2) \wedge (P_1 \vee P_3)$$

generally fail.

Example 15.4. In $\mathbb{C}^2$ consider the orthogonal projections P_1, P_2, and P_3 onto the first and second coordinate axes and the diagonal, respectively. Then $P_1 \vee P_2 = I$, $P_1 \wedge P_3 = P_2 \wedge P_3 = 0$, and

$$(P_1 \vee P_2) \wedge P_3 = P_3, \quad (P_1 \wedge P_3) \vee (P_2 \wedge P_3) = 0.$$

15.2 States and Observables in Quantum Mechanics

From now on H is a *separable complex Hilbert space.*

15.2.a States

Upon replacing indicator functions of measurable sets by orthogonal projections in H, one is led to the idea to define a *state* as a mapping $v : \mathscr{P}(H) \to [0,1]$ that satisfies $v(0) = 0$, $v(I) = 1$, and is *countably additive* in the sense that

$$\sum_{n \geqslant 1} v(P_n) = v(P)$$

whenever $(P_n)_{n \geqslant 1}$ is a (finite or infinite) sequence of pairwise disjoint orthogonal projections and P is their least upper bound, that is, P is the orthogonal projection onto the closure of the span of the ranges of P_n, $n \geqslant 1$. Here, two orthogonal projections are called *disjoint* if their ranges are mutually orthogonal, and a family of projections said to be *disjoint* if every two distinct members of this family are disjoint.

Although this definition is quite satisfactory in many ways, it suffers from the defect that it does not present an obvious way to extend v to nonnegative linear combinations of pairwise disjoint orthogonal projections. In the classical picture, the expected value of a nonnegative simple function $f = \sum_{n=1}^{N} c_n \mathbf{1}_{B_n}$ in state v is given by its integral $\int_X f \, \mathrm{d}v = \sum_{n=1}^{N} c_n v(B_n)$. The desideratum

$$v\left(\sum_{n=1}^{N} c_n P_n\right) = \sum_{n=1}^{N} c_n v(P_n) \tag{15.1}$$

can be thought of as a quantum analogue of this, and constitutes the first step towards defining the expected value for more general classes of observables. However, if one attempts to take (15.1) as a definition, a problem of well-definedness arises (that such a problem indeed may arise is demonstrated by the example at the end of this section).

The next definition proposes a way around this difficulty. Recall that the *convex hull* of a subset S of a vector space V is the smallest convex set in V containing S and is denoted by $\mathrm{co}(S)$.

Definition 15.5 (Affine mappings). Let S be a subset of a vector space V. A mapping $v : S \to [0,1]$ is called *affine* if it extends to a mapping $v : \mathrm{co}(S) \to [0,1]$ satisfying

$$v\left(\sum_{n=1}^{N} \lambda_n v_n\right) = \sum_{n=1}^{N} \lambda_n v(v_n)$$

for all $N \geqslant 1$, $v_1, \ldots, v_N \in S$, and scalars $\lambda_1, \ldots, \lambda_N \geqslant 0$ satisfying $\sum_{n=1}^{N} \lambda_n = 1$.

Let us denote by $\mathscr{P}_{\mathrm{fin}}(H)$ the set of all *finite rank projections* in $\mathscr{P}(H)$, that is, the set of all projections with finite-dimensional ranges. To prepare for the definition of a state, we prove the following result.

Proposition 15.6. *Let* $v : \mathscr{P}_{\mathrm{fin}}(H) \to [0,1]$ *be affine and satisfy* $v(0) = 0$. *Then there exists a unique positive trace class operator* T *on* H *such that*

$$v(P) = \mathrm{tr}(PT), \quad P \in \mathscr{P}_{\mathrm{fin}}(H).$$

It satisfies

$$\mathrm{tr}(T) = \sup_{P \in \mathscr{P}_{\mathrm{fin}}(H)} v(P).$$

Conversely, if T *is a positive trace class operator on* H, *then*

$$v(P) := \mathrm{tr}(PT), \quad P \in \mathscr{P}(H),$$

defines an affine mapping $v : \mathscr{P}(H) \to [0,1]$ *satisfying* $v(0) = 0$ *and*

$$\sup_{P \in \mathscr{P}_{\mathrm{fin}}(H)} v(P) = \sup_{P \in \mathscr{P}(H)} v(P) = v(I) = \mathrm{tr}(T).$$

Moreover, v *countably additive.*

Proof To prove uniqueness, suppose that $T, \widetilde{T} \in \mathscr{L}(H)$ are such that $\mathrm{tr}(PT) = \mathrm{tr}(P\widetilde{T})$ for all $P \in \mathscr{P}_{\mathrm{fin}}(H)$. Taking P to be the rank one projection $h \otimes h : x \mapsto (x|h)h$, with $h \in H$ of norm one, gives $(Th|h) = (\widetilde{T}h|h)$. By scaling, this identity extends to arbitrary $h \in H$, and it implies $T = \widetilde{T}$ by Proposition 8.1.

The existence proof proceeds in several steps.

Step 1 – Throughout this step it is important to keep in mind that, when considering general convex or nonnegative-linear combinations of projections $P_1, \ldots, P_N$ in $\mathscr{P}_{\mathrm{fin}}(H)$, the projections P_n need not be mutually orthogonal and the same projection may be used multiple times.

Fix orthogonal projections $P_1, \ldots, P_N \in \mathscr{P}_{\mathrm{fin}}(H)$ and scalars $0 \leqslant c_1, \ldots, c_N \leqslant 1$ satisfying $\sum_{n=1}^{N} \leqslant 1$. With $c_{N+1} := 1 - \sum_{n=1}^{N} c_n$ and $P_{N+1} := 0$, the affinity assumption implies

$$v\Big(\sum_{n=1}^{N} c_n P_n\Big) = v\Big(\sum_{n=1}^{N+1} c_n P_n\Big) = \sum_{n=1}^{N+1} c_n v(P_n) = \sum_{n=1}^{N} c_n v(P_n),$$

where we used that $v(0) = 0$. Also, if an operator admits two such representations, say

$$\sum_{n=1}^{N} c_n P_n = \sum_{n=1}^{N'} c_n' P_n',$$

then by the same argument the affinity of v implies that

$$\sum_{n=1}^{N} c_n v(P_n) = \sum_{n=1}^{N'} c_n' v(P_n').$$

Consider next the case of scalars $c_1, \ldots, c_N \geqslant 0$, and consider an operator of the form $S = \sum_{n=1}^{N} c_n P_n$ with projections $P_n \in \mathscr{P}_{\mathrm{fin}}(H)$. Fix an arbitrary integer $k \geqslant \sum_{n=1}^{N} c_n$. Then, by what we just proved, the number

$$kv(\tfrac{1}{k}S) = kv\Big(\sum_{n=1}^{N} \frac{c_n}{k} P_n\Big) = k \sum_{n=1}^{N} \frac{c_n}{k} v(P_n) = \sum_{n=1}^{N} c_n v(P_n)$$

is independent of k. Hence we may define an extension of v, again denoted by v, by

$$v(S) := kv(\tfrac{1}{k}S) = \sum_{n=1}^{N} c_n v(P_n).$$

If S admits two such representations, say $S = \sum_{n=1}^{N} c_n P_n = \sum_{n=1}^{N'} c_n' P_n'$, then by taking $k \geqslant$

$\max\{\sum_{n=1}^{N} c_n, \sum_{n=1}^{N'} c'_n\}$ and using the well-definedness in the case already considered we obtain that $v(S)$ is well defined.

The extension just defined is finitely additive on the set of operators S of the form just described. Indeed, this follows by induction from the fact that if $S = \sum_{n=1}^{N} c_n P_n$ and $S' = \sum_{n=N+1}^{N'} c_n P_n$ are two such operators, then

$$v(S+S') = v\left(\sum_{n=1}^{N'} c_n P_n\right) = \sum_{n=1}^{N'} c_n v(P_n) = \sum_{n=1}^{N} c_n v(P_n) + \sum_{n=N+1}^{N'} c_n v(P_n) = v(S) + v(S').$$

Shifting the index in the expression for S' is justified since no restrictions are imposed on the projections occurring in the expressions for S and S' other than their membership of $\mathscr{P}_{\mathrm{fin}}(H)$; cf. the remark at the beginning of the proof.

Step 2 – Consider now an operator of the form $S = \sum_{n=1}^{N} c_n P_n$ with coefficients $c_n \in \mathbb{R}$ and projections $P_n \in \mathscr{P}_{\mathrm{fin}}(H)$. Then we may write $S = S_+ - S_-$, where $S_\pm$ are nonnegative-linear combinations of projections in $\mathscr{P}_{\mathrm{fin}}(H)$ as in Step 1, and define

$$v(S) := v(S_+) - v(S_-).$$

To see that this is well defined, let $S = S_+ - S_- = S'_+ - S'_-$ be two such representations. By the finite additivity proved in Step 1,

$$v(S_+) + v(S'_-) = v(S_+ + S'_-) = v(S'_+ + S_-) = v(S'_+) + v(S_-),$$

so $v(S_+) - v(S_-) = v(S'_+) - v(S'_-)$ as desired. Similarly it is checked that $cv(S) = v(cS)$ for all $c \in \mathbb{R}$ and that $v(S + S') = v(S) + v(S')$.

If $S = \sum_{n=1}^{N} c_n P_n$ with coefficients $c_n \in \mathbb{C}$ and projections $P \in \mathscr{P}_{\mathrm{fin}}(H)$, we set

$$v(S) := \frac{1}{2} v(S + S^\star) + \frac{1}{2i} v(i(S - S^\star)). \tag{15.2}$$

Then v is easily seen to be additive and real-linear, and from

$$v(iS) = \frac{1}{2} v(iS - iS^\star) + \frac{1}{2i} v(i(iS + iS^\star))$$

$$= i\left(\frac{1}{2i} v(iS - iS^\star) - \frac{1}{2} v(-(S + S^\star))\right) = iv(S)$$

it follows that v is in fact complex-linear.

Step 3 – Let $S \in \mathscr{K}(H)$ be any finite rank operator. We may represent S as $\sum_{n=1}^{N} c_n P_n$ with $c_1, \ldots, c_N \in \mathbb{C}$ and mutually orthogonal projections $P_1, \ldots, P_N \in \mathscr{P}_{\mathrm{fin}}(H)$. In doing so, we obtain

$$|v(S)| = \left| v\left(\sum_{n=1}^{N} c_n P_n \right) \right| = \left| \sum_{n=1}^{N} c_n v(P_n) \right| \leqslant \max_{1 \leqslant n \leqslant N} |c_n| \sum_{n=1}^{N} v(P_n)$$

$$= \max_{1 \leqslant n \leqslant N} |c_n| v\left(\sum_{n=1}^{N} P_n \right) \leqslant \max_{1 \leqslant n \leqslant N} |c_n| = \left\| \sum_{n=1}^{N} c_n P_n \right\| = \|S\|.$$

(15.3)

Here we used that $v(\sum_{n=1}^{N} P_n) \leqslant 1$ since $\sum_{n=1}^{N} P_n$ is an orthogonal projection.

Step 4 – By the spectral theorem (Theorem 9.1), every compact selfadjoint operator $S \in \mathscr{K}(H)$ can be approximated, in the norm of $\mathscr{L}(H)$, by a sequence of finite rank operators S_n. The estimate (15.3), applied to their differences, entails that the limit

$$v(S) := \lim_{n \to \infty} v(S_n)$$

exists. If the finite rank operators S'_n form another approximating sequence, then by what has been proved before we have

$$|v(S_n) - v(S'_n)| = |v(S_n - S'_n)| \leqslant \|S_n - S'_n\| \to 0.$$

This shows it that the number $v(S)$ is independent of the choice of approximating sequence.

For general compact operators $S \in \mathscr{L}(H)$ we define $v(S)$ by (15.2) and find

$$|v(S)| \leqslant \frac{1}{2} \|S + S^\star\| + \frac{1}{2} \|i(S - S^\star)\| \leqslant 2\|S\|.$$

Repeating previous arguments, this extension is again seen to be linear.

Step 5 – The argument of Step 4 proves that we may identify v with an element in $(\mathscr{K}(H))^*$, the dual of the space $\mathscr{K}(H)$ of compact operators on H. By trace duality (Theorem 14.29) there exists a unique trace class operator $T \in \mathscr{L}_1(H)$ such that for all $S \in \mathscr{K}(H)$ we have $v(S) = \text{tr}(ST)$. By considering the orthogonal projection $P = h \bar{\otimes} h$ onto the span of the norm one vector h, we obtain $(Th|h) = \text{tr}(PT) = v(P) \geqslant 0$. This implies that T is positive.

If P_n is an increasing sequence of finite rank projections converging to the identity operator strongly, then

$$\text{tr}(T) = \lim_{n \to \infty} \text{tr}(P_n T) = \lim_{n \to \infty} v(P_n) \leqslant \sup_{P \in \mathscr{P}_{\text{fin}}(H)} v(P).$$

In the opposite direction, for any $P \in \mathscr{P}_{\text{fin}}(H)$ we have

$$v(P) = \text{tr}(PT) = \text{tr}(TP) \leqslant \text{tr}(T).$$

Taking the supremum over all $P \in \mathscr{P}_{\text{fin}}(H)$ we obtain $\sup_{P \in \mathscr{P}_{\text{fin}}(H)} v(P) \leqslant \text{tr}(T)$. This proves the identity $\text{tr}(T) = \sup_{P \in \mathscr{P}_{\text{fin}}(H)} v(P)$. thereby completing the proof of the first assertion of the theorem.

Step 6 – We now turn to the converse statement. Let T be a positive trace class operator on H and define $v(S) := \text{tr}(ST) = \text{tr}(TS)$ for $S \in \mathscr{L}(H)$. Its restriction to $\mathscr{P}(H)$, which we shall denote by v again, is obviously affine and satisfies $v(0) = 0$. To prove countable additivity, let $(P_n)_{n \geq 1}$ be a sequence of disjoint orthogonal projections and let P be the orthogonal projection onto the closure of the span of their ranges. If $(h_j^{(n)})_{j \geq 1}$ is an orthonormal basis for the range of P_n, then the union of these sequences can be relabelled into an orthonormal basis $(h_k)_{k \geq 1}$ for the range of P. Then,

$$v(P) = \text{tr}(TP) = \sum_{k \geq 1} (Th_k | h_k) = \sum_{n \geq 1} \left(\sum_{j \geq 1} (TP_n h_j^{(n)} | h_j^{(n)}) \right) = \sum_{n \geq 1} \text{tr}(TP_n) = \sum_{n \geq 1} v(P_n),$$

the fourth identity being justified by the nonnegativity of the summands.

Step 7 – Let $P \in \mathscr{P}(H)$ be arbitrary and choose an orthonormal basis $(h_n)_{n \geq 1}$ for $(\text{R}(P))^{\perp}$. Denoting the coordinate projections by P_n, by the countable additivity of v we have $v(P) \leq v(P) + \sum_{n \geq 1} v(P_n) = v(I)$. This being true for any $P \in \mathscr{P}(H)$ it follows that

$$\sup_{P \in \mathscr{P}(H)} v(P) \leq v(I).$$

On the other hand, if $(h_n)_{n \geq 1}$ is an orthonormal basis for H, the countable additivity of v gives $\lim_{N \to \infty} v(\sum_{n=1}^{N} P_n) = v(I)$. Since $\sum_{n=1}^{N} P_n \in \mathscr{P}_{\text{fin}}(H)$ this implies

$$\sup_{P \in \mathscr{P}_{\text{fin}}(H)} v(P) \geq v(I).$$

In combination with the second part of step Step 5, which shows that for any positive trace class operator T on H we have $\text{tr}(T) = \sup_{P \in \mathscr{P}_{\text{fin}}(H)} \text{tr}(PT)$, this proves the identities in the second part of the theorem. $\square$

In what follows we denote by $\mathscr{S}(H)$ the convex set of all positive trace class operators with unit trace on H. We will see below (Proposition 15.14) that this set is the closed convex hull of its set of extreme points and that these extreme points are precisely the orthogonal rank one projections in H.

A functional $\phi : \mathscr{L}(H) \to \mathbb{C}$ is called *positive* if $\phi(T) \geq 0$ for every positive $T \in \mathscr{L}(H)$, and *normal* if

$$\sum_{n \geq 1} \phi(P_n) = \phi(P)$$

whenever $(P_n)_{n \geq 1}$ is a sequence of disjoint orthogonal projections in H and P is their least upper bound. The same terminology applies to functionals $\phi : \mathscr{K}(H) \to \mathbb{C}$.

Theorem 15.7. *The following six sets are in one-to-one correspondence:*

(1) *affine mappings* $v : \mathscr{P}_{\text{fin}}(H) \to [0,1]$ *satisfying* $v(0) = 0$ *and* $\sup_{P \in \mathscr{P}_{\text{fin}}(H)} v(P) = 1$;

(2) *affine mappings* $v : \mathscr{P}(H) \to [0,1]$ *satisfying* $v(0) = 0$ *and* $v(I) = 1$;

(3) *positive trace class operators* T *on* H *satisfying* $\text{tr}(T) = 1$, *via*

$$v(P) = \text{tr}(PT), \quad P \in \mathscr{P}_{\text{fin}}(H);$$

(4) *positive trace class operators* T *on* H *satisfying* $\text{tr}(T) = 1$, *via*

$$v(P) = \text{tr}(PT), \quad P \in \mathscr{P}(H);$$

(5) *positive functionals* $\phi : \mathscr{K}(H) \to \mathbb{C}$ *satisfying* $\sup_{P \in \mathscr{P}_{\text{fin}}(H)} \phi(P) = 1$, *via*

$$\phi(S) = \text{tr}(ST), \quad S \in \mathscr{K}(H);$$

(6) *positive normal functionals* $\phi : \mathscr{L}(H) \to \mathbb{C}$ *satisfying* $\phi(I) = 1$, *via*

$$\phi(S) = \text{tr}(ST), \quad S \in \mathscr{L}(H).$$

Proof For $m, n = 1, 2, 3, 4$ we write $(m) \Rightarrow (n)$ to express that every object in the set described by (m) uniquely defines an element in the set described by (n).

(1)$\Leftrightarrow$(3): This one-to-one correspondence is contained in Proposition 15.6.

(3)$\Rightarrow$(6): Let T be a positive trace class operator with unit trace and define $\phi : \mathscr{L}(H) \to \mathbb{C}$ as in (6). Then $\phi(I) = \text{tr}(T) = 1$. To prove the positivity of ϕ, let $S \geqslant 0$. If $(h_n)_{n \geqslant 1}$ is an orthonormal basis for H, the positivity of T implies

$$\phi(S) = \text{tr}(ST) = \text{tr}(S^{1/2}TS^{1/2}) = \sum_{n \geqslant 1} (TS^{1/2}h_n | S^{1/2}h_n) \geqslant 0.$$

The normality of ϕ follows from the countable additivity of the mapping $P \mapsto \text{tr}(PT)$ proved in the second part of Proposition 15.6.

(6)$\Rightarrow$(5): This inclusion follows from Step 7 of the proof of Proposition 15.6.

(5)$\Rightarrow$(1): The restriction $v := \phi|_{\mathscr{P}_{\text{fin}}(H)}$ is affine, takes values in $[0,1]$, and satisfies $v(0) = 0$ and $\sup_{P \in \mathscr{P}_{\text{fin}}(H)} v(P) = 1$.

(6)$\Rightarrow$(2)$\Rightarrow$(1): The first inclusion is obtained in the same way and the second again follows from Step 7 of the proof of Proposition 15.6.

(1)$\Rightarrow$(4)$\Rightarrow$(3): These inclusions are also contained in Proposition 15.6. $\square$

We my now define a *state* as either one of these six sets. For the sake of definiteness we take the sixth:

Definition 15.8 (States). A *state* is a positive normal functional $\phi : \mathscr{L}(H) \to \mathbb{C}$ satisfying $\phi(I) = 1$.

This definition captures what is generally called a *normal state* in the mathematical literature on Quantum Mechanics; the term *state* is usually reserved for general positive functionals $\phi : \mathscr{L}(H) \to \mathbb{C}$ satisfying $\phi(I) = 1$. The small abuse of terminology committed by omitting the adjective 'normal' from our terminology may be excused by the third item in the above list, which does not involve normality.

Remark 15.9 (Density functions). In the Physics literature, the positive trace class operator T with unit trace associated with state ψ is called the *density operator* associated with ϕ.

As the following example shows, a countably additive mapping $v : \mathscr{P}(\mathbb{C}^2) \to [0,1]$ satisfying $v(0) = 0$ and $v(I) = 1$ need not be affine (and therefore need not define a state).

Example 15.10 (Failure of affineness in two dimensions). Let $H = \mathbb{C}^2$ and let S denote its unit sphere. Let $f : S \to [0,1]$ be a function with the following two properties:

(i) $f(h_1) = f(h_2)$ whenever $\mathrm{span}(h_1) = \mathrm{span}(h_2)$;
(ii) $f(h_1) + f(h_2) = 1$ whenever $h_1 \perp h_2$.

Apart from these restrictions, f can be completely arbitrary.

Define $v : \mathscr{P}(H) \to [0,1]$ by $v(0) := 0$, $v(I) = 1$, and

$$v(P_h) := f(h), \quad h \in S,$$

where P_h is the orthogonal projection onto $\mathrm{span}(h)$. It is clear that v is countably additive: if the orthogonal projections $P_1, P_2, \ldots$ are pairwise disjoint, then all but at most two must be zero. If there are zero or one nonzero projections, then countable additivity is trivial, and if there are two nonzero projections they must be of the form P_{h_1} and P_{h_2} with $h_1 \perp h_2$; in that case countable additivity follows from

$$v(P_{h_1}) + v(P_{h_2}) = f(h_1) + f(h_2) = 1 = v(I) = v(P_{h_1} + P_{h_2}).$$

If there exists a positive operator T on H with unit trace such that for all $P \in \mathscr{P}(H)$ we have $v(P) = \mathrm{tr}(PT)$, then

$$f(h) = v(P_h) = \mathrm{tr}(P_h T) = (Th|h)$$

depends continuously on h. It is, however, easy to construct discontinuous functions f satisfying the conditions (i) and (ii). Indeed, once the value of f at a given point $h_0 \in S$ is fixed, the conditions (i) and (ii) fix the values of f only on the points $e^{i\theta} h_0$ and all points orthogonal to them. If we identify S with the unit sphere S^3 in $\mathbb{R}^4$, these points define a 'great circle' incident with h_0 and an 'equator' relative to the 'north pole' h_0. Therefore, in a sufficiently small neighbourhood of h_0, f is only determined on a submanifold of dimension 1. This leaves enough room to construct functions f satisfying (i) and (ii) but discontinuous at h_0.

If v were affine we could represent it by a positive operator T. This would contradict the discontinuity of f.

It is not a coincidence that this counterexample lives in two dimensions: A celebrated theorem due to Gleason asserts that if $\dim(H) \geqslant 3$, then every countably additive mapping $v : \mathscr{P}(H) \to [0, 1]$ is affine and hence defines a state.

15.2.b Pure States

Theorem 15.7 establishes four equivalent ways of looking at the convex set of all states. Since the correspondences between them preserve convex combinations and hence extreme points, the following definition makes sense from each of these points of view:

Definition 15.11 (Pure states). A *pure state* is an extreme point of the convex set of states.

Proposition 15.12. *A state* $\phi : \mathscr{L}(H) \to \mathbb{C}$ *is pure if and only if it is a* vector state, *that is, there exists a unit vector $h \in H$ such that*

$$\phi(S) = (Sh|h), \quad S \in \mathscr{L}(H).$$

This unit vector is unique up to a scalar multiple of modulus one.

The first assertion can be equivalently stated as saying that the extreme points of the set of all positive trace class operators with unit trace are precisely the orthogonal projections of rank one.

Proof 'Only if': Let ϕ be a state and let T be the associated positive trace class operator on H with unit trace. By the singular value decomposition (Theorem 14.15) we have $T = \sum_{n \geqslant 1} \lambda_n h_n \bar{\otimes} h_n$ for some orthonormal basis $(h_n)_{n \geqslant 1}$ of H and a nonnegative scalar sequence $(\lambda_n)_{n \geqslant 1}$ such that $\sum_{n \geqslant 1} \lambda_n = \text{tr}(T) = 1$. This allows us to write T as a convex combination of distinct states unless all but one λ_n vanish, in which case we have $T = h_v \bar{\otimes} h_v$ for some unit vector $h_v \in H$ and $v(P) = \text{tr}(P \circ (h_v \bar{\otimes} h_v)) = (Ph_v|h_v)$ for all orthogonal projections $P \in \mathscr{P}(H)$.

'If': If ϕ is a vector state, then the associated positive trace class operator is of the form $T = h \bar{\otimes} h$ with $\|h\| = 1$. If $T = (1 - \lambda)T_0 + \lambda T_1$ is a convex combination of positive trace class operators T_0 and T_1 with unit trace, then the unit vector $h = Th = (1 - \lambda)T_0 h + \lambda T_1 h$ is a convex combination of two vectors of norm at most one. Hence we must have either $h = (1 - \lambda)T_0 h$ or $h = \lambda T_1 h$. Since T_0 and T_1 are contractive, this is only possible if $\lambda = 0$ (in the first case) or $\lambda = 1$ (in the second case). This means that either $T = T_0$ or $T = T_1$, so T is an extreme point of the convex set of positive trace class operator on H with unit trace. Since the correspondence between states and the associated positive trace class operators preserves convex combinations, it follows that ϕ is an extreme point of the convex set of states.

The uniqueness assertion follows by observing that for all $\theta \in \mathbb{R}$ and $h \in H$ we have

$$(e^{i\theta}h) \bar{\otimes} (e^{i\theta}h) = h \bar{\otimes} h.$$

$$\square$$

Remark 15.13 (Bras, kets, superpositions, mixed states). In the Physics literature, the pure state corresponding to a unit vector $h \in H$ is commonly denoted by $|h\rangle$ and referred to as the *ket* or *wave function* associated with h; often, $|h\rangle$ is identified with h.

The addition in H can be used to define, for orthogonal unit vectors $h_1, h_2 \in H$ and scalars $\alpha_1, \alpha_2 \in \mathbb{C}$ satisfying $|\alpha_1|^2 + |\alpha_2|^2 = 1$, the pure state

$$\alpha_1 |h_1\rangle + \alpha_2 |h_2\rangle := |\alpha_1 h_1 + \alpha_2 h_2\rangle.$$

Such states are referred to as (coherent) *superpositions* of the states $|h_1\rangle$ and $|h_2\rangle$. Such states should be carefully distinguished from states that can be built by using the addition of $\mathscr{L}_1(H)$. Indeed, for $h_1, h_2 \in H$ are linearly independent unit vectors in H and $\lambda \in [0, 1]$ the convex combination $(1 - \lambda)h_1 \bar{\otimes} h_1 + \lambda h_2 \bar{\otimes} h_2$, or, in Physics notation,

$$(1 - \lambda)|h_1\rangle\langle h_1| + \lambda |h_2\rangle\langle h_2|$$

defines a state in $\mathscr{L}_1(H)$. Such states, which are not pure unless $\lambda = 0$ or $\lambda = 1$, are called *mixed states* or, more precisely, *mixtures* of the states $|h_1\rangle$ and $|h_2\rangle$.

We recall that $\mathscr{S}(H)$ denotes the convex set of all positive trace class operators with unit trace on H. As we have seen in Theorem 15.7, the elements of this set are in one-to-one correspondence with states. By Proposition 15.12, the extreme points of $\mathscr{S}(H)$ are the rank one projections of the form $h \bar{\otimes} h$ with $h \in H$ of norm one.

Proposition 15.14. *The set $\mathscr{S}(H)$ is the closed convex hull of its extreme points. The extreme points of this set are precisely the rank one projections of the form $h \bar{\otimes} h$ with $h \in H$ of norm one.*

Proof By the singular value decomposition of Theorem 14.15, every element of $T \in \mathscr{S}(H)$ is of the form $T = \sum_{n \geqslant 1} \lambda_n h_n \bar{\otimes} h_n$, with convergence in trace norm, with $(h_n)_{n \geqslant 1}$ an orthonormal basis in H and $(\lambda_n)_{n \geqslant 1}$ a nonnegative sequence satisfying $\sum_{n \geqslant 1} \lambda_n = 1$. This gives the first assertion. The second follows from Theorem 15.7, which informs us that the operators of the form $h \bar{\otimes} h$ with $h \in H$ of norm one are in one-to-one correspondence with the vector states, which are the extreme points of the convex set of all states by Proposition 15.12. $\square$

15.2.c Observables

Let $(\Omega, \mathcal{F})$ be a measurable space. Classically, an Ω-valued observable on the state space $(X, \mathcal{X})$ is a measurable function $f : X \to \Omega$. By definition of measurability, f induces a mapping from $\mathcal{F}$ to $\mathcal{X}$ given by

$$F \mapsto f^{-1}(F), \quad F \in \mathcal{F},$$

and this mapping is countably additive, in the sense that if the sets $F_n \in \mathcal{F}$ are pairwise disjoint, then $f^{-1}(\bigcup_{n \geqslant 1} F_n) = \bigcup_{n \geqslant 1} f^{-1}(F_n)$. Identifying sets in $\mathcal{X}$ by their indicator functions and replacing them by orthogonal projections in a Hilbert space H, we arrive at the following definition of an observable in Quantum Mechanics.

Definition 15.15 (Observables). Let $(\Omega, \mathcal{F})$ be a measurable space and H a Hilbert space. An Ω-*valued observable* is a countably additive mapping $P : \mathcal{F} \to \mathscr{P}(H)$ satisfying $P(\Omega) = I$. An *elementary observable* is a $\{0, 1\}$-valued observable.

By Corollary 9.18, the elementary observables are precisely the orthogonal projections. This should be compared to the classical situation where elementary observables are given as the indicator functions of measurable sets.

Observables defined in this way are sometimes called *sharp observables*, as opposed to *unsharp observables* which will be introduced in Section 15.3.b.

Following notation introduced in Chapter 9 we write $P_F := P(F)$ for $F \in \mathcal{F}$. For vectors $h \in H$, we denote by P_h the nonnegative probability measure on Ω given by

$$P_h(F) := (P_F h | h), \quad F \in \mathcal{F}.$$

In the language of Chapter 9 a real-valued observable is nothing but a projection-valued measure on $\mathbb{R}$, and by the spectral theorem (Theorem 10.56) we can associate a unique selfadjoint operator A with P determined by

$$D(A) = \left\{ h \in H : \int_{\mathbb{R}} |\lambda|^2 \, dP_h(\lambda) < \infty \right\}$$

and, for $h \in D(A)$,

$$(Ah | h) = \int_{\mathbb{R}} \lambda \, dP_h(\lambda)$$

(see Theorem 10.50). In the converse direction, the spectral theorem asserts that every selfadjoint operator A arises from a projection-valued measure on $\mathbb{R}$ in this way and hence defines an observable.

Thus we arrive at the conclusion that *real-valued* observables are in one-to-one correspondence with selfadjoint operators. In most treatments of Quantum Mechanics this is simply taken as a postulate. In a sense, the present treatment provides the deeper motivation for this postulate, in that this correspondence appears as a consequence of the point of view that, on the mathematical level, the classical-to-quantum transition is simply the transition from the Boolean algebra of subsets of measurable space to the lattice of orthogonal projections on a Hilbert space. A further advantage of the present approach is that, in the same vein, the spectral theorem for normal operators can be reinterpreted as establishing a one-to-one correspondence between complex-valued observables and normal operators, and between observables with values in the unit circle and unitary operators.

We return to the abstract setting of observables $P : \mathscr{F} \to \mathscr{P}(H)$ with values in Ω. If $\phi : \mathscr{L}(H) \to \mathbb{C}$ is a pure state represented by the unit vector $h \in H$, then

$$\phi(P_F) = (P_F h | h) = P_h(F), \quad F \in \mathscr{F},$$

so the assignment $F \mapsto \phi(P_F)$ defines a probability measure. The following proposition is an immediate consequence of the fact that states are normal and that the least upper bound of a sequence of disjoint orthogonal projections is given by their sum. It is the mathematical counterpart of the so-called Born rule in Quantum Mechanics and allows us to interpret the number $\phi(P_F)$ as "the probability that measuring P results in a value contained in $F \in \mathscr{F}$ when the system is in state ϕ".

Proposition 15.16 (Born rule). *If $\phi : \mathscr{L}(H) \to \mathbb{C}$ is a state and $P : \mathscr{F} \to \mathscr{P}(H)$ an Ω-valued observable, the mapping*

$$F \mapsto \phi(P_F), \quad F \in \mathscr{F},$$

defines a probability measure on $(\Omega, \mathscr{F})$.

If P is a real-valued (or complex-valued) observable represented by a selfadjoint (or, more generally, a normal) operator A, then, as a projection-valued measure, P is supported on the spectrum $\sigma(A)$ and therefore P can be thought of as an $\sigma(A)$-valued observable. The physical interpretation is that "with probability one, a measurement of A produces a value belonging to $\sigma(A)$".

15.2.d The Uncertainty Principle

If P is a real-valued observable represented by a bounded selfadjoint operator A, the *expected value of P in state ϕ* is defined as the number

$$\langle A \rangle_\phi := \phi(A).$$

If $\phi = |h\rangle$ is a pure state associated with a unit vector $h \in H$ contained in $\mathrm{D}(A)$, we have

$$\langle A \rangle_{|h\rangle} = (Ah | h).$$

In this situation, for $h \in \mathrm{D}(A)$ we can define the *variance* by

$$\mathrm{var}_{|h\rangle}(A) := \left\langle (A - \langle A \rangle_{|h\rangle})^2 \right\rangle_{|h\rangle} = \| (A - (Ah | h)) h \|^2.$$

The *uncertainty* of A in state $|h\rangle$ is defined by

$$\Delta_{|h\rangle}(A) := (\mathrm{var}_{|h\rangle}(A))^{1/2}.$$

Theorem 15.17 (Uncertainty principle). *Let $|h\rangle$ be a pure state associated with the unit vector $h \in H$, and consider two real-valued observables with associated selfadjoint operators A and B. If $h \in \mathsf{D}([A,B]) := \{h \in \mathsf{D}(A) \cap \mathsf{D}(B) : Ah \in \mathsf{D}(B), Bh \in \mathsf{D}(A)\}$ and $[A,B]h := ABh - BAh$, then*

$$\Delta_{|h\rangle}(A)\Delta_{|h\rangle}(B) \geqslant \frac{1}{2}|([A,B]h|h)|.$$

Proof The operators $\widetilde{A} := A - (Ah|h)$ and $\widetilde{B} := B - (Bh|h)$ with domains $\mathsf{D}(\widetilde{A}) = \mathsf{D}(A)$ and $\mathsf{D}(\widetilde{B}) = \mathsf{D}(B)$ are selfadjoint. In particular we note that $\widetilde{A}h \in \mathsf{D}(\widetilde{B})$, $\widetilde{B}h \in \mathsf{D}(\widetilde{A})$, and we have $[\widetilde{A}, \widetilde{B}]h = [A,B]h$. The Cauchy–Schwarz inequality implies

$$\Delta_{|h\rangle}(A)\Delta_{|h\rangle}(B) = \|\widetilde{A}h\|\|\widetilde{B}h\| \geqslant |(\widetilde{A}h|\widetilde{B}h)| \geqslant |\mathrm{Im}(\widetilde{A}h|\widetilde{B}h)|$$
$$= \frac{1}{2}|(\widetilde{A}h|\widetilde{B}h) - (\widetilde{B}h|\widetilde{A}h)| = \frac{1}{2}|([\widetilde{A}, \widetilde{B}]h|h)| = \frac{1}{2}|([A,B]h|h)|.$$

$\square$

The physical interpretation of the next result is that a measurement of A in a pure state $|h\rangle$ gives the expected value $(Ah|h)$ with probability one if and only if the representing unit vector h is an eigenvector of A, and in this case the eigenvalue equals $(Ah|h)$.

Proposition 15.18. *Let P be a real-valued observable, represented by the selfadjoint operator A, and let $h \in \mathsf{D}(A)$ satisfy $\|h\| = 1$. The following assertions are equivalent:*

(1) *A has zero uncertainty in the state $|h\rangle$;*
(2) *h is an eigenvector for A.*

If these equivalent conditions hold, then for the corresponding eigenvalue λ we have

$$\lambda = (Ah|h) \quad and \quad (P_{\{\lambda\}}h|h) = 1.$$

Proof (1)$\Rightarrow$(2): If $\mathrm{var}_{|h\rangle}(A) = 0$, then $Ah = (Ah|h)h$, so h is an eigenvector of A with eigenvalue $\lambda = (Ah|h)$.

(2)$\Rightarrow$(1): If $Ah = \lambda h$, then

$$\mathrm{var}_{|h\rangle}(A) = \|(A - (Ah|h))h\|^2 = \|(A - \lambda)h\|^2 = 0.$$

If the equivalent conditions hold, then by Corollary 10.59 for all measurable functions $f : \sigma(A) \to \mathbb{C}$ we have $f(A)h = f(\lambda)h$ and consequently

$$\int_{\sigma(A)} f\,\mathrm{d}P_h = (f(A)h|h) = f(\lambda).$$

This forces $P_h = \delta_{\{\lambda\}}$ and therefore $(P_{\{\lambda\}}h|h) = \int_{\sigma(A)} \mathbf{1}_{\{\lambda\}}\,\mathrm{d}P_h = \mathbf{1}_{\{\lambda\}}(\lambda) = 1.$ $\square$

15.2.e The Qubit

It is instructive to take a closer look at the simplest genuinely quantum mechanical system, the qubit. It is the quantum version of the *bit* $\{0,1\}$, which we think of as equipped with the counting measure μ giving mass 1 to each of the two elements of $\{0,1\}$. Physically, the qubit models a spin $\frac{1}{2}$ particle. We write $\mathbf{1}_{\{0\}}$ and $\mathbf{1}_{\{1\}}$ for the unit basis vectors of the Hilbert space $L^2(\{0,1\})$ and denote the pure states associated with them by $|0\rangle$ and $|1\rangle$. Every pure state is then of the form $\alpha|0\rangle + \beta|1\rangle$ with $\alpha, \beta \in \mathbb{C}$ satisfying $|\alpha|^2 + |\beta|^2 = 1$. Since pure states are defined up to a complex number of modulus one, every pure state can be uniquely written in the form

$$\cos(\theta/2)|0\rangle + e^{i\varphi}\sin(\theta/2)|1\rangle \tag{15.4}$$

for suitable $0 \leqslant \theta \leqslant \pi$ and $0 \leqslant \varphi < 2\pi$. In spherical coordinates, the variables θ and φ uniquely determine a point

$$(\sin\theta\cos\varphi,\ \sin\theta\sin\varphi,\ \cos\theta) \tag{15.5}$$

on the unit sphere S^2 of $\mathbb{R}^3$. This representation of pure states is frequently referred to as the *Bloch sphere*.

In what follows we identify $L^2(\{0,1\})$ isometrically identify with $\mathbb{C}^2$. Under this identification, linear operators on $L^2(\{0,1\})$ correspond to 2×2 matrices with complex coefficients. States can be identified with points in the closed unit ball of $\mathbb{R}^3$ as follows. Any selfadjoint operator $T = (t_{ij})_{i,j=1}^2$ on $\mathbb{C}^2$ with unit trace $\mathrm{tr}(T) = t_{11} + t_{22} = 1$ is of the form

$$T = \frac{1}{2}\begin{pmatrix} 1+c_3 & c_1 - ic_2 \\ c_1 + ic_2 & 1 - c_3 \end{pmatrix} \tag{15.6}$$

The Bloch sphere
(Source: Wikipedia)

with $c_1, c_2, c_3 \in \mathbb{R}$. The vector $c = (c_1, c_2, c_3) \in \mathbb{R}^3$ is called the *Bloch vector* of T. It is easily checked that the eigenvalues of T are $\frac{1}{2}(1 \pm |c|)$. From this we see that $T \geqslant 0$ if and only if $|c| \leqslant 1$.

A routine computation shows that the pure state $|h\rangle = \cos(\theta/2)|0\rangle + e^{i\varphi}\sin(\theta/2)|1\rangle$ corresponds to the operator

$$T = h \bar{\otimes} h = \frac{1}{2}\begin{pmatrix} 1+\cos\theta & e^{-i\varphi}\sin\theta \\ e^{i\varphi}\sin\theta & 1 - \cos\theta \end{pmatrix} \tag{15.7}$$

with Bloch vector $(\sin\theta\cos\varphi,\ \sin\theta\sin\varphi,\ \cos\theta)$. Thus the Bloch sphere representation of the pure state $|h\rangle$ equals the Bloch vector of the associated operator $h \bar{\otimes} h$.

Equation (15.6) can be written as

$$T = \frac{1}{2}\begin{pmatrix} 1 & 0 \\ 0 & 1 \end{pmatrix} + \frac{c_1}{2}\begin{pmatrix} 0 & 1 \\ 1 & 0 \end{pmatrix} + \frac{c_2}{2}\begin{pmatrix} 0 & -i \\ i & 0 \end{pmatrix} + \frac{c_3}{2}\begin{pmatrix} 1 & 0 \\ 0 & -1 \end{pmatrix}$$

$$= \frac{1}{2}(I + c_1\sigma_1 + c_2\sigma_2 + c_3\sigma_3),$$

where

$$\sigma_1 = \begin{pmatrix} 0 & 1 \\ 1 & 0 \end{pmatrix}, \quad \sigma_2 = \begin{pmatrix} 0 & -i \\ i & 0 \end{pmatrix}, \quad \sigma_3 = \begin{pmatrix} 1 & 0 \\ 0 & -1 \end{pmatrix}$$

are the three *Pauli matrices*. These matrices are selfadjoint and their spectra equal $\{\pm 1\}$. Therefore they are associated with ± 1 valued observables, also denoted by σ_1, σ_2, and σ_3. The corresponding eigenstates of σ_j are called the *spin up/spin down states along the jth axis*. Every selfadjoint operator A on $\mathbb{C}^2$ is of the form

$$A = \begin{pmatrix} a & c - id \\ c + id & b \end{pmatrix} = \begin{pmatrix} c_0 + c_3 & c_1 - ic_2 \\ c_1 + ic_2 & c_0 - c_3 \end{pmatrix} = c_0 I + c_1\sigma_1 + c_2\sigma_2 + c_3\sigma_3$$

for certain $a, b, c, d, c_0, c_1, c_2, c_3 \in \mathbb{R}$ with $a = c_0 + c_3$, $b = c_0 - c_3$, $c = c_1$, and $d = c_2$. It follows that the quadruple $\{I, \sigma_1, \sigma_2, \sigma_3\}$ is a basis for the real-linear vector space of selfadjoint operators on $\mathbb{C}^2$.

15.2.f Entanglement

The natural choice for the state space of a system of N classical point particles in $\mathbb{R}^3$ is $\mathbb{R}^{3N} \times \mathbb{R}^{3N}$, the idea being that six coordinates are needed (three for position, three for momentum) to describe the state of each particle. In Quantum Mechanics, the natural choice of Hilbert space is $L^2(\mathbb{R}^{3N})$. Labelling the points of $\mathbb{R}^{3N}$ as $x = (x_j^{(n)})_{j,n=1}^{3,N}$, this choice suggests the following natural definition of observables $\widehat{x}_j^{(n)}$ describing the jth coordinate of the nth particle:

$$\widehat{x}_j^{(n)} f(x) := x_j^{(n)} f(x), \quad f \in D(\widehat{x}_j^{(n)}), \quad x \in \mathbb{R}^{3N},$$

where $D(\widehat{x}_j^{(n)}) = \{f \in L^2(\mathbb{R}^{3N}) : x_j^{(n)} f \in L^2(\mathbb{R}^{3N})\}$. Later we will see that the corresponding momentum operators are given by

$$\widehat{p}_j^{(n)} f := \frac{1}{i}\frac{\partial}{\partial x_j^{(n)}} f, \quad f \in D(\widehat{p}_j^{(n)})$$

with their natural domains.

The space $L^2(\mathbb{R}^{3N})$ is isometric in a natural way to the N-fold Hilbert space tensor product (see Definition 14.30 and the discussion following it):

$$L^2(\mathbb{R}^{3N}) \simeq \underbrace{L^2(\mathbb{R}^3) \otimes \cdots \otimes L^2(\mathbb{R}^3)}_{N \text{ times}}.$$

This suggests that if the Hilbert spaces $H_1, \ldots, H_N$ describe the states of N quantum mechanical systems, then their Hilbert space tensor product

$$H_1 \otimes \cdots \otimes H_N$$

serves to describe the system composed of these N subsystems. In what follows we focus on the case $N = 2$, but everything we say extends to general N without difficulty.

Let H and K be Hilbert spaces, and let $H \otimes K$ be their Hilbert space tensor product. For unit vectors $h \in H$ and $k \in K$ we write $|h\rangle$ and $|k\rangle$ for the pure states in H and K represented by these vectors, and

$$|h\rangle|k\rangle := |h \otimes k\rangle$$

for the pure state represented by the unit vector $h \otimes k$ in $H \otimes K$.

Suppose now that orthonormal vectors $h_1, h_2 \in H$ and orthonormal vectors $k_1, k_2 \in K$ are given. Then the unit vectors $h_1 \otimes k_1$ and $h_2 \otimes k_2$ are orthogonal in $H \otimes K$. Hence, for scalars $\alpha_1, \alpha_2 \in \mathbb{C}$ satisfying $|\alpha_1|^2 + |\alpha_2|^2 = 1$, the superposition $\alpha_1 h_1 \otimes k_1 + \alpha_2 h_2 \otimes k_2$ defines a unit vector in $H \otimes K$. To this unit vector corresponds the pure state

$$\alpha_1|h_1\rangle|k_1\rangle + \alpha_2|h_2\rangle|k_2\rangle := |\alpha_1 h_1 \otimes k_1 + \alpha_2 h_2 \otimes k_2\rangle.$$

Unless $\alpha_1 = 0$ or $\alpha_2 = 0$, such states cannot be written in the form $|h\rangle|k\rangle$ and are called *entangled states*.

The partial trace (see Section 14.4) can be used to define states of subsystems starting from the state of a composite system. More concretely, suppose that $T \in \mathscr{L}_1(H \otimes K)$ is a positive trace class operator with unit trace. Then the operators $\operatorname{tr}_K(T)$ and $\operatorname{tr}_H(T)$ are positive trace class operators with unit trace in $\mathscr{L}_1(H)$ and $\mathscr{L}_1(K)$, respectively. If we think of T as describing the state of a system with Hilbert space $H \otimes K$, $\operatorname{tr}_K(T)$ and $\operatorname{tr}_H(T)$ can be thought of as describing the states of the two constituent subsystems with Hilbert spaces H and K, respectively. For example, by the result of Example 14.32, if the operator T corresponds to a unit vector $h \otimes k$ in $H \otimes K$, the states corresponding to $\operatorname{tr}_K(T)$ and $\operatorname{tr}_H(T)$ are the pure states $|h\rangle$ and $|k\rangle$, that is,

$$\operatorname{tr}_K(T) = |h\rangle\langle h|, \quad \operatorname{tr}_H(T) = |k\rangle\langle k|.$$

15.3 Positive Operator-Valued Measures

We discuss next a natural extension of the notion of an observable.

15.3.a Effects

As a warm-up we show:

Proposition 15.19. *Let $(X, \mathscr{X})$ be a measurable space. The closed convex hull in $B_{\mathrm{b}}(X)$ of the set of elementary observables $\{\mathbf{1}_B : B \in \mathscr{X}\}$ equals*

$$\mathscr{E}(X) := \{f \in B_{\mathrm{b}}(X) : 0 \leqslant f \leqslant 1 \text{ pointwise}\}.$$

The extreme points of $\mathscr{E}(X)$ are precisely the elementary observables $\mathbf{1}_B$, $B \in \mathscr{X}$.

Proof Denote by $E(X)$ the closed convex hull of the set of elementary observables.

The inclusion $E(X) \subseteq \mathscr{E}(X)$ is trivial. To prove the inclusion $\mathscr{E}(X) \subseteq E(X)$, let $f \in \mathscr{E}(X)$ be given. Given $\varepsilon > 0$, select a simple function $g = \sum_{j=1}^{k} c_j \mathbf{1}_{B_j}$ such that $\|f - g\|_\infty < \varepsilon$; this function may be chosen in such a way that the measurable sets B_j are disjoint and the coefficients satisfy $0 \leqslant c_j \leqslant 1$. After relabelling we may assume that $0 \leqslant c_1 \leqslant \cdots \leqslant c_k \leqslant 1$.

If $k = 1$, then $g = (1 - c_1)\mathbf{1}_\varnothing + c_1 \mathbf{1}_{B_1}$ belongs to $E(X)$. If $k \geqslant 2$ we set

$$L_0 := \varnothing \quad \text{and} \quad L_i := \bigcup_{j=i}^{k} B_j \quad (j = 1, \ldots, k)$$

and

$$\lambda_0 := 1 - c_k, \quad \lambda_1 := c_1, \quad \text{and} \quad \lambda_i := c_i - c_{i-1} \quad (i = 2, \ldots, k).$$

Then $0 \leqslant \lambda_i \leqslant 1$, $\sum_{i=0}^{k} \lambda_i = 1$, and

$$g = \sum_{j=1}^{k} c_j \mathbf{1}_{B_j} = \sum_{i=0}^{k} \lambda_i \mathbf{1}_{L_i}.$$

It follows that g belongs to $E(X)$. Since $\varepsilon > 0$ was arbitrary, this proves that $f \in E(X)$.

If $g \in \mathscr{E}(X)$ is an elementary observable and $g = \lambda f_0 + (1 - \lambda)f_1$ with $0 < \lambda < 1$ and $0 \leqslant f_j \leqslant 1$ for $j = 0, 1$, then $0 = g(\xi) = \lambda f_0(\xi) + (1 - \lambda)f_1(\xi)$ implies $f_0(\xi) = f_1(\xi) = 0$ and $1 = g(\xi') = \lambda f_0(\xi') + (1 - \lambda)f_1(\xi')$ implies $f_0(\xi') = f_1(\xi') = 1$, that is, $f_0 = f_1 = g$ pointwise. It follows that every elementary observable is an extreme point of $\mathscr{E}(X)$. If $g \in \mathscr{E}(X)$ is not an elementary observable, then the set $\{\varepsilon \leqslant g \leqslant 1 - \varepsilon\}$ is nonempty for sufficiently small $\varepsilon > 0$, and then it is easy to produce measurable $f_0 \neq f_1$ satisfying $0 \leqslant f_j \leqslant 1$ for $j = 0, 1$ and $g = \frac{1}{2}f_0 + \frac{1}{2}f_1$. It follows that g is not an extreme point of $\mathscr{E}(X)$. $\qquad \square$

The quantum mechanical counterpart of the elementary observables are the orthogonal projections. In analogy to the above result we now characterise the closed convex hull of $\mathscr{P}(H)$ in $\mathscr{L}(H)$. We write $S \leqslant T$ to express that $T - S$ is a positive operator.

Proposition 15.20. *The closed convex hull in $\mathscr{L}(H)$ of $\mathscr{P}(H)$ equals*

$$\mathscr{E}(H) := \{E \in \mathscr{L}(H) : 0 \leqslant E \leqslant I\}.$$

The extreme points of $\mathscr{E}(H)$ are precisely the orthogonal projections.

Proof Every element of the convex hull of $\mathscr{P}(H)$ belongs to $\mathscr{E}(H)$, and this passes on to the closed convex hull.

Since elements of $\mathscr{E}(H)$ are positive and hence selfadjoint with $\sigma(E) \subseteq [0,1]$, every $E \in \mathscr{E}(H)$ admits a representation as

$$E = \int_{[0,1]} \lambda \, dP(\lambda)$$

where P is the projection-valued measure of E supported in $\sigma(E)$.

Let

$$f_n := \sum_{j=0}^{2^n-1} \frac{j}{2^n} \mathbf{1}_{I_j},$$

where $I_0 := [0, \frac{1}{2^n}]$ and $I_j := (\frac{j}{2^n}, \frac{j+1}{2^n}]$ for $1 \leqslant j \leqslant 2^n - 1$. Set

$$E_n := \int_{[0,1]} f_n(\lambda) \, dP(\lambda) = \frac{1}{2^n} \left(P_{[0,1/2^n]} + \sum_{j=1}^{2^n-1} P_{(j/2^n,1]} \right).$$

Then E_n is in the convex hull of $\mathscr{P}(H)$ and

$$\lim_{n\to\infty} \|E - E_n\| \leqslant \lim_{n\to\infty} \sup_{\lambda \in [0,1]} |\lambda - f_n(\lambda)| = 0.$$

This proves that E is in the closed convex hull of $\mathscr{P}(H)$.

If $E \in \mathscr{E}(H)$ is an orthogonal projection and $E = \lambda E_0 + (1-\lambda)E_1$ with $0 < \lambda < 1$ and $0 \leqslant E_j \leqslant I$ for $j = 0, 1$, then for all $x \in \mathsf{N}(E)$ we have $\lambda(E_0 x|x) + (1-\lambda)(E_1 x|x) = 0$ with $(E_i x|x) \geqslant 0$ for $i = 0, 1$, and this is possible only if $(E_0 x|x) = (E_1 x|x) = 0$. For all norm one vectors $x \in \mathsf{R}(E)$ we have $(Ex|x) = (x|x) = 1$ and consequently $\lambda(E_0 x|x) + (1-\lambda)(E_1 x|x) = 1$. Since $(E_i x|x) \leqslant 1$ for $i = 0, 1$, this is possible only if $(E_0 x|x) = (E_1 x|x) = 1$. It follows that $E_0 = E_1 = 0$ on $\mathsf{N}(E)$ and $E_0 = E_1 = I$ on $\mathsf{R}(E)$, and therefore $E_0 = E_1 = E$. It follows that E is an extreme point of $\mathscr{E}(H)$.

If $E \in \mathscr{E}(H)$ is not an orthogonal projection, then the spectral theorem for bounded selfadjoint operators implies that the spectrum $\sigma(E)$ cannot be equal to $\{0,1\}$. Since $\sigma(E)$ is contained in $[0,1]$ it follows that $[\varepsilon, 1-\varepsilon] \cap \sigma(E)$ is nonempty for all sufficiently small $\varepsilon > 0$ and then, again by the spectral theorem, it is easy to produce operators $E_0 \neq E_1$ in $\mathscr{E}(H)$ such that $E = \frac{1}{2}E_0 + \frac{1}{2}E_1$. It follows that E is not an extreme point of $\mathscr{E}(H)$. $\qquad\square$

Definition 15.21 (Effects). An *effect* is an element of the set $\mathscr{E}(H)$.

Effects are selfadjoint, and it follows from Theorem 8.11 that a selfadjoint operator on H is an effect if and only if its spectrum is contained in the unit interval $[0,1]$. If T is an arbitrary nonzero positive operator, then for all $0 \leqslant c \leqslant \|T\|^{-1}$ the operator cT is an effect. Indeed, this is clear for $c = 0$, and if $c > 0$ the operator $c^{-1}I - T$ is positive since it is selfadjoint and has positive spectrum.

A mapping $v : \mathscr{E}(H) \to [0,1]$ is said to be *finitely additive* if

$$\sum_{n=1}^{N} v(E_n) = v(E)$$

whenever $E_1, \ldots, E_N, E \in \mathscr{E}(H)$ satisfy $E_1 + \cdots + E_N = E$.

Theorem 15.22 (Busch). *Every finitely additive mapping $v : \mathscr{E}(H) \to [0,1]$ satisfying $v(I) = 1$ restricts to an affine mapping $v : \mathscr{P}(H) \to [0,1]$ and hence defines a state.*

Proof By assumption we have $v(I) = 1$ and from $1 = v(I) = v(I+0) = v(I) + v(0) = 1 + v(0)$ it follows that $v(0) = 0$. By additivity, the restriction of v to $\mathscr{P}(H)$ to $[0,1]$ is affine. Now the result follows from Proposition 15.6.

$\square$

15.3.b Positive Operator-Valued Measures

The next definition generalises the notion of a projection-valued measure by replacing the role of orthogonal projections by effects.

Definition 15.23 (Positive operator-valued measures). A *positive operator-valued measure* (POVM) on a measurable space $(\Omega, \mathscr{F})$ is a mapping $Q : \mathscr{F} \to \mathscr{E}(H)$ that assigns to every set $F \in \mathscr{F}$ an effect $Q_F := Q(F) \in \mathscr{E}(H)$ with the following properties:

(i) $Q_{\Omega} = I$;
(ii) for all $x \in H$ the mapping

$$F \mapsto (Q_F x | x), \quad F \in \mathscr{F},$$

defines a measure Q_x on $(\Omega, \mathscr{F})$.

The measure defined by (ii) is denoted by Q_x. Thus, for all $F \in \mathscr{F}$ and $x \in H$, by definition we have

$$(Q_F x | x) = Q_x(F) = \int_{\Omega} \mathbf{1}_F \, dQ_x.$$

Note that

$$Q_x(\Omega) = (Q_{\Omega} x | x) = (x | x) = \|x\|^2.$$

This shows that the measures Q_x are finite.

Every projection-valued measure is a POVM. In the converse direction we have the following simple result.

Proposition 15.24. *A POVM $Q : \mathscr{F} \to \mathscr{E}(H)$ is a projection-valued measure if and only if $Q_F Q_{F'} = Q_{F \cap F'}$ for all $F, F' \in \mathscr{F}$.*

Proof The 'only if' part has already been established in Section 9.2. The 'if' part is evident from $Q_F^2 = Q_{F \cap F} = Q_F$, which shows that each Q_F is a projection. Since Q_F is also positive, it is an orthogonal projection. □

A POVM which is not projection-valued is sometimes called an *unsharp observable*. An example will be discussed in Section 15.3.d.

We have seen in Proposition 15.16 that if $P : \mathscr{F} \to \mathscr{P}(H)$ is a projection-valued measure, then for every state ϕ the mapping

$$F \mapsto \phi(P_F), \quad f \in \mathscr{F},$$

is probability measure on $(\Omega, \mathscr{F})$. This sets up an affine mapping from $\mathscr{S}(H)$ to the convex set $M_1^+(\Omega)$ of probability measures on $(\Omega, \mathscr{F})$; we recall that $\mathscr{S}(H)$ denotes the convex set of all positive trace class operators with unit trace on H. As we have seen in Proposition 15.14, this set is the closed convex hull of its extreme points, which are precisely the rank one projections of the form $h \bar{\otimes} h$ with $h \in H$ of norm one.

Inspection of this argument shows that it extends to POVMs. The following proposition shows that in the converse direction, every POVM arises in this way.

Theorem 15.25 (POVMs as unsharp observables). *Let $(\Omega, \mathscr{F})$ be a measurable space. If $\Phi : \mathscr{S}(H) \to M_1^+(\Omega)$ is an affine mapping, then there exists a unique POVM $Q : \mathscr{F} \to \mathscr{E}(H)$ such that for all $T \in \mathscr{S}(H)$ we have*

$$(\Phi(T))(F) = \mathrm{tr}(Q_F T), \quad F \in \mathscr{F}.$$

Proof The proof consists of two steps.

Step 1 – We claim that Φ extends to a bounded operator from $\mathscr{L}_1(H)$ into $M(\Omega)$. The proof of this claim is accomplished in three steps. First, we set $\Phi(0) := 0$ and, for an arbitrary nonzero positive operator $T \in \mathscr{L}_1(H)$,

$$\Phi(T) := \|T\|_1 \Phi(T/\|T\|_1),$$

where $\|T\|_1 = \mathrm{tr}(T) > 0$ since T is positive and nonzero. Note that for all $c \geqslant 0$ we have

$$\Phi(cT) = c\Phi(T).$$

The identity

$$S + T = (\|S\|_1 + \|T\|_1)\Big(\lambda \frac{S}{\|S\|_1} + (1 - \lambda)\frac{T}{\|T\|_1}\Big),$$

where $\lambda = \|S\|_1/(\|S\|_1 + \|T\|_1)$, implies that if $S, T \in \mathscr{L}_1(H)$ are positive, then

$$\Phi(S+T) = \Phi\Big((\|S\|_1 + \|T\|_1)\Big(\lambda \frac{S}{\|S\|_1} + (1-\lambda)\frac{T}{\|T\|_1}\Big)\Big)$$

$$= (\|S\|_1 + \|T\|_1)\Phi\Big(\lambda \frac{S}{\|S\|_1} + (1-\lambda)\frac{T}{\|T\|_1}\Big)$$

$$= (\|S\|_1 + \|T\|_1)\Big(\lambda\Phi\Big(\frac{S}{\|S\|_1}\Big) + (1-\lambda)\Phi\Big(\frac{T}{\|T\|_1}\Big)\Big)$$

$$= \|S\|_1\Phi\Big(\frac{S}{\|S\|_1}\Big) + \|T\|_1\Phi\Big(\frac{T}{\|T\|_1}\Big) = \Phi(S) + \Phi(T),$$

where we used the assumption that Φ is affine. Applying this to aS and bT with $a, b \geqslant 0$ we find that

$$\Phi(aS + bT) = \Phi(aS) + \Phi(bT) = a\Phi(S) + b\Phi(T).$$

Next, for an arbitrary selfadjoint $T \in \mathscr{L}_1(H)$ write $T = T_1 - T_2$ with T_1, T_2 positive operators in $\mathscr{L}_1(T)$. Such decompositions always exist; one could take for instance $T_1 = \frac{1}{2}(T + |T|)$ and $T_2 = T_1 - T$. We then set

$$\Phi(T) := \Phi(T_1) - \Phi(T_2).$$

To see that this is well defined, suppose that we also have $T = T_1' - T_2'$ with T_1', T_2' positive operators in $\mathscr{L}_1(T)$. Then $T_1 + T_2' = T_2 + T_1'$ and hence, by what we just proved,

$$(\Phi(T_1) - \Phi(T_2)) - (\Phi(T_1') - \Phi(T_2')) = \Phi(T_1 + T_2') - \Phi(T_2 + T_1') = 0.$$

As in the proof of Theorem 15.7 it is checked that Φ is real-linear.

Finally, for an arbitrary $T \in \mathscr{L}_1(T)$ we set

$$\Phi(T) := \Phi(A) + i\Phi(B),$$

where $A := \frac{1}{2}(T + T^\star)$ and $B := \frac{1}{2i}(T - T^\star)$ are the unique selfadjoint operators such that $T = A + iB$. As in the proof of Theorem 15.7 it is checked that Φ is linear.

Step 2 – We now turn to the proof of the theorem. Using the extension provided by Step 1, for every fixed $F \in \mathscr{F}$ the mapping $T \mapsto (\Phi(T))(F)$ defines a bounded functional on $\mathscr{L}_1(H)$ and therefore by Theorem 14.29 it defines a bounded operator $Q_F \in \mathscr{L}(H)$ such that

$$(\Phi(T))(F) = \mathrm{tr}(TQ_F), \quad T \in \mathscr{L}_1(H).$$

For all norm one vectors $h \in H$ we have

$$(Q_F h | h) = \mathrm{tr}((h \bar{\otimes} h) \circ Q_F) = (\Phi(h \bar{\otimes} h))(F) \in [0, 1],$$

which gives the operator inequality $0 \leqslant Q_F \leqslant I$, that is, we have $Q_F \in \mathscr{E}(H)$.

It is clear that $Q_\Omega = I$, and for every norm one vector $h \in H$ the measure

$$F \mapsto (Q_F h|h) = \Phi(h \bar\otimes h)(F), \quad F \in \mathscr{F},$$

is a probability measure. This proves that $Q : F \mapsto Q_F$ is a POVM.
Uniqueness is clear since $\mathrm{tr}(TQ_F) = 0$ for all $T \in \mathscr{L}_1(H)$ implies $Q_F = 0$. $\square$

Remark 15.26. The assumption that Φ should be affine is a reasonable one in the light of the following argument. Suppose we have two quantum mechanical systems at our disposal, represented by the operators T_1 and T_2 in $\mathscr{S}(H)$ describing their states. We use a classical coin to decide which state is going to be observed: if, with probability p, 'heads' comes up we observe the system corresponding to T_1; otherwise we observe the system corresponding to T_2. This experiment can be described as observing the state corresponding to the convex combination $pT_1 + (1 - p)T_2$. If Φ is the observable to be measured, we expect the probability distribution of the outcomes, $\Phi(pT_1 + (1 - p)T_2)$, to be given by $p\Phi(T_1) + (1 - p)\Phi(T_2)$.

POVMs admit a bounded functional calculus, but an important difference with the bounded functional calculus for projection-valued measures of Theorem 9.8 is that the calculus for POVMs fails to be multiplicative (see, however, (15.10) for a partial result on multiplicativity).

Proposition 15.27 (Bounded functional calculus for POVMs). *Let $Q : \mathscr{F} \to \mathscr{L}(H)$ be a POVM. There exists a unique linear mapping $\Psi : B_b(\Omega) \to \mathscr{L}(H)$ satisfying*

$$\Psi(\mathbf{1}_F) = Q_F, \quad F \in \mathscr{F},$$

and

$$\|\Psi(f)\| \leqslant \|f\|_\infty, \quad f \in B_b(\Omega).$$

It satisfies

$$\Psi(f)^\star = \Psi(\overline{f}), \quad f \in B_b(\Omega).$$

Proof For $x, y \in H$ consider the complex measure $Q_{x,y}$ defined by

$$Q_{x,y}(F) := (Q_F x|y), \quad F \in \mathscr{F}.$$

That this indeed defines a measure follows by a polarisation argument from the countable additivity of the measures Q_x, $x \in H$. For any measurable partition $\Omega = F_1 \cup \cdots \cup F_k$ we have, by the Cauchy–Schwarz inequality applied twice,

$$\sum_{j=1}^k |Q_{x,y}(F_j)| = \sum_{j=1}^k |(Q_{F_j} x|y)| \leqslant \sum_{j=1}^k (Q_{F_j} x|x)^{1/2} (Q_{F_j} y|y)^{1/2}$$

$$= \sum_{j=1}^k Q_x(F_j)^{1/2} Q_y(F_j)^{1/2}$$

$$\leqslant \Big(\sum_{j=1}^{k} Q_x(F_j) \Big)^{1/2} \Big(\sum_{j=1}^{k} Q_y(F_j) \Big)^{1/2}$$

$$= Q_x(\Omega)^{1/2} Q_y(\Omega)^{1/2} = \|x\| \|y\|,$$

from which it follows that $Q_{x,y}$ has finite variation $|Q_{x,y}|(\Omega) \leqslant \|x\| \|y\|$.
For $f \in B_b(\Omega)$ define

$$\mathfrak{a}_f(x,y) := \int_\Omega f \, dQ_{x,y}, \quad x,y \in H.$$

The form $\mathfrak{a}$ is sesquilinear and bounded and defines a bounded operator $\Psi(f)$ on H by
Proposition 9.15. It is clear that $\Psi(\mathbf{1}_F) = Q_F$ for all $F \in \mathscr{F}$ and

$$|(\Psi(f)x|y)| = \Big| \int_\Omega f \, dQ_{x,y} \Big| \leqslant \int_\Omega |f| \, d|Q_{x,y}| \leqslant \|f\|_\infty \|x\| \|y\|.$$

The identity $(\Psi(f))^\star = \Psi(\overline{f})$ is a consequence of $\overline{Q_{y,x}} = Q_{x,y}$, from which it follows
that

$$((\Psi(f))^\star x|y) = (x|\Psi(f)y) = \overline{(\Psi(f)y|x)} = \overline{\mathfrak{a}_f(y,x)}$$

$$= \overline{\int_\Omega f \, dQ_{y,x}} = \int_\Omega \overline{f} \, dQ_{x,y} = \mathfrak{a}_{\overline{f}}(x,y) = (\Psi(\overline{f})x|y).$$

Uniqueness is clear from the fact that $\Psi(\mathbf{1}_F) = Q_F$ and the simple functions are dense
in $B_b(\Omega)$. □

15.3.c Naimark's Theorem

If J is an isometry from H into another Hilbert space $\widetilde{H}$ and $\widetilde{P}$ is an orthogonal projection
in $\widetilde{H}$, then $J^\star \widetilde{P} J$ is an effect in H: for all $x \in H$ we have

$$0 \leqslant (\widetilde{P}Jx|Jx) = \|\widetilde{P}Jx\|^2 \leqslant \|x\|^2 = (x|x)$$

and therefore $0 \leqslant J^\star \widetilde{P} J \leqslant I$. This gives a method of producing POVMs from projection-
valued measures:

Proposition 15.28 (Compression). *Let J be an isometry from H into another Hilbert
space $\widetilde{H}$. If $\widetilde{P} : \mathscr{F} \to \mathscr{P}(\widetilde{H})$ is a projection-valued measure, then $Q := J^\star \widetilde{P} J : \mathscr{F} \to \mathscr{E}(H)$ is a POVM.*

Proof By what we just observed, Q maps sets $F \in \mathscr{F}$ to elements of $\mathscr{E}(H)$. It is clear
that $Q_\Omega = J^\star J = I$. To see that Q is a POVM, it remains to observe that for all $x \in H$
and $F \in \mathscr{F}$ we have

$$Q_x(F) = (Q_F x|x) = (\widetilde{P}_F Jx|Jx) = \widetilde{P}_{Jx}(F),$$

from which it follows that Q_x is a finite measure on $(\Omega, \mathscr{F})$. □

The main result of this section is *Naimark's theorem*, which asserts that, conversely, every POVM arises in this way.

Theorem 15.29 (Naimark). *Let $(\Omega, \mathscr{F})$ be a measurable space and let $Q : \mathscr{F} \to \mathscr{E}(H)$ be a POVM. There exists a Hilbert space $\widetilde{H}$, a projection-valued measure $\widetilde{P} : \mathscr{F} \to \mathscr{P}(\widetilde{H})$, and an isometry $J : H \to \widetilde{H}$ such that*

$$Q_F = J^* \widetilde{P}_F J, \quad F \in \mathscr{F}.$$

To motivate the proof of this theorem we consider first the special case $\Omega = \mathbb{T}$ and $\mathscr{F} = \mathscr{B}(\mathbb{T})$ its Borel σ-algebra. If $Q : \mathscr{B}(\mathbb{T}) \to \mathscr{E}(H)$ is a POVM, the operator

$$T := \int_{\mathbb{T}} z \, \mathrm{d}Q(z)$$

is a contraction on H by Proposition 15.27. By the Sz.-Nagy dilation theorem (Theorem 8.36) there exist a Hilbert space $\widetilde{H}$, a unitary operator $U \in \mathscr{L}(\widetilde{H})$, and an isometry $J : H \to \widetilde{H}$ such that

$$T^n = J^* U^n J, \quad n \in \mathbb{N}.$$

Using the spectral theorem for bounded normal operators, let $\widetilde{P} : \mathscr{B}(\mathbb{T}) \to \mathscr{P}(\widetilde{H})$ be its associated projection-valued measure. Then, by the properties of the bounded functional calculus of U,

$$T^n = J^* U^n J = J^* \left(\int_{\mathbb{T}} z^n \, \mathrm{d}\widetilde{P}(z) \right) J = \int_{\mathbb{T}} z^n \, \mathrm{d}Q(z), \quad n \in \mathbb{N}.$$

We claim that $\widetilde{P}$ has the desired properties. Indeed, for all $x \in H$ we have

$$\int_{\mathbb{T}} \lambda^n \, \mathrm{d}Q_x(\lambda) = (T^n x | x) = (U^n Jx | Jx) = \int_{\mathbb{T}} \lambda^n \, \mathrm{d}\widetilde{P}_{Jx}(\lambda).$$

This means that the nonnegative Fourier coefficients of the probability measures Q_x and $\widetilde{P}_{Jx}$ agree. Hence $Q_x = \widetilde{P}_{Jx}$ by Theorem 5.31 and the observation following it. But this implies, for all Borel subsets $B \in \mathscr{B}(\mathbb{T})$,

$$(Q_B x | x) = Q_x(B) = \widetilde{P}_{Jx}(B) = (\widetilde{P}_B Jx | Jx) = (J^* \widetilde{P}_B Jx | x).$$

This being true for all $x \in \widetilde{H}$, we conclude that $Q_B = J^* \widetilde{P}_B J$.

This argument cannot be extended to cover the general case, but it does suggest a proof strategy for Theorem 15.29, namely, to adapt the proof of the Sz.-Nagy dilation theorem.

Proof of Theorem 15.29   Let

$$S := \mathscr{F} \times H = \{(F, x) : F \in \mathscr{F}, x \in H\}$$

and consider the function $\widetilde{Q} : S \times S \to \mathbb{C}$ by

$$\widetilde{Q}(p, p') := (Q_{F \cap F'} x | x') \text{ for } p = (F, x), \ p' = (F', x').$$

We claim that this function is *positive definite* in the sense that for all finite choices of $p_1, \ldots, p_N \in S$ and $z_1, \ldots, z_N \in \mathbb{C}$ we have

$$\sum_{n,m=1}^{N} \widetilde{Q}(p_n, p_m) z_n \bar{z}_m \geqslant 0. \tag{15.8}$$

First assume that $p_n = (F_n, x_n)$ with the sets F_n disjoint. In that case,

$$\sum_{n,m=1}^{N} \widetilde{Q}(p_n, p_m) z_n \bar{z}_m \sum_{n,m=1}^{N} (Q_{F_n \cap F_m} x_n | x_m) z_n \bar{z}_m = \sum_{n=1}^{N} (Q_{F_n} z_n x_n | z_n x_n) \geqslant 0$$

by the positivity of the operators Q_{F_n}. For general $F_1, \ldots, F_N \in \mathscr{F}$ we write their union $\bigcup_{n=1}^{N} F_n$ as a union of 2^N disjoint sets C_σ in $\mathscr{F}$, indexed by the elements $\sigma \in 2^N$, the power set of $\{1, \ldots, N\}$, as follows. For $\sigma \in 2^N$ we set

$$C_\sigma := \bigcap_{n \in \sigma} F_n \setminus \bigcup_{m \notin \sigma} F_m.$$

It is straightforward to check that the sets C_σ are pairwise disjoint and that for all $n = 1, \ldots, N$ we have

$$F_n = \bigcup_{\substack{\sigma \in 2^N \\ n \in \sigma}} C_\sigma, \quad F_n \cap F_m = \bigcup_{\substack{\sigma \in 2^N \\ \{n,m\} \subseteq \sigma}} C_\sigma.$$

Then, by the additivity of Q and the positivity of the operators Q_{C_σ},

$$\begin{aligned}
\sum_{n,m=1}^{N} \widetilde{Q}(p_n, p_m) z_n \bar{z}_m &= \sum_{n,m=1}^{N} (Q_{F_n \cap F_m} x_n | x_m) z_n \bar{z}_m \\
&= \sum_{n,m=1}^{N} \Big(\sum_{\substack{\sigma \in 2^N \\ \{n,m\} \subseteq \sigma}} Q_{C_\sigma} x_n \Big| x_m \Big) z_n \bar{z}_m \\
&= \sum_{\sigma \in 2^N} \sum_{\substack{1 \leqslant n,m \leqslant N \\ \{n,m\} \subseteq \sigma}} (Q_{C_\sigma} x_n | x_m) z_n \bar{z}_m \\
&= \sum_{\sigma \in 2^N} \Big(Q_{C_\sigma} \sum_{\substack{1 \leqslant n \leqslant N \\ n \in \sigma}} z_n x_n \Big| \sum_{\substack{1 \leqslant m \leqslant N \\ m \in \sigma}} z_m x_m \Big) \geqslant 0.
\end{aligned}$$

This completes the proof of (15.8).

Let V be the vector space of finitely supported complex-valued functions defined on S. The elements of V are functions $h : S \to \mathbb{C}$ such that $f(p) = 0$ for all but at most finitely many pairs $p = (F, x) \in S$. The function $v \in V$ that maps $p \in S$ to the complex number z and is identically zero otherwise will be denoted as $v = z \mathbf{1}_p$. For two functions $v, v' \in V$, say $v = \sum_{n=1}^{N} z_n \mathbf{1}_{p_n}$ and $v' = \sum_{n=1}^{N} z_n' \mathbf{1}_{p_n}$ (allowing some of the z_n and z_n' to be zero) we define

$$(v|v') := \sum_{n,m=1}^{N} \widetilde{Q}(p_n, p_m) z_n \overline{z'_m}. \tag{15.9}$$

Arguing as in the proof of Theorem 8.34, this uniquely defines a sesquilinear mapping from $V \times V$ to $\mathbb{C}$ which satisfies $(v|v') = \overline{(v'|v)}$ for all $v, v' \in V$ and $(v|v) \geqslant 0$ for all $v \in V$, and

$$N = \{ v \in V : (v|v) = 0 \}$$

is a subspace of V. It follows that (15.9) induces an inner product on the vector space quotient V/N. Let $\widetilde{H}$ denote the Hilbert space completion $\widetilde{H}$ of V/N with respect to this inner product.

Consider elements in $\widetilde{H}$ of the form $p + N = (\Omega, x) + N$ and $p' + N = (\Omega, x') + N$ with $x, x' \in H$. Then

$$(p + N | p' + N)_{\widetilde{H}} = (Q_\Omega x | x') = (x | x').$$

Taking $x' = x$, in particular we may identify $x \in H$ isometrically with the element $p + N$ in $\widetilde{H}$, where $p = (\Omega, x)$. In this way we obtain an isometric embedding J of H into $\widetilde{H}$.

To ease up notation we use the notation $p = (F, x)$ for general elements of $\widetilde{H}$, rather than the more precise notation $p + N = (F, x) + N$. With this notation, $Jx = (\Omega, x)$.

The mapping $\pi : \widetilde{H} \to \widetilde{H}$ defined by

$$\pi(F, x) := (\Omega, Q_F x)$$

satisfies $\pi^2(F, x) = \pi(\Omega, Q_F x) = (\Omega, Q_\Omega Q_F x) = (\Omega, Q_F x) = \pi(F, x)$. We extend π by linearity and check that this results in a selfadjoint, hence orthogonal, projection in $\widetilde{H}$ whose range equals H. From

$$(\pi(F, x)|x')_{\widetilde{H}} = (Q_F x | x')_H = ((F, x)|(\Omega, x'))_{\widetilde{H}} = ((F, x)|Jx')_{\widetilde{H}}$$

it follows that $\pi = J^*$ as mappings from $\widetilde{H}$ to H.

Finally set

$$\widetilde{P}_F(F', x) := (F \cap F', x).$$

Again it is routine to check that $\widetilde{P}_F$ is an orthogonal projection in $\widetilde{H}$. By the properties (i) and (ii) in Definition 15.23 the mapping $\widetilde{P} : \mathscr{F} \to \mathscr{L}(\widetilde{H})$ is a projection valued measure. Finally, from

$$J^* \widetilde{P}_F Jx = \pi \widetilde{P}_F(\Omega, x) = \pi(\Omega \cap F, x) = \pi(F, x) = Q_F x$$

we conclude that $Q_F = J^* \widetilde{P}_F J$. □

The argument given after the statement of Theorem 5.31 works for general contractions:

Theorem 15.30. *For every contraction* $T \in \mathcal{L}(H)$, *there exists a unique POVM* $Q :$ $\mathcal{B}(\mathbb{T}) \to \mathcal{E}(H)$ *such that*

$$T^n = \frac{1}{2\pi} \int_{-\pi}^{\pi} e^{in\theta} \, dQ(\theta), \quad n \in \mathbb{N}.$$

If Q is a POVM with the above property, then T is unitary if and only if Q is a projection-valued measure.

Proof Existence follows follows by following the lines just mentioned: if U is a unitary dilation of T and P is its projection-valued measure, the compression Q of P has the required properties.

To prove uniqueness, suppose that for all $x \in H$ we have

$$(T^n x | x) = \int_{\mathbb{T}} z^n \, dQ_x(z) = \int_{\mathbb{T}} z^n \, d\widetilde{Q}_x(z), \quad n \in \mathbb{N},$$

where $\widetilde{Q}$ is another POVM on $\mathbb{T}$. This means that the nonnegative Fourier coefficients of the probability measures Q_x and $\widetilde{Q}_x$ agree. Now Theorem 5.31 (and the observation following it) can be applied to see that $Q_x = \widetilde{Q}_x$.

For the final statement it only remains to prove the 'only if' part. But this follows from uniqueness, for if $T := \int_{\mathbb{T}} z \, d\widetilde{Q}(z)$ is unitary for some POVM $\widetilde{Q}$ on $\mathbb{T}$, then we may also represent T in terms of its associated projection-valued measure P, that is, $T = \int_{\mathbb{T}} z \, dP(z)$. By uniqueness, $\widetilde{Q} = P$. □

If Q is as in Theorem 15.30, then for all trigonometric polynomials $f \in C(\mathbb{T})$ of the form $f(z) = \sum_{n=0}^{N} c_n z^n$ we have

$$\Psi(f) = \int_{\mathbb{T}} f \, dQ = f(T),$$

where $T = \int_{\mathbb{T}} z \, dQ(z)$. By the continuity of the bounded functional calculus with respect to the supremum norm, this identity persists for functions f in the *disc algebra* $A(\mathbb{D})$, the Banach space of all functions $f \in C(\mathbb{T})$ which have continuous extension to $\overline{\mathbb{D}}$ which is holomorphic on $\mathbb{D}$; these are precisely the functions belonging to the closure in $f \in C(\mathbb{T})$ of the trigonometric polynomials of the form just considered. An easy consequence is that the bounded functional calculus of a POVM Q on $\mathbb{T}$ is multiplicative on the disc algebra, that is,

$$\Psi(f)\Psi(g) = \Psi(fg), \quad f, g \in A(\mathbb{D}). \tag{15.10}$$

15.3.d The Phase/Number Pair

A convenient model for the *number operator*, the selfadjoint operator in Quantum Optics that corresponds to the observable of counting the number of photons, can be given on the Hardy space $H^2(\mathbb{D})$ considered in Section 7.3.d. Recall that $H^2(\mathbb{D})$ is the Hilbert space of all holomorphic functions on $\mathbb{D}$ of the form $f(z) = \sum_{n \in \mathbb{N}} c_n z^n$ with

$$\|f\|^2 := \sum_{n \in \mathbb{N}} |c_n|^2 < \infty.$$

As we have seen in that section, the mapping

$$\sum_{n \in \mathbb{N}} c_n z_n \mapsto \sum_{n \in \mathbb{N}} c_n e_n,$$

where $z_n(z) := z^n$ and $e_n(\theta) := e^{in\theta}$, sets up an isometry from $H^2(\mathbb{T})$ onto the closed subspace of $L^2(\mathbb{T})$ consisting of all functions whose negative Fourier coefficients vanish. This allows us to identify $H^2(\mathbb{D})$ with the range of the Riesz projection $\sum_{n \in \mathbb{Z}} c_n e_n \mapsto \sum_{n \in \mathbb{N}} c_n e_n$ on $L^2(\mathbb{T})$.

In $H^2(\mathbb{D})$ we consider the unbounded selfadjoint operator N with domain

$$D(N) = \left\{ f = \sum_{n \in \mathbb{N}} c_n z_n \in H^2(\mathbb{D}) : \sum_{n \in \mathbb{N}} n^2 |c_n|^2 < \infty \right\},$$

given by

$$N z_n = n z_n, \quad n \in \mathbb{N}.$$

The sequence $(z_n)_{n \in \mathbb{N}}$ is an orthonormal basis of eigenvectors for N and accordingly we have $\mathbb{N} \subseteq \sigma(N)$. On the other hand, if $\lambda \in \mathbb{C} \setminus \mathbb{N}$, then for every $f = \sum_{n \in \mathbb{N}} c_n z_n$ in $H^2(\mathbb{D})$ the equation $(\lambda - N)u = f$ is uniquely solved by $u = \sum_{n \in \mathbb{N}} \frac{c_n}{\lambda - n} z_n \in H^2(\mathbb{D})$. This implies that $\lambda \in \rho(N)$. We conclude that

$$\sigma(N) = \mathbb{N}. \tag{15.11}$$

(This is a special case of Proposition 10.32, but the proof could be simplified here because we have precise information about the domain of the operator.) We think of the eigenfunctions z_n on N as the pure states describing the n-photon states of an electromagnetic field. In this interpretation, (15.11) tells us that the number of photons observed is a nonnegative integer.

The projection-valued measure N associated with N is given by $N_{\{n\}} = \pi_n$, the orthogonal projection in $H^2(\mathbb{D})$ onto the one-dimensional subspace spanned by z_n, so that

$$(Nf|f) = \int_{\mathbb{N}} n \, dN_f(n) = \sum_{n \in \mathbb{N}} n(N_{\{n\}} f|f), \quad f \in D(N).$$

To define *phase* as a $\mathbb{T}$-valued unsharp observable in the sense of POVMs we proceed as follows. Let S be the 'left shift' on $H^2(\mathbb{D})$, that is,

$$S \sum_{n \in \mathbb{N}} c_n z_n := \sum_{n \in \mathbb{N}} c_{n+1} z_n.$$

In the language of Section 7.3.d, S is the Toeplitz operator T_ϕ with symbol $\phi(z) = z$.

Identifying $H^2(\mathbb{D})$ with the range of the Riesz projection in $L^2(\mathbb{T})$, a unitary dilation of S is given by the 'left shift' $\widetilde{S}$ on $L^2(\mathbb{T})$,

$$\widetilde{S} \sum_{n \in \mathbb{Z}} c_n e_n := \sum_{n \in \mathbb{Z}} c_{n+1} e_n,$$

with $e_n(\theta) = e^{in\theta}$ as before. The projection-valued measure $P : \mathscr{B}(\mathbb{T}) \to \mathscr{P}(L^2(\mathbb{T}))$ associated with $\widetilde{S}$ is easily checked to be given by

$$P_B f = \mathbf{1}_B f, \quad B \in \mathscr{B}(\mathbb{T}), \ f \in L^2(\mathbb{T}). \tag{15.12}$$

Its compression to $H^2(\mathbb{D})$ is a POVM $\Phi : \mathbb{T} \to \mathscr{E}(H^2(\mathbb{D}))$, which is called the *phase observable*. It satisfies

$$S^n = \int_{\mathbb{T}} z^n \, d\Phi(z), \quad n \in \mathbb{N}.$$

The covariance property expressed in the following theorem identifies the POVM Φ as the "complementary unsharp observable" to the number observable N. The notions of covariance and complementarity will be developed in more detail in Section 15.5.

Theorem 15.31 (Covariance of phase). *The phase observable Φ is covariant with respect to the unitary C_0-group generated by $-iN$, that is, for all Borel subsets $B \subseteq \mathbb{T}$ we have*

$$U(t)\Phi_B U^\star(t) = \Phi_{e^{it}B}, \quad t \in \mathbb{R},$$

where $e^{it}B = \{e^{it}z : z \in B\}$ is the rotation of B over t.

Proof Since the POVM Φ is the compression of the projection-valued measure P given by (15.12), for all $m, n \in \mathbb{N}$ we have

$$(\Phi_B U^\star(t)e_n|e_m) = (P_B J U^\star(t)e_n|Je_m) = e^{int}(\mathbf{1}_B e_n|e_m),$$

while at the same time, with $A = \{\theta \in (\pi, \pi] : e^{i\theta} \in B\}$,

$$(U^\star(t)\Phi_{e^{it}B}e_n|e_m) = (P_{e^{it}B}Je_n|JU(t)e_m) = e^{imt}(\mathbf{1}_{e^{it}B}e_n|e_m)$$

$$= e^{itm}\frac{1}{2\pi}\int_A e^{i(n-m)(\eta+t)} \, d\eta = e^{int}\frac{1}{2\pi}\int_A e^{i(n-m)\eta} \, d\eta = e^{int}(\mathbf{1}_B e_n|e_m).$$

Since the functions e_n, $n \in \mathbb{N}$, have dense span in $H^2(\mathbb{D})$, this completes the proof. $\square$

As will be explained in Section 15.6.e, N can be thought of as the Hamiltonian of the quantum harmonic oscillator. In a sense made precise in Problem 15.13(f), N and Φ are complementary in the sense of satisfying a Heisenberg-type commutation relation. For physical reasons, this means that Θ can be thought as a time variable. As such, it is a special case of a more general state-dependent notion of 'thermal time'. This idea will be revisited in the Notes to this chapter.

15.4 Hidden Variables

The one-to-one correspondence of Theorem 15.25 between the set of POVMs and the set of affine mappings from $\mathscr{S}(H)$ to $M_+^1(\Omega)$ is particularly satisfying from a philosophical point of view, as it characterises unsharp observables in an operational way: an unsharp observable is nothing but a rule of assigning probability distributions to states in such a way that convex combinations are respected. The rationale of this assumption has been discussed in Remark 15.26.

Thinking of unsharp observables as affine mappings from $\mathscr{S}(H)$ to $M_1^+(\Omega)$, analogously we can define classical unsharp observables as affine mappings from $M_1^+(X)$ to $M_1^+(\Omega)$, where $(X, \mathscr{X})$ is the state space of the classical system. Indeed, in Section 15.2 we have defined an observable as a measurable function from X to Ω, and such a function f induces an affine mapping from $M_1^+(X)$ to $M_1^+(\Omega)$ by sending μ to its image measure $f(\mu) = \mu \circ f^{-1}$. In this way, every classical observable defines a classical unsharp observable.

The following theorem shows that every family of quantum observables with values in a locally compact Hausdorff space admits a classical model, in the sense made precise in the formulation of the theorem. As before, we use the notation $|h\rangle$ for the pure state $h \bar\otimes h \in \mathscr{S}(H)$ with $h \in H$ of norm one. The theorem is phrased in terms of *countably generated* locally compact Hausdorff spaces. By definition, these are locally compact Hausdorff spaces whose topology generated by a countable family of open sets. On such a space Ω, using Urysohn functions (cf. Proposition 4.3) it is not hard to see that the indicator of every open set can be approximated pointwise by a non-increasing sequence of continuous functions $f_n \in C_0(\Omega)$; this fact will be used in the proof.

Theorem 15.32 (Hidden variables). *Let Ω be a locally compact Hausdorff space whose topology is countably generated, and suppose that $\Phi^{(i)} : \mathscr{S}(H) \to M_1^+(\Omega)$, $i \in I$, are the affine mappings associated with a family of unsharp quantum mechanical observables. Then there exists a locally compact Hausdorff space X and a family of affine maps $\phi^{(i)} : M_1^+(X) \to M_1^+(\Omega)$, $i \in I$, such that the following conditions hold:*

(1) *the elements of X are the equivalence classes of the pure states $|h\rangle$ modulo indiscernibility under $\Phi^{(i)}$, $i \in I$; here, two pure states $|h_1\rangle$ and $|h_2\rangle$ are said to be indiscernible under $\Phi^{(i)}$, $i \in I$, if*

$$\Phi^{(i)}|h_1\rangle = \Phi^{(i)}|h_2\rangle, \quad i \in I;$$

(2) *the quotient mapping sending a pure $|h\rangle$ state to its equivalence class $[h]$ in X is continuous;*
(3) *the classical unsharp observables $f^{(i)}$ are related to the unsharp observables $\Phi^{(i)}$ by*

$$\phi^{(i)}(\delta_{[h]}) = \Phi^{(i)}|h\rangle,$$

where $\delta_{[h]} \in M_1^+(X)$ *is the Dirac measure supported on* $[h] \in X$.

Proof We start by observing that each $\Phi^{(i)}$ induces a bounded operator $S^{(i)} : C_0(\Omega) \to \mathscr{L}(H)$ by the presciption

$$S^{(i)}(f) := \int_\Omega f \, dQ^{(i)},$$

where $Q^{(i)}$ is the POVM underlying $\Phi^{(i)}$ as in Theorem 15.25; the boundedness of this operator follows from Proposition 15.27.

We endow the set $\text{Extr}(\mathscr{S}(H))$ of pure states of $\mathscr{S}(H)$ with the coarsest topology τ such that all mappings $T \mapsto \int_\Omega f \, d\Phi^{(i)}(T)$ with $i \in I$ and $f \in C_0(\Omega)$ are continuous. As a subset of the closed unit ball of $\mathscr{L}_1(H)$, $\text{Extr}(\mathscr{S}(H))$ is relatively compact in the closed unit ball of $(\mathscr{L}_1(H))^{**} = (\mathscr{L}(H))^*$, using trace duality (Theorem 14.29) to identify the dual of $\mathscr{L}_1(H)$ isometrically with $\mathscr{L}(H)$. By the Banach–Alaoglu theorem, $\text{Extr}(\mathscr{S}(H))$ is relatively compact with respect to the weak* topology inherited from the closed unit ball of $(\mathscr{L}_1(H))^{**}$. We have

$$\int_\Omega f \, d\Phi^{(i)}(T) = \langle T, S^{(i)} f \rangle,$$

using the duality between $\mathscr{L}_1(H)$ and $\mathscr{L}(H)$ on the right-hand side. This identity implies that the topology τ is coarser than the weak* topology inherited from the closed unit ball of $(\mathscr{L}_1(H))^{**}$. As a result, the weak*-closure of $\text{Extr}(\mathscr{S}(H))$ is relatively τ-compact and therefore the topological space $(\text{Extr}(\mathscr{S}(H)), \tau)$ is locally compact.

As mentioned in the statement of the theorem, we define

$$X := \text{Extr}(\mathscr{S}(H))/\sim,$$

where $\sim$ is the equivalence relation of indiscernibility under $\Phi^{(i)}$, $i \in I$. We endow this space with the quotient topology $\tau/\sim =: \upsilon$, that is, we declare a subset of X to belong to υ if its pre-image under the quotient mapping $q : |h\rangle \mapsto q|h\rangle =: [h]$ belongs to τ. This topology renders the quotient mapping from $\text{Extr}(\mathscr{S}(H))$ to X continuous. As a result, the space X is a locally compact space with respect to υ. It is also Hausdorff, for if $x_1 \neq x_2$ in X, we have $x_1 = [h_1]$ and $x_2 = [h_2]$ with $|h_1\rangle \not\sim |h_2\rangle$ in $\text{Extr}(\mathscr{S}(H))$, so there is an $i \in I$ such that

$$\Phi^{(i)} |h_1\rangle \neq \Phi^{(i)} |h_2\rangle . \tag{15.13}$$

This means that $\Phi^{(i)} |h_1\rangle$ and $\Phi^{(i)} |h_2\rangle$ can be separated by open sets of the weak* topology of the closed unit ball of $(\mathscr{L}_1(H))^{**}$. We claim that they can actually be separated by open sets of τ. Indeed, suppose for a contradiction, that

$$\int_\Omega f \, d\Phi^{(i)} |h_1\rangle = \int_\Omega f \, d\Phi^{(i)} |h_2\rangle , \quad f \in C_0(\Omega).$$

This translates into

$$\int_\Omega f \, dQ_{h_1}^{(i)} = \int_\Omega f \, dQ_{h_2}^{(i)}, \quad f \in C_0(\Omega).$$

By the observation preceding the statement of the theorem, this implies that $Q_{h_1}^{(i)}(U) = Q_{h_2}^{(i)}(U)$ for all open sets $U \subseteq X$. Since these generate the topology of X, Dynkin's lemma E.4 then implies that $Q_{h_1}^{(i)} = Q_{h_2}^{(i)}$. It follows that for every $B \in \mathscr{B}(\Omega)$ we have $(Q_B^{(i)} h_1 | h_1) = (Q_B^{(i)} h_2 | h_2)$, and this in turn implies $\Phi^{(i)} |h_1\rangle (B) = \Phi^{(i)} |h_2\rangle (B)$. This being true for all $B \in \mathscr{B}(\Omega)$, we conclude that $\Phi^{(i)} |h_1\rangle = \Phi^{(i)} |h_2\rangle$, contradicting (15.13). This completes the proof that the topology υ is Hausdorff on X.

Observing that the singletons $\{[h]\}$ belong to $\mathscr{B}(X)$, the Dirac measures $\delta_{[h]}$ belong to $M_1^+(X)$. We extend the mapping

$$\delta_{[h]} \mapsto \Phi^{(i)} |h\rangle = (\Phi_i(\cdot)h | h)$$

to their convex hull in $M_1^+(X)$ by convexity:

$$\phi^{(i)} \left(\sum_{n=1}^N \lambda_n \delta_{[h_n]} \right) := \sum_{n=1}^N \lambda_n (\Phi_i(\cdot)h_n | h_n)$$

for scalars $0 \leqslant \lambda_n \leqslant 1$ such that $\sum_{n=1}^N \lambda_n = 1$. Clearly, each $\phi^{(i)}$ preserves convex combinations. The functions $\phi^{(i)}$ are continuous with respect to the weak* topologies of $M_1^+(X)$ and $M_1^+(\Omega)$. To see this, by identifying the elements of $M_1^+(X)$ as bounded functionals on $C_b(X)$, we observe that the mapping $\phi^{(i)}$ is the restriction of the adjoint of the bounded operator $R^{(i)}$ from $C_0(\Omega)$ to $C_b(X)$ given by

$$(R^{(i)}f)([h]) := \int_\Omega f \, dQ_h^{(i)},$$

where $Q^{(i)}$ is the POVM associated with $\Phi^{(i)}$ as in Theorem 15.25. Note that $R^{(i)}f$ is well defined pointwise as a function on X, for if $|h_1\rangle \sim |h_2\rangle$, then $\int_\Omega 1_B \, dQ_{h_1}^{(i)} = \int_\Omega 1_B \, dQ_{h_2}^{(i)}$ for all Borel sets B in Ω, and therefore $\int_\Omega f \, dQ_{h_1}^{(i)} = \int_\Omega f \, dQ_{h_2}^{(i)}$ by linearity and a limiting argument. Also note that $R^{(i)}f$ is continuous on (X, υ); this follows from the identity

$$(R^{(i)}f)([h]) = \int_\Omega f \, dQ_h^{(i)} = \int_\Omega f \, d\Phi^{(i)} |h\rangle$$

and the definition of υ as the quotient topology associated with τ.

Since the convex hull of the Dirac measures is dense in $M_1^+(X)$ with respect to the weak* topology inherited from $(C_b(X))^*$, these functions $\phi^{(i)}$ admit unique weak*-continuous extensions to all of $M_1^+(X)$, and these extensions again preserve convex combinations. $\qquad\square$

The set $\mathrm{Extr}(\mathscr{S}(H))$ can also be viewed as a subset of the image of closed unit ball of H under the quotient mapping $H \mapsto H/\mathbb{C}$ which identifies two vectors h and h' whenever $h = ch'$ for some $|c| = 1$, that is, when $|h\rangle = |h'\rangle$. As such, it is locally compact with respect from the quotient topology induced by the weak topology of H, by the reflexivity of H (see Example 4.57). This would only obscure the proof, however, as it hides the trace duality underlying the main idea of the proof.

It is of some interest to work through the details of this construction for the qubit. Accordingly let $H = \mathbb{C}^2$. As we have seen in Section 15.2.e, the set of extreme points of $\mathscr{S}(H)$ then corresponds to the Bloch sphere S^2 in $\mathbb{R}^3$. Under this correspondence the Bloch vector $\xi = (\sin\theta\cos\varphi, \sin\theta\sin\varphi, \cos\theta) \in S^2$ corresponds to the operator $T_\xi \in \mathscr{S}(\mathbb{C}^2)$ given in matrix form as

$$T_\xi = \frac{1}{2}\begin{pmatrix} 1+\cos\theta & e^{-i\varphi}\sin\theta \\ e^{i\varphi}\sin\theta & 1-\cos\theta \end{pmatrix}.$$

Let us take the Pauli matrices as the set of observables of interest:

$$\{\sigma_1, \sigma_2, \sigma_3\}.$$

Let P_j be the projection-valued measure associated with σ_j. For example, $(P_1)_{\{1\}}$ and $(P_1)_{\{-1\}}$ are the orthogonal projections onto the one-dimensional subspaces spanned by the eigenvectors corresponding to the eigenvalues 1 and -1 of $\sigma_1 = \begin{pmatrix} 0 & 1 \\ 1 & 0 \end{pmatrix}$:

$$(P_1)_{\{1\}} = \frac{1}{2}\begin{pmatrix} 1 & 1 \\ 1 & 1 \end{pmatrix}, \quad (P_1)_{\{-1\}} = \frac{1}{2}\begin{pmatrix} 1 & -1 \\ -1 & 1 \end{pmatrix}.$$

We view these as unsharp observables with values in $\Omega := \{\pm 1\}$. Since $\{P_1, P_2, P_3\}$ separates the points of S^2, the space X constructed in the proof of Theorem 15.32 can be identified with S^2. The corresponding family classical variables $\phi = \{\phi_1, \phi_2, \phi_3\}$ is given by the mappings $\phi_j : M_1^+(S^2) \to M_1^+(\{\pm 1\})$,

$$(\phi_1(\delta_\xi))(\{1\}) = (\Phi(T_\xi))(\{1\}) = \mathrm{tr}((P_1)_{\{1\}}T_\xi)$$
$$= \mathrm{tr}\left(\frac{1}{2}\begin{pmatrix} 1 & 1 \\ 1 & 1 \end{pmatrix} \circ \frac{1}{2}\begin{pmatrix} 1+\cos\theta & e^{-i\varphi}\sin\theta \\ e^{i\varphi}\sin\theta & 1-\cos\theta \end{pmatrix}\right)$$
$$= \frac{1}{4}\mathrm{tr}\begin{pmatrix} 1+\cos\theta+e^{i\varphi}\sin\theta & 1-\cos\theta+e^{-i\varphi}\sin\theta \\ 1+\cos\theta+e^{i\varphi}\sin\theta & 1-\cos\theta+e^{-i\varphi}\sin\theta \end{pmatrix}$$
$$= \frac{1}{4}(1+\cos\theta+e^{i\varphi}\sin\theta+1-\cos\theta+e^{-i\varphi}\sin\theta)$$
$$= \frac{1}{2}(1+\cos\varphi\sin\theta) = \frac{1}{2}(1+\xi_1)$$

and likewise

$$(\phi_1(\delta_\xi))(\{-1\}) = \frac{1}{2}(1 - \xi_1).$$

Similar computations give

$$(\phi_2(\delta_\xi))(\{\pm 1\}) = \frac{1}{2}(1 \pm \xi_2), \quad (\phi_3(\delta_\xi))(\{\pm 1\}) = \frac{1}{2}(1 \pm \xi_3).$$

By considering convex combinations of Dirac measures and a limiting argument, it follows that the classical unsharp observables $\phi_j : M_1^+(S^2) \to M_1^+(\{\pm 1\})$ are given by

$$d\phi_j(\mu) = \left(1 + p \int_{S^2} \xi_j \, d\mu(\xi)\right) dp, \quad j = 1, 2, 3,$$

where dp is the probability measure on $\{\pm 1\}$ giving each point mass $\frac{1}{2}$.

15.5 Symmetries

In order to motivate our definition of a *symmetry* we introduce some notation and terminology. The *adjoint* of a conjugate-linear mapping $T : H \to H$ (that is, a mapping satisfying $T(x+y) = T(x) + T(y)$ and $T(cx) = \bar{c}x$) is the unique conjugate-linear mapping $T^\star : H \to H$ defined by

$$(x|T^\star y) = (Tx|y), \quad x, y \in H.$$

A mapping $T : H \to H$ is called *antiunitary* if it is conjugate-linear and satisfies $TT^\star = T^\star T = I$.

It is straightforward to check that if $U : H \to H$ is unitary or antiunitary, then the mapping

$$\mathscr{U}(T) := UTU^\star$$

is well defined as a mapping from $\mathscr{S}(H)$ to $\mathscr{S}(H)$ and satisfies

$$\mathrm{tr}(\mathscr{U}(T_1)\mathscr{U}(T_2)) = \mathrm{tr}(T_1 T_2), \quad T_1, T_2 \in \mathscr{S}(H). \tag{15.14}$$

Here, as before, $\mathscr{S}(H)$ denotes the set of all positive trace class operators on H with unit trace. As we have seen, the extreme points of this set are precisely the rank one projections $h \bar{\otimes} h$ with $h \in H$ of norm one. The physical intuition of (15.14) is that $\mathscr{U}$ *preserves transition probabilities* between pure states. Indeed, if $|g\rangle$ and $|h\rangle$ are pure states, then

Eugene Wigner, 1902–1995

$$\mathrm{tr}((g \bar{\otimes} g) \circ (h \bar{\otimes} h)) = ((g \bar{\otimes} g)h|h) = |(g|h)|^2$$

is the expected value of the observable $g \bar{\otimes} g$ in state $|h\rangle$. In Physics parlance, this is the probability of "finding a system with state $|h\rangle$ in state $|g\rangle$ when measuring it against an orthonormal basis containing g", that is, the transition probability between $|h\rangle$ and $|g\rangle$.

A remarkable theorem due to Wigner provides a converse to (15.14):

Theorem 15.33 (Wigner). *If $\mathscr{U} : \mathscr{S}(H) \to \mathscr{S}(H)$ is a bijection with the property that*

$$\mathrm{tr}(\mathscr{U}(T_1)\mathscr{U}(T_2)) = \mathrm{tr}(T_1 T_2), \quad T_1, T_2 \in \mathscr{S}(H),$$

there exists a mapping $U : H \to H$ which is either unitary or antiunitary such that

$$\mathscr{U}(T) = UTU^\star, \quad T \in \mathscr{S}(H).$$

This mapping is unique up to a complex scalar of modulus one.

We sketch a proof of the theorem only for the case of the qubit, that is, for $H = \mathbb{C}^2$, and refer to the Notes for some missing details and references to the general case.

Sketch of the proof of Theorem 15.33 for the qubit We begin by recalling (15.4) and (15.7), which state that if we write a pure state $|h\rangle$ as $\cos(\theta/2)|0\rangle + e^{i\varphi}\sin(\theta/2)|1\rangle$ with $0 \leqslant \theta \leqslant \pi$ and $0 \leqslant \varphi < 2\pi$, then the rank one projection $h \bar{\otimes} h$ in $\mathbb{C}^2$ onto the span of h is given as a matrix by

$$h \bar{\otimes} h = \frac{1}{2} \begin{pmatrix} 1 + \cos\theta & e^{-i\varphi}\sin\theta \\ e^{i\varphi}\sin\theta & 1 - \cos\theta \end{pmatrix}.$$

By elementary computation,

$$\mathrm{tr}((h \bar{\otimes} h)(h' \bar{\otimes} h')) = |(h|h')|^2 = \frac{1}{2}(1 + x_h \cdot x_{h'}),$$

where, as in (15.5),

$$x_h = (\sin\theta\cos\varphi, \sin\theta\sin\varphi, \cos\theta)$$

is the Bloch vector of h. Under the bijective correspondence $h \bar{\otimes} h \leftrightarrow x_h$ between the elements of $\mathscr{S}(\mathbb{C}^2)$ and the points of the unit sphere S^2 of $\mathbb{R}^3$, the assumption of the theorem implies that $\mathscr{U}$ induces a mapping $\mathscr{R} : S^2 \to S^2$ satisfying $\frac{1}{2}(1 + \mathscr{R}x_h \cdot \mathscr{R}x_{h'}) = \frac{1}{2}(1 + x_h \cdot x_{h'})$, that is,

$$\mathscr{R}x_h \cdot \mathscr{R}x_{h'} = x_h \cdot x_{h'}.$$

This identity implies that the 3×3 matrix R defined by

$$R_{ij} = \mathscr{R}u_i \cdot u_j, \quad i, j \in \{1, 2, 3\},$$

with u_1, u_2, u_3 the standard unit vectors of $\mathbb{R}^3$, is orthogonal. Now we use the algebraic fact, taken for granted here, that for every orthogonal 3×3 matrix R with real coefficients there exists a mapping $U : \mathbb{C}^2 \to \mathbb{C}^2$ which is either unitary or antiunitary, and which is unique up to a complex scalar of modulus one, such that

$$U(x \cdot \sigma)U^\star = (Rx) \cdot \sigma, \quad x \in \mathbb{R}^3,$$

where $x \cdot \sigma := x_1 \sigma_1 + x_2 \sigma_2 + x_3 \sigma_3$, where $x \in \mathbb{R}^3$ and $\sigma_1, \sigma_2, \sigma_3$ are the Pauli matrices. The mapping U has the required properties. $\qquad\square$

Informally speaking, Wigner's theorem tells us that symmetries $\mathscr{U}$ of quantum mechanical systems are given by operators U acting on the underlying Hilbert space that are either unitary or antiunitary. In practice one is primarily interested in one-parameter groups of symmetries indexed by time. Suppose $(\mathscr{U}(t))_{t \in \mathbb{R}}$ is such a group. By the uniqueness part of Wigner's theorem, for all $s, t \in \mathbb{R}$ the identity $\mathscr{U}(t)\mathscr{U}(s) = \mathscr{U}(t+s)$ implies the existence of a scalar $c(t,s)$ of modulus one such that the corresponding (anti)unitary operators satisfy

$$U(t)U(s) = c(t,s)U(t+s).$$

From the associative law $U(t)(U(s)U(r)) = (U(t)U(s))U(r)$ we obtain the *cocycle identity*

$$c(s,r)c(t,s+r) = c(t,s)c(t+s,r).$$

In this situation, a theorem of Bargmann implies that there exists function d, taking values in the scalars of modulus one, such that

$$c(t,s) = \frac{d(t)d(s)}{d(t+s)}$$

and the operators $V(t) := d(t)^{-1}U(t)$ are unitary. They satisfy

$$(\mathscr{U}(t))(T) = V(t)^\star TV(t), \quad V(t)V(s) = V(t+s).$$

The unitary group $(V(t))_{t \in \mathbb{R}}$ can be shown to be strongly continuous. Hence, by Stone's theorem, it follows that there exists a selfadjoint operator $\mathscr{H}$, the *Hamiltonian* associated with the family $\mathscr{U}$, such that $V(t) = e^{it\mathscr{H}}$ for $t \in \mathbb{R}$. The action of the unitary C_0-group $(e^{it\mathscr{H}})_{t \in \mathbb{R}}$ on pure states is given by $\mathscr{U}(t)(h \bar\otimes h) = V(t)h \bar\otimes V(t)h$. The equation

$$\frac{\mathrm{d}}{\mathrm{d}t}V(t)h = i\mathscr{H}V(t)$$

is an abstract version of the Schrödinger equation (13.33) (which corresponds to the special case $H = L^2(\mathbb{R}^d, m)$ and $\mathscr{H} = -\Delta + \text{potential}$). These considerations motivate the following definition.

Definition 15.34 (Symmetry, of a Hilbert space). A *symmetry* of H is a unitary operator on H.

For later use we also introduce the following classical counterpart of this definition.

Definition 15.35 (Symmetry, of a measure space). A *symmetry* of the measure space $(\Omega, \mathscr{F}, \mu)$ is a measurable bijective mapping $g : \Omega \to \Omega$ with measurable inverse that leaves μ invariant, that is,

$$(g(\mu))(F) := \mu(g^{-1}(F)) = \mu(F), \quad F \in \mathscr{F}.$$

If g is a symmetry of $(\Omega, \mathscr{F}, \mu)$, then for all $F \in \mathscr{F}$ the set $g(F)$ is measurable and $\mu(F) = \mu(g^{-1}(g(F))) = \mu(g(F))$, that is, μ is also invariant under g^{-1}.

15.5.a Covariance

Definition 15.36 (Conservation and covariance). Let $(\Omega, \mathscr{F}, \mu)$ be a measure space and let U be a symmetry of a Hilbert space H.

(i) An observable $P : \mathscr{F} \to \mathscr{P}(H)$ is said to be *conserved* under U if $U P_F U^\star = P_F$ for all $F \in \mathscr{F}$.

(ii) An observable $P : \mathscr{F} \to \mathscr{P}(H)$ is said to be *covariant* under the pair (g, U), where g is a symmetry of $(\Omega, \mathscr{F}, \mu)$, if $U P_F U^\star = P_{g(F)}$ for all $F \in \mathscr{F}$, that is, if the following diagram commutes:

$$
\begin{array}{ccc}
\mathscr{F} & \xrightarrow{\ F \mapsto g(F)\ } & \mathscr{F} \\
\big\downarrow{\scriptstyle P} & & \big\downarrow{\scriptstyle P} \\
\mathscr{P}(H) & \xrightarrow{\ P_F \mapsto U P_F U^\star\ } & \mathscr{P}(H)
\end{array}
$$

As we will see shortly, position is covariant with respect to translation (an object at position x appears at position $x - x'$ if the origin is translated over x') and momentum is covariant with respect to boosts (cf. Section 15.5.a; an object with momentum p appears with momentum $p - p'$ if a boost of size p' is applied, that is, if the origin 'in momentum space' is translated over p').

Locally Compact Abelian Groups An interesting special case arises when observables take values in a locally compact abelian (LCA) group G. Our treatment borrows some results from the theory of LCA groups that will not be proved here. For the special cases $\mathbb{R}^d$ and $\mathbb{T}$ the presentation is self-contained, as all missing details can be filled in with the help of the results of Chapter 5. A fuller treatment of symmetries should also cover the case of (noncommutative) Lie groups such as $SO(3)$ and $SU(2)$, but this would take us too far afield.

Every LCA group G admits a *Haar measure*, that is, a Borel measure μ such that $\mu(B) = \mu(g^{-1}(B))$ for all $B \in \mathscr{B}(G)$ and $g \in G$. This measure is unique up to a scalar multiple. With respect to a Haar measure μ, every $g \in G$ induces a symmetry on G by

$$g : g' \mapsto gg', \quad g' \in G.$$

This induces a symmetry U_g on $L^2(G) := L^2(G, \mu)$ given by

$$U_g f = f \circ g^{-1}.$$

We refer to U_g as the *translation over g*. We have

$$U(g_1)U(g_2)f = (f \circ g_2^{-1}) \circ g_1^{-1} = f \circ (g_1 g_2)^{-1} = U(g_1 g_2)f,$$

so $U(g_1)U(g_2) = U(g_1 g_2)$. This means that the mapping $U : G \to \mathscr{L}(L^2(G))$, $g \mapsto U_g$, is multiplicative and hence defines a unitary representation. It is easily checked that this representation is strongly continuous.

A *character* of G is a continuous group homomorphism $\gamma : G \to \mathbb{T}$. The set Γ of all characters of G is called the *Pontryagin dual of G*. It has the structure of a locally compact abelian group in a natural way by endowing it with the weak* topology inherited from $L^\infty(G)$ (local compactness being a consequence of the fact that $\Gamma \cup \{0\}$ is a weak* closed subset of the closed unit ball $\overline{B}_{L^\infty(G)}$ which is weak* compact by the Banach–Alaoglu theorem). It follows that Γ carries a Haar measure which is again unique up to a normalisation. Every $g \in G$ defines a character $\gamma \mapsto \gamma(g)$ on Γ, and the *Pontryagin duality theorem* asserts that these are the only ones and the Pontryagin dual of Γ equals G both as a set and as an LCA group.

Every character $\gamma \in \Gamma$ induces a symmetry on $L^2(G)$ via

$$V_\gamma f(g') = \gamma(g')f(g'), \quad g' \in G, \ f \in L^2(G).$$

We refer to V_γ as the *boost over γ*. In the language of Chapter 5, V_γ is the pointwise multiplier with γ.

It is immediate from the above definitions that the so-called *Weyl commutation relation* holds:

Proposition 15.37 (Weyl commutation relation). *For all $g \in G$ and $\gamma \in \Gamma$ we have*

$$V_\gamma U_g = \gamma(g) U_g V_\gamma.$$

Proof For $f \in L^2(G)$ and $g' \in G$ we have

$$V_\gamma U_g f(g') = \gamma(g') U_g f(g') = \gamma(g')f(g^{-1}g')$$

and

$$\gamma(g) U_g V_\gamma f(g') = \gamma(g) V_\gamma f(g^{-1}g') = \gamma(g)\gamma(g^{-1}g')f(g^{-1}g') = \gamma(g')f(g^{-1}g').$$

$\square$

We now turn to the definition of a pair of canonical observables that can be associated with LCA groups. For the statement of the second part of the theorem we need the *Plancherel theorem for LCA groups*, which asserts that the Fourier–Plancherel transform $\mathscr{F} : L^1(G) \to L^\infty(\Gamma)$ defined by

$$\mathscr{F}f(\gamma) := \int_G f(g)\overline{\gamma(g)}\,\mathrm{d}\mu(g), \quad \gamma \in \Gamma,$$

where μ is a Haar measure on G, maps $L^1(G) \cap L^2(G)$ into $L^2(\Gamma)$ and there is a unique normalisation of the Haar measure of Γ such that $\mathscr{F}$ extends to a unitary operator from $L^2(G)$ onto $L^2(\Gamma)$. For later reference we note that $\overline{\gamma(g)} = \gamma(g^{-1})$ and hence

$$\mathscr{F}U_g f(\gamma) = \int_G f(g^{-1}g')\overline{\gamma(g')}\,\mathrm{d}\mu(g') = \int_G f(g')\overline{\gamma(gg')}\,\mathrm{d}\mu(g') = \gamma(g^{-1})\mathscr{F}f(\gamma),$$
$$(15.15)$$

where the second identity follows by substitution and invariance of μ.

Theorem 15.38 (Position and momentum). *Let G be an LCA group, let Γ be its Pontryagin dual, and let $\mathscr{B}(G)$ and $\mathscr{B}(\Gamma)$ be their Borel σ-algebras. Then:*

(1) *there exists a unique G-valued observable $X : \mathscr{B}(G) \to \mathscr{P}(L^2(G))$ such that for all $\gamma \in \Gamma$ we have*

$$V_\gamma = \int_G \gamma\,\mathrm{d}X,$$

and it is given by $X_B f = \mathbf{1}_B f$ for $f \in L^2(G)$ and $B \in \mathscr{B}(G)$;

(2) *there exists a unique Γ-valued observable $\Xi : \mathscr{B}(\Gamma) \to \mathscr{P}(L^2(G))$ such that for all $g \in G$ we have*

$$U_g = \int_\Gamma g\,\mathrm{d}\Xi,$$

and it is given by $\Xi_B f = \mathscr{F}^{-1}\mathbf{1}_B \mathscr{F}f$ for $f \in L^2(G)$ and $B \in \mathscr{B}(\Gamma)$.

Proof (1): Consider the G-valued observable $X : \mathscr{B}(G) \to \mathscr{L}(L^2(G))$ defined by

$$X_B f := \mathbf{1}_B f, \quad B \in \mathscr{B}(G), \; f \in L^2(G).$$

For Borel sets $B \subseteq G$ the operator $T_{\mathbf{1}_B} := \int_G \mathbf{1}_B \,\mathrm{d}X$ on $L^2(G)$ is the pointwise multiplier $T_{\mathbf{1}_B} f = \mathbf{1}_B f$. By linearity, for μ-simple functions ϕ on G the operator $T_\phi := \int_G \phi\,\mathrm{d}X$ on $L^2(G)$ is the pointwise multiplier $T_\phi f = \phi f$. By approximation, the operator $T_\gamma := \int_G \gamma\,\mathrm{d}X$ is the pointwise multiplier $T_\gamma f = \gamma f = V_\gamma f$. This proves existence.

We only sketch the proof of uniqueness; for the special cases $G = \mathbb{R}^d$ and $G = \mathbb{T}$ the missing details are easily filled in by using the properties of the Fourier transform proved in Chapter 5. If $\widetilde{X}$ is an observable satisfying $V_\gamma = \int_G \gamma\,\mathrm{d}\widetilde{X}$, then for all $f \in L^2(G)$ we have $\int_G \gamma\,\mathrm{d}\widetilde{X}_f = \int_G \gamma\,\mathrm{d}X_f$. This can be interpreted as saying that the finite Borel measures

$\widetilde{X}_f$ and X_f have the same Fourier transforms. Their equality therefore follows from the injectivity of the Fourier transform as a mapping from $M(G)$ to $L^\infty(\Gamma)$.

(2): We begin with the existence part. Applying the construction of the preceding part to Γ we obtain the dual position operator $X^\Gamma : \mathscr{B}(\Gamma) \to \mathscr{L}(L^2(\Gamma))$ given by

$$X_B^\Gamma \phi := \mathbf{1}_B \phi, \quad B \in \mathscr{B}(\Gamma), \ \phi \in L^2(\Gamma).$$

By conjugation with the Fourier–Plancherel transform it induces an observable $\Xi : \mathscr{B}(\Gamma) \to \mathscr{L}(L^2(G))$:

$$\Xi_B f := \mathscr{F}^{-1} X_B^\Gamma \mathscr{F} f, \quad B \in \mathscr{B}(\Gamma), \ f \in L^2(G).$$

This observable has the desired property. Uniqueness is proved in the same way as in part (1). □

Definition 15.39 (Position and momentum). The G-valued observable X and the Γ-valued observable Ξ of the theorem are called the *position* and *momentum* observables of G, respectively.

The special cases where $G = \mathbb{R}^d$ and $G = \mathbb{T}$ will be discussed in Sections 15.5.b and 15.5.c, respectively.

Proposition 15.40. *Let G be an LCA group and let X and Ξ be the position and momentum observables of G.*

(1) *X is covariant with respect to every (g, U_g) and conserved under every V_γ,*

$$U_g X_B U_g^\star = X_{gB}, \quad V_\gamma X_B V_\gamma^\star = X_B.$$

(2) *Ξ is conserved under every U_g and covariant with respect to every (γ, V_γ),*

$$U_g \Xi_B U_g^\star = \Xi_B, \quad V_\gamma \Xi_B V_\gamma^\star = \Xi_{\gamma B}.$$

Proof (1): For all $g, g' \in G$, $f \in L^2(G)$, and $B \in \mathscr{B}(G)$ we have

$$U_g X_B U_g^\star f(g') = [\mathbf{1}_B U_{g^{-1}} f](g^{-1} g') = \mathbf{1}_B(g^{-1} g') f(g') = X_{gB} f(g'),$$

proving covariance with respect to (g, U_g). Conservation under V_γ is even simpler:

$$V_\gamma X_B V_\gamma^\star f = \gamma(\cdot) \mathbf{1}_B(\cdot) \gamma^{-1}(\cdot) f(\cdot) = \mathbf{1}_B f = X_B f.$$

The proof of (2) is entirely similar. □

Remark 15.41 (Complementarity). In the Physics literature, the 'duality' between the position and momentum observables is referred to as *complementarity*. As we will see in the next two sections, this captures the complementarity of the position and momentum observables in $\mathbb{R}^d$ as well as that of the angle and angular momentum observables in $\mathbb{T}$.

15.5.b The Case $G = \mathbb{R}^d$: Position and Momentum

We now specialise to $G = \mathbb{R}^d$ with normalised Lebesgue measure $dm(x) = (2\pi)^{-d/2} dx$ as the Haar measure. Every character $\gamma : \mathbb{R}^d \to \mathbb{T}$ is of the form

$$\gamma(x) = e_\xi(x) := e^{ix \cdot \xi}$$

for some $\xi \in \mathbb{R}^d$. Under the identification $\gamma \leftrightarrow e_\xi$ we have $\Gamma = \mathbb{R}^d$, the Haar measure being the normalised Lebesgue measure m. To distinguish $G = \mathbb{R}^d$ from its dual $\Gamma = \mathbb{R}^d$ we use roman letters for elements of G and greek letters for elements of its dual Γ.

For every $y \in \mathbb{R}^d$, the translation $y : x \mapsto x + y$ is a symmetry of $\mathbb{R}^d$. The induced symmetry U_y on $L^2(\mathbb{R}^d, m)$ is right translation over y:

$$U_y f(x) = f(x - y), \quad x \in \mathbb{R}^d, \ f \in L^2(\mathbb{R}^d, m).$$

For every $\xi \in \mathbb{R}^d$ the boost V_ξ on $L^2(\mathbb{R}^d, m)$ is given by

$$V_\xi f(x) = e^{ix \cdot \xi} f(x), \quad x \in \mathbb{R}^d, \ f \in L^2(\mathbb{R}^d, m).$$

The Weyl commutation relation takes the form

$$V_\xi U_x = e^{ix \cdot \xi} U_x V_\xi, \quad x, \xi \in \mathbb{R}^d.$$

The position observable $X : \mathscr{B}(\mathbb{R}^d) \to \mathscr{P}(L^2(\mathbb{R}^d, m))$ and momentum observable $\Xi : \mathscr{B}(\mathbb{R}^d) \to \mathscr{P}(L^2(\mathbb{R}^d, m))$ of Theorem 15.42 can be described as selfadjoint operators as follows. For $1 \leqslant j \leqslant d$ we define $X_j : \mathscr{B}(\mathbb{R}) \to \mathscr{P}(L^2(\mathbb{R}^d, m))$ by

$$(X_j)_B f := \mathbf{1}_{\mathbb{R} \times \cdots \times \mathbb{R} \times B \times \mathbb{R} \cdots \times \mathbb{R}} f, \quad f \in L^2(\mathbb{R}^d, m),$$

with the Borel set $B \subseteq \mathbb{R}$ at the jth position. This projection-valued measure is interpreted as giving the jth position coordinate. We will prove that the selfadjoint operator A_j in $L^2(\mathbb{R}^d, m)$ associated with X_j equals $\widehat{x}_j$, where

$$\widehat{x}_j f(x) := x_j f(x), \quad x \in \mathbb{R}^d, \tag{15.15a}$$

for $f \in D(\widehat{x}_j) := \{f \in L^2(\mathbb{R}^d) : x \mapsto x_j f(x) \in L^2(\mathbb{R}^d, m)\}$. Indeed, for Borel sets $B \subseteq \mathbb{R}$ and $f \in L^2(\mathbb{R}^d)$ we have

$$\int_\mathbb{R} \mathbf{1}_B \, d(X_j)_f = ((X_j)_B f | f) = (\mathbf{1}_{\mathbb{R} \times \cdots \times \mathbb{R} \times B \times \mathbb{R} \cdots \times \mathbb{R}} f | f) = \int_{\mathbb{R}^d} \mathbf{1}_B(x_j) |f(x)|^2 \, dm(x).$$

By linearity and a limiting argument, for $f \in D(\widehat{x}_j)$ this implies $f \in D(A_j)$, where

$$D(A_j) = \left\{ f \in L^2(\mathbb{R}^d) : \int_\mathbb{R} |\lambda|^2 \, d(X_j)_f(\lambda) < \infty \right\}$$

and

$$(A_j f | f) = \int_\mathbb{R} \lambda \, d(X_j)_f(\lambda) = \int_{\mathbb{R}^d} x_j |f(x)|^2 \, dm(x) = (\widehat{x}_j f | f).$$

This proves the inclusion $\widehat{x}_j \subseteq A_j$. Since both A_j and $\widehat{x}_j$ are self-adjoint, it follows from Proposition 10.49 that $A_j = \widehat{x}_j$ with equal domains.

Likewise, the selfadjoint operator in $L^2(\mathbb{R}^d)$ associated with the jth momentum coordinate $\Xi_j : \mathscr{B}(\mathbb{R}) \to \mathscr{P}(L^2(\mathbb{R}^d, m))$,

$$(\Xi_j)_B f := \mathscr{F}^{-1} \mathbf{1}_{\mathbb{R} \times \cdots \times \mathbb{R} \times B \times \mathbb{R} \cdots \times \mathbb{R}} \mathscr{F} f, \quad f \in L^2(\mathbb{R}^d, m),$$

is given by

$$\widehat{\xi}_j f(x) := \frac{1}{i} \frac{\partial f}{\partial x_j}(x), \quad x \in \mathbb{R}^d, \tag{15.15b}$$

for $f \in D(\widehat{\xi}_j) := \{f \in L^2(\mathbb{R}^d) : x \mapsto \frac{1}{i} \frac{\partial f}{\partial x_j}(x) \in L^2(\mathbb{R}^d, m)\}$.

Applying the Weyl commutation relation for X and Ξ to functions in $C_c^1(\mathbb{R}^d)$ and differentiating this relation with respect to x_j and ξ_k, we obtain the *Heisenberg commutation relation*

$$\widehat{x}_j \widehat{\xi}_k - \widehat{\xi}_k \widehat{x}_j = i \delta_{jk} I, \tag{15.16}$$

the rigorous interpretation being that for all $f \in C_c^1(\mathbb{R}^d)$ the equality $\widehat{x}_j \widehat{\xi}_k f - \widehat{\xi}_k \widehat{x}_j f = i \delta_{jk} f$ holds in $L^2(\mathbb{R}^d, m)$. Of course (15.16) could also be derived directly from (15.15a) and (15.15b). Note that $C_c^1(\mathbb{R}^d)$ is dense in the commutator domain $D([\widehat{x}_j, \widehat{\xi}_k])$ for all $1 \leqslant j, k \leqslant d$, and (15.16) extends to functions in this domain.

For pure states ϕ represented by a norm one function $f \in L^2(\mathbb{R}^d)$ such that $f \in D(\widehat{x}_j) \cap D(\widehat{\xi}_j)$ and $\widehat{x}_j f, \widehat{\xi}_j f \in D(\widehat{x}_j) \cap D(\widehat{\xi}_j)$, the uncertainty principle of Theorem 15.17 takes the form

$$\Delta_\phi(\widehat{x}_j) \Delta_\phi(\widehat{\xi}_j) \geqslant \frac{1}{2}.$$

It follows from Proposition 15.40 that the position observable X is covariant with respect to translations and conserved under boosts, and that the momentum observable Ξ is conserved under translations and covariant with respect to boosts. This essentially characterises these observables:

Theorem 15.42 (Covariance characterisation of position and momentum). *Up to conjugation with a translation, respectively a boost, position and momentum are characterised by their covariance and conservation properties. More precisely, denoting by X and Ξ the position and momentum observables, the following assertions hold.*

(1) *if the observable $P : \mathscr{B}(\mathbb{R}^d) \to \mathscr{P}(L^2(\mathbb{R}^d, m))$ is covariant with respect to all pairs (x, U_x), $x \in \mathbb{R}^d$, and conserved under all boosts V_ξ, $\xi \in \mathbb{R}^d$, then there exists a unique $y \in \mathbb{R}^d$ such that*

$$P_B = U_y X_B U_y^\star, \quad B \in \mathscr{B}(\mathbb{R}^d);$$

(2) *if the observable* $P : \mathscr{B}(\mathbb{R}^d) \to \mathscr{P}(L^2(\mathbb{R}^d, m))$ *is invariant under all translations* U_x, $x \in \mathbb{R}^d$, *and covariant with respect to all pairs* (ξ, V_ξ), $\xi \in \mathbb{R}^d$, *then there exists a unique* $\eta \in \mathbb{R}^d$ *such that*

$$P_B = V_\eta \Xi_B V_\eta^\star, \quad B \in \mathscr{B}(\mathbb{R}^d).$$

Proof Let the projection-valued measure $P : \mathscr{B}(\mathbb{R}^d) \to \mathscr{P}(L^2(\mathbb{R}^d, m))$ be covariant with respect to all pairs (x, U_x) and conserved under all boosts V_ξ. The boost invariance means that every projection P_B commutes with pointwise multiplication with every trigonometric exponential $x \mapsto \exp(ix \cdot \xi)$. By Lemma 5.33, this implies that P_B is a pointwise multiplier of the form $P_B f(x) = m_B(x) f(x)$ with $m_B \in L^\infty(\mathbb{R}^d, m)$. Since P_B is a projection, m_B must be an indicator function, say of the set C_B:

$$P_B f = \mathbf{1}_{C_B} f.$$

Substituting this into the covariance with respect to translations, we arrive at the identity $\mathbf{1}_{C_B}(x - t) = \mathbf{1}_{C_{B+t}}$ as elements of $L^\infty(\mathbb{R}^d, m)$, that is, we have

$$C_B + t = C_{B+t}$$

up to a null set. Similarly one sees that $C_{\mathbb{R}^d} = \mathbb{R}^d$ and $C_{\complement B} = \complement C_B$ up to null sets. Finally, if B and B' are disjoint, then so are P_B and $P_{B'}$, and therefore C_B and $C_{B'}$ are disjoint up to a null set. It follows that the mapping $B \mapsto C_B$ commutes up to null sets with translations and the Boolean set operations.

Let $B := [0,1)^d$ be the half-open unit cube. Suppose $x, y \in \mathbb{R}^d$ are Lebesgue points of $\mathbf{1}_{C_B}$ satisfying $\max_{1 \leqslant j \leqslant d} |x_j - y_j| > 1$. Then C_B and $C_B + y - x$ intersect in a set of positive measure. This is only possible if B and $B + y - x$ intersect in a set of positive measure, but these sets are disjoint. This contradiction proves that all Lebesgue points x, y of $\mathbf{1}_{C_B}$ satisfy $\max_{1 \leqslant j \leqslant d} |x_j - y_j| \leqslant 1$. Since almost every point of $\mathbf{1}_{C_B}$ is a Lebesgue point, it follows that up to a null set, C_B is contained in the rectangle $\prod_{j=1}^d [a_j, b_j]$, where

$$a_j := \inf\{x_j : x \text{ is a Lebesgue point of } \mathbf{1}_{C_B}\},$$

$$b_j := \sup\{x_j : x \text{ is a Lebesgue point of } \mathbf{1}_{C_B}\}.$$

In particular, since $0 \leqslant b_j - a_j \leqslant 1$, up to a null set we have $C_B \subseteq \prod_{j=1}^d [a_j, a_j + 1) = a + B$, where $a = (a_1, \ldots, a_d)$.

We next claim that up to a null set we have equality

$$C_B = a + B.$$

Indeed, the sets $k + B$ with $k \in \mathbb{Z}^d$ are pairwise disjoint and their union is $\mathbb{R}^d$. Hence, up to null sets, the sets $C_{k+B} = k + C_B$ are disjoint and their union is $\mathbb{R}^d$. This is only possible if $(a + B) \setminus C_B$ is a null set. This proves the claim.

Let $n \in \mathbb{N}$ and consider the set $B^{(n)} := [0, 2^{-n})^d$. The same argument as above proves that there exists an $a^{(n)} \in \mathbb{R}^d$ such that $C_{B^{(n)}} = a^{(n)} + B^{(n)}$. Now B is the disjoint union of 2nd translates $k + B^{(n)}$, $k \in \{j2^{-n} : j = 0, 1, \ldots 2^n - 1\}^d$. Therefore, up to a null set, $C_B = a + B$ is the disjoint union of the 2nd sets $C_{k+B^{(n)}} = a^{(n)} + k + B^{(n)}$. This union equals $a^{(n)} + B$. This shows that $a^{(n)} = a$ for all $n \in \mathbb{N}$.

Summarising what we have proved, we find that for all sets B of the form $y + B^{(n)}$ with $y \in \mathbb{R}^d$ and $n \in \mathbb{Z}$, we have

$$P_B f = \mathbf{1}_{a+B} f.$$

Equivalently, this can be expressed as

$$P_B = U_a X_B U_{-a} = U_a X_B U_a^\star.$$

This proves the first part of the theorem. To prove the second part, suppose that the projection-valued measure $P : \mathscr{B}(\mathbb{R}^d) \to \mathscr{P}(L^2(\mathbb{R}^d, m))$ is conserved under all U_y and covariant with respect to all pairs (η, V_η). Then $\widetilde{P} := \mathscr{F}^{-1} P_B \mathscr{F}$ defines a projection-valued measure that is covariant with respect to all pairs (y, U_y) and conserved under all V_η. It follows from the previous step that $\widetilde{P} = U_a X U_a^\star$ for some $a \in \mathbb{R}^d$, and therefore $P = \mathscr{F} \widetilde{P} \mathscr{F}^{-1} = V_a \Xi V_a^\star$ for some $a \in \mathbb{R}^d$ by (15.15). □

15.5.c The Case $G = \mathbb{T}$: Angle and Angular Momentum

The results of the preceding section have natural analogues for the unit circle $\mathbb{T}$. We identify $\mathbb{T}$ with the unit circle of $\mathbb{C}$ and take the normalised Lebesgue measure on $\mathbb{T}$ as the Haar measure. Every character $\gamma : \mathbb{T} \to \mathbb{T}$ is of the form

$$\gamma(z) = z^k, \quad z \in \mathbb{T},$$

for some $k \in \mathbb{Z}$. Under this identification we have $\Gamma = \mathbb{Z}$, its normalised Haar measure being the counting measure.

For every $w \in \mathbb{T}$ the rotation $z \mapsto wz$ is a symmetry of $\mathbb{T}$. The induced symmetry U_w on $L^2(\mathbb{R}^d, m)$ is given by

$$U_w f(z) = f(w^{-1} z), \quad z \in \mathbb{T}, \ f \in L^2(\mathbb{T}). \tag{15.17}$$

For every $k \in \mathbb{Z}$ the boost V_k on $L^2(\mathbb{T})$ is given by

$$V_k f(z) = z^k f(z), \quad z \in \mathbb{T}, \ f \in L^2(\mathbb{T}). \tag{15.18}$$

The Weyl commutation relation takes the form

$$V_k U_z = z^k U_z V_k, \quad z \in \mathbb{T}, \ k \in \mathbb{Z}.$$

The position and momentum observables in $\mathbb{T}$ associated with the symmetries U_z and V_k are denoted by Θ and L and are called the *angle* and (orbital) *angular momentum* observables. They take values in $\mathbb{T}$ and $\mathbb{Z}$ respectively; in particular, angular momentum can only assume discrete values. In the Physics literature one speaks about 'quanta' of angular momentum.

Remark 15.43. By viewing $\mathbb{Z}$ as a subset of the real line, we may identify L with a real-valued observable and thus associate with L a selfadjoint operator $\widehat{l}$ on $L^2(\mathbb{R})$. There is no natural way, however, to do the same with Θ. One could identify $\mathbb{T}$ with the interval $(-\pi, \pi]$ contained in the real line and thus identify Θ with a real-valued observable. The choice of the interval $(-\pi, \pi]$ is somewhat arbitrary, however, and entails a nonuniqueness issue that cannot be resolved satisfactorily. The associated selfadjoint operator $\widehat{\theta}$ appears not to be very useful. For instance, it does not satisfy the 'continuous variable' Weyl commutation relation

$$e^{is\widehat{\theta}}e^{it\widehat{l}} = e^{ist}e^{it\widehat{l}}e^{is\widehat{\theta}}.$$

This will be further discussed in Problem 15.13.

By Proposition 15.40, Θ is covariant under every pair (z, U_z) and conserved under every V_k, and L is conserved under every U_z and covariant under every pair (k, V_k). Repeating the proof of Theorem 15.42 almost *verbatim* we arrive at the following result.

Theorem 15.44 (Covariance characterisation of angle and angular momentum). *Up to conjugation with a translation, respectively a boost, angle and angular momentum are characterised by their covariance and conservation properties. More precisely, denoting by Θ and L the position and momentum operators associated with the rotations U_z and boosts V_k given by* (15.17) *and* (15.18), *the following assertions hold.*

(1) *if the observable $P : \mathscr{B}(\mathbb{T}) \to \mathscr{P}(L^2(\mathbb{T}))$ is covariant with respect to all pairs (z, U_z), $z \in \mathbb{T}$, and conserved under all V_k, $k \in \mathbb{Z}$, then there exists a unique $w \in \mathbb{T}$ such that*

$$P_B = U_w \Theta_B U_w^\star, \quad B \in \mathscr{B}(\mathbb{T});$$

(2) *if the observable $P : \mathscr{B}(\mathbb{Z}) \to \mathscr{P}(L^2(\mathbb{T}))$ is conserved under all U_z, $z \in \mathbb{T}$, and covariant with respect to all pairs (k, V_k), $k \in \mathbb{Z}$, then there exists a unique $j \in \mathbb{Z}$ such that*

$$P_B = V_j L_B V_j^\star, \quad B \in \mathscr{B}(\mathbb{Z}).$$

15.5.d The Stone–Von Neumann Theorem

We have seen in Theorem 15.42 that the $\mathbb{R}^d$-valued position and momentum observables are uniquely determined, up to conjugation with a translation and a boost respectively, by their transformation properties under translations and boosts. It is interesting to observe that both the covariance relation for position

$$U_x X_B U_x^\star f = X_{xB} f, \quad x \in \mathbb{R}^d, \quad f \in L^2(\mathbb{R}^d, m),$$

and the covariance relation for momentum

$$V_\xi \Xi_B V_\xi^* f = \Xi_{\xi B} f \quad \xi \in \mathbb{R}^d, \quad f \in L^2(\mathbb{R}^d, m),$$

imply the Weyl commutation relation. Here, as before, $dm(x) = (2\pi)^{-d/2} dx$ is the normalised Lebesgue measure. Indeed, approximating $e^{ix \cdot \xi}$ by simple functions, as in the proof of Theorem 15.38 we find that the covariance relation for position implies the identity $V_\xi U_x = e^{ix \cdot \xi} U_x V_\xi$ for $x, \xi \in \mathbb{R}^d$, which is the Weyl commutation relation. In the same way the covariance relation for momentum implies the Weyl commutation relation. In view of this it is reasonable to ask to what extent position and momentum are determined by the Weyl commutation relation. The answer to this question is provided by a theorem due to Stone and von Neumann (Theorem 15.48 and its corollary). Proving this theorem is the main objective of the present section.

We start with some preparation. Suppose that $\widetilde{U}, \widetilde{V} : \mathbb{R}^d \to \mathscr{L}(H)$ are strongly continuous unitary representations of $\mathbb{R}^d$ on a Hilbert space H such that the Weyl commutation relation holds, that is,

$$\widetilde{V}_\xi \widetilde{U}_x = e^{ix \cdot \xi} \widetilde{U}_x \widetilde{V}_\xi, \quad x, \xi \in \mathbb{R}^d. \tag{15.19}$$

The relation (15.19) states that $\widetilde{U}$ and $\widetilde{V}$ 'commute up to a multiplicative scalar of modulus one'. This suggests to interpret (15.19) as a 'projective' unitary representation of $\mathbb{R}^d \times \mathbb{R}^d$ on H. There is a quick way to extend (15.19) to a unitary representation as follows. Consider the unitary operators

$$\widetilde{W}(x, \xi) := e^{\frac{1}{2} ix \cdot \xi} \widetilde{U}_x \widetilde{V}_\xi = e^{-\frac{1}{2} ix \cdot \xi} \widetilde{V}_\xi \widetilde{U}_x, \quad x, \xi \in \mathbb{R}^d. \tag{15.20}$$

The operators $\widetilde{W}(x, \xi)$ defined by (15.20) satisfy

$$\begin{aligned}
\widetilde{W}(x, \xi) \widetilde{W}(x', \xi') &= e^{-\frac{1}{2} i(x \cdot \xi + x' \cdot \xi')} \widetilde{V}_\xi \widetilde{U}_x \widetilde{V}_{\xi'} \widetilde{U}_{x'} \\
&= e^{-\frac{1}{2} i(x \cdot \xi + x' \cdot \xi') - ix \cdot \xi'} \widetilde{V}_\xi \widetilde{V}_{\xi'} \widetilde{U}_x \widetilde{U}_{x'} \\
&= e^{\frac{1}{2} i(x' \cdot \xi - x \cdot \xi')} e^{-\frac{1}{2} i(x+x')(\xi+\xi')} \widetilde{V}_{\xi+\xi'} \widetilde{U}_{x+x'} \\
&= e^{\frac{1}{2} i(x' \cdot \xi - x \cdot \xi')} \widetilde{W}(x+x', \xi+\xi').
\end{aligned} \tag{15.21}$$

From this it follows that the unitary operators defined by

$$\widetilde{W}(x, \xi, t) := e^{it} \widetilde{W}(x, \xi) \tag{15.22}$$

satisfy

$$\begin{aligned}
\widetilde{W}(x, \xi, t) \widetilde{W}(x', \xi', t') &= e^{i(t+t')} \widetilde{W}(x, \xi) \widetilde{W}(x', \xi') \\
&= e^{i(t+t' + \frac{1}{2}(x' \cdot \xi - x \cdot \xi'))} \widetilde{W}(x+x', \xi+\xi') \\
&= \widetilde{W}((x, \xi, t) \circ (x', \xi', t')),
\end{aligned} \tag{15.23}$$

where

$$(x,\xi,t) \circ (x',\xi',t') := \left(x+x',\ \xi+\xi',\ t+t'+\frac{1}{2}(x'\cdot\xi - x\cdot\xi') \right).$$

One easily checks that the operation $\circ$ turns $\mathbb{H}^d := \mathbb{R}^d \times \mathbb{R}^d \times \mathbb{R}$ into a group:

Definition 15.45 (Heisenberg group). The *Heisenberg group* in dimension d is the group $\mathbb{H}^d := \mathbb{R}^d \times \mathbb{R}^d \times \mathbb{R}$ with composition law

$$(x,\xi,t) \circ (x',\xi',t') := \left(x+x',\xi+\xi',t+t'+\frac{1}{2}(x'\cdot\xi - x\cdot\xi') \right).$$

The identity 15.23 informs us that $\widetilde{W}$ defines a unitary representation of the Heisenberg group $\mathbb{H}^d$ on H. It is strongly continuous and it satisfies

$$\widetilde{W}(0,0,t) = e^{it}I, \quad t \in \mathbb{R}. \tag{15.24}$$

Definition 15.46 (Schrödinger representation). The *Schrödinger representation* is the unitary representation $W : \mathbb{H}^d \to \mathscr{L}(L^2(\mathbb{R}^d,m))$ arising in the special case where unitary representations $U,V : \mathbb{R}^d \to \mathscr{L}(L^2(\mathbb{R}^d,m))$ are given by translations and boosts, respectively.

Explicitly, the Schrödinger representation is given by

$$W(x,\xi,t)f(x') = e^{it}e^{-\frac{1}{2}ix\cdot\xi}e^{ix'\cdot\xi}f(x'-x). \tag{15.25}$$

Proposition 15.47. *The Schrödinger representation is* irreducible, *that is, the only closed subspaces of $L^2(\mathbb{R}^d,m)$ invariant under the action of W are the trivial subspaces $\{0\}$ and $L^2(\mathbb{R}^d,m)$.*

The proof of this proposition will be given at the end of the section, for it uses elements of the proof of the following theorem which says that the Schrödinger representation is essentially the only irreducible unitary representation of $\mathbb{H}^d$ satisfying (15.24):

Theorem 15.48 (Stone–von Neumann). *Let $\widetilde{W} : \mathbb{H}^d \to \mathscr{L}(H)$ be a strongly continuous unitary representation of $\mathbb{H}^d$ on a separable Hilbert space H. If $\widetilde{W}$ is irreducible and satisfies $\widetilde{W}(0,0,t) = e^{it}I$ for all $t \in \mathbb{R}$, then $\widetilde{W}$ is unitarily equivalent to the Schrödinger representation W. More precisely, there exists a unitary operator $S : L^2(\mathbb{R}^d,m) \to H$ such that*

$$\widetilde{W}(x,\xi,t) = SW(x,\xi,t)S^\star, \quad (x,\xi,t) \in \mathbb{H}^d.$$

The operator S is unique up to a multiplicative scalar of modulus one.

Here, we use the term *unitary operator* for an operator $S : H \to K$, where H and K are Hilbert spaces, such that $S^\star S = I$ (the identity operator on H) and $SS^\star = I$ (the identity operator on K).

We have the following immediate corollary for representations arising from pairs of unitary representations satisfying the Weyl commutation relation.

Corollary 15.49. *Let* $\widetilde{U},\widetilde{V} : \mathbb{R}^d \to \mathscr{L}(H)$ *be strongly continuous unitary representations on a separable Hilbert space H satisfying the Weyl commutation relation*

$$\widetilde{V}_\xi \widetilde{U}_x = e^{ix\cdot\xi}\widetilde{U}_x\widetilde{V}_\xi, \quad x,\xi \in \mathbb{R}^d.$$

Suppose furthermore that the family $\{\widetilde{U}_x,\widetilde{V}_\xi : x \in \mathbb{R}^d, \xi \in \mathbb{R}^d\}$ acts irreducibly on H in the sense that the only closed subspace invariant under all operators U_x and $\widetilde{V}_\xi$, $x \in \mathbb{R}^d$, $\xi \in \mathbb{R}^d$, are the trivial subspaces $\{0\}$ and H. Then there exists a unitary operator $S : L^2(\mathbb{R}^d,m) \to H$ such that

$$\widetilde{U}_x = SU_xS^\star, \quad x \in \mathbb{R}^d,$$
$$\widetilde{V}_\xi = SV_\xi S^\star, \quad \xi \in \mathbb{R}^d,$$

where U and V are the translation and boost representations on $L^2(\mathbb{R}^d,m)$, respectively. The operator S is unique up to a multiplicative constant of modulus one.

Proof Defining $\widetilde{W} : \mathbb{H}^d \to \mathscr{L}(H)$ by (15.20) and (15.22), the irreducibility assumption of the corollary translates into the irreducibility of the representation $\widetilde{W}$. $\quad\square$

We now fix a strongly continuous unitary representation $\widetilde{W} : \mathbb{H}^d \to \mathscr{L}(H)$ and define

$$\widetilde{U}_x := W(x,0,0), \quad \widetilde{V}_\xi := W(0,\xi,0), \quad \widetilde{W}(x,\xi) := \widetilde{W}(x,\xi,0).$$

Then (15.19)–(15.23) hold again. We write m for both the normalised Lebesgue measures on $\mathbb{R}^d$ and $\mathbb{R}^{2d}$.

Definition 15.50 (Weyl transform). For $a \in L^1(\mathbb{R}^{2d},m)$ we define the operator $\widetilde{W}(a) \in \mathscr{L}(H)$ by

$$\widetilde{W}(a)h := \int_{\mathbb{R}^{2d}} a(x,\xi)\widetilde{W}(x,\xi)h\,dm(x)\,dm(\xi), \quad h \in H,$$

where the integral is a Bochner integral in H.

The next two lemmas state some properties for the Weyl transform associated with the Schrödinger representation $W : \mathbb{H}^d \to \mathscr{L}(L^2(\mathbb{R}^d,m))$.

Lemma 15.51. *For all $a \in L^1(\mathbb{R}^{2d},m) \cap L^2(\mathbb{R}^{2d},m)$ the operator $W(a)$ is Hilbert–Schmidt on $L^2(\mathbb{R}^d,m)$ and*

$$\|W(a)\|_{\mathscr{L}_2(L^2(\mathbb{R}^d,m))} = \|a\|_{L^2(\mathbb{R}^{2d},m)}.$$

Proof For the Schrödinger representation we have the explicit formula (15.25),

$$W(x,\xi)f(x') = e^{-\frac{1}{2}ix\cdot\xi}e^{ix'\cdot\xi}f(x'-x), \quad f \in L^2(\mathbb{R}^d,m),$$

where $W(x,\xi) := W(x,\xi,0)$ as in (15.22). By a change of variables and Fubini's theorem we obtain

$$W(a)f = \int_{\mathbb{R}^{2d}} a(x,\xi) e^{-\frac{1}{2}ix\cdot\xi} e^{i(\cdot)\cdot\xi} f(\cdot - x) \, dm(x) \, dm(\xi)$$

$$= \int_{\mathbb{R}^d} \left(\int_{\mathbb{R}^d} a(\cdot - x, \xi) e^{-\frac{1}{2}i(\cdot - x)\cdot\xi} e^{i(\cdot)\cdot\xi} \, dm(\xi) \right) f(x) \, dm(x)$$

$$:= \int_{\mathbb{R}^d} k(x,\cdot) f(x) \, dm(x),$$

where

$$k(x,x') = \int_{\mathbb{R}^d} a(x' - x, \xi) e^{\frac{1}{2}i(x+x')\cdot\xi} \, dm(\xi).$$

By Plancherel's theorem,

$$\int_{\mathbb{R}^{2d}} \left| k\left(\frac{y-z}{2}, \frac{y+z}{2}\right) \right|^2 dm(y)\,dm(z) = \int_{\mathbb{R}^{2d}} \left| \int_{\mathbb{R}^d} a(z,\xi) e^{\frac{1}{2}iy\cdot\xi} \, dm(\xi) \right|^2 dm(y)\,dm(z)$$

$$= 2^d \int_{\mathbb{R}^{2d}} |a(z,y)|^2 \, dm(y)\,dm(z) = 2^d \|a\|_2^2$$

and hence

$$\int_{\mathbb{R}^{2d}} |k(x,x')|^2 \, dm(x)\,dm(x') = \frac{1}{2^d} \int_{\mathbb{R}^{2d}} \left| k\left(\frac{y-z}{2}, \frac{y+z}{2}\right) \right|^2 dm(y)\,dm(z) = \|a\|_2^2.$$

The result now follows from Example 14.3, which says that an integral operator with square integrable kernel is Hilbert–Schmidt, with Hilbert–Schmidt norm equal to the L^2-norm of the kernel. $\qquad\square$

Since $L^1(\mathbb{R}^{2d},m) \cap L^2(\mathbb{R}^{2d},m)$ is dense in $L^2(\mathbb{R}^{2d},m)$, the lemma implies that the mapping $W : a \mapsto W(a)$ has a unique extension to an isometry from $L^2(\mathbb{R}^{2d},m)$ into the space of Hilbert–Schmidt operators $\mathscr{L}_2(L^2(\mathbb{R}^d,m))$. This extension is again denoted by W.

A special role is played by the functions

$$a_0(x,\xi) := \exp\left(-\frac{1}{4}(|x|^2 + |\xi|^2)\right), \quad x,\xi \in \mathbb{R}^d,$$

$$\phi_0(x) := 2^{d/4} \exp\left(-\frac{1}{2}|x|^2\right), \quad x \in \mathbb{R}^d.$$

Note that $\|a_0\|_{L^2(\mathbb{R}^{2d},m)} = \|\phi_0\|_{L^2(\mathbb{R}^d,m)} = 1$.

Lemma 15.52. *The operator $W(a_0)$ equals the rank one projection $\phi_0 \bar{\otimes} \phi_0$.*

Proof Using (15.25) and the elementary identity (which follows from Lemma 5.19)

$$\int_{\mathbb{R}^d} e^{-\frac{1}{2}ix\cdot\xi} e^{iy\cdot\xi} \exp\left(-\frac{1}{4}|\xi|^2\right) dm(\xi) = 2^{d/2} \exp\left(-\left|y - \frac{1}{2}x\right|^2\right)$$

we obtain, for $f \in L^2(\mathbb{R}^d, m)$,

$$W(a_0)f = \int_{\mathbb{R}^{2d}} \exp\left(-\frac{1}{4}|x|^2\right) \exp\left(-\frac{1}{4}|\xi|^2\right) e^{-\frac{1}{2}ix\cdot\xi} e^{i(\cdot)\cdot\xi} f(\cdot - x)\, dm(x)\, dm(\xi)$$

$$= 2^{d/2} \int_{\mathbb{R}^d} \exp\left(-|(\cdot) - \frac{1}{2}x|^2\right) \exp\left(-\frac{1}{4}|x|^2\right) f(\cdot - x)\, dm(x)$$

$$= 2^{d/2} \exp\left(-\frac{1}{2}|\cdot|^2\right) \int_{\mathbb{R}^d} \exp\left(-\frac{1}{2}|x|^2\right) f(x)\, dm(x) = (\phi_0 \bar{\otimes} \phi_0)f.$$

$\square$

Returning to a general strongly continuous unitary representation $\widetilde{W} : \mathbb{H}^d \to \mathscr{L}(H)$, we note the following important multiplicativity property.

Lemma 15.53. *For all $a, b \in L^1(\mathbb{R}^{2d}, m)$ we have*

$$\widetilde{W}(a)\widetilde{W}(b) = \widetilde{W}(a\#b),$$

where the so-called twisted convolution $a\#b \in L^1(\mathbb{R}^{2d}, m)$ *is defined by*

$$a\#b(x, \xi) := \int_{\mathbb{R}^{2d}} e^{\frac{1}{2}i(x'\cdot\xi - x\cdot\xi')} a(x - x', \xi - \xi') b(x', \xi')\, dm(x')\, dm(\xi').$$

Young's inequality implies that $a\#b$ does indeed belong to $L^1(\mathbb{R}^d, m)$.

Proof Fix $h \in H$. By (15.21), a change of variables, and Fubini's theorem,

$$\widetilde{W}(a)\widetilde{W}(b)h = \int_{\mathbb{R}^{4d}} a(x, \xi) b(x', \xi') \widetilde{W}(x, \xi) \widetilde{W}(x', \xi') h\, dm(x)\, dm(\xi)\, dm(x')\, dm(\xi')$$

$$= \int_{\mathbb{R}^{4d}} e^{\frac{1}{2}i(x'\cdot\xi - x\cdot\xi')} a(x, \xi) b(x', \xi')$$

$$\times \widetilde{W}(x + x', \xi + \xi') h\, dm(x)\, dm(\xi)\, dm(x')\, dm(\xi')$$

$$= \int_{\mathbb{R}^{4d}} e^{\frac{1}{2}i(x'\cdot\xi - x\cdot\xi')} a(x - x', \xi - \xi') b(x', \xi')$$

$$\times \widetilde{W}(x, \xi) h\, dm(x)\, dm(\xi)\, dm(x')\, dm(\xi')$$

$$= \int_{\mathbb{R}^{2d}} a\#b(x, \xi) \widetilde{W}(x, \xi) h\, dm(x)\, dm(\xi) = \widetilde{W}(a\#b)h.$$

$\square$

This lemma is used to establish the following technical fact.

Lemma 15.54. *We have*

$$\widetilde{W}(a_0)\widetilde{W}(x, \xi)\widetilde{W}(a_0) = a_0(x, \xi)\widetilde{W}(a_0), \quad x, \xi \in \mathbb{R}^d.$$

Proof Repeating the steps in the proof of Lemma 15.53, for all $h \in H$ we obtain

$$\widetilde{W}(x,\xi)\widetilde{W}(a_0)h = \int_{\mathbb{R}^{2d}} a_0(x',\xi')\widetilde{W}(x,\xi)\widetilde{W}(x',\xi')h \, dm(x') \, dm(\xi')$$

$$= \int_{\mathbb{R}^{2d}} e^{\frac{1}{2}i(x'\cdot\xi - x\cdot\xi')} a_0(x-x',\xi-\xi')\widetilde{W}(x,\xi)h \, dm(x') \, dm(\xi') \quad (15.26)$$

$$= \widetilde{W}(a_{x,\xi})h,$$

where $a_{x,\xi}(x',\xi') := e^{\frac{1}{2}i(x'\cdot\xi - x\cdot\xi')} a_0(x-x',\xi-\xi')$. Hence, by Lemma 15.53, the lemma is equivalent to the statement that

$$\widetilde{W}(a_0 \# a_{x,\xi}) = a_0(x,\xi)\widetilde{W}(a_0).$$

For this it suffices to show that

$$a_0 \# a_{x,\xi} = a_0(x,\xi)a_0.$$

By the injectivity of the Schrödinger representation W, which follows from Lemma 15.51, this in turn is equivalent to showing that

$$W(a_0)W(x,\xi)W(a_0) = W(a_0 \# a_{x,\xi}) = a_0(x,\xi)W(a_0).$$

The verification of this identity proceeds by explicit calculation. By Lemma 15.52,

$$W(a_0)W(x,\xi)W(a_0)f = (\phi_0 \bar\otimes \phi_0)W(x,\xi)(\phi_0 \bar\otimes \phi_0)f$$

$$= (W(x,\xi)\phi_0|\phi_0)(f|\phi_0)\phi_0 = (W(x,\xi)\phi_0|\phi_0)W(a_0)f.$$

Moreover, by (15.25) and an elementary computation,

$$(W(x,\xi)\phi_0|\phi_0) = e^{-\frac{1}{2}ix\cdot\xi}(e^{i(\cdot)\cdot\xi}\phi_0(\cdot - x)|\phi_0)$$

$$= 2^{d/2}e^{-\frac{1}{2}ix\cdot\xi}\int_{\mathbb{R}^d} e^{iy\cdot\xi}\exp\left(-\frac{1}{2}|y-x|^2\right)\exp\left(-\frac{1}{2}|y|^2\right) dm(y)$$

$$= \exp\left(-\frac{1}{4}|x|^2\right)\exp\left(-\frac{1}{4}|\xi|^2\right) = a_0(x,\xi).$$

$\square$

We are now ready for the proof of the Stone–von Neumann theorem.

Proof of Theorem 15.48 We split the proof into three steps.

Step 1 – We begin by showing that $\widetilde{W}(a_0)$ is a rank one projection. By Lemmas 15.52 and 15.53 (applied to W),

$$W(a_0 \# a_0) = W(a_0)W(a_0) = (\phi_0 \bar\otimes \phi_0)^2 = \phi_0 \bar\otimes \phi_0 = W(a_0).$$

By the injectivity of W (which follows from Lemma 15.51), this implies that $a_0 \# a_0 = a_0$. Another application of Lemma 15.53, this time to $\widetilde{W}$, gives

$$\widetilde{W}(a_0)\widetilde{W}(a_0) = \widetilde{W}(a_0 \# a_0) = \widetilde{W}(a_0).$$

This means that $\widetilde{W}(a_0)$ is a projection. We will use the assumption of irreducibility of $\widetilde{W}$ to prove that this projection has rank one.

We begin by showing that $\widetilde{W}(a_0) \neq 0$. Indeed, if we had $\widetilde{W}(a_0) = 0$, then for all $x, \xi \in \mathbb{R}^d$ and $h, h' \in H$ we would have, by (15.21) and (15.26),

$$
\begin{aligned}
0 &= (\widetilde{W}(x,\xi)\widetilde{W}(a_0)\widetilde{W}(-x,-\xi)h|h') \\
&= (\widetilde{W}(a_{x,\xi})\widetilde{W}(-x,-\xi)h|h') \\
&= \int_{\mathbb{R}^{2d}} e^{\frac{1}{2}i(x'\cdot\xi - x\cdot\xi')} a_0(x - x', \xi - \xi') \\
&\qquad\qquad \times (\widetilde{W}(x',\xi')\widetilde{W}(-x,-\xi)h|h')\, dx'\, d\xi' \\
&= \int_{\mathbb{R}^{2d}} e^{-\frac{1}{2}i(x'\cdot\xi - x\cdot\xi')} a_0(x',\xi') \\
&\qquad\qquad \times (\widetilde{W}(x - x', \xi - \xi')\widetilde{W}(-x,-\xi)h|h')\, dx'\, d\xi' \\
&= \int_{\mathbb{R}^{2d}} e^{-i(x'\cdot\xi - x\cdot\xi')} a_0(x',\xi')(\widetilde{W}(-x',-\xi')h|h')\, dx'\, d\xi' \\
&= \int_{\mathbb{R}^{2d}} e^{i(x'\cdot\xi - x\cdot\xi')} a_0(x',\xi')(\widetilde{W}(x',\xi')h|h')\, dx'\, d\xi'.
\end{aligned}
$$

This being true for all $x, \xi \in \mathbb{R}^d$, the Fourier inversion theorem would then imply that $(\widetilde{W}(x',\xi')h|h') = 0$ for almost all $x', \xi' \in \mathbb{R}^d$. Since $h, h' \in H$ were arbitrary, it would follow that $W(x',\xi') = 0$ for almost all $x', \xi' \in \mathbb{R}^d$, contradicting the fact that all these operators are unitary.

Fix any nonzero $h \in R(\widetilde{W}(a_0))$; this is possible by the preceding argument. Let $\widetilde{Y}_h$ be the closed linear span of the set $\{\widetilde{W}(x,\xi)h : x, \xi \in \mathbb{R}^d\}$. From (15.21) we see that $\widetilde{Y}_h$ is invariant under each operator $\widetilde{W}(x,\xi)$ and hence under the representation $\widetilde{W}$. Since $\widetilde{W}$ is assumed to be irreducible it follows that $\widetilde{Y}_h = H$.

By (15.19) and (15.20),

$$
\widetilde{W}(x,\xi)^* = e^{\frac{1}{2}ix\cdot\xi}\widetilde{U}_x^*\widetilde{V}_\xi^* = e^{\frac{1}{2}ix\cdot\xi}\widetilde{U}_{-x}\widetilde{V}_{-\xi} = e^{-\frac{1}{2}ix\cdot\xi}\widetilde{V}_{-\xi}\widetilde{U}_{-x} = \widetilde{W}(-x,-\xi).
$$

Hence if $h, h' \in R(\widetilde{W}(a_0))$, say $h = \widetilde{W}(a_0)g$ and $h' = \widetilde{W}(a_0)g'$, then

$$
\begin{aligned}
(\widetilde{W}(x,\xi)h|\widetilde{W}(x',\xi')h') &= (\widetilde{W}(-x',-\xi')\widetilde{W}(x,\xi)\widetilde{W}(a_0)g|\widetilde{W}(a_0)g') \\
&= e^{\frac{1}{2}i(x'\cdot\xi - x\cdot\xi')}(\widetilde{W}(x - x', \xi - \xi')\widetilde{W}(a_0)g|\widetilde{W}(a_0)g') \\
&= e^{\frac{1}{2}i(x'\cdot\xi - x\cdot\xi')} a_0(x - x', \xi - \xi')(\widetilde{W}(a_0)g|g') \\
&= e^{\frac{1}{2}i(x'\cdot\xi - x\cdot\xi')} a_0(x - x', \xi - \xi')(h|h'),
\end{aligned} \tag{15.27}
$$

using that $\widetilde{W}(a_0)$ is an orthogonal projection and $(\widetilde{W}(a_0)g|g') = (\widetilde{W}(a_0)g|\widetilde{W}(a_0)g') = (h|h')$. In particular, if $h \perp h'$ with $h' \in R(\widetilde{W}(a_0))$, then $\widetilde{Y}_h \perp \widetilde{Y}_{h'}$. Since $\widetilde{Y}_h = H$ this implies $\widetilde{Y}_{h'} = \{0\}$, which, by the previous reasoning, is only possible if $h' = 0$. This proves that $R(\widetilde{W}(a_0))$ equals the one-dimensional span of h.

Step 2 – Define

$$S \sum_{n=1}^{N} c_n W(x_n, \xi_n) \phi_0 := \sum_{n=1}^{N} c_n \widetilde{W}(x_n, \xi_n) h_0,$$

where $h_0 \in \mathsf{R}(\widetilde{W}(a_0))$ has norm $\|h_0\| = 1 = \|\phi_0\|$. It follows from (15.27) (applied to both W and $\widetilde{W}$) that S is well defined and isometric on the linear span of the functions $W(x, \xi)\phi_0$, $x, \xi \in \mathbb{R}^d$, and hence extends to an isometry from Y_{ϕ_0} onto $\widetilde{Y}_{h_0}$, the former being defined as the closed linear span of the functions $W(x, \xi)\phi_0$, $x, \xi \in \mathbb{R}^d$. But $\widetilde{Y}_{h_0} = H$, and by applying this to W we see that likewise $Y_{\phi_0} = L^2(\mathbb{R}^d, m)$. This proves that S is isometric from $L^2(\mathbb{R}^d, m)$ onto H, and hence unitary. Since $S\phi_0 = h_0$, this proves that S has the desired properties.

Step 3 – If $T : L^2(\mathbb{R}^d, m) \to H$ is another unitary operator with the property that $\widetilde{W}(x, \xi, t) = TW(x, \xi, t)T^\star$ for all $(x, \xi, t) \in \mathbb{H}^d$, then $S^\star T W(x, \xi) = W(x, \xi)S^\star T$ for all $x, \xi \in \mathbb{R}^d$. From this it follows that $S^\star T$ commutes with $W(a_0)$, and therefore it maps the one-dimensional range of this operator onto itself. This implies that $S^\star T f = e^{i\theta} f$ for some $\theta \in \mathbb{R}$ and all $f \in \mathsf{R}(W(a_0))$. Then,

$$T \sum_{n=1}^{N} c_n W(x_n, \xi_n) f = SS^\star T \sum_{n=1}^{N} c_n W(x_n, \xi_n) f$$

$$= S \sum_{n=1}^{N} c_n W(x_n, \xi_n) S^\star T f = e^{i\theta} S \sum_{n=1}^{N} c_n W(x_n, \xi_n) f$$

and therefore $T = e^{i\theta} S$. $\qquad\square$

Proof of Proposition 15.47 Reasoning by contradiction, suppose that Y is a nontrivial closed subspace invariant under W and let $Y^\perp$ be its orthogonal complement. The identity $W(x, \xi)^\star = W(-x, -\xi)$ implies that $Y^\perp$ is invariant under W as well. By restriction we thus obtain two strongly continuous unitary representations $W_Y : \mathbb{H}^d \to \mathscr{L}(Y)$ and $W_{Y^\perp} : \mathbb{H}^d \to \mathscr{L}(Y^\perp)$, and they satisfy

$$W_Y(0, 0, t) = e^{it} I_Y, \quad W_{Y^\perp}(0, 0, t) = e^{it} I_{Y^\perp}.$$

Lemma 15.52 and Step 1 of the proof of Theorem 15.48 imply that $W(a_0)$, $W_Y(a_0)$, and $W_{Y^\perp}(a_0)$ are orthogonal projections of rank one in $L^2(\mathbb{R}^d, m)$, Y, and $Y^\perp$, respectively. This leads to the contradiction

$$y_0 \bar{\otimes} y_0 = W(a_0) = W_Y(a_0) + W_{Y^\perp}(a_0),$$

as it represents the rank one projection $y_0 \bar{\otimes} y_0$ as a sum of two disjoint rank one projections. $\qquad\square$

The final result of this section describes the Ornstein–Uhlenbeck semigroup in terms of the Weyl calculus. Let us first recall some notation from Theorem 13.56 (where a different normalisation of Lebesgue measure was used). The multiplication E,

$$Ef(x) := \exp\left(-\frac{1}{4}|x|^2\right)f(x)$$

is unitary from $L^2(\mathbb{R}^d,\gamma)$ to $L^2(\mathbb{R}^d,m)$, and the dilation D,

$$Df(x) := 2^{d/4}f(\sqrt{2}x)$$

is unitary on $L^2(\mathbb{R}^d,m)$. Consequently the operator

$$U := D \circ E \tag{15.28}$$

is unitary from $L^2(\mathbb{R}^d,\gamma)$ to $L^2(\mathbb{R}^d,m)$.

Theorem 15.55. *For all $t > 0$ we have, with $s := \frac{1-e^{-t}}{1+e^{-t}}$,*

$$OU(t) = (1+s)^d\, U^\star W(\widehat{a_s})U,$$

where W is the Schrödinger representation and

$$a_s(x,\xi) := \exp\left(-s(|x|^2+|\xi|^2)\right), \quad x,\xi \in \mathbb{R}^d.$$

Proof Let $a \in L^1(\mathbb{R}^{2d},m) \cap L^2(\mathbb{R}^{2d},m)$. A formal calculation, using the definition of the Weyl transform, the identity (15.25), and a change of variables, gives

$$W(\widehat{a})f = \int_{\mathbb{R}^{2d}}\left(\int_{\mathbb{R}^{2d}}a(u,v)e^{-i(u\cdot x+v\cdot\xi)}\,dm(u)\,dm(v)\right)W(x,\xi)f\,dm(x)\,dm(\xi)$$

$$= \int_{\mathbb{R}^{3d}}\left(\int_{\mathbb{R}^d}e^{-i(v+\frac{1}{2}x-(\cdot))\cdot\xi}\,dm(\xi)\right)a(u,v)e^{-iux}f(\cdot-x)\,dm(u)\,dm(v)\,dm(x)$$

$$= \int_{\mathbb{R}^{3d}}\delta_{v+\frac{1}{2}x-(\cdot)}a(u,v)e^{-iux}f(\cdot-x)\,dm(u)\,dm(v)\,dm(x)$$

$$= \int_{\mathbb{R}^{3d}}\delta_{v-\frac{1}{2}(\cdot+x)}a(u,v)e^{-iu(\cdot-x)}f(x)\,dm(v)\,dm(u)\,dm(x)$$

$$= \int_{\mathbb{R}^{2d}}a\left(u,\frac{1}{2}(x+\cdot)\right)e^{iu(x-\cdot)}f(x)\,dm(u)\,dm(x).$$

This computation can be made rigorous by replacing the use of the physicist's δ-function by a mollifier argument as in the proof of Theorem 5.20.

By the definition of U, this gives the explicit formula

$$U^\star W(\widehat{a})Uf(y) = \int_{\mathbb{R}^{2d}}a\left(u,\frac{1}{2}(x+\frac{y}{\sqrt{2}})\right)$$

$$\times \exp\left(iu(x-\frac{y}{\sqrt{2}})\right)\exp\left(-\frac{1}{2}|x|^2+\frac{1}{4}|y|^2\right)f(x\sqrt{2})\,dm(u)\,dm(x)$$

$$= \frac{1}{2^{d/2}}\int_{\mathbb{R}^{2d}}a\left(u,\frac{x+y}{2\sqrt{2}}\right)$$

$$\times \exp\left(iu(\frac{x-y}{\sqrt{2}})\right) \exp\left(-\frac{1}{4}|x|^2 + \frac{1}{4}|y|^2\right) f(x)\, dm(u)\, dm(x)$$

$$=: \int_{\mathbb{R}^d} K_a(y,x) f(x)\, dm(x)$$

with

$$K_a(y,x) = \frac{1}{2^{d/2}} \exp\left(-\frac{1}{4}|x|^2 + \frac{1}{4}|y|^2\right) \int_{\mathbb{R}^d} a\left(u, \frac{x+y}{2\sqrt{2}}\right) \exp\left(iu(\frac{x-y}{\sqrt{2}})\right) dm(u).$$

Applying this to the function a_s we obtain

$$K_{a_s}(y,x) = \frac{1}{2^{d/2}} \exp\left(-\frac{1}{4}|x|^2 + \frac{1}{4}|y|^2\right)$$

$$\times \int_{\mathbb{R}^d} \exp\left(-s(|u|^2 + \frac{1}{8}|x+y|^2)\right) \exp\left(iu(\frac{x-y}{\sqrt{2}})\right) dm(u)$$

$$= \frac{1}{2^{d/2}} \exp\left(-\frac{s}{8}|x+y|^2\right) \exp\left(-\frac{1}{4}|x|^2 + \frac{1}{4}|y|^2\right)$$

$$\times \int_{\mathbb{R}^d} \exp\left(-s|u|^2 + iu(\frac{x-y}{\sqrt{2}})\right) dm(u)$$

$$= \frac{1}{2^{d/2}} \exp\left(-\frac{1}{8s}|x-y|^2\right) \exp\left(-\frac{s}{8}|x+y|^2\right) \exp\left(-\frac{1}{4}|x|^2 + \frac{1}{4}|y|^2\right)$$

$$\times \int_{\mathbb{R}^d} \exp\left(-s|\eta|^2\right) dm(\eta)$$

$$= \frac{1}{2^d s^{d/2}} \exp\left(-\frac{1}{8s}|x-y|^2\right) \exp\left(-\frac{s}{8}|x+y|^2\right) \exp\left(-\frac{1}{4}|x|^2 + \frac{1}{4}|y|^2\right)$$

$$= \frac{1}{2^d s^{d/2}} \exp\left(-\frac{1}{8s}(1-s)^2(|x|^2 + |y|^2) + \frac{1}{4}(\frac{1}{s} - s)xy\right) \exp\left(-\frac{1}{2}|x|^2\right).$$

Therefore, with $s = \frac{1-e^{-t}}{1+e^{-t}}$,

$$(1+s)^d U^\star W(\hat{a}_s) U f(y)$$

$$= (1+s)^d \int_{\mathbb{R}^d} K_{a_s}(y,x) f(x)\, dm(x)$$

$$= \frac{(1+s)^d}{2^d (2\pi s)^{d/2}} \int_{\mathbb{R}^d} \exp\left(-\frac{1}{8s}(1-s)^2(|x|^2 + |y|^2) + \frac{1}{4}(\frac{1}{s} - s)xy\right) f(x) \exp\left(-\frac{1}{2}|x|^2\right) dx$$

$$= \frac{1}{2^d (2\pi)^{d/2}} \left(\frac{2}{1+e^{-t}}\right)^d \left(\frac{1+e^{-t}}{1-e^{-t}}\right)^{d/2}$$

$$\times \int_{\mathbb{R}^d} \exp\left(-\frac{1}{2} \frac{e^{-2t}}{1-e^{-2t}}(|x|^2 + |y|^2) + \frac{e^{-t}}{1-e^{-2t}} xy\right) \exp\left(-\frac{1}{2}|x|^2\right) f(x)\, dx$$

$$= \frac{1}{(2\pi)^{d/2}} \left(\frac{1}{1-e^{-2t}}\right)^{d/2} \int_{\mathbb{R}^d} \exp\left(-\frac{1}{2} \frac{|e^{-t}y - x|^2}{1-e^{-2t}}\right) f(x)\, dx$$

$$= \int_{\mathbb{R}^d} M_t(y,x) f(x)\, dx = OU(t) f(y),$$

where

$$M_t(y,x) = \frac{1}{(2\pi)^{d/2}} \left(\frac{1}{1 - e^{-2t}} \right)^{d/2} \exp\left(-\frac{1}{2} \frac{|e^{-t}y - x|^2}{1 - e^{-2t}} \right)$$

is the Mehler kernel; the last step used the Mehler formula (13.31) for $OU(t)$.   □

15.6 Second Quantisation

Up to this point we have been concerned with the problem of *first quantisation*, namely, how to define the quantum analogues of classical observables. In order to arrive at a version of Quantum Mechanics that is consistent with Special Relativity, one must be able to describe systems with a variable number of particles. This is due to the fact that the equivalence of mass and energy makes it possible that particles are created and annihilated. If one uses a Hilbert space H to describe the pure states of a single particle, one postulates that the n-fold Hilbert space tensor product

$$H^{\otimes n} := \underbrace{H \otimes \cdots \otimes H}_{n \text{ times}}$$

describes the pure states of a system of n such particles. As explained in Appendix B, we have a direct sum decomposition

$$H^{\otimes n} = \Gamma^n(H) \oplus \Lambda^n(H)$$

into symmetric and antisymmetric tensor products. A *boson* is a particle whose n-particle states are given by elements of $\Gamma^n(H)$ and a *fermion* is a particle whose n-particle states are given by elements of $\Lambda^n(H)$. We will discuss the bosonic theory only; the fermionic theory requires deeper tools from noncommutative analysis that would take us too far afield. The bosonic theory, moreover, has interesting connections to several other topics covered in this work.

The elements of the Hilbert space direct sum

$$\Gamma(H) := \bigoplus_{n \in \mathbb{N}} \Gamma^n(H)$$

correspond to superpositions of states carrying different numbers of bosons. The process of passing from H to $\Gamma(H)$ is called *(bosonic, or symmetric) second quantisation*. The observation that every contraction T on H extends to a contraction $\Gamma(T)$ on $\Gamma(H)$ (see Section 15.6.c) allows us to establish a beautiful connection, for the special case $H = \mathbb{K}^d$, with the Ornstein–Uhlenbeck semigroup discussed in Section 13.6.e, namely,

$$OU(t) = \Gamma(e^{-t}I)$$

(Theorem 15.68). Under this correspondence, the negative generator $-L$ of this semigroup corresponds to the number operator of Section 15.3.d (where a unitarily equivalent model of it was studied). Our study will also uncover a deep connection between second quantisation and the Fourier transform: over the complex scalars, the Fourier–Plancherel transform is unitarily equivalent to the second quantisation of the operator $-iI$ (Theorem 15.70). Taken together, these facts connect the Fourier–Plancherel transform to the Ornstein–Uhlenbeck semigroup. Some connections of second quantisation with Number Theory will be discussed in the Notes.

For simplicity we will limit ourselves to the case where the Hilbert space describing the pure states of a single particle is finite-dimensional. *Mutatis mutandis*, the theory generalises to arbitrary Hilbert spaces H if one replaces the Gaussian measure by a so-called H-isonormal process, a central object in Malliavin calculus. Although this generalisation does not pose any mathematical difficulties we will not pursue it, as it adds a layer of abstraction that would only obscure the various connections just described.

Unless otherwise stated the scalar field $\mathbb{K}$ is allowed to be either real or complex.

15.6.a The Wiener–Itô Chaos Decomposition

For $h \in \mathbb{R}^d$ we define $\phi_h \in L^2(\mathbb{R}^d, \gamma) = L^2(\mathbb{R}^d, \gamma; \mathbb{K})$, where $\mathbb{K} = \mathbb{R}$ or $\mathbb{K} = \mathbb{C}$ by

$$\phi_h(x) := (x|h) = x \cdot h = \sum_{j=1}^{d} x_j h_j, \quad x \in \mathbb{R}^d.$$

Let $(H_n)_{n \in \mathbb{N}}$ be the sequence of Hermite polynomials introduced in Section 3.5.b. For $n \in \mathbb{N}$ we define

$$\mathscr{H}_n := \overline{\mathrm{span}} \{ H_n(\phi_h) : h \in \mathbb{R}^d, |h| = 1 \},$$

the closure being taken in $L^2(\mathbb{R}^d, \gamma)$. Here, $(H_n(\phi_h))(x) := H_n(\phi_h(x)) = H_n((x|h))$ for $x \in \mathbb{R}^d$. The space $\mathscr{H}_n$ is sometimes referred to as the *Gaussian chaos* of order n. Note that $\mathscr{H}_0 = \mathbb{K}\mathbf{1}$ is the one-dimensional space of constant functions.

We are going to prove that the subspaces $\mathscr{H}_n$ are pairwise orthogonal and induce a direct sum decomposition

$$\bigoplus_{n \in \mathbb{N}} \mathscr{H}_n = L^2(\mathbb{R}^d, \gamma).$$

In a second step we will identify orthonormal bases for the summands $\mathscr{H}_n$.

To start our analysis, for $h \in \mathbb{R}^d$ we define the functions

$$K_h := \exp\left(\phi_h - \frac{1}{2} |h|^2 \right).$$

From $\exp(\phi_h) \in L^2(\mathbb{R}^d, \gamma)$ we see that K_h is well defined as an element of $L^2(\mathbb{R}^d, \gamma)$.

From (3.15) we see that

$$K_h = \exp\left(|h|\phi_{h/|h|} - \frac{1}{2}|h|^2\right) = \sum_{n \in \mathbb{N}} \frac{|h|^n}{n!} H_n(\phi_{h/|h|}), \quad h \in \mathbb{R}^d. \tag{15.29}$$

In particular,

$$K_h = \sum_{n \in \mathbb{N}} \frac{1}{n!} H_n(\phi_h), \quad |h| = 1. \tag{15.30}$$

Lemma 15.56. *The functions K_h, $h \in \mathbb{R}^d$, span a dense subspace in $L^2(\mathbb{R}^d, \gamma)$.*

Proof Suppose that $f \in L^2(\mathbb{R}^d, \gamma)$ is such that $(f|K_h) = 0$ for all $h \in \mathbb{R}^d$. Then $\int_{\mathbb{R}^d} f \exp(\phi_h) \, d\gamma = 0$ for all $h \in \mathbb{R}^d$. Taking $h := \sum_{j=1}^d c_j e_j$, with e_j the jth standard unit vector of $\mathbb{R}^d$, we see that

$$\int_{\mathbb{R}^d} f(x) \exp\left(\sum_{j=1}^d c_j x_j\right) d\gamma(x) = 0$$

for all $c_1, \ldots, c_d \in \mathbb{R}$. By analytic continuation we obtain that the same holds for all $c_1, \ldots, c_d \in \mathbb{C}$. Taking $c_j = -iy_j$ with $y_j \in \mathbb{R}$, this implies that the Fourier transform of the function $x \mapsto f(x) \exp\left(-\frac{1}{2}|x|^2\right)$ vanishes. By the injectivity of the Fourier transform (Theorem 5.20) we conclude that $f(x) \exp\left(-\frac{1}{2}|x|^2\right) = 0$ for almost all $x \in \mathbb{R}^d$, that is, $f(x) = 0$ for almost all $x \in \mathbb{R}^d$. $\quad\square$

Theorem 15.57 (Wiener–Itô decomposition). *We have the orthogonal decomposition*

$$L^2(\mathbb{R}^d, \gamma) = \bigoplus_{n \in \mathbb{N}} \mathscr{H}_n.$$

Proof Fix $h, h' \in \mathbb{R}^d$ with $|h| = |h'| = 1$ and $s, t \in \mathbb{R}$. Repeating the steps of (3.16), for all $s, t \in \mathbb{R}$ we have

$$(H(s, \phi_h)|H(t, \phi_{h'})_{L^2(\mathbb{R}^d, \gamma)} = \exp\left(st \left(h|h'\right)\right).$$

Substituting $H(r, \cdot) = \sum_{k=0}^\infty \frac{r^k}{k!} H_k$ and $\exp(st(h|h')) = \sum_{k=0}^\infty \frac{(st)^k}{k!} ((h|h'))^k$, and taking the partial derivative $\frac{\partial^{n+m}}{\partial s^m \partial t^n}$ at $s = t = 0$ on both sides of the resulting identity, we obtain

$$(H_m(\phi_h)|H_n(\phi_{h'}))_{L^2(\mathbb{R}^d, \gamma)} = \delta_{mn} n! (h|h')^n.$$

Noting that $\delta_{mn} n! = \delta_{mn} \sqrt{m!} \sqrt{n!}$, this can be equivalently stated as

$$\left(\frac{H_m(\phi_h)}{\sqrt{m!}} \Big| \frac{H_n(\phi_{h'})}{\sqrt{n!}}\right) = \delta_{mn}(h|h')^n. \tag{15.31}$$

For $m \neq n$ this implies $\mathscr{H}_m \perp \mathscr{H}_n$.

If $f \perp \mathscr{H}_n$ for all $n \in \mathbb{N}$, then $(f|H_n(\phi_h))_{L^2(\mathbb{R}^d, \gamma)} = 0$ for all $h \in \mathbb{R}^d$ with $|h| = 1$, and therefore (15.29) implies that $(f|K_h)_{L^2(\mathbb{R}^d, \gamma)} = 0$ for all $h \in \mathbb{R}^d$, and therefore $f = 0$ by Lemma 15.56. $\quad\square$

The next result shows that the Wiener–Itô decomposition diagonalises the Ornstein–Uhlenbeck semigroup OU on $L^2(\mathbb{R}^d, \gamma)$ introduced in Section 13.6.e:

Theorem 15.58. *The following identities hold:*

(1) *for all $h \in \mathbb{R}^d$ and $t \geqslant 0$,*

$$OU(t)K_h = K_{e^{-t}h};$$

(2) *for all $n \in \mathbb{N}$, $F \in \mathcal{H}_n$, and $t \geqslant 0$,*

$$OU(t)F = e^{-nt}F.$$

Proof Completing squares in the exponential, for all $h \in \mathbb{R}^d$ and $t \geqslant 0$ we have

$$\int_{\mathbb{R}^d} \exp\left(\sqrt{1-e^{-2t}}(y|h)\right) d\gamma(y) = \prod_{j=1}^{d} \frac{1}{\sqrt{2\pi}} \int_{\mathbb{R}} \exp\left(\sqrt{1-e^{-2t}}y_j h_j - \frac{1}{2}|y_j|^2\right) dy_j$$

$$= \prod_{j=1}^{d} \exp\left(\frac{1}{2}(1-e^{-2t})|h_j|^2\right) = \exp\left(\frac{1}{2}(1-e^{-2t})|h|^2\right).$$

Hence, from the definitions of $OU(t)$, ϕ_h, and K_h,

$$OU(t)K_h(x) = \int_{\mathbb{R}^d} \exp\left(\phi_h(e^{-t}x + \sqrt{1-e^{-2t}}y) - \frac{1}{2}|h|^2\right) d\gamma(y)$$

$$= \exp\left(e^{-t}(x|h) - \frac{1}{2}|h|^2\right) \int_{\mathbb{R}^d} \exp\left(\sqrt{1-e^{-2t}}(y|h)\right) d\gamma(y)$$

$$= \exp\left(e^{-t}(x|h) - \frac{1}{2}|h|^2\right) \exp\left(\frac{1}{2}(1-e^{-2t})|h|^2\right)$$

$$= \exp\left((x|e^{-t}h) - \frac{1}{2}|e^{-t}h|^2\right) = K_{e^{-t}h}(x).$$

If $|h| = 1$ and $s \geqslant 0$, it follows from (15.29) and the preceding calculation that

$$OU(t) \sum_{n \in \mathbb{N}} \frac{s^n}{n!} H_n(\phi_h) = OU(t)K_{sh} = K_{e^{-t}sh} = \sum_{n \in \mathbb{N}} \frac{s^n e^{-nt}}{n!} H_n(\phi_h).$$

Taking nth derivatives in s and evaluating at $s = 0$, we obtain the identity

$$OU(t)H_n(\phi_h) = e^{-nt}H_n(\phi_h).$$

By linearity and taking limits, this gives (2). □

Part (2) of the theorem implies that each summand $\mathcal{H}_n$ is contained in $D(L)$, where L is the generator of $(OU(t))_{t \geqslant 0}$, and

$$LF = -nF, \quad F \in \mathcal{H}_n, \, n \in \mathbb{N}.$$

Over the complex scalars, Proposition 10.32 can be applied and we obtain:

Corollary 15.59. $\sigma(-L) = \mathbb{N}$.

15.6.b The Wiener–Itô Isometry

Our next aim is to find an orthonormal basis for each summand $\mathscr{H}_n$. This will be achieved in Theorem 15.60 by means of multivariate Hermite polynomials.

For $\mathbf{n} = (n_1, \ldots, n_k) \in \mathbb{N}^k$ we write

$$|\mathbf{n}| := \sum_{j=1}^{k} n_j, \quad \mathbf{n}! := \prod_{j=1}^{k} n_j!.$$

For orthonormal systems $\mathbf{h} = (h_j)_{j=1}^{k}$ in $\mathbb{R}^d$ and $\mathbf{n} \in \mathbb{N}^k$ we define

$$H_{\mathbf{n}}(\phi_{\mathbf{h}}) := \prod_{j=1}^{k} H_{n_j}(\phi_{h_j}).$$

Theorem 15.60. *Let* $\mathbf{h} = (h_j)_{j=1}^{d}$ *be an orthonormal basis for* $\mathbb{R}^d$. *For each* $n \in \mathbb{N}$ *the family*

$$\left\{ \frac{1}{\sqrt{\mathbf{n}!}} H_{\mathbf{n}}(\phi_{\mathbf{h}}) : \mathbf{n} \in \mathbb{N}^d, |\mathbf{n}| = n \right\}$$

defines an orthonormal basis for $\mathscr{H}_n$. *As a consequence, the family*

$$\left\{ \frac{1}{\sqrt{\mathbf{n}!}} H_{\mathbf{n}}(\phi_{\mathbf{h}}) : \mathbf{n} \in \mathbb{N}^d \right\}$$

defines an orthonormal basis for $L^2(\mathbb{R}^d, \gamma)$.

Proof The proof is divided into three steps.

Step 1 – First we prove that the family $\{ \frac{1}{\sqrt{\mathbf{n}!}} H_{\mathbf{n}}(\phi_{\mathbf{h}}) : \mathbf{n} \in \mathbb{N}^d \}$ is an orthonormal system in $L^2(\mathbb{R}^d, \gamma)$. Let $\mathbf{m}, \mathbf{n} \in \mathbb{N}^d$. By separation of variables and (15.31),

$$\left(\frac{1}{\sqrt{\mathbf{m}!}} H_{\mathbf{m}}(\phi_{\mathbf{h}}) \Big| \frac{1}{\sqrt{\mathbf{n}!}} H_{\mathbf{n}}(\phi_{\mathbf{h}}) \right) = \prod_{i=1}^{d} \prod_{j=1}^{d} \left(\frac{1}{\sqrt{m_i!}} H_{m_i}(\phi_{h_i}) \Big| \frac{1}{\sqrt{n_j!}} H_{n_j}(\phi_{h_j}) \right)$$

$$= \prod_{i=1}^{d} \prod_{j=1}^{d} \delta_{m_i n_j}(h_i | h_j)^{n_j} = \prod_{j=1}^{d} \delta_{m_j n_j} = \delta_{\mathbf{m}\mathbf{n}}. \tag{15.32}$$

Step 2 – Next we prove completeness of this system in $L^2(\mathbb{R}^d, \gamma)$. Suppose $f \in L^2(\mathbb{R}^d, \gamma)$ is such that $(f | H_{\mathbf{n}}(\phi_{\mathbf{h}})) = 0$ for all $\mathbf{n} \in \mathbb{N}^d$. Fix an arbitrary $h \in \mathbb{R}^d$ and put $g_k := \sum_{j=0}^{k} \frac{1}{j!} \phi_h^j$. Then $\lim_{k \to \infty} g_k = \exp(\phi_h)$ in $L^2(\mathbb{R}^d, \gamma)$ by dominated convergence. By writing $h = \sum_{j=1}^{d} c_j h_j$ we see that each g_k is a polynomial in $\phi_{h_1}, \ldots, \phi_{h_d}$, and such polynomials are linear combinations of the functions $H_{\mathbf{n}}(\phi_{\mathbf{h}})$ for appropriate multi-indices $\mathbf{n} \in \mathbb{N}^d$. It follows that $(f | g_k) = 0$ for all $k \in \mathbb{N}$. Passing to the limit $k \to \infty$ it follows that $(f | \exp(\phi_h)) = 0$, and therefore $(f | K_h) = 0$. Since $h \in \mathbb{R}^d$ was arbitrary, Lemma 15.56 implies that $f = 0$. Together with Step 1, this proves that $\{ H_{\mathbf{n}}(\phi_{\mathbf{h}}) : \mathbf{n} \in \mathbb{N}^d \}$ is an orthonormal basis of $L^2(\mathbb{R}^d, \gamma)$.

Step 3 – The final step is to prove that $\{\frac{1}{\sqrt{\mathbf{n}!}}H_{\mathbf{n}}(\phi_{\mathbf{h}}) : \mathbf{n} \in \mathbb{N}^d, |\mathbf{n}| = n\}$ is an orthonormal basis for $\mathscr{H}_n$. Denote by $\mathscr{G}_n$ the closed linear span of the set $\{H_{\mathbf{n}}(\phi_{\mathbf{h}}) : \mathbf{n} \in \mathbb{N}^d, |\mathbf{n}| = n\}$. By Step 1, $\mathscr{G}_m \perp \mathscr{G}_n$ if $m \neq n$. If $h = \sum_{j=1}^d c_j h_j \in \mathbb{K}^d$ and $0 \leqslant m \leqslant n$, then $H_m(\phi_h) = H_m(\sum_{j=1}^d c_j \phi_{h_j})$ is a linear combination of polynomials of the form $H_{\mathbf{k}}(\phi_{\mathbf{h}})$ with $|\mathbf{k}| \leqslant \mathbf{m}$, and therefore $H_m(\phi_h) \in \bigoplus_{j=1}^m \mathscr{G}_j$. In particular this implies that $\mathscr{H}_m \subseteq \bigoplus_{j=1}^m \mathscr{G}_j \subseteq \bigoplus_{j=1}^n \mathscr{G}_j$ and therefore

$$\bigoplus_{j=1}^n \mathscr{H}_j \subseteq \bigoplus_{j=1}^n \mathscr{G}_j.$$

Also, by Step 1,

$$\mathscr{H}_n \perp \bigoplus_{j=1}^{n-1} \mathscr{G}_j.$$

It follows that $\mathscr{H}_n \subseteq \mathscr{G}_n$. This being true for all $n \in \mathbb{N}$, by Step 2 it follows that

$$L^2(\mathbb{R}^d, \gamma) = \bigoplus_{n \in \mathbb{N}} \mathscr{H}_n \subseteq \bigoplus_{n \in \mathbb{N}} \mathscr{G}_n = L^2(\mathbb{R}^d, \gamma).$$

From this we infer that $\mathscr{H}_n = \mathscr{G}_n$ for all $n \in \mathbb{N}$. □

The orthogonal projection in $L^2(\mathbb{R}^d, \gamma)$ onto $\mathscr{H}_n$ will be denoted by J_n.

Corollary 15.61. *For all $n \in \mathbb{N}$ and $h \in \mathbb{R}^d$ with $|h| = 1$ we have*

$$J_n(\phi_h^n) = H_n(\phi_h).$$

More generally, if $(h_j)_{j=1}^k$ is orthonormal in $\mathbb{R}^d$ and $\mathbf{n} = (n_1, \ldots, n_k) \in \mathbb{N}^k$, then

$$J_{|\mathbf{n}|}(\phi_{h_1}^{n_1} \cdots \phi_{h_k}^{n_k}) = H_{n_1}(\phi_{h_1}) \cdots H_{n_k}(\phi_{h_k}).$$

Proof We have

$$H_{n_1}(\phi_{h_1}) \cdots H_{n_k}(\phi_{h_k}) = \phi_{h_1}^{n_1} \cdots \phi_{h_k}^{n_k} + q(\phi_{h_1}, \ldots, \phi_{h_k}), \tag{15.33}$$

where q is a polynomial of degree strictly less than $|\mathbf{n}|$. Hence Theorem 15.60 implies that $q(h_1, \ldots, h_k) \in \bigoplus_{j=0}^{|\mathbf{n}|-1} \mathscr{H}_j$, and the corollary follows by projecting (15.33) onto $\mathscr{H}_{|\mathbf{n}|}$. Since $H_{n_1}(\phi_{h_1}) \cdots H_{n_k}(\phi_{h_k}) \in \mathscr{H}_{|\mathbf{n}|}$, the left-hand side remains unchanged; the right-hand side is mapped to $J_{|\mathbf{n}|}(\phi_{h_1}^{n_1} \cdots \phi_{h_k}^{n_k})$. □

Corollary 15.62. *For all $h \in \mathbb{R}^d$ with $|h| = 1$,*

$$K_h = \sum_{n \in \mathbb{N}} \frac{1}{n!} J_n(\phi_h^n).$$

Proof This follows from the previous corollary and (15.30). □

Proposition 15.63. *For all $g_1, \ldots, g_n \in \mathbb{R}^d$ and $h_1, \ldots, h_n \in \mathbb{R}^d$ we have*

$$(J_n(\phi_{g_1} \cdots \phi_{g_n}) | J_n(\phi_{h_1} \cdots \phi_{h_n})) = \sum_{\sigma \in S_n} (g_1 | h_{\sigma(1)}) \cdots (g_n | h_{\sigma(n)}),$$

where S_n is the group of permutations of $\{1, \ldots, n\}$.

Proof   Let $(e_j)_{j=1}^d$ be the standard basis of $\mathbb{R}^d$. Choose nonnegative integers $\ell_1, \ldots, \ell_j$ and $m_1, \ldots, m_k$ such that $\ell_1 + \cdots + \ell_j = m_1 + \cdots + m_k = n$. By adding extra zeroes to the shortest of these two sequences we may assume that $j = k$. By Corollary 15.61,

$$J_n(\phi_{e_{i_1}}^{\ell_1} \cdots \phi_{e_{i_k}}^{\ell_k}) = H_{\ell_1}(\phi_{e_{i_1}}) \cdots H_{\ell_k}(\phi_{e_{i_k}}),$$

$$J_n(\phi_{e_{i_1}}^{m_1} \cdots \phi_{e_{i_k}}^{m_k}) = H_{m_1}(\phi_{e_{i_1}}) \cdots H_{m_k}(\phi_{e_{i_k}}).$$

Hence, by (15.32),

$$(J_n(\phi_{e_{i_1}}^{\ell_1} \cdots \phi_{e_{i_k}}^{\ell_k}) | J_n(\phi_{e_{i_1}}^{m_1} \cdots \phi_{e_{i_k}}^{m_k})) = \mathbf{m}! \delta_{\ell_1, m_1} \cdots \delta_{\ell_k, m_k}.$$

On the other hand, with

$$g_1 = \cdots = g_{\ell_1} := e_{i_1}, \quad \ldots, \quad g_{\ell_1 + \cdots + \ell_{k-1} + 1} = \cdots = g_{\ell_1 + \cdots + \ell_k} = e_{i_k},$$
$$h_1 = \cdots = h_{m_1} := e_{i_1}, \quad \ldots, \quad h_{m_1 + \cdots + m_{k-1} + 1} = \cdots = m_{\ell_1 + \cdots + \ell_k} = e_{i_k}, \tag{15.34}$$

we have

$$\sum_{\sigma \in S_n} (g_1 | h_{\sigma(1)}) \cdots (g_n | h_{\sigma(n)}))$$

$$= \sum_{\sigma \in S_n} (e_{i_1} | h_{\sigma(1)}) \cdots (e_{i_1} | h_{\sigma(\ell_1)}) \cdots$$

$$\cdots (e_{i_k} | h_{\sigma(\ell_1 + \cdots + \ell_{k-1} + 1)}) \cdots (e_{i_k} | h_{\sigma(\ell_1 + \cdots + \ell_{k-1} + \ell_k)}))$$

$$= m_1! \cdots m_k! \cdot \delta_{\ell_1, m_1} \cdots \delta_{\ell_k, m_k} = \mathbf{m}! \delta_{\ell_1, m_1} \cdots \delta_{\ell_k, m_k}.$$

This proves the corollary in the special case (15.34). By n-linearity, the corollary then also follows if each of the functions f and g are finite linear combinations of such expressions, and finally the general case follows by density. $\square$

The main result of this section relates the spaces $\mathcal{H}_n$ to the n-fold symmetric tensor product of $\mathbb{K}^d$. The n-fold tensor product

$$(\mathbb{K}^d)^{\otimes n} := \underbrace{\mathbb{K}^d \otimes \cdots \otimes \mathbb{K}^d}_{n \text{ times}}$$

is a Hilbert space with respect to the inner product

$$\left(\sum_{j=1}^{\ell} g_1^{(j)} \otimes \cdots \otimes g_n^{(j)} \,\Big|\, \sum_{k=1}^m h_1^{(k)} \otimes \cdots \otimes h_n^{(k)} \right) := \sum_{j=1}^{\ell} \sum_{k=1}^m (g_1^{(j)} | h_1^{(k)}) \cdots (g_n^{(j)} | h_n^{(k)}).$$

We identify $(\mathbb{K}^d)^{\otimes 0}$ with the scalar field $\mathbb{K}$. From Appendix B we recall that the *n-fold symmetric tensor product* of $\mathbb{K}^d$, denoted by $\Gamma^n(\mathbb{K}^d)$, is defined as the range of the orthogonal projection $P_{\Gamma_n} \in \mathscr{L}((\mathbb{K}^d)^{\otimes n})$ given by

$$P_{\Gamma_n}(h_1 \otimes \cdots \otimes h_n) := \frac{1}{n!} \sum_{\sigma \in S_n} h_{\sigma(1)} \otimes \cdots \otimes h_{\sigma(n)}, \quad h_1, \ldots, h_n \in \mathbb{K}^d,$$

and extended by linearity, where S_n is the group of permutations of $\{1, \ldots, n\}$. Equivalently, $\Gamma^n(\mathbb{K}^d)$ is the subspace of those elements of $(\mathbb{K}^d)^{\otimes n}$ that are invariant under the action of S_n.

For the formulation of the next theorem we introduce the following notation. Let $\mathbf{h} = (h_j)_{j=1}^k$ be a finite sequence in $\mathbb{K}^d$. For $\mathbf{n} \in \mathbb{N}^k$ with $|\mathbf{n}| = n$ let

$$\mathbf{h}^{\otimes \mathbf{n}} := h_1^{\otimes n_1} \otimes \cdots \otimes h_k^{\otimes n_k},$$

where $h_j^{\otimes n_j} = h_j \otimes \cdots \otimes h_j$ (n_j times) with the convention that terms of the form $h_j^{\otimes 0}$ are omitted. Similarly we let

$$\phi_{\mathbf{h}}^{\otimes \mathbf{n}} := \phi_{h_1}^{n_1} \cdots \phi_{h_k}^{n_k}.$$

Theorem 15.64 (Wiener–Itô isometry). *There exists a unique isometric isomorphism*

$$W : \Gamma(\mathbb{K}^d) \to L^2(\mathbb{R}^d, \gamma; \mathbb{K})$$

with the following property: For every $1 \leqslant k \leqslant d$, every orthonormal system $\mathbf{h} = (h_j)_{j=1}^k$ in $\mathbb{R}^d$, every $n \in \mathbb{N}$, and every multi-index $\mathbf{n} \in \mathbb{N}^k$ with $|\mathbf{n}| = n$,

$$W(P_{\Gamma_n}(\mathbf{h}^{\otimes \mathbf{n}})) = \frac{1}{\sqrt{n!}} H_{\mathbf{n}}(\phi_{\mathbf{h}}), \quad \mathbf{n} \in \mathbb{N}^k.$$

This mapping W will be referred to as the *Wiener–Itô isometry*.

Proof For complex scalars, the result follows from the real case by complexification. We may therefore assume that $\mathbb{K} = \mathbb{R}$.

Uniqueness being clear, we concentrate on existence. Let $\mathbf{e} = (e_j)_{j=1}^d$ be the standard basis in $\mathbb{R}^d$. For every integer $n \in \mathbb{N}$ consider the linear mapping $W_n : \Gamma^n(\mathbb{R}^d) \to \mathscr{H}_n$ defined by

$$W_n : P_{\Gamma_n}(\mathbf{e}^{\otimes \mathbf{n}}) \mapsto \frac{1}{\sqrt{n!}} H_{\mathbf{n}}(\phi_{\mathbf{e}}), \quad \mathbf{n} \in \mathbb{N}^d, \ |\mathbf{n}| = n,$$

and extended by linearity. We begin by showing that if $\mathbf{h} = (h_i)_{i=1}^k$ is any orthonormal system in $\mathbb{R}^d$, then

$$W_n(P_{\Gamma_n}(\mathbf{h}^{\otimes \mathbf{n}})) = \frac{1}{\sqrt{n!}} H_{\mathbf{n}}(\phi_{\mathbf{h}}), \quad \mathbf{n} \in \mathbb{N}^k, \ |\mathbf{n}| = n. \tag{15.35}$$

As a second step we show that W_n is an isometry from $\Gamma^n(\mathbb{R}^d)$ onto $\mathscr{H}_n$. In view of the Wiener–Itô decomposition, these two facts prove the theorem.

Step 1 – Let $h_i = \sum_{j=1}^d c_{ij} e_j$, $i = 1, \ldots, k$, be the expansion in terms of the standard basis. Let $\mathbf{n} \in \mathbb{N}^k$ be a multi-index satisfying $|\mathbf{n}| = n$. Then,

$$
\begin{aligned}
W_n(P_{\Gamma_n}(\mathbf{h}^{\otimes \mathbf{n}})) &= W_n\left(P_{\Gamma_n}\left(\Big(\sum_{j=1}^d c_{1j} e_j\Big)^{\otimes n_1} \otimes \cdots \otimes \Big(\sum_{j=1}^d c_{kj} e_j\Big)^{\otimes n_k} \right) \right) \\
&= W_n\Big(P_{\Gamma_n} \sum_{|\mathbf{m}|=n} a_\mathbf{m} \mathbf{e}^{\otimes \mathbf{m}} \Big) = W_n\Big(\sum_{|\mathbf{m}|=n} a_\mathbf{m} P_{\Gamma_n} \mathbf{e}^{\otimes \mathbf{m}} \Big) \\
&= \sum_{|\mathbf{m}|=n} a_\mathbf{m} \frac{1}{\sqrt{n!}} H_\mathbf{m}(\phi_\mathbf{e}) = \frac{1}{\sqrt{n!}} \sum_{|\mathbf{m}|=n} a_\mathbf{m} \prod_{j=1}^d H_{m_j}(\phi_{e_j}) \\
&= \frac{1}{\sqrt{n!}} \sum_{|\mathbf{m}|=n} a_\mathbf{m} J_n\Big(\prod_{j=1}^d \phi_{e_j}^{m_j} \Big) = \frac{1}{\sqrt{n!}} J_n\Big(\sum_{|\mathbf{m}|=n} a_\mathbf{m} \prod_{j=1}^d \phi_{e_j}^{m_j} \Big) \\
&= \frac{1}{\sqrt{n!}} J_n\Big(\sum_{|\mathbf{m}|=n} a_\mathbf{m} \phi_\mathbf{e}^\mathbf{m} \Big) = \frac{1}{\sqrt{n!}} J_n\Big(\prod_{\ell=1}^k \Big(\sum_{j=1}^d c_{\ell j} \phi_{e_j}\Big)^{n_\ell} \Big) \\
&= \frac{1}{\sqrt{n!}} J_n\Big(\prod_{\ell=1}^k \phi_{h_\ell}^{n_\ell} \Big) = \frac{1}{\sqrt{n!}} \prod_{\ell=1}^k H_{n_\ell}(\phi_{h_\ell}) = \frac{1}{\sqrt{n!}} H_\mathbf{n}(\phi_\mathbf{h}),
\end{aligned}
$$

where the coefficients $a_\mathbf{m}$, $\mathbf{m} \in \mathbb{N}^d$, are determined by the identity

$$
\sum_{|\mathbf{m}|=n} a_\mathbf{m} \xi^\mathbf{m} = \prod_{\ell=1}^k \Big(\sum_{j=1}^d c_{\ell j} \xi_j \Big)^{n_\ell}
$$

in the formal variables $\xi_1, \ldots, \xi_d$ and $\xi^\mathbf{m} = \xi_1^{m_1} \cdots \xi_d^{m_d}$. This establishes (15.35).

Step 2 – In this step we show that the mappings W_n are isometric from $\Gamma^n(\mathbb{R}^d)$ onto $\mathscr{H}_n$. First let $h_1, \ldots, h_n \in \mathbb{R}^d$ be arbitrary. By Proposition 15.63 we have

$$
\begin{aligned}
\|J_n(\phi_{h_1} \cdots \phi_{h_n})\|^2 &= \sum_{\sigma \in S_n} (h_1 | h_{\sigma(1)}) \cdots (h_n | h_{\sigma(n)}) \\
&= \sum_{\sigma \in S_n} (h_1 \otimes \cdots \otimes h_n | h_{\sigma(1)} \otimes \cdots \otimes h_{\sigma(n)}) \\
&= n!(h_1 \otimes \cdots \otimes h_n | P_{\Gamma_n}(h_1 \otimes \cdots \otimes h_n)) = n! \|P_{\Gamma_n}(h_1 \otimes \cdots \otimes h_n)\|^2,
\end{aligned}
$$

the projection $P_{\Gamma_n} = P_{\Gamma_n}^2$ being orthogonal and hence selfadjoint. Specialising to the standard basis of $\mathbb{R}^d$ and using Corollary 15.61, we obtain

$$
\begin{aligned}
\|P_{\Gamma_n}(\mathbf{e}^{\otimes \mathbf{n}})\| &= \|P_{\Gamma_n}(e_1^{\otimes n_1} \otimes \cdots \otimes e_d^{\otimes n_d})\| = \frac{1}{\sqrt{n!}} \|J_n(\phi_{e_1}^{n_1} \cdots \phi_{e_d}^{n_d})\| \\
&= \frac{1}{\sqrt{n!}} \|H_{n_1}(\phi_{e_1}) \cdots H_{n_d}(\phi_{e_d})\| = \frac{1}{\sqrt{n!}} \|H_\mathbf{n}(\phi_\mathbf{e})\|.
\end{aligned}
$$

This identity extends to finite linear combinations by the Pythagorean identity, noting that both on the left and on the right the contributing terms in the sums are orthogonal. This proves that the mapping in the statement of the proposition is an isometry. Since the multivariate Hermite polynomials of degree n form an orthonormal basis in $\mathcal{H}_n$, this isometry is surjective. □

As a special case, note that for all $h \in \mathbb{R}^d$ with $|h| = 1$,

$$W(h^{\otimes n}) = \frac{1}{\sqrt{n!}} H_n(\phi_h). \tag{15.36}$$

15.6.c Second Quantised Operators

For a linear operator T on H we obtain a linear operator $T^{\otimes n}$ on $H^{\otimes n}$ by

$$T^{\otimes n}(h_1 \otimes \cdots \otimes h_n) := (Th_1 \otimes \cdots \otimes Th_n)$$

and linearity. For $n = 0$ we understand that $H^{\otimes n} = \mathbb{K}$ and $T^{\otimes 0} = I_{\mathbb{K}}$, where $\mathbb{K}$ is the scalar field.

Proposition 15.65. *If T is a bounded operator on H, then $T^{\otimes n}$ is a bounded operator on $H^{\otimes n}$ of norm*

$$\|T^{\otimes n}\| = \|T\|^n.$$

Proof By a scaling argument it suffices to show that if $\|T\| = 1$, then $\|T^{\otimes n}\| = 1$. The inequality $\|T^{\otimes n}\| \geqslant 1$ being obvious from the definition of the inner product on $H^{\otimes n}$, we prove the inequality $\|T^{\otimes n}\| \leqslant 1$.

If the scalar field is real we denote by $H_{\mathbb{C}}$ the complexification of H. Endowed with the norm $\|h + ih'\|_{H_{\mathbb{C}}}^2 := \|h\|^2 + \|h'\|^2$, this is a complex inner product space. If T is a contraction on H, then $T_{\mathbb{C}}(h + ih') := Th + iTh'$ defines a contraction on $H_{\mathbb{C}}$ of the same norm. Noting that $(T_{\mathbb{C}})^{\otimes n} = (T^{\otimes n})_{\mathbb{C}}$, it suffices to prove the proposition in the case of complex scalars.

If T is unitary, then

$$\left\| T^{\otimes n} \sum_{j=1}^k c_j h_1^{(j)} \otimes \cdots \otimes h_n^{(j)} \right\|^2 = \sum_{i=1}^k \sum_{j=1}^k c_i \bar{c}_j \prod_{m=1}^n (Th_m^{(i)} | Th_m^{(j)})$$

$$= \sum_{i=1}^k \sum_{j=1}^k c_i \bar{c}_j \prod_{m=1}^n (h_m^{(i)} | h_m^{(j)}) = \left\| \sum_{j=1}^k c_j h_1^{(j)} \otimes \cdots \otimes h_n^{(j)} \right\|^2$$

and therefore $T^{\otimes n}$ is an isometry on $H^{\otimes n}$. The corresponding result for contractions follows from the fact that every contraction T on H can be represented as a convex combination of four unitaries by Lemma 14.25. □

Restricting $T^{\otimes n}$ to the symmetric part $\Gamma_n(H)$ of $H^{\otimes n}$, we obtain well-defined contractions $\Gamma_n(T)$ on $\Gamma_n(H)$. By taking direct sums,

$$\Gamma(T) := \bigoplus_{n \in \mathbb{N}} \Gamma^n(T)$$

defines a contraction on $\bigoplus_{n \in \mathbb{N}} \Gamma^n(H)$.

Definition 15.66 (Symmetric second quantisation). The Hilbert space completion $\Gamma(H)$ of $\bigoplus_{n \in \mathbb{N}} \Gamma^n(H)$ is called the *symmetric Fock space* over H. When T is a contraction on H, the contraction $\Gamma(T)$ on $\Gamma(H)$ is called the *symmetric second quantisation* of T.

Antisymmetric second quantisation can be defined similarly but will not be studied here. Because of this, we will omit the adjective 'symmetric' from now on and simply talk about *second quantisation*.

If S and T are contractions on H, their second quantisations satisfy

$$\Gamma(I) = I, \quad \Gamma(ST) = \Gamma(S)\Gamma(T), \quad \Gamma(T^\star) = (\Gamma(T))^\star. \tag{15.37}$$

In what follows we take again $H = \mathbb{K}^d$. If T is a contraction on $\mathbb{K}^d$, via the Wiener–Itô isometry (Theorem 15.64) the operator $\Gamma(T)$ induces a contraction on $L^2(\mathbb{R}^d, \gamma) = L^2(\mathbb{R}^d, \gamma; \mathbb{K})$ which, by a slight abuse of notation, will be denoted by $\Gamma(T)$ as well. It is easily checked that (15.37) holds again.

Lemma 15.67. *If T is a contraction on $\mathbb{K}^d$, then for all $h \in \mathbb{R}^d$,*

$$\Gamma(T)K_h = K_{Th}.$$

Proof Denoting by W the Wiener–Itô isometry, for all $h \in \mathbb{R}^d$ with $Th \neq 0$ we have

$$K_{Th} = \sum_{n \in \mathbb{N}} \frac{|Th|^n}{n!} H_n(\phi_{Th/|Th|}) = \sum_{n \in \mathbb{N}} \frac{|Th|^n}{n!} W((Th/|Th|)^{\otimes n})$$

$$= \sum_{n \in \mathbb{N}} \frac{1}{n!} W((Th)^{\otimes n}) = W\Big(\sum_{n \in \mathbb{N}} \frac{1}{n!} \Gamma_n(T) h^{\otimes n}\Big)$$

$$= W\Big(\Gamma(T) \sum_{n \in \mathbb{N}} \frac{1}{n!} h^{\otimes n}\Big) = \Gamma(T)\Big(W\Big(\sum_{n \in \mathbb{N}} \frac{1}{n!} h^{\otimes n}\Big)\Big)$$

$$= \Gamma(T) \sum_{n \in \mathbb{N}} \frac{1}{n!} H_n(\phi_h) = \Gamma(T)K_h,$$

where the first identity follows from (15.29) and the second and penultimate steps follow from (15.36). If $Th = 0$, then $K_{Th} = K_0 = \mathbf{1} = \Gamma(T)K_0$. $\qquad\square$

As a special case, for the Ornstein–Uhlenbeck semigroup we obtain:

Theorem 15.68 (Ornstein–Uhlenbeck semigroup and second quantisation). *Under the Wiener–Itô isometry, for all $t \geqslant 0$ we have*

$$OU(t) = \Gamma(e^{-t}I).$$

Proof This follows from Theorem 15.58 and Lemma 15.67, which give

$$OU(t)K_h = K_{e^{-t}h} = \Gamma(e^{-t}I)K_h,$$

and the density of the span of the functions K_h in $L^2(\mathbb{R}^d, \gamma)$ shown in Lemma 15.56. $\square$

Over the real scalars we have the following positivity result.

Theorem 15.69 (Positivity). *If T is a contraction on $\mathbb{R}^d$, then $\Gamma(T)$ is a positivity preserving contraction on $L^2(\mathbb{R}^d, \gamma)$.*

Proof By Lemma 15.67, for all $h \in \mathbb{R}^d$ we have $\Gamma(T)K_h = K_{Th} \geqslant 0$. Moreover, for all $c \in \mathbb{R}$,

$$\begin{aligned}
\Gamma(T)(\exp(c\phi_h)) = \Gamma(T)(\exp(\phi_{ch})) &= \exp(\tfrac{1}{2}c^2|h|^2)\Gamma(T)K_{ch} \\
&= \exp(\tfrac{1}{2}c^2|h|^2)K_{cTh} \\
&= \exp\Big(c\phi_{Th} + \tfrac{1}{2}c^2(|h|^2 - |Th|^2)\Big).
\end{aligned}$$

By analytic continuation this identity extends to arbitrary $c \in \mathbb{C}$.

Let $0 \leqslant f \in \mathscr{F}^2(\mathbb{R}^d) = \{f \in L^1(\mathbb{R}^d) \cap L^2(\mathbb{R}^d) : \widehat{f} \in L^1(\mathbb{R}^d) \cap L^2(\mathbb{R}^d)\}$. By Fourier inversion,

$$\begin{aligned}
\Gamma(T)f &= \frac{1}{(2\pi)^{d/2}} \int_{\mathbb{R}^d} \widehat{f}(\xi_1, \ldots, \xi_n)\Gamma(T)\exp(i\phi_\xi)\,d\xi \\
&= \frac{1}{(2\pi)^{d/2}} \int_{\mathbb{R}^d} \widehat{f}(\xi_1, \ldots, \xi_n)\exp\Big(i\phi_{T\xi} - \tfrac{1}{2}(|\xi|^2 - |T\xi|^2)\Big)\,d\xi.
\end{aligned}$$

In view of $\phi_{T\xi}(x) = (x|T\xi) = (T^*x|\xi) = T^*x \cdot \xi$, by dominated convergence we have $\Gamma(T)f(\cdot) = \lim_{\varepsilon \downarrow 0}\breve{F}_\varepsilon(T^*\cdot)$, where

$$F_\varepsilon(\xi) := \widehat{f}(\xi)g_\varepsilon(\xi) \quad \text{with} \quad g_\varepsilon(\xi) := \exp\Big(-\tfrac{1}{2}(|\xi|^2 - (1-\varepsilon)|T\xi|^2)\Big).$$

By taking inverse Fourier transforms and applying Lemma 5.19, we conclude that $(2\pi)^{d/2}\breve{F}_\varepsilon = f * \breve{g}_\varepsilon$. If we can prove that the $\breve{g}_\varepsilon$ is nonnegative almost everywhere on $\mathbb{R}^d$, it follows that $\Gamma(T)f \geqslant 0$ almost everywhere on $\mathbb{R}^d$.

Since T is a contraction we may write

$$|\xi|^2 - (1-\varepsilon)|T\xi|^2 = ((I - (1-\varepsilon)T^*T)\xi|\xi) = |D_\varepsilon\xi|^2,$$

where $D_\varepsilon := (I - (1 - \varepsilon)T^\star T)^{1/2}$ is invertible. Hence,

$$\breve{g}_\varepsilon(x) = \frac{1}{(2\pi)^{d/2}} \int_{\mathbb{R}^d} \exp\left(-\frac{1}{2}|D_\varepsilon \xi|^2\right) \exp(ix \cdot \xi) \, d\xi.$$

After a change of variables, the right-hand side can be evaluated as a Fourier transform of a Gaussian and is therefore strictly positive on $\mathbb{R}^d$.   □

15.6.d The Segal–Plancherel Transform

In this section we discuss a Gaussian analogue of the Fourier–Plancherel transform $\mathscr{F}$, the so-called *Segal–Plancherel transform* $\mathscr{W}$ on $L^2(\mathbb{R}^d, \gamma)$. We work over the complex scalars. As before we denote by

$$dm(x) := \frac{1}{(2\pi)^{d/2}} \, dx$$

the normalised Lebesgue measure. If we reinterpret the Fourier transform as an operator from $L^1(\mathbb{R}^d, m)$ to $L^\infty(\mathbb{R}^d, m)$,

$$\mathscr{F}f(\xi) = \int_{\mathbb{R}^d} \exp(-ix \cdot \xi) f(x) \, dm(x), \quad \xi \in \mathbb{R}^d, \ f \in L^1(\mathbb{R}^d, m),$$

its restriction to $L^1(\mathbb{R}^d, m) \cap L^2(\mathbb{R}^d, m)$ extends to an isometry on $L^2(\mathbb{R}^d, m)$. In the present section, the term Fourier–Plancherel transform will refer to this operator.

As in Section 15.5.d we let $U := D \circ E$, where $D : L^2(\mathbb{R}^d, m) \to L^2(\mathbb{R}^d, m)$ and $E : L^2(\mathbb{R}^d, \gamma) \to L^2(\mathbb{R}^d, m)$ are the unitary operators

$$Df(x) := 2^{d/4} f(\sqrt{2}x), \quad Ef(x) := e(x)f(x),$$

with $e(x) := \exp\left(-\frac{1}{4}|x|^2\right)$.

Theorem 15.70. *The mapping* $\mathscr{W} : f \mapsto \mathscr{W}f$, *defined for multivariate polynomials* $f : \mathbb{R}^d \to \mathbb{C}$ *and analytic continuation by*

$$\mathscr{W}f(x) := \int_{\mathbb{R}^d} f(-ix + \sqrt{2}y) \, d\gamma(y), \quad x \in \mathbb{R}^d,$$

extends to a unitary operator on $L^2(\mathbb{R}^d, \gamma)$ *and we have*

$$\mathscr{W} = U^\star \circ \mathscr{F} \circ U. \tag{15.38}$$

Proof   Since D, E, and $\mathscr{F}$ are unitary, the unitarity of $\mathscr{W}$ will follow from the operator identity (15.38). To prove this identity, we substitute $\eta = \sqrt{2}y$ and $\xi = -ix + \eta$ to obtain

$$\mathscr{W}f(x) = \frac{1}{(4\pi)^{d/2}} \int_{\mathbb{R}^d} f(-ix + \eta) \prod_{j=1}^{d} \exp\left(-\frac{1}{4}\eta_j^2\right) d\eta$$

$$= \frac{1}{(4\pi)^{d/2}} \int_{-ix+\mathbb{R}^d} f(\xi) \prod_{j=1}^{d} \exp\left(-\frac{1}{4}(\xi_j + ix_j)^2\right) d\xi$$

$$\overset{(*)}{=} \frac{1}{(4\pi)^{d/2}} \int_{\mathbb{R}^d} f(\xi) \prod_{j=1}^{d} \exp\left(-\frac{1}{4}(\xi_j + ix_j)^2\right) d\xi$$

$$= \frac{1}{(4\pi)^{d/2}} \int_{\mathbb{R}^d} f(\xi) \exp\left(-\frac{1}{4}(|\xi|^2 - \frac{1}{2}i\xi \cdot x + \frac{1}{4}|x|^2)\right) d\xi.$$

To justify $(*)$ it suffices, by writing f as a linear combination of monomials and separating variables, to show that for any $k \in \mathbb{N}$ and $x \in \mathbb{R}$,

$$\int_{-ix+\mathbb{R}} \xi^k \exp\left(-\frac{1}{4}(\xi + ix)^2\right) d\xi = \int_{\mathbb{R}} \xi^k \exp\left(-\frac{1}{4}(\xi + ix)^2\right) d\xi.$$

But this is clear by Cauchy's integral formula and a limiting argument using the decay at infinity. Hence,

$$(E \circ \mathscr{W} \circ E^\star) f(x) = \frac{1}{(4\pi)^{d/2}} \exp\left(-\frac{1}{4}|x|^2\right)$$

$$\times \int_{\mathbb{R}^d} \exp\left(\frac{1}{4}|\xi|^2\right) f(\xi) \exp\left(-\frac{1}{4}|\xi|^2 - \frac{1}{2}i\xi \cdot x + \frac{1}{4}|x|^2\right) d\xi$$

$$= \frac{1}{(4\pi)^{d/2}} \int_{\mathbb{R}^d} f(\xi) \exp\left(-\frac{1}{2}i\xi \cdot x\right) d\xi$$

$$= \frac{1}{(2\pi)^{d/2}} \int_{\mathbb{R}^d} 2^{d/2} f(2\xi) \exp(-i\xi \cdot x) \, dm(\xi)$$

$$= (\mathscr{F} \circ D^2) f(x) = (D^\star \circ \mathscr{F} \circ D) f(x),$$

where we used that $\mathscr{F} \circ D = D^\star \circ \mathscr{F}$. $\qquad \square$

We have the following representation of $\mathscr{W}$ in terms of second quantisation:

Theorem 15.71 (Segal). $\mathscr{W} = \Gamma(-iI)$.

Proof Let $f : \mathbb{R}^d \to \mathbb{C}$ be a multivariate polynomial. Mehler's formula and Theorem 15.68 tell us that for all $t > 0$,

$$\Gamma(e^{-t}I) f = OU(t) f = \int_{\mathbb{R}^d} f(e^{-t}(\cdot) + \sqrt{1 - e^{-2t}} y) \, d\gamma(y).$$

By analytic continuation we may replace $t > 0$ by any $\operatorname{Re} z > 0$. Now let $z \to \frac{1}{2}\pi i$. $\qquad \square$

The preceding two theorems combine to the following result. Recall the definition of the unitary operator U of (15.28).

Corollary 15.72. *The Fourier–Plancherel transform is unitarily equivalent to* $\Gamma(-iI)$. *More precisely, we have*

$$U^\star \circ \mathscr{F} \circ U = \Gamma(-iI).$$

Here, U is as in (15.28). This gives a neat "explanation" of the identity $\mathscr{F}^4 = I$: by the multiplicativity of second quantisation it follows from the identity $(-i)^4 = 1$!

15.6.e Creation and Annihilation

For $h \in \mathbb{R}^d$ and $n \in \mathbb{N}$ the *(bosonic) creation operator* $a_n^\dagger(h) : \Gamma^n(\mathbb{R}^d) \to \Gamma^{n+1}(\mathbb{R}^d)$ is defined by

$$a_n^\dagger(h) \sum_{\sigma \in S_n} h_{\sigma(1)} \otimes \cdots \otimes h_{\sigma(n)}$$

$$:= \frac{1}{\sqrt{n+1}} \sum_{\sigma \in S_n} \sum_{j=1}^{n+1} h_{\sigma(1)} \otimes \cdots \otimes h_{\sigma(j-1)} \otimes h \otimes h_{\sigma(j)} \otimes \cdots \otimes h_{\sigma(n)},$$

and the *(bosonic) annihilation operator* $a_{n+1}(h) : \Gamma^{n+1}(\mathbb{R}^d) \to \Gamma^n(\mathbb{R}^d)$ by $a_0(h) := 0$ and

$$a_{n+1}(h) \sum_{\sigma \in S_{n+1}} h_{\sigma(1)} \otimes \cdots \otimes h_{\sigma(n+1)}$$

$$:= \frac{1}{\sqrt{n+1}} \sum_{\sigma \in S_{n+1}} \sum_{j=1}^{n+1} (h_{\sigma(j)}|h) \, h_{\sigma(1)} \otimes \cdots \otimes \widehat{h_{\sigma(j)}} \otimes \cdots \otimes h_{\sigma(n+1)}$$

using the notation $\widehat{}$ to express that this term is omitted. These operators are well defined and bounded, and their operator norms are bounded by

$$\|a_n^\dagger(h)\|_{\mathscr{L}(\Gamma^n(\mathbb{R}^d), \Gamma^{n+1}(\mathbb{R}^d))} = \|a_{n+1}(h)\|_{\mathscr{L}(\Gamma^{n+1}(\mathbb{R}^d), \Gamma^n(\mathbb{R}^d))} \leqslant C_n |h|$$

with constants C_n depending on n only. The first equality follows from the duality

$$a_n^{\dagger\star}(h) = a_{n+1}(h).$$

Furthermore, a straightforward computation gives the commutation relations

$$a_n(h)a_{n+1}^\dagger(h) - a_n^\dagger(h)a_{n+1}(h) = |h|^2 I.$$

Let $(e_j)_{j=1}^d$ denote the standard basis of $\mathbb{R}^d$. Using the Wiener–Itô isometry we may define operators A and $A^\dagger$ as densely defined operators from $L^2(\mathbb{R}^d, \gamma) = L^2(\mathbb{R}^d, \gamma; \mathbb{K})$ to $L^2(\mathbb{R}^d, \gamma; \mathbb{K}^d)$ and from $L^2(\mathbb{R}^d, \gamma; \mathbb{K}^d)$ to $L^2(\mathbb{R}^d, \gamma)$, respectively, by putting

$$AW(x) := \big(W(a_n(e_1)x), \ldots, W(a_n(e_d)x)\big), \quad x \in \Gamma^n(\mathbb{R}^d), \ n \in \mathbb{N},$$

and

$$A^\dagger \big(W(x_1), \ldots, W(x_d)\big) := a_n^\dagger(e_1)x_1 + \cdots + a_n^\dagger(e_d)x_d, \quad x_1, \ldots, x_d \in \Gamma^n(\mathbb{R}^d), \ n \in \mathbb{N}.$$

The operators A and $A^\dagger$ are dual to each other in the sense that

$$(Af|g) = (f|A^\dagger g), \quad f \in \mathscr{H}_n, \ g \in \mathscr{H}_n^d. \tag{15.39}$$

The identity (15.39) easily implies that A and $A^\dagger$ are closable. From now on we denote by A and $A^\dagger$ their closures and by $D(A)$ and $D(A^\dagger)$ the domains of their closures.

Let $\nabla = (\partial_1, \ldots, \partial_d)$ be the gradient, viewed as a densely defined closed operator from $L^2(\mathbb{R}^d, \gamma)$ to $L^2(\mathbb{R}^d, \gamma; \mathbb{K}^d)$ with its natural domain $D(\nabla) = H^1(\mathbb{R}^d, \gamma)$, the Hilbert space of all functions in $L^2(\mathbb{R}^d, \gamma)$ admitting a weak derivative belonging to $L^2(\mathbb{R}^d, \gamma)$.

Lemma 15.73. *The space* $\mathrm{Pol}(\mathbb{R}^d)$ *of polynomials in the real variables* $x_1, \ldots, x_d$ *is dense* $H^1(\mathbb{R}^d, \gamma)$.

Proof We sketch the main line of argument and leave some tedious details to the reader (cf. Problem 15.16). As we have seen in Section 13.6.e, the densely defined closed operator associated with the sesquilinear form

$$\mathfrak{a}_{OU}(f, g) = \int_{\mathbb{R}^d} \nabla f \cdot \overline{\nabla g} \, d\gamma(x), \quad f, g \in D(\mathfrak{a}_{OU}),$$

with $D(\mathfrak{a}_{OU}) = H^1(\mathbb{R}^d, \gamma)$, equals $-L$, where L is the generator of the Ornstein–Uhlenbeck semigroup OU on $L^2(\mathbb{R}^d, \gamma)$. We claim that $D(L)$ is dense in $D(\mathfrak{a}_{OU}) = H^1(\mathbb{R}^d, \gamma)$. This is a special case of a general density result mentioned in the Notes to Chapter 13 but can be proved directly as follows. Since the Ornstein–Uhlenbeck semigroup is analytic (Theorem 13.55), for all $f \in L^2(\mathbb{R}^d, \gamma)$ and $t > 0$ we have $OU(t) \in D(L)$ by Theorem 13.31. In particular this implies $OU(t)f \in D(\mathfrak{a}_{OU}) = H^1(\mathbb{R}^d, \gamma)$ and it suffices to prove that

$$\lim_{t\downarrow 0} \|OU(t)f - f\|_{H^1(\mathbb{R}^d, \gamma)} = 0$$

for all $f \in H^1(\mathbb{R}^d, \gamma)$. For this, in turn, it suffices to check that for all such f we have

$$\lim_{t\downarrow 0} \|\nabla OU(t)f - \nabla f\|_{L^2(\mathbb{R}^d, \gamma; \mathbb{K}^d)} = 0.$$

Writing $g_j := \partial_j f$, this follows by differentiating under the integral in the definition of the Ornstein–Uhlenbeck semigroup:

$$\partial_j OU(t) = \int_{\mathbb{R}^d} \partial_j f(e^{-t}(\cdot) + \sqrt{1 - e^{-2t}} y) \, d\gamma(y)$$

$$= e^{-t} \int_{\mathbb{R}^d} g_j(e^{-t}(\cdot) + \sqrt{1 - e^{-2t}} y) \, d\gamma(y) = e^{-t} OU(t) \partial_j f.$$

The definition of the domain of the operator associated with a form, combined with a straightforward computation, shows that $\mathrm{Pol}(\mathbb{R}^d)$ is contained in $D(L)$. We will show that $\mathrm{Pol}(\mathbb{R}^d)$ is invariant under the Ornstein–Uhlenbeck semigroup. Once this has been established, Proposition 13.5 implies that $\mathrm{Pol}(\mathbb{R}^d)$ is dense in $D(L)$.

To prove the invariance of the space $\mathrm{Pol}(\mathbb{R}^d)$ under the Ornstein–Uhlenbeck semigroup, first let f be a monomial of the form

$$f(x) = x_1^{k_1} \cdot \cdots \cdot x_d^{k_d}, \quad x \in \mathbb{R}^d, \tag{15.40}$$

with $k_j \in \mathbb{N}$ for all $j = 1, \ldots, d$. For γ-almost all $x \in \mathbb{R}^d$ we have

$$OU(t)f(x) = \int_{\mathbb{R}^d} f(e^{-t}x + \sqrt{1 - e^{-2t}}\,y)\,d\gamma(y) =: F_t(x).$$

Substituting the expression (15.40), by direct evaluation we see that $F_t \in \mathrm{Pol}(\mathbb{R}^d)$. The desired invariance follows by taking linear combinations.   □

Proposition 15.74. *We have* $A = \nabla$ *with equality of domains.*

Proof First let $f = \frac{1}{\sqrt{n!}}H_n(\phi_h)$ with $n \in \mathbb{N}$ and $h \in \mathbb{R}^d$ with $|h| = 1$. For $n = 0$ the identity $Af = \nabla f$ is trivial, so in what follows we take $n \geqslant 1$. The function f is the image under the Wiener–Itô isometry of the element $h^{\otimes n} \in \Gamma^n(\mathbb{R}^d)$ and we have

$$(Af)_j = (AW(h^{\otimes n}))_j = W\big(a_n(e_j)\underbrace{h \otimes \cdots \otimes h}_{n \text{ times}}\big)$$

$$= \frac{1}{\sqrt{n}}\sum_{\ell=1}^{n}(h|e_j)W\big(\underbrace{h \otimes \cdots \otimes h}_{n-1 \text{ times}}\big)$$

$$= \sqrt{n}(h|e_j)W\big(\underbrace{h \otimes \cdots \otimes h}_{n-1 \text{ times}}\big) = \frac{\sqrt{n}}{\sqrt{(n-1)!}}(h|e_j)H_{n-1}(\phi_h).$$

On the other hand, since $H_n' = nH_{n-1}$ and $\partial_j\phi_h = (e_j|h)$,

$$(\nabla f)_j = \frac{1}{\sqrt{n!}}\partial_j H_n(\phi_h) = \frac{\sqrt{n}}{\sqrt{(n-1)!}}(e_j|h)H_{n-1}(\phi_h).$$

Since $(h|e_j) = (e_j|h)$ (keeping in mind that $h \in \mathbb{R}^d$), this proves that $Af = \nabla f$ for all $f \in \mathscr{H}_n$ and $n \in \mathbb{N}$. Since the linear span of functions $f \in \mathscr{H}_n$, $n \in \mathbb{N}$, is dense in $\mathsf{D}(A)$ by definition, and since ∇ is closed, this gives the inclusion $A \subseteq \nabla$.

To prove equality $A = \nabla$ it remains to be shown that the linear span of functions $f \in \mathscr{H}_n$, $n \in \mathbb{N}$, is also dense in $H^1(\mathbb{R}^d, \gamma)$. For this purpose we recall from the proof of Theorem 15.60, in the special case of the standard unit basis of $\mathbb{R}^d$, that for each $n \in \mathbb{N}$ the linear span of the polynomials $H_n(\phi_h)$, $|h| = 1$, equals the space of all polynomials of the form $x \mapsto H_{n_1}(x_1) \cdot \cdots \cdot H_{n_d}(x_d)$ with $n_1 + \cdots + n_d = n$. Their linear span when n ranges over $\mathbb{N}$ equals the space $\mathrm{Pol}(\mathbb{R}^d)$ introduced above. This space is dense in $H^1(\mathbb{R}^d, \gamma)$ by Lemma 15.73.   □

In what follows we work over the complex scalars. For $j = 1, \ldots, d$ and $f \in \mathrm{Pol}(\mathbb{R}^d)$ we let

$$a_j f := (Af)_j, \quad a_j^\dagger f := A^\dagger(0, \ldots, 0, f, 0, \ldots)$$

with f in the jth place. By (15.39) these operators are dual to each other, in the sense that with respect to the inner product of $L^2(\mathbb{R}^d, \gamma)$ we have

$$(a_j f|g) = (f|a_j^\dagger g), \quad f, g \in \mathrm{Pol}(\mathbb{R}^d).$$

Define the *position operator* $Q = (q_1, \ldots, q_d)$ by

$$q_j := \frac{1}{\sqrt{2}}(a_j + a_j^\dagger).$$

The choice of the normalising constant $1/\sqrt{2}$ in the definition of q_j may appear unnatural. The reason for this choice will become apparent in (15.41), (15.42), and (15.45).

Viewed as an operator in $L^2(\mathbb{R}^d, \gamma)$ with dense initial domain $\mathrm{Pol}(\mathbb{R}^d)$, this operator is symmetric and therefore closable. We claim that its closure, which we denote by q_j again, is selfadjoint. First we claim that, for almost all $x \in \mathbb{R}^d$,

$$q_j f(x) = \frac{1}{\sqrt{2}} x_j f(x), \quad f \in \mathrm{Pol}(\mathbb{R}^d).$$

Indeed, since by Proposition 15.74 we have $a_j = \partial_j$, the directional derivative in the direction of e_j, it follows that

$$\sqrt{2}(q_j f | g) = (f | \partial_j g) + (\partial_j f | g).$$

Suppose now that $f, g \in \mathrm{Pol}(\mathbb{R}^d)$. Then, with $x = (x_1, \ldots, x_d)$,

$$(\partial_j^\star f | g) = \frac{1}{(2\pi)^{d/2}} \int_{\mathbb{R}^d} f(x)\overline{\partial_j g(x)} \exp\left(-\frac{1}{2}|x|^2\right) dx$$

$$= \frac{1}{(2\pi)^{d/2}} \int_{\mathbb{R}^d} [x_j f(x) - \partial_j f(x)]\overline{g(x)} \exp\left(-\frac{1}{2}|x|^2\right) dx = -(\partial_j f | g) + (x_j f | g).$$

It follows that $(\partial_j f + \partial_j^\star) f(x) = x_j f(x)$. This proves the claim. The asserted selfadjointness of q_j is an easy consequence of this claim.

In a similar way we define the *momentum operator* $P = (p_1, \ldots, p_j)$ by

$$p_j := \frac{1}{i\sqrt{2}}(a_j - a_j^\dagger).$$

Again this operator, initially defined on functions in $\mathrm{Pol}(\mathbb{R}^d)$, is symmetric and hence closable, and its closure is selfadjoint.

The identities below are understood in the sense that they hold when applied to functions in $\mathrm{Pol}(\mathbb{R}^d)$, Some additional details are addressed in Problems 15.17–15.20. From the commutation relation $[a_j, a_j^\dagger] = I$ we have

$$[p_j, q_j] = \frac{1}{i}(a_j a_j^\dagger - a_j^\dagger a_j) = \frac{1}{i} I \tag{15.41}$$

as well as the identity

$$\frac{1}{2}(p_j^2 + q_j^2) = \frac{1}{2}(a_j^\dagger a_j + a_j a_j^\dagger) = a_j^\dagger a_j + \frac{1}{2}[a_j, a_j^\dagger] = a_j^\dagger a_j + \frac{1}{2} I. \tag{15.42}$$

As is checked by an easy computation, in terms of the annihilation and creation operators, the Ornstein–Uhlenbeck operator is given by

$$-L = \nabla^\star \nabla = A^\dagger A = \sum_{j=1}^{d} a_j^\dagger a_j, \tag{15.43}$$

so that by (15.42),

$$-L = -\frac{d}{2} + \frac{1}{2}(P^2 + Q^2) = -\frac{d}{2} + \sum_{j=1}^{d} \frac{1}{2}(p_j^2 + q_j^2), \tag{15.44}$$

again in the sense that these identities hold when the operators are applied to functions in $\mathrm{Pol}(\mathbb{R}^d)$. The operators P and Q intertwine with the momentum operator $D = \frac{1}{i}\nabla$ and the position operator X, in the sense that

$$U \circ q_j \circ U^\star = x_j, \quad U \circ p_j \circ U^\star = \frac{1}{i}\partial_j, \tag{15.45}$$

with U the unitary operator of Sections 15.5.d and 15.6.d. These relations are easy to check by explicit computation and justify the terminology 'position' and 'momentum' for q_j and p_j. In this way we recover the unitary equivalence, established in Theorem 13.56, of $-L + \frac{d}{2}$ with the quantum harmonic oscillator.

Problems

15.1 Prove the assertions about orthogonal projections in Section 15.1.c.

15.2 Prove that if P and Q are orthogonal projections on a Hilbert space such that $\mathrm{R}(P)$ is contained in $\mathrm{R}(Q)$, then $Q = P \vee (Q \wedge \neg P)$.

15.3 Let $\phi : \mathscr{L}(H) \to \mathbb{C}$ be a state, let $(h_n)_{n \geqslant 1}$ be an orthonormal basis for H, and let P_n denote the orthogonal projection onto the span of the set $\{h_1, \ldots, h_n\}$. Prove that for all $T \in \mathscr{L}(H)$ we have

$$\phi(T) = \lim_{n \to \infty} \phi(P_n T).$$

Hint: Apply the Cauchy–Schwarz inequality to the mapping $(T, U) \mapsto \phi(TU^\star)$ and take $U := I - P_n$.

15.4 Consider a qubit in state $\alpha|0\rangle + \beta|1\rangle$, where $\alpha, \beta \in \mathbb{C}$ satisfy $|\alpha|^2 + |\beta|^2 = 1$. Compute the probabilities that upon measuring the spin in direction $j \in \{1,2,3\}$ we find 'up', respectively 'down'.

15.5 We take a closer look at the Pauli matrices σ_1, σ_2, σ_3.

(a) Show that the complex exponentials of the Pauli matrices are given by

$$\exp(i\theta\sigma_j) = (\cos\theta)I + i(\sin\theta)\sigma_j, \quad j = 1,2,3.$$

(b) Show that if v is a unit vector in $\mathbb{R}^3$, then for all $n \in \mathbb{N}$ we have

$$(v \cdot \sigma)^n = \begin{cases} I, & n \text{ even,} \\ v \cdot \sigma, & n \text{ odd.} \end{cases}$$

Use this to prove the identity

$$\exp(i\theta v \cdot \sigma) = (\cos\theta)I + i(\sin\theta)v \cdot \sigma.$$

Furthermore show that $\det(\exp(i\theta v \cdot \sigma)) = 1$.

(c) Conclude that

$$\{\exp(i\theta v \cdot \sigma) : \theta \in [0, 2\pi]\} = SU(2),$$

the group of unitary matrices acting on $\mathbb{C}^2$ with determinant 1.

15.6 Prove that if U is a symmetry of H, then the mapping $\tau_U : \mathscr{P}(H) \to \mathscr{P}(H)$ given by $\tau_U(P) := U^\star PU$ enjoys the following properties:

  (i) $\tau_U(I) = I$;
  (ii) for all $P \in \mathscr{P}(H)$ we have $\tau_U(\neg P) = \neg\tau_U(P)$;
  (iii) for all $P_1, P_2 \in \mathscr{P}(H)$ we have

$$\tau_U(P_1 \wedge P_2) = \tau_U(P_1) \wedge \tau_U(P_2);$$
$$\tau_U(P_1 \vee P_2) = \tau_U(P_1) \vee \tau_U(P_2).$$

15.7 Show that position and momentum are covariant with respect to rotations R_ρ on $L^2(\mathbb{R}^d, m)$ given by $R_\rho f(x) = f(\rho^{-1}x)$, where $\rho \in SO(d)$, the group of orthogonal transformations on $\mathbb{R}^d$ with determinant 1.

15.8 For $G = \mathbb{Z}/2\mathbb{Z}$, determine the position and momentum operators on $L^2(G) \simeq \mathbb{C}^2$.

15.9 Find the projection-valued measures associated with the selfadjoint operators $\widehat{x}_j$ and $\widehat{\xi}_j$ discussed in Section 15.5.b.

15.10 In this problem we prove *Wintner's theorem*: There exists no pair of *bounded* operators $S, T \in \mathscr{L}(H)$ satisfying the Heisenberg commutation relation $ST - TS = iI$. We may absorb the imaginary constant i into one of the two operators and consider the identity $ST - TS = I$ instead. Assuming that $S, T \in \mathscr{L}(H)$ satisfy $ST - TS = I$, obtain a contradiction by completing the following steps.

  (a) Show that for all $n = 1, 2, \ldots$ we have $S^nT - TS^n = nS^{n-1}$.
  (b) Deduce that $S^{n-1} \neq 0$ and $n\|S^{n-1}\| \leqslant 2\|S^{n-1}\|\|S\|\|T\|$.

15.11 The aim of this problem is to prove that for a linear mapping $\phi : \mathscr{L}(H) \to \mathbb{C}$ the following assertions are equivalent:

  (1) $\phi(T) = \sum_{j=1}^k (Tx_j|y_j)$ for suitable $k \geqslant 1$ and $x_1, \ldots, x_k, y_1, \ldots, y_k \in H$;
  (2) ϕ is continuous with respect to the weak topology of $\mathscr{L}(H)$;
  (3) ϕ is continuous with respect to the strong topology of $\mathscr{L}(H)$.

(a) Prove the implications (1)$\Rightarrow$(2)$\Rightarrow$(3).

The remainder of the problem is devoted to a proof of the implication (3)$\Rightarrow$(1).

(b) Show that (3) implies that there exist $x_1, \ldots, x_k \in H$ such that

$$|\phi(T)| \leqslant \max_{1 \leqslant j \leqslant k} \|Tx_j\|, \quad T \in \mathscr{L}(H).$$

(c) Let K be the closure of the subspace $\{(Tx_1, \ldots, Tx_k) \in H^k : T \in \mathscr{L}(H)\}$ in H^k. Show that the linear mapping $\psi : K \to \mathbb{C}$ defined by

$$\psi(Tx_1, \ldots, Tx_k) := \phi(T)$$

is well defined and bounded.

(d) Using the Riesz representation theorem, show that ϕ is of the form as in (1).

15.12 Prove that if $\phi : \mathscr{L}(H) \to \mathbb{C}$ is linear, the following assertions are equivalent:

(1) there exist sequences $(x_n)_{n \geqslant 1}$ and $(y_n)_{n \geqslant 1}$ satisfying

$$\sum_{n \geqslant 1} \|x_n\|^2 < \infty \quad \text{and} \quad \sum_{n \geqslant 1} \|y_n\|^2 < \infty$$

such that for all $T \in \mathscr{L}(H)$ we have $\phi(T) = \sum_{\geqslant 1}(Tx_n|y_n)$;

(2) ϕ is continuous on $\overline{B}_{\mathscr{L}(H)}$ with respect to the weak topology of $\mathscr{L}(H)$;

(3) ϕ is continuous on $\overline{B}_{\mathscr{L}(H)}$ with respect to the strong topology of $\mathscr{L}(H)$;

(4) ϕ is normal.

If ϕ is positive and satisfies $\phi(I) = 1$, these conditions are equivalent to:

(5) there exists an orthogonal sequence $(x_n)_{n \geqslant 1}$ satisfying $\sum_{n \geqslant 1} \|x_n\|^2 = 1$ such that for all $T \in \mathscr{L}(H)$ we have $\phi(T) = \sum_{n \geqslant 1}(Tx_n|x_n)$.

15.13 Using the functional calculus for projection-valued measures on $\mathbb{T}$ we may define

$$\widehat{\theta} := \int_{\mathbb{T}} \arg(z) \, d\Theta(z),$$

where the projection-valued measure $\Theta : \mathscr{B}(\mathbb{T}) \to \mathscr{P}(L^2(\mathbb{T}))$ is the angle observable of Section 15.5.c. There is some ambiguity here as to how to take the argument; for the sake of definiteness we take it in $(-\pi, \pi]$.

(a) Show that $\widehat{\theta}$ is bounded and selfadjoint on $L^2(\mathbb{T})$.

(b) Show that for all $f, g \in L^2(\mathbb{T})$ we have

$$(\widehat{\theta}f|g) = \frac{1}{2\pi} \int_{-\pi}^{\pi} \theta f(e^{i\theta})\overline{g(e^{i\theta})} \, d\theta.$$

Define the *angular momentum operator* as the selfadjoint operator $\widehat{l}$ defined by the angular momentum observable $L : \mathscr{B}(\mathbb{Z}) \to \mathscr{P}(L^2(\mathbb{T}))$ of Section 15.5.c,

$$\widehat{l} := \sum_{n \in \mathbb{Z}} nL_{\{n\}}.$$

(c) Show that, with an appropriate choice of domain, $\widehat{l}$ is selfadjoint on $L^2(\mathbb{T})$.

(d) Prove that $\widehat{\theta}$ and $\widehat{l}$ satisfy the Heisenberg commutation relation

$$\widehat{l}\,\widehat{\theta} - \widehat{\theta}\,\widehat{l} = il$$

on $\mathsf{D}(\widehat{l\theta}) \cap \mathsf{D}(\widehat{\theta l})$ and show that this domain is dense in $L^2(\mathbb{T})$.

The operator $\widehat{\theta}$ appears to be of little use in Physics. This is related to the failure of the 'continuous variable' Weyl commutation relation for $\widehat{\theta}$ and $\widehat{l}$:

(e) Show that there exists no bounded operator T on $L^2(\mathbb{T})$ such that the following identity holds for all $s, t \in \mathbb{R}$:

$$e^{isT}e^{it\widehat{l}} = e^{ist}e^{it\widehat{l}}e^{isT}. \tag{15.46}$$

Show that the same conclusion holds if we assume that T is a (possibly unbounded) selfadjoint operator.

Hint: Show that if an $s \in \mathbb{R}$ exists such that the identity in (15.46) holds for all $t \in \mathbb{R}$, then $s \in \mathbb{Z}$.

(f) Prove a similar result for the phase operator of Section 15.3.d.

15.14 Show that if T is a contraction on $\mathbb{R}^d$, then for every $1 \leqslant p < \infty$ the second quantised operator $\Gamma(T)$ extends to a contraction on $L^p(\mathbb{R}^d, \gamma)$.

15.15 Show that if U is an isometry on $\mathbb{R}^d$, then for all $f \in L^2(\mathbb{R}^d, \gamma)$ we have

$$\Gamma(U)f(x) = f(U^\star x)$$

for almost all $x \in \mathbb{R}^d$.

15.16 Complete the details of the proof of Lemma 15.73.

15.17 Complete the details of the proofs that the position and momentum operators q_j and p_j are selfadjoint on $L^2(\mathbb{R}^d, \gamma)$.

15.18 Prove the commutation relation $[a_j, a_j^\dagger] = I$ used in the proof of (15.41). Also prove that if $j \neq k$, then $[a_j, a_k^\dagger] = 0$ and $[p_j, q_k^\dagger] = 0$.

15.19 Show that the operator $-\frac{d}{2} + \frac{1}{2}(P^2 + Q^2)$, considered in (15.45) as a densely defined operator in $L^2(\mathbb{R}^d, \gamma)$ with domain $\mathrm{Pol}(\mathbb{R}^d)$, is closable, with closure $-L$.

15.20 Show that the position and momentum operators q_j and p_j introduced in Section 15.6.e satisfy the relations

$$q_j \circ \mathscr{W} = \mathscr{W} \circ p_j, \quad p_j \circ \mathscr{W} = -\mathscr{W} \circ q_j,$$

consistent (modulo the difference in normalisations of the Fourier transform) with the relations $x_j \circ \mathscr{F} = \mathscr{F} \circ (\frac{1}{i}\partial_j)$ and $(\frac{1}{i}\partial_j) \circ \mathscr{F} = -\mathscr{F} \circ x_j$ for position and momentum operators of Section 15.5.b.

Appendix A
Zorn's Lemma

Zorn's lemma provides a sufficient condition for the existence of maximal elements in partially ordered sets. Its formulation uses some terminology which we introduce first. A *relation* on a set S is a subset R of the cartesian product $S \times S$. Instead of $(x, y) \in R$ we often write $x R y$.

Definition A.1 (Partially ordered sets). A *partially ordered* set is a pair $(S, \leqslant)$, where S is a set and $\leqslant$ is a relation on S such that for all $x, y, z \in S$ we have:

   (i) (reflexivity) $x \leqslant x$;
  (ii) (antisymmetry): if $x \leqslant y$ and $y \leqslant x$, then $x = y$;
 (iii) (transitivity): if $x \leqslant y$ and $y \leqslant z$, then $x \leqslant z$.

A *totally ordered* set is a partially ordered set $(S, \leqslant)$ with the property that for all $x, y \in S$ we have $x \leqslant y$ or $y \leqslant x$ (or both, in which case $x = y$).

Definition A.2 (Maximal elements, upper bounds). Let $(S, \leqslant)$ be a partially ordered set. An element $x \in S$ is said to be *maximal* if $x \leqslant y$ implies $y = x$. An element $x \in S$ is said to be an *upper bound* for the subset $S' \subseteq S$ if $x' \leqslant x$ holds for all $x' \in S'$.

Assuming the Axiom of Choice, one has the following result.

Theorem A.3 (Zorn's lemma). *If $(S, \leqslant)$ is a nonempty partially ordered set with the property that each of its totally ordered subsets has an upper bound in S, then S has a maximal element.*

Appendix B
Tensor Products

Let V and W be vector spaces and let $\mathscr{B}(V,W)$ denote the vector space of all bilinear mappings from $V \times W$ into the scalar field $\mathbb{K}$, that is, all mappings $\phi : V \times W \to \mathbb{K}$ satisfying

$$\phi(cv,w) = \phi(v,cw) = c\phi(v,w)$$

for all $c \in \mathbb{K}$, $v \in V$, and $w \in W$, and

$$\phi(v+v',w) = \phi(v,w) + \phi(v',w),$$
$$\phi(v,w+w') = \phi(v,w) + \phi(v,w')$$

for all $v,v' \in V$ and $w,w' \in W$.

For all $v \in V$ and $w \in W$, the mapping

$$v \otimes w : \phi \mapsto \phi(v,w)$$

defines an element of $\mathscr{B}(V,W)^\dagger$, the vector space of all linear mappings from $\mathscr{B}(V,W)$ to $\mathbb{K}$. Note that

$$c(v \otimes w) = (cv) \otimes w = v \otimes (cw)$$

for all $c \in \mathbb{K}$, $v \in V$, and $w \in W$, and

$$(v+v') \otimes w = v \otimes w + v' \otimes w$$
$$v \otimes (w+w') = v \otimes w + v \otimes w'$$

for all $v,v' \in V$ and $w,w' \in W$.

Definition B.1 (Algebraic tensor product). The (algebraic) *tensor product*

$$V \otimes W$$

of V and W is the linear span in $\mathscr{B}(V,W)^\dagger$ of the set $\{v \otimes w : v \in V, w \in W\}$.

We have natural isomorphisms of vector spaces

$$\mathbb{K} \otimes V \simeq V \otimes \mathbb{K} \simeq V.$$

By the above definition and the identities preceding it, every element of $V \otimes W$ admits a representation as a finite sum $\sum_{j=1}^{k} v_j \otimes w_j$.

If the (finite or infinite) sets $\{v_i : i \in I\}$ and $\{w_i : i \in I\}$ are both linearly independent, so is the set $\{v_i \otimes w_i : i \in I\}$. Indeed, suppose that $\sum_{j=1}^{k} c_j v_{i_j} \otimes w_{i_j} = 0$ for certain $k \geqslant 1$ and scalars $c_1, \ldots, c_k$. For $\phi \in V^\dagger$ and $\psi \in W^\dagger$ the mapping $\zeta : (v,w) \mapsto \phi(v)\psi(w)$ belongs to $\mathscr{B}(V,W)$ and accordingly

$$0 = \Big(\sum_{j=1}^{k} c_j v_{i_j} \otimes w_{i_j} \Big)(\zeta) = \sum_{j=1}^{k} c_j \zeta(v_{i_j}, w_{i_j}) = \sum_{j=1}^{k} c_j \phi(v_{i_j}) \psi(w_{i_j}) = \psi \Big(\sum_{j=1}^{k} c_j \phi(v_{i_j}) w_{i_j} \Big).$$

This being true for all $\psi \in W^\dagger$, the linear independence of $\{w_1, \ldots, w_k\}$ implies that $\phi(c_j v_{i_j}) = c_j \phi(v_{i_j}) = 0$ for all $\phi \in V^\dagger$ and $j = 1, \ldots, k$. But this implies that $c_j v_{i_j} = 0$ for all $j = 1, \ldots, k$. The linear independence of $\{v_1, \ldots, v_k\}$ implies that $v_{i_j} \neq 0$ for all $j = 1, \ldots, k$, so we must have $c_j = 0$ for all $j = 1, \ldots, k$. This proves our claim.

Remark B.2. The above argument relies on the availability of sufficiently many linear functionals. This issue can be avoided by observing that if there is a linear dependence relation in $V \otimes W$, then there exist finite-dimensional subspaces $V' \subseteq V$ and $W' \subseteq W$, containing the finitely many elements v_i and w_i of V and W involved in the linear dependence, such that this linear dependence also exists in $V' \otimes W'$. Running the argument in $V' \otimes W'$, we may test against the coordinate functionals of bases for V' and W' containing the v_i and w_i, respectively.

Remark B.3. A similar remark can be made with regard to the very construction of the tensor product $V \otimes W$ presented here: this space is nontrivial only if a sufficient supply of bilinear mappings from $V \times W$ to $\mathbb{K}$ can be guaranteed. This can be done by using Zorn's lemma, which allows one to find algebraic bases for V and W. With such bases at hand, one may use the associated coordinate functionals to construct nontrivial bilinear mappings. Although an alternative construction of the tensor product can be given which circumvents this issue, the present approach has the advantage of connecting in a direct and intuitive way with the various functional analytic settings where tensor products are employed. In the main text, V and W will always be Hilbert spaces and the required supply of bilinear and linear functionals is guaranteed through the inner product.

As a corollary to this observation we obtain that if V and W are finite-dimensional, with bases $(v_i)_{i=1}^{d_V}$ and $(w_j)_{j=1}^{d_W}$, then $(v_i \otimes w_j)_{i,j=1}^{d_V, d_W}$ is a basis for $V \otimes W$. In particular,

$$\dim(V \otimes W) = \dim(V)\dim(W).$$

For vector spaces U, V, W, the mapping

$$(u \otimes v) \otimes w \mapsto u \otimes (v \otimes w)$$

uniquely extends to an isomorphism of vector spaces

$$(U \otimes V) \otimes W \simeq U \otimes (V \otimes W).$$

Stated differently, taking tensor products is associative. This allows us to define the tensor product

$$U \otimes V \otimes W$$

as either one of the spaces in this isomorphism; for the sake of concreteness we will use the space on the left-hand side. With this in mind we can define the tensor product $V_1 \otimes \cdots \otimes V_N$ of vector spaces V_n, $n = 1, \ldots, N$, inductively by

$$V_1 \otimes \cdots \otimes V_N := (V_1 \otimes \cdots \otimes V_{N-1}) \otimes V_N.$$

Alternatively one could define $V_1 \otimes \cdots \otimes V_N$ in terms of functionals on the space of N-linear mappings; the resulting space is isomorphic in a natural way to the one just defined. In what follows we write $V^{\otimes n} := V \otimes \cdots \otimes V$ for the n-fold tensor product of V.

The *n-fold symmetric tensor product* $\Gamma^n(V)$ is defined as the range of the projection P_Γ on $V^{\otimes n}$ given by

$$P_\Gamma : v_1 \otimes \cdots \otimes v_n \mapsto \frac{1}{n!} \sum_{\sigma \in S_n} v_{\sigma(1)} \otimes \cdots \otimes v_{\sigma(n)}, \quad v_1, \ldots, v_n \in V,$$

where S_n is the group of permutations of $\{1, \ldots, n\}$. Likewise one defines the *n-fold antisymmetric product* $\Lambda^n(V)$, also known as the *n-fold exterior product*, of a vector space V as the range of the projection on $V^{\otimes n}$ given by

$$P_\Lambda : v_1 \otimes \cdots \otimes v_n \mapsto \frac{1}{n!} \sum_{\sigma \in S_n} \mathrm{sign}(\sigma) v_{\sigma(1)} \otimes \cdots \otimes v_{\sigma(n)}, \quad v_1, \ldots, v_n \in V.$$

In connection with second quantisation, these spaces are sometimes denoted by $V^{\circledS n}$ and $V^{\circledast n}$, respectively.

We conclude with the observation that if S and T are linear operators on the vector spaces V and W, respectively, then

$$(S \otimes T) : v \otimes w \mapsto Sv \otimes Tw$$

uniquely defines a linear operator $S \otimes T$ on the tensor product $V \otimes W$. To see that this operator is well defined, suppose that an element in $V \otimes W$ admits two representations

$$\sum_{n=1}^{N} c_n v_n \otimes w_n = \sum_{n=1}^{N'} c'_n v'_n \otimes w'_n.$$

If $\phi : V \times W \to \mathbb{K}$ is bilinear, then the mapping $\phi_{S,T} : V \times W \to K$ given by

$$\phi_{S,T}(v,w) := \phi(Sv, Tw)$$

is bilinear and

$$\left(\sum_{n=1}^{N} c_n Sv_n \otimes Tw_n \right)(\phi) = \sum_{n=1}^{N} c_n \phi(Sv_n, Tw_n)$$

$$= \sum_{n=1}^{N} c_n \phi_{S,T}(v_n, w_n) = \left(\sum_{n=1}^{N} c_n v_n \otimes w_n \right)(\phi_{S,T})$$

and, by the same argument,

$$\left(\sum_{n=1}^{N'} c'_n Sv'_n \otimes Tw'_n \right)(\phi) = \left(\sum_{n=1}^{N'} c'_n v'_n \otimes w'_n \right)(\phi_{S,T}).$$

It follows that

$$\left(\sum_{n=1}^{N} c_n Sv_n \otimes Tw_n \right)(\phi) = \left(\sum_{n=1}^{N'} c'_n Sv'_n \otimes Tw'_n \right)(\phi).$$

This being true for all bilinear $\phi : V \times w \to \mathbb{K}$, it follows that

$$\sum_{n=1}^{N} c_n Sv_n \otimes Tw_n = \sum_{n=1}^{N'} c'_n Sv'_n \otimes Tw'_n.$$

Appendix C
Topological Spaces

This appendix offers a brief treatment of topological spaces. Only those notions are covered that find their way into the main text. Several others will only be needed in the more concrete setting of metric spaces and will be discussed in that context.

Definition and General Properties

Definition C.1 (Topological spaces). A *topological space* is a pair (X, τ), where X is a set and τ is a *topology* on X, that is, τ is a collection of subsets of X with the following properties:

(i) $\varnothing \in \tau$ and $X \in \tau$;
(ii) τ is closed under taking arbitrary unions;
(iii) τ is closed under taking finite intersections.

A subset U of X is said to be *open* if $U \in \tau$, and *closed* if its complement is open. The *interior* S° of a subset S is the union of all open subsets U in X contained in S. The *closure* $\overline{S}$ of a subset S is the intersection of all closed subsets F of X containing S. Note that S° is the largest open subset of X contained in S and $\overline{S}$ is the smallest closed subset of X containing S. A set S is *dense* in a closed set S' if $\overline{S} = S'$.

If $\mathscr{C}$ is a collection of subsets of X, the *topology generated by* $\mathscr{C}$ is the intersection of all topologies on X containing $\mathscr{C}$. The topology generated by a collection $\mathscr{C}$ is the smallest topology containing every element of $\mathscr{C}$.

In what follows we often omit the topology τ from our notation and write X instead of the more cumbersome (X, τ) to denote topological spaces, except in those situations where confusion could arise. In such situations we may speak of τ-*open* and τ-*closed* sets instead of open and closed sets in order to emphasise the role of τ.

A topological space X is said to be *Hausdorff* if for every two distinct points $x_1, x_2 \in X$ there exist disjoint open sets $U_1, U_2 \in \tau$ such that $x_1 \in U_1$ and $x_2 \in U_2$.

Proposition C.2. *Finite subsets of a Hausdorff topological space are closed.*

Proof Since finite unions of closed sets are closed it suffices to prove that every singleton $\{x\}$ in a Hausdorff space X is closed. For any $y \in X \setminus \{x\}$ choose an open set U_y such that $y \in U_y$ and $x \notin U_y$. This is possible by the Hausdorff assumption (and actually uses less than that). We then have $\complement\{x\} = \bigcup_{y \in X \setminus \{x\}} U_y$, and this set is open since τ is closed under taking arbitrary unions. It follows that $\{x\}$ is closed. $\square$

An important class of Hausdorff topological space is the class of metric spaces; they are discussed in more detail in Section D. Further examples relevant to Functional Analysis are Banach spaces with their weak topology, dual Banach spaces with their weak* topology, and spaces of bounded operators acting between Banach spaces with their strong and weak operator topologies. For their definitions we refer to the main text.

Continuity

Let (X, τ_X) and (Y, τ_Y) be topological spaces and consider a mapping $f : X \to Y$.

Definition C.3. We call f *continuous at the point* $x_0 \in X$ if for every open set $V \in \tau_Y$ containing $f(x_0)$ there exists an open set $U \in \tau_X$ containing x_0 such that $f(U) \subseteq V$. We call f *continuous* if f is continuous at every point of X.

As an immediate consequence of the definition we note that if (X, τ_X), (Y, τ_Y), (Z, τ_Z) are topological spaces and $f : X \to Y$ is continuous at the point $x_0 \in X$ and $g : Y \to Z$ is continuous at the point $f(x_0) \in Y$, then the composition $g \circ f : X \to Z$ is continuous at the point $x_0 \in X$. In particular, the composition of two continuous mappings is continuous.

Proposition C.4. *Let (X, τ_X) and (Y, τ_Y) be topological spaces. For a mapping $f : X \to Y$ the following assertions are equivalent:*

(1) *f is continuous;*
(2) *$f^{-1}(V)$ is open for every open subset V of Y;*
(3) *$f^{-1}(F)$ is closed for every closed subset F of Y.*

Proof (1)$\Rightarrow$(2): Suppose that f is continuous and let V be an open set in Y. Let $x \in f^{-1}(V)$ be arbitrary. Using the definition of continuity we choose an open subset $U_x \in \tau_X$ containing x such that $f(U_x) \subseteq V$. This means that $U_x \subseteq f^{-1}(V)$. It follows that $f^{-1}(V) = \bigcup_{x \in f^{-1}(V)} U_x$, and this set is open since τ is closed under taking arbitrary unions. This shows that $f^{-1}(V)$ is open in X.

(2)$\Rightarrow$(3): Suppose $f^{-1}(V)$ is open for every open $V \subseteq Y$. Let $F \subseteq Y$ be closed. Then its complement $\complement F$ is open in Y, hence by our assumption $f^{-1}(\complement F)$ is open. From $f^{-1}(F) = \complement f^{-1}(\complement F)$ it follows that $f^{-1}(F)$ is closed.

(3)$\Rightarrow$(2): This is proved in the same way, interchanging the roles of 'open' and 'closed'.

(2)$\Rightarrow$(1): Let $x \in X$ be arbitrary and let $V \subseteq Y$ be open and contain $f(x)$. The set $U = f^{-1}(V)$ is open in X by assumption, x is an element of this set, and we have $f(U) \subseteq V$. Thus f is continuous at the point x. $\qquad \square$

Compactness

Let X be a topological space and let S be a subset of X. A collection $\mathcal{U}$ of open subsets of X is called an *open cover* of S if $S \subseteq \bigcup_{U \in \mathcal{U}} U$. A *subcover* is a cover $\mathcal{U}'$ of S contained in $\mathcal{U}$. The set S is called *compact* if every open cover of S has a finite subcover. A set is called *relatively compact* if its closure is compact.

Proposition C.5. *Let X be a topological space. Then:*

(1) *every closed subset of X contained in a compact subset of X is compact;*
(2) *if X is Hausdorff, then every compact subset of X is closed.*

Proof (1): Let the closed set F be contained in the compact subset S of X. Let $\mathcal{U}_F$ be an open cover of F, and extend it to an open cover $\mathcal{U}$ of S by adjoining the open set $\complement F$. The resulting cover of S has a finite subcover, and this subcover also covers F. Removing the set $\complement F$ from this subcover, we are left with a finite subcover of $\mathcal{U}$ for S. It follows that F is compact.

(2): Let S be a compact subset of the Hausdorff space X. We first claim that for every $x \in \complement S$ there is an open set U_x containing x and disjoint from S. Indeed, for every $y \in S$, the Hausdorff property provides us with two disjoint open sets $U_{x,y}$ and $V_{x,y}$ such that $x \in U_{x,y}$ and $y \in V_{x,y}$. The open cover $\mathcal{V}_x = \{V_{x,y} : y \in S\}$ of S has a finite subcover, say $\mathcal{V}'_x = \{V_{x,y_1}, \ldots, V_{x,y_{k_x}}\}$, where $k_x \geq 1$ is an integer depending on x. The set $U_x := \bigcap_{j=1}^{k_x} U_{x,y_j}$ is open, contains x, and is disjoint from S. This proves the claim. But now we see that $\complement S = \bigcup_{x \in \complement S} U_x$, so $\complement S$ is open and S is closed. $\qquad \square$

A collection of subsets of a topological space has the *finite intersection property* if every finite subcollection has nonempty intersection.

Proposition C.6. *A nonempty closed subset S of a topological space X is compact if and only if every collection of closed subsets of S with the finite intersection property has nonempty intersection.*

Proof 'Only if': Let $\mathscr{C}$ be a collection of closed subsets of S having the finite intersection property. If we had $\bigcap_{C \in \mathscr{C}} C = \varnothing$, then $\mathscr{U} := \{\complement C : C \in \mathscr{C}\}$ is an open cover of S without a finite subcover. For if $\complement C_1, \ldots, \complement C_k$ were to cover S, then $C_1 \cap \cdots \cap C_k = \varnothing$. It follows that S is not compact.

'If': Reasoning by contradiction, assume that every collection of closed subsets of S with the finite intersection property has nonempty intersection and assume that there exists an open cover $\mathscr{U}$ of S without finite subcover. Then for any finite choice of sets $U_1, \ldots, U_k \in \mathscr{U}$ we have $S \setminus \bigcup_{j=1}^{k} U_j \neq \varnothing$. It follows that $\bigcap_{j=1}^{k} (S \cap \complement U_j) \neq \varnothing$. From the assumption on S we infer that $\bigcap_{U \in \mathscr{U}} (S \cap \complement U) \neq \varnothing$. But then $\mathscr{U}$ does not cover S and we have arrived at a contradiction. $\square$

Compactness is preserved under continuous mappings:

Proposition C.7. *Let X and Y be topological spaces. Let $f : X \to Y$ be a continuous mapping. If S is a compact subset of X, then $f(S)$ is compact in Y.*

Proof Let $\mathscr{U}$ be an open cover of $f(S)$. Then $\{f^{-1}(U) : U \in \mathscr{U}\}$ is an open cover of S by Proposition C.4. Since S is compact, it has a finite subcover $\{f^{-1}(U_1), \ldots, f^{-1}(U_n)\}$. The collection $\{U_1, \ldots, U_n\}$ is then a finite subcover of $f(S)$. $\square$

Every continuous function $f : [a, b] \to \mathbb{R}$ has a global maximum and a global minimum on $[a, b]$. More generally we have:

Theorem C.8 (Global maxima and minima). *Let X be a compact topological space and let $f : X \to \mathbb{R}$ be continuous. Then f attains a global maximum and a global minimum.*

Proof We prove that f attains a global maximum; by applying this to the continuous function $-f$ it follows that f also attains a global minimum.

For $n \geqslant 1$ let $U_n = \{x \in X : f(x) < n\}$. The collection $\mathscr{U} = \{U_n : n \geqslant 1\}$ is an open cover of X and has, thanks to the compactness of X, a finite subcover. From this it follows that the range of f is bounded above. Let $m := \sup\{f(x) : x \in X\}$.

Suppose that there is no $x \in X$ such that $f(x) = m$; we show that then X cannot be compact. The assumption just made implies that the collection $\mathscr{V} = \{V_n : n \geqslant 1\}$ is an open cover of X, where $V_n := \{x \in X : f(x) < m - \frac{1}{n}\}$. Since for every $n \geqslant 1$ there is an $x \in X$ such that $f(x) \geqslant m - \frac{1}{n}$ (this follows from the definition of the supremum) $\mathscr{V}$ has no finite subcover. $\square$

A topological space is called *normal* if for any two disjoint closed subsets F and G there exist disjoint open subsets U and V such that $F \subseteq U$ and $G \subseteq V$.

Proposition C.9. *Every compact Hausdorff space X is normal.*

Proof Let F and G be disjoint nonempty closed subsets of the compact Hausdorff space X. Then F and G are compact by Proposition C.5. Fix a point $x \in F$. Since X is

Hausdorff, for all $y \in G$ there exist disjoint open subsets $U_{x,y}$ and $V_{x,y}$ such that $x \in U_{x,y}$ and $y \in V_{x,y}$. By letting y range over all points of G and using compactness we find an open cover $V_{x,y_1}, \ldots, V_{x,y_{k_x}}$ of G. Set $U_x := \bigcap_{j=1}^{k_x} U_{x,y_j}$ and $V_x := \bigcup_{j=1}^{k_x} V_{x,y_j}$. Then $x \in U_x$, $G \subseteq V_x$, and $U_x \cap V_x = \varnothing$. Letting x range over F and using compactness we find an open cover $U_{x_1}, \ldots, U_{x_\ell}$ of F. The sets $U := \bigcup_{j=1}^{\ell} U_{x_j}$ and $V := \bigcap_{j=1}^{\ell} V_{x_j}$ are open and satisfy $F \subseteq U$, $G \subseteq V$, and $U \cap V = \varnothing$. $\qquad\square$

In Appendix D we will see that also every metric space is normal.

Corollary C.10. *Let X be a normal space. If $F \subseteq U \subseteq X$ with F compact and U open, then there exists an open set V such that*

$$F \subseteq V \subseteq \overline{V} \subseteq U.$$

Proof By normality there exist disjoint open sets W and W' such that $F \subseteq W$ and $\complement U \subseteq W'$. Since $\complement W$ is closed and $W' \subseteq \complement W \subseteq \complement F$, we have $\overline{W'} \subseteq \complement F$ and therefore

$$F \subseteq \complement \overline{W'} \subseteq \complement W' \subseteq U.$$

The set $V := \complement \overline{W'}$ satisfies $F \subseteq V \subseteq \overline{V} \subseteq \complement W' \subseteq U$, where the third inclusion holds since $\complement W'$ is a closed set containing V. $\qquad\square$

Urysohn's Lemma

In normal spaces, disjoint closed sets can be separated by continuous functions. This is the content of the next result.

The *support* of a continuous function $f : X \to \mathbb{K}$, where X is a topological space, is defined as the complement of the largest open set $U \subseteq X$ such that $f \equiv 0$ on U or, equivalently, as the set closure of the set $\{x \in D : f(x) \neq 0\}$. The support of f is denoted by $\mathrm{supp}(f)$.

Proposition C.11 (Urysohn's lemma). *Let X be a normal space. If $F \subseteq U \subseteq X$ with F closed and U open. Then there exists a continuous function $f : X \to [0,1]$ such that $f \equiv 1$ on F and $\mathrm{supp}(f) \subseteq U$.*

Proof A rational number $q \in [0,1]$ is called *dyadic* if it is of the form $\frac{k}{2^n}$, where $k, n \in \mathbb{N}$ and $0 \leqslant k \leqslant 2^n$. We will construct, for every dyadic $q \in [0,1]$, an open set U_q such that

$$F \subseteq U_q \subseteq \overline{U_q} \subseteq U$$

and, for all dyadic $r \in [0,1]$,

$$q > r \text{ implies } \overline{U_q} \subseteq U_r.$$

By Corollary C.10 (applied twice) there exist open sets U_0 and U_1 such that

$$F \subseteq U_1 \subseteq \overline{U_1} \subseteq U_0 \subseteq \overline{U_0} \subseteq U.$$

Reasoning by induction, suppose that for some $n \in \mathbb{N}$ and $k = 0, \ldots, 2^n$ the open sets $U_{\frac{k}{2^n}}$ have been chosen such that $q > r$ implies $\overline{U_q} \subseteq U_r$. Using Corollary C.10, for all $k = 0, \ldots, 2^n - 1$ we find an open set $U_{\frac{2k+1}{2^{n+1}}}$ such that

$$\overline{U_{\frac{k+1}{2^n}}} \subseteq U_{\frac{2k+1}{2^{n+1}}} \subseteq \overline{U_{\frac{2k+1}{2^{n+1}}}} \subseteq U_{\frac{k}{2^n}}.$$

Then $\overline{U_q} \subseteq U_r$ holds for all dyadic $q > r$ of the form $\frac{k}{2^{n+1}}$ with $0 \leqslant k \leqslant 2^{n+1}$.
Now define

$$f_q(x) := \begin{cases} q, & \text{if } x \in U_q, \\ 0, & \text{otherwise,} \end{cases} \qquad g_r(x) := \begin{cases} 1, & \text{if } x \in \overline{U_r}, \\ r, & \text{otherwise,} \end{cases}$$

and put $f(x) := \sup_q f_q(x)$ and $g(x) := \inf_r g_r(x)$. Then f is lower semicontinuous, g is upper semicontinuous, $0 \leqslant f \leqslant 1$, $f \equiv 1$ on F, and $\mathrm{supp}(f) \subseteq U$. To conclude the proof we show that $f = g$.

If $f_q(x) > g_r(x)$, then we must have $q > r$, $x \in U_q$, and $x \notin \overline{U_r}$. But $q > r$ implies $U_q \subseteq U_r$. This contradiction shows that $f_q(x) \leqslant g_r(x)$ for all dyadic $q, r \in [0, 1]$ and $x \in X$. This implies $f \leqslant g$.

If $f(x) < g(x)$, there are dyadic numbers $q, r \in [0, 1]$ such that $f(x) < r < q < g(x)$. But $f(x) < r$ implies that $x \notin U_r$ and $g(x) > q$ implies $x \in \overline{U_q}$. This contradicts the fact that $q > r$ implies $\overline{U_q} \subseteq U_r$. It follows that $f = g$. $\qquad\square$

As an application we prove:

Theorem C.12 (Partition of unity). *Let X be a normal space and let*

$$F \subseteq U_1 \cup \cdots \cup U_k,$$

where F is compact and the sets U_j are open in X for all $j = 1, \ldots, k$. Then there exist nonnegative continuous functions $f_j : X \to [0, 1]$ with support in U_j, $j = 1, \ldots, k$, such that

$$f_1 + \cdots + f_k \equiv 1 \text{ on } F.$$

The same result holds if X is a locally compact Hausdorff space.

Proof Every $x \in F$ is contained in at least one of the sets U_j, and applying normality to the closed sets $\{x\}$ and $\complement U_j$ we find an open subset V containing x and whose closure is contained in U_j. Letting x range over F and using that F is compact, it follows that we can cover F with finitely many open sets $V_1, \ldots, V_n$ such that for all $m = 1, \ldots, n$ we have $\overline{V_m} \subseteq U_{j_m}$ for some $1 \leqslant j_m \leqslant k$. Set

$$F_j := \bigcup_{m: \overline{V_m} \subseteq U_j} \overline{V_m}, \quad j = 1, \ldots, k.$$

This set is closed and contained in U_j. By Urysohn's lemma we can find continuous functions $g_j : X \to [0, 1]$ topologically supported in U_j such that $g_j \equiv 1$ on F_j. Put $f_1 := g_1$ and

$$f_j := (1 - g_1) \cdots (1 - g_{j-1}) g_j, \quad j = 2, \ldots, k.$$

The support of f_j is contained in U_j and an easy induction argument shows that

$$f_1 + \cdots + f_k = 1 - (1 - g_1) \cdots (1 - g_k). \tag{C.1}$$

If $x \in F$, then $g_j(x) = 1$ for at least one $j = 1, \ldots, k$ and therefore (C.1) implies that $f_1 + \cdots + f_k \equiv 1$ on F.

The case of locally compact Hausdorff spaces may be reduced to the case of compact Hausdorff spaces by the same argument as in Proposition 4.3. $\qquad\square$

We conclude with a useful extension theorem. Its proof makes use of the fact, mentioned in Section 2.2.a, that the space $C_b(X)$ of all bounded continuous functions on a topological space X is complete as a normed space endowed with the supremum norm. The proof of this elementary fact is direct and does not introduce any circularity.

Theorem C.13 (Tietze extension theorem). *Let F be a closed subset of a normal space X and let $f : F \to [0, 1]$ be continuous. Then there exists a continuous function $g : X \to [0, 1]$ such that $g|_F = f$.*

Proof The sets $A := \{x \in F : f(x) \in [0, \frac{1}{3}]\}$ and $B := \{x \in F : f(x) \in [\frac{2}{3}, 1]\}$ are disjoint and closed in F, and hence closed in X (since F is closed in X). By Urysohn's lemma there exists a continuous function $g_1 : X \to [0, 1/3]$ such that $g_1 \equiv 0$ on A and $g_1 \equiv \frac{1}{3}$ on B. This function satisfies $0 \leqslant f - g_1 \leqslant \frac{2}{3}$ pointwise on F. Proceeding inductively, we construct continuous functions $g_k : X \to [0, 2^{k-1}/3^k]$, $k \geqslant 1$, such that for every $k \geqslant 1$ we have

$$g_k \equiv 0 \text{ on the set } \left\{ x \in F : f(x) - \sum_{j=1}^{k-1} g_j(x) \leqslant 2^{k-1}/3^k \right\}$$

and

$$g_k \equiv 2^{k-1}/3^k \text{ on the set } \left\{ x \in F : f(x) - \sum_{j=1}^{k-1} g_j(x) \geqslant 2^k/3^k \right\}.$$

We then have $0 \leqslant f - \sum_{j=1}^k g_j \leqslant 2^k/3^k$ pointwise on F; the lower bound is clear from the construction and the upper bound follows by induction.

Set $g := \sum_{k \geqslant 1} g_k$. The partial sums of this sum converge uniformly and therefore g is continuous, by the completeness of $C_b(X)$. On the set F we have

$$0 \leqslant f - g \leqslant f - \sum_{j=1}^{k} g_j \leqslant 2^k/3^k$$

for every $k \geqslant 1$, forcing that $f = g$ on F. □

Tychonov's Theorem

Let I be a nonempty set and suppose that for every $i \in I$ a topological space (X_i, τ_i) is given. The *cartesian product* of the family $(X_i)_{i \in I}$ is the set $X = \prod_{i \in I} X_i$ whose elements are the mappings $x : I \to \bigcup_{i \in I} X_i$ with the property that $x(i) \in X_i$ for all $i \in I$. For each $i \in I$ we define the *coordinate mapping* $p_i : X \to X_i$ by $p_i(x) := x(i)$. The *product topology* of $X = \prod_{i \in I} X_i$ is the topology generated by the sets $p_i^{-1}(U_i)$, where U_i ranges over all open sets in X_i and i ranges over I. It is the smallest topology $\tau = \prod_{i \in I} \tau_i$ with the property that all coordinate mappings $p_i : x \mapsto x(i)$ are continuous as mappings from X into X_i. If $I = \{i_1, \ldots, i_k\}$ is finite, the topology of $X = \prod_{j=1}^{k} X_{i_j}$ coincides with the topology generated by the sets of the form $U = U_{i_1} \times \cdots \times U_{i_k}$ with U_{i_j} open in X_j for all $j = 1, \ldots, k$.

Theorem C.14 (Tychonov). *The product of any family of compact spaces is a compact space. If each one of the spaces is Hausdorff, then so is its product.*

Proof Let $X = \prod_{i \in I} X_i$, where (X_i, τ_i) is a compact topological space for each $i \in I$. If $X_i = \varnothing$ for some $i \in I$ we have $X = \varnothing$ and there is nothing to prove. We may therefore assume that $X_i \neq \varnothing$ for all $i \in I$.

Fix a collection $\mathscr{C}$ of closed subsets of X with the finite intersection property. We wish to prove that $\bigcap_{C \in \mathscr{C}} C \neq \varnothing$. Once this has been proved, Proposition C.6 implies that X is compact.

Let **D** be the set of all collections $\mathscr{D}$ of subsets of X which have the finite intersection property and contain $\mathscr{C}$ as a subcollection. The set **D** is nonempty (it contains $\mathscr{C}$) and can be partially ordered by set inclusion, that is, we declare $\mathscr{D} \leqslant \mathscr{D}'$ to mean that $\mathscr{D} \subseteq \mathscr{D}'$. Note that we do not insist on closedness of the sets in the collections $\mathscr{D}$.

Let $\mathbf{T} \subseteq \mathbf{D}$ be a *totally ordered* subset, that is, a subset with the property that for all $\mathscr{T}_1, \mathscr{T}_2 \in \mathbf{T}$ we have either $\mathscr{T}_1 \subseteq \mathscr{T}_2$ or $\mathscr{T}_2 \subseteq \mathscr{T}_1$. We claim that $\bigcup_{\mathscr{T} \in \mathbf{T}} \mathscr{T}$ belongs to **D**. For this it suffices to check that this union has the finite intersection property. To this end suppose that $T_1, \ldots, T_k \in \bigcup_{\mathscr{T} \in \mathbf{T}} \mathscr{T}$, say $T_j \in \mathscr{T}_j \in \mathbf{T}$ for $j = 1, \ldots, k$. Since **T** is totally ordered, after relabelling we may assume that $\mathscr{T}_1 \subseteq \cdots \subseteq \mathscr{T}_k$. Then every T_j belongs to $\mathscr{T}_k$ and therefore the finite intersection property of $\mathscr{T}_k$ implies that $\bigcap_{j=1}^{k} T_j \neq \varnothing$. This proves that $\bigcup_{\mathscr{T} \in \mathbf{T}} \mathscr{T}$ has the finite intersection property.

Evidently, the union $\bigcup_{\mathcal{T} \in \mathbf{T}} \mathcal{T}$ is an upper bound for $\mathbf{T}$ in $\mathbf{D}$. We may therefore apply Zorn's lemma and obtain that $\mathbf{D}$ has a maximal element. We denote it by $\mathcal{M}$. For each $i \in I$ consider the collection

$$\mathcal{X}_i := \{\overline{p_i(M)} : M \in \mathcal{M}\},$$

where $p_i : x \mapsto x(i)$ are the coordinate mappings. It consists of closed subsets of X_i and has the finite intersection property since $\mathcal{M}$ has it. Since X_i is compact, the set $Y_i := \bigcap_{M \in \mathcal{M}} \overline{p_i(M)}$ is nonempty by Proposition C.6. For every $i \in I$ choose a $y_i \in Y_i$ and let $x \in X$ be defined by $x(i) := y_i$, $i \in I$. We will prove in two steps that $x \in \overline{M}$ for all $M \in \mathcal{M}$.

Step 1 – If U_i is open in X_i and contains $x(i) = y_i$, the fact that $x(i) \in \overline{p_i(M)}$ for all $M \in \mathcal{M}$ implies that $x \in p_i(M) \cap U_i \neq \varnothing$ and hence $x \in M \cap p_i^{-1}(U_i) \neq \varnothing$ for all $M \in \mathcal{M}$. It follows that the collection $\mathcal{M} \cup \{p_i^{-1}(U_i) : i \in I\}$ has the finite intersection property and belongs to $\mathbf{D}$. By maximality, this collection equals $\mathcal{M}$. Therefore $p_i^{-1}(U_i) \in \mathcal{M}$ for all $i \in I$.

Step 2 – Let U be an open set in X containing x. By the definition of the product topology there are indices $i_1, \ldots, i_k \in I$ and open sets $U_{i_j} \in \tau_{i_j}$ for $j = 1, \ldots, k$ such that $x \in \bigcap_{j=1}^{k} p_{i_j}^{-1}(U_{i_j}) \subseteq U$. By the result of Step 1 and the fact that $\mathcal{M}$ has the finite intersection property we have $M \cap p_{i_1}^{-1}(U_{i_1}) \cap \cdots \cap p_{i_k}^{-1}(U_{i_k}) \neq \varnothing$ for all $M \in \mathcal{M}$. In particular, $M \cap U \neq \varnothing$ for all $M \in \mathcal{M}$. This being true for all open sets U containing x, it follows that $x \in \overline{M}$ for all $M \in \mathcal{M}$.

It now follows that

$$x \in \bigcap_{M \in \mathcal{M}} \overline{M} \subseteq \bigcap_{C \in \mathcal{C}} \overline{C} = \bigcap_{C \in \mathcal{C}} C,$$

where we used that $\mathcal{C} \subseteq \mathcal{M}$ and the fact that the elements of $\mathcal{C}$ are closed sets. Therefore $\bigcap_{C \in \mathcal{C}} C \neq \varnothing$ and we conclude that X is compact.

Suppose now that each space X_i is Hausdorff. If $x, x' \in X$ and $x \neq x'$, then for some $i \in I$ we must have $x(i) \neq x'(i)$ and since X_i is Hausdorff there are disjoint open sets U_i and U_i' in X_i containing $x(i)$ and $x'(i)$, respectively. Their inverse images under π_i are open and disjoint in X. $\qquad \square$

Appendix D
Metric Spaces

We now introduce an important class of Hausdorff spaces, namely, the class of metric spaces. All results of the previous appendix apply to metric spaces, but in order to make the present appendix independently readable some proofs are repeated. In addition, our treatment of metric spaces includes a number of additional topics.

Definition and General Properties

Definition D.1 (Metric spaces). A *metric space* is a pair (X,d), where X is a set and d a *metric* (or *distance function*) on X, that is, a function $d : X \times X \to [0,\infty)$ such that for all x, y, z in X the following conditions are satisfied:

  (i) $d(x,y) = 0 \Leftrightarrow x = y$;
  (ii) $d(x,y) = d(y,x)$;
  (iii) $d(x,z) \leqslant d(x,y) + d(y,z)$ (the *triangle inequality*).

In what follows we omit the distance function d from the notation and write X instead of the more cumbersome (X,d) to denote metric spaces, except in those situations where confusion could arise.

Let X be a metric space, let $x \in X$, and let $r > 0$. The set

$$B(x;r) := \{y \in X : d(x,y) < r\}$$

is called the *open ball* with centre x and radius r. The set

$$\overline{B}(x;r) := \{y \in X : d(x,y) \leqslant r\}$$

is called the *closed ball* with centre x and radius r.

A subset S of a metric space X is called *open* if for all $x \in S$ there exists an $r > 0$ such that $B(x;r) \subseteq S$. A subset S of a metric space X is called *closed* if its complement $\complement S = X \setminus S$ is open. It is an easy consequence of the triangle inequality that every open ball $B(x;r)$ is open and every closed ball $\overline{B}(x;r)$ is closed; this justifies the terminology 'open ball' and 'closed ball'.

The *interior* of a subset S of a metric space X is the union of all open subsets of X contained in S and is denoted by S°. It is the largest open subset of X contained in S. The *closure* of a subset S of a metric space X is the intersection of all closed subsets of X containing S and is denoted by $\overline{S}$. It is the smallest closed subset of X containing S.

The closure $\overline{B(x;r)}$ of an open ball is always contained in the closed ball $\overline{B}(x;r)$, but this inclusion may be strict. For example, take $X = \mathbb{Z}$ with distance function $d(m,n) = |n - m|$. The open ball $B(0;1) = \{0\}$ is also closed, so its closure equals $\overline{B(0;1)} = \{0\}$. On the other hand, $\overline{B}(0;1) = \{-1,0,1\}$.

For any metric space (X,d), the collection of its open sets is a Hausdorff topology, the so-called *Borel topology* of X. More is true:

Proposition D.2. *Every metric space is normal.*

Proof Let F and G be disjoint closed sets in a metric space X. We must find disjoint open set U and V such that $F \subseteq U$ and $G \subseteq V$. We may assume that F and G are both nonempty, since otherwise the result is trivial.

The function

$$f(x) := \frac{d(x,F)}{d(x,F) + d(x,G)}$$

is well defined, continuous, takes values in $[0,1]$, and satisfies $f \equiv 0$ on F and $f \equiv 1$ on G. The sets $U := \{x \in E : f(x) < \frac{1}{2}\}$ and $V := \{x \in E : f(x) > \frac{1}{2}\}$ are open and have the desired properties. $\square$

Convergence

A sequence $(x_n)_{n\geqslant 1}$ in a metric space X is called *convergent* if there exists an $x \in X$ such that for all $\varepsilon > 0$ there is an index $N \geqslant 1$ with the property $d(x_n,x) < \varepsilon$ for all $n \geqslant N$. We then write

$$\lim_{n\to\infty} x_n = x$$

and call x a *limit* of the sequence $(x_n)_{n\geqslant 1}$. It is clear that $\lim_{n\to\infty} x_n = x$ if and only if $\lim_{n\to\infty} d(x_n,x) = 0$. Limits are unique, for if $\lim_{n\to\infty} x_n = x$ and $\lim_{n\to\infty} x_n = y$, then for all indices n the triangle inequality gives $0 \leqslant d(x,y) \leqslant d(x,x_n) + d(x_n,y) \to 0 + 0 = 0$ and therefore $d(x,y) = 0$.

A subset S of a metric space X is called *sequentially closed* if the limit of every sequence in S that converges in X belongs to S.

Proposition D.3. *For a subset S in a metric space X, the following assertions are equivalent:*

(1) *S is closed;*

(2) *S is sequentially closed.*

Proof (1)$\Rightarrow$(2): Let $(x_n)_{n\geqslant 1}$ be a sequence in S, convergent in X with limit x. We must show that $x \in S$. Assume the contrary. Then $x \in \complement S$, and since $\complement S$ is open there is an $\varepsilon > 0$ such that $B(x;\varepsilon) \subseteq \complement S$. On the other hand, since $(x_n)_{n\geqslant 1}$ converges to x there is an index $N \geqslant 1$ such that $d(x_n,x) < \varepsilon$ for all $n \geqslant N$, that is, $x_n \in B(x;\varepsilon)$ (and hence $x_n \in \complement S$) for all $n \geqslant N$. For these indices we obtain the contradiction $x_n \in S \cap \complement S$.

(2)$\Rightarrow$(1): We must show that $\complement S$ is open. Choose $x \in \complement S$ arbitrarily. We must show that there is an $\varepsilon > 0$ such that $B(x;\varepsilon) \subseteq \complement S$. Suppose that such an $\varepsilon > 0$ does not exist. Then for every $n \geqslant 1$ we can find an $x_n \in B(x;\frac{1}{n}) \cap S$. The resulting sequence $(x_n)_{n\geqslant 1}$ is contained in S and satisfies $d(x_n,x) < \frac{1}{n}$ for all $n \geqslant 1$, that is, we have $\lim_{n\to\infty} x_n = x$. Since S is sequentially closed we conclude that $x \in S$, in contradiction with the assumption that $x \in \complement S$. $\square$

As a corollary we have the following useful criterion for determining which elements belong to the closure of a given set:

Proposition D.4. *For a subset S in a metric space X and a point $x \in X$, the following assertions are equivalent:*

(1) $x \in \overline{S}$;

(2) $S \cap B(x;\varepsilon) \neq \varnothing$ *for all $\varepsilon > 0$;*

(3) *there exists a sequence $(x_n)_{n\geqslant 1}$ in S with $\lim_{n\to\infty} x_n = x$.*

In particular, a set S is dense in the closed set S' if and only every $s' \in S$ is the limit of a sequence in S.

Proof (1)$\Rightarrow$(2): If $S \cap B(x;\varepsilon) = \varnothing$ for some $\varepsilon > 0$, then $S \subseteq \complement B(x;\varepsilon)$. Since $\complement B(x;\varepsilon)$ is closed, this implies $\overline{S} \subseteq \complement B(x;\varepsilon)$ and therefore $x \notin \overline{S}$.

(2)$\Rightarrow$(3): For every $n \geqslant 1$ we choose $x_n \in S \cap B(x;\frac{1}{n})$. In this way we obtain a sequence $(x_n)_{n\geqslant 1}$ in S converging to x.

(3)$\Rightarrow$(1): If $x \notin \overline{S}$, then $x \in \complement\overline{S}$ and this set is open. Hence there exists an $\varepsilon > 0$ such that $B(x;\varepsilon) \subseteq \complement\overline{S}$. In particular it holds that $d(x,y) \geqslant \varepsilon$ for all $y \in S$. This implies that no sequence in S can converge to x. $\square$

Completeness

A sequence $(x_n)_{n\geqslant 1}$ in a metric space X is called a *Cauchy sequence* if for all $\varepsilon > 0$ there is an index $N \geqslant 1$ such that $d(x_n, x_m) < \varepsilon$ for all $n, m \geqslant N$.

Proposition D.5. *In a metric space X the following assertions hold:*

(1) *every convergent sequence is a Cauchy sequence;*
(2) *every Cauchy sequence with a convergent subsequence is convergent.*

Proof (1): Let x be the limit of the convergent sequence $(x_n)_{n\geqslant 1}$ and let $\varepsilon > 0$ be arbitrary. Choose N so large that $d(x_k, x) < \varepsilon$ for all $k \geqslant N$. By the triangle inequality, for all $n, m \geqslant N$ it follows that $d(x_n, x_m) \leqslant d(x_n, x) + d(x, x_m) < \varepsilon + \varepsilon = 2\varepsilon$.

(2): Let $(x_n)_{n\geqslant 1}$ be a Cauchy sequence with a subsequence $(x_{n_k})_{k\geqslant 1}$ convergent to x. We check that $(x_n)_{n\geqslant 1}$ converges to x. Choose $\varepsilon > 0$ arbitrarily and let $N \geqslant 1$ be such that $d(x_n, x_m) < \varepsilon$ for all $m, n \geqslant N$. Let $K \geqslant 1$ be such that for all $k \geqslant K$ we have both $n_k \geqslant N$ and $d(x_{n_k}, x) < \varepsilon$. Now choose $k \geqslant K$ arbitrarily. Then for all $n \geqslant N$ we have $d(x_n, x) \leqslant d(x_n, x_{n_k}) + d(x_{n_k}, x) < \varepsilon + \varepsilon = 2\varepsilon$. $\qquad\square$

A subset S of a metric space X is called *complete* if every Cauchy sequence contained in S converges to a limit in S. In particular, X is complete if every Cauchy sequence in X is convergent in X. very complete sets S is sequentially closed and hence closed, and conversely if X is complete, then every closed subset S is complete.

Theorem D.6 (Completion). *If X is a metric space, there exists a complete metric space $(\overline{X}, \overline{d})$ and a mapping $i : X \to \overline{X}$ with the following properties:*

Augustin-Louis Cauchy,
1789–1857

(i) *i is isometric, that is, $\overline{d}(ix, iy) = d(x, y)$ for all $x, y \in X$;*
(ii) *i has dense range, that is, $i(X)$ is dense in $\overline{X}$.*

Moreover, if $(\overline{\overline{X}}, \overline{\overline{d}})$ is another complete metric space and $i' : X \to \overline{\overline{X}}$ is a mapping satisfying (i) *and* (ii)*, then the identity mapping on X has a unique extension to an isometry from $\overline{X}$ onto $\overline{\overline{X}}$.*

A more precise way of stating the last assertion is that there exists a unique isometry j from $\overline{X}$ onto $\overline{\overline{X}}$ which satisfies $j(ix) = i'x$ for all $x \in X$.

Proof On the set of Cauchy sequences in X we define an equivalence relation by declaring the Cauchy sequences $(x_n)_{n\geqslant 1}$ and $(x'_n)_{n\geqslant 1}$ equivalent if $\lim_{n\to\infty} d(x_n, x'_n) = 0$. Let $\overline{X}$ be the set of all equivalence classes. The elements of $\overline{X}$ are denoted by $\overline{x}$ On $\overline{X}$ we define a metric $\overline{d}$ by

$$\overline{d}(\overline{x}, \overline{y}) := \lim_{n\to\infty} d(x_n, y_n),$$

where $(x_n)_{n\geqslant 1}$ is a Cauchy sequence representing $\overline{x}$. Using the triangle inequality, it is readily checked that this limit indeed exists and is independent of the choice of sequences $(x_n)_{n\geqslant 1}$ and $(y_n)_{n\geqslant 1}$ representing x and y.

If $x \in X$, the constant sequence $(x)_{n\geqslant 1}$ is Cauchy and therefore defines an element of $\overline{X}$ which we shall denote by $[x]$. We thus obtain a mapping $i : X \to \overline{X}$ by declaring $ix := [x]$. From $\overline{d}(ix, iy) = \overline{d}([x], [y]) = \lim_{n\to\infty} d(x, y) = d(x, y)$ we see that this mapping is isometric.

This gives property (i). To prove property (ii) let $\overline{x} \in \overline{X}$, and let $(x_n)_{n\geqslant 1}$ be a Cauchy sequence in X representing $\overline{x}$. For any $\varepsilon > 0$ we may choose N so large that $d(x_n, x_m) < \varepsilon$ for all $m, n \geqslant N$. Then $\overline{d}(ix_N, \overline{x}) = \lim_{n\to\infty} d(x_N, x_n) \leqslant \varepsilon$. This shows that $i(X)$ is dense in $\overline{X}$.

Next we prove that $\overline{X}$ is complete. Suppose $(\overline{x}_n)_{n\geqslant 1}$ is a Cauchy sequence in $\overline{X}$. By the density of X in $\overline{X}$ we may pick elements $x_n \in X$ such that $\overline{d}(\overline{x}_n, [x_n]) < \frac{1}{n}$. From

$$d(x_n, x_m) = \overline{d}([x_n], [x_m]) \leqslant \overline{d}([x_n], \overline{x}_n) + \overline{d}(\overline{x}_n, \overline{x}_m) + \overline{d}(\overline{x}_m, [x_m])$$

$$\leqslant \frac{1}{n} + \overline{d}(\overline{x}_n, \overline{x}_m) + \frac{1}{m}$$

and the fact that the right-hand side tends to 0 as $n, m \to \infty$, we infer that $(x_n)_{n\geqslant 1}$ is a Cauchy sequence in X. Let $\overline{x} \in \overline{X}$ be its equivalence class. As was shown in the proof of density, we have $\lim_{n\to\infty} \overline{d}([x_n], \overline{x}) = 0$. Then

$$\overline{d}(\overline{x}_n, \overline{x}) \leqslant \overline{d}(\overline{x}_n, [x_n]) + \overline{d}([x_n], \overline{x}) < \frac{1}{n} + \overline{d}([x_n], \overline{x})$$

shows that $\lim_{n\to\infty} \overline{d}(\overline{x}_n, \overline{x}) = 0$. This proves the completeness of $\overline{X}$.

Let $I : X \to X$ denote the identity mapping on X, and let $(\overline{\overline{X}}, \overline{\overline{d}})$ and $i' : X \to \overline{\overline{X}}$ satisfy (i) and (ii). We obtain an isometry $j : \overline{X} \to \overline{\overline{X}}$ by putting

$$j\overline{x} := i' \lim_{n\to\infty} x_n,$$

where $(x_n)_{n\geqslant 1}$ is a Cauchy sequence representing $\overline{x}$ and the limit on the right-hand side is taken in $\overline{\overline{X}}$. This limit exists because the sequence $(i'x_n)_{n\geqslant 1}$ is Cauchy in the complete space $\overline{\overline{X}}$. The resulting mapping j is an isometry from $\overline{X}$ onto $\overline{\overline{X}}$, whose inverse is obtained by applying the same procedure with the roles of $\overline{X}$ and $\overline{\overline{X}}$ interchanged.

If j' is another isometry from $\overline{X}$ onto $\overline{\overline{X}}$ extending the identity mapping on X, then

$j'x = \lim_{n\to\infty} j'ix_n = \lim_{n\to\infty} jix_n = jx$ for all $x \in \overline{X}$ represented by the Cauchy sequence $(x_n)_{n\geqslant 1}$ in X. This gives the uniqueness of j. □

A complete metric space $(\overline{X}, \overline{d})$ is called a *completion* of (X, d) if there exists a mapping $i : X \to \overline{X}$ with properties (i) and (ii). The second part of the theorem asserts that completions are "unique up to isometry". To avoid this minor ambiguity we agree to call the metric space $(\overline{X}, \overline{d})$ constructed in the above proof "the" completion of (X, d).

Continuity

Let X and Y be metric spaces and consider a mapping $f : X \to Y$.

Definition D.7. A mapping $f : X \to Y$ is *continuous at the point* $x_0 \in X$ if and only if for every $\varepsilon > 0$ there exists a $\delta > 0$ such that for all $x \in X$ with $d_X(x, x_0) < \delta$ we have $d_Y(f(x), f(x_0)) < \varepsilon$. We call f *continuous* if f is continuous at every point of X.

This definition is consistent with the one in the previous appendix; this is clear from the fact that every open set U in X containing x_0 contains the open balls $B(x_0; \delta)$ for sufficiently small $\delta > 0$ and every open set V in Y containing $f(x_0)$ contains the open balls $B(f(x_0); \varepsilon)$ for sufficiently small $\varepsilon > 0$.

A mapping $f : X \to Y$ is called *sequentially continuous at the point* $x_0 \in X$ if for every sequence $(x_n)_{n\geqslant 1}$ in X with $\lim_{n\to\infty} x_n = x_0$ we have $\lim_{n\to\infty} f(x_n) = f(x_0)$. We call f *sequentially continuous* if f is sequentially continuous at every point of X.

Proposition D.8. *For a mapping $f : X \to Y$ between the metric spaces X and Y the following assertions are equivalent:*

(1) *f is continuous at the point $x_0 \in X$;*
(2) *f is sequentially continuous at the point $x_0 \in X$.*

In particular, f is continuous if and only if f is sequentially continuous.

Proof (1)$\Rightarrow$(2): Suppose that $\lim_{n\to\infty} x_n = x_0$ in X. Choose $\varepsilon > 0$ arbitrarily and choose, using the continuity of f at x_0, a $\delta > 0$ such that $d_Y(f(x), f(x_0)) < \varepsilon$ for all $x \in X$ with $d_X(x, x_0) < \delta$. Since $\lim_{n\to\infty} d_X(x_n, x_0) = 0$ we can find an index $N \geqslant 1$ such that $d_X(x_n, x_0) < \delta$ for all $n \geqslant N$. For all $n \geqslant N$ it then holds that $d_Y(f(x_n), f(x_0)) < \varepsilon$.

(2)$\Rightarrow$(1): Suppose that there is an $\varepsilon > 0$ such that no $\delta > 0$ can be found with the property that $d_Y(f(x), f(x_0)) < \varepsilon$ for all $x \in X$ with $d_X(x, x_0) < \delta$. Then for every $n \geqslant 1$ we can find $x_n \in X$ with $d_X(x_n, x_0) < \frac{1}{n}$ and $d_Y(f(x_n), f(x_0)) \geqslant \varepsilon$. But this implies that f is not sequentially continuous. □

A function $f : X \to Y$ is *uniformly continuous* if for every $\varepsilon > 0$ there exists a $\delta > 0$ such that whenever $x, y \in X$ satisfy $d_X(x,y) < \delta$, then $d_Y(f(x), f(y)) < \varepsilon$. Every uniformly continuous function is continuous, but the converse is false even for bounded functions: the function $f : (0,1) \to [-1,1]$, $f(x) = \sin(1/x)$, is continuous but not uniformly continuous. Below we prove that if X is compact, then every continuous function from X into another metric space is uniformly continuous.

Uniformly continuous functions have the following extension property:

Proposition D.9. *Let X and Y be metric spaces, with Y complete. If $f : X \to Y$ is uniformly continuous, there exists a unique uniformly continuous function $\overline{f} : \overline{X} \to Y$ extending f.*

Proof Let $x \in \overline{X}$ and choose a sequence $x_n \to x$ with each x_n in X. Then $(x_n)_{n \geqslant 1}$ is Cauchy in X, and the uniform continuity of f implies that $(f(x_n))_{n \geqslant 1}$ is Cauchy in Y. Since Y is complete, this sequence converges to a limit, say y. We set $\overline{f}(x) := y$. We need to check that $\overline{f}$ is well defined (that is, $\overline{f}(x)$ does not depend on the choice of the approximating sequence) and is uniformly continuous.

If $\widetilde{x}_n \to x$ with each $\widetilde{x}_n$ in X, then $\lim_{n \to \infty} d_X(x_n, \widetilde{x}_n) = 0$. By the uniform continuity of f it follows that $\lim_{n \to \infty} d_Y(f(x_n), f(\widetilde{x}_n)) = 0$, and therefore $\lim_{n \to \infty} f(\widetilde{x}_n) = \lim_{n \to \infty} f(x_n)$. This proves that $\overline{f}$ is well defined.

To prove that $\overline{f}$ is uniformly continuous let $\varepsilon > 0$ and choose $\delta > 0$ as in the definition of uniform continuity of f. If $\overline{d}_X(x,y) < \delta$ in $\overline{X}$, and if $(x_n)_{n \geqslant 1}$ and $(y_n)_{n \geqslant 1}$ are approximating sequences in X, then $(f(x_n))_{n \geqslant 1}$ and $(f(y_n))_{n \geqslant 1}$ are approximating sequences for $\overline{f}(x)$ and $\overline{f}(y)$ in Y, and for large enough n we have $d_X(x_n, y_n) < \delta$ and $d_Y(f(x_n), f(y_n)) < \varepsilon$. From this we obtain $d_Y(\overline{f}(x), \overline{f}(y)) = \lim_{n \to \infty} d_Y(f(x_n), f(y_n)) \leqslant \varepsilon$.

It is clear from the construction that $\overline{f}$ extends f. $\qquad\square$

Compactness

Let (X,d) be a metric space. We recall that a subset S of a metric space X is *compact* if every open cover of S has a finite subcover, and *relatively compact* if its closure $\overline{S}$ is compact. In order to characterise compactness in terms of sequences we introduce the following terminology. A subset S of a metric space X is called *sequentially compact* when every sequence in S has a convergent subsequence with limit in S.

A subset S of a metric space X is called *totally bounded* if for every $r > 0$ there is a finite cover of S with balls of radius r.

Theorem D.10 (Compactness and total boundedness). *For a subset S of a metric space X the following assertions are equivalent:*

(1) *S is compact;*

(2) *S is sequentially compact;*

(3) *S is complete and totally bounded.*

Proof (1)$\Rightarrow$(2): Suppose that $(x_n)_{n\geqslant 1}$ is a sequence in S not containing any subsequence converging to an element of S. We will construct an open cover of S without finite subcover.

The assumption entails that for every $x \in S$ there is an $\varepsilon(x) > 0$ with the property that the open ball $B(x;\varepsilon(x))$ contains at most finitely many terms of the sequence $(x_n)_{n\geqslant 1}$. Let $\mathscr{U} = \{B(x;\varepsilon(x)) : x \in S\}$. This is an open cover of S without finite subcover, as every $U \in \mathscr{U}$ contains at most finitely many terms of the sequence $(x_n)_{n\geqslant 1}$. But this sequence has infinitely many distinct terms: otherwise we could immediately pick a convergent subsequence.

(2)$\Rightarrow$(3): Suppose that S is not totally bounded. We will construct a sequence in S without any subsequence converging to an element of S.

By our assumption there exists an $\varepsilon > 0$ such that S has no finite cover with ε-balls with centres in S. Choose $x_1 \in S$ arbitrarily. The collection $\{B(x_1;\varepsilon)\}$ does not cover S, so there is an $x_2 \in S$ with $x_2 \notin B(x_1;\varepsilon)$. Note that $d(x_1,x_2) \geqslant \varepsilon$.

The collection $\{B(x_1;\varepsilon), B(x_2;\varepsilon)\}$ is not a cover of S, so there is an $x_3 \in S$ with $x_3 \notin B(x_1;\varepsilon) \cup B(x_2;\varepsilon)$. Note that $d(x_1,x_3) \geqslant \varepsilon$ and $d(x_2,x_3) \geqslant \varepsilon$.

Continuing this way we obtain a sequence $(x_n)_{n\geqslant 1}$ with the property that $d(x_n,x_m) \geqslant \varepsilon$ for all choices of n and m. This sequence has no Cauchy subsequence, and therefore no convergent subsequence.

Next we prove that S is complete. Suppose that $(x_n)_{n\geqslant 1}$ is a Cauchy sequence in S. Since S is sequentially compact this sequence has a convergent subsequence with limit x in S. But then $(x_n)_{n\geqslant 1}$ itself converges to x. This shows that S is complete.

(3)$\Rightarrow$(1): Suppose, for a contradiction, that S is complete and totally bounded but not compact.

Since S is not compact, there is an open cover $\mathscr{U}$ of S without finite subcover. Since S is totally bounded, for every $n \geqslant 1$ we can find a finite cover $\mathscr{B}_n$ of S consisting of $\frac{1}{n}$-balls with centres in S.

There is a ball $B_1 \in \mathscr{B}_1$ such that $S \cap B_1$ cannot be covered by finitely many open sets in $\mathscr{U}$. In the same way there is a ball $B_2 \in \mathscr{B}_2$ such that $S \cap B_1 \cap B_2$ cannot be covered by finitely many open sets in $\mathscr{U}$. Continuing in this way we find a sequence of balls $B_k \in \mathscr{B}_k$ such that $S \cap B_1 \cap \cdots \cap B_k$ cannot be covered by finitely many open sets in $\mathscr{U}$.

The sequence of centres $(x_n)_{n\geqslant 1}$ of these balls is a Cauchy sequence in S. To see this, we note that for all $n,m \geqslant 1$ the intersection $B_n \cap B_m$ is nonempty. If x_{mn} is an element in the intersection, with the triangle inequality we find that

$$d(x_n,x_m) \leqslant d(x_n,x_{nm}) + d(x_{nm},x_m) \leqslant \tfrac{1}{n} + \tfrac{1}{m}.$$

In view of $\lim_{n,m\to\infty} \frac{1}{n} + \frac{1}{m} = 0$ our assertion follows.

By completeness, the sequence $(x_n)_{n\geqslant 1}$ converges to a limit which belongs to S. Choose $U \in \mathscr{U}$ such that $x \in U$ and choose $r > 0$ such that $B(x;r) \subseteq U$. Choose N so large that $\frac{1}{N} < \frac{1}{2}r$ and $d(x_N,x) < \frac{1}{2}r$ for all $n \geqslant N$. Then

$$B_1 \cap \cdots \cap B_N \subseteq B_N = B(x_N; \tfrac{1}{N}) \subseteq B(x_N; \tfrac{1}{2}r) \subseteq B(x;r) \subseteq U.$$

But this means that $B_1 \cap \cdots \cap B_N$ is covered by the finite subcollection $\{U\}$ of $\mathscr{U}$. This contradiction concludes the proof. □

The equivalence of (1) and (3) implies that in a complete metric space, a subset is compact if and only if it is closed and totally bounded, and relatively compact if and only if it is totally bounded. The 'only if' parts are trivial, and for the 'if' parts we note that the closure of a totally bounded set is totally bounded; for if S can be covered with finitely many balls of radius $B(x_n; \varepsilon)$ with centres in S, the balls $B(x_n; 2\varepsilon)$ cover $\overline{S}$. Now it remains to observe that a closed subset of a complete metric space is complete.

Theorem D.11 (Bolzano–Weierstrass). *A subset of $\mathbb{R}^d$ is compact if and only if it is closed and bounded.*

Proof We have seen that, in any metric space, compact sets are always closed and bounded. Suppose, conversely, that the set S is closed and bounded.

Step 1 – We prove the theorem for $d = 1$ and the interval $[a,b]$. Let $\mathscr{U}$ be a cover for $[a,b]$ with open subsets of $\mathbb{R}$. We must show that $\mathscr{U}$ contains a finite subcover.

Let us call a point $x \in [a,b]$ *reachable from a* if there is a finite subcollection of $\mathscr{U}$ covering $[a,x]$. Let S be the set of all points that are reachable from a. We must show that $b \in S$.

First we observe that S is nonempty: clearly we have $a \in S$. Since S is bounded above (by b) we may put $p := \sup S$. Choose $U \in \mathscr{U}$ such that $p \in U$. Since U is open, there is an $\varepsilon > 0$ with $(p - \varepsilon, p + \varepsilon) \subseteq U$. Since $p = \sup S$ we can find an $x \in S$ with $p - \varepsilon < x \leqslant p$. Choose a finite subcollection $\mathscr{U}'$ of $\mathscr{U}$ covering $[a,x]$. The collection $\mathscr{U}'' = \mathscr{U}' \cup \{U\}$ is a finite subcollection of $\mathscr{U}$ covering $[a,p]$. We conclude that $p \in S$. We can also conclude that $p = b$. Indeed, if we had that $p < b$, then we could find a $y \in [a,b] \cap (p, p+\varepsilon)$. Then $\mathscr{U}''$ also covers interval $[a,y]$, and it follows that $y \in S$. This contradicts the fact that $p = \sup S$.

Step 2 – Suppose now that $S \subseteq \mathbb{R}^d$ is closed and bounded. Since S is bounded, we can find an $r > 0$ such that $S \subseteq [-r,r]^d$. We claim that $[-r,r]^d$ is compact. Once this has been shown, it follows that S, being a closed subset of the compact set $[-r,r]^d$, is compact.

To prove that $[-r,r]^d$ is compact we show that $[-r,r]^d$ is sequentially compact. Let $(x_n)_{n\geqslant 1}$ be a sequence in $[-r,r]^d$. The d coordinate sequences are sequences in the interval $[-r,r]$, which is sequentially compact by the Bolzano–Weierstrass theorem. By taking d consecutive subsequences we arrive at a subsequence $(x_{n_k})_{k\geqslant 1}$ all of whose coordinate sequences converge in $[-r, r]$. The sequence $(x_{n_k})_{k \geqslant 1}$ then converges in $\mathbb{R}^d$, with a limit in $[-r, r]^d$. $\qquad \square$

Theorem D.12. *Let (X, d_X) and (Y, d_Y) be metric spaces with X compact. Every continuous mapping $f : X \to Y$ is uniformly continuous.*

Proof Let $\varepsilon > 0$ be arbitrary. For every $x \in X$ we can find a $\delta(x) > 0$ such that for all $x' \in X$ with $d_X(x, x') < \delta(x)$ we have $d_Y(f(x), f(x')) < \frac{1}{2}\varepsilon$. The collection

$$\mathscr{U} = \left\{ B(x; \tfrac{1}{2}\delta(x)) : x \in X \right\}$$

is an open cover of X, and therefore has a finite subcover, say

$$\mathscr{U}' = \left\{ B(x_j; \tfrac{1}{2}\delta(x_j)) : j = 1, \ldots, n \right\}.$$

Let $\delta = \min\{\frac{1}{2}\delta(x_j) : j = 1, \ldots, n\}$.

Suppose now that $x, x' \in X$ satisfy $d_X(x, x') < \delta$. We have $x \in B(x_j; \frac{1}{2}\delta(x_j))$ for some $1 \leqslant j \leqslant n$ (since $\mathscr{U}'$ covers X). Then $d_X(x, x_j) < \frac{1}{2}\delta(x_j)$ and

$$d_X(x', x_j) \leqslant d_X(x', x) + d_X(x, x_j) < \delta + \tfrac{1}{2}\delta(x_j) \leqslant \delta(x_j).$$

Consequently, $d_Y(f(x), f(x')) \leqslant d_Y(f(x), f(x_j)) + d_Y(f(x_j), f(x')) < \frac{1}{2}\varepsilon + \frac{1}{2}\varepsilon = \varepsilon$. $\quad \square$

In many applications, the following simple special case of Tychonov's theorem (Theorem C.14) suffices.

Proposition D.13. *If $K_1, \ldots, K_n$ are compact metric spaces, then their cartesian product $K := K_1 \times \cdots \times K_n$ is a compact metric space with respect to the* product metric

$$d(s, t) := \sum_{j=1}^{n} d_j(s_j, t_j), \quad s, t \in K.$$

Proof Given $\varepsilon > 0$, for $j = 1, \ldots, n$ choose finitely many open d_j-balls of radius ε/n to cover K_j. Their cartesian products are open, contained in d-balls of radius ε, and cover K. Since $\varepsilon > 0$ was arbitrary, this shows that K is totally bounded. Since the completeness of the spaces K_j implies that K is complete, this proves the compactness of K. $\qquad \square$

Definition D.14 (Separability). A metric space is called *separable* if it contains a dense countable subset.

Proposition D.15. *Every compact metric space is separable.*

Proof For each $n = 1, 2, \ldots$ cover the metric space with finitely many open balls of radius $\frac{1}{n}$, say $B_1^{(n)}, \ldots, B_{N_n}^{(n)}$. Together, the centres of all these balls form a dense subset. Indeed, any nonempty open set U contains an open ball B, say of radius $r > 0$, and this ball must contain at least one of the balls $B_j^{(n)}$ for each $n \geqslant 1$ such that $\frac{1}{n} < \frac{1}{3}r$, for otherwise the sets $B_1^{(n)}, \ldots, B_{N_n}^{(n)}$ cannot cover B. The centres of such balls are in U. $\quad \square$

Appendix E

Measure Spaces

This appendix reviews the basic elements of Measure Theory.

σ-Algebras

Let Ω be a set.

Definition E.1 (σ-Algebras). A *σ-algebra* in Ω is a collection $\mathscr{F}$ of subsets of Ω with the following properties:

(i) $\Omega \in \mathscr{F}$;

(ii) $F \in \mathscr{F}$ implies $\complement F \in \mathscr{F}$;

(iii) $F_1, F_2, \cdots \in \mathscr{F}$ implies $\bigcup_{n \geqslant 1} F_n \in \mathscr{F}$.

Here, $\complement F = \Omega \setminus F$ is the complement of F.

A *measurable space* is a pair $(\Omega, \mathscr{F})$, where Ω is a set and $\mathscr{F}$ is a σ-algebra in Ω. The sets in $\mathscr{F}$ are often referred to as the *measurable subsets* of Ω.

These properties express that $\mathscr{F}$ is nonempty, closed under taking complements, and closed under taking countable unions. From

$$\bigcap_{n \geqslant 1} F_n = \complement \left(\bigcup_{n \geqslant 1} \complement F_n \right)$$

it follows that $\mathscr{F}$ is closed under taking countable intersections. Clearly, $\mathscr{F}$ is closed under finite unions and intersections as well.

Example E.2. When $\mathscr{C}$ is any collection of subsets of Ω, the *σ-algebra generated by* $\mathscr{C}$ is defined as the intersection of all σ-algebras in Ω containing $\mathscr{C}$, and is denoted by $\sigma(\mathscr{C})$. It is the smallest σ-algebra containing $\mathscr{C}$. Such σ-algebras arise in a variety of situations:

(1) When (X, τ) is a topological space, the σ-algebra $\mathscr{B}(X)$ generated by τ is called the *Borel σ-algebra* of (X, τ).

(2) Let $(\Omega_1, \mathscr{F}_1)$, $(\Omega_2, \mathscr{F}_2)$, ... be a sequence of measurable spaces. On the cartesian product $\prod_{n \geqslant 1} \Omega_n$, the *product σ-algebra* $\prod_{n \geqslant 1} \mathscr{F}_n$ is the σ-algebra generated by all sets of the form

$$F_1 \times \cdots \times F_N \times \Omega_{N+1} \times \Omega_{N+2} \times \cdots$$

with $N = 1, 2, \ldots$ and $F_n \in \mathscr{F}_n$ for $n = 1, \ldots, N$.

The product of finitely many measurable spaces $(\Omega_1, \mathscr{F}_1)$, ... $(\Omega_N, \mathscr{F}_N)$ is defined similarly; here one takes the σ-algebra in $\Omega_1 \times \cdots \times \Omega_N$ generated by all sets of the form $F_1 \times \cdots \times F_N$ with $F_n \in \mathscr{F}_n$ for $n = 1, \ldots, N$. By way of example, the reader may check that

$$(\mathbb{R}^d, \mathscr{B}(\mathbb{R}^d)) = \prod_{n=1}^{d} (\mathbb{R}, \mathscr{B}(\mathbb{R})).$$

For a proof one may use that every open set in $\mathbb{R}^d$ is a countable union of open rectangles of the form $(a_1, b_1) \times \cdots \times (a_d, b_d)$.

(3) Let Ω and Ω' be sets and let $f : \Omega \to \Omega'$ be any function. When $\mathscr{F}'$ is a σ-algebra in Ω', for $F' \in \mathscr{F}'$ we define

$$\{f \in F'\} := \{\omega \in \Omega : f(\omega) \in F'\}.$$

The collection

$$\sigma(f) = \{\{f \in F'\} : F' \in \mathscr{F}'\}$$

is a σ-algebra in Ω, the *σ-algebra generated by f*. The σ-algebra generated by a family of functions is defined similarly.

Measures

Let $(\Omega, \mathscr{F})$ be a measurable space.

Definition E.3 (Measures). A *measure* on $(\Omega, \mathscr{F})$ is a mapping $\mu : \mathscr{F} \to [0, \infty]$ with the following properties:

  (i) $\mu(\varnothing) = 0$;
  (ii) for all disjoint sets $F_1, F_2, \ldots$ in $\mathscr{F}$ we have $\mu(\bigcup_{n \geqslant 1} F_n) = \sum_{n \geqslant 1} \mu(F_n)$.

A triple $(\Omega, \mathscr{F}, \mu)$, with μ a measure on a measurable space $(\Omega, \mathscr{F})$, is called a *measure space*.

A measure space $(\Omega, \mathscr{F}, \mu)$ is called *finite* if μ is a *finite* measure, that is, if $\mu(\Omega) < \infty$. If $\mu(\Omega) = 1$, then μ is called a *probability measure* and $(\Omega, \mathscr{F}, \mu)$ is called a *probability space*. In probability theory, it is customary to use the symbol $\mathbb{P}$ for a probability measure. A measure space $(\Omega, \mathscr{F}, \mu)$ is called σ-*finite* if there exist $F_1, F_2, \ldots$ in $\mathscr{F}$ such that $\bigcup_{n \geq 1} F_n = \Omega$ and $\mu(F_n) < \infty$ for all $n \geq 1$. A *Borel measure* on a topological space (X, τ) is measure $\mu : \mathscr{B}(X) \to [0, \infty]$, where $\mathscr{B}(X)$ is the Borel σ-algebra of X.

The following properties of measures are easily checked:

(i) if $F_1 \subseteq F_2$ in $\mathscr{F}$, then $\mu(F_1) \leq \mu(F_2)$;

(ii) if $F_1, F_2, \ldots$ in $\mathscr{F}$, then

$$\mu\left(\bigcup_{n \geq 1} F_n\right) \leq \sum_{n \geq 1} \mu(F_n);$$

(iii) if $F_1 \subseteq F_2 \subseteq \ldots$ in $\mathscr{F}$, then

$$\mu\left(\bigcup_{n \geq 1} F_n\right) = \lim_{n \to \infty} \mu(F_n);$$

(iv) if $F_1 \supseteq F_2 \supseteq \ldots$ in $\mathscr{F}$ and $\mu(F_1) < \infty$, then

$$\mu\left(\bigcap_{n \geq 1} F_n\right) = \lim_{n \to \infty} \mu(F_n).$$

In (iii) and (iv), the limits (in $[0, \infty]$) exist by monotonicity.

Dynkin's Lemma

Lemma E.4 (Dynkin's lemma). *Let μ_1 and μ_2 be two finite measures defined on a measurable space $(\Omega, \mathscr{F})$. Let $\mathscr{A} \subseteq \mathscr{F}$ be a collection of sets with the following properties:*

(i) *$\Omega \in \mathscr{A}$;*

(ii) *$\mathscr{A}$ is closed under finite intersections;*

(iii) *the σ-algebra generated by $\mathscr{A}$, equals $\mathscr{F}$.*

If $\mu_1(A) = \mu_2(A)$ for all $A \in \mathscr{A}$, then $\mu_1 = \mu_2$.

Proof Let $\mathscr{D}$ denote the collection of all sets $D \in \mathscr{F}$ with $\mu_1(D) = \mu_2(D)$. Then $\mathscr{A} \subseteq \mathscr{D}$ and $\mathscr{D}$ is a *Dynkin system*, that is,

- $\Omega \in \mathscr{D}$;
- if $D_1 \subseteq D_2$ with $D_1, D_2 \in \mathscr{D}$, then also $D_2 \setminus D_1 \in \mathscr{D}$;
- if $D_1 \subseteq D_2 \subseteq \ldots$ with all $D_n \in \mathscr{D}$, then also $\bigcup_{n \geq 1} D_n \in \mathscr{D}$.

By assumption we have $\mathscr{D} \subseteq \mathscr{F} = \sigma(\mathscr{A})$, the σ-algebra generated by $\mathscr{A}$; we will show that $\sigma(\mathscr{A}) \subseteq \mathscr{D}$. To this end let $\mathscr{D}_0$ denote the smallest Dynkin system in $\mathscr{F}$ containing $\mathscr{A}$. We will show that $\sigma(\mathscr{A}) \subseteq \mathscr{D}_0$. In view of $\mathscr{D}_0 \subseteq \mathscr{D}$, this proves the lemma.

Let $\mathscr{C} = \{D_0 \in \mathscr{D}_0 : D_0 \cap A \in \mathscr{D}_0 \text{ for all } A \in \mathscr{A}\}$. This is a Dynkin system and $\mathscr{A} \subseteq \mathscr{C}$ since $\mathscr{A}$ is closed under taking finite intersections. It follows that $\mathscr{D}_0 \subseteq \mathscr{C}$, since $\mathscr{D}_0$ is the smallest Dynkin system containing $\mathscr{A}$. But obviously, $\mathscr{C} \subseteq \mathscr{D}_0$, and therefore $\mathscr{C} = \mathscr{D}_0$.

Now let $\mathscr{C}' = \{D_0 \in \mathscr{D}_0 : D_0 \cap D \in \mathscr{D}_0 \text{ for all } D \in \mathscr{D}_0\}$. This is a Dynkin system and the fact that $\mathscr{C} = \mathscr{D}_0$ implies that $\mathscr{A} \subseteq \mathscr{C}'$. Hence $\mathscr{D}_0 \subseteq \mathscr{C}'$, since $\mathscr{D}_0$ is the smallest Dynkin system containing $\mathscr{A}$. But obviously, $\mathscr{C}' \subseteq \mathscr{D}_0$, and therefore $\mathscr{C}' = \mathscr{D}_0$.

It follows that $\mathscr{D}_0$ is closed under taking finite intersections. But a Dynkin system with this property is a σ-algebra. Thus, $\mathscr{D}_0$ is a σ-algebra, and now $\mathscr{A} \subseteq \mathscr{D}_0$ implies that also $\sigma(\mathscr{A}) \subseteq \mathscr{D}_0$.   □

Outer Measures

Let S be a set. The *power set* of S, that is, the set of all subsets of S, is denoted by 2^S.

Definition E.5 (Outer measures). A mapping $\nu : 2^S \to [0, \infty]$ is called an *outer measure* if

   (i) $\nu(\varnothing) = 0$;
   (ii) $A \subseteq B$ implies $\nu(A) \leqslant \nu(B)$;
   (iii) for all $A_1, A_2, \ldots \in 2^S$ we have

$$\nu\left(\bigcup_{n \geqslant 1} A_n\right) \leqslant \sum_{n \geqslant 1} \nu(A_n).$$

Lemma E.6. *Let $\mathscr{C} \subseteq 2^S$ satisfy $\varnothing \in \mathscr{C}$ and suppose that $\mu : \mathscr{C} \to [0, \infty]$ satisfies $\mu(\varnothing) = 0$. For subsets $A \subseteq S$ define*

$$\mu^*(A) := \inf\left\{\sum_{j \geqslant 1} \mu(C_j) : A \subseteq \bigcup_{j \geqslant 1} C_j, \text{ where } C_j \in \mathscr{C} \text{ for all } j \geqslant 1\right\} \qquad \text{(E.1)}$$

with the convention that $\mu^*(A) = \infty$ if the above set is empty. Then μ^* is an outer measure.

Proof The mapping $\mu^* : 2^S \to [0, \infty]$ clearly satisfies the conditions (i) and (ii) in Definition E.5. In order to check condition (iii) let $A_1, A_2, \ldots$ be subsets of S and let $\varepsilon > 0$ be arbitrary. If $\mu^*(A_n) = \infty$ for some $n \geqslant 1$, then (iii) trivially holds. We may therefore assume that $\mu^*(A_n) < \infty$ for all $n \geqslant 1$. By the definition of μ^*, for each fixed $n \geqslant 1$ we can find $C_{n,j} \in \mathscr{C}$ such that

$$A_n \subseteq \bigcup_{j \geqslant 1} C_{n,j} \quad \text{and} \quad \sum_{j \geqslant 1} \mu(C_{n,j}) \leqslant \mu^*(A_n) + 2^{-n}\varepsilon.$$

Then $\bigcup_{n \geqslant 1} A_n \subseteq \bigcup_{n,j=1}^{\infty} C_{n,j}$, and, again by the definition of μ^*,

$$\mu^*\left(\bigcup_{n \geqslant 1} A_n\right) \leqslant \sum_{n,j=1}^{\infty} \mu(C_{n,j}) \leqslant \sum_{n \geqslant 1} (\mu^*(A_n) + 2^{-n}\varepsilon) = \varepsilon + \sum_{n \geqslant 1} \mu^*(A_n).$$

Since $\varepsilon > 0$ was arbitrary, this proves the required estimate. $\qquad\square$

Let $\mu : 2^S \to [0,\infty]$ be a mapping which satisfies $\mu(\varnothing) = 0$. A set $A \subseteq S$ is called *μ-measurable* if

$$\mu(Q) = \mu(Q \cap A) + \mu(Q \cap \complement A) \quad \text{for all } Q \in 2^S.$$

The collection of all μ-measurable sets is denoted by $\mathscr{M}_\mu$.

Theorem E.7 (Measures from outer measures). *If $\nu : 2^S \to [0,\infty]$ is an outer measure, then $\mathscr{M}_\nu$ is a σ-algebra and ν is a measure on $(S, \mathscr{M}_\nu)$.*

For the proof of the theorem we need the following terminology. A *ring* in S is a subset $\mathscr{R}$ of 2^S with the following properties:

(i) $\varnothing \in \mathscr{R}$;
(ii) $A, B \in \mathscr{R}$ implies $A \setminus B \in \mathscr{R}$;
(iii) $A, B \in \mathscr{R}$ implies $A \cup B \in \mathscr{R}$.

If $\mathscr{R}$ is a ring, the identity $A \cap B = A \setminus (A \setminus B)$ implies that if $A, B \in \mathscr{R}$, then $A \cap B \in \mathscr{R}$.

Proof of Theorem E.7 We proceed in two steps.

Step 1 – We begin by checking that if $\mu : 2^S \to [0,\infty]$ is any mapping which satisfies $\mu(\varnothing) = 0$, then $\mathscr{M}_\mu$ is a ring and μ is additive on $\mathscr{M}_\mu$.

It is clear that $\varnothing \in \mathscr{M}_\mu$. In order to check that $\mathscr{M}_\mu$ is a ring we check the following:

(a) $A \in \mathscr{M}_\mu$ implies $\complement A \in \mathscr{M}_\mu$;
(b) $A, B \in \mathscr{M}_\mu$ implies $A \cap B \in \mathscr{M}_\mu$.

Given these properties it is straightforward to check that $\mathscr{M}_\mu$ is a ring. Indeed, this follows from the formulas $B \setminus A = B \cap \complement A$ and $A \cup B = \complement(\complement A \cap \complement B)$.

Property (a) is clear. To check (b) let $A, B \in \mathscr{M}_\mu$ and set $C := A \cap B$. Let $Q \in 2^S$ be arbitrary. Observing that $A \cap \complement B = \complement C \cap A$ and $\complement A = \complement C \cap \complement A$, and making repeated use of the definition of $\mathscr{M}_\mu$, we have

$$\mu(Q) = \mu(Q \cap A) + \mu(Q \cap \complement A)$$

$$= \mu(Q \cap A \cap B) + \mu(Q \cap A \cap \complement B) + \mu(Q \cap \complement A)$$
$$= \mu(Q \cap C) + \mu(Q \cap \complement C \cap A) + \mu(Q \cap \complement C \cap \complement A)$$
$$= \mu(Q \cap C) + \mu(Q \cap \complement C).$$

Therefore, $A \cap B = C \in \mathcal{M}_\mu$.

To check that μ is additive on $\mathcal{M}_\mu$ fix two disjoint sets $A, B \in \mathcal{M}_\mu$ and let $Q := A \cup B$. Then $Q \cap A = A$ and $Q \cap \complement A = B$. Since $A \in \mathcal{M}_\mu$, we find

$$\mu(A \cup B) = \mu(Q) = \mu(Q \cap A) + \mu(Q \cap \complement A) = \mu(A) + \mu(B).$$

Step 2 – We now turn to the proof of the theorem. From Step 1 we know that $\mathcal{M}_\nu$ is a ring and ν is additive on $\mathcal{M}_\nu$. In view of property (a) it remains to check that for any disjoint sequence $(A_n)_{n \geqslant 1}$ in $\mathcal{M}_\mu$,

$$A := \bigcup_{n \geqslant 1} A_n \in \mathcal{M}_\nu \quad \text{and} \quad \nu(A) = \sum_{n \geqslant 1} \nu(A_n). \tag{E.2}$$

Let $B_n = \bigcup_{j=1}^n A_j$ for each $n \geqslant 1$. Fix an arbitrary subset Q of S. By Step 1, for all $n \geqslant 1$ we have $\complement A \subseteq \complement B_n$, $B_n \in \mathcal{M}_\nu$, and

$$\sum_{j=1}^n \nu(Q \cap A_j) + \nu(Q \cap \complement A) = \nu(Q \cap B_n) + \nu(Q \cap \complement A)$$
$$\leqslant \nu(Q \cap B_n) + \nu(Q \cap \complement B_n) = \nu(Q).$$

Using the σ-subadditivity of ν and then passing to the limit $n \to \infty$, we infer

$$\nu(Q \cap A) + \nu(Q \cap \complement A) \leqslant \sum_{j \geqslant 1} \nu(Q \cap A_j) + \nu(Q \cap \complement A) \leqslant \nu(Q). \tag{E.3}$$

On the other hand, by subadditivity also the converse inequality $\nu(Q) \leqslant \nu(Q \cap A) + \nu(Q \cap \complement A)$ holds. This shows that $A \in \mathcal{M}_\nu$ and that the inequalities in (E.3) are in fact equalities. Now (E.2) follows by taking $Q = A$ in (E.3). $\quad\square$

Carathéodory's Extension Theorem

For additive functions $\mu : \mathcal{R} \to [0, \infty]$ one has the following result.

Lemma E.8. *Let $\mathcal{R}$ be a ring and $\mu : \mathcal{R} \to [0, \infty]$ be additive, that is,*

$$\mu\left(\bigcup_{j=1}^n A_n\right) = \sum_{j=1}^n \mu(A_j)$$

holds for all disjoint sets $A_1, \ldots, A_n \in \mathcal{R}$. The following assertions hold:

(1) *if $A, B \in \mathcal{R}$ and $A \subseteq B$, then $\mu(A) \leqslant \mu(B)$;*

(2) *if* $A_1, A_2, \cdots \in \mathscr{R}$ *and* $\bigcup_{j \geqslant 1} A_j \in \mathscr{R}$ *and* μ *is countably additive on* $\mathscr{R}$, *then*

$$\mu\Big(\bigcup_{j \geqslant 1} A_j\Big) \leqslant \sum_{j \geqslant 1} \mu(A_j).$$

Proof (1): Writing $B = A \cup (B \setminus A)$, we see that

$$\mu(B) = \mu(A \cup (B \setminus A)) = \mu(A) + \mu(B \setminus A) \geqslant \mu(A).$$

(2): The sets $B_1 := A_1$, $B_2 := A_2 \setminus A_1$, $B_3 := A_3 \setminus (A_1 \cup A_2)$, ... are disjoint and we have $\bigcup_{j \geqslant 1} B_j = \bigcup_{j \geqslant 1} A_j$. Therefore, by the countable additivity of μ,

$$\mu\Big(\bigcup_{j \geqslant 1} A_j\Big) = \mu\Big(\bigcup_{j \geqslant 1} B_j\Big) = \sum_{j \geqslant 1} \mu(B_j) \leqslant \sum_{j \geqslant 1} \mu(A_j).$$

$\square$

Theorem E.9 (Carathéodory's extension theorem). *Let* $\mathscr{R}$ *be a ring in S and suppose that* $\mu : \mathscr{R} \to [0, \infty]$ *is countably additive on* $\mathscr{R}$ *and satisfies* $\mu(\varnothing) = 0$. *Let* μ^* *be the associated outer measure. Then:*

(1) *the outer measure* μ^* *restricts to a measure on* $\sigma(\mathscr{R})$ *extending* μ;
(2) *if* μ^* *is* σ-*finite on* $\sigma(\mathscr{R})$ *and if* ν *is another* σ-*finite measure on* $\sigma(\mathscr{R})$ *extending* μ, *then* $\mu^* = \nu$.

Proof By Theorem E.7, μ^* is a measure on the σ-algebra $\mathscr{M}_{\mu^*}$. We prove that it has the following properties:

(i) $\mathscr{R} \subseteq \mathscr{M}_{\mu^*}$;
(ii) $\mu^*(A) = \mu(A)$ for all $A \in \mathscr{R}$.

Clearly, part (1) of the theorem follows from the claim, which actually shows that there is a further extension to the possibly larger σ-algebra $\mathscr{M}_{\mu^*}$.

Step 1 – In this step we prove (i). Let $A \in \mathscr{R}$ and $Q \subseteq S$ be given. The subadditivity of μ^* gives $\mu^*(Q) \leqslant \mu^*(Q \cap A) + \mu^*(Q \cap \complement A)$. The converse estimate $\mu^*(Q \cap A) + \mu^*(Q \cap \complement A) \leqslant \mu^*(Q)$ trivially holds if $\mu^*(Q) = \infty$. If $\mu^*(Q) < \infty$, choose $B_1, B_2, \cdots \in \mathscr{R}$ such that $Q \subseteq \bigcup_{n \geqslant 1} B_n$. Then $B_n \cap A$ and $B_n \cap \complement A = B_n \setminus A$ belong to $\mathscr{R}$ for all $n \geqslant 1$, and

$$Q \cap A \subseteq \bigcup_{n \geqslant 1} B_n \cap A \quad \text{and} \quad Q \cap \complement A \subseteq \bigcup_{n \geqslant 1} B_n \cap \complement A.$$

Using first the definition of μ^* and then the additivity of μ on $\mathscr{R}$, we find

$$\mu^*(Q \cap A) + \mu^*(Q \cap \complement A) \leqslant \sum_{n \geqslant 1} \mu(B_n \cap A) + \sum_{n \geqslant 1} \mu(B_n \cap \complement A) = \sum_{n \geqslant 1} \mu(B_n).$$

Taking the infimum over all admissible sequences $B_1, B_2, \ldots$ as specified above, we obtain $\mu^*(Q\cap A)+\mu^*(Q\cap\complement A)\leqslant\mu^*(Q)$. Combining both estimates, we conclude that $A\in\mathscr{M}_{\mu^*}$.

Step 2 – In this step we prove (ii). Let $A\in\mathscr{R}$. It is clear that $\mu^*(A)\leqslant\mu(A)$; this follows by taking $B_1=A$ and $B_n=\varnothing$ for $n\geqslant 2$ in (E.1). The converse estimate $\mu(A)\leqslant\mu^*(A)$ trivially holds if $\mu^*(A)=\infty$. If $\mu^*(A)<\infty$, choose $B_1,B_2,\cdots\in\mathscr{R}$ such that $A\subseteq\bigcup_{n\geqslant 1}B_n$. Then, by Lemma E.8, where part (1) is applied to the inclusion $A\cap B_n\subseteq B_n$ and part (2) to the union $A=\bigcup_{n\geqslant 1}A\cap B_n$,

$$\mu(A)\leqslant\sum_{n\geqslant 1}\mu(A\cap B_n)\leqslant\sum_{n\geqslant 1}\mu(B_n).$$

Taking the infimum over all admissible sequences $B_1,B_2,\ldots$ as specified above, we obtain $\mu(A)\leqslant\mu^*(A)$.

Let now the assumptions of part (2) be satisfied and choose pairwise disjoint sets $S_n\in\sigma(\mathscr{R})$ such that $S=\bigcup_{n\geqslant 1}S_n$ and $\mu^*(S_n)<\infty$ and $\nu(S_n)<\infty$. Then the restrictions of μ^* and ν agree on the σ-algebras $\{F\cap S_n:F\in\sigma(\mathscr{R})\}$ in S_n by Dynkin's lemma (which can be applied, noting that the collections $\mathscr{R}_n:=\{R\cap S_n:R\in\mathscr{R}\}$ are rings in S_n and hence are closed under finite intersections). By countable additivity, this in turn implies that μ^* and ν agree on $\sigma(\mathscr{R})$. $\qquad\square$

To verify the countable additivity condition in Carathéodory's result one may use the following sufficient condition.

Proposition E.10. *Let $\mathscr{R}$ be a ring in a set S and let $\mu:\mathscr{R}\to[0,\infty]$ be an additive map with the property that $\mu(\varnothing)=0$. If for each nonincreasing sequence $(A_n)_{n\geqslant 1}$ in $\mathscr{R}$ with $\bigcap_{n\geqslant 1}A_n=\varnothing$ we have $\lim_{n\to\infty}\mu(A_n)=0$, then μ is countably additive on $\mathscr{R}$.*

Proof Let $(B_j)_{j\geqslant 1}$ be a disjoint sequence in $\mathscr{R}$ with $B:=\bigcup_{j\geqslant 1}B_j\in\mathscr{R}$. We need to show that

$$\mu(B)=\sum_{j\geqslant 1}\mu(B_j). \tag{E.4}$$

Let $A_n=\bigcup_{j\geqslant n}B_j=B\setminus(B_1\cup\cdots\cup B_{n-1})$. Then $A_n\in\mathscr{R}$ and $\bigcap_{n\geqslant 1}A_n=\varnothing$, and therefore $\mu(A_n)\to 0$ by assumption. On the other hand,

$$\mu(B)=\mu(A_n\cup B_1\cup B_2\cup\cdots\cup B_{n-1})=\mu(A_n)+\sum_{j=1}^{n-1}\mu(B_j).$$

Therefore, $0\leqslant\mu(B)-\sum_{j=1}^{n-1}\mu(B_j)=\mu(A_n)\to 0$ as $n\to\infty$, and (E.4) follows. $\qquad\square$

Lebesgue Measure

As a first application of Carathéodory's theorem we construct the *Lebesgue measure*.

For $a = (a_1,\ldots,a_d)$ and $b = (b_1,\ldots,b_d)$ such that $a_j \leqslant b_j$ for $j = 1,\ldots,d$ we write

$$(a,b] := \{x \in \mathbb{R}^d : a_j < x_j \leqslant b_j,\ j = 1,\ldots,d\}.$$

The collection $\mathscr{I}^d$ of all finite unions of half-open rectangles is a ring and every set in $\mathscr{I}^d$ can be written as a finite union of *disjoint* half-open rectangles.

For $I = (a,b]$ let

$$|I| := \prod_{j=1}^{d} (\beta_j - \alpha_j).$$

For $A \in \mathscr{I}^d$ of the form $A = I_1 \cup \cdots \cup I_n$, with disjoint $I_j \in \mathscr{I}^d$, define $\lambda_d : \mathscr{I}^d \to [0,\infty]$ by

$$\lambda_d(A) := \sum_{j=1}^{n} |I_j|.$$

We must check that this number is well defined. To this end suppose that $A = (a_1,b_1] \cup \cdots \cup (a_m,b_m] = (c_1,d_1] \cup \cdots \cup (c_n,d_n]$ are two representations of A as unions of disjoint half-open rectangles. Then $I_{ij} = (a_i,b_i] \cap (c_j,d_j]$ is either empty or a nonempty half-open rectangle, and we have

$$\bigcup_{i=1}^{m} I_{ij} = (c_j,d_j] \text{ and } \bigcup_{j=1}^{n} I_{ij} = (a_i,b_i].$$

From the definition and the disjointness of the sets I_{ij} we obtain

$$\sum_{i=1}^{m} \lambda_d((a_i,b_i]) = \sum_{i=1}^{m} \lambda_d\Big(\bigcup_{j=1}^{n} I_{ij}\Big) = \sum_{i=1}^{m}\sum_{j=1}^{n} \lambda_d(I_{ij})$$
$$= \sum_{j=1}^{n}\sum_{i=1}^{m} \lambda_d(I_{ij}) = \sum_{j=1}^{n} \lambda_d\Big(\bigcup_{i=1}^{m} I_{ij}\Big) = \sum_{j=1}^{n} \lambda_d((c_j,d_j]),$$

which proves the asserted well-definedness.

When the dimension d is fixed and there is no danger of confusion we write λ for λ_d.

Lemma E.11. *The function* $\lambda : \mathscr{I}^d \to [0,\infty]$ *is countably additive on* $\mathscr{I}^d$.

Proof By Proposition E.10 it suffices to prove that for each nonincreasing sequence $(A_n)_{n\geqslant 1}$ in $\mathscr{I}^d$ satisfying $\bigcap_{n\geqslant 1} A_n = \varnothing$ we have $\mu(A_n) \to 0$. Fix such a sequence $(A_n)_{n\geqslant 1}$ and let $\varepsilon > 0$. We have to find $N \in \mathbb{N}$ such that $\lambda(A_n) < \varepsilon$ for all $n \geqslant N$.

Step 1 – For each $n \in \mathbb{N}$ choose a $B_n \in \mathscr{I}^d$ such that $\overline{B_n} \subseteq A_n$ and $\lambda(A_n \setminus B_n) \leqslant 2^{-n}\varepsilon$. Since $\overline{B_n} \subseteq A_n$, we also have $\bigcap_{n\geqslant 1} \overline{B_n} = \varnothing$. It follows that the complements of the sets $\overline{B_n}$ form an open cover of the set $\overline{A_1}$, which is compact by the Bolzano–Weierstrass theorem. Therefore, there exists an N such that $\overline{A_1} \subseteq \bigcup_{n=1}^{N} \complement\overline{B_n}$. It follows that $\bigcap_{n=1}^{N} \overline{B_n} \subseteq \complement A_1$. Since $\overline{B_n} \subseteq A_1$ for all $n \geqslant 1$, we must have that $\bigcap_{n=1}^{N} \overline{B_n} = \varnothing$.

Step 2 – Let $C_n = \bigcap_{j=1}^n B_j$ for $n \geqslant 1$. For every $n \geqslant 1$, $A_n \setminus C_n = \bigcup_{j=1}^n (A_n \setminus B_j) \subseteq \bigcup_{j=1}^n (A_j \setminus B_j)$. Therefore, using Lemma E.8 (part (1) in $(*)$ and part (1) for finite unions in $(**)$) we find

$$\lambda(A_n \setminus C_n) \overset{(*)}{\leqslant} \lambda\left(\bigcup_{j=1}^n (A_j \setminus B_j)\right) \overset{(**)}{\leqslant} \sum_{j=1}^n \lambda(A_j \setminus B_j) \leqslant \sum_{j=1}^n 2^{-j}\varepsilon < \varepsilon.$$

Since $C_n = \varnothing$ for all $n \geqslant N$, we conclude that $\lambda(A_n) = \lambda(A_n \setminus C_n) < \varepsilon$ for all $n \geqslant N$. $\quad\square$

Theorem E.12 (Lebesgue measure). *There exists a unique σ-finite Borel measure λ on $\mathbb{R}^d$ satisfying*

$$\lambda(I) = |I|$$

for all $I \in \mathscr{I}^d$. Moreover, for all $h \in \mathbb{R}^d$ and $A \in \mathscr{B}(\mathbb{R}^d)$ we have

$$\lambda(A + h) = \lambda(A)$$

where $A + h := \{x + h : x \in A\}$.

Proof In Lemma E.11 we have shown that λ is countably additive on the ring $\mathscr{I}^d$. Therefore, by Theorem E.9, λ admits a unique extension to a σ-finite measure on $\sigma(\mathscr{I}^d) = \mathscr{B}(\mathbb{R}^d)$.

To prove translation invariance, fix $h \in \mathbb{R}^d$. We claim that for every $A \in \mathscr{B}(\mathbb{R}^d)$ the set $A + h$ belongs to $\mathscr{B}(\mathbb{R}^d)$. To see this, let $\mathscr{A}_h = \{A \in \mathscr{B}(\mathbb{R}^d) : A + h \in \mathscr{B}(\mathbb{R}^d)\}$. This is a σ-algebra contained in $\mathscr{B}(\mathbb{R}^d)$. For each open set A, the set $A + h$ is open and hence belongs to $\mathscr{B}(\mathbb{R}^d)$. It follows that $\mathscr{A}_h$ contains all open sets. Since $\mathscr{B}(\mathbb{R}^d)$ is the smallest σ-algebra containing all open sets, it follows that $\mathscr{B}(\mathbb{R}^d) \subseteq \mathscr{A}_h$. Since also $\mathscr{A}_h \subseteq \mathscr{B}(\mathbb{R}^d)$ we obtain equality $\mathscr{A}_h = \mathscr{B}(\mathbb{R}^d)$. This proves the claim.

For $A \in \mathscr{B}(\mathbb{R}^d)$ set $\mu_h(A) := \lambda(A + h)$. Then μ_h is a σ-finite Borel measure on $\mathscr{B}(\mathbb{R}^d)$ and for any half-open rectangle I, $\mu_h(I) = |I + h| = |I| = \lambda(I)$. By uniqueness, we find that $\mu_h(A) = \lambda(A)$ for all $A \in \mathscr{B}(\mathbb{R}^d)$. $\quad\square$

Product Measures

As a second application of Carathéodory's theorem we prove the existence of product measures.

Theorem E.13 (Product measures). *Let $(\Omega_j, \mathscr{F}_j, \mu_j)$, $j = 1, \ldots, n$, be σ-finite measure spaces. Then there exists a unique σ-finite measure $\mu = \prod_{j=1}^n \mu_n$ on the product σ-algebra $\mathscr{F} = \prod_{j=1}^n \mathscr{F}_j$ which satisfies*

$$\mu(F_1 \times \cdots \times F_n) = \prod_{j=1}^n \mu_j(F_j)$$

whenever the sets $F_j \in \mathscr{F}_j$ satisfy $\mu_j(F_j) < \infty$ for $j = 1, \ldots, n$.

The measure μ is called the *product* of $\mu_1, \ldots, \mu_n$.

Proof Let $\mathscr{R}$ be the ring consisting of all finite unions of measurable rectangles of finite measure, that is, sets of the form $\prod_{j=1}^{n} F_j$ with $F_j \in \mathscr{F}_j$ satisfying $\mu_j(F_j) < \infty$ for $j = 1, \ldots, n$. Since the intersection of finitely many measurable rectangles of finite measure is a measurable rectangle of finite measure, every $R \in \mathscr{R}$ can be written as a finite union of *disjoint* measurable rectangles of finite measure, say $R = R^{(1)} \cup \cdots \cup R^{(k)}$ and we may define

$$\mu(R) := \sum_{j=1}^{k} \mu(R^{(j)}),$$

where each $\mu(R^{(j)})$ is given by the product formula in the statement of the theorem. The proof that $\mu(R)$ is well defined follows the lines of the proof for the Lebesgue measure. It is clear that μ is additive on $\mathscr{R}$. We claim that μ is countably additive on $\mathscr{R}$. Once we know this, the existence of a unique σ-finite product measure follows from Carathéodory's theorem.

A quick proof of the claim is obtained by applying Proposition E.10 in combination with the dominated convergence theorem. The reader may check that no circularity is introduced by borrowing this result at this stage. Thus let $(A_j)_{j \geqslant 1}$ be a nonincreasing sequence of sets in $\mathscr{R}$ satisfying $\bigcap_{j \geqslant 1} A_j = \varnothing$. We must show that $\lim_{j \to \infty} \mu(A_j) = 0$. We have

$$\mu(A_j) = \int_{\Omega_n} \cdots \int_{\Omega_1} \mathbf{1}_{A_j} \, d\mu_1 \cdots d\mu_n,$$

using that A_j is the finite union of measurable rectangles of finite measure and that the identity holds for such sets by definition. The asserted convergence now follows by applying dominated convergence n times consecutively. $\qquad\square$

Example E.14. The Lebesgue measure on $(\mathbb{R}^d, \mathscr{B}(\mathbb{R}^d))$ is the product measure of d copies of the Lebesgue measure on $(\mathbb{R}, \mathscr{B}(\mathbb{R}))$.

Borel Measures on Metric Spaces

Definition E.15 (Regularity). A Borel measure μ on a topological space X is called:

- *inner regular*, if for all Borel subsets B of X and all $\varepsilon > 0$ there is a closed set F in X such that $F \subseteq B$ and $\mu(B \setminus F) < \varepsilon$;
- *outer regular*, if for all Borel subsets B of X and all $\varepsilon > 0$ there is an open set U in X such that $B \subseteq U$ and $\mu(U \setminus B) < \varepsilon$;
- *regular*, if it is inner regular and outer regular.

Proposition E.16 (Regularity). *Every finite Borel measure on a metric space is regular.*

Proof Let μ be a finite Borel measure on a metric space X. Denote by $\mathscr{A}(X)$ the collection of all Borel sets A in X which have the property that for all $\varepsilon > 0$ there exist a closed set F and an open set U such that $F \subseteq A \subseteq U$ and $\mu(U \setminus F) < \varepsilon$. We must prove that $\mathscr{A}(X) = \mathscr{B}(X)$, the Borel σ-algebra of X.

Claim 1: $\mathscr{A}(X)$ *is a σ-algebra.* It is clear that $\varnothing \in \mathscr{A}(X)$ and that $\mathscr{A}(X)$ is closed under taking complements. To see that $\mathscr{A}(X)$ is closed under taking countable unions, let $(A_n)_{n \geqslant 1}$ be sets in $\mathscr{A}(X)$. Let $\varepsilon > 0$ be given and let $(F_n)_{n \geqslant 1}$ and $(U_n)_{n \geqslant 1}$ be sequences of closed and open sets such that $F_n \subseteq A_n \subseteq U_n$ and $\mu(U_n \setminus F_n) < \frac{\varepsilon}{2^n}$. The set $U = \bigcup_{n \geqslant 1} U_n$ is open. In view of

$$\mu\left(U \setminus \bigcup_{n \geqslant 1} F_n\right) \leqslant \sum_{n \geqslant 1} \mu(U_n \setminus F_n) < \varepsilon$$

there exists an index N such that

$$\mu\left(U \setminus \bigcup_{n=1}^{N} F_n\right) < \varepsilon.$$

The set $F := \bigcup_{n=1}^{N} F_n$ is closed, satisfies $F \subseteq \bigcup_{n \geqslant 1} A_n \subseteq U$, and $\mu(U \setminus F) < \varepsilon$.

Claim 2: $\mathscr{A}(X)$ *contains all closed subsets of X.* To see this, let F be a closed subset of X and define, for $k \geqslant 1$, $U_k := \{x \in X : d(x,F) < \frac{1}{k}\}$. Then each U_k is open and we have $\bigcap_{k \geqslant 1} U_k = F$. Hence, $\lim_{n \to \infty} \mu(U_k) = \mu(F)$ and the claim follows.

Combining the two claims we see that $\mathscr{A}(X) = \mathscr{B}(X)$. $\square$

In order to state the theorem we need the following terminology.

Definition E.17 (Tightness). A finite Borel measure μ on a topological space X is called *tight* if for every $\varepsilon > 0$ there exists a compact set K in X such that $\mu(X \setminus K) < \varepsilon$.

The following proposition gives a sufficient condition for tightness.

Proposition E.18. *Every finite Borel measure μ on a separable complete metric space X is tight.*

Proof Let $(x_n)_{n \geqslant 1}$ be a dense sequence in X and fix $\varepsilon > 0$. For each integer $k \geqslant 1$, the closed balls $\overline{B}(x_n; \frac{1}{k})$ cover X, and therefore there exists an index $N_k \geqslant 1$ such that

$$\mu\left(\bigcup_{n=1}^{N_k} \overline{B}(x_n; \tfrac{1}{k})\right) \geqslant \mu(X) - \frac{\varepsilon}{2^k}.$$

The set

$$K := \bigcap_{k \geqslant 1} \bigcup_{n=1}^{N_k} \overline{B}(x_n; \tfrac{1}{k})$$

is closed and totally bounded. Since X is assumed to be complete, K is compact by Theorem D.10. Moreover,

$$\mu(\complement K) \leqslant \sum_{k \geqslant 1} \frac{\varepsilon}{2^k} = \varepsilon.$$

$\square$

For separable complete metric spaces, this result implies the following improvement to the inner regularity of Borel measures provided by Proposition E.16.

Corollary E.19. *Let μ be a finite Borel measure on a separable complete metric space X. Then for all Borel subsets B in X and all $\varepsilon > 0$ there is a compact set K in X such that $K \subseteq B$ and $\mu(B \setminus K) < \varepsilon$.*

Proof The measure μ is inner regular by Proposition E.16, so for every Borel set B there is a closed set F in X such that $F \subseteq B$ and $\mu(B \setminus F) < \frac{1}{2}\varepsilon$. By Proposition E.18 there is a compact set G in X such that $\mu(X \setminus G) < \frac{1}{2}\varepsilon$. Then $K := F \cap G$ is compact, contained in B, and satisfies $\mu(B \setminus K) < \varepsilon$. $\square$

Definition E.20. A finite Borel measure μ on a topological space X is called *Radon*, if for every Borel subset B of X and all $\varepsilon > 0$ there is a compact set K in X and an open set U in X such that $K \subseteq B \subseteq U$ and $\mu(U \setminus K) < \varepsilon$.

Proposition E.21. *Every finite Borel measure μ on a separable complete metric space X is Radon.*

Proof The measure μ is outer regular by Proposition E.16, so for every Borel set B there is an open set U in X such that $B \subseteq U$ and $\mu(U \setminus B) < \frac{1}{2}\varepsilon$. By Corollary E.19 there is a compact set K in X such that $K \subseteq B$ and $\mu(B \setminus K) < \varepsilon$. This gives the result. $\square$

Appendix F
Integration

In this appendix we review the Lebesgue integral.

Measurable Functions

Let $(\Omega_1, \mathscr{F}_1)$ and $(\Omega_2, \mathscr{F}_2)$ be measurable spaces. A function $f : \Omega_1 \to \Omega_2$ is said to be *measurable* if $\{f \in F\} \in \mathscr{F}_1$ for all $F \in \mathscr{F}_2$. Clearly, compositions of measurable functions are measurable. If $\mathscr{C}_2$ is a subset of $\mathscr{F}_2$ with the property that $\sigma(\mathscr{C}_2) = \mathscr{F}_2$, then a function $f : \Omega_1 \to \Omega_2$ is measurable if and only if

$$\{f \in C\} \in \mathscr{F}_1 \text{ for all } C \in \mathscr{C}_2.$$

Indeed, just notice that $\{F \in \mathscr{F}_2 : \{f \in F\} \in \mathscr{F}_1\}$ is a sub-σ-algebra of $\mathscr{F}_2$ containing $\mathscr{C}_2$.

If $f : \Omega_1 \to \Omega_2$ is measurable, then

$$f_*(\mu_1)(F_2) := \mu_1\{f \in F_2\}, \quad F_2 \in \mathscr{F}_2,$$

defines a measure $f_*(\mu_1)$ on $(\Omega_2, \mathscr{F}_2)$. This measure is called the *image* of μ_1 under f. Measurable functions f from a probability space $(\Omega, \mathscr{F}, \mathbb{P})$ to another measurable space are called *random variables* and the image of the probability measure $\mathbb{P}$ under f is called the *distribution* of f under $\mathbb{P}$.

In most applications we are concerned with measurable functions f from a measurable space $(\Omega, \mathscr{F})$ to $(\mathbb{K}, \mathscr{B}(\mathbb{K}))$. Such functions are said to be *Borel measurable*. Since open sets are Borel, continuous functions are Borel measurable.

In what follows we summarise some measurability properties of Borel measurable functions. The adjective 'Borel' will be omitted except when confusion is likely to arise. By the observation made earlier, a function $f : \Omega \to \mathbb{R}$ is measurable if and only if $\{f > a\} \in \mathscr{F}$ for all $a \in \mathbb{R}$, and a function $f : \Omega \to \mathbb{C}$ is measurable if and only if

$\{\operatorname{Re} f > a, \operatorname{Im} f > b\} \in \mathscr{F}$ for all $a, b \in \mathbb{R}$. From this it follows that linear combinations, products, and quotients (if defined) of measurable functions are measurable. For example, if the real-valued functions f and g are measurable, then $f + g$ is measurable since

$$\{f + g > a\} = \bigcup_{q \in \mathbb{Q}} \{f > q\} \cap \{g > a - q\}.$$

The measurability of the sum of two complex-valued measurable functions is proved in the same way.

If $f : \Omega \to \mathbb{K}$ and $g : \Omega \to \mathbb{K}$ are measurable, then $fg = \frac{1}{2}[(f + g)^2 - (f^2 + g^2)]$ is measurable.

If $f : \Omega \to \mathbb{C}$ is measurable, then its complex conjugate $\overline{f}$ is measurable, and therefore $\operatorname{Re} f = \frac{1}{2}(f + \overline{f})$ and $\operatorname{Im} f = \frac{1}{2i}(f - \overline{f})$ are measurable. Conversely, if $\operatorname{Re} f$ and $\operatorname{Im} f$ are measurable, so is $f = \operatorname{Re} f + i \operatorname{Im} f$.

If $f = \sup_{n \geqslant 1} f_n$ pointwise and each $f_n : \Omega \to \mathbb{R}$ is measurable, then f is measurable since

$$\{f > a\} = \bigcup_{n \geqslant 1} \{f_n > a\}.$$

It follows from $\inf_{n \geqslant 1} f_n = -\sup_{n \geqslant 1}(-f_n)$ that the pointwise infimum of a sequence of measurable functions is measurable as well. From this we get that the pointwise limit superior and limit inferior of measurable functions are measurable, since

$$\limsup_{n \to \infty} f_n = \lim_{n \to \infty} \left(\sup_{k \geqslant n} f_k \right) = \inf_{n \geqslant 1} \left(\sup_{k \geqslant n} f_k \right)$$

and $\liminf_{n \to \infty} f_n = -\limsup_{n \to \infty}(-f_n)$. It follows that the pointwise limit $\lim_{n \to \infty} f_n$ of a sequence of measurable functions $f_n : \Omega \to \mathbb{R}$ is measurable. By considering real and imaginary parts separately, the latter extends to pointwise limits of functions $f_n : \Omega \to \mathbb{C}$.

In the above considerations involving suprema, infima, and limits it is implicitly assumed that these suprema and infima exist and are finite pointwise. This restriction can be lifted by considering functions $f : \Omega \to [-\infty, \infty]$. Such functions are said to be *Borel measurable* if the sets $\{f \in B\}$, $B \in \mathscr{B}(\mathbb{R})$, as well as the sets $\{f = \infty\}$ and $\{f = -\infty\}$ are in $\mathscr{F}$.

A *simple function* is a function $f : \Omega \to \mathbb{K}$ that can be represented in the form

$$f = \sum_{n=1}^{N} c_n \mathbf{1}_{F_n}$$

with coefficients $c_n \in \mathbb{K}$ and disjoint sets $F_n \in \mathscr{F}$ for all $n = 1, \dots, N$.

Proposition F.1. *A function $f : \Omega \to \mathbb{K}$ is measurable if and only if it is the pointwise limit of a sequence of simple functions $f_n : \Omega \to \mathbb{K}$. This sequence may be chosen to satisfy $0 \leqslant |f_n| \uparrow |f|$ pointwise. If f is bounded, it may in addition be arranged that the* *convergence is uniform. If f is nonnegative, we may furthermore arrange that* $0 \leqslant f_n \uparrow f$ *pointwise, and uniformly if f is bounded.*

Proof   There is no loss of generality in taking $\mathbb{K} = \mathbb{C}$.

The 'if' part is clear from the fact that measurability is preserved under taking pointwise limits. It remains to prove the 'only if' part.

To prove the first assertion, for $j, k \in \mathbb{Z}$ and $n \in \mathbb{N}$ consider the rectangles $R_{jk}^{(n)} = [\frac{j}{2^n}, \frac{j+1}{2^n}) + i[\frac{k}{2^n}, \frac{k+1}{2^n})$ in the complex plane. Let $c_{jk}^{(n)}$ be the unique point in the closure of $R_{jk}^{(n)}$ with minimum modulus. Then the simple functions

$$f_n = \sum_{j,k=-2^{2n}}^{2^{2n}} c_{jk}^{(n)} \mathbf{1}_{\{f \in R_{jk}^{(n)}\}}$$

satisfy $f_n \to f$ and $0 \leqslant |f_n| \uparrow |f|$ pointwise. If f is bounded, the convergence is uniform.

To prove the second assertion, for $j \in \mathbb{N}$ and $n \in \mathbb{N}$ consider the intervals $I_j^{(n)} = [\frac{j}{2^n}, \frac{j+1}{2^n})$ in the nonnegative real line. Then the simple functions

$$f_n = \sum_{j=0}^{2^{2n}} \frac{j}{2^n} \mathbf{1}_{\{f \in I_j^{(n)}\}}$$

satisfy $0 \leqslant f_n \uparrow f$ pointwise. If f is bounded, the convergence is uniform. $\square$

The Lebesgue Integral

The construction of the Lebesgue integral proceeds in two stages: in the first step, the Lebesgue integral of an arbitrary nonnegative measurable function is defined (allowing the value ∞); in the second step, the notion of integrability is introduced and the Lebesgue integral of an integrable function is defined.

Let $(\Omega, \mathscr{F}, \mu)$ be a measure space. For a nonnegative simple function $f = \sum_{n=1}^{N} c_n \mathbf{1}_{F_n}$ we define

$$\int_{\Omega} f \, d\mu := \sum_{n=1}^{N} c_n \mu(F_n).$$

We allow $\mu(F_n)$ to be infinite; this causes no problems because the coefficients c_n are nonnegative (we use the convention $0 \cdot \infty = 0$). It is easy to check that this integral is well defined, in the sense that it does not depend on the particular representation of f as a simple function. Also, the integral is linear with respect to addition and multiplication with nonnegative scalars,

$$\int_{\Omega} af + bg \, d\mu = a \int_{\Omega} f \, d\mu + b \int_{\Omega} g \, d\mu,$$

and monotone in the sense that if $0 \leqslant f \leqslant g$ pointwise, then

$$\int_\Omega f \, \mathrm{d}\mu \leqslant \int_\Omega g \, \mathrm{d}\mu.$$

In what follows, a *nonnegative function* is a function with values in $[0, \infty]$. Recall that such a function is said to be *(Borel) measurable* if all sets $\{f \in B\}$ with $B \in \mathscr{B}(\mathbb{R})$, as well as the set $\{f = \infty\}$, are in $\mathscr{F}$.

For a nonnegative measurable function f we choose a sequence of simple functions $0 \leqslant f_n \uparrow f$ (see Proposition F.1) and define

$$\int_\Omega f \, \mathrm{d}\mu := \lim_{n \to \infty} \int_\Omega f_n \, \mathrm{d}\mu.$$

The following lemma implies that this definition does not depend on the approximating sequence.

Lemma F.2. *For a nonnegative measurable function f and nonnegative simple functions f_n and g such that $0 \leqslant f_n \uparrow f$ and $g \leqslant f$ pointwise we have*

$$\int_\Omega g \, \mathrm{d}\mu \leqslant \lim_{n \to \infty} \int_\Omega f_n \, \mathrm{d}\mu.$$

Proof   First consider the case $g = \mathbf{1}_F$. Fix $\varepsilon > 0$ arbitrary and let $F_n := \{\mathbf{1}_F f_n \geqslant 1 - \varepsilon\}$. Then $F_1 \subseteq F_2 \subseteq \dots$ and $\bigcup_{n \geqslant 1} F_n = F$, and therefore $\mu(F_n) \uparrow \mu(F)$. Since $\mathbf{1}_F f_n \geqslant (1 - \varepsilon)\mathbf{1}_{F_n}$,

$$\lim_{n \to \infty} \int_\Omega f_n \, \mathrm{d}\mu \geqslant \lim_{n \to \infty} \int_\Omega \mathbf{1}_F f_n \, \mathrm{d}\mu \geqslant (1 - \varepsilon) \lim_{n \to \infty} \mu(F_n)$$

$$= (1 - \varepsilon)\mu(F) = (1 - \varepsilon)\int_\Omega g \, \mathrm{d}\mu.$$

This proves the lemma for $g = \mathbf{1}_F$. The general case follows by linearity. $\square$

The integral is linear and monotone on the set of nonnegative measurable functions. Indeed, if f and g are such functions and $0 \leqslant f_n \uparrow f$ and $0 \leqslant g_n \uparrow g$, then for $a, b \geqslant 0$ we have $0 \leqslant af_n + bg_n \uparrow af + bg$ and therefore

$$\int_\Omega af + bg \, \mathrm{d}\mu = \lim_{n \to \infty} \int_\Omega af_n + bg_n \, \mathrm{d}\mu$$

$$= a \lim_{n \to \infty} \int_\Omega f_n \, \mathrm{d}\mu + b \lim_{n \to \infty} \int_\Omega g_n \, \mathrm{d}\mu = a \int_\Omega f \, \mathrm{d}\mu + b \int_\Omega g \, \mathrm{d}\mu.$$

If such f and g satisfy $f \leqslant g$ pointwise, then from $0 \leqslant f_n \leqslant \max\{f_n, g_n\} \uparrow g$ we see that

$$\int_\Omega f \, \mathrm{d}\mu = \lim_{n \to \infty} \int_\Omega f_n \, \mathrm{d}\mu \leqslant \lim_{n \to \infty} \int_\Omega \max\{f_n, g_n\} \, \mathrm{d}\mu \leqslant \int_\Omega g \, \mathrm{d}\mu.$$

Let us now take a closer look at the role of null sets. We begin with a simple observation.

Proposition F.3. *If f is a nonnegative measurable function, then:*

(1) *if $\int_\Omega f \, d\mu < \infty$, then $\mu\{f = \infty\} = 0$;*
(2) *if $\int_\Omega f \, d\mu = 0$, then $\mu\{f \neq 0\} = 0$.*

Proof For all $c > 0$ we have $0 \leqslant c \mathbf{1}_{\{f=\infty\}} \leqslant f$ and therefore

$$0 \leqslant c\mu\{f = \infty\} \leqslant \int_\Omega f \, d\mu.$$

The first result follows from this by letting $c \to \infty$. For the second, note that for all $n \geqslant 1$ we have $\frac{1}{n}\mathbf{1}_{\{f \geqslant \frac{1}{n}\}} \leqslant f$ and therefore

$$\frac{1}{n}\mu\Big\{f \geqslant \frac{1}{n}\Big\} \leqslant \int_\Omega f \, d\mu = 0.$$

It follows that $\mu\{f \geqslant \frac{1}{n}\} = 0$. Now note that $\{f > 0\} = \bigcup_{n \geqslant 1}\{f \geqslant \frac{1}{n}\}$. $\qquad\square$

The Monotone Convergence Theorem

The next theorem is the cornerstone of Integration Theory.

Theorem F.4 (Monotone Convergence Theorem). *Let $0 \leqslant f_1 \leqslant f_2 \leqslant \ldots$ be a sequence of nonnegative measurable functions converging pointwise to a function f. Then,*

$$\lim_{n \to \infty} \int_\Omega f_n \, d\mu = \int_\Omega f \, d\mu.$$

Proof First note that f is nonnegative and measurable. For each $n \geqslant 1$ choose a sequence of simple functions $0 \leqslant f_{nk} \uparrow_k f_n$. Set

$$g_{nk} := \max\{f_{1k}, \ldots, f_{nk}\}.$$

For $m \leqslant n$ we have $g_{mk} \leqslant g_{nk}$. Also, for $k \leqslant l$ we have $f_{mk} \leqslant f_{ml}$, $m = 1, \ldots, n$, and therefore $g_{nk} \leqslant g_{nl}$. We conclude that the functions g_{nk} are monotone in both indices.

From $f_{mk} \leqslant f_m \leqslant f_n$, $1 \leqslant m \leqslant n$, we see that $f_{nk} \leqslant g_{nk} \leqslant f_n$, and we conclude that $0 \leqslant g_{nk} \uparrow_k f_n$. From

$$f_n = \lim_{k \to \infty} g_{nk} \leqslant \lim_{k \to \infty} g_{kk} \leqslant f$$

we deduce that $0 \leqslant g_{kk} \uparrow f$. Recalling that $g_{kk} \leqslant f_k$ it follows that

$$\int_\Omega f \, d\mu = \lim_{k \to \infty} \int_\Omega g_{kk} \, d\mu \leqslant \lim_{k \to \infty} \int_\Omega f_k \, d\mu \leqslant \int_\Omega f \, d\mu.$$

$\square$

Example F.5. We have the following *substitution formula*. For any measurable f : $\Omega_1 \to \Omega_2$ and nonnegative measurable $\phi : \Omega_2 \to \mathbb{R}$,

$$\int_{\Omega_1} \phi \circ f \, \mathrm{d}\mu = \int_{\Omega_2} \phi \, \mathrm{d}f_*(\mu).$$

To prove this, note that this is trivial for simple functions $\phi = \mathbf{1}_F$ with $F \in \mathscr{F}_2$. By linearity, the identity extends to nonnegative simple functions ϕ, and by monotone convergence (using Proposition F.1) to nonnegative measurable functions ϕ.

From the monotone convergence theorem we deduce the following useful corollary.

Theorem F.6 (Fatou's Lemma)**.** *Let* $(f_n)_{n \geqslant 1}$ *be a sequence of nonnegative measurable functions on* $(\Omega, \mathscr{F}, \mu)$*. Then*

$$\int_\Omega \liminf_{n \to \infty} f_n \, \mathrm{d}\mu \leqslant \liminf_{n \to \infty} \int_\Omega f_n \, \mathrm{d}\mu.$$

Proof   From $\inf_{k \geqslant n} f_k \leqslant f_m$, $m \geqslant n$, we infer

$$\int_\Omega \inf_{k \geqslant n} f_k \, \mathrm{d}\mu \leqslant \inf_{m \geqslant n} \int_\Omega f_m \, \mathrm{d}\mu.$$

Hence, by the monotone convergence theorem,

$$\int_\Omega \liminf_{n \to \infty} f_n \, \mathrm{d}\mu = \int_\Omega \lim_{n \to \infty} \inf_{k \geqslant n} f_k \, \mathrm{d}\mu = \lim_{n \to \infty} \int_\Omega \inf_{k \geqslant n} f_k \, \mathrm{d}\mu$$

$$\leqslant \lim_{n \to \infty} \inf_{m \geqslant n} \int_\Omega f_m \, \mathrm{d}\mu = \liminf_{n \to \infty} \int_\Omega f_n \, \mathrm{d}\mu.$$

$\square$

The Dominated Convergence Theorem

A measurable function $f : \Omega \to \mathbb{K}$ is called *integrable* if

$$\int_\Omega |f| \, \mathrm{d}\mu < \infty.$$

Clearly, if f and g are measurable and $|g| \leqslant |f|$ pointwise, then g is integrable if f is integrable. In particular, if f is integrable, then the nonnegative functions f^+ and f^- are integrable, and we define

$$\int_\Omega f \, \mathrm{d}\mu := \int_\Omega f^+ \, \mathrm{d}\mu - \int_\Omega f^- \, \mathrm{d}\mu.$$

For a set $F \in \mathscr{F}$ we write

$$\int_F f \, \mathrm{d}\mu := \int_\Omega \mathbf{1}_F f \, \mathrm{d}\mu,$$

noting that $1_F f$ is integrable. The monotonicity and additivity properties of this integral carry over to this more general situation, provided we assume that the functions we integrate are integrable.

The next result is among the most useful in all of Analysis.

Theorem F.7 (Dominated Convergence Theorem). *Let $(f_n)_{n \geqslant 1}$ be a sequence of integrable functions such that $\lim_{n \to \infty} f_n = f$ pointwise. If there exists an integrable function g such that $|f_n| \leqslant |g|$ for all $n \geqslant 1$, then*

$$\lim_{n \to \infty} \int_\Omega |f_n - f| \, d\mu = 0.$$

In particular,

$$\lim_{n \to \infty} \int_\Omega f_n \, d\mu = \int_\Omega f \, d\mu.$$

Proof We make the preliminary observation that if $(h_n)_{n \geqslant 1}$ is a sequence of nonnegative measurable functions such that $\lim_{n \to \infty} h_n = 0$ pointwise and h is a nonnegative integrable function such that $h_n \leqslant h$ for all $n \geqslant 1$, then by the Fatou lemma

$$\int_\Omega h \, d\mu = \int_\Omega \liminf_{n \to \infty} (h - h_n) \, d\mu \leqslant \liminf_{n \to \infty} \int_\Omega h - h_n \, d\mu = \int_\Omega h \, d\mu - \limsup_{n \to \infty} \int_\Omega h_n \, d\mu.$$

Since $\int_\Omega h \, d\mu$ is finite, it follows that $0 \leqslant \limsup_{n \to \infty} \int_\Omega h_n \, d\mu \leqslant 0$ and therefore

$$\lim_{n \to \infty} \int_\Omega h_n \, d\mu = 0.$$

The theorem follows by applying this to $h_n = |f_n - f|$ and $h = 2|g|$. $\qquad \square$

If f is integrable and $\mu\{f \neq 0\} = 0$, then

$$\int_\Omega f \, d\mu = 0.$$

Indeed, by considering f^+ and f^- separately we may assume f is nonnegative. Choose simple functions $0 \leqslant f_n \uparrow f$. Then $\mu\{f_n > 0\} \leqslant \mu\{f > 0\} = 0$ and therefore $\int_\Omega f_n \, d\mu = 0$ for all $n \geqslant 1$. The claim follows from this. Consequently in the main results of the previous section, in particular in the monotone convergence theorem (Theorem F.4) and the dominated convergence theorem (Theorem F.7), *we may replace pointwise convergence by pointwise convergence μ-almost everywhere*, where the latter means that we allow an exceptional set of μ-measure zero in the assumptions. For instance, in the monotone convergence theorem it suffices to assume that $0 \leqslant f_n \uparrow f$ pointwise μ-almost everywhere, and similarly in the dominated convergence theorem it suffices to assume that $\lim_{n \to \infty} f_n = f$ pointwise μ-almost everywhere and $|f_n| \leqslant |g|$ pointwise μ-almost everywhere for all n.

Fubini's Theorem

Proposition F.8. *Let* $(\Omega_1, \mathscr{F}_1)$ *and* $(\Omega_2, \mathscr{F}_2)$ *be measurable spaces and let* $f : \Omega_1 \times \Omega_2 \to \mathbb{K}$ *be measurable with respect to the product σ-algebra* $\mathscr{F}_1 \times \mathscr{F}_2$. *Then:*

(1) *for all* $\omega_1 \in \Omega_1$ *the function* $\omega_2 \mapsto f(\omega_1, \omega_2)$ *is measurable;*
(2) *for all* $\omega_2 \in \Omega_2$ *the function* $\omega_1 \mapsto f(\omega_1, \omega_2)$ *is measurable.*

Proof   The collection $\mathscr{F}$ of all sets $F \in \mathscr{F}_1 \times \mathscr{F}_2$ such that (1) and (2) hold for $f = \mathbf{1}_F$ is a σ-algebra containing every set of the form $F_1 \times F_2$ with $F_1 \in \mathscr{F}_1$ and $F_2 \in \mathscr{F}_2$. Since $\mathscr{F}_1 \times \mathscr{F}_2$ is the smallest σ-algebra containing these sets, it follows that $\mathscr{F} = \mathscr{F}_1 \times \mathscr{F}_2$.
   This proves that the proposition holds for all indicator functions $\mathbf{1}_F$ with $F \in \mathscr{F}_1 \times \mathscr{F}_2$. By taking linear combinations, the result extends to simple functions. The result for arbitrary measurable functions then follows by pointwise approximation with simple functions. □

Theorem F.9 (Fubini, first version). *Let* $(\Omega_1, \mathscr{F}_1, \mu_1)$ *and* $(\Omega_2, \mathscr{F}_2, \mu_2)$ *be σ-finite measure spaces. If* $f : \Omega_1 \times \Omega_2 \to \mathbb{K}$ *is nonnegative and measurable with respect to the product σ-algebra* $\mathscr{F}_1 \times \mathscr{F}_2$, *then:*

(1) *the nonnegative function* $\omega_2 \mapsto \int_{\Omega_1} f(\omega_1, \omega_2) \, d\mu_1(\omega_1)$ *is measurable;*
(2) *the nonnegative function* $\omega_1 \mapsto \int_{\Omega_2} f(\omega_1, \omega_2) \, d\mu_2(\omega_2)$ *is measurable;*
(3) *we have*

$$\int_{\Omega_1 \times \Omega_2} f \, d(\mu_1 \times \mu_2) = \int_{\Omega_2} \int_{\Omega_1} f \, d\mu_1 \, d\mu_2 = \int_{\Omega_1} \int_{\Omega_2} f \, d\mu_2 \, d\mu_1.$$

Proof   First suppose that $\mu_1(\Omega_1) = \mu_2(\Omega_2) = 1$ and let $\mathscr{F}$ be the collection of all sets $F \in \mathscr{F}_1 \times \mathscr{F}_2$ such that (1)–(3) hold for $f = \mathbf{1}_F$. We claim that $\mathscr{F}$ is a σ-algebra. Indeed, (1)–(3) are trivial for $f = \mathbf{1}_\varnothing = 0$. If (1)–(3) hold for a set $F \in \mathscr{F}_1 \times \mathscr{F}_2$, then $\mathbf{1}_{\complement F}(\omega_1, \omega_2) = 1 - \mathbf{1}_F(\omega_1, \omega_2)$ implies that (1) and (2) hold for $\complement F$, and furthermore

$$\int_{\Omega_1 \times \Omega_2} \mathbf{1}_{\complement F} \, d(\mu_1 \times \mu_2) = \int_{\Omega_1 \times \Omega_2} 1 - \mathbf{1}_F \, d(\mu_1 \times \mu_2)$$

$$= 1 - \int_{\Omega_1 \times \Omega_2} \mathbf{1}_F \, d(\mu_1 \times \mu_2) = 1 - \int_{\Omega_2} \int_{\Omega_1} \mathbf{1}_F \, d\mu_1 \, d\mu_2$$

$$= \int_{\Omega_2} \int_{\Omega_1} 1 - \mathbf{1}_F \, d\mu_1 \, d\mu_2 = \int_{\Omega_2} \int_{\Omega_1} \mathbf{1}_{\complement F} \, d\mu_1 \, d\mu_2$$

and similarly for the other repeated integral, so (3) holds for F. If (1)–(3) hold for disjoint sets $\mathbf{1}_{F_1}, \mathbf{1}_{F_2}, \cdots \in \mathscr{F}_1 \times \mathscr{F}_2$, the monotone convergence theorem implies that (1)–(3) hold for $\mathbf{1}_F$ with $F = \bigcup_{n \geqslant 1} F_n$. This proves the claim. It is clear that (1)–(3) hold for all rectangles $F_1 \times F_2$ with $F_1 \in \mathscr{F}_1$ and $F_2 \in \mathscr{F}_2$. Since $\mathscr{F}_1 \times \mathscr{F}_2$ is the smallest σ-algebra containing these rectangles, it follows that $\mathscr{F} = \mathscr{F}_1 \times \mathscr{F}_2$.
   If μ_1 and μ_2 are finite, we apply the preceding step to the normalised measures

$\mu_1/\mu_1(\Omega_1)$ and $\mu_2/\mu_2(\Omega_2)$ and again find that (1)–(3) hold for all $F \in \mathscr{F}_1 \times \mathscr{F}_2$. This extends to the σ-finite case by approximation and monotone convergence.

By taking linear combinations, the result extends to nonnegative simple functions. The result for arbitrary nonnegative measurable functions then follows by another application of monotone convergence. □

A variation of the Fubini theorem holds if we replace 'nonnegative measurable' by 'integrable':

Theorem F.10 (Fubini, second version). *Let* $(\Omega_1, \mathscr{F}_1, \mu_1)$ *and* $(\Omega_2, \mathscr{F}_2, \mu_2)$ *be* σ-*finite measure spaces. If* $f : \Omega_1 \times \Omega_2 \to \mathbb{K}$ *is integrable with respect to the product measure* $\mu_1 \times \mu_2$, *then:*

(1) *the function* $\omega_2 \mapsto \int_{\Omega_1} f(\omega_1, \omega_2)\, d\mu_1(\omega_1)$ *is integrable with respect to* μ_2;
(2) *the function* $\omega_1 \mapsto \int_{\Omega_2} f(\omega_1, \omega_2)\, d\mu_2(\omega_2)$ *is integrable with respect to* μ_1;
(3) *we have*

$$\int_{\Omega_1 \times \Omega_2} f\, d(\mu_1 \times \mu_2) = \int_{\Omega_2} \int_{\Omega_1} f\, d\mu_1\, d\mu_2 = \int_{\Omega_1} \int_{\Omega_2} f\, d\mu_2\, d\mu_1.$$

Proof By splitting into real and imaginary parts and then into positive and negative parts, we may assume that f is nonnegative. Hence (3) holds by Theorem F.9, with a finite left-hand side. It follows that the two repeated integrals are finite. Since an integral with respect to a measure μ of a nonnegative function is finite only if the integrand is finite μ-almost everywhere, assertions (1) and (2) follow as well. □

Appendix G

Notes

Historical perspectives on Functional Analysis are presented in Dieudonné (1981); Monna (1973); Pietsch (2007). Among the many excellent textbooks on Functional Analysis, our favourites include Bressan (2013); Brezis (2011); Conway (1990); Dunford and Schwartz (1988a); Einsiedler and Ward (2017); Lax (2002); Rudin (1991); Schechter (2002); Werner (2000); Yosida (1980).

Chapter 1

Exhaustive treatments of the theory of Banach spaces and the Bochner integral are given in Albiac and Kalton (2006); Diestel (1984); Dunford and Schwartz (1988a); Li and Queffélec (2004), and in Diestel and Uhl (1977); Dunford and Schwartz (1988a); Hytönen et al. (2016), respectively.

Chapter 2

The proofs of Propositions 2.19 and 2.23 are taken from Haase (2007). Our presentation of Proposition 2.34 and Section 2.3.d follows Hytönen et al. (2016), where more detailed information on this topic can be found. The classical reference is Stein (1970). The Fréchet–Kolmogorov compactness theorem is usually stated for bounded subsets of $L^p(\mathbb{R}^d)$. That boundedness follows from the assumptions (i) and (ii) was observed later by Sudakov; the simple proof presented here is from Hanche-Olsen et al. (2019). The presentation of Section 2.3.d follows Hytönen et al. (2016).

Our treatment of Theorems 2.45 and 2.46 follows Bogachev (2007b).

Comprehensive treatments of vector lattices, Banach lattices, and positive operators are given in Aliprantis and Burkinshaw (1985); Luxemburg and Zaanen (1971); Meyer-Nieberg (1991); Schaefer (1974); Zaanen (1997).

The problem of proving the boundedness of $T \otimes I_X$ on $L^p(\Omega; X)$ for bounded operators T on $L^p(\Omega)$ discussed in Problem 2.28 is highly nontrivial. An interesting complement to the results mentioned in the problem is the following result of Paley and Marcinkiewicz–Zygmund: If T is bounded on $L^p(\Omega)$ with $1 \leqslant p < \infty$ and X is a Hilbert space, then $T \otimes I_X$ admits a unique extension to a bounded operator on $L^p(\Omega; X)$ and its norm equals $\|T\|$. The proof uses properties of Gaussian random variables. It was shown by Kwapień that the Fourier–Plancherel transform (see Section 5.5) extends to a bounded operator on $L^2(\mathbb{R}^d; X)$ if and only if X is isomorphic to a Hilbert space (and as such is unitary if X is a Hilbert space); by results of Bourgain and Burkholder, the Hilbert transform (see Section 5.6) extends to a bounded operator on $L^p(\mathbb{R}; X)$ for some $1 < p < \infty$ if and only if it extends to a bounded operator on $L^p(\mathbb{R}; X)$ for all $1 < p < \infty$ if and only if X has the so-called *UMD property*; this abbreviation stands for "unconditionality of martingale differences". Proofs of these results and their ramifications can be found in Hytönen et al. (2016); Pisier (2016). The UMD property also characterises the boundedness of the vector-valued extension of the Itô stochastic integral of Problem 3.25; see van Neerven et al. (2015) and the references given therein.

Chapter 3

Theorem 3.13 characterises Hilbert spaces up to isomorphism. More precisely, the following deep theorem has been proved in Lindenstrauss and Tzafriri (1971): A Banach space X is isomorphic to a Hilbert space if and only if every closed subspace of X is the range of a bounded projection in X.

The proof of the Radon–Nikodým theorem outlined in Problem 3.22 is due to von Neumann and follows Rudin (1987). The construction, in Problem 3.24, of a linear operator on ℓ^2 which fails to be bounded depends on the existence of an algebraic basis in ℓ^2 (see Problem 3.23). This, in turn, is deduced with the help of Zorn's lemma. The latter being equivalent to the Axiom of Choice, this raises the question whether a constructive example of an unbounded operator can be given. Within Zermelo–Fraenkel Set Theory (ZF) the answer is negative: it is consistent with ZF that every linear operator on a Banach space is bounded. In fact, it is a theorem in ZF extended with the so-called Axiom of Determinacy and the Countable Axiom of Choice that every linear operator on a Banach space is bounded (Fremlin, 2015, Theorem 567H (c)).

It can be quite hard to decide whether a given subspace is dense in a given Hilbert space. The following example may illustrate this. Let H denote the Hilbert space of all scalar sequences $c = (c_n)_{n \geqslant 1}$ for which the norm

$$\|c\|_H^2 := \sum_{n \geqslant 1} \frac{1}{n^2} |c_n|^2$$

is finite. For $x \in \mathbb{R}$ let $\{x\}$ denote its fractional part, that is, the unique real number in $[0, 1)$ such that $x = k + \{x\}$ for some integer $k \in \mathbb{Z}$. For $m = 1, 2, 3, \ldots$ let $c^{(m)} := (\{\frac{n}{m}\})_{n \geqslant 1}$ and note that these sequences belong to H. It is a theorem of Nyman and Baez–Duarte that the linear span of the sequence $(c^{(m)})_{m \geqslant 1}$ is dense in H if and only if the Riemann hypothesis holds. This result, as well as several related ones, is surveyed in Baghi (2006). The Riemann hypothesis is considered by many mathematicians as one of the most important open problems in all of Mathematics.

Chapter 4

Our proof of Theorem 4.2 combines ideas of Folland (1999) and Ruzhansky and Turunen (2010). One should be aware that different authors use slightly different definitions of Radon measures.

The real version of the Hahn–Banach theorem is due to Banach; the extension to complex scalars was added a decade later by Hahn. Banach also proved the sequential version of the Banach–Alaoglu theorem; the general version is due to Alaoglu. A detailed survey of the Hahn–Banach theorem is given in Buskes (1993).

The weak and weak* topologies are special cases of so-called *locally convex* topologies. For systematic introductions to this subject we recommend Aliprantis and Burkinshaw (1985); Conway (1990); Rudin (1991). Theorem 4.34 is a special case of the so-called *principle of local reflexivity*. Its full formulation can be found, for example, in Albiac and Kalton (2006).

Our proof of Theorem 4.63 is closely related to that presented in Bogachev (2007a), where more refined versions of the theorem can be found.

The result of Problem 4.10 is discussed in Phelps (1960). The converse also holds: If every functional on a closed subspace on X has a unique Hahn–Banach extension of the same norm, then X^* is strictly convex; see Foguel (1958); Taylor (1939).

The result of Problem 4.32 is due to Sobczyk (1941).

Chapter 5

General references on the theory of bounded operators include Beauzamy (1988); Gohberg et al. (2003, 2013); Nikolski (2002).

The proof of the uniform boundedness theorem sketched in Problem 5.3 is taken from Pietsch (2007), where it is credited to Lebesgue.

Most treatments of the Fourier–Plancherel transform use the Schwartz space $\mathscr{S}(\mathbb{R}^d)$ of rapidly decreasing smooth functions instead of our $\mathscr{F}^2(\mathbb{R}^d)$. Our treatment of the

Fourier transform and the Hilbert transform follows that of Hytönen et al. (2016). The L^p-boundedness of the Hilbert transform is classical; we follow Grafakos (2008).

The theory of Fourier multiplier operators can be meaningfully extended to the L^p-setting, where it becomes a powerful tool in the Calderón–Zygmund theory of singular integral operators. The prime example of such an operator is the Hilbert transform. Detailed treatments of the Hilbert transform and singular integral operators in the L^p-setting are given in Grafakos (2008); Stein (1970, 1993). The theory of the Hilbert transform extends to higher dimensions where analogous statements hold for the *Riesz transforms*, defined as the Fourier multipliers operators associated with the functions $m_j \in L^\infty(\mathbb{R}^d)$ defined by

$$m_j(\xi) := \frac{\xi_j}{|\xi|}, \quad j = 1,\ldots,d.$$

An exhaustive treatment of these matters belongs to the realm of Harmonic Analysis; see Grafakos (2008); Stein (1970) and Chapter 5 of Hytönen et al. (2016).

The proof of the Riesz–Thorin theorem 5.38 presented here is taken from Hytönen et al. (2016), where also the argument proving $\|T_{\mathbb{C}}\| = \|T\|$ can be found. In this reference, the proof of the Clarkson inequalities sketched in Problem 5.26 is attributed to Jürgen Voigt. It is a famous result of Beckner (1975) that the constant 1 in the Hausdorff–Young inequality

$$\|\mathscr{F}f\|_{L^q(\mathbb{R}^d,m)} \leqslant \|f\|_{L^p(\mathbb{R}^d,m)}$$

for the Fourier transform with respect to the normalised Lebesgue measure m, where $1 \leqslant p \leqslant 2$ and $\frac{1}{p} + \frac{1}{q} = 1$, can be improved to

$$\|\mathscr{F}f\|_{L^q(\mathbb{R}^d,m)} \leqslant C_p^d \|f\|_{L^p(\mathbb{R}^d,m)}$$

with $C_p = (p^{1/p}/q^{1/q})^{1/2}$. In the same paper, Beckner proved the improvement to the Young inequality mentioned in the main text and showed that both results are sharp. The proofs rely on (but go beyond) the techniques developed in Section 15.6. Counterexamples to (5.7) in the range $q < p$ can be found in Riesz (1926) and Maligranda (1997).

The Marcinkiewicz interpolation theorem is of fundamental importance in the theory of singular integrals; see Grafakos (2008); Stein (1970) and the forthcoming work Hytönen et al. (2022+). Our treatment follows Hytönen et al. (2016). The L^p-boundedness of the Hilbert transform, here derived as a consequence of the Riesz–Thorin theorem, can also be derived from the Marcinkiewicz interpolation theorem; the required weak L^1-bound is due to Kolmogorov; see Duoandikoetxea (2001).

The result of Problem 5.16 is due to Pettis (1938). It is no coincidence that the counterexample for $p = 1$ in part (b) lives in the space c_0: part (a) extends to $p = 1$ for all Banach spaces not containing a closed subspace isomorphic to c_0. A proof of this fact is given in Diestel and Uhl (1977); further results on the Pettis integral can be found in van Dulst (1989); Musiał (2002); Talagrand (1984).

Chapter 6

There are many excellent treatments of spectral theory, such as the monumental classic Dunford and Schwartz (1988b) and the monographs Arveson (2002); Aupetit (1991); Müller (2007). A discussion of the conditions (i) and (ii) for contours in Cauchy's theorem can be found in Rudin (1987). The result of Problem 6.15 is due to Gelfand (1941); the proof outlined here is due to Allan and Ransford (1989).

Chapter 7

Our proofs of Proposition 7.27 and Theorems 7.29 and 7.31 follow Schechter (2002), Bleecker and Booß Bavnbek (2013), and Böttcher and Silbermann (2006), respectively. The proof of Theorem 7.33 follows Coburn (1966, 1967); see also Arveson (2002). Another proof is given in Douglas (1998), where also the results from the theory of Hardy spaces alluded to in the proof can be found. Toeplitz operators are covered in more depth in Arveson (2002); Böttcher and Silbermann (2006); Douglas (1998); Nikolski (2002).

The winding number of a continuous closed curve in $\mathbb{C} \setminus \{0\}$ parametrised by the function $\phi : [0, 1] \to C \setminus \{0\}$, $t \mapsto \phi(e^{2\pi i t})$ can be computed as follows. One shows that there exists a continuous function $g : [0, 1] \to \mathbb{C}$ such that

$$\phi(e^{2\pi i t}) = e^{2\pi i g(t)}, \quad t \in [0, 1].$$

The identity $e^{2\pi i g(1)} = \phi(1) = e^{2\pi i g(0)}$ implies that $g(1) - g(0) \in \mathbb{Z}$. This integer equals the winding number of ϕ. Proofs and some easy consequences can be found in Arveson (2002).

The result of Problem 7.4 has the following interesting complement, due to Pitt: For all $1 \leqslant p < q < \infty$, every bounded operator from ℓ^q to ℓ^p is compact. An immediate consequence is that the spaces ℓ^p and ℓ^q are not isomorphic. The proof of Pitt's theorem requires some effort; see for instance Albiac and Kalton (2006); Ryan (2002). The result of Problem 7.8 is due to Terzioğlu (1971).

By using some elementary C^*-algebra techniques it is possible to derive Theorem 7.31 as a simple corollary to Theorem 7.33. We begin by introducing some terminology. A *Banach algebra* is a Banach space $\mathscr{A}$ endowed with a composition mapping $(x, y) \mapsto xy$ such that

$$\|xy\| \leqslant \|x\| \|y\|$$

holds for all $x, y \in \mathscr{A}$. A *unital Banach algebra* is a Banach algebra $\mathscr{A}$ with a unit element, that is, an element $e \in \mathscr{A}$ such that

$$ex = xe = x, \quad x \in \mathscr{A}.$$

The *spectrum* $\sigma_{\mathscr{A}}(x)$ of an element x of a unital Banach algebra $\mathscr{A}$ is the set of all $\lambda \in \mathbb{C}$ for which no $y \in \mathscr{A}$ can be found such that $(\lambda - x)y = y(\lambda - x) = e$, and most of the spectral theory contained in Chapter 6 can be routinely extended to this situation.

A $C^\star$-*algebra* is a Banach algebra $\mathscr{A}$ with an *involution*, that is, a mapping $x \mapsto x^\star$ on $\mathscr{A}$ satisfying

$$(x + y)^\star = x^\star + y^\star, \quad (cx)^\star = \overline{c}x^\star, \quad (xy)^\star = y^\star x^\star,$$

as well as

$$\|x\| = \|x^\star\|, \quad \|x^\star x\| = \|x\| \|x^\star\|$$

for all $x, y \in \mathscr{A}$ and $c \in \mathbb{C}$. According to the *Gelfand–Naimark theorem*, every $C^\star$-algebra $\mathscr{A}$ is $\star$-isometric to a closed $\star$-subalgebra of $\mathscr{L}(H)$ for a suitably chosen Hilbert space H (a $\star$-*subalgebra* being a subalgebra closed under taking involutions and a $\star$-*isometric isomorphism* being an isometric isomorphism U with the additional properties that $U(xy) = UxUy$ and $U(x^\star) = (Ux)^\star$ for all $x, y \in \mathscr{A}$). This theorem connects the abstract definition of $C^\star$-algebras given here with the concrete approach taken in Section 9.5.

The following generalisation of Proposition 8.19 holds (see (Folland, 2016, Proposition 1.23) or (Rudin, 1991, Theorem 11.29)): If $\mathscr{A}$ is a closed unital $\star$-subalgebra of a unital $C^\star$-algebra $\mathscr{B}$, then

$$\sigma_{\mathscr{A}}(x) = \sigma_{\mathscr{B}}(x), \quad x \in \mathscr{A}. \tag{G.1}$$

The proof follows Proposition 8.19, except for the fact that $x = x^\star$ implies $\sigma_{\mathscr{B}}(x^\star x) \subseteq \mathbb{R}$; this fact can be proved by combining the Gelfand–Naimark theorem and Proposition 8.19. An elementary proof can be given as follows. If $u \in \mathscr{B}$ is unitary, that is, $uu^\star = u^\star u = e$, the argument suggested in Problem 8.2 proves that $\sigma_{\mathscr{B}}(u) \in \mathbb{T}$. Then, the argument suggested in Problem 8.3 proves that if $x = x^\star$, then $\sigma_{\mathscr{B}}(x) \in \mathbb{R}$.

Using (G.1), let us now give a simple alternative proof of Theorem 7.31 based on Theorem 7.33 (cf. Corollary 1 in (Arveson, 2002, Section 4.3)). The results of Problem 7.12 prove that for any Banach space X the Calkin $\mathscr{L}(X)/\mathscr{K}(X)$ is a unital Banach algebra. If H is a Hilbert space, then $\mathscr{L}(H)/\mathscr{K}(H)$ is a unital $C^\star$-algebra.

In what follows we take $H := H^2(\mathbb{D})$. Since $\mathscr{K}(H)$ is contained in the Toeplitz algebra $\mathscr{T}$ it is meaningful to consider the quotient space $\mathscr{T}/\mathscr{K}(H)$. This space is a unital $\star$-subalgebra of $\mathscr{L}(H)/\mathscr{K}(H)$. By Coburn's theorem the mapping $T_\phi + K \mapsto \phi$ sets up a $\star$-isometry from $\mathscr{T}$ onto $C(\mathbb{T})$ and we have the commuting diagram where π are the quotient mappings and j is the composition of the $\star$-isometry from

$$0 \longrightarrow \mathscr{K}(H)) \longrightarrow \mathscr{T} \xrightarrow{\ \pi\ } C(\mathbb{T}) \longrightarrow 0$$
$$\downarrow{=} \qquad \downarrow{\subseteq} \qquad \downarrow{j}$$
$$0 \longrightarrow \mathscr{K}(H) \longrightarrow \mathscr{L}(H) \xrightarrow{\ \pi\ } \mathscr{L}(H)/\mathscr{K}(H) \longrightarrow 0$$

$C(\mathbb{T})$ onto $\mathscr{T}/\mathscr{K}(H)$ provided by Theorem 7.33 and the natural inclusion mapping from $\mathscr{T}/\mathscr{K}(H)$ into $\mathscr{L}(H)/\mathscr{K}(H)$. As such, j is injective.

Now suppose that $\phi \in C(\mathbb{T})$ is such that $T_\phi \in \mathscr{T}$ is Fredholm. By Atkinson's theorem there exists an $S \in \mathscr{L}(H^2(\mathbb{D}))$ such that $I - T_\phi S$ and $I - S T_\phi$ are compact. This means that T_ϕ defines an invertible element in $\mathscr{L}(H)/\mathscr{K}(H)$. As an application of (G.1), T_ϕ defines an invertible element in $\mathscr{T}/\mathscr{K}(H)$. A moment's reflection reveals that this implies that $S \in \mathscr{T}$, say $S = T_\psi + K$ with $\psi \in C(\mathbb{T})$ and $K \in \mathscr{K}(H)$. It then follows that

$$j\phi\psi = \pi(T_\psi T_\phi) = \pi(S T_\phi) = \pi(I) = j\mathbf{1},$$

and the injectivity of j implies $\phi\psi = \mathbf{1}$. This is only possible if ϕ is zero-free.

Chapter 8

Our proof of Theorem 8.18 is taken from Rudin (1991). A proof of Runge's theorem, which was used in the proof of part (i) of Theorem 8.20, may be found in Rudin (1987). The clever proof of Proposition 8.21 is taken from Whitley (1968). The proof of Theorem 8.36 is taken from Davies (1980).

The proof of the Toeplitz–Hausdorff theorem proposed in Problem 8.9 is due to Li (1994). More about numerical ranges can be found in Gustafson and Rao (1997).

Chapter 9

Most treatments of the spectral theorem for normal operators proceed via the theory of $C^\star$-algebras; see, for example, Arveson (2002); Rudin (1991). This permits a concise abstract proof, but has the drawback that this theory depends on the existence of maximal ideals, a well-known consequence of Zorn's lemma. Our approach avoids the use of Zorn's lemma. The idea to use Proposition 9.12 to prove that the projection-valued measure is concentrated on the spectrum is from Haase (2018). Our treatment of the von Neumann bicommutant theorem and the result stated in Problem 15.11 are taken from Pedersen (2018). The presentation of Section 9.6 follows Koelink (1996).

In Heuser (2006) a direct, albeit tricky, proof is given of the result of Problem 9.13 which relies solely on the continuous functional calculus for selfadjoint operators.

Chapter 10

References for this chapter include Akhiezer and Glazman (1981a); Birman and Solomjak (1987); Dunford and Schwartz (1988b); Edmunds and Evans (2018b); Kato (1995); Reed and Simon (1975); Schmüdgen (2012). Our proof of the spectral theorem combines elements of Rudin (1991) and Schmüdgen (2012), and is elementary in that it avoids the use of C^*-algebra techniques. For selfadjoint operators a more direct construction of the measurable calculus can be given; see, for example, Rudin (1991), where it is used to give a simpler proof of the existence and uniqueness of square roots for positive selfadjoint operators.

The proof of the inclusion '$\subseteq$' of Theorem 9.28 presented in Section 10.3.b is taken from Sz.-Nagy (1967).

Chapter 11

The connections between Functional Analysis and the theory of partial differential equations are emphasised in Bressan (2013); Brezis (2011); Jost (2013). The results of this chapter barely scratch the surface of what can be said in this context.

Sobolev spaces are treated in detail in Adams and Fournier (2003); Evans (2010). Some of our proofs are modelled after those presented in these references. Our presentation of Propositions 11.5 and 11.16 follows Hytönen et al. (2016). The proof of Theorem 11.12 follows an idea of Krylov (2008).

Extension operators are treated in Adams and Fournier (2003); Evans (2010). The proof of Step 1 of Theorem 11.28 is from Adams and Fournier (2003). Our proof of Theorem 11.27 is based on unpublished lecture notes by Mark Veraar. The theorem, which asserts the density of $C^\infty(\overline{D})$ in $W^{k,p}(D)$ for bounded C^k-domains D, actually gives the stronger result that for any $f \in W^{k,p}(D)$ there exists a sequence of functions $f_n \in C^\infty(\mathbb{R}^d)$ whose restrictions to D satisfy $\lim_{n\to\infty} \|f_n - f\|_{W^{k,p}(D)} = 0$. In this connection it is worth mentioning that if D is a bounded C^k-domain, then every function $f \in C^k(\overline{D})$ is the restriction of a function in $C^k(\mathbb{R}^d)$; the analogous result holds for functions in $C^\infty(\overline{D})$ when D is a bounded C^∞-domain. In both cases, the extensions can be realised through a linear mapping. This result is due to Seeley (1964).

The proof of Theorem 11.24 follows Arendt and Urban (2010) and Brezis (2011). The C^1-conditions of the second part of the theorem can be relaxed; see Biegert and Warma (2006). If D is bounded and has C^1-boundary ∂D, then for $1 \leqslant p < \infty$ the mapping $f \mapsto f|_{\partial D}$ for $f \in C^\infty(\overline{D})$ admits a unique extension to a bounded operator T, the *trace operator*, from $W^{1,p}(D)$ to $L^p(\partial D)$. Here, we think of ∂D as being equipped with its surface measure. Moreover, for a function $f \in W^{1,p}(D)$ one has $f \in W_0^{1,p}(D)$ if and only if $Tf = 0$. The details can be found in Adams and Fournier (2003); Brezis (2011); Evans (2010).

It is not true in general that the weak solution of the Poisson problem $-\Delta u = f$ with $f \in L^2(D)$ subject to Dirichlet boundary conditions belongs to $H^2(D)$; a counterexample can be found in Satz 6.89 of Arendt and Urban (2010). Proofs of the H^2-regularity result mentioned in Remark 11.38 and its analogue for Neumann boundary conditions can be found in Chapter 6 of Evans (2010).

Systematic treatments of the finite element method are presented in the monographs Atkinson and Han (2009); Brenner and Scott (2008).

Problems 11.23 and 11.24 are taken from Krylov (2008) and reproduce Sobolev's original proof of the inequality named after him. The outline of the proof, in Problem 11.26, that for $f \in C_c(D)$ no solution in $C^2(D) \cap C(\overline{D})$ to the Poisson problem may exist, is taken from Arendt and Urban (2010). Problems 11.28 and 11.29 are modelled after the same reference.

Chapter 12

Excellent references for the theory of forms are Kato (1995); Ouhabaz (2005). Some of our proofs follow the latter reference. For the spectral theory of differential operators the reader is referred to Edmunds and Evans (2018a,b) and the references given therein, and, for variational methods, Henrot (2006). More complete treatments of Dirichlet and Neumann Laplacians can be found in the lecture notes Arendt (2006) and the survey papers Arendt (2004); Grebenkov and Nguyen (2013), where further references to the literature are given. A standard reference for the theory of elliptic second-order differential operators is Gilbarg and Trudinger (2001).

Our proof of Theorem 12.12 follows Arendt (2006). Further results along the lines of this theorem and its corollary can be found there and in Ouhabaz (2005). Among other things, under the assumptions of the corollary, $D(A)$ is dense in $D(\mathfrak{a})$.

In some of the results about the Neumann Laplacian, the C^1 assumption on the boundary can be relaxed. For example, the Neumann Laplacian has point spectrum if D has continuous boundary and D lies 'on one side' of it. The steps are as follows: If D a has continuous boundary, then $W^{1,2}(D)$ is compactly embedded in $L^2(D)$ according to Theorem V.4.17 of Edmunds and Evans (2018b) and the Neumann Laplacian has a compact resolvent. The assumption on the boundary in Theorem 12.26(2) and Theorem 12.27 may be weakened accordingly.

Kac's question "Can one hear the shape of a drum?" was asked in Kac (1966) and answered to the negative in Gordon et al. (1992). Our presentation of Weyl's theorem follows Higson (2004). An example of a Jordan curve of positive area is given in Osgood (1903).

The inequality $\mu_n \leqslant \lambda_n$ of Corollary 12.28 comparing the Dirichlet eigenvalues λ_n and the Neumann eigenvalues μ_n admits a significant improvement, due to Friedlander (1991) who proved that for all $n \geqslant 1$ we have

$$\mu_{n+1} \leqslant \lambda_n.$$

After reducing to smooth domains, an important step in the proof is the *spectral flow inequality*

$$N_{\text{Neum}}(\lambda) - N_{\text{Dir}}(\lambda) = n(\lambda),$$

for $\lambda > 0$ satisfying $\lambda \notin \sigma(-\Delta_{\text{Dir}}) \cup \sigma(-\Delta_{\text{Neum}})$, where

$$N_{\text{Dir}}(\lambda) = \#\{\lambda_n \in \sigma(-\Delta_{\text{Dir}}) : \lambda_n < \lambda\},$$
$$N_{\text{Neum}}(\lambda) = \#\{\lambda_n \in \sigma(-\Delta_{\text{Neum}}) : \lambda_n < \lambda\},$$

and $n(\lambda)$ is the number of negative eigenvalues of the *Dirichlet-to-Neumann operator* R_λ, counting multiplicities throughout. This is the operator on $L^2(D)$ which maps a function $f \in L^2(\partial D)$ to $\frac{\partial u}{\partial \nu}|_{\partial D} \in L^2(D)$, where $u \in H^1(D)$ is the unique solution of the problem

$$\begin{cases} -\Delta u = \lambda u & \text{on } D, \\ \quad u = f & \text{on } \partial D. \end{cases}$$

It is selfadjoint, bounded below, and has compact resolvent.

A simpler proof of Friedlander's theorem, based on a variant of the Courant–Fischer theorem, was obtained by Filonov (2005). Levine and Weinberger (1986) obtained the inequality

$$\mu_{n+d} \leqslant \lambda_n$$

for bounded convex domains D in $\mathbb{R}^d$, with strict inequality when ∂D is smooth.

Weyl's theorem has been extended to other types of boundary conditions, including Neumann boundary conditions, and positive selfadjoint elliptic operators. For more details the reader is referred to Safarov and Vassiliev (1997). Such extensions are nontrivial even for the Laplace operator because the domain monotonicity for Dirichlet eigenvalues of Lemma 12.30 generally fails for boundary conditions other than Dirichlet. This is demonstrated by the following example, taken from Funano (2022). We use the notation $a \lesssim b$ to express that $a \leqslant Cb$ for a universal constant C.

For $1 \leqslant p \leqslant 2$ let $B_{\ell_d^p}$ denote the open unit ball of ℓ_d^p, the space $\mathbb{K}^d$ endowed with the norm given by $\|x\|_p^p = \sum_{j=1}^d |x_j|^p$. If the positive real number $r_{d,p}$ is defined by the condition $\text{vol}(r_{d,p} B_{\ell_d^p}) = 1$, then $r_{d,p} \asymp d^{1/p}$. The smallest Neumann eigenvalue for the Laplace operator on $D' := r_{d,p} B_{\ell_d^p}$ can be shown to satisfy

$$\mu_{2,D'} \gtrsim 1$$

(keep in mind our convention that $\mu_{1,D'} = 0$). Approximating the segment in D' connecting the origin and the point $(r_{d,p}, 0, 0, \ldots, 0)$ by a convex C^1-domain $D \subseteq D'$, it can be shown that

$$\mu_{2,D} \approx \frac{1}{r_{d,p}^2} \approx \frac{1}{d^{2/p}}.$$

In the positive direction, in the same reference the following monotonicity result is proved: If $D, D' \subseteq \mathbb{R}^d$ bounded convex sets with C^1 boundaries and if $D \subseteq D'$, then for all $n \geqslant 1$ we have

$$\mu_{n,D'} \lesssim d^2 \mu_{n,D}, \quad n \geqslant 1.$$

The counterexample (upon letting $p \downarrow 1$) shows that the factor d^2 is essentially optimal. Problems 12.2–12.4 are taken from Arendt (2006).

Chapter 13

Excellent introductions to the theory of C_0-semigroups include the monographs Applebaum (2019); Davies (1980); Engel and Nagel (2000); Pazy (1983). For a discussion of the examples in Section 13.6 we refer to these sources. The monumental 1957 treatise Hille and Phillips (1957) is freely available online.

Parts of Sections 13.1–13.4 and Figure 13.1, as well as Figure 10.1 in Chapter 10, are taken from Appendix G of Hytönen et al. (2017), which in turn is based on the corresponding material in the author's lecture notes for the 2006/07 Internet Seminar "Stochastic Evolution Equations", available on the author's webpage.

Theorem 13.11 is due to Phillips (1955). Theorem 13.16 is a special case of a result of Jorgensen (1982). The idea to use this result to prove Wiener's tauberian theorem is from van Neerven (1997). Theorem 13.17 was obtained independently by Hille and Yosida near the end of the 1940s. An extension to arbitrary C_0-semigroups, which is somewhat more technical to state, was found soon afterwards. A detailed account of Theorem 13.17 and its history is given in Engel and Nagel (2000). The intimate connections between semigroups and the theory of Laplace transforms are emphasised in Arendt et al. (2011).

Fuller treatments of the abstract Cauchy problem are given in Amann (1995); Arendt (2004); Tanabe (1979).

Analytic semigroups are treated in detail in Lunardi (1995). Maximal regularity for bounded analytic C_0-semigroups on Hilbert spaces was first proved in De Simon (1964). The result remains valid if L^2 is replaced by L^p with $1 < p < \infty$ throughout; this follows from rather deep extrapolation arguments for singular integral operators and falls outside our scope. For a full treatment as well as references to the extensive literature on the subject the reader is referred to Hytönen et al. (2022+) whose treatment we follow. The method of applying maximal regularity to solving time-dependent problems of Section 13.4.d goes back to Clément and Li (1993/94) and has been extended to cover a wealth of other nonlinear problems.

In the light of Example 13.39 (which is revisited at the end of this section) it is of some interest to mention that, in the converse direction, every generator $-A$ of an analytic C_0-semigroup of contractions on a complex Hilbert space H can be represented in divergence form, in the following precise sense: There exists a Hilbert space $\mathscr{H}$, a closed operator $V : \mathsf{D}(V) \subseteq H \to \mathscr{H}$ with dense domain and dense range, and a bounded coercive operator $B \in \mathscr{L}(\mathscr{H})$, that is, we have $(Bx|x)_{\mathscr{H}} \geqslant \beta \|x\|_{\mathscr{H}}$ for some $\beta > 0$ and all $x \in \mathscr{H}$, such that

$$A = V^* B V.$$

More precisely, there exists a densely defined, closed, sectorial form $\mathfrak{a}$ in H with domain $\mathsf{D}(\mathfrak{a}) = \mathsf{D}(V)$ such that A is the operator associated with $\mathfrak{a}$ and

$$a(g,h) = (BVg|Vh), \quad g,h \in \mathsf{D}(V).$$

A proof of this result can be found in Maas and van Neerven (2009), where it is also shown that this representation is essentially unique.

Our proofs of Theorem 13.48 and Lemma 13.52 are taken form Arendt (2006).

The Ornstein–Uhlenbeck semigroup has many interesting properties, for which the reader is referred to Hytönen et al. (2023+); Janson (1997); Nualart (2006). Probabilistically, up to a time scaling it arises as the transition semigroup associated with the solution $(u_x(t))_{t \geqslant 0}$ of the stochastic differential equation

$$\mathrm{d}u(t) = -\frac{1}{2}u(t)\,\mathrm{d}t + \mathrm{d}B_t, \quad t \geqslant 0,$$

with initial condition $u(0) = x$; the driving process $(B_t)_{t \geqslant 0}$ is a standard Brownian motion in $\mathbb{R}^d$. More precisely, for all $t \geqslant 0$ and $f \in L^2(\mathbb{R}^d, \gamma)$ one has

$$OU(t/2)f(x) = \mathbb{E}(f(u_x(t)))$$

for almost all $x \in \mathbb{R}^d$.

The domain identification $\mathsf{D}(L) = W^{2,p}(\mathbb{R}^d, \gamma)$ for the Ornstein–Uhlenbeck operator L in $L^p(\mathbb{R}^d, \gamma)$ with $1 < p < \infty$ is due, in a more general formulation, to Metafune et al. (2002). This paper also contains references to earlier papers on this subject, in particular regarding the special case $p = 2$. The Ornstein–Uhlenbeck semigroup extends to an analytic C_0-contraction semigroup in $L^p(\mathbb{R}^d, \gamma)$ for $1 < p < \infty$, with optimal angle $\theta - p$ given by

$$\cos \theta_p = \left| \frac{2}{p} - 1 \right|.$$

This result is due to Epperson (1989), who also showed that the exact domain of holomorphy is the set

$$E_p := \{z = x + iy \in \mathbb{C} : |\sin(y)| < \tan(\theta_p)\sinh(x)\}.$$

A simpler proof of the latter was given in van Neerven and Portal (2018).

The L^p-to-L^q bound (13.35) for the free Schrödinger group $(S(t))_{t \in \mathbb{R}}$ in Section 13.6.g is an example of a so-called *dispersive* estimate. It informs us that initial data in $L^1(\mathbb{R}^d) \cap L^2(\mathbb{R}^d)$ are mapped instantaneously (in forward and backward time) to $L^\infty(\mathbb{R}^d)$ and decay to 0 as $|t| \to \infty$ with respect to the norm of $L^q(\mathbb{R}^d)$ for all $2 < q \leqslant \infty$. This bound lies at the basis of a class of deep estimates, named after Strichartz who proved an analogous estimate for the wave group, the simplest of which gives a bound for the $L^p(\mathbb{R}; L^q(\mathbb{R}^d))$ norm of $S(\cdot)f$ for initial data $f \in L^2(\mathbb{R}^d)$ and suitable exponents p, q. Such estimates, in turn, are the key to solving certain important classes of nonlinear Schrödinger equations. For a detailed treatment of these matters the reader is referred to the lecture notes Hundertmark et al. (2013) and the references cited there; an elementary introduction is presented in Stein and Shakarchi (2011).

The argument in the first part of Section 13.6.h is taken from Hundertmark et al. (2013).

The example in Problem 13.17 is due to Arendt (1995).

If A is a densely defined operator acting in a Banach space with the property that $(-\infty, 0) \subseteq \rho(A)$ and

$$\sup_{\lambda \in (0,\infty)} |\lambda| \|(\lambda + A)^{-1}\| < \infty,$$

then there exists a unique densely defined closed operator $A^{1/2}$ such that

$$(A^{1/2})^2 = A.$$

Moreover, $D(A)$ is dense in $D(A^{1/2})$ and

$$A^{1/2}x = \frac{1}{\pi} \int_0^\infty \lambda^{-1/2}(\lambda + A)^{-1} Ax \, d\lambda, \quad x \in D(A).$$

A proof of this result can be found in Section 3.8 of Arendt et al. (2011). In particular it applies to $A = -B$ whenever B is the generator of a uniformly bounded C_0-semigroup on X. The result should be compared to Proposition 10.60, where it was shown that if A is a positive selfadjoint operator acting in a Hilbert space, then A admits a unique positive square root $A^{1/2}$.

Let us now apply this to the divergence form operator $A_a := -\text{div}(a\nabla)$ of Example 13.39 associated with the sesquilinear form

$$\mathfrak{a}_a(f,g) := \int_{\mathbb{R}^d} a\nabla f \overline{\nabla g}, \quad f, g \in H^1(\mathbb{R}^d),$$

where the $d \times d$ matrix $a = (a_{ij})_{i,j=1}^{d}$ is assumed to have bounded measurable real-valued coefficients satisfying the uniform ellipticity condition stated in the example. Since $-A_a$ generates an analytic C_0-semigroup of contractions, by the above discussion the square root $A_a^{1/2}$ is well defined. The *Kato square root problem* is to decide whether the domain equality

$$\mathsf{D}(A_a^{1/2}) = \mathsf{D}(\mathfrak{a}_a) = H^1(\mathbb{R}^d)$$

holds, with equivalence of homogeneous norms

$$\|A_a^{1/2} f\| \eqsim \|\nabla f\|$$

for all functions f in this common domain. Starting with the papers Kato (1961) and McIntosh (1982), this problem has witnessed a long and interesting history. It was finally resolved in the affirmative, in the generality stated here, in Auscher et al. (2002). This paper also contains references to the various special cases that had been obtained before. An alternative proof based on the theory of bisectorial operators was obtained subsequently in Axelsson et al. (2006).

Chapter 14

The results of Sections 14.1–14.3 are standard. The results of Section 14.4 are taken from Attal (2013).

The argument in Step 1 of Proposition 14.21 follows (Dunford and Schwartz, 1988b, Lemma XI.6.21). Our proof of Lidskii's theorem (Theorem 14.34) is due to Simon (1977), whose arguments we follow here. A survey of the connections between determinants and traces, containing a proof of MacMahon's formula as well as a treatment of Fredholm determinants, is Cartier (1989). A proof of Lemma 14.43 based on a theorem of Borel and Carathéodory, is given in Simon (1977). For much more on this topic the reader may consult Simon (2005).

For positive kernel operators, Theorem 14.45 is due to Mercer (1909). The extension to general integral operators of trace class is taken from Birman (1989). In that paper it is also shown how to extend the result to general measure spaces as long as its L^2 space is separable. Further interesting results on this topic can be found in Brislawn (1988). It is of interest to note that not every integral operator with continuous kernel is of trace class; a classical counterexample can be found in Carleman (1916).

The proof of Theorem 14.46 is taken from Murphy (1994). Theorem 14.47 is from Helton and Howe (1973). The derivation of Euler's formula from the trace of the Dirichlet Laplacian on the interval is taken from Grieser (2007).

The proof outlined in Problem 14.7 is taken from Arendt (2006), where it is attributed to Markus Haase. Problem 14.15 is taken from Helton and Howe (1973), Problems

14.16 and 14.17 are from Murphy (1994), and (14.18) is taken from Connes and Consani (2021).

Chapter 15

Historical aspects of the interaction between Functional Analysis and Quantum Mechanics are well recorded in Landsman (2019). An excellent modern introduction to Quantum Mechanics from the mathematician's point of view is Hall (2013). More advanced treatments are offered in Landsman (1998, 2017); Mackey (1968); Parthasarathy (2005); Takhtajan (2008).

The mathematical formulation of Quantum Mechanics using the language of Hilbert space theory is due to von Neumann (1968). Ever since the publication of this work in 1932, physicists, mathematicians, and philosophers have wondered as to why Nature made that choice by looking for deeper criteria characterising Hilbert spaces. A first important step in this direction was taken in Piron (1964) and Amemiya and Araki (1966/1967) in the 1960s, who proved that a complex inner product space is a Hilbert space if and only if it is orthomodular. By definition, an inner product space H is *orthomodular* if $Y + Y^\perp = H$ for every closed subspace of H. The theorem of Piron and Amemiya–Araki was extended by several mathematicians to inner product spaces over $\mathbb{R}$, $\mathbb{C}$, and $\mathbb{H}$, the field of quaternions. The definitive result in this direction was proved in Solèr (1995). In order to state her result we need the following terminology.

Let H be a vector space over a field $\mathbb{K}$. A *Hermitian form* on H is a mapping $(\cdot|\cdot) : H \times H \to \mathbb{K}$ satisfying the axioms of an inner product except the requirement that $(x|x) = 0$ should imply $x = 0$. A *Hermitian vector space* is a vector space endowed with a Hermitian form. A subspace Y of a Hermitian vector space H is called *closed* if $Y^{\perp\perp} = Y$, orthogonal complements being defined in the obvious way using the Hermitian form. A Hermitian vector space H is called *orthomodular* if $Y + Y^\perp = H$ for every closed subspace Y of H. A field $\mathbb{K}$ is called a *⋆-field* if it admits an involution, that is, a mapping $c \mapsto c^\star$ from $\mathbb{K}$ onto itself satisfying $(c_1 + c_2)^\star = c_1^\star + c_2^\star$, $(c_1 c_2)^\star = c_2^\star c_1^\star$, and $c^{\star\star} = c$ for all $c_1, c_2, c \in \mathbb{K}$).

Now we are ready to state Solèr's theorem: If H is a Hermitian vector space over a ⋆-field $\mathbb{K}$ admitting an infinite orthonormal sequence (orthonormality being defined in the obvious way using the Hermitian form), then:

- $\mathbb{K}$ equals $\mathbb{R}$, $\mathbb{C}$, or $\mathbb{H}$;
- the Hermitian form is an inner product;
- H is a Hilbert space over $\mathbb{K}$.

A survey of Solèr's theorem is given in Holland (1995). Very recently, the theorem was used in Heunen and Kornell (2022) to give a characterisation of the category of Hilbert spaces as the unique category (in the sense of category theory) satisfying certain natural category theoretical axioms.

Modern treatments of the foundations of Quantum Mechanics replace the language of operator theory on Hilbert spaces by that of C^*-algebras. By a theorem of Gelfand, Naimark, and Segal (see, for example, Rudin (1991)), every closed $*$-subalgebra can be represented as a $*$-subalgebra of $\mathscr{L}(H)$ for an appropriate Hilbert space H, so not much seems to be gained. The advantage of this approach, however, is that it covers both the classical and the quantum settings: by a theorem due to Gelfand, every *commutative* C^*-algebra can be represented as a space $C_0(\Omega)$ for some locally compact Hausdorff space Ω, and by $C(K)$ for some compact Hausdorff space K if the C^*-algebra has a unit. In this precise sense, the 'classical world' is commutative, while the 'quantum world' in noncommutative. Comprehensive treatments of C^*-algebras and states defined on them are offered in Blackadar (2006); Bratteli and Robinson (1987); Pedersen (2018); Takesaki (2002).

Proofs of Gleason's theorem mentioned at the end of Section 15.2.a can be found in Landsman (2017); Parthasarathy (2005).

Our proof of Theorem 15.29 is taken form Akhiezer and Glazman (1981b). Another proof can be derived from Stinespring's dilation theorem. This approach is presented in Han et al. (2014), which may be consulted for more on (positive) operator-valued measures. Older references on the subject are Berberian (1966); Davies (1976); Holevo (2011); Landsman (1998). For a detailed discussion and examples of unsharp observables the reader is referred to Busch et al. (1995), which is also the source for the results of Section 15.3.d. The phase POVM Φ introduced in this section was studied in Garrison and Wong (1970).

Let us now sketch an elegant proof of Naimark's theorem based on Stinespring's theorem. We leave out some details which can be found in Stinespring (1955); see also Paulsen (2002). Let Q be a POVM on $(\Omega, \mathscr{F})$ and let $\Psi_Q : B_b(\Omega) \to \mathscr{L}(H)$ be the bounded functional calculus of Proposition 15.27. The crucial observation is that every bounded operator $\Psi : B_b(\Omega) \to \mathscr{L}(H)$ is *completely positive*, that is, for all $n = 1, 2, \ldots$ and all $f_1, \ldots f_n \in B_b(\Omega)$ and $h_1, \ldots, h_n \in H$ we have

$$\sum_{j,k=1}^{n} (\Psi(f_j \overline{f_k}) h_j | h_k) \geqslant 0.$$

By Gelfand's theorem there is no loss of generality in assuming that Ω is a compact Hausdorff space and that $\mathscr{F}$ is its Borel σ-algebra. Fixing an integer $n \geqslant 1$, by the Riesz representation theorem we find a finite Borel measure μ on Ω such that

$$\sum_{j=1}^{n} (\Psi(g) h_j | h_j) = \int_{\Omega} g \, d\mu, \quad g \in C(\Omega).$$

By the Radon–Nikodým theorem there exist functions $h_{jk} \in L^1(\Omega, \mu)$ such that

$$(\Psi(g)h_j | h_k) = \int_\Omega g h_{jk} \, d\mu, \quad g \in C(\Omega).$$

One then checks that the matrix $(h_{jk})_{j,k=1}^n$ is positive μ-almost everywhere. Also the matrix $(f_j \overline{f_k})_{j,k=1}^n$ is positive μ-almost everywhere. It follows that $\sum_{j,k=1}^n f_j \overline{f_k} h_{jk} \geqslant 0$ μ-almost everywhere and therefore

$$\sum_{j,k=1}^n (\Psi(f_j \overline{f_k}) h_j | h_k) = \int_\Omega \sum_{j,k=1}^n f_j \overline{f_k} h_{jk} \, d\mu \geqslant 0,$$

as was to be shown. Now (a special case of) *Stinespring's theorem* asserts that every completely positive bounded mapping $\Psi : B_b(\Omega) \to \mathscr{L}(H)$ satisfying $\|\Psi(\mathbf{1})\| = 1$ is of the form

$$\Psi(f) = J^\star \Pi(f) J,$$

where J is an isometry from H to a Hilbert space $\widetilde{H}$ and $\Pi : B_b(\Omega) \to \mathscr{L}(\widetilde{H})$ is a $\star$-homomorphism. Applying this to Ψ_Q and restricting Π to indicator functions, Proposition 15.24 gives us the desired projection-valued measure.

Physically, the qubit corresponds to the 2-dimensional irreducible unitary representation of $SU(2)$ and as such it models a spin-$\frac{1}{2}$ particle. For every $n \in \mathbb{N}$, $SU(2)$ admits an irreducible representation which acts on $\mathbb{C}^{n+1}$ and represents a spin-$\frac{1}{2}n$ particle (which is a boson if n is even and a fermion if n is odd). More on this topic can be found in Sternberg (1994); Woit (2017).

A complete proof of Theorem 15.33, including a proof of the algebraic fact that was used in our proof for the qubit case, is given in Landsman (2017). Our proof for the qubit case is extracted from it. Bargmann's theorem mentioned in the text is in Bargmann (1954); a complete proof is also found in Parthasarathy (2005).

Theorem 15.32 is a straightforward generalisation of the hidden variable result of Holevo (2011), where also the resulting hidden variable model for the qubit is derived. The existence of hidden variables for the qubit was first observed by Bell (1966). There is an extensive literature on the *nonexistence* of hidden variables, but such results usually work with more restrictive notions of hidden variables. A discussion of these results can be found in Landsman (2017).

For introductions to Lie groups and LCA groups we recommend Folland (2016). More complete treatments of covariance lead to the notion of *systems of imprimitivity* studied in Mackey (1968). For in-depth discussions of covariance and the way it pins down observables we recommend Parthasarathy (2005); Varadarajan (1985). A discussion from the physicist point of view is given in Busch et al. (1995). Theorem 15.38 is a special case of a generalisation of Stone's theorem (Theorem 13.44) for arbitrary strongly continuous unitary representations of G; see Theorem 4.5 in Folland (2016).

The presentation of the Stone–von Neumann theorem follows Folland (1989) and Hall (2013). The theorem admits a generalisation to LCA groups, essentially due to Mackey. For modern references to the literature on this subject the reader is referred to the survey article Rosenberg (2004). The formula for the Ornstein–Uhlenbeck semigroup in Theorem 15.55 goes back, at least, to Unterberger (1979); in its present form it is taken from van Neerven and Portal (2018).

Treatments of second quantisation can be found in Janson (1997); Parthasarathy (1992); Simon (1974). For a discussion from the Physics perspective we recommend Talagrand (2022). The proof of Theorem 15.69 is taken from Simon (1974). Theorems 15.70 and 15.71 are due to Segal (1956). Our discussion of the position and momentum operators follows Parthasarathy (1992), except that we use different normalisations designed to arrive at the physicist's identities (15.43) and (15.44) for the quantum harmonic oscillator. Proposition 15.74 can be found in the notes to Chapter 1 of Nualart (2006). As mentioned in the text, most results in Section 15.6 generalise to infinite dimensions if one replaces the Gaussian measure γ on $\mathbb{R}^d$ by a so-called H-isonormal process defined on a probability space $(\Omega, \mathbb{P})$, where H is a real Hilbert space taking the role of $\mathbb{R}^d$. The resulting theory has deep connections with the theory of stochastic integration; see, for example, Nualart (2006).

Let us finish with describing an interesting connection with Number Theory. Roughly speaking it says that, spectrally, the positive integers are precisely the second quantised primes. The starting point to make this into a rigorous statement one is a theorem of Brown and Pearcy (1966) that if T is a bounded operator on a Hilbert space H, then the spectrum of its n-fold tensor product $T^{\otimes n}$ acting on the Hilbert space $H^{\otimes n}$ equals

$$\sigma(T^{\otimes n}) = \{\lambda_1 \cdots \lambda_n : \lambda_j \in \sigma(T), \ j = 1, \ldots, n\}.$$

If $\|T\| < 1$, by taking direct sums one arrives at the formula

$$\sigma\left(\bigoplus_{n \in \mathbb{N}} T^{\otimes n}\right) = \overline{\bigcup_{n \in N} \{\lambda_1 \cdots \lambda_n : \lambda_j \in \sigma(T) \text{ for } 1 \leqslant j \leqslant n; \ n \geqslant 1\}}$$

with contribution 1 for the spectrum of $T^{\otimes 0} := I$. Now let $\mathbb{P} = \{2, 3, 5, 7, 11, \ldots\}$ be the set of primes and consider the Hilbert space $\ell^2(\mathbb{P})$. Denoting the standard unit basis vectors of this space by $e_2, e_3, e_5, \ldots$ we consider the contraction

$$T : e_p \mapsto \frac{1}{p} e_p, \quad p \in \mathbb{P}.$$

Then $\|T\| = \frac{1}{2}$,

$$\sigma(T) = \{\frac{1}{p} : p \in \mathbb{P}\} \cup \{0\},$$

and accordingly

$$\sigma\Big(\bigoplus_{n\in\mathbb{N}} T^{\otimes n}\Big) = \Big\{\frac{1}{n} : n \in \mathbb{N}, n \geqslant 1\Big\} \cup \{0\},$$

with each point $\frac{1}{n}$ being a simple pole, thanks to the uniqueness of prime factorisation. This observation (which extends to the symmetric second quantisation of T), as well as deeper connections, can be found in Bost and Connes (1995); Connes (1994).

Appendices

Most of the material is standard; some proofs are taken from Folland (1999); Kallenberg (2002); Ryan (2002). Zorn's lemma is equivalent with the Axiom of Choice. A proof of this fact and further equivalences can be found in Jech (1973, 2003); Rubin and Rubin (1970). The proof of Tychonov's theorem follows Ruzhansky and Turunen (2010). The treatment of Carathéodory's treatment is based on lecture notes by Mark Veraar.

Credits

John von Neumann's picture is reprinted with permission from George Karger/The Chronicle Collection/Getty Images. All other pictures are in the public domain.

References

Adams, R. A., and Fournier, J. J. F. 2003. *Sobolev spaces*. 2nd edn. Pure and Applied Mathematics (Amsterdam), vol. 140. Elsevier/Academic Press, Amsterdam. 680, 681

Akhiezer, N. I., and Glazman, I. M. 1981a. *Theory of linear operators in Hilbert space. Vol. I.* Monographs and Studies in Mathematics, vol. 9. Pitman, Boston, Mass.-London. 680

Akhiezer, N. I., and Glazman, I. M. 1981b. *Theory of linear operators in Hilbert space. Vol. II.* Monographs and Studies in Mathematics, vol. 10. Pitman, Boston, Mass.-London. 688

Albiac, F., and Kalton, N. J. 2006. *Topics in Banach space theory*. Graduate Texts in Mathematics, vol. 233. New York: Springer. 673, 675, 677

Aliprantis, C. D., and Burkinshaw, O. 1985. *Positive operators*. Pure and Applied Mathematics, vol. 119. Academic Press, Inc., Orlando, FL. 673, 675

Allan, G. R., and Ransford, T. J. 1989. Power-dominated elements in a Banach algebra. *Studia Math.*, **94**(1), 63–79. 677

Amann, H. 1995. *Linear and quasilinear parabolic problems, Vol. I: Abstract linear theory*. Monographs in Mathematics, vol. 89. Birkhäuser Boston, Inc., Boston, MA. 683

Amemiya, I., and Araki, H. 1966/1967. A remark on Piron's paper. *Publ. Res. Inst. Math. Sci. Ser. A*, **2**, 423–427. 687

Applebaum, D. 2019. *Semigroups of linear operators*. London Mathematical Society Student Texts, vol. 93. Cambridge University Press, Cambridge. 683

Arendt, W. 1995. Spectrum and growth of positive semigroups. Pages 21–28 of: *Evolution equations (Baton Rouge, LA, 1992)*. Lecture Notes in Pure and Appl. Math., vol. 168. Dekker, New York. 685

Arendt, W. 2004. Semigroups and evolution equations: functional calculus, regularity and kernel estimates. Pages 1–85 of: *Evolutionary equations. Vol. I.* Handb. Differ. Equ. North-Holland, Amsterdam. 681, 683

Arendt, W. 2006. *Heat kernels*. Lecture notes of the 9th Internet Seminar 2005/06, available at www.uni-ulm.de/mawi/iaa/members/arendt/. 681, 683, 684, 686

Arendt, W., and Urban, K. 2010. *Partielle Differentialgleichungen: Eine Einführung in analytische und numerische Methoden*. Spektrum. 680, 681

Arendt, W., Batty, C. J. K., Hieber, M., and Neubrander, F. 2011. *Vector-valued Laplace transforms and Cauchy problems*. 2nd edn. Monographs in Mathematics, vol. 96. Birkhäuser/Springer Basel AG, Basel. 683, 685

Arveson, W. 2002. *A short course on spectral theory*. Graduate Texts in Mathematics, vol. 209. Springer-Verlag, New York. 677, 678, 679

Atkinson, K., and Han, W. 2009. *Theoretical numerical analysis. A functional analysis framework*. 3rd edn. Texts in Applied Mathematics, vol. 39. Springer, Dordrecht. 681

Attal, S. 2013. *Tensor products and partial traces*. Lecture notes, available at math.univ-lyon1.fr/~attal/. 686

Aupetit, B. 1991. *A primer on spectral theory*. Universitext. Springer-Verlag, New York. 677

Auscher, P., Hofmann, S., Lacey, M., McIntosh, A., and Tchamitchian, Ph. 2002. The solution of the Kato square root problem for second order elliptic operators on $\mathbb{R}^n$. *Ann. of Math. (2)*, **156**(2), 633–654. 686

Axelsson, A., Keith, S., and McIntosh, A. 2006. Quadratic estimates and functional calculi of perturbed Dirac operators. *Invent. Math.*, **163**(3), 455–497. 686

Baghi, B. 2006. On Nyman, Beurling and Baez–Duarte's Hilbert space reformulation of the Riemann hypothesis. *Proc. Indian Acad. Sci. (Math. Sci.)*, **116**, 137–146. 675

Bargmann, V. 1954. On unitary ray representations of continuous groups. *Ann. of Math. (2)*, **59**, 1–46. 689

Beauzamy, B. 1988. *Introduction to Operator Theory and Invariant Subspaces*. North-Holland Mathematical Library. North-Holland. 675

Beckner, W. 1975. Inequalities in Fourier analysis. *Ann. of Math. (2)*, **102**(1), 159–182. 676

Bell, J. S. 1966. On the problem of hidden variables in quantum mechanics. *Rev. Modern Phys.*, **38**, 447–452. 689

Berberian, S. K. 1966. *Notes on spectral theory*. Van Nostrand Mathematical Studies, No. 5. D. Van Nostrand Co., Inc., Princeton, NJ-Toronto, Ont.-London. 2nd edition by the author available at web.ma.utexas.edu/mp_arc/c/09/09-32.pdf. 688

Biegert, M., and Warma, M. 2006. Regularity in capacity and the Dirichlet Laplacian. *Potential Anal.*, **25**(3), 289–305. 680

Birman, M. Sh. 1989. *A proof of the Fredholm trace formula as an application of a simple embedding for kernels integral operators of trace class in $L^2(\mathbb{R}^m)$*. Report LIT-MATK-R-89-30, Department of Mathematics, Linköpign University. 686

Birman, M. Sh., and Solomjak, M. Z. 1987. *Spectral theory of selfadjoint operators in Hilbert space*. Mathematics and its Applications (Soviet Series). D. Reidel Publishing Co., Dordrecht. 680

Blackadar, B. 2006. *Operator algebras*. Encyclopaedia of Mathematical Sciences, vol. 122. Springer-Verlag, Berlin. Theory of C^*-algebras and von Neumann algebras, Operator Algebras and Non-commutative Geometry, III. 688

Bleecker, D. D., and Booß Bavnbek, B. 2013. *Index theory—with applications to mathematics and physics*. International Press, Somerville, MA. 677

Bogachev, V. I. 2007a. *Measure theory. Vol. I*. Springer-Verlag, Berlin. 675

Bogachev, V. I. 2007b. *Measure theory. Vol. II*. Springer-Verlag, Berlin. 673

Bost, J.-B., and Connes, A. 1995. Hecke algebras, type III factors and phase transitions with spontaneous symmetry breaking in number theory. *Selecta Math. (N.S.)*, **1**(3), 411–457. 691

Böttcher, A., and Silbermann, B. 2006. *Analysis of Toeplitz operators*. 2nd edn. Springer Monographs in Mathematics. Springer-Verlag, Berlin. 677

Bratteli, O., and Robinson, D. W. 1987. *Operator algebras and quantum statistical mechanics. Vol. 1*. 2nd edn. Texts and Monographs in Physics. Springer-Verlag, New York. 688

Brenner, S. C., and Scott, L. R. 2008. *The mathematical theory of finite element methods*. 3rd edn. Texts in Applied Mathematics, vol. 15. Springer, New York. 681

Bressan, A. 2013. *Lecture notes on functional analysis*. Graduate Studies in Mathematics, vol. 143. Amer. Math. Soc., Providence, RI. 673, 680

Brezis, H. 2011. *Functional analysis, Sobolev spaces and partial differential equations*. Universitext. Springer, New York. 673, 680, 681

Brislawn, C. 1988. Kernels of trace class operators. *Proc. Amer. Math. Soc.*, **104**(4), 1181–1190. 686

Brown, A., and Pearcy, C. 1966. Spectra of tensor products of operators. *Proc. Amer. Math. Soc.*, **17**, 162–166. 690

Busch, P., Grabowski, M., and Lahti, P. J. 1995. *Operational quantum physics*. Lecture Notes in Physics. New Series m: Monographs, vol. 31. Springer-Verlag, Berlin. 688, 689

Buskes, G. 1993. The Hahn-Banach theorem surveyed. *Dissertationes Math.*, **327**, 49. 675

Carleman, T. 1916. Über die Fourierkoeffizienten einer stetigen Funktion. *Acta Math.*, **41**(1), 377–384. 686

Cartier, P. 1989. A course on determinants. Pages 443–557 of: *Conformal invariance and string theory (Poiana Braşov, 1987)*. Perspect. Phys. Academic Press, Boston, MA. 686

Clément, Ph., and Li, S. 1993/94. Abstract parabolic quasilinear equations and application to a groundwater flow problem. *Adv. Math. Sci. Appl.*, **3** (Special Issue), 17–32. 684

Coburn, L. A. 1966. Weyl's theorem for nonnormal operators. *Michigan Math. J.*, **13**, 285–288. 677

Coburn, L. A. 1967. The C^*-algebra generated by an isometry. *Bull. Amer. Math. Soc.*, **73**, 722–726. 677

Connes, A. 1994. *Noncommutative geometry*. Academic Press, Inc., San Diego, CA. 691

Connes, A., and Consani, C. 2021. The scaling Hamiltonian. *J. Operator Theory*, **85**(1), 257–276. 687

Conway, J. B. 1990. *A course in functional analysis*. 2nd edn. Graduate Texts in Mathematics, vol. 96. Springer-Verlag, New York. 673, 675

Davies, E. B. 1976. *Quantum theory of open systems*. Academic Press, London-New York. 688

Davies, E. B. 1980. *One-parameter semigroups*. London Mathematical Society Monographs, vol. 15. Academic Press, Inc., London-New York. 679, 683

De Simon, L. 1964. Un'applicazione della teoria degli integrali singolari allo studio delle equazioni differenziali lineari astratte del primo ordine. *Rend. Sem. Mat. Univ. Padova*, **34**, 205–223. 683

Diestel, J. 1984. *Sequences and series in Banach spaces*. Graduate Texts in Mathematics, vol. 92. Springer-Verlag, New York. 673

Diestel, J., and Uhl, Jr., J. J. 1977. *Vector measures*. Providence, RI: Amer. Math. Soc. 673, 677

Dieudonné, J. 1981. *History of functional analysis*. North-Holland Mathematics Studies, vol. 49. North-Holland Publishing Co., Amsterdam-New York. 673

Douglas, R. G. 1998. *Banach algebra techniques in operator theory*. 2nd edn. Graduate Texts in Mathematics, vol. 179. Springer-Verlag, New York. 677

van Dulst, D. 1989. *Characterizations of Banach spaces not containing ℓ^1*. CWI Tract, vol. 59. Amsterdam: CWI. 677

Dunford, N., and Schwartz, J. T. 1988a. *Linear operators. Part I: General theory*. Wiley Classics Library. John Wiley & Sons, Inc., New York. Reprint of the 1958 original. 673

Dunford, N., and Schwartz, J. T. 1988b. *Linear operators. Part II: Spectral theory, selfadjoint operators in Hilbert space*. Wiley Classics Library. John Wiley & Sons, Inc., New York. Reprint of the 1963 original. 677, 680, 686

Duoandikoetxea, J. 2001. *Fourier analysis*. Graduate Studies in Mathematics, vol. 29. Amer. Math. Soc., Providence, RI. Translated and revised from the 1995 Spanish original. 676

Edmunds, D. E., and Evans, W. D. 2018a. *Elliptic differential operators and spectral analysis*. Springer Monographs in Mathematics. Springer, Cham. 681

Edmunds, D. E., and Evans, W. D. 2018b. *Spectral theory and differential operators*. 2nd edn. Oxford Mathematical Monographs. Oxford University Press, Oxford. 680, 681

Einsiedler, M., and Ward, Th. 2017. *Functional analysis, spectral theory, and applications*. Graduate Texts in Mathematics, vol. 276. Springer, Cham. 673

Engel, K.-J., and Nagel, R. 2000. *One-parameter semigroups for linear evolution equations*. Graduate Texts in Mathematics, vol. 194. New York: Springer-Verlag. 683

Epperson, J. B. 1989. The hypercontractive approach to exactly bounding an operator with complex Gaussian kernel. *J. Funct. Anal.*, **87**(1), 1–30. 685

Evans, L. C. 2010. *Partial differential equations*. 2nd edn. Graduate Studies in Mathematics, vol. 19. Amer. Math. Soc., Providence, RI. 680, 681

Filonov, N. 2005. On an inequality between Dirichlet and Neumann eigenvalues for the Laplace operator. *St. Petersburg Math. J.*, **16**(2), 413–416. 682

Foguel, S. R. 1958. On a theorem by A. E. Taylor. *Proc. Amer. Math. Soc.*, **9**, 325. 675

Folland, G. B. 1989. *Harmonic analysis in phase space*. Annals of Mathematics Studies, vol. 122. Princeton University Press, Princeton, NJ. 690

Folland, G. B. 1999. *Real analysis*. 2nd edn. Pure and Applied Mathematics (New York). New York: John Wiley & Sons Inc. 675, 691

Folland, G. B. 2016. *A course in abstract harmonic analysis*. 2nd edn. Textbooks in Mathematics. CRC Press, Boca Raton, FL. 678, 689

Fremlin, D. H. 2015. *Measure theory. Vol. 5. Set-theoretic measure theory. Part I*. Torres Fremlin, Colchester. Corrected reprint of the 2008 original. 674

Friedlander, L. 1991. Some inequalities between Dirichlet and Neumann eigenvalues. *Arch. Ration. Mech. Anal.*, **116**(2), 153–160. 682

Funano, K. 2022. A note on domain monotonicity for the Neumann eigenvalues of the Laplacian. *arXiv:2202.03598*. 682

Garrison, J. C., and Wong, J. 1970. Canonically conjugate pairs, uncertainty relations, and phase operators. *J. Mathematical Phys.*, **11**, 2242–2249. 688

Gelfand, I. M. 1941. Zur Theorie der Charaktere der Abelschen topologischen Gruppen. *Mat. Sbornik N. S.*, **51**, 49–50. 677

Gilbarg, D., and Trudinger, N. S. 2001. *Elliptic partial differential equations of second order*. Classics in Mathematics. Springer-Verlag, Berlin. Reprint of the 1998 edition. 681

Gohberg, I., Goldberg, S., and Kaashoek, M. 2003. *Basic classes of linear operators*. Birkhäuser. 675

Gohberg, I., Goldberg, S., and Kaashoek, M. A. 2013. *Classes of linear operators*. Operator Theory: Advances and Applications, vol. 63. Birkhäuser. 675

Gordon, C., Webb, D. L., and Wolpert, S. 1992. One cannot hear the shape of a drum. *Bull. Amer. Math. Soc. (N.S.)*, **27**(1), 134–138. 681

Grafakos, L. 2008. *Classical Fourier analysis*. 2nd edn. Graduate Texts in Mathematics, vol. 249. New York: Springer. 676

Grebenkov, D. S., and Nguyen, B.-T. 2013. Geometrical structure of Laplacian eigenfunctions. *SIAM Review*, **55**(4), 601–667. 681

Grieser, D. 2007. Über Eigenwerte, Integrale und $\frac{\pi^2}{6}$: die Idee der Spurformel. *Math. Semester-ber.*, **54**(2), 199–217. 686

Gustafson, K. E., and Rao, D. K. M. 1997. *Numerical range.* Universitext. Springer-Verlag, New York. 679

Haase, M. H. A. 2007. Convexity inequalities for positive operators. *Positivity*, **11**(1), 57–68. 673

Haase, M. H. A. 2018. *Functional calculus.* Lecture notes of the 21st Internet Seminar, available at www.math.uni-kiel.de/isem21. 679

Hall, B. C. 2013. *Quantum theory for mathematicians.* Graduate Texts in Mathematics, vol. 267. Springer, New York. 687, 690

Han, D., Larson, D. R., Liu, B., and Liu, R. 2014. Operator-valued measures, dilations, and the theory of frames. *Mem. Amer. Math. Soc.*, **229**(1075). 688

Hanche-Olsen, H., Holden, H., and Malinnikova, E. 2019. An improvement of the Kolmogorov–Riesz compactness theorem. *Expositiones Math.*, **37**(1), 84–91. 673

Helton, J. W., and Howe, R. E. 1973. Integral operators: commutators, traces, index and homology. Pages 141–209. Lecture Notes in Math., Vol. 345 of: *Proceedings of a Conference Operator Theory (Dalhousie Univ., Halifax, N.S., 1973).* 686

Henrot, A. 2006. *Extremum problems for eigenvalues of elliptic operators.* Frontiers in Mathematics. Birkhäuser Verlag, Basel. 681

Heunen, C., and Kornell, A. 2022. Axioms for the category of Hilbert spaces. *Proceedings of the National Academy of Sciences*, **119**(9), e2117024119. 687

Heuser, H. 2006. *Funktionalanalysis.* 4th edn. Mathematische Leitfäden. B. G. Teubner, Stuttgart. 679

Higson, N. 2004. The local index formula in noncommutative geometry. Pages 443–536 of: *Contemporary developments in algebraic K-theory.* ICTP Lect. Notes, XV. Abdus Salam Int. Cent. Theoret. Phys., Trieste. 681

Hille, E., and Phillips, R. S. 1957. *Functional analysis and semi-groups.* Amer. Math. Soc. Colloq. Publ., vol. 31. Providence, RI: Amer. Math. Soc. rev. ed. 683

Holevo, A. 2011. *Probabilistic and statistical aspects of quantum theory.* 2nd edn. Quaderni/ Monographs, vol. 1. Edizioni della Normale, Pisa. 688, 689

Holland, Jr., S. S. 1995. Orthomodularity in infinite dimensions; a theorem of M. Solèr. *Bull. Amer. Math. Soc. (N.S.)*, **32**(2), 205–234. 687

Hundertmark, D., Machinek, L., Meyries, M., and Schnaubelt, R. 2013. *Operator semigroups and dispersive equations.* Lecture notes of the 16th Internet Seminar 2012/13, available at isem.math.kit.edu. 685

Hytönen, T. P., van Neerven, J. M. A. M., Veraar, M. C., and Weis, L. W. 2016. *Analysis in Banach spaces. Vol. I: Martingales and Littlewood-Paley theory.* Ergebnisse der Mathematik und ihrer Grenzgebiete. 3. Folge, vol. 63. Springer, Cham. 673, 674, 676, 680

Hytönen, T. P., van Neerven, J. M. A. M, Veraar, M. C., and Weis, L. W. 2017. *Analysis in Banach spaces. Vol. II: Probabilistic methods and operator theory.* Ergebnisse der Mathematik und ihrer Grenzgebiete. 3. Folge, vol. 67. Springer, Cham. 683

Hytönen, T. P., van Neerven, J. M. A. M., Veraar, M. C., and Weis, L. W. 2022+. *Analysis in Banach spaces. Vol. III: Harmonic analysis and operator theory.* In preparation. 676, 683

Hytönen, T. P., van Neerven, J. M. A. M., Veraar, M. C., and Weis, L. W. 2023+. *Analysis in Banach spaces. Vol. IV: Stochastic analysis.* In preparation. 684

Janson, S. 1997. *Gaussian Hilbert spaces.* Cambridge Tracts in Mathematics, vol. 129. Cambridge: Cambridge University Press. 684, 690

Jech, Th. J. 1973. *The axiom of choice*. North-Holland, Amsterdam-London. Studies in Logic and the Foundations of Mathematics, Vol. 75. 691

Jech, Th. J. 2003. *Set theory*. Springer Monographs in Mathematics. Springer-Verlag, Berlin. Third edition. 691

Jorgensen, P. 1982. Spectral theory for infinitesimal generators of one-parameter groups of isometries: The min-max principle and compact perturbations. *J. Math. Anal. Appl.*, **90**, 343–370. 683

Jost, J. 2013. *Partial differential equations*. 3rd edn. Graduate Texts in Mathematics, vol. 214. Springer, New York. 680

Kac, M. 1966. Can one hear the shape of a drum? *Amer. Math. Monthly*, **73**(4, part II), 1–23. 681

Kallenberg, O. 2002. *Foundations of modern probability*. 2nd edn. Probability and its Applications (New York). Springer-Verlag, New York. 691

Kato, T. 1961. Fractional powers of dissipative operators. *J. Math. Soc. Japan*, **13**, 246–274. 686

Kato, T. 1995. *Perturbation theory for linear operators*. Classics in Mathematics. Springer-Verlag, Berlin. Reprint of the 1980 edition. 680, 681

Koelink, H. T. 1996. 8 Lectures on quantum groups and q-special functions. *arXiv:q-alg/9608018*. 679

Krylov, N. V. 2008. *Lectures on elliptic and parabolic equations in Sobolev spaces*. Graduate Studies in Mathematics, vol. 96. Amer. Math. Soc., Providence, RI. 680, 681

Landsman, N. P. 1998. *Mathematical topics between classical and quantum mechanics*. Springer Monographs in Mathematics. Springer-Verlag, New York. 687, 688

Landsman, N. P. 2017. *Foundations of quantum theory*. Fundamental Theories of Physics, vol. 188. Springer, Cham. 687, 688, 689

Landsman, N. P. 2019. Quantum theory and functional analysis. *arXiv:1911.06630*. 687

Lax, P. D. 2002. *Functional Analysis*. Wiley. 673

Levine, H. A., and Weinberger, H. F. 1986. Inequalities between Dirichlet and Neumann eigenvalues. *Arch. Ration. Mech. Anal.*, **94**(3), 193–208. 682

Li, C.-K. 1994. *C*-numerical ranges and *C*-numerical radii. *Linear and Multilinear Algebra*, **37**(1-3), 51–82. 679

Li, D., and Queffélec, H. 2004. *Introduction à l'étude des espaces de Banach*. Cours Spécialisés, vol. 12. Société Mathématique de France, Paris. 673

Lindenstrauss, J., and Tzafriri, L. 1971. On the complemented subspaces problem. *Israel J. Math.*, **9**, 263–269. 674

Lunardi, A. 1995. *Analytic semigroups and optimal regularity in parabolic problems*. Progress in Nonlinear Differential Equations and their Applications, vol. 16. Birkhäuser, Basel. 683

Luxemburg, W. A. J., and Zaanen, A. C. 1971. *Riesz spaces*. North-Holland. 673

Maas, J., and van Neerven, J. M. A. M. 2009. Boundedness of Riesz transforms for elliptic operators on abstract Wiener spaces. *J. Funct. Anal.*, **257**(8), 2410–2475. 684

Mackey, G. W. 1968. *Induced representations of groups and quantum mechanics*. W. A. Benjamin, Inc., New York-Amsterdam; Boringhieri, Turin. 687, 689

Maligranda, L. 1997. On the norms of operators in the real and the complex case. Pages 67–71 of: *Seminar on Banach spaces and related topics: 14/12/1997-14/12/1997*. 676

McIntosh, A. 1982. On representing closed accretive sesquilinear forms as $(A^{1/2}u, A^{*1/2}v)$. Pages 252–267 of: *Nonlinear partial differential equations and their applications. Collège de France Seminar, Vol. III (Paris, 1980/1981)*. Res. Notes in Math., vol. 70. Pitman, Boston, Mass.-London. 686

Mercer, J. 1909. Functions of positive and negative type, and their connection to the theory of integral equations. *Phil. Trans. Royal Soc. London. Series A*, **209**(441-458), 415–446. 686

Metafune, G., Prüss, J., Rhandi, A., and Schnaubelt, R. 2002. The domain of the Ornstein–Uhlenbeck operator on an L^p-space with invariant measure. *Ann. Scuola Norm.-Sci*, **1**(2), 471–485. 684

Meyer-Nieberg, P. 1991. *Banach lattices*. Universitext. Berlin: Springer-Verlag. 673

Monna, A. F. 1973. *Functional analysis in historical perspective*. John Wiley & Sons, New York-Toronto, Ont. 673

Müller, V. 2007. *Spectral theory of linear operators and spectral systems in Banach algebras*. 2nd edn. Operator Theory: Advances and Applications, vol. 139. Birkhäuser, Basel. 677

Murphy, G. J. 1994. Fredholm index theory and the trace. *Proc. Roy. Irish Acad. Sect. A*, **94**(2), 161–166. 686, 687

Musiał, K. 2002. Pettis integral. Pages 531–586 of: *Handbook of measure theory, Vol. I, II*. Amsterdam: North-Holland. 677

van Neerven, J. M. A. M. 1997. Elementary operator-theoretic proof of Wiener's Tauberian Theorem. *Rendic. Istit. Matem. Univ. Trieste, Suppl. Vol. XXVIII*, 281–286. 683

van Neerven, J. M. A. M., and Portal, P. 2018. The Weyl calculus with respect to the Gaussian measure and restricted L^p-L^q boundedness of the Ornstein-Uhlenbeck semigroup in complex time. *Bull. Soc. Math. France*, **146**(4), 691–712. 685, 690

van Neerven, J. M. A. M., Veraar, M. C., and Weis, L. W. 2015. Stochastic integration in Banach spaces – a survey. In: *Stochastic analysis: A series of lectures*. Progress in Probability, vol. 68. Birkhäuser Verlag. 674

Nikolski, N. K. 2002. *Operators, functions, and systems: an easy reading, Vol. 1*. Mathematical Surveys and Monographs, vol. 92. Amer. Math. Soc., Providence, RI. 675, 677

Nualart, D. 2006. *The Malliavin calculus and related topics*. 2nd edn. Probability and its Applications (New York). Springer-Verlag, Berlin. 684, 690

Osgood, W. F. 1903. A Jordan curve of positive area. *Trans. Amer. Math. Soc.*, **4**(1), 107–112. 681

Ouhabaz, E. M. 2005. *Analysis of heat equations on domains*. London Mathematical Society Monographs Series, vol. 31. Princeton University Press, Princeton, NJ. 681

Parthasarathy, K. R. 1992. *An introduction to quantum stochastic calculus*. Modern Birkhäuser Classics. Birkhäuser/Springer Basel AG, Basel. 2012 reprint of the 1992 original. 690

Parthasarathy, K. R. 2005. *Mathematical foundations of quantum mechanics*. Texts and Readings in Mathematics, vol. 35. New Delhi: Hindustan Book Agency. 687, 688, 689

Paulsen, V. 2002. *Completely bounded maps and operator algebras*. Cambridge Studies in Advanced Mathematics, vol. 78. Cambridge University Press, Cambridge. 688

Pazy, A. 1983. *Semigroups of linear operators and applications to partial differential equations*. Applied Math. Sciences, vol. 44. New York: Springer-Verlag. 683

Pedersen, G. K. 2018. *C^*-algebras and their automorphism groups*. Pure and Applied Mathematics (Amsterdam). Academic Press, London. 679, 688

Pettis, B. J. 1938. On integration in vector spaces. *Trans. Amer. Math. Soc.*, **44**(2), 277–304. 676

Phelps, R. R. 1960. Uniqueness of Hahn-Banach extensions and unique best approximation. *Trans. Amer. Math. Soc.*, **95**, 238–255. 675

Phillips, R. S. 1955. The adjoint semi-group. *Pac. J. Math.*, **5**, 269–283. 683

Pietsch, A. 2007. *History of Banach spaces and linear operators*. Boston, MA: Birkhäuser Boston Inc. 673, 675

Piron, C. 1964. Axiomatique quantique. *Helv. Phys. Acta*, **37**, 439–468. 687

Pisier, G. 2016. *Martingales in Banach spaces*. Cambridge Studies in Advanced Mathematics, vol. 155. Cambridge University Press. 674

Reed, M., and Simon, B. 1975. *Methods of modern mathematical physics. Vol. II: Fourier analysis, self-adjointness*. New York: Academic Press. 680

Riesz, M. 1926. Sur les maxima des formes bilinéaires et sur les fonctionnelles linéaires. *Acta Math.*, **49**(3-4), 465–497. 676

Rosenberg, J. 2004. A selective history of the Stone-von Neumann theorem. Pages 331–353 of: *Operator algebras, quantization, and noncommutative geometry*. Contemp. Math., vol. 365. Amer. Math. Soc., Providence, RI. 690

Rubin, H., and Rubin, J. E. 1970. *Equivalents of the axiom of choice*. Studies in Logic and the Foundations of Mathematics. North-Holland Publishing Co., Amsterdam-London. 691

Rudin, W. 1987. *Real and complex analysis*. 3rd edn. McGraw-Hill Book Co., New York. 674, 677, 679

Rudin, W. 1991. *Functional analysis*. 2nd edn. International Series in Pure and Applied Mathematics. McGraw-Hill, Inc., New York. 673, 675, 678, 679, 680, 688

Ruzhansky, M., and Turunen, V. 2010. *Pseudo-differential operators and symmetries*. Pseudo-Differential Operators. Theory and Applications, vol. 2. Birkhäuser Verlag, Basel. 675, 691

Ryan, R. A. 2002. *Introduction to tensor products of Banach spaces*. Springer Monographs in Mathematics. Springer-Verlag London, Ltd., London. 677, 691

Safarov, Yu., and Vassilev, D. 1997. *The asymptotic distribution of eigenvalues of partial differential operators*. Transl. Math. Monogr., vol. 155. Amer. Math. Soc., Providence, RI. 682

Schaefer, H. H. 1974. *Banach lattices and positive operators*. Grundlehren der Mathematischen Wissenschaften, vol. 215. New York: Springer-Verlag. 673

Schechter, M. 2002. *Principles of functional analysis*. 2nd edn. Graduate Studies in Mathematics, vol. 36. Amer. Math. Soc., Providence, RI. 673, 677

Schmüdgen, K. 2012. *Unbounded self-adjoint operators on Hilbert space*. Graduate Texts in Mathematics, vol. 265. Springer, Dordrecht. 680

Seeley, R. T. 1964. Extension of C^∞ functions defined in a half space. *Proc. Amer. Math. Soc.*, **15**, 625–626. 680

Segal, I. E. 1956. Tensor algebras over Hilbert spaces. I. *Trans. Amer. Math. Soc.*, **81**, 106–134. 690

Simon, B. 1974. *The $P(\phi)_2$ Euclidean (quantum) field theory*. Princeton Series in Physics. Princeton University Press, Princeton, NJ. 690

Simon, B. 1977. Notes on infinite determinants of Hilbert space operators. *Adv. Math.*, **24**(3), 244–273. 686

Simon, B. 2005. *Trace ideals and their applications*. 2nd edn. Mathematical Surveys and Monographs, vol. 120. American Mathematical Soc. 686

Sobczyk, A. 1941. Projection of the space (m) on its subspace (c_0). *Bull. Amer. Math. Soc.*, **47**, 938–947. 675

Solèr, M. P. 1995. Characterization of Hilbert spaces by orthomodular spaces. *Comm. Algebra*, **23**(1), 219–243. 687

Stein, E. M. 1970. *Singular integrals and differentiability properties of functions*. Princeton Mathematical Series, No. 30. Princeton University Press, Princeton, NJ. 673, 676

Stein, E. M. 1993. *Harmonic analysis: real-variable methods, orthogonality, and oscillatory integrals*. Princeton University Press. 676

Stein, E. M., and Shakarchi, R. 2011. *Functional analysis*. Princeton University Press, Princeton, NJ. Volume 4 of Princeton Lectures in Analysis. 685

Sternberg, S. 1994. *Group theory and physics*. Cambridge University Press, Cambridge. 689

Stinespring, W. F. 1955. Positive functions on C^*-algebras. *Proc. Amer. Math. Soc.*, **6**, 211–216. 688

Sz.-Nagy, B. 1967. *Spektraldarstellung linearer Transformationen des Hilbertschen Raumes*. Ergebnisse der Mathematik und ihrer Grenzgebiete, Band 39. Springer-Verlag, Berlin-New York. 680

Takesaki, M. 2002. *Theory of operator algebras. I*. Encyclopaedia of Mathematical Sciences, vol. 124. Springer-Verlag, Berlin. Reprint of the first (1979) edition, Operator Algebras and Non-commutative Geometry, 5. 688

Takhtajan, L. A. 2008. *Quantum mechanics for mathematicians*. Graduate Studies in Mathematics, vol. 95. Amer. Math. Soc., Providence, RI. 687

Talagrand, M. 1984. Pettis integral and measure theory. *Mem. Amer. Math. Soc.*, **51**(307). 677

Talagrand, M. 2022. *What is a quantum field theory?* Cambridge University Press. 690

Tanabe, H. 1979. *Equations of evolution*. Monographs and Studies in Mathematics, vol. 6. Pitman, Boston, Mass.-London. 683

Taylor, A. E. 1939. The extension of linear functionals. *Duke Math. J.*, **5**, 538–547. 675

Terzioğlu, T. 1971. A characterization of compact linear mappings. *Arch. Math. (Basel)*, **22**, 76–78. 677

Unterberger, A. 1979. Oscillateur harmonique et opérateurs pseudo-différentiels. *Ann. Inst. Fourier (Grenoble)*, **29**(3), xi, 201–221. 690

Varadarajan, V. S. 1985. *Geometry of quantum theory*. 2nd edn. New York: Springer-Verlag. 689

von Neumann, J. 1968. *Mathematische Grundlagen der Quantenmechanik*. Grundlehren der mathematischen Wissenschaften, vol. 38. Springer-Verlag, Berlin-New York. Reprint of the first edition, 1932. 687

Werner, D. 2000. *Funktionalanalysis*. 3rd edn. Springer-Verlag, Berlin. 673

Whitley, R. 1968. The spectral theorem for a normal operator. *Amer. Math. Monthly*, **75**, 856–861. 679

Woit, P. 2017. *Quantum theory, groups and representations*. Springer, Cham. An introduction. 689

Yosida, K. 1980. *Functional analysis*. 6th edn. Grundlehren der Mathematischen Wissenschaften, vol. 123. Springer-Verlag, Berlin-New York. 673

Zaanen, A. C. 1997. *Introduction to operator theory in Riesz spaces*. Springer-Verlag, Berlin. 673

Index

Printed in the United States
by Baker & Taylor Publisher Services